THE SEA

Ideas and Observations on Progress in the Study of the Seas

THE SEA

Ideas and Observations on Progress in the Study of the Seas

Edited by

**E. D. GOLDBERG
I. N. McCAVE
J. J. O'BRIEN
J. H. STEELE**

VOLUME 6

Marine Modeling

A Wiley-Interscience Publication

JOHN WILEY & SONS · **New York** · **London** · **Sydney** · **Toronto**

Library of Congress Cataloging in Publication Data:

Main entry under title:
The Sea: ideas and observations on progress in the
  study of the seas.

   General editor, v. 1–3: M. N. Hill; v. 4, pts. 1–
A. E. Maxwell; v. 5–6: E. D. Goldberg.
   Vol. 4, pts. 1–   have imprint: New York, Wiley-Inter-
sciences; v. 5: New York, Wiley.
   Includes bibliographies and indexes.
   CONTENTS: v. 1. Physical oceanography.—v. 2. The
composition of sea-water. Comparative and descriptive
oceanography.—v. 3. The earth beneath the sea. History.
—v. 4. New concepts of sea floor evolution. pt. 1.
General observations. pt. 2. Regional observations,
concepts.
   1. Oceanography. I. Hill, Maurice Neville, 1919–
ed. II. Maxwell, Arthur Eugene, 1925–   ed.
III. Goldberg, Edward D., ed.

GC11.S4   551.4′6   62-18366

ISBN 0-471-31091-3(V.6)

Printed in the United States of America

10 9 8 7 6 5 4 3 2 1

CONTRIBUTORS TO VOLUME 6

GRAEME BONHAM-CARTER, Sharpe House, Wiveliscombe, Somerset, England

G. T. CSANADY, Woods Hole Oceanographic Institution, Woods Hole, Massachusetts 02543

KENNETH L. DENMAN, Marine Ecology Laboratory, Bedford Institute of Oceanography, Dartmouth, Nova Scotia, Canada

DOMINIC M. DI TORO, Environmental Engineering and Science Program, Manhattan College, New York, New York 10471; and Hydroscience, Inc., Westwood, New Jersey

R. C. DUGDALE, Bigelow Laboratory of Marine Science, West Boothbay Harbor, Maine 04575

K. R. DYER, Institute of Oceanographic Sciences, Taunton, England

JOHN W. HARBAUGH, Department of Applied Earth Sciences, Stanford University, Stanford, California 94305

MYRL C. HENDERSHOTT, Scripps Institution of Oceanography, University of California at San Diego, La Jolla, California 92037

WILLIAM R. HOLLAND, Geophysical Fluid Dynamics Laboratory, Princeton University, Princeton, New Jersey 08450

ALAN D. JASSBY, Marine Ecology Laboratory, Bedford Institute of Oceanography, Dartmouth, Nova Scotia, Canada

PAUL D. KOMAR, School of Oceanography, Oregon State University, Corvallis, Oregon 97331

P. H. LEBLOND, Department of Oceanography, University of British Columbia, Vancouver, Canada

ABRAHAM LERMAN, Department of Geological Sciences, Northwestern University, Evanston, Illinois 60201

FRED T. MACKENZIE, Department of Geological Sciences, Northwestern University, Evanston, Illinois 60201

JOHN L. MANCINI, Hydroscience, Inc., Westwood, New Jersey

V. V. MENSHUTKIN, Academy of Sciences of the USSR, Leningrad, USSR

FRANK J. MILLERO, Rosenstiel School of Marine and Atmospheric Science, University of Miami, Miami, Florida 33149

D. W. MOORE, Nova University, Fort Lauderdale, Florida 33314

MICHAEL M. MULLIN, Scripps Institution of Oceanography, University of California at San Diego, La Jolla, California 92037

L. A. MYSAK, Department of Oceanography, University of British Columbia, Vancouver, Canada

PEARN P. NIILER, School of Oceanography, Oregon State University Corvallis, Oregon 97331

DONALD J. O'CONNOR, Environmental Engineering and Science Program, Manhattan College, New York, New York 10471; and Hydroscience, Inc., Westwood, New Jersey

M. W. OWEN, Hydraulics Research Station, Wallingford, England

S. G. H. PHILANDER, Princeton University, Princeton, New Jersey 08450

TREVOR PLATT, Marine Ecology Laboratory, Bedford Institute of Oceanography, Dartmouth, Nova Scotia, Canada

R. O. REID, Department of Oceanography, Texas A & M University, College Station, Texas

PETER B. RHINES, Woods Hole Oceanographic Institution, Woods Hole, Massachusetts 02543

A. S. SARKISYAN, P. P. Shirshov Institute of Oceanology, Academy of Sciences of the USSR, Moscow, USSR

J. DUNGAN SMITH, Department of Oceanography, University of Washington, Seattle, Washington 98195

JOHN H. STEELE, Department of Agriculture and Fisheries for Scotland, Marine Laboratory, Aberdeen, Scotland

P. A. TAYLOR, Department of Oceanography, University of Southampton, England

ROBERT V. THOMANN, Environmental Engineering and Science Program, Manhattan College, New York, New York 10471; and Hydroscience, Inc., Westwood, New Jersey

A. C. Vastano, Department of Oceanography, Texas A & M University, College Station, Texas

George Veronis, Department of Geology and Geophysics, Yale University, New Haven, Connecticut

M. E. Vinogradov, Institute of Oceanology, Moscow, USSR

J. J. Walsh, Division of Oceanographic Sciences, Brookhaven National Laboratory, Upton, New York 11973

J. J. Wanstrath, Department of Oceanography, Texas A & M University, College Station, Texas

R. E. Whitaker, Department of Oceanography, Texas A & M University, College Station, Texas

Roland Wollast, Institute of Industrial Chemistry, University of Brussels, Brussels, Belgium

DEDICATION

This volume is dedicated to Bostwick H. Ketchum, one of the earlier and more successful modelers of both physical and biological systems. A student of A. C. Redfield, Buck Ketchum has always shared his mentor's curiosity and intuitive understanding of the relationships between marine organisms and their physical and chemical environment. Although much of his lifetime devotion to problems of plankton ecology has not involved modeling per se, his approach to the subject as a dynamic process involving the physical and biochemical cycling of nutrients has provided both the intellectual and factual basis for subsequent synthesis by the modeling fraternity. More than anyone else, he was responsible for the development of a large and diversified group of plankton ecologists at the Woods Hole Oceanographic Institution, a team that has over the past 30 years gathered many of the basic data leading to our present knowledge of the ecology and population dynamics of plankton communities. The pioneer works in plankton modeling by G. A. Riley and later by J. H. Steele have a substantial part of their roots in the earlier Woods Hole plankton studies of Ketchum and his colleagues.

In his studies during the early 1950's of the fate of enteric bacteria in waste water discharged to the sea, Ketchum was among the first to recognize that both biological factors (predation and mortality) and physical processes (advection and dispersion) must be considered. These concepts were subsequently extended to other pollutants, including radionuclides, and have provided a basis for much of the current work on the distribution and fate of wastes in the marine environment.

Perhaps one of the most neglected areas of physical oceanography is that dealing with the circulation and flushing of estuaries. Most of the definitive work in this area was done some 20 years ago by D. W. Pritchard, working in the James River estuary, and K. F. Bowden, in the Mersey River estuary, together with a handful of others. It is interesting that Buck Ketchum, a biologist, figures prominently in this group. As a natural outgrowth of his work on the dispersion of pollutants and the behavior of plankton populations, Ketchum developed one of the early models of estuarine flushing, which he then applied to studies of the Raritan River, Passamaquoddy Bay, Barnstable Harbor, the Hudson River, and other locales in the northeastern United States. Although lacking in the mathematical sophistication of later efforts, Ketchum's model and the information obtained through its application are still both valid and useful.

Extrapolation of his study of river mouths and harbors to coastal waters led Ketchum to the concept of the entire continental shelf as an estuarine system, and to the calculation of rates and annual cycles of mixing and flushing of its accumulated river water. The numerical treatment of this phenomenon is one of the factors that has influenced modern advances in coastal circulation studies and has contributed to the evolution from rather vague concepts of such processes as coastal upwelling to the more refined quantitative approaches of Dugdale, Walsh, and others.

JOHN RYTHER

Woods Hole Oceanographic Institution
December 1976

PREFACE

The recent expansive growth in oceanography has stimulated the formation of strong subdisciplinary groups—biological oceanography, marine chemistry, ecology, physical oceanography, and geophysics, among others. At the larger oceanographic institutions, these groups are often housed separately. They have their own journals and produce their own reference and textbooks. Many small departments, particularly in the biological sciences, do marine work, but not within a broad oceanographic context. As a consequence oceanography, as practiced in many places, has lost much of its interdisciplinary flavor. The oceanographic colloquium that attracts an audience from most of the subdisciplines has become a rarity. Students associate themselves with the subdiscipline rather than the major field.

There are two strong counteracting forces to this tendency toward overspecialization in oceanography. The first has been the birth and survival of small oceanographic centers in which there are usually no more than several members of each subdiscipline. At such places the mood of oceanography as it was practiced a quarter of a century ago persists. Chemists talk to and listen to biologists. The seminar by the physicist attracts the geologists.

The development of computer modeling of marine systems has been a second force to counteract the trend from the generalist to the specialist approach in the study of the oceans. Although the first simulations of natural situations involved only the parameters of a given subdiscipline, the rapid progress in computer science has allowed more complex problems to be attacked, and these more complex problems often involve inputs from a variety of subdisciplines. The prediction of climatic trends requires data from meteorologists, physical oceanographers, chemists, and geologists to formulate mathematical models. Models of biological productivity may demand similar types of information as well as data on organisms and their interactions. The construction of some models by oceanographers cannot help but be interdisciplinary for all but the simplest systems. Perhaps the production of this volume is an indication of how a group of specialists sought to expose their colleagues to the possibilities of more general approaches to marine problems.

Still, many of the presentations are within the scope of a single subdiscipline. For coherence, they have been grouped within four categories: physics, chemistry, biology, and geology. Perhaps there should have been a fifth category in applied oceanography or ocean engineering for which we have one contribution. The largest section involves the physical aspects of modeling, a not unexpected situation. The physicists usually have a greater facility with mathematics than their colleagues in other subdisciplines of oceanography. As a consequence they have recognized earlier, and with greater dedication to their use, the tools of mathematic modeling. The numbers of contributions in the other areas reflect most probably the activity of computer modeling in the area as well as the abilities of the editor in the subdiscipline to collect manuscripts.

The idea of the volume came about at a Conference on Modeling of Marine Systems held in Ofir, Portugal in June 1973 under the auspices of the NATO Science Committee. All four editors were in attendance. There was a clear recognition of the substantial advances in understanding the dynamics of ocean systems as well as in the ability to pose soluble problems previously unattackable, through the use of computers. Whereas the conference provided a state-of-the-art assessment of modeling (*Modeling of Marine Systems*, Jacques C. J. Nihoul, ed., Elsevier, 1975, 272 pp.), the guiding concept of this volume is an elaboration of those ideas and techniques that will allow important advances in our knowledge of the ocean environment. This volume is intended, as were its predecessors, to provide a springboard for research in this field of marine science.

THE EDITORS

CONTENTS

MARINE MODELING

Part III.   Chemistry

Part IV.   Biology

Marine Modeling

I. PHYSICS

1. OCEANIC GENERAL CIRCULATION MODELS

WILLIAM R. HOLLAND

1. Introduction

Despite substantial progress during the past few decades in understanding the large-scale ocean circulation, the description of the general circulation of the ocean is far from complete. There are several reasons for this. First, the observational basis for constructing models is wanting, and new observations are continually modifying our choice about which physical phenomena are important in the large-scale circulation. For example, recent field programs (e.g., MODE) have suggested that horizontal eddy mixing due to mesoscale eddies is likely to be important to the general circulation. In addition, we do not know observationally the detailed behavior of heat and momentum fluxes into the ocean (the boundary conditions) for all relevant space and time scales. A second reason why progress has been slow is that although the physical laws governing the system (the Navier–Stokes equations) are well known, valid approximations necessary for model studies are still unknown. Much of our understanding of large-scale currents has come from analytical studies of highly idealized model oceans. These have had simple forcing and simple geometry and usually have included simple turbulent viscosity hypotheses that are at best a temporary expediency. A third reason is that even a minimum set of physical laws necessary for a valid description of the oceanic circulation is quite complicated. It is apparent that advection of momentum, heat, and salt is important, and that as a result the problem is inherently nonlinear, at least in certain regions of the flow.

These difficulties in constructing a valid picture of the oceanic circulation have led to the development of numerical models that are an extension of the earlier analytic theories. Such models are capable of including a complexity in terms of relevant physical laws and processes that is not possible in a purely analytical approach. In addition, these models allow the investigator to isolate a few physical processes and determine, one by one, the influence of various factors on results. This is the rationale for the construction of numerical general circulation models at present. Such models are beginning to play an important part in planning field programs as well as in testing hypotheses about ocean dynamics. Eventually when models with correct dynamical processes are developed as shown by comparison of results with observation, they will also be useful for prediction purposes, climate studies, and studies of the fate of man-made and natural pollutants in the ocean.

What do we mean by the "general circulation" of the ocean? Lorentz (1967) has discussed this question for the atmosphere, and much of his discussion is directly applicable to the oceans as well. In this chapter we are primarily concerned with the large-scale, time-averaged, three-dimensional structure of ocean currents and the related temperature and salinity fields. We discuss the attempts with both simple and

3

TABLE I

Some Numerical Studies of Ocean Circulation

Study	Forcing	Density	Dissipation
A. Early ocean models			
Bryan (1963)	Wind	Homogeneous	Lateral friction
Veronis (1966a, b)	Wind	Homogeneous	Bottom friction
Blandford (1971)	Wind	Homogeneous	Lateral and bottom friction
Beardsley (1973)	Wind	Homogeneous	Lateral and bottom friction
Bryan and Cox (1967)	Wind + thermohaline	Baroclinic	Lateral friction
Bryan and Cox (1968a, b)	Wind + thermohaline	Baroclinic	Lateral friction
B. Geographical and process studies			
Holland (1967)	Wind	Homogeneous	Lateral friction
Schulman and Niiler (1970)	Wind	Homogeneous	Bottom friction
Holland and Hirschman (1972)	Wind + thermohaline	Baroclinic (diagnostic)	Lateral friction
Holland (1973)	Wind + thermohaline	Baroclinic	Lateral friction
Gill and Bryan (1971)	Wind + thermohaline	Baroclinic	Lateral friction
Holland (1975b)	Wind + thermohaline	Baroclinic	Lateral friction
C. Realistic basin and global ocean studies			
Friedrich (1970)	Wind + thermohaline	Baroclinic	Lateral and bottom friction
Cox (1970)	Wind + thermohaline	Baroclinic	Lateral friction
Semtner (1973)	Wind + thermohaline	Baroclinic	Lateral friction
Bryan and Cox (1972)	Wind	Homogeneous	Lateral friction
Cox (1975)	Wind + thermohaline	Baroclinic	Lateral friction
D. Eddy-resolving general circulation models			
Holland and Lin (1975a, b)	Wind	Baroclinic	Lateral friction

comprehensive numerical models to unravel the essential physical laws that govern this mean circulation. In this regard we are largely concerned with closed basins in which complete vorticity and energetic balances can be (but often have not been) discussed. We examine studies with coarse resolution as well as studies that make use of fine resolution to study physical processes in important geographical regions such as the equatorial and western boundary regions that are needed for quantitative prediction of the large-scale circulation.

Limited area models of transient currents and diagnostic models of large-scale currents (in which the density field is prescribed by observations) are considered in other chapters in this volume (Chapters 7 and 9, respectively). Therefore, although there has been a natural important feedback between these model results and results from the prognostic, large-scale models that are considered here, only passing reference to them is made.

In the sections that follow we briefly discuss the early analytical theories of the wind-driven ocean circulation which set the stage for the development of numerical ocean models. Then numerical modeling results are described. Several ways of organizing the numerous studies are possible. They could be classified according to increasing complexity of either physical laws (i.e., homogeneous to baroclinic) or geometry (i.e., rectangular basin with constant depth to global oceans with realistic topography), or they could be arranged chronologically. None of these approaches is strictly followed; instead we try to show how each study developed out of earlier ideas and related to previous work. Therefore, in Section 3 the early numerical studies of homogeneous oceans which were direct extensions of analytical work are discussed, and then the first attempts at calculating the three-dimensional, baroclinic structure of the circulation are described. The studies included there are listed in Table IA. Following that, in Section 4, various studies of physical processes in closed basins (effects of bottom topography, baroclinity, and wind and thermohaline driving) and of special geographical regions (the Southern Ocean, the Equatorial Region) are described. These studies are summarized in Table IB. Finally, in Section 5 the recent results of numerical experiments in realistic ocean basins (the North Atlantic, the Indian Ocean) and the results from global calculations with ocean models and with joint ocean–atmosphere models are presented. Table IC shows these works. In Section 6 very recent results from eddy-resolving general circulation models are presented to suggest that the earlier parameterizations of horizontal mixing of heat, salt, and momentum are likely to be incorrect in the light of recent observations, and in fact a major physical process in the oceanic general circulation, mesoscale eddies, must be taken into account properly.

2. Early Analytical Models

A brief review of some of the analytical studies that laid the foundations for our understanding of the large-scale wind-driven ocean circulation is undertaken here. A more comprehensive review of the formulation of these mathematical models and their relationship to observations can be found in Stommel's *The Gulf Stream* (1965). A collection of reprinted papers on the wind-driven circulation has been published (Robinson, 1963). In addition, there are several important review articles on the theoretical aspects of the dynamics of ocean circulation. These are contained in Stommel (1957), Fofonoff (1962), Robinson (1965), Veronis (1973), Bryan (1975), and Welander (1975).

Analytical models of ocean circulation are most easily discussed when separated into two categories, wind-driven (horizontal mass transport) theories and thermocline (interior vertical structure) theories. This division is somewhat arbitrary and, in fact, it has been one of the tasks of numerical models to bring these two aspects of the oceanic circulation into a single framework. There is clearly an important (not yet understood) coupling of these problems as evidenced, for example, by the numerical studies on the joint effect of bottom topography and baroclinity discussed below.

One class of analytical, wind-driven ocean models is concerned with the horizontal transport in a homogeneous, rectangular ocean basin of constant depth. When lateral friction is included by a simple eddy viscosity hypothesis (Munk, 1950), and vertical friction is included by means of Ekman layers at the top and bottom, a single vorticity equation governing the mass transport stream function ψ includes virtually all the models and encompasses all the results of the early steady, wind-driven theories. That equation is

$$\frac{1}{\rho H} \frac{\partial(\psi, \nabla^2 \psi)}{\partial(x, y)} + \beta \psi_x = \text{curl}_z \, \tau - \frac{1}{H} \left(\frac{K_m f}{2}\right)^{1/2} \nabla^2 \psi + A_m \nabla^4 \psi \tag{1}$$

where x and y are the eastward and northward coordinates, β is the northward gradient of the Coriolis parameter f, H is the depth, and τ is the wind stress. The vertical and horizontal coefficients of eddy viscosity are K_m and A_m, respectively. This equation can be written, with some simplifications, in the nondimensional form,

$$Ro \frac{\partial(\psi', \nabla^2 \psi')}{\partial(x', y')} + \psi'_x = \text{curl}_z \, \tau' - Ek_v \nabla^2 \psi' + Ek_h \nabla^4 \psi' \tag{2}$$

where Ro is the Rossby number, and Ek_v and Ek_h are the vertical and horizontal Ekman numbers. In some studies the Reynolds number, $Re = Ro/Ek_h$, is introduced. This nondimensionalization is accomplished by the scaling $(x, y) = L(x', y')$ and $\psi = T_0 \psi'/\beta$ where L is the length scale of the basin and T_0 is the amplitude of the wind stress. Then

$$Ro = \frac{T_0}{\rho H \beta^2 L^3} \tag{3}$$

$$Ek_v = \left(\frac{K_m f}{2}\right)^{1/2} (\beta H L) \tag{4}$$

$$Ek_h = \frac{A_m}{\beta L^3} \tag{5}$$

The classical work of Ekman (1905) showed how the wind stress, acting at the sea surface, could put momentum into the ocean in a thin boundary layer at the sea surface and how bottom friction could remove it in a thin layer at the bottom. The theory is simple, and the vertically integrated effects of these *Ekman layers* are included in equation 1 as the first and second terms on the right-hand side. A more elaborate treatment of the surface mixed layer is probably necessary to reproduce details of the vertical momentum flux; in the simple Ekman theory the influence of stratification is ignored, and a simple prescription of the vertical mixing process is chosen. At this time much observational and theoretical work is underway on this problem. Here, though, we stick to the simple Ekman model that has been used in nearly all of the numerical models, but it is clear that a proper parameterization of the vertical momentum (and heat) flux near the sea surface will have to be incorporated into future large-scale ocean models.

Sverdrup (1947) proposed that a simple balance in the vorticity equation 1 between the curl of the wind stress and the β effect (due to the southward transport of planetary vorticity) would be applicable in the central ocean far from coasts. There the inertial terms, lateral friction, and bottom friction would be negligible. This simple but far-reaching idea has been implicit in almost all analytical models of the wind-driven and thermocline circulations. Sverdrup found a reasonable agreement between theory and observations in the eastern tropical Pacific, but the idea has not been critically tested. In fact, there has been a long-standing discrepancy between observations of Gulf Stream transport and the interior transport calculated from mean wind-stress values by Sverdrup's technique. This lack of agreement may be caused in two ways (at least): (1) the wind-stress observations are in error (the Sverdrup balance is in fact correct); (2) the Sverdrup balance is in error due to important inertial recirculations (Veronis, 1966b), bottom pressure torques (Holland, 1973), or transient eddies acting in the interior (Holland and Lin, 1975a, b). As we shall see, numerical models have begun to examine these possibilities but definitive conclusions have not yet been reached.

Stommel (1948) showed that, if a simple bottom friction was added to Sverdrup's balance, a closed basin circulation could be achieved. Moreover, this model successfully showed that westward intensification was due to the β effect and thus provided a qualitative explanation for the presence of western boundary currents. For the purposes of this chapter, this model is also symbolically important as the first oceanic "general circulation model" (GCM); that is, it has closed momentum, energy, and vorticity budgets, including both interior and boundary flows. It is the prototype for all future GCM's, including the numerical ones. It should be emphasized here that most closed basin circulation models, both analytical and numerical, have computed the steady circulation resulting from steady driving. Energy and vorticity budgets in such cases can help enormously in sorting out the relevant physical processes going on (see Holland, 1973, 1975a). In experiments in which transient processes due to instabilities of the flow are explicitly included in general circulation calculations (Holland and Lin, 1975a, b) it is even more important to understand fully the way in which energy and vorticity are put in, redistributed, and lost in various regions of the flow field.

Munk (1950) suggested that lateral friction, rather than bottom friction, was the relevant dissipative mechanism in the ocean (last term on right-hand side of equation 1). The advective term was neglected as in previous studies. This work, like that of Sverdrup, dealt with vertical averages of equations applicable to a general stratified ocean. With the assumption that the deep ocean is motionless, however, and the non-linear terms unimportant, the transport equation is identical to that for a homogeneous, constant-depth ocean. The results of Munk's study are very similar to Stommel's (Figs. 1a, 1b). However, the boundary conditions result in a fundamental difference in the way vorticity is lost from the basin. In Stommel's case it is lost through the bottom, largely in the intense western boundary current. In Munk's case the vorticity must diffuse to the lateral boundaries, mainly the western one. Although this results in only slight differences in the flow patterns in the linear models, rather important differences occur when inertial effects and nonsteady (unstable) cases are examined. This point is discussed below.

It was realized early that the inertial terms could not be ignored in the region of the western boundary current. Munk et al. (1950) examined a slightly nonlinear extension of the Munk (1950) model by a perturbation technique and showed that the nonlinear terms tended to concentrate the boundary current into the northwest

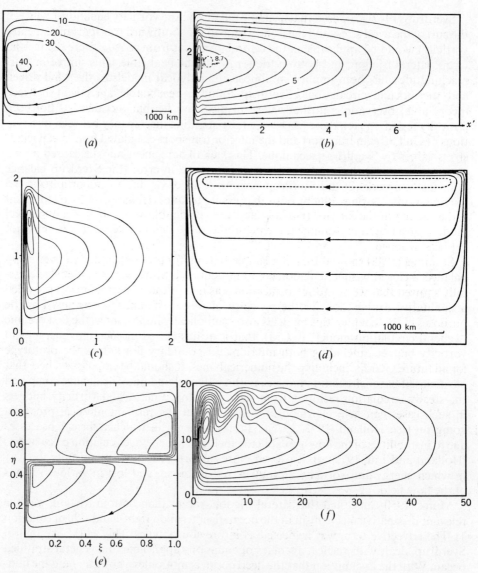

Fig. 1. Analytic models of the wind-driven ocean circulation. Patterns of the horizontal mass transport streamfunction from (a) Stommel (1948); (b) Munk (1950); (c) Munk, Groves, and Carrier (1950); (d) Fofonoff (1952); (e) Carrier and Robinson (1962); and (f) Moore (1963).

corner (Fig. 1c). There was no tendency for the flow to separate near the maximum of the wind-stress curl. Fofonoff (1954) ignored the wind-driving and the friction terms in equation 1 and found a class of solutions in closed basins (Fig. 1d). In this case the β effect produces a north–south asymmetry rather than an east–west one. The solutions are not unique and an eastward jet can be put at any latitude. The fundamental questions about how energy and vorticity are put into and taken out of the ocean are not addressed by this model. Numerical experiments by Veronis (1966b) and an analysis by Niiler (1966) help answer these questions.

The nature of an inertial western boundary layer has been examined by Charney (1955) and Morgan (1956), but these studies are valid only for the formation region of the current. Dissipation and wind driving are assumed to occur elsewhere in the ocean. Carrier and Robinson (1962) tried to extend the inertial case to the forced, small viscosity limit. They suggested that the western boundary current should separate at the latitude of maximum wind-stress curl (Fig. 1e). However, as Welander (1975) points out, the problem again seems nonunique, for the eastward jet can be placed at any latitude since the interior flow is everywhere westward as in Fofonoff's result. Moore (1963) made an approximate analytical treatment of the problem and suggested that the flow in the northern half of the basin need not have a boundary layer character. His result suggests a standing Rossby wave pattern in the northern half of the basin (Fig. 1f).

The difficulties of analytical treatment of inertial models of the wind-driven circulation, in particular the difficulty in piecing together the various dynamical regions to give a coherent picture of the general circulation, illustrates the way in which numerical models can support and extend analytical work. Section 3 describes the work aimed at settling some of the questions raised by these analytical treatments of the nonlinear wind-driven problem.

The thermocline theories have had, in a certain sense, a similar history to the wind-driven theories. Linear models of an ocean with thermohaline forcing were first solved with vertical friction terms only (Lineykin, 1955; Stommel and Veronis, 1957) and then extended to include horizontal diffusion of momentum and heat (Pedlosky, 1969). However, these models were unrealistic in that either the bottom boundary condition of no heat flux could not be met or statically unstable regions had to occur. In addition, they were linear in the heat equation, which by scale estimates can be shown to be incorrect for oceanic motions. When advection terms were included in the heat equation (Robinson and Stommel, 1959; Welander, 1959a; Robinson and Welander, 1963; Blandford, 1965; Welander, 1971), the problem became inherently nonlinear and only some quite special solutions have been found. It has not been possible yet to piece together the various parts of the flow (including boundary layers), and it has been quite difficult to decide how applicable these studies are to actual ocean circulations. [See Welander (1975) for a discussion of the difficulties encountered in trying to generalize the thermocline theories.]

As with the wind-driven theories, numerical modeling could be quite helpful in sorting out the various thermohaline regimes, but as yet, in contrast to the horizontal transport problem, little has been done. Bryan and Cox (1967) compared results from a numerical model with an exact solution found by Blandford (1965). The heat balances in the interior subtropical gyre are quite similar, lending strength to the thesis that numerical models can be helpful in furthering thermocline theory.

3. Early Numerical Models

There are many important contributions by numerical modelers to understanding the dynamics of the ocean. It has not been possible to include all these; a few numerical results have been chosen to present the thread of the development and some of the important physical results. The selection has been restricted to those results designed to address the general circulation question, and I have used those papers with which I am most familiar or about which I have special knowledge. This is by no means intended to be a balanced or unbiased critique of the many excellent works in the field.

A. Homogeneous Ocean Models

Several numerical studies have been carried out (Table IA) that are direct extensions of the analytical wind-driven circulation models discussed in the preceding section. It was pointed out that the development of inertial general circulation models presented serious difficulties that were not overcome until Bryan (1963) and later Veronis (1966a, b) examined the problem numerically. Their results showed the power and usefulness of even simple numerical models and presented new problems to be solved by the analytical worker (Niiler, 1966).

Bryan's study was carried out to determine how the linear solution of Munk (1950) was altered by the inclusion of the inertial terms (equation 1 without bottom friction).

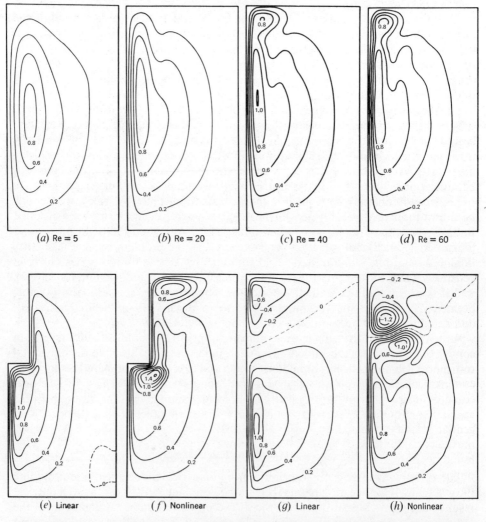

(a) Re = 5   (b) Re = 20   (c) Re = 40   (d) Re = 60

(e) Linear   (f) Nonlinear   (g) Linear   (h) Nonlinear

Fig. 2.   Numerical experiments on the wind-driven ocean circulation by Bryan (1963). (a)–(d) The effect on the mass transport streamfunction of decreasing the horizontal Ekman number while keeping the Rossby number fixed. (e) and (f) The effect on the transport pattern of a change in the direction in the western wall. (g) and (h) The effect on the transport pattern due to a more complex wind-stress distribution.

That question stemmed directly from the proposal by Carrier and Robinson (1962) that the western boundary current would leave the western boundary at the maximum of the wind-stress curl (Fig. 1e), whereas the perturbation approach of Munk, Groves, and Carrier (1950) suggested that the circulation would intensify toward the northern boundary of the gyre (Fig. 1c). These results were connected to the question of the mechanism by which the western boundary current in actual ocean basins leaves the coast to turn eastward as an intense eastward flowing current.

Some of Bryan's results are shown in Fig. 2. The solutions in a rectangular basin with simple wind-stress distribution are shown (Fig. 2a–d) as a function of decreasing lateral viscosity. The Ekman number was decreased while the Rossby number was held fixed. The important result was found that the boundary current intensified near the northwest corner and the boundary current tended to extend all the way to the northern boundary, supporting the Munk, Groves, and Carrier (1950) result rather than the Carrier and Robinson (1962) one. In cases where the current reached the northern boundary, a northern boundary current tended to form but soon turned southward. Bryan also explored the effects of irregular lateral boundaries (Fig. 2e and f) and more complex wind-stress distributions (Fig. 2g and h) to show that in more inertial cases sharp changes of direction in the lateral boundary did not cause boundary current separation and that the current tended to leave shore only near the zero of the wind-stress curl. Finally Bryan found that the more nonlinear cases were inherently time dependent owing (apparently) to a barotropic instability in the boundary current. This effect, and the possibility of mixed barotropic–baroclinic instabilities in baroclinic models (Orlanski and Cox, 1973) may in fact be an important oceanic process that has been largely ignored until recently.

The numerical study by Veronis (1966a, b) complements that of Bryan. Veronis used bottom friction as the dissipative mechanism rather than lateral friction and was able to explore rather more nonlinear effects. In fact the results (Fig. 3) were able to trace the connection between Stommel's linear model and Fofonoff's inertial one. With Niiler's (1966) analysis, the piecing together of this simplest of highly inertial general circulation models was nearly complete.

The Veronis results, however, raised another question: what are the proper boundary conditions for general circulation calculations? Bryan used no-slip conditions on the western and eastern boundaries (the tangential velocity is zero there), whereas the bottom friction model requires no condition on the tangential velocity component at the boundary. Stewart (1964) addressed this question and pointed out that when the tangential velocity vanishes at the boundary (the no-slip case) eastern boundary currents, such as those found in the Fofonoff and Veronis cases, could not exist. This result is based upon the fact that the average relative vorticity and the transport of relative vorticity vanish in a boundary layer in which the tangential velocity vanishes on each side. Because the shear stress generated by the movement of the fluid along the boundary must oppose the movement, Stewart (1964) found, a current moving with a boundary on its right can generate at the boundary only clockwise vorticity, and a current moving with a boundary on its left can generate only counterclockwise vorticity. Therefore a boundary current flowing northward, with its requirement that there be a source of counterclockwise vorticity (to match the increasing planetary vorticity), must be on the western side of the basin. By the same argument a southward flowing boundary current must also be on the western side of the basin. Note that a northern boundary current with no-slip conditions is also excluded by this argument. Since the boundary supplies some counterclockwise

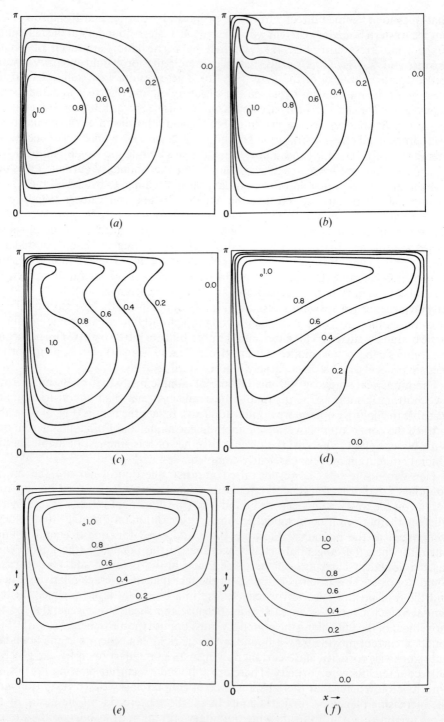

Fig. 3.   Numerical experiments on the wind-driven ocean circulation by Veronis (1966b). The respective Rossby and vertical Ekman numbers for each case are (a) $Ro = 1 \times 10^{-6}$, $Ek_v = 0.05$; (b) $Ro = 2.5 \times 10^{-3}$, $Ek_v = 0.05$; (c) $Ro = 1 \times 10^{-2}$, $Ek_v = 0.05$; (d) $Ro = 1 \times 10^{-2}$, $Ek_v = 0.025$; (e) $Ro = 1.2 \times 10^{-2}$, $Ek_v = 0.018$; (f) $Ro = 4 \times 10^{-2}$, $Ek_v = 0.025$.

vorticity to the current that is not balanced by a planetary vorticity tendency, that current also cannot exist.

Blandford (1971) performed a series of experiments to illustrate this argument and to clarify this important difference between the Bryan (1963) and Veronis (1966a, b) models. Indeed these results are of crucial importance in understanding all the numerical studies that followed. Figure 4 shows some of Blandford's results. Figure 4a shows the transport pattern without lateral viscosity (i.e., $A_m = 0$ in equation 1). It is analogous to Veronis' results. Figure 4b shows the transport pattern including both bottom and lateral friction but with slip (stress free) boundary conditions. Finally, Fig. 4c shows the transport pattern results when no-slip boundary conditions are imposed on all boundaries. The northwest corner is highly transient in this last case. [Note that Bryan (1963) imposed no-slip conditions on the eastern and western walls but slip conditions on the northern and southern walls. Thus he *could* have a northern boundary current in his results.]

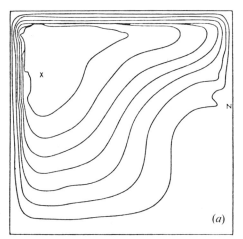

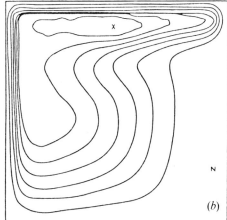

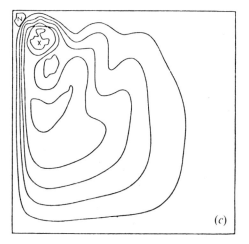

Fig. 4.   Numerical experiments on the wind-driven ocean circulation by Blandford (1971). The cases are (a) no lateral viscosity, (b) small lateral viscosity with slip boundary conditions, and (c) small lateral viscosity with no-slip boundary conditions. Case (c) is highly transient while (a) and (b) are steady.

These results show the importance of Stewart's argument. It is not only the nature of the frictional dissipation (i.e., lateral versus bottom friction) that is important in these cases but also the nature of the boundary conditions (slip versus no-slip). This factor must be taken into account in trying to explain results in more complex numerical general circulation calculations.

It should be pointed out that Blandford's results depend somewhat on the fact that bottom friction dominates lateral friction. As Niiler (1966) has discussed, if there were no bottom friction in a strongly inertial case with slip boundary conditions, transient behavior should be expected. This is because the vorticity induced by the wind must be diffused outward across every bounding streamline. However, if the interior is bounded by an inertial boundary current region, the vorticity gradient is the wrong sign to allow vorticity to diffuse to the boundary. Thus every streamline must pass through a viscous region which, it would seem, can be accomplished only by the presence of transient eddies and their accompanying Reynolds stresses.

In laboratory and numerical experiments in a sliced cylinder, Beardsley (1969, 1973) showed that the instability found by the numerical models is indeed physical and that the numerical results quite accurately reproduce that behavior. Therefore numerical modelers can have some confidence (an especially important ingredient to numerical modeling activities!) that future, more complex situations in which transient eddies spontaneously arise can be handled by numerical techniques as long as sufficient attention is paid to horizontal resolution requirements.

B. Baroclinic Ocean Models

The pioneering work in development of prognostic, wind- and thermohaline-driven ocean circulation models has been carried out in the Soviet Union by Sarkisyan (1962) and in the United States by Bryan and Cox (1967). These models seek to reproduce the three-dimensional density distribution, that is, the temperature and salinity distributions, as well as the three-dimensional flow field within the basin. The model equations are based upon the Navier–Stokes equations and conservation equations for heat and salt suitably averaged over the grid scale in space and a corresponding time interval. These averaged equations then contain Reynolds stress terms which account for a diffusion of momentum, heat, and salt by small-scale processes not directly included in the model. In the calculations to be discussed, a simple turbulent viscosity hypothesis has been used. For momentum mixing, it is identical to that of Munk (1950). A similar assumption is made for eddy mixing of heat and salt.

A correct representation (or closure hypothesis) for these eddy fluxes remains one of the major roadblocks to further progress in ocean modeling. The work described in Section 6 represents a beginning at *explicitly* including some of the small-scale (mesoscale) effects that the eddy parameterization is supposed to represent. The results there suggest that fluxes of momentum, heat, and salt proportional to the gradient (and down gradient) are not an adequate parameterization, and in fact it is not clear that such a parameterization can even be found. If that turns out to be the case, general circulation calculations will have to explicitly include mesoscale phenomena to achieve valid results.

Our discussion is based upon the ocean model developed by Bryan and co-workers (see Bryan, 1969). Most prognostic, three-dimensional models contain similar physics, and in fact many of the numerical experiments to be discussed have been carried out with variants of that basic model. The equations of motion are based upon two assumptions in addition to the eddy viscosity hypothesis already mentioned: (a) the Boussinesq approximation is made in which density variations are ignored except in buoyancy terms; and (b) a hydrostatic balance is assumed. These approximations are valid for large-scale oceanic motions but, as we shall see, certain aspects of the circulation (e.g., convective overturning) must be treated implicitly because they are inherently small-scale.

Let $\mathbf{V}$ be the horizontal velocity vector (with eastward and northward components u and v) and let w be the vertical velocity. Then the equations of motion and continuity are

$$\mathbf{V}_t + \mathbf{V} \cdot \nabla V + w \mathbf{V}_z + \mathbf{f} \times \mathbf{V} = -\frac{1}{\rho_0} \nabla p + \mathbf{F} + K_m \mathbf{V}_{zz} \tag{6}$$

$$\rho g = -p_z \tag{7}$$

$$\nabla \cdot \mathbf{V} + w_z = 0 \tag{8}$$

Here $\mathbf{f}$ is the vertical component of the Coriolis parameter (equal to $2\omega \sin \phi \ \mathbf{k}$, where $\mathbf{k}$ is a unit vector in the vertical direction z), p is the pressure, and ρ is the density. K_m is the vertical coefficient of eddy viscosity and $\mathbf{F}$ is a horizontal body force caused by lateral friction. In Bryan's model, the latitudinal and longitudinal components of $\mathbf{F}$ are given by

$$F^\lambda = A_m \left[\nabla^2 u + \frac{(1 - \tan^2 \phi)}{a^2} u - \frac{2 \tan \phi}{a^2 \cos \phi} v_\lambda \right] \tag{9}$$

$$F^\phi = A_m \left[\nabla^2 v + \frac{(1 - \tan^2 \phi)}{a^2} v + \frac{2 \tan \phi}{a^2 \cos \phi} u_\lambda \right] \tag{10}$$

where λ and ϕ are longitude and latitude, a is the radius of the earth, and A_m is the coefficient of lateral mixing of momentum. These terms represent the simple closure hypothesis which parameterizes the effects of Reynolds stresses due to scales of motion (subgrid scale) not explicitly included in the model.

The equation of state relating density to potential temperature θ, salinity S, and pressure is complicated, but can be expressed by various analytic approximations or in tables. In general

$$\rho = \rho(\theta, S, p) \tag{11}$$

Equations for the conservation of heat and salt are

$$(\theta, S)_t + \mathbf{V} \cdot \nabla(\theta, S) + w(\theta, S)_z = A_h \nabla^2(\theta, S) + K_h(\theta, S)_{zz} \tag{12}$$

where it has been assumed that the eddy mixing coefficients A_h and K_h are constant.

In many studies of the dynamics of ocean circulation, a simple equation of state (equation 11) has been used; that is,

$$\rho = \rho_0(1 - \alpha T) \tag{13}$$

where T is an apparent temperature and α is the coefficient of thermal expansion. T is then calculated from equation 12; that is,

$$T_t + \mathbf{V} \cdot \nabla T + w T_z = A_h \nabla^2 T + K_h T_{zz} \tag{14}$$

The appropriate surface boundary conditions for T are then either that the total buoyancy flux is given (T_z given), or that values of T based upon distributions of θ and S are given (T given), or some mix of these two conditions.

In the two-dimensional analytical and numerical studies of a homogeneous ocean it was shown that three nondimensional numbers were needed to describe the flow, the Rossby number, and the horizontal and vertical Ekman numbers. Here the set of equations 6–10 and 14 can be put in dimensionless form (following Bryan, 1975) by letting

$$(x, y) = L(x', y')$$ (15)

$$z = Dz'$$ (16)

$$t = \frac{L}{V^*} t'$$ (17)

$$p = (2\Omega V^* L \rho_0) p'$$ (18)

$$T = \frac{2\Omega V^* L}{g\alpha D} T'$$ (19)

$$w = \frac{V^* D}{L} w'$$ (20)

$$\mathbf{V} = V^* \mathbf{V}'$$ (21)

Here x and y are the eastward and northward coordinates, equal to $a(\lambda - \lambda_0) \cos \phi$ and $a(\phi - \phi_0)$, respectively, L and D are the horizontal and vertical length scales of the basin under consideration, and V^* is the velocity scale defined in terms of the surface boundary conditions. Then

$$Ro[\mathbf{V}'_t + \mathbf{V}' \cdot \nabla \mathbf{V}' + w' V'_z] + \sin \theta\, k x \mathbf{V}' = -\nabla p' + Ek_h \mathbf{F} + Ek_v V'_{zz}$$ (22)

$$T' = p'_z$$ (23)

$$\nabla \cdot \mathbf{V}' + w' = 0$$ (24)

$$T'_t + \mathbf{V} \cdot \nabla T' + w' T'_z = Pe_h^{-1} \nabla^2 T' + Pe_v^{-1} T'_{zz}$$ (25)

where

$$Ro = \frac{V^*}{2\Omega L}$$ (26)

$$Ek_h, Ek_v = \frac{A_m}{2\Omega L^2}, \frac{K_m}{2\Omega D^2}$$ (27)

$$Pe_h^{-1}, Pe_v^{-1} = \frac{A_h}{V^* L}, \frac{K_h}{V^* L}$$ (28)

Here Ro is the Rossby number, Ek_h and Ek_v are the horizontal and vertical Ekman numbers, and Pe_h and Pe_v are the horizontal and vertical Peclet numbers.

These nondimensional numbers can be related to four other nondimensional numbers that are associated with the width of the western boundary layer, $L_f/L = (Ek_h)^{1/3}$, $L_i/L = Ro^{1/2}$, $L_s/L = Ek_v^{1/2}$, and $L_m/L = Pe_h^{-1}$. These correspond to boundary layers in which there are important effects caused by lateral friction (Munk, 1950), inertial effects (Fofonoff, 1954), bottom friction (Stommel, 1948), and lateral diffusion of density (Bryan and Cox, 1968), respectively.

The vertical Ekman and Peclet numbers are measures of the depth to which the surface boundary conditions on wind stress and apparent temperature are felt for simple linear balances. The ratio of the Ekman layer depth D_e to the depth of the fluid is given by $Ek_v^{1/2}$ (Ekman, 1905), and the ratio of the thermocline depth D_t to the total depth is given by $Pe_v^{-1}L^2/D^2$ (Munk, 1966).

Observations can be used to establish order of magnitude estimates of the values of these various nondimensional numbers and length scales for the ocean. Typical basin values, $L = 5000$ km and $D = 5000$ m, are used together with (a) an estimate of the depth of the bottom Ekman layer [according to Wimbush and Munk (1970) the equivalent bottom Ekman depth is a few meters], (b) an estimate of the lateral diffusion coefficient [$A_m < 10^6$ cm^2/sec (Stommel, 1955)], (c) an estimate of the width of the western boundary current (about 50 km), and (d) an estimate of the depth of the thermocline ($D_t = 1000$ m according to Munk, 1966).

From (a):

$$\frac{L_s}{L} = Ek_v^{1/2} = 4 \times 10^{-4}$$

$$Ek_v = 1.6 \times 10^{-7}$$

$$L_s = 2 \text{ km}$$

This requires a value of K_m at the ocean bottom of 6 cm^2/sec. *From (b)*, using $A_m = 10^6$ cm^2/sec:

$$Ek_h = \frac{A_m}{2\Omega L^2} = 3 \times 10^{-8}$$

$$L_f = (Ek_h)^{1/3}L = 15 \text{ km}$$

These values are smaller if A_m is smaller. *From (c)*, we conclude that

$$L_i = 50 \text{ km}$$

That is, the boundary current is essentially inertial in character (we show below that $L_m < L_i$, so that the only possible length scale to match the observed width is L_i). Then

$$Ro = 10^{-4}$$

and

$$V^* = 7 \text{ cm/sec}$$

This is consistent with observed current speeds in the upper part of the open ocean. Using this value of V^* and assuming $A_h = A_m = 10^6$ cm^2/sec,

$$Pe_h^{-1} = \frac{A_h}{V^*L} = 3 \times 10^{-4}$$

$$L_m = 1.5 \text{ km}$$

Note that these numbers will be smaller if A_h is smaller. *From (d)*, $D_t = 1000$ m so that

$$Pe_v^{-1} = 2 \times 10^{-7}$$

These approximate values of the nondimensional numbers governing the oceanic circulation, together with the various length scales associated with the western boundary current, are shown in Table II. Note, however, that the simple formulation of eddy mixing that led to the nondimensional numbers Ek_h, Ek_v, Pe_h^{-1}, and Pe_v^{-1} may be unrealistic and the values of these numbers irrelevant. It seems likely that mixing (by mesoscale eddies, for example) cannot be parameterized in this simple fashion.

TABLE II

Nondimensional Numbers	Western Boundary Current Widths (km)
$Ro = 1 \times 10^{-4}$	$L_i = 50$
$Ek_h = 2 \times 10^{-8}$	$L_f = 15$
$Ek_v = 2 \times 10^{-7}$	$L_s = 2$
$Pe_h^{-1} = 3 \times 10^{-4}$	$L_m = 2$
$Pe_v^{-1} = 2 \times 10^{-7}$	

These considerations make clear the basic difficulty in understanding the oceanic circulation problem as a problem in geophysical fluid dynamics. In addition to variable factors such as the geometry of ocean basins and wind and temperature distributions at the sea surface, there is a five-dimensional parameter space to explore! Since only a few dozen large-scale numerical studies have been done, there is considerable work to do.

The prognostic, general circulation models to be discussed have, for the most part, solved the equations as an initial value calculation subject to given boundary conditions (given wind stress and temperature or heat flux). One of the difficulties in carrying out such calculations is the very long time scale for the ocean to come to equilibrium when the calculation is begun from arbitrary initial conditions. The early studies (and some of the later ones) were not carried out to equilibrium because of this long spin-up time, which was shown by Bryan and Cox (1968) to be well predicted by the vertical diffusion time, $\Delta t = Z^2/K_h$. If Z is the depth of the thermocline ($K_h = 1$ cm^2/sec according to Munk, 1966), then the spin-up time is a few hundred years. Though the adjustment time for the boundary layers is much shorter, that for the deep ocean is even longer. Thus each experiment represents a considerable investment in computer time. This situation has been made better by the fact that computer technology has progressed rapidly over the last decade (about two orders of magnitude increase in speed during the last 10 years according to Leith, 1975). Unfortunately our ambitions in attempting to model the ocean have grown even faster.

We conclude this section with a discussion of the results of a single numerical experiment by Bryan and Cox (1968). Starting from initial conditions of uniform stratification and complete rest, an extensive integration in time was carried out until near equilibrium was reached in all layers. The parameters were chosen so that the solution was quite nonlinear in contrast to the earlier study by Bryan and Cox (1967). It is clear, however, that the ocean was still too diffusive, both for momentum and heat, but for the first time in a baroclinic case nonlinearity was beginning to dominate.

Although the results of the earlier study (Bryan and Cox, 1967) showed that the topography of the thermocline agreed better with observations when both wind and thermal driving were taken into account, the follow-up study treated a more nonlinear case in which a comparison with the previous wind-driven calculations in homogeneous oceans could be made. Some of the results are shown in Fig. 5. Figures 5a and b compare the horizontal mass transport for the nonlinear case with the linear (Munk) solution. Note the similarity in the way that the nonlinear terms enhance the flow in the subtropical gyre and shift the maximum northward, as happened in the homogeneous case (see Fig. 2). Figures 5c and 5d show something of the vertical structure of the flow. Figure 5c shows the total vertical overturning, that is, the mass

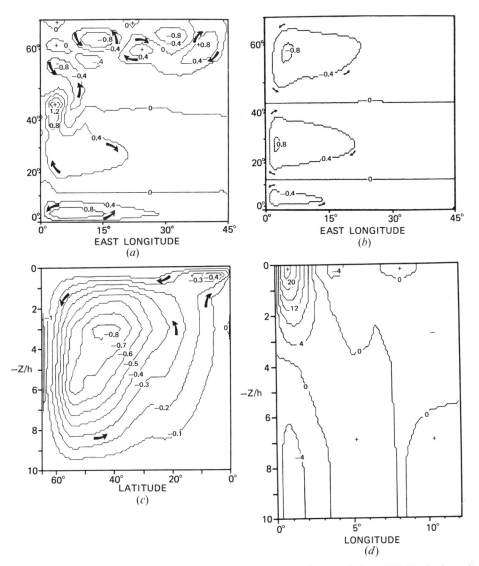

Fig. 5.   The results of a baroclinic (6 level) numerical experiment by Bryan and Cox (1968). The horizontal mass transport streamfunctions are shown for (a) the fully nonlinear and (b) the linear (Munk) solutions. The meridional transport streamfunction is shown in (c) and an east–west section across the western boundary current is shown in (d).

transport in the meridional plane. Note that almost as much water circulates in this vertical plane as in the horizontal plane. Figure 5d shows an east–west section across the western boundary current. A swift northward surface current and an important (southward) return flow are found, in agreement with observations (Swallow and Worthington, 1961).

Many observed features of the structure of the thermocline and circulation were reproduced in this calculation, and further numerical studies were stimulated by this work. From this point such models were used to ask (and partly to answer) a number of important questions that had to be solved before realistic general circulation models could be evolved. These questions included the following:

1. What is the influence of bottom topography?
2. What effect does transient (seasonal) forcing have?
3. What influence does the lateral geometry have?
4. Is the physics included in these early models adequate to study more general problems?
5. Are the parameterizations (both horizontal and vertical) of eddy mixing of heat, salt, and momentum adequate?
6. What role is played by naturally occurring transients due to barotropic and/or baroclinic instability?

These and other questions have been examined in a number of numerical experiments during the last decade and are discussed in the next sections.

4. Physical Process Studies in Closed Basins

The preceding section described the simplest of the homogeneous, wind-driven models and the baroclinic, wind- and thermohaline-driven models. As pointed out there, even those models have a wide variety of behaviors depending upon basic nondimensional parameters, geometry, distribution of wind stress and thermal forcing, and boundary conditions. In this section we look into the effect of complications in geometry (both bottom topography and lateral shape of the basin) necessary to describe the circulation in actual ocean basins and present results from studies of idealized models of special regions (the Southern Ocean, the Equatorial Ocean).

A. Bottom Topography in a Homogeneous Ocean

The influence of bottom topography has been examined in a number of homogeneous and baroclinic (both diagnostic and prognostic) studies. In a simple extension of the constant-depth model of Bryan (1963), Holland (1967) introduced variable depth into the homogeneous case. The appropriate nondimensional equations are

$$-\psi_y'\left(\frac{Ro\zeta' + f'}{h'}\right)_x + \psi_x'\left(\frac{Ro\zeta' + f'}{h}\right)_y = +\operatorname{curl}_z\left(\frac{\tau'}{h'}\right) + Ek_h\nabla^2\zeta' \qquad (29)$$

where h' is the nondimensional depth and ζ' is the nondimensional vorticity equal to $\zeta' = \nabla \cdot (\nabla\psi'/h')$. In this study it was found that when a simple sloping region was included leading away from the western boundary, the western boundary current could be induced to leave the coast and follow topographic contours (actually f'/h' contours) seaward (Fig. 6). As the Rossby number was increased, steady meanders in the free current began to be important. Thus the results of Bryan (1963) were extended to incorporate the ideas suggested by Warren (1963) concerning the meandering nature of the path of the Gulf Stream after it had left Cape Hatteras and turned seaward.

It is of interest to ask how these rather frictional results behave in the highly inertial limit. As Veronis (1966b) and Niiler (1966) showed, the Fofonoff (1954) solution gives a good representation of the inertial limit of Stommel's (1948) model with nonlinear terms added. Therefore it is useful to extend Fofonoff's model to

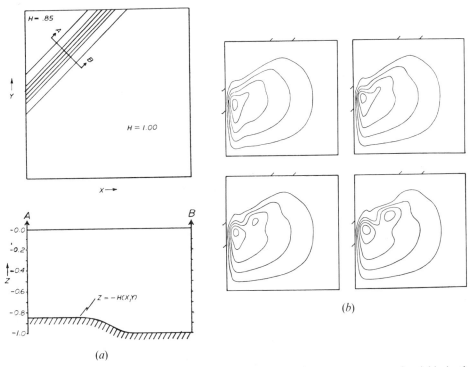

Fig. 6. Numerical experiments on the wind-driven circulation in a homogeneous ocean of variable depth (Holland, 1967). (a) The topography; (b) several streamfunction patterns with increasing nonlinearity. Note the tendency for the flow to follow f/h contours in the frictional limit and for the flow to meander about f/h contours in the more nonlinear cases.

the variable-depth case. The vorticity equation then would be

$$\frac{D}{Dt}\left(\frac{\zeta + f}{h}\right) = 0 \qquad (30)$$

where

$$\zeta = \nabla \cdot \frac{\nabla \psi}{h} \qquad (31)$$

Then, following Fofonoff, we look for special solutions such that

$$\frac{\zeta + f}{h} = A + B\psi \qquad (32)$$

The second-order, elliptic differential equation is easily solved by relaxation for any $h(x, y)$. Figure 7 shows inertial solutions corresponding (a) to the flat bottom case, and (b) to the topography used by Holland (1967). Maps of h and f/h used in the topographic case are also shown. Note that, in the inertial limit, the western boundary current overshoots the topographic step at the western boundary and forms a northern boundary current. However, there is a significant recirculation with rather strong southwestward flow along the f/h contours.

Another study of topographic effects in the homogeneous case is that of Schulman and Niiler (1970). In their model bottom friction is used rather than the lateral friction used by Holland (1967). The basic results are rather similar to Holland's, showing

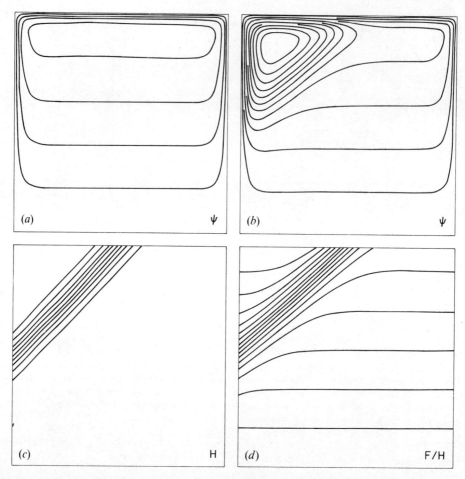

Fig. 7.   Steady flow in a frictionless homogeneous ocean. (*a*) Inertial flow in a constant depth ocean (Fofonoff, 1954); (*b*) inertial flow in a variable depth ocean; (*c*) the topography; (*d*) the f/h contours.

the very strong tendency for the flow to follow f/h lines. In this regard an analytic study by Welander (1968) examines several special cases and shows that frictional effects must be important even in the interior when closed f/h lines are present. This is easily established by examining the circulation around a closed circuit following an f/h line. This circulation theorem shows that in the interior the tendency for wind to drive the flow around a closed f/h line must be balanced by frictional effects; the topographic and planetary vorticity terms cannot export any net vorticity. Holland (1973) discusses this constraint on the flow for the baroclinic model.

B. Bottom Topography in a Baroclinic Ocean

The importance of the influence of bottom topography on the general circulation has long been in doubt. The early thinking suggested that, owing to the adjustment of the density field, the deep currents which could "feel" bottom topography would be negligibly small. This phenomenon was known as compensation. The idea that compensation was complete was inherent in the horizontal mass transport model of

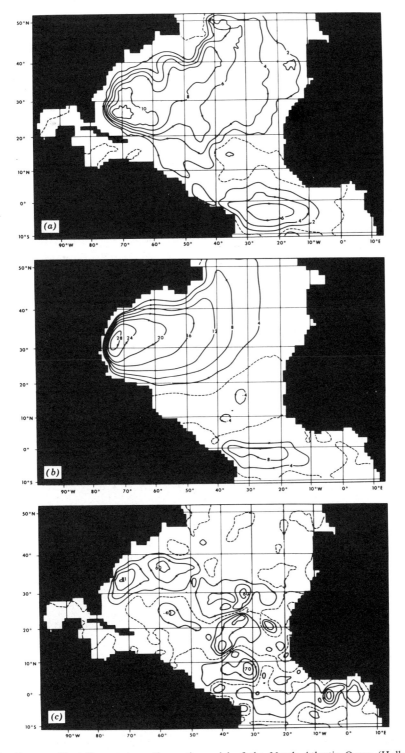

Fig. 8. Topographic influences in a diagnostic model of the North Atlantic Ocean (Holland and Hirschman, 1972). (a) A homogeneous ocean of variable depth; (b) a baroclinic ocean of constant depth; the bottom artificially placed at a depth of 1273 meters; (c) a baroclinic ocean of variable depth. The maximum transports in the model Gulf Streams are respectively 14, 28, and 81 million m³/sec.

23

Munk (1950). It was based upon the observational evidence that a level of no motion in the deep water gave an acceptable representation of surface currents. However, even a small barotropic component, when integrated through the entire depth of the water column could contribute strongly to the mass transport. Welander (1959b) discussed this situation and showed that even a small lack of compensation would result in an important topographic effect. Schulman (1975) reviewed the situation and concluded that bottom relief should play an important role in the general circulation.

The joint effect of baroclinity and bottom relief has been examined in a number of numerical calculations. Most of these have been *diagnostic* studies (especially by Sarkisyan and co-workers in the Soviet Union) that showed a very strong influence of bottom relief. Compensation, as defined by Welander (1959b), was not complete. Holland and Hirschman (1972) did such a calculation for the North Atlantic Ocean using the entire NODC data files to construct the observed density field in that basin. The results are summarized by showing three cases (Fig. 8): (*a*) the mass transport streamfunction for a homogeneous ocean but with complete bottom topography; (*b*) the mass transport streamfunction in the baroclinic (diagnostic density) case but with a flat bottom artificially added at a depth of 1273 m; and (*c*) the mass transport streamfunction with complete (diagnostic) density and topography. Note that the second case is essentially a Munk (1950) calculation, identical to assuming "complete" compensation, whereas the first shows the result of topographic relief in a homogeneous ocean and the third shows the result of topographic relief in a baroclinic

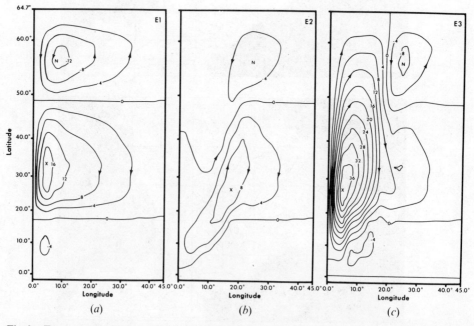

Fig. 9.   Topographic influences in a prognostic baroclinic model of the ocean (Holland, 1973). The basin is 45° of longitude wide and stretches from the equator (with symmetry conditions there) to 64.7°N. There are 19 vertical levels in the model. The mass transport streamfunction is shown for three cases: (*a*) a constant-depth, baroclinic ocean, (*b*) a variable-depth, homogeneous ocean, and (*c*) a variable-depth, baroclinic ocean. Note the large enhancement when both variable depth and variable density are present.

ocean. Note that the relief *reduces* the transport by the boundary current in the homogeneous case (from 29 to 14 $\times$ 10^6 m^3/sec) and *enhances* the transport in the baroclinic case (from 28 to 81 $\times$ 10^6 m^3/sec). In addition the joint effect of baroclinity and bottom relief tends to break up the transport pattern considerably, creating a number of isolated gyres that are no longer simply related to the wind-stress pattern. Worthington (1975) has recently suggested the presence of such gyres in the North Atlantic Ocean.

This kind of diagnostic study, like the homogeneous ocean studies, posed a question that required an answer. Both kinds of studies indicated that topography would play an important role in determining the transport pattern. However, the nature and degree of compensation was still rather obscure. The flat bottom prognostic models (Bryan and Cox, 1967, 1968) showed important deep mean currents associated with the meridional overturning. These deep currents were especially large near the western boundary. Therefore, it might be expected that in prognostic models with bottom relief the transport pattern might be seriously altered.

Holland (1973) examined this question and showed that the horizontal transports could be considerably enhanced by bottom torques associated with currents near bottom slopes. Figure 9 shows three cases: (a) constant-depth baroclinic, (b) variable-depth homogeneous, and (c) variable-depth baroclinic. Note the large enhancement in transport in the last case. Holland showed that these effects were due to bottom pressure forces acting against bottom relief, entirely analogous to the mountain torques well known to the meteorologist. In the vorticity equation governing the horizontal mass transport, these pressure forces act as pressure torques which contribute vorticity to the vertically averaged flow.

C. The Energetics of Baroclinic Oceans

An analysis of the energetics of such flows (Holland, 1975a) clarifies how the energy associated with the horizontal mass transport can be enhanced in the presence of bottom topography and baroclinity. If the kinetic energy equations are formed from equations 6–14 it is found

$$\frac{dE}{dt} = A + G + B + W + D \qquad (33)$$

where E is the kinetic energy per unit area (erg/cm^2) equal to

$$\frac{1}{A_s} \iiint \rho_0 \frac{\mathbf{V} \cdot \mathbf{V}}{2} \, dx \, dy \, dz$$

Here A_s is the surface area of the basin. The kinetic energy in a fixed volume is changed in several ways:

1. By the advection of kinetic energy into the region across the boundary ($V_n =$ inward normal velocity):

$$A = \frac{1}{A_s} \iint \rho_0 \frac{\mathbf{V} \cdot \mathbf{V}}{2} V_n \, dS \qquad (34)$$

This is a redistribution of kinetic energy.

2. By the performance of work by pressure forces at the boundary as a consequence of flow across the boundary:

$$G = \frac{1}{A_s} \iint p V_n \, dS \tag{35}$$

This is also a redistribution of kinetic energy. Note that $A = G = 0$ if there is no flow across the boundary.

3. By the performance of work done at the sea surface by the wind stress:

$$W = \frac{1}{A_s} \iint \mathbf{V} \cdot \boldsymbol{\tau}^w \, dx \, dy \tag{36}$$

4. By the performance of work done by buoyancy forces in the fluid:

$$B = -\frac{1}{A_s} \iiint g\rho w \, dx \, dy \, dz \tag{37}$$

The kinetic energy increases if ρ and w are negatively correlated (ρ small in upward motion, large in downward motion) and decreases if ρ and w are positively correlated. This represents an exchange between potential and kinetic energy in the basin.

5. By the loss of energy due to dissipation distributed throughout the fluid and due to work done by tangential stresses at the sides and bottom of the basin:

$$D = \frac{1}{A_s} \iiint (\mathbf{V} \cdot \mathbf{F} - \rho_0 K_m \mathbf{V}_z \cdot \mathbf{V}_z) \, dx \, dy \, dz - \frac{1}{A_s} \iint \mathbf{V} \cdot \boldsymbol{\tau}^b \, dx \, dy \tag{38}$$

If the region under consideration is a closed basin, with no flow in and out, the governing energy equation becomes

$$\frac{dE}{dt} = B + W + D \tag{39}$$

To understand the problem of the horizontal mass transport in the ocean, it is convenient to divide the energy per unit area, E, into two parts

$$\bar{E} = \frac{1}{A_s} \iiint \rho_0 \frac{\bar{\mathbf{V}} \cdot \bar{\mathbf{V}}}{2} \, dx \, dy \, dz \tag{40}$$

$$\hat{E} = \frac{1}{A_s} \iiint \rho_0 \frac{\hat{\mathbf{V}} \cdot \hat{\mathbf{V}}}{2} \, dx \, dy \, dz \tag{41}$$

where the overbar and caret indicate a vertical average and the deviation from the vertical average. $\bar{E}$ is the kinetic energy associated with the horizontal mass transport and $\hat{E}$ is the kinetic energy associated with the vertical shear. The rate of change of these quantities can then be written

$$\bar{E}_t = N_e + B_e + W_e + D_e \tag{42}$$

$$\hat{E}_t = N_i + B_i + W_i + D_i \tag{43}$$

where

$$N_e = -\frac{1}{A_s} \iiint \rho_0 \overline{\mathbf{V}} \cdot (\mathbf{V} \cdot \nabla \mathbf{V} + w\mathbf{V}_z)\, dx\, dy\, dz \tag{44}$$

$$B_e = -\frac{1}{A_s} \iiint \overline{\mathbf{V}} \cdot \nabla p\, dx\, dy\, dz \tag{45}$$

$$W_e = \frac{1}{A_s} \iint \overline{\mathbf{V}} \cdot \tau^w\, dx\, dy \tag{46}$$

$$D_e = -\frac{1}{A_s} \iint \overline{\mathbf{V}} \cdot \tau^b\, dx\, dy + \frac{1}{A_s} \iiint \overline{\mathbf{V}} \cdot \overline{\mathbf{F}}\, dx\, dy\, dz \tag{47}$$

and

$$N_i = A - N_e \tag{48}$$

$$B_i = G + B - B_e \tag{49}$$

$$W_i = W - W_e \tag{50}$$

$$D_i = D - D_e \tag{51}$$

In a mechanically closed basin, where $A = G = 0$,

$$N_e + N_i = 0 \tag{52}$$

$$B_e + B_i = B \tag{53}$$

Equation 52 shows that the nonlinear advective terms provide an energy exchange between the energy associated with the horizontal mass transport and that associated with the vertical shear, but they do not lead to a net increase in kinetic energy. This exchange does provide a mechanism by which the vertically averaged component of flow can be enhanced, however.

Equation 53 states that in general there is a conversion of potential energy to (or from) the kinetic energies associated with the horizontal mass transport and the vertical shear by the action of buoyancy forces. From equation 45, however, it can be shown that

$$B_e = -\frac{1}{A_s} \iiint (\bar{p} - p_b) \frac{\overline{\mathbf{V}} \cdot \nabla H}{H}\, dx\, dy\, dz \tag{54}$$

and hence that B_e is zero for a closed basin if either the bottom is flat ($H = $ constant) or the ocean is homogeneous ($\bar{p} = p_b$). Here $\bar{p}$ is the vertically averaged pressure and p_b is the bottom pressure. Thus the energy associated with the horizontal mass transport can be enhanced only if the ocean has variable density *and* variable topography.

Holland (1975a) compared the energetics of the three cases shown in Fig. 9. The energy diagrams for these cases are shown in Fig. 10. In the first of these, B_e is zero because the depth is constant; the energy associated with the horizontal mass transport is maintained by direct wind driving only (N_e and N_i are negligible for the parameters chosen in these experiments). In the homogeneous case, B_e is again zero because the density is constant. In the baroclinic case with variable depth, B_e is no longer zero and in fact supplies energy to the horizontal mass transport mode at a rate

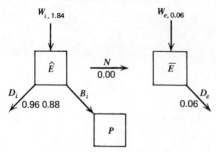

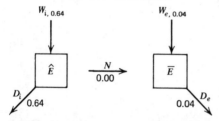

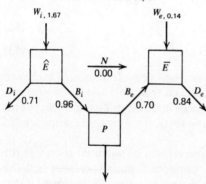

Fig. 10. The energetics of the three cases shown in Fig. 9. Here $\hat{E}$ and $\bar{E}$ are the total kinetic energies associated with the vertical shear and the vertical mean respectively, and P is the potential energy. Note that in the first two models there is no effective link (since N_i is negligible in these calculations) between $\bar{E}$ and the other energy sources, while in the third case baroclinic processes in the presence of topographic relief lead to a significant energy flow into the barotropic mode.

five times larger than the work by the wind stress. The average kinetic energy of this vertically averaged flow is an order of magnitude larger than in the other cases. Thus the joint effect of baroclinity and bottom topography can be an important factor in determining the horizontal mass transport pattern in the ocean.

D. Wind and Thermohaline Forcing

An examination of these kinds of energetic balances also reveals the nature of the two basic driving mechanisms for the general circulation of the ocean. The ocean is forced directly by the wind stress and by thermohaline processes acting at the sea surface. Which of these is more important is yet to be determined, but it is clear that both are necessary to give realistic results. Bryan and Cox (1968) showed that in a highly nonlinear baroclinic case with a flat bottom intense western boundary currents occur both with and without wind driving. The inertial terms enhance the vertically averaged flow. Holland (1975a) also compares cases with and without wind driving but in model oceans with variable depth and with negligible inertial effects (Fig. 11).

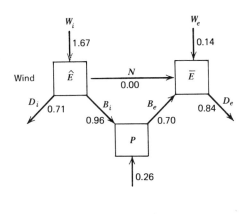

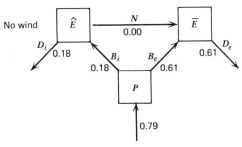

Fig. 11. The energetics for two baroclinic, variable-depth cases with and without wind forcing. In the first case wind and thermohaline forcing compete, and the wind tends to supply energy to the potential energy of the system. In the second case thermohaline forcing maintains the potential energy, and baroclinic processes supply energy to both forms of kinetic energy.

In the case in which wind is acting (same as case C, Fig. 10) the horizontal mass transport is driven both by wind ($W_e = 0.14$ erg/cm^2/sec) and by bottom pressure forces ($B_e = 0.70$ erg/cm^2/sec). Energy flows from kinetic to potential energy; that is, the variable density field is maintained by the wind in the face of diffusion processes. In the case with no wind forcing, the only source of energy is the thermohaline driving which maintains the potential energy of the system. The bottom pressure forces again act to maintain a horizontal mass transport in the face of dissipation. If the depth were constant, however, B_e would be zero and $\bar{E}$ would be very small. Only the vertical shear energy $\hat{E}$ would be significant.

E. The Dynamics of a Southern Hemisphere Ocean

The combined effects of lateral basin shape and bottom topography were explored by Gill and Bryan (1971). The model was distinct from earlier studies in that a gap in the meridional boundaries was inserted to simulate the Drake Passage. Two important effects were found. Firstly the meridional circulation was strongly affected by the presence of the gap (Fig. 12a), and secondly the horizontal mass transport was enhanced by the presence of a topographic ridge across the bottom of the gap (Fig. 12b). The first result arises because a pressure gradient can build up against a meridional wall, but cannot where there is a gap in the wall. The second result follows from the considerations given above regarding topographic effects; that is, the horizontal mass transport streamfunction can be enhanced in the presence of baroclinity and bottom relief.

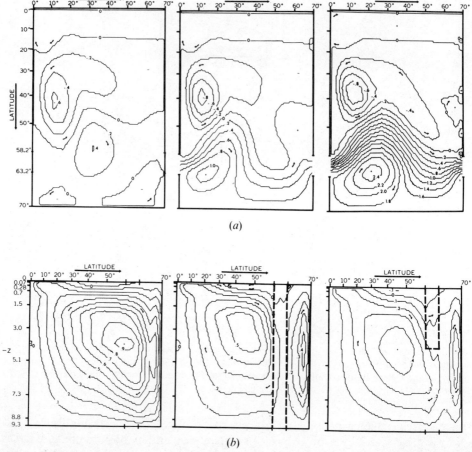

Fig. 12. Numerical experiments with a baroclinic model of the Southern Ocean (Gill and Bryan, 1961). Three cases show the closed basin circulation, the circulation with a full gap in the meridional boundary, and the circulation with a gap which reaches only halfway to the bottom where (*a*) shows the meriodional transport streamfunctions and (*b*) shows the horizontal transport streamfunctions.

F. The Equatorial Undercurrent

Early baroclinic models of the general circulation (Bryan and Cox, 1967, 1968) showed a shallow equatorial circulation including a broad and weak equatorial undercurrent. The countercurrents north and south of the equator probably could be simulated rather well if sufficient realism and complexity in the wind pattern were included in the model, but it was not clear what elements were important in establishing a realistic undercurrent. A number of analytic studies have looked into the problem (see the survey article by Philander, 1973), and several numerical calculations have included the equatorial region with more or less success. Some of these are discussed later for global and real basin models. Holland (1975b), however, performed a series of three-dimensional calculations with high horizontal and vertical resolution at the equator designed to examine the dynamical nature of the undercurrent. In contrast to earlier basin models, the resolution was sufficient near the equator (about

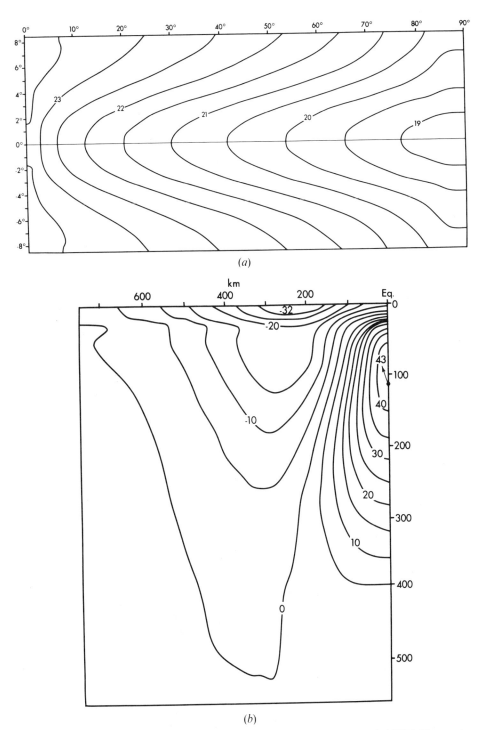

Fig. 13. The steady equatorial circulation in a fine resolution model (Holland, 1975b). The northern boundary is at 20°N and the southern boundary at the equator, where a symmetry boundary condition is imposed. The model is driven by a steady zonal (westward) wind stress and by heat flux boundary conditions. (a) The sea surface temperature pattern predicted by the model; (b) the zonal (positive eastward) velocity pattern at the equator showing the strong equatorial undercurrent and countercurrents at a mid-longitude.

25 km north–south; 10 grid points in the upper 500 m) so that the structure of the current could be realistically determined.

The surface boundary conditions were extremely simple in this model. The wind stress was purely zonal and uniform with latitude, with $\tau = 0.3$ dyne/cm^2. The thermal boundary condition was a specified heat flux taken from observations so that the surface temperature, an important element to be predicted, could be determined by the thermodynamics of the model.

Figure 13 shows some results. The predicted surface temperature pattern is shown in Fig. 13a and a meridional section of zonal velocity is shown in Fig. 13b. These results are in general agreement with observations except that the maximum under-current velocity is somewhat smaller than observed. It seems clear, however, that we are approaching quite realistic simulations when sufficient resolution and small enough lateral eddy viscosity and diffusion coefficients are used.

5. Single Ocean and Global Ocean Studies

The preceding studies were highly idealized for the purpose of studying basic physical processes in isolation. In this section we examine attempts to model real ocean basins with models that include most of the physical factors thought to be important. Some of these are single oceanic basins like the North Atlantic and the Indian Ocean, and, as such, cannot be closed oceans with no flow through the lateral boundary. However, because it is important to attempt to model actual basins in order to verify how good the models are, we relax our restriction about considering only closed basins. Thus lateral boundary conditions at the open boundaries may lead to complications in interpretation. The only way to get around this difficulty is to perform calculations for the entire world ocean. The initial attempts to accomplish that task are also discussed here. Table IC lists the studies used to illustrate these various attempts to simulate real oceans. Many others could equally well have been included.

Friedrich (1970) examined the circulation in the North Atlantic Ocean between 57.5°N and 17.5°S using a 14-layer model with 3° horizontal resolution. The actual topography and coastline shape were included. The surface boundary conditions were the climatological fields of temperature, salinity, and wind stress including the annual variation. Starting from an initial state of rest and uniform stratification, the equations were integrated for 80 yr on a 5° grid, then 70 yr more on a 3° grid to reach a near equilibrium.

The results show that the larger-scale features of the temperature and salinity fields agree with observations (Defant, 1936), although Friedrich points out that the ocean is probably too diffusive. The influence of the annual variation of wind and thermohaline driving on the mean flow, if any, has not been explicitly examined, but the annual march of temperature in the upper layers is fairly realistic. The horizontal transport pattern (Fig. 14a) shows considerable topographic control of the circulation, the topographic term contributing importantly to the vorticity balance for the vertically averaged flow. The meridional transport (Fig. 14b) shows the important net overturning of water in high northern latitudes which produces North Atlantic Deep Water. The lack of symmetry about the equator, in which surface water flows northward and deep water southward, is reproduced. These results are encouraging, leading one to believe that if sufficient resolution (1° or less) and smaller horizontal diffusivities could be reached, many of the observed features of the large-scale circulation would be reproduced.

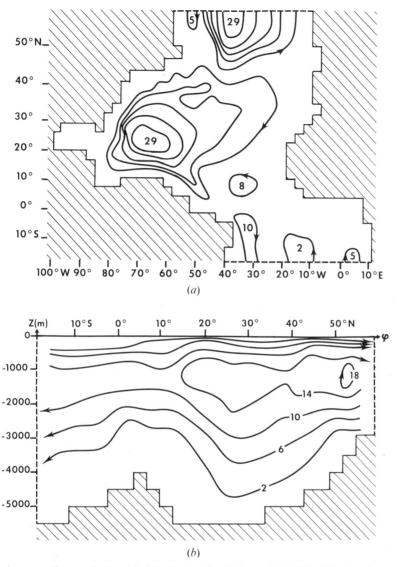

Fig. 14. A prognostic numerical model of the North Atlantic Ocean (Friedrich, 1970). Actual topography and coastline shape are included, and the ocean is driven by climatological wind, temperature, and salinity data at the sea surface. (a) The horizontal transport streamfunction, and (b) the meridional transport streamfunction for the North Atlantic.

Cox (1970) undertook the study of the circulation in the Indian Ocean driven by the very strong annual variation in wind forcing due to the coming and going of the monsoons. His study, like Friedrich's, sought to calculate the currents and water mass properties and compare them with observations. Particular emphasis was placed on the annual response of the Somali Current to the monsoonal wind field. The problem was again solved by an extended numerical integration (~ 200 yr) of the time-dependent equations on successively finer horizontal grids, finishing with $1°$ resolution. There were seven layers in the vertical in the final stage of the calculation.

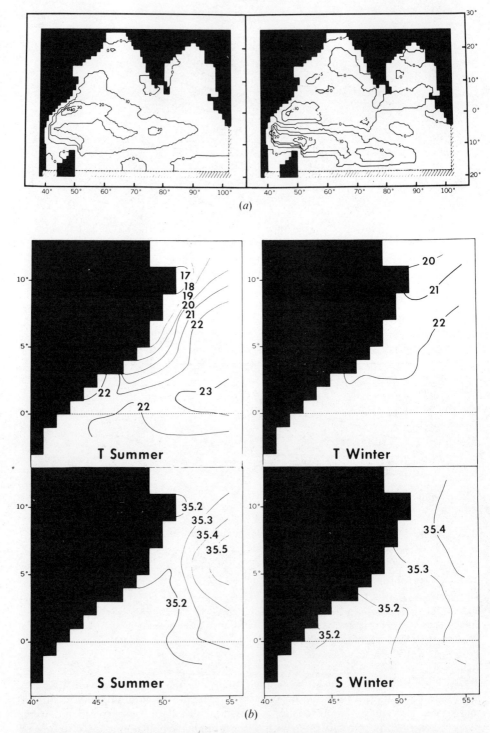

Fig. 15. A prognostic numerical model of the Indian Ocean under monsoonal wind and thermal driving (Cox, 1970). (a) The horizontal transport streamfunction in summer (left) and winter (right). Note the strongly developed Somali Current during the summer (Southwest) monsoon. (b) The temperature and salinity patterns, at the second level in the model, showing the important effects of upwelling during the Southwest monsoon.

34

The model currents and water mass characteristics compare favorably with observations, given the rather coarse vertical resolution. [See Cox (1970) for a thorough discussion of the observations.] The time-dependent behavior of the Somali Current in the model is particularly interesting and shows fair agreement with observations (Fig. 15a). The current is very well developed during the summer (the Southwest Monsoon) and practically absent in the winter. The predicted upwelling along the Somali coast (Fig. 15b) is found to be controlled by local winds blowing parallel to the coastline, and Cox suggests that the rapid response of the Somali Current to the monsoons is, in part, a local effect and not due to the changes in large-scale forcing distributed across the whole basin.

Semtner (1973) examined the circulation in the Arctic Ocean. This region of the world ocean is particularly difficult to model for two reasons: (1) the effect of the ice pack on appropriate surface boundary conditions is not well understood, and (2) the water mass characteristics are largely determined by inflow from the North Atlantic Ocean, that is, the open boundary is a controlling factor on the distribution of water mass properties. Despite these difficulties, Semtner found considerable success in simulating the currents and water mass characteristics in the Arctic Basin to the extent that they are known. In addition, he made a number of idealized runs (a homogeneous case, a constant-depth case) to compare with the full baroclinic case in order to begin to sort out the relevant physics in the Arctic Basin.

As has been mentioned, the single basin models suffer from the fact that the basin is not closed, and the lateral open boundary conditions give rise to special difficulties in interpretation of results. The way around this problem is to treat the entire world ocean. A beginning has been made on this kind of model, notably the homogeneous models of Ilyin et al. (1969), Sag (1969), Bye and Sag (1972), and Bryan and Cox (1972), and the baroclinic model calculations by Bryan and co-workers (Cox, 1975; Bryan, Manabe, and Pacanowski, 1975). Since the homogeneous cases are really of interest only when compared to their more realistic baroclinic counterparts, only the companion·cases of Bryan and Cox (1972) and Cox (1975) are discussed. The same model has been used in the climate study by Bryan et al. (1975) and Manabe et al. (1975).

The mass transport patterns in the world ocean are shown in Fig. 16 for four cases: (a) a homogeneous ocean of constant depth, (b) a homogeneous ocean of variable depth, (c) a diagnostic baroclinic ocean of variable depth, and (d) a fully baroclinic ocean. The first three cases require only a few months of ocean time to come to equilibrium since only the barotropic mode is involved in the adjustment, but case (d) would require several hundred years to fully adjust to the boundary conditions. Here only the initial adjustment of the model over a short period (2.3 yr) has been carried out, but Cox (1975) suggests that the patterns already are in much better agreement with the observed dynamic topography than the diagnostic model.

Note the important topographic steering found in the homogeneous cases by comparing the cases shown in Figs. 16a and 16b. The results are similar to those discussed earlier for idealized models; that is, variable depth considerably reduces the transports in homogeneous oceans. Note that the Antarctic Circumpolar Current transport changes from 600×10^6 to 22×10^6 m^3/sec. The diagnostic model gives more realistic results, showing the enhancement of peak transports in subtropical gyres as well as in the Circumpolar Current (186×10^6 m^3/sec) over the homogeneous case, in agreement with basin results. Finally the prognostic model (after 2.3 yr) shows considerable smoothing of the fields found in the diagnostic results (note the loss of the strong gyre in the South Atlantic). This may result either from too strong diffusivity in the prognostic model (the diagnostic results are better) or from lack of

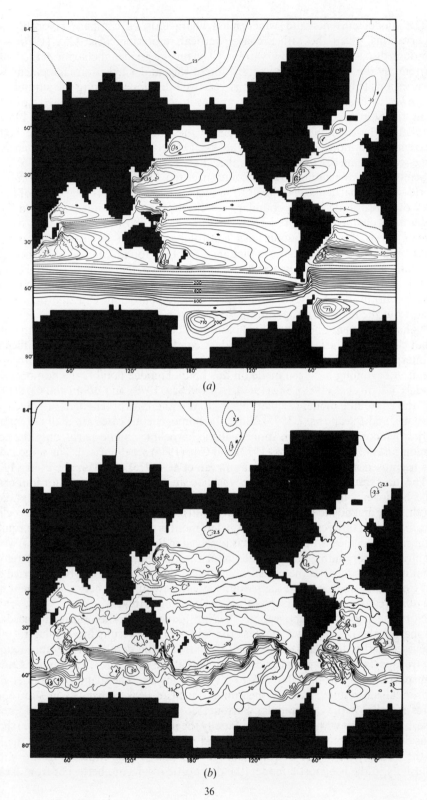

(a)

(b)

36

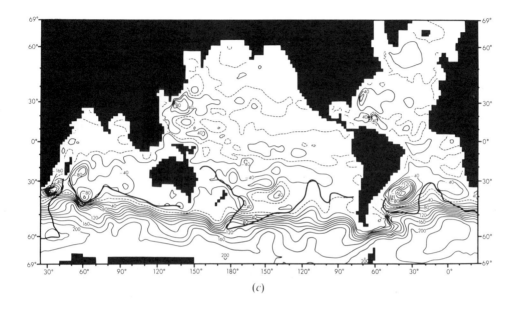

(c)

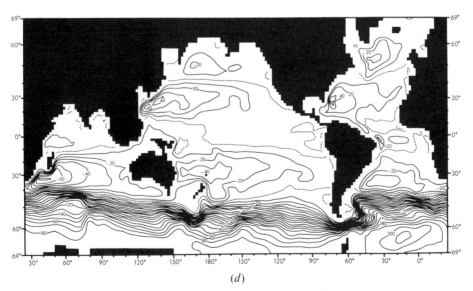

(d)

Fig. 16. The mass transport patterns in the world ocean. (a) A homogeneous ocean of constant depth; (b) a homogeneous ocean with a realistic depth configuration (both from Bryan and Cox, 1972); (c) a baroclinic (diagnostic) ocean, and (d) a baroclinic (prognostic) ocean (both from Cox, 1975).

37

good density data and their adjustment to bottom relief in the diagnostic model (the prognostic results are better). In reality both difficulties are probably important, and we must wait for longer, higher resolution, smaller diffusivity prognostic runs to give realistic results.

6. Eddy-Resolving General Circulation Models

The equations of motion derived above included terms that parameterized the effects of Reynolds stresses (or subgrid scale motions) in a simple way. A proper way to parameterize the effects of Reynolds stresses and eddy heat fluxes due to small-scale motions is as yet unknown. In fact the simple linear law used in nearly all ocean model studies is known to be incorrect in some regions, based upon limited Gulf Stream data (Webster, 1961).

One way to examine this problem is to perform numerical experiments with models that explicitly include fine-scale motions. When very fine resolution is included in such models, the eddies previously left out of the calculation (except in their parameterized effect) can now be physically resolved and their influence determined. The difficulty is of course that the computation requirements increase rapidly with resolution, a factor of about eight when the horizontal resolution is doubled.

Several early experiments (Bryan, 1963; Bryan and Cox, 1968; Blandford, 1971) did find naturally occurring transients in the models. Because at the time the ubiquitous nature of mesoscale eddies was not realized, not much attention was paid to the transients. During the last few years, however, a new look has been taken at the nature of eddies in the ocean. The Polygon and the Mid-Ocean Dynamics Experiments (MODE) have focused attention on mesoscale phenomena. New looks at old data as well as results from these recent large observational programs seem to show that mesoscale eddies, with horizontal wavelengths of 250–500 km and periods of 50–200 days are an exceedingly important phenomena in mid-ocean. It now seems likely that the energy associated with transient eddies exceeds that associated with the mean flow in the central ocean by at least an order of magnitude. A proper parameterization of the effects of these eddies, or, alternatively, explicitly including them in general circulation calculations, would indeed seem necessary if we are to understand the mechanisms which govern the distribution of currents, temperature, and salinity in the oceanic general circulation.

Recently Holland and Lin (1975a, b) have begun a series of investigations in which mesoscale eddy motions are explicitly included in a simple general circulation model. They examined a rectangular, wind-driven ocean with two layers, the simplest extension to the baroclinic case of the homogeneous ocean studies of Bryan (1963) and Veronis (1966a, b). Fine enough resolution was included to resolve horizontal scales on the order of the internal radius of deformation, which is the preferred scale for the production of baroclinic eddies.

An example of one case is shown in Fig. 17. The model ocean is spun up from rest. During the initial phase the energy put in by the wind increases the kinetic energy of the upper layer and the available potential energy of the system (Fig. 17a). At about 500 days the flow becomes baroclinically unstable and the available potential energy falls rapidly to some lower equilibrium value. During that time the kinetic energy of the lower layer rapidly increases as mesoscale eddies, with horizontal wavelengths of 400 km and periods of 64 days, appear in the region of instability. After some adjustment time the ocean reaches a statistically steady state in which the eddies are superimposed upon a mean circulation. The energetics of the mean and

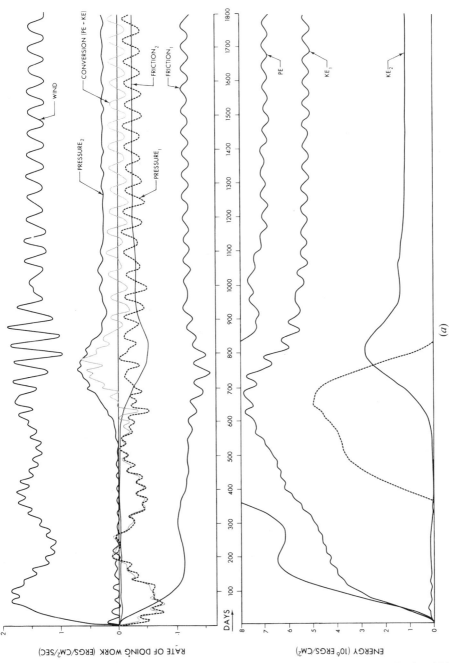

Fig. 17.   Results from a high resolution, two-layer ocean showing mesoscale eddies (Holland and Lin, 1975a, b). (a) The time history of the spatially averaged energies and energy rates. The upper panel shows various rates of doing work as well as the rate of energy conversion between potential and kinetic. The lower panel shows the kinetic energy in each layer and the potential energy of the system. (b) The energy-box diagram for the final, statistically steady state; (c, d, e) the instantaneous, mean, and eddy fields for the final statistically steady state.

39

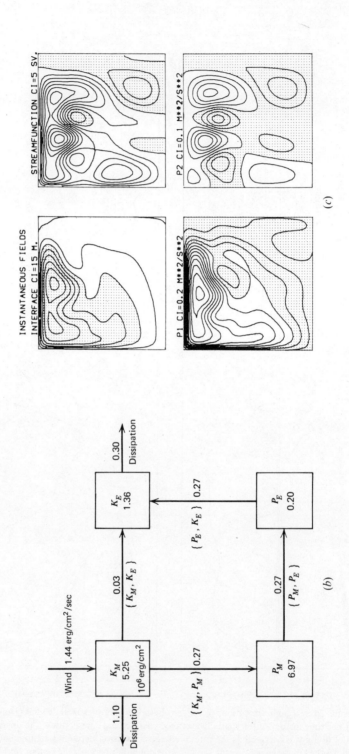

40

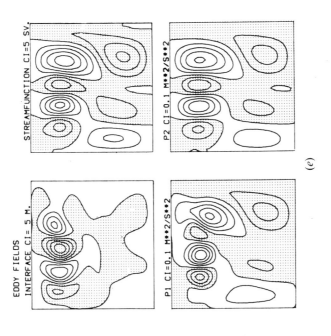

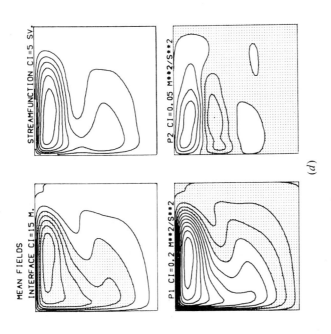

41

eddy flows are shown in Fig. 17*b*. Note the large drain of energy from the total mean energy to the eddies. Figs. 17*c*, 17*d*, and 17*e* show the instantaneous, mean, and eddy fields. Note the important result that the baroclinic eddies set up a secondary circulation in the deep water, entirely driven by the Reynolds stresses. Such a result would not be possible in steady models which parameterize the Reynolds stresses as in earlier studies since here the eddy viscosity is *negative*. This driving of abyssal currents by eddy processes may account for some of the observational discrepancies between water mass analysis and geostrophic dynamics (Worthington, 1976).

These very preliminary experiments do show that mesoscale processes alter the general circulation in the ocean just as transient disturbances do in the atmosphere. The difficulty for the ocean modeler is that these oceanic eddies have very small horizontal scales and thus are difficult to include in models of entire ocean basins. It is not yet known whether some parameterization of these eddy processes in terms of large-scale flow properties is possible. It is clear that the simple closure laws previously used are not adequate.

Where do we go from here? Firstly, new observational programs designed to reveal the nature and particularly the distribution of eddies in various parts of the world ocean are needed. The joint Soviet Union–United States POLYMODE program is one step in that process, but other parts of the ocean (e.g., the Circumpolar Current, the Pacific, etc.) need exploration as well. Little information is available about the variability of eddy energy from ocean to ocean. Secondly, numerical modeling efforts to unravel the role of mesoscale eddies in establishing the characteristics of the large-scale circulation must continue and expand if we are to improve our capability to realistically model the world ocean.

References

Beardsley, R. C., 1969. A laboratory model of the wind-driven ocean circulation. *J. Fluid Mech.*, **38**, 255–271.

Beardsley, R. C., 1973. A numerical model of the wind-driven ocean circulation in a circular basin. *Geophys. Fluid Dyn.*, **4**, 211–241.

Blandford, R., 1965. Notes on the theory of the thermocline. *J. Mar. Res.*, **23**, 18–29.

Blandford, R., 1971. Boundary conditions in homogeneous ocean models. *Deep-Sea Res.*, **18**, 739–751.

Bryan, K., 1963. A numerical investigation of a nonlinear model of a wind-driven ocean. *J. Atmos. Sci.*, **20**, 594–606.

Bryan, K., 1969. A numerical method for the study of the circulation of the world ocean. *J. Comp. Phys.*, **3**, 347–376.

Bryan, K., 1975. Three-dimensional numerical models of the ocean circulation. *Proc. Symposium on Numerical Models of Ocean Circulation*, Durham, N.H. National Academy of Sciences, Washington, D.C.

Bryan, K. and M. D. Cox, 1967. A numerical investigation of the oceanic general circulation. *Tellus*, **19**, 54–80.

Bryan, K. and M. D. Cox, 1968a. A nonlinear model of an ocean driven by wind and differential heating: Part I. Description of the three-dimensional velocity and density fields. *J. Atmos. Sci.*, **25**, 945–967.

Bryan, K. and M. D. Cox, 1968b. A nonlinear model of an ocean driven by wind and differential heating: Part II. An analysis of the heat, vorticity and energy balance. *J. Atmos. Sci.*, **25**, 968–978.

Bryan, K. and M. D. Cox, 1972. The circulation of the world ocean: A numerical study. Part I. A homogeneous model. *J. Phys. Oceanogr.*, **2**, 319–335.

Bryan, K., S. Manabe, and R. C. Pacanowski, 1975. A global ocean-atmosphere climate model: Part II. The oceanic circulation. *J. Phys. Oceanogr.*, **5**, 30–46.

Bye, J. A. T. and T. W. Sag, 1972. A numerical model for circulation in a homogeneous world ocean. *J. Phys. Oceanogr.*, **2**, 305–318.

Carrier, G. F. and A. R. Robinson, 1962. On the theory of the wind-driven ocean circulation. *J. Fluid Mech.*, **12**, 49–80.

Charney, J. G., 1955. The Gulf Stream as an inertial boundary layer. *Proc. Natl. Acad. Sci.*, **41**, 731–740.

Cox, M. D., 1970. A mathematical model of the Indian Ocean. *Deep-Sea Res.*, **17**, 45–75.

Cox, M. D., 1975. A baroclinic numerical model of the world ocean: preliminary results. *Proc. Symposium on Numerical Models of Ocean Circulation*, Durham, N.H. National Academy of Sciences, Washington, D.C.

Defant, A., 1936. Schichtung und Zirkulation des Atlantischen Ozeans. *Wiss. Ergeb. Dtsch. Atl. Exped. "Meteor"*, **6**, Atlas.

Ekman, V. W., 1905. On the influence of the earth's rotation on ocean currents, *Ark. Mat. Astron. Fys.*, **2**, 1–53.

Fofonoff, N. P., 1954. Steady flow in a frictionless homogeneous ocean. *J. Mar. Res.*, **13**, 254–262.

Fofonoff, N. P., 1962. The sea, ideas and observations. In *Dynamics of Ocean Currents*. Vol. 1. Pergamon Press, New York, pp. 323–396.

Friedrich, H. J., 1970. Preliminary results from a numerical multilayer model for the circulation in the North Atlantic. *Dtsch. Hydrogr. Z.*, **23**, 145–164.

Gill, A. E. and K. Bryan, 1971. Effects of geometry on the circulation of a three-dimensional southern-hemisphere ocean model. *Deep-Sea Res.*, **18**, 685–721.

Holland, W. R., 1967. On the wind-driven circulation in an ocean with bottom topography. *Tellus*, **19**, 582–600.

Holland, W. R., 1973. Baroclinic and topographic influences on the transport in western boundary currents. *Geophys. Fluid Dyn.*, **4**, 187–210.

Holland, W. R., 1975a. Energetics of baroclinic oceans. *Proc. Symposium on Numerical Models of Ocean Circulation*, Durham, N.H. National Academy of Sciences, Washington, D.C.

Holland, W. R., 1975b. On the dynamics of the equatorial ocean. In preparation.

Holland, W. R. and A. D. Hirschman, 1972. A numerical calculation of the circulation in the North Atlantic Ocean. *J. Phys. Oceanogr.*, **2**, 336–354.

Holland, W. R. and L. B. Lin, 1975a. On the generation of mesoscale eddies and their contribution to the oceanic general circulation. I. A preliminary numerical experiment. *J. Phys. Oceanogr.*, **5**, 642–657.

Holland, W. R. and L. B. Lin, 1975b. On the generation of mesoscale eddies and their contribution to the oceanic general circulation. II. A parameter study. *J. Phys. Oceanogr.*, **5**, 658–669.

Ilyin, A. M. and V. M. Kamenkovich, 1963. On the influence of friction on ocean current. *Dokl. Akad. Nauk SSSR*, **150**, 1274–1277.

Ilyin, A. M., V. M. Kamenkovich, T. G. Zhugrina, and M. M. Silkina, 1969. On the calculation of the complete circulation in the world ocean (stationary problem). *Izv. Acad. Sci. USSR Atmos. Oceanic Phys.*, **5**, 1160–1171 (English, 668–674).

Kamenkovich, V. M., 1962. On the theory of the Antarctic Circumpolar Current. *Trans. Inst. Okeanol., Acad. Sci. USSR*, **56**, 291–306 (English, 245–301).

Leith, C. E., 1975. Future computing machine configurations and numerical models. *Proc. Symposium on Numerical Models of Ocean Circulation*, Durham, N.H. National Academy of Sciences, Washington, D.C.

Lineykin, P. C., 1955. On the determination of the thickness of the baroclinic layer of fluid heated uniformly above and non-uniformly from below. *Dolk. Akad. Nauk SSSR*, **101**, 461.

Lorentz, E. N., 1967. *The Nature and Theory of the General Circulation of the Atmosphere*. World Meteorological Organization, 161 pp.

Manabe, S., K. Bryan, and M. J. Spelman, 1975. A global ocean–atmosphere climate model: Part I. The atmospheric circulation. *J. Phys. Oceanogr.*, **5**, 3–29.

Marchuk, G. I., A. S. Sarkisyan, and V. P. Kochergin, 1973. Calculations of flows in a baroclinic ocean: numerical methods and results. *Geophys. Fluid Dyn.*, **5**, 89–100.

Moore, D. W., 1963. Rossby waves in ocean circulation. *Deep-Sea Res.*, **10**, 735–747.

Morgan, G. W., 1956. On the wind-driven ocean circulation. *Tellus*, **8**, 301–320.

Munk, W. H., 1950. On the wind-driven ocean circulation. *J. Meteorol.*, **7**, 79–93.

Munk, W. H., 1966. Abyssal recipes. *Deep-Sea Res.*, **13**, 707–730.

Munk, W. H., G. W. Groves, and G. F. Carrier, 1950. Note on the dynamics of the Gulf Stream. *J. Mar. Res.*, **9**, 218–238.

Munk, W. H. and E. Palmén, 1951. Note on the dynamics of the Antarctic Circumpolar Current. *Tellus*, **3**, 53–56.

Niiler, P. P., 1966. On the theory of wind-driven ocean circulation. *Deep-Sea Res.*, **13**, 597–606.

Orlanski, I. and M. Cox, 1973. Baroclinic instability in ocean currents. *Geophys. Fluid Dyn.*, **4**, 297–332.

Pedlosky, J., 1969. Linear theory of the circulation of a stratified ocean. *J. Fluid Mech.*, **35**, 185–205.

Philander, S. G. H., 1973. Equatorial undercurrent: Measurements and theories. *Rev. Geophys. Space Phys.*, **11**, 513–570.

Robinson, A. R., ed., 1963. *On the Wind-Driven Ocean Circulation*. Blaisdell, New York, 161 pp.

Robinson, A. R., 1965. Oceanography. In *Research Frontiers in Fluid Dynamics*. R. J. Seeger and G. Temple, eds. Wiley-Interscience, New York, pp. 504–533.

Robinson, A. R. and H. Stommel, 1959. The oceanic thermocline and the associated thermohaline circulation. *Tellus*, **11**, 295–308.

Robinson, A. R. and P. Welander, 1963. Thermal circulation on a rotating sphere, with application to the oceanic thermocline. *J. Mar. Res.*, **21**, 25–38.

Sag, T. W., 1969. A numerical model for wind-driven circulation of the world's ocean. *Phys. Fluids Suppl.*, **2**, 177–183.

Sarkisyan, A. S. and V. F. Ivanov, 1971. Joint effect of baroclinicity and bottom relief as an important factor in the dynamics of sea currents. *Izv. Acad. Sci. USSR Atmos. Oceanic Phys.*, **7**, 173–188 (English, 116–124).

Sarkisyan, A. S., 1962. On the dynamics of the origin of wind currents in the baroclinic ocean. *Okeanologie*, **11**, 393–409.

Sarkisyan, A. S. and V. F. Ivanov, 1972. Comparison of various methods of calculating currents in a baroclinic ocean. *Izv. Acad. Sci. USSR Atmos. Oceanic Phys.*, **8**, 403–418 (English, 230–237).

Sarkisyan, A. S. and V. P. Keondzhian, 1972. Calculation of the surface level and total mass transport function for the North Atlantic. *Izv. Acad. Sci. USSR Atmos. Oceanic Phys.*, **8**, 1202–1215 (English, 701–708).

Schulman, E. E., 1975. Numerical models of ocean circulation: a study of topographic effects. *Proc. Symposium on Numerical Models of Ocean Circulation*, Durham, N.H. National Academy of Sciences, Washington, D.C.

Schulman, E. E. and P. P. Niiler, 1970. Topographic effects on the wind-driven ocean circulation. *Geophys. Fluid Dyn.*, **1**, 439–462.

Semtner, A. J., 1973. A numerical investigation of Arctic Ocean circulation. Ph.D. thesis, Princeton University, 251 pp.

Stewart, R. W., 1964. The influence of friction on inertial models in oceanic circulation. In *Studies in Oceanography*. K. Yoshida, ed. University of Tokyo Press, pp. 3–9.

Stommel, H., 1948, The westward intensification of wind-driven ocean currents. *Trans. Am. Geophys. Union*, **29**, 202–206

Stommel, H., 1957. A survey of ocean current theory. *Deep-Sea Res.*, **4**, 149–184.

Stommel, H., 1965. *The Gulf Stream*. 2nd ed. University of California Press.

Stommel, H., 1955. Lateral eddy viscosity in the Gulf Stream system. *Deep-Sea Res.*, **3**, 88–90.

Stommel, H. and G. Veronis, 1957. Steady convective motion in a horizontal layer of fluid heated uniformly above and non-uniformly from below. *Tellus*, **9**, 401–407.

Sverdrup, H. U., 1947, Wind-driven currents in a baroclinic ocean, with application to the equatorial currents of the eastern Pacific. *Proc. Natl. Acad. Sci. US.* **33**, 319–326.

Swallow, J. C. and L. V. Worthington, 1961. An observation of a deep countercurrent in the western North Atlantic. *Deep-Sea Res.*, **8**, 1–19.

Veronis, G., 1966a. Wind-driven ocean circulation: Part I. Linear theory and perturbation analysis. *Deep-Sea Res.*, **13**, 17–29.

Veronis, G., 1966b. Wind-driven ocean circulation: Part II. Numerical solutions of the non-linear problem. *Deep-Sea Res.*, **13**, 31–55.

Veronis, G., 1969. On theoretical models of the thermocline circulation. *Deep-Sea Res., Suppl.*, **16**, 301–323.

Veronis, G., 1973. Large scale ocean circulation. In *Advances in Applied Mechanics*. Vol. 13. Academic Press, New York, pp. 1–92.

Warren, B. A., 1963. Topographic influences on the path of the Gulf Stream. *Tellus*, **15**, 167–183.

Webster, F., 1961. The effect of meanders on the kinetic energy balance of the Gulf Stream. *Tellus*, **13**, 392–401.

Welander, P., 1959a. An advective model of the ocean thermocline. *Tellus*, **11**, 309–318.

Welander, P., 1959b. On the vertically integrated mass transport in the oceans. In *The Atmosphere and the Sea in Motion*. B. Bolin, ed. Rockefeller Institute Press and Oxford University Press, pp. 95–101.

Welander, P., 1968. Wind-driven circulation in one and two-layer oceans of variable depth. *Tellus*, **20**, 1–15.

Welander, P., 1971. Some exact solutions to the equations describing an ideal-fluid thermocline. *J. Mar. Res.*, **29**, 60–68.

Welander, P., 1975. Analytical modeling of the oceanic circulation. *Proc. Symposium on Numerical Models of Ocean Circulation*, Durham, N.H. National Academy of Sciences, Washington, D.C.

Wimbush, M. and W. Munk, 1970. The benthic boundary layer. In *The Sea*. Vol. 4, Part 1. Wiley, New York, pp. 731–758.

Worthington, L. V., 1976. *On the North Atlantic Circulation*. The Johns Hopkins Press, Baltimore. To be published.

2. NUMERICAL MODELS OF OCEAN TIDES

Myrl C. Hendershott

1. Introduction

The incorporation, by Laplace in 1776, of Newton's second law into a continuum approximation suitable for the description of shallow water waves completed the framework within which the theoretical study of ocean tides has proceeded up to the present-day global numerical efforts discussed below. Results obtained from what was essentially analysis of the eigenvalue spectrum of Laplace's tidal equations (LTE), as the customary modern statement of the governing approximate continuum equation is usually called, played an important role in the evolution of geophysical fluid mechanics. However, technical difficulties in solving LTE historically prevented success sufficient for practical prediction of sea level directly from the astronomically derived tide generating forces, and tidal prediction proceeded by extrapolation of observations rather than theoretically.

Modern computational techniques offer hope of further progress in the predictive solution of LTE at a time when new scientific and engineering goals further motivate efforts to provide such solutions. Renewed scientific interest in theoretical prediction of ocean tides stems from the pervasiveness, both observed and suspected, of tidal signals and of tidal effects in solid earth and fluid ocean geophysics. Tidal signals in records of gravity, of solid earth tilt and strain, of oceanic and terrestrial electric and magnetic fields, and of precisely measured satellite altitude are, in part, of oceanic origin. Elucidation of the geophysical information inherent in these records requires an accurate model of ocean tides. Provision of a suitable model requires further progress in specifying oceanic sinks of tidal energy and in solving LTE with these sinks properly incorporated. Renewed engineering interest in the theoretical prediction of ocean tides stems from present-day ability to significantly alter the tidal characteristics of sizable bays and rivers by construction of dams (sometimes for tidal power) or causeways. Economic and environmental considerations require that such alterations be predictable in advance of construction; numerical solution of LTE with appropriate boundary conditions is the only rational predictive scheme capable of correctly responding to changes in geometry in advance of subsequent observation of the (altered) tides.

This chapter summarizes the achievements and lacunae of modern finite difference attempts to solve LTE for domains spanning the global ocean or a major ocean basin. On account of this intentionally limited scope, it is important to cite recent works of a more general nature. Platzman (1971) recapitulates the basic dynamical foundations. Hendershott and Munk (1970) and Hendershott (1973) together attempt an up-to-date summary of theoretical understanding of ocean tides. Voit (1967) reviews the Soviet literature. Wunsch (1974) summarizes and augments understanding of the important related problem of ocean internal tides. Miles (1974) points out that work subsequent to Laplace has essentially been concerned with the solution of LTE rather than with extension of the dynamical theory, and critically examines the limit process inherent in obtaining LTE.

The contents and bibliographies of these articles cannot be recapitulated here. Citations of subsequent work are representative rather than exhaustive. The discussion concentrates upon the structure of a theoretical model of ocean tides and upon

47

the methodology existing and needed but not yet developed, for solving such a model by finite differences.

The most general model of ocean tides would also include all effects of solid earth and atmospheric tides as well as effects of direct thermal forcing of the oceans. The model outlined here is far simpler. Atmospheric pressure effects on oceanic tides are excluded altogether, as are the effects of direct thermal forcing of the oceans. For prediction of ocean tides (but *not* of meteorologically induced surges) the resulting error is small (Wunsch, 1972, displays the atmospheric effect at Bermuda; Munk and Cartwright, 1966, estimate the thermal effects) and it appears unlikely to be energetically significant (Munk and McDonald, 1960, p. 204). However, effects of solid earth tides on ocean tides are of order one (Fig. 1) and must be included in the ocean model. Here, this is done in the Love number approximation appropriate to a static and dissipationless elastic solid earth. The appropriate refinements of this procedure have been developed by Farrell (1972a, b); it appears to be of an accuracy far superseding that of our ability to treat the fluid portion of the problem. Its adaptation rules out solid earth dissipation ab initio—the high Q's of low-mode, seismic, free oscillations of the earth and the success (Farrell, 1972b) of this approach in removing the ocean effect from solid earth (gravity) tides are invoked in its justification—and we may correspondingly draw only indirect conclusions about solid earth dissipation from any of the model solutions (although the proposed formalism could in principle allow for complex Love numbers).

The fluid part of the model, a modified version of LTE with appropriate boundary conditions, presents difficulties because of the closeness to resonance of the problem to be made discrete. Although strongly conditioned by dissipation, ocean tides appear to be of sufficiently high Q that the admittance of realistic models may well *not* be essentially flat over tidal bands (Hendershott, 1973). Numerically, this is likely to mean that some degree of the extreme sensitivity to variation of details of discretization encountered in dissipationless global tide calculations will persist in more realistically dissipative ones, especially in application of the models more demanding than simple reproduction of the major features of the elevation distribution.

The exposition of the model begins by recapitulating LTE (Section 2) with emphasis upon a formalism for inclusion of solid earth deformation and ocean self-attraction via the Love number representation. Inclusion of these effects significantly modifies

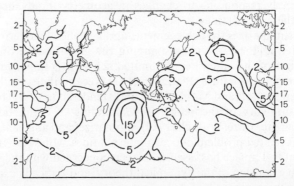

Fig. 1.   Co-range lines (cm) of the self potential of ocean tides plus the potential due to the earth's deformation under the weight of the ocean tidal column. Numbers on margins are amplitudes (cm) of astronomical potential plus potential due to solid earth deformation by astronomical potential (all potentials divided by g).

the usual statement of energy conservation (Section 4). From the global point of view, boundary conditions effectively parameterize the trapping, scattering, and dissipation of tidal energy occurring over shelves and in marginal seas. Formulation (Section 3) of appropriate boundary conditions hinges upon the conceptual separation of deep-sea and shallow-sea problems. Once suitable bathymetry is prepared, the differential part of the gobal tide problem may be discretized straightforwardly and either time stepped or solved for individual tidal harmonics to obtain the forced tide. The numerical apparatus required is discussed in Section 5; it is fundamental in studies (Platzman, 1972a, 1974; Christensen, 1972, 1973) of ocean normal modes as well, but the further development required to extract the normal modes is not reviewed here. Neglect of solid earth loading and of ocean self-attraction is a first-order failing common to all published global solutions of LTE. A procedure for their inclusion is proposed and results of its application in a simple geometry are summarized. Finally (Section 6) results of various global computations are compared.

2. The Model Equations

LTE are

$$u_t - (2\Omega \sin \theta)v = -\frac{g(\zeta - \Gamma/g)_\phi}{a \cos \theta} + \frac{F^\phi}{\rho D} \tag{1}$$

$$v_t + (2\Omega \sin \theta)u = -\frac{g(\zeta - \Gamma/g)_\theta}{a} + \frac{F^\theta}{\rho D} \tag{2}$$

$$(\zeta_t - \delta_t) + \frac{1}{a \cos \theta}[(uD)_\phi + (vD \cos \theta)_\theta] = 0 \tag{3}$$

Here (ϕ, θ) are longitude and latitude (*not* colatitude) with corresponding velocity components (u, v), ζ and δ are geocentric ocean and solid earth tides, Γ is the total tide producing potential, g is the (constant) downward acceleration of gravity at the spherical surface of the earth (of radius a), ρ is the density of the oceans, D is the (variable) depth of the ocean, and Ω is the earth's angular rate of rotation. Subscripts ϕ, θ, and t refer, respectively, to differentiation by longitude, latitude, and time; the superscripts ϕ and θ denote the zonal and meridional components of dissipative stress F. Note that most authors work in the conventional spherical coordinates, in which θ represents colatitude rather than latitude, so that their equations have a slightly different appearance. Equations 1–3 formally neglect any vertical variation of the velocities u and v and hence from the outset exclude the internal tides associated with the stable stratification of the real oceans. Consequently, although they may form an adequate basis for discussion of the elevation field, they are not directly suitable for velocity computations except in vertically well mixed marginal seas.

The geocentric tides ζ and δ are not directly observable. Rather, the observed ocean tide recorded by a gauge fixed to the surface of the earth is

$$\zeta_0 = \zeta - \delta \tag{4}$$

The total tide-generating potential Γ contains not only an astronomical contribution U computable from the mass of the tide generating bodies and their motions; there are also contributions arising from the elastic yielding of the solid earth, both to the astronomical component of tide generating force as well as to the weight of the oceanic tidal column, and from the Newtonian self-attraction of the oceans

(Hendershott, 1972). If the tide-generating potentials Γ and U, the observed ocean tide ζ_0, and the solid earth tide δ are decomposed into spherical harmonic components (denoted by subscript n^1) then the usual Love numbers k_n, h_n, k'_n, and h'_n for body forces and surface loading (Munk and McDonald, 1960) allow expression of the total potential as

$$\Gamma_n = (1 + k_n)U_n + (1 + k'_n)g\alpha_n\zeta_{0_n} \tag{5}$$

and the geocentric earth tide is

$$\delta_n = \left(h_n\frac{U_n}{g}\right) + (h'_n\alpha_n\zeta_{0_n}) \tag{6}$$

Here $\alpha_n \equiv [3/(2n + 1)]$ (water density/solid earth density).

These express the potential and the solid earth tide in terms of the astronomical potential plus the unknown "observed" ocean tide ζ_0. It is convenient to use them to rewrite equations 1–3 so that the only independent variables appearing are u, v, and ζ_0. We have, from equations 5 and 6,

$$\zeta - \frac{\Gamma}{g} = \zeta_0 + \left(\delta - \frac{\Gamma}{g}\right) = \zeta_0 - \sum_n \frac{(1 + k_n - h_n)U_n}{g} - \sum_n (1 + k'_n - h'_n)\alpha_n\zeta_{0_n}$$

in which the ultimate term is evidently the convolution of the observed tide ζ_0 with a certain Green's function $G(\theta', \phi'|\theta, \phi)$ which has been constructed by Farrell (1972c) for a dissipationless, static, perfectly elastic earth:

$$\sum_n (1 + k'_n - h'_n)\alpha_n\zeta_{0_n} = \iint d\theta' \, d\phi' \, \cos\theta' \, G(\theta', \phi'|\theta, \phi)\zeta_0(\theta', \phi')$$

Equations 1–3 thus become

$$u_t - (2\Omega \sin\theta)v = -\frac{g}{a\cos\theta}\frac{\partial}{\partial\phi}\left[\zeta_0 - \sum_n \frac{(1 + k_n - h_n)U_n}{g}\right.$$
$$\left. - \iint d\theta' \, d\phi' \, \cos\theta' \, G(\theta', \phi'|\theta, \phi)\zeta_0(\theta', \phi')\right] + \frac{F^\phi}{\rho D} \tag{7}$$

$$v_t + (2\Omega \sin\theta)u = -\frac{g}{a}\frac{\partial}{\partial\theta}\left[\zeta_0 - \sum_n \frac{(1 + k_n - h_n)U_n}{g}\right.$$
$$\left. - \iint d\theta' \, d\phi' \, \cos\theta' \, G(\theta', \phi'|\theta, \phi)\zeta_0(\theta', \phi')\right] + \frac{F^\theta}{\rho D} \tag{8}$$

$$\zeta_{0_t} + \frac{1}{a\cos\theta}[(uD)_\phi + (vD\cos\theta)_\theta] = 0 \tag{9}$$

and are clearly integrodifferential in nature on account of the effects of ocean loading and of ocean self-attraction summarized in the convolution. It is further convenient to let

$$Z = \zeta - \frac{\Gamma}{g} = \zeta_0 + \left(\delta - \frac{\Gamma}{g}\right) \tag{10}$$

---

[1] The astronomical potential is almost entirely of order $n = 2$.

That is, the departure Z of the geocentric ocean tide from equilibrium with the full tide-generating potential Γ replaces ζ_0 as independent variable and so

$$u_t - (2\Omega \sin \theta)v = -\frac{g}{a \cos \theta} Z_\phi + \frac{F^\phi}{\rho D} \tag{11}$$

$$v_t + (2\Omega \sin \theta)u = -\frac{g}{a} Z_\theta + \frac{F^\theta}{\rho D} \tag{12}$$

$$Z_t + \frac{1}{a \cos \theta} [(uD)_\phi + (vD \cos \theta)_\theta] = -\sum_n \frac{(1 + k_n - h_n)U_{n_t}}{g}$$
$$- \iint d\theta' \, d\phi' \cos \theta' \, G(\theta', \phi'|\theta, \phi)\zeta_{0_t}(\theta', \phi') \tag{13}$$

Many authors prefer to work in Mercator coordinates. In negative Mercator coordinates (with the additional minus sign prefixed to equation 14 in order to keep north Mercator latitude positive), the Mercator latitude τ and the geographical spherical latitude θ are related by

$$\tau = -\ln \tan\left(\frac{\pi}{4} - \frac{\theta}{2}\right) \tag{14a}$$

$$\theta = \frac{\pi}{2} - 2 \arctan(\exp - \tau) \tag{14b}$$

Mercator velocity components $\mathscr{U}$ and $\mathscr{V}$ and stresses $\mathscr{F}^\phi$ and $\mathscr{F}^\tau$ are related to their spherical counterparts by

$$\begin{array}{ll}
\mathscr{U} = u \text{ sech } \tau & \mathscr{V} = v \text{ sech } \tau \\
\mathscr{F}^\phi = F^\phi \text{ sech } \tau & \mathscr{F}^\tau = F^\theta \text{ sech } \tau
\end{array} \tag{15}$$

so that equations 11–13 become

$$\mathscr{U}_t - (2\Omega \tanh \tau)\mathscr{V} = -\frac{gZ_\phi}{a} + \frac{\mathscr{F}^\phi}{\rho D} \tag{16}$$

$$\mathscr{V}_t + (2\Omega \tanh \tau)\mathscr{U} = -\frac{gZ_\tau}{a} + \frac{\mathscr{F}^\tau}{\rho D} \tag{17}$$

$$Z_t + \frac{1}{a \text{ sech}^2 \tau} [(\mathscr{U}D)_\phi + (\mathscr{V}D)_\tau] = -\sum_n \frac{(1 + k_n - h_n)U_{n_t}}{g}$$
$$- \iint d\tau' \, d\phi' \text{ sech}^2 \tau' \, G(\phi', \tau'|\phi, \tau)\zeta_{0_t}(\phi', \tau') \tag{18}$$

The quasi-planar form of these equations is attractive for numerical work.

LTE are hyperbolic in x, y, and t and are natural candidates for numerical solution by time stepping. Alternatively, we may note that they are linear provided the dissipative stresses may be linearized. Consequently, the time dependence of their solutions is that of the potential U; harmonic at discrete frequencies σ calculable from astronomy (this is observed to be the case to a remarkable degree). Each tidal harmonic obeys an equation elliptic in x and y; the ensemble of elliptic problems may be solved frequency by frequency and the results linearly superposed.

We thus introduce the complex forms of all independent variables, writing, for example

$$\text{observed tide } \zeta_0(\phi, \theta, t) = Re[\text{complex elevation } \zeta_0(\phi, \theta)e^{-i\sigma t}] \tag{19}$$

Note, however, that many tidal theorists take $e^{+i\sigma t}$. We use the same symbol to denote either the real, time-dependent part of a quantity or its complex, spatially variable form. In the latter notation, equations 16–18 become

$$-i\sigma \mathcal{U} - (2\Omega \tanh \tau)\mathcal{V} = -\frac{gZ_\phi}{a} + \frac{\mathcal{F}^\phi}{\rho D} \tag{20}$$

$$-i\sigma \mathcal{V} + (2\Omega \tanh \tau)\mathcal{U} = -\frac{gZ_\tau}{a} + \frac{\mathcal{F}^\tau}{\rho D} \tag{21}$$

$$-i\sigma Z_t + \frac{1}{a \operatorname{sech}^2 \tau}[(\mathcal{U}D)_\phi + (\mathcal{V}D)_\tau] = +i\sigma \sum_n \frac{(1 + k_n - h_n)U_n}{g}$$

$$+ i\sigma \iint d\tau' \, d\phi' \operatorname{sech}^2 \tau' \, G(\phi', \tau' | \phi, \tau)\zeta_0(\phi', \tau') \tag{22}$$

Now equations 20 and 21 may be rewritten as

$$\mathcal{U} = -\frac{g}{2\Omega a}\left[\frac{isZ_\phi - \tanh \tau \, Z_\tau}{(s^2 - \tanh^2 \tau)} + \frac{1}{\rho D}\frac{- is\mathcal{F}^\phi + \tanh \tau \, \mathcal{F}^\tau}{(s^2 - \tanh^2 \tau)}\right]$$

$$\mathcal{V} = -\frac{g}{2\Omega a}\left[\frac{isZ_\tau + \tanh \tau \, Z_\phi}{(s^2 - \tanh^2 \tau)} + \frac{1}{\rho D}\frac{- is\mathcal{F}^\tau - \tanh \tau \, \mathcal{F}^\phi}{(s^2 - \tanh^2 \tau)}\right] \tag{23}$$

where $s = \sigma/2\Omega$. Substitution of equation 23 into 22 now gives a single eliptic equation for Z, namely,

$$[(QHZ_\phi)_\phi + (QHZ_\tau)_\tau] - \frac{i}{s}[(QH \tanh \tau \, Z_\phi)_\tau - (QH \tanh \tau \, Z_\tau)_\phi]$$

$$+ \varepsilon^2 \operatorname{sech}^2 \tau \, Z = \frac{1}{\rho D_0}[(Q\mathcal{F}^\phi)_\phi + (Q\mathcal{F}^\tau)_\tau] - \frac{i}{\rho D_0 s}$$

$$\times [(Q \tanh \tau \, \mathcal{F}^\phi)_\tau - (Q \tanh \tau \, \mathcal{F}^\tau)_\phi] - \varepsilon^2 \operatorname{sech}^2 \tau\left[\sum_n \frac{(1 + k_n - h_n)U_n}{g}\right.$$

$$\left. + \iint d\tau' \, d\phi' \operatorname{sech}^2 \tau' \, G(\phi', \tau' | \phi, \tau)\zeta_0(\phi, \tau)\right] \tag{24}$$

where $D = D_0 H(\phi, \tau)$ divides the relief into its mean D_0 and normalized variation $H(\phi, \tau)$, $Q = 1/(s^2 - \tanh^2 \tau)$, and $\varepsilon^2 = 4\Omega^2 a^2/gD_0$.

Further cross-differentiation of the left-hand side of equation 24 produces a slightly simpler equation

$$QH\nabla^2 Z + Z_\phi\left[(QH)_\phi - \frac{i}{s}(QH \tanh \tau)_\tau\right]$$

$$+ Z_\tau\left[(QH)_\tau + \frac{i}{s}(QH \tanh \tau)_\phi\right] + \varepsilon^2 \operatorname{sech}^2 \tau \, Z = \cdots \tag{24a}$$

If the dissipative stresses are now taken linear in the velocities, that is,

$$\mathscr{F}^\phi = -\mathscr{U}D_0\rho\sigma r$$
$$\mathscr{F}^\tau = -\mathscr{V}D_0\rho\sigma r \tag{25}$$

where r is a dimensionless but spatially variable bottom friction coefficient (not the γ of equation 26, then equations 20 and 21 may be rewritten

$$-i\sigma\left(1 + i\frac{r}{H}\right)\mathscr{U} - (2\Omega\tanh\tau)\mathscr{V} = -\frac{gZ_\phi}{a}$$

$$-i\sigma\left(1 + i\frac{r}{H}\right)\mathscr{V} + (2\Omega\tanh\tau)\mathscr{U} = -\frac{gZ_\tau}{a}$$

and equation 24 becomes the rather simpler

$$\left[\left(Q'\left(1 + i\frac{r}{H}\right)HZ_\phi\right)_\phi + \left(Q'\left(1 + i\frac{r}{H}\right)HZ_\tau\right)_\tau\right]$$

$$-\frac{i}{s}[(Q'H\tanh\tau\ Z_\phi)_\tau - (Q'H\tanh\tau\ Z_\tau)_\phi]$$

$$+ \varepsilon^2\operatorname{sech}^2\tau\ Z = -\varepsilon^2\operatorname{sech}^2\tau\left[\sum_n \cdots\right] \tag{24b}$$

where

$$Q' = \frac{1}{s^2\left(1 + i\dfrac{r}{H}\right)^2 - \tanh^2\tau}$$

Up to this point, we have sidestepped the question of a representation for the bottom stresses **F**. Unfortunately, no definitive recommendation may be made. The linearization (equation 25) is based upon the bottom drag law

$$\frac{\mathbf{F}}{\rho D} = -\gamma|u|\mathbf{u} \qquad \gamma \sim 0.002 \tag{26}$$

customary and reasonably satisfactory in shallow-water calculations. The work of Kagan (1972) bears upon possible improvement of this. In the deep ocean, complete expressions for **F** should include internal wave conversion; the work of Wunsch (1969), Baines (1974), and Bell (1973) might be taken as the starting point. In numerical work an eddy viscosity term is often added to some form of equation 26 to insure stability (see equation 57).

Aside from this difficulty, numerical solution of LTE by time stepping or frequency by frequency would be relatively straightforward (with the possible exception of the convolution terms) for oceans covering the entire globe. However, the existence of continents raises the question of boundary conditions.

3. The Boundary Condition

The boundary between land and ocean is not smooth at any imaginable global scale of discretization. Indeed at the global scales of discretization currently computationally feasible (1–6°), single grid points may represent entire shallow sea entrances or hundreds of kilometers of open coastline. At the former, the seemingly natural coastal condition of vanishing normal flow

$$\mathbf{u} \cdot \hat{n} = 0 \quad \text{at coasts} \tag{27}$$

is not appropriate whereas at the latter, the velocity field must be well resolved. But the shoal relief characteristic of continental shelves may well force the horizontal scales of the tides there to be locally smaller than the global scale of discretization so that the nearshore variation is not resolved at all.

Employment of equation 27 everywhere raises the additional requirement that the dissipative stresses $F^\phi/\rho D$, $F^\theta/\rho D$ appearing in LTE be accurately described as part of the solution. Now even if all contributions to these stresses except the bottom friction were negligible, so that the customary drag law $\rho\gamma\mathbf{u}_b|u_b|$ with γ typically of order 0.002 were available, LTE do not have in them the value $\mathbf{u}_b$ of tidal flow velocity at the bottom so that evaluation of the drag expression requires the assumption, not necessarily valid where stratification is appreciable, that $\mathbf{u}_b$ is only the barotropic component. Even if we retreat to vertically well mixed shallow water, it is unlikely that LTE discretized at a typical global scale will there resolve $\mathbf{u}$ adequately to estimate $\rho\gamma\mathbf{u}|u|$ with accuracy. In spite of these objections, a number of reasonably realistic global solutions take equation 27 as sole boundary condition (Pekeris and Accad, 1969; Zahel, 1970, 1973; Gordeev, Kagan, and Rivkind, 1973). Use of a graded net to overcome some of these difficulties appears impractical in a finite difference formulation because of the extreme irregularity of boundary shape. Consequently alternative boundary conditions are required. They must reflect both dissipation and phase delay associated with small-scale (usually subgrid) tidal features occurring over shoal relief and in relatively small marginal seas.

One alternative approach is to suppose (as is not inconsistent with the existing inadequate data) that the action of dissipative stresses is confined primarily to shallow water regions over shelves and in marginal seas. If observations ζ_{os} of tidal elevation at the seaward edge of all such regions were available, then equation 27 could be replaced by

$$\zeta_0 = \zeta_{os} \quad \text{seaward of shelves and marginal seas} \tag{28}$$

and the solutions would be relieved of the requirement that they reproduce the dissipative stresses point by point. There would be, at any instant, generally nonzero flow of water onto or off the shallow regions treated by equation 28 and hence the energy flux into these regions would be computable once the deep sea solution had been obtained.

In fact, the required observations ζ_{os} are all but nonexistent. Nevertheless, supposing that elevations at the seaward edge of shelves are not greatly different from those measured at the coasts themselves, solutions have been obtained with essentially

$$\zeta_0 = \zeta_{oc} \tag{29}$$

seaward of shelves, where ζ_{oc} is the observed coastal elevation *landward* of these. This procedure is not entirely arbitrary, as the following discussion indicates, but it ought to be modified appreciably for certain types of offshore relief and at the mouths of marginal seas. It has proved useful in obtaining fairly realistic numerical solutions for deep-sea elevations (Bogdanov and Magarik, 1967, 1969), solutions which further yield realistic global dissipation (Hendershott, 1972).

A third approach, originally due to Proudman (1944) requires specification of an albedo α_p (possibly complex) relating normal velocity and elevation seaward of a dissipative region.

$$\mathbf{u} \cdot n = \alpha_p \zeta \quad \text{seaward of dissipative regions} \tag{30}$$

For long, narrow rectangular gulfs, the Kelvin wave fits of Hendershott and Speranza (1971) to gulf tides suggest a way to approximate α_p. The formulation of Garrett (1974) gives precise meaning to the albedo in terms of certain response functions characterizing the deep sea and marginal seas of arbitrary shape. For extended marginal sea openings or for large shelves characterized by coastally trapped energy-propagating modes, α_p is dependent upon the energetics of the entire shoal region and is not simply interpretable in a purely local sense. In large-scale application, α_p has been determined phenomenologically (Gohin, 1961).

Needless to say, any global mixture of equations 27, 29, and 30 may also be dealt with numerically. Equations 29 and 30 have the disadvantage that the resulting global solutions need not conserve mass at every instant; that is, in general,

$$\partial_t \int_{\text{oceans}} d\theta \, d\phi \, \cos\theta \, \zeta_0(\phi, \theta) \neq 0$$

This occurs because those solutions do not encompass the entire ocean. The neglected loading raises difficulties when evaluating the convolutions

$$\iint d\theta' \, d\phi' \, \cos\theta' \, G(\phi', \theta' | \phi, \theta) \zeta_0(\phi', \theta')$$

(Farrell, 1972c) and hence constitutes a problem to be overcome before meaningful solutions of LTE allowing for self-attraction and solid earth loading may be obtained with these boundary conditions.

In order to suggest improvements in the alternative boundary conditions it is necessary to discuss the properties of tides over shelf regions and tides in marginal seas. Besides providing the link between coastal observations and deep-sea boundary conditions, these problems (especially the latter) are miniature versions of the deep-sea problem and experience with them provides useful guidance as to how to proceed with the global calculation.

The shoaling relief of long straight continental shelves is capable of appreciable local modification of motions which would occur at a vertical coastline fronting deep water and of giving rise to a number of additional locally trapped modes (Munk, Snodgrass, and Wimbush, 1970). The former comprise the usual rotationally trapped Kelvin wave as well as a continuum of Poincaré waves which may be thought of as the nearshore representation of basin-wide modes having appreciable (possibly even greatest) amplitudes of elevation variation in the deep sea. The latter include an infinite family of refractively trapped Stokes high-frequency edge waves and an infinite family of quasi-geostrophic low-frequency shelf waves. The Stokes edge waves exist as trapped modes only for longshore wavelengths sufficiently short that total internal reflection may occur at shelf's edge. For a nonrotating shelf of width a, depth D_s, and offshore depth D_o, the resulting minimal frequencies σ_c for Stokes waves are $a^2 \sigma_c^2 = n^2 \pi^2 g D_o D_s / (D_o - D_s)$, $n = 0, 1, 2, \ldots$. Rotation increases these for $n \neq 0$. One $n = 0$ wave becomes the usual Kelvin wave as $\sigma \to 0$; the other is cut off by rotation at a frequency slightly exceeding the Coriolis parameter.

At mid-latitudes over relatively narrow shelves ($a = 150$ km, $D_s = 0.5$ km, $D_o = 4$ km; $\sigma_c \sim 20n$ c.p.d.) such as the Californian or Chilean shelves, neither edge nor shelf modes are resonantly excited and the cross-relief variation of elevation associated with the remaining Kelvin wave and Poincaré continuum is only slightly altered by the relief (Stock, personal communication). The transition from coastal observations ζ_{oc} to elevation ζ_{os} seaward of the shelf is thus quite gentle and construction of ζ_{os} from ζ_{oc} for use in equation 28 appears straightforward. Even equation 29 is not unreasonable. In these cases, the energy flow is parallel to the shelf so that dissipation is small in spite of the shoal shelf depth.

Over a wider and more shoal shelf such as the Patagonian shelf ($a = 1000$ km, $D_s = 0.1$ km, $D_o = 4$ km; $\sigma_c \sim 2n$ c.p.d.), Stock (personal communication) points out that the Kelvin mode has virtually decayed to zero by shelf's edge so that seaward extrapolation of coastal observation requires accurate delineation of the Poincaré contribution. Resonant excitation of the first refractively trapped edge mode (traveling opposite the Kelvin wave) cannot be ruled out. Stock further notes that the coastal shape and shelf width are not smooth over the (rather small) long-coast wavelength of these modes, so that the mode-fitting procedure of Munk, Snodgrass, and Wimbush (1970) cannot be successfully applied here, even if the formalism were modified to include the significant dissipation (Miller, 1966) occurring over this shelf. What we do learn is that the small-scale tidal features observed over the shelf do not likely extend far enough into the deep sea to be reflected in a deep-sea boundary condition such as equation 28 just as in the numerical calculation of von Trepka (1967) (Fig. 2) for an idealized shelf and basin. The apparent dominance of trapped modes means that a local numerical model of Patagonian shelf tides could probably be constructed which would be relatively insensitive to conditions seaward of the shelf. Unfortunately, it probably also means that seaward extrapolation of coastal observation to obtain

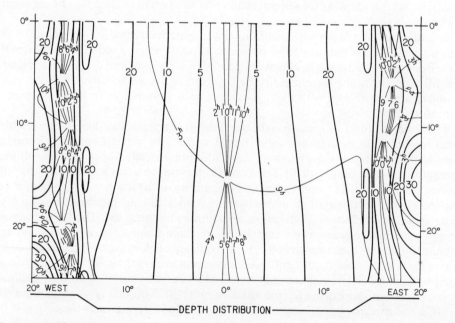

Fig. 2.   Co-tidal and co-range lines for a rectangular ocean with continental shelves at east and west coasts (Von Trepka, 1967).

ζ_{os} will be unreliable here. The alternative to full resolution of tides over the shelf is an admittance type of condition (equation 30). We momentarily defer discussion of this possibility.

At high latitudes, diurnal tides may excite one or more quasi-geostrophic shelf modes. This is not likely to be of direct importance for the elevations but means that large diurnal tidal currents may appear over shelves. Cartwright (1969) reports observations suggesting this possibility.

Shelf regions such as the east coast of the United States which are characterized by almost simultaneous (semidiurnal) tidal rise and fall have received less theoretical attention, although Redfield (1958) points out a noteworthy correlation between shelf width and local coastal tidal properties.

The relationship between deep-sea tides and tides in marginal seas has been elucidated by Garrett (1974). The discussion is useful here not only because of the problem of representing marginal seas correctly in the global models but more generally because of the light which it throws upon the procedure advocated in equations 29 and 30 above, of solving part rather than all of the essentially elliptic boundary value problem posed (for harmonic time dependence) by LTE with equation 27 at boundaries.

Consider a gulf G opening on the deep ocean O with mouth M (across which distance is measured by s). The usual approach, described by Defant (1961) to solving for gulf tides ζ_G is to solve LTE twice, (a) first with

$$\mathbf{u} \cdot \hat{n} = 0 \qquad \text{at gulf coast } B$$

$$\zeta_{(a)} = 0 \qquad \text{at gulf mouth } M$$

$$\Gamma \neq 0 \qquad \text{in gulf (gravitational forcing)}$$

and (b) secondly with

$$\mathbf{u} \cdot \hat{n} = 0 \qquad \text{at gulf coast } B$$

$$\zeta_{(b)} = \zeta_M(s) \quad \text{the elevation observed across the mouth } M$$

$$\Gamma = 0 \qquad \text{in gulf (no forcing)}$$

thus obtaining, respectively, $\zeta_{(a)}(\mathbf{x})$ and $\zeta_{(b)}(\mathbf{x})$. The full solution $\zeta_G = \zeta_{(a)} + \zeta_{(b)}$ satisfies LTE with potential Γ and observed variation $\zeta_M(s)$ of elevation across the mouth. Its resonances are those obtained by seeking natural solution of the homogeneous gulf problem, that is, (b) with $\zeta_M(s) = 0$, but they are not the resonances that would actually be observed in the gulf.

Garrett (1974) considers the ocean and deep-sea problems simultaneously so that $\zeta_M(s)$ is determined as part of the solution rather than fixed a priori. To the extent that the result is insensitive to the distant details of the deep-sea problem, the procedure is a viable local one for gulf tides. It also provides the correct way of summarizing the properties of those gulf tides into a deep-sea boundary condition such as equation 30. To this end, let $\zeta_G^{(1)}$ and $\zeta_0^{(1)}$ be the gulf and deep-sea tides obtained by solving LTE (with forcing) with $\mathbf{u} \cdot \hat{n} = 0$ at M, that is, with the mouth closed by a thin membrane. Furthermore, let the responses of gulf and deep sea to the flux $D\mathbf{u} \cdot \hat{n} = \delta(s - \sigma)$ at the mouth be, respectively, $\zeta_G(\mathbf{x}) = K_G(\mathbf{x}, \sigma)$ and $\zeta_0(\mathbf{x}) = -K_0(\mathbf{x}, \sigma)$. Then the elevations $\zeta_G^{(2)}$ and $\zeta_0^{(2)}$ obtained by solving LTE without forcing but with an arbitrarily specified flux $F(s)$ across the mouth are respectively

$$\int_M K_G(\mathbf{x}, \sigma)F(\sigma)\,d\sigma \quad \text{and} \quad -\int_M K_0(\mathbf{x}, \sigma)F(\sigma)\,d\sigma$$

The sums $\zeta_G^{(1)} + \zeta_G^{(2)}$ and $\zeta_0^{(1)} + \zeta_0^{(2)}$ now solve the forced LTE for both deep sea and gulf provided that $F(s)$ is chosen to make them continuous at M, that is, such that it satisfies the integral equation

$$\zeta_G^{(1)}(s) + \int_M K_G(s, \sigma)F(\sigma)\,d\sigma = \zeta_0^{(1)}(s) - \int_M K_0(s, \sigma)F(\sigma)\,d\sigma = \zeta_M(s) \qquad (31)$$

This is a relationship closely analogous to the Proudman boundary condition (equation 30); if $\zeta_G^{(1)}(s) \ll \zeta_M(s)$ (i.e., if the closed gulf is far from resonance) and, if the mouth is small enough that the integrals over it may be replaced by simple averages, then the first of equation 31 becomes

$$K_G D_M u_M = \zeta_M$$

which is just the form of equation 30. However, equation 31 is far more general.

Garrett discusses the various approximate balances of equation 31 made possible by differing geometries of gulf and ocean and he gives details of the case of a very narrow gulf opening onto a semi-infinite plane ocean. Especially noteworthy is the fact that the resonances of the full solution are *not* those of the gulf with $\mathbf{u} \cdot \hat{n} = 0$. The principal resonance is near $k_G d = \pi/2$ (where $k_G{}^2 = \sigma^2/gD_G$, D_G is the depth of the gulf), that is, near but not precisely equal to the resonance of the Defant decomposition, and it is bounded even when the entire problem is dissipation free. This corresponds to radiative damping of the oscillation in the gulf, a possibility altogether excluded by the usual imposition of observed elevations across the mouth, and is an especially appropriate point of view for short-period (and hence short-wavelength) gulf or harbor seiches. If the ocean is taken of finite size rather than as semi-finite, then the combined system of dissipationless gulf plus dissipationless ocean can of course display unbounded resonances. Some of these will be close to quarter-wave resonances of the gulf if the ocean is large or deep, but locating them more precisely requires knowledge of $\int_M K_0(\mathbf{x}, \sigma)F(\sigma)\,d\sigma$. The difference between predictions obtained in this manner and those obtained in the usual manner may be of considerable practical significance in model assessment of the effects upon gulf tides of proposed large-scale engineering alterations of gulf geometry (Garrett, 1972).

The two alternative boundary conditions (equations 29 and 30) already utilized for the global problem are entirely analogous to the two procedures of Defant (1961) and Garrett (1974) outlined above. The simpler equation 29 may well prove adequate for dynamical interpolation of elevations provided that appropriate observations become available, and it may be useful for estimates of global dissipation, but it cannot yield a correct frequency response and may well be overly sensitive to details of discretization because its resonances are damped only by open ocean dissipation included in the model. Inclusion of the convolution terms of (13) or (18) requires particular attention to global mass conservation. A discrete version of equation 31 offers a natural way to join deep-sea tidal models to more detailed, local models of gulfs or shallow seas, provided that the latter models are linear. The models may be used to obtain values of $K_G(s, \sigma)$ and $K_0(s, \sigma)$ at mesh points; subsequent solution of a discrete analogue of equation 31 would then give $\zeta_M(s)$ so that the global solution could follow from equation 28. However, many details remain to be worked out.

4. Energetics of the Model

Before proceeding to numerical details, we summarize the energetics of the model. Further details are given in Hendershott (1972). Multiplication of equations 1 and 2 by $\rho u D$ and $\rho v D$, respectively, yields

$$\frac{1}{2}\rho D(u^2 + v^2)_t = -\rho g D \mathbf{u} \cdot \mathbf{V}\left(\zeta - \frac{\Gamma}{g}\right) + \mathbf{u} \cdot \mathbf{F} \tag{32}$$

where $\mathbf{V}$ is the two-dimensional gradient over the surface of the sphere. Use of

$$D\mathbf{u} \cdot \mathbf{V}\left(\zeta - \frac{\Gamma}{g}\right) = \mathbf{V} \cdot \left(\mathbf{u}D\left(\zeta - \frac{\Gamma}{g}\right)\right) - \left(\zeta - \frac{\Gamma}{g}\right)\mathbf{V} \cdot \mathbf{u}D$$

plus continuity (3) to eliminate $\mathbf{V} \cdot \mathbf{u}D$ yields, with addition of $\rho g(-\delta\delta_t + \delta_t D)$ to both sides of equation 32,

$$(KE + PE)_t + \mathbf{V} \cdot \mathbf{P} = W + \mathbf{u} \cdot \mathbf{F} \tag{33}$$

where

$$KE = \frac{1}{2}\rho D(u^2 + v^2) \tag{34}$$

$$PE = \frac{1}{2}\rho g(\zeta^2 - \delta^2 + 2\delta D) = \frac{1}{2}\rho g(\zeta_0{}^2 + 2\zeta_0\delta + \delta D) \tag{35}$$

$$\mathbf{P} = \rho g \mathbf{u}D\zeta = \rho g \mathbf{u}D(\zeta_0 + \delta) \tag{36}$$

and

$$W = \rho\zeta_{0_t}\Gamma + \rho g(\zeta_0 + D)\delta_t + \rho\mathbf{V} \cdot (\mathbf{u}D\Gamma) \tag{37}$$

are, respectively, kinetic and potential energies, power flux, and the rate of working on the oceans by tide-generating forces and by the tidally heaving ocean floor.

These expressions are valid for an arbitrarily specified total potential Γ and solid earth tide δ. They appear more familiar if specialized to the case of a rigid earth covered by a non-self-attracting ocean:

$$KE' = \frac{1}{2}\rho D(u^2 + v^2) \tag{38}$$

$$PE' = \frac{1}{2}\rho g\zeta_0{}^2 \tag{39}$$

$$\mathbf{P}' = \rho g \mathbf{u}D\zeta_0 \tag{40}$$

$$W' = \rho\zeta_{0_t}U + \rho\mathbf{V} \cdot (\mathbf{u}DU) \tag{41}$$

In this form, they are the basis for a number of studies of coastal tidal energy loss initiated by Taylor (1919) and continuing through Miller (1966). However, it is not clear that the terms thus neglected are insignificant. Garrett (1974) points out that the ultimate term of equation 41 has incorrectly been excluded from discussions of marginal sea dissipation.

Equations 33–37 or their (well chosen!) numerical equivalents are the logical starting point for a discussion of the energetics of semidiurnal and diurnal global tides (at long periods, the rotation introduces ambiguity into the formulation of the energy equation; Thompson, 1974). When discussing the dominant M2 tide, these equations or their traditional versions (equations 38–41) are often averaged over one tidal period (the average being denoted by $\langle \ \rangle$) to obtain (with the asterisk denoting complex conjugation of the complex quantities defined as in equation 19)

$$\nabla \cdot \langle \mathbf{P} \rangle = \langle W_t \rangle + \langle \mathbf{u} \cdot \mathbf{F}^* \rangle \tag{42}$$

$$\langle \mathbf{P} \rangle = \frac{1}{2} \operatorname{Re}[\rho g D(\langle \mathbf{u} \zeta_0^* \rangle + \langle \mathbf{u} \delta^* \rangle)] \tag{43}$$

$$\langle W \rangle = \frac{1}{2} \operatorname{Re}[\rho \langle \zeta_{0_t} \Gamma^* \rangle + \rho g \langle \zeta_{0_t} \delta_t^* \rangle + \rho \nabla \cdot (\mathbf{u} D \Gamma^*)] \tag{44}$$

or less general counterparts based upon equations 38–41.

Although the actual tide contains many other harmonic constituents (the semi-diurnal S2 and N2 and the diurnal K1 and 01 being the largest and comprising with M2 the most significant components of ocean tidal variation), it is instructive to proceed as though the only harmonic were M2; the results may be used directly to discuss tidal friction with somewhat better than order of magnitude accuracy. Integrated over an ocean at whose coasts $\mathbf{u} D \cdot \hat{n} = 0$, equation 42 becomes

$$\langle W^0 \rangle = \frac{1}{2} \operatorname{Re} \int_{\text{oceans}} \langle W \rangle \, da = -\frac{1}{2} \operatorname{Re} \int_{\text{oceans}} \langle \mathbf{u} \cdot \mathbf{F}^* \rangle \, da$$

that is, the total rate of energy loss $\langle W^0 \rangle$ in the oceans may be obtained either by direct evaluation of dissipative stresses, possible only if the model accurately resolves $\mathbf{u}$ in the regions of intense dissipation, or else from basin-wide integration of equation 44

$$\langle W^0 \rangle = \frac{1}{2} \operatorname{Re} \int_{\text{oceans}} [\rho \langle \zeta_{0_t} \Gamma^* \rangle + \rho g \langle \zeta_0 \delta_t^* \rangle] \, da \tag{45}$$

Equation 45 is valid for arbitrarily specified total tide generating potential Γ and solid earth tide δ, regardless of whether or not the solid earth is dissipative. The rate of energy loss $\langle W^e \rangle$ within the normally stressed solid earth is

$$\frac{1}{2} \operatorname{Re} \int_{\text{ocean, land}} [\rho_e \langle \delta_t \Gamma^* \rangle - \langle P \delta_t^* \rangle] \, da$$

where ρ_e is the surface density of the earth (assumed incompressible) and P is the surface pressure given by the weight $\rho g \zeta_0$ of the ocean tidal column. Thus

$$\langle W^e \rangle = \frac{1}{2} \operatorname{Re} \left[\int_{\text{ocean}} [\rho_e \langle \delta_t \Gamma^* \rangle - \rho g \zeta_0 \delta_t^* \rangle] \, da + \int_{\text{land}} \rho_e \langle \delta_t \Gamma^* \rangle \, da \right] \tag{46}$$

where, as before, Γ and δ are entirely general.

We now introduce the spherical harmonic expansions for the ocean tide ζ_0 and astronomical potential U

$$\zeta_0 = \sum_{n, m} (a_{n, m} \cos m \, \phi + b_{n, m} \sin m \, \phi) P_n^{m}(\cos \theta) \tag{47}$$

$$U = \sum_N U_N = \sum_{N, M} (A_{N, M} \cos M \, \phi + B_{N, M} \sin M \, \phi) P_N^{M}(\cos \theta) \tag{48}$$

and require the solid earth to be quasi-static and dissipation free, thus allowing ourselves the use of the (real) Love number expansions 5 and 6:

$$\Gamma = \sum_N (1 + k_N)U_N + \sum_{n,m} (1 + k'_n)\alpha_n(a_{n,m} \cos m\phi + b_{n,m} \sin m\phi)P_n{}^m(\cos\theta) \quad (49)$$

$$\delta = \sum_N \frac{h_N U_N}{g} + \sum_{n,m} h'_n \alpha_n(a_{n,m} \cos m\phi + b_{n,m} \sin m\phi)P_n{}^m(\cos\theta) \quad (50)$$

In terms of these, ocean dissipation $\langle \overline{W}^0 \rangle$ (the overbar reminds us that vanishing solid earth dissipation has been assumed if the Love numbers are real) is given by equation 45 as

$$\langle \overline{W}^0 \rangle = \frac{1}{2} \text{Re}\left[-i\sigma\rho \sum_{N,M} (1 + k_N - h_N)(a_{N,M} A^*_{N,M} + b_{N,M} B^*_{N,M}) \int Y_N{}^{M2} da \right] \quad (51)$$

where the orthogonality of spherical harmonics $Y_N{}^M$ and $\text{Re}\langle -i\sigma AA^* \rangle = 0$ have been employed.

Thus even in the case where solid earth loading and ocean self-attraction are significant, the total rate of working on the ocean by tide-generating forces depends only upon the amplitude and phase of those spherical harmonic components of the ocean tide also represented in the tide generating potential. For the particular case of semidiurnal M2 tides, this rate of working is given by

$$\langle \overline{W}^0 \rangle = \frac{1}{2} \text{Re} \int -i\sigma\rho(1 + k_2 - h_2)\zeta_0 U^*_2 \, da \quad (52)$$

This is the *form* of the expression which would also obtain if solid earth loading and ocean self-attraction had been neglected entirely, but we must remember that their inclusion will modify ζ_0 itself.

5. The Numerical Structure of Global Tidal Models

A. The Choice of Bottom Relief

Regardless of the particular form of LTE chosen for numerical solution, the variable ocean depth appears in the equations as a set of variable coefficients which must be estimated empirically. Two tabulations of global bathymetry have been explicitly acknowledged in the literature on global solutions of LTE:

1. Dishon (1964) estimates ocean depths averaged over areas bounded by 1° of longitude and 1° of Mercator latitude, lying between about 70°NS. Most of the Arctic and Antarctic are thus excluded. Ship's soundings at irregular intervals are interpolated to grid points by adjusting the grid point values to minimize a local measure of the error of reproduction of the irregularly spaced original data by bilinear interpolation from the grid. A part of the iterative process is the application of a smoothing operator (Shuman, 1957) to suppress "small wavelengths of the order of about two grid lengths while leaving larger wavelengths (associated with the overall bottom-formation) substantially unchanged."

The resulting bathymetric map is shown in Fig. 3. Pekeris and Accad (1969) refer to Dishon's work but indicate a slightly different method of data processing—

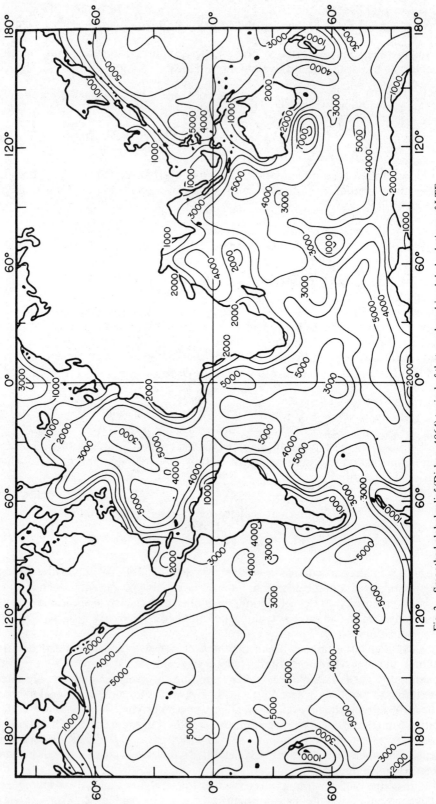

Fig. 3.   Smoothed global relief (Dishon, 1964) typical of that employed in global solutions of LTE.

averaging all available observations over a $1°$ square and then smoothing the resulting grided values by

$$D(\mathbf{r}) = \frac{\displaystyle\sum_{\substack{\text{points } \mathbf{d} \\ \text{near } \mathbf{r}}} W(\mathbf{d})D(\mathbf{r} + \mathbf{d})}{\displaystyle\sum W(\mathbf{d})}$$

where   (53)

$$W(\mathbf{d}) = \frac{R^2 - d^2}{R^2 + d^2}$$

and R is a smoothing radius. They choose $R = 12°$ and give a slightly different and more detailed bathymetric map.

2. Smith, Menard, and Sharman (1965) estimate average ocean depths and land elevations from contour charts, subjectively (but repeatedly, by different workers) averaging over areas bounded by $1°$ of latitude and by the following longitude intervals:

Latitude (°)	Longitude (°)
0–49	1
50–69	2
70–79	5
80–89	10

The estimates are available in computer-compatible form without further smoothing. Gates and Nelson (1973) have revised this tabulation.

It is not always obvious how the relief used in a given computation has been derived. Nevertheless, smoothing sufficient that estimates of horizontal depth gradients are not badly biased by detail at scales subgrid to the global mesh is required. Often this is obtained simply by averaging to the mesh scale chosen although further smoothing is sometimes employed. Thus Pekeris and Accad (1969), computing on a 1 or $2°$ global grid, note that their calculations are based on a bathymetry smoothed with $R = 12°$ and that, with $R = 5°$, the frictionless computations "become unstable."

Since most computations are carried out over global nets with mesh spacings of $1–6°$, this smoothing virtually obliterates such short wavelength features as the Hawaiian Island Arc, and it notably broadens continental shelf and slope regions. Thus from the beginning, certain aspects of the actual ocean tide are not accessible to a global model. Larsen (private communication) points out that semidiurnal tides on the southwestern side of the Hawaiian Island Arc lag those on the northeastern side by about an hour. This effect is absent from all global computations. Whether it is local and does not much alter North Pacific tides or whether it is indicative of far-reaching scattering of the semidiurnal wave is a matter for conjecture (although numerical experimentation might help to resolve the point).

The artificial widening of continental shelves means that the world's long but rather narrow shelf regions with their steep seaward slopes are replaced in the models by regions resembling the Patagonian shelf in width but having a much more gentle seaward slope. This degree of shelf broadening may well decrease the cutoff frequencies of refractively trapped Stokes–Kelvin edge waves to near tidal values, and the possibility of spurious tidal edge waves thus arises in choosing a global computational mesh and in interpreting the model solutions.

B. *Discretization and Solution of the Differential Problem*

Finite difference (FD) methods for solving LTE fall into two classes: time stepping (TS) and harmonic methods. The former envisage solution of LTE as an initial value problem in which, after several tidal cycles, dissipative effects eliminate all memory of the initial conditions (which are therefore arbitrary) and the solution becomes periodic. The latter assume ab initio a time dependence harmonic at one of the tidal lines and require solution of a FD equivalent of equation 24a or 24b. These are elliptic boundary value problems. All unknowns of their FD analogues are found simultaneously by solving a large system of coupled algebraic equations. They are thus entirely implicit whereas TS procedures are explicit or at most only partly implicit.

Questions of accuracy and stability are foremost. The choice of spatial discretization is to some extent independent of whether TS or harmonic solutions are sought. In both cases, it is advantageous to choose FD equivalents of LTE which are quasi-conservative, that is, in which the FD representations of mass and energy are precisely conserved (in the absence of forcing or dissipation) when the equations are made discrete in space but not in time. This guarantees harmonic solutions which conserve energy precisely and eliminates at least one possibility of instability of a TS procedure. In regularly shaped basins with constant depth or analytically defined relief, some sophistication in the choice of space discretization is possible and could be advantageous [thus Accad and Pekeris (1964) find fourth-order differences in space far superior to second-order ones in the rectangular sea problem of Rossiter (1958)], but the irregular shape and relief of the real oceans have forced adoption of nothing more complex than second-order space differences in actual global computations.

The stability of TS schemes is usually discussed "in the large," that is, by constructing plane wave solutions of the difference equations specialized to a plane ocean of constant depth and requiring the dispersion relation to yield at most damped waves or, equivalently, by requiring the amplification matrix (i.e., equation 66) to have eigenvalues of unit magnitude or smaller. This approach is usually not able to take into account the manner in which boundary conditions are imposed. They may well destabilize a scheme apparently stable "in the large," although this is often discovered only by actual trial.

In order to discuss numerical schemes in a simple way, we consider the FD analogues of equations 1–3 in plane coordinates although actual global computations must, of course, take into account the spherical geometry. If Mercator coordinates are adopted, the global equations 16–18 differ only slightly in form from the plane equations. We first discuss numerical methods of treating the differential part of the problem, that is, planar equations 1–3, but with neglect of solid earth loading and ocean self-attraction. Γ and δ are thus specifiable from the beginning. In this section we thus replace $\zeta - \Gamma/g$ by ζ to finally obtain the convenient planar forms

$$u_t - fv = -g(\zeta)_x + \frac{F^x}{\rho D} \tag{54}$$

$$v_t + fu = -g(\zeta)_y + \frac{F^y}{\rho D} \tag{55}$$

$$(\zeta)_t + (uD)_x + (fD)_y = -\left(\frac{\Gamma}{g} - \delta\right)_t \tag{56}$$

In actual global computations, the dissipative stresses have usually been taken to be of the form

$$\frac{\mathbf{F}}{\rho D} = -r(u, D)\mathbf{u} + A\nabla^2\mathbf{u} \tag{57}$$

The most obvious explicit scheme, defining u, v, and ζ at every mesh point of a rectangular grid and simultaneously updating all variables by forward time stepping, is unsuitable not only on account of the basic instability of forward differencing but also because the scheme is redundant. The FD problem can break into two independent problems if the FD equivalents of equations 54 and 55 contain no lateral eddy viscosity term, and into four independent problems if additionally the Coriolis parameter f vanishes. Not only is this uneconomical of computing resources, but also it leads to grid point oscillations as the independent solutions diverge from one another. These are apt to be especially troublesome in the global tide problem on account of the nearness to resonance of the frictionless system.

These difficulties are avoided in the widely used discretization, equations 58–61:

$$\frac{u^{n+1}(x, y) - u^{n-1}(x, y)}{2\Delta t} - f\bar{v}_u^{n-1}(x, y) = -g(\zeta^n(x, y))_{\bar{x}} - ru^{n-1}(x, y) + A\nabla_u^2 u^{n-1}(x, y) \tag{58}$$

$$\frac{v^{n+1}(x, y) - v^{n-1}(x, y)}{2\Delta t} + f\bar{u}_v^{n-1}(x, y) = -g(\zeta^n(x, y))_{\bar{y}} - rv^{n-1}(x, y) + A\nabla_v^2 v^{n-1}(x, y) \tag{59}$$

$$\frac{\zeta^{n+2}(x, y) - \zeta^n(x, y)}{2\Delta t} + (u^{n+1}(x, y)D(x, y))_{\bar{x}} + (v^{n+1}(x, y)D(x, y))_{\bar{y}}$$

$$= -\left(\frac{\Gamma^n(x, y)}{g} - \delta^n(x, y)\right)_t \tag{60}$$

in which u, v, and ζ are defined on the lattice of equation 61 with velocities computed at time steps $(n - 1)\Delta t$, $(n + 1)\Delta t$, etc., and elevations computed at time steps $n\Delta t$, $(n + 2)\Delta t$, etc., so that the order of updating is

$$\tag{61}$$

In the FD equations 58–60 the subscripts $\bar{x}$ and $\bar{y}$ denote space differencing; that is, $(\zeta(x, y))_{\bar{x}} = (\zeta(x + \Delta x, y) - \zeta(x - \Delta x, y))/2\Delta x$, while

$$\nabla_u^2 u(x, y) = \frac{u(x + 2\Delta x, y) + u(x - 2\Delta x, y) - 2u(x, y)}{4\Delta x^2}$$

$$+ \frac{u(x, y + 2\Delta y) + u(x, y - 2\Delta y) - 2u(x, y)}{4\Delta y^2} \tag{62}$$

and

$$\bar{v}_u(x, y) = \frac{\begin{aligned}v(x + \Delta x, \Delta y + y) + v(x - \Delta x, y + \Delta y)\\ + v(x + \Delta x, y - \Delta y) + v(x - \Delta x, y - \Delta y)\end{aligned}}{4} \qquad (63)$$

hold in equation 58 with similar replacements in equation 59.

This scheme is entirely explicit. Its stability analysis "in the large" is facilitated by the replacements

$$U = uD, \qquad V = vD, \qquad \Phi = -\sqrt{gD}\,\zeta \qquad (64)$$

We now suppose constancy of depth D, Coriolis parameter f, and bottom drag coefficient r (equation 57), and then cast the FD solutions in plane wave form

$$\begin{pmatrix} U(x, y, t) \\ V(x, y, t) \\ \Phi(x, y, t) \end{pmatrix} = \begin{pmatrix} U^{n+1} \\ V^{n+1} \\ \Phi^n \end{pmatrix} \exp(ik_2 x + ik_2 y) \qquad (65)$$

Proceeding as in Fischer (1959) by substituting equation 65 into equations 58–60 leads to

$$\begin{pmatrix} U^{n+2} \\ V^{n+2} \\ \Phi^{n+1} \end{pmatrix} = \begin{pmatrix} 1 - c\Delta t & f'\Delta t & ia \\ -f'\Delta t & 1 - c\Delta t & ib \\ ia(1 - c\Delta t) - ibf'\Delta t & ib(1 - c\Delta t) + iaf'\Delta t & 1 - (a^2 + b^2) \end{pmatrix} \begin{pmatrix} U^n \\ V^n \\ \Phi^{n-1} \end{pmatrix}$$

$$(66)$$

where, in the *amplification matrix* of equation 66,

$$a = \sqrt{gD}\left(\frac{2\Delta t}{\Delta s}\right)\sin k_1 \Delta s$$

$$b = \sqrt{gD}\left(\frac{2\Delta t}{\Delta s}\right)\sin k_2 \Delta s$$

$$\qquad (67)$$

$$c = 2r + \frac{A(2 - \cos 2k_1\Delta s - \cos 2k_2 \Delta s)}{\Delta s^2}$$

$$f' = f(\cos k_1 \Delta s - \cos k_2 \Delta s)$$

and we have taken $\Delta x = \Delta y = \Delta s$ for simplicity.

For computational stability, the eigenvalues λ of the amplification matrix must be less than 1 in absolute value. If they are of unit amplitude, the scheme is neutrally stable. If any are less than 1 in absolute value, it is damped. Now if

$$c\Delta t \ll 1, \qquad f't \ll 1 \qquad (68)$$

that is, if the damping is not too large, then the eigenvalues of the amplification matrix are approximately

$$\lambda_1 = 1 - c\Delta t$$

$$\lambda_{2,3} = \left[1 - \frac{c\Delta t}{2} - \frac{(a^2 + b^2)}{2}\right] \pm \frac{1}{2}[(a^2 + b^2 + c\Delta t)^2 - 4(a^2 + b^2 + f'^2\Delta t^2)]^{1/2}$$

$$(69)$$

For

$$4(a^2 + b^2 + f'^2\Delta t^2) \geq (a^2 + b^2 + c\Delta t)^2 \tag{70}$$

we have

$$|\lambda_1| = (1 - c\Delta t) \tag{71a}$$

$$|\lambda_{2,3}| = (1 - c\Delta t + f'^2\Delta t^2)^{1/2} \tag{71b}$$

If rotation and dissipation are both neglected, that is, if $f = r = A = 0$, then $c = f' = 0$ and equations 71 become $|\lambda| = 1$ provided equation 70 holds, that is, provided that

$$\Delta t < \frac{\Delta s}{\sqrt{2gD}} \tag{72}$$

which is the Courant–Friederichs–Lewy (CFL) criterion for stability. Without dissipation, the scheme of equations 58–60 is unstable to rotation, even if the CFL criterion is satisfied, since with $f \neq 0$ but $r = A = 0$ equation 71b yields $|\lambda_{2,3}| > 1$. Evidently some dissipation $(c > 0)$ is required for stability. Equation 71b requires that

$$\Delta t < \frac{c}{f'^2} \tag{73}$$

which, for the case of bottom drag only, becomes

$$\Delta t < \frac{r}{f^2} \tag{73a}$$

Because of their entirely explicit treatment, both the damping terms in fact require Δt to be appreciably smaller than the CFL criterion would indicate. Supposing the (generally weaker)

$$\Delta t < c^{-1}, \qquad \Delta t < \frac{c}{f'^2}$$

which follow from equations 71 for a stable computation, we must also require equation 70, which may be rewritten

$$\Delta t^2 < \frac{\Delta s^2}{8gD}(2 - c\Delta t + 2\sqrt{1 - c\Delta t + f'^2\Delta t^2}) \tag{74}$$

This is often a considerable decrease from the CFL criterion. Thus Zahel (1970) notes that with a (spherical) grid of horizontal mesh spacing $2\Delta\phi = 4°$ (i.e., about 64 km at the highest latitude 82°N of the domain), a bottom drag coefficient (equation 57) or $r = 0.003$ and a lateral eddy viscosity of 10^7 m^2/sec, the most rapid step possible on the basis of an analogue of equation 74 is 155 sec, whereas the CFL criterion (with depth 3000 m) would lead to about 280 sec. (By allowing the mesh spacing to grow toward the poles, Zahel is in fact able to use a step of 435 sec.) In shallower water, the problem is more severe, and in extremely shallow water (a few meters bottom drag may become the limiting term.

This FD scheme has been outlined in some detail because it is simple, widely used, and provides a convenient point of reference for explanation of other TS schemes. Platzman (1963) discusses its close relationship to the ideas of Richardson (1922) and points out the sense in which it is one of an entire family of difference schemes for the linearized shallow-water equations (essentially LTE) without rotation or lateral diffusivity. Consideration of the various possible arrangements of u, v, and ζ points similar to the Richardson lattice (equation 61) leads him to the concept of conjugate Richardson lattices. A particularly elegant discretization for the shallow-water equations with rotation involves simultaneous updating of a pair of conjugate Richardson lattices:

The lattice pair of equations 75 are coupled by the Coriolis terms; in the absence of rotation they would be entirely independent unless coupled by boundary conditions. This procedure is inherently stable to rotation but requires doubling of computer storage vis-à-vis equation 61. If it is desired to retain the economy of storage of equation 61, Platzman (1972a) points out that replacement of the simple space average of equation 63 used to obtain the Coriolis acceleration in equation 58 by

$$\overline{\overline{V}}_u(x, y) = [(q(x, y) + q(x + \Delta x, y + \Delta y))v(x + \Delta x, y + \Delta y)$$
$$+ (q(x, y) + q(x - \Delta x, y + \Delta y))v(x - \Delta x, y + \Delta y)$$
$$+ (q(x, y) + q(x + \Delta x, y - \Delta y))v(x + \Delta x, y - \Delta y)$$
$$+ (q(x, y) + q(x - \Delta x, y - \Delta y))v(x - \Delta x, y - \Delta y)]/8q(x, y) \quad (76)$$

where $q(x, y) = f(y)/D(x, y)$, with a similar replacement of $\bar{u}_v(x, y)$ in equation 59, makes the space discretization (equations 58–60) quasi-conservative. Stability of the time stepping procedure of these equations, without dissipation, to rotation may be achieved by rewriting equations 58 and 59 as

$$\frac{u^{n+1} - u^{n-1}}{2\Delta t} - f\bar{\bar{v}}_u{}^{n-1} = \cdots \quad (77)$$

$$\frac{v^{n+1} - v^{n-1}}{2\Delta t} + f\bar{\bar{u}}_v{}^{n+1} = \cdots \quad (78)$$

(Sielecki, 1968) and then always advancing the variables in the order ζ, u, v using, respectively, equations 60, 77, and 78. Thus executed, the computation remains fully explicit.

An alternative to the Richardson lattice is a staggered one on which both velocity components are defined at the same points:

$$
\begin{array}{ccc}
\zeta & & \zeta \\
\zeta \quad \zeta & -\ uv\ - & \zeta \quad \zeta \\
\zeta \quad \zeta \quad \zeta \quad \longrightarrow \quad & -\ uv\ -\ uv\ - \quad \longrightarrow \quad & \zeta \quad \zeta \quad \zeta \\
\zeta \quad \zeta & -\ uv\ - & \zeta \quad \zeta \\
\zeta & - & \zeta \\
& & \\
n & n+1 & n+2
\end{array}
\tag{79}
$$

The FD equations are still equations 58–60, but now $\bar{v}_u(x, y) = v(x, y)$, etc. The lateral diffusivity terms $A\nabla^2 u(x, y)$ may be evaluated either as in equation 62 or, with slightly better accuracy, by instead using the four velocities located one diagonal space from x, y. If both f and A are zero, the system of equations 58–60 decouple into two independent sets.

The stability analysis differs only slightly from that above and will not be repeated. The scheme is already quasi-conservative. The TS may be made stable to rotation either by the procedure (equations 77, 78) of Sielecki (1968) or else by making the Coriolis terms implicit, that is, by writing

$$
\frac{u^{n+1} - u^{n-1}}{2\Delta t} - f\left(\frac{v^{n+1} + v^{n-1}}{2}\right) = \cdots
\tag{80}
$$

$$
\frac{v^{n+1} - v^{n-1}}{2\Delta t} + f\left(\frac{u^{n+1} + u^{n-1}}{2}\right) = \cdots
\tag{81}
$$

For practical purposes, even this scheme is explicit since the 2×2 set of coupled equations for u^{n+1}, v^{n+1} may be solved algebraically.

The application of boundary conditions of equation 27, $\mathbf{u} \cdot \hat{n} = 0$ at coasts, and equation 28, $\zeta_0 = \zeta_{os}$ at open-water boundaries, is discussed by von Trepka (1967) for the Richardson lattice (equation 61) and by Heaps (1969) for the staggered lattice (equation 79). The basic idea is to approximate the irregular coasts by a connected series of north–south line elements passing through u points and east–west segments passing through v points, while the open-water boundary is made up of ζ points. The coast is thus replaced by a series of right-angle "steps" at which either u or v vanishes. At open-water boundaries, ζ is specified. There the additional condition that the normal derivative of normal velocity be zero is often imposed in order to evaluate the Coriolis terms at open-water velocity points according to equation 63 or 76 when using a Richardson lattice (although this is not necessary; a one-sided average may be taken by omitting half the terms in equation 76 and dividing by 4 instead of 8). The staggered lattice requires no further condition unless lateral diffusivity is included. In that case, the foregoing conditions are usually augmented by requiring the vanishing of the normal gradient of tangential velocity at the coasts.

The literature dealing with applications of these explicit schemes is vast. Hansen (1962, 1966) summarizes extensive application in the North Sea and other marginal seas of equations 58–60 with Richardson lattice (61), a procedure which he and his co-workers call the "HN procedure." Zahel (1970, 1973) applies the method to M2 and K1 global tides. Lauwerier (1962) and Hollsters (1962) discuss marginal sea applications of the staggered lattice (79) and Heaps (1969) adopts it (with 45° rotation) to forecast storm surges around Britain and in the North Sea. Gohin (1961a) employs a similar scheme in his computation of the M2 tide in the Atlantic and Indian Oceans (Gohin, 1961b). Ueno's (1964) global M2 calculation makes use of the staggered lattice with implicit Coriolis terms (equations 80 and 81).

In all these explicit schemes, the smallness of the largest stable time step imposed by conditions like equation 74 is a severe restriction. Various implicit treatments of the dissipative terms and/or the elevation gradient terms have been proposed with an eye toward relaxing equation 74. In extremely shallow water, the bottom drag limits the time step, and in this case useful improvement results from making the drag terms implicit:

$$\frac{u^{n+1} - u^{n-1}}{2\Delta t} - fv_{n-1} = \cdots - ru^{n+1} \cdots \text{etc.} \tag{82}$$

[the nonlinear dependence of r upon u requires further consideration (Hollsters, 1962)]. In this simplest case, the computation may still be carried out explicitly. However, when lateral diffusivity and/or elevation gradients are made implicit, each time step requires solution of the finite difference analogue of a certain elliptic boundary value problem (hopefully one more simple than equations 24!).

If only dissipative terms are made implicit, the CFL criterion will effectively fix the greatest stable step. If the elevations are additionally made implicit, the calculation becomes altogether stable. In other kinds of problems, for example, those in which one wishes to forecast the evolution of a quasi-geostrophic flow without either subjecting the equations to a filtering approximation or being forced to proceed at a time step limited by irrelevant gravity waves, an implicit representation of the elevation field may result in great computational savings. Here the case is less clear-cut because the gravity waves themselves must be resolved adequately. Making the elevation field implicit in the most straightforward manner results in a two-dimensional elliptic problem not a great deal easier than equations 24 to be solved at each step. A useful economization is achieved by a splitting procedure.

We illustrate these last remarks with a typical (but not necessarily optimal) scheme for the dissipationless, rotating shallow-water problem displaying implicit treatment of the elevation field (Elvius and Sundstrom, 1973):

$$\frac{u^{n+1} - u^{n-1}}{2\Delta t} - fv^{n-1} = -\frac{g(\zeta^{n+1} + \zeta^{n-1})_{\bar{x}}}{2}$$

$$\frac{v^{n+1} - v^{n-1}}{2\Delta t} + fu^{n-1} = -\frac{g(\zeta^{n+1} + \zeta^{n-1})_{\bar{y}}}{2} \tag{83}$$

$$\frac{\zeta^{n+1} - \zeta^{n-1}}{2\Delta t} + D\left[\frac{(u^{n+1} + u^{n-1})_{\bar{x}}}{2} + \frac{(v^{n+1} + v^{n-1})_{\bar{y}}}{2}\right] = 0$$

Eliminating u^{n+1} and v^{n+1} (and abbreviating, for simplicity, $u^{n+1} = u^+, u^{n-1} = u^-$, etc.) yields

$$gD(\zeta^+_{\bar{x}\bar{x}} + \zeta^+_{\bar{y}\bar{y}}) - \frac{\zeta^+}{2\Delta t^2} = -gD(\zeta^-_{\bar{x}\bar{x}} + \zeta^-_{\bar{y}\bar{y}}) - \frac{\zeta^-}{2\Delta t^2} + [(fv^-)_{\bar{x}} - (fu^-)_{\bar{y}} + u^-_{\bar{x}} + v^-_{\bar{y}}]/2 \tag{84}$$

Now each time step requires solution of equation 84, which is a FD analogue of the elliptic boundary value problem

$$gD\nabla^2\zeta - \frac{\zeta}{2\Delta t^2} = \cdots \tag{84a}$$

The CFL condition (72) no longer limits the stable step length, but each step has become almost as complicated as solving a discrete version of equations 24, that is, LTE with harmonic time variation assumed. For tides, which *are* predominantly harmonic, little has been gained. Leendertse (1967) therefore employs a splitting method. In the notation of equation 83 it is

$$\frac{u^{n+1} - u^{n-1}}{\Delta t} - \frac{f}{2} v^n = -g\zeta_{\bar{x}}^{n+1} - ru^n \tag{85a}$$

$$\frac{\zeta^{n+1} - \zeta^n}{\Delta t} + D[u_{\bar{x}}^{n+1} + v_{\bar{y}}^n] = 0 \tag{85b}$$

$$\frac{v^{n+1} - v^n}{\Delta t} + \frac{f}{2} u^{n+1} = -g\zeta_{\bar{y}}^n - rv^{n+1} \tag{85c}$$

$$\frac{v^{n+2} - v^{n+1}}{\Delta t} = \frac{f}{2} u^{n+1} = -q\zeta_{\bar{y}}^{n+2} - rv^{n+1} \tag{85d}$$

$$\frac{\zeta^{n+2} - \zeta^{n+1}}{\Delta t} + D[u_{\bar{x}}^{n+1} + v_{\bar{y}}^{n+2}] = 0 \tag{85e}$$

$$\frac{u^{n+2} - u^{n+1}}{\Delta t} - \frac{f}{2} v^{n+2} = -q\zeta_{\bar{x}}^{n+1} - ru^{n+2} \tag{85f}$$

Now equations 85a and 85b are implicit in u^{n+1}, ζ^{n+1}; at each step we must solve the one-dimensional

$$gD\zeta_{\bar{x}\bar{x}}^{n+1} - \frac{\zeta^{n+1}}{\Delta t^2} = \cdots \tag{86a}$$

Having done this, equation 85c yields v^{n+1} and we go to equations 85d and 85e, implicit in v^{n+2}, ζ^{n+2} so that we must solve

$$gD\zeta_{\bar{y}\bar{y}}^{n+2} - \frac{\zeta^{n+2}}{\Delta t^2} = \cdots \tag{86b}$$

Finally equation 85f gives u^{n+2} and the step is complete. This splitting procedure allows us to replace solution of the two-dimensional equation 84a by equations 86a and 86b of one-dimensional problems without sacrificing the improvement in stability (Leendertse, 1967; Elvius and Sundstrom, 1973).

Leendertse (1967) has employed equations 85 with Coriolis, bottom drag, tidal forcing, *and* nonlinear advective terms. The variables are discretized on a Richardson lattice (61) and the Coriolis terms are obtained by local averaging (equation 63, etc.). At each half step the Coriolis terms are treated as in equations 77 and 78. Leendertse (1967) reports application of the model to the southern part of the North Sea and to a river estuary of the Rhine. Satisfactory elevations are obtained with time steps in excess of five times the CFL limit.

Gordeev, Kagan, and Rivkind (1973) have carried out a global time-stepping calculation for the M2 tide making use of a scheme (Marchuk, Kagan and Tamsalu, 1969) a slight variant of which (Marchuk, Gordeev, Rivkind, and Kagan, 1973) they characterize as "explicit for equations of gradient of the level and implicit for equations for complete flows." In the latter reference, bottom drag and lateral diffusivity are dealt with implicitly by a splitting procedure closely analogous to that of equations

85 but the elevations are not implicit in both half steps. In their calculations of North and Arctic sea tides, this procedure enables stable time-stepping at virtually the CFL limit.

It is of interest to compare the explicit calculation of Zahel (1970) with the dissipative implicit one of Gordeev, Kagan, and Rivkind (1973). Both employ the same bottom drag and eddy diffusivity coefficients, $r = 0.003$ and $A = 10^7$ cm^2/sec. Zahel (1970) employs a 4° spherical net graded toward the poles to escape the small step that would be imposed by the narrowing of a uniform mesh at high latitudes. For a 5° grid, extending as far as 82°N poleward without grading, he estimates, from an analogue of equation 74, that the maximum stable step would be 220 sec. This is slightly more than half of the (dissipationless) CFL criterion for the ungraded 5° mesh. The Soviet calculation is on a 5° grid extending only as far poleward as 70°S, thus slightly increasing the CFL limit, and was carried out with a time step of 360 sec. This is evidently very close to the CFL limit.

To judge from Leendertse's (1967) experience in shallow-water calculations, the additional labor of making the elevations implicit would allow one to use time steps five times as great or even greater. But even if the calculation remains stable for such long steps, they are beginning to get long enough in comparison with a tidal period that overall accuracy would suffer.

Much of the development of time-stepping schemes has been for purposes of storm surge prediction. There, because of the irregular time variation of meteorological forcing, they offer the only alternative to statistical input-response techniques (Cartwright, 1968). For tides, however, the assumption of harmonic time variation offers an alternate approach.

The assumption that all quantities have time dependence $e^{-i\sigma t}$ makes LTE (with linearized bottom drag) elliptic in space. Discretization of the resulting three partial differential equations for u, v, and ζ may either follow or precede elimination of one or more of the dependent variables.

It is of course not necessary to eliminate any of the independent variables. Thus Platzman (1974) discretizes the unforced, dissipationless LTE on a Richardson lattice to obtain equations analogous to equations 58–60,

$$-i\sigma \begin{pmatrix} uD \\ vD \\ \zeta \end{pmatrix} = \begin{pmatrix} 0 & -f\Box & gD\Delta_x \\ f\Box & 0 & gD\Delta_y \\ \Delta_x & \Delta_y & 0 \end{pmatrix} \begin{pmatrix} uD \\ vD \\ \zeta \end{pmatrix} \qquad (87)$$

in which $\Delta_x \zeta \equiv \zeta_{\bar{x}}$ and $\Box v$ is equation 76, which is preferable to the simpler space averaging of equation 63 because the matrix of coefficients is then Hermitian and the natural frequencies σ of the discrete problem are real i.e., the discrete problem conserves energy. Platzman (1974) uses this as the starting point for a computation of normal modes of the Atlantic and Indian oceans.

Pekeris and Accad (1969) start with equations 1–3 without solid earth tides or ocean self-gravitation and with a dissipative term of the form

$$\frac{\mathbf{F}}{\rho D} = \alpha \sigma \left(\frac{D(\phi, \tau)}{D_0} \right)^n \mathbf{u}$$

where D_0 is a reference depth (1000 m) characteristic of coastal regions in the digitized relief and α and n are constants chosen to make dissipation small away from coasts. Values of α employed ranged from 0 (no dissipation) to 0.5, and n was 1, 2, or 3. The equations are written in Mercator coordinates (equations 14a and 14b), and the elevation is eliminated to obtain a pair of coupled partial differential equations in the quantities

$$F = iuD \frac{\cos \theta}{\sigma a}, \qquad G = ivD \frac{\cos \theta}{\sigma a}$$

These are solved for the M2 tide "by finite differences, imposing the boundary conditions of the vanishing of F along a coastal parallel, and of G along a meridional segment of the coastline."

This procedure is the most natural one if the coastal boundary condition is that of vanishing normal velocity (equation 27). If one prefers instead the specification (equation 29) of boundary elevations, then it is natural to eliminate the velocity components (either before or after discretization) and solve for the elevation. Bogdanov and Magarik (1967, 1969) and Hendershott (1972) proceed in this manner, first eliminating the velocity components to obtain equation 24b and then discretizing and solving. Details of the Soviet calculational procedure are given in Bogdanov, Kim, and Magarik (1964).

The prototype problem is readily obtained from equations 54–56 by choosing $\mathbf{F}/\rho D = -r\mathbf{u}$ and eliminating u and v to obtain a (beta[1]) plane version of equation 24b:

$$\left[\left(\frac{H\zeta_x}{\sigma'^2 - f^2} \right)_x + \left(\frac{H\zeta_y}{\sigma'^2 - f^2} \right)_y \right] - \frac{i}{\sigma'} \left[\left(\frac{-fH\zeta_y}{\sigma'^2 - f^2} \right)_x + \left(\frac{fH\zeta_x}{\sigma'^2 - f^2} \right)_y \right]$$

$$+ \frac{\sigma}{\sigma'} \frac{\zeta}{gD_0} = \frac{\sigma}{\sigma'} \frac{(\Gamma/g - \delta)}{gD_0} \qquad (88)$$

where $\sigma' = \sigma + ir$ and $D(x, y) = D_0 H(x, y)$. Further differentiation within equation 88 produces

$$\nabla^2 \zeta + \left[\left(\frac{H}{\sigma'^2 - f^2} \right)_x - \frac{i}{\sigma'} \left(\frac{fH}{\sigma'^2 - f^2} \right)_y \right] \left(\frac{\sigma'^2 - f^2}{H} \right) \zeta_x$$

$$+ \left[\left(\frac{H}{\sigma'^2 - f^2} \right)_y + \frac{i}{\sigma} \left(\frac{fH}{\sigma'^2 - f^2} \right)_x \right] \left(\frac{\sigma'^2 - f^2}{H} \right) \zeta_y$$

$$+ \frac{\sigma}{\sigma'} \frac{\sigma'^2 - f^2}{gD_0 H} \zeta = \frac{\sigma}{\sigma'} \frac{\sigma'^2 - f^2}{gD_0 H} \left(\frac{\Gamma}{g} - \delta \right) \qquad (89)$$

It will be convenient to abbreviate these as

$$(A\zeta_x)_x + (A\zeta_y)_y - i[(-B\zeta_y)_x + (B\zeta_x)_y] + C\zeta = D \qquad (88a)$$

and

$$\nabla^2 \zeta + a\zeta_x + b\zeta_y + c\zeta = d \qquad (89a)$$

respectively.

[1] That is, the set of plane equations but with $f = \beta y$ modeling qualitatively the latitude dependence of $2\Omega \sin \theta$.

Bogdanov, Kim, and Magarik (1964) and Hendershott (1972) now replace the derivatives in equation 89a by second-order space differences to obtain

$$\left(1 + \frac{a_p\Delta s}{2}\right)\zeta_e + \left(1 - \frac{a_p\Delta s}{2}\right)\zeta_w + \left(1 + \frac{b_p\Delta s}{2}\right)\zeta_n + \left(1 - \frac{b_p\Delta s}{2}\right)\zeta_s$$

$$+ (c_p\Delta s^2 - 4)\zeta_p = d_p\Delta s^2 \quad (90a)$$

(where we have introduced the notation $\zeta_p = \zeta(x, y)$, $\zeta_e = \zeta(x + \Delta x, y)$, $\zeta_{ee} = \zeta(x + 2\Delta x, y)$, $\zeta_{ne} = \zeta(x + \Delta x, y + \Delta y)$, etc.) at every point x, y of a square mesh of spacing $\Delta x = \Delta y = \Delta s$. Specification of boundary values of ζ completes the numerical problem.

This is the simplest discretization, since it leads to a "five-point star"; that is, the difference equation at x, y involves only nearest north $(x, y + \Delta y)$, south $(x, y - \Delta y)$, east $(x + \Delta x, y)$, and west $(x - \Delta x, y)$ neighbors. Others are possible and sometimes preferable. One alternative is to discretize equation 88a by using second differences. The result is

$$[A_e(\zeta_{ee} - \zeta_p) - A_w(\zeta_p - \zeta_{ww}) + A_n(\zeta_{nn} - \zeta_p) - A_s(\zeta_p - \zeta_{ss})]$$

$$- i[-B_e(\zeta_{ni} - \zeta_{se}) + B_w(\zeta_{nw} - \zeta_{sw}) + B_n(\zeta_{ne} - \zeta_{nw}) - B_s(\zeta_{se} - \zeta_{sw})]$$

$$+ c_p\Delta s^r\zeta_p = D_p\Delta s^r \quad (90b)$$

The difference equations 90b now occupy a "nine-point" star composed of the ζ points sketched in the staggered lattice (79). In fact, the discretization 90b may also be obtained by discretizing LTE 54–56 on that staggered lattice and then eliminating u, v. That discretization of LTE is quasi-conservative; hence equation 90b is guaranteed to conserve energy (the FD analogues of the energy equation 33 are readily constructed since FD forms of LTE *before* elimination of u, v are available) but it has the undesirable feature that, if B is constant, the FD problem breaks into two independent problems and grid point oscillations arise (Stock, personal communication). An alternative energy conserving discretization is got by rewriting equation 88a as

$$(A\zeta_x)_x + (A\zeta_y)_y - i[-(B_x\zeta)_y + (B_y\zeta)_x] + C\zeta = D \quad (88b)$$

and again replacing derivatives by second-order differences. The "star" is now a five-point one and, with proper boundary conditions, the FD matrix of coefficients corresponding to equation 88a is Hermitian so that real natural frequencies are guaranteed (see discussion below) but no simple explicit analog of the energy equation exists.

The most commonly used method of solving such elliptic difference equations is some form of Young's (1954) technique of sequential overrelaxation (SOR). The basic relaxation procedure is to scan the mesh repeatedly, updating mesh point values according to, for example

$$\zeta_p^{i+1} = (1 - \alpha)\zeta_p^i + \alpha\left\{d_p\Delta s^2 - \left[\left(1 + \frac{a_p\Delta s}{2}\right)\zeta_e^i + \left(1 - \frac{a_p\Delta s}{2}\right)\zeta_w^{i+1}\right.\right.$$

$$\left.\left. + \left(1 + \frac{b_p\Delta s}{2}\right)\zeta_n^{i+1} + \left(1 - \frac{b_p\Delta s}{2}\right)\zeta_s^i\right]\right\}/(c_p\Delta s^2 - 4) \quad (90c)$$

which reduces to equation 90a if the overrelaxation coefficient α is unity. In this example, the scan is left to right, top to bottom so that when updating ζ_p, the upper and left neighboring values have just been updated and the eastern and lower neighboring values have yet to be updated. The behavior of this iteration is understood in many cases (Forsythe and Wasow, 1960) and it is easily employed with highly irregular boundaries. Values of α greater than one but less than two often speed the convergence of the iteration dramatically.

Garabedian (1956) has given an illuminating heuristic deviation of Young's results in which the course of the relaxation is shown to be approximately governed by a certain hyperbolic equation. Garabedian lets

$$\zeta(x, y, \text{ith iteration}) = \psi(x, y, s(x, y, t))$$

where t is a timelike variable increasing linearly with the number of iterations and, with $s = (x - y + 2\rho t)/2$, finds for ψ essentially

$$\left(\frac{4 - \alpha}{2\alpha}\right)\psi_{ss} + 2\left(\frac{2 - \alpha}{\alpha\delta}\right)\psi_s = L(\psi) - d \tag{91}$$

Here the subscript s denotes partial differentiation with respect to the timelike variable s, ρ is the interval of s separating adjacent iterations, δ is the spatial mesh spacing Δs of equation 90c, $L = \nabla^2 + a\partial/\partial x + b\partial/\partial y + c$ is a second-order differential operator having constant coefficients a, b, c and

$$\begin{aligned} L(\zeta_0) &= d & x, y \in \text{region } R \\ \zeta_0 &= 0 & x, y \in \text{boundary } C \text{ of region } R \end{aligned} \tag{92}$$

is the original problem which is to be solved by overrelaxation with overrelaxation coefficient α. Clearly the behavior of ψ during the course of the relaxation depends on the spectrum of eigenvalues of L defined by

$$\begin{aligned} L(u_j) &= \gamma_j u_j & x, y \in R \\ u_j &= 0 & x, y \in C \end{aligned} \tag{93}$$

for if we write

$$\psi = \zeta_0 + \sum_j a_j^{\pm} \mu_j \exp(sp_j^{\pm}) \tag{94}$$

we find

$$(p_j^{\pm})^2\left(\frac{4 - \alpha}{2\alpha}\right) + (p_j^{\pm})2\left(\frac{2 - \alpha}{\alpha\delta}\right) - \gamma_j = 0$$

That is,

$$p_j^{\pm} = -\frac{2}{\delta}\left(\frac{2 - \alpha}{4 - \alpha}\right) \pm \left[\frac{4}{\delta^2}\left(\frac{2 - \alpha}{4 - \alpha}\right)^2 + \left(\frac{2\alpha}{4 - \alpha}\right)\gamma_j\right]^{1/2} \tag{95}$$

The relaxation is seen to converge only if $\text{Re } p_j^{\pm} < 0$ for all j.

If the γ_j are all negative, as they are if $L = \nabla^2$, then the square root is either imaginary or less in absolute value than $|(2/\delta)(2 - \alpha)/(4 - \alpha)|$ for $0 < \alpha < 2$ so that the iteration does always converge for α in this range, although the convergence may be oscillatory rather than monotone.

Now if $L = \nabla^2 + \sigma^2/c_0^2$, the case of oscillations of a frictionless membrane across which waves propagate at speed c_0, then the eigenvalues of L are those of ∇^2 shifted along the positive real axis by σ^2/c_0^2. Convergence is assured only if these shifted eigenvalues are all negative, that is, only if the driving frequency is *less* than the frequency of the grave mode of the membrane. If $L = \nabla^2 + i\beta/\sigma\partial_x$, the case of divergence-free planetary waves of frequency σ, then the eigenvalues of L are those of ∇^2 shifted along the positive real axis by $\beta^2/4\sigma^2$. Now convergence is assured only if the driving frequency is *greater* than the natural frequency of the most rapid planetary wave mode. Inasmuch as LTE allow both gravitational and rotational free oscillations, it appears likely that SOR applied to the dissipationless global tidal problem will converge only within a rather narrow range of frequencies if at all.

Computational experience, published and unpublished, indicates difficulties with SOR in the global tide problem. Bogdanov, Kim, and Magarik (1964) report that "Seidel–Nekrasov" iteration of a dissipationless analogue of equation 90c in the Pacific shows a fluctuating instability. In the global calculations of Bogdanov and Magarik (1967, 1969), the dissipationless problem is posed but, in its solution by Seidel's method with optimal overrelaxation (effectively, SOR), a term of the form $1 - i\Delta$, $\Delta \ll 1$, is made to multiply the penultimate term of equation 90a to ensure convergence.

To avoid these difficulties, Pekeris and Accad (1969) and Hendershott (1972) prefer a method of solution which, in the absence of errors, would solve the difference equation exactly in a finite number of operations. Hendershott (1972) embeds the irregular ocean basins in a rectangular region which is uniformly covered by the mesh so that there are MN mesh points ordered from left to right and top to bottom by M horizontal lines each line containing N points. At ocean ζ points equation 90a or 90b is the difference equation, whereas at boundary points, elevations are specified ($\zeta_p = \zeta_{obs}$). At land points, the dummy equation $\zeta_p = 0$ is satisfied. The solution X of the difference problem $AX = B$ is a column vector (equation 96) with grid point values of ζ following one another in television raster order. The matrix A of coefficients of the finite difference problem is block tridiagonal when the mesh points are ordered in this fashion (other orderings are also possible), and the difference problem $AX = B$ may be solved efficiently by the triangularization method of Fox (1962).

The difference equations have the general form

$$\begin{pmatrix} A_1 & C_1 & & & & \\ D_1 & A_2 & C_2 & & & \\ & D_1 & A_3 & C_3 & & \\ & & D_{M-2} & A_{M-1} & C_{M-1} \\ & & & D_{M-1} & A_M \end{pmatrix} \begin{pmatrix} X_1 \\ X_2 \\ X_3 \\ X_{M-1} \\ X_M \end{pmatrix} = \begin{pmatrix} B_1 \\ B_2 \\ B_3 \\ B_{M-1} \\ B_M \end{pmatrix} \qquad (96)$$

where D_i and C_i are, at most, tridiagonal matrices of order N and the A_i are at most pentadiagonal (unless the region is periodically connected, as in the case of a canal-shaped region encircling a sphere). X_0 and B_i are column vectors of length N. The solution proceeds as in Fox (1962):

$$\left. \begin{aligned} E_0 &= 0 \\ E_{j-1} &= (D_{j-1}F_{j-1}^{-1}), & F_j &= (A_j - E_{j-1}G_{j-1}) \\ G_j &= C_j, & Y_j &= (B_j - E_{j-1}Y_{j-1}) \end{aligned} \right\} \quad j = 1, M \qquad (97)$$

$$\left. \begin{aligned} G_M &= 0 \\ X_j &= F_j^{-1}(Y_j - G_j X_{j+1}) \end{aligned} \right\} \quad j = M, 1$$

and the principal computation is the inversion of the M (usually dense) matrices F_i each of order N.

If, in the course of the computation, the E_j, F_j^{-1}, and G_j have been stored (on tape), we then have essentially a "Green's function" for solving $AX = B$ very rapidly with many right-hand sides B':

$$\left.\begin{array}{l} E_0 = 0 \\ Y_j = B_j' - E_{j-1}Y_{j-1} \end{array}\right\} \quad j = 1, M$$

$$\left.\begin{array}{l} G_M = 0 \\ X_j' = F_j^{-1}(Y_j - G_j X_{j+1}') \end{array}\right\} \quad j = M, 1 \tag{98}$$

If the discretization 90a is used, the submatrices A_i are tridiagonal and the C_i, D_i are diagonal. If equation 90b is used, A is pentadiagonal and the method of equation 97 appears inappropriate. However, a reordering of mesh points corresponds to a 45° rotation of the mesh brings A into tridiagonal form.

With Dirichlet boundary conditions, a FD matrix corresponding to equation 90a or 90b may be rewritten as

$$A'(\sigma) + I\sigma$$

where I is the unit matrix. In this form, it is clear that finite difference free oscillations have natural frequencies which are those eigenvalues λ of $A'(\sigma)$ such that $\lambda = -\sigma$. If $A'(\sigma)$ is Hermitian for real σ, the λ are real and consequently the FD system is energy preserving. Equation 90b and the straightforward discretization of equation 88b result in Hermitian A' but equation 90a does not. For global solutions of LTE, equation 90a nevertheless appears to yield reasonable solutions for forced tides. The question is of more acute procedural importance in calculations of normal modes (Platzman, 1972a, 1974).

It is feasible to impose other boundary conditions such as equation 27 or equation 30 even when working with elevations only. The velocities are expressed in terms of ζ by solving equations 54 and 55 to obtain

$$\mathbf{u} \cdot \hat{n} = \alpha\zeta_x + \beta\zeta_y$$

where α and β depend upon the Coriolis parameter and the strike of the boundary. The general condition

$$\alpha\zeta_x + \beta\zeta_y + \gamma\zeta = \delta \tag{99}$$

has equations 27 and 28 or 29 and 30 as limiting cases.

Unless the FD boundary conditions are composed with care, they may destroy the Hermitian property of the matrix of coefficients. The following implementations of equation 99 are not guaranteed to preserve the Hermitian property.

At boundaries paralleling mesh lines, equation 99 is most easily implemented by discretizing both the boundary condition 99 and the elevation equation 88a or 88b with centered second-order differences at each boundary point. Both discretizations occupy a "five-point star," one member of which is a ficticious mesh point falling outside the boundary. Reference to this point may be eliminated between the two discretizations centered at the boundary point to produce one difference equation, of second-order accuracy, as the FD boundary condition. This procedure is feasible along boundaries composed of (not necessarily colinear) mesh point intersections, although where the boundary is curved, two fictitious points are required and then centered space differences must be abandoned in one direction. Nonreentrant corners are an exception since they are stagnation points and there are effectively two boundary conditions (corresponding to the vanishing of the two components of velocity normal to the two sides of the corner) rather than only one. Truly reentrant corners in the physical boundary pose special problems (Forsythe and Wasow, 1960).

Ignoring the difference between the location of the actual physical boundary and a computational boundary made up of mesh point intersections results in only first-order accuracy of FD representation of boundary conditions. It is possible to apply the foregoing procedure with additional Taylor expansion of the elevation away from the actual physical boundary to obtain a discretization of equation 99 which is formally of second-order accuracy. For basins (i.e., circles) whose actual physical boundaries are smooth at the mesh scale, Hendershott (unpublished) thus obtains a modest increase in accuracy, but in the geophysical case the actual physical boundary is never smooth at the mesh scale and the technique offers no clear-cut advantage.

Harmonic solutions for the global M2 tide have been obtained by Bogdanov and Magarik (1967) on a 5° spherical coordinate grid and by Hendershott (1972) on a 6° Mercator grid using dissipationless LTE and coastal specification equation 3 of elevations. Bogdanov and Magarik (1967, 1969) similarly solved for global S2, K1, and 01. Pekeris and Accad (1969) display various solutions on 1° and 2° Mercator grids with and without linearized bottom drag, all obtained by requiring equation 1, the vanishing of the component of velocity normal to a computational coastline made up of straight-line segments of meridians and parallels.

With TS solutions, it is possible to solve the global problem with the full tide-generating force (rather than just one harmonic) and, at the end, to analyze the solution for individual components. All nonlinearities, especially the nonlinear bottom friction, may thus be included in a straightforward manner if the mesh has sufficient resolution. The harmonic method requires modification to include nonlinearities.

If only one harmonic, say M2, is considered in isolation, then the ability (equation 98) to solve the problem repeatedly with different forcing terms allows one to treat the nonlinear bottom drag in an iterative manner. In the strongly dissipative Gulf of California, Stock (personal communication) reports rapid convergence of this procedure. However, calculations of weaker harmonics must take into account the fact that each harmonic is influenced by the others through the nonlinearity of the bottom drag terms. Jeffreys (1952) discusses the expansion of this term for the case of a multi-component tide. Sidjebat (1970) has proposed and tested, with good results, an iterative scheme which incorporates Jeffrey's expansion into a set of coupled elliptic equations, one for each tidal constituent. Sidjebat's method works well in the very shallow (a few meters) Bight of Abaco, where nonlinear drag effects are strong. It would be directly applicable to the deep-sea problem were it not for the fact that global models fail to resolve shallow-water velocities with anything like the accuracy needed to properly evaluate the bottom drag.

C. Inclusion of the Convolution Terms

All of the foregoing fails to allow for the effects of solid earth loading and of ocean self-attraction, although they are appreciable on a global scale (Hendershott, 1972). Naïvely retaining these effects, but neglecting dissipation, we might imagine either solving equations 16–18 by time stepping or else eliminating u, v between equations 20 and 22 to obtain[1]

$$\mathscr{L}(\zeta_0) = \mathscr{L}'\left(\zeta + \iint G(\theta|\theta')\zeta_0(\theta')\,d\theta'\right) \qquad (100)$$

---

[1] The variables of integration are abbreviated in an obvious manner throughout this discussion.

Here $\mathscr{L}$ is the elliptic operator of equation 24, $\mathscr{L}' = \mathscr{L} - \varepsilon^2 \sec^2 \tau$,

$$\bar{\zeta} = \sum_n \frac{(1 + k_n - h_n)U_n}{g}$$

is the equilibrium tide relative to the solid earth neglecting the convolution terms, and $G(\theta|\theta')$ is the Green's function of Farrell (1972c). We would then solve for individual tidal harmonics. In either case, the computational task is prohibitively long; we must either reevaluate the global convolution integral at each time step or else solve a set of elliptic finite difference equations with a dense matrix of coefficients.

Hendershott (1972) suggests iterative solution of equation 100, that is,

$$\mathscr{L}(\zeta_0^{i+1}) = \mathscr{L}'\left(\bar{\zeta} + \iint G(\theta|\theta')\zeta_0^{i}(\theta')\, d\theta'\right) \tag{101}$$

but his iteration appears to be getting off to a bad start. That it need *not* converge in general is easily seen by considering the nonrotating, constant depth, frictionless spherical problem for which

$$\mathscr{L} = \nabla^2 + \frac{\sigma^2 a^2}{gD}$$

$$\mathscr{L}' = \nabla^2$$

If we let

$$\bar{\zeta} = \sum_n \frac{1 + k_u - h_n}{g} \sum_m (A_{n,m} \cos m\phi + B_{n,m} \sin m\phi)P_n^m(\sin\theta)$$

and $\qquad\qquad\qquad\qquad\qquad\qquad\qquad\qquad\qquad\qquad\qquad\qquad (102)$

$$\zeta_0 = \sum_n \sum_m (a_{n,m} \cos m\phi + b_{n,m} \sin m\phi)P_n^m(\sin\theta)$$

then the solution of equation 100 is

$$\begin{pmatrix} a_{n,m} \\ b_{n,m} \end{pmatrix} = \begin{pmatrix} A_{n,m} \\ B_{n,m} \end{pmatrix} \frac{-n(n+1)(1 + k_n - h_n)/g}{\sigma^2 a^2/gD - n(n+1)(1 - \alpha_n(1 + k_n' - h_n'))} \tag{103}$$

If we iterate as in equation 101

$$\left(-n(n+1) + \frac{\sigma^2 a^2}{gD}\right)\begin{pmatrix} a_{n,m}^{i+1} \\ b_{n,m}^{i+1} \end{pmatrix} = (-n(n+1)(1 + k_n' - h_n')\alpha_n \begin{pmatrix} a_{n,m}^{i} \\ b_{n,m}^{i} \end{pmatrix}$$

$$- \frac{n(n+1)(1 + k_n - h_n)}{g}\begin{pmatrix} A_{n,m} \\ B_{n,m} \end{pmatrix} \tag{104}$$

whence

$$\frac{a_{n,m}^{i+1} - a_{n,m}^{i+1}}{a_{n,m}^{i+1} - a_{n,m}^{i}} = \frac{n(n+1)(1 + k_n' - h_n')\alpha_n}{n(n+1) - \sigma^2 a^2/gD} \tag{105}$$

As $n \to \infty$, this ratio $\to 0$ like n^{-1} but, for frequencies close to a resonance $\sigma^2 a^2 = gDn(n+1)$ of the problem without the convolution terms, it may diverge. If the potential is rich in spherical harmonics U_n, most of them will converge as the iteration proceeds, but not those very near this resonance.

Although we cannot analyze the behavior of the iteration in the general case ($\mathscr{L}$ given by equation 24), the foregoing suggests that we there let an estimate of ζ_0 be

$$\hat{\zeta}_0 = \sum_{i=1}^{N} l_i \zeta_0^{(i)} \tag{106}$$

where the $\zeta_0^{(i)}$ are obtained by iterating equation 101 N times, even if divergently, and then find the l_i by minimizing

$$E = \iint \left| \mathscr{L}(\hat{\zeta}_0) - \mathscr{L}'\left(\tilde{\zeta} + \iint G(\theta|\theta') \hat{\zeta}_0(\theta')\, d\theta' \right) \right| d\theta \tag{107}$$

that is, by solving the N equations.

$$\frac{\partial E}{\partial l_i} = 0, \qquad i = 1, N \tag{108}$$

If the individual iterates $\zeta_0^{(i)}$ are made to satisfy homogeneous boundary conditions, the $\hat{\zeta}_0$ likewise satisfy them. If the boundary conditions are inhomogeneous, then the minimization (equation 108) of E should be constrained in such a way that $\hat{\zeta}_0$ satisfies the inhomogeneous conditions. For example, if each iterate is made to satisfy the boundary condition $\zeta_0(i) = \zeta_{obs}$ for convenience, then equation 108 should be subject to the constraint $\sum l_i = 1$.

Parke and Hendershott (unpublished) have studied the quality of solutions $\hat{\zeta}_0$ obtained as above for the M2 tide in the rectangular, flat bottom sea of Rossiter (1958), at whose edges the normal component of fluid velocity vanishes. With a depth of 3930 m, the iterates $\zeta_0^{(i)}$ appear to be slowly converging; at the central mesh point Re $\zeta_0^{(i)}$ takes on the values 1.002, 1.699, 2.011, 2.216, 2.330, 2.401, 2.442, 2.468 and 2.483 for $i = 1, 9$. Table I shows the weights l_i for $N = 2, 4, 6$ and the resulting relative errors $(E/\iint |(\hat{\zeta}_0)|^2 \, d\theta)^{1/2}$. These errors decrease rapidly and the estimate 106 appears simply to accelerate an already convergent iteration. With a depth of 3200 m, Re $\zeta_0^{(i)}$ at the same central point takes on the values 4.413×10, -2.626×10^2, 1.537×10^3, -8.971×10^3, 5.249×10^4, -3.072×10^5, 1.797×10^6, and -1.052×10^9 for $i = 1, 8$. Now the problem is clearly near resonant and the basic iteration diverges. Nevertheless, the estimate $\hat{\zeta}_0$ is stable and the relative error rapidly decreases (Table II).

Further development is required to apply this technique to solutions obtained with other boundary conditions 28–31. If its use with them is feasible, solution by the harmonic method using a solution algorithm such as equation 98, which generates an efficient representation of the application of the inverse operator A^{-1} during the first iteration, appears far more economical than time-stepping.

TABLE I

N	l_1	l_2	l_3	l_4	l_5	l_6	Relative Error
2	0.005	1.106					1.55×10^{-1}
4	0.115	-0.270	1.377	2.537			1.08×10^{-2}
6	-0.000	-0.008	0.136	-0.247	-1.509	2.628	3.3×10^{-5}

TABLE II

N	l_1	l_2	l_3	l_4	l_5	l_6	Relative Error
2	0.737	0.123					2.86×10^{-1}
4	-0.042	-0.168	1.026	0.180			3.32×10^{-3}
6	-0.034	-0.147	0.817	0.325	0.035	0.001	2.43×10^{-3}

6. Model Results and Recommendations

The results of more than a decade of numerical effort to solve LTE for global tides are summarized in Table III and in the cotidal–corange charts of Figs. 4 and 5 for semidiurnal (M2) and diurnal (K1) tides, respectively. Both semidiurnal and diurnal maps are characterized by the appearace of amphidromes (points at which the range vanishes and around which crests rotate) and regions that we may call anti-amphidromes (within which the range is maximum and the phase varies only slowly). None of the calculations include the effects of ocean self-attraction or of solid earth loading, yet taken together they qualitatively capture the overall global tide as reconstructed empirically by, for example, Dietrich (1944).

Hendershott (1973) previously compared a number of M2 calculations. The calculation of Gordeev, Kagan, and Rivkind (1972) only subsequently became available and has been included in Fig. 4. The majority of the M2 calculations agree in the prediction of both a North and a South Atlantic amphidrome, although they vary widely in their resolution and prediction of coastal tides. None are able to resolve the tides over the shallow Patagonian shelf. The central Indian Ocean is consistently characterized by a pronounced anti-amphidrome, although again, coastal details differ widely. The models show the greatest divergence in the Pacific. In the southeast Pacific the two Soviet calculations yield a low-amplitude confluence of cotidal lines where the remaining models put an amphidrome. No data exist to resolve the point empirically. The northwest Pacific is different in every model, and all differ from the empirical reconstruction of Dietrich (1944). Those computations taking vanishing normal velocity as their coastal boundary condition consistently reverse the direction of propagation of the M2 tide along the west coast of North America. The computational broadening of the continental shelf, the neglect of Hawaiian Island arc scattering, and the possible misrepresentation of north Pacific edge dissipation all come to mind as possible reasons for this difficulty but we have no clear explanation for it.

The two diurnal computations (Fig. 5) generally show more gentle space variation than the semidiurnal ones (Fig. 4), but they also disagree significantly in the Pacific.

Hendershott (1972) and Pekeris and Accad (1969) explicitly note the proximity to resonance of their calculations in the absence of dissipation. Platzman (1974) has studied the dissipationless normal modes of a basin composed of the Atlantic and Indian Oceans. His results do much to clarify the various calculations of Table III. He finds 37 free oscillations with periods between 8 and 100 hr, and a number of these have periods not far removed from the tidal bands (the closest to tidal resonance are 29.4, 23.5, 20.9, ..., 12.8, 12.2, 11.3, ... hr). The proximity to resonance of the various global calculations is thus to be expected. Indeed, the fact that the various global calculations all appear to qualitatively capture the overall global tide, even though they are all based upon a seriously incomplete potential, suggests that their response to tidal forcing is dominated, at least in overall form, by their own internal resonances.

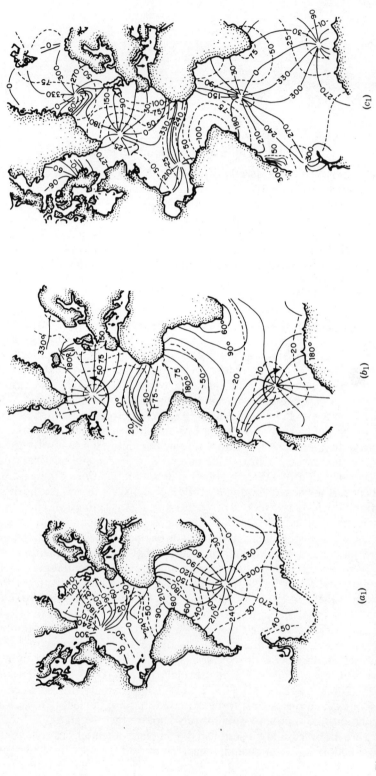

(a_1) (b_1) (c_1)

Fig. 4. Respectively, Atlantic Ocean $(a_1)-(f_1)$, Indian Ocean $(a_2)-(f_2)$, and Pacific Ocean $(a_3)-(f_3)$ numerical solutions of LTE for the M2 tide. $(a_1)-(a_3)$ due to Bogdanov and Magarik (1967); $(b_1)-(b_3)$ due to Hendershott (1973); $(c_1)-(c_3)$ due to Zahel (1970); $(d_1)-(d_3)$ due to Gordeev, Kagan and Rivkind (1972); $(e_1)-(e_3)$ due to Dietrich (empirical map, 1944); $(f_1)-(f_3)$ due to Pekeris and Accad (1969).

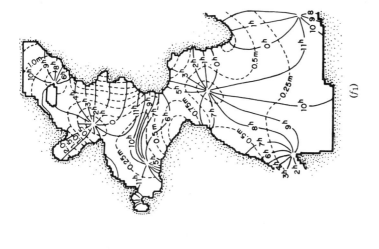

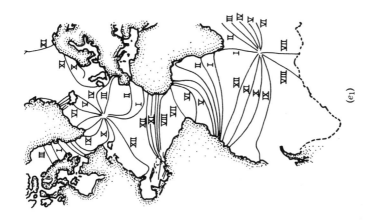

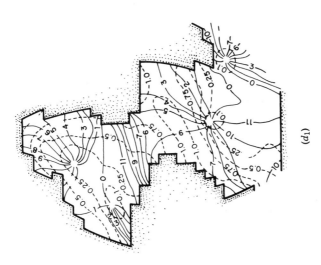

Fig. 4. (*Continued*)

83

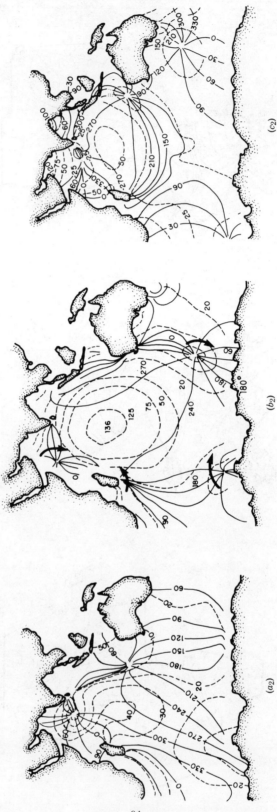

84

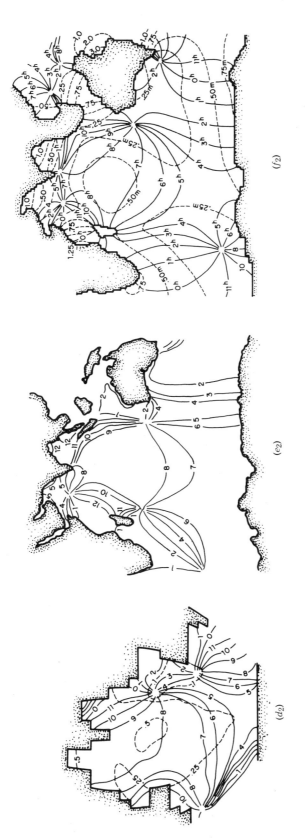

Fig. 4. (Continued)

85

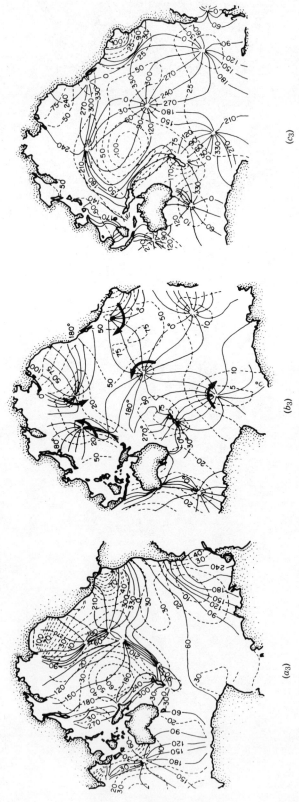

(c_3)

(b_3)

(a_3)

86

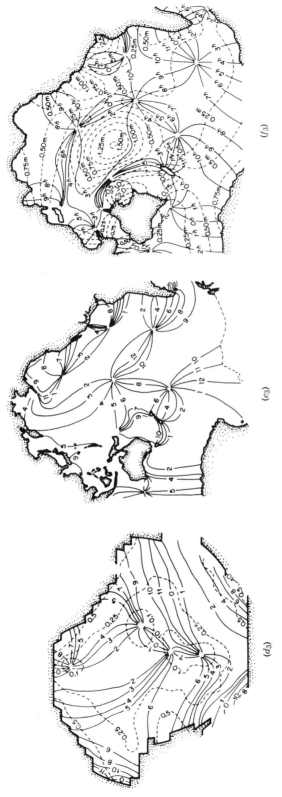

Fig. 4. (*Continued*)

(d_3)

(e_3)

(f_3)

87

TABLE III

Summary of Large-Scale Tidal Models

Investigator	Type of Model	Boundary Condition	Dissipation	Earth Tide
Zahel (1970, 1972)	M2, K1 tides for globe	Impermeable coast	Bottom stress in shallow water	None
Pekeris and Accad (1969)	M2 tide for globe	Impermeable coast	Linearized artificial bottom stress in shallow water	None
Hendershott (1972)	M2 tide for globe	Coastal elevation specified	At coast only	See Section 5
Bogdanov and Magarik (1967, 169)	M2, S2, K1, 01 tides for globe	Coastal elevation specified	At coast only	None
Tiron et al. (1967)	M2, S2, K1, 01 tides for a plane model	Coastal elevation specified where data exist, otherwise impermeable coast	At coast only	None
Gohin (1971a, b)	M2 tide for Atlantic, Indian Oceans	Coastal impedance specified	At coast only	Yielding to astronomical force only
Platzman (1972a, b, 1974)	Normal modes of major basins, especially Atlantic	Adiabatic boundaries	Zero	—
Munk et al. (1970)	Normal mode representation of California coastal tides	Impermeable coast	Zero, although energy flux parallel to coast may reflect dissipation up coast	Yielding to astronomical force only
Cartwright (1971)	β plane representation for South Atlantic tides	Island values specified	Zero, although energy flux may reflect distant dissipation	Yielding to astronomical force only
Ueno (1964)	M2 tide for globe	Impermeable coast	None	None
Gordeev, Kagan, and Rivkind (1972)	M2 tide for globe	Impermeable coast	Linearized bottom stress in shallow water	None

88

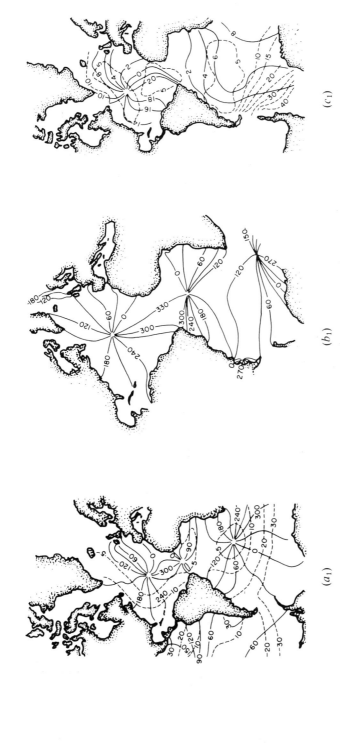

(a_1)

(b_1)

(c_1)

Fig. 5. Same as Fig. 4 except of the K1 tide. (a_1)–(a_3) due to Zahel (1972), (b_1)–(b_3) due to Bogdanov and Magarik (1969), and (c_1)–(c_3) due to Dietrich (1944).

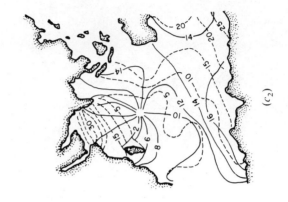

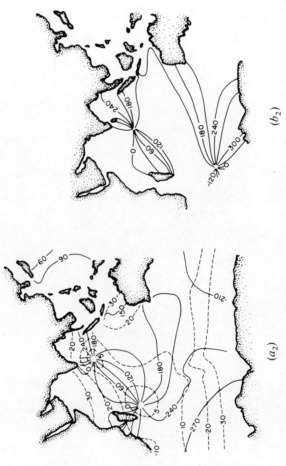

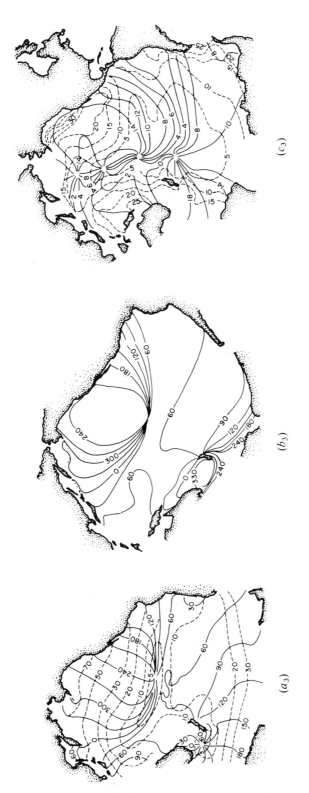

(a_3)

(b_3)

(c_3)

Fig. 5. (*Continued*)

91

Qualitatively, the several modes of near tidal period strongly resemble the computed cotidal–corange maps. For example, several of the Indian Ocean modes with periods not far from semidiurnal show an anti-amphidrome in apparent correspondence with the one occurring in the various M2 calculations and delineated by island tidal data in the Indian Ocean. Platzman (1974) notes this resemblance as well as the correspondence between the computed modes and the Atlantic tide (see also Platzman, 1972b).

It is likely that a number of features of the global tide will find relatively simple explanation in a normal mode synthesis of the forced tide. One of the difficulties standing in the way of such a synthesis is the inclusion of dissipation. We do not know how strongly these modes are damped or how much their form is changed by the (possibly strongly localized) damping. What information we do have is as follows.

The total amount of energy dissipated in lunar ocean tides is estimable from astronomy (Munk and McDonald, 1960, give 2.7×10^{19} erg/sec, but see also Muller, 1974) and directly, if with wide uncertainty, from coastal tide observations (Miller, 1966, finds 0.7×10^{19}–2.5×10^{19} erg/sec at the outside edges of his range of uncertainty, and Hendershott, 1972, estimates 3.04×10^{19} erg/sec by dynamical interpolation of coastal observations, that is, by solving LTE with coastal boundary condition 29). These estimates are for coasts, shelves, and marginal seas only. Wunsch (1974) argues against the possibility that internal tides might significantly augment these figures by invoking their low group velocity.

Although dissipative, ocean tides are not so strongly damped that no evidence of resonance remains. Hendershott (1972) estimated an overall global Q of 34 for M2 (which would be reduced to 20 if the near equipartition of potential and kinetic energy characterizing the gravitational normal modes of Platzman is assumed to hold for the M2 tide), a number in rough accord with the value 25 estimated by Garrett and Munk (1971) from the age of the tide. Wunsch (1972) found that tidal admittances at Bermuda and in the Azores vary by multiplicative factors of nearly four over the semidiurnal band alone, and he identified a 14.8-hr resonant period apparently corresponding to the 14.4-hr resonance computed by Platzman (1972b, 1974). Gallet reported similar behavior in the northeast Pacific during the First GEOP Conference [see conference summary accompanying Hendershott (1973)].

In order to substantially improve upon the calculations of Table III, a number of things must be done. First of all, the complete potential must be employed, including the effects of ocean self-attraction and of solid earth loading. The difficulties involved are primarily numerical. If they are resolved in the manner suggested in Section 5, the normal modes of Platzman (1974) are attractive candidates for inclusion in the sequence of approximating functions used in equation 100.

Once this has been done, the model has two kinds of parameters which may be regarded as adjustable in order to optimize the elevation prediction or interpolation. One set of parameters is the set of free mode periods very close to tidal lines. Especially if a near tidal mode largely escapes effective damping by shallow-water friction, as might well happen on account of the highly localized nature of the dissipation as reconstructed by Miller (1966), its amplitude may have to be fixed by requiring that the model have realistic frequency response (i.e., a realistic behavior of the admittance as compared with stations in the region where the mode is appreciable). The other set of parameters is the representation of dissipation itself in the model. It is not clear from any of the present solutions how sensitive the elevation predictions are to variation in the distribution of dissipation. We may speculate that the ability of the solutions to resolve different distributions of dissipation will be limited by the number of free modes contributing effectively to the solution but the point requires detailed study.

With the advent of satellite altimetry measurements of open ocean tides and/or of areal averages of tidal elevation obtained by inverting gravity observations, it will be feasible to adjust these two sets of free parameters to obtain optimal estimates of ocean tidal dissipation and of ocean tidal elevation. Shoal marginal sea regions may best be included by constructing separate (linearized) models and coupling them to the global calculation in the manner of Garrett (Section 3). Appreciable progress beyond this point, if required, is likely to be won only at the cost of a complete reformulation of the problem with improved parameterizations of the dissipative processes and a much finer computational grid.

References

Accad, Y. and C. L. Pekeris, 1964. The K2 tide in oceans bounded by meridians and parallels. *Proc. Roy. Soc. London, A*, **278**, 110–128.

Baines, P. G., 1974. The generation of internal tides over steep continental slopes. *Phil. Trans. Roy. Soc. London, Ser. A.*, **277**, 27–58.

Bell, T. N., 1973. Internal wave generation by deep ocean flows over abyssal topography. PhD Thesis, Johns Hopkins University, Baltimore.

Bogdanov, K. T. and V. A. Magarik, 1967. A numerical solution of the problem of semidiurnal tidal wave distribution (M2 and S2) in the ocean. *Dokl. Akad. Nauk. USSR*, **172**, 6 (in Russian).

Bogdanov, K. T. and V. A. Magarik, 1969. A numerical solution of the problem of tidal wave propagation in the world ocean. *Izv. Atmos. Oceanic Phys.*, **5** (12), 1309–1317 (in Russian).

Bogdanov, K. T., K. Kim and A. Magarik, 1964. A numerical solution of the hydrodynamical equations of tides on the BESM-2 computer for the world ocean area. *Trans. Inst. Oceanogr. Acad. Sci.* **75**, (in Russian).

Cartwright, D. E., 1968. A unified analysis of tides and surges round north and east Britain. *Phil. Trans. Roy. Soc., London, Ser. A*, **263**, 1–55.

Cartwright, D. E., 1969. Extraordinary tidal currents near St. Kilda. *Nature*, **223**, 938–932.

Cartwright, D. E., 1971. Tides and waves in the vicinity of St. Helena. *Phil. Trans. Roy. Soc., London, Ser. A*, **270**, 603–649.

Christensen, Niels, Jr., 1972. Numerical simulation of free oscillations of enclosed basins on a rotating earth. *Hawaii Inst. Geophys. Rept. HIG-72-9*, 181 pp.

Christensen, Niels, Jr., 1973. On free modes of oscillation of a hemispherical basin centered on the equator. *J. Mar. Res.*, **31**, 168–174.

Defant, A., 1961. *Physical Oceanography. Vol 2*. Pergamon Press, New York.

Dietrich, G., 1944. Die Schwingungssysteme der halb- und eintägigen Tiden in den Ozeanen. *Veroeffentl. Inst. Meereskd. Univ. Berlin, A*, **41**, 7–68.

Dishon, M., 1964. Determination of the average ocean depths from bathymetric data. *Int. Hydro. Rev.*, **41** (2), 77–90.

Dronkers, J. J., 1964. *Tidal Computations in Rivers and Coastal Waters*. North-Holland, Amsterdam.

Elvius, T. and A. Sundstrom, 1973. Computationally efficient schemes and boundary conditions for a fine-mesh barotropic model based on the shallow water equations. *Tellus*, **25** (2), 132–156.

Farrell, W. E., 1972a. Deformation of the earth by surface loads. *Rev. Geophys. Space Phys.*, **10**, 761–797.

Farrell, W. E., 1972b. Global calculations of tidal loading. *Nature*, **238**, 43.

Farrell, W. E. 1972c. Earth tides, ocean tides and tidal loading. *Phil. Trans. Roy. Soc. London, Ser. A*, **274**, 45.

Fischer, G., 1959. Windstau und Gezeiten in Randmeeren. *Tellus*, **11**, 60–76.

Forsythe, G. E. and W. R. Wasow, 1960. *Finite Difference Method for Partial Differential Equations*. Wiley, New York.

Fox, L. N., 1962. *Numerical Solution of Ordinary and Partial Differential Equations*. Addison-Wesley, Reading, Mass.

Garabedian, P. R., 1956. Estimation of the relaxation factor for small mesh size. *Math. Tables Aids Comput.*, **10**, 183–185.

Garrett, C. J. R., 1972. Tidal resonance in the Bay of Fundy and Gulf of Maine. *Nature*, **238**, 441–443.

Garrett, C. J. R., 1974. Tides in gulfs. *Deep-Sea Res.*, **22**, 23–35.

Garrett, C. J. R. and W. H. Munk, 1971. The age of the tide and the Q of the oceans. *Deep-Sea Res.*, **18**, 493–503.

Gates, L. R. and A. B. Nelson, n.d. *A new Tabulation of the Scripps Topography on a 1° Global Grid. Part I, Terrain Heights, R-1276; Part II, Ocean Depths, R-1277.* The Rand Corporation, Santa Monica, Calif., Part I, 133 pp.; Part II, 136 pp.

Gohin, F., 1961a. Etude des marées oceaniques à l'aide de modèles máthematiques. *Proc. Symp. Math.- Hydrodyn. Methods Phys. Oceanogr.*, *Sept. 1961.* Institüt Meereskunde, University of Hamburg, pp. 179–198.

Gohin, F., 1961b. Determination des denivellations et des còurants de maree. *Proc. Seventh Conf. Coastal Eng.*, The Hague, Netherlands, August 1960, Vol. 2, pp. 485–509.

Gordeev, R. G., B. A. Kagan, and V. Ya. Rivkind, 1973. Numerical solution of the equations of tidal dynamics in the world ocean. *Dokl. Akad. Nauk USSR*, **209** (2), 340–343 (in Russian).

Hansen, W., 1962. Tides. In *The Sea*, I. M. Hill, ed. Vol. 1. Wiley, New York.

Hansen, W., 1966. The reproduction of motion in the sea by means of hydrodynamical and numerical methods. *Mitt. Inst. Meereskd., Univ. Hamburg*, **25**.

Heaps, N. S., 1969. A two-dimensional numerical sea model. *Phil. Trans. Roy. Soc., A.* **265**, 93–137.

Hendershott, M. C., 1972. The effects of solid earth deformation on global ocean tides. *Geophys. J. Roy. Astron. Soc.*, **29**, 389–402.

Hendershott, M. C., 1973. Ocean tides. *EOS*, **54**, 76–86.

Hendershott, M. C. and W. H. Munk. 1970. Tides. *Ann. Rev. Fluid Mech.*, **2**, 205–224.

Hendershott, M. C. and A. Speranza, 1971. Co-oscillating tides in long, narrow bays: The Taylor problem revisited. *Deep-Sea Res.*, **18**, 959–980.

Hollsters, M., 1962. Remarques sur la stabilité dans les calculs de marée. *Proc. Symp. Math.-Hydrodyn. Methods Phys. Oceanogr.*, *Sept. 1961.* Institüt Meereskunde, University of Hamburg, pp. 211–225.

Jeffreys, M., 1970. *The Earth.* Cambridge University Press.

Kagan, B. A., 1972. On the resistance law for the tidal flow. *Izv. Atmos. Oceaňic Phys.*, **8** (5), 533–542 (in Russian).

Lauwerier, H., 1962. Some recent work of the Amsterdam Mathematical Centre on the hydrodynamics of the North Sea. *Proc. Symp. Math.-Hydrodyn. Methods Phys. Oceanogr.*, *Sept. 1961.* Institüt Meereskunde, University of Hamburg, pp. 13–24.

Leendertse, J. J., 1967. Aspects of a computational model for long-period water-wave computation. *Memo. RM-5294-PR.* The Rand Corporation, Santa Monica, Calif., 165 pp.

Marchuk, G. I., R. G. Gordeev, V. Ya. Rivkind, and B. A. Kagan, 1973. A numerical method for the solution of tidal dynamics equations and the results of its application. *J. Comp. Phys.*, **13**, 15–34.

Marchuk, G. I., B. A. Kagan, and R. E. Tamsalu, 1969. A numerical method of calculating tidal motion in border seas. *Izv. Atmos. Oceanic Phys.*, **5** (7), 694–703.

Miller, G., 1966. The flux of tidal energy out of the deep oceans. *J. Geophys. Res.*, **71** (4), 2485–2489.

Miles, J., 1974. On Laplace's tidal equations. *J. Fluid Mech.*, **66**, 241–260.

Muller, P. M. 1974. Determination of the rate of change of G from solar system observation (in preparation).

Munk, W. H. and G. MacDonald, 1960. *The Rotation of the Earth.* Cambridge University Press, London.

Munk, W. H. and D. Cartwright, 1966. Tidal spectroscopy and prediction. *Phil. Trans. Roy. Soc., London, Ser. A.*, **259**, 533–581.

Munk, W. H., F. Snodgrass, and M. Wimbush, 1970. Tides off-shore: Transition from California coastal to deep-sea waters. *Geophys. Fluid Dyn.*, **1**, 161–235.

Pekeris, C. L. and Y. Accad, 1969. Solution of Laplace's equation for the M2 tide in the world oceans. *Phil. Trans. Roy. Soc., London, Ser. A*, **265**, 413–436.

Platzman, G. W., 1963. The dynamical prediction of wind tides on Lake Erie. *Meteorol. Monogr.*, **4** (26), 44 pp.

Platzman, G. W., 1971, Ocean tides and related waves. In *Lectures on Applied Mathematics.* Vol. 14. American Mathematics Society, Providence, R.I.

Platzman, G. W., 1972a. Two-dimensional free oscillations in natural basins. *J. Phys. Oceanogr.*, **2**, 117–138.

Platzman, G. W., 1972b. North Atlantic Ocean: Preliminary description of normal modes. *Science*, **178**, 156–157.

Platzman, G. W., 1974. Normal modes of the Atlantic and Indian Oceans. *J. Phys. Oceanogr.* (in press).

Proudman, J. 1944. The tides of the Atlantic Ocean. *Mon. Not. Roy. Astron. Soc.*, **104**, 244–256.

Redfield, A., 1958. Influence of the continental shelf on tides of the Atlantic Coast of the United States. *J. Mar. Res.*, **17**, 432–448.

Richardson, L. F., 1922. *Weather Prediction by Numerical Methods.* Cambridge University Press, London; Dover, New York, 1965.

Rossiter, J. R., 1958. Application of relaxation methods to oceanic tides. *Proc. Roy. Soc., A*, **248**, 482–498.

Shuman, F. G., 1957. Numerical methods in weather prediction. *Mon. Weather Rev.*, **85**, 357–361.

Sidjebat, M., 1970. The numerical modeling of tides in a shallow semienclosed basin by a modified elliptic method. Ph.D. thesis, University of Miami, Coral Gables, Fla.

Sielecki, Anita, 1968. An energy-conserving difference scheme for the storm surge equations. *Mon. Weather Rev.*, **96**, 150–156.

Smith, Stuart M., H. M. Menard, and George Sharman, 1966. World-wide ocean depths and continental elevations averaged for areas approximating one degree squares of latitude and longitude. *Ref.* 65-8, Scripps Institution of Oceanography, La Jolla, Calif., 14 pp.

Taylor, G. I., 1919. Tidal friction in the Irish Sea. *Phil. Trans. Roy. Soc., London, A*, **220**, 1–93.

Thompson, R. E. 1974. The energy and energy flux of planetary waves in an ocean of variable depth. *Geophys. Fluid. Dyn.*, **5** (4), 385–399.

Tiron, K. D., Y. Sergeev, and A. Michurin, 1967. Tidal charts for the Pacific, Atlantic and Indian Oceans (transl.). *Vestn. Leningrad Univ.*, **24**, 123–135.

Ueno, T., 1964. Theoretical studies on tidal waves travelling over the rotating globe, I, II. *Oceanogr. Magazine*, **15** (2), 99–101; **16** (3), 47–54.

Villain, C., 1952. Cartes des lignes cotidals dans les océans. *Ann. Hydrog. (Paris)*, **3**, 269–388.

Voit, S. S., 1967. Tides in seas and oceans. *Fis. Atmos. Okeana*, **3** (6), 579–591 (in Russian).

Von Trepka, L., 1967. Anwendung des Hydrodynamischen—Numerischen Verfahrens zur Emmittlung des Schelfeinflusses auf die Gezeiten in Modellkanälen und Modellozeanen. *Mitt. Inst. Meereskd. Univ. Hamburg*, **9**.

Wunsch, C., 1969. Progressive internal waves on slopes. *J. Fluid Mech.*, **35**, 131–144.

Wunsch, C., 1972. Bermuda sea level in relation to tides, weather, and baroclinic fluctuations. *Rev. Geophys.*, **10** (1), 1–49.

Wunsch, C., 1974. Internal tides in the ocean. *Rev. Geophys. Space Phys.*, **13**, 167–182.

Young, D. M., 1954. Iterative methods for solving partial difference equations of elliptic type. *Trans. Am. Math. Soc.*, **76**, 92–111.

Zahel, W., 1970. Die Reproduktion Gezeitenbedingter Bewegungsvorgänge im Weltozean Mittels des Hydrodynamisch-numerischen Verfahrens, *Mitt. Inst. Meereskd. Univ. Hamburg*, **17**.

Zahel, W., 1973. The diurnal K1 tide in the world ocean—a numerical investigation. *Pure Appl. Geophy.*, **109** (8), 1819–1825.

3. ONE-DIMENSIONAL MODELS OF THE SEASONAL THERMOCLINE

Pearn P. Niiler

1. Introduction

The seasonal heat exchange between the atmosphere and the oceans produces large horizontal-scale changes of the temperature structure in the surface layers of the oceans. This seasonal cycle of the heat content and temperature of the ocean surface water column has been well documented from B.T. and X.B.T. observations both in the North Pacific (Bathen, 1970; Robinson, 1973) and North Atlantic (Schroeder, 1963). A variety of continuous records also exists at ocean weather ship stations for the years 1948–1973. Excellent descriptions of the seasonal cycle at ocean station Papa are given by Tully and Giovando (1963) and Tabata (1965). Estimates of the seasonal heat exchange with the atmosphere have also been made (e.g., Budyko, 1956; Wyrtki, 1966). At specific locations such as ocean weather stations, the estimates by different investigators agree, whereas over wide areas, these computations suffer from lack of data on net solar radiation absorbed in the ocean. An attempt to provide a detailed review and a geographic summary of these observations is beyond the scope of this presentation; however, several general descriptive aspects that lie at the base of theories and models of the seasonal variability in the ocean are noteworthy.

In the early northern spring, the daily solar warming of the ocean surface becomes larger than the nighttime cooling, and layers of successively warmer and lighter water are formed over the cold abyss. Over a period of two months of intense warming, this cyclic process builds a strong vertical thermal gradient near the surface, a seasonal thermocline. If summer winds are sufficiently strong, a homogeneous column, the mixed layer, persists on top of the seasonal thermocline. In early fall, the atmospheric storms become more frequent, and a net cooling of the ocean surface sets in. The heavy convective elements now aid in the vigorous vertical stirring of the surface column, and the erosion of the summer thermocline begins. This erosion is accomplished by a slowly penetrating, vertically homogeneous layer, whereas, in contrast, buildup is a result of a series of individual, daily events (see Tully and Giovando, 1963). The processes by which heat is distributed into the ocean in summer are observed to be quite different than those that remove this stored thermal energy in winter. The maximum depth of the wintertime erosion marks the top of the permanent thermocline. At each location in mid ocean, a well defined level can be found at which the seasonal temperature and salinity variations are no longer perceptible (Wunsch, 1972).

There are strong observational and theoretical indications that the seasonal thermocline locally stores the heat that is given up by the atmosphere in summer and released by the ocean in winter. Gill and Niiler (1973) have shown this concept to emerge from considering large horizontal-scale (1000 km latitudinal and 3000 km longitudinal) seasonal forcing of the ocean circulation by the atmosphere. Both vertical and horizontal advection of heat can be shown to be much smaller when averages over such large-scale areas are considered. Gill (1974) and Bernstein (1974) have shown that the mesoscale eddy field can produce strong, local advection of the surface patterns; however, simple recipes such as horizontal averaging over many eddy scales or using a variable (advective) depth of seasonal penetration produce the repeatable, smooth pattern of local, seasonal heat storage which is balanced by the atmospheric input. Mid-oceanic surface front formation (Voorhis, 1969), heat advection by western boundary currents (Niiler and Richardson, 1973; Montgomery, 1974), and tropical seasonal patterns (Yoshida, 1967) are notable exceptions to this balance because these phenomena contain small, nonperiodic, horizontal-scale features.

Finally, it is observed that the vertical shape of the seasonal thermocline and the maximum depth of the winter penetration vary considerably over the mid ocean. The geographic variability of the heating cycle cannot account for such differences because contours of constant seasonal heating are not mirrored by contours of the constant sea-surface temperature cycle. Although the general surface circulation of the oceans might not affect the total heat storage, it does play an important role in determining the vertical distribution of that storage and the shape of the seasonal thermocline. The vertical turbulent thermal diffusion in the ocean strongly depends upon the vertical shear of the near-surface currents.

A strong hydrodynamic similarity exists between the atmospheric boundary layer and the oceanic layers under ice and the ocean bottom (Smith and McPhee, 1972; Munk and Wimbush, 1970). The surface layer mixing processes in the ocean, however, are somewhat different because the influence of surface waves is different from that of a solid boundary. This review does not cover modeling of the turbulent boundary layer in the atmospheric parameter ranges. Also, it is useful at the outset to note that although the models of the seasonal time scale evolution are presented, this cycle is a composite of very short-lived events, and of necessity, the modeling of the seasonal cycle must begin by modeling these events.

The central modeling problem of the planetary boundary layer, or the ocean surface layer, is parameterization of the turbulent processes that transfer heat and momentum vertically. Two somewhat distinct viewpoints on this matter are found in the literature. The earliest concepts were to parameterize the turbulent transfer of heat and momentum as functionals of the mean gradients. The lowest-order parameterization in this scheme is a familiar one in which gradients of temperature and velocity govern the transfer mechanism (Munk and Anderson, 1949). In more recent developments, second-order closure schemes for turbulence are employed to render a set of simultaneous differential equations for the mean fields and the vertical turbulent fluxes (Lumley and Khajeh-Nouri, 1974). In contrast, the models presented here parameterize the observed features of vertical structure of the mean distributions and develop a dynamically consistent set of constraints for the turbulent fluxes. To close the system, a variety of hypotheses are advanced which are based on laboratory models of penetrative turbulent mixing and convection.

In Section 2, the equations of the model are developed, and the time-dependent response of the system is investigated under typical oceanic conditions in Section 3. In Section 4, the application of the model to development of the seasonal thermocline is reviewed, and results of comparison with oceanic conditions are discussed. Some conclusions are drawn in Section 5.

2. The Conservation Equations

The ocean surface water column under seasonal influence can be divided into two principal dynamic regimes. Below an active wind–wave zone, a mixed layer is observed in which the density and velocity field are uniform with depth (Halpern, 1974; Pollard and Millard, 1970; Van Leer, et al., 1974). An entrainment interface marks the

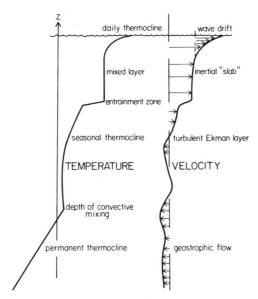

Fig. 1.   Schematic of the dynamic regimes in a one-dimensional model of the seasonal thermocline.

lower boundary of the mixed layer and the upper boundary of the seasonal thermocline. During summer and fall storms, a strong vertical shear of horizontal currents is found across the entrainment zone; in laboratory experiments, a sharp gradient of the turbulent intensity exists across this zone as well (Linden and Turner, 1975). The seasonal thermocline is a region of high internal wave activity in which patchy turbulent regions appear; billow or shear turbulences play an important role in the vertical transfer of heat and momentum (Woods and Wiley, 1972).

Figure 1 displays the schematic of this single-dimensional model. The thermal structure is seen to closely follow the conditions summarized by Tully and Giovando (1963) at ocean station Papa.

The conservation equations of mixed layer models have appeared in a variety of publications (e.g., Niiler, 1975), as have the equations for a weakly sheared and strongly stratified thermocline (Zilitinkevich et al., 1967). The mixed layer models are differentiated by the physical assumptions about the conditions under which a vertical homogeneous column is allowed to exist. In the simplest scheme, a mixed layer is formed only during a surface cooling process, a vertical convective adjustment (Warren, 1972; Thompson, 1974). In the thermocline, a variety of the parameterizations of vertical turbulent fluxes in terms of the mean fields is used. The simplest scheme postulates that the vertical turbulent flux of momentum is proportional to the vertical gradient of the horizontal currents, and the vertical turbulent flux of heat is proportional to the vertical gradient of the temperature. The constants of proportionality are the vertical austausch coefficients of momentum and heat. Here a derivation of the coupled equations is presented for the case in which both the entrainment interface and the surface wave zone are thinner than the mixed layer or the seasonal thermocline.

The one-dimensional conservation equation for heat (or density) is

$$C_p \rho_0 \left(\frac{\partial T}{\partial t} + \frac{\partial \overline{W'T'}}{\partial z} \right) = 0 \qquad (1)$$

and for momentum it is

$$\rho_0\left(\frac{\partial \mathbf{V}}{\partial t} + \hat{f} \times \mathbf{V} = \frac{-\partial \overline{W'\mathbf{V}'}}{\partial z}\right) \tag{2}$$

Here T and $\mathbf{V}$ are the mean temperature and velocity, respectively, and primes designate the fluctuating quantities; the overbar designates the Reynolds' ensemble average. Henceforth the subscript s is used to refer to quantities within the mixed layer of depth h, and plus signs designate quantites at the top of the seasonal thermocline.

In the mixed layer, T and $\mathbf{V}$ are to be independent of depth; therefore, $\rho_0 C_p(\overline{W'T'})_s$ and $\rho_0(\overline{W'\mathbf{V}'})_s$ are linear functions of depth. At the ocean surface, there is a pre-scribed heat flux Q_0 and a momentum flux τ_0. As the entrainment interface erodes into the top of the seasonal thermocline there is a rapid change of both quantities due to an entrainment heat flux of $\rho_0 C_p[T_s - T(z = -h^+)](\partial h/\partial t)$ and entrainment momentum flux of $\rho_0[\mathbf{V}_s - \mathbf{V}(z = -h^+)](\partial h/\partial t)$ (for a rigorous derivation see Niiler, 1975). Thus if q_+ is the heat flux at the top of the seasonal thermocline, and τ_+ is the momentum flux,

$$-\rho_0 C_p(\overline{W'T'})_s = Q_0 + \frac{z}{h}\left\{Q_0 - C_p\rho_0(T_s - T_+)\frac{\partial h}{\partial t} - q_+\right\} \tag{3}$$

$$-\rho_0(\overline{W'\mathbf{V}'})_s = \tau_0 + \frac{z}{h}\left\{\tau_0 - \rho_0(\mathbf{V}_s - \mathbf{V}_+)\frac{\partial h}{\partial t} - \tau_+\right\} \tag{4}$$

These expressions hold from the base of the mixed layer to the wind–wave zone at the ocean surface. In writing equations 3 and 4 it has been assumed that τ and Q do not change through the wave zone. Upon substituting from equations 3 and 4 into 1 and 2, there results,

$$\frac{\partial T_s}{\partial t} + (T_s - T_+)\frac{\partial h}{\partial t} = \frac{1}{\rho_0 C_p h}(Q_0 - q_+) \tag{5}$$

$$\frac{\partial \mathbf{V}_s}{\partial t} + \hat{f} \times \mathbf{V}_s = \frac{\tau_0}{\rho_0 h} - \frac{1}{h}(\mathbf{V}_s - \mathbf{V}_+)\frac{\partial h}{\partial t} - \frac{\tau_+}{\rho_0 h} \tag{6}$$

In this model, the kinetic energy equation for the turbulent components is now used to determine the conditions under which sufficient mechanical energy is generated to keep the layer well mixed. The total balance, as expressed in an integral form, is

$$\int_{-h^+}^{0}\left\{\frac{1}{2}\frac{\partial \overline{C'^2}}{\partial t} + \frac{\partial}{\partial z}\overline{W'\left(\frac{C'^2}{2} + \frac{p'}{\rho_0}\right)} + \overline{W'\mathbf{V}'}\cdot\frac{\partial \mathbf{V}}{\partial z}\right\}dz = \alpha g\int_{-h^+}^{0}\overline{W'T'}\,dz - \int_{-h^+}^{0}\varepsilon\,dz \tag{7}$$

The integral is taken from directly beneath the entrainment interface, h^+, to the ocean surface; in the above, $\overline{C'^2}/2$ is the turbulent kinetic energy density per unit volume, $\alpha = \partial\rho/\partial T$, and ε, a positive quantity, is the dissipation rate of turbulent energy per unit volume.

At the ocean surface, there is a wind–wave driven current of magnitude $\mathbf{U}_*$ and across the entrainment zone there is a rapid velocity change of $(\mathbf{V}_s - \mathbf{V}_+)$. The expressions for $\overline{W'\mathbf{V}'}$ and $\overline{W'T'}$ within the mixed layer are obtained from equations 3 and 4; in the entrainment zone, equations 1 and 2 are integrated in z to obtain these expressions for the case where $\mathbf{V}$ and T are rapid functions of z (see Niiler, 1975, for a derivation of these latter expressions).

The following further assumptions are now made.

1. The kinetic energy density $\bar{C}'^2/2$ is vertically uniform within the mixed layer, and its intensity is rapidly reduced across the entrainment interface. The rate of accumulation of turbulent energy is approximately equal to

$$\int_{-h^+}^{0} \frac{1}{2} \frac{\partial \bar{C}'^2}{\partial t} \, dz \simeq c_e \frac{U_*^2}{2} \frac{\partial h}{\partial t}; \quad U_* \text{ is the frictional velocity within the layer} \quad (7a)$$
$$\text{and } \rho_0 U_*^2 = |\tau_0|.$$

2. The vertical energy flux, $\overline{W'[(C'^2/2) + (p'/\rho_0)]}$, within the seasonal thermocline is much smaller than in the mixed layer. This flux is produced at the ocean surface by the shear of the wind–wave driven flow ($\tau_0 \cdot \partial \mathbf{U}_*/\partial z$). Within the entrainment zone additional production takes place which is proportional to $|\mathbf{V}_s - \mathbf{V}_+|^2(\partial h/\partial t)$. A portion of this flux is dissipated at a rate ε_s, and the excess is written as

$$\int_{-h^+}^{0} \overline{W'\mathbf{V}'} \cdot \frac{\partial \mathbf{V}}{\partial z} \, dz + \overline{W'\left(\frac{C'^2}{2} + \frac{p'}{\rho_0}\right)}\Big|^{z=0} + \int_{-h^+}^{0} \varepsilon_s \, dz = -m_0 U_*^3$$

$$-\frac{1}{2}|\mathbf{V}_s - \mathbf{V}_+|^2 \frac{\partial h}{\partial t} - \frac{1}{2\rho_0}\tau_+ \cdot (\mathbf{V}_s - \mathbf{V}_+) \quad (7b)$$

3. There is an excess of background dissipation. Its amplitude is not directly related to wind-driven turbulence (the U_* dissipation) and simply

$$\int_{-h^+}^{0} (\varepsilon - \varepsilon_s) \, dz \simeq \varepsilon_0 h \quad (8)$$

Upon substituting from equations 5–8 and 3 into 7, the following is obtained:

$$\frac{1}{2}\frac{\partial h}{\partial t}\{(c_e U_*^2) + \alpha g h(T_s - T_+) - |\mathbf{V}_s - \mathbf{V}_+|^2\} = m_0 U_*^3 - \frac{\alpha g}{2C_p \rho_0} Q_0 h$$

$$- \varepsilon_0 h + \frac{\tau_+}{\rho_p} \cdot (\mathbf{V}_s - \mathbf{V}_+) - \frac{\alpha g}{2C_p \rho_0} q_+ \quad (9)$$

In general, three adjustable constants, c_e, m_0, and ε_0 appear, and the parameterization of the fluxes into the seasonal thermocline must also be specified. However, for oceanic values of U_*, Q_0 (and reasonable choice of ε_0), each of these variables has been shown to be related to *different time scale* development of the surface layer structure and can be evaluated from relatively independent data sets (Niiler, 1975; deSzoeke and Rhines, 1975). In the next section, the deepening of the oceanic mixed layer during onset of a strong wind is reviewed with emphasis on a determination of the parameters in equation 9.

Equations 5, 6, and 9 apply only during the time when the mixed layer is deepening ($\partial h/\partial t > 0$) and there is entrainment of colder and less rapidly moving fluid across its lower interface. This is always the case during the formation of the seasonal thermocline, as successively warmer mixed layers are formed each day at the ocean surface and rapidly entrain into the deep homogeneous water column left behind the preceding winter. However, as will be shown, successively warmer layers penetrate to successively shallower depths, and it appears as if the mixed layer depth as defined over a week's interval is decreasing. For all practical purposes, during the time that successively shallower layers are formed, $\partial h/\partial t = 0$ in equations 5, 6, and 9, the momentum and heat balance is determined without the entrainment effects, and h_m is determined from the solution to the right-hand side of equation 9 [h_m is the Monin–Obukhov length (Monin and Yaglom, 1971) for the mixed layer].

In the region of the seasonal thermocline, the vertical heat and momentum flux are parameterized as

$$-\rho_0 C_p \overline{W'T'} = \rho_0 C_p l_T{}^2 \left|\frac{\partial \mathbf{V}}{\partial z}\right| \frac{\partial T}{\partial z} \qquad (10)$$

$$-\rho_0 \overline{W'\mathbf{V}'} = \rho_0 l_m{}^2 \left|\frac{\partial \mathbf{V}}{\partial z}\right| \frac{\partial \mathbf{V}}{\partial z} \qquad (11)$$

Here l_T and l_m are the length scales for the energy-containing eddies for heat and momentum, respectively. Because a momentum flux can be carried by vertically propagating internal waves, whereas heat cannot, l_m, in general, need not equal l_T. Within the scope of this presentation, adequate review of turbulent transfer theory in stratified shear flow cannot be given (see, Zilitinkevich et al., 1967). Various hypotheses for the turbulent l_T and l_m as functionals of $\partial \mathbf{V}/\partial z$ and $\alpha g(\partial T/\partial z)$, however, differentiate one theory from another, and integration of the equations with a few of the simplest formulations have been carried out in this presentation.

It should be noted that direct radiative heating is observed to penetrate into the ocean so that all the radiative heat is not absorbed within the surface wave zone (as has been assumed here). Denman and Miyake (1973) and Thompson (1974) have made a parameter study of this effect, and for simplicity, we have not considered it here. Also, a buoyancy flux must be computed in the turbulent energy balance, and here only the temperature effect on the density change has been considered. In general, a pycnocline should be substituted for a thermocline, and a mixing process for salt also must be introduced below the mixed layer. A self-consistent development of the combined thermohaline problem is straightforward, and since only temperature effects are compared with observations, such additional formalism seems a bit cumbersome for this presentation. Clearly, salinity changes would be important in describing the seasonal variability in the tropical oceans and in parameterizing the main thermocline in the subarctic regions, and should be included when quantitative comparisons with observations are made.

3. The Time-Dependent Response at Onset of a Wind Event

Consider a quiescent ocean that is stratified, with wind stress beginning to act on the ocean surface. The simplest model of response of the surface layer, without surface heating, is obtained by setting $\varepsilon_0 = l_T = l_m = 0$ (or $\tau_+ = q_+ = 0$ in equations 5, 6, and 9). Phillips (1966), Niiler (1975), and deSzoeke and Rhines (1975) have shown that four distinct physical regimes of dynamics follow which can be characterized best in terms of the turbulent kinetic energy equation (with N the Brunt–Väisälä frequency of the stratified column).

$$\frac{1}{2}\frac{\partial h}{\partial t}\left(c_e U_*^2 + \frac{N^2 h^2}{2} - |\mathbf{V}_s|^2\right) = m_0 U_*^3 \qquad (12)$$

$$\quad (A) \qquad\quad (B) \qquad (C) \qquad\quad (D)$$

A = storage rate of turbulent energy within the mixed layer
B = rate of increase of potential energy of the mixed layer due to entrainment of cold water from the thermocline
C = rate of production of turbulent mechanical energy from the entrainment stress
D = rate of production of turbulent mechanical energy by surface wave processes

1. Initially, there is a balance between A and D and $h \simeq 2(m_0/c_e)U_* t$; the rate of increase of perturbation kinetic energy within the water column is proportional to the rate at which it is produced at the ocean surface. The depth of the mixed layer grows to a few meters over a fraction of Brunt–Väisälä period. This balance holds for $t \simeq 100$ sec, for $U_* = $ 2–3 cm/sec, $c_e = 2$, $m_0 = 1.2$.

2. A balance between B and D is attained, and $h \simeq (12m_0)^{1/3}(U_*/N)(Nt)^{1/3}$. The perturbation kinetic energy which is produced at the surface goes into increasing the potential energy of the mixed layer by entrainment of cold water from below. In the ocean, the mixed layer grows to 10–15 m within an hour. This regime breaks down after $Nt \sim 10$.

3. Meanwhile, the mean flow has been accelerating, and the production of turbulent energy at the base of the mixed layer becomes significant. If $|V_s|$ develops more rapidly than $t^{2/3}$, then $h \simeq \sqrt{2}|V_s|/N$, and a balance between B and C is attained in equation 12. The turbulent energy produced at the base of the mixed layer is used to increase the potential energy of the homogeneous water column. Since $|V_s|$ is primarily an inertial current (see equation 6), its amplitude is characteristically of magnitude $(|\tau_0|/fh)$. Within a quarter pendulum day, the mixed layer erodes to $h_* \simeq 2(|\tau_0|/\rho_0 Nf)^{1/2}$ ($\simeq$ 17–35 m). Subsequently, the intensity of the current decreases, as does the production rate of turbulent energy at the base of the layer, and the surface energy production again dominates the turbulent energy production.

4. Finally, with the decrease of the first inertial oscillation, a slow erosion continues as a balance between B and D. This is identical to the balance described in step 2. There is a net input of energy from surface processes (D), and the potential energy of the column slowly increases by entrainment of cold water from the thermocline (B). In the absence of dissipation ($\varepsilon = 0$), and with cyclic heating and cooling, this erosion of the thermocline will prevail over many cycles. This is due to the fact that over many cycles $U_*^3 = |\tau_0/\rho_0|^{2/3}$ represents a net production of turbulent energy.

Figure 2 displays these regimes from an actual numerical integration of equation 12.

These event-initiated behaviors are important because the coefficients c_e and m_0 have been determined by a best fit to data in a number of laboratory experiments and field observations. Lundgren and Wang (1973) have estimated that for homogeneous turbulence, $c_e = 0.56$; Kato and Phillips (1969), in a laboratory study of entrainment (our dynamic region 2), and Denman and Miyake (1973), in a field study of slow erosion at ocean station Papa (our dynamic region 4), find that $m_0 = 1.0$–1.3.

We have obtained numerical solutions to the equations for the case where a diffusive leakage of momentum (and heat) is allowed to the region below the mixed layer. This is a test of whether a quantitative change of m_0 needs to be made in a more complex system. The partial differential equations below the mixed layer are finite differenced on a vertical grid of depth d, and these are integrated forward in time, together with the equations for the mixed layer fields, h, T_s, V_s, using a fourth-order Adams–Moulton modification of the Runga–Kutta scheme. A relative error of 10^{-4} after each integration step is allowed. This scheme is accurate but inefficient and has a

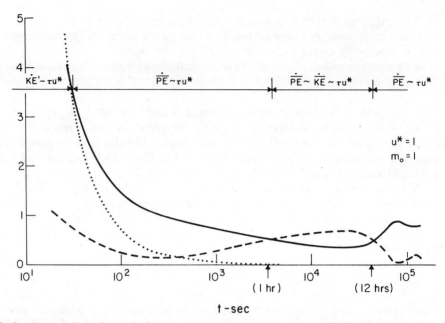

Fig. 2. A comparison of terms in the numerical integration of equation (12), when $|V_s|$ is the solution of equation (6) with $\tau^+ = 0$. Dotted line is the ratio (A/D); dashed line is the ratio (B/D); solid line is the ratio (C/D). $U_* = 1$ cm/sec, $N = 10^{-2}$ sec^{-1}, $m_0 = 1$ (from deSzoeke and Rhines, 1975).

variable time step, which is required to describe some pathological behaviors revealed by the analytic solutions discussed above. Also, a variable number of grid layers is used below the mixed layer as these are needed. We have found good numerical stability in the system if an additional diffusive grid is added when the shear across the last moving layer exceeds 10^{-3}/sec. If an arbitrary (large) number of diffusive grid layers is specified beforehand, numerical instability can appear. The mixed layer is allowed to erode into the diffusive grids in a fashion so that piecewise continuous transition of the viscous stress and entrainment stress between the two uppermost grids is effected. Hence as the mixed layer "swallows up" a diffusive grid, no discontinuous jumps appear in $(V_s - V_+)$, τ_0, or τ_+ in the turbulent energy equation or momentum equations. The integration scheme leaks a little momentum and heat out of the lowest grid; however, we keep track of this leakage numerically and have presented runs in which, over the integration time, these fluxes can account for only 1–2% of the total storage. In effect, the surface fluxes of heat and momentum are stored in the column [i.e., $(\partial/\partial t) \int_{-H}^{0} T\,dz \simeq (Q_0/\rho_0 C_p)$ and $(\partial/\partial t) \int_{-H}^{0} V\,dz + \hat{f}x \int_{-H}^{0} V \simeq (\tau_0/\rho_0)$], over the maximum depth of seasonal penetration, H.

Figure 3 compares time-dependent evolution of the mixed layer depth at the onset of wind ($Q_0 = 0$) for the cases where $l_T = 0$, no heat is allowed to diffuse below the mixed layer, and the mixing for momentum is a constant, $l_m = 0$, 3.0 m, 0.3 m; also on the same graph are the solutions where $l_m = 0.3h$ and $0.03h$. No detectable difference can be graphed from these solutions on the $h(t)$ display. Momentum diffusion will not act to stabilize the erosion process because h will grow without bound. Figures 4 and 5 show the corresponding graphs for $(U^2 + V^2)^{1/2}$ as a function of z. It is apparent that when the mixing length is proportional to h, the diffusion is larger, and the solutions become stationary spirals sooner. In these cases, the evolution of

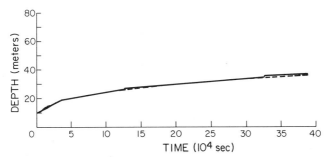

Fig. 3.   Initial deepening of a mixed layer into a stratified column with momentum diffusion to lower layer (———). Dashed line is Niiler's (1975) model without diffusion ($\tau^+ = 0$). The parameters are $m_0 = 1.2$, $\tau_0 = (1, 0)$ dynes/cm^2, $N^2 = 0.8 \times 10^{-4}$ sec^{-2}. For diffusive scales, see text.

the mixed layer is initially governed by process 3 and later by 4; in both cases, the additional turbulent, mechanical energy generated by the viscous stress at the base of the mixed layer is small, and the vertical shear produced over the initial half pendulum day is not effectively reduced by the viscous momentum diffusion. Note, however, in later times, momentum has diffused well below the mixed layer, and in this model, this is the region where heat can also begin to diffuse vertically through a crudely parameterized shear turbulence. This latter phenomenon is important for seasonal scale response, because it is observed that the seasonal thermocline formed

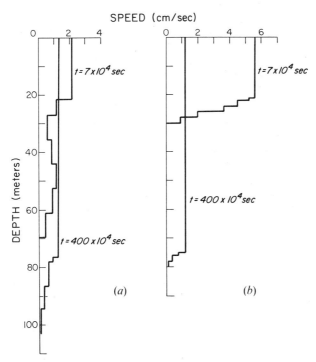

Fig. 4.   Speed profiles of integration of Figure 3 for $l_T = 0$, $l_m = 3.0$ m (a), 0.3 m (b). The discontinuities in the graphs mark the limits of integration grids.

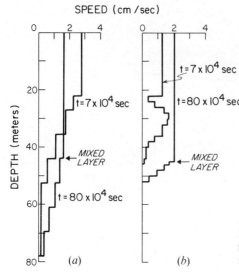

Fig. 5. Speed profiles for integration of Figure 3 for $l_T = 0$, $l_m = 0.3h$ (a), $0.03h$ (b). The discontinuities in the graphs mark the limits of integration grids.

by early June is modified considerably by early October, when the erosion process begins. In this model, such an adjustment can happen only if heat can be transferred through the mixed layer to the seasonal thermocline.

4. Models of the Cyclic Thermocline

Munk and Anderson (1949) present a calculation for the stationary structure of the upper layers of the ocean under heating conditions. Over the entire water column, they use the formulas

$$-\rho_0 \overline{W'V'} = \{A_0(1 + \beta_v R)^{-1/2}\} \frac{\partial V}{\partial z}, \qquad \beta_v = 10 \tag{13}$$

$$-\rho_0 \overline{W'T'} = \{A_0(1 + \beta_T R)^{-3/2}\} \frac{\partial T}{\partial z}, \qquad \beta_T = 3.33 \tag{14}$$

and solve the momentum equations for the case of constant heat flux through the system, $-C_p\rho_0 \overline{W'T'}(z) = Q_0$. The Richardson number is $R = (\alpha g \, (\partial T/\partial z)/|\partial V/\partial z|^2)$. They show that with the above conditions, the depth of the minimum stress also corresponds to the depth of the maximum temperature gradient. A computation of this depth was made for different values of A_0, for the summertime conditions of the North Pacific. Principally, it was found that in a model where the turbulent momentum heat flux depends upon the Richardson number there is an equilibrium state for the temperature field which has a nearly homogeneous layer near the surface with a sharp thermocline in the region of minimum stress. The relevance of this model and the North Pacific observations has not been tested adequately with the modern data bank of hydrographic and meteorological observations. This model has been used in a variety of calculations for oceanic and lake circulation studies (see, for example, Robinson, 1960; Bennett, 1974). Munk and Anderson note that they were not able to parameterize the case of thermal convection (surface cooling) adequately.

In studies of the formation of the daily thermocline, which is overlaid by the atmospheric boundary layer, Pandolfo (1969) and Pandolfo and Jacobs (1972) use a formalism expressed in equations 13 and 14, except the "austausch" coefficient is given by

$$A_0 = l^2 \left\{ \left| \frac{\partial \mathbf{V}}{\partial z} \right| + S_w \right\}$$

(15)

where S_w is the shear of the average orbital velocity of the gravity wave pattern; the mixing length for momentum and heat are identical, $l_{T,m}^2 = 0.02$ cm^2. During nighttime cooling, the turbulent transfer is parameterized in a manner identical to the atmospheric boundary layer (Kitaigorodskiy, 1960). The daily cycle is investigated in a series of initial value problems where the finite difference form of the conservation equations are solved for ocean station Echo (33°N, 47°W) and BOMEX 1969 data. A 4-day to 1-week simulation of the sea–surface temperature compares quite well with the oceanic bucket temperatures, as does the model heat storage resemble the observed storage.

The Pandolfo and Jacobs study is characterized by a state of strong heating and light winds. Under such conditions, a mixed layer begins to form as in steps 1–3. However, the entrainment can be very quickly arrested if the right-hand side of equation 9 vanishes for $h \leqslant h_*$. This happens when the layer depth has reached the Monin–Obukhov depth (with $l_T = l_m = 0$) $h_m = (2m_0 U_*^3 C_p \rho_0 / \alpha g Q_0)$ (see Kim, 1975, for a discussion of the equilibration time scale during heating). For the BOMEX conditions discussed by Pandolfo and Jacobs (1972) $U_* \simeq 0.8$ cm/sec, $Q_0 \simeq 1.5 \times 10^{-2}$ cal/cm^2 sec; hence $h_m < 5$ m, and the daily storage of heat, which is observed to penetrate to 30 m, can be accomplished within a diffusive daily thermocline. Under such conditions, the general model does behave as if a diffusive transfer operates over the entire water column, a stratified Ekman layer.

Kraus and Turner (1967) show how their model (with $\mathbf{V} = l_m = l_T = 0$) can parameterize a seasonal cycle. They begin an integration of the equations over a homogeneous water column, from which $t = 0$, $h(0) = \infty$. During the heating cycle, $\partial h / \partial t$ is set equal to zero in the equations, and the decreasing h_m represents the mixed layer depth. A seasonal thermocline is laid down which does not change until the winter erosion removes it. As the heating intensity decreases, the mixed layer begins to erode the seasonal thermocline owing to wind stirring, and when cooling begins, this entrainment is considerably enhanced. As the heat is removed from the water column, the mixed layer again increases without bound to the homogeneous conditions. The authors recognize that such a treatment would not be valid if such a process were to operate over a permanent thermocline. Each year, there would be a *net* amount of entrainment (increase of potential energy) because the wind does a

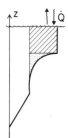

Fig. 6. Miropol'skiy (1970) model of the seasonal thermocline. The region below the mixed layer is smoothly joined to the top of the main thermocline during both heating and cooling periods.

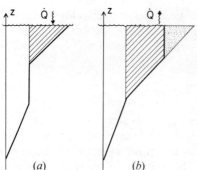

Fig. 7. Warren's (1972) model of the seasonal thermocline. During spring and summer, a linear seasonal thermocline expands to receive the heat (a). During cooling, heat is removed by "convective adjustment" (b).

net amount of work on the system. The deep winter convection would erode the main thermocline over a number of years.

Two deficiencies are apparent in the Kraus–Turner model in relation to observations. The oceanic seasonal thermocline does become warmed over the course of the four months from June to October, and there is a well-defined limit to the seasonal penetration.

Miropol'skiy (1970) presents a similar seasonal calculation to Kraus and Turner (1967). A seasonal thermocline, a cubic polynomial shaped form, is fit between the mixed layer and the top of the permanent thermocline (Figure 6). The temperature profile is continuous, and the perturbation energy equation is again used to relate the net increase of potential energy through the seasonal cycle to the kinetic energy produced by the shear of the wind-driven (or wave-driven) shear of near surface currents. Miropol'skiy presents only one annual cycle in his paper. Similarly, Thompson (1974) uses a variety of models to integrate a year's worth of oceanic response with ocean station November (March, 1967–March, 1968) meteorological conditions, but the behavior of his recipes over a long period is not clear. These equations possess the same character as the Kraus–Turner model and can lead to unstable solutions over many years of integration. Gill and Turner (1974) suggest that the Kraus–Turner (1967) model be modified by reducing the entrainment due to thermal convection, and present numerical evidence that such a model which overlays a homogeneous water column gives a stable seasonal temperature and potential energy cycle. We have found this to be true; nevertheless, the mixed layer depth still continues the erosive behavior during the wintertime cooling (see Figure 9).

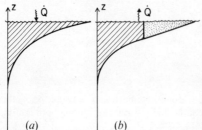

Fig. 8. Cox's (1970) model of the seasonal behavior. During heating the water is warmed through linear diffusion (a). During cooling, heat is removed by "convective adjustment" (b).

A Dissipative Stability Criterion

Consider a constant, cyclic surface heating of amplitude $+Q_0$ for time t_0 and $-Q_0$ for the remaining half year. The surface stirring is a constant, U_*^3, throughout the seasons. There is a constant "background" dissipation rate ε_0; $V = 0$. The equations that describe the warming period are

$$h_m = \frac{2m_0 U_*^3}{(\alpha g Q_0 / C_p \rho_0) + \varepsilon_0} \tag{16}$$

and

$$\frac{dT_s}{dt} = \frac{Q_0}{h_m C_p \rho_0} \tag{17}$$

At the time heating begins, $T_s = T_0$, the homogeneous temperature of the water column. After the time t_0,

$$T_s = T_0 + \frac{Q_0 t_0}{h_m C_p \rho_0} \tag{18}$$

When cooling begins,

$$\frac{d}{dt}[(T_s - T_0)h] = -\frac{Q_0}{C_p \rho_0} \tag{19}$$

and since at $t = t_0, h = h_m$, it follows from equation 18 and integration of equation 19 that

$$(T_s - T_0)h = \frac{Q_0}{C_p \rho_0}(2t_0 - t) \tag{20}$$

The mixed layer depth during cooling period is derived from equation 9,

$$h(T_s - T_0)\frac{dh}{dt} = 2m_0 U_*^3 - \left(\varepsilon_0 - \frac{\alpha g}{C_p \rho_0} Q_0\right)h \tag{21}$$

or using equations 20 and 16,

$$(2t_0 - t)\frac{dh}{dt} = (2 + \delta)h_m - \delta h \tag{22}$$

where

$$\delta = \frac{\varepsilon_0 C_p \rho_0}{\alpha g Q_0} - 1$$

Equation 22 is cast into

$$\frac{d}{dt}\{(2t_0 - t)^{-\delta}h\} = (2 + \delta)h_m(2t_0 - t)^{-\delta - 1} \tag{23}$$

and integrating equation 23, with $h = h_m$ at $t = t_0$,

$$\frac{h}{h_m} = \frac{(2 + \delta)}{\delta} - \frac{2}{\delta}\left(2 - \frac{t}{t_0}\right)^\delta \tag{24}$$

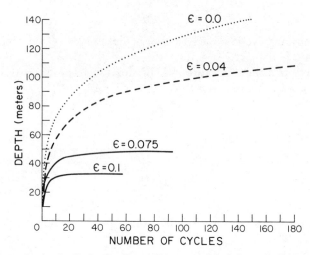

Fig. 9.   Mixed layer depth of a dissipative Kraus–Turner model of the cyclic thermocline as a function of the number of cycles of forcing. The heating cycle of half-amplitude 200 cal/day is 180° out of phase with an average seasonal wind stress of half-amplitude of 1.0 dynes/cm^2; there is no mean heating, and the mean wind stress is 0.25 dynes/cm^2. $m_0 = 1.2$; N^2 of main thermocline is 0.8×10^{-4} sec^{-2}. The critical non-dimensional dissipation rate is $\varepsilon = 0.05$, with instability for lower values. The dotted curve, $\varepsilon = 0.0$, is the most severely dissipating solution of Gill and Turner's model with $\delta = 0$.

1. If $\delta > 0$ (or $\varepsilon_0 > \alpha g Q_0/C_p \rho_0$), $h \to h_m(2 + \delta)/\delta$ as $t \to 2t_0$, and the erosion process will be halted by dissipation at a finite depth; $(T_s - T_0) \to 0$ as $t \to 2t_0$. (Note that in this model h is a discontinuous function of time at $t = t_0^+$). A critical level of dissipation is needed to ensure stability of the mixed layer depth cycle. Figures 9 and 10 display integration of the Kraus–Turner model with constant dissipation, which overlays a permanent thermocline. Both the wind stress and the heating are cyclic, and 180° out of phase. For such a system, we have found an apparent numerical stability, *provided that* $(\varepsilon_0 + \alpha g Q_0/C_p \rho_0) > 0$ *for the entire cycle.* A minimum depth of the mixed layer is typically given by h_m, and the maximum depth is dissipation limited.

2. Gill and Turner's (1974) model in this simple cyclic model corresponds to the case $-1 < \delta \leqslant 0$. In that case, we see from equation 24 that h becomes unbounded as $t \to 2t_0$. However, the surface temperature, T_s, heat storage, $(T_s - T_0)h$, and potential energy, $(T_s - T_0)(h^2/2)$, all have a bounded cycle. As presented by Gill and Turner (1974), a plot of heat storage and potential energy as a function of the surface temperature would present a closed hysteresis in a solution over a homogeneous water column (in the Kraus–Turner model, $\delta = -1$, and the potential energy does *not* form a closed cycle). Figure 9 also shows an integration of the Gill and Turner model over a main thermocline with cyclic heating and wind stress, and the erosion of the mixed layer into the main thermocline is apparent.

Finally, Warren (1972) presents a calculation which appears to have stable characteristics over a long integration period. Figure 7 shows the parameterization in Warren's model; the shape of the seasonal thermocline is specified during heating, and convective adjustment is used only during cooling (no entrainment due to wind stirring or cooling is allowed). The heat exchange with the atmosphere is a function of the surface temperature. Warren reports good numerical stability and shows a remarkable skill in predicting the seasonal temperature cycle and depth of maximum

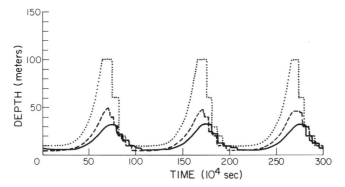

Fig. 10.   Mixed layer depth of a dissipative Kraus–Turner model of the cyclic thermocline as a function of time. The parameters are the same as in Figure 9. The dissipation $\varepsilon_0 = 1.6 \times 10^{-3}$ cm^2/sec^3 ($\cdots\cdots$) (unstable); 3.0×10^{-3} cm^2/sec^3 ($- - - - -$) (stable), 4.0×10^{-3} cm^2/sec^3 (——) (stable).

seasonal penetration in the western portion of the North Atlantic and North Pacific. Cox (1970) computes the large-scale seasonal circulation in the Indian Ocean with a diffusive, convectively adjusted model (Figure 8); however, his vertical resolution of the temperature does not permit an adequate description of the surface layer structures that we have attempted to provide here.

5. Conclusions

We have carried out integrations of the stable models through many seasonal cycles with "climatological" ocean conditions in the North Pacific averaged over a 5° latitude, 15° longitude region centered at 27.5°N, 152.5°W. Seckel (1970) uses bulk formulas to compute the monthly mean heat exchange and wind stress over a 2-year period, 1963–1965, and we have fit an annual and biannual harmonic to this data. The monthly mean stirring function, U_*^3 was computed from the formula $\rho_0 U_*^3 = (\bar{\tau}_x^2 + \bar{\tau}_x'^2 + \bar{\tau}_y^2 + \bar{\tau}_y'^2)^{3/2}$, where $\bar{\tau}_x$, $\bar{\tau}_y$ are the monthly mean wind-stress components, and $\bar{\tau}_x'$, $\bar{\tau}_y'$ are the standard deviations of these components. The integration is carried out by initially prescribing that the observed main thermocline reaches the surface, and the analytic, climatological functions are used to energize the model; only thermal effects were considered. Figure 11 displays these data, together with our cyclic fit.

Figure 12 displays results of an integration with a dissipative, stable Kraus–Turner model ($l_T = l_m = 0$). The model temperature profile in late August is shown, together with the observations in June and October. For any given heating and wind cycle, the minimum depth of the mixed layer and the maximum summer temperature can be fixed by selecting m_0 ($m_0 = 1.0$, which is the value used by Denman and Miyake, 1973). The maximum stable depth of seasonal penetration and the minimum winter temperature can be fixed by selecting ε_0 ($\varepsilon_0 = 3.4 \times 10^{-3}$ cm^2/sec^3). Hence excellent skill can be demonstrated in reproducing the seasonal range of observed sea-surface temperatures. However, a conclusion that such a model is now relevant to describing the seasonal oceanic conditions is both misleading and erroneous. The *shape* of the observed seasonal thermocline is not reproduced. Furthermore, commonly published values of Q_0 (Wyrtki, 1966) computed from climatological data are low by a factor of 1.5–2 when compared with the observed seasonal storage rate over large areas of the North Pacific (Bathen, 1970). As shown by Gill and Niiler (1973), horizontal

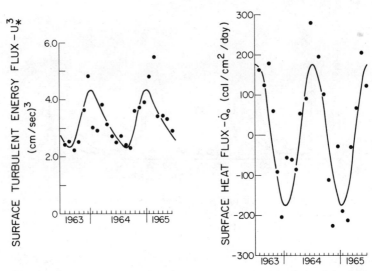

Fig. 11. Climatological forcing data for the perturbation kinetic energy equation at 27.5°N, 152.5°W. The circles are estimates by Seckel (1970), and the solid line is the harmonic fit that was used in the model.

advection cannot correct this imbalance over a seasonal time scale. Two unrealistic factors combine in a way by which a third realistic result is found. Note also from Fig. 12 that the observations indicate a warming of the seasonal thermocline (below the mixed layer) between July and October. This phenomenon cannot be produced by a Kraus–Turner model.

Figure 13 displays the speed and temperature profiles of the combined model outlined in Section 2. The mixing length for momentum is proportional to the depth

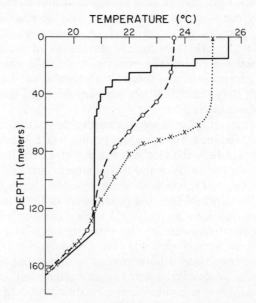

Fig. 12. The September temperature profile of a stable, dissipative, Kraus–Turner model, with forcing of Figure 11 $m_0 = 1.2$, $\varepsilon_0 = 3.4 \times 10^{-3}$ cm^2/sec^3. Open circles are from hydrographic data from July–August. The crosses are from hydrographic data from October–November period.

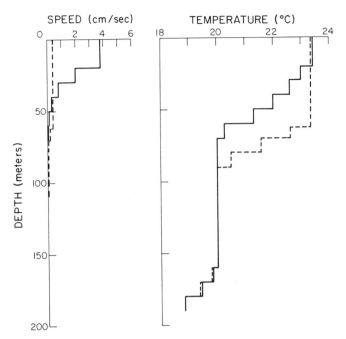

Fig. 13.   Speed and temperature profiles of the combined dissipative mixed layer and diffusive thermocline. All parameters are the same as in Figure 12 except the m_0, ε_0 and $Q_{0\,max}$ have been increased by a
factor of two. The profiles are drawn after three years of integration with cyclic forcing. Solid line, $l_T =$
0.1 m, $l_m = 0.03h$. Dashed line, $l_T = 0.5$ m, $l_m = h$.

of the mixed layer, and the mixing length for heat is a constant. The surface heating
rate, the dissipation, and the wind stirring have been increased by a factor of two over
the conditions on Fig. 12. Although a direct comparison with data is premature
(and not within the scope of this paper), it is apparent that diffusion below the mixed
layer does act to warm the seasonal thermocline and strongly influences the vertical
distribution of the heat storage toward the observed behavior on the seasonal time
scale.

It does appear that a weakly dissipative mixed layer overlying a diffusive seasonal
thermocline is needed to model the principal observed scales within the seasonal
thermocline. At present, we lack observational data on dissipation rates within the
mixed layer, and crucial tests against many years of oceanic observations within the
seasonal pycnocline have yet to be carried out. Modeling the seasonal thermocline
in the oceans is at an infant stage compared to modeling the planetary boundary
layer under the atmospheric circulation.

Acknowledgment

I wish to express my gratitude to Ms. P. Stabeno for carrying out the long series of
computations; Ms. N. Walker has been the stoic critic of this manuscript. This
research is sponsored by N00014-75-C-0299, NR 083-315.

References

Bathen, K. H., 1970. Heat storage and advection in the North Pacific Ocean. *Hawaii Inst. Geophys. Rept.*
        HIG-70-6, 122 pp., 55 figs.
Bennett, J., 1974. Personal communication.

Bernstein, R., 1974. Baroclinic mesoscale eddies: interaction with sea surface temperature. *Norpax Highlights*, **2** (2), 1–3.

Budyko, M. I., 1956. The heat balance of the earth's surface. In *Gidrometeorol* (trans. from Russian by N. A. Stepanova, U.S. Weather Bureau, Washington, D.C., 1958).

Cox, M., 1970. A mathematical model of the Indian Ocean. *Deep-Sea Res.*, **17**, 47–75.

Denman, K. L. and M. Miyake, 1973. Upper layer modification at Ocean Station Papa: observations and simulations. *J. Phys. Oceanogr.*, **3**, 185–196.

deSzoeke, R. and P. B. Rhines, 1975. Asymptotic regimes in mixed-layer deepening. *J. Mar. Res.*, in press.

Gill, A. E., 1974. The relationship between heat content of the upper ocean and the sea surface temperature. *Norpax Highlights*, **2** (3), 1–4.

Gill, A. E. and P. P. Niiler, 1973. The theory of the seasonal variability in the ocean. *Deep-Sea Res.*, **20**, 14–177.

Gill, A. E. and J. S. Turner, 1974. Mixing models for the seasonal thermocline. *Norpax Highlights*, **2** (5), 9–12.

Halpern, D., 1974. Observations of deepening of the wind-mixed layer in the northeast Pacific Ocean. *J. Phys. Oceanogr.*, **4** (3), 454–466.

Kato, H., and O. M. Phillips, 1969. On the penetration of a turbulent layer into stratified fluid. *J. Fluid Mech.*, **37**, 643–655.

Kim, Jeong-Woo, 1975. Oceanic mixed layer parameterization: II. Theory. *Rand Corp. Rept. WN-8901-ARPA*, Santa Monica, Calif.

Kitaigorodskiy, S. A., 1960. On the computation of the thickness of the wind-mixing layer in the ocean. *Bull. Acad. Sci. USSR Geophys. Ser.*, **3**, 284–287.

Kraus, E. B. and J. S. Turner, 1967. A one-dimensional model of the seasonal thermocline, II. *Tellus*, **19**, 98–105.

Linden, P. F. and J. S. Turner, 1975. Small-scale mixing in stably stratified fluids: a report on Euromech 51. *J. Mar. Res.*, **67**, part 1, 1–16.

Lumley, J. L. and Bejan Khajeh-Nouri, 1974. Computational modeling of turbulent transport. In *Proc. Turbulent Diffusion in Environmental Pollution, International Symposium, 1973*. Academic Press, New York, pp. 169–192.

Lundgren, T. S. and F. C. Wang, 1973. Eddy viscosity models for free turbulent flows. *Phys. Fluids*, **16** (2), 174–178.

Miropol'skiy, Yu. Z., 1970. Nonstationary model of the wind-convection mixing layer in the ocean. *Izv. Atmos. Oceanic Phys.*, **6** (12), 1284–1294.

Monin, A. S. and A. M. Yaglom, 1971. *Statistical Fluid Mechanics: Mechanics of Turbulence*. MIT Press, Cambridge, Mass.

Montgomery, R. B., 1974. Comments on "Seasonal variability of the Florida current," by Niiler and Richardson. *J. Mar. Res.*, **32** (3), 533–535.

Munk, W. H. and E. A. Anderson, 1949. Notes on a theory of the thermocline. *J. Mar. Res.*, **7**, 276–295.

Munk, W. H. and M. Wimbush, 1970. The benthic boundary layer. In *The Sea*. Vol. 4. M. N. Hill, ed. pp. 731–758.

Niiler, P. P., 1975. Deepening of the wind-mixed layer. *J. Mar. Res.*, **33**, pp. 405–422.

Niiler, P. P. and W. S. Richardson, 1973. Seasonal variability of the Florida Current. *J. Mar. Res.*, **31** (3), 144–167.

Pandolfo, J. B., 1969. Motions with inertial and diurnal period in a numerical model of the navifacial boundary layer. *J. Mar. Res.*, **26**, 301–317.

Pandolfo, J. P. and C. A. Jacobs, 1972. Numerical simulations of the tropical air-sea planetary boundary layer. *Boundary-Layer Meteorol.*, **3**, 15–46.

Phillips, O. M., 1966. *The Dynamics of the Upper Ocean*. Cambridge University Press, 261 pp.

Pollard, R. T. and R. C. Millard, Jr., 1970. Comparison between observed and simulated wind-generated inertial oscillations. *Deep-Sea Res.*, **17**, 813–821.

Robinson, A. R., 1960. The general thermal circulation in equatorial regions. *Deep-Sea Res.*, **6** (4), 311–317.

Robinson, M., 1973. Mean sea surface and sub-surface temperature and depth of the top of the thermocline, North Pacific Ocean, May 1971. *Atlas Monthly*. Published by Fleet Numerical Weather Control, Monterey, Calif.

Schroeder, E. H., 1963. North Atlantic Temperatures at a depth of 200 meters. *Serial Atlas of the Marine Environment*, Folio 2.

Seckel, G. R., 1970. The trade wind zone oceanography pilot study, part IX: wind stress values July 1963 to June 1965. *U.S. Fish Wildlife Serv. Spec. Sci. Rept. Fisheries*, **620**.

Smith, J. D. and M. G. McPhee, 1972. An experimental investigation of the boundary layer under pack ice. *Aidjex Bull.* (University of Washington), **14**, 40–43.

Tabata, Susumu, 1965. Variability of oceanographic conditions at Ocean Station "P" in the northeast Pacific Ocean. *Trans. Roy. Soc. Can. Sect. III*, Ser. 4, **3**, 367–418.

Thompson, R. O. R. Y., 1974. Predicting the character of the well-mixed layer. *Woods Hole Oceanogr. Inst. Tech. Rept. WHOI-74-82*, Woods Hole, Mass.

Tully, J. P. and L. F. Giovando, 1963. Seasonal temperature structure in the eastern subarctic Pacific Ocean. In *Roy. Soc. Can. Spec. Publ.* **5**, 10–36 (M. J. Dunbar, ed.).

Van Leer, J. C., W. R. Johnson, and E. Mehr, 1974. Cyclesonde data report from the winter, 1973. NSF Continental Shelf Dynamics Program. *UN-RSMAS 74033*.

Voorhis, A. D., 1969. The horizontal extent and persistence of thermal fronts in the Sargasso Sea. *Deep-Sea Res.*, **16** (supplement), 331–338.

Warren, B. A., 1972. Insensitivity of subtropical mode water characteristics to meteorological fluctuations. *Deep-Sea Res.*, **19**, 1–20.

Woods, J. D. and R. L. Wiley, 1972. Billow turbulence and ocean micro-structure. *Deep-Sea Res.*, **19**, 87–121.

Wunsch, C., 1972. Bermuda sea level in relation to tides, weather and baroclinic fluctuations. *Rev. Geophys. Space Phys.*, **10**, 1–50.

Wyrtki, K., 1966. Seasonal variation of heat exchange and surface temperature in the North Pacific. *Hawaii Inst. Geophys. Rept. HIG-66-3*, 8 pp., 72 figs.

Yoshida, K., 1967. Circulation in the eastern tropical oceans with special reference to undercurrents and upwelling. *Jap. J. Geophys.*, **4**, 1–75.

Zilitinkevich, S. S., D. L. Laikhtman, and A. S. Monin, 1967. Dynamics of the atmospheric boundary layer—a review. *Izv. Atmos. Oceanic Phys.*, **3** (3), 297–333.

4. THE COASTAL JET CONCEPTUAL MODEL IN THE DYNAMICS OF SHALLOW SEAS

G. T. CSANADY

1. Introduction

In the atmospheric sciences we deal with fluid motions of great complexity depending on a very large number of external and internal parameters. Moreover, in most problems we are interested in enough of the details of those motions so that their statistics alone do not satisfy us. Therefore the two usual approaches of physical theory become inappropriate: we cannot isolate a limited number of independent parameters which fully determine a phenomenon of interest (in the manner of a controlled laboratory experiment), nor is it fruitful to consider many realizations of the phenomenon concerned and confine attention to its average properties alone (in the manner of the kinetic theory of gases). One response to this challenge that has emerged in the atmospheric sciences is a theoretical tool that may be described as the "conceptual model." As we shall understand it, such a model is deliberately incomplete and aims only at reproducing predominant characteristics of some distinctly identifiable, observable phenomenon. One good example of a "distinct" phenomenon in oceanography is the Gulf Stream, noting at once that its distinctness is clear enough in a gross way, but a precise definition of its boundaries would be problematic. It is also clear that the Stream interacts in a complex way with its environment, and yet there is no question of the usefulness of regarding it as a separate entity. An example of the conceptual model is Stommel's (1948) theory of westward intensification, which has allowed us a great deal of insight into the dynamics of the Stream. The key steps in gaining this insight have been: (a) the distillation from complex experience of a clearly identifiable, distinct flow structure called the Gulf Stream; and (b) the construction of a model ocean of very simple properties, but containing just enough complexity to produce a western boundary current of gross characteristics quantitatively very similar to the Gulf Stream. It is likely that in the atmospheric sciences dynamical understanding of complex phenomena will forever be based on such incomplete conceptual models, even if our models become more and more sophisticated as a given branch of the science advances. This chapter is devoted to the discussion of one such conceptual model which has provided insight into the dynamics of nearshore water movements in shallow seas.

To be more specific, the tidal equations have some aperiodic solutions that provide simple but realistic models of certain wind-driven flow problems. The first such model was proposed by Charney (1955) in a discussion of geostrophic adjustment near a coastline, and was named by him the "coastal jet." Although intended originally to elucidate Gulf Stream dynamics, the model has proved useful in connection with directly (locally) wind-driven coastal currents (which the Gulf Stream is not) in the Great Lakes and over continental shelves. A grossly simplified summary of the dynamical principles involved is as follows. We understand wind-driven flow in the deep ocean in terms of the conceptual model known as "Ekman drift," in which the Coriolis force associated with crosswind water transport balances the stress exerted by the wind over the water surface. A coastline parallel to the wind, however, prevents Ekman drift. In a conceptual model without bottom or side friction, the wind stress then accelerates the water within some nearshore region and produces a "coastal

117

jet." In a basin of constant depth the width of the jet turns out to be proportional to the Rossby radius of deformation. Especially relevant to observation are internal mode (baroclinic) coastal jets, which simulate clearly identifiable wind-driven coastal currents.

As may be expected, various complications arise due to the finite length and width of basins (lakes or continental shelves) and to their variable depths. In the following sections we first give an account of the coastal jet conceptual model, beginning with the crudest approach and making it gradually more realistic, and then illustrate some observed coastal currents which the coastal jet model is intended to simulate.

2. The Simplest Coastal Jet Model: Semi-infinite Ocean

Consider a semi-infinite ocean bounded by a straight coastline coincident with the y axis, with the x axis pointing out to sea. Assume that the water depth is a constant $H = h + h'$, a slightly lighter top layer of equilibrium depth h lying over dense water of depth h'. The fractional density defect $\varepsilon = (\rho' - \rho)/\rho$ is small, of order 10^{-3}. Let a wind begin to blow at time $t = 0$, exerting a stress $\tau_0 = \rho F$ at the water surface, directed parallel to the coast, constant in space and in time for $t > 0$.

For a sufficiently short period the motions generated in the water will be slow enough for nonlinear accelerations to be neglected. Also, friction at the stable interface and at the bottom will for a time be unimportant, as will frictional stresses in vertical planes. Therefore we can elucidate some important properties of the initial motion with the aid of the linearized equations of momentum and mass balance integrated over depth separately for top and bottom layers. The pressure is assumed to be hydrostatic, and the pressure gradients expressed in terms of surface and thermocline elevations above equilibrium, ζ and ζ', both supposed small compared to equilibrium depths. The resulting well-known transport equations are as follows:

Top layer:

$$\frac{\partial U}{\partial t} - fV = -gh\frac{\partial \zeta}{\partial x}$$

$$\frac{\partial V}{\partial t} + fU = -gh\frac{\partial \zeta}{\partial y} + F$$

$$\frac{\partial U}{\partial x} + \frac{\partial V}{\partial y} = -\frac{\partial}{\partial t}(\zeta - \zeta')$$

Bottom layer: (1)

$$\frac{\partial U'}{\partial t} - fV' = -gh'\frac{\partial \zeta}{\partial x} + gh'\varepsilon\frac{\partial}{\partial x}(\zeta - \zeta')$$

$$\frac{\partial V'}{\partial t} + fU' = -gh'\frac{\partial \zeta}{\partial y} + gh'\varepsilon\frac{\partial}{\partial y}(\zeta - \zeta')$$

$$\frac{\partial U'}{\partial x} + \frac{\partial V'}{\partial y} = -\frac{\partial \zeta'}{\partial t}$$

where U, V, U', and V' are depth-integrated velocities or volume transports (separately for top and bottom layers) along x and y.

Boundary conditions at the coast $x = 0$ are such that the normal transports U and U' vanish. At infinity, $x \to \infty$, we postulate vanishing surface and interface elevations.

The above sets of equations for top and bottom layers are clearly coupled through the presence of both ζ and ζ' in either set. It is possible, however, to produce two linear combinations of the two sets in such a way that the resulting sets of equations are uncoupled *normal mode* equations of the form:

$$\frac{\partial U_n}{\partial t} - f V_n = -g h_n \frac{\partial \zeta_n}{\partial x}$$

$$\frac{\partial V_n}{\partial t} + f U_n = -g h_n \frac{\partial \zeta_n}{\partial y} + F_n \tag{2}$$

$$\frac{\partial U_n}{\partial x} + \frac{\partial V_n}{\partial y} = -\frac{\partial \zeta_n}{\partial t}$$

where h_n is an appropriate "equivalent depth," although it may also be regarded a factor modifying gravity. These equations describe motions of a *homogeneous* fluid of depth h_n. Equations 2 arise from adding a constant α_n times the first set of three equations of equation 1 to the second set of three, so that the combined variables are

$$U_n = \alpha_n U + U'$$
$$V_n = \alpha_n V + V'$$
$$\zeta_n = \alpha_n(\zeta - \zeta') + \zeta' \tag{3}$$
$$F_n = \alpha_n F$$

These linear transformations, when introduced into equations 1 result in a set of equations of the form of equation 2, provided that α_n ($n = 1, 2$) is a root of Stokes's equation:

$$\alpha^2 + \left(\frac{h'}{h} - 1\right)\alpha - \frac{h'}{h}(1 - \varepsilon) = 0 \tag{4}$$

The corresponding equivalent depths are then

$$h_n = h'\varepsilon(1 - \alpha_n)^{-1} \tag{5}$$

The above separation of the problem into normal modes is exact under the postulated conditions. The physical properties of the normal mode equations are best exhibited, however, using approximate values of the roots $\alpha_{1,2}$, valid for small values of the density defect ε. The two roots are, expanded in powers of ε, as follows:

$$\alpha_1 = 1 - \frac{\varepsilon h'}{h + h'} + O(\varepsilon^2)$$

$$\tag{6}$$

$$\alpha_2 = -\frac{h'}{h} + \frac{\varepsilon h'}{h + h'} + O(\varepsilon^2)$$

with corresponding equivalent depths, to the lowest order only:

$$h_1 = h' + h + 0(\varepsilon)$$

$$h_2 = \frac{\varepsilon h h'}{h + h'} + 0(\varepsilon^2)$$

(7)

To lowest order in ε, the normal mode equations (to be called "surface" or "barotropic" and "internal" or "baroclinic" modes) are now written in terms of the original variables, according to equation 3, but multiplied by a constant factor in the case of the internal mode:

Surface mode:

$$\frac{\partial(U + U')}{\partial t} - f(V + V') = -g(h + h')\frac{\partial \zeta}{\partial x}$$

$$\frac{\partial(V + V')}{\partial t} + f(U + U') = -g(h + h')\frac{\partial \zeta}{\partial y} + F$$

$$\frac{\partial(U + U')}{\partial x} + \frac{\partial(V + V')}{\partial y} = -\frac{\partial \zeta}{\partial t}$$

Internal mode:

(8)

$$\frac{\partial U'}{\partial t} - fV' = -g\varepsilon \frac{hh'}{h + h'}\frac{\partial \zeta'}{\partial x}$$

$$\frac{\partial V'}{\partial t} + fU' = -g\varepsilon \frac{hh'}{h + h'}\frac{\partial \zeta'}{\partial y} - \frac{h'}{h + h'}F$$

$$\frac{\partial U'}{\partial x} + \frac{\partial V'}{\partial y} = -\frac{\partial \zeta'}{\partial t}$$

The approximate surface mode equations are exactly those that apply to a homogeneous fluid of depth $h + h'$, acted upon by a wind stress ρF, along the y axis. In this mode (and to zeroth order) the fluid does not "feel" the small density defect of the top layer. The internal mode equations are identical in form to the surface mode ones, when applied to motions of the *bottom* layer of fluid alone, that is, as if the thermocline were a free surface, but with gravity reduced by a factor of $\varepsilon h/(h + h')$. Also the effective force acting on the bottom layer is *opposite* in direction to the applied wind stress, and its magnitude is reduced by the factor $h'/(h + h')$. The reduction in gravity is by a large factor, whereas the effective stress is not much different from that acting at the surface. One may therefore at once suspect that very large thermocline displacements may be produced. The opposite direction of the effective force in the internal mode has the consequence of producing opposite surface and thermocline displacements.

The full solution of the problem we posed (suddenly imposed wind) contains some inertial oscillations and an aperiodic part describing a "coastal jet" structure near shore and Ekman drift far offshore. Full details are given by Crépon (1967); here we concern ourselves only with the aperiodic part of the response. Solving the approximate normal mode equations 8 separately, and adding the results, we arrive at the particular solution (correct to order $\varepsilon^{1/2}$):

$$u = \frac{U}{h} = \frac{F}{f(h + h')} \left\{ 1 - \exp\left(\frac{-x}{R_1}\right) + \frac{h'}{h}\left[1 - \exp\left(\frac{-x}{R_2}\right)\right]\right\}$$

$$v = \frac{V}{h} = \frac{Ft}{h + h'} \left[\exp\left(\frac{-x}{R_1}\right) + \frac{h'}{h} \exp\left(\frac{-x}{R_2}\right)\right]$$

$$\zeta = -\frac{Ft}{fR_1} \left[\exp\left(\frac{-x}{R_1}\right) + \frac{h'}{h}\frac{R_2}{R_1} \exp\left(\frac{-x}{R_2}\right)\right]$$

$$u' = \frac{U'}{h'} = \frac{F}{f(h + h')} \left\{ 1 - \exp\left(\frac{-x}{R_1}\right) - \left[1 - \exp\left(\frac{-x}{R_2}\right)\right]\right\}$$ (9)

$$v' = \frac{V'}{h'} = \frac{Ft}{h + h'} \left[\exp\left(\frac{-x}{R_1}\right) - \exp\left(\frac{-x}{R_2}\right)\right]$$

$$\zeta' = \frac{Ft}{fR_2}\frac{h'}{h + h'} \left[\exp\left(\frac{-x}{R_2}\right) - \frac{R_2}{R_1} \exp\left(\frac{-x}{R_1}\right)\right]$$

Here u, v, etc. are velocities averaged over the two layers separately and the distances $R_{1,2}$ are surface and internal radii of deformation:

$$R_1 = f^{-1}\sqrt{g(h + h')}$$
$$R_2 = f^{-1}\sqrt{g\varepsilon hh'/(h + h')}$$ (10)

These distances give the widths of coastal jets: as equations 9 show, within bands of scale width R_1 and R_2 longshore velocities, surface, and thermocline elevations increase in direct proportion to time or, more descriptively, to the impulse Ft of the wind stress. The onshore–offshore velocity reduces to zero on approaching the coast within a band of scale width R_1 and R_2, referring to the contribution of the surface and internal mode, respectively.

Very close to shore, $x \ll R_2$ (note that $R_2 \ll R_1$, by equation 10) the onshore velocities u and u' in either layer are negligible. The longshore velocities are, on the other hand, to a high degree of approximation:

$$v = \frac{Ft}{h} \quad (x \ll R_2)$$ (11)

$$v' = 0$$

In other words, the impulse of the wind stress is distributed over the top layer only. The physical reason is contained in the fifth of equations 1: in the absence of pressure gradients along y (and of friction at the interface), bottom layer longshore velocity can be generated only by Coriolis force due to onshore–offshore flow. Because of the shore constraint this is zero; hence the water of the bottom layer must remain stagnant.

At intermediate values of x, $R_2 \ll x \ll R_1$, onshore–offshore velocities are approximately

$$u = \frac{F}{f(h + h')}\frac{h'}{h}$$

$$\qquad\qquad (R_2 \ll x \ll R_1) \qquad\qquad (12)$$

$$u' = -\frac{F}{f(h + h')}$$

The total offshore transport is thus zero, $uh + u'h' = 0$, outflow in the top layer (if F is positive) being balanced by inflow in the bottom layer. Longshore velocities are in this same intermediate region:

$$v = \frac{Ft}{h + h} = v' \qquad (R_2 \ll x \ll R_1) \qquad\qquad (13)$$

The impulse of the wind stress has been distributed evenly here over the entire water column. Within the bottom layer this has happened through the medium of the Coriolis force, because by the fifth of equations 1,

$$V' = -\int_0^t f U' \, dt = \frac{Ft}{h + h'} h' \qquad\qquad (14)$$

bottom layer onshore transport having been substituted from equation 12. The total onshore displacement in the bottom layer is thus just sufficient to produce the required longshore transport through the action of the Coriolis force.

Far from shore, $x \ll R_1$, longshore velocities are zero, the bottom layer is completely at rest, and Ekman drift in the top layer balances wind stress.

Turning now to surface and thermocline elevations, we see that these increase in proportion to time, or rather to wind-stress impulse, and are present at any distance not large compared to R_1. Because $R_2 \ll R_1$, thermocline elevations are large, but only close to shore, $x \ll R_2$. It is easily shown that the total accumulation of bottom layer water near shore, beneath the rapidly rising part of the thermocline, is exactly what is flowing inshore in the "intermediate" region (equation 12). It is also easily checked that this water is transferred from a region of scale width R_1 (where the thermocline is being slowly depressed) to an interior region of scale width R_2 (where it is rapidly rising). Corresponding to the much larger extension of the R_1 region, the depression of the thermocline at any instant is small compared to its elevation over the R_2 region.

The surface elevation is depressed (by a positive F) over both R_1 and R_2 regions, to the total extent required by the Ekman drift far offshore. At the shore, however, the depression is greater than if the fluid were homogeneous. Even if R_2/R_1 is small, this extra elevation may be detectable, because the combined factor $h'R_2/hR_1$ may not be much smaller than unity.

We also note that, given time-independent onshore–offshore flow, the Coriolis force of longshore flow at any instant is balanced by onshore–offshore pressure gradients. The momentum balance along x is thus geostrophic, with both longshore flow and offshore pressure gradient increasing linearly in time.

The above results are best appreciated if one considers "typical" magnitudes of the quantities involved. Table I lists plausibly assumed values of the independent parameters characterizing shallow seas. The ratio of the radii of deformation is seen to be typically about 50, and the magnitude of R_1 is so great that $x \gg R_1$ is not a

TABLE I

"Typical" Data of Shallow Seas

Total depth $h + h'$	10^4 cm	Wind force $\tau_0/\rho = F$	1 cm^2/sec^2
Top layer h	2.10^3 cm	Surface mode radius of deformation R_1	316 km
Bottom layer h'	8.10^3 cm		
Effective gravity εg	2 cm/sec^2	Internal mode radius of deformation R_2	5.66 km
Coriolis parameter f	10^{-4}/sec		

realistic part of the model: no shallow sea is more than 1000 km wide. However, the nearshore and intermediate regions may be realistically described by this model. The validity of our results must be restricted to an initial period. We make calculations for $t = 2.10^4$ sec, or just under 6 hr.

Equations 9 show that the velocities are scaled by $F/f(h + h')$ and $Ft/(h + h')$: these become 1 and 2 cm/sec, respectively, with the above assumptions. Surface elevations are proportional to $Ft/fR_1 = 6.32$ cm, thermocline elevations to $[Ft/fR_2][h'/(h + h')] = 283$ cm. The most conspicuous effect of the wind is clearly the movement of the thermocline, which rapidly outgrows the linearizing assumption of $|\zeta'| \ll h$. Calculated characteristics are listed in Table II. The coastal jet is "baroclinic" near shore (confined to the top layer, and associated with an inclined thermocline), and "barotropic" in the intermediate region (evenly distributed over the depth, and associated with a surface slope). Figure 1 is an attempt to illustrate the properties of a two-layer coastal jet structure with a wind tending to produce upwelling. The opposite wind produces an opposite jet and a downwelled thermocline, of otherwise the same characteristics.

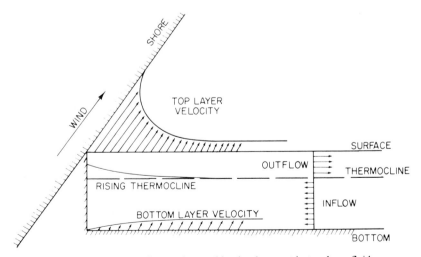

Fig. 1. Schematic picture of coastal jet development in two-layer fluid.

TABLE II

Some Characteristics of Two-Layer Coastal Jet

Variable	Close to Shore $x \ll R_2$	Intermediate $R_2 \ll x \ll R_1$	Far $x \gg R_1$
u (cm/sec)	0	4	5
v (cm/sec)	10	2	0
ζ (cm)	-6.46	-6.32	0
u' (cm/sec)	0	-1	0
v' (cm/sec)	0	2	0
ζ' (cm)	275	-5.06	0

3. Opposing Shores

The most important inadequacy of the preceding section's simple model is that a region "far" from shore does not exist in reality, yet the boundary conditions imposed there play a certain role in determining the structure of the solution. The simplest "fix" for this problem is to assume a second parallel shore at $x = a$, that is, to use a canal model in place of a semi-infinite ocean.

The full solution of the initial value problem for the canal model is also available (Crépon, 1969) and is not very different from the previous case. If one takes the y axis to be at the center of the canal, the aperiodic part of the solution becomes symmetrical or antisymmetrical, containing hyperbolic functions in place of the exponentials in equation 9. To order $\varepsilon^{1/2}$ the solution is

$$u = \frac{F}{f(h + h')}\left[1 - \frac{\cosh{(x/R_1)}}{\cosh{(a/2R_1)}} + \frac{h'}{h}\left(1 - \frac{\cosh{(x/R_2)}}{\cosh{(a/2R_2)}}\right)\right]$$

$$v = \frac{Ft}{h + h'}\left(\frac{\cosh{(x/R_1)}}{\cosh{(a/2R_1)}} + \frac{h'}{h}\frac{\cosh{(x/R_2)}}{\cosh{(a/2R_2)}}\right)$$

$$\zeta = -\frac{Ft}{fR_1}\left(\frac{\sinh{(x/R_1)}}{\sinh{(a/2R_1)}} + \frac{h'}{h}\frac{R_2}{R_1}\frac{\sinh{(x/R_2)}}{\sinh{(a/2R_2)}}\right)$$

$$u' = \frac{F}{f(h + h')}\left[1 - \frac{\cosh{(x/R_1)}}{\cosh{(a/2R_1)}} - \left(1 - \frac{\cosh{(x/R_2)}}{\cosh{(a/2R_2)}}\right)\right] \tag{15}$$

$$v' = \frac{Ft}{h + h'}\left(\frac{\cosh{(x/R_1)}}{\cosh{(a/2R_1)}} - \frac{\cosh{(x/R_2)}}{\cosh{(a/2R_2)}}\right)$$

$$\zeta' = \frac{Ft}{fR_2}\frac{h'}{(h + h')}\left(\frac{\sinh{(x/R_1)}}{\sinh{(a/2R_1)}} - \frac{R_2 \sinh{(x/R_2)}}{R_1 \sinh{(a/2R_2)}}\right)$$

The case of interest is when $a \ll R_1$; in this case the "intermediate" region $R_2 < x < R_1$ occupies most of the canal, outside boundary layers of scale width R_2. At one shore the thermocline is rising; at the other it is sinking. Fluid in the top layer is transferred effectively from one shore to the other, to the right of the wind, while fluid in the bottom layer compensates. Over most of the canal the wind-imparted momentum is distributed evenly over the water column. In the bottom layer this is

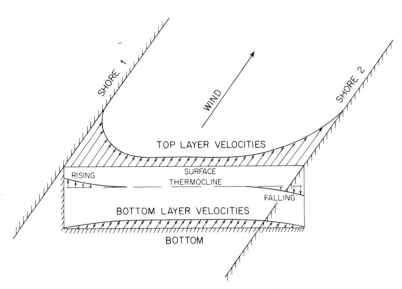

Fig. 2.   Coastal jets in canal model.

brought about by the Coriolis force due to cross flow. The same effect in the top layer *subtracts* the appropriate amount of momentum from the wind input. The surface elevation distribution is nearly

$$\zeta = -\frac{Ft}{fR_1} \cdot \frac{x}{2a} \tag{16}$$

the slight constant cross-canal slope balancing the Coriolis force of longshore flow.

Within the coastal boundary layers longshore flow is confined to the top layer and conditions are in every way exactly as in the case of the semi-infinite ocean. Figure 2 illustrates the canal model of coastal jets.

4. End Walls

Another unrealistic feature of the simple coastal jet model is the infinite extension of the coast along the y axis. Closing the ends at $y = \pm b/2$ produces a more realistic model of a closed basin at the expense of some rather drastic modifications of the dynamical response. Although no part of the response of a closed basin is strictly linear in time, the coastal jet type of response emerges again as the initial behavior of low-frequency long waves, with some important additional features and qualifications.

The response of a rectangular, two-layer basin of constant depth to a suddenly imposed constant wind stress consists of a time-independent (static) particular solution and a series of free oscillations. Most of the free oscillations have the character of the well-known seiches (Wilson, 1972) or Poincaré waves (Mortimer, 1963), the frequencies of which are higher than inertial. From our point of view, these oscillatory solutions applying to a simplified basin constitute conceptual models fundamentally different from coastal jets; we are not concerned with them here. However, in sufficiently large basins some wavelike modes have a frequency much less than f and may be combined with the static solution to provide a *quasi-static* response pattern of the kind we are discussing here.

The solution is most conveniently discussed in connection with equations 2, that is, separately for the surface and internal modes. In either mode, $n = 1$ or 2, the *static* solution is

$$\zeta_n = \frac{F_n}{f^2 R_n^2} \left\{ y \frac{\cosh (x/R_n)}{\cosh (a/2R_n)} - b \sum_{j=1}^{\infty} (-1)^j \frac{4R_j^2/R_n^2 \sinh (y/R_j)}{(2j-1) \sinh (b/2R_j)} \cos \left[(2j-1) \frac{\pi x}{a} \right] \right\}$$

(17)

where

$$\frac{1}{R_j^2} = \frac{1}{R_n^2} + \frac{(2j-1)^2 \pi^2}{a^2} \qquad (j = 1, 2, 3, \ldots)$$

(17a)

and R_n have the same definitions as in equation 10. The corresponding equivalent transport components are

$$U_n = \frac{\partial \psi_n}{\partial y} \qquad V_n = -\frac{\partial \psi_n}{\partial x}$$

(18)

where

$$\psi_n = f R_n^2 \zeta_n - \frac{F_n}{f} y$$

(18a)

so that ψ_n is volume-transport stream function in the "equivalent" basin.

The properties of this solution depend on the size of the equivalent basin in comparison with the radius of deformation. For a small basin, $a/R_n, b/R_n \to 0$, the terms of the trigonometric series in the brackets are all negligible and the hyperbolic cosines are unity, so that, approximately

$$\zeta_n = \frac{F_n}{f^2 R_n^2} y \qquad \left(\frac{a}{R_n}, \frac{b}{R_n} \to 0 \right)$$

(19)

Equation 18 now shows that the volume transports are zero to the same approximation. This solution is known as a static "setup," the surface slope of the water just balancing the wind stress everywhere.

In the large basin limit $(a/R_n, b/R_n \to \infty)$ the hyperbolic functions in the denominators in equation 17 are very large, so that ζ_n is nearly zero everywhere, except where at least some of the numerators are also large, that is, within distances of order R_n from boundaries. Along the side walls, $x = \pm a/2$, at least several R_n from the end walls, the elevation distribution is exactly as given by equation 19, but this drops to zero within a coastal boundary layer of scale width R_n. At the center of the basin the elevation is zero, and the transport is Ekman transport balancing wind stress. At the end walls the elevations are $\pm F_n b/2f^2 R_n^2$, that is, also as in the small basin, but again the drop along a line parallel to the y-axis is not gradual but sudden and is accomplished within a distance of order R_n. The equivalent volume transports U_n, V_n are as follows:

$$U_n = \frac{F_n}{f}, \qquad V_n = 0 \qquad \text{(center portion)}$$

$$U_n = \pm \frac{b}{2R_n} \frac{F_n}{f}, \qquad V_n = 0 \qquad \text{(end walls)}$$

(20)

$$U_n = 0, \qquad V_n = \pm \frac{y}{R_n} \frac{F_n}{f} \qquad \text{(side walls)}$$

by the setup (longshore pressure gradient), which opposes the wind stress. The same pressure gradient reduces the longshore velocity of the surface layer by the factor $h'/(h + h')$, as compared to the simple coastal jet model without setup. Outside the coastal boundary layers of scale width R_2 there is again no flow and little thermocline displacement (more precisely, ζ' is of order ζ), and a surface elevation gradient balances the wind stress.

We may sum up this discussion of a finite, constant-depth basin in the observation that the longshore pressure gradient produced by the end walls establishes return flow below the thermocline within the coastal boundary layer, where along an infinite shore the water would remain stagnant. There is also another effect: the slow propagation around the basin of the stagnation point, in a cyclonic sense (Csanady and Scott, 1974). This follows from the propagation of the Kelvin waves we already mentioned but is outside the scope of the coastal jet conceptual model.

5. Depth Variations

Next we would like to extend the coastal jet concept to still more realistic models by allowing the depth of the basin to vary. To the accuracy of linearized theory, and of the hydrostatic approximation, equations 1 remain valid if h and h' depend on x and y, except of course where h or h' tend to zero (in the linearization process h and h' are assumed large compared to ζ and ζ'). The coastal jet concept is only useful when there is a long and straight coast, so that the most general model we need consider is a long and narrow basin with arbitrary depth distribution in the cross section, as illustrated in Fig. 5. The basin has end walls, but these are supposed far away and the depth is only a function of the x-coordinate. Where h or h' tends to zero, equations 1 have singularities that lead to some physically unrealistic features of the solutions. These do not prove to be unduly disturbing (the physical reasons for the anomalies being clear enough) so that we shall not concern ourselves with the corrections that, strictly speaking, would have to be made near these singularities.

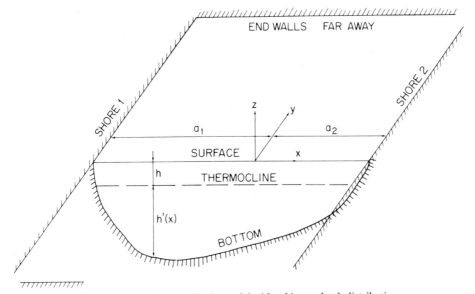

Fig. 5.   Long and narrow basin model with arbitrary depth distribution.

A thermocline is present only where the total depth is greater than the top layer equilibrium depth h_0. Within this part of the basin all six of equations 1 must be used, outside the $h + h' = h_0$ locus only the first three (but setting $\zeta' = 0$). Within the two-layer portion, and sufficiently far from the end walls, a coastal jet type of response may be expected to be described by a solution of equations 1 of the following form:

$$
\begin{aligned}
U &= U(x) & U' &= U'(x) \\
V &= B(x)t & V' &= B'(x)t \\
\zeta &= W(x)t + Sy & \zeta' &= W'(x)t + S'y
\end{aligned}
\tag{22}
$$

Here B and B' are depth-integrated longshore accelerations and W and W' are vertical velocities, all functions of x only. Longshore gradients of surface elevation, S, and of thermocline elevation S', are supposed constant and they represent the only influence we allow the end walls. The magnitudes of these longshore gradients may be thought to be determined by nondimensional basin dimensions, length divided by appropriate radii of deformation, as suggested by our earlier results.

Substituted into equations 1, equations 22 lead to the relationships:

$$
W = -\frac{d}{dx}(U + U')
$$

$$
B = \frac{gh}{f}\frac{dW}{dx} = -\frac{gh}{f}\frac{d^2}{dx^2}(U + U')
$$

$$
W' = -\frac{dU'}{dx}
\tag{23}
$$

$$
B' = \frac{gh'}{f}\frac{dW}{dx} + \frac{g\varepsilon h'}{f}\frac{d}{dx}(W' - W)
$$

$$
= \frac{gh'}{f}\frac{d^2}{dx^2}(U + U') + \frac{g\varepsilon h'}{f}\frac{d^2 U}{dx^2}
$$

These allow longshore accelerations and vertical velocities to be calculated once the distribution of onshore–offshore transports U and U' is known. The latter are subject to (still from equations 1 and 22, eliminating W, B, W', and B' with the aid of equations 23):

$$
-\frac{gh}{f}\frac{d^2}{dx^2}(U + U') + fU = -ghS + F
$$

$$
-\frac{gh'}{f}\frac{d^2}{dx^2}(U + U') + \frac{g\varepsilon h'}{f}\frac{d^2 U}{dx^2} + fU' = -gh'S - g\varepsilon h'(S' - S)
\tag{24}
$$

Our previous results lead us to suspect that the latter two equations have solutions consisting of a sum of "surface" and "internal" modes. For small proportionate density defect ε these models should be characterized by (1) a vertically uniform velocity, and (2) a vanishing total transport. We therefore seek solutions consisting of two additive components, indexed 1 and 2, each subject to constraints as follows:

$$
U = U_1 + U_2 \qquad U' = U'_1 + U'_2
$$

$$
\frac{U'_1}{h'} - \frac{U_1}{h} = 0\left(\varepsilon\frac{U_1}{h_0}\right)
\tag{25}
$$

$$
U_2 + U'_2 = 0(\varepsilon U_2)
$$

We also anticipate that the two solutions will be scaled by very different character-istic lengths, so that when equations 25 are substituted into equations 24, index 1 and index 2 quantities should balance separately. Dividing the first of equations 24 by h, the second by h', and subtracting, we find

$$- \frac{g\varepsilon}{f} \frac{d^2 U}{dx^2} + f\left(\frac{U}{h} - \frac{U'}{h'}\right) = \frac{F}{h} + g\varepsilon(S' - S) \tag{24a}$$

Substituting now the form of solution postulated in equations 25 we obtain

$$- \frac{g\varepsilon}{f} \frac{d^2 U_2}{dx^2} + f\left(\frac{1}{h} + \frac{1}{h'}\right)U_2 = \frac{F}{h} + g\varepsilon(S' - S)$$

$$+ \left[\left(\frac{fU_1'}{h'} - \frac{fU_1}{h}\right) + \frac{g\varepsilon}{f} \frac{d^2 U_1}{dx^2} - \frac{U_2 + U_2'}{h'}\right] \tag{26}$$

By hypothesis the last square-bracketed term on the right of this equation is of order ε. The second term on the right is also of order ε, unless either S or S' is much larger than F/gh, a possibility we disallow far from the end walls. The first term on the left also contains ε, but it is not negligible if (as expected) the length scale or variations of U_2 is of order $(g\varepsilon h_0)^{1/2} f^{-1}$. Dropping the order ε terms from equation 26 we have after some cross multiplications:

$$\frac{d^2 U_2}{dx^2} - \frac{f^2}{g\varepsilon}\left(\frac{1}{h} + \frac{1}{h'}\right)U_2 = - \frac{fF}{g\varepsilon h} \tag{27}$$

Within the two-layer portion of our system $h = h_0 = $ constant, but h' is variable. Equation 27 then constitutes a relatively simple second-order equation from which U_2 may be determined. For h' also constant the solution is exactly what we already found for the *internal mode* in a constant-depth canal. Boundary conditions are $U_2' = 0 \cong -U_2$ at both locations where the thermocline intersects the bottom.

Adding the two equations 24 leads to

$$\frac{g}{f}(h + h') \frac{d^2}{dx^2}(U + U') - f(U + U') = g(h + h')S - F + g\varepsilon h'(S' - S) + \frac{1}{f} \frac{d^2 U}{dx^2} \tag{24b}$$

Substituting the "Ansatz" of equation 25 and dropping undifferentiated terms multiplied by ε there is left:

$$\frac{g}{f}(h + h') \frac{d^2}{dx^2}(U_1 + U_1') - f(U_1 + U_1') = g(h + h')S - F + \frac{g\varepsilon h'}{f} \frac{d^2 U_2}{dx^2}$$

$$- \frac{g}{f}(h + h') \frac{d^2}{dx^2}(U_2 + U_2') \tag{28}$$

Although $U_2 + U_2'$ is by hypothesis of order εU_2, the last term on the right is not negligible because its horizontal variation is scaled by $(g\varepsilon h_0)^{1/2} f^{-1}$, as is the variation of the second-last term, both according to equation 27. Because the other terms in the equation are scaled differently, these two terms must balance between themselves, a condition from which the precise variation of $(U_2 + U_2')$ may be calculated. This is an order ε quantity of no special interest and we shall not further concern ourselves with it. The remaining terms in the equation leave

$$\frac{d^2}{dx^2}(U_1 + U_1') - \frac{f^2}{g(h + h')}(U_1 + U_1') = fS - \frac{fF}{g(h + h')} \qquad (29)$$

The same equation applies outside the two-layer portion, but of course with $h' = 0$, $U' = 0$.

The solution of this equation yields the response in the surface or barotropic mode, which is (to order one) independent of stratification. With constant $h + h'$ our earlier results are again recovered. As one easily demonstrates, equations 27 and 29 possess solutions for at least some well behaved h' distributions, so that they constitute the extension of the coastal jet model to the variable depth case, the conjectures incorporated in equations 22 and 25 having been substantiated.

6. The Pattern of Barotropic Flow

Equation 29 describes the aperiodic response of a variable-depth basin in the surface or barotropic mode to suddenly impose longshore wind. As we have already pointed out, we expect the coastal jet conceptual model to be useful along a long, straight coast; hence we concentrate our attention on a basin much longer than it is wide, illustrated schematically in Fig. 5.

A scale analysis of the problem shows the second term on the left of equation 29 to be small, if the basin width is small compared to an appropriately defined radius of deformation. Let the maximum depth of the basin be H, and let us nondimensionalize the terms of equation 29 according to the following scheme:

$$R^2 = \frac{gH}{f^2}$$

$$x^* = \frac{x}{R} \qquad h^* = \frac{h + h'}{H} \qquad (30)$$

$$U^* = \frac{f(U_1 + U_1')}{F} \qquad S^* = \frac{gHS}{F}$$

Let the shores of the basin be at $x = -a_1, a_2$, and assume that

$$a^*_{1,2} = \frac{a_{1,2}}{R} \ll 1 \qquad (31)$$

Equation 29 becomes with the nondimensional quantities introduced:

$$\frac{d^2 U^*}{dx^{*2}} = \frac{U^*}{h^*} + S^* - \frac{1}{h^*} \qquad (29a)$$

The nondimensional cross flow U^* is defined as the fraction of Ekman transport and we may expect this to be at most of order unity. The nondimensional pressure gradient S^* is equal to one if the pressure gradient exactly balances the wind stress in the deepest part of the basin. We may expect this to be also of order unity; in a sufficiently long basin it could even be of smaller order, according to our previous results. Thus unless the singularity at the shore (where h^* tends to zero) causes difficulty, U^* obtained by a double integration of equation 29a should be of order $a^{*2}_{1,2}$, that is, small. An ordinary sloping beach, where h^* varies linearly with $(x \pm a_{1,2})$

does *not* cause a problem and the conclusion regarding the order of magnitude of U^* stands, unless the bottom slope becomes unrealistically small.

One may further verify the truth of this assertion using some specific model. Figure 6 illustrates an idealized cross section with simple, symmetrical sloping beaches ($a_1 = a_2 = a$) and a constant-depth center. The dimensions of this cross section approximate those of Lake Ontario. Over the sloping parts the solution of equation 29a is

$$U^* = C_1 \lambda I_1(\lambda) - \lambda K_1(\lambda) - S^* h^* + 1 \tag{32}$$

where $I_1(\ )$, $K_1(\ )$ are modified Bessel functions, C_1 is an integration constant, and

$$\lambda = \sqrt{\frac{4|x \pm a|H}{sR}}$$

with s the bottom slope $s = d(h + h')/dx$. Over the center part, on the other hand:

$$U^* = 1 - S^* + C_2 \cosh x^* \tag{33}$$

with C_2 another integration constant. These solutions already satisfy one boundary condition each, zero offshore flow at the shore, and symmetry of the U^* distribution. Matching U^* and dU^*/dx (i.e., transport and surface elevation) at the point where the two sections join, one may determine the constants. Given the quantitative data of Fig. 6, the center (maximum) value of U^* turns out to be of order 10^{-3}. The dimensional quantity $U_1 + U'_1$ is thus three orders of magnitude less than the Ekman flux F/f.

The depth-integrated longshore acceleration in the surface mode is, using equations 23 and 29:

$$B_1 + B'_1 = -\frac{g}{f}(h + h')\frac{d^2}{dx^2}(U_1 + U'_1)$$

$$= -g(h + h')S + F - f(U_1 + U'_1) \tag{34}$$

As we have just seen, the last term in this equation is negligible, so that the longshore momentum balance (which equation 34 represents) is between wind stress, pressure gradient, and acceleration, Coriolis force due to cross flow being of subordinate importance.

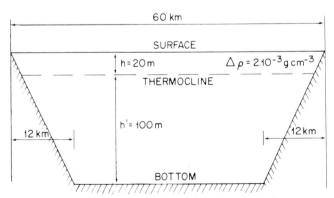

Fig. 6.   Idealized model of Lake Ontario cross section.

With F and S constant, one term on the right of equation 34 is constant and the other varies as $h + h'$, so that the acceleration $B_1 + B'_1$ also varies. In the constant-depth case, for a sufficiently short basin, we have seen that $B_1 + B'_1$ was negligible everywhere and the wind stress was balanced by the pressure gradient. The nearest analogue in the present case is that the cross-sectional integrals of the two balance, that is,

$$gS \int_{-a_1}^{a_2} (h + h')\, dx = F(a_2 + a_1) \tag{35}$$

This also implies then:

$$\int_{-a_1}^{a_2} (B_1 + B'_1)\, dx = 0 \tag{36}$$

so that no fluid accumulates windward of the cross section considered.

It is tempting to impose equation 35 (or equivalently, 36) as a condition and call it part of the coastal jet conceptual model for an appropriately small basin. However, in order to maintain reasonable correspondence with what might be observable coastal jets, we would have to include all low-frequency modes of motion that contribute to coastal currents. To the extent that the effects of the Coriolis force are entirely negligible, it is easy enough to justify equation 36, because any net accelerations are then associated with the relatively high seiche frequencies. In a rotating basin of variable depth, on the other hand, there may exist "second class motions" (Ball, 1965), analogous to "shelf waves" or "topographic Rossby waves." The combination of these waves with the steady-state flow pattern may give rise to a completely different resultant, much as the static elevation plus Kelvin waves produce a coastal jet in a *large* constant-depth basin unlike either of its constituents. This question needs further mathematical investigation (and, of course, corresponding further scrutiny of experimental evidence). For the time being, we accept equations 35 and 36 on an intuitive basis, as probably reasonable for describing the aperiodic barotropic flow in a variable-depth basin of modest size, at least for a period after a wind-stress impulse, before the Coriolis force can develop some more complex effects. [For further remarks on this question see a recent discussion by Simons (1974) and my reply.]

Bennett (1974) seems to have been the first to explore the interesting implications of equation 34, with equation 36 imposed as a condition. The integral on the left of equation 35 is the total fluid-filled area of the canal cross section, which we may use to define a mean depth:

$$h_m = \frac{1}{a_2 + a_1} \int_{-a_1}^{a_2} (h + h')\, dx \tag{37}$$

The longshore pressure gradient is then $S = F/gh_m$, and equation 34 gives zero depth-integrated acceleration $B_1 + B'_1$ for those points in the cross section where $h + h' = h_m$. In shallower water the wind stress overcomes the pressure gradient and accelerates the water downwind. Indeed in *very* shallow water, where $h + h' \ll h_m$, the effect of the pressure gradient becomes negligible. This should hold independently of the truth of the intuitively imposed equation 36, as long as the pressure gradient S is of order F/gh_m. In water deeper than h_m the pressure gradient overwhelms the wind stress and the developing volume transport is against the wind.

Fig. 7.   Schematic pattern of wind-driven total transport in long and narrow basins.

We have arrived at these results for a given basin cross section. Their implication for a long and narrow basin is a sort of double-gyre flow pattern, illustrated in Fig. 7, the streamlines shown indicating depth-integrated volume transport. The velocities are the transports divided by water depth, that is, relatively large in shallow water. We conclude that a pronounced barotropic coastal jet should be present in a variable-depth basin *outside* the mean depth contour, a very different result from the constant-depth case. In water much shallower than the cross-sectional mean depth h_m the momentum gain is nearly equal to the wind stress impulse divided by total depth. This is exactly the same result as we found earlier for an infinite coast (without end walls), the reason for the correspondence being that the effects of longshore gradients in shallow water are unimportant.

7. Baroclinic Flow Over Variable Depth

We return now to equation 27 and inquire how internal mode coastal jets are affected by depth variations. This equation of course applies only where the bottom layer equilibrium depth h' is nonzero and where the top layer depth $h = h_0$ is constant. A solution of equation 27 yields the top layer baroclinic onshore–offshore transport U_2, the bottom layer transport, U'_2, being equal and opposite to order ε. Other quantities of interest may then be found from equations 23, especially thermocline movement and longshore acceleration:

$$W'_2 = \frac{dU_2}{dx}$$

(38)

$$B_2 = -\frac{g\varepsilon hh'}{f(h + h')}\frac{d^2 U_2}{dx^2}$$

and $B'_2 = -B_2$. The surface elevation (or rather its rate of movement, W) is a quantity of order $\varepsilon^{1/2}$.

Equation 27 is a second-order nonhomogeneous equation of relatively simple structure which, for simple h' distributions, reduces to the hypergeometric equation. Specifically for linearly varying h' ($h' = xs$, with s the bottom slope) the equation transforms into

$$\frac{d^2 U_*}{dx_*^2} - \left(\frac{1}{4} + \frac{k}{x_*}\right)U_* = -\frac{1}{4}$$

(39)

where

$$U_* = \frac{fU_2}{F}$$

$$x_* = \frac{2x}{L}$$

$$k = \frac{h_0}{2sL}$$

with

$$L = \frac{(g\varepsilon h_0)^{1/2}}{f}$$

This is Whittaker's equation with a constant nonhomogeneous term, the solutions of which are generalized Struve functions (Babister, 1967). The solutions reduce to a simpler form for special values of the slope parameter k. The character of the solution may be illustrated using $k = 0.5$, in which case one finds (Csanady, 1971):

$$U_2 = C\frac{Fx}{fL}\left[I_0\left(\frac{x}{L}\right) + I_1\left(\frac{x}{L}\right)\right] - \frac{Fx}{fL}\left[1 + \frac{\pi}{2}L_0\left(\frac{x}{L}\right) + \frac{\pi}{2}L_1\left(\frac{x}{2}\right)\right] \quad (40)$$

where I_0 and I_1 are modified Bessel functions L_0 and L_1 are modified Struve functions, and C is an integration constant. The solution already satisfies the boundary condition of zero normal transport at the coast. The constant C may be determined by patching the solution 40 to a constant-depth model at some specific $x = x_0$, with continuous U_2 and dU_2/dx.

In a quantitative illustration the following "typical" values may be used:

$$s = 3.10^{-3}$$
$$\varepsilon = 2.10^{-3}$$
$$h_0 = 18 \text{ m}$$
$$f = 10^{-4}/\text{sec}$$

These result in $k = 0.5$, $L = 6$ km. Transition to constant depth may be supposed to occur at $x = 30$ km, resulting in a maximum bottom layer depth of 90 m, total water depth of 108 m. All these data are representative of conditions in shallow seas, especially in the Great Lakes or over the eastern seaboard continental shelf, although the slope chosen is on the low side.

When the transition to constant depth takes place at a distance much larger than L, the value of the constant C becomes approximately $\pi/2$ and U_2 tends with large x/L to the constant value F/f (the total Ekman transport) already over the sloping part of the bottom. In this special case the depth-integrated longshore acceleration is

$$B_2 = F\frac{x}{L}\left\{\left(1 + \frac{x}{L}\right)^{-1} + 1 - \frac{\pi}{2}\left[I_0\left(\frac{x}{L}\right) - L_0\left(\frac{x}{L}\right)\right] - \frac{\pi}{2}\left[I_1\left(\frac{x}{L}\right) - L_1\left(\frac{x}{L}\right)\right]\right\} \quad (41)$$

This vanishes at $x/L = 0$ and $x/L \to \infty$, the variation in between being illustrated in Fig. 8. Notable is the relatively small maximum value of B_2/F: in a constant-depth model the maximum of this quantity is $h'/(h + h')$ (see equation 9), which is usually close to unity. In the present example, the maximum B_2 occurs where $h'/(h + h')$ is

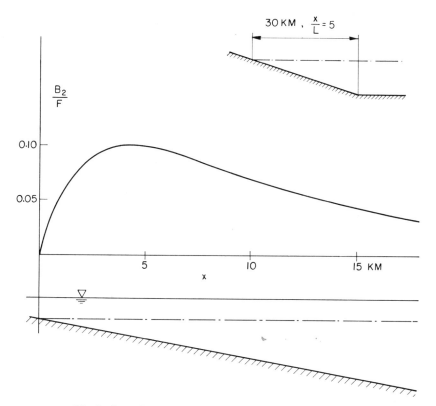

Fig. 8.   Longshore acceleration of top layer over sloping beach.

approximately $\frac{1}{2}$, but by the time this depth is reached, the distance from the thermocline–bottom intersection is of order L, so that the baroclinic acceleration has decayed appreciably.

The top layer longshore acceleration is B_2 divided by the constant top layer depth h_0 and therefore also varies as B_2/F (Fig. 8). However, the bottom layer longshore acceleration is different, $-B_2/sx$, which has the nonzero value of $-(F/h_0)(2-\pi/2)$ at the shore.

The upward velocity of the thermocline may be calculated to be, from equations 38 and 40, with $C = \pi/2$:

$$W_2' = \frac{F}{fL}\left\{\frac{\pi}{2}\left(1+\frac{x}{L}\right)\left[I_0\left(\frac{x}{L}\right)-L_0\left(\frac{x}{L}\right)\right]+\frac{\pi}{2}\frac{x}{L}\left[I_1\left(\frac{x}{L}\right)-L_1\left(\frac{x}{L}\right)\right]-\left(1+\frac{x}{L}\right)\right\}$$

(42)

This also vanishes at large x/L, while its value at $x/L = 0$ is $(\pi/2 - 1)F/fL$, or a little over half of the value which applies in a constant-depth model with a relatively large h'/h ratio (see again equation 9).

We may now combine the barotropic and baroclinic components of the flow to obtain the total response over a sloping beach. We assume for simplicity that the length scale of baroclinic jets, L, is small compared to basin dimensions, and that even the beach of constant slope s extends much further from shore than L. This also implies that the depth at a distance of order L from the thermocline–bottom intersection is small compared to basin average depth, so that the effect of any longshore pressure gradient on the barotropic flow is small within this same range.

The longshore barotropic acceleration is then (by equation 34) $F/(h + h')$. At the point where the thermocline intersects the bottom this gives longshore velocities $v = Ft/h_0$, $v' = (\pi/2 - 1)Ft/h_0$ (the latter with the results of the above typical example). The depth of the bottom layer is here of course zero, so that the wind-induced momentum is all in the top layer. The bottom layer, however, is *not* at rest now (in contrast with the constant-depth model, at the shore). The physical reason for the difference is easily discovered if we recall that the thermocline is rising in our typical beach model with a velocity of $W' = (\pi/2 - 1)F/fL$. Because of the presence of a bottom slope, this requires an inflow velocity $u' = U'/h'$ in the bottom layer of a magnitude equal to vertical velocity divided by the slope, $u' = -W'/s = -(\pi/2 - 1)F/fh_0$ (noting that in our typical example with $k = 0.5$, $L = h_0/s$). The Coriolis force acting on an inward moving parcel of fluid then produces the longshore acceleration $fu' = (\pi/2 - 1)F/h_0$, explaining our earlier result.

At distances long compared to L the baroclinic contribution to B and B' decays to zero and the longshore acceleration pattern is what we described in the preceding section, evenly distributed over the total depth. There is, however, equal and opposite onshore–offshore flow in the top and bottom layers. The momentum of the wind stress is transferred to the bottom layer by the medium of the Coriolis force, as we discussed in greater detail in connection with the constant-depth model.

To sum up this discussion of depth effects, we may conclude that depth variations have an important influence on the pattern of barotropic flow basinwide (this may be pursued beyond the scope of the coastal jet conceptual model; see, e.g., Csanady 1973, 1974). In shallow water the net effect is that the momentum imparted by the wind simply accelerates the water column, the longshore pressure gradient having a negligible influence. In this respect, the variable-depth shore zone behaves as an infinitely long coast in the simplest coastal jet model. The baroclinic contribution still rearranges the total momentum input in favor of the top layer, within a nearshore band of scale width L, which is very nearly the same as the baroclinic radius of deformation R_2 in a constant-depth basin of large bottom layer to top layer depth ratio. However, the monopoly of the top layer on the windward momentum is not complete over a sloping beach, because the bottom layer acquires windward velocity even very close to the thermocline–bottom intersection. In our "typical" example (which was actually closer to the flattest beach extreme) the bottom layer velocity near $h' = 0$ was about half the top layer one in the same location. A larger bottom slope would tend to reduce this toward the zero value appropriate for a vertical beach. The top layer longshore acceleration is influenced relatively little by the baroclinic contribution for two reasons: (1) this contribution decays with scale length L, and (2) the bottom layer grows only slowly in depth within this distance range, and can absorb only small amounts of wind-imparted momentum. The net result, however, is still that within a band extending a distance of order L beyond the thermocline–bottom intersection the momentum input of the wind is mainly reflected in a momentum gain by the top layer.

It is anticlimactic to arrive at the result that the simplest, infinite shore, constant-depth coastal jet model predicted much the same behavior as the variable-depth, closed basin one (albeit a long and narrow basin). The differences in detail are not as important practically as the similarities, although they are essential from the point of view of a dynamical understanding of the phenomena involved.

8. Application to Observed Coastal Currents

During and in preparation for the International Field Year on the Great Lakes (IFYGL, carried out during 1972) a systematic series of observations have been carried out on the coastal boundary layer of Lake Ontario, defined as the band extending from 0 to about 10 km from shore. Lake Ontario is much longer than it is wide (by about a factor of 5) and it possesses some relatively straight, uncompli-cated shorelines (Fig. 9). Both westerly and easterly storms are moderately frequent, and are sometimes preceded and followed by quiescent periods, allowing a study of isolated longshore wind-stress impulses. During the summer months a resonably sharp thermocline separates the top *epilimnion* from a colder and slightly denser *hypolimnion*, producing conditions to which the coastal jet conceptual model should apply. The IFYGL observations have indeed produced flow phenomena very much with characteristics as discussed above; some of these are illustrated in Figs. 10–12b.

All these illustrations show examples of coastal jets in their *initial* development, that is, more or less immediately following a shore-parallel wind stress impulse acting upon a relatively quiescent lake. Their predominant features are as follows:

1. High velocities in the jets are confined to the warm layer and to a nearshore band of the order of 5 km width. Water below the thermocline moves slowly (with a velocity of a few cm/sec) in the same direction as above.

2. A thermocline upwelling or downwelling occurs in the region of the jet. A downwelling accompanies flow leaving the shore to the right, upwelling an opposite current. The velocity of the warm water relative to the cold is geostrophically balanced by the thermocline slope, to a good approximation.

3. During IFYGL the wind-stress impulse Ft could be estimated from over-the-lake wind observations. The total depth-integrated momentum at the core of the jet

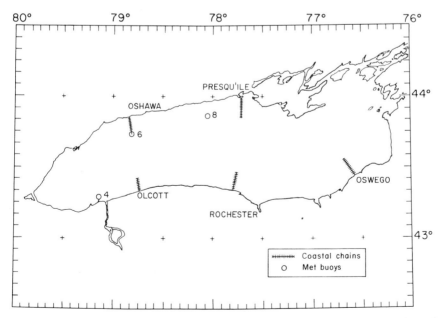

Fig. 9.   L. Ontario with locations of coastal current studies.

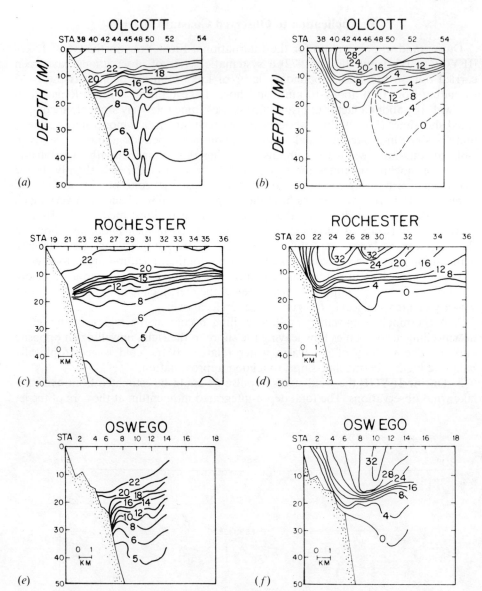

Fig. 10.   Example of coastal jet, 23 July 1972, south shore showing isotherms (°C, left) and constant longshore velocity contours (cm sec^{-1}). Direction of flow is such as to leave shore to the *right*.

was always of the same order of magnitude as the wind-stress impulse (although somewhat less, typically 70 % of *Ft*).

4. The distribution of depth-integrated momentum with distance from shore (Fig. 11) showed a peak at 6–7 km from shore, presumably owing to strong frictional influences in water less than 30 m deep. In somewhat deeper water the depth-integrated momentum was close to the wind-stress impulse (as already pointed out) but it then decreased to zero roughly where the depth became equal to the cross-sectional average.

On the basis of this evidence it appears legitimate to regard flow structures of the kind depicted in Figs. 10–12 as distinct observable phenomena, the principal features of which are in good accord with the coastal jet conceptual model. There are other characteristics of these nearshore currents which are more complex: for instance, the reversal of flow upon the passage of a Kelvin wave (Csanady and Scott, 1974), or the observed preference for a flow direction leaving the shore to the right (Blanton, 1974),

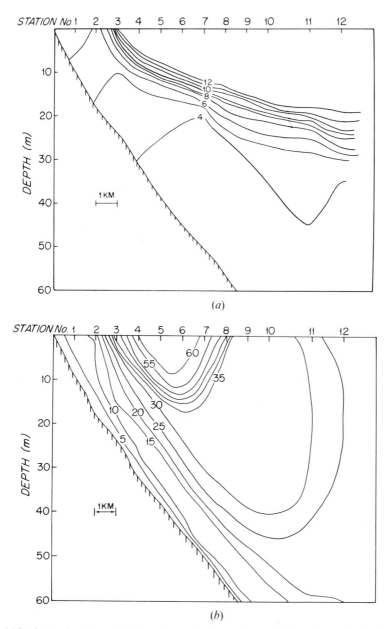

Fig. 11. (a) isotherms, °C; (b) constant longshore velocity contours at Oshawa immediately after major storm (10 Oct. 1972). Flow now leaves shore to the *left*.

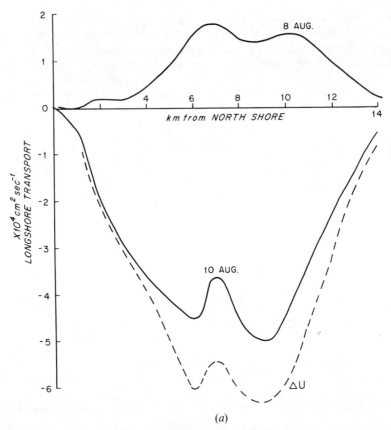

Fig. 12. (a) Depth-integrated transport distribution at Oshawa, Aug. 8–10, 1972, following two opposing wind stress impulses. Lake depth is equal to the average for this cross section at 16 km from shore.

not to speak of the effects of friction clearly evident in Fig. 11. Some of these complexities may be accounted for by an extension of the coastal jet conceptual model, whereas others will no doubt require recourse to different conceptual models. It seems appropriate at this point to let the case rest.

In conclusion, a few words may be said regarding the application of the same conceptual model to flow phenomena over the continental shelves. Given the depths and widths of these, there can be little doubt that very much the same phenomena as documented in detail on the Great Lakes will occur within 10 km or so of oceanic coasts. We have carried out a pilot study of coastal currents south of Long Island (so far unpublished) which confirms this expectation. Fragmented reports of similar phenomena have surfaced from time to time (see, e.g., the review of Bumpus, 1973, referring to upwelling along the Maine coast) but have not been systematically assembled. In studying upwelling along the Oregon coast, O'Brien and his collaborators have explicitly invoked the coastal jet conceptual model (see O'Brien and Hurlburt, 1972; Thompson and O'Brien, 1973). It appears, however, that the Oregon upwelling may be affected by some nonlocal influences, which produce, among other effects, a poleward undercurrent. Possibly the coastal jet conceptual model will prove useful

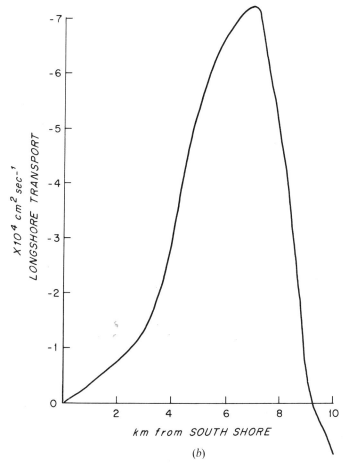

Fig. 12. (b) Depth-integrated transport distribution at Olcott, Aug. 10, 1972, following two opposing wind stress impulses. Lake depth is equal to the average for this cross section at 9 km from shore.

in describing short-term response phenomena in this situation, becoming one of several tools in a larger kit to which one may have to have recourse to understand fully a more complex total picture.

Acknowledgment

This work was supported by the Great Lakes Environmental Research Laboratory of NOAA under contract 03-5-022-26 and by the Brookhaven National Laboratory under contract 325373S.

References

Babister, A. W., 1967. *Transcendental Functions.* Macmillan, New York, 414 pp.
Ball, F. K., 1965. Second-class motions of a shallow liquid. *J. Fluid Mech.*, **23**, 545–561.
Bennett, J. R., 1974. On the dynamics of wind-driven lake currents. *J. Phys. Oceanogr.*, **4**, 400–414.
Blanton, J. O., 1974. Some characteristics of nearshore currents along the North shore of Lake Ontario. *J. Phys. Oceanogr.*, **4**, 415–424.

Bumpus, D. F., 1973. A description of the circulation on the continental shelf of the east coast of the United States. In *Progress in Oceanography*. Vol. 6. Pergamon Press, New York, 111–159 pp.

Charney, J. G., 1955. Generation of oceanic currents by wind. *J. Mar. Res.*, **14**, 477–498.

Crépon, M., 1967 and 1969. Hydrodynamique marine en regime impulsionnel. *Cah. Oceanogr.*, **19**, 847–880; **21**, 333–353, 863–877.

Csanady, G. T., 1971. Baroclinic boundary currents and long edge-waves in basins with sloping shores. *J. Phys. Oceanogr.*, **1**, 92–104.

Csanady, G. T., 1972. The coastal boundary layer in Lake Ontario *J. Phys. Oceanogr.*, **2**, 168–176.

Csanady, G. T., 1973. Wind induced barotropic motions in long lakes. *J. Phys. Oceanogr.*, **3**, 429–438.

Csanady, G. T., 1974. Barotropic currents over the continental shelf. *J. Phys. Oceanogr.*, **4**, 357–371.

Csanady, G. T., and J. T. Scott, 1974. Baroclinic coastal jets in Lake Ontario during IFYGL. *J. Phys. Oceanogr.*, **4**, 524–541.

Mortimer, C. H., 1963. Frontiers in physical limnology with particular reference to long waves in rotating basins. *Great Lakes Res. Div. Publ.* (University of Michigan) **10**, 9–42.

O'Brien, J. J. and Hurlburt, H. E., 1972. A numerical model of coastal upwelling. *J. Phys. Oceanogr.*, **2**, 14–26.

Simons, T. J., 1974. Comments on "Wind Induced Barotropic Motions in Long Lakes." *J. Phys. Oceanogr.*, **4**, 270–271; Reply, 271–273.

Stommel, H. M., 1948. The westward intensification of wind-driven ocean currents. *Trans. Am. Geophys. Union*, **29**, 202–206.

Thompson, J. D. and O'Brien, J. J., 1973. Time-dependent coastal upwelling. *J. Phys. Oceanogr.*, **3**, 33–46.

Wilson, B. W., 1972. *Seiches Advan. Hydrosci.*, **8**, 1–94.

5. EXPERIMENTS IN STORM SURGE SIMULATION

R. O. Reid, A. C. Vastano, R. E. Whitaker, and J. J. Wanstrath

1. Introduction

The general system of equations that represent storm surges has been treated numerically with success for the past 20 years. In various expressions, these systems may model one- or two-spatial dimensions, usually in terms of vertically integrated differential equations, and extend in complexity to include the effect of the earth's rotation and the field accelerations. Initial numerical experiments with one-dimensional algorithms such as the study of forced and free surges by Reid (1957) and the method of characteristics (Freeman, 1951) encouraged the expansion of efforts, in keeping with computer capacity and speed, toward the relatively complex systems that are presently available to predict surge characteristics on the continental shelf (Jelesnianski, 1972; Hansen, 1956) and the more detailed computations of flow and water elevation in adjacent estuarys and low-lying regions (Reid and Bodine, 1968; Leendertse, 1970). A comprehensive survey of the basic methods of finite difference approximations in surge dynamics was carried out by Welander (1961). This work is a guide to explicit methods for integrating the equations of motion. More recently, two-dimensional implicit schemes have been discussed by Dronkers (1969) and a three-dimensional system has been investigated by Heaps (1971).

Though methods for integrating the equations of motion have been improved to produce fast and accurate algorithms, the general predictive capabilities of surge algorithms have approached a limit inherently established by the physical mechanisms represented in the models. An increasing activity in observation and analysis and the current focus on man's relation to nearshore processes have emphasized the need for refinement of surge estimates. Present research focuses on the need for better physical representations of boundary conditions.

The investigation of a storm-induced surge is often a two-stage computational process requiring prediction on the continental shelf and a more spatially and temporally detailed study of a particular nearshore region. The latter depends on shelf surge computations to provide the oceanic forcing function on the seaward boundaries. In view of this relation, the present elaboration of estuarine models coincides with better shelf surge prediction.

2. Shelf Surges in Transformed Coordinates

Algorithms used in shelf surge prediction carry out the integration of the equations of motion in large spatial regions that generally extend over the continental shelf to the shore. An effective means for achieving greater accuracy in these computations is a more precise representation of the natural shoreline and seaward boundaries of the model. The problem is to find a transformation involving mapping relations $\xi = \xi(x, y)$, $\eta = \eta(x, y)$ that preserve orthogonality and in which ξ and η are continuous monotonic functions of x and y. Further, the transformation must map the coastline and seaward boundaries as isolines of the curvilinear coordinate, η. A point (x, y) on the z plane will be transformed to (ξ, η) in a rectangular region on the

ζ plane and satisfy these conditions if the mapping relation is conformal, such that

$$\xi + i\eta = F(x + iy) \tag{1}$$

or

$$x + iy = G(\xi + i\eta) \tag{2}$$

where F and G are single-valued real functions.

The function $G(\zeta)$ may be represented by a truncated Fourier series

$$G(\zeta) = P_0 + Q_0\zeta + \sum_{n=1}^{N} (P_n \cos n\zeta + Q_n \sin n\zeta) \tag{3}$$

where $\zeta = \xi + i\eta$ and the coefficients P_n and Q_n are complex constants. The real and imaginary parts of this series yield x and y, respectively, in terms of ξ and η. The coefficients are taken such that for a constant value, $\eta = \beta$, the resulting x and y as a function of ξ will map out the reach of the coastline under study. As an additional constraint, the seaward boundary, taken as the 200 m depth contour, is represented by $\eta = -\beta$.

An iterative procedure is used for evaluating the coefficients P_n and Q_n so as to give a best fit in a least squares sense. Let

$$P_n = A_n + iB_n$$
$$Q_n = C_n + iD_n$$

where A_n, B_n, C_n, and D_n $(n = 1, 2, \ldots, N)$ are real constants. Equation 3 then yields

$$\chi = A_0 + C_0\xi - D_0\eta + \sum_{n=1}^{N} (A_n \cosh n\eta - D_n \sinh n\eta) \cos n\xi$$

$$+ \sum_{n=1}^{N} (B_n \sinh n\eta + C_n \cosh n\eta) \sin n\xi \tag{4}$$

and

$$y = B_0 + C_0\eta + D_0\xi + \sum_{n=1}^{N} (B_n \cosh n\eta + C_n \sinh n\eta) \cos n\xi$$

$$+ \sum_{n=1}^{N} (D_n \cosh n\eta - A_n \sinh n\eta) \sin n\xi \tag{5}$$

for a range of ξ from $-\pi$ to π. The parametric representations of these curves are

Seaward:

$$x = x_s(\xi, -\beta)$$
$$y = y_s(\xi, -\beta)$$

Coastline: (6)

$$x = x_c(\xi, +\beta)$$
$$y = y_c(\xi, +\beta)$$

respectively, where

$$x_s(\xi, -\beta) = A_0 + C_0\xi + D_0\beta + \sum_{n=1}^{N}(C_n \cosh n\beta - B_n \sinh n\beta) \sin n\xi$$

$$+ \sum_{n=1}^{N}(A_n \cosh n\beta + D_n \sinh n\beta) \cos n\xi \qquad (7)$$

$$y_s(\xi, -\beta) = B_0 - C_0\beta + D_0\xi + \sum_{n=1}^{N}(B_n \cosh n\beta - C_n \sinh n\beta) \cos n\xi$$

$$+ \sum_{n=1}^{N}(A_n \sinh n\beta + D_n \cosh n\beta) \sin n\xi \qquad (8)$$

$$x_c(\xi, \beta) = A_0 + C_0\xi - D_0\beta + \sum_{n=1}^{N}(B_n \sinh n\beta + C_n \cosh n\beta) \sin n\xi$$

$$+ \sum_{n=1}^{N}(A_n \cosh n\beta - D_n \sinh n\beta) \cos n\xi \qquad (9)$$

$$y_c(\xi, \beta) = B_0 + C_0\beta + D_0\xi + \sum_{n=1}^{N}(B_n \cosh n\beta + C_n \sinh n\beta) \cos n\xi$$

$$+ \sum_{n=1}^{N}(D_n \cosh n\beta - A_n \sinh n\beta) \sin n\xi \qquad (10)$$

The periodic range of 2π represented by $x_s(\xi, -\beta)$ or $x_c(\xi, +\beta)$ corresponds to the distance 2λ in Fig. 1, and this implies that

$$C_0 = \frac{\pi}{\lambda} \quad \text{and} \quad A_n = D_n = 0, \quad n = 0, 1, \ldots, N \qquad (11)$$

B_0 is calculated as the mean distance between the coast and seaward boundary curves.

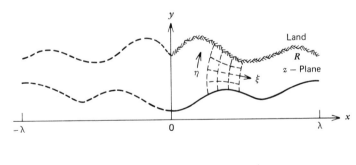

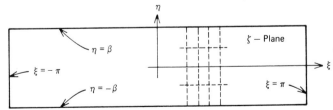

Fig. 1. The coast curves and seaward boundary to be mapped from (x, y) space to straight line boundaries in (ξ, η) space.

The range for ξ and the scale factor C_0 are free parameters in the transformation relation. However, β must be determined along with the coefficients in relations 7–10 by a curve fitting technique.

The procedure for evaluating β and the coefficients is similar to that employed by Reid and Vastano (1966). This set is to be computed for a given value of N such that equations 7–10 give a suitable approximation to the curves X_s, Y_s, X_c, and Y_c in the sense of minimizing a mean square error. These curves are the parametric representation of the given coastline and seaward boundary curves, respectively, expressed as a function of arc length. Since the latter functions are not known directly in terms of ξ, but rather in terms of arc length, the technique is an iterative one beginning with an initial approximation of arc length in terms of ξ for each curve. A convergent algorithm has been based on minimizing the error function

$$E = \frac{1}{2\pi} \int_{-\pi}^{\pi} [(Y_s - y_s)^2 + (X_s - x_s)^2 + (Y_c - y_c)^2 + (X_c - x_c)^2] \, d\xi \qquad (12)$$

An application of the transformation procedure has been made for the continental shelf regions of the Gulf of Mexico and the eastern seaboard of the United States. However, before surge computations are made, the composition of the orthogonal curvilinear grid system is stretched in both the shoreward and longshore directions to distribute the grid resolution. In this manner an economy is achieved in terms of the number of grid points required by the program while providing the highest resolution in regions of specific interest. Figure 2a shows a curvilinear mesh in (x, y) space which is a direct result of the transformation application. The corresponding mesh in (ξ, η) space is shown in Fig. 2b. In order to provide a computing grid of constant grid increments, a transformation to (S^*, T^*) space is made in two steps. The functional relationship

$$S^* = S^*(S_p(\xi)) \qquad (13)$$

is used as the basis for generating a constant ΔS^* spacing where S_p is arc length distance along the transform generated coast. Pragmatically, this relationship is generated by a choice of ΔS^* that yields the desired longshore resolution. In the shoreward direction, the travel time, T^*, for a long wave to proceed from the seaward to shore boundary along a ξ line is subdivided into a selected number of increments. This is represented by

$$T^* = T^*(S_n(\eta)) \qquad (14)$$

where S_n is distance along the ξ line and T^* is given by

$$T^* = \int_{S_n(\eta)} \frac{ds}{\sqrt{gD}} \qquad (15)$$

and $\sqrt{gD}$ is the local wave celerity. Choices of ΔS^* and ΔT^* generate the (S^*, T^*) grid schematic shown in Fig. 2c. Figure 3a presents the transformed coordinate system expressed in (x, y) space for a portion of the shelf region in the western Gulf of Mexico and Fig. 3b is the associated (S^*, T^*) grid utilized in the storm surge computations (Wanstrath, 1975). The shoreline and seaward boundaries are now identified by constant values of the coordinate T^*, while constant values of S^* identify the lateral boundaries.

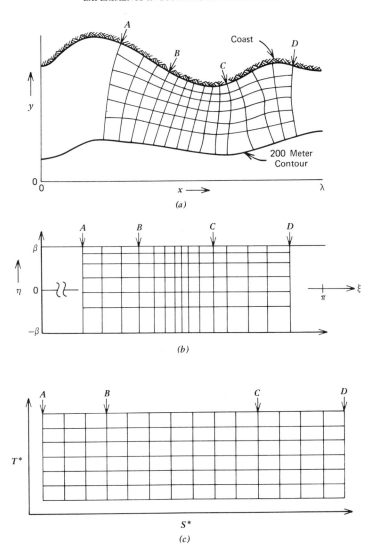

Fig. 2. (a) (x, y) space to be mapped by the transformation relation given by equations 4 and 5. (b) The transformed system represented in (ξ, η) space. (c) (ξ, η) space transformed by equations 13 and 14 to (S^*, T^*) space.

The vertically integrated equations appropriate for the surge in the (S^*, T^*) system are

$$\frac{\partial Q_{S^*}}{\partial t} - f Q_{T^*} + \frac{gD}{F\mu} \frac{\partial}{\partial S^*} (H - H_B) = \tau_{S^*} - \sigma_{S^*} \tag{16}$$

$$\frac{\partial Q_{T^*}}{\partial t} + f Q_{S^*} + \frac{gD}{Fv} \frac{\partial}{\partial T^*} (H - H_B) = \tau_{T^*} - \sigma_{T^*} \tag{17}$$

$$\frac{\partial H}{\partial t} + \frac{1}{F^2} \left[\frac{1}{\mu} \frac{\partial}{\partial S^*} F Q_{S^*}) + \frac{1}{v} \frac{\partial}{\partial T^*} (F Q_{T^*}) \right] = 0 \tag{18}$$

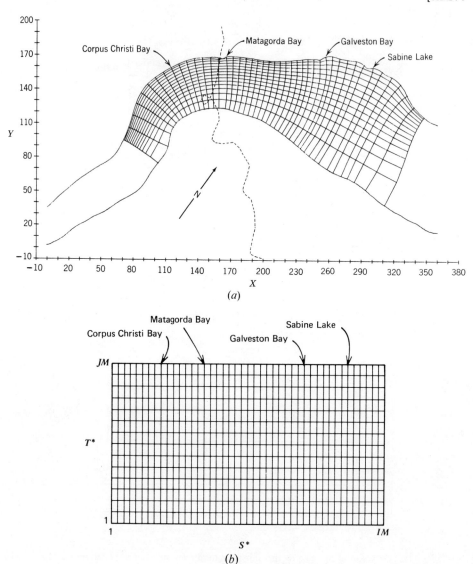

Fig. 3.   (a) A portion of the shelf region of the western Gulf of Mexico showing a grid of (ξ, η) expressed in (x, y) space. (b) The final (S^*, T^*) grid system employed for storm surge computations in this region.

where Q is the volume transport per unit width, τ is the kinematic wind stress, and σ is the bottom resistance stress, f is the Coriolis parameter, g is the local acceleration of gravity, D is the depth of the water, H is the sea-surface elevation relative to mean sea level, and H_B represents the hydrostatic elevation of the sea surface corresponding to the atmospheric pressure anomaly. The term F is the curvilinear coordinate system scale factor given by

$$ F = \left[\left(\frac{\partial x}{\partial \xi} \right)^2 + \left(\frac{\partial y}{\partial \xi} \right)^2 \right]^{1/2} \tag{19} $$

and the terms μ and v are further scale factors representing the transformation to the (S^*, T^*) system such that

$$\mu = \frac{\partial \xi}{\partial S_p} \frac{\partial S_p}{\partial S^*} \tag{20}$$

and

$$v = \frac{\partial \dot{\eta}}{\partial S_n} \frac{\partial S_n}{\partial T^*} \tag{21}$$

The relation between the kinematic wind stress and wind speed at an elevation of 10 m above the surface, W_{10}, is

$$\tau = \frac{\rho_a C_D W_{10}^2}{\rho_w} \tag{22}$$

where ρ_a is the air density, ρ_w is the water density, and C_D is a nondimensional drag coefficient with values given by

$$\frac{\rho_a}{\rho_w} C_D = K = \begin{cases} K_1 & \text{if } W_{10} < 7 \text{ m/sec} \\ K_1 + \left(1 - \frac{7}{W_{10}}\right)^2 K_2 & \text{if } W_{10} \geqslant 7 \text{ m/sec} \end{cases} \tag{23}$$

where $K_1 = 1.1 \times 10^{-6}$ and $K_2 = 2.5 \times 10^{-6}$. The bed resistance terms are

$$\sigma_{S^*} = K_0 \frac{QQ_{S^*}}{D^2} \tag{24}$$

$$\sigma_{T^*} = K_0 \frac{QQ_{T^*}}{D^2} \tag{25}$$

where K_0 is a nondimensional drag coefficient taken as 2.5×10^{-3}.

The wind-stress components in (x, y) space at a distance r from the hurricane center are given by

$$\tau_x = K V_R^2 (-y \cos \phi - x \sin \phi) G \tag{26}$$
$$\tau_y = K V_R^2 (x \cos \phi - y \sin \phi) G \tag{27}$$

where

$$G = \frac{r^2}{R^3} \qquad \text{if } r < R$$
$$= \frac{r^2}{R} \qquad \text{if } r \geqslant R \tag{28}$$

which is the analytic representation given by Jelesnianski (1965). The distance from storm center to the region of maximum winds is R, V_R is the maximum wind speed, and ϕ is the ingress angle. The movement of the storm is accounted for in this model by vector addition in the manner prescribed by Jelesnianski. The stress components in (S^*, T^*) space are given by

$$\tau_{S^*} = \tau_x \cos \theta + \tau_y \sin \theta \tag{29}$$
$$\tau_{T^*} = -\tau_x \sin \theta + \tau_y \cos \theta \tag{30}$$

where

$$\theta = \tan^{-1}\left(\frac{\partial y/\partial \xi}{\partial x/\partial \xi}\right) \tag{31}$$

The surface wind fields given by equations 26–28 were altered in the presence of land to provide deformed fields in the nearshore region (Wanstrath, 1975) for the surge computations. Figure 4a presents the symmetric wind field representation of Hurricane Carla given by the above equations. Isovel contours are shown in meters per second. The distorted wind field, which analytically conforms to the representation given by the Hydrometeorological Section of the National Weather Service, is shown in Fig. 4b.

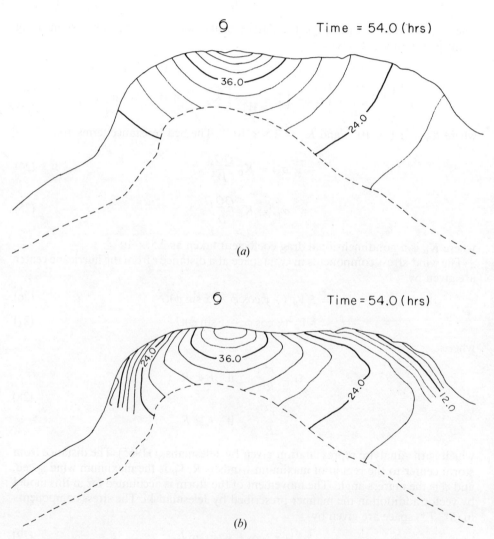

Fig. 4.   (a) Isovel values of wind speed for a symmetric model of Hurricane Carla winds. (b) Isovel values of wind speed for the deformed model of Hurricane Carla winds used in the surge computations.

The atmospheric surface pressure field was analytically represented by

$$P = P_0 + (P_\infty - P_0)e^{-R/r},\tag{32}$$

where P_0 is the central pressure of the storm and P_∞ is the associated far-field pressure. The pressure field was not deformed in the nearshore region to reflect the wind field deformation.

The distinct advantage of the (S^*, T^*) grid system lies in the inherent accuracy of the boundary conditions in curvilinear coordinates. The boundary conditions at the sea boundaries of the (S^*, T^*) grid system are similar to those selected by Jelesnianski (1965). For the lateral boundaries the gradient of the S^*-directed volume transport is assumed to vanish,

$$\frac{\partial Q_{S^*}}{\partial S^*} = 0 \tag{33}$$

and on the seaward boundary the sea-surface elevation is taken as the hydrostatic equivalent of the atmospheric pressure anomaly,

$$H = H_B \tag{34}$$

At the shoreline boundary the conditions applied to the system are based on the assumption of an infinitely high wall that is represented by no transport in the shoreward direction

$$Q_{T^*} = 0 \tag{35}$$

The finite difference algorithm is an explicit, centered-difference, leapfrog scheme (Alvarez, 1973). The computing lattice provides transport components at the same location and staggered in space and time with respect to the water level anomaly. The finite difference analogues of the conservation relations are

$$Q_{S^*}(i, j, n + 1) = \frac{G_1 G_2 + f\Delta t G_3}{G_1{}^2 + (f\Delta t)^2} \tag{16a}$$

and

$$Q_{T^*}(i, j, n + 1) = \frac{G_1 G_3 - f\Delta t G_2}{G_1{}^2 + (f\Delta t)^2} \tag{17a}$$

and

$$H(i, j, n + 1) = H(i, j, n - 1) - \frac{\Delta t}{F(i, j)^2}$$

$$\times \left[\frac{F(i + 1, j)Q_{S^*}(i + 1, j, n) - F(i - 1, j)Q_{S^*}(i - 1, j, n)}{\mu(i)\Delta S^*} \right.$$

$$\left. + \frac{F(i, j + 1)Q_{T^*}(i, j + 1, n) - F(i, j - 1)Q_{T^*}(i, j - 1, n)}{\nu(j)\Delta T^*} \right] \tag{18a}$$

where i, j, and n are indices related to S^*, T^*, and time, respectively, Δt is the time step, and

$$G_1 = \frac{1 + [2K_0 \Delta t Q(i, j, n - 1)]}{\bar{D}^2} \tag{36}$$

$$G_2 = Q_{S*}(i, j, n - 1) + f \Delta t Q_{T*}(i, j, n - 1) - \frac{g \Delta t \bar{D}}{\mu(i) F(i, j) \Delta S^*}$$

$$\times [H(i + 1, j, n) - H_B(i + 1, j, n) - H(i - 1, j, n) + H_B(i - 1, j, n)] + 2\Delta t \tau_{S*} \tag{37}$$

$$G_3 = Q_{T*}(i, j, n - 1) - f \Delta t Q_{S*}(i, j, n - 1) - \frac{g \Delta t \bar{D}}{v(j) F(i, j) \Delta T^*}$$

$$\times [H(i, j + 1, n) - H_B(i, j + 1, n) - H(i, j - 1, n) + H_B(i, j - 1, n)] + 2\Delta t \tau_{T*} \tag{38}$$

$$\bar{D} = \frac{D(i + 1, j, n) + D(i - 1, j, n) + D(i, j + 1, n) + D(i, j - 1, n)}{4} \tag{39}$$

and

$$Q = [Q_{S*}(i, j, n - 1)^2 + Q_{T*}(i, j, n - 1)^2]^{1/2}$$

The lateral boundary condition analogues are given by

$$Q_{S*}(1, j, n + 1) = Q_{S*}(3, j, n + 1)$$
$$Q_{S*}(IM, j, n + 1) = Q_{S*}(IM - 2, j, n + 1) \tag{33a}$$

for odd j where IM is the total number of S^* grid points. The fluxes Q_{S*} along $i = 2$ and $i = IM - 1$ for even j are obtained from an average of the four neighboring Q_{S*} values. The seaward boundary condition is

$$H(i, 1, n + 1) = H_B(i, 1, n + 1) \tag{34a}$$

for even values of i. The values of H for odd i along $j = 2$ are obtained using a four-point average of neighboring surge heights. A shoreline boundary condition analogue, consistent with equation 35, is

$$Q_{T*}(i, JM, n) = 0 \tag{35a}$$

taken at odd values of the index i, where JM is the total number of T^* grid points. At alternate values of i, the required water elevations are computed on the basis of volume conservation by specifying the value $Q_{T*}(i, JM + 1, n)$ as the negative of $Q_{T*}(i, JM - 1, n)$.

The dashed line in Fig. 3a is the track of Hurricane Carla which crossed the Texas coastline near Port O'Connor on September 11, 1961. The relevant general characteristics of this storm are a central pressure of 940 mbar, a far-field pressure of 1016 mbar, maximum winds of 51.5 m/s, an ingress angle of 30°, and a radius to maximum winds of 42.6 km. The forward speed of the storm along its track varies from 3.1 to 4.5 m/s. The numerical model was employed for a time prototype span of 66 hr with a time step of 180 sec. Two examples of open-coast hydrographs are shown in Figs. 5 and 6 for Pleasure Pier at Galveston and Sabine Pass. The solid line represents the computed shelf surge. The observations are extracted from the U.S. Army Corps of Engineers

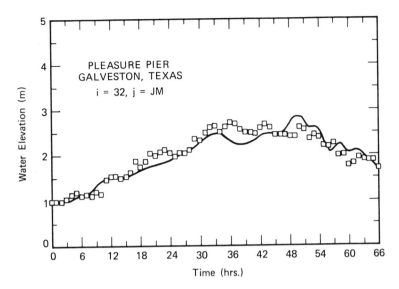

Fig. 5. Hydrograph for the shelf surge at Pleasure Pier, Galveston, Texas. The solid line represents the numerical computations and open squares represent observed data.

Hurrican Carla report (Galveston District, 1963) and are shown as squares. The computed maximum surge was 3.8 m approximately 46.3 km northeast of the entrance to Matagorda Bay, and the observed coastal high water measured by the U.S. Army Corps of Engineers was 3.7 m. Although the computations agree fairly well with those observed for the peak coastal surge and for Galveston, there is some discrepancy for regions far from the storm track such as Sabine Pass. This may be attributed to

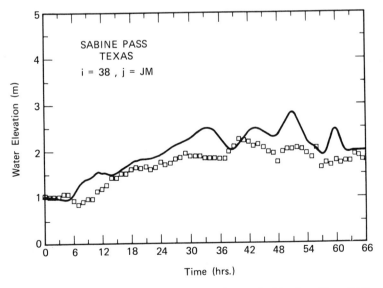

Fig. 6. Hydrograph for the shelf surge at Mud Bayou southwest of Sabine Pass. The solid line represents the numerical computations and open squares represent observed data.

the problem of proper deformation of the wind field (both speed and direction) to reflect the influence of land. This is a subject of continuing research.

The simulation of this hurricane constitute the initial verification of surge computations in transformed coordinates. The advantages of formulating the finite difference equations for the (S^*, T^*) grid arise from the ease and accuracy in applying the boundary conditions. Additionally, this is accompanied by a significant economy of program logic and computer resources. Further verifications and applications are planned for the Gulf of Mexico and the eastern seaboard of the United States.

3. Boundary Conditions in One Dimension

In view of the fact that accurate shelf surge calculations are necessary to provide the proper dynamic time history on the seaward boundary of an estuarine model, the effects of the seaward and shoreline boundary conditions on such solutions are under study with a time-dependent, quasi-one-dimensional surge algorithm. The model is similar to the one in Section 2,

$$\frac{\partial Q_x}{\partial t} - fQ_y = \tau_x - \tau\sigma_x \tag{40}$$

$$\frac{\partial Q_y}{\partial t} + fQ_x + gD\frac{\partial}{\partial y}(H - H_B) = \tau_y - \sigma_y \tag{41}$$

$$\frac{\partial H}{\partial t} + \frac{\partial Q_y}{Qy} = 0 \tag{42}$$

where x is directed longshore. An assumption of uniformity has been applied to Q and H in this direction, eliminating the associated gradients in equations 40 and 42. An explicit algorithm which is a reduced version in (x, y) space of the one selected for the computations in Section 2 provides for a leapfrog computation of H and Q along the traverse shown in Figs. 7 and 8. The underwater topography along this line generally represents the bottom profile from a depth of 2 km offshore across a continental shelf and into a bay along the Gulf of Mexico. Within the bay the numerical computations are reduced to the simultaneous solution of equations 41 and 42 with the omission of the Coriolis term. The storm characteristics are typical of a hurricane which might strike the northwestern coastline of the Gulf of Mexico. This storm is employed throughout the series of experiments and is represented by the symmetric model given by equations 26–28 and 32.

In order to establish a storm surge case that will be a standard for comparisons, the condition

$$H = 0 \tag{43}$$

is chosen for the seaward boundary, and a coupling condition is implied at the shoreline in the traverse shown in Fig. 7. The quasi-two-dimensional algorithm equations 40–42 are joined with the one-dimensional bay by continuous computations of H and Q_y. The seaward boundary condition, equation 43, implies that a node in the water elevation exists at this point of the traverse. Figure 9 represents the dynamic conditions in this case immediately after the storm has crossed the shoreline. The hydrographs shown in Fig. 10 pertain to the two positions indicated by ① and ② in Fig. 9. These curves appear as dashed lines on the hydrographs of succeeding cases. A distinct characteristic of this solution is the drawdown in water elevation caused by the convergence of water in the vicinity of the hurricane eye.

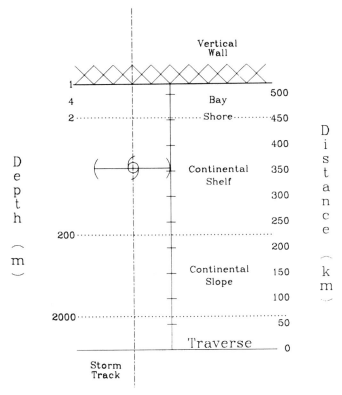

Fig. 7.   The coordinate system for quasi-one-dimensional shelf surge computations. The storm and radius to maximum winds is shown by the figure ←↺→.

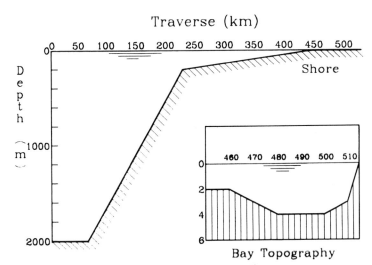

Fig. 8.   Underwater topography along the traverse line shown in Figure 7.

Wind Stress Vectors

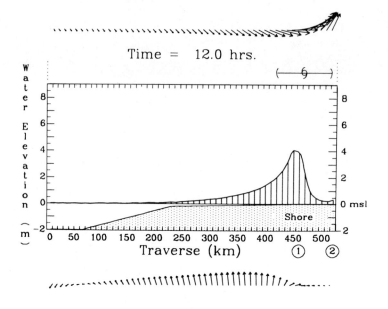

Flow Vectors

Fig. 9.   Water elevation, wind and flow (transport) vectors for case 1a as the storm crosses the coastline.

Hydrograph Records

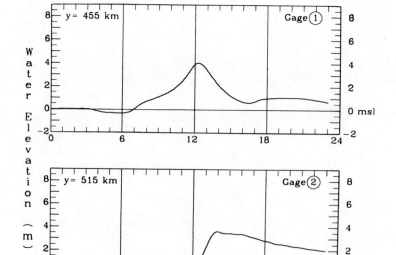

Elapsed Time (hrs.)

Fig. 10.   Hydrographs, at positions ① and ② indicated in Figure 9, for case 1a.

A series of numerical experiments have been carried out to examine the effect of boundary conditions on the storm surge. Two shoreline boundary conditions are considered here: (1) a one-dimensional bay and (2) a wall. The seaward boundary conditions employed are (*a*) a node in the water elevation and (*b*) a radiation condition accounting for a barometric forcing function.

Case 2*a* represents an alteration in the shoreline boundary relative to 1*a*. The condition

$$Q_y = 0 \qquad\qquad (44)$$

replaces the coupling to the bay and implies a wall of infinite height that permits no flow into the bay. The boundary condition in this case is the most prevalent one in surge models. Figure 11 corresponds to Fig. 9 of case 1*a* and indicates an increased buildup of water at the shoreline wall, and Fig. 12 compares the hydrographs for 1*a* and 2*a*. The drawdown is greater in 1*a* since the volume of water in the bay is not available. The implication of this experiment is that computations with a wall at the shoreline boundary result in significant overestimates of the maximum water elevation in topographic situations where low-lying regions are shoreward of the coastline. More importantly, the second-stage computations discussed in the introduction should be forced with water elevation and flow time histories rather than simple elevations. This difference may affect the amount and phasing of inundation in second-stage computations.

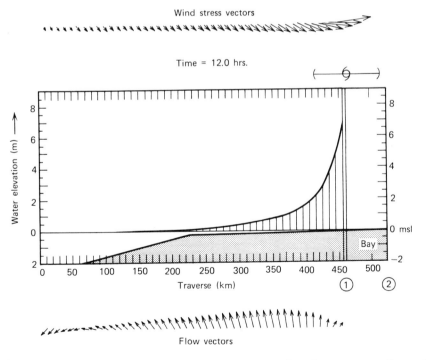

Fig. 11.   Water elevations, wind and flow (transport) vectors for case 2*a* as the storm crosses the coastline.

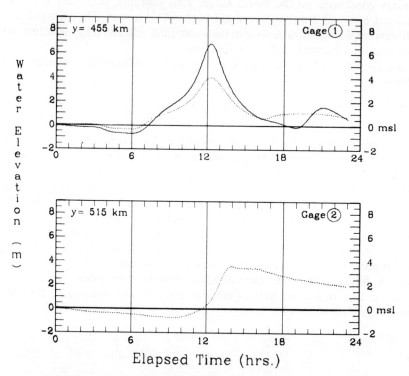

Fig. 12.   Hydrographs, at positions ① and ② indicated in Figure 9, for case 2*a*.

Case 1*b* introduces a condition at the seaward boundary that accounts for the barometric term, H_B. This condition specifies the water elevation at the boundary

$$H = H_B - Q_y/\sqrt{gD} \tag{45}$$

In addition to the boundary condition, the water elevation corresponding to the atmospheric pressure anomaly was applied to the entire traverse as an initial condition. The hydrographs for this case are shown in Fig. 13. The water elevation at the shoreline is quite similar to case 1*a* in the region of maximum surge and in the recession stage. The slight differences are related to the initial conditions. In the early stages of the storm a significant difference exists because no drawdown occurs seaward of the shoreline. The presence of the initial setup of the water surface is primarily responsible for eliminating the drawdown in case 1*b*. The hydrographs for the final case, 2*b*, are shown in Fig. 14 and bear out this statement. In this case, the volume of water from the bay is eliminated from the computations and the character of the shoreline hydrograph is similar to that of 1*b* in the early stages. The implication is clear: to realistically counteract offshore winds in the early stages of numerical surge computations, the effect of the atmospheric pressure anomaly must be taken into account. Further experiments are being carried out with this simple one-dimensional model to investigate nonlinear shoreline boundary conditions that represent flooding.

Hydrograph Records

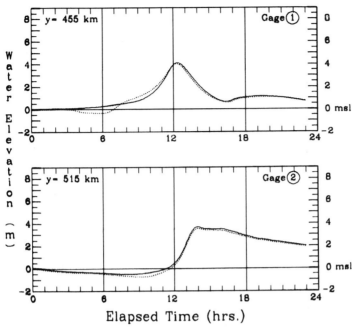

Fig. 13.   Hydrographs, at positions ① and ② indicated in Figure 9, for case 1*b*.

Hydrograph Records

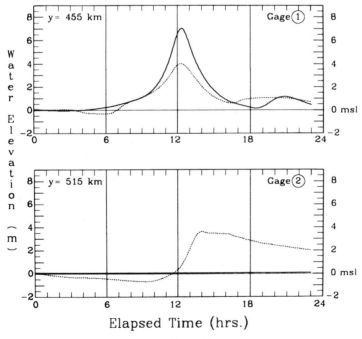

Fig. 14.   Hydrographs, at positions ① and ② indicated in Figure 9, for case 2*b*.

4. Simulation of Subgrid Scale Obstacles

Applying numerical models to bays and estuaries poses difficult dynamic problems, particularly when dense vegetative canopies extend over large portions of the wetted area. A numerical model has been developed which simulates flooded marshes as an ensemble of subgrid scale obstacles (Reid and Whitaker, 1976). If the obstructions are submerged, the model treats the wind-driven canopy flow as a two-layer system with the interfacial stress formulated in terms of a coupling coefficient and the flow differential. The resistance offered by the individual canopy elements is parameterized in terms of a drag coefficient and the dimensional properties of the obstacles. Additionally, the dynamic influence of the canopy elements extending into the air is introduced through a sheltering factor when any portion of the canopy is above water.

The obstructions are assumed to be rigid elements of specified width w and height b oriented normal to the flow and distributed evenly over the bottom. When the depth D is greater than b, the integrated equation of motion for the lower layer is

$$\frac{\partial \mathbf{U}_1}{\partial t} + gb\nabla H = \tau_c - \tau_B - \mathbf{F}_c \tag{46}$$

while, for the region above the obstructions,

$$\frac{\partial \mathbf{U}_2}{\partial t} + g(D - b)\nabla H = \tau_s - \tau_c \tag{47}$$

where the subscripts 1 and 2 refer to the lower and upper layers, respectively.

Equations 46 and 47 may be combined to give

$$\frac{\partial \mathbf{U}}{\partial t} + gD\nabla H = \tau_s - \tau_B - \mathbf{F}_c \tag{48}$$

where $\mathbf{U}$ is the combined flux $\mathbf{U}_1 + \mathbf{U}_2$.

The continuity requirement for the total water column is

$$\frac{\partial H}{\partial t} + \nabla \cdot \mathbf{U} = 0 \tag{49}$$

As in the preceding sections, the stresses are taken in quadratic form. Explicitly,

$$\tau_B = f_1|\mathbf{u}_1|\mathbf{u}_1 \tag{50}$$

and

$$\tau_c = f_2|\mathbf{u}_2 - \mathbf{u}_1|(\mathbf{u}_2 - \mathbf{u}_1) \tag{51}$$

where $\mathbf{u}_1$ and $\mathbf{u}_2$ are depth mean velocities for the respective layers. The coefficients f_1 and f_2 are nondimensional. The resistance per unit horizontal area afforded the flow in the lowest layer is

$$\mathbf{F}_c = C_D wbN|\mathbf{u}_1|\mathbf{u}_1 \tag{52}$$

where N is the number of obstructions per unit horizontal area, and C_D is a nondimensional drag coefficient. Clearly, the interfacial stress τ_c vanishes and b is replaced by D in equation 52 if $D \leqslant b$.

A sheltering coefficient S modifies the ordinary quadratic kinematic wind stress relation to model the dynamic influence of the canopy elements when $D < b$.

Requiring that the windstress be continuous at the elevation b and assuming a quadratic form of the resistance to the wind presented by the individual elements, the nondimensional coefficient S takes the form

$$S = \frac{1}{1 + C_D N w h_c / K} \tag{53}$$

where $h_c = b - D$ if $D < b$, or $h_c = 0$ if $D \geqslant b$ and K is the wind stress coefficient (Reid and Whitaker, 1976). Equation 53 is valid if $D < b$, and as required, S approaches unity as $N w h_c$ vanishes.

The numerical integration of the analogue of the scalar conservation equations is performed on a single Richardson lattice (Platzman, 1963; Reid and Bodine, 1968).

The computation yields values of H at odd-half steps of x and y, and even-half time steps. The x-directed transports U and U_1 are evaluated for even-half steps of x, odd-half steps of y, and odd-half time steps. The y-directed fluxes V and V_1 are obtained for the odd-half steps of x, even-half steps of y, and odd-half time steps.

The bottom elevations Z, relative to mean water level, are specified as permanent storage for the same locations as H to facilitate the required evaluations of $D = Z - h$, and the thickness of the layer above the canopy. In addition, the wind-stress components are provided at spatial locations consistent with U, U_1, and V, V_1, respectively, but at even-half time steps.

We obtain the recursion equations for U, U_1, V, V_1, and H by taking the scalar form of 46, 48, and 49, and approximating the spatial and time derivatives with centered differences:

$$U'(j, k + \tfrac{1}{2}) = \frac{[(1 + \gamma_{1x})\overline{(D - b)_x} + \gamma_{2x}\overline{D}_x]U^* - \gamma_{1x}\overline{(D - b)_x}U_1^*}{[(1 + \gamma_{1x})\overline{(D - b)_x} + \gamma_{2x}\overline{D}_x + \gamma_{1x}\gamma_{2x}b]} \tag{54}$$

$$U_1'(j, k + \tfrac{1}{2}) = \frac{\gamma_{2x}bU^* + \overline{(D - b)_x}U_1^*}{[(1 + \gamma_{1x})\overline{(D - b)_x} + \gamma_{2x}\overline{D}_x + \gamma_{1x}\gamma_{2x}b]} \tag{55}$$

$$V'(j + \tfrac{1}{2}, k + 1) = \frac{[(1 + \gamma_{1y})\overline{(D - b)_y} + \gamma_{2y}\overline{D}_y]V^* - \gamma_{1y}\overline{(D - b)_y}V_1^*}{[(1 + \gamma_{1y})\overline{(D - b)_y} + \gamma_{2y}D_y + \gamma_{1y}\gamma_{2y}b]} \tag{56}$$

$$V_1'(j + \tfrac{1}{2}, k + 1) = \frac{\gamma_{2y}bV^* + \overline{(D - b)_y}V_1^*}{[1 + \gamma_{1y})\overline{(D - b)_y} + \gamma_{2y}\overline{D}_y + \gamma_{1y}\gamma_{2y}b]} \tag{57}$$

and

$$H'(j - \tfrac{1}{2}, k + \tfrac{1}{2}) = H(j - \tfrac{1}{2}, k + \tfrac{1}{2}) + [U'(j, k + \tfrac{1}{2}) - U'(j - 1, k + \tfrac{1}{2})$$

$$+ V'(j - \tfrac{1}{2}, k + 1) - V'(j - \tfrac{1}{2}, k)]\frac{\Delta t}{\Delta x} \tag{58}$$

where in equations 54 and 55

$$U_1^* \equiv U_1(j, k + \tfrac{1}{2}) - gb[H(j + \tfrac{1}{2}, k + \tfrac{1}{2}) - H(j - \tfrac{1}{2}, k + \tfrac{1}{2})]\frac{\Delta t}{\Delta x} \tag{59}$$

and

$$U^* \equiv U(j, k + \tfrac{1}{2})$$

$$+ \left\{ T_{sx}(j, k + \tfrac{1}{2}) + g\bar{D}_x[H(j + \tfrac{1}{2}, k + \tfrac{1}{2}) - H(j - \tfrac{1}{2}, k + \tfrac{1}{2}] \frac{1}{\Delta x} \right\} \Delta t \quad (60)$$

The primed quantities are at the new time level and overbars denote the spatial averages,

$$\bar{D}_x = \tfrac{1}{2}[D(j + \tfrac{1}{2}, k + \tfrac{1}{2}) + D(j - \tfrac{1}{2}, k + \tfrac{1}{2})] \quad (61)$$

and

$$\overline{(D - b)}_x = \tfrac{1}{2}[(D - b)(j + \tfrac{1}{2}, k + \tfrac{1}{2}) + (D - b)(j - \tfrac{1}{2}, k + \tfrac{1}{2})] \quad (62)$$

Additionally,

$$\gamma_{1x} \equiv (f_1 + C_D NA) \frac{|\mathbf{u}_1|}{b} \Delta t \quad (63)$$

and

$$\gamma_{2x} \equiv f_2 \frac{|\mathbf{u}_2 - \mathbf{u}_1|}{b} \Delta t \quad (64)$$

The expressions for V_1^*, V^*, $\bar{D}_y$, $\overline{(D - b)}_y$, γ_{1y}, and γ_{2y} used in equations 56 and 57 are counterparts to those given in equations 59–64. Equations 54–58 constitute the canopy analogue.

As a verification, a one-dimensional version of the model was applied to a laboratory study by Tickner (1957). Tickner reported the steady-state, wind-induced, water surface profile at six points along a laboratory channel. For the first set of observations, the channel bottom was smooth. Subsequent sets of measurements were made with the bottom roughened by placing strips of ordinary wire screen normal to the channel axis.

The parameters characterizing the obstacles required for numerical simulation of Tickner's results were either provided by Tickner or were measured directly. The empirical coefficients f_2, K, and C_D were evaluated by calibrating the model to the results obtained by Tickner (Reid and Whitaker, 1976). The coefficient f_1 was taken as 0.005 and calibration yielded a C_D of 1.77 and an f_2 of 0.00146. The coefficient K was expressed as a function of the mean water depth and wind speed. Generally, K varies as the depth and wind speed over the range 2×10^{-6} to 4×10^{-6}.

Figure 15 gives the comparison between the computed steady-state profile (solid line) and the observed profiles (circles) with a wind speed of 9.15 m/sec, an obstacle spacing of 0.061 m, and height of 0.035 m (horizontal dotted line), and three mean depths. The dashed line indicates the surface profile observed over a smooth bottom. The agreement obtained between the observed and simulated surface profiles indicates the model is sufficiently calibrated. The associated longitudinal distribution of the depth mean speeds in the lower (u_1) and upper (u_2) layer are shown in Fig. 16.

The first two-dimensional application of the model was to a flat bottom, enclosed,

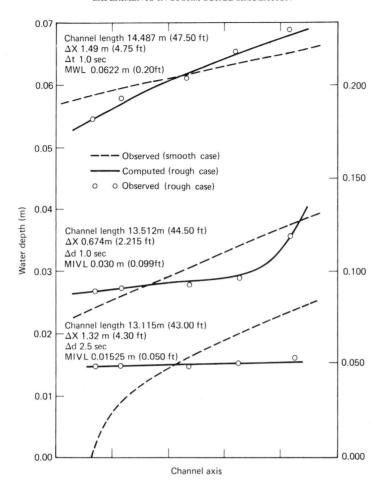

Fig. 15. Comparison of computed and observed surface profiles for the rough bottom channel with a screen spacing of 0.061 m and a wind speed of 9.15 m/sec. The dashed line represents the smooth channel case.

rectangular basin driven by a uniform wind. A field of vertical elements representing dense vegetation was specified over half the basin. For one case investigated, the wind was directed toward the modeled marsh whereas in another instance an opposite wind was utilized. The last case assumed the wind to be parallel to the marsh open water boundary. Figures 17a and 17b are examples of the surface topography and flow regime obtained with the obstacle height greater than the mean water depth. The wind is directed toward the top of the figure with the simulated marsh on the left-hand side of the basin. The small anticyclonic cell depicted in Fig. 17b developed slowly from the general cyclonic circulation as the simulated vegetation in the upper end of the basin was submerged. Interestingly, a cyclonic flow can also be obtained by applying a uniform wind to a laterally sloping bottom basin free of vegetation.

Another two-dimensional application of the subgrid scale obstacle model concerns a transient phenomenon. During an attempt to model the October 1950 storm

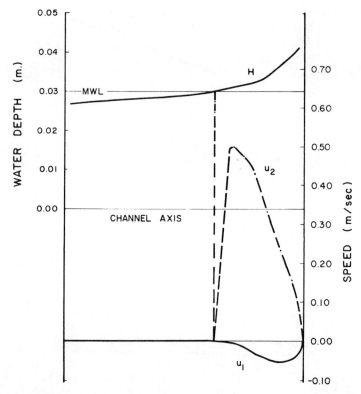

Fig. 16.   Surface profile and current speeds in lower (u_1) and upper (u_2) layers obtained with a screen spacing of 0.061 m and wind speed of 9.15 m/sec.

surge in Lake Okeechobee, Florida, with an existing surge algorithm, it became apparent that the dense marsh in the southwest quadrant of the lake exerted considerable influence on the circulation (Whitaker, Reid, and Vastano, 1973). Adjusting the bottom friction factor for the vegetated region proved to be inadequate. Attempts to obtain reasonable agreement between computed and observed water levels at a hydrometeorological station (HGS-1) situated on the southwestern periphery of the vegetated region were particularly futile. The center of the October 1950 hurricane moved from south to north across the lake. Hence HGS-1 experienced prolonged high winds coming off long reaches of the marsh.

The surge model applied to Lake Okeechobee was modified to interrupt the normal sequence of instructions for those computational points located in the marsh. Computations at these exceptional points were made with the canopy routine. Figure 18 shows the observed water levels at HGS-1 and the computed water levels obtained with and without the simulated marsh. The difference in the results obtained from the two computational modes is significant.

The applications of the canopy model have been encouraging. However, further research is required to define those parameters related to the physical characteristics of realistic subgrid scale features. In particular, the technique may hold promise for surge computations with man-made obstacles such as buildings during surge flood stages.

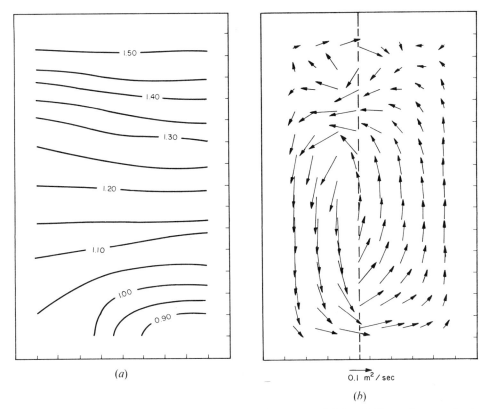

Fig. 17. (a) The surface topography at time level 330 with a canopy height of 1.28 m and a mean depth of 1.22 m, contour interval 0.05 m. (b) The transport vector field at time level 329.5 with a canopy height of 1.28 m and a mean depth of 1.22 m.

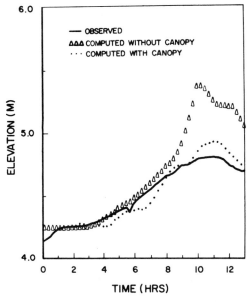

Fig. 18 Comparison of observed and computed water levels in Lake Okeechobee. The observed levels were recorded during the October 1950 hurricane at the leeward edge of a densely vegetated area.

Acknowledgments

The majority of the research presented in this chapter was sponsored by the Civil Engineering Research Center, U.S. Army Corps of Engineers, through the Texas A & M Research Foundation. The National Center for Atmospheric Research (sponsored by the National Science Foundation) provided grants for the computations. The authors gratefully acknowledge the assistance of the computer facility staff at NCAR, and thank Dr. Alan Cline and Mr. Tom Wright of NCAR in particular for their contributions and interest.

References

Alvarez, J. A. 1973. Numerical prediction of storm surges in the Rio De la Plata area. Ph.D. Thesis, The University of Buenos Aires, 70 pp. plus tables and figures.

Dronkers, J. J., 1969. Tidal computations for rivers, coastal areas, and seas. *J. Hydraul. Div., Proc. Am. Soc. Civ. Eng.*, **95**, (HY1), 29–77.

Freeman, J. C., 1951. The solution of nonlinear meteorological problems by the method of characteristics. In *Compendium of Meteorology*, American Meteorological Society, pp. 421–433.

Galveston District, U.S. Army Corps of Engineers, 1963. *Hurricane Carla.*

Hansen, W., 1956. Theorie zur Errechnung des Wasserstandes und der Strömungen in Randmeeren Nebst Anwendungen. *Tellus*, **8**, 287–300.

Harris, D. L. 1963. Characteristics of the hurrican storm surge. *U.S. Weather Bur. Tech. Paper No. 48.*

Heaps, N. S., 1971. On the numerical solution of the three-dimensional hydrodynamical equations for tides and storm surges. *Mém. Soc. Sci. Liège*, **1**, 143–180.

Jelesnianski, C. P., 1965. A numerical calculation of storm tides induced by a tropical storm impinging on a continental shelf. *Mon. Weather Rev.*, **93**, 343–358.

Jelesnianski, C. P., 1972. SPLASH (special program to list amplitudes of surges from hurricanes). I. Storms reaching land. *Natl. Weather Serv. Ref. TDL-46.*

Leendertse, J. J., 1970. A water-quality simulation model for well-mixed estuaries and coastal seas: Volume I, principles of computation. *Memo. RM-6230-RC*, The Rand Corporation, Santa Monica, Calif.

Platzman, G. W., 1963. The dynamical prediction of wind tides on Lake Erie. *Am. Meteorol. Soc.*, **26** (4), 1–99.

Reid, R. O., 1957. Forced and free surges in a narrow basin of variable depth and width: a numerical approach. *A & M College of Texas, Dept. Oceanogr. Tech. Rept. Ref. 57-25T.*

Reid, R. O. and B. R. Bodine, 1968. Numerical model for storm surges in Galveston Bay. *J. Waterw. Harbors Div., Proc. Am. Soc. Civ. Eng.*, **94**, (WW1), 33–57.

Reid, R. O. and A. C. Vastano, 1966. Orthogonal coordinates for analysis of long gravity waves near islands. *Santa Barbara Speciality Conference in Coastal Engineering, Proc. Am. Soc. Civ. Eng.*, pp. 1–20.

Reid, Robert O. and Robert E. Whitaker, 1976. Wind-driven flow of water influenced by a canopy. *J. of the Waterways Harbors and Coastal Engineering Division, Am. Soc. Civ. Eng.*, Vol. 102, No. WW1, Proc. Paper 11926, Feb. 1976, pp. 61–77.

Tickner, E. G., 1957. Effect of bottom roughness on wind tide in shallow water. *U.S. Army Beach Erosion Board Tech. Mem. 95.*

Wanstrath, J. J., 1975. Storm surge simulation in transformed coordinates. Ph.D. Thesis, Texas A & M University.

Welander, P., 1961. Numerical prediction of storm surges. *Advan. Geophys.*, **8**, 316–379.

Whitaker, R. E., R. O. Reid, and A. C. Vastano, 1973. Drag coefficient at hurricane wind speeds as deduced from numerical simulation of dynamical water level changes in Lake Okeechobee. *Texas A & M Univ., Dept. Oceanogr. Tech. Rept. 73-13-T.*

6. USE OF TRACERS IN CIRCULATION STUDIES

GEORGE VERONIS

1. Introduction

Apart from wave studies most investigations in dynamical oceanography involve the use of parametric representations of processes with time and space scales that lie outside the range appropriate to the phenomenon under study. This is especially true of investigations involving the distribution of tracers such as salinity, temperature, dissolved oxygen, nutrients, and stable and radioactive chemicals. When the models are steady, as they often are for studies of large-scale flows, parametric treatment of transport processes is unavoidable. Hence the assessment of models involving the joint tracer-circulation problem should rest as strongly on what one learns about transport processes as it does on the agreement between observed and deduced distributions. The situation will remain so until the nature of transport mechanisms is much better understood than it is at the present time. Therefore, one can look upon tracer-circulation models as complements to the more mechanistic explorations of transport processes. It is this aspect of the problem that is stressed here.

Early studies of tracer distribution were motivated by the desire to deduce the circulation by indirect methods because mean velocities in the open ocean are too slow to be obtained by direct measurements. Thus Wüst (1928), Möller (1929), and Sverdrup (1930) used the observed distributions of salinity and temperature to identify continuous tongues or cores of water of similar characteristics. They associated these tongues with mean flows over large horizontal distances and gradually developed a picture of the mean circulation, particularly in the deeper layers of the ocean. Investigations of this type have continued to the present. A recent example is a very careful study by Reid and Lynn (1971) of the penetration of deep water from the polar regions of the Atlantic to the Indian and Pacific Oceans.

The significant contribution of such investigations is the identification of water of a given type at a point far removed from the source. Clearly the water has somehow penetrated to a distant point. However, the path of a water particle and the processes important for the redistribution cannot be determined from the observed tracer distribution alone. Thus the presence of Mediterranean water in the western North Atlantic does not necessarily imply a westward mean velocity from Gibraltar. It may be that outflow from the Mediterranean turns northward along the Portuguese coast and is diffused westward along isopycnic surfaces. The actual process can be determined only from models that include dynamically consistent velocities and other suitably constrained mechanisms. A few such studies are described in the following sections.

The last decade or so has seen a great deal of activity in the use of vertical distributions of tracers as a measure of vertical transfer processes. Basic to these models is the neglect of horizontal processes and a balance between vertical convection and diffusion. Depending on the tracer, radioactive decay and consumption or production terms may also be included. A key parameter that emerges is the scale height, K/W, where K is the vertical diffusion coefficient and W is the vertical velocity. (Both these parameters are normally assumed constant.) The vertical tracer profile provides a measure of the scale height and then some additional process allows a determination of K and W separately.

Under the assumed conditions and with no dynamical constraints this procedure is largely an exercise in rationalizing the data, because none of the parameters are externally determined. Furthermore, the results are not easily interpreted because the scale height varies both horizontally and vertically. Nevertheless, this approach has resulted in several careful studies of data and it has provided some impetus toward a search for situations that are more appropriate to the basic assumptions of the model. Also, a number of physical, chemical, and biological processes pertinent to the model have been subjected to careful scrutiny in order to provide more reliable figures for required balances. Section 2 contains a discussion of some of the general features of this approach.

This is followed by a brief description of the balances assumed in studies of the thermohaline circulation. These thermocline models involve geostrophic balance and usually include vertical diffusion in the heat equation. The aim is to deduce the processes important for the maintenance of the thermocline. A scaling analysis, more or less summarizing the results of the different approaches to the problem, is outlined. The conclusion is that thermocline models in their present form have served to provide dynamically consistent balances for the thermocline processes but they have not narrowed the allowable range of parameters to the point where they provide a really significant test of the different processes. Nevertheless, it is possible to distinguish between regions where diffusion appears to be necessary and those where advective processes suffice.

A recent study by Rooth and Östlund (1972), summarized in Section 4, has made use of observed vertical distributions of temperature and tritium to account for the dynamics and, by a clever manipulation of the advective-diffusive equations, they have obtained an upper bound for the vertical mixing coefficient in the water below the $18°$ water in the Sargasso Sea. This upper bound of K (≈ 0.2 cm^2/sec) is smaller than most estimates by other methods. Rooth and Östlund also point out that horizontal processes probably dominate the distributions of temperature and tritium in deeper water but that the use of a vertical advective-diffusive balance leads to the usual magnitude for K (≈ 1 cm^2/sec).

Section 5 summarizes the results of the horizontal advective-diffusive models for determining the oxygen distribution in abyssal waters when Stommel's (1958) abyssal circulation model is used for the velocity field. Arons and Stommel (1967) made the first such calculation for a relatively crude model of the North Atlantic. By considering many different combinations of the various parameters they obtained a best fit with the observed oxygen distribution for a horizontal mixing coefficient in the range 6.0×10^6–7.4×10^6 cm^2/sec and an oxygen consumption coefficient in the range 2×10^{-3}–2.5×10^{-3} ml/l/yr. More exact calculations by Kuo and Veronis (1970, 1973) for a world model lead to the same best-fit values. An important conclusion that emerges from the latter study is that the horizontal distribution of oxygen is strongly dependent on lateral mixing and that a naïve use of the oxygen gradient to deduce a mean flow direction yields a velocity of the wrong sign.

Because of the importance of numerical methods when more extensive calculations are made, the method of approximating convective derivatives by weighted differences is described briefly in Section 6. The approach taken by Kuo and Veronis (1973) is a special form of a more general treatment used by Fiadeiro (1975). The general procedure is outlined for a one-dimensional form of the advective-diffusive equation.

Fiadeiro (1975) carried out a three-dimensional calculation for the distribution of tracers, using Stommel's abyssal circulation model for the velocities. His model requires observed values of the tracer along the upper boundary of the abyssal layer, and in this way it is more constrained by observations than the Kuo–Veronis calculation. The two calculations show very similar distributions of oxygen in abyssal waters and thereby provide an upper bound for the intensity of vertical mixing.

The chapter concludes with a critique of Holland's (1971) three-dimensional model for a complete dynamical system with passive tracers as part of the calculation. It is shown that an important feature of Holland's calculation, namely, sinking of water through the thermocline in most of the open ocean, is a direct result of his use of a large diffusion coefficient along geopotential surfaces. Along the western boundary current where density surfaces slope sharply this mixing forces an extremely strong upwelling, which requires that abyssal water be supplied by sinking of surface water over most of the rest of the ocean. Hence the use of lateral mixing along isopycnic, rather than geopotential, surfaces is strongly suggested. The tracer distribution obtained in Holland's calculation is also strongly controlled by horizontal diffusion, as can be seen by comparison with the results of Kuo and Veronis for a strongly diffusive case.

2. Vertical Advective-Diffusive Models with no Dynamical Balance

Most of the investigations involving tracers have been oriented either toward deducing gross features of the circulation from the tracer distribution or toward obtaining gross features of the tracer distribution from assumed simple flow patterns. An example of the former approach is Wüst's (1935) proposed core method in which he traced high-salinity Mediterranean water to distant regions of the North Atlantic. Examples of the latter include studies by Defant (1929) and Thorade (1931), in which a distinctive feature of a property is introduced into a region by horizontal advection and is eroded downstream by vertical diffusion. This role of vertical diffusion or mixing came to be accepted more or less as a basic mechanism after Sverdrup et al. (1942, pp. 143–146) demonstrated how T–S curves for observed soundings could be explained very plausibly by vertical mixing. In particular, they improved upon Wüst's observation by showing how the decreasing salinity anomaly with distance from the Mediterranean could be rationalized by the vertical mixing argument.

More recent studies have focused on vertical exchange processes in order to determine the magnitude of the pertinent parameters. Koczy (1958) used geochemical tracers to obtain an estimate of the vertical velocity. Wyrtki (1962) calculated vertical profiles of oxygen and compared them with observed profiles to determine the vertical distribution of oxygen consumption. He concluded that oxygen minima are due to biochemical processes but that the vertical position of the minimum is related to the circulation. Munk (1966) estimated the scale depth of salinity and temperature in the water layer below the thermocline and then made use of radioactive tracer distributions to determine the magnitudes of the vertical velocity and vertical mixing coefficient.

Craig (1969) incorporated most of the important features of these vertical exchange models and made several important observations. Ignoring all horizontal processes and assuming a steady state, he started with the following equation for the concentration, C, of a tracer substance

$$KC'' + J = WC' + \lambda C \qquad (1)$$

where K is the vertical mixing coefficient, W is the vertical velocity, λ is the radioactive decay rate, J is the production or consumption rate associated with particulate flux or biochemical processes, and primes denote vertical (z) derivatives. The parameters K, J, λ, and W are assumed constant. If the concentration is given by constant values C_0 and C_m at the levels $z = 0$ and $z = z_m$, respectively, the solution is

$$C = \frac{J}{\lambda}$$

$$+ \frac{\left(C_m - \dfrac{J}{\lambda}\right)\exp\left[\dfrac{z - z_m}{2z^*}\right]\sinh\left(\dfrac{Az}{2z^*}\right) + \left(C_0 - \dfrac{J}{\lambda}\right)\exp\left(\dfrac{z}{2z^*}\right)\sinh\left[\dfrac{A(z_m - z)}{2z^*}\right]}{\sinh\left(\dfrac{Az_m}{2z^*}\right)}$$

(2)

where

$$A = \left(1 + \frac{4z^*\lambda}{W}\right)^{1/2}, \qquad z^* = \frac{K}{W}$$

Wyrtki (1962) was concerned with the oxygen distribution so that he chose $\lambda = 0$ but he considered the case with J as a function of z. For radioactive tracers Munk (1966) ignored the production term J. Since equation 2 includes all the special cases, it can be used to recover the earlier results.

The depth, z^*, is obtained from the balance $KC'' = WC'$ and represents the scale depth of the system. Its value can be determined from observed temperature and salinity profiles. Wyrtki, Munk, and Craig used vertical profiles in subthermocline and abyssal waters to obtain values for z^* ranging between 0.8 km and 1.3 km for the Pacific.

For stable, nonconservative tracers ($\lambda = 0$, $J \neq 0$) equation 2 simplifies to

$$C = C_0 + (C_m - C_0)f(z) + \frac{J}{W[z - z_m f(z)]}$$

(3)

where

$$f(z) = \frac{\exp(z/z^*) - 1}{\exp(z_m/z^*) - 1}$$

and the calculated profile for C can be compared with observations to determine the ratio J/W. From various observed profiles of total dissolved carbon dioxide ($\sum CO_2$) Craig estimated $2.6\ \mu M/kg/km \leqslant J/W \leqslant 3.7\ \mu M/kg/km$. He then used an estimate by Li, Takahashi, and Broecker (1969) of a $CaCO_3$ flux of $0.15\ M/m^2/yr$ to obtain a rough value of J for $\sum CO_2$ of $0.21\ \mu M/kg/yr$. With a mean value of $J/W = 3.5$ $\mu M/kg/km$ for the Pacific he then deduced $W \approx 6 \pm 3\ m/yr$ and with $z^* = 1$ km, he found $K \approx 2\ cm^2/sec$. These values for W and K are roughly the same as those obtained by Wyrtki and Munk as well as by other people.

As a check on the values listed above, Craig used an estimate made by Redfield et al. (1963) of the ratio of O_2 consumption to CO_2 production expected for oxidation of planktonic organic matter to arrive at the relation $J(O_2) = -1.3J(CO_2)$. Then modifying the ratio, because the solution of $CaCO_3$ adds to the total CO_2, he obtained $J(O_2) \approx J(CO_2) \approx 0.21\ \mu M/kg/yr$. The latter value is about twice the estimate by Riley (1951) for $J(O_2)$, but it falls in the range given by Munk (1966) and by Wyrtki (1962). In the course of his study Craig also emphasized the point that nonconservative tracers will be linearly correlated only if the differences $(C_m - C_0)$ of the boundary concentration for the two substances have the same ratio as the J values. This point is clear from the formulation of the problem but it is often overlooked.

Radiocarbon has been used as a tracer by various investigators because the known decay rate can serve as a clock for the circulation. Craig made two important contributions to this area. He pointed out that earlier workers had compared calculations of ^{14}C with observed profiles of $^{14}C/^{12}C$. Thus if ^{12}C has a local maximum it would be interpreted incorrectly as a ^{14}C minimum. By obtaining results for $^{14}C/CO_2$, Craig was able to identify such spurious extrema and to obtain profiles for the corrected distribution of ^{14}C. Specifically, he showed that in the deep waters of the Pacific ^{14}C is essentially independent of z and that the apparent mid-depth minimum of $^{14}C/^{12}C$ is due to the particulate flux of ^{12}C. Because of radioactive decay ^{14}C should show a subsurface minimum if there is no production of ^{14}C. Thus Craig showed that radioactive decay of ^{14}C is just balanced by the particulate flux, that is, $J \approx \lambda C$. It is an unfortunate coincidence that the model calculations for ^{14}C with $J = 0$ produce a vertical profile that agrees with that for the observed $^{14}C/^{12}C$ ratio. Thus one can obtain the "correct" profile with incorrect physical assumptions.

The foregoing analyses are based on the assumption that horizontal processes can be ignored. Wyrtki (1962) suggested that in the Indian and Pacific oceans such vertically balanced models may be realistic for deep water, that is, for water below the thermocline but above the bottom layer (≈ 4000 m depth), which is supplied by Antarctic bottom water. However, even an order-of-magnitude analysis casts doubt on the assumption. For example, horizontal advection ($v \, \partial C/\partial y$) is negligible in equation 1 if it is small compared to all other terms. Thus a necessary condition is $v \ll J/(\partial C/\partial y)$. Two of the vertical profiles that Wyrtki used are ocean stations Carnegie 154 and 158, which are about 1500 km apart and at 2000 m show a difference of dissolved oxygen concentration of 22 $\mu M/kg$. For oxygen with $J \approx 0.22 \ \mu M/kg/yr$, therefore, horizontal advection is negligible if $v \ll 0.05$ cm/sec. Mean horizontal velocities in mid-ocean regions are too small to be determined by direct observation but most indirect estimates belie the restriction on v. By a similar argument horizontal diffusion cannot be ruled out as an important mechanism either. If horizontal exchange processes are, in fact, as important as vertical processes, a problem arises when trying to interpret the results of the vertical exchange models. For example, horizontal advection can serve much the same function as local production or consumption does in altering the vertical structure. But the latter mechanism has little to do with the circulation per se, whereas the former is an implicit part of the circulation. Therefore, the need is for a more rigidly constrained system, one that would provide a more crucial test of the roles of the various processes.

Although horizontal advection has been excluded from these vertical exchange models in the water above the bottom layer, it has often been included in an equation for the vertically integrated bottom layer in order to provide an estimate of the horizontal velocity there when horizontal gradients are known from observations. The direction of the horizontal velocity is thus dependent on the direction of the tracer gradient. However, horizontal eddy diffusion may control the horizontal tracer distribution nearly completely and, if it does, the tracer distribution cannot be used to determine the direction of the mass velocity.

One final criticism of the vertical exchange models is that the calculations lead to different scale depths in different regions of the ocean. The characteristic depth in the thermocline is substantially smaller than that for deeper water. Most of the models using tracers have been restricted to deep water and have avoided this problem.

However, even in deep water the scale depth changes from one station to another. How is one to interpret such a result?

In spite of these objections the investigations employing vertical models must be considered as important contributions to the overall tracer problem. A good deal of strenuous effort and thought has gone into the data analysis and the interpretation of the results, and consequently several important questions have been answered. Even the derived magnitudes of the parameters may be roughly correct. The main point is that they have provided one framework within which parametric estimates can be made; the derived values can serve as a guide for more complete models.

Perhaps the most obvious shortcoming of these simple models is the total absence of a dynamical constraint. Including complete dynamical balance would require a solution of the general circulation problem before a consistent tracer distribution can be obtained. However, before trying to cope with that problem it is advisable to consider somewhat simpler systems, incorporating at least some dynamical constraints.

3. Thermocline Models

The aim of thermocline models is to obtain a consistent dynamical system that includes a realistic density structure. Although these models are not specifically oriented toward tracer distributions, almost all of them involve vertical diffusion and advective processes, and the required dynamical balance provides an added constraint for limiting the range of magnitude of the parameters K and W.

Many of the articles on thermocline models are concerned with the mathematical solutions of the system of equations. Since the purpose here is to determine the effect of the dynamical constraint on the parameters important to the tracer problem, only a simple scale analysis of the equations is presented. The reader is referred to Veronis (1969) for a more detailed review.

Since the thermocline problem involves global space scales and describes essentially a steady state phenomenon, the dynamical equations reduce to geostrophic and hydrostatic balance

$$2\Omega v \rho_0 a \cos \varphi \sin \varphi = \frac{\partial p}{\partial \lambda} \tag{4}$$

$$2\Omega u \rho_0 a \sin \varphi = -\frac{\partial p}{\partial \varphi} \tag{5}$$

$$\frac{\partial p}{\partial z} = -g\rho \tag{6}$$

$$\frac{\partial u}{\partial \lambda} + \frac{\partial (v \cos \varphi)}{\partial \varphi} + \frac{a \cos \varphi}{\partial z} \frac{\partial w}{\partial z} = 0 \tag{7}$$

The Boussinesq approximation has been used with ρ_0 as a constant reference density; ρ is the perturbation density and the equations are written in spherical coordinates (λ, φ, z) corresponding to longitude, latitude, and vertical with velocity components (u, v, w). Use of equation 6 in equations 4 and 5 leads to the thermal wind equations

$$2\Omega \rho_0 a \cos \varphi \sin \varphi \frac{\partial v}{\partial z} = -g\frac{\partial \rho}{\partial \lambda} \tag{8}$$

$$2\Omega \rho_0 a \sin \varphi \frac{\partial u}{\partial z} = g\frac{\partial \rho}{\partial \varphi} \tag{9}$$

Cross-differentiating equations 4 and 5 to eliminate the pressure and making use of equation 7 yields the planetary divergence equation

$$v = a \tan \varphi \, \frac{\partial w}{\partial z} \tag{10}$$

The equation for the density is

$$\frac{u}{a \cos \varphi} \frac{\partial \rho}{\partial \lambda} + \frac{v}{a} \frac{\partial \rho}{\partial \varphi} + W \frac{\partial \rho}{\partial z} = K \frac{\partial^2 \rho}{\partial z^2} \tag{11}$$

where only vertical diffusion is taken into account. Equations 8–11 together with appropriate boundary conditions comprise the thermocline problem as it has most often been formulated.

The system is nonlinear and only two rather special analytical solutions have been found, one with an exponential decay of density with depth and the other with an inverse power law in z for the density. In the former the vertical velocity is composed of two separate parts, $W = w_a + w_d$, which are associated with two different balances (Veronis, 1969)

$$\frac{u}{a \cos \varphi} \frac{\partial \rho}{\partial \lambda} + \frac{v}{a} \frac{\partial \rho}{\partial \varphi} + w_a \frac{\partial \rho}{\partial z} = 0 \tag{12}$$

$$w_d \frac{\partial \rho}{\partial z} = K \frac{\partial^2 \rho}{\partial z^2} \tag{13}$$

Here the subscripts a and d correspond to balances by advective and diffusive processes, respectively. Because of this division of the vertical convective process into two disjoint parts, the exponential solution provides no real test of the relative roles of advection and diffusion. The inverse power law solution does not involve such a simple separation, but its use is limited because it is a similarity solution and only special forms of the surface boundary conditions can be satisfied.

To some extent, therefore, the analytical results from the thermocline models suffer from the same inadequacies as the simple vertical exchange models. Only a limited number of processes are taken into account and the solutions that can be obtained restrict the possible balances even further. These simplifications are convenient and even necessary to make the problem tractable, but nature may require quite a different set of balances.

Welander (1971) discussed the thermocline problem in some detail. The following is a simple scale analysis that reproduces some of his results by exploring the balances between different terms.

To begin with it is helpful to distinguish between the effects of anticyclonic and cyclonic wind stresses. In the former case the net effect of the wind is to induce a local downwelling by means of Ekman pumping, W_e, and it is not possible to obtain the balance given by equation 13 because with positive K and negative w the density would have to decrease with depth. Hence within the limiting assumptions of the model, it is necessary that the effect of Ekman pumping be balanced by advection as in equation 12. The scaling obtained from equation 12 is then the same as that given by the continuity equation 7 or the planetary divergence equation 10 provided that horizontal scales are global, that is, given by the radius of the earth, a. Then with the scaling $v \sim u \sim V$, $z \sim H$, $w \sim W_e$ for midlatitudes ($\cos \varphi \sim \sin \varphi \sim 1$) equations 8 or 9 and 10 yield

$$\frac{2\Omega V}{H} \sim \frac{g\rho}{a} \tag{14}$$

$$\frac{V}{a} \sim \frac{W_e}{H} \tag{15}$$

Eliminating V yields an expression for H in terms of the remaining parameters

$$H_a \sim \left(\frac{2\Omega W_e a^2}{g\rho}\right)^{1/2} \tag{16}$$

where subscript a emphasizes the advective balance that is expressed. In cgs units the Ekman pumping velocity, W_e, is order 10^{-4} and $\rho \sim 10^{-3}$. Hence with $g \sim 10^{-3}$, $2\Omega \sim 10^{-4}$, $a \sim 6 \times 10^8$, one obtains

$$H_a \sim 600 \text{ m} \tag{17}$$

This value is smaller than the depth scale obtained from subthermocline water (Munk, 1966), but the latter is normally associated with diffusion and a mean upwelling rather than with the Ekman pumping velocity.

At the base of this advective layer in anticyclonic gyres the vertical velocity will have been reduced and will most likely reverse sign because of the effect of diffusive processes. In this region and in cyclonic gyres the balance given by equation 13 can be valid. Then

$$H_d \sim \frac{K}{w_d} \tag{18}$$

where subscript d refers to a diffusive layer and corresponds to a deep upwelling velocity. Equation 18 gives the scale depth for subthermocline waters. Of the three quantities involved, H_d is the only observable one, and the expression is normally used to evaluate the ratio K/w_d. Vertical sounding of geochemical tracers determine K and w_d separately, as in Section 2.

The remaining possibility is that diffusion and advection contribute equally. Then equations 14, 15, and 18 form a single system and the scale depth is given in terms of K by

$$H \sim K^{1/3} \left(\frac{2\Omega a^2}{g\rho}\right)^{1/3} \tag{19}$$

The remaining variables, V and W, are

$$V \sim ak^{1/3} \left(\frac{g\rho}{2\Omega a^2}\right)^{2/3} \tag{20}$$

$$W \sim K^{2/3} \left(\frac{g\rho}{2\Omega a^2}\right)^{1/3} \tag{21}$$

In this region of overlap the vertical velocity cannot be identified with only Ekman pumping or only upwelling. Hence dropping the subscripts from W in equations 16 and 18 makes it possible to write equation 19 in terms of H_a and H_d as

$$H \sim H_d^{1/3} H_a^{2/3} \tag{22}$$

Hence the scale depth for the combined balance takes on an intermediate value between the two pure cases, as might be expected.

It is advisable, of course, to use the best observables (usually H and ρ) to obtain estimates for the remaining quantities. For deep water $H \sim 10^5$ cm and $\rho \sim 10^{-4}$ so that $K \sim 2.5$ cm^2/sec, $V \sim 0.15$ cm/sec and $W \sim 2.5 \times 10^{-5}$ cm/sec. These results are consistent with Craig's (1969). For water at the base of the thermocline $H \sim 4 \times 10^4$ cm and $\rho \sim 2 \times 10^{-4}$ yielding $K \sim 0.3$ cm^2/sec, $V \sim 0.11$ cm/sec, and $W \sim 0.75 \times 10^{-5}$ cm/sec. Using a numerical analysis of this thermocline model Stommel and Webster (1962) concluded that K lies in the range 0.1–1 cm^2/sec in the thermocline layer.

The discussion and the estimates given here indicate that different dynamical and thermodynamical balances may be expected in different regions of the ocean. Welander (1971) emphasized this point and proposed a rough division of the ocean into advective and diffusive regions. However, although some progress has been made with these thermocline models they have not served to narrow the range of allowable values of the turbulent mixing parameters. The models are still too simple to provide an effective test of those processes that involve unmeasurable parameters.

4. A Semiquantitative Model

Rooth and Östlund (1972) have presented a novel and informative, alternative approach to the tracer problem, based on the distribution of tritium. Almost all of the measurable tritium in the oceans was injected during the periods 1956–1957 and 1961–1962 when hydrogen bombs were tested in the atmosphere. With a half-life of 12.3 yr tritium is a good tracer for processes that have a time scale of several decades. The so-called 18° water in the Sargasso Sea is formed seasonally and, since it contains a high concentration of tritium because of its recent contact with the atmosphere, it serves as a source of tritium for the water below.

The argument by Rooth and Östlund is based on the observed relation between tritium and temperature. Data from several sampling stations made during cruises P-6808 and P-6810 by R/V *Pillsbury* in October 1969 over a region in the Sargasso Sea from about 20°N to 30°N give rise to a vertical temperature profile that is nearly exponential with a scale height of 440 m in the temperature range $7°C \leqslant T - 2.3°C \leqslant 15°C$. The tritium distribution is also exponential with depth in this range and a plot of tritium versus temperature on logarithmic scales is closely represented by the line

$$\phi = \mu\theta + \text{const} \tag{23}$$

where μ is constant, $\phi = \ln$ (tritium), and $\theta = \ln (T - 2.3°C)$. When tritium is expressed in tritium units (10^{18} times the mole fraction of tritium in total hydrogen), the points cluster about the line with $\mu = 5$ for higher temperature.

If temperature, T, and tritium concentration, C, are determined by advective-diffusive equations with anisotrophic diffusive processes governed by the coefficients K (vertical) and K_H (horizontal), the equations are

$$C_t + \mathbf{v} \cdot \nabla C + \lambda C = KC_{zz} + K_z C_z + K_H \nabla_H^2 C + \nabla_H (K_H \cdot \nabla_H C) \tag{24}$$

$$\mathbf{v} \cdot \nabla T = KT_{zz} + K_z T_z + K_H \nabla_H^2 T + \nabla_H (K_H \cdot \nabla_H T) \tag{25}$$

where subscripts z and t correspond to partial derivatives, λ is the decay rate for tritium, and ∇_H^2 is the horizontal Laplacian. The temperature is taken to be steady but the tritium concentration is decaying from its initial value injected during the period 1956–1962.

Substituting $\phi = \ln C$ and $\theta = \ln(T - 2.3°C)$ into equations 24 and 25 yields

$$\phi_t + \mathbf{v} \cdot \nabla \phi + \lambda = K[(\phi_z)^2 + \phi_{zz}] + K_z\phi_z + K_H[(\nabla_H\phi)^2 + \nabla_H^2\phi] + \nabla_H(K_H \cdot \nabla_H\phi) \tag{26}$$

$$\mathbf{v} \cdot \nabla \theta = K[(\theta_z)^2 + \theta_{zz}] + K_z\theta_z + K_H[(\nabla_H\theta)^2 + \nabla_H^2\theta] + \nabla_H(K_H \cdot \nabla_H\theta) \tag{27}$$

Now the use of equation 23 in equations 26 and 27 leads to

$$\phi_t + \lambda = (\mu^2 - \mu)[K(\theta_z)^2 + K_H(\nabla_H\theta)^2] \tag{28}$$

or, representing derivatives of θ by scales characteristic for the water below 18°C, that is, $\theta_z \sim H^{-1}, \nabla_H\theta \sim L^{-1}$, gives

$$\phi_t + \lambda = (\mu^2 - \mu)[KH^{-2} + K_HL^{-2}] \tag{29}$$

Hence a weighted diffusion time scale is expressed in terms of the local rate of change and the decay rate of the tracer. The dynamics and boundary conditions of the problem are effectively prescribed by the power law (equation 23). By making use of this observed relation, Rooth and Östlund have avoided the difficulties of the dynamical analysis but have accounted for the dynamical effects.

From analysis of samples taken in 1971, Rooth and Östlund have concluded that the transient term, ϕ_t, is at most 30% of the decay rate for tritium. The remaining parameters of the analysis are known. With $\lambda = 1.8 \times 10^{-9}$/sec, $H = 440$ m, and the fact that K_HL^{-2} is positive, an upper bound for K can be determined from equation 29.

$$K < 0.2 \text{ cm}^2/\text{sec} \tag{30}$$

A corresponding upper bound for the magnitude of the upwelling velocity, W, follows from

$$W = \frac{K}{H} < 0.45 \times 10^{-5} \text{ cm/sec} = 1.4 \text{ m/yr} \tag{31}$$

These values of K and W are at the lower extremes of the ranges given by earlier workers.

Two additional results that Rooth and Östlund have obtained have direct ramifications for studies of the mixing processes. The first is based on the fact that for temperature below 14°C the data points from nearly all the *Pillsbury* stations fall on a line with $\mu = 3$ in equation 23; that is, the tritium values are larger than those on the $\mu = 5$ line. These colder waters appear to have sources peripheral to the Sargasso Sea and the higher tritium values are brought in by horizontal exchange processes in the main thermocline region below the Sargasso Sea. Thus lateral exchange processes appear to dominate over vertical exchange processes. Formal substitution of $\mu = 3$ in equation 29 leads to the values $K = 0.75$ cm^2/sec and $W = 5.2$ m/yr. These values are closer to the ones obtained from vertical exchange models. The implication is that these "acceptable" values for W and K are arrived at from analyses that omit the important, and presumably dominant, process of lateral exchange. Hence even though such an analysis may lead to reasonable values of K and W, the vertical advective-diffusive model may not be at all pertinent.

The second point is that with a horizontal length scale, L, of 1000 km in equation 29 horizontal diffusion will dominate over vertical exchange for $K_H \approx 10^6$ cm^2/sec.

However, Rooth and Östlund point out that lateral exchange processes act predominantly along isopycnic rather than horizontal surfaces. From their data they conclude that for exchange along isopycnic surfaces the lateral diffusivity required to negate the significance of vertical diffusion in equation 29 is $K_H = 10^8$ cm^2/sec. This value is on the high side of the range that has been used by other investigators. However, nearly all quantitative studies involving lateral exchange processes assume the exchange to take place along geopotential rather than isopycnic surfaces. Although this distinction is not significant in regions where the two surfaces are nearly parallel, it is of utmost importance when the density surfaces have a significant slope. In these latter cases, "horizontal" and "vertical" should refer to directions parallel and normal to isopycnics. Until that distinction is made, the evidence obtained from models making use of K_H along geopotential surfaces is bound to be ambiguous and misleading.

5. Horizontal Advective-Diffusive Models with Given Abyssal Circulation

Mean velocities in the deep ocean are generally too weak to be observable and estimates of the velocity magnitudes are based on indirect evidence, often from the distribution of tracers. In Section 2 the point was made that the direction of velocity cannot be deduced unambiguously from the direction of the tracer gradient because lateral diffusion may dominate the tracer distribution. Therefore, it is important to analyze combined circulation-tracer models to obtain distributions based on dynamically consistent flows.

The formal problem is simply stated. Given the mean velocities, from either a theoretical model or observations, and either flux or concentration values of the tracer at all boundaries, find the tracer distribution in the region under consideration. The problem would have a unique solution if all parameters were known. However, none of the parameters, except for the decay constant of radiochemicals, can be determined with any precision. Hence it is necessary to find solutions for different values of the parameters and then choose the set that fits the observed tracer distribution best. It is by no means certain that this problem has a unique solution. Indeed, with the relatively sparse network of observations even the distribution against which the solutions are tested is only an approximation. The problem is further complicated by the fact that the "given" circulation is also one of the unknowns. If one had total freedom to choose a circulation, the problem would almost certainly have multiple solutions. However, the dynamical constraint on the circulation is actually quite restrictive and there is some hope that the problem is well posed and that the adjustable parameters are those associated with tracer mixing and consumption.

Several calculations have been made along these lines and all have made use of Stommel's (1958) abyssal circulation model for the velocity distribution. This model assumes a homogeneous abyssal ocean extending upward 3 km from the (flat) bottom, $z = 0$, to the base of the thermocline, $z = H$. The flow is geostrophic everywhere except in boundary layers (required to satisfy mass continuity) along the western sides of the ocean basins and along the northern boundaries of the Pacific and Indian oceans. The circulation is driven by an assumed uniform upwelling at $z = H$. Abyssal water is supplied at sources of equal strength in the North Atlantic and in the Weddell Sea. The magnitude of the upwelling (or of the sources) is a parameter of the problem.

The circulation pattern for an idealized geometry of the oceans is shown in Fig. 1. A recirculation, R, around Antarctica of arbitrary magnitude can be appended to the one shown. Details of the calculation are given by Stommel and Arons (1960) but

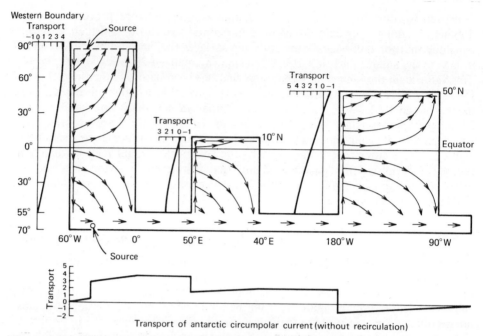

Fig. 1. An idealized world ocean basin bounded by parallels of longitude and latitude as marked. Particle trajectories for the abyssal circulation in the idealized world ocean are shown in the interior and in the boundary layers. A uniform recirculation around Antarctica is indicated by arrows. Sources of equal strength are located in the Atlantic at 90°N and at 45°W, 70°S and supply the abyssal water which is subsequently lost to the surface layer via uniform upwelling of magnitude w_0. Transport is in units of $w_0 a^2$. The lower part of the figure shows the transport required in the Antarctic to supply the western boundary currents and to absorb the poleward interior flow.

it should be noted that the eastward and northward velocities, u and v, are independent of z, whereas the vertical velocity, w, is a linear function of z.

The equation determining the distribution of a tracer is

$$\nabla \cdot (\mathbf{v}C^*) = K_v \frac{\partial^2 C^*}{\partial z^2} + K_H \nabla_H{}^2 C^* + J^* - \lambda C^* \tag{32}$$

where K_v and K_H are the vertical and horizontal diffusion coefficients and C^* is the tracer concentration. A vertical average $(H^{-1} \int_0^H (\ ) \, dz)$ of equation 32 yields

$$\nabla \cdot (vC) + (wC^*)_{z=H} = \left(K_v \frac{\partial C^*}{\partial z} \right)_{z=H} + K_H \nabla^2 C + J - \lambda C \tag{33}$$

where the unstarred variables are now vertical averages and ∇ is the horizontal gradient. The boundary conditions $w|_{z=0} = 0$ and $K_v(\partial C^*/\partial z)_{z=0} = 0$ have been used. The terms evaluated at $z = H$ are sources or sinks of tracer and have been ignored (more or less arbitrarily) in all of the two-dimensional (horizontal) calculations to be discussed. Then equation 33 reduces to

$$\nabla \cdot (\mathbf{v}C) = K\nabla^2 C + J - \lambda C \tag{34}$$

where $K \equiv K_H$. For a stable tracer, such as dissolved oxygen, λ vanishes. In the calculations that have been made the advection term has been used in the form $\mathbf{v} \cdot \nabla C$.

However, the form $\nabla \cdot (\mathbf{v}C)$ is preferable for numerical purposes. It should be noted that the two are not identical because the horizontal velocity field is divergent.

The initial calculation, by Arons and Stommel (1967), was applied to the North Atlantic only and was not fully two-dimensional. The latitude bands 0 to 30°N and 30°N to 60°N were treated as strips with north–south variation determined by western boundary values for each strip. Because of its restricted nature the calculation is not described in detail, but it is important to point out that Arons and Stommel made their calculations for a wide range of parameters, a necessary procedure for these models where so little is known about the various processes. They concluded that for the oxygen distribution a best fit with observations is obtained for K in the range 6.0×10^6 to 7.4×10^6 cm²/sec and J in the range -0.088 to -0.11 μM/kg/yr. These values are roughly the same as the best-fit values obtained with considerably more sophisticated models.

Kuo and Veronis (1970, 1973) have treated the two-dimensional problem for the oxygen distribution ($\lambda = 0$) with the geometry of Fig. 1. The circulation is as shown with sources of fluid of known concentration in the polar regions of the Atlantic. Except for these source regions there is no flux of oxygen ($\partial C/\partial n = 0$) at lateral boundaries. Then integrating equation 34 over horizontal space yields

$$\left(v_n C - K \frac{\partial C}{\partial n} \right)_{\text{sources}} = JA \tag{35}$$

where subscript n corresponds to the normal component at the sources and A is the area of the domain. Hence the sources supply the oxygen that is consumed in the main body of the ocean. Had the fluxes at the top boundary not been ignored, they could provide an additional source or sink that could affect the balance of equation 35.

In the Kuo–Veronis calculation the advection term is written as $\mathbf{v} \cdot \nabla C$ instead of $\nabla \cdot (\mathbf{v}C)$. Furthermore, since the vertical velocity has magnitude w_0 (the upwelling amplitude), the horizontal velocities are proportional to aw_0/H, where a is the radius of the earth. Writing $\nabla = \nabla'/a$ where ∇' is nondimensional yields the nondimensional form of equation 34 for the Kuo–Veronis calculation

$$P\mathbf{v} \cdot \nabla C = \nabla^2 C + j \tag{36}$$

where $P = w_0 a^2/HK$ is the Peclét number, $j = J/a^2K$, and all quantities are non-dimensional. The recirculation, R, around Antarctica is made dimensionless by the definition $R = (KH\delta\phi)\rho$, where $\delta\phi (=15°)$ is the latitudinal extent of the Circumpolar Current.

The principal difficulties encountered in the solution have to do with the relaxation of the finite-difference analogue of equation 36 when the Peclét number exceeds unity, that is, when advection dominates. The difficulties were resolved in the second Kuo–Veronis paper by taking weighted differences for the advection term, that is,

$$\frac{\partial C}{\partial x} = \frac{1+\sigma}{2} \frac{C_{n+1} - C_n}{h} + \frac{1-\sigma}{2} \frac{C_n - C_{n-1}}{h} \tag{37}$$

where h is the increment in x and $-1 \leqslant \sigma \leqslant 1$. For $\sigma = 0$ equation 37 reduces to the standard centered-difference scheme. For $\sigma < 0$ the left-hand difference dominates and for $\sigma > 0$ the right-hand difference dominates. The optimal choice of σ as a function of the advective velocity was determined empirically from the solution of a simple one-dimensional problem and the values of σ so determined were then used at each

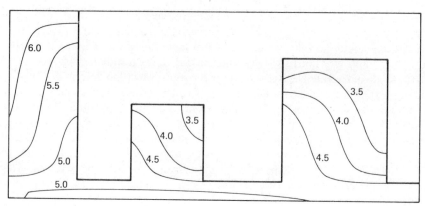

Fig. 2. Contours of dissolved oxygen in the idealized abyssal world ocean as obtained by Kuo and Veronis (1973). The parameters that provide the best fit with the observations are $w_0 = 1.5 \times 10^{-5}$ cm^2/sec, $K = 6 \times 10^6$ cm^2/sec, $R = 35 \times 10^6$ m^3/sec, $J = -2 \times 10^{-3}$ ml/l/yr.

point and in each direction in the two-dimensional relaxation. The relaxation method converged for all values of P.

Calculations were made for wide ranges of values of P, j, and ρ and a best fit to the observed oxygen distribution at 4 km depth was obtained for $P = 3.4, j = -4.3$, $\rho = 75$. The observed and calculated distributions are shown in Figs. 2 and 3, respectively. The parameters P, j, and ρ are combinations of the dimensional parameters. In order to obtain the optimum values of w_0, J, and R, the solutions were examined for individual characteristics that agreed with observations. The extent of the 5.0 ml/l curve in the Antarctic Circumpolar Current served to specify R. Then the optimal value of ρ determined K, P determined w_0, and j determined J. These optimal values are $K = 6 \times 10^6$ cm^2/sec, $w_0 = 1.5 \times 10^{-5}$ cm/sec, $J = -8.8 \times 10^{-2}$ μM/kg/yr, and $R = 3.5 \times 10^7$ m^3/sec.

The general features of observed and calculated distributions agree reasonably well. There is a gradual decrease of concentration from southwest to northeast in the Indian and Pacific oceans and highest concentrations occur in the vicinity of the western boundaries. A particularly important point is that the concentration decreases northward in the Pacific in spite of the southward flow in the southern hemisphere. The distribution is clearly dominated by lateral mixing. Thus a naïve determination of meridional velocity from the tracer gradient would lead to a deduced flow in the wrong direction.

The calculation also served as a test of Stommel's abyssal circulation model. The main conclusion here is that there is an obvious need for a circulation model that includes the effect of bottom topography since the greatest disparity between observation and calculation can be traced to topographic influence.

6. Finite-Difference Approximation for the Advective-Diffusive Equation

In the preceding section the difficulties encountered in solving the advective-diffusive equation were mentioned. The original Kuo–Veronis calculation by relaxation made use of centered differences and the iterations converged only for sufficiently small values of the ratio w_0/K. The weighted-difference scheme (equation 37) with σ determined empirically for the one-dimensional problem made it possible to obtain convergence for all values of w_0/K.

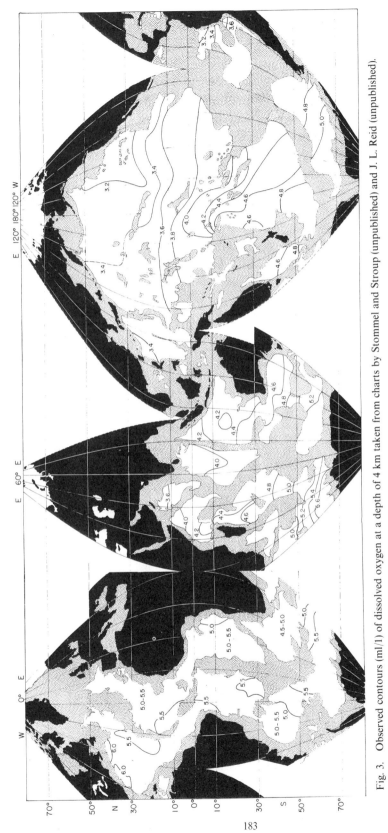

Fig. 3. Observed contours (ml/l) of dissolved oxygen at a depth of 4 km taken from charts by Stommel and Stroup (unpublished) and J. L. Reid (unpublished).

The convergence properties of the finite-difference methods are discussed in detail by Fiadeiro (1975), who also developed an analogous procedure for an approximation to the continuous problem when the domain is divided into cells at each of which the concentration is represented by the value averaged over the cell. The latter approach was suggested by Keeling and Bolin (1967), who made use of transfer functions that had been used by Welander (1959) for a study of air–sea interaction. Fiadeiro (1975) analyzed the several methods of approximation used for this "cell" approach to the problem and has shown that a weighted form for the value of the concentration at the cell boundaries includes the other forms as special cases. In his thorough treatment Fiadeiro discusses convergence of the iteration scheme in terms of the properties of the matrices which represent the discrete system.

In most analyses of numerical schemes in hydrodynamics the advection terms must be written in divergent form in order that the discrete system retain conservative properties such as mass, energy, and vorticity in the absence of dissipation. Use of the divergent form of advection also reduces the finite-difference equations to the same set of equations as the cell formulation provided that the same approximations are used for derivatives.

These two formulations have characteristic properties that are similar. For both of them it is possible to derive a set of equations equivalent to a centered-difference approximation. In this case convergent iteration schemes are possible only for sufficiently fine grids or small cells. A weighted scheme analogous to the one used by Kuo and Veronis can be formulated for the "cell" approach also and convergence can be achieved for a system with larger cells. With either formulation the weighted scheme can be shown to be equivalent to the centered-difference approximation with a virtual diffusion coefficient that depends on the velocity and is therefore a function of position. For the finite-difference scheme discussed in Section 5 this virtual diffusion coefficient takes the form $K(1 + uh\sigma/2K)$ and for errors of $O(h^2)$ the value of σ is $uh/6K$. For the cell approach the coefficient is $K[1 + (1 + u^2h^2/K^2)^{1/2}]/2$ to the same order of approximation. In either case, the price of obtaining convergence is an effectively larger diffusion. Hence the physical system is not exactly equivalent to the continuous one.

Although the points mentioned here are important for an understanding of the results of the numerical models, the mathematical details would take us too far afield. Hence the reader is referred to Fiadeiro's work for the detailed treatment.

7. A Three-Dimensional Advective-Diffusive Model with Given Abyssal Circulation

Fiadeiro (1975) has extended the Kuo–Veronis type of calculation to three dimensions, again using Stommel's abyssal circulation model for the velocity field. The calculation is confined to the Pacific Ocean and the boundary conditions include observed concentration values along the inflow and outflow parts of the southern boundary and at the top of the abyssal layer (i.e., at the base of the thermocline). Along the remaining lateral (solid) boundaries the concentration flux vanishes. Hence this three-dimensional model incorporates much more observational data as a constraint than did the two-dimensional Kuo–Veronis calculation which made use of observed concentrations only in the source regions.

Data from the GEOSECS[1] program made it possible for Fiadeiro to work with

---

[1] GEOSECS is the acronym given to the program Geochemical Sections, sponsored by the National Science Foundation as one of six major observational programs in the International Decade of Ocean Exploration. In the GEOSECS program surface-to-bottom measurements of many chemical and physical quantities were taken along certain tracks in the Atlantic and Pacific Oceans.

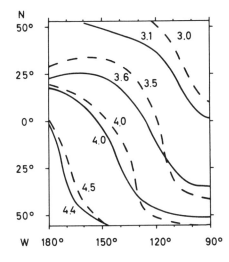

Fig. 4. Calculated contours of dissolved oxygen (ml/1) at a depth of 4 km in an idealized Pacific basin as obtained by Fiadeiro (solid curves) and by Kuo and Veronis (dashed curves) for values of the parameters given in the text.

several different tracers. A model for the salinity distribution was used to obtain best-fit values for the horizontal and vertical eddy diffusion coefficients and for the magnitude of the recirculation around Antarctica. These values were then used in the model for dissolved oxygen to obtain a best fit for the oxygen consumption parameter, J. Similar calculations were made for nutrients (phosphates and nitrates), silicates, and other tracers.

When this chapter was being written, Fiadeiro's work had not yet been published and only a preliminary result is quoted. In his calculation of dissolved oxygen he obtained a distribution at a depth of 4 km, which can be compared with one obtained by Kuo and Veronis. The results of the two calculations are shown in Fig. 4. For the Kuo–Veronis calculation the parametric values that were used are $R = 15 \times 10^6$ m^3/sec, $w_0 = 5$ m/yr, $K_H = 6 \times 10^6$ cm^2/sec, $J = -0.088$ $\mu M/kg/yr$. Fiadeiro's values for R and w_0 are the same but he has $K_H = 7.5 \times 10^6$ cm^2/sec, $J = -0.059$ $\mu M/kg/yr$. His remaining parameters and his actual best-fit values will be available when his work is published. The results given here are intended to exhibit the remarkably good agreement between the two models. In view of the more stringent observational constraints in Fiadeiro's calculation the two could have shown rather different distributions. Of course, the simpler calculation sheds no light on the vertical mixing process or the value of the vertical diffusion coefficient. These aspects of the more complete problem are included in Fiadeiro's work. However, the agreement in the abyss implies an upper bound to the intensity of vertical mixing since the distribution at the base of the thermocline is very different from the deep one and consequently could not have penetrated down by vertical diffusion. Fiadeiro finds that about half of the abyssal oxygen is lost by vertical diffusion, so his required value of J is smaller than the Kuo–Veronis value.

8. A Three-Dimensional Predictive Calculation

A more complete study of the combined circulation-tracer problem should contain a determination of the circulation as well as the tracer distribution from given external conditions. This approach is the most general one, and it is the one that should ultimately provide the complete solution to the problem. Since it effectively incorporates the general circulation as part of the overall study, it is the most difficult of the several approaches that have been undertaken and it requires a numerical treatment and a large amount of computer time. The only calculation of this type that has been published is one by Holland (1971), who treated a rectangular basin corresponding to the North Atlantic. A critique of the more important aspects of Holland's calculation is given elsewhere (Veronis, 1975), so this discussion is confined to the issues that have been raised earlier in this article.

Extensive numerical calculations of the type required for three-dimensional circulation models put heavy demands on computing machines, and it is often necessary to use rather coarse grid networks in order that the amount of computer time be kept within reasonable limits. Since nearly all the calculations make use of centered differences, the larger the grid increment the larger the diffusion coefficients necessary for numerical stability. Hence for reasons that have nothing to do with the physics, it is often necessary to use unrealistically large diffusion coefficients in these three-dimensional models.

One of the consequences of using large diffusion coefficients is that the width of boundary layers is exaggerated. In fact, that is one of the sought-after features, because the width must be sufficiently large to be resolved by the grid network. The hope is that the essential qualitative character of the system can be captured and that subsequent refinements will enable one to draw closer to the actual situation.

Another possibility is that the calculated flow is qualitatively different from the real flow. For example, a more dissipative system may be stable, whereas one with small diffusion coefficients may not. This possibility is always present, but one is aware of it; with subsequent refinements the stability can be examined. More serious cases are those in which basic physical balances are different from the actual ones or those in which the results do not agree with preconceptions based on physically plausible simple models, thereby casting doubt on the continued use of the latter.

The large numerical models of general circulation (e.g., Bryan, 1969, and earlier references in that paper) contain a result also obtained by Holland (1971), which contradicts the basic assumption in Stommel's (1958) abyssal circulation model. Stommel noted that, since the observed thermal structure in the ocean has warmer water overlying colder, turbulent processes would transport heat downward, thus heating the lower layers. For a statistically steady system the excess heat must be removed by some mechanism and the specific assumption is that the required process is an upwelling of cold water. The cold water that upwells through the base of the thermocline drives the abyssal circulation of Stommel's model. One of the results of the numerical calculations is a downwelling through the thermocline over most of the ocean. Hence the distribution of tracers obtained by Holland (1971) is associated with a circulation quite different from Stommel's abyssal flow field. The latter was used by both Kuo and Veronis (1973) and by Fiadeiro (1975) to obtain tracer distributions that are in reasonable agreement with observations.

The reason for the downwelling that emerges from the numerical models can be traced to the use of the large diffusion coefficient (5×10^7 cm^2/sec) along geopotential surfaces. As long as the isopycnics are essentially horizontal, lateral mixing along geopotential surfaces serves mainly to distribute properties along those surfaces. However, as Stommel pointed out to me, when isopycnics have a substantial slope, as they do in the western boundary current of Holland's model, mixing along geopotential surfaces induces mixing across isopycnics that is effectively vertical mixing. In fact, when Holland's results are used to determine the effective mixing across isopycnals, one calculates a vertical velocity of the order of 10^{-3} cm/sec balanced by vertical diffusion with a coefficient of the order of 50 cm^2/sec. Such intense upwelling in the western boundary current requires an inordinately large supply of abyssal water, which is provided by downwelling in the interior. Thus upwelling under the western boundary current more or less "drives" the abyssal circulation.

In a subsequent (as yet unpublished) calculation with a lateral diffusion coefficient of 10^7 cm^2/sec, Holland has obtained a circulation with a much reduced upwelling in the western boundary current and a slow upwelling in the interior. Hence the numerical calculation leads to a flow pattern more like that of Stommel's abyssal circulation. The tracer distribution associated with this later calculation was not available when this chapter was being written, so the result cannot be shown. In Holland's (1971) earlier calculation the tracer distribution is dominated by lateral diffusions and is similar to the one obtained by Kuo and Veronis for a very diffusive system.

The results of the numerical calculations reinforce the point made earlier in connection with Rooth and Östlund's discussion of the role of vertical diffusion in the region below the 18°C water in the Sargasso Sea. Horizontal and vertical diffusion should be associated with directions parallel and perpendicular to isopycnic surfaces. Only in this way will the intensity and significance of these processes be subject to a controlled test by deductive models.

Acknowledgment

This summary is based on work supported by National Science Foundation grants GA 40705-X and DES 73-00424 A01.

References

Arons, A. B. and H. Stommel, 1967. On the abyssal circulation of the world ocean. III. An advective-lateral mixing model of the distribution of tracer property in an ocean basin. *Deep-Sea Res.*, **14**, 441–457.

Bryan, K., 1969. A numerical method for the study of ocean circulation. *J. Comp. Phys.*, **4**, 347–376.

Craig, H., 1969. Abyssal carbon and radiocarbon in the Pacific. *J. Geophys. Res.*, **74**, 5491–5506.

Defant, A., 1929. Stabile Lagerung oceanischer Wasserkörper und dazu gehörige Stromsysteme. *Berlin Univ. Inst. Meereskd. Veroeff, NF., A Georgr.-Naturwiss. Reihe*, **19**, 33 pp.

Fiadeiro, M. E. M., 1975. The modelling of the tracer distributions in the deep Pacific Ocean. Ph.D. thesis, University of California, San Diego.

Holland, W. R., 1971. Ocean tracer distributions: Part I. A preliminary numerical experiment. *Tellus*, **23**, 371–392.

Keeling, C. D. and B. Bolin, 1967. The simultaneous use of chemical tracers in oceanic studies. *Tellus*, **20**, 566–581.

Koczy, F. F., 1958. Natural radium as a tracer in the ocean. *Proc. 2nd U.N. Int. Conf. Peaceful Uses At. Energy*, **18**, 351–357.

Kuo, H. H. and G. Veronis, 1970. Distribution of tracers in the deep oceans of the world. *Deep-Sea Res.*, **17**, 29–46.

Kuo, H. H. and G. Veronis, 1973. The use of oxygen as a test for an abyssal circulation model. *Deep-Sea Res.*, **20**, 871–888.

Li, Y. H., T. Takahashi, and W. S. Broecker, 1969. Degree of saturation of $CaCO_3$ in the oceans. *J. Geophys. Res.*, **74**, 5507–5525.

Möller, L., 1929. Die Zirkulation des Indisches Ozeans. *Berlin Veroeff. Inst. Meereskd.*, A, **21**.

Munk, W. H., 1966. Abyssal recipes. *Deep-Sea Res.*, **13**, 707–730.

Redfield, A. C., B. H. Ketchum, and F. A. Rickards, 1963. The influence of organisms on the composition of seawater. In *The Sea*. Vol. 2. M. N. Hill, ed., pp. 26–77.

Reid, J. L. and R. J. Lynn, 1971. On the influence of the Norwegian-Greenland and Weddell Seas upon the bottom waters of the Indian and Pacific oceans. *Deep-Sea Res.*, **18**, 1063–1088.

Riley, G. A., 1951. Oxygen, phosphate and nitrate in the Atlantic Ocean. *Bull. Bingham Oceanogr. Coll.*, **13**, 1–126.

Rooth, C. and G. Östlund, 1972. Penetration of tritium into the Atlantic thermocline. *Deep-Sea Res.*, **19**, 481–492.

Stommel, H., 1958. The abyssal circulation. *Deep-Sea Res.*, **5**, 80–82.

Stommel, H. and A. B. Arons, 1960. On the abyssal circulation of the world ocean. II. An idealized model of the circulation pattern and amplitude in oceanic basins. *Deep-Sea Res.*, **6**, 217–233.

Stommel, H. and J. Webster, 1962. Some properties of the thermocline equations in a subtropical gyre. *J. Mar. Res.*, **20**, 42–56.

Sverdrup, H. U., 1930. The origin of the deep-water of the Pacific Ocean as indicated by the oceanographic work of the *Carnegie, Gerlands Beitr. Geophys.*, **29**, 95–105.

Sverdrup, H. U., M. W. Johnson, and R. H. Fleming, 1942. *The Oceans*. Prentice-Hall, New York, 1087 pp.

Thorade, H., 1931. Strömmung und zungenförmige Ausbreitung des Wassers. *Gerlands Beitr. Geophys.*, **34**, (3), 57–76.

Veronis, G., 1969. On theoretical models of the thermohaline circulation. *Deep-Sea Res.*, **16** (Suppl.), 301–323.

Veronis, G., 1975. The role of models in tracer studies. *Numerical Models of Ocean Circulation*. Nat. Acad. Sci., Washington D.C.

Welander, P., 1959. On the frequency response of some different models describing the exchange of matter between the atmosphere and the sea. *Tellus*, **11**, 348–354.

Welander, P., 1971. The thermocline problem. *Phil. Trans. Roy. Soc. London A*, **270**, 69–73.

Wüst, G., 1928. Der Ursprung der Atlantischen Tiefenwässer. In *Berlin ZS. Ges. Erdk. Sonderband zur Hundertjahrfeier der Gesellschaft*, 506–534.

Wüst, G., 1935. Schichtung und Zirkulation des Atlantischen Ozeans. Die Stratosphäre. Dtsch. Atlant. Exped. Meteor, 1925–27. *Wiss. Ergeb.*, **6**, (1).

Wyrtki, K., 1962. The oxygen minima in relation to ocean circulation. *Deep-Sea Res.*, **9**, 11–23.

7. THE DYNAMICS OF UNSTEADY CURRENTS

PETER B. RHINES

1. Summary

This is an account of the dynamics of quasi-geostrophic flows of a wavelike or turbulent nature. An *historical review*, Section 2, is followed by a section on *kinematics*, Section 3, which illustrates the characteristically different appearance of the fields of pressure, velocity, vorticity, and density, even though they are linearly related. The spectral breadth is the important quantity distinguishing them. Their different natural weighting with respect to length scale leads to simple experimental tests of the (ω, k) (i.e., frequency, wave number) relation from single moorings. These make use of an empirical "turbulent" dispersion relation, $\omega \propto Uk$, where U is the root-mean-square (rms) fluid velocity. The spectra and correlation functions useful for nondivergent fields are reviewed.

Section 4 is a review of the *lowest-order dynamics* of a wedge-shaped analogue of a β plane ocean, stressing the nearly geometrical nature of Sverdrup flow and long, baroclinic waves. An impedance is defined to quantify the vertical stiffness of rotating fluids, and these classical flows occur when forcing is so gentle that fluid columns can resist vertical stretching and compression.

Section 5 is a review of *topographic Rossby waves*, which occur when the potential vorticity balance is linear, for small wave steepness, ε/ω, where ε is the Rossby number and ω the wave frequency divided by the Coriolis frequency. Attention centers on (1) barotropic, (2) fast baroclinic, and (3) slow baroclinic waves; type 2 relies on the slope of the bottom. All three are important to the nonlinear dynamics that follow. The partition of initially prescribed flow among the types, and a steady flow, is demonstrated.

Oceanic observations are given in support of the gross properties, particularly westward propagation (types 1 and 3), intensification near the bottom (type 2), and the inverse nature of the dispersion relation (type 1).

Section 6 gives a new treatment of *nonlinear cascades* that occur in a *flat-bottom ocean* when ε/ω is not small. For oceanic energy levels it is shown that energy may travel faster through wave number space than physical space, in the sense that significant horizontal and vertical eddy–eddy interactions can occur before propagation has moved the energy a single wavelength.

The cascades carry barotropic (depth-independent) energy toward large scale. However, geophysical flows find many ways to counter the lateral expansion of eddies, for example, Rossby-wave propagation may take hold, while developing persistent anisotropy that favors zonal currents. (This is related to the induction of mean circulation by eddies; see Section 8.)

Baroclinic energy (currents with vertical shear) moves toward the Rossby deformation scale from either smaller or larger scales; there the eddies above and below the thermocline lock together, producing a barotropic state with surprising efficiency. A proof of the migration from baroclinic toward barotropic flow is given. Once this transformation has occurred, the theory for a simple homogeneous fluid applies.

This chain of events means that, for example, a slowly propagating, long baroclinic Rossby wave with modest currents fragments into deformation scale eddies (in a

189

generalized baroclinic instability), jumps to the barotropic mode, expands to larger scale again, and propagates away much more quickly, as a barotropic Rossby wave, whose currents are much swifter. A Gulf Stream meandering experiment is described, which involves the same cascades. The meteorological analogue is discussed.

Section 7 adds to these "primary" cascades the effect, crucial to the oceans, of a *rough bottom* and *coastal boundaries*. Both act (via simple formulas given here) as sources of enstrophy which can grossly alter the energy cascades. In addition, they suggest the relatively small-scale flows found in the deep water and near western boundaries.

Sea-floor roughness (at scales greater than the deformation radius) is found to be essential in preserving the vertical shear of currents found in the oceans; without this topography, the cascade toward depth-independent flow would operate within a few months [with the caveat that disequilibrium (forcing or damping) of the field, or severe spatial intermittency, such as found in Gulf Stream rings, can also preserve baroclinity]. Computer experiments show this control of the vertical structure to occur when $\varepsilon/\delta \lesssim 0.5$, where δ is the rms topographic height, say, in a 500-km region of interest. Baroclinic instability of large-scale flows is altered by bottom topography, which severely inhibits the "occlusion" stage. Topography generally whitens the wave number spectrum, whereas the nonlinear cascades, alone, tend to sharpen it.

Over topography the energy develops a patchy distribution, even after time averaging. Such "fine structure" in the intensity has recently been discovered at sea. At the same time a steady component of flow develops spontaneously, about f/h contours in the deep water (f is the Coriolis frequency and h is the depth).

Linear wave theory with a rough bottom is reviewed, and a simple baroclinic "double Kelvin wave" derived, which is trapped both horizontally and vertically. The wave theory lends insight to the behavior of the nonlinear, rough-bottom ocean, for the linear waves themselves exhibit spatial intermittency, a cascade toward small scale, a predilection for vertical shear, and an inability to carry energy efficiently in the horizontal.

In all, the oceanic case contrives to make horizontal energy flux in eddies triply dependent on the energy level. Only the more intense baroclinic flows succeed in switching from baroclinic to barotropic modes, and expanding in the horizontal. Both such changes act to increase the group velocity, and the concurrent release of potential energy increases the kinetic energy being transported. Horizontal propagation experiments are shown as illustration. Baroclinic energy, in the nonlinear, rough-bottom case, moves principally westward from its source (less than 1 km/day in other directions), qualitatively as in linear theory.

The fluid tends to adjust toward states near the transition between waves and geostrophic turbulence. Ideas of linear propagation still have value; combined with knowledge of the nonlinear changes in structure, they predict qualitatively the movement of energy about these model oceans. Yet the transfer spectra verify that quasi-equilibrium of an eddy field involves a continuing conflict among the cascades due to advection and topography.

Section 8, a discussion of *mean flows*, begins with a formula (related to an early model of Kolmogorov) for the difference between ensemble-averaged Eulerian and Lagrangian mean flow, which is

$$\langle u_i \rangle = \frac{\partial}{\partial x_j} \kappa_{ij}$$

where $\kappa_{ij} = \int_0^t R_{ji}(\tau)\, d\tau$, is the diffusivity integral of the Lagrangian correlation function. Particles on the average are attracted to regions of intense eddies (large κ_{ij}); oceanic evidence is given for important gradients in eddy intensity.

Holland and Lin's (1975) simulation of eddy interaction with the mean ocean flow is described. Then a simple vorticity-flux theory is given for mean-flow generation by turbulent eddies (or waves), which accounts for f/h-contour-following currents found in the laboratory and computer simulations. In the simplest case, isolated forcing of a barotropic β plane fluid, a westward zonally ($\overline{}$) averaged flow,

$$\overline{u} = -\frac{\beta}{2}\,\overline{\eta^2}$$

develops in the far field, where η is the displacement, north and south, of particles from their point of origin. This applies to either wavelike or turbulent inviscid flow, and leads to a prediction of both the eastward jet and westward-flowing far field found experimentally by Whitehead (1975). With weak bottom friction, coefficient D, a nondiffusive ($\kappa_{22} = 0$), wavelike field induces flow with speed ranging from the above value to twice it, depending on the correlation time of the velocities. If, instead, particles wander freely in latitude, then

$$u = -\frac{\beta\kappa_{22}}{D}$$

The Lagrangian drift is compared with these Eulerian values.

Generalizations are given, and related to Green's and Welander's work. The time-dependent theory here involves both positive and negative diffusivity of potential vorticity. The curl of these stresses, $(\partial/\partial x_i)(\kappa_{ij}\partial Q/\partial x_j)$ where Q is a slowly varying mean potential vorticity distribution, is a likely source of surface and abyssal circulation, both in regions immediate to intense currents, and also in random eddy fields. The argument combines with others to predict, especially, elongated bands of zonal, or f/h-following currents in the oceans. The gradients of diffusivity, κ_{ij}, may in many regions provide the dominant driving.

Section 9 is a further discussion of *recent observations*, of *sources* and of *sinks*.

This paper is arranged, first, with flows that would be driven by gentle and slowly varying forces and then, successively, by quickly varying and vigorous forcing. The set of parameters that determines the evolution of a field of quasigeostrophic eddies is: L/L_ρ, the ratio of the horizontal scale of the dominant eddies, and L_ρ, the Rossby deformation scale; $\beta L^2/U$ where U is the vertical-average rms velocity; ε/δ, the ratio of Rossby number and topographic height variations found between scales L and L_ρ; and P/K, the ratio of available potential and kinetic energies in the eddy field. In addition, the level of external forcing, the strength of mean currents and their associated potential energy, the bottom drag, internal-wave interaction rate, and degree of intermittency may occur, if they are not negligible. In some interesting cases the turbulent dynamics alter the parameters toward $L/L_\rho \sim 1$, then $P/K \to 0$, $\beta L^2/U \to 1$.

2. Historical Introduction

The kinetic energy of the oceans takes four dominant forms: surface waves, inertial (or near-inertial) and tidal oscillations, the climatological mean circulation, and unsteady currents of period greater than one-half pendulum day. The latter, particularly the nearly geostrophic eddies of roughly 200 km diameter, are the concern of this chapter. I have sought to make this a self-contained account, and yet to include recent results on the nonlinear dynamics of eddies. Doing so has severely limited its breadth, and in particular many important investigations of Rossby wave propagation have had to be omitted. Sections 2–5 are in part a review of classical ideas, whereas Sections 6–9 are new or recently published material. A broader discussion of ocean variability is given by Monin, Kamenkovich and Kort, 1974.

A. Observational History

The beginnings of concentrated investigation of transient ocean currents came in the 1950s with a series of cruises led by Fuglister to map the instantaneous form of the Gulf Stream and the rings of current that are thrown off it, to the north and south. Earlier, those who looked closely at other parts of the Stream (e.g., Bache, 1846; Pillsbury, 1890) found unsteadiness and spatial complexity to be the rule. Turbulence in concentrated streams (recorded in sketches by Leonardo da Vinci) at one extreme, and seasonal reversals of the Somali Current, at the other, were familiar long ago. But eddy motion was generally related in the observer's mind to nearby intense currents. The notion that the ocean interior is populated by chaotic, variable currents is recent (excepting ancient speculations, like those of Plato, about bodies of water oscillating deep within the earth, feeding the seas, and about the occasional, malign Charybdis).

Perhaps self-protection causes the human mind to imagine unknown regions to be simply structured, or structureless. Thus in the early nineteenth century, before even the mid-oceanic depth had been determined, it was held that below the first few hundred meters the ocean was stagnant and lifeless. The notion of inert abyssal regions was reinforced by the incorrect observation of frequent 4°C temperatures (measured with unprotected thermometers), combined with the incorrect notion that seawater has its maximum density at this temperature.

The zoologist, in search of the beginning of this lifeless (azoic) zone, gradually pushed downward the known limits of both life and dynamic activity. Finally, benthic animals were dredged from the sea floor in increasing quantities. Broader scientific interest in the abyssal ocean grew when Darwin suggested it might contain a living record of early animal evolution, and technical interest came with the laying of telegraph cables.

Physical oceanography often bootlegged on geographical or biological cruises. Complexity and variability of the deep-temperature field quickly became apparent. Thomson and Carpenter on the *Lightning* (1868), for example, measured temperatures to 1000 m near the Faeros Bank: "it had been shown that there are great masses of water at different temperatures moving about, each in its particular course, maintaining a remarkable system of ocean circulation, and yet keeping so distinct from one another that an hour's sail may be sufficient to pass from the extreme of heat to the extreme of cold." The use of density structure to infer horizontal current came with gradual appreciation for the intensity of the Coriolis force. Routine dynamic computations followed Bjerknes' circulation theorem at the beginning of this century.[1]

The dominant concern of the hydrographic work that followed was the climatological mean circulation. Fuglister and Worthington's (1951) explorations were the first to capture the instantaneous picture of the transient meanders and eddies. The enormity of the task is obvious, requiring vast numbers of ships or moored instruments. Reproduced in Fig. 1 is a map of the mean temperature in the upper 200

---

[1] Two interesting histories of the general subject are those of Deacon (1972) and Schlee (1973).

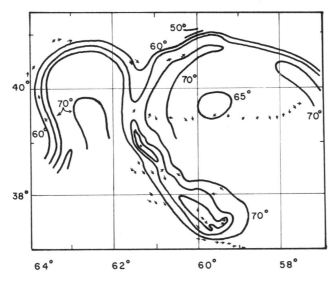

Fig. 1. A Gulf Stream ring at the moment of detachment (Fuglister and Worthington, 1951). Mean temperature in the upper 200 m, in degrees Fahrenheit. Current directions from towed electrodes (GEK). See also Figs. 28, 57.

m, from a part of the 1950 survey. A 500-km-long meander has begun to detach and form an autonomous, cold, cyclonic ring. The effect of this nucleus of energy, both potential and kinetic, on the adjoining ocean is clear, for example in the long sections of Fuglister's (1960) atlas. But typical isopycnal slopes decrease rapidly as one moves away from the Gulf Stream, and we are left to wonder what activity there is in the open ocean.

Stommel established in 1954 a monitor station a few miles southeast of Bermuda (32°N, 65°W). It was a novel prospect to record a long hydrographic time series at a point far from the regions of greatest activity. But the temperature series clearly resolved long-period (~150-day) eddies in the thermocline height, well below the level of direct seasonal penetration (Fig. 2). Schroeder and Stommel (1969) showed the dynamic height variations to be coherent with the sea level at Bermuda (corrected for atmospheric pressure) and thus eliminated the possibility that they were poorly sampled internal waves. The only uncertainty is the degree to which Bermuda itself affects these stations, conceivably casting off eddies into passing currents.

The *Aries* cruises in 1955 seemed to be the conclusive step. Swallow, Crease, and Stommel organized them to measure directly the currents below the thermocline. The search for a slowly moving, steady ocean interior ended as surely as did the nineteenth-century belief in an azoic zone, when neutrally buoyant floats moved away rapidly, altering direction every few weeks.

The Swallow floats showed currents exceeding 10 cm/sec (at the 2000 and 4000 m levels, southwest of Bermuda) and crudely identified their length scale (~70 km separation before velocity coherence is lost). The size of these deep eddies was comparable with those in Fuglister's and Stommel's shallow data, but the time scale and intensity were far smaller. In the succeeding sections we hope to show convincing dynamic reasons for the generic differences and similarities among these eddies.

Little was known before this decade about the geographic distribution of eddy energy. It is easy to appreciate that wiggliness of isotherms in classical sections

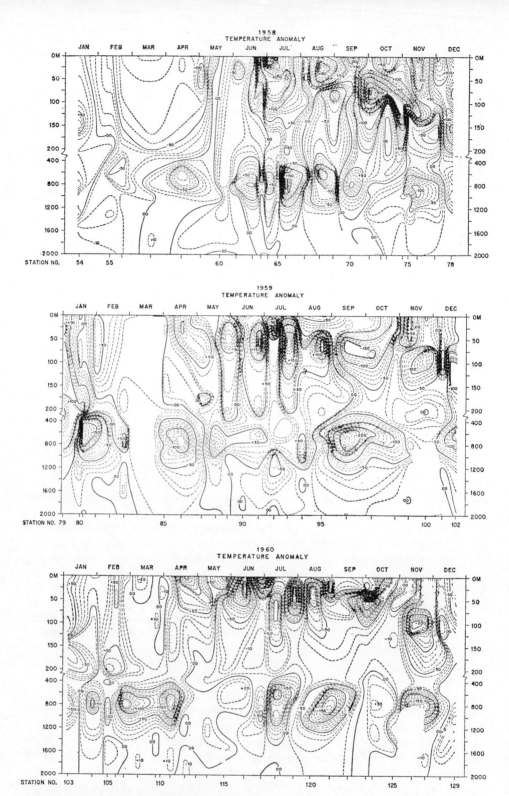

Fig. 2.   Time series of temperature anomaly (relative to 8-yr average seasonal cycle) from the *Panulirus* monitor site, 15 km southeast of Bermuda (Schroeder and Stommel, 1969), for the years 1958–1960. The associated dynamic height is coherent with the Bermuda tide gauge at low frequencies, suggesting that these are geostrophic eddies.

194

diminishes away from coasts and intense currents, but it is difficult to produce satisfying maps of perturbation energy from them. Parker (1971) used the bathy-thermograph archives to produce an interesting map of the distribution of shallow, intense thermal features in the western North Atlantic. Again, the decrease in numbers as one leaves the Gulf Stream is clear, but the spatial irregularity of the data base makes further interpretation difficult.

Only now is the proliferation of moored current meters and thermistors yielding a quantitative picture of the geography of eddies and mean flow. We have come to realize that really very little is known about the broad pattern of these quantities; even the dynamic topography of the time-mean circulation in a region so widely traveled as the North Atlantic is unreliable, beyond verifying the western-boundary activity, the "polar" front, and equatorial currents, and a hint of the nature of the broad, shallow return flow.

B. Dynamic Background

It is the fault of eddies that the general circulation is hard to observe, but they are more than just observational noise. Stommel (1957) discussed how precarious was the dependence of linear, steady models on the Sverdrup interior flow. It may be that such circulation models, together with their time-dependent counterpart, the linear Rossby wave, are sufficiently inspired to hold true beyond their strict limits of validity. But there is increasing evidence that deeper understanding will come from nonlinear, interactive models of eddies and mean flow.

The atmospheric circulation gives a precedent. There, one would like to have dismissed fronts, internal waves, cyclones, and monsoon and orographic circulation to find a simple theory of the maintenance of the zonal winds, but Jeffreys (1926) and Starr (e.g., 1968) showed some of these to be essential in redistributing mean angular momentum. The physical source and nature of the "eddies" that produce the relevant Reynolds' stresses are complex and include both classical baroclinic instability and large-scale orographic and thermal contrasts.

The ocean contains many similar elements; the energy-containing eddies, though less than 1/10 as big as those in the atmosphere, are roughly the same size if rescaled by the density structure (the Rossby radius). Oceanic flow speeds are perhaps 1/25 as big, yet again appear the same, relative to the ratio of the beta effect and square of the Rossby radius. But the absence of unobstructed paths for zonal flow (everywhere but, possibly, in the Southern Ocean) is likely to make very different the driving or retarding of the mean circulation by eddies.

The ocean basins, in addition, are perhaps 20–60 eddy diameters wide, whereas the atmospheric domain is rather small, measured against the size of cyclones. This, together with the confinement of the intense mean flow to a far smaller region in the oceans, suggests that inhomogeneity is the greater, and that the nature of lateral influences is more crucial there, than in the atmosphere.

There is, then, a competition between the response as local flow to local external forcing by winds and heating, on the one hand, and the necessity that unusual clumps of energy and density anomalies spread themselves about, on the other. The classical mean circulation presumes that lateral influence is powerful, and the constitutive relation of the fluid somehow allows forces excited, say, on the Labrador Sea to be felt off Cape Hatteras. A picture of rubbery seas, propagating energy about, is incomplete without actual circulation. As in striking an elastic wheel, which is free to spin about its hub, the signaling process distributes angular momentum until internal friction dissipates all but the new rigid-body rotation. An important parameter of this analogue is the travel time for waves, relative to the time characteristic of the forcing. If the forcing is slowly varying, measured in these terms, the transients are insignificant and the body effectively rigid. But in the other extreme, a flaccid material will have gross undulations superimposed on the steady mode. It is this latter case that appropriately describes the oceanic response to weather and the seasons: the speed of travel for the energy-containing eddies is probably less than 5 km/day.

Unfortunately, however, these strongly excited transients (and those generated by internal instability) are so energetic, and the medium itself is so irregular, that linear superposition fails. We must then consider carefully the nature of geostrophic and topographic turbulence in altering the horizontal transports, and driving circulation wherever they are found.

3. Kinematics of Eddy Fields

The difficulties posed by complex, variable fields of current, density, and pressure are not merely analytical and instrumental. One's judgment and memory of the simplest qualities of an eddy field can fail, and two observers rarely seem to agree upon, say, the dominant scale or period of a flow, however perfectly it has been measured. The problems of description lead to vagueness in all later stages of analysis. The object of this section, therefore, is to review some of the descriptors of eddies, and some of their immediate application to observations.

A. Space

Consider an example, Fig. 3, of instantaneous maps of some artificially produced eddies. Each of the three major boxes, Figs. 3a–3c, shows four fields corresponding to a two-dimensional, nondivergent flow; counterclockwise from the lower right, these are ψ (stream function or pressure), $\partial\psi/\partial x$, $-\partial\psi/\partial y$ (velocity components), and $\nabla^2\psi$ (vorticity). The maxim, "integration smooths, differentiation roughens," applies. Observations using current meters would identify a smaller dominant length scale than those from pressure gauges, and a vorticity meter would be the most confusing of all.

The Fourier coefficients $\hat{\psi}_{\mathbf{K}}$ of the instantaneous stream function are defined by

$$\psi(\mathbf{x}) = \sum_n \sum_m \hat{\psi}_{\mathbf{K}} e^{i\mathbf{k}\cdot\mathbf{x}} \qquad (\hat{\psi}_{\mathbf{K}} = \hat{\psi}^*_{-\mathbf{K}})$$

where $\mathbf{k} = 2m\pi/L, 2n\pi/L$; $n, m = 1, \infty$, periodic over a large domain, width L. Let L become large, and imagine an equivalent continuum of wave numbers obtained by blurring the discrete k over a fixed, small interval (or see Batchelor, 1953, p. 30). Then the power spectra of the various fields, with respect to vector and scalar wave number, are, for geostrophic flows,

$$P(\mathbf{k}) = \rho_0 f_0 |\psi_{\mathbf{K}} \psi_{\mathbf{K}}^*| \qquad \mathscr{P}(k) = \int_0^{2\pi} P(\mathbf{k})k\, d\theta \quad \text{(pressure)}$$

$$E(\mathbf{k}) = \frac{1}{2}\frac{|\mathbf{k}|^2 P(\mathbf{k})}{\rho_0^2 f_0^2} \qquad \mathscr{E}(k) = \frac{1}{2}\frac{k^2 \mathscr{P}(k)}{\rho_0^2 f_0^2} \qquad \text{(kinetic energy)}$$

$$\Omega(\mathbf{k}) = \frac{1}{2}\frac{|\mathbf{k}|^4 P(\mathbf{k})}{\rho_0^2 f_0^2} \qquad \hat{\Omega}(\mathbf{k}) = \frac{1}{2}\frac{k^4 \mathscr{P}(k)}{\rho_0^2 f_0^2} \qquad \text{(enstrophy)}$$

$$k = |\mathbf{k}|, \qquad \tan^{-1}\theta = \frac{\mathbf{k}\cdot\text{north}}{\mathbf{k}\cdot\text{east}}$$

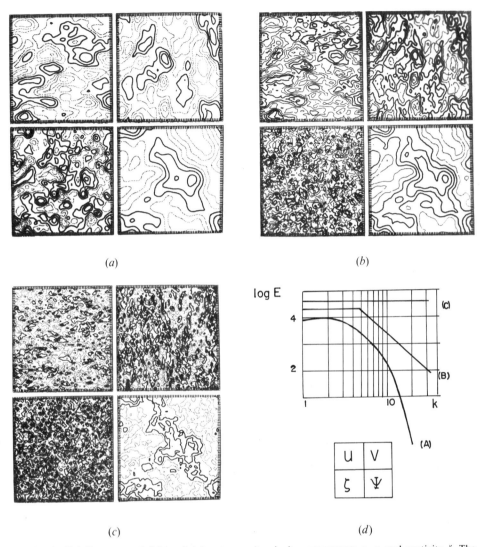

Fig. 3. Artificially generated fields of ψ (or pressure), velocity components u, v, and vorticity ζ. The azimuthal wave number dependence is fixed. (a) For scalar-wave number spectrum $E = ke^{-2/3k}$; (b) for $E = 1.0$ ($k \leqslant 5$), $E = (k/5)^{-3}$ ($k > 5$); (c) for $E = \text{const}$. The discrimination between associated fields ψ, u, v, ζ increase with spectral bandwidth.

where ρ_0 is density, f_0 is Coriolis frequency. Successive differentiation in space corresponds to high-pass filtering of the spectrum. The *degree* to which this filtering affects the picture depends of course on the breadth of the spectrum. A monochromatic field will be unaltered in character by differentiation. The three realizations in Fig. 3 illustrate this. Each one has the same directional makeup in its spectrum, yet with different dependence on k, Fig. 3d. As one proceeds from relatively narrow to white spectra, differentiation has more and more of an effect. The dominant scale of the vorticity for a white velocity spectrum is infinitely small (limited here only by the 64×64 grid). The distinction is experimentally useful because arrays of current meters and pressure gauges, used together, can give a measure of spectral breadth beyond that estimated from either kind of instrument alone. The "breadth," which may be defined by

$$k_2{}^2 = \int |k - k_1|^2 \mathscr{E}(k)\, dk \bigg/ \int \mathscr{E}\, dk$$

clearly controls the sensitivity of the picture to the measuring device. The centers of mass of the spectra, k_1, are 3.1, 4.8, and 16, respectively.

The correlation functions of such fields carry the same information, but in different forms. If the velocity correlation tensor is

$$R_{ij}(\mathbf{r}) = L^{-2} \iint u_i(\mathbf{x})u_j(\mathbf{x} + \mathbf{r})\, d\mathbf{x}$$

(the data being defined over a very large square of area L^2), the spectral tensor, the Fourier transform of R_{ij},

$$\Phi_{ij}(\mathbf{k}) = \frac{1}{4\pi^2} \iint R_{ij}(\mathbf{r})e^{-i\mathbf{k}\cdot\mathbf{r}}\, d\mathbf{r}$$

has as its trace $2E(\mathbf{k})$. For two-dimensional, isotropic, nondivergent flow the correlation may be written as

$$R_{ij}(\mathbf{r}) = \frac{f(r) - g(r)}{r^2} r_i r_j + g(r)\delta_{ij} \qquad r = |\mathbf{r}|$$

where $g(r) = \overline{u_2(\mathbf{x})u_2(\mathbf{x} + r_1)}$, for example, is the lateral covariance and $f(r) = \overline{u_1(\mathbf{x})u_1(\mathbf{x} + r_1)}$ is the longitudinal covariance. The bar is a probability average or, for homogeneous fields, a spatial average.

Now the ψ field has a correlation

$$C(r) = \overline{\psi(\mathbf{x})\psi(\mathbf{x} + \mathbf{r})}$$

By differentiating we find

$$u_1(\mathbf{x})u_1(\mathbf{x} + r_1) \equiv f(r) = \frac{-C'}{r}$$

and

$$u_2(\mathbf{x})u_2(\mathbf{x} + r_1) \equiv g(r) = -C'' = (rf)'$$

Here the basic ψ field and velocity fields have correlation functions which themselves are related by differentiation. The lateral correlation $g(r)$, being a differentiated function of the longitudinal correlation, $f(r)$, generally has a smaller dominant scale. This distinction is made clear by the anisotropy in the maps (Fig. 3) of the velocity components. The spectrum with respect to scalar wave number is related to the covariances by various expressions,

$$\mathscr{E}(k) = \tfrac{1}{4}\pi k \iint (f + g)e^{-i\mathbf{k}\cdot\mathbf{x}}\, d\mathbf{x}$$

$$= \tfrac{1}{2}\pi k \int_0^\infty r(f + g)J_0(kr)\, dr$$

$$= \tfrac{1}{2}\pi k^3 \int_0^\infty rCJ_0(kr)\, dr$$

for isotropic fields. The inverses of these relations are

$$f(r) = \frac{2}{r} \int_0^\infty k^{-1} \mathcal{E}(k) J_1(kr) \, dk$$

$$g(r) = \frac{2}{r} \int_0^\infty \mathcal{E}(k) \left[J_0(kr) - \frac{1}{kr} J_1(kr) \right] dk$$

$$C(r) = 2 \int_0^\infty k^{-2} \mathcal{E}(k) J_0(kr) \, dk$$

$$= (\rho_0 f_0)^{-2} \int_0^\infty \mathcal{P}(k) J_0(kr) \, dk$$

A useful model energy spectrum,

$$\mathcal{E}(k) = \frac{1}{2} \frac{k^3}{[r_0^{-2} + k^2]^{\mu+1}}$$

corresponds to a scalar correlation function

$$C(r) = C(0) \frac{2^{1-\mu}}{\Gamma(\mu)} \left(\frac{r}{r_0} \right)^\mu K_\mu \left(\frac{r}{r_0} \right) \sim r^{\mu-1/2} e^{-r} \quad \text{for } r \to \infty$$

[and corresponding lateral and longitudinal velocity correlations, $g(r) = -C''$, $f(r) = -r^{-1}C'$]. For instance, a spectrum with a peak at r_0^{-1} and asymptotic tail $\sim k^{-3}$ has a scalar correlation

$$C(r) = \frac{1}{2} C(0) \left(\frac{r}{r_0} \right)^2 K_2 \left(\frac{r}{r_0} \right) \qquad (\mu = 2)$$

The zero crossing of $g(r)$ occurs at $r = 1.34 r_0$.

B. Scales

Above we have used the centroid, k_1, of the wave number spectrum as a measure of dominant scale. It seems wise to use the entire spectrum in such a way, to increase confidence in the result. Measures like the position of the spectral maximum will tend to be less stable. However, when comparisons of k are being made, one must take care that the range of wave numbers being considered, particularly at the low end, is the same.

The correlation function for scalar fields has associated with it the integral scale,

$$r_1 = \int_0^\infty C(r) \, dr$$

and the "microscale," $r_2 = [C''(0)]^{-1/2}$. In terms of isotropic spectra,

$$r_1 = (\rho_0 f_0)^{-2} \int_0^\infty \mathcal{P}(k) \, dk$$

and $r_2^2 = (\rho_0 f_0)^{-2} \int_0^\infty k^2 \mathcal{P}(k) \, dk$; these are very different scales from k_1^{-1}. For the velocity correlations of classical turbulence theory, the corresponding microscale is $[f''(0)]^{-1/2}$; $f''(0) = \frac{1}{10} \int_0^\infty k^2 \mathcal{E}(k) \, dk$ (proportional to the total dissipation) weights heavily the smaller scales.

Thus it is a rather different picture one gets from spectra on the one hand, by remembering the centroid k_1, breadth k_2, and perhaps asymptotic tail $N(\mathscr{E}(k) \sim k^{-N})$, than from correlations on the other hand, where the scale r_1, r_2, and perhaps the position of the first zero crossing are kept track of. The use of spectra has the advantage that, if linear waves are present, a dispersion equation may associate particular wave vectors, $\mathbf{k}$, of spatial Fourier components with frequencies, ω, of temporal Fourier components. If, instead, the flow is turbulent, there may still be a theory predicting the shape and dynamical role of the spectral tail. Neither kind of theory is readily applied to correlation functions.

C. Time

In individual time series similar remarks hold, except that the temporal correlation functions are simple one-dimensional Fourier transforms of the spectra. For example, if $C_\psi(t, \tau) = \overline{\psi(t)\psi(t + \tau)}$, $C_u = \overline{u(t)u(t + \tau)}$, then the spectrum of ψ, is

$$P_1(\omega) = \int C_\psi(t, \tau)e^{-i\omega\tau} \, d\tau$$

and the spectrum of the velocity is $\mathscr{E}_1(\omega) = \int C_u(t, \tau)e^{-i\omega\tau} \, d\tau$. Centroid frequencies, ω_1, breadth $[\overline{(\omega - \omega_1)^2}]^{1/2}$, spectral tails ω^{-N}, integral time scales $t_1 = \int_0^\infty C \, d\tau$, and microscales $t_2 = [C''(0)]^{-1/2}$ exist in analogy with their spatial counterparts. Here $t_2{}^2$ is just twice the mean-square acceleration $\overline{((\partial u/\partial t)^2)}$ and t_1 is $\lim_{\omega \to 0} \mathscr{E}(\omega)$ (this must not be confused with the mean, which is presumed to have been removed).

In temporal records there arise two natural reference frames: purely Eulerian records measured at a point fixed in space, and Lagrangian records measured from points moving with fluid parcels. For later reference we define the Lagrangian data as $\mathbf{u}^L(t; \mathbf{x}_0)$, depending on time and the initial particle position, $\mathbf{x}_0$, and Eulerian data $\mathbf{u}^E(t, \mathbf{x})$, depending on the fixed observation point $\mathbf{x}$. The corresponding spectra are $\mathscr{E}_{ij}^L(\omega; \mathbf{x}_0) = \int R_{ij}(\tau, \mathbf{x}_0)e^{-i\omega\tau} \, d\tau$ and $\mathscr{E}^E(\omega; \mathbf{x}) = \int R_{ij}^E(\tau; \mathbf{x})e^{-i\omega\tau} \, d\tau$.

D. Dynamics

Some simple dynamic remarks are in order. First, if the complete space–time correlations or wave number–frequency spectra [corresponding to $\mathbf{u}(\mathbf{x}, t)$, say] can be measured, then a simple dynamic system might reveal itself by the appearance of spectral energy only along certain dispersion surfaces, $\omega = \omega(\mathbf{k})$. But usually this is neither observationally possible nor theoretically expected. Grosser dynamic comparisons can be made, however. For instance, the breadth or asymptotic tail of a wave number spectrum tells us whether or not there is dissipation occurring at small scales, that is, dissipation strong enough to damp actively and critically the energy-containing eddies. Violently energetic, three-dimensional turbulence has broad enough spectra (e.g., $k^{-5/3}$) to do this critical damping, but many flows such as geostrophic turbulence and plume convection have spectra too sharp (e.g., k^{-3}) to be dominated in this way by friction.

One of many ways to distinguish waves from turbulence is by the strength of the diffusivity of marked fluid particles. The diffusivity at long times, for a homogeneous, stationary field, is

$$\kappa_{ji} = \overline{u_i u_j} t_{ij}; \qquad t_{ij} = \int_0^\infty R_{ij}(\tau) \, d\tau$$

(see Section 8), and one sees that a wave field in which particles do not stray far from their origins will necessarily yield a correlation function with deep negative lobes capable of making the integral scale, t_{ij}, vanishingly small, while diffusive turbulence will have a substantial t_{ij}.

Another wave/turbulence comparison can be made using the above discussion of spectral breadth. Suppose, first, that turbulence is endowed with a crude dispersion relation $\omega \sim U|\mathbf{k}|$, where U is the root mean square (rms) particle speed. This is not Taylor's hypothesis, which is a more precise consequence of a strong mean current, but merely suggests that turbulent eddies advect one another past a fixed sensor in the advective time $(|\mathbf{k}|U)^{-1}$. If this is so, then our remarks about the spatial filtering action of pressure sensors relative to current meters (in a geostrophic flow) apply also to temporal records: a pressure record at a single point should be of character- istically longer period than a velocity record at the same point. Figure 4, a computer experiment with two-dimensional turbulence, shows this to be true: the larger spatial scale of ψ or p, compared with velocity, maps onto a longer period, and the *extent* of the difference gives a measure of spectral breadth. If, instead, the motion had obeyed the dispersion relation of planetary waves, $\omega \propto k^{-1}$, the reverse would have occurred: longer waves would have the higher frequency, and a pressure time series would have

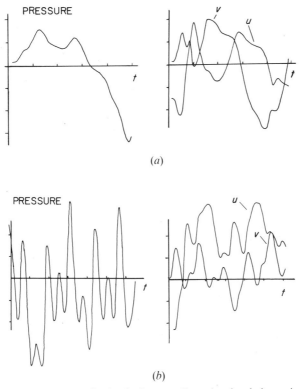

Fig. 4.   (a) Pressure and velocity at a fixed point in a two-dimensional turbulence simulation (see Table I for details of computer experiment). The velocities are more rapidly changing than the pressure; this verifies that turbulence maps large wave-numbers on to large frequencies. (b) As in (a), but with a field of linear Rossby waves (for which $\sigma = -\beta \cos \theta/|k|$, where θ is the angle of $\mathbf{k}$ from the east). The inverse dispersion relation causes velocities to change *less* rapidly than pressure.

been more quickly varying than current (Fig. 4b). A crude test of the dispersion relation is thus possible with a single pair of records from these different sensors.

Time series at fixed (or drifting) points seem to be easier to obtain than space series (although one cannot claim that the *Panulirus* time series preceded the discovery of eddies in classical sections). Further use of the advective dispersion relation thus may be of interest. For instance, Eulerian frequency spectra characteristically drop down steeply at frequencies above those of the dominant eddies, roughly like ω^{-3}. A deep spectral valley separates the eddies from the inertial-tidal-internal wave band. Now the energetic eddies must sweep smaller eddies past a fixed sensor, and thus a wave number spectrum falling off something like k^{-3}, beyond *its* maximum, is expected, as is a real dearth of energy at scales (wavelength/2) $\sim$ 2–10 km (corresponding typically to periods of 2–10 days, for $U \sim 5$ km/day). The implication of this spectral valley (Rhines, 1973) is simply that no local cascade of energy is occurring from the eddies to small scales. If such a transfer happens at all, it must be jumping the valley via boundary turbulence, intermittent internal turbulence, or nonlocal cascade into internal waves. The rate of energy dissipation in the eddies is crucial, and at this point unknown.

What about temperature and salinity data, used in conjunction with current records? The thermal wind equation, $f(\partial \mathbf{u}/\partial z) = -\mathbf{g} \times \nabla\rho/\rho$, admits two possibilities. If eddies are geometrically similar, broader ones penetrating more deeply, the density and velocity wave number spectra should have the same shape, and the baroclinic velocity measurements are redundant with the density data (assuming geostrophy). However, if the eddies have the same depth penetration regardless of breadth (say, the baroclinic energy occupies only the first vertical mode), then the density field acts like pressure, with spectrum $k^{-2}\mathscr{E}(k)$ [$\mathscr{E}(k)$ being the spectrum of baroclinic velocity]. In this case the apparent scale of the density eddies should exceed that of the velocity field, by an amount proportional to the spectral breadth. Again, time series of density and velocity at a point allow a crude estimate of the corresponding dispersion relation.

D. Record Length

In the measurement of mid-ocean eddies with period, T, 50–300 days, scales $\sim$ 40–100 km, we are never likely to have enough data to make the stability of spectral estimates really satisfactory. A measurement with record length, τ, of more than 10 periods would be ambitious, and the estimates settle down only as $(\tau/T)^{1/2}$, even assuming the spectrum to be rather narrow. It is thus doubly important to focus on important dynamic regions and to be pragmatic, looking at those regions where the environmental time constants are smaller, for instance, on the continental rise, where waves of period 5–15 days are important. In all, it seems to be more useful to invent crude dynamic tests that can withstand the paucity of data, rather than make a monolithic drive to determine the full wave number—frequency spectrum of the eddies. By itself, such a result might even prove to be unenlightening, for many theories produce the same spectrum.

4. Dynamics of the Gentlest Kind

It is well to describe the ideal response of a gently driven rotating fluid, before examining the flows caused by stronger forcing, and in the presence of less simple boundaries and more complex density stratification. Indeed it is the linear Sverdrup solution that is the heart of modern circulation theories, whether thermohaline or wind-driven, and its strength or weakness must be noted.

Here we describe a model that attempts to show in more intuitive form the basic solutions described by Stommel (1957). It reproduces a number of the known principal features of spherical geometry, yet in a wedge-shaped container, Fig. 5. The primitive Boussinesq equations for the velocity, $\mathbf{u}$, and pressure, p, are

$$\underset{1}{}\quad\underset{2}{}\quad\underset{3}{}\quad\underset{4}{}\quad\underset{5}{}\quad\underset{6}{}$$

$$\frac{\partial \mathbf{u}}{\partial t} + \mathbf{u}\cdot\nabla\mathbf{u} + f\hat{\mathbf{k}}\times\mathbf{u} = -\frac{\nabla p}{\rho_0} + g\rho' + \nu\Delta\mathbf{u} \tag{1}$$

$$\nabla\cdot\mathbf{u} = 0 \qquad\qquad \frac{\partial \rho'}{\partial t} + \nabla\cdot\rho'\mathbf{u} = 0$$

where $\rho \equiv \rho_0(1 + \rho')$ is the potential density, f is the Coriolis frequency, 2Ω, $\hat{\mathbf{k}}$ is a vertical unit vector, $\mathbf{g}$ is the gravity field, and ν is the kinematic viscosity. For flows describable by a single horizontal length scale L, vertical length scale H, time scale T, and velocity scale U, the ratio of typical sizes of terms 1 and 3 is $(fT)^{-1} \equiv \omega$, a scaled frequency; the ratio $2{:}3$ is $U/fL \equiv \varepsilon$, the Rossby number; and $6{:}3$ $\nu(H^{-2} + L^{-2})/f \equiv E$, the Ekman number.

Basic geostrophy occurs when each of ω, ε, and E is small, whence

$$f\hat{\mathbf{k}}\times\mathbf{u} = -\frac{\nabla p}{\rho_0} + O(\omega) + O(\varepsilon) + O(E) \tag{2}$$

If we now neglect the density stratification, 5, the Taylor–Proudman approximation follows:

$$f(\hat{\mathbf{k}}\cdot\nabla)\mathbf{u} = 0 \tag{3}$$

to the same order; such flows involve no vertical shear of the velocities, and ideal vertical dye lines remain vertical. The Coriolis force acts only normal to $\hat{\mathbf{k}}$, and hence equation 3 relies in addition on the hydrostatic balance in the vertical.

Considerable intuition for these approximations may be found in linear internal wave theory, just as sound waves give us insight into "incompressible" fluids. Plane-wave solutions of equation 1 with $E \to 0$ pass smoothly to the limit $\omega \to 0$ where they take on the character of Taylor–Proudman flows, even while the energetics of wave theory still applies. The frequency over f is just $\omega = \cos\theta$, where θ is the angle of the wave number vector, $\mathbf{k}$, from the vertical. The group velocity lies along the wave crests, normal to $\mathbf{k}$, with magnitude $f|\mathbf{k}|^{-1}\sin\theta$. Energy propagates, therefore, ever closer to the vertical as $\omega \to 0$, painting out Taylor columns induced, say, by a

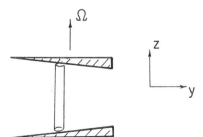

Fig. 5. Wedge-shaped container filled with homogeneous fluid uniformly rotating. y is analogous to north on a β plane.

disturbance below. Rigidity develops in fluid columns owing to the rapidity of this energy propagation, the group velocity approaching fL [and the time for signals to penetrate vertically through a distance H approaching $f^{-1}(H/L)$] even as the frequency vanishes. In order that the fluid not be strained vertically, free motions in a container like that in Fig. 5 must be directed along the depth contours, $h = \text{constant}$.

A. Hough, Goldsbrough, Sverdrup Flow

The lowest-order constraints due to strong rotation alone provide the solution to simple interior flows in the wedge-shaped model of the β plane. Imagine the response of the fluid to gentle downward motion, w_0, imposed at the upper surface (whether by Ekman convergence, fluid sources, or simply mechanical motion of the lid itself). Since equation 3 implies $\nabla_H \cdot \mathbf{u} = 0$, the only recourse for the fluid is to move toward greater depth, avoiding any vertical compression (the vertical velocity equaling w_0 throughout the column). The "southward" velocity, $-v$, is given by

$$v = \frac{w_0}{\alpha} \tag{4}$$

where $\alpha/2$ is the semi-angle of the wedge. In the case of an imposed stress, τ, quasi-steady Ekman layers along the rigid boundaries produce interior vertical velocity, $w_0 = \hat{\mathbf{k}} \cdot \nabla \times \tau/\rho_0 f$, and then renaming α as $\beta H/f$, where H is the mean depth, equation 4 becomes

$$\beta v = \frac{\nabla \times \tau|_z}{\rho_0 H} \tag{5}$$

The horizontal flow is approximately nondivergent, $\nabla_H \cdot \mathbf{u} = 0$; hence $u = \int v_y \, dx$, the limits of integration as yet unspecified.

If the wedge geometry is complicated by adding rough-bottom topography, free geostrophic flows must still follow $h(x, y)$ contours, and equation 5 generalizes to

$$hu \cdot \nabla\left(\frac{f}{h}\right) = \frac{\nabla \times \tau|_z}{\rho_0 H} \tag{6}$$

Either equation 5 or 6 puts severe constraints on the forcing pattern and basic geometry if violations of this level of geostrophy are to be avoided.

B. Stratification

The principal effects of stratification can be included without loss of simplicity using two homogeneous layers of slightly different density, $\Delta\rho$, and mean depths at $y = 0$ of H_1 and H_2, respectively, in the same geometry (Fig. 6a). Depth gradients occur in both layers, in the rest state. (Provided $f^2 L^2/gH \ll 1$, the interface at $z = \eta(\mathbf{x}, t)$ is practically flat in the absence of flow.) Equation 2 applies above and below the interface, yet a jump in velocity across it is allowed by Margules' relation, $g'\hat{\mathbf{k}} \times \nabla\eta = -f[\mathbf{u}_1 - \mathbf{u}_2]$ (g' is the reduced gravity, $g\Delta\rho/\rho_0$), which follows from the expression (equation 2) for upper-layer velocity, $\mathbf{u}_1$, and lower-layer velocity, $\mathbf{u}_2$, with the hydrostatic relation $p_2 - p_1 = g'\eta$ for the perturbation pressure fields, respectively.

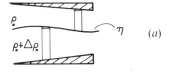

(a)

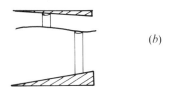

(b)

(c)

Fig. 6. Two-layer analogues of the β plane. (a) Equal mean depths of the fluid; (b) with a thin upper layer, the bottom sloping more steeply than the lid, for a uniform potential vorticity gradient; (c) on a rotating sphere, with columns remaining parallel to Ω (yet the effect of continuous stratification in the oceans makes this less than a perfect idealization).

Now the same driving by a vertical velocity at the top causes, by the constraint of stiffness in each layer, a north–south transport

$$\tfrac{1}{2}(v_1 + v_2) = \frac{w_0}{\alpha} \quad \text{for } H_1 = H_2$$

This is the analogue of the stratified Sverdrup relation. For the flow to be steady, $\partial\eta/\partial t = 0$, v_2 must vanish, leaving the transport entirely confined above the thermocline, which tilts, preventing the upper-layer pressure gradients from reaching the depths.

However, the time required to set up such a flow is great, so that either unsteadiness of the winds or diffusive effects may modify the vertical structure. The latter lead to thermocline theories, which do not concern us here.

C. Time Dependence

A more realistic resolution of the vertical structure with stratification, and a description of the setup time for steady flow, come from the approach to steady circulations found when the frequency of an oscillatory wind-stress pattern is led to vanish. Then linear wave theory gives a complete solution.

Consider the free motions possible ($w_0 = 0$) beyond that of flow along geostrophic contours. Time and space variations are allowed, but they must be gradual. In the equal-depth geometry of Figure 6a north–south motion is possible without stretching fluid columns, only if $v_2 = -v_1$ and if the interface moves vertically in response. Combining with the equation for geostropic thermocline tilts,

$$\frac{\partial \eta}{\partial x} = \frac{f}{g'}(v_2 - v_1) = \frac{-2f}{g'}v_1$$

and continuity,

$$\frac{\partial \eta}{\partial t} = v_1 \alpha$$

we find

$$\frac{\partial \eta}{\partial t} + \frac{\alpha g'}{2f}\frac{\partial \eta}{\partial x} = 0.$$

Where the thermocline is locally depressed, the upper flow is northward on the western side and southward to the east. This flow into and out of the wedge forces the thermocline downward in the west, upward in the east. It follows that an arbitrary pattern of interface displacements, varying in both x and y, moves "westward" without dispersion, at speed $\alpha g'/2f \equiv \beta c_0^2/f^2$ where c_0 is the speed of long internal waves without rotation, or equivalently, βL_ρ^2, where L_ρ is Rossby's internal deformation radius, c_0/f, and in doing so obeys the most trivial, nearly kinematic rules of Taylor–Proudman flow. This internal Rossby wave of Veronis and Stommel (1956) describes the baroclinic adjustment to unsteady winds. Its phase speed, $\lesssim 4$ km/day at mid-latitudes, shows that many years must pass before signals can cross the ocean and complete the spin-up of a baroclinic circulation from rest.

It is now a simple matter to imagine the response to a wind stress of large scale, say, a steady pattern turned on at time zero. Let the imposed vertical velocity be

$$w_0(x, y, t) = \frac{\nabla \times \tau|_z}{\rho_0 f} = f_1(x)f_2(y) \qquad t \geq 0$$

$$= 0 \qquad t < 0$$

Including unequal layer depths H_1 and H_2 (Fig. 6b), the equivalent $-\beta$ slopes are now αH_1 and αH_2, where $\beta = f\nabla h/h = f\alpha$. Then,

$$\alpha H_1 v_1 = w_0 - \frac{\partial \eta}{\partial t}$$

$$\alpha H_2 v_2 = \frac{\partial \eta}{\partial t}$$

$$\frac{\partial \eta}{\partial x} = \frac{f}{g'}(v_2 - v_1)$$

and we find

$$\frac{\partial \eta}{\partial x} - c_0^{-1}\frac{\partial \eta}{\partial t} = -\gamma w_0 = -\gamma f_1(x)f_2(y)$$

where

$$\gamma = \frac{f}{g'\alpha H_1}$$

$$c_0 = \frac{\alpha g' H_1 H_2}{f(H_1 + H_2)} = \beta L_\rho^2$$

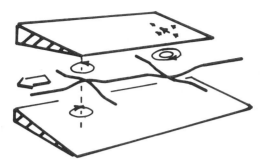

Fig. 7. Currents induced by steady localized wind stress turned on at $t = 0$ (yet slowly enough that barotropic signals are negligible). The transient, circular baroclinic wave moves off to the west, leaving a steady Sverdrup gyre. This occurs without stretching or compression of fluid columns.

The solution is

$$\eta = -\gamma f_2(y) \int_x^\infty [f_1(x' + c_0 t) - f_1(x')] \, dx$$

If, for example, the stress curl has a dipole nature, say, $w_0 = -x \exp[-(x^2 + y^2)]$, the interface is

$$\eta = \tfrac{1}{2}\gamma \exp(-y^2)[\exp[-(x + c_0 t)^2] - \exp(-x^2)]$$

which is a steady, downward depression in the thermocline just beneath the stress, and its mirror image, propagating westward along $y = 0$ (Fig. 7). The currents are initially barotropic, with an anticyclonic vortex at the origin. But as time progresses, this stationary circulation intensifies above the thermocline, and vanishes below it as the transient vortex moves off to the west. The currents in the transient vortex themselves are concentrated in the thin upper layer, $v_2/v_1 = H_1/H_2$. Lighthill (1969) has described in detail such events, particularly as applied to equatorial wave modes excited by the arrival of monsoon winds. Our major point here is that they fall within the range of interesting effects whose dynamics are so gentle that there is no vortex stretching; this is possible even near the equator, when the typical "impedance" of a rotating fluid with respect to vertical stretching, shown below to be $f^2 L^2 T/H$, becomes large.

It might seem that motions with such slow propagation would find difficulty in remaining linear, but the expected principal violation, the advection of density $u \cdot \nabla \eta$, vanishes identically. We may thus expect westward propagation of thermocline eddies of rather large amplitude.

If broad topography is added to the bottom, beyond that which is here imitating β, the free-wave equation becomes

$$\frac{\partial \eta}{\partial x} - \frac{f^2}{g'} \left(\frac{1}{\beta H_2 + \delta f} + \frac{1}{\beta H_1} \right) \frac{\partial \eta}{\partial t} = 0$$

where δ is the true north–south bottom slope, here taken to be uniform. A strong slope thus *increases* the baroclinic wave speed to

$$\frac{f^2}{g'} \left(\frac{\beta H_1(\beta H_2 + \delta f)}{\beta(H_1 + H_2) + \delta f} \right)$$

which approaches $1 + (H_1/H_2)$ times the speed without topography. The change arises, as above, when the steep slopes make very little deep, horizontal motion necessary to yield a given vertical velocity, with the result that this slowly varying circulation is virtually confined to the upper layer, as $\partial f/\beta H_2 \to \infty$. Rhines (1970) has suggested, and these heuristic models support, that over generally rough topography (with slopes greater than 10^{-3}) we should similarly expect the baroclinic waves to propagate faster, with the lower layer more nearly at rest. Rooth (private communication) remarks that strong bottom friction, $E^{1/2} \gg \beta L_\rho^2/fL$, does just the same thing, bringing the deep water to rest and speeding the propagation up, while in the limit causing vanishingly small dissipation.

D. Spherical Geometry

These same arguments provide a local description of mid-latitude flows in a two-layer spherical ocean, (Fig. 6c) with rigidity imparted to the fluid parallel to the rotation axis. The wedge geometry is provided, in effect, by the northward gradient of the axial projection of the layer depths, equivalent to gradients of f/h_i where $f = 2\Omega \sin(\text{latitude})$. The inclusion of continuous stratification, the final link to the real oceans, is a difficult step. The integrity of fluid columns parallel to $\mathbf{\Omega}$ is then lost and scale analysis gives support to the "traditional approximation," the neglect of locally horizontal components of the earth's rotation, which makes our wedge geometry an analogue rather than a replica of the complete spherical ocean.

Finally we mention that other, more distant analogues may be of value to the intuition. For example, the nonrotating Boussinesq thermal convection of a fluid with strong basic temperature stratification, $\bar{T}(z)$, obeys

$$\frac{d\bar{T}}{dz} w = Q(x, z)$$

$$\frac{\partial u}{\partial x} + \frac{\partial w}{\partial z} = 0$$

for a gentle internal heat source distribution, $Q(x, z)$. This is a Hadley circulation in (x, z) fundamental to atmospheric modeling, and is formally the same as Sverdrup flow in (x, y) with $\beta \equiv d\bar{T}/dz$, $\nabla \times \tau|_z/\rho_0 f_0 \equiv Q$. Fluid that is heated rises slightly to a new equilibrium level, without significantly altering the mean state. Simple gyres in (x, z) are possible in confined geometry, if only $\int Q \, dx = 0$ at each level. Otherwise conductive boundary layers form at the side walls to close the vertical circulation.

By establishing that Sverdrup flow and long internal Rossby waves involve nearly inextensible translation of fluid columns in the wedge analogue, we can gauge the transition to more vigorous currents. To do so, we quantify the stiffness by solving for the flow induced by simple forcing with $\beta = 0$; imposed vertical velocity must then produce relative vertical vorticity, ζ, according to $\partial \zeta/\partial t = -f(\partial w/\partial z)$. Suppose $w_0 = w_0 \sin(x/L) \sin(t/T)$; the resulting pressure field is

$$P = \frac{p}{\rho} = \frac{f^2 T w_0 L^2}{H} \sin \frac{x}{L} \cos \frac{t}{T}.$$

The oscillatory currents have an *impedance* with respect to vertical stretching,

$$\left| \frac{P}{w_0} \right| = \frac{f^2 L^2 T}{H},$$

here a reactance. The increased vertical stiffness at large scale or large rotation rate is evident.

The average kinetic energy produced here by pressure work, $\overline{Pw_0}$, below the lid is $f^2 w_0^2 T^2 L^2 / 16H$, which may be compared to the kinetic energy in a pure Sverdrup flow ($v = w_0/\alpha$) driven by the same distribution of vertical velocity (but with sloping geometry), which averages to $Hw_0^2/8\alpha^2$. Now if the forcing becomes so rapidly changing as to make the ratio of these energies, $(\alpha f TL/2H)^2$ $[\equiv (\beta LT/2)^2]$, less than unity, the least work is expended if the fluid begins to yield to vertical stretching and compression, with less motion to the north and south. The less energetic choice is made, and Sverdrup flow gives way to Rossby waves, involving essential relative vorticity, when βLT falls below unity. If the forcing is more intense than this, so that nonlinearity is significant, nonnegligible vorticity appears when $\alpha f L^2/HU < 1$ (i.e., $\beta L^2/U < 1$), and two-dimensional turbulence appears, instead of linear waves, to replace the classical circulation. These same conclusions follow from a more conventional scale analysis of the potential vorticity equation, which is developed in the next section.

5. Linear Potential-Vorticity Waves

The earliest explorations of long-period waves ("linearized eddies") on a rotating sphere followed the train of thought found in classical elastic oscillations of solids and membranes, and in the short-period tides, that is, dominance of grave modes of large scale. This seemed appropriate because the assumed forcing effect, the winds and sun, act on a large scale, and also because the natural frequencies and propagation speeds turned out to be the *largest* for the grave modes (with all "higher" modes contained at lower frequency). With the intellectual focus moving more or less continuously downward through the frequency spectrum, the grave modes were the first oscillations to be found at periods greater than one-half pendulum day. LaPlace's tidal equation, which reduces to the spherical form of the potential vorticity equation at small ω, thus was divided by Hough (1898) into first- and second-class waves according to whether ω was greater or less than unity. Longuet-Higgins (1964 et seq.) has produced the most thorough discussion of the second-class planetary waves on a sphere, as did Ball (1963 et seq.) for smoothly shaped rotating basins.

It was imagined that, in combination with classical circulation theory, the planetary waves would provide the complete solution to the currents forced by the real, unsteady winds and heating. In fact, the possibility of nonlinear interaction in a weak sense was recognized, and some calculations of weak rectified flows (wave–current interaction) and wave–wave resonant interactions were produced (e.g., Pedlosky, 1965; Veronis, 1970; Kenyon, 1964; Longuet-Higgins and Gill, 1966).

Now these grave-mode planetary waves may be relevant to oceanwide seiches of periods less than a month once the modifications owing to variable ocean depth are included (Rhines, 1969; Platzmann, 1974). However, despite recent measurements supporting their existence, they are most unenergetic, accounting for currents not exceeding 1 cm/sec or so. The evidence is now overwhelming that the scale of the dominant energy-containing eddies is 100 km or less, corresponding more closely with the Rossby internal radius of deformation, L_ρ, than with the geometry of either the basin or the external forcing effects.[2]

Linear waves are worthy of study, even though the linear approximation is doubly bad at the mesoscale (the currents being far stronger there, and the phase speeds far slower than with the basinwide planetary modes). However, it turns out that vestiges of the linear theory apply well into the nonlinear range, even when energy transformations are becoming violent. In addition, powerful intuition about the spin-up of steady circulations (and the resolution of degeneracies in them) is provided by the group velocity of low-frequency waves, and this intuition provides rough upper bounds to the rate at which influence can propagate laterally, above and beyond any fluid advection velocity. Veronis and Stommel (1956) and Lighthill (1967, 1969) have emphasized this use of wave theory, and produced a rich picture of the regions of influence and rates of communication of currents forced on a β plane.

A. Derivation of Equations

By centering attention on the $O(100 \text{ km})$ scales, we can be less uneasy about the mid-latitude β plane approximation, which is essentially an expansion of the problem in terms of L/R_e, R_e being the earth's radius multiplied by tan(latitude). The traditional approximation is also taken. This neglects the upward Coriolis force $2\Omega \cos \lambda u$ (due to overwhelming static stability) and the eastward Coriolis force $2\Omega \cos \lambda w$ (due to the smallness of vertical velocity, w, in low-frequency motion of small aspect ratio, H/L; see Miles (1974) and Needler and LeBlond (1973). The momentum and mass conservation equations for free adiabatic motions are

$$\frac{\partial \mathbf{u}}{\partial t} + (\mathbf{u} \cdot \nabla)\mathbf{u} + \mathbf{f} \times \mathbf{u} = -\frac{\nabla p}{\rho} - \mathbf{g} + \nu \Delta \mathbf{u} \tag{7}$$

$$\frac{\partial \rho}{\partial t} + \nabla \cdot \rho \mathbf{u} = 0$$

Here (x, y, z), $(\hat{\mathbf{i}}, \hat{\mathbf{j}}, \hat{\mathbf{k}})$ are Cartesian coordinates and unit vectors eastward, northward, and locally upward, t is time, $\mathbf{u}$ is the velocity, $f = 2\Omega \sin \lambda$ where λ is latitude, $\mathbf{g}$ is the gravity vector, plus centrifugal acceleration, p is the pressure, and v is the kinematic viscosity. The Boussinesq approximation applies when the compressibility is slight enough that the density scale height far exceeds the fluid depth, and when temperature- and salinity-induced density variations are also small. For the oceans the pressure effect on density is not really negligible (fractional range of in situ density is $\sim 3\%$) and deep-ocean vertical density gradients far exceed the gradients of potential density, $\partial \rho_p/\partial z = (\partial \rho/\partial z) - (\rho g/c^2)$ (the dynamically relevant quantity), where c is the speed of sound. However, we henceforth ignore such effects, which mainly affect the calculation of static stability from vertical soundings.

---

[2] This chapter exclusively treats mid-latitude regions. Within a band of a few degrees north and south of the equator the waves are, on the whole, more rapidly propagating, particularly in the baroclinic modes due essentially to the reduced impedance of the fluid with respect to vertical stretching. The ocean may turn out to be rather like the atmosphere, in being largely nonlinear at middle latitudes, yet full of energetic linear waves near the equator. The zone of distinct equatorial dynamics is $(gH)^{1/2}/\beta \sim \pm 2.5°$ for the oceans (H = equivalent depth ~ 0.75 m), whereas for the atmosphere it is much greater, $\sim \pm 33°$ ($H \sim 10$ km). (Higher vertical modes, trapped within $10°$ of the virtual equator, are also of meteorological interest.)

Take the following nondimensionalization:

$$[x, y] = L, \qquad [z] = H, \qquad [t] = (f\omega)^{-1}, \qquad [u, v] = U, \qquad [w] = \frac{UH}{L},$$

$$[\zeta] = \frac{U}{L}, \qquad [p'] = fUL\rho_0, \qquad [\rho'] = \frac{\rho_0 fUL}{gH}, \qquad [\zeta_H] = \frac{U}{H}$$

where $(\zeta_H \cdot \hat{\mathbf{i}}, \zeta_H \cdot \hat{\mathbf{j}}, \zeta)$ is the vorticity, $\nabla \times \mathbf{u}$. The departures from mean hydrostatic values $\bar{p}$ and $\bar{\rho}$ of the pressure and density are p' and ρ', respectively: $p = \bar{p} + p'$, $\rho = \bar{\rho} + \rho'$. Let $f = f_0[1 + (L/R)y]$, where $R =$ earth radius $\times \tan$ (mean latitude). The vorticity equation, the curl of equation 7, is then

$$\omega\zeta_t + \varepsilon\mathbf{u}_H \cdot \nabla\zeta + \varepsilon\frac{H}{L}w\zeta_z - \left(\varepsilon\zeta + y\frac{L}{R} + 1\right)w_z + \varepsilon(\zeta_H \cdot \nabla)w + \frac{L}{R}v$$

$$= \varepsilon\frac{\rho_0^2}{\rho^2}\frac{f^2L^2}{gH}(p'_x\rho'_y - p'_y\rho'_x) \quad (8)$$

for the vertical component and

$$\omega\zeta_{H,t} + \varepsilon\mathbf{u}_H \cdot \nabla\zeta_H + \varepsilon\frac{H}{L}w\zeta_{H,z} - \varepsilon(\zeta_H \cdot \nabla)\mathbf{u}_H - (1 + yL/R + \varepsilon\zeta)\mathbf{u}_{H,z}$$

$$= \frac{\bar{\rho}\rho_0}{\rho^2}\left(1 + \frac{\varepsilon\rho_0 f_0^2 L^2}{\bar{\rho}gH}p'_z\right)\hat{\mathbf{k}} \times \nabla\rho' + \frac{\rho_0 H}{\rho H_s}\hat{\mathbf{k}} \times \nabla p' \quad (9)$$

$H_s^{-1} = -(1/\bar{\rho})\bar{\rho}_z \equiv N^2/g$, the density scale height, for the horizontal. With oceanic scaling the right side of equation 8, the creation of vertical vorticity by twisting, $\nabla p' \times \nabla\rho'$, is of relative order $10^{-3}\varepsilon$, and will be ignored. The term $(H/H_s)\hat{\mathbf{k}} \times \nabla p'$ in equation 9 is negligible ($\sim 10^{-2}$ at most) relative to $\hat{\mathbf{k}} \times \nabla\rho'$.

The vertical momentum and continuity equations are

$$\omega w_t + \varepsilon\mathbf{u} \cdot \nabla w = \frac{L^2}{H^2}(-p'_z - \rho') \quad (10)$$

$$\omega\rho'_t + \varepsilon(\mathbf{u} \cdot \nabla\rho) = \delta B^2 w \frac{\bar{\rho}}{\rho_0} \quad (11)$$

$$\nabla \cdot \mathbf{u} = 0 \quad (12)$$

where $B = NH/f_0 L$.

Now expand in ε, with $\omega = O(\varepsilon)$ formally, and $H/H_s \leqslant O(\varepsilon)$, $f_0^2 L^2/gH \leqslant O(\varepsilon)$. $L/R = O(\varepsilon), (\bar{p} - \rho_0)/\rho_0 \leqslant O(\varepsilon^2) (u, v, w, p', \ldots) = (u^0 + \varepsilon u' + \cdots, v^0 + \varepsilon v' + \cdots, \ldots)$ $O(\varepsilon^0)$:

$$w_z^0 = 0 \quad (13a)$$

$$u_z^0 = \rho_y^0, v_z^0 = -\rho_x^0 \quad (13b)$$

$$u_x^0 + v_y^0 = 0 \quad (13c)$$

$$p_z^0 = -\rho^0 \quad (13d)$$

$$B^2 w^0 = 0 \quad (13e)$$

This says that the scaling was mistaken in one respect: $w \sim \varepsilon(H/L)U$ rather than UH/L. The typical inclination of the velocity vector from horizontal is far smaller than the aspect ratio, H/L, would alone suggest. At this order, then, there exists a stream function for the horizontal velocities, $u^0 = \hat{\mathbf{k}} \times \nabla\psi$, which is simply proportional to the pressure field:

$$-\hat{\mathbf{k}} \times \mathbf{u}^0 = -\nabla p^0$$

The next order balance is of interest, not so much in giving small corrections, but in revealing the slow evolution of the dominant fields.

$$O(\varepsilon): w'_z = \frac{\omega}{\varepsilon}\zeta^0_t + \mathbf{u}_H \cdot \nabla\zeta^0 + \frac{L}{R}v^0 \tag{14a}$$

$$\mathbf{u}'_{H,z} + \hat{\mathbf{k}} \times \nabla\rho' = \frac{\omega}{\varepsilon}\boldsymbol{\zeta}^0_{H,t} + (\mathbf{u}_H \cdot \nabla)\boldsymbol{\zeta}_H{}^0 - (\boldsymbol{\zeta}_H{}^0 \cdot \nabla)\mathbf{u}_H{}^0 \tag{14b}$$

$$u'_x + v'_y = 0 \tag{14c}$$

$$p'_z = -g\rho' \tag{14d}$$

$$\frac{D^0\rho^0}{Dt} = \delta B^2 w' \tag{14e}$$

where

$$\frac{D^0}{Dt} = \frac{\omega}{\varepsilon}\frac{\partial}{\partial t} + \mathbf{u}_H{}^0 \cdot \nabla$$

The left side of the vertical vorticity equation 14a is eliminated with equations 14e and 13d:

$$\frac{D^0}{Dt}(\zeta^0 + (B^{-2}p_z{}^0)_z) + \frac{L}{\varepsilon R}v^0 = 0$$

In terms of the stream function,

$$\frac{D^0}{Dt}\left(\nabla^2\psi^0 + (B^{-2}\psi_z)_z + \frac{\omega}{\varepsilon}f\right) = O\left(\varepsilon + \frac{H}{H_s} + \varepsilon\frac{f^2L^2}{gH}\right) \tag{15}$$

with $(D^0/Dt)\psi_z{}^0 = (\delta B^2/\varepsilon)\nabla\psi^0 \times \nabla h \cdot \hat{\mathbf{k}}$ on $z = -H$, $(D^0/Dt)\psi_z{}^0 = 0$ on $z = 0$. Here the fluid is confined between rigid flat top, $z = 0$, and rigid bottom, $z = -H(1 + \delta h(\mathbf{x}))$, $\delta \lesssim \varepsilon$. This is the conservation of geostrophic potential vorticity, following nearly horizontal particle trajectories. In other situations more exact conservation laws exist; for instance, the barotropic potential vorticity law,

$$\frac{D}{Dt}\left(\frac{h\nabla \cdot h^{-1}\nabla\psi + f}{h}\right) = O\left[\varepsilon^2\left(\frac{H}{L}\right)^2\right] \tag{16}$$

obeyed by uniform density fluid in a rapidly rotating container of depth h.[3] Here the vertical stretching term imposed by horizontal flow across an uneven bottom, $f(D/Dt)h^{-1}$, replaces the stretching of fluid between two, infinitesimally separated, isopycnal surfaces, $(D^0/Dt)(B^{-2}\psi_z)_z$, in equation 15. Ertel's (1942) relation,

$$\frac{D}{Dt}\left(\frac{(\nabla \times \mathbf{u} + \mathbf{f}) \cdot \nabla\rho}{\rho}\right) = 0$$

is the most exact of all, giving a potential vorticity that is conserved even in strongly ageostrophic, nondiffusive flows. For a more thorough account, see Phillips (1963).

B. Linear Waves

The range of phenomena governed by equation 15 is vast. Taking the limit $\varepsilon/\omega \to 0$, we recover the purely linear wave regime. It cannot be overstressed that nonlinearities are rarely negligible in this sense, even though the momentum equation is quite accurately geostrophic [$O(\varepsilon)$]. But linear theory is important just as, in the study of surface gravity waves in a full gale, the linear propagation theory would still hold in a gross sense for the dominant waves (and accurately for much longer waves). The smallness of the terms in equation 15 requires that we not forget for long diffusive effects and interaction with small-scale processes like internal waves.

Three distinct kinds of wave can be identified in the linear solutions of equation 15. To illustrate, take the boundary conditions to be free in the horizontal, rigid at the sea surface, $z = 0$, and at the sea floor, which slopes uniformly in the north–south direction:

$$\psi_z = 0 \qquad \text{on } z = 0 \tag{17}$$

$$\omega B^{-2}\psi_{zt} = \delta\psi_x \qquad \text{on } z = -1 \tag{18}$$

where the zero superscript has been dropped, and the bottom slope, of magnitude $\delta L/H$, is small $\delta \sim \omega$. The interior equation is

$$(\nabla_H^2\psi + (B^{-2}\psi_z)_z)_t + \left(\frac{L}{\omega R}\right)\psi_x = 0 \tag{19}$$

---

[3] In a case of particular interest, waves in homogeneous fluid on a thin spherical shell, care is required in applying equation 16. Though it is appropriate to discuss more gentle motions (as in Fig. 6c) in which fluid columns retain their integrity *parallel to the rotation axis*, such a model gives grossly incorrect results when used to develop a wave theory. This is because, on a sphere, the error in applying the Taylor–Proudman approximation is of order $\omega H/L$ [whereas on a flat β plane the error is $O(\omega H/L)^2$ only]. If, instead, the entire vector vorticity equation is expanded in ω, rather than assuming away parts of it, the correct wave speeds result, and fluid columns retain their alignment with the *local* vertical. The solution agrees in the short-wave limit with the usual β plane formula. The mistake of assuming the Taylor–Proudman approximation to be exact seems to be common. Another illustration is the later interpretation of the wave pattern produced in experiments of Fultz and Frenzen (1955). When an obstacle was dragged westward about a latitude circle in the spherical shell, a distinct train of waves was produced that obeyed the planetary wave dispersion relation. When the obstacle was caused to move in the opposite (eastward) direction, no planetary waves should have been possible, yet there appeared a peculiar, periodic wave train with cusp-like particle paths. Though elaborate planetary solutions of equation 16 have been produced to explain the pattern, it is very likely that they were simply standing inertial oscillations (which have cycloidal particle paths). Once again, the *possibility* of β plane, Taylor–Proudman dynamics applying has spirited away the simpler f plane, nonhydrostatic flows from our minds. (The experiments are shown by Greenspan, 1968, p. 266.)

The problem is separable into vertical and horizontal parts. For uniform density stratification, plane-wave solutions are

$$\psi = e^{ikx + ly - \hat{\omega}t} \begin{Bmatrix} \cosh \mu z \\ \cos mz \end{Bmatrix}$$

already satisfying equation 17. Substituting into equations 18 and 19, we have

$$\hat{\omega} = \frac{-kL/R\omega}{k^2 + l^2 + B^{-2}m^2}$$

and

$$m \tan m = \frac{\delta R}{L} [(k^2 + l^2)B^2 + m^2] \qquad (20)$$

or

$$\mu \tanh \mu = - \frac{\delta R}{L} [(k^2 + l^2)B^2 - \mu^2]$$

The solutions divide into those with vanishing vertical shear, those with oscillatory structure, and those which are evanescent in the vertical.

Type 1. Fast Barotropic

For $k^2 + l^2 \ll B^{-2}$ a solution with $m^2 = O[(\delta R/L)(k^2 + l^2)B^2] \ll 1$ is (returning to dimensional variables)

$$\sigma = \frac{-k(\beta - f_0 \alpha/H)}{k^2 + l^2} = \frac{(\beta - f_0 \alpha/H) \cos \theta}{|\mathbf{k}|} \qquad (21)$$

$$\beta = \frac{df}{d \text{ (latitude)}}, \qquad \sigma = \hat{\omega}\omega f_0, \qquad \alpha = \nabla h = \frac{\delta H}{L}$$

These familiar topographic Rossby waves involve only slight $[O(\delta)]$ density perturbations, and are the same shearfree modes found on an unstratified β plane. Both the frequency and propagation speeds rise with wavelength, indefinitely in this model.

Type 2. Fast Baroclinic

The above solution ceases to exist at scales as small as the Rossby radius, NH/f. There arises to replace it a wave confined within a layer of thickness fL/N above the sloping bottom. Its purest form has $k^2 + l^2 \gg B^{-2}$ and $\beta = 0$. Then, in dimensional form, the dispersion relation is

$$\sigma = -N\alpha \sin \phi \qquad (22)$$

where ϕ is the angle of the wave vector (k, l) from Oy. The wave field, also in dimensional form, is

$$\psi = e^{i(kx + ly - \sigma t)} \cosh\left(\frac{zN}{f_0 L}\right)$$

This is a harmonic solution of $\Delta\psi = 0$, the rather passive form which equation 19 takes in the absence of β. The spatial scaling $B \sim 1$ is ubiquitous to rotating, stratified fluids. The dispersion relation is that of a simple buoyancy oscillation. In common with other trapped solutions like Kelvin waves, it is independent of the very property,

f, that causes its trapping. But this mode has a kind of dual nature. If we replace h in equation 21 by the penetration height, fL/N (with $\beta = 0$), the dispersion relations 21 and 22 become the same. We may thus consider this to be a topographic Rossby wave, where density stratification provides a lid for vortex stretching.

Type 3. Slow Baroclinic

At all scales, modes exist that are oscillatory in the vertical, in the presence of β. For small slopes (relative to H/R, $\sim 10^{-2}$) these obey equation 20 $\tan m \approx 0 \to$ $m \approx n\pi$, while for larger slopes, $m \to (n + \frac{1}{2})\pi$. The most quickly varying such wave (for $\delta \to 0$), $m = \pi$, has a period of about 1 yr and horizontal scale $L_\rho \sim 40$ km. At larger scales, $|\mathbf{k}|^2 B \ll 1$, the dispersion relation becomes

$$\hat{\omega} = -\frac{kL/R\omega}{\pi^2 B^{-2}} \qquad \left(\frac{\delta R}{L} \to 0, m \approx \pi\right)$$

$$= \frac{4kL/R\omega}{B^{-2}\pi^2} \qquad \left(\frac{\delta R}{L} \to \infty, m \approx \frac{\pi}{2}\right)$$

This is the nondispersive, purely westward propagating mode derived in a simpler fashion in Section 4. The increase in frequency and propagation speed over a sloping bottom by a factor of four becomes somewhat less dramatic when realistic, nonuniform density stratification of the ocean is included.

The dispersion relation (Fig. 8) shows each of the three limiting types. Sketched on the figure is the corresponding vertical structure of the current. With realistic stratification, which is strongest in the upper ocean, the energy of type 3 waves is increasingly confined there. A fraction, roughly $(L/L_\rho)^2$, of the energy is potential. This normally exceeds unity in cases of interest, so it is appropriate to call these baroclinic motions "thermocline eddies."

The fast baroclinic waves (type 2) are, conversely, confined near the sloping bottom. Their inclusion here is of course schematic for those regions of ocean with complex topography. But it makes clear what was omitted from earlier wave models, that vertical shear (and density perturbations) with periods far shorter than 1 yr can exist under linear quasi-geostrophic dynamics.

How great must the bottom slope be for type 2 waves to occur? It turns out (Rhines, 1970, 1971a) that one may imagine there to be a competition for the highest frequency, among the basic planetary, topography, and bouyancy waves. The β effect alone provides a frequency $\sim fL/R$. Simple vortex stretching by the topography without stratification causes a frequency $\sim f|\nabla h|L/H$. Motion forced up a slope, disturbing the density field, suggests the frequency $\sim N|\nabla h|$. Predominance of bottom-trapped waves thus occurs if

$$\frac{NH}{fL} > 1 + \frac{H}{R|\nabla h|}$$

on the basis of scale analysis. If, therefore, the slope exceeds that equivalent to β ($\sim 10^{-3}$), the type 2 wave may be expected at scales of order the Rossby internal radius, $L_\rho \sim NH/f$, and smaller. At slopes so large that $\omega \gtrsim 1$ ageostrophic theory is required, yet the same, trapped buoyancy oscillation occurs.

The fast barotropic wave, type 1, is of course a β plane representation of the planetary wave on a sphere. In a forced problem, with wind blowing across the ocean surface, say, this mode is encountered alone at large frequency. The response, undiminished through the water column, is of crucial importance, and is one of the

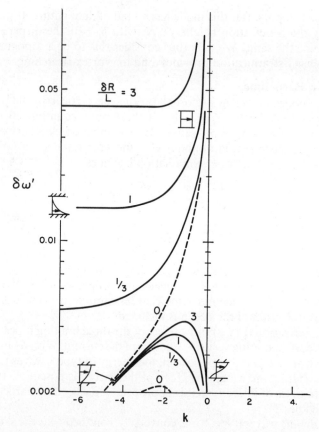

Fig. 8.   Dispersion relation for linear topographic-Rossby waves (β plane, uniform upslope to the north), for several values of slope. Rossby waves, $\delta = 0$, are plotted with dashes. The sketches show the vertical structure. The three classes of structure are referred to as 1, 2, and 3 in the text. $R = 6000$ km (43° latitude), wave number normalized by $NH/f = 120$ km.

striking predictions of linear theory, for "weatherlike" time scales over horizontal scales greater than L_ρ. Veronis and Stommel (1956) established the result for an ocean of constant depth, and it holds here with a sloping bottom, unless $L \lesssim L_\rho$.

Perspective views of the dispersion surface are shown in Figs. 9 and 10, and $\delta > 0$ (upslope to the north). The double arrows give the direction and relative size of the horizontal group velocity, $\omega f L(\partial \hat{\omega}/\partial \mathbf{k})$. The transition from classical type 1 waves to bottom-trapped type 2 waves carries one from a situation in which the group velocity can point in any direction (yet the phase velocity always has a westward component) to one in which both group and phase velocities must have a westward component. For arbitrary orientation of the depth gradient, type 2 waves tend to move energy and phase to the left facing shallow water, while type 1 is a topographic Rossby wave, relative to a "pseudo-westward" direction along f/h contours, and type 3 tends still to favor actual westward propagation. The larger group velocities occur near the origin of Fig. 9, and these have a westward component. Conversely, those components, generated by a point excitation, say, which do radiate eastward have large amplitude, if anything like equal energy flux occurs with respect to direction.

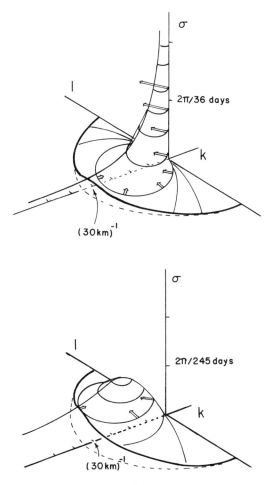

Fig. 9.   Perspective view of dispersion surface, $\sigma(k, l)$ (constant depth). Equal-σ curves are shown. Double arrows indicate direction and (schematically) magnitude of group velocity. Upper portion is the barotropic mode, lower portion the baroclinic. Note different frequency scales. N/f is uniform, $= 30$.

Another immediate consequence of the linear theory is that thermocline eddies (type 3), already very slowly propagating, will be doubly inefficient at carrying energy north or south from a source. The greatest north–south component of group velocity is $0.23\beta c_0^2/f_0^2$ (for $\delta = 0$), whereas the westward velocity approaches $\beta c_0^2/f_0^2$ in the entire set of long waves. For oceanic conditions these velocities are roughly 0.9 cm/sec and 3 cm/sec, respectively

Lighthill (1967, 1969) has emphasized the application of wave theory in the limit of vanishing frequency to the development of forced, steady flows. For instance, the Taylor column in a simple rotating fluid may be visualized as a region of influence of inertial waves of small frequency. In that limit the group velocity remains large ($\sim f_0 L$, where L is the horizontal scale of the "source") and acts to resolve a degenerate flow without the immediate need of invoking viscosity. Here the degenerate, free flows are those along geostrophic contours (for barotropic flow) or zonally (for large-scale baroclinic flow), or along depth contours (for small-scale deep baroclinic flow).

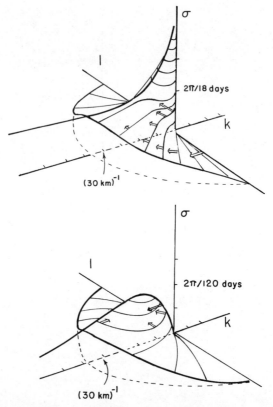

Fig. 10. Perspective view of $\sigma(k, l)$, with a bottom slope of 1.7×10^{-3} (up to the north). $(\delta R/L = 2)$. Both types have increased frequency and propagation rate, relative to $\delta = 0$. The constant-frequency loci show an absence of eastward group velocities with short, bottom-intensified waves.

Types 1, 3, and 2 show that the signal from a slowly oscillating forcing function propagates westward or pseudo-westward at rates $(\beta + \alpha f_0)L^2$, βL_ρ^2, and $N|\nabla H|/L$, respectively, and in doing so carries with it the developing flow. This justifies the classical procedure of calculating Sverdrup flow by zeroing the disturbance to the east of the wind stress (Lighthill, 1967).

C. Initial-Value Problems

The wave pattern (Fig. 11) arising from the dispersion of an initial Gaussian vortex shows some of these tendencies. There is no bottom slope, and the currents are infinitesimal and entirely barotropic. This computer realization shows the characteristic penetration of east–west crested waves due westward from the origin, while slower, shorter, stronger north–south crested waves appear east of the origin. The general picture agrees with far-field ray theory, and with the Green's function for a steadily oscillating delta-function source,[4]

$$\nabla^2\psi_t + \beta\psi_x = \delta(\mathbf{x})\exp(-i\omega_0 t)$$

[4] This canonical problem is slightly misleading, in that the right side is not the result of purely local forces; the stress, τ, itself obeys $\oint \tau \cdot d\mathbf{l} = 1$ about any circuit enclosing the origin. The difference in wave pattern between this and local forcing is probably small, but the difference in amplitude structure may not be.

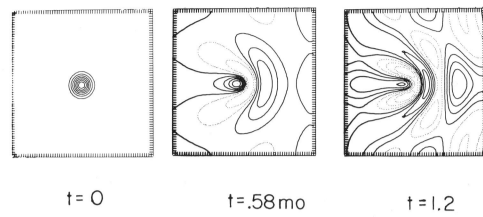

$$t = 0 \qquad\qquad t = .58\,\text{mo} \qquad\qquad t = 1.2$$

Fig. 11.   Linear, barotropic Rossby waves dispersing from an initially Gaussian current pattern; for more than infinitesimal currents, the efficiency of the radiation is severely altered. In this and later computer simulations the contours are solid for *negative* ψ, dashed for positive ψ (choosing ψ to be proportional to pressure). The boundary conditions are periodic, and the numerical scheme is the spectral code (cutoff wave number 32, corresponding to 64^2 spatial map) developed by Dr. Orszag.

which is known to be

$$\psi = \exp\left(-ik_3 x\right) H_0^{(2)}(k_3 r) \qquad k_3 \equiv \frac{\beta}{2\omega_0}$$

This particular Hankel function is chosen so that the group velocity points outward at large radius [while the phase of $H_0^{(2)}(k_3 r)$ spirals inward]. The wave crests, in the stationary phase approximation, obey

$$r = \frac{\text{const}}{\cos^2(\tfrac{1}{2}\theta)}$$

(r, θ) being polar coordinates relative to the origin. Longuet-Higgins and Gill (1966) have given the response to a space–time delta function of stress-curl, which also resembles these patterns.

Considering now the stratified case, the wave modes derived here, together with free geostrophic flow, completely describe linearized motions which develop from given currents and density. This is true provided that the initial data is itself sufficiently close to a state of geostrophy. The appearance of the frequency in the lower boundary condition renders the set of equations 20 nonorthogonal, but this may be circumvented as in the problem of a vibrating string with a nonrigid support (Morse and Feshbach, 1953, p. 1343).

To illustrate the role of the bottom-trapped type 2 we consider (with $\beta = 0$) an initial current of arbitrary vertical structure, but limited to a single horizontal Fourier component,

$$\psi(\mathbf{x}, 0) = e^{i(kx + ly)} F(z)$$

The solution to equations 15, 17, and 18, derived by Suarez (1971), is

$$\psi = e^{i(kx + ly)} \left[(e^{-i\sigma t} - 1) \frac{\cosh \mu z}{\cosh \mu} F(-1) + F(z) \right]$$

where $\sigma/f_0 = -kB^2/\mu \tanh \mu$. It is the simple sum of a wave (type 1 or 2) whose velocity at the bottom is equal to that of the initial data, and a steady geostrophic flow which vanishes there and elsewhere accounts for the remainder of the initial data. By summing such solutions about the azimuth, one may construct eddylike patterns; for instance, with

$$\psi(\mathbf{x}, 0) = J_0(k_0 r) \qquad r^2 = x^2 + y^2$$

the solution is

$$\psi(\mathbf{x}, t) = J_0(k_0 r^1)\frac{\cosh \mu z}{\cosh \mu} + \left(1 - \frac{\cosh \mu z}{\cosh \mu}\right)J_0(k_0 r)$$

$$r^1 = ((x - ct)^2 + y^2)^{1/2} \qquad c = \frac{-B^2}{\mu \tanh \mu}$$

This represents a steady circular flow at the origin, which decreases toward the bottom, as a bottom-trapped circular pattern creeps out from beneath it, along the depth contours to the "left." The difficulty with such solutions is that they are not truly local, and rely on inward radiation from infinity to defeat the natural dispersive nature of the waves.

The role of a weak β effect would be to cause the "steady" component to propagate itself, as a type 3 wave. Howard and Siegmann (1969) have discussed in general terms the possible steady geostrophic flows in such cases.

D. Observations

The observations made in the introduction fixed the gross time and length scales of the energy-containing eddies, both deep and shallow. We are rapidly obtaining refined measurements to improve upon this picture, which bring observations and theory somewhat closer together.

Type 1.   Fast Barotropic

The Mid-Ocean Dynamics Experiment (MODE-I) covered a $(200 \text{ km})^2$ region centered on $28°\text{N}$, $69°40'\text{W}$, in the Sargasso Sea, Spring 1973. Rossby and Webb's neutrally buoyant SOFAR floats drifted with the currents at 1500 m. Characteristics of the floats were unusual accuracy (roughly 1 km absolute, $<\frac{1}{2}$ km relative position fixing) and longevity (the observations in this area began in September 1972 and continue at the time of this writing, May 1975).

There were often more than 10 floats present in the area, and they allowed rather accurate, objective stream function maps to be produced (in real time by Bretherton, and subsequently by Freeland and Gould). Figure 12a from Freeland, Rhines, and Rossby (1975), is an x–t plot of the stream function at $28°\text{N}$; Fig. 12b is a y–t plot taken along $70°\text{W}$. There appears a persistent tendency for phase lines to move westward with time, but no striking north–south propagation. Thus the nearly ubiquitous property of westward phase propagation, found in the linear theory of Rossby waves, actually appears in currents below the main thermocline.

The phase speed ranges from 2 to 12 cm/sec in the figure, averaging about 5 cm/sec. This slightly exceeds the rms current speed, which is 4 cm/sec in the region, and far exceeds the mean flow. The horizontal length scale is also visible; the transverse spatial correlation function first crosses zero at 50 km. For linear theory, flat-bottom waves (equation 21) have a westward phase speed equal to $\beta/|\mathbf{k}|^2$. The estimate

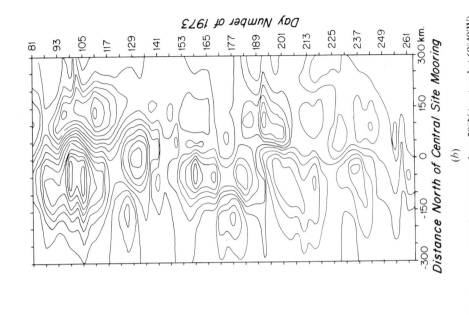

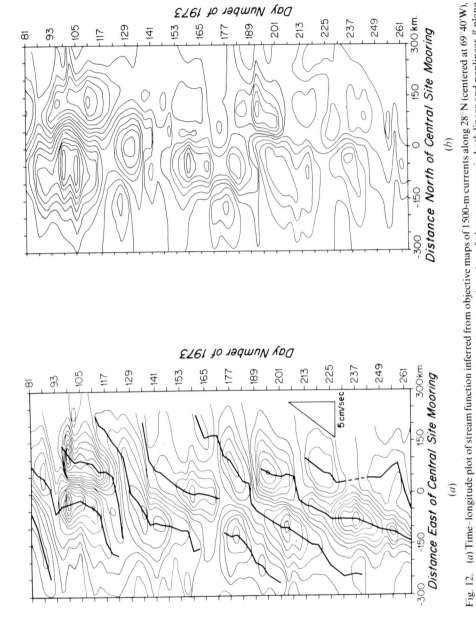

Fig. 12. (*a*) Time–longitude plot of stream function inferred from objective maps of 1500-m currents along 28°N (centered at 69°40'W), by Freeland, Rhines, and Rossby (1975). There is evidence of westward motion of phase, as occurs in both linear and nonlinear β plane theory (compare Fig. 20*b*). The large-scale f/h contours in the region lie roughly east and west. The data came from neutrally buoyant SOFAR floats (Figs. 64–66). The currents at this depth are not dominated by thermocline eddies, but are more representative of the deep ocean. (*b*) As in (*a*), but time–latitude plot along 69°40'W.

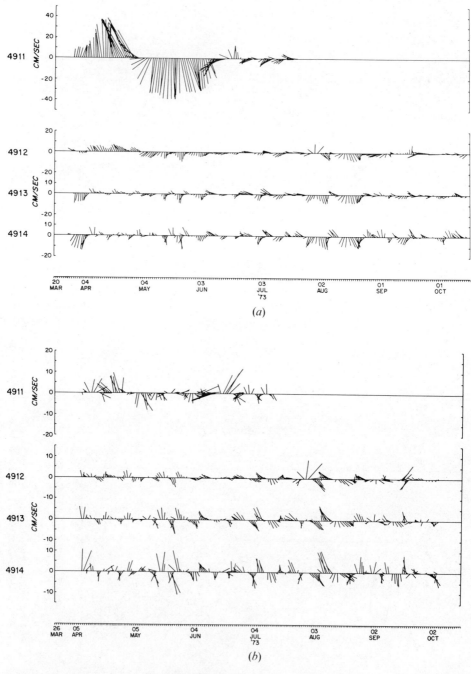

Fig. 13. (a) Currents at Site D (39°10'N, 70°W), at levels 205 m, 1019 m, 2030 m, and 2550 m. The ocean depth is 2650 m. A thermocline eddy dominates the upper flow, followed by an interesting, rapid oscillation. (b) High-pass filtered version of (a). The deep westward mean flow and upper-level thermocline eddy are thus removed. The deeper layers are dominated by fast oscillations with episodes of clear bottom intensification and polarization ($\overline{uv} < 0$).

222

5 cm/sec implies a rational scale, $|\mathbf{k}|^{-1}$, of 50 km. More is said about these exceptional measurements below.

Type 2.   Fast Baroclinic

Cursory examination of current meter records frequently shows the existence of vertical shear at high frequencies, relative to the predicted, $O(1 \text{ yr})$, cutoff for baroclinic Rossby waves (yet less than f_0). This is particularly true as one approaches the ocean floor. Figure 13 shows current meter records from site D $(39°10'\text{N}, 70°\text{W})$ taken by Luyten, Schmitz, and Thompson of the Woods Hole Oceanographic Institution. The location is some 100 km north of the mean Gulf Stream axis and 50 km south of the continental shelf. There is a persistent northward shoaling with slope 8×10^{-3}. First, observe the agreement in general terms of the character of the vertical structure with the linear picture. The shallow level is dominated by a strong current which varies with a 3-month period. The deeper currents are weaker, yet more quickly varying.

High-pass frequency filtering (by Dr. Luyten) gives a rather definite character to the deep oscillations. Below 1000 m, they tend to increase in energy with depth, and develop a strong polarization, with u and v negatively correlated. All these features are expected of linear waves of type 2; in particular, the natural period $\geqslant 2\pi/N|\nabla h|$ ≈ 6 days here, and the independence of the period from the wavelength allows the testing of the dispersion relation by a *single* vertical string of current meters (Rhines, 1970, 1971a). In addition the vertical scale observed is directly proportional to the horizontal scale, which may thus be inferred. Thompson and Luyten (1976) and Rhines (1971b) have made just such comparisons, using spectral analysis. Figure 14

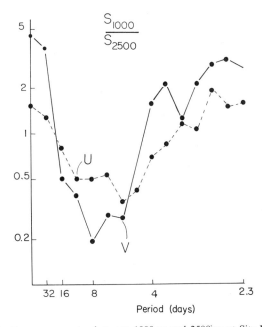

Fig. 14.   Ratio of kinetic energy spectra between 1000 m and 2500 m at Site D versus frequency, from Thompson and Luyten (1976). The records occupied 10 months in 1972. Both u and v spectra are shown. Most of the energy in the v spectrum lies between 8 and 20 days. Energy in a significant band is bottom-intensified. The wavelength inferred from linear theory ranges from 90 to 160 km.

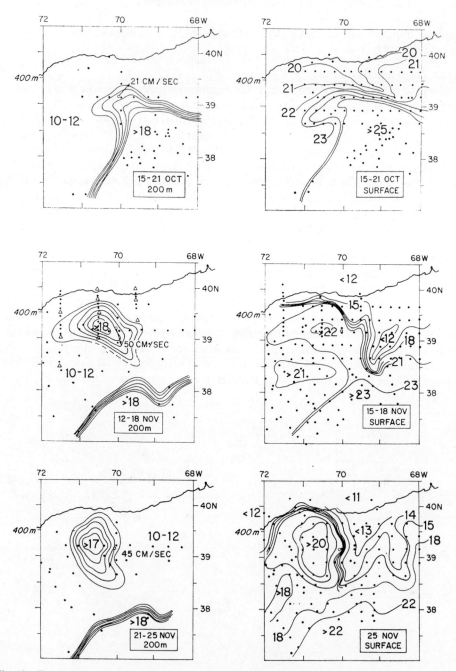

Fig. 15.   Temperature (°C) maps showing a warm anticyclonic ring cast northward by the Gulf Stream (Saunders, 1971). The current vector originates at Site D. The surface flow entrains cold water from the shallow continental shelf while the deep field remains relatively simple. Such an event probably caused the 200-m flow in Fig. 13a. Although the ring moved westward, it would be an oversimplification to charac-terize it as a baroclinic Rossby wave.

shows how, in an intermediate band of frequencies, the spectral energy increases downward, as predicted.

What is particularly exciting is not only that the "waves" exist with apparently correct local behavior, but that their $\overline{uv}$ polarization implies group velocity with a component to the north. Thompson (1971) recognized that the nearby Gulf Stream was a likely source, and that this polarization would be expected with type 1 Rossby waves. Fortunately, the same behavior arises with type 2 waves; there is probably enough breadth to the distribution of scales that a range encompassing types 1 and 2 waves occurs.

Type 3. Slow Baroclinic

These current data are a good example of the many levels of understanding that apply to ocean observations. Some features of the deep flow agree with type 1 Rossby wave theory, yet the addition of type 2 waves fills out the picture. The strong current at the upper level might be interpreted as a type 3 baroclinic Rossby wave, yet in fact it has the signature of cutoff Gulf Stream eddies. An event very like this one was documented by Saunders (1971), Fig. 15 (see also Fig. 57). That particular eddy in fact moved westward, at roughly the speed of the climatological mean flow, and rejoined the Stream. With this mixture of advection and propagation present, it would be unwise to say that type 3 waves had been seen, so for better evidence we turn to the open ocean, far from intense currents. The POLYGON group (Brek-hovskikh, Federov, Fomin, Koshlyakov, and Yampolsky, 1971), Bernstein and White (1975), and the MODE group have each witnessed the slow westward movement of thermocline eddies. The MODE observations are summed up in Fig. 16, and x–t

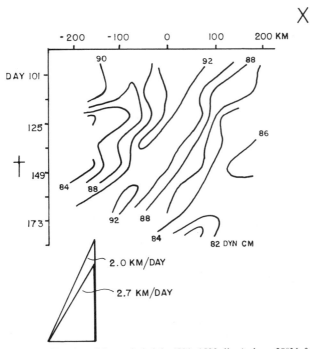

Fig. 16.   Time–longitude diagram of dynamic height (501–1500 dbar) along 28°N, from the Mid-Ocean Dynamics Experiment. There is increasing evidence of such westward phase propagation in thermocline eddies, at a rate comparable with the linear speed (see also Fig. 41); note that this is significantly slower than the observed phase motion in 1500-m currents (Fig. 12a).

diagram, again along 28°N, of the dynamic height anomaly across the main thermo-cline. This measure of the density field shows a clear westward movement at a rate 2 cm/sec. It is less likely than Gulf Stream eddies to have involved net westward movement of the water, although this is not yet certain. In fact, it is the very slowness of the propagation of thermocline eddies that makes it hard to resolve the mean flow (for one must average over many eddy periods).

Type 1.   Fast Barotropic (Ultralong)

It is shown in Section 6 that the use of linear theory is marginal, at best, in studying the above observations. But a "benthic" group in MODE-I provided observations of sea-floor pressure that are much more likely to relate to linear waves. It was discovered that the pressure records were surprisingly coherent (>0.95) over 200-km separations in the MODE region (Fig. 17; Brown, Munk, Snodgrass, Mofjeld, and

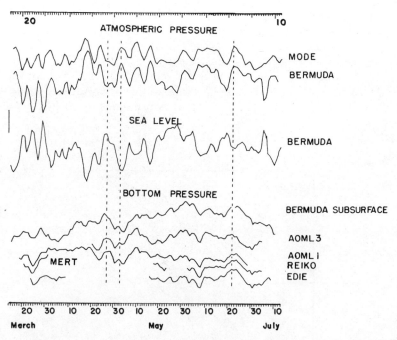

Fig. 17.   Time series of bottom pressure in MODE (Brown et al., 1975). The gauges centered on 28°N, 69°40'W are remarkably coherent, despite separations $O(180$ km). Furthermore, the pressure-corrected sea level at Bermuda (~ 650 km distant) (labeled "Bermuda Subsurface") is as well. This is strong evidence for fast, type 1 planetary waves.

Zetler, 1975). Such separations are well beyond the coherence distance for the energy-containing eddies. Then, looking at the record of the Bermuda tide gauge 700 km away, they found that it too was remarkably coherent ($\gtrsim 0.8$, once the inverted barom-eter effect of atmospheric pressure had been subtracted). This established that the pressure just below the sea surface is coherent both vertically with the sea floor, and also horizontal, over 700 km.

It may be the first observation of "ultralong" planetary waves, of the kind anticipated many decades ago.[5] To see whether the observed large scale is indeed consistent with wave theory, we converted the observed $P(\sigma)$ frequency spectrum to a wave number spectrum, assuming a model dominated by a single orientation (say, $k = l$). Let

$$\hat{P}(k) = \frac{P(\sigma)\,d\sigma}{dk}$$

where $\sigma = -\beta/2k$, $d\sigma/dk = -\beta/2k^2$. Although the frequency spectrum is "red," increasing toward small frequency, the estimated wave number spectrum is flatter, with a central peak at $k \sim 2\pi/7000$ km. The large scale is sufficient to explain the highly coherent records, with phase lag of less than a day. This is particularly so because it gives only a rough upper bound to the phase lag. The sense of the lag, though marginally determined, seems to have Bermuda leading, suggesting westward propagation. The atmospheric pressure patterns, conversely, move eastward across the region. The inferred energy spectrum $[\propto k^2 P(k)]$ is very "blue" and suggests that, indeed, the kinetic energy in the deep-ocean peaks near the 50–100 km (rational) scale.

This example verifies the almost inescapable result, discussed in Section 3, that pressure gauges see longer scales than current meters, but we could not have anticipated how great the difference would be. The observed association of higher frequencies dominantly with large scales gives support to the planetary "inverse" dispersion relation, $\omega \propto k^{-1}$, for ultralong waves as against the turbulent relation $\omega \propto Uk$, described in Section 3. The periods and scales of these waves are closer to the scales of moving weather systems, which may favor the setting up of planetary seiches by both the atmospheric pressure and winds. The result should motivate a systematic search of island-based sea-surface records, and give impetus to the refinement of the sea-floor pressure recorders.

Energy Flux

Are the large-scale motions energetically important? True, their energy density is slight relative to the small-scale eddies, but their energy flux is in fact much larger. Given that the group velocity for planetary waves has magnitude $\sigma/k = -\beta/|\mathbf{k}|^2$, one can define an energy *flux* spectrum,

$$\mathscr{F}(k) = \beta k^{-2} \mathscr{E}(k) = \frac{\beta}{\rho_0 f^2}\, \hat{P}(k) \qquad (k = |\mathbf{k}| \text{ here}),$$

giving the density with respect to scalar wave number of the kinetic energy flux, without regard to direction. $\mathscr{F}(k)$ is thus simply proportional to $\hat{P}(k)$ and is dominated by the ultralong waves.

Scale Transformations

This remark exaggerates the importance of the longer waves, unless mechanisms exist for the conversion from large to small scale. Bottom roughness provides one such mechanism, but a simpler, classical feature works in a flat bottom, linear ocean.

---

[5] Thus the Bermuda tide gauge has helped in establishing both type 1 waves, and at longer periods type 3 waves, in collaboration with the *Panulirus* stations. In either case there is significant concern over such use of island stations, for local island dynamics could conceivably have dominated the sea-level records. Happily, this result shows that the higher-frequency components are indeed of large scale. This was predicted by Rhines (1969) for the linear theory, but he gave no reasons why nonlinear, unsteady "island wakes" might not add significant noise.

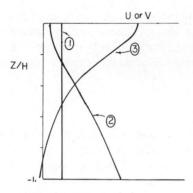

Fig. 18.   Typical vertical structure of the linear waves, types 1, 2, and 3 for exponential stratification $N \propto \exp (2z/H)$.

This is the striking fact (Pedlosky, 1965; Phillips, 1966) that the reflection laws for Rossby waves cause fast, long, weak, westward-propagating waves to reflect at a western boundary into slow, short, strong, eastward-propagating waves. It can be verified, in Fig. 9, by finding two waves of the same frequency and north–south wave number. The ratio of $|\mathbf{k}|$ of the reflected wave to that of the incident wave exceeds unity and the ratio of energy densities is the square of this ratio. Any reasonable model of the dissipation then leaves the western region filled with energetic eddies fed by a nearly invisible, ocean wide source. This occurred clearly in the laboratory (Phillips and Ibbetson, 1967). We return to the topic in Section 7.

Synthesis

Using linear building blocks we can produce current and density fields that contain a number of observed features. McWilliams and Flierl (1975) gave done so, to produce an optimal linear description of the fields in MODE-I. It should already be clear that the decrease in characteristic time scale with depth and the great decrease in kinetic energy with depth, followed by a gradual increase, are consistent with a superposition of wave types 1, 2, and 3, such that type 3 dominates the flow above the thermocline, type 1 the flow at mid-depth, and type 2 the flow in the lowest kilometer. Figure 18 shows this schematically, using an exponential stratification and a bottom slope of 8×10^{-3}. The thermocline eddies are disposed this way by the concentration of the density gradients in the upper ocean. Away from the western regions of the oceans, particularly, where the 18° water provides a deep thermocline, the type 3 waves would be strongly confined near to the ocean surface, where they could intermingle with the directly driven wind-mixed layer.

The picture may be refined by allowing the modes to interact weakly, and by including a large-scale mean current. But this might deflect us from the central issue: the thermocline eddies, even in a quiet part of the ocean like the MODE area, are sufficiently strong to blow the tops off barotropic waves, trying to propagate through them. This is the subject of the next section.

6. Nonlinear Waves and Turbulence: Primary Cascades

A classical ocean filled with linear waves would present us with a problem in spectral discrimination and synthesis. All that would remain would be the identification of sources and sinks. The basic gradient of mean potential vorticity dominates in that case, and provides a smooth restoring force. Based on typical mid-ocean conditions, however, none of the three principal wave types has a steepness that is small. Take an upper ocean velocity of 10 km/day and deep-ocean velocity of 4 km/day; the particle excursion during a wave period $\div 2\pi$, or ratio of particle speed to theoretical phase speed, $\hat{\varepsilon} \equiv \varepsilon/\omega$, is,

$$\hat{\varepsilon}_1 = \frac{2Uk^2}{\beta} = 2 \qquad\qquad (\text{deep-ocean } U, k = k_\rho \equiv 1/45 \text{ km})$$

$$= 5 \qquad\qquad (\text{shallow } U, k = k_\rho)$$

$$\hat{\varepsilon}_2 = \frac{2Uk}{N\alpha} = 0.8 \qquad (N = 2.6 \times 10^{-3}, \alpha = 5 \times 10^{-3}, k = 1/15 \text{ km})$$

$$\hat{\varepsilon}_3 = \frac{3Uk_\rho^2}{\beta} = 7$$

for the barotropic, fast baroclinic, and slow baroclinic types, respectively. The numerical factors allow the criteria to apply to the average wave speed, rather than an extreme value. Although special solutions can be found whose nonlinear advection is far smaller than these numbers suggested, the interactions between different Fourier components must, on the average, be strong, particularly in regions of more vigorous currents than these. Such a situation is remarkable, when it is remembered that for deep-ocean velocities any less than these, the current-meter rotors would be stalled a significant fraction of the time. However, all the forces acting to accelerate the fluid are weak and a state of horizontal turbulence is readily reached. A persuasive picture of the wave steepness (Fig. 64) is the ensemble of tracks of the SOFAR float experiment in MODE, Rossby, Voorhis, and Webb (1975). Even at 1500 m, the slowest level in the ocean, the tracks show particles not to be confined near latitude lines, but chaotically traveling distances far greater than the dominant length scale, say 50 km, of the flow.

A. The Diagram

Figure 19 shows schematically the nonlinear cascades that occur in a flat-bottom ocean. We have plotted turbulent states on this dispersion diagram by endowing turbulent fluid of dominant wave number k and rms particle speed U with a frequency kU. There is some reality in this, (Section 3), as a crude representation of the concentrated frequency–wave number spectrum. The vertical axis is a crude two-point representation of the vertical wave number. Thus baroclinic modes appear on the upper plane, barotropic modes on the lower plane. For linear waves, $U < \omega(k)/k$, the exact dispersion relation $\omega(k)$ applies instead. This delineates the wave regime to the left of the solid curves. As suggested by the shortness of trajectory a, interactions in the linear region are slow. A cursory look at the other trajectories shows a general movement toward the deformation radius on the baroclinic plane, downward to the barotropic plane, and then to the left. Taken literally it would say that in a freely evolving ocean vertical shear would disappear, and the horizontal size of eddies would at first approach the deformation radius, and then increase. We first describe the individual elements of this statement, and then in the succeeding section modify it to account for the real, rough-bottom ocean.

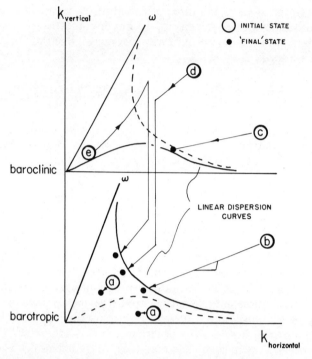

Fig. 19. Schematic representation of freely evolving, nonlinear eddy fields in a *flat-bottom* ocean. Turbulent states are plotted on this ω, **k** diagram by ascribing to them a frequency kU, where k is the centroid of the energy spectrum, and U the rms particle speed. The position of these states relative to the linear dispersion relations (below solid curves) is crucial. Energy-preserving changes of scale occur, within baroclinic eddies, toward the deformation radius from either side. This same scale is an aperture through which energy passes, downward from baroclinic to barotropic states, followed by cascade to small wave number. Wherever energy meets a dispersion region it tends to stagnate.

B. Barotropic Cascade, Path b

The behavior on the lower plane is just the same as if the ocean were unstratified; under the Boussinesq approximation, arbitrary barotropic motion itself satisfies the equations. This case is described in detail by Rhines (1973, 1975). The essential result (Batchelor, 1953) follows from the integral constraints on total energy and squared relative vorticity, or enstrophy, in purely two-dimensional flow:

$$\frac{\partial}{\partial \tau} \int_0^\infty \mathscr{E} \, dk = -2v \int k^2 \mathscr{E} \, dk - 2R \int \mathscr{E} \, dk < 0$$

$$\frac{\partial}{\partial \tau} \int_0^\infty k^2 \mathscr{E} \, dk = -2v \int k^4 \mathscr{E} \, dk - 2R \int k^2 \mathscr{E} \, dk \tag{23}$$

There is no forcing and an Ekman drag, $R\mathbf{u}$, acts at the base of the fluid. As in Section 3, the domain is taken to be periodic, of very large size, and Fourier coefficients are blurred into a continuous function. If initial values $\psi_0(\mathbf{x})$ are specified and then allowed to evolve freely, both right sides are negative. For a finite initial scale of motion they both vanish with R, v (for all finite time), leaving $(\partial/\partial t) \int \mathscr{E} \, dk = 0$, $(\partial/\partial t) \int k^2 \mathscr{E} \, dk = 0$.

The argument proceeds by assuming that an initially narrow spectrum spreads in time, about its mean wave number,

$$\frac{\partial}{\partial \tau} \int (k - k_1)^2 \mathscr{E} \, dk > 0$$

where $k_1 = \int k\mathscr{E} \, dk / \int \mathscr{E} \, dk$. It follows immediately that

$$\frac{\partial k_1}{\partial t} < 0 \qquad\qquad (24)$$

This "red" cascade toward small wave numbers may be appreciated by imagining possible redistributions of a "mass" $\mathscr{E}(k)$ along the k axis, such as to conserve the moment of inertia of $\mathscr{E}$ about $k = 0$. If a unit amount of $\mathscr{E}$ is taken from a narrow peak at $k = k_0$ to twice k_0, then the equivalent of *four* units of ε must move to $\frac{1}{2}k_0$ to compensate. Given an initially narrow concentration, no more than one fourth of the total energy can move to $2k_0$, for the entire remainder must then move to $k = 0$ to preserve the second moment.

The transfer of energy out of small eddies into larger ones, when 3D vortex stretching is suppressed, is a feature having links with many kinds of fluid flows. The transfer from cyclones to the zonal-average winds in the atmosphere may be considered a special case, and in terms of westerly momentum, appears as a negative eddy viscosity (Starr, 1968). The corresponding spectral transfer function $T_\psi(k)$ shows the nonlinear contribution to $\partial \mathscr{E}(k)/\partial t$ at each wave number; more complex flows with stratification and topography continue to have a "red" $T_\psi(k)$ (Figs. 31, 38), but the velocity components normal to isopleths of density and topography yield competing transfer spectra, that may reverse this trend.

The absence of vortex-stretching in two dimensions (2D) prevents high-Reynolds number turbulence from dissipating energy at all efficiently. Whereas the e-folding time for energy decay in 3D is approximately L/U regardless of how fine the dissipation scale, here a viscous time of order $(\nu/L^2 + R)^{-1}$ is required. The presence of strong eddy motions cannot, therefore, be taken as an indication that dissipation is occurring, in the gross sense encountered, for instance, in a laboratory jet.

Enstrophy, $k^2\mathscr{E}$, on the other hand, must in the mean be carried to small scale to balance the expansion in size of the dominant eddies. The notion that an inertial subrange will form at large wave number, to carry enstrophy, dominates the literature (e.g., Kraichnan, 1967). However, the similarity of this process to the energy-carrying Kolmogorov inertial range of 3D turbulence breaks down when rates are considered: 3D energy can apparently reach infinitesimal scale during a single revolution of an eddy ($\sim L/U$). Yet the time required for 2D enstrophy to do so is logarithmically infinite, as the Reynolds number increases without bound. This is because the time for enstrophy to double its characteristic wave number is of order $(\int_0^k k^2\mathscr{E} \, dk)^{-1/2}$, which is strictly bounded in two dimensions. A high-Reynolds number, 2D fluid is thus inviscid for all finite time, given energy initially at finite scale. The fact that the effective Reynolds number *is* large is plausible once one views the wealth of fine structure in, say, sea-surface temperature photographs. In spite of blurring of temperature gradients by exchange with the atmosphere, strong contrasts persist down to a scale of 1 km and less. Clearly there is a wide gap between the energy-containing scale, say, k_1^{-1}, and the scale of significant thermal dissipation. Less is known about the velocity fine structure, but current records on fixed moorings, low-passed to remove internal waves, do show frequent rapid accelerations as if a corresponding jaggedness in the velocity distribution were present.

The removal of viscosity from the dominant dynamics at large Reynolds number suggested to Batchelor (1969) a similarity solution for the time evolution of the energy-containing eddies. If details of the initial conditions are eventually forgotten in the evolution the only external parameter is the rms particle speed, U. This suggests a shape-preserving solution,

$$\mathscr{E}(k, t) = \tfrac{1}{2}U^3 tg(Ukt) \tag{25}$$

where g is a normalized, unknown shape function ($\int_0^\infty g(\xi)\,d\xi = 1$). Solution 25 exhibits both rightward enstrophy flux and leftward energy flux. The dominant scale, k_1^{-1}, of this eddy field expands according to

$$\frac{dk_1^{-1}}{dt} = TU \tag{26}$$

where

$$T^{-1} = \int_0^\infty \xi g\,d\xi, \text{ a constant}$$

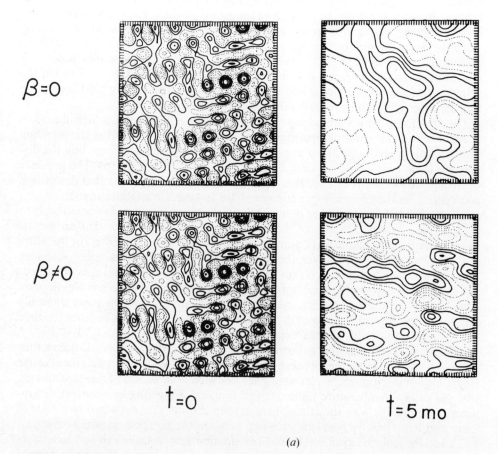

Fig. 20. (a) Streamlines in barotropic evolution experiments with and without the β effect. Beside keeping the scale small, β tends to produce a striated pattern of shifting, principally zonal, flows. (a) Contour interval 0.11 at $t = 0$, 0.16 at $t = 5$.

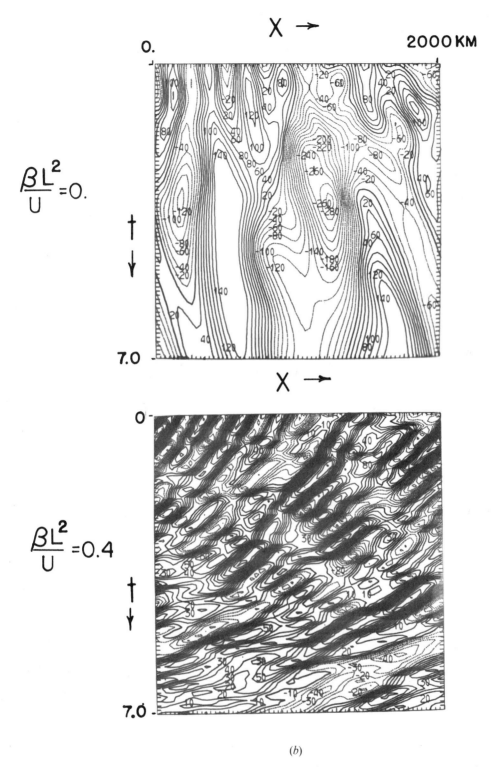

Fig. 20. (b) contour interval 0.11 at $t = 0$, 0.06 at $t = 5$. (See Table I for parameters.) (b) Time–longitude diagrams for ψ, from the barotropic runs in (a). The cascade to large scales occurs in either case; with beta, it leads to increasingly fast propagation, until being halted at $\kappa \sim \kappa_\beta$.

(c)

Fig. 20.   (c) Vorticity field in a high resolution barotropic turbulence experiment. With a strong β effect this vorticity would be radiated rather than sheared out.

Experimental values for T are given by Rhines (1975); $T = 3.0 \times 10^{-2}$ for initially narrow spectra. Thus we are closer to having useful results about the evolution of the energy-containing eddies in two than in three dimensions, and numerical computation is far more economical in two dimensions, so we have the added benefit of many simulations.

If these results are applied directly to the ocean, say, the 200-km-wide thermocline eddy viewed in MODE-I, equation 26 gives about 100 days for the time required to double its size (taking $U \sim 10$ cm/sec). But this certainly did not occur, for reasons that will soon become clear.

A numerical simulation of pure, barotropic 2D turbulence is shown in Fig. 20a. The computation uses a 64^2 degree-of-freedom spectral scheme (see Rhines, 1975). The dominant scale increased rapidly, and its evolution may also be seen in the x–t plot of stream function (along a single latitude line), Fig. 20b. The corresponding wave number spectrum is sharp, with an inertial range steeper than k^{-3}. This narrowness of the spectrum lends confidence to the use of single scales L, τ, and U in describing the field, which is the basis of Fig. 19. The experimental parameters for this and succeeding runs are given in Table I.

C. Obstacles to the Red Cascade

Evidence from mid-ocean observations is that the eddies are closely packed and not very intermittent. But the mechanism of thin-jet instability makes the field more erratic nearby the Gulf Stream or other intense currents. The eddies followed

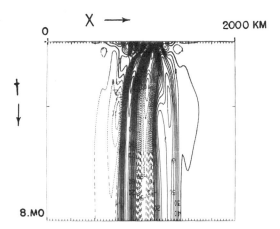

Fig. 21.   The effect of isolating a patch of eddies with surrounding, still fluid (time–longitude plot). Inter-action ceases when there are too few eddies to act as turbulence (roughly four).

by Fuglister were distinct and isolated, both in their foreign water-mass character-istics, and in their unusual energy content.

Figure 21 shows how isolation can affect the simple 2D cascade. A initial cluster of eddies, surrounded by quiet fluid, cascades to longer scale, until the energetic patch contains too few eddies to act as turbulence. At this point in the experiment further evolution ceased, as is clear in the x–t diagram. A number of integral properties of such an isolated distribution must be conserved: the total vorticity, the center of vorticity $\iint \mathbf{x}\zeta\, d\mathbf{x}^2 / \iint \zeta\, d\mathbf{x}^2$, and the dispersion $\iint |\mathbf{x}|^2\zeta\, d\mathbf{x}^2 / \iint \zeta\, d\mathbf{x}^2$ about the center (Batchelor 1967, p. 528). These guarantee that the "size" of a cluster of turbulence remain about the same as it evolves, if the net vorticity is significantly nonzero. In the present experiment, however, the net vorticity vanishes identically, and there is no guarantee that the energetic region will remain confined. In some other realizations, in fact, vortex dipoles (which are self-propelled) managed to escape the cluster. Regardless of the ultimate detail, the initially isolated cluster must eventually cease to act as turbulence.

A second, and more subtle, obstacle to the 2D cascade is the restoring force provided by the β effect, or its topographic equivalent (owing to gradients of ambient potential vorticity). The energy and enstrophy integrals, equation 23, are unaltered (in an unbounded domain) by β, indicating that the direction of the cascade should be the same as before. Scale analysis of the equation,

$$\frac{D}{Dt}\nabla^2\psi + \beta\psi_x = 0$$

however, suggests that eventually the red cascade (path b in Fig. 19) will carry the flow into the regime of linear waves, no matter how intense or how small the initial eddies. The relative strength of nonlinearity $\hat{\varepsilon}_1 \equiv 2Uk_1{}^2/\beta$, decreases like $2/\beta Ut^2T^2$, according to equation 26. When it passes through unity, radiation of vorticity begins to supplant advection, and interactions between different wave numbers demand frequency resonance obeying the dispersion relation of type 1 waves if they are to develop fully. This requirement of physical coincidence plus spectral resonance

TABLE I

Experimental Parameters[a]

Figure	t	K_1	K_2	P	k_ρ	H_2/H_1	dt	v
20a	0	0.80	—	—	—	—	0.004	0.004
	4.0	0.17	—	—	—	—	—	—
20b	0	0.80	—	—	—	—	0.004	0.004
	4.0	0.10	—	—	—	—	—	—
21	0	0.036	—	—	—	—	0.004	0.0025
	2.5	0.0082	—	—	—	—	—	—
22a	0	0.23	0.23	0.12	5.7	1	0.016	0.003
	4.8	0.10	0.11	0.023	—	—	—	—
22b	0	0.40	0	0.16	5.7	1	0.008	0.002
	4.8	0.22	0.054	0.056	—	—	—	—
23, 24	0	0.62	0	5.6	8.0	3.57	0.004	0.004
	6.2	0.64	1.5	1.97	—	—	—	—
25	0	1.0	0.33	20.4	8.0	3.57	0.006	0.004
	3.0	5.0	8.4	3.3	—	—	—	—
27–29	0	0.24	0	2.11	5.0	3.17	0.008	0.006
	3.2	0.26	0.064	1.95	—	—	—	—
	8.5	0.46	0.52	0.74	—	—	—	—
31	0	1.2	0.70	1.8	8.0	3.57	0.008	0
	5.1	0.57	1.64	0.24	—	—	—	—
34	0	1.1	0	1.9	8.0	3.57	0.008	0.004
	3.0	0.53	0.40	0.81	—	—	—	—
	8.8	0.13	0.11	0.29	—	—	—	—
35, 41	0	1.2	0.30	2.4	8.0	3.57	0.008	0
	7.0	0.33	1.0	0.48	—	—	—	—
39, 40	0	0.25	0	3.1	8.0	3.57	0.008	0.004
	2.6	0.48	0.80	1.7	—	—	—	—
	5.1	0.38	0.61	0.95	—	—	—	—
39	0	0.25	0	3.1	8.0	3.57	0.008	0.004
	2.6	0.43	0.64	2.1	—	—	—	—
	5.1	0.50	0.87	0.88	—	—	—	—
39	0	0.25	0	3.1	8.0	3.57	0.008	0.004
	2.6	0.20	0.05	3.1	—	—	—	—
	5.1	0.33	0.41	2.5	—	—	—	—
39	0	2.6×10^{-11}	0	3.1×10^{-10}	8.0	3.57	0.008	0.004
	5.1	3.7×10^{-11}	2.0×10^{-11}	5.0×10^{-11}	—	—	—	—
42	0	0.48	0	0.16	6.0	1.0	0.008	0.003
	3.5	0.16	0.016	0.096	—	—	—	—
43b, e	0	0.49	0	0.79	8.0	3.57	0.008	0.004
	8.8	0.046	0.035	0.16	—	—	—	—
43c	0	0.50	1.5	2.0	8.0	3.57	0.008	0
	7.0	0.21	0.85	0.30	—	—	—	—
43d	0	0.60	0.15	1.21	8.0	3.57	0.008	0
	7.0	0.19	0.44	0.30	—	—	—	—
45	0	1.1×10^{-4}	0	3.0×10^{-4}	8.0	3.57	0.008	0.006
	7.0	2.8×10^{-5}	6.0×10^{-6}	1.6×10^{-4}	—	—	—	—

[a] In "computer" units such that the box width is 2π. In the later simulations this corresponds to 2000 km, with the time unit 1 month, and velocity unit 10 cm/sec (then $\beta = 17.8$ and $f = 200$). Energies (K_1, K_2, P) are in velocity units; multiply by H_1 to get energy. k_ρ is the inverse Rossby deformation radius, $(F_1 + F_2)^{1/2}$. v, R, and Q are lateral, bottom, and k^4 friction coefficients; δ is the rms topographic height/H_2; $k_1^{(1)}$ and $k_1^{(2)}$ are

R	Q	δ	β	$k_1^{(1)}$	$k_1^{(2)}$	U_1	U_2	Experiment
0	0	0	0	10.4	—	1.3	—	2DT5
—	—	—	—	3.5	—	0.58	—	
0	0	0	52.0	10.4	—	1.3	—	βT5
—	—	—	—	5.7	—	0.45	—	
0	0	0	0	14.0	—	0.27	—	Entrain-1
—	—	—	—	—	—	0.13		
0	0	0	0	8.5	8.5	0.68	0.68	2LT-2
—	—	—	—	4.6	3.7	0.45	0.47	
0	0	0	0	6.5	6.4	0.89	0	2LT-1
—	—	—	—	—	—	—	—	
0	0	0	20.0	2.7	—	1.1	0	BE
—	—	—	—	4.6	3.9	1.1	1.7	
0	0	0	20.0	2.2	2.9	1.4	0.43	BCRW
—	—	—	—	6.3	4.6	3.2	2.2	
0	0	0	0	—	—	0.69	0	GS3-2
—	—	—	—	2.6	3.8	0.71	0.20	
—	—	—	—	3.0	2.4	0.96	0.57	
0.04	10^{-5}	0	17.8	6.5	6.3	1.5	0.65	MODE 4B
—	—	—	—	4.7	4.2	1.1	0.96	
0.04	0	0.053	17.8	5.5	—	1.5	0	MODE 3B
—	—	—	—	6.3	7.2	1.0	0.47	
—	—	—	—	5.8	7.0	0.51	0.24	
0.04	10^{-5}	0.053	17.8	5.6	5.6	1.5	0.41	MODE 4A
—	—	—	—	6.4	6.2	0.81	0.76	
0	0	0.053	20.0	2.0	—	0.71	0	MODE 2_1
—	—	—	—	4.9	5.5	0.98	0.67	
—	—	—	—	5.5	6.2	0.87	0.58	
0	0	0.027	20.0	2.0	—	0.71	0	MODE 2_2
—	—	—	—	4.6	4.5	0.92	0.60	
—	—	—	—	5.4	5.0	1.0	0.70	
0	0	0	20.0	2.0	—	0.71	0	MODE 2_3
—	—	—	—	2.1	2.6	0.64	0.17	
—	—	—	—	4.9	5.1	0.81	0.48	
0	0	0.053	20.0	2.0	—	7.2×10^{-6}	0	MODE 2_3
—	—	—	—	3.5	6.0	6.3×10^{-6}	5.3×10^{-6}	
0	0	0.11	0	7.5	—	0.98	0	2LT-5
—	—	—	—	6.2	8.2	0.56	0.18	
0.04	0	0.053	17.8	5.7	5.7	0.99	0	HORIZ-1
—	—	—	—	5.6	6.9	0.30	0.14	
0.04	2×10^{-5}	0.053	17.8	5.8	5.7	1.0	0.92	HORIZ-2
—	—	—	—	5.8	5.9	0.64	0.69	
0.04	1×10^{-5}	0.053	17.8	5.7	5.7	1.1	0.29	HORIZ-2
—	—	—	—	6.2	6.7	0.62	0.50	
0	0	0.053	17.8	5.5	—	0.015	0	MODE 3A
—	—	—	—	5.1	7.3	0.0074	0.0018	

the first moments of the upper and lower kinetic energy spectra, respectively. U_1 and U_2 are rms velocities. Note that in the experiments with regions of empty ocean these will be less than the typical velocities.

The early runs, using lateral friction, were heavily damped by friction, but these have all been repeated with the more effective (less devastating) k^4 friction.

reduced the cascade rate T by a factor of five in the numerical simulations. The result was a slowly evolving Rossby wave field of scale given by

$$k_\beta^2 = \frac{\beta}{2U}$$

and steepness $\hat{\varepsilon}_1$ slightly below unity. The action of turbulence as a source of wave motion (with no loss of energy to dissipation) is one of the ironies of geostrophic flow.

The numerical experiment is shown in Fig. 20. For comparison, the ψ field at an intermediate time, 5.1 months, taking $U \sim 5$ cm/sec and a domain 2000 km wide, with and without β, is shown. Within the time of a single eddy revolution the differences become apparent. The growth of energy in the smallest wave numbers is inhibited by β, as is the leftward migration of the spectral maximum. The x–t plots in Fig. 20b show how the initial clustering occurs as before, but both the slower advection time of the bigger eddies $[\sim(k_1 U)^{-1}]$ and the ever greater frequency of Rossby waves of scale k_1^{-1} continue to make the transition to wave motion a quick one. The westward phase propagation, at a rate comparable with U, takes over, and further changes of scale occur only gradually. For this reason, the constant energy trajectory b in Fig. 19 is shown to terminate at the threshold of the wavelike region. The vorticity field, (Fig. 20c) in pure 2D turbulence shows contours elongated by the shear, in exactly the pattern of a passive dye trace. In more complex experiments below, this shearing action shows reliably when horizontal advection dominates the dynamics.

D. Anisotropy

An interesting feature of 2D turbulence with β is the development of anisotropy in which eddies are elongated along latitude lines. This preference for flow along geostrophic contours is a common thread running through the remainder of this chapter, including the consideration of mean-flow generation by eddies. It may be anticipated by realizing that a general feature of weak-wave interactions seems to be that they cascade predominantly toward small frequency, proofs of the initial tendency having been given by Hasselmann (1967) for single triads. Combined with the 2D cascade to small wave number, this rules out continuing isotropy, for the type 1 dispersion relation associates small frequencies with *large* wave number, if we fix the mix of propagation directions. The end state of this cascade was speculated (Rhines, 1975) to be a nearly steady pattern of zonal current, alternating with scale k_β^{-1}. There is some controversey about whether the tendency proceeds this far, but at the intermediate times of interest for the ocean, it is clearly effective. When energetic eddies do not show this tendency (for instance, those observed along 70°W in the deep water are anisotropic in the opposite sense) other constraints (e.g., the proximity of a continental margin) must be suspected.

For typical deep-ocean velocities (5 cm/sec) the scale of these waves of unit steepness is ~ 70 km (wavelength 440 km), which is comparable with estimates from the observed correlation functions. Although there is much more to the story, it appears that the planetary restoring force contributes to the smallness of mesoscale eddies.

E. Baroclinic Cascades; Path c

The remaining initial-value problems involve baroclinity. For comparison with experiments we approximate the full potential vorticity equation 15 by a two-layer model, similar to that used by Phillips in early atmospheric models and theory. The equations for the upper and lower layer in dimensional variables, are

$$\frac{D_1}{Dt}[\nabla^2\psi_1 + F_1(\psi_2 - \psi_1)] + \beta\psi_{1,x} = \nu\nabla^4\psi_1 + Q\nabla^6\psi_1$$

$$\frac{D_2}{Dt}[\nabla^2\psi_2 + F_2(\psi_1 - \psi_2)] + \beta\psi_{2,x} = \nu\nabla^4\psi_2 + R\nabla^2\psi_2 + Q\nabla^6\psi_2;$$

(27)

$D_i/Dt \equiv \partial/\partial t + J(\psi_i, \ )$, $F_i = f_0^2/g'H_i$, $g' = g\Delta\rho/\rho$. Here the interface height is $f_0(\psi_2 - \psi_1)/g'$. (In an n-layer model, the interior layers have thickness $\propto \psi_{n+1} - 2\psi_n + \psi_{n-1}$, from the hydrostatic relation. This approaches ψ_{zz}, equation 15, in the limit.) Hydrostatic, quasi-geostrophic motion is assumed. A distinct change in dynamics occurs between scales on either side of the Rossby deformation radius, $(F_1 + F_2)^{-1/2} \equiv k_\rho^{-1}$. When $k_\rho L \gtrsim 1$ the layers are strongly coupled by interfacial motion and the attendant vortex stretching [the term $F_i D_i(\psi_2 - \psi_1)/Dt$]. At scales far smaller than k_ρ^{-1}, on the other hand, the coupling terms are negligible, and the interface is effectively rigid. In later runs a high-order friction ($Q\nabla^6\psi$) is used.

Thus the initial behavior of a field of eddies of scale $k_\rho L \ll 1$ is just that of two decoupled layers of 2D turbulence. The cascade labeled c in Fig. 19 has this character, following a constant-energy trajectory until it meets the edge of the wave regime, where it stagnates. The waves themselves, in this extreme ($\beta/2Uk_\rho^2 \gg 1$), have the same nature, being effectively two decoupled layers of barotropic Rossby waves; that they may be expressed as baroclinic modes of odd vertical symmetry obscures this independence.

F. Path d

Imagine now, the more interesting case, Fig. 19, path d, in which the energy level is greater, so that the initially small eddies reach the wave number k_ρ before feeling the β effect. As they expand toward the deformation scale the pressure perturbations build up (for the eddies are geostrophic and U remains fixed) and the layers begin to communicate. Where there is a strong cyclonic vortex in the upper layer, the interface is elevated by the low pressure, and vorticity of the same sign is induced in the fluid below. This tendency was noted long ago by Prandtl (see Prandtl, 1952, p. 386) and by meteorologists studying the development and occlusion of mature cyclones in the atmosphere (e.g., Wiin–Nielsen, 1962).

But what is striking about the experiments (Fig. 22), is the totality of the process. Very quickly, eddies of like sign, above and below the thermocline lock together to produce a barotropic field: if there are none present initially, in one layer, they will soon be produced by this process. In doing so the cascade goes far beyond the "equivalent barotropic" state familiar in the atmosphere. The 2D results again become relevant, for once the solutions have dropped to the barotropic plane in Fig. 19, they are indistinguishable from experiment b.

The necessity of these events can be established for path d by noting that *no state other than near-barotropy* is consistent with these initial values of potential vorticity, for energy that reaches small wave number. For example in the experiment with eddies of small scale ($k_0 \gg k_\rho$) initially confined above the thermocline, the lower layer potential vorticity is nearly zero:

$$\nabla^2\psi_2 - \frac{f\eta}{H_2} = O\left(\frac{k_\rho}{k_0}\right)$$

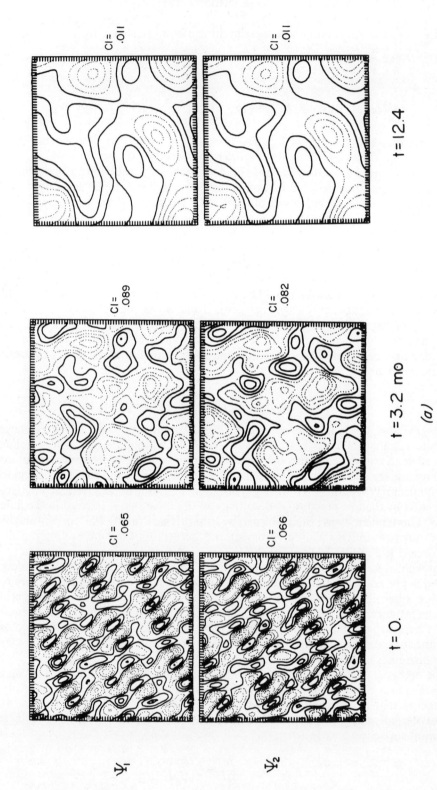

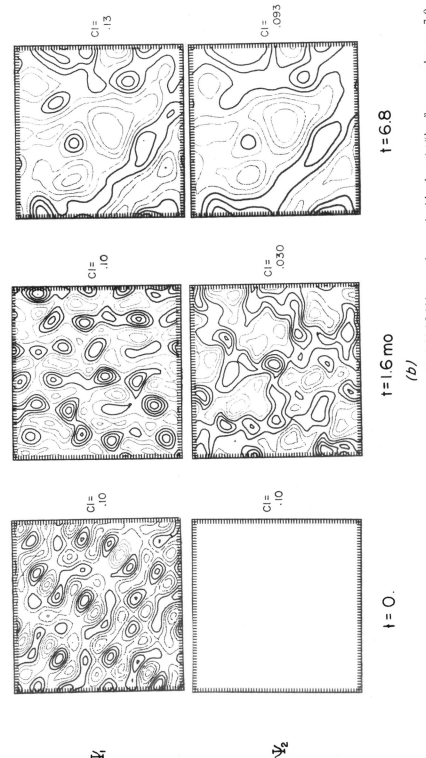

Fig. 22. (a) Sequence of two-layer streamline fields in a free spin-down experiment. The initial field has equal energy in either layer (at "box" wave number $\kappa = 7$–9, near $\kappa_p = 7$) but with *random* phase relation. The box width is 2000 km, and $\beta = 0$. After locking together in the vertical, the eddies behave as 2D turbulence. CI means contour interval. (b) As in (a), but with the lower layer initially at rest. Energy is now communicated downward as a part of the cascade toward barotropy and large horizontal scale.

where η is the thermocline height. The Fourier transform of this expression yields

$$-k^2\hat{\psi}_2 = \frac{f\hat{\eta}}{H_2} \equiv (\hat{\psi}_2 - \hat{\psi}_1) + O\left(\frac{k_\rho}{k_0}\right)\hat{k}_\rho^{\,2}; \qquad \hat{k}_\rho^{\,2} \equiv k_\rho^{\,2}\,\frac{H_1}{H_1 + H_2}$$

or

$$\left(\frac{k^2}{\hat{k}_\rho^{\,2}} + 1\right)\hat{\psi}_2 = \hat{\psi}_1 + O\left(\frac{k_\rho}{k_0}\right)$$

Later, therefore, energy that is delivered to the Fourier components of small wave number $k \ll k_\rho$, has $\hat{\psi}_2 \to \hat{\psi}_1$, while at large k, $\hat{\psi}_2 \ll \hat{\psi}_1$. The ratio of potential to kinetic energy, though often biased towards large values at small k, behaves like $(k/k_\rho)^2$, which itself becomes small.

In fact, under conditions far more general than $(k_0/k_\rho) \gg 1$, the same events are found to occur. Evidence for this is the analogous expression for initial eddies of any size:

$$\left(\frac{k^2}{\hat{k}_\rho^{\,2}} + 1\right)\hat{\psi}_2 = \hat{\psi}_1 + \hat{\psi}_1^{\,0}$$

where $\hat{\psi}_1^{\,0}|\mathbf{k}|$ is the Fourier transform of the initial ψ_1 pattern *distorted as a passive tracer* by the subsequent deep flow. Now the nature of the horizontal enstrophy cascade is to elongate a passive tracer into thin sheets (Fig. 20c) typically with spectrum $\propto k^{-1}$ weighting heavily the large k. If $k_0 \sim k_\rho$, this argues against any significant cascade of $\hat{\psi}_1^{\,0}$ to smaller k. In addition, the conservation of energy during the cascade implies that $\hat{\psi}_2$ and $\hat{\psi}_1$, proportional to the pressures, will increase vastly with time. We are therefore left with the same result: barotropy is a necessary consequence of a "red" horizontal cascade and is intrinsically related to the transfer of enstrophy to large wave number.[6] For initially large-scale eddies (path e below), the cascade of potential enstrophy to large k/k_ρ is still more crucial to the development of barotropy. Cases appear in which the cascade is suppressed, and total barotropy is not achieved.

These results complement the geostrophic turbulence theory of Charney (1971), which applies to eddies far from the top and bottom boundaries, of large vertical wave number (the above considerations are readily extended to this, continuously stratified situation); then the full, continuously stratified equations tend to exhibit a cascade to large scales, both horizontal and vertical. This is seen by writing the conservation of total energy and potential enstrophy, from equation 15 (with $\beta = 0$):

$$\frac{d}{dt}\iiint |\nabla\psi|^2 \, dx\,dy\,dz = 0$$

$$\frac{d}{dt}\iiint |\Delta\psi|^2 \, dx\,dy\,dz = 0 \qquad (28)$$

for an inviscid system, where z has been rescaled by fL/N. The additional terms $|\psi_z|^2$ and $|\psi_{zz}|^2$ represent, respectively, potential energy and the contribution to potential enstrophy from the variations in isopycnal layer thickness. In the absence of boundaries, ψ has a three-dimensional Fourier expansion, and the analogues of equation 23 give, as before, $d|\mathbf{k}_1|/dt < 0$ if the dispersion of the wave number spectrum

---

[6] The barotropic state, however, may not be total if the eddies are very intermittent (as with a few distant Gulf Stream rings), if there is disequilibrium (strong forcing or bottom friction), or if there is a very shallow thermocline or strong bottom topographic roughness (see Section 7).

about its mean is to increase. This is a self-similar cascade, with ellipsoidal eddies expanding in size, yet keeping the same eccentricity. The red cascade again suggests the similarity solution 25 for the evolving wave number spectrum. The theory provides basic reasons for the predominance in the ocean and atmosphere of rather grave vertical modes.

We have shown that when top and bottom boundaries are encountered, the interior fluid will still try to expand its vertical scale, and thus develop into depth-independent flow. The continuing equipartition of energy between the potential and two kinetic components, in Charney's cascade, must then be lost, for the cascade d efficiently converts potential energy to kinetic. Once d reaches the barotropic plane, the system again enters into the wave-turbulence conflict described for b. The predicted spectra are very different from Charney's k^{-3} (for both velocity and temperature); here the uncorrelated nature of the layers at $k \gg k_\rho$ suggests $K(k) \propto k^{-3}$ and $P(k) \propto k^{-5}$ roughly, whereas in $k < k_\rho$ the barotropy suggests a red $K(k)$, yet blue or white $P(k)$.

G. Path e

The final possibility, path e, for these initial value experiments is a field of baroclinic eddies far larger in scale than the deformation radius. At small amplitude these states would simply propagate as type 3 baroclinic Rossby waves, due westward at speed βk_ρ^{-2}. In fact the waves are unstable and quickly break down into eddies of scale $\sim k_\rho^{-1}$.

This is the first violation of the "red" cascade so far encountered. The constraints 23 are now relaxed by production of relative enstrophy. The analogues of equation 23 follow from conservation of total energy

$$\tfrac{1}{2}[H_1(\Delta\psi_1)^2 + H_2(\nabla\psi_2)^2 + H_1F_1(\psi_2 - \psi_1)^2]$$

[whose spectra are K_1 (upper kinetic), K_2 (lower kinetic), and P (potential)], and potential enstrophies

$$H_1[\nabla^2\psi_1 + F_i(\psi_2 - \psi_1)]^2, \qquad H_2[\nabla^2\psi_2 + F_2(\psi_1 - \psi_2)]^2$$

With $\beta = 0$ for simplicity, and $F_1 = F_2 = F$, we have, analogous to equation 23

$$\frac{d}{dt}\int(K_1 + K_2 + P)\,dk = 0 \qquad (29)$$

$$\frac{d}{dt}\int[k^2(K_1 + K_2) + 2(F + k^2)P]\,dk = 0 \qquad (30)$$

$$\frac{d}{dt}\int(k^2 + 2F)(K_1 - K_2)\,dk = 0 \qquad (31)$$

When $F \to 0$, $P \to 0$ we recover equations 23. Otherwise, equations 29 and 30 yield

$$\frac{\partial}{\partial t}\int E\,dk = 0$$

$$\frac{\partial}{\partial t}\int k^2 E\,dk = -\int(k^2 + 2F)\frac{\partial P}{\partial t}\,dk \qquad (32)$$

$$E = K_1 + K_2 + P$$

Comparing with equations 23, we find that the second moment of $E(k)$ can increase if potential energy is released, $\partial P/\partial t < 0$, in the weighted sense of the integral. A wave number region of significantly growing eddies with $\partial P/\partial t > 0$ cannot occur far above $k = k_\rho$, for it would tend to change the right side back to negative. This amounts to an integral statement, valid for turbulence, of the favoring of wave numbers $k^2 \lesssim 2F$ for the growing eddies. The energy and enstrophy invariants thus permit a "blue" cascade, in which the dominant wave number, say, $\int kE\,dk/\int E\,dk$, increases spontaneously. Path e follows, as do a to d, if particles tend to separate with time.

Equations 29 and 30 are the two-layer analogues of the invariants 28 for the continuously stratified model. They can be rewritten as

$$\frac{\partial}{\partial t}\int \sum_{i=1}^{2}(k^2 + m_i^2)\tilde{\psi}_i^2\,d\mathbf{x} = 0$$

$$\frac{\partial}{\partial t}\int \sum_{i=1}^{2}(k^2 + m_i^2)^2\tilde{\psi}_i^2\,d\mathbf{x} = 0$$

where $m_0 = 0$, $m_1 = (2F)^{1/2}$ appear as the vertical wave numbers and $\tilde{\psi}_1$, $\tilde{\psi}_2$ are the Fourier transforms of the barotropic and baroclinic modes, $\psi_1 + \psi_2$ and $\psi_1 - \psi_2$, respectively. This version emphasizes that to violate the red cascade in the horizontal sense with baroclinic instability, the motion must evolve toward larger vertical scales to compensate for the growth of smaller horizontal scales. For $d|\mathbf{k}_1|/dt < 0$ still, based on the total wave number (k, m_i). Bass (1974) has discussed the n-layer case.

Though baroclinic instability is well-known in the study of nearly zonal mean flows, its appearance as a turbulent cascade from big eddies to small, is less familiar. Yet it is the *primary* nonlinear effect in large baroclinic eddies. The cascade rate can crudely be inferred from the linearized stability theory, which yields growth rates $\sim Uk_\rho$, U being the large-scale baroclinic velocity. If we scale the small eddy velocities with U as well (implying a well-developed cascade), the energy flux to them is roughly U^3k_ρ.

An experiment following path e (Fig. 23) involves a set of big eddies (of widths ~ 200–1000 km), initially above the thermocline. The successive streamline contours show meanders developing locally, where the flow is particularly intense. This first occurs in the region of southwestward flow (Fig. 23b). The shallowness of the upper layer makes westward currents, U_W, more unstable than eastward currents, U_E, if $|U_W|/|U_E| > H_1/H_2$ (see Section 8). The situation changes in time owing to westward propagation. Later the band of east-northeastward current near the north end of the box breaks down in a well-defined instability, developing four wavelengths along the current (Fig. 23c).

At the end stages, the meanders collapse and produce an intensified band of zonal flow (Fig. 23c). By this time a considerable amount of barotropic, nearly zonal flow has developed. Arguments for the production of anisotropy and zonal-averaged east-west flow, which varies only slowly in time, were given in the purely barotropic case above. Here the generalized baroclinic instability augments these effects, producing even more striking "mean" flow. This occurs, first, owing to the greater instability of meridional flows, which break down into preferentially zonal eddies (the perturbation currents are then unopposed by β), and second, owing to the finite amplitude eddy–eddy induction of zonal flow, to be discussed in Section 8. The strength and persistence of the zonal velocities may be seen in a time–latitude plot of stream function (Fig. 24). The upper-level flow in the north central region intensifies and endures throughout the 9-month experiment. Ocean observations (Section 9) hint at such a banded current structure.

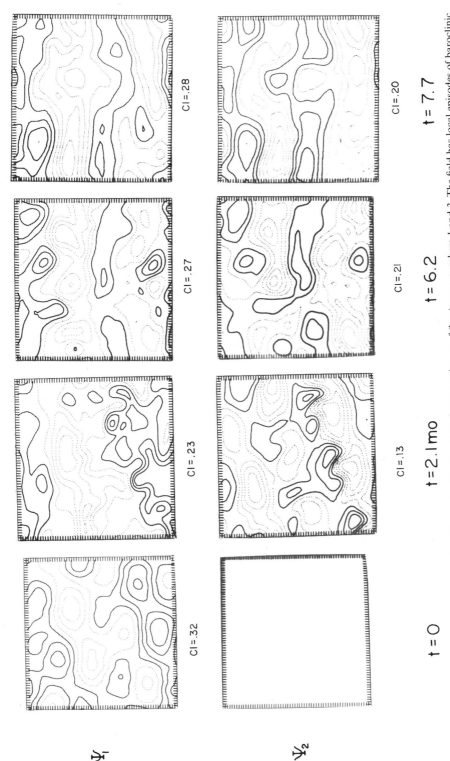

$t = 0$ $t = 2.1$mo $t = 6.2$ $t = 7.7$

$CI = .32$ $CI = .23$ $CI = .27$ $CI = .28$

$CI = .13$ $CI = .21$ $CI = .20$

ψ_1

ψ_2

Fig. 23. Evolution of large-scale eddies, initially confined to the upper layer, a mixture of (box) wave numbers 1 and 3. The field has local episodes of baroclinic instability that in turn reduce the scale, develop deep flow, and increase the scale again. The final, banded zonal pattern of flow is evident.

245

LATITUDE →

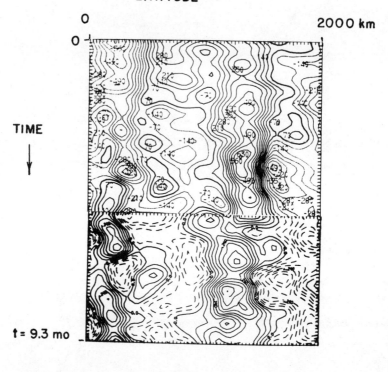

Fig. 24.   Time–latitude plot of upper-layer ψ (actually $-\psi$) (or pressure), at mid-longitude. The persistence and growth in strength of the zonal flow are evident. (At the end viscosity finally damps the current.)

The end state of this unforced flow has in it some uncertainty: though the barotropy that developed in experiments with smaller eddies, with a lesser store of potential energy, was often total, here we cannot be sure that all the large-scale potential energy will be tapped (the ratio of potential to kinetic energy started at 9.0, and dropped to 0.83 after 8.7 months). If a zonal configuration develops soon enough, vertical shear ΔU can remain, and is stable, if $\beta L_\rho^2/\Delta U >$ about 1 (the classical result). The zonal configuration of the currents, in other words, can defeat the potential enstrophy cascade which is a prerequisite for barotropy. Indeed, this seems to be happening in Fig. 23d. But the nature of an equilibrium or near equilibrium end-state in these, or in steady forced experiments, is uncertain and must depend critically on the method of damping and driving the flow. The rms velocities certainly grow during the instability, but in a range of such experiments, they rarely exceeded the speeds of the initial flow by a factor of three. Even this required rather special, vulnerable flows, such as in the succeeding two examples. The rate of cascade toward k_ρ must be studied in detail, and compared with the other transfer spectra, and dissipation, to see how much of the energy completes the journey along path e.

These events amount to a statement that the classical baroclinic Rossby wave, so often used to describe oceanic adjustment processes, is unstable. To isolate the simplest case, I looked at a single Fourier mode (wavelength 1000 km) combined with

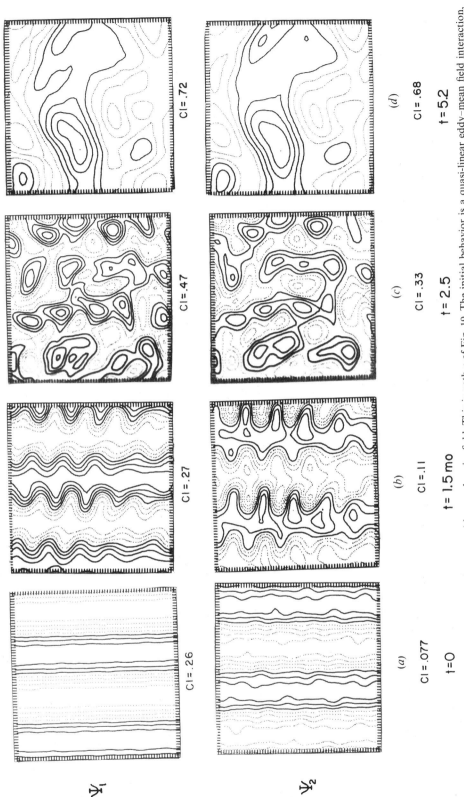

Ψ_1

CI = .26

CI = .27

CI = .47

CI = .72

Ψ_2

(a)

CI = .077

t=0

(b)

CI = .11

t = 1.5 mo

(c)

CI = .33

t = 2.5

(d)

CI = .68

t = 5.2

Fig. 25. Instability of a single baroclinic Rossby wave with a weak noise field. This is path e of Fig. 19. The initial behavior is a quasi-linear eddy–mean field interaction, yet mutual eddy interactions quickly take command.

247

a weak field of random noise. It is a somewhat bizarre experiment, for it so clearly passes through all the stages of path e, and hence demonstrates *all* the primary cascades!

The streamlines (Fig. 25a, b) show sharply tuned growth of energy at a scale of $(1.3-1.6)L_\rho$. After reaching large steepness (with upper layer velocities ~ 30 cm/sec) the eddies interact laterally while locking together in the vertical (Fig. 25c). This yields a final surge of kinetic energy, the upper layer exceeding 35 cm/sec (as against 14 cm/sec initially), when the last of the potential energy drains away. Thenceforth the dynamics are just those described earlier, for purely barotropic flow: the red cascade continues until transforming into barotropic mode 1 waves, after which anisotropy builds up, favoring nearby zonal flow of scale $\sim k_\beta^{-1}$ (Fig. 25d). The entire affair occupied 6 months, during which the ratio of potential to kinetic energies dropped from 15.7 to 0.03, the wave field switched modes from baroclinic to barotropic, and the initial meridional flow gave way to nearly zonal flow. A good summary is the time–longitude plots of pressure and temperature (Fig. 26), in which the westward propagation speed rises abruptly from 2.6 to 50 cm/sec. This has implications for the lateral propagation of energy which survive in the more oceanically relevant experiments (Section 7). The flow patterns that develop spontaneously in these experiments may seem unreal to a reader with a well-developed picture of the ocean in his mind; they are certainly incomplete, yet represent oceans not terribly distant in nature from ours.

The instability is tractable analytically, and Kim's (1975) treatment is a relevant model for the ocean. It has been common in the past to apply baroclinic instability theory based on "meteorological" flows which are steady, zonal, and often of infinite horizontal extent. In its most general form the Rossby-wave model develops both baroclinic and barotropic instabilities and transfer toward k_ρ from both sides. It shows β to inhibit all but zonal perturbations and to depress the "red" cascade when $k < k_\beta$. If extended to finite amplitude, it will model much of Fig. 19 rather well.

H. Meanders in a Two-Layer Gulf Stream

These broad, homogeneous fields of eddies have an air of remoteness about them that, I believe, can be dispelled by describing a highly structured flow, in which these same interactions are found to occur. Consider a thin, zonal jet flowing eastward at the surface, through the periodic box. Its thinness (half-width $\sim L_\rho$) makes it more a model of the Gulf Stream than of the atmospheric westerlies (at least in their classical conception). The initial field (Fig. 27) contains a weak, broad-band noise. There is also a slight counterflow such that no y-averaged transport passes through the section. The exact form chosen is $\psi_1 = \tanh{(y/a)} - y/L + \text{noise}$, where $a = 78$ km, and L is the half-width of the box. This is now a smaller domain, $2L = 1250$ km, with the thermocline at 960 m, and the total depth 4000 m. The deformation scale is 40 km, $\beta = 0$, and $U_1 \sim 50$ cm/sec, averaged across the stream (by rescaling the time, this is equivalent to a very fast flow in the presence of β, $\beta L_\rho^2/U \ll 1$).

In less than 20 days, the random noise organizes into a meandering instability with a regular set of elliptical eddies below the thermocline. The deep motion and its phase shift (leading the perturbations overhead) are known to be essential to baroclinic instability. At the early stages, the deep currents are predominantly north–south, and the whole pattern moves downstream as it intensifies.

After the meander steepness exceeds unity, the independently growing eddies begin to interact with one another; here the turbulence theory becomes relevant. A symptom of the eddy–eddy interactions beginning to compete with eddy-mean flow

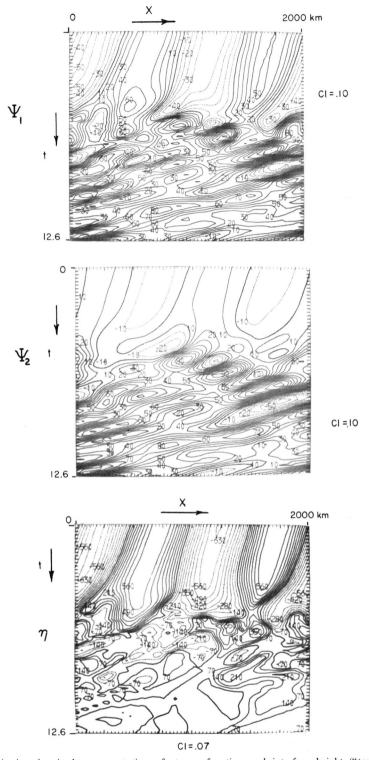

Fig. 26. A time–longitude representation of stream function and interface height ("temperature") shows the rapid breakdown of the field, which jumps from baroclinic to barotropic modes, increasing its propagation speed by a factor of 20. The temperature structure vanishes.

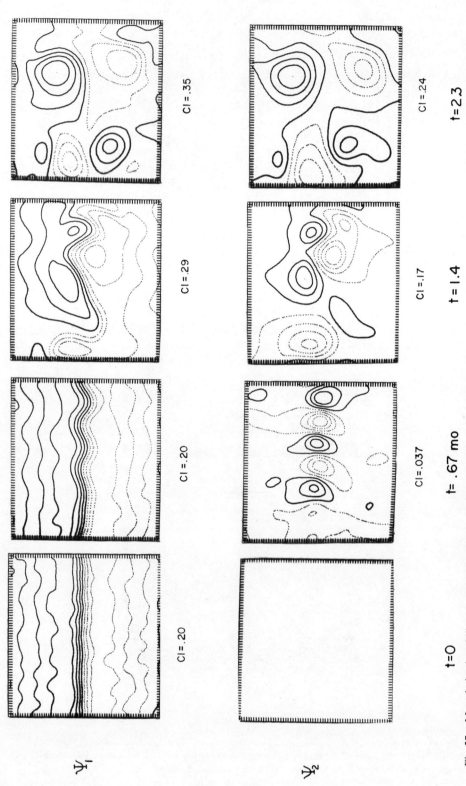

Fig. 27. Meandering instability of a thin zonal jet, initially flowing in the upper layer, passes through the stage of path e, Fig. 19. After linear growth, which involves immediately the deep water (producing predominantly north–south flow there), the eddy field interacts, expanding horizontally and "occluding" in the vertical. The box width is 1250 km, $\beta = 0$.

interactions is the stretching of vorticity contours by the horizontal shear (as in 2D turbulence) (Fig. 20c). At $t = 42$ days the deep eddies of like sign begin to coalesce. This pairing requires a north–south displacement which automatically creates a zonal-mean abyssal flow, in the same sense as the upper level stream; the transport increases with time.

Lateral turbulence also causes the horizontal scale to expand; the initial wavelength, 330 km, increases as the meanders begin to break up ($t = 42$ days). At this stage there is a weak lateral convergence of eastward upper-level momentum into the mean stream. Soon thereafter, the stream almost vanishes from the ψ fields, seeming to appear and disappear at various longitudes, as the cascade toward

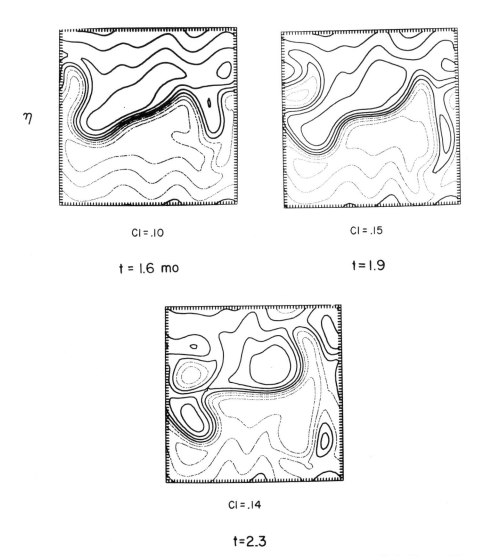

η

CI = .10   CI = .15

t = 1.6 mo   t = 1.9

CI = .14

t = 2.3

Fig. 28.   None of the velocity events in Fig. 27 are visible in the "hydrographers'" Gulf Stream. There the jet remains identifiable for longer. An elongated meander detached to form a closed ring. Some of the flavor of Fig. 1 is thus reproduced in a hydrostatic two-layer model.

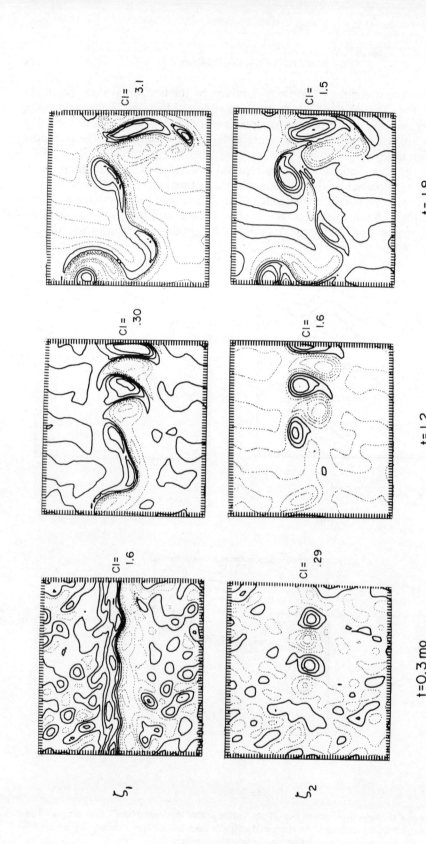

Fig. 29. The vorticity fields for the meandering experiment show simply shaped eddies emerging from the noise field, but at the stage of horizontal inter-action, elongating in the manner of Fig. 20c.

barotropy takes hold, beginning to lock together the upper- and lower-level eddies ($t = 68$ days). The "hydrographer's" Gulf Stream, however, is, still present in the density field (Fig. 28) long after it is disguised in the velocity fields by the growing barotropic turbulence. The sequence in fact catches the growth of a very deep "sock," and the birth of a detached ring, reminiscent of Fig. 1. The end result, a field of very large eddies with a strong, deep flow, may suggest the nature of the North Atlantic east of the Grand Banks. The temporal instability may best be related to spatially growing meanders in the ocean.

The example of jet instability contains numerous dynamic similarities with homogeneous turbulence (path e of Fig. 19). It also reminds one of the accounts of oceanographers like Fuglister and Luyten who, on their return from a Gulf Stream tracking expedition' paint a beautiful picture, but one also of frightening complexity. Also, the figures demonstrate how much of the flow field is lost if one has access only to the hydrographic field. Barotropic and baroclinic modes are both essential to the dynamic picture. One regrets that the barotropic field was for so long obscured as a "level of no motion." The vorticity field (Fig. 29) is a sensitive indication of lateral eddy–eddy interactions. After initially linear growth, the shearing begins to work as in 2D turbulence.

Other experiments, with the important additions of β and topography, appear in the literature. The simulations by Orlanski and Cox (1973), appropriate to the Gulf Stream in shallow water south of Cape Hatteras, are also of interest.

I. Energy-Transfer Spectra

The events in Fig. 19 may be viewed in more rigorous fashion in wave number space by computing the Fourier transforms of the various advective terms in the potential vorticity equation. Thus we decompose the contributions to the change in energy at the scalar wave number, k:

$$\frac{\partial(K + P)}{\partial t} = T_\psi(k) + T_p(k) + T_H(k) - D(k)$$

where

$$T_\psi = H_1 \, \mathrm{Re}\, [\hat{\psi}_1^* \hat{J}_1(\psi_1, \nabla^2\psi_1)] + H_2 \, \mathrm{Re}\, [\hat{\psi}_2^* \hat{J}(\psi_2, \nabla^2\psi_2)],$$

$$T_P(k) = \mathrm{Re}\, [F_1 H_1(\hat{\psi}_1^* - \hat{\psi}_2^*)\hat{J}(\psi_1, \psi_2)],$$

is the result of density advection, $T_H(k) = \mathrm{Re}\, [f_0^2 \hat{\psi}_2^* \hat{J}(\psi_2, h_2)]$ involves the bottom topography, and $D(k) = (R + vk^2 + Qk^4)K(k)$ involves lateral, bottom and "k^4" friction. Here ($\hat{\ }$) = Fourier transform. This relation is shown in Fig. 30 along with

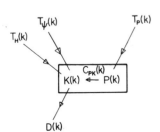

Fig. 30. Kinetic plus potential energy $(K + P)$ at a single scalar wave number, k, changes due to advection of vorticity (T_ψ), advection of density (T_P), flow across bottom topography (T_H), and dissipation (D). The internal conversion C_{PK} from P to K is unique if one defines T_P to be the only "direct" source of P.

the conversion between kinetic and potential energy at a single wave number,

$$C_{PK}(k) = T_P - \frac{\partial P}{\partial t}$$

[Note $\int_0^\infty T_i(k)\, dk \equiv 0$.] A similar format is used in meteorology (e.g., Smagorinsky, 1963) for analysis of general circulation dynamics. Here we recognize the signature of 2D turbulence (Fig. 31) as being a drain of energy by $T_\psi(k)$ from the wave numbers at which it is concentrated, predominantly toward small k. The production of "barotropy" from moderate-scale eddies appears as a positive C_{PK}, conversion from potential to kinetic energy in the vicinity of the deformation radius together with the cascade of potential energy toward that wave number by $T_p(k)$. (The baroclinic instability of initially large-scale eddies is an approach to k_ρ from a larger reservoir

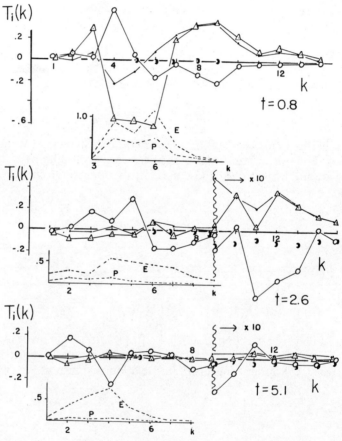

Fig. 31. Transfer spectra for a flat-bottom experiment similar to Fig. 22b, initial energy at box wave numbers $k = 4\text{--}6$, ($U_1 = 14$, deep layer at rest). At first T_P carries energy away from the spectral maximum, to the right. Near κ_ρ ($= 8$) it is converted to kinetic energy, whereupon T_ψ carries it back to the left. Barotropy is well-developed by $t = 5.1$ (total potential energy, E and P, are shown); then T_ψ is dominant everywhere. The large-wave number "tail," $k = 10\text{--}14$, is magnified. The dynamics there is initially as above, with T_P cascading to the right, T_ψ to the left. Dissipation dominates only greater wave numbers than these, $k > 15$. See also Steinberg (1973). $\bigcirc = T$, $\cdot = C_{PK}$, $\triangle = T_P$, $) = -D$.

of energy at small k.) Near $k = k_\rho$, C_{PK} produces kinetic energy which finally cascades back toward small k via T_ψ, a more exact description of path e.

The region beyond the spectral maximum (magnified) is of interest. The balance there is a (nearly) statistically steady microcosm of path e, with energy carried to the right by T_P, converted by C_{PK}, and then carried back to the left by T_ψ, lateral vorticity advection. The dissipation is small in $K < 15$. As Steinberg suggested from his more viscous runs, the dynamics is far different from a 2D spectral tail.

There is a strong analogy with the meteorologists' concept of cyclone growth, followed by occlusion (vertical uplifting of frontal density surfaces) which causes the upper-level flow to fall into an equivalent barotropic state (similar perturbation streamlines, yet less energy aloft). The transfer of kinetic energy, at the late stages, back to the zonal-average winds, is the analogue of the final red cascade by T_ψ, and its attendant anisotropy. The sequential nature of these events is related to the observed "index cycle." Steadily forced computer simulations by Steinberg (1973) and Barros and Wiin-Nielsen (1974) form an interesting complement to our spin-down experiments, and some of the energy cycles are very similar. The emphasis in these papers is the nature of the viscous-inertial subrange, which carries potential enstrophy toward dissipation at large wave number.

The character of the transfer spectra is rather insensitive to the physical configuration for given wave number spectrum; this agrees with the appearance of the same cascades in both homogeneous turbulence and an unstable Gulf Stream. An interesting line of theoretical research involves the prediction of transfer spectra from instability analysis, or with turbulent closure models.

The vertical eddy–eddy interactions found here are striking, but they must now be examined in a model with better vertical resolution. A reassuring sign is the recent work of Bass (1974), who has simulated baroclinic instability with channel geometry. His resolution varies from 2 to 16 levels, yet in all cases the development of barotropy may be seen at interior levels. The fluid nearest the rigid top and bottom contains chaotic, energetic "pseudo"-fronts. Bass's experiment, though framed in a meteorological setting, shows some of the richness of vertical structure that we may anticipate in the ocean. Bretherton and Owens (private communication) have also been making multilevel simulations for oceanic turbulence, which will be interesting to compare with the present work.

7. Basins and Bottom Topography

The picture developed to this point of the free geostrophic–turbulence cascade would be simple enough to suggest ocean experiments, and to fit directly into a theory that contained forcing, mean flows, and dissipation; however, there are effects of ocean-basin geometry that work against the homogeneity and "narrow-band" evolution of those models. As an example, the vertical structure in the primary experiments developed toward a barotropic state far more rapidly than is consistent with the ocean. We demonstrate here that topography and side walls act to counter several of the nonlinear cascades, and to alter the nature of horizontal propagation.

A. Coastal Boundaries

First we describe, from the aspect of turbulence theory, the role of idealized slippery coasts. For a single-fluid layer on a β plane the energy invariant remains, yet the relative enstrophy is no longer conserved (although potential enstrophy, $[\nabla^2 \psi + f]^2$, is). Instead (Rhines, 1975), integration over the basin yields

$$\frac{\partial}{\partial t} \int \mathscr{E} \, dk = \frac{1}{2} \frac{\partial}{\partial t} \iint |\nabla \psi|^2 \, dx \, dy = 0 \qquad (33)$$

$$\frac{\partial}{\partial t} \int k^2 \mathscr{E} \, dk = \frac{1}{2} \frac{\partial}{\partial t} \iint |\nabla^2 \psi|^2 \, dx \, dy = -\frac{1}{2} \beta \oint_{\mathscr{C}} |\nabla \psi|^2 \sin \theta_b \, ds \qquad (34)$$

where θ_b is the angle from east of a positive unit vector tangent to the boundary, $\mathscr{C}$, upon which $\psi = 0$. The source term allows the second moment of $E(k)$ to increase wherever shoreline lies to the west of moving fluid, and conversely. The "red" cascade, besides being blocked by Rossby wave propagation, can thus be reversed near a western boundary, where eddies or circulation of large scale are transformed to small scale. It amounts to a generalization to unsteady, nonlinear flow of classical arguments for western intensification; with linear Rossby waves, western-wall reflection converts long waves to short, increasing the enstrophy. Combined with the knowledge that small-scale [large-enstrophy/energy] motions propagate slowly, this favors the concentration of both energy and enstrophy in the west, and their removal from the eastern ocean.

In the linear frictional "Gulf Stream" the enstrophy produced according to equation 34 maintains its narrowness while, in a steady state, being dissipated by bottom friction (yet such a current at the eastern side would lose its enstrophy to the coast). The generation of enstrophy by wind stress is negligible in this case. Finally, in the nonlinear, frictionless, free gyres of Fofonoff (1954) a steady solution is made possible by east–west symmetry, enstrophy inflow and outflow just canceling one another.

With two-layer stratification (equations 29–31) the result becomes

$$\frac{\partial}{\partial t} \int (X_1{}^2 + X_2{}^2) \, dx \, dy = -\beta \oint_{\mathscr{C}} (\hat{K} + \hat{P}) \sin \theta_b \, ds$$

where

$$X_i = \nabla^2 \psi_i + F(\psi_j - \psi_i), \qquad j = 3 - i$$

and

$$\hat{K} = \frac{H_1}{2} |\nabla \psi_1|^2 + \frac{H_2}{2} |\nabla \psi_2|^2, \qquad \hat{P} = \frac{1}{2} \frac{f_0{}^2}{g'} |\psi_1 - \psi_2|^2$$

Energy of either form at the coast alters the net "nonplanetary" enstrophy. The cascade arguments, equations 29 and 30, become

$$\frac{\partial}{\partial t} \int E \, dk = 0$$

$$\frac{\partial}{\partial t} \int k^2 E \, dk = \frac{-\partial}{\partial t} \int (k^2 + 2F) P \, dk - \beta \oint_{\mathscr{C}} \hat{E} \sin \theta_b \, ds$$

$$\hat{E}(\mathbf{x}) = \hat{K}(\mathbf{x}) + \hat{P}(\mathbf{x}), \qquad E(k) = K_1(k) + K_2(k) + P(k)$$

The combined effect is clear: both potential-energy release within the body of the fluid and energy at its western periphery can increase the second moment of E, and move the center of mass of $E(k)$ to larger wave number.

B. Rough Bottoms

The final and most difficult topic in the dynamics of eddies is the effect of an irregular sea floor. This works through vertical vortex stretching produced by currents flowing across slopes. The sensitivity to this effect is clear in the intrinsic smallness of the vertical velocity ($\sim U\varepsilon H/L$ or $U\omega H/L$) under geostrophic scaling, and by the picture of vertical stiffness developed in Section 3. The effect is easiest to see in a hydrostatic, quasi-geostrophic layered model, where equation 16 becomes the conservation law for potential vorticity,

$$\frac{D}{Dt}\left(\frac{h_i\nabla(h_i^{-1}\nabla\psi) + f}{h_i}\right) = O\left(\varepsilon\frac{H}{L}\right)^2$$

for the ith layer. The two-layer model (equation 27) becomes, in the dimensional variables,

$$\frac{D_1}{Dt}[\nabla^2\psi_1 + F_1(\psi_2 - \psi_1)] + \beta\psi_{1,x} = \nu\nabla^4\psi$$

$$\frac{D_2}{Dt}\left(\nabla^2\psi_2 + F_2(\psi_1 - \psi_2) - \frac{\delta f_0\hat{h}_2}{H}\right) + \beta\psi_{2,x} = \nu\nabla^4\psi_2 - R\nabla^2\psi_2$$

where

$$F_i = \frac{f_0^2}{g'H_i}, \qquad \frac{D_i}{Dt} = \frac{\partial}{\partial t} + J(\psi_i, \ ), \qquad h_2(\mathbf{x}) = H_2 + \delta\hat{h}_2$$

The scaling has typical topographic heights ($\sim H_2\delta$) of order ε or ω, whichever is larger, and this leads to the retention of depth variations only in the vortex stretching terms (neglecting for example, $(D/Dt)(\Delta h_2^{-1}\cdot\nabla\psi_2)$). As before, $\delta \sim L/R$ in order that the dynamics include both planetary and topographic waves. In order to consider islands and continental margins, $\delta \sim 1$, the depth should be allowed to vary throughout the equations.

C. Effect on the Primary Nonlinear Cascades

The topography can generate relative enstrophy, fragmenting large eddies into small ones, and thus counter the red cascade of 2D turbulence, The process with weak currents resembles wave scattering by a random medium, but this gives a convergent result only when $\varepsilon \ll \delta \ll L/R$, plane Rossby waves being the first approximation. But neither inequality is valid for the energy-containing eddies. Figure 32 is a schematic diagram of the parameter space (ε, $\delta R/L_H$, kL_H) for a *single*-layer fluid with topography of dominant horizontal scale L_H. Near the base plane are found linear solutions, with weakly scattered Rossby waves at the left (I) (Thompson, 1975; Rhines, 1970) and topographic Rossby waves (geometrical optics, short waves) (Smith, 1971) at the far right (III). The range II, $\varepsilon \ll \delta \gtrsim L/R$, $kL_H \sim 1$, represents linear oscillations in a very irregular medium, in the worst case being a kind of topographic "turbulence" (Rhines and Bretherton, 1974).

Above the base plane nonlinearity acts, approaching 2D turbulence above $\varepsilon \sim \delta + (kR)^{-1}$. The evolution of solutions is, as in Fig. 19, to the left in the turbulent region (arrows), until meeting the transition surface where they tend to stagnate. Yet below, solutions tend to move to the right by scattering and refraction. The representation of a wave number spectrum by a single length scale is less convincing here, however, than

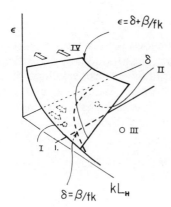

$\epsilon = \delta + \beta/fk$

$\delta = \beta/fk$ kL_H

Fig. 32. The role of rough-bottom topography (for simplification, with single scale L_H) in *homogeneous* fluid motions. Sufficiently intense flows ($\epsilon \gg \delta + \beta/k$), region IV, act as 2D turbulence, cascading to the left (arrows). Gentler flows are increasingly fragmented by roughness and migrate to the right. $\beta = f_0/R$.

in Section 6. For spectral broadening develops from interaction of energy at wave number **k** and topography at wave number **μ**, producing energy at wave number **k** ± **μ**. Typically, topography has a rather flat spectrum, $H(k) \sim k^{-1.5}$ or k^{-2}, which quickly "whitens" the energy spectrum, although favoring transfer to large k (in fact, $H \propto k^{-2}$ is a white *slope* spectrum).

In the stratified problem, topography may affect the structure in the vertical as well as the horizontal. Though the nonlinear effects direct energy toward large vertical scale in the interior, the coup de grace, the destruction of vertical shear (path d and e in Fig. 19) is no longer so likely; the disappearance of the barotropic wave type 1 when $kL_\rho \gtrsim 1$, $\epsilon \ll \delta$ is evidence. In its place arise baroclinic waves 2 and 3, of both large and small frequency.

I carried out a series of spin-down experiments to explore the rough bottom eddies, and these have revealed a number of properties of relevance to the ocean. The topography, of rms amplitude 200 m (Fig. 33a) is randomly generated with a $k^{-1.5}$ scalar-wave number spectrum, cut off at $k > 8$ (relative to the domain width of 2π). This is just at the deformation scale, $L_\rho = (F_1 + F_2)^{-1/2}$, set to be 40 km, with the "ocean" width 2000 km. The contours of $f/h_2(x)$ (Fig. 33b) show the potential vorticity gradients due to β and h_2 to be comparable, with occasional closed contours appearing. The degree of openness of these contours is crucial to the processes of horizontal propagation and mean-flow induction (Section 8). As in many of the earlier experiments, an initially narrow-band field of eddies was allowed to evolve freely for 200–500 days, with mild damping, and without driving. Here the evolution of deformation-scale eddies, very large initial eddies, and linear waves are separately described, as is another series of inhomogeneous runs in which horizontal propagation is important.

D. Deformation-Scale Eddies

A typical case (Fig. 34) starts with eddies slightly bigger than the deformation radius above the thermocline and still fluid below. In this instance the currents are not strong, ($U_1 \sim 15$ cm/sec, $U_2 = 0$ at $t = 0$) but still significantly nonlinear, representative of the western Atlantic at 30°N. The initial adjustment phase occupies about 3 months $[\sim (U_{rms}k_\rho)^{-1}]$, during which the deep ocean is set into motion. But unlike the flat-bottom case, this development is halted by the irregularity of f/h (even though the "islands" of f/h are few). The layers become detuned by the topography so that the barotropic mode is a stable vertical structure only at large

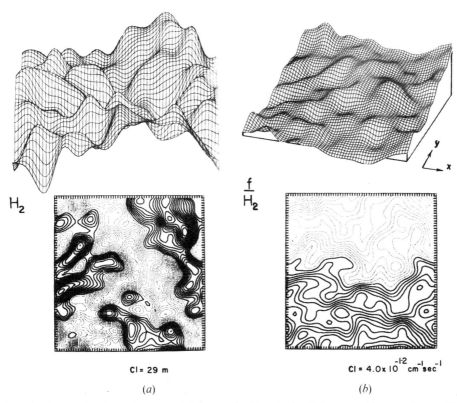

Fig. 33.   Perspective plot and contours of topography (a) and f/h_2 (b) for succeeding experiments. The depth has standard deviation 200 m ($\delta = 0.053$), or a typical bottom slope equivalent to β. (Actually (a) shows minus the topography; a ridge runs southeast through the central region.) Later, flows will be shown responding to the large rise seen at the western end of $y = 0$. The box is 2000 km wide. The domain is 2000 km across. Topography of scale smaller than 40 km was omitted, even though it may be important.

scales, far from the activity at k_ρ. If in the theory, Section 6, demonstrating the necessary increase in barotropy, we add a fine-grained pattern of topographic potential vorticity with $\delta \gtrsim \varepsilon$, the argument is destroyed.

To grasp the gross sensitivity of these results to energy level, compare the evolution in Figs. 34 and 35. The sole difference is the addition of an initially weak eddy field below the thermocline, in Fig. 35. The more energetic run has rms $|U_1| = 15.3$ cm/sec, $|U_2| = 4.1$ cm/sec, and no great reservoir of potential energy. It is still a plausible intensity for the oceans. Yet the eddies succeed in interacting vertically, tending toward barotropy, expanding horizontally, and developing stronger anisotropy and f/h contour flow, practically oblivious to the roughness.

As it is, in the weakly energized flow (Fig. 34) one can spy the primary cascade working in limited regions for limited times, over unusually flat f/h topography, or in a region of unusually strong current (see also Fig. 37). A remarkable feature of this flow, and those seen earlier, is the persistence of westward phase propagation well into the nonlinear regime (Fig. 41). This is particularly so in the thermocline eddies, type 3, which dominate the upper level currents and the thermocline height. There the phase

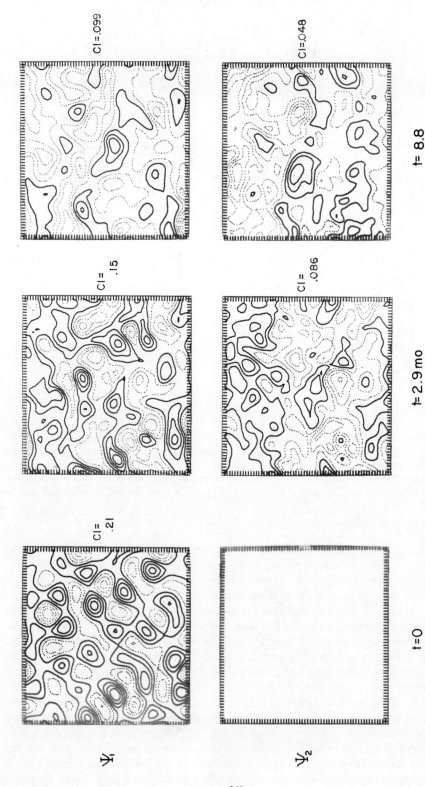

Fig. 34. Spin-down of deformation-scale eddies, $U \sim 15$, deep layer initially at rest. The uneven sea bed so detunes the vertical structure that barotropy (path e, Fig. 19) is now prevented. Note the anticyclone trapped over a predominant rise in the deep water, right center. After 7 months, $K/P = 0.85$.

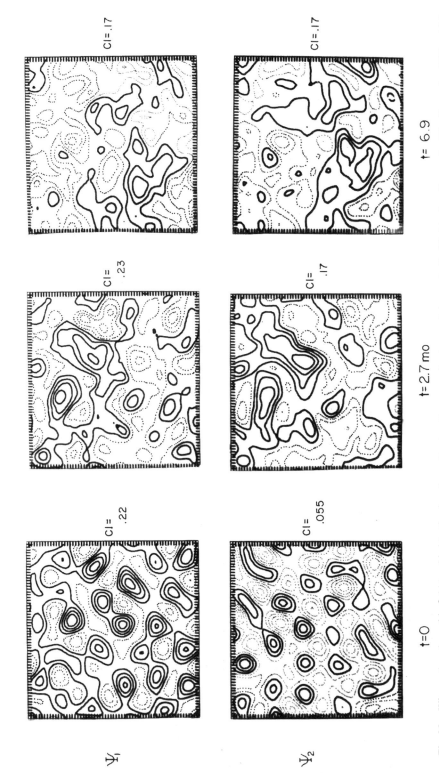

Ψ_1

CI = .22

CI = .23

CI = .17

Ψ_2

CI = .055

CI = .17

CI = .17

t=0

t=2.7 mo

t= 6.9

Fig. 35.   When a 4-cm/sec eddy field is added to the deep layer, it raises the energy level enough to regain the "flat-bottom" cascade. After 7 months, $K/P = 12.3$.

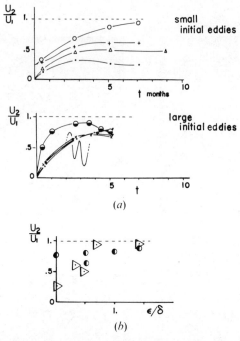

Fig. 36. Downward penetration of energy, as measured by the vertical structure, U_2/U_1 for a variety of values of ε/δ. (*a*) Growth from small-scale highly baroclinic initial flows toward an equilibrated value. Values of ε/δ are (upper): $\bigcirc$, 0.70; $+$, 0.34; $\triangle$, 0.51; $\cdot$, 0; (lower): $\ominus$, 0; $\times$, 0.50; $\triangledown$, 1.0; $\cdot$, 0.50. With large initial eddies the energy penetrates more easily. The wavy curve ($\varepsilon = 0, \delta = 0$) reminds us that linear effects can periodically alter U_2/U_1, here by a beating between vertical modes. Here ε is based on the vertical r.m.s. value of U at $t = 0$, and the length-scale L_ρ. The values of $\beta L_2{}^2/U_2$ are, (upper); $\bigcirc$, 0.6; $+$, 0.8; $\triangle$, 0.8; $\cdot$, 0.8; (lower): $\ominus$, ∞; $\times$, 0.9; $\triangledown$, 1.4; $\cdot$, 0.6. A triangle was inadvertently left off the final graph, ($U_2/U_1 = 0.31$, $\varepsilon/\delta = 0.17$). (*b*) The equilibrated vertical structure as a function of ε/δ. $\triangleright$, L_ρ-scale initial eddies; $\mathbb{O}$, large-scale initial eddies.

speed is not far from the linear prediction. The deep flow shows westward propagation in regions of favorable topography, but not elsewhere (Fig. 41). The dual nature of geostrophic eddies, exhibiting properties of both waves and turbulence even in the most complex of cases, promises a growing theoretical understanding of them.

For a given configuration at $t = 0$, the developed vertical structure (which evolves very slowly after the initial phase) depends upon ε/δ and $\beta L_\rho{}^2/U \equiv \varepsilon L_\rho/R$. A gross measure of the structure is U_2/U_1, based on rms currents without regard to their relative phase. The time evolution of U_2/U_1 (Fig. 36*a*) for a number of experiments (with fixed β) shows a clear dependence on topographic heights and energy level.[7] The equilibrated structure (Fig. 36*b*) rises with ε/δ, providing a plausible ocean structure, say, $U_2/U_1 = \frac{1}{3}$, at $\varepsilon/\delta \lesssim 0.5$. But as a warning, points are also shown corresponding with initial eddies of large scale; these clearly penetrate more easily to the depths.

---

[7] The dependence on $\beta L_\rho{}^2/U$ has not been thoroughly explored. When it is large, the fluid will be unable to develop barotropy even in the absence of topography. Also, the boundary conditions (periodic in ψ) imply nonzero large-scale pressure gradients. If, instead, these are taken to vanish, a large-scale westward flow develops in response to topographic drag (Bretherton and Karweit, 1975). The choice of boundary conditions (periodic ψ or periodic velocity) is moot.

At this point we should compare fields predicted in the models of Sections 5–7 from the point of view of the seagoing experimentalist. Figures 37a–d show time series measured at a single mooring in a linear, flat-bottom, a nonlinear flat-bottom, a nonlinear rough-bottom, and a linear, rough-bottom ocean, respectively. In each case the initial configuration of the current is the same as in Fig. 34. First, with weak currents the subsequent oscillations are purely linear Rossby waves. The long period of the baroclinic mode shows in the "temperature" (the interface elevation), and, with this modal mix, dominates the upper-level currents. The average vertical structure obeys $U_2/U_1 = H_1/H_2 \sim 0.28$, and evolves as soon as the two modes separate from one another, roughly one-half barotropic period.

Second (Fig. 37b), the bottom is flat but the currents are stronger. The deep and shallow flows begin to move in apparently unrelated ways, but suddenly, after 4 months the layers lock together. The temperature field exhibits much faster oscillations than are possible in linear theory, owing to horizontal advection. [Sharp temperature features are found at sea, even well below the surface (Fig. 63) with time scales far less than those of type 3 waves.] The time scale of the currents decreases after the switch of modes.

The third comparison run (Fig. 37c) comes from an ocean with $\varepsilon/\delta \approx 0.4$. This most complicated case shows periods during which the layers begin to lock together, but

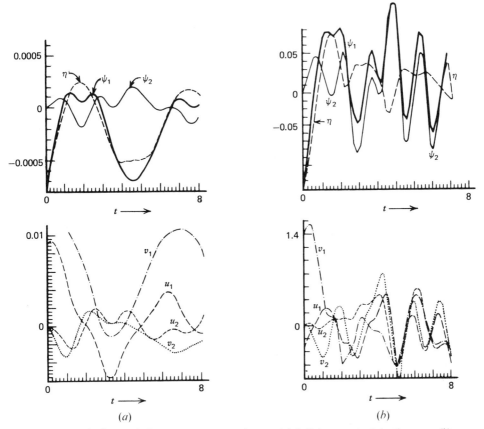

Fig. 37. "Mooring" records from computer experiments. (a) A linear constant-depth ocean; (b) nonlinear, constant depth.

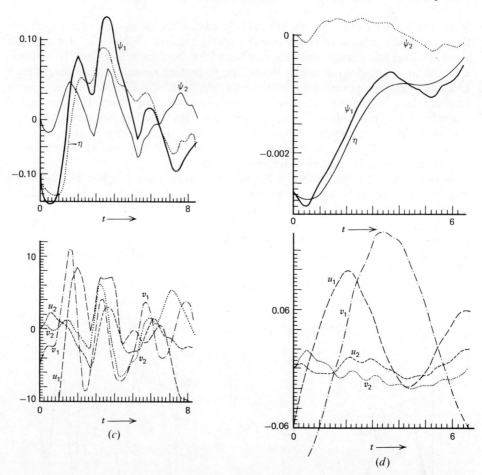

Fig. 37. (c) nonlinear, rough bottom; (d) linear, rough bottom. Upper graphs give ψ_1, ψ_2, η; lower graphs give velocities (cm/sec). Note different abscissa in (d). Although there are crude similarities with the linear solution, nonlinearity adds shorter time scales to the density signals, redistributes energy in the vertical, and alters the horizontal scale; topography confines fast oscillations to the deep layer, whitens the horizontal spectra, and counters the downward flow of energy from shallow water. At any given moment, however, the linear theory has some partial validity, particularly after a comfortable vertical and horizontal structure has been achieved.

the records soon diverge again, and the vertical shear returns. The topography has a noticeable effect on the time scale of the deep flow; here it is rather short, but this is not always true.

Finally (Fig. 37d), a *linear* rough-bottom ocean responds with fast oscillations in the deep water. Yet for these initial conditions the vast majority of the energy remains in and above the thermocline ($U_2/U_1 \sim 0.33$). The trapping of fast waves in the deep water and slow waves in shallow water occurs just as in the linear theory for simpler geometry, in types 2 and 3.

The spectral transfer functions (Fig. 38), now have an added member $T_H(k)$, the contribution of $U_2 \nabla \cdot h_2$ to $(\partial/\partial t)E(k)$. $T_H(k)$ characteristically removes energy from the spectral maximum of $K(k)$, sending it predominantly to large k. The dynamics of the spectral tail is different from the flat-bottom ocean (Fig. 31). Now T_H replaces

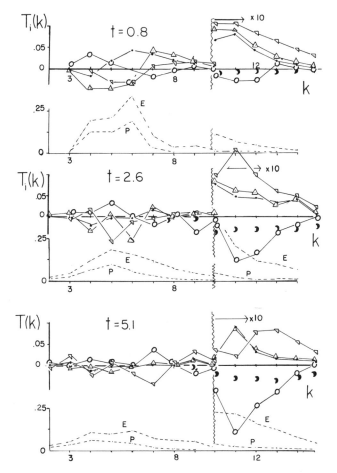

Fig. 38. Transfer spectra with topography, for an experiment with L_p-sized eddies, initial $U_2 = 8.0$, $U_1 = 2.7$ cm/sec. Compare Fig. 31 (here ▽ is T_H). The basic shapes of T_ψ, T_P, and C_{PK} are as before, yet they are relatively smaller, and barotropy fails to develop (compare E and P spectra). Topographic scattering broadens the spectra, relative to Fig. 31, sending energy to large wave number. The spectral "tail" (magnified) now has totally different dynamics: T_H, rather than T_P, carries the energy to the right and T_ψ returns it to the left. Dissipation is small throughout $K < 14$.

T_P as the carrier of energy to large k, yet T_ψ still carries it back to the left, allowing a quasi-steady "topographic" subrange. It reduces the amount of energy carried to small k (by T_ψ), and seems to reduce both the magnitude of T_P and the conversion C_{PK} to kinetic energy. In this manner, the progress of the fluid along the paths in Fig. 19 is impeded, and sometimes reversed. The flow evolves slowly after initial adjustment, but the persistent broadening due to T_H suggests that a true equilibrium spectrum will be impossible without forcing and dissipation.

E. Initially Large-Scale Eddies

Those baroclinic eddies with diameter greater than about 150 km contain an excess of potential energy. Conversion into deformation-scale flow is the dominant nonlinear process, followed, with bottom flat, by further conversion to kinetic energy as barotropy develops, and finally the reverse cascade of 2D turbulence. What is the effect of bottom roughness in the process? We anticipate two competing effects: first, the horizontal scattering by topography which, if it proceeds to large k, must release potential energy (for small eddies cannot possibly contain it); second, and in opposition, roughness has been shown to prevent barotropy, which develops in the late stage of baroclinic instability.

The net conversion C_{PK} is shown for a number of comparison runs in Fig. 39. The initial release is indeed augmented, and the rougher the bottom, the more so. But in the second phase the conversion drops far below the flat-bottom case, as suggested. The time-integrated release of potential energy is severely reduced by the inhibition of barotropy. To aid in the comparison, a calculation shown in the figure was stopped at $t = 2$ months, the bottom denuded of topography, and the run allowed to proceed. After a brief shock, the conversion accelerates with this new freedom, as the deep layer comes up to speed.

In making "rough–smooth" comparisons it must be remembered that the smooth-bottom baroclinic instability develops at a rate initially depending on the amount of noise specified at scales near L_ρ. For random "big eddies," no noise is necessary for the energy cascade to develop, but with initially simple, rectilinear patterns (here $\psi_1 \propto \sin 2x$ at $t = 0$), noise is essential. Thus there is an unnatural delay in the development of C_{PK} without topography.

It should be clear that potential energy release does not rely solely on nonlinearity $(\mathbf{u} \cdot \nabla \eta)$ in this problem, for the scale transformation by topographic scattering brings about a similar end. A purely linear run (infinitesimal currents) is plotted in Fig. 39 (with its own normalization) and it exhibits this conversion. The vertical structure resulting late in these runs was included in Fig. 36. The instability process penetrates the deep water rather more easily than did smaller eddies of comparable energy, thus making U_2/U_1 a function of both initial configuration, ε/δ, and of $\beta L_\rho^2/U$.

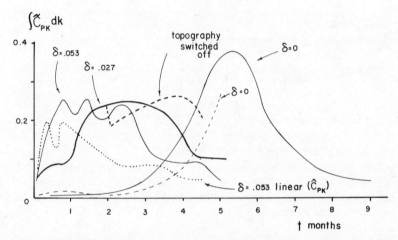

Fig. 39.   Total conversion from potential to kinetic energy versus time, for three values of topographic height, δ; large-scale initial eddies. The stronger the topography, the greater the initial conversion (due to deep scattering) yet the less the total conversion (the area beneath the curves). C_{PK} is a nonlinear effect in a flat-bottom ocean, but here occurs also with purely linear motions. $\tilde{C}_{PK} = C_{PK} P_0^{-3/2} k_\rho k_0^{-2}$, $(P_0 = $ initial potential energy), $(k_0 = $ initial wave number) $\hat{C}_{PK} = C_{PK} P_0^{-1}$.

Fig. 40. Flow patterns for rough-bottom baroclinic instability, $\delta = 0.027$, $\epsilon/\delta \simeq 1.0$. The large scale of the developing flow (relative to Fig. 25, flat bottom) comes as scattering carries energy continuously toward large k rather than jumping to $k = k_\rho$. f/h-Contour flow and a significant amount of deep-water energy develops ($U_2/U_1 \rightarrow 0.7$).

A visual impression of the streamline patterns (Fig. 40) in these runs is one of large-scale flow, strikingly different from the sharply defined L_ρ-size eddies in pure baroclinic instability (compare Fig. 25). This is the result of spectral broadening by the roughness, which carries energy continuously along the k axis rather than having it leap to the deformation radius. Examination of the wave number spectra verify that the process of baroclinic instability is *intrinsically different with rough topography*, even though the f/h contours are still "open." As above, its new nature appears when $\varepsilon/\delta \lesssim 1$ so that vigorous flows like the Gulf Stream (Section 6) are probably less affected by small-scale roughness (although large-scale bottom slopes will still be important, for they alter the mean cross-stream potential–vorticity gradient).

The vertical structure developing in these flows (Fig. 36) shows greater barotropy, on the average, than with initially small eddies (for given ε/δ). This emphasizes that the oceanic vertical profile of current depends not only on levels of energy and topography, but also on the *manner of supplying the energy*. The streamlines (Fig. 40), show a degree of barotropy at large scale. The huge apparent scale of ψ is due in part to the spectral broadening by topography. In fact, the energy-containing eddies as measured by k_1 are not strikingly different in scale from k_ρ. Note the strongly zonal character to the well-developed flow.

F. Fine Structure, Anisotropy, Mean Flows

The experiments described above were meant to be homogeneous in space; both topography and initial fields have randomly generated spectra, with specified scalar-wave number shape. Without uneven bottom topography the flow remains reasonably homogeneous, at least in its energy density.

A striking spatial intermittency, however, develops over a rough bottom. The result that the vertical adjustment of the water column occurs in a very few months, is suggestive of this, in that spatial propagation of energy during this period is small (less than 200 km in 2 months, say). The fluid can thus respond locally to its topographic environment, which Fig. 36 shows to be a decisive control over the vertical structure. Beyond this, the purely linear waves over a rough bottom become rather intermittent themselves, owing to local "seamount" resonance. Linear, flat-bottom waves, of course, tend to be of an extreme opposite nature, propagating with unchanging vertical structure, and gradual horizontal evolution, in the far field.

The kind of heterogeneity that develops in the current is shown in Fig. 41, a time–longitude plot of the deep pressure (or ψ) field from the experiment in Fig. 34. Compare it with the smoothness of thermocline-height propagation, also shown. Over geostrophic contours that are bunched, yet still run east–west, there are fast oscillations and a clear westward phase propagation (the eastern half of the region; see Fig. 33*b*). Conversely, in the west there develops a low-pressure circulation that remains nearly steady, above the dominant 300-km-wide rise (Fig. 33*b*). The flow field is variable in energy density, as well as character, as is consistent with recently observed fine-structure of eddy energy (Section 9). Another example of topographically induced spottiness in the deep eddies may be seen by looking ahead to Fig. 43*f*. Whether or not one is interested in the erratic statistical distribution of the eddies per se, it presents a severe aliasing problem for ocean experiments.

The generation of mean currents by eddy vorticity-flux is discussed in Section 8. An example with $\beta = 0$ (Fig. 42) shows a weak ($U_2 \sim 2$ cm/sec, $\varepsilon/\delta \sim 0.14$) deep-current field to begin more and more to resemble the topographic contours (Fig. 33*a*) with time. G. Holloway (private communication) has shown this condition to be

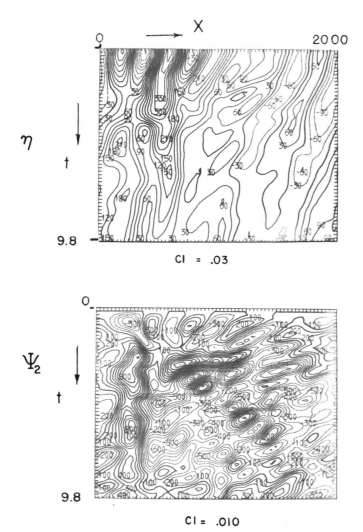

Fig. 41.   Time–longitude plots of ψ_2 and η, from the southernmost latitude shown in Fig. 33. The topography causes a natural inhomogeneity in the deep water: unusually fast oscillations in one area, yet a strong anticyclone trapped above a ridge in another. The thermocline eddies are less disturbed, and propagate westward with occasional episodes of vertical interaction.

reliable in a single-layer fluid, and it seems to be the generalization to f/h flow of the zonal currents appearing spontaneously throughout Section 6. He points out that, with $\overline{(\zeta + fH/h)^2}$ conserved for at least the initial period, an increase in relative enstrophy, $\overline{\zeta^2}$, implies an initially growing negative correlation between ζ (and hence ψ) and the geostrophic contours, $(\partial/\partial\tau)\overline{\zeta f/h} < 0$ (the overbar is integration throughout space). In fact, we have shown $\overline{\zeta^2}$ to increase owing to either topographic scattering or the presence of energy at a western boundary. This suggests both the appearance of persistent gyres about small seamounts and, on the large, appearance of basin-scale circulation with $\overline{\zeta y} < 0$ ($f = f_0 + \beta y$); see Section 8.

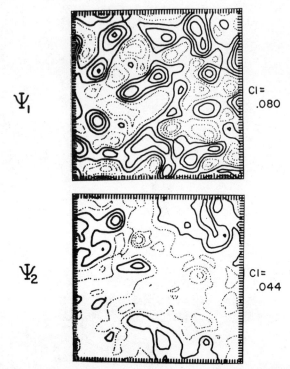

Fig. 42. ψ fields late in an experiment with $\beta = 0$. The rather weak eddy field is nearly held above the thermocline by the topography, $U_1{}^0 = 9.3$, $\delta = 0.11$, $\varepsilon/\delta = 0.4$, $U_2/U_1 \rightarrow 0.3$. The weak abyssal current develops toward a contour circulation (see Fig. 33a) with anticyclones above high ground, and conversely.

G. Lateral Propagation

One motivation for this kind of work was to develop improved ideas about the horizontal transports of energy by unsteady currents, for both their intrinsic interest and their role in the general circulation. This is of particular relevance to the oceans owing to their great size, measured against that of the $O(L_\rho)$ energetic eddies, and to the sparse distribution of really active energy sources. Estimates of travel time from a conjectured source of energy are often made from linear, flat-bottom theory. But what the experiments have shown most conclusively is that nonlinearity and topography can act to transform these scales, and the mix of vertical modes, in a rapid fashion.

We have initiated some further experiments with oceans initially one-half full of eddies, to provide information about the energy velocity and to suggest what the appearance of eddies arriving from a distant source should be. The energy was initially placed above the thermocline in deformation-scale eddies, occupying 1000-km bands separated by equal areas of quiet ocean. Time–latitude sections, (Figs. 43a–c) show the intrusion of energy from the north and south. Figure 43a is linear, with a flat bottom; the tilted contours indicate northward phase propagation, hence southward group velocity. After 250 days the barotropic mode has grought some energy into the quiet region, but the baroclinic mode has made only a meager contribution (the largest north–south group velocity of the baroclinic mode is $\beta/4(F_1 + F_2) \sim 0.8$ cm/sec).

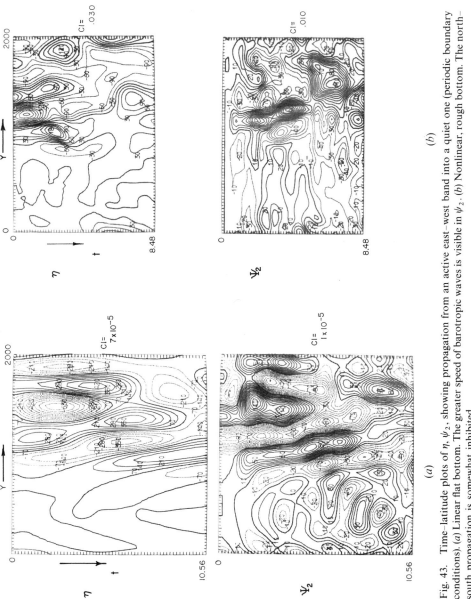

Fig. 43. Time–latitude plots of η, ψ_2, showing propagation from an active east–west band into a quiet one (periodic boundary conditions). (a) Linear flat bottom. The greater speed of barotropic waves is visible in ψ_2. (b) Nonlinear, rough bottom. The north–south propagation is somewhat inhibited.

271

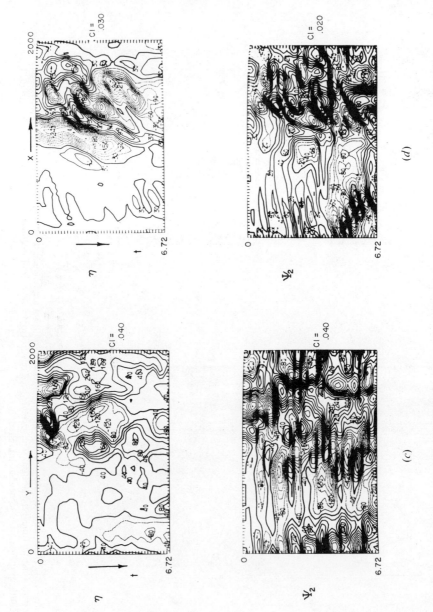

Fig. 43. (c) As above, yet a greater initial energy level. The nonlinear switch from baroclinic to barotropic modes, horizontal expansion, and C_{PK} release all combine to increase the energy flux into the quiet region. (d) Time–longitude maps of nonlinear, rough-bottom, *east–west* propagation from meridional bands of energy. The linear prediction survives, of greater flux east–west than north–south. Note westward phase propagation, even for energy moving eastward in ψ_2.

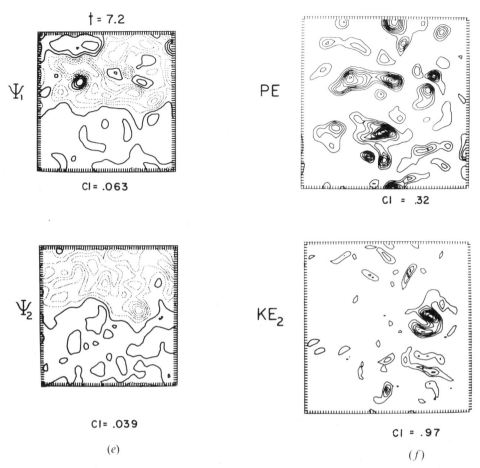

Fig. 43.   (e) ψ Fields for experiment (b). Southward propagation has been slight after 7 months. There is a zonal-average Eulerian zonal flow in both layers, related to the eddy-flux of potential vorticity (Section 8). (f) Maps of potential and deep-kinetic energy in the east–west experiment in (d), $t = 6.9$. A 250-km-wide seamount has trapped 95% of the deep energy in a single eddy (it also affects (e)).

The same configuration of currents was set off over a rough bottom, with rms upper-level speed initially of 7.4 cm/sec (Fig. 43b). The propagation is less orderly, as the usual spectral broadening occurs. The thermocline eddies are at least as slow as in the linear case. In the deep water, rather less energy reaches the empty region than before, due to topographic backscatter and distortion. The topography works in subtle ways, affecting both vertical and horizontal scale of the fluid, as well as causing backscatter and refraction.

The third comparison run (Fig. 43c) illustrates the crucial influence that nonlinear scale transformations can exert. Here the current speed is initially 14 cm/sec in the upper ocean, 13 cm/sec below, so that the topographic resistance to barotropy is ineffective. The eddies switch modes and in doing so augment greatly their group velocity. The empty region is filled within 200 days, suggesting an effective north–south group velocity of about 5 cm/sec.

A fourth experiment (Fig. 43*d*) shows the relatively rapid propagation in the east–west direction, from alternating meridional bands of energetic and quiet ocean. The energy level is rather weak, yet the spatial field for Fig. 43*c* is shown in Fig. 43*e*. After 7.2 months, little energy has reached the initially empty fluid. The deep flow is again intermittent, with much of the energy trapped in a single anticyclone. This same rise in the sea floor trapped the majority of the deep energy in the east–west run (Fig. 43*d*); see Fig. 43*f*.

If the ocean were subjected to intermittent, strong episodes of forcing by winds or meandering of boundary currents, energy could burst across the domain with ease. The nonlinear effects contrive doubly to increase the group velocity, by switching to the fastest vertical mode, and then expanding in horizontal scale. This suggests rather intermittent far-field energy levels, bearing in mind that periods of greatest group velocity occur when the greatest energy density is to be transported. Is it possible that far-reaching spasms of activity occur in the oceans, if only we could see them?

H. The Linear-Wave Problem and Small-Scale Topography

It has proved useful to imagine the properties of eddies near the transition between wave motion and turbulence, for in the region of overlap, both theories can contribute. The parameter ε/δ, which discriminated between "rough" and "smooth" nonlinear cascades, in fact also represents the steepness of topographic waves. For topography of a single dominant horizontal scale, L_H, if significant, tends to induce fluid motions of scales near L_H, and frequency $\sim \delta f$, whence ε/δ is the ratio of current speed to the phase speed. In addition, ε/δ measures the extent to which fluid crosses an entire topographic feature during a wave period ($\div 2\pi$); linear waves apply when the excursions thus measured are small, and one approaches quasi-steady Taylor column- and Taylor "cone" problems when they are large.

This extreme, $\varepsilon/\delta \gg 1$, holds especially for small-scale (10–30 km) seamounts, which have been filtered from our model. Hogg (1973), and McCartney (1975) are among the recent investigators of the purely steady limit, and Huppert and Bryan (1975) have looked at the crucially important "start-up" problem. Recent ocean measurements (Section 9) emphasize that the larger-scale flow shifts sufficiently often in the deep water that the steady, potential-vorticity conserving deflections are a poor description of the effects of topography, even at the smallest scales relevant to geostrophic flow (~ 10 km). The transient-flow problem is complex, and it is essential to decide whether flow, starting from rest, causes just a single starting vortex to be swept from above the bottom feature (leaving behind the steady, bound vortex), or whether a continual train of shed vortices is created, as in a classical cylinder wake. (The author believes he has seen vortex shedding from a Taylor column in the laboratory, and there are numerous satellite photographs of vortex streets in the lee of islands, occupying either fluid.) The parameterization of small-scale topography, and its wave drag (including internal waves) on the mesoscale eddies is crucial; our present practice of removing all topography with $L_H < L_\rho$, and replacing it with a linear drag, may be severely in error.

We return to the larger scales, where $\varepsilon/\delta \lesssim 1$ frequently (if only by inference from the observed baroclinity of ocean currents). There the wave theory has already suggested some of the turbulent cascade results. First, linear scattering of long waves provides a model for $T_H(k)$, the topographic energy transfer spectrum, in the general case. This may allow a quantitative estimate of spectral broadening and fragmentation, of even nonlinear eddies. Second, the occurrence of the bottom-trapped type 2 is consistent with high-frequency oscillations found in the deep water (but not above the thermocline) in the simulations. Third, the disappearance of the barotropic type 1 at $L \lesssim L_\rho$ suggests, in the turbulent runs, the sustenance of vertical shear (the defeat of barotropy), for it is the scales near L_ρ that form the aperture through which the different levels communicate. Fourth, the prediction from theory of horizontally trapped waves over rough topography is suggestive of the immobility of energy found in the nonlinear cases. This argues further that spatial intermittency of energy and "local" equilibrium of the eddies should develop in both linear and nonlinear oceans.

The linear-wave theory becomes difficult in the most relevant case of topographic "turbulence," $kL_H \sim 1$, $\varepsilon \ll \delta \gtrsim L_H/R$. Some closed-form solutions have been found but there is much to be done. A relevant idealization is that of isolated bottom features, which occur when f/h contours are packed together (as at the continental margins and Mid-Atlantic Ridge), or when they form closed "islands."

The simplest such wave, which we derive for illustration, is that found at a near-discontinuity in the depth, say, along $y = 0$, $h_2 = H_2$ for $x < 0$, $h = H_2(1 + \delta)$ for $x > 0$. This provides a delta function of the restoring effect, the slope. The two-layer equations (linear, $\beta = 0$) become, in dimensional variables,

$$[\nabla^2 \psi_1 + F_1(\psi_2 - \psi_1)]_t = 0$$

$$[\nabla^2 \psi_2 + F_2(\psi_1 - \psi_2)]_t - \frac{f}{H_2} h_{2,y} \psi_{2,x} = 0$$

for $\delta \ll 1$. Trapped waves exist of the form

$$\psi_1 = (a_1 e^{-k|y|} + a_2 e^{-\kappa|y|}) e^{i(kx - \sigma t)}$$

$$(1 + \gamma)\psi_2 = (a_1 e^{-k|y|} - a_2 e^{i\kappa|y|}) e^{i(kx - \sigma t)}$$

They satisfy the exterior equations if $\kappa^2 = k^2 + F_1 + F_2$, $\gamma = H_2/H_1 = F_1/F_2$. The interface height ($\propto e^{-\kappa|y|}$) is tent-shaped, with its scale being the smaller of k^{-1} and the deformation radius k_ρ^{-1}. At $y = 0$ we must match normal flux and pressure; integration of the lower-level equation across the step determines the discontinuity in $\psi_{2,y}$:

$$[\psi_1] = 0 = [\psi_2] = [\psi_{1,y}]$$

$$[\psi_{2,y}] = \frac{f}{\sigma} \delta \psi_{2,y} \qquad (y = 0)$$

An oscillatory vortex sheet occurs in the lower layer, driven by the upslope velocity (and has this strength even when the particle excursions are rather great). The matching conditions yield the dispersion relation

$$\frac{\sigma}{f} = -\frac{\delta}{2} \frac{\gamma}{1 + \gamma} \left(\frac{k}{\kappa \gamma} + 1 \right)$$

Large scales, $k^2 \ll F_1 + F_2$ are barotropic "double Kelvin waves" (Rhines, 1969) with $\sigma/f = -(\delta/2)(\gamma/(1 + \gamma)) = -\frac{1}{2} \times$ fractional change in total depth across the step, with $\psi_1 = \psi_2$. Waves shorter than the deformation scale become increasingly confined to the lower layer. In this limit they feel the interface as a rigid lid, and hence $\sigma/f \to -\delta/2$, $|\psi_1/\psi_2| \to k_\rho|k \ll 1$. The topography provides trapping in vertical and horizontal directions. For the simplest case the group velocity, $\partial\omega/\partial k$, vanishes at both extremes of k, but is significant near $k = k_\rho$. Phase and energy each move to the left, facing shallow water. Simple solutions also may be found for seamounts and islands, where the phase progresses clockwise (the analogue of westward-propagating Rossby waves).

The eigenvalue problem for sinusoidal depth variations (Suarez, 1971; McWilliams, 1974; Rhines and Bretherton, 1974) provides an interesting model of "roughness"

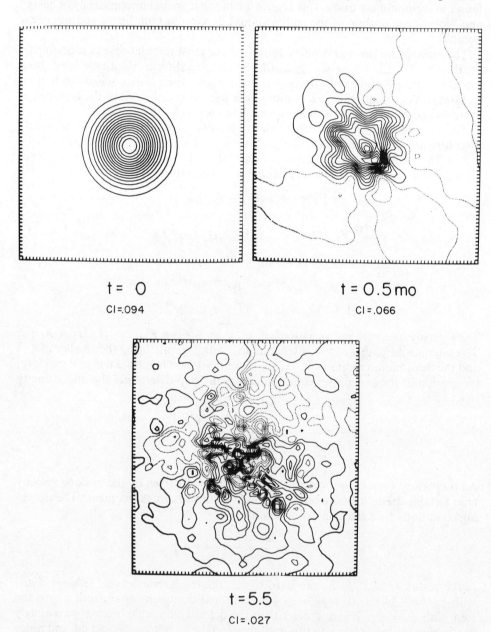

Fig. 44.   Linear-wave dissolution of a circular vortex in a rough-bottom ocean (homogeneous fluid). The topography is confined to wave number 6–11. Compare Fig. 11. In spite of the spatial complexity, the pressure at a fixed point varies rather sinusoidally in time, as in a trapped "seamount" oscillation.

waves, the frequencies again being $O(\delta)$. For arbitrary depth in a one-layer ocean enclosed by a coast, upper bounds can be established for the frequency, and are of this same order:

$$\sigma \le \left| f - \frac{f_0 h}{h_0} \right|_{\max}$$

where f_0 and h_0 are the mean values of f and h, respectively. If $\delta R/L \ll 1$ this essentially sets the greatest planetary-wave frequency at $|f - f_0|_{\max}$.

If we fix the total rms derivation of f/h from its mean in a model basin, the average energy propagation tends to be faster, the more smoothly the topography is distributed. This is not quite the same as saying that the addition of rough topography to a smooth β plane must reduce the group velocity, for the group velocity eventually scales up with δ, and hence must become large if δ does.

Numerical experiments with a single-layer fluid exhibit some of the interesting horizontal-cascade effects. Figure 44, taken from a movie sequence, shows the fragmentation of a large Gaussian vortex by narrow-band bottom roughness. (See also Rhines, 1973, figure 5.) The energy gradually percolates outward from seamount to seamount, for there is no large-scale β effect to support fast, long waves. A natural fine structure builds up, even with these linear dynamics.

A linear experiment in the two-layer stratified ocean (Fig. 45) uses initial conditions and topography very like the nonlinear runs (Figs. 34, 35). The energy reaching down to the lower layer is rapidly "whitened," and again becomes severely intermittent

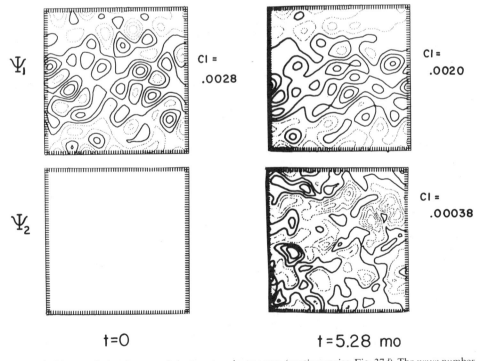

Fig. 45.   Linear solution for a rough-bottom two-layer ocean (see time series, Fig. 37d). The wave number spectrum is very "white" in the deep water, although the finer scales are not evident in ψ_2.

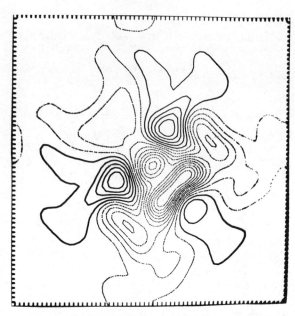

Fig. 46.   The linear solution, after a Gaussian vortex is placed over a tessellated bottom, $h \propto \sin x \sin y$.

in space. The time series at fixed moorings, shown in Fig. 37d, are dominated by slightly modified baroclinic Rossby waves in the upper currents and temperature, yet fast topographic waves in the deep flow. In this instance, the average vertical structure (included on Fig. 36) remains strongly baroclinic, $U_2/U_1 \simeq 0.33$.

Surely the topographic waves best documented in nature are those trapped in the coastal wave guide formed by the continental slope. These are particularly quickly propagatory because $\delta \sim 1$, providing periods as short as a few days; see Chapter 10.

But the convenience of acquiring coastal data should not deter us from looking elsewhere: the possibilities for lateral trapping and for unanticipated kinds of fluid dynamics in mid-ocean are numerous. Within so complex a domain as the sea, we may yet find Cthulhu (Fig. 46).

8. Mean-Flow Interaction

The energy-containing eddies are themselves worthy of attention, and in addition can affect the time- or space-averaged flow in which they are embedded. The inter-action takes many forms. A purely steady forcing effect, if sufficiently strong, leads to circulations that become unsteady. There can then be feedback of the eddies onto the mean. Conversely, a purely oscillatory forcing frequently causes a rectified flow. Examples of both extremes are given below.

It is important to realize how many different ways there are to define "mean flow." Meteorologists, possessing a simple geometry at the large scale, favor the zonal, Eulerian mean of instantaneous wind velocity. Their eddy field contains the meanders of the mid-latitude jet streams, nearly steady monsoon circulations and orographic deflections, as well as detached cyclones. Owing to the generalized Stokes drift, the velocity averaged about fixed control surfaces does not accurately describe the average paths of fluid parcels, either in a meridional plane or a level surface. But oceanographers, who are interested in the life history of salts, heat, and chemical tracers, as well as momentum, must pay more attention to such Lagrangian means, averaging over an ensemble of realizations. In the presence of coastal boundaries it is particularly difficult to find fixed control surfaces that yield useful overall statements about linear or angular momentum, or vorticity. The oceanographer's interest in the paths of fluid parcels is in part pragmatic: in much of the ocean the time-averaged current at a fixed point is nearly unmeasurable by direct means. (An illustrative example is given in Section 9.)

A simple calculation of the difference between Eulerian and Lagrangian mean flow can be made for a field of geostrophic turbulence. Imagine, as a model, that the dispersal of particles obeys a diffusion equation,

$$\frac{\partial C}{\partial t} = \nabla \cdot (\kappa \nabla C)$$

where $C(\mathbf{x}, t)$ is the spatial concentration of fluid markers, and α the diffusivity. Multiplying by $\mathbf{x}$ and integrating, we find that the center of mass obeys

$$\frac{\partial \overline{\mathbf{x}}}{\partial t} = \overline{\nabla \kappa}$$

where $\overline{(\ )} = \int (\ ) C\, d\mathbf{x} / \int C\, d\mathbf{x}$. If C is a delta function, this gives the most likely flow of a single particle. The particles move preferentially toward regions of large diffusivity, even in the absence of an Eulerian mean flow. The spread of the probability distribution about the expected path occurs, to a first approximation at rate $\overline{\kappa}$. This model was investigated by Kolmogorov (see Monin and Yaglom, 1972, p. 610).

It is very likely that, for a stationary, turbulent field of slowly varying statistics (relative to the excursion of particles during one eddy-period), the appropriate generalization is

$$\frac{\partial \langle \overline{x}_i \rangle}{\partial t} = \frac{\partial \overline{\kappa_{ij}}}{\partial x_j} + \langle u_i(\mathbf{x}) \rangle$$

where

$$\kappa_{ij} = \int_0^t R_{ji}(\tau | \mathbf{x})\, d\tau, \qquad R_{ij} = \langle u_i(t|\mathbf{x}) u_j(t + \tau | \mathbf{x}) \rangle$$

This uses Taylor's formula for the diffusivity in terms of the Lagrangian correlation function, $R_{ij}(\tau | \mathbf{x})$ for particles released at the point $\mathbf{x} \equiv (x_1, x_2)$ at $t = 0$. The brackets are an ensemble average, and $\langle u_i(\mathbf{x}) \rangle$ is the Eulerian mean flow. In the limit of small-amplitude waves this becomes the classical Stokes drift. The difference between the two mean flows will be most significant when the eddy intensity varies greatly on the scale of the eddies themselves. In surface gravity waves, for example, the difference is $\partial x / \partial t = \partial \kappa_{xz} / \partial z$, and below the troughs the Eulerian average flow completely vanishes.

To get a feel for the result imagine the dispersion of neutrally buoyant particles in a turbulent boundary layer, say, at the base of the atmosphere. The boundary exerts the same effect as a strong diffusivity gradient (which also exists in the air itself). It is intuitive that eventually the center of mass of a marked region will rise away from the boundary, even though the Eulerian-average vertical velocity vanishes everywhere.

Taylor's (1921) formulation of turbulent diffusion (essentially the identity that $(\partial / \partial t)\langle x_i x_j \rangle = \int_{-t}^{t} R_{ij}(\tau)\, d\tau$ in a *homogeneous* field) emphasizes also that in a given region there are many different Lagrangian drifts and many different diffusivities, depending on the subset of particles being counted, and on the recent history of the field.[8] At small times a delta-function cloud disperses with $\langle x_i^2 \rangle \propto \langle u_i^2 \rangle t^2$ (initially zero diffusivity) yet after the initial velocities are forgotten, the expansion slows toward a random walk, $\langle x_i^2 \rangle \to 2(t \int_0^\infty R_{ii} \, d\tau - \int_0^\infty \tau R_{ii} \, d\tau)$. Dyed patches of fluid of different sizes (marked with different colors) thus spread at rates inconsistent with a single diffusivity and their centroids move with different velocities.

This result suggested itself after neutrally buoyant SOFAR floats in MODE were observed to behave rather erratically in their mean drift, at times acting very unlike the mean flow seen by current meters moored nearby. In the same region, we found precipitous, permanent gradients in eddy intensity.

A. A Whole-Gyre Model

The dynamical studies in Sections 5–7 focused for simplicity on homogeneous fields, without boundaries. But we know the ocean to be heterogeneous, and an independent line of attack includes an explicit source, here the meanders of a wind-driven circulation, which may in turn radiate to the central ocean. A number of investigators have been experimenting with such models reminiscent of Stommel's single-gyre circulation, yet with stratification and explicit eddies present. One such calculation, by Holland and Lin (1975), has reached an advanced stage,[9] and we describe their results in some detail.

Holland and Lin drive their ocean, in which there is simple lateral friction, by spinning up from rest with a steady wind stress, sinusoidally varying with latitude. At moderate Reynolds number the mean flow is highly inertial. The Gulf Stream (with free-slip boundary conditions) turns along the northern wall and then decelerates as fluid returns to the interior (Fig. 47). This is typical of inertial gyres; only at smaller Reynolds number, or with no-slip walls, do the more classical western boundary layer patterns return.

After 2 yr of driving by the wind (of amplitude 1 dyne/cm²) the circulation spontaneously begins to meander, and closed cells of transport move throughout the basin. Unlike our free initial-value experiments, the mean state here is continuously maintained and a statistically steady, fully interactive state is reached after about 3 yr (Fig. 47).

The time-mean and perturbation fields are at first sight surprising. The perturbations are strongest, not in the vicinity of the Gulf Stream, but in the westward return flow. There the upper layer meanders lag those in the lower layer by roughly 60°. This is a tilt of the phase of pressure in the xz plane opposite to the sense of the mean velocity, a familiar signature of baroclinic instability. The dominance of the open-ocean return flow is in part due to the stabilization provided by the northern wall, but more strongly due to the narrowness of the westward flow. For the intensity of the eastward-flowing Gulf Stream is about 20 cm/sec (averaged over its 80-km width), whereas that of the return flow is about one-third as great. Simple instability theory Section 6, which ignores horizontal shear, suggests that the eastward flow, to be unstable, must obey $|U_E| > g'H_2\beta/f$ (in order that the mean potential vorticity gradient $\overline{Q}_y$ have opposite signs in the two layers). The westward flow can more easily

[8] The identity in general involves $\int_{t_0-t}^0 (R_{ij}(\tau, t|\mathbf{x}, t_0) + R_{ji}(\tau, t|\mathbf{x}, t_0)) \, d\tau$. If the statistics are stationary but inhomogeneous, however, this diffusivity is non-stationary.

[9] Others currently investigating this area include Haidvogel, Mintz, Robinson, and co-workers.

INSTANTANEOUS FIELDS
INTERFACE CI=15 M.   STREAMFUNCTION CI=5 SV.

P1 CI=0.2 M**2/S**2   P2 CI=0.1 M**2/S**2

Fig. 47. Typical flow pattern from Holland and Lin's wind-driven ocean. P1 and P2 are ψ_1 and ψ_2 in our notation, PSI is $\psi_1 + \psi_2$, the total transport.

reverse the gradient in the thin upper layer, if only $|U_W| > g'H_1\beta/f$.[10] Thus if $U_W/U_E > H_1/H_2$, the return flow will be the more vulnerable, as it is here: $H_1/H_2 \sim 0.2$, $U_W/U_E \sim 0.3$ (averaging horizontally over the Rossby radius ~ 50 km). When Holland doubled the north–south extent of the basin, so that the maximum westerly winds blew at its middle latitude, a two-gyre circulation occurred (Fig. 49). With the restraint of the rigid wall removed, the separated Gulf Stream was weakly unstable, but the predominant energy conversion occurred, as before, in the return flow.

There is in the deep layer of Holland and Lin's ocean a time-mean flow (Fig. 48) including gyres both co- and counterrotating relative to the upper flow; here the eddies drive an abyssal circulation. Above the thermocline, however, Holland demonstrates that the flow is significantly weaker with eddies than without. Averaging over the water column, there is a net transfer of kinetic energy from mean to perturbations at a rate $\frac{1}{9}$ the conversion from potential to kinetic energy. This remarkable braking action is suggested by Thompson's (1971a) qualitative argument, and our analysis below, that spontaneous wave radiation will intensify an eastward jet, yet weaken a westward jet. It is likely that, given a more realistic intensity of the separated Gulf Stream (which exceeds 100 cm/sec averaged over the upper kilometer) and of the return flow, which cannot far exceed 10 cm/sec, the center of energy release will move

[10] The wedge model of a two-layer β plane ocean (Fig. 6) shows this asymmetry simply, for the slope equivalent to β is greater in the thicker lower layer. Hence a rather steep uptilt of the thermocline to the north is required to cause dh/dy to have opposite signs in the two layers, yet a milder tilt in the opposite sense will manage to do so.

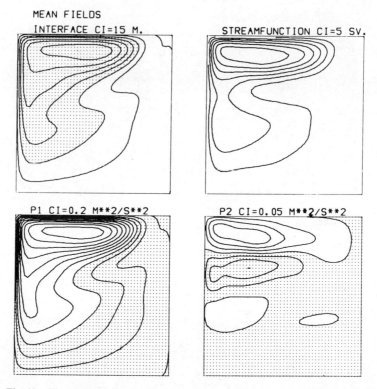

MEAN FIELDS

INTERFACE CI=15 M.   STREAMFUNCTION CI=5 SV.

P1 CI=0.2 M**2/S**2   P2 CI=0.05 M**2/S**2

Fig. 48.   Time-averaged flow, showing an abyssal circulation driven by the eddies.

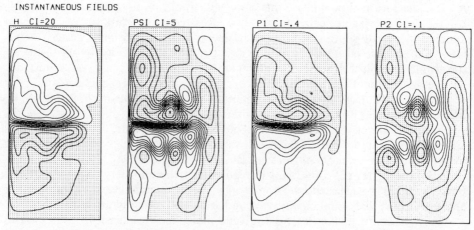

INSTANTANEOUS FIELDS

H CI=20   PSI CI=5   P1 CI=.4   P2 CI=.1

Fig. 49.   As above, but in an elongated basin with the maximum eastward stress at mid-latitude. The separated Gulf Stream is now unstable, and becomes more so, the larger the Reynolds number.

282

to the Stream itself, which will meander more like the model in Fig. 27, and the eddies will then act, as in meteorological flows, to intensify the circulation in both layers.[11] The experiment nevertheless shows the sensitivity of westward flows to baroclinic instability (relying on the upper layer being thinner than the lower layer) which theoreticians have remarked upon (e.g., Gill, Green, and Simmons, 1974). It may be that in the real ocean the region of eddy production analogous to Holland's case is farther south in the equatorial currents. Dominance of the tight non-Sverdrup gyre of north–south scale $\sim L_\rho$, which is not observed in the oceans, may lessen when reasonable meandering instabilities of the Gulf Stream are included, by raising the Reynolds number (Holland, private communication).

The distinction between instability of eastward and westward currents is particularly important in considering the north–south heat flux, for baroclinically unstable westward streams will mix heat, equatorward, against the overall global gradients of temperature.

The distant eddy field radiated from the north in Holland's experiment is virtually barotropic. Neither mean advection nor relatively slow baroclinic propagation was able to carry thermocline eddies to the south (in fact, the basin-average eddy kinetic energy was 6.8 times the eddy potential energy). This dearth of strong thermocline eddies is in disagreement with the observed ocean, where the ratio of eddy potential to eddy kinetic energy probably exceeds unity. Yet the disparity is just the same one found in the flat-bottom cascades (Section 6). It is very likely that addition of realistic topography and reduction of the lateral damping will allow a baroclinic far field to develop. It is also an illustration that long-term experiments, involving a balance between forcing and dissipation, may be rather sensitive to the nature of the friction, more so than the short-term evolution in initial-value experiments. Heterogeneous experiments like this one also require accurate modeling of horizontal fluxes of energy, which in turn are themselves very dependent on both friction and bottom topography.

Despite these intricacies the calculations seem to be the first to include the entire list of ingredients (excepting rough topography) needed to understand the mean circulation.

B. Rectified Circulation on a Homogeneous β Plane

The problem of eddy–mean flow interaction needs the focus provided by simplified geometry and forcing, in addition to calculations like Holland's showing its role in complex ocean models. A laboratory experiment by Whitehead (1975) exhibits succinctly the rectified flow generated by localized forcing on a homogeneous β plane. The 2-m Woods Hole tank was rotated with its surface free (covered by a plastic "windscreen"), yielding a paraboloidal β plane. A circular disk was mounted in a horizontal plane, at mid-depth and mid-latitude. Forced vertical oscillation of the disk produced a mixture of waves and turbulence in the otherwise still fluid. Radial dye streaks revealed a persistent zonal circulation (Fig. 50) which was prograde ("eastward") at the latitude of the forcing and retrograde ("westward") elsewhere. Circulation in this sense occurred equally well when the disk was replaced by a source of small air bubbles. E. Firing, G. Williams, and E. Lorenz (private communications) have reported the analogous result from truncated numerical calculations on a rotating sphere covered by homogeneous fluid.

---

[11] In this case the eddies are likely to have a decisive effect on the upper-layer circulation, as well as abyssal flow, rather than altering it only slightly.

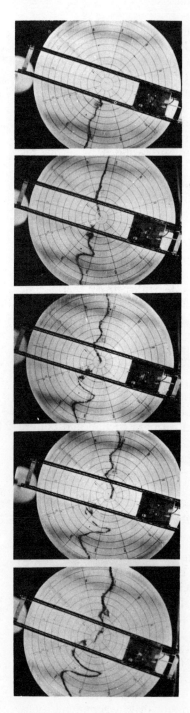

Fig. 50.   Mean circulation induced by an isolated disturbance (beneath the black square) on a polar β plane (Whitehead, 1975). The dye streaks deforming with time show a prograde (eastward) jet at the forcing latitudes, with westward flow elsewhere.

C. Inviscid Theory

It is of interest to write an analytical expression for this mean flow which holds for both waves and geostrophic turbulence. Consider a polar β plane like Whitehead's but, for simplicity, with constant depth and Coriolis frequency, f, decreasing linearly away from the center (Fig. 51). Neglecting at first forcing and dissipation, we have simple conservation of barotropic potential vorticity,

$$q \equiv \beta y + \zeta \qquad [\zeta \equiv (\nabla \times \mathbf{u})_z]$$

$$\frac{Dq}{Dt} = 0$$

(35)

Here y, v are the (inward) radial coordinate and velocity. Integrate over a region within a fixed latitude circle, $\mathscr{C}$. The Eulerian, zonally averaged u velocity is then given by

$$\frac{\partial}{\partial t} \bar{u}_e = \overline{qv}$$

where $(\overline{}) = \oint_{\mathscr{C}} ()\, dx$ is the integral about the latitude circle. Now a fluid column which would have zero relative vorticity at latitude y_0 has potential vorticity $q = \beta y_0 \equiv \beta y - \beta(y - y_0)$. With $v = Dy/Dt$, $\bar{v} = 0$, and defining $\eta = y - y_0$, it follows that

$$\frac{\partial \bar{u}_e}{\partial t} = -\beta(y - y_0)\overline{\frac{D(y - y_0)}{Dt}}$$

$$= -\frac{1}{2}\beta \overline{\frac{D}{Dt}\eta^2}$$

(36)

which is an exact relation, regardless of the intensity or nature of the fluid motion. If the convective part of the right side is small, equation 36 becomes

$$\bar{u}_e = -\tfrac{1}{2}\beta\overline{\eta^2}$$

(37)

for an initial state of rest. The neglect of the convective terms leading to equation 37 is not so severe as to require linear wave motion. It implies $\overline{\mathbf{u} \cdot \nabla \eta^2} \equiv (\partial/\partial y)\overline{v\eta^2} \ll (\partial\overline{\eta^2}/\partial t)$ or $\gamma(U/c)[(\overline{\eta^2})^{1/2}/L] \ll 1$ where γ is a correlation coefficient between v and η^2, U is a scale particle speed, c a scale-phase speed, and L a length scale defining the

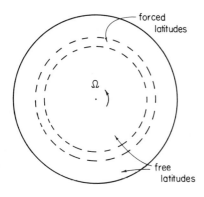

forced
latitudes

free
latitudes   Fig. 51.   Geometry for the circulation experiment.

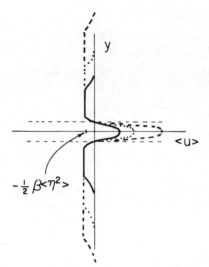

Fig. 52. Sketch of the inviscid solution for the Eulerian circulation due to random forcing confined to the dashed band. At successive times the jet strengthens, as the westward flow (of constant strength) fills an ever-greater region.

north–south envelope of variation of $\overline{\eta^2}$. Equation 37 is thus valid for either weak disturbances ($U/c \ll 1$) or nonlinear fields in which the scale of variation L of the intensity is much greater than the typical north–south particle displacement.

Random trading of fluid particles across the latitude circle $\mathscr{C}$ systematically decreases the net relative vorticity within $\mathscr{C}$, yielding an Eulerian mean circulation. This westward momentum appears at "free" latitudes for which equation 35 holds. At forced latitudes, where the wave maker was located in Whitehead's experiment, eastward momentum is left behind as a prograde jet whose strength depends upon the nature of the wavemaker: net angular momentum vanishes if the source (for example, the air bubbles used in some of Whitehead's runs) provides none. In the steadily excited inviscid case the westward circulation at a given latitude builds up to its asymptotic value $-\frac{1}{2}\beta\overline{\eta^2}$ as soon as the disturbance arrives. The westward jet continues to accelerate indefinitely, to compensate for the presence of westward flow in an ever larger region. The solution is sketched in Fig. 52. The central result, which holds in more general circumstances, is that the depth-averaged flux of potential vorticity equals the force, plus the momentum influx, exerted along $\mathscr{C}$ by eddies on the fluid instantaneously occupying the fixed contour, $\mathscr{C}$. Green (1970) made use of the steady form of this relation. Below, it is shown to shed light also on the eddy-forcing of flow, averaged in time at a point in space, rather than about a geostrophic contour.

D. Viscous Theory

Add to the problem a simple bottom drag, $-D\mathbf{u}$, which would be provided by a linear Ekman layer. The vorticity equation,

$$\left(\frac{D}{Dt} + D\right)\zeta = -\beta v$$

has the formal solution, following a fluid parcel,

$$\zeta = -\beta \int_0^t v(t')e^{D(t'-t)}\,dt' + \zeta_0 e^{-Dt}$$

with initial value ζ_0 at $t = 0$. The northward flux of vorticity becomes

$$\overline{\zeta v} = -\beta v \overline{\int_0^t v e^{D(t'-t)}\, dt'} + \overline{\zeta_0 e^{-Dt} v}$$

which is simplified by defining $\xi = \int_0^t v e^{D(t'-t)}\, dt'$, to

$$\overline{\zeta v} = -\beta \overline{\xi\left(\frac{D\xi}{Dt} + D\xi\right)} + e^{-Dt}\overline{\zeta_0 v}$$

The exact relation analogous to equation 36 is

$$\left(\frac{\partial}{\partial t} + D\right)\bar{u}_e = -\beta\overline{\left(\frac{1}{2}\frac{D}{Dt} + D\right)\xi^2} + e^{-Dt}\overline{\zeta_0 v} \tag{38}$$

Here, ξ is the north–south particle displacement weighted over the previous spin-up time, which expresses the fading memory that fluid has for its initial latitude. The second right side term gives the decaying dependence on initial relative vorticity. Again, if

$$\frac{U}{c}\frac{(\overline{\eta^2})^{1/2}}{L} \ll 1$$

the advective part of D/Dt may be neglected, and the explicit solution is

$$\bar{u}_e = -\tfrac{1}{2}\beta\left(\overline{\xi^2} + D\overline{\int_0^t \xi^2 e^{-D(t-t')}\, dt'}\right) + e^{-Dt}\overline{\zeta_0 \int_0^t v\, dt}$$

There are two interesting limits, the inviscid, which yields equation 36, or

$$\bar{u}_e = -\tfrac{1}{2}\beta\overline{\xi^2} \equiv -\tfrac{1}{2}\beta\overline{\eta^2} \tag{39}$$

(if $\zeta_0 = 0$), and the steady, which yields

$$\bar{u}_e = -\left(\frac{\beta}{D}\right)\int_0^\infty R_{22}(\tau)e^{-D\tau}\, d\tau \tag{40}$$

If the spinup time, D^{-1}, far exceeds the time scale of the eddies, this simplifies to: (a) $u_e = -\beta\kappa_{22}/D$ if the field is diffusive ($\kappa_{22} \neq 0$); (b) $u_e = -\tfrac{1}{2}\beta\overline{\eta^2}$, if η^2 is bounded ($\kappa_{22} = 0$) and the correlation R_{22} falls to zero rapidly, relative to D^{-1}; or (c) $u_e = -\beta\overline{\eta^2}$ (twice that of (b)) if $\kappa_{22} = 0$ and R_{22} has long memory, relative to the time D^{-1}. Note that (a) resembles (b) if we replace η by the average excursion of particles in one spinup time. This illustrates the difference between wavelike and turbulent flow, as according to whether $\overline{\eta^2}$ is bounded or not. Neither expression depends explicitly on the time or length scales of the eddies. In the solution, (c), the circulation reaches $-\beta\xi^2/2$ as soon as the disturbance is established but then, gradually if D is small, the circulation continues to increase to twice this value after a few spin-up times. This would be a feature to look for in an experiment.

At forced latitudes the flow depends, as before, on the exact nature of the source. If no time-averaged forces are exerted on the fluid, the regions of positive and negative angular momentum sum to zero: each is not finite, being limited by friction.

This analysis contributes to earlier arguments for anisotropy, favoring zonal or f/h currents on a β plane, even without external forcing. Random increase or decrease in $\overline{\eta^2}$ will occur due to the eddy motion itself. The slowness of energy propagation north–south, relative to east–west, may enhance the effect by maintaining north–south gradients, yet smoothing out those east and west.

E. Taylor's Formula

The derivation was suggested by a result in Taylor's remarkable 1915 paper which considered, among other things, the stability of a plane, nonrotating, inviscid flow in a channel. He finds (in our notation)

$$\frac{\partial U}{\partial t} = \frac{1}{2} U''(y) \frac{\partial \overline{\eta^2}}{\partial t} \tag{41}$$

as a consequence of conservation of relative vorticity. (See also Dickenson, 1969.)

Here $U(y) \equiv \bar{u}_e$; Taylor's formulation is centered on slight deviations from a strong, parallel flow. In this case, unless U'' is a constant, the displacements, η, and perturbation velocities must be assumed small. Equation 41 in effect equates the vorticity flux to the divergence of the momentum flux. When integrated across the channel the left side vanishes, there being no sources of momentum. This gives Rayleigh's criterion, that $U''(y)$ must change sign somewhere for the spontaneous growth of disturbances, based on the novel definition of instability, that $\overline{\eta^2}$ increase everywhere. [In our analogous β plane application, the integration of equation 37 (where β is a positive constant) across a zonal channel bounded by rigid walls (or, without walls, to distant latitudes which are quiescent) shows that $\int \overline{\eta^2} \, dy$ must be constant in the absence of forcing or dissipation: on an *unforced* β plane random motion is ultimately limited in north–south excursion. External agents are required to mix the potential vorticity.]

Now a positive value of $\frac{1}{2}\partial(\overline{\eta^2})\partial t$, which we henceforth call κ, acts (though not exactly) like a positive viscosity, reducing the momentum where U'' is positive, and conversely (Fig. 53). This redistribution of momentum, on the whole, reduces the energy of the mean flow.

Two situations exist, however, in which the perturbations sharpen the jet and increase its energy. First, stable perturbations, $\kappa < 0$, can exist in a potentially unstable flow and they will increase U wherever the curvature U'' is negative. Second, the flow may be absolutely stable, $U'' \neq 0$ everywhere, and then the integral of equation 41 shows that κ must take on both signs within the fluid. Imagine, for example, a parabolic inviscid flow, with imposed initial values of $\overline{\eta^2}$ that are large near its axis, and vanish towards the edges. It is most plausible that the disturbance will decrease where it is large and increase where it is small. If this is so, the flow will be intensified, with a flux of x momentum against its own gradient.

Starr (1968), in describing "negative viscosity," frequently invoked two-dimensional thought problems. Here we have shown in detail how such flows may redistribute momentum aginst its gradient, but only when, in some part of the flow, the eddy diffusivity of *vorticity*, κ, is negative.

Consider now the combined situation of a zonal current, $U(y)$, on a β plane. The equivalent of equations 36 and 41 is

$$\frac{\partial U}{\partial t} = (U'' - \beta) \frac{1}{2} \frac{\partial \overline{\eta^2}}{\partial t} \equiv (U'' - \beta)\kappa \tag{42}$$

The *unstable case* requires that the potential vorticity gradient vanish somewhere for $\overline{\eta^2}$ to increase everywhere. By sketching the curvature of a typical jetlike profile, it becomes clear that westward jets are less stable than eastward jets of the same shape. During the instability the role of β is to cause westward accelerations of the mean flow in a broader band, and eastward accelerations in a narrower band, than in the comparison problem with $\beta = 0$.

In the *stable case*, $U'' - \beta \neq 0$, κ must once again take both signs, and singling out the case where $\kappa < 0$ near the jet axis and $\kappa > 0$ elsewhere, we find a dramatic difference in the momentum redistribution. The stable, outward moving disturbance sharpens an eastward jet (in the sense that $\partial U/\partial t > 0$ where $\kappa < 0$), as it did with $\beta = 0$, but, by equation 42, a westward jet is now instead decelerated at the center. The westward momentum moves out to its flanks, where $\kappa > 0$.

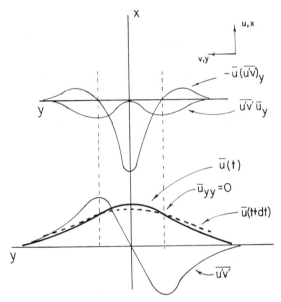

Fig. 53.   Sketch of energy transformation terms and Reynolds' stress in a simple "laboratory" parallel jet. The perturbation field spreads the jet, reducing its energy. The conversion between x-mean and perturbation flow is clear, in an integrated sense, yet its local evaluation is obscured by spatial fluxes.

This shows the interplay between mean shear and β in deciding the redistribution of momentum. β takes the dominant role when $\beta a^2/U > 1$, a being the horizontal scale of U. The contribution of mean shear is in a sense included in the original derivation (equation 36) where it appears by altering the reference latitudes, y_0, of the fluid particles.

The descriptions show simply how a variety of up- and down-gradient momentum fluxes may occur in parallel flows; they are meant to contrast linear stability theory on the one hand (e.g., Howard and Drazin, 1964), with countergradient fluxes that may appear in a stable flow, on the other. It is significant that in unforced, stable flows the north–south diffusivity of fluid parcels, κ, must take both signs and average to zero.

F. Energy Conversion to and from the Mean Flow

We have discussed the redistribution of momentum by eddies, without writing down the corresponding energetic relations. For two-dimensional inviscid channel flow, homogeneous in x, the usual equations for mean (in x) and perturbation kinetic energy are

$$\frac{\partial \overline{K}_M}{\partial t} = -U \frac{\partial}{\partial y} \overline{u'v'} \tag{43}$$

$$\frac{\partial \overline{K'}}{\partial t} + \frac{\partial}{\partial y} [\overline{v'(K' + p')}] = \overline{u'v'} \frac{\partial U}{\partial y} \tag{43a}$$

where

$$u = U + u', \ldots, K' = \tfrac{1}{2}(u'^2 + v'^2), \qquad K_M = \tfrac{1}{2}U^2,$$

and

$$\overline{()} = \lim_{L \to \infty} \frac{1}{2L} \int_{-L}^{L} () \, dx$$

Mean energy increases wherever downstream momentum converges, and perturbation energy changes owing to fluxes of energy, pressure work, or conversion from mean energy.

One must not be too facile in describing the conversion terms, however. At the outer edges of an unstable, classical jet, (Fig. 53) for example, the perturbations are growing, with the right side of equation 43a positive; yet the mean flow is also growing, and the right side of equation 43 is also positive! The two conversion terms differ by a divergence, $(\partial/\partial y)(\overline{u'v'U})$, which vanishes upon integration over the domain. It may be appropriate to rewrite equation 43a in the form

$$\frac{\partial \overline{K'}}{\partial t} + \frac{\partial}{\partial y} [\overline{v'(K - K_M + p')}] = U \frac{\partial}{\partial y} \overline{u'v'}$$

$$K \equiv \tfrac{1}{2}[(u' + U)^2 + v'^2]$$

Now the conversions are equal and opposite, and the flux of perturbation energy is altered: the naming of $v'(K - K_M)$ as perturbation energy flux seems appropriate, since $K = K_M + K' + u'U$, the final cross-term vanishing only after x averaging. The spatial flux terms are thus difficult to distinguish from the conversions between mean and perturbations. After integration over the entire domain the fluxes vanish, but it is not easy to ascribe the conversion to a particular region in space. Conclusions drawn from incomplete measurements, in only one part of the flow, may thus be ambiguous.

G. Webster's Experiment

Apparently the first measurement of these fluxes and conversions in a large-scale ocean current was by Webster (1965). The Gulf Stream between Miami and Cape Hatteras frequently exhibits billowlike undulations on its inshore side. These are now a familiar sight, thanks to satellite-borne infrared photography (Fig. 58), but were once known only from sparse point measurements. The importance of turbulent-viscous theories of the Gulf Stream motivated this experiment in which towed electrodes established the surface values of u', v', and U, where Ox is downstream and time averages replace x averages in many experiments (although not in Fig. 54).

A typical cross-stream profile of near-surface values of the flux and conversion terms is shown in Fig. 54, from near Cape Hatteras. The persistently negative value of $-U(\partial/\partial y)\overline{u'v'}$ nearest the shore, in the absence of other fluxes, indicates a retardation of the time-average flow there, whereas its positive value at the core indicates an intensification. This action may be related to the rather sharp inshore edge of the Stream. The net conversion in the region of measurement is directed into the mean kinetic energy, agreeing in sense with the computer simulation of Orlanski and Cox (1973), and the theoretical speculations of Starr (1968) and Green (1970). In each of these works, the energy source is the density structure below the surface.

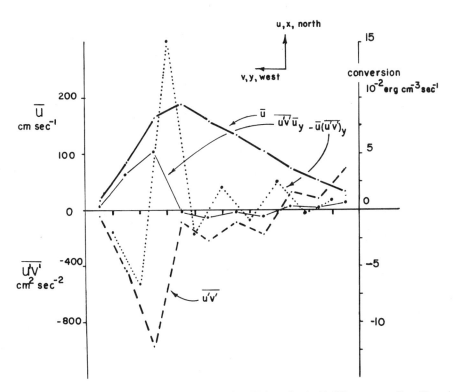

Fig. 54.   The quantities analogous to Fig. 53 estimated by Webster for the Gulf Stream near Cape Hatteras. The energetic terms are very different, and both indicate net transfer from eddies to mean flow at the ocean surface. As above, the *local* evaluation of conversion is precarious.

This example reminds one of the difficulty in making energetic analyses when the measurements do not extend out to vanishing values of $\overline{u'v'}$ or U. The integrated conversion terms between mean and perturbation kinetic energy are not then equal and opposite, when written as in equation 43. The common practice of using $\overline{u'v'}(\partial U/\partial y)$ to sum up the mutual interaction of mean with fluctuations is therefore incomplete.

H. Linearized Theory: Momentum Transport in Rossby Waves

Returning to the theory of zonal-flow generation, we concentrate on the role of β. We have shown that outward-moving disturbances in a basically stable flow necessarily sharpen an eastward jet while broadening a westward jet. But β also hastens the redistribution of $\overline{\eta^2}$ itself by allowing waves to propagate. Thompson (1971b) first suggested this qualitative result on the basis of far-field, linear theory. As a consequence of their dispersion relation, equation 21, barotropic Rossby waves generated along an east–west line have crests arranged in a herringbone pattern which points to the east. The correlation of east and north velocities implied by motion to and fro along these crests represents a flux of westward momentum, uv, away from the source. The eastward momentum left behind in a force free environment would augment an eastward jet like the Gulf Stream beyond Cape Hatteras (see also Figs. 11, 55).

A similar pattern of u–v correlation may be seen in the wave-crests generated by a point source oscillating at a single frequency (Section 5). The Green's function is

$$\psi = H_0^{(2)}\left(\frac{\beta r}{2\omega_0}\right) \exp\left[-i\left(\frac{\beta x}{2\omega_0} + \omega_0 t\right)\right]$$

$$r^2 = x^2 + y^2, \qquad \tan\theta = \frac{y}{x}$$

In the far field the crests, $r = \text{const}/\cos^2(\frac{1}{2}\theta)$, look somewhat like the waves excited by a fishline in a moving stream.

Thompson (1971a) verified that, indeed, currents at a site mooring (Site D) north of the Gulf Stream at 39°N, 70°W, are polarized with significantly negative $\overline{uv}$. This region, though not in the far field of the stream, is sufficiently removed that a persistent westward countercurrent occurs, which is very like that suggested by the theory; see also Green (1970). The role of such countercurrents in the general circulation is significant; at shallow levels they seem to form closed, climatological-mean gyres by reconnecting with the stream, whereas at deep levels they contribute to the equatorward flow of North Atlantic Deep Water.

Reinforced by general results like equation 37, and powerful qualitative arguments like Thompson's, we are encouraged to write down a simple, nearly linear result which demonstrates more exactly the westward momentum contained in Rossby waves.[12]

Consider an unbounded, barotropic, β plane. The zonal, inviscid momentum equation,

$$\frac{\partial u}{\partial t} + u\frac{\partial u}{\partial x} + v\frac{\partial u}{\partial y} - fv = -\frac{\partial p}{\partial x}$$

yields upon averaging zonally, and expansion of the fields in powers of the non-linearity, $\gamma \sim \varepsilon/\omega$,

$$\frac{\partial \overline{u^{(1)}}}{\partial t} = -\frac{\partial}{\partial y}\overline{u^{(0)}v^{(0)}}$$

where $u = u^{(0)} + \gamma u^{(1)} + \cdots$. This gives the $O(\gamma^2)$ induced Eulerian circulation due to the $O(\gamma)$ wave field. Now imagine exciting the fluid with a moving corrugated wall

---

[12] Dr. Stern has recently reported a similar calculation (private communication).

at $y = 0$. A convenient artifice, the term $e^{\alpha t}(\alpha \ll \gamma)$, will be used to show the uniqueness of the wave-induced flow [otherwise, one can add an arbitrary steady zonal flow, $\gamma^2 U(y)$].

The equation and boundary conditions, at lowest order, are

$$\frac{\partial}{\partial t}\nabla^2\psi^{(0)} + \beta\frac{\partial\psi^{(0)}}{\partial x} = 0$$

$$\frac{\partial\psi^{(0)}}{\partial x}\bigg|_{y=0} = A\cos k(x - c_x t)e^{\alpha t}, \qquad \psi^{(0)} \to 0 \qquad \text{as} \quad \begin{cases} t \to -\infty \\ y \to \infty \end{cases}$$

with

$$(u^{(0)}, v^{(0)}) = \left(-\frac{\partial\psi^{(0)}}{\partial y}, \frac{\partial\psi^{(0)}}{\partial x}\right)$$

The general solution, $\psi^{(0)} = \mathscr{R}\{B\exp[ik(x - c_x t) + ily + \alpha t]\}$ is constrained by the boundary conditions to have $B = -iA/k$, and by the equation to have

$$\beta ik - (k^2 + l^2)(ikc_x + \alpha) = 0$$

Now specifying k and c_x to be real, we find l is complex,

$$l^2 = -k^2 + \frac{\beta ik}{ikc_x + \alpha}$$

For small α, let $l = l_0 + i\alpha\delta$, yielding

$$l_0 = \left(-k^2 + \frac{\beta}{c_x}\right)^{1/2}$$

$$\alpha\delta = \tfrac{1}{2}l_0^{-1}\frac{\beta\alpha}{kc_x^2}$$

The complete linear solution,

$$\psi^{(0)} = \frac{A}{k}\sin[k(x - c_x t) + l_0 y]e^{\alpha(t - \delta y)} + O(\gamma) + O(\alpha)$$

automatically reveals the energy velocity by the rate of northward propagation of the envelope, $e^{\alpha(t - \delta y)}$. The rate is the same as that predicted by group-velocity theory. The second-order zonal flow, $\overline{u^{(1)}}$, is found from

$$\frac{\partial\overline{u^{(1)}}}{\partial t} = \frac{\partial}{\partial y}\overline{\left(\frac{\partial\psi^{(0)}}{\partial y}\frac{\partial\psi^{(0)}}{\partial x}\right)}$$

$$= \frac{\partial}{\partial y}\left(\frac{A^2 l_0}{k}\overline{\cos^2[k(x - c_x t) + l_0 y]}e^{2\alpha(t - \delta y)}\right)$$

or

$$\overline{u^{(1)}} = -\frac{\delta l_0}{2k}A^2 e^{2\alpha(t - \delta y)}$$

The solution to this order includes an outward propagating wave, growing everywhere because the boundary forcing is increasing, and a westward flow induced by it. For this problem the Eulerian mean flow varies only slightly over a wavelength, and hence the Lagrangian (particle-drift) mean is identical, to $O(\alpha)$. Here, unlike the free jet models, there is an increase in total x momentum, owing to the external agent.

This total westward momentum may be found from an argument used by R. W. Stewart for the case of internal gravity waves (see Bretherton, 1969). The force, say, F, exerted westward to maintain the motion of the corrugated wall creates kinetic energy in the fluid at a rate Fc_x, and creates momentum (or more generally, impulse) at a rate F. Thus the average (in x and y directions) energy density E and momentum density M must obey

$$M = \frac{E}{c_x}$$

where $c_x = -\beta/K^2$ for Rossby waves of total wave number K.[13] With $E = \frac{1}{2}\beta^2\overline{\eta^2}/K^2$, this becomes a special case of our result, equation 37.

I. Rossby Waves in a Shear Flow

To this point we have omitted the classic eddy–mean flow interaction, that of linear waves moving through a slowly varying zonal flow, for reasons of appropriateness. Nevertheless the results of the theory are interesting. First, wave action, $E/(\omega - \mathbf{U} \cdot \mathbf{k})$, is a conservative property for a packet of waves. Here E is the integrated wave energy, measured by an observer moving with the mean flow, $\mathbf{U} = U\hat{\mathbf{i}}$, and ω and $\mathbf{k}$ are the frequency and wave number in an absolute frame of reference. Multiplied by $\mathbf{k} \cdot \hat{\mathbf{i}}$, this is the total westward momentum of the waves, which is invariant even though energy is being exchanged with the mean flow (a consequence of the constant wave momentum being carried up the gradient of ambient mean flow).

Now it seems to be less than well-known that this conservation property holds for more extreme, geophysically more interesting, variation in U with y. Imagine the case when U takes on two uniform values, U_1 and U_2, on either side of a vortex sheet lying along $y = 0$. A steady train of Rossby waves approaches from the south. Ignoring possible instability, we can calculate the partial reflection that occurs at the interface. The fluid boundary between the two regions can be treated rather like the corrugated wall considered above; an observer in region 2, moving with speed U_2, sees the corrugations move steadily westward at speed $c = \omega/\mathbf{k} \cdot \hat{\mathbf{i}}$, and records that that material surface is doing work on region 2 at a rate $-F(U_2 - c)$ where F is the x force exerted by fluid in region 1, on the interface. Similarly, an observer in region 1, moving with speed U_1, reports that the upper fluid is doing work at a rate $F(U_1 - c)$ on the lower. The evidence implies that wave-energy fluxes $\mathscr{F}_1$ and $\mathscr{F}_2$, measured in either case by observers riding on the mean flow, obey

$$\frac{\mathscr{F}_1}{\omega - kU_1} = \frac{\mathscr{F}_2}{\omega - kU_2}$$

by elimination of the force, F. Thus the wave action is conserved also in this *rapidly* varying medium, and is now partitioned among incident, reflected, and transmitted waves.[14] The argument applies to a large class of waves, including some with non-

---

[13] The work of Bretherton (1969) and McIntyre (1971) suggests extreme caution in the application of ideas of wave–momentum density; the mean flows associated with wave packets radiate far away from them. Here the x averaging conceals this.

[14] Yet an observer at rest, including as "wave energy" both the oscillatory and induced mean flow due to the waves, reports from the same argument that wave *energy* flux is conserved.

trivial structure normal to the propagation plane (e.g., short surface waves propagating among currents). It is related to the result of Eliassen and Palm (1960) for steady wave trains in a density-stratified atmosphere with horizontal .winds of arbitrary vertical structure, that the momentum flux $\overline{uw}$ is independent of height.[15]

These results add further light to the discussion of jetlike mean flows. For waves propagating outward from the center of an eastward jet, where $U - c \,(\propto E)$ is large (recall that $c < 0$), will transport westward momentum outward from the jet core, thus increasing the x-mean energy there, while E decreases. If $U - c > 0$ everywhere in the corresponding westward jet, outward-propagating waves will still carry westward momentum, now at the *expense* of the zonal-average energy of the core as E increases. When the theory of critical-layer absorption is added in, covering flows where $U - c$ vanishes at some latitude, the interactions can be calculated in detail, and are of relevance particularly to atmospheric flows (Dickenson, 1970).

J. Topographic Effects: Nonlinear Theory

The theory of circulation induced by eddies given in most general form in equations 36 and 38 applies as well when the potential vorticity gradient is not simply a constant. The geostrophic contours of f/h, where h is the depth of a barotropic model, are generalizations of latitude circles, and the induced circulation obeys

$$\frac{\partial \overline{\mathbf{u}}_E}{\partial t} = -\overline{\frac{h}{2}\left(\nabla\left(\frac{f}{h}\right) \times \hat{\mathbf{k}}\right)\frac{D\eta^2}{Dt}}$$

for inviscid flow; a similar generalization of equation 38 follows with Ekman friction. Here η denotes the particle displacement normal to the contours, and the averaging bar is a line integral about a complete contour. The variations of $\nabla(f/h)$ in space, over distances $\sim L_h$, however, impose the additional restriction that $\eta \ll L_h$.

We may expect to see anticyclonic circulation above seamounts and ridges in the deep ocean, when random forcing acts at a distance so that the region in question is "free." Such circulation has indeed been found in my numerical experiments (Section 7), and Holloway (private communication) has discovered similar contour currents in his barotropic model (Section 7).

An illustrative example from my early barotropic experiments (Fig. 55a) shows particle trajectories in finite-amplitude "roughness" waves above a sinusoidal bottom. The oscillations, in the mean, cause the fluid over ridges to have negative relative vorticity, and conversely. This yields contour currents of both Eulerian and Lagrangian flow, which are fully as strong as the currents associated with the primary wave.

An extreme example from the sea is the persistent anticyclonic current found above the Great Meteor Seamount by Meincke (1971), which occurred in the presence of fluctuations at tidal period and longer. The results apply not only to the deep ocean; the continental rise and shelf provide a systematic, strong potential-vorticity gradient, and shelf waves and turbulence are present in abundance. The above theory, which applies when lateral boundaries do not block the geostrophic contours, provides an alternative to direct wind-generated longshore currents.

---

[15] A further consequence of this formula is that, if the phase speed, c, lies between U_1 and U_2, $\mathscr{F}_1$ and $\mathscr{F}_2$ have opposite signs. With no incoming energy from $+\infty$, then, $\mathscr{F}_1 < 0$, and net energy must flow away from the interface on both sides of it. This is known as "overreflection" of an incident wave.

Small-scale roughness on the bottom acts in a distinct way on mesoscale flows, probably providing an augmented drag owing to both its small-scale geostrophic wake and lee Rossby-wave and internal-wave generation. Bretherton and Karweit (1975) have emphasized this role of the roughness.

K. Applications

The theory has been presented in an idealized form, but its extensions bear on the oceanic case with stratification, topography, and lateral boundaries. There the eddy flux of vorticity into a fixed, elemental region yields an average stress curl which can then drive a large-scale Eulerian circulation in the classical manner. The result analogous to equation 36 for a single-layer ocean, with mean potential vorticity $Q(x, y)$ is that the potential vorticity flux-divergence may be rewritten to give

$$\frac{\partial}{\partial t} \nabla^2 \langle \psi \rangle + J(\langle \psi \rangle, Q) = \frac{\partial}{\partial x_i} \left(\kappa_{ik} \frac{\partial Q}{\partial x_k} \right)$$

where $\kappa_{ij} = \int_0^t R_{ji}(\tau | \mathbf{x}) \, d\tau$, $R_{ij} = \langle u_i(t)u_j(t + \tau) \rangle$, $u_i = (\partial \psi / \partial x_2, -\partial \psi / \partial x_1)$. Here $\langle \psi \rangle$ is the ensemble-averaged flow, and $Q(x, y)$, κ_{ij} are assumed slowly varying in space, relative to the particle excursions. If κ_{ij} is identified with the eddy diffusivity, then this becomes Welander's (1970) formulation, which was based heuristically on a down-gradient diffusion of vorticity. Unlike Welander (1970), Green (1970), and Rossby (1947), however, we suggest that the time-dependent theory, with κ_{ij} allowed to vary wildly in space, is likely to be of interest for the ocean. Over times greater than an eddy period, but less than the climatological time scale, a pulsing of eddy energy can drive temporary "mean" flows with the diffusivity either positive or negative; in free initial-value problems described earlier, in fact, the area-averaged diffusivity vanished.

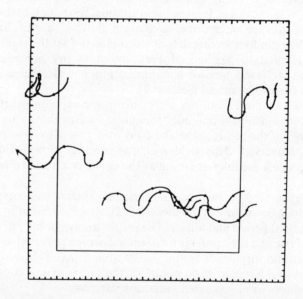

Fig. 55.   Two examples of eddy-induced mean circulation in a homogeneous model. (*a*) The Lagrangian paths of particles in finite-strength waves above a sinusoidally corrugated bottom lie east–west along depth contours.

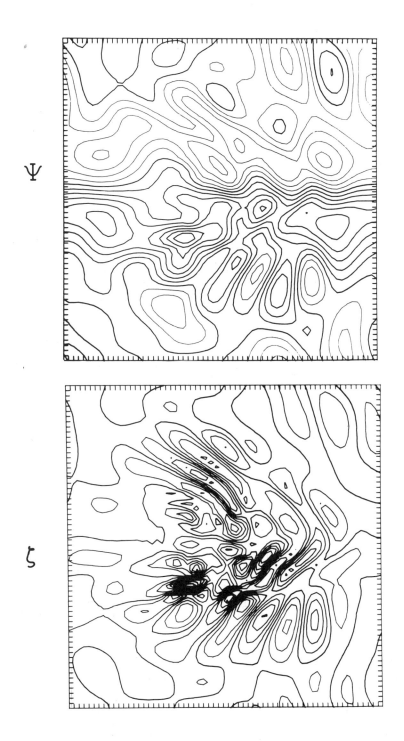

Fig. 55. (b) an initial cluster of eddies near the center interacts and radiates, developing a large-scale eastward jet in the center, westward flow at the periphery (full lines are positive ψ here). The vorticity field shows the small-scale tilted troughs, the vehicle of vorticity flux.

As an example (Fig. 55b) witness the large-scale Eulerian flow induced in a simple β plane spin-down experiment. The fluid is homogeneous. Initially a small cluster of eddies is prescribed at the center of the region. As these interact and radiate into the surrounding, quiet fluid, the intensity decreases at the center and increases elsewhere. Direct application of equation 36 promises an eastward zonal-mean flow at mid-latitudes ($\kappa < 0$) and westward flow elsewhere ($\kappa > 0$).

The eddy-stress term is related to the expression for Lagrangian mean flow, given at the beginning of this section; combination of that result with the above turbulent vorticity equation shows that, if κ_{ij} is symmetric, then

$$\left\langle \frac{\partial \zeta^E}{\partial t} \right\rangle + \left\langle \frac{D\zeta^L}{Dt} \right\rangle = -2\beta \langle v^E \rangle.$$

The Eulerian and Lagrangian rates of change of vorticity are equal but *opposite* for ensembles which have $\langle v^E \rangle = 0$. Over brief periods, floats and current meters will tend to register the same vorticity tendency, but over long times, the opposite tendency.

As an example, in the configuration of Whitehead's rotating paraboloid, it turns out that the circulation integral about a *moving* fluid contour is equal and opposite to the Eulerian circulation about a fixed latitude circle (which itself often is equal to the Lagrangian particle drift). This may be appreciated by realizing that the area enclosed by a dyed contour, initially lying on a latitude circle, must decrease if the fluid is displaced in any fashion. Kelvin's theorem then yields *eastward* circulation about this moving contour, but all particle motion (in free latitudes), and the Eulerian momentum, are directed westward.

In addition to the time-dependent flows driven by eddy vorticity flux, the discovery of quasi-permanent fine structure in the eddy intensity both in the models and the ocean, tremendously increases the stress curl, $(\partial/\partial x_i)[\kappa_{ik}(\partial Q/\partial x_k)]$.

In a stratified ocean these ideas apply to vertically averaged vorticity flux. The vigorous diffusivity of the upper-level flow may, in this case, drive abyssal circulation, as in Holland's model. The required vertical flux of horizontal momentum has been discussed by Bretherton and Karweit (1975). An experiment yielding mean circulation in this way was shown in Fig. 43e; the advancing front of eddy energy in this "propagation" run yielded a positive diffusivity with strong gradient, and created westward zonal-average flow in both shallow and deep layers. In other regions being drained of their eddy energy, the diffusivity was *negative* and the mean flow eastward.

The vorticity-flux theory gives dynamical significance to the observed diffusion of water-mass properties, for instance, the silicate in Antarctic Bottom Water, and the salt in the Mediterranean outflow. In addition, the direct measurement of Lagrangian diffusion can now be made with neutrally buoyant SOFAR floats (Freeland, Rhines, and Rossby, 1975); the spreading of the cluster with time (Fig. 64) gives a first estimate of 1500-m-level diffusivity, $\simeq 8 \times 10^6$ cm^2/sec.

9. Observational Notes

Observations of unsteady currents are now widespread, and we may anticipate a rapidly improving picture of their geographical distribution. For a sample, Hamon (1968), and Boland and Hamon (1970) have recorded eddies and pulsations in the East Australia Current; Mazeika (1973) and Koshlyakov and Grachev (1973) have described eddies in the North Equatorial Current; Bernstein and White (1975) have produced time sequences of thermocline eddies in the eastern North Pacific; Swallow and Bruce (1966) and Bruce (1973), describe a "separation bubble" in the Somali

Fig. 56.   Sea-surface temperature patterns along the west coast of the United States (from the NOAA-II satellite, courtesy of NOAA-National Environmental Satellite Service). Darker areas are warm, light areas cold. Sept. 11, 1974, after an intense period of coastal upwelling.

299

Current; Düing, Katz, and the GATE group (private communication) have found the Atlantic Equatorial Undercurrent to be oscillating on a large scale; Foster (1972) has seen gross irregularities in the Antarctic Circumpolar Current at the Drake Passage (where the net inferred transport was westward during the current meter experiment!); the MEDOC Group (1970) found eddies to occur after violent, meso-scale, deep convection in the western Mediterranean; and Swallow and Hamon (1960) and Gould (1971) report variable currents in the eastern North Atlantic and Bay of Biscay. As in the atmosphere, the meandering and pulsation of intense currents appears as eddy energy if time averaging is used, yet there is some distinction between such dynamics, and those of detached, freely moving eddies and waves in the ocean interior.

Surely the most dramatic evidence for eddies is the infrared photographs of the sea-surface temperature field, from earth satellites. Figure 56, from the NOAA series of satellites, shows billowlike patterns at the edge of cold coastal water lying off Oregon and California. Can these be the instabilities of the cold, southward coastal current? The temperature structure of the separated Gulf Stream (Fig. 57)

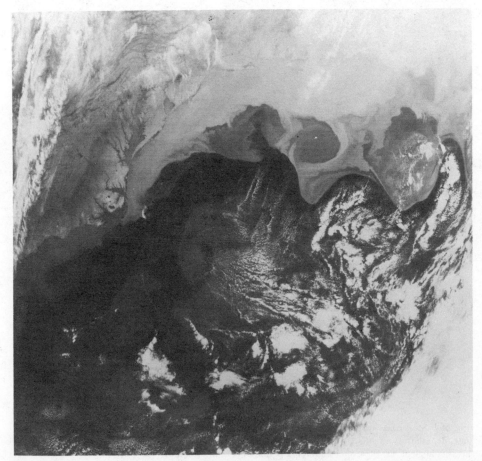

Fig. 57. The Gulf Stream near Cape Hatteras from a NOAA satellite, April 28, 1974. Woods Hole Oceanographic Institution's Site D is found to the left side of the prominent eddy just north of the Stream (black dot).

shows the distinctness of transition between shelf, slope, Gulf Stream, and Sargasso Sea water masses, giving life to these classical water-mass divisions. Active entrainment of cold water by warm eddies appears at the shelf edge (cf. Fig. 15) and a cold eddy seaward of the Gulf Stream entrains warm water from the stream near Cape Hatteras. Site D, the source of current records (Fig. 13), is indicated. It is remarkable that, amidst this chaos, linear wave theory continues to have qualitative truth (Sections 4 and 8). At Hatteras, the southward flowing shelf water is also entrained into the Gulf Stream, and but occasionally penetrates the crescent-shaped bays farther south. The crispness of these patterns reemphasizes the inability of geostrophic flow to cascade energy to small scales, in the efficient sense of 3D turbulence. Even though surface frontogenesis is no doubt occurring, and sharpening the temperature gradients, the picture lacks the fuzziness that we see in a laboratory turbulent jet, rich in energy over a broad range of scales.

Finally (Fig. 58), surface traces of eddies have appeared in mid-ocean. Dr. Legechis of NOAA has enhanced the signals (the satellites were designed for meteorology, not

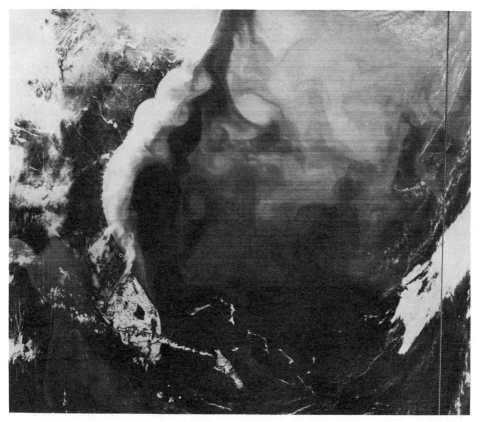

Fig. 58.   NOAA-satellite view of the Sargasso Sea, April 1, 1974, enhanced by Dr. Legechis. Eddies are visible (at least in the original) not only near the Florida Current, but in open ocean. The MODE experiment occurred beneath the prominent north–south tongue of warm water in the right-center. Observed structure suggests that thermocline eddies should be made visible on the surface by the action of their velocities, more than their own temperatures. A cold, cyclonic Gulf Stream ring is visible, entraining warm water from the Stream. The gray shades cover a 9.1°C temperature range (darker = warmer), while black is 2.1°C range above this, and white represents all other temperatures, whether cold or hot.

oceanography), and finds warm and cold tongues hundreds of kilometers in scope with a clear imprint of horizontal advection. This picture reveals the organic reality of the mid-ocean north–south temperature gradient (which on climatological maps is so smooth). It is suggestive of down-gradient mixing by thermocline eddies. The connection with the deeper density field is unknown, as is the strength of the subtropical front suggested by the picture. Here at the top of the mixed layer, one can imagine the white imprint of cold windstorms continually being distorted by lateral stirring of the thermocline eddies. Elsewhere, entrainment appears by a cold eddy (the same as in Fig. 57) near the Gulf Stream, and billows occur on the shoreward side of the stream (see Rao, Strong, and Koffler, 1971), which were associated with momentum convergences by Webster (Section 8). A great deal of activity, perhaps with pulse-like variations of the Florida Current, occurs on the Blake Plateau.

A. Sources

With improved observation, the sources and sinks of eddy energy will become more and more apparent. Our present list of sources includes direct wind generation, violent instabilities of intense currents and radiation from them, slower instability of the gentle mid-ocean currents, flow past rough topography and irregular coastline, occasional sinking of cold water, enhancement by western-boundary reflection, and possible driving by internal-wave stresses. The analogy of the dynamics to those in the atmosphere appears rather weak here, for the sources of energy seem to be more sparsely distributed in the ocean, and the domain itself is far bigger, measured by deformation radii and propagation rates, than the atmosphere.

B. Dissipation

A crucial, unknown aspect of the long-term distribution of energy is the dissipation process. For example, the nature of interaction between mesoscale eddies and internal waves is uncertain even as to sign (Müller, 1974, predicts that internal waves drain energy from eddies, whereas other, more deterministic theories like critical-layer absorption may suggest the opposite). Again the ocean and atmosphere are very dissimilar, internal waves being far weaker than the large-scale flows in the atmosphere. The reason may be the absorptive nature of the stratosphere (owing to the effect of decreasing density in the kinetimatic viscosity), and less efficient generation at the ground. The ocean bottom, on the other hand, probably has an albedo of at least $\frac{1}{2}$, and turbulence and wind waves at the surface are a potentially strong source.

Dissipation by lateral friction is far weaker than classical eddy coefficients would suggest. The very nature of geostrophic turbulence, by its inability to extend vortex lines indefinitely, is to avoid such dissipation. The deep valley in frequency spectra between periods of a day and a few weeks (e.g., Rhines, 1973) attests to the lack of a homogeneous cascade, local in (ω, k) space, between geostrophic and ageostrophic flow. Bottom friction, from conventional drag laws, is very slight, a few percent of 1 dyne/cm^2 in mid-ocean, giving a spin-down time exceeding 500 days. [Lee-wave drag exerted by topography of a few kilometers' lateral extent, may be far more significant (Bell, 1975).]

Once again the analogy between atmosphere and ocean breaks down in comparisons of the boundary drag and dissipation. The atmospheric lower boundary layer is $O(1 \text{ km})$ thick, fully 10 % of the depth of the troposphere. Perhaps one-half of the energy dissipation occurs there (Kung, 1967). This suggests a rapid spin-down of atmospheric energy (e folding, say, in 3–6 days compared with an inertial time scale $L/U \sim 1$ day for $L \sim 1000$ km, $U \sim 10$ m/sec). In the ocean, on the other hand, active three-dimensional turbulence does not seem to exist in regions thicker than $O(10$ m), or 0.2 % of the fluid depth.

Yet a crude estimate of overall dissipation time for oceanic kinetic energy is the ratio of vertically integrated kinetic energy density, to the rate of working by wind stress, τ, or $\rho \int |\mathbf{u}|^2 \, dz/\tau U_s$, where U_s is a subsurface downwind current. For $U_s = 10$ cm/sec, $\tau = 1$ dyne/cm, and currents of 10 cm/sec above the thermocline and 4 cm/sec below, this yields a time of 19 days: a very short time in view of the paucity of three-dimensional turbulence, and one comparable with the inertial time scale L/U (~ 5–20 days). The presence of severe intermittency may make such an estimate meaningless, but it suggests the need to search for the sinks as well as the sources of energy.

C. MODE; Velocity

Some further data from the MODE experiment, discussed by Schmitz et al. (1975), is of more than casual interest here. Two site moorings were maintained for longer than 2 yr: MODE "east" (28°10′N, 68°35′W) over hilly topography and, 100 km to the west, MODE "center" (28°00′N, 69°40′W), above the Hatteras abyssal plain. Daily current vectors at three levels (filtered of internal waves) (Figs. 59, 60) show the usual decrease in time scale with depth. The energy level also decreases downward across the thermocline, yet rises slightly below 1500 m. The series are quite regular, yet there are occasional bursts of unusual activity. The 1500-m and 4000-m levels are visually coherent in the vertical, but not the horizontal. There appears, particularly at 1500 m, an eastward decrease in eddy-energy density. At the 4000-m eastern site, an unusually strong mean current flows at 2 cm/sec to the south–southwest. The presence, some 15 km to the southeast, of a dominant ridge topography (35 km × 5 km × 500 m high) may not be incidental. This deep mean flow is known from other measurements to be of small lateral extent. The appearance of these noticeable gradients in intensity over small lateral separations, and of small-scale, deep, mean flows was a dominant feature of the rough bottom simulations (Section 7).

D. Zonal Bands

At 500 m (note the rotated coordinates, Fig. 60), on several occasions, a strong burst of zonal flow appears at both moorings. This coherence over 100 km is unusual (it does not appear below the thermocline) and is suggestive of the anisotropy found to be so persistent in the computer experiments, when they were sufficiently energetic (Figs. 20b, 23d, 25d). There is some indication of a sympathetic pulse of energy in the deep water at these times.

There is evidence in other forms of zonally banded currents, from a variety of sources; P. Richardson and collaborators have recently followed SOFAR floats, depths 700–1100 m, in the region 32°N to 36°N, attempting to lay them within Gulf Stream rings. The trajectories, as well as showing some loops, followed long zonal excursions, westward and occasionally eastward, at an average velocity of about 5 cm/sec. Yoshida (1970) and Bryan (private communication) suggest that tropical regions exhibit banded currents, more extensive than the usual equatorial system. (Yoshida describes five or six distinct jets in the Pacific, 20°N to 30°N.) One must of course be wary of contours drawn from north–south sections widely spaced in longitude. Recalling Whitehead's experiment, we remark that any permanence in the

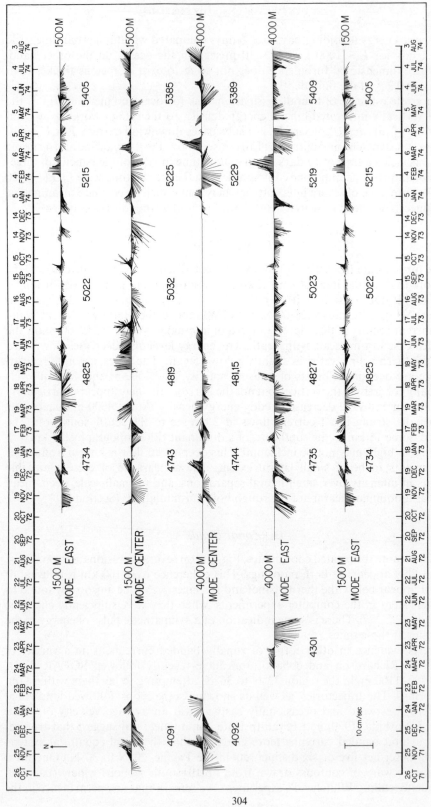

Fig. 59. Current meter records from two site moorings in the MODE experiment, 100 km apart, at 28°N, 69°40′W and 28°10′N, 68°35′W, courtesy of Dr. Schmitz. The two levels are below the main thermocline. The records are visually coherent in the vertical but not the horizontal. Note the east–west variation in intensity and the persistent, deep mean flow in the rough, eastern area. Dr. Schmitz notes that minor changes will be made in this and the four succeeding figures, for final publication.

304

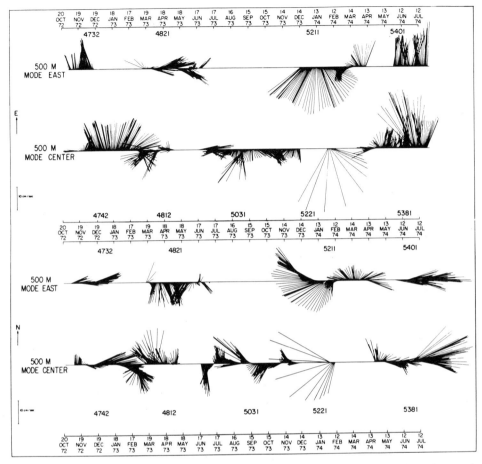

Fig. 60.   Records from site moorings as in Fig. 59, 500-m level, plotted two ways (east = up, then north = up). Bursts of strong, zonal energy occur, coherent over 100 km (recall banded structures which recur in the numerical models, e.g., Figs. 23, 40).

forcing pattern will yield permanent bands, whereas in our free spin-down experiments a banded appearance occurred by chance, with the growth or decay of $\langle \eta^2 \rangle$ in a given region. G. Williams' (private communication) experiments very likely involve the same physics, in a model of Jupiter's atmosphere.

A shorter sequence with dense vertical coverage is from the intensive period of the experiment (Fig. 61). The upper portion is excerpted from the record at MODE-east, in the rough area. It shows the kind of waxing and waning of vertical coherence that appeared, with nonlinear "capture" followed by topographic scattering, in the simulations, Fig. 37c. A period of southward flow (the eastern side of a warm thermocline eddy) begins at almost all levels; yet it gives way, below, to the shorter time scales natural to the deep water. The deepest level, 100 m above the sea floor, is grossly out of step; we suggest that it is dominated by fine-scale topographic oscillations, frontal activity (where thermal gradients intersect the boundary), and perhaps occasional intrusion of the bottom mixed layer.

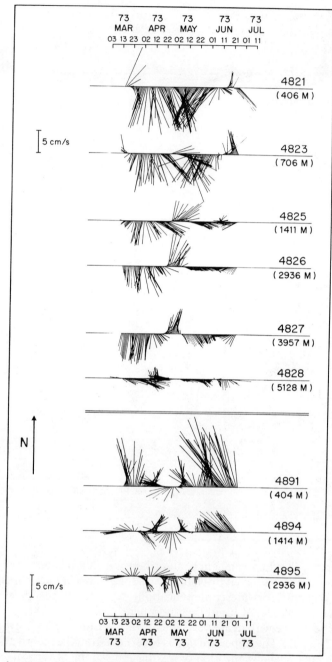

Fig. 61. Two short records from MODE showing well the vertical current structure, mooring 482 (28°9'N, 68°39'W) and 489 (29°35'N, 70°W). The more quickly varying yet weaker currents in the deep water and brief episodes of great vertical coherence are commonplace.

Bryden (1975) has demonstrated the value of such dense vertical sampling of currents for dynamical studies. The thermal-wind equation may be manipulated to give

$$|\mathbf{u}_H|^2 \frac{\partial \theta}{\partial z} = \frac{g}{f\rho_0} \mathbf{u}_H \cdot \nabla \rho$$

where θ is the angle of the horizontal current, $\mathbf{u}_H$, from the east. The right side is difficult to measure directly. Its importance in the adiabatic density equation,

$$\frac{\partial \rho}{\partial t} + \mathbf{u}_H \cdot \nabla \rho = -w \frac{\partial \rho}{\partial z}$$

relative to the nearly linear, right term, gives a direct comparison of the advective and propagative nature of the eddy field. [Moored temperature recorders and instrumented SOFAR floats measure $\partial \rho / \partial t$ and $\partial \rho / \partial t + \mathbf{u}_H \cdot \nabla \rho$, respectively (given a

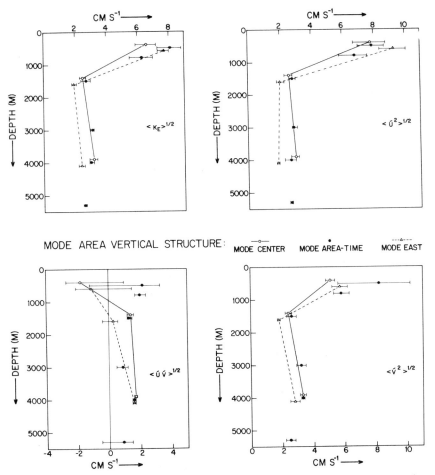

Fig. 62. Three representations of the average vertical current structure, from the two site moorings and the brief, spatially intense MODE experiment.

T–S relation). In all, they provide three determinations of the three terms in the equation.] Bryden concluded that, for a few weeks in MODE-I, the horizontal advection of density was well-correlated with, and comparable in magnitude to, the local time-derivative of density. This is consistent with the nonlinear picture of an eddy field.

The lower section of Fig. 61 shows perhaps a more classical structure of a slowly varying thermocline eddy, superimposed upon fast oscillations which are themselves more highly coherent in the vertical. This record was taken 170 km north of the MODE center.

In the time-averaged vertical structure (Fig. 62), the minimum in speed at 1500 m is visible, with a slight increase toward greater depth. If we adopt this profile to a

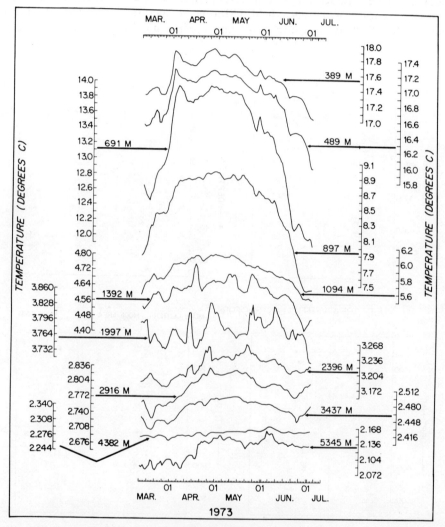

Fig. 63. Temperature records from the densely instrumented central mooring of MODE, courtesy of Drs. Schmitz and Wunsch. Time scales far shorter than those of linear, classical Rossby waves are present here.

two-layer model, $U_2/U_1 \simeq 0.4$, and ε/δ, based on 5 cm/sec rms current (the vertical average) and 200-m rms topographic height, yields $\varepsilon \simeq 0.02$, $\delta \simeq 0.05$, $\varepsilon/\delta \simeq 0.4$. Referring to Fig. 36, this agrees satisfactorily with the simulated U_2/U_1. The dependence of the simulated structure on the manner of supplying the energy, however, requires study. The computer experiments, despite their spontaneous fine structure, suggest that the Hatteras abyssal plain is too narrow (~ 150 km) at this latitude to allow the fluid to respond to it; one cannot apply Fig. 36 to such a small region.

D. Temperature

The MODE center was heavily instrumented, also, for temperature (Schmitz et al., 1975; Wunsch, 1975). Figure 63, the pressure-corrected time series, shows a rapid warming, extending nearly to the ocean bottom, followed by a more gradual cooling. The noisy signal at mid-depth is characteristic of salty Mediterranean water, undergoing active mixing with its surroundings. There is unusual noise just above the bottom, where the currents were also observed to be special. It is easy to imagine a simple, sharp-edged thermocline eddy from this figure, but in fact the temperature rise seems to be a frontlike feature embedded in the more gradual and extensive western side of the eddy. During the arrival of this front, the currents veered sharply at all levels. These kinds of quickly varying, interlocked fields of temperature and velocity are an example of "mobile" fine structure, which linear dynamics is incapable of producing.

E. Particle Paths

The superposition of all tracks (October 1972–March 1974) of the 1500-m SOFAR floats appears in Fig. 64 (Rossby, Voorhis, and Webb, 1975). The floats were released at various times to June 1973, in the neighborhood of the MODE center (28°N, 69°40'W). The paths, although at the *least* energetic depth of the ocean, show most clearly the "steepness" of the field of unsteady currents. Rather than executing slight oscillations about latitude lines, added to slow Sverdrup drift, they undergo wild hooks and spirals, and also an occasional long, straight track. The particle excursion during the time characteristic of the deep eddies (say, 10–20 days) is fully equal to their length scale, and this sets the wave steepness near unity. Yet within this pattern (Fig. 12) there was clear westward propagation of phase; a ubiquitous feature of the nonlinear simulations was in fact the westward phase progression, except where the large-scale ($L > L_\rho$) f/h contours were badly distorted (Figs. 20, 24).

The spacing of daily fixes indicates the speeds, which have strong geographical variation. Averaged plots of the kinetic energy (Fig. 65) agree with this impression; the energy level increases by a factor of three in the 500 km south of the center, and decreases by one-half in 300 km east of center. Later data seem to support these profiles. The intensity of eddies thus varies on scales far smaller than those of the ocean basins.

The orientation of the tracks seems also to have a persistent geographical variety. Regions of predominantly zonal tracks abut regions of frequent north–south flow. Farther to the north and south, the paths seem to trace out the shape of the large-scale topography (the Blake Outer Ridge and Escarpment and in the south, the Antilles).

These geographical gradations of particle trajectories and energy were familiar in the rough bottom numerical experiments. They may in addition give us clues to the source of the deep eddies. One sees the nearby Florida Current (Fig. 58), and the western boundary undercurrent (which has been recorded directly beneath the looping

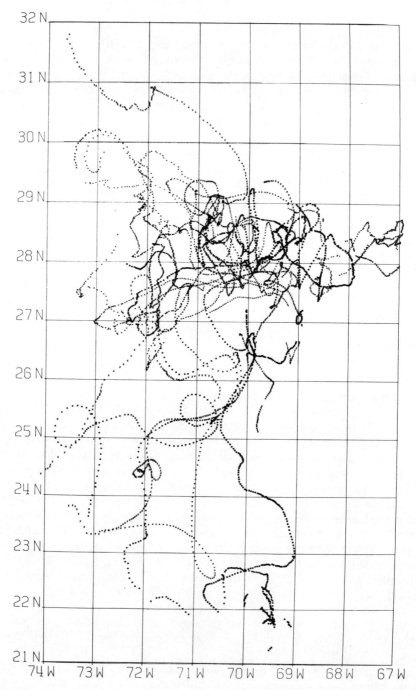

Fig. 64. The superposition of all trajectories, October 1972–March 1974 from neutrally buoyant SOFAR floats at 1500 m (from Rossby et al., 1975), daily positions plotted. A geographical variation in energy, as well as in the turbulent character of the flow, is evident.

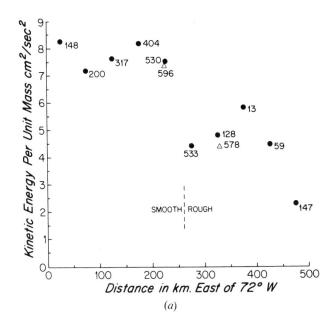

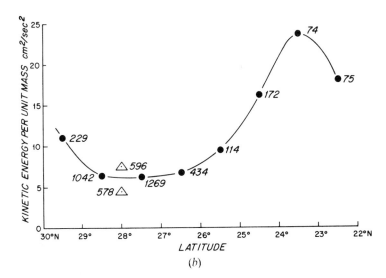

Fig. 65. Average 1500-m kinetic energy seen by the floats as a function of (a) longitude and (b) latitude. This permanent $O(1)$ variation in eddy intensity over scales no bigger than the eddies themselves forms a fixed "fine structure," and hints at a dearth of energy farther east, and an abundance to the north, west, and south. Topography of both large and small scale can lead to such gradients (e.g., Figs. 41, 43*f*), but so also could highly structured driving agents.

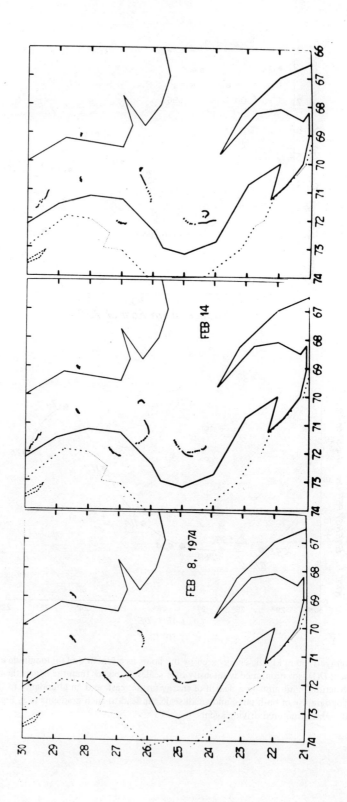

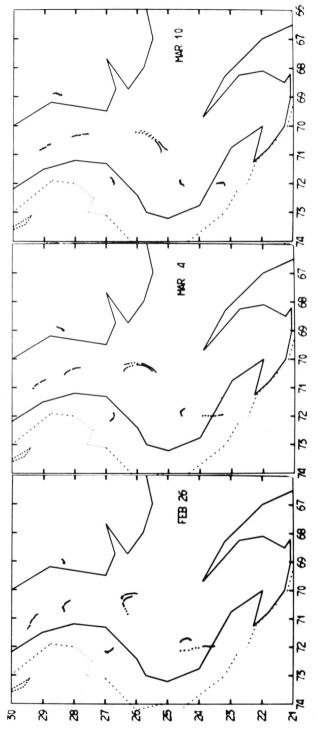

Fig. 66. Sequence of 1500-m float tracks above the Hatteras abyssal plain (5400-m isobath, solid; 5000-m, dashed; 4500-m, tip of Blake Outer Ridge at northwest corner.

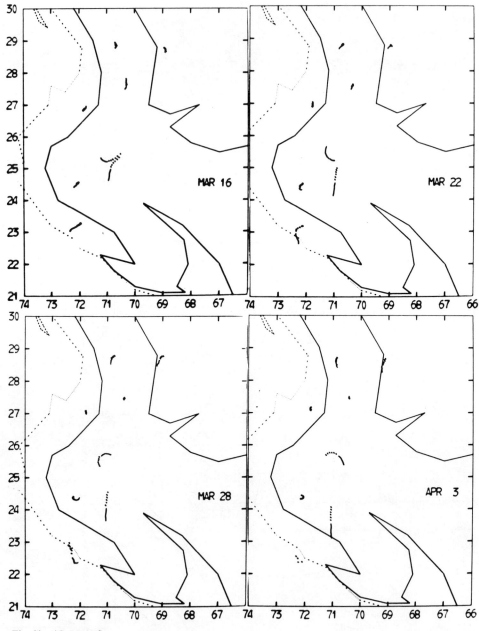

Fig. 66.   (*Continued*)

tracks at 30°N, 73°W), and the abundance of Gulf Stream energy to the north. In addition the region is near a "corner" in the large-scale topography, which probably forms a jagged, dynamic western wall for the subthermocline eddies.

When, soon, we have similar trajectories from amidst the thermocline eddies (say, at 500 m), their behavior will be very different. Given both the longer Eulerian time scale of upper-level flow (say, by a factor of 2–3), and its greater speed (a factor of 2–5), the floats may be expected to race around the eddies, mapping them while they change slowly with time. In this sense, they may be particularly "cost effective." Furthermore, the particle tracks in these westward-moving thermocline patterns will provide the most direct resolution of their dynamics, subtly apportioned between advection and propagation.

The development of dynamic theory has rested on the conservation of quasi-geostrophic potential vorticity, with an assumed, weak dissipation. If these quasi-Lagrangian floats, following close to constant pressure surfaces, are sufficiently like fluid particles, their tracks can provide immediate statements of dynamics; it is attractive to imagine testing the potential vorticity relation using float clusters. One sees in Fig. 64 an excursion of 550 km to the south in 120 days. If the change in planetary vorticity $(0.23f)$ from start to finish were to appear entirely as relative vorticity, it would be 10 times the typical value in the field $(0.02f)$! Nor can a change in isopycnal thickness of this degree be expected. (For coherent motion of the subthermocline water, it would imply an 880-m difference in thermocline height from start to finish.) One is left to suggest that rather rapid dissipation of enstrophy can occur, if the quasi-geostrophic theory is to be believed.

The final figure, Fig. 66, shows the tracks of some such southward journeys of the floats. It describes, better than my commentary, the flow of a nonlinear ocean.

Acknowledgments

This work had its early beginnings in the teaching of F. P. Bretherton, the papers of M. J. Lighthill, and the survey by N. P. Fofonoff in Volume 1 of this series. They may, however, find it difficult to recognize the result. I am grateful to the National Science Foundation, grant GX-36342, now IDO73-09737, for support. The computer experiments were done at the National Center for Atmospheric Research, Boulder, Colorado. This is contribution number 3371 from the Woods Hole Oceanographic Institution, and number 51 from the MODE project. Special thanks to MLLR and BPG.

References

Bache, A. D., 1846. *Am. J. Sci. Arts*, **30** (90) (1860), 313–329.

Baer, F., 1972. An alternate scale representation of atmospheric energy spectra. *J. Atmos. Sci.*, **29**, 649–669.

Ball, F. C., 1963. Some general theorems concerning the finite motion of a shallow rotating liquid lying on a paraboloid. *J. Fluid Mech.*, **17**, 2.

Barros, V. R. and A. Wiin-Nielsen, 1974. On quasi-geostrophic turbulence: a numerical experiment. *J. Atmos. Sci.*, **31**, 609–621.

Bass, A., 1974. Pseudospectral numerical study of geostrophic turbulence. Ph.D. thesis, Department of Meteorology, M.I.T., Cambridge, Mass.

Batchelor, G. K., 1953. *Homogeneous Turbulence.* Cambridge University Press.

Batchelor, G. K., 1967. *An Introduction to Fluid Dynamics.* Cambridge University Press.

Batchelor, G. K., 1969. Computation of the energy spectrum in homogeneous two-dimensional turbulence. *Phys. Fluids*, **12**, II, 233–238.

Bell, T. H., 1975. Topographically generated internal waves in the open ocean. *J. Geophys. Res.*, **80**, 320–327.

Bernstein, R. L. and W. B. White, 1975. Time and length scales of baroclinic eddies in the central North Pacific Ocean. *J. Phys. Oceanogr.*, **4**, 613–624.

Boland, F. M. and B. V. Hamon, 1970. The East Australian Current, 1965–1968. *Deep-Sea Res.*, **17**, 777–794.

Brekhovskikh, L., K. Federov, L. Fomin, M. Koshlyakov, and A. Yampolsky, 1971. Large-scale multi-buoy experiment in the tropical Atlantic. *Deep-Sea Res.*, **18**, 1189–1206.

Bretherton, F. P., 1969. On the mean motion induced by internal gravity waves. *J. Fluid Mech.*, **36**, 785–803.

Bretherton, F. P., and M. Karweit, 1975. Mid-ocean mesoscale modelling. *Proc. Durham Conf. Numer. Methods Oceanogr.* In press.

Brown, W., W. Munk, F. Snodgrass, H. Mofjeld, and B. Zetler, 1975. MODE bottom experiment. *J. Phys. Oceanogr.*, **5**, 75–85.

Bruce, J. G., 1973. Large-scale variation of the Somali Current during the southwest monsoon, 1970. *Deep-Sea Res.*, **20**, 837–846.

Bryden, H., 1975. Horizontal advection of temperature for low-frequency motions, *Deep-Sea Res.*, submitted.

Charney, J. G., 1971. Geostrophic turbulence. *J. Atmos. Sci.*, **28**, 1087–1095.

Crease, J., 1962. Velocity measurements in the deep water of the western North Atlantic. *J. Geophys. Res.*, **67**, 3173–3176.

Deacon, M., 1972. *Scientists and the Sea, 1650–1900, A study of Marine Science.* Academic Press, New York.

Dickenson, R. E., 1969. Theory of planetary-wave zonal flow interaction. *J. Atmos. Sci.* **26**, 73–81.

Dickenson, R. E., 1970. Development of a Rossby-wave critical level. *J. Atmos. Sci.*, **27**, 627–633.

Elliasen, A., and E. Palm, 1960. On the transfer of energy in stationary mountain waves. *Geof. Publ.*, **22** (3), 1–23.

Fofonoff, N. P., 1954. Steady flow in a frictionless homogeneous ocean. *J. Mar. Res.*, **13**, 517–537.

Foster, L., 1972. Current measurements in the Drake Passage. Ph.D. thesis, Dalhousie University.

Freeland, H., P. Rhines, and T. Rossby, 1975. Statistical observations of the trajectories of neutrally buoyant floats in the North Atlantic. *J. Mar. Res.*, **34**, 69–92.

Fuglister, F. C., 1960. *Atlantic Ocean Atlas of temperature and salinity profiles and data from the International Geophysical Year of 1957–58.* Woods Hole Oceanographic Institution Atlas Series. Vol. I. Woods Hole, Mass., 209 pp.

Fuglister, F., and L. V. Worthington, 1951. Some results of a multiple ship survey of the Gulf Stream. *Tellus*, **3** (1), 1–14.

Fultz, D., and P. Frenzen, 1955. A note on certain interesting ageostrophic motions in a rotating hemispherical shell. *J. Meteorol.*, **12**, 332–338.

Gill, A. E., J. S. A. Green, and A. J. Simmons, 1974. Energy partition in the large-scale ocean circulation and the production of mid-ocean eddies. *Deep-Sea Res.*, **21**, 499–528.

Gould, W. J., 1971. Spectral characteristics of some deep current records from the eastern North Atlantic. *Phil. Trans. Roy. Soc., London A*, **270**, 437–450.

Green, J. S. A., 1970. Transfer properties of the large-scale eddies and the general circulation of the atmosphere. *Quart. J. Roy. Meteorol. Soc.*, **96**, 157–185.

Greenspan, H. P., 1968. *The Theory of Rotating Fluids.* Cambridge Univ. Press.

Hamon, B. V., 1968. Spectrum of sea level at Lord Howe Island in relation to circulation. *J. Geophys. Res.*, **73**, 6925–6927.

Hasselmann, K., 1967. A criterion for nonlinear wave stability. *J. Fluid Mech.*, **30**, 737–739.

Hogg, N. G., 1973. On the stratified Taylor column. *J. Fluid Mech.*, **58**, 517–537.

Holland, W. R. and L. B. Lin, 1975. On the generation of mesoscale eddies and their contribution to the oceanic general circulation, I and II. *J. Phys. Oceanogr.*, **5**, 642–669.

Hough, S. S., 1898. On the application of harmonic analysis to the dynamical theory of the tides, II. On the general integration of Laplace's dynamical equations. *Phil. Trans. Roy. Soc., A*, **191**, 139.

Howard, L. N. and P. G. Drazin, 1964. On instability of a parallel flow of inviscid fluid in a rotating system with variable Coriolis parameter. *J. Math. Phys.*, **18**, 83–99.

Howard, L. N. and W. Siegmann, 1969. On the initial-value problem for rotating stratified flow. *Stud. Appl. Math.*, **48**, 153–169.

Huppert, H. E. and K. Bryan, 1975. Topographically generated eddies, preprint.

Jeffreys, H., 1926. On the dynamics of geostrophic winds. *Quart. J. Roy. Meteorol. Soc.*, **52**, 85–104.

Kenyon, L., 1964. *Nonlinear Rossby Waves.* Summer G.F.D. Program, Woods Hole Oceanographic Institution.

Kim, K., 1975. Instability and energetics in a baroclinic ocean. Ph.D. thesis, W.H.O.I./M.I.T. Joint Program in Oceanography.

Koshlyakov, M. N. and Y. M. Grachev, 1973. Mesoscale currents at a hydrophysical polygon in the tropical Atlantic. *Deep-Sea Res.*, **20**, 507–526.

Kung, E. C., 1967. Diurnal and long-term variations of the kinetic energy generation and dissipation for a five-year period. *Mon. Weather Rev.*, **95**, 593–606.

Kraichnan, R. H., 1967. Inertial ranges in two-dimensional turbulence. *Phys. Fluids*, **10**, 1417–1428.

Lighthill, M. J., 1967. On waves generated in dispersion systems by travelling forcing effects with application to the dynamics of rotating fluids. *J. Fluid Mech.*, **27**, 4.

Lighthill, M. J., 1969. Dynamic response of the Indian Ocean to west of the southwest monsoon. *Phil. Trans. Roy. Soc. A*, **265**, 45–92.

Longuet-Higgins, M. S., 1964. Planetary waves on a rotating sphere. *Proc. Roy. Soc.*, A, **279**, 446–473.

Longuet-Higgins, M. S., 1975. The response of a stratified ocean to stationary or moving wind systems. *Deep-Sea Res.*, **12**, 923–973.

Longuet-Higgins, M. S. and A. E. Gill, 1966. Resonant interactions between planetary waves. *Proc. Roy. Soc.*, A, **299**, 1–145.

Mazeika, P. A., 1973. Circulation and water masses east of the Lesser Antilles. *Dtsch. Hydrogr. Z.*, **26**, 51–73.

McCartney, M. S., 1975. Inertial Taylor columns on a beta-plane. *J. Fluid Mech.*, **68**, 71–95.

McWilliams, J., 1974. Forced transient flow and small-scale topography. *Geophys. Fluid Dyn.*, **6**, 49–79.

McWilliams, J. and G. Flierl, 1975. Optimal quasi-geostrophic wave analysis of MODE array data, preprint.

MEDOC Group, 1970. Observation of formation of deep water in the Mediterranean Sea, 1969. *Nature*, **227**, 1037–1040.

Meincke, J., 1971. Observation of an anticyclonic vortex trapped above a seamount. *J. Geophys. Res.*, **76**, 7432–7436.

Miles, J., 1974. On Laplace's tidal equation. *J. Fluid Mech.*, **66**, 241–260.

Monin, A., V. M. Kamenkovich, and V. G. Kort, 1974. *The Variability of the Ocean*, Gidrometeosdat Leningrad (in Russian).

Monin, A., and A. M. Yaglom, 1972. *Statistical Hydromechanics*. M.I.T. Press.

Morse, P., and H. Feshbach, 1953. *Methods of Theoretical Physics*. McGraw-Hill, New York.

Müller, P., 1974. On the interaction between short internal waves and larger scale motion in the ocean. *Hamb. Geophys. Einzelschr.*, **23**.

Needler, G. and P. H. LeBlond, 1973. On the influence of the horizontal component of the earth's rotation on long-period wave. *Geophys. Fluid Dyn.*, **5**, 23–46.

Orlanski, I. and C. Cox, 1973. Baroclinic instability in ocean currents. *Geophys. Fluid Dyn.*, **4**, 297–332.

Parker, C. E., 1971. Gulf Stream rings in the Sargasso Sea. *Deep-Sea Res.*, **18**, 981–993.

Pedlosky, J., 1965. A note on the western intensification of the oceanic circulation. *J. Mar. Res.*, **23**, 207–209.

Pedlosky, J., 1965. A study of time-dependent ocean circulation. *J. Atmos. Sci.*, **22**, 267–272.

Phillips, N. A., 1963. Geostrophic motion. *Rev. Geophys.*, **1**, 123–171.

Phillips, N. A., 1966. Large-scale eddy motion in the western Atlantic. *J. Geophys. Res.*, **71**, 3883–3891.

Phillips, N. A. and A. Ibbetson, 1967. Some laboratory experiments on Rossby waves with application to the ocean. *Tellus*, **19**, 81–88.

Pillsbury, J. E., 1890. The Gulf Stream—a description of the methods employed in the investigation and the results of the records, App. 10. Rep. *Supt. U.S. Coast Geod. Surv. 1890*, pp. 461–620.

Platzmann, G., 1974. Normal mode of the Atlantic and Indian Oceans. *Univ. Chicago Dept. Geophys. Sci. Tech. Rept. 25*.

Prandtl, L., 1952. *Essentials of Fluid Dynamics*. Hatner, New York.

Rao, P. K., A. Strong, and R. Koffler, 1971. Gulf Stream meanders and eddies as seen in satellite infrared imagery. *J. Phys. Oceanogr.*, **1**, 237–239.

Rhines, P. B., 1969. Slow oscillations in an ocean of varying depth, parts I, II. *J. Fluid Mech.*, **37**, 161–205.

Rhines, P. B., 1970. Edge-, bottom- and Rossby waves in a rotating, stratified fluid. *Geophys. Fluid Dyn.*, **1**, 273–302.

Rhines, P. B., 1971a. A comment on the Aries observations. *Phil. Trans. Roy. Soc.*, A, **270**, 461–463.

Rhines, P. B., 1971b. A note on long-period motions at Site D. *Deep-Sea Res.*, **18**, 21–26.

Rhines, P. B., 1973. Observations of the energy-containing oceanic eddies, and theoretical models of waves and turbulence. *Boundary Layer Meteorol.*, **4**, 345–360.

Rhines, P. B., 1975. Waves and turbulence on a β-plane. *J. Fluid Mech.*, **69**, 417–443.

Rhines, P. B. and F. P. Bretherton, 1974. Topographic Rossby-waves in a rough-bottomed ocean. *J. Fluid Res.*, **61**, 583–607.

Rossby, C. G., 1947. On the distribution of angular velocity in gaseous envelopes under the influence of large-scale mixing processes. *Bull. A.M.S.*, **28**, 53–68.

Rossby, T., A. Voorhis, and D. Webb, 1975. A quasi-Lagrangian study of mid-ocean variability using long-range SOFAR floats, *J. Mar. Res.*, **33**, 355–382.

Saunders, P. M., 1971. Anticyclonic eddies formed from shoreward-meanders of the Gulf Stream. *Deep-Sea Res.*, **18**, 1207–1219.

Schlee, S., 1973. *The Edge of an Unfamiliar World, a History of Oceanography*. E. P. Dutton, New York.

Schmitz, W. J., Jr., J. R. Luyten, and W. Sturges, 1975. On the spatial variation of low-frequency fluctuations in the western North Atlantic: I. Along 70° west longitude. Submitted to *J. Mar. Res.*

Schroeder, E. and H. M. Stommel, 1969. How representative is the series of Panulirus stations of monthly mean conditions off Bermuda? *Progr. Oceanogr.*, **5**, 31–40.

Smagorinsky, J., 1963. General circulation experiments with the primitive equations, I, the basic experiment. *Mon. Weather Rev.*, **91**, 99–167.

Smith, R., 1971. Ray theory for topographic Rossby waves. *Deep-Sea Res.*, **18**, 477–483.

Starr, V. P., 1953. Note concerning the nature of the large-scale eddies in the atmosphere. *Tellus*, **5**, 494–498.

Starr, V. P., 1968. *Negative Viscosity Phenomena*. McGraw-Hill, New York.

Steinberg, L., 1973. Numerical simulation of quasi-geostrophic turbulence. *Tellus*, **25**, 233–246.

Stommel, H. M., 1957. A survey of ocean current theory. *Deep-Sea Res.*, **4**, 149–184.

Suarez, A. A., 1971. The propagation and generation of topographic oscillations in the ocean. Ph.D. thesis, Department of Meteorology, M.I.T.

Swallow, J. C. and B. V. Hamon, 1960. Some measurements of deep currents in the eastern North Atlantic. *Deep-Sea Res.*, **6**, 155–168.

Swallow, J. C. and J. G. Bruce, 1966. Current measurements off the Somali coast during the southwest monsoon of 1964. *Deep-Sea Res.*, **13**, 861–888.

Taylor, G. I., 1915. Eddy motion in the atmosphere. *Phil. Trans. Roy. Soc.*, A, **240**, 1–26.

Taylor, G. I., 1921. Diffusion by continuous movements. *Proc. London Math. Soc.*, **2** (XX), 196–212.

Thompson, R., 1971a. Topographic Rossby waves at a site north of the Gulf Stream. *Deep-Sea Res.*, **18** (1), 1–19.

Thompson, R., 1971b. Why there is an intense eastward current in the North Atlantic, but not the South Atlantic. *J. Phys. Oceanogr.*, **1**, 235–237.

Thompson, R. and J. R. Luyten, 1976. Evidence for bottom-trapped topographic Rossby waves from single moorings. *Deep-Sea Res.*, in press.

Thomson, C. W., 1868. *The Depths of the Sea*. 2d ed. McMillan and Co., London.

Thompson, R. E., 1975. Propagation of planetary waves over random bottom topography. *J. Fluid Mech.*, **70**, 267–286.

Veronis, G., 1970. Effect of fluctuating winds on ocean circulation. *Deep-Sea Res.*, **17**, 421–434.

Veronis, G., and H. M. Stommel, 1956. The action of variable wind stresses on a stratified ocean. *J. Mar. Res.*, **15**, 43–45.

Webster, F., 1965. Measurements of eddy fluxes of momentum in the surface layer of the Gulf Stream. *Tellus*, **17**, 239–245.

Welander, P., 1970. Lateral friction in the ocean as an effect of potential vorticity mixing. *Geophys. Fluid Dyn.*, **5**, 101–120.

Whitehead, J. A., Jr., 1975. Mean-flow driven by circulation in a β-plane. *Tellus*, **27**, 358–364.

Wiin-Nielsen, A., 1962. On transformation of kinetic energy between the vertical shear flow and the vertical mean flow in the atmosphere. *Mon. Weather Rev.*, **90**, 311–323.

Wunsch, C., 1975 private communication.

Yoshida, K., 1970. Subtropical countercurrents: band structures revealed from CSK data. In *The Kuroshio—A Symposium on the Japan Current*. J. C. Moor, ed. East–West Center Press, Honolulu, pp. 197–204.

8. MODELING OF THE TROPICAL OCEANIC CIRCULATION

DENNIS W. MOORE AND S. G. H. PHILANDER

1. Introduction

The oceanic circulation in the tropical oceans is distinguished by the predominantly zonal currents in the oceanic interior (see Fig. 1) and the intense boundary currents near the coasts. This circulation is in response to the stress exerted upon the ocean surface by the trade winds over the Atlantic and Pacific, and the monsoons over the Indian Ocean. The models that have been developed thus far to study this oceanic response fall into two groups. The first group, which consists of linear, inviscid, time-dependent models, attempts to explain how the currents are generated in a stably stratified ocean that is initially at rest. The results are relevant for a short period after the sudden onset of the winds. If the winds should remain steady, then the oceanic circulation will ultimately be time-independent (unless the currents are unstable). The second group of models describes such a steady-state circulation and can thus account for only some of the features of the mean circulation because the trade winds and monsoons are subject to fluctuations that cause the currents to be variable. The amplitude of this variability is largest in the Indian Ocean where the seasonal reversal of the monsoons causes the Somali Current to reverse its direction, and causes the alternate generation and destruction of the Equatorial Undercurrent. The first group of models should describe the circulation in the Indian Ocean during the early stages of the generation of these currents.

There were as of March 1975 no models that describe the response of the tropical current system to variable winds. Because of the mathematical complexity of the problem, such models will have to be numerical. The design of these models should be based on results from simpler models such as those mentioned above. We already have considerable information available about the spatial and temporal resolution necessary in numerical models. There is a need for further studies with simple models. For example, before incorporating the actual bathymetry of the ocean floor into a numerical model, it will be valuable to know how, in an idealized model, topographic features influence equatorial waves.

Another factor that will affect the development of numerical models is the increasing availability of measurements that provide information about variability in the tropical oceans. There are, at the moment, insufficient data to test the results of linear inviscid models, but the situation will soon be remedied by projects such as the one of Knox (1974), and by experiments such as the GARP Atlantic Tropical Experiment (GATE), conducted during the summer of 1974. Similar experiments in different ocean basins will reveal the effect of a change in geometry, and forcing function, on the oceanic circulation. The ultimate test for future numerical models of the tropical oceanic circulation will be their ability to simulate the differences and similarities between the circulation patterns observed in the different ocean basins.

We confine our attention here to phenomena with a frequency much lower than 1/day. We therefore do not concern ourselves with the precise manner in which momentum is transferred from the atmosphere to the ocean. The high-frequency turbulent processes that are involved in this transfer are parameterized with a coefficient of eddy viscosity in the steady-state models. This permits the specification

319

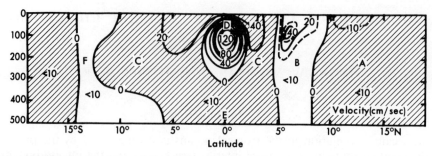

Fig. 1.   Meridional cross-section of the zonal component of the velocity (cm/sec) in the central Pacific. Shaded regions indicate westward flow. Near the equator, velocities are based on direct observations; at higher latitudes, geostrophic velocities (relative to the flow at 1000 m) were computed (after Knauss, 1963). A, North Equatorial Current; B, North Equatorial Countercurrent; C, South Equatorial Current; D, Equatorial Undercurrent; E, Intermediate Equatorial Current; F, South Equatorial Countercurrent.

of a stress boundary condition at the surface. The models are thus capable of simulating laboratory flows in which laminar boundary layers such as Ekman layers occur, but they can cope with only limited aspects of the interaction between the ocean and atmosphere. The linear inviscid models mentioned earlier assume that the wind stress at the surface gives rise to a body force in the mixed upper layer of the ocean. The question of how this mixed layer comes into existence and how it is maintained is not addressed. There is, as yet, no explanation for the unusually deep, mixed surface layers in the western part of the Pacific and Indian Oceans, or for the absence of such mixed surface layers in the eastern part of the Pacific (see Fig. 2) and Atlantic. (In the latter two ocean basins, there are, on the other hand, mixed layers below the core of the Equatorial Undercurrent; in the Pacific Ocean, this deep mixed layer coincides approximately with the region between the 12 and 14°C isotherms in Fig. 2.) The numerical models that simulate the tropical oceanic circulation are dependent on a better understanding of, and an improved parameterization for, the turbulent processes that determine the structure of these mixed layers.

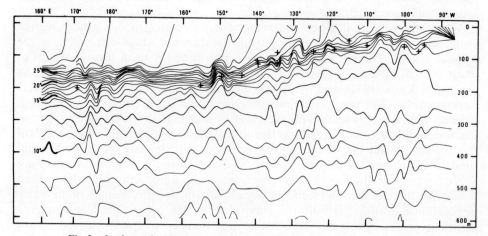

Fig. 2.   Isotherms in an equatorial plane in the Pacific (after Colin et al., 1971).

2. Equations of Motion

Consider motion in an ocean basin of width L and depth H, that spans the equator. Let r, θ, and ϕ be spherical coordinates positive in the radial, latitudinal, and longitudinal directions, respectively. Let R be the radius of the earth, and introduce a β plane coordinate system (Veronis, 1973) by setting

$$Lx = R\phi, \qquad Ly = R\theta, \qquad Hz = r - R.$$

The nondimensional coordinates x, y, and z and the corresponding velocity components $U_0 u$, $U_0 v$, and $(HU_0/L)w$ are positive eastward, northward (from the equator), and upward (from the ocean floor), respectively. (U_0 is a typical horizontal velocity.) Assume the following expressions for the density and pressure:

$$\text{Density} = \rho_0 + \Delta\rho_v \bar{\rho}(z) + \Delta\rho_H \rho(x, y, z, t), \tag{1}$$

$$\text{Pressure} = \rho_0 g H(1 - z) + g\Delta\rho_v H \int_z^1 \bar{\rho}(z')\, dz'$$

$$+ \left(2\Omega U_0 \rho_0 \frac{L^2}{R}\right) p(x, y, z, t), \tag{2}$$

where ρ_0, the mean density of the ocean, is constant; $\Delta\rho_v$ is the density difference between the ocean surface and floor; $\Delta\rho_H$ is the density difference between points a horizontal distance L apart; Ω is the rate of rotation of the earth; g is the gravitational acceleration; and $t(R/2\Omega L)$ denotes (dimensional) time. We assume that the Boussinesq approximation is valid and that the motion is hydrostatic (except for the possible importance of horizontal Coriolis terms). The equations expressing the conservation of momentum, mass, and heat may then be written as follows[1]:

$$\frac{Du}{Dt} + \gamma w - yv + \frac{\partial p}{\partial x} = E_v \frac{\partial}{\partial z}\left(v(z)\frac{\partial u}{\partial z}\right) + E_H \nabla_H^2 u, \tag{3a}$$

$$\frac{Dv}{Dt} + yu + \frac{\partial p}{\partial y} = E_v \frac{\partial}{\partial z}\left(v(z)\frac{\partial v}{\partial z}\right) + E_H \nabla_H^2 v, \tag{3b}$$

$$-\gamma u + \frac{\partial p}{\partial z} + \frac{1}{RoFr}\rho = 0, \tag{3c}$$

$$\frac{\partial u}{\partial x} + \frac{\partial v}{\partial y} + \frac{\partial w}{\partial z} = 0, \tag{3d}$$

and

$$\frac{D\rho}{Dt} + Ro\frac{\Delta\rho_v}{\Delta\rho_H}\frac{\partial\bar{\rho}}{\partial z}w = \Gamma_1 \frac{\partial}{\partial z}\left(k(z)\frac{\partial\rho}{\partial z}\right) + \Gamma_2 \nabla_H^2 \rho, \tag{3e}$$

where

$$\frac{D}{Dt} \equiv \frac{\partial}{\partial t} + Ro\left(u\frac{\partial}{\partial x} + v\frac{\partial}{\partial y} + w\frac{\partial}{\partial z}\right),$$

$$\nabla_H^2 = \frac{\partial^2}{\partial x^2} + \frac{\partial^2}{\partial y^2},$$

[1] In subsequent sections, subscripts x, y, z, and t are used for partial derivatives; for example, $u_x = \partial u/\partial x$.

Rossby number $\qquad Ro = \dfrac{U_0 R}{2\Omega L^2},$ $\hfill$ (4a)

Froude number $\qquad Fr = \dfrac{4\Omega^2 L^4}{R^2(\Delta\rho_H/\rho_0)gH},$ $\hfill$ (4b)

Aspect ratio $\qquad \gamma = \dfrac{HR}{L^2},$ $\hfill$ (4c)

Ekman numbers $\qquad E_v = \dfrac{v_v R}{2\Omega H^2 L},$ $\hfill$ (4d)

$$E_H = \frac{v_H}{v_v}\frac{H^2}{L^2}E_v,$$ $\hfill$ (4e)

Inverse Peclet numbers $\qquad \Gamma_1 = \dfrac{K_v R}{2\Omega L H^2},$ $\hfill$ (4f)

$$\Gamma_2 = \frac{K_v}{K_H}\frac{H^2}{L^2}\Gamma_1.$$ $\hfill$ (4g)

Here $v_v v(z)$, v_H, $K_v k(z)$, and K_H denote vertical and horizontal coefficients of eddy viscosity and thermal diffusivity.

In Sections 3–5, we discuss solutions to equations 3 that are valid for different ranges of the parameters given in equations 4. The steady-state motion to be discussed in Section 4 is driven by the surface wind stress $\tau_0\boldsymbol{\tau}$ [where the nondimensional $\boldsymbol{\tau} = (\tau^x, \tau^y)$ is 0(1)], and by horizontal density gradients. The density differences $\Delta\rho_v$ and $\Delta\rho_H$ in equation 1 are assumed to be of the same order of magnitude. Without loss of generality, we thus assume $\Delta\rho_v = 0$, and refer to $\Delta\rho_H$ as $\Delta\rho$. If we choose the velocity scale U_0 to be $\tau_0(R/2\Omega L v_v)^{1/2}$, then the surface boundary conditions may be written

$$E_v^{1/2}\frac{\partial u}{\partial z} = \tau^x, \qquad E_v^{1/2}\frac{\partial v}{\partial z} = \tau^y, \qquad w = 0$$

and $\hfill$ (5)

$$\rho = \rho_s(x, y) \qquad \text{at } z = 1.$$

In Section 3, where linear, time-dependent, inviscid models are discussed, the initial vertical stratification of the ocean [$\Delta\rho_v\bar{\rho}(z)$ in equation 1] is assumed to be given. We also assume $\Delta\rho_H = Ro\Delta\rho_v$; the distinction between $\Delta\rho_v$ and $\Delta\rho_H$ is important. The boundary condition at the ocean bottom is $w = 0$ at $z = 0$. At the sea surface, the appropriate boundary conditions in the absence of a surface wind stress are that the sea surface is a locus of particles and that the sea-surface pressure equals the atmospheric pressure, taken to be constant. These two conditions may be combined to give the single boundary condition

$$w = \frac{\Delta\rho_H}{\rho_0}Fr\frac{\partial p}{\partial t} \qquad \text{at } z = 1,$$

the nominal position of the free surface. The dimensional sea-surface displacement from its nominal position is given by

$$\zeta = \frac{2\Omega L^2}{R}\frac{u_0}{g}p(x, y, 0, t).$$

These kinematic boundary conditions also hold in the presence of surface wind-stress forcing. The incorporation of such a surface wind stress into this model is discussed in Section 3.B.

3. Linear, Time-Dependent Models

A. Free Modes

Vertical Structure

The equations for the linear inviscid model are

$$u_t + \gamma w - yv + p_x = 0, \tag{6a}$$

$$v_t + yu + p_y = 0, \tag{6b}$$

$$u_x + v_y + w_z = 0, \tag{6c}$$

$$-\gamma u + p_z + \frac{1}{RoFr}\rho = 0, \tag{6d}$$

and

$$\rho_t + Ro\frac{\Delta\rho_v}{\Delta\rho_H}\frac{d\bar\rho}{dz}w = 0. \tag{6e}$$

Horizontal density differences arise only by advection of the basic state density field $\bar\rho(z)$, so we choose the scale $\Delta\rho_H = Ro\Delta\rho_v$, and no parameters appear in equation 6e. The parameter $1/RoFr$ in equation 6d is then equal to $(gHR^2/4\Omega^2L^4)(\Delta\rho_v/\rho_0)$, and the surface boundary condition is

$$w = \frac{4\Omega^2L^4}{gHR^2}p_t \quad \text{at } z = 1.$$

Now if we set the parameter γ equal to zero in equations 6a and 6d, the resulting equations admit solutions for which the vertical (z) dependence is separable. Formally, we regard this as the zeroth order of an ordinary perturbation expansion in the parameter γ. The vertical separation proceeds as follows. Define two new parameters, $\varepsilon = \Delta\rho_v/\rho_0$ and $\Theta = gHR^2/4\Omega^2L^4$. The equations we wish to solve are

$$u_t - yv + p_x = 0, \tag{7a}$$

$$v_t + yu + p_y = 0, \tag{7b}$$

$$u_x + v_y + w_z = 0, \tag{7c}$$

$$p_z + \varepsilon\Theta\rho = 0, \tag{7d}$$

and

$$\rho_t - ws(z) = 0, \tag{7e}$$

where

$$s(z) = -\frac{d\bar{\rho}}{dz}$$

is nonnegative for static stability.

The boundary conditions are $w = 0$ at $z = 0$ and $p_t = \Theta w$ at $z = 1$. The density, ρ, may be eliminated from equations 7d and 7e to give

$$p_{zt} = -\varepsilon\Theta w s(z). \tag{7f}$$

If we assume

$$u = F(z)\boldsymbol{u}(x, y, t), \tag{8a}$$

$$v = F(z)\boldsymbol{v}(x, y, t), \tag{8b}$$

$$p = F(z)\boldsymbol{p}(x, y, t), \tag{8c}$$

and

$$w = G(z)\boldsymbol{w}(x, y, t), \tag{8d}$$

and substitute in equations 7a, 7b, 7c, and 7f, we find that $F(z)$ is a common factor in each term of equations 7a and 7b, and equation 7f may be separated into

$$\boldsymbol{w} = -\boldsymbol{p}_t \quad \text{and} \quad F_z = \varepsilon\Theta s(z)G. \tag{9a}$$

Equation 3.2d then separates if

$$G_z = -\lambda F, \tag{9b}$$

where λ is a separation constant. The boundary conditions on equations 9a and 9b are $G = 0$ at $z = 0$ and $F + \Theta G = 0$ at $z = 1$. Combining equations 9a and 9b, we find

$$G_{zz} + \lambda\varepsilon\Theta s(z)G = 0, \tag{9c}$$

with $G = 0$ at $z = 0$ and $G_z = \lambda\Theta G$ at $z = 1$.

Let us now investigate the possibility of a solution to equation 9c for which $\lambda\Theta$ is $O(1)$, and $\varepsilon = \Delta\rho_v/\rho_0$ is formally treated as small. That is to say, we assume

$$G = G_0 + \varepsilon G_1 + \varepsilon^2 G_2 + \cdots,$$

and

$$\lambda = \lambda_0 + \varepsilon\lambda_1 + \varepsilon^2\lambda_2 + \cdots.$$

Then the equation to leading order is

$$G_{0_{zz}} = 0,$$

with boundary conditions $G_0 = 0$ at $z = 0$ and $G_{0_z} = \lambda_0\Theta G_0$ at $z = 1$. Clearly, $G_0 = az$, $\lambda_0\Theta = 1$, with a an arbitrary constant is the only solution. The corresponding F is a constant. In this case, the horizontal velocities do not vary in z (to leading order in ε), and the vertical velocity is linear in z. The stratification $s(z)$ does not enter the problem to leading order. This is the barotropic mode.

The other solutions of equation 9c have eigenvalues λ for which $\varepsilon\lambda = \lambda'$ is $O(1)$, and to leading order in ε the boundary condition on G is $G = 0$ at both $z = 0$ and $z = 1$.

Equation 9c is then

$$G_{zz} + \lambda'\Theta s(z)G = 0,$$

and clearly the solutions G and eigenvalue λ' depend on the details of the stratification $s(z)$. If $s(z)$ is everywhere positive, the eigenvalues λ' are all nonnegative and the corresponding vertical models are the denumerably infinite set of baroclinic modes. The simple case of uniform stratification $[s(z) \equiv 1]$ gives $G = \sin n\pi z$, with $\lambda'\Theta = n^2\pi^2$ to leading order in ε, with n a positive integer.

The equations governing the horizontal and time dependence of the u, v, and p fields corresponding to a given vertical mode are as follows:

$$u_t - yv + p_x = 0, \tag{10a}$$

$$v_t + yu + p_y = 0, \tag{10b}$$

and

$$\lambda p_t + u_x + v_y = 0, \tag{10c}$$

where λ is the separation constant for the mode being considered. Clearly, $1/\sqrt{\lambda}$ is a dimensionless measure of a horizontal phase speed for this system of equations. The corresponding dimensional phase speed turns out to be $\sqrt{gH}$ for the barotropic mode ($\lambda_0 = 1/\Theta$). If λ_n is the eigenvalue corresponding to the nth baroclinic mode, the quantity $H'_n = (\lambda_0/\lambda_n)H$ is called the equivalent depth for that particular baroclinic mode. The corresponding dimensional phase speed for the nth baroclinic mode is $\sqrt{gH'_n}$. Since all the λ_n are $0(\varepsilon^{-1})$, the equivalent depths for the baroclinic modes are $0(\varepsilon H)$. Typical values for the equivalent depths of the first five baroclinic modes in the equatorial Atlantic Ocean are 60, 20, 8, 4, and 2 cm (E. Katz, personal communication).

The results on the vertical separation may be easily understood in terms of the equivalent depths as follows. The quantity $2\Omega/R$, which appears in many places in this chapter, is the equatorial value of Rossby's (1939) parameter β. It has dimensions $[L^{-1}T^{-1}]$. If we pick the length scale L and time scale T such that $\beta = 1/LT$ and $L/T = \sqrt{gH}$, where H is the actual depth in the case of the barotropic mode or the equivalent depth in the case of a baroclinic mode, we obtain the natural time and length scale for the particular vertical mode we are considering. These choices for L and T are equivalent to setting λ equal to 1 in equation 10c. Then equations 10 (with $\lambda = 1$) provide a canonical system governing the horizontal and time dependence for each baroclinic mode.

We have seen that for the barotropic mode, the phase speed that emerges is $L/T = \sqrt{gH}$, where H is the actual ocean depth. For an ocean 5 km deep, the corresponding time scale is about 4 hr and the length scale about 3000 km. This means that for the barotropic mode, the equatorial region cannot be treated as isolated from the rest of the ocean basin. The barotropic problem is global, not local.

However, the baroclinic modes can be studied in isolation. For the equivalent depths for the first five baroclinic modes as given above, we find the time and length scales given in Table I.

For a velocity scale $U_0 = 1$ cm/sec, the unit of sea-surface elevation corresponding to the first baroclinic mode ($\beta L^2 U_0/g$) is equal to 0.25 cm. Therefore, there is a measurable sea-surface displacement associated with baroclinic motions, at least in the first mode.

TABLE I

n	H' (cm)	$\sqrt{gH'}$ (cm/sec)	L (km)	T (days)
1	60	240	325	1.5
2	20	140	247	2.0
3	8	88	197	2.6
4	4	63	165	3.1
5	2	44	139	3.6

Dispersion Relation

We now restrict our attention to baroclinic modes. Equations 10 with $\lambda = 1$ may be combined to give a single equation for v, which is

$$v_{xxt} + v_{yyt} + v_x - y^2 v_t - v_{ttt} = 0. \tag{11}$$

Before investigating the solutions to this equation, let us consider possible motions for which $v \equiv 0$. In this case, equations 10 become

$$u_t + p_x = 0, \tag{12a}$$

$$-yu + p_y = 0, \tag{12b}$$

and

$$p_t + u_x = 0. \tag{12c}$$

The general solution of this system is

$$u = \pm p = F(x \mp t)\varepsilon^{\mp y^2/2}. \tag{13}$$

We choose the upper sign since the solutions for the other choice are unbounded for large y. This solution represents an equatorially trapped Kelvin wave traveling at unit speed ($\sqrt{gH'}$ in dimensional units) from west to east without change of shape. That is to say, the wave is nondispersive.

All other solutions of equations 10 must have a nonzero v field satisfying equation 11. The separable solutions of equation 11 which are bounded for large y are of the form

$$v = e^{i(Kx - \omega t)}\psi_m(y), \tag{14}$$

where $\psi_m(y)$ is given by

$$\psi_m = \frac{\exp(-y^2/2)H_m(y)}{\sqrt{2^m m! \sqrt{\pi}}}, \tag{15}$$

where H_m is the mth Hermite polynomial. We call $\psi_m(y)$ a Hermite function. It satisfies the equation

$$\frac{d^2\psi_m}{dy^2} + (2m + 1 - y^2)\psi_m = 0, \tag{16}$$

where m is a nonnegative integer. It is easy to see from equation (16) that $\psi_m(y)$ is oscillatory on the interval $|y| \le (2m + 1)^{1/2}$, and monotonic (exponentially decaying) for $|y| > (2m + 1)^{1/2}$. $|y| = (2m + 1)^{1/2}$ is the turning latitude for the mode being considered. These motions are described as being "equatorially trapped." The functions $\psi_m(y)$ defined by equation 15 have been normalized so that

$$\int_{-\infty}^{\infty} \psi_m(y)\psi_n(y) \; dy = \delta_{mn}.$$

By substituting the solution given by equation 14 into equation 11, we obtain the dispersion relation

$$K^2 + \frac{K}{\omega} - \omega^2 + 2m + 1 = 0, \tag{17}$$

which may be solved for K as a function of m and ω to obtain

$$K = -\frac{1}{2\omega} \pm \sqrt{\omega^2 + \frac{1}{4\omega^2} - (2m + 1)}. \tag{18}$$

The special case $m = 0$ deserves comment. In this case, the two roots for K are

$$K = \omega - \frac{1}{\omega}, \quad \text{and} \quad K = -\omega.$$

The solution with

$$v = \exp\left\{i\left[\left(\omega - \frac{1}{\omega}\right)x - \omega t\right]\right\}\psi_0(y) \tag{19}$$

is called a Yanai wave, or mixed Rossby gravity wave. The corresponding u and p fields are

$$u = p = \frac{i\omega}{\sqrt{2}} \exp\left\{i\left[\left(\omega - \frac{1}{\omega}\right)x - \omega t\right]\right\}\psi_1(y), \tag{20}$$

and involve only a single Hermite function.

The other root, $K = -\omega$, is not acceptable on an equatorial β plane because although the v field is bounded everywhere, the corresponding u and p fields grow exponentially in y for y large (Matsuno, 1966; Moore, 1968).

For $m \ge 1$, the east–west wave number K is real for frequencies ω as long as the quantity under the radical in equation 18 is positive. This implies either $|\omega| \ge [(m + 1)/2]^{1/2} + (m/2)^{1/2}$ (gravity waves), or $|\omega| \le [(m + 1)/2]^{1/2} - (m/2)^{1/2}$ (planetary waves). For $[(m + 1)/2]^{1/2} + (m/2)^{1/2} > |\omega| > [(m + 1)/2]^{1/2} - (m/2)^{1/2}$, the wave numbers K are complex, with real part $K_{Re} = -1/2\omega$ corresponding to westward phase propagation. The imaginary part is $K_{Im} = \pm [2m + 1 - \omega^2 - (1/4\omega^2)]^{1/2}$, corresponding to exponential decay toward the east or west, depending on choice of sign. The possible existence of these boundary layer type solutions is important in the oceans owing to the presence of continental boundaries, and distinguishes the oceanic equatorial waves from atmospheric ones, for which only real wave number solutions periodic around the entire globe are admissible.

For $m \geq 1$, the $\boldsymbol{u}$ and $\boldsymbol{p}$ fields corresponding to $\boldsymbol{v}$ given by equation 14 are

$$\boldsymbol{u} = \frac{i}{2} e^{i(Kx - \omega t)} \left\{ \frac{\sqrt{2(m+1)}\, \psi_{m+1}(y)}{\omega - K} + \frac{\sqrt{2m}\, \psi_{m-1}(y)}{\omega + K} \right\}, \tag{21}$$

and

$$\boldsymbol{p} = \frac{i}{2} e^{i(Kx - \omega t)} \left\{ \frac{\sqrt{2(m+1)}\, \psi_{m+1}(y)}{\omega - K} - \frac{\sqrt{2m}\, \psi_{m-1}(y)}{\omega + K} \right\}. \tag{22}$$

Thus the structure of the $\boldsymbol{u}$ and $\boldsymbol{p}$ fields involves Hermite functions of one order higher and one order lower than the corresponding $\boldsymbol{v}$ field.

Figure 3 shows the dispersion curves for the Kelvin and Yanai waves. Figures 4 through 6 show the dispersion curves for $m = 1, 2,$ and 3.

Energy Density and Energy Flux

From the original linearized equations 7, before any vertical separation is done, it is possible to obtain an energy equation of the form

$$\frac{1}{2} \frac{\partial}{\partial t} \left(u^2 + v^2 + \frac{\Theta}{s(z)} \rho^2 \right) = -(up)_x - (vp)_y - (wp)_z. \tag{23}$$

The left side is the local time rate of change of the energy, where the $u^2 + v^2$ part represents kinetic energy due to the horizontal motion, and the $[\Theta/s(z)]\rho^2$ term is the available potential energy. The right side represents the work done by pressure

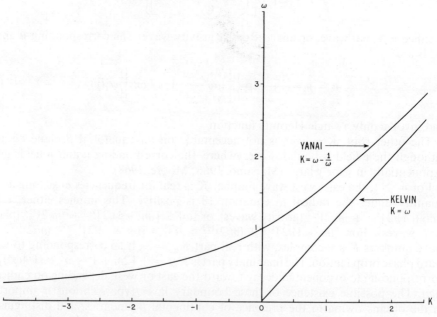

Fig. 3.   Dispersion curves for Kelvin and Yanai waves. The wave number K is real for all real frequencies ω, and the group velocity ($\partial \omega / \partial K$) is eastward at all frequencies.

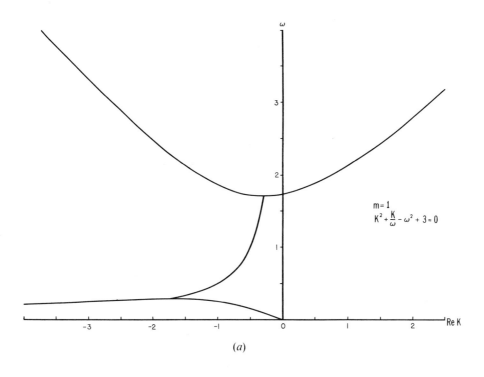

(a)

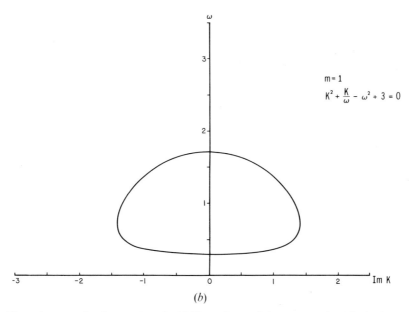

(b)

Fig. 4.   Dispersion curve for the $m = 1$ mode. (a) The real part of the wave number (abscissa) versus the frequency (ordinate). The lower branch (planetary wave) is connected to the upper branch (gravity wave) by the curve Re $K = -(1/2\omega)$. (b) The imaginary part of K in this transition frequency range.

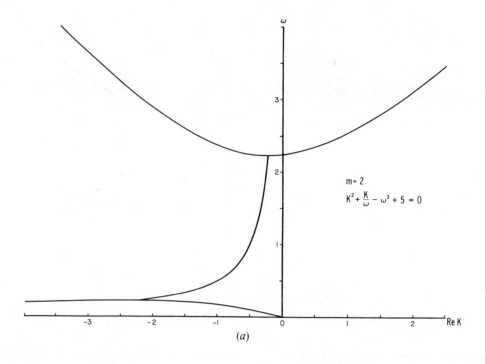

(a)

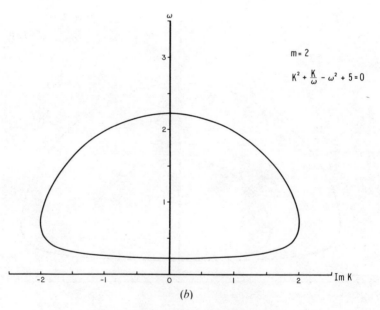

(b)

Fig. 5. Same as Fig. 4, but with $m = 2$.

330

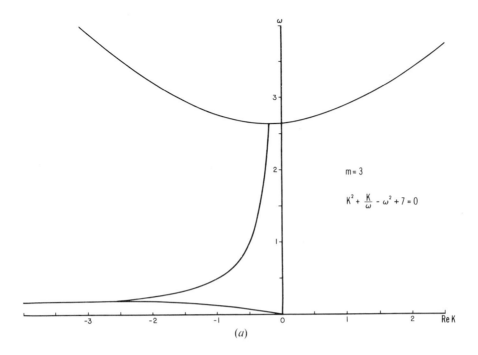

(a)

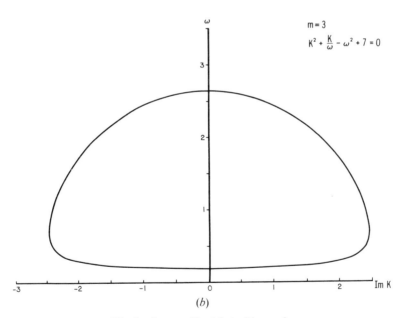

(b)

Fig. 6. Same as Fig. 4, but with $m = 3$.

forces. If we substitute for ρ in terms of p from equation 7d and integrate with respect to z from $z = 0$ to $z = 1$, we obtain

$$\frac{\partial}{\partial t}\left(\frac{\boldsymbol{u}^2 + \boldsymbol{v}^2 + \boldsymbol{p}^2}{2}\right) = -(\boldsymbol{u}\boldsymbol{p})_x - (\boldsymbol{v}\boldsymbol{p})_y, \tag{24}$$

the last term on the right side of equation 23 contributing nothing since $w = 0$ at $z = 0$ and $z = 1$. Now we define the energy density

$$E = \int_{-\infty}^{\infty} \overline{\frac{\boldsymbol{u}^2 + \boldsymbol{v}^2 + \boldsymbol{p}^2}{2}}\, dy, \tag{25}$$

and the energy flux

$$F = \int_{-\infty}^{\infty} \overline{\boldsymbol{u}\boldsymbol{p}}\, dy, \tag{26}$$

where the overbar denotes a time average.

For a single equatorial wave mode with complex amplitude A (i.e., $\boldsymbol{v} = $ real part of $Ae^{i(Kx - \omega t)}\psi_m(y)$, etc.), we find

$$E = \frac{|A|^2}{4}\left[1 + \frac{m+1}{(\omega - K)^2} + \frac{m}{(\omega + K)^2}\right], \tag{27}$$

and

$$F = \frac{|A|^2}{4}\left[\frac{m+1}{(\omega - K)^2} - \frac{m}{(\omega + K)^2}\right], \tag{28}$$

where both ω and K are assumed to be real. Then it is easy to prove that

$$F = EC_g, \tag{29}$$

where the group velocity C_g may be found from the dispersion relation (equation 17) to be

$$C_g = \frac{\partial \omega}{\partial K} = \frac{2K + (1/\omega)}{2\omega + (K/\omega^2)}. \tag{30}$$

Furthermore, if we superpose a number of waves at a given frequency ω with different real wave numbers K, they do not interfere with each other. That is to say, if we set

$$v = \mathrm{Re}\left\{\sum_{\substack{m=0 \\ s=\pm}}^{\infty} A_{ms} \exp\left[i(K_{ms}x - \omega t)\right]\psi_m(y)\right\} \tag{31}$$

etc., where all the $K_{m\pm} = -1/2\omega \pm [\omega^2 + (1/4\omega^2) - (2m + 1)]^{1/2}$ are real, we find

$$F = \overline{\boldsymbol{u}\boldsymbol{p}} = \sum_{\substack{m=0 \\ s=\pm}}^{\infty} \frac{|A_{ms}|^2}{4}\left\{\frac{m+1}{(\omega - K_{ms})^2} - \frac{m}{(\omega + K_{ms})^2}\right\}. \tag{32}$$

Reflections at Western Boundaries

In Figures 3–6 we distinguish waves whose group velocity is to the west from waves whose group velocity is to the east. If $C_g = \partial\omega/\partial K$ is positive (negative), the group velocity is eastward (westward). Thus the Kelvin and Yanai waves have group velocity to the east for all ω. For the gravity waves and planetary waves, the curve $K = -(1/2\omega)$ separates the eastward from the westward group velocities. If $K > -(1/2\omega)$, gravity waves have eastward group velocity and planetary waves have westward group velocity. If $K < -(1/2\omega)$, gravity waves have westward group velocity and planetary waves have eastward group velocity. The existence of the equatorially trapped Kelvin and Yanai waves with group velocity always to the east result in very asymmetric reflection properties at north–south boundaries. The reflection at a western boundary ($x = 0$, say) of an incoming wave (i.e., one with real wave number K and group velocity toward the west) is the simplest problem to treat. This reflection problem is solvable in closed form, as follows.

To be explicit, consider an incoming gravity wave at a given frequency ω, with meridional velocity given by

$$v_{\mathrm{Inc}} = \exp\left[i(K_{M-}x - \omega t)\right]\psi_M(y). \tag{33}$$

The wave number of the incoming wave is

$$K_{M-} = -\frac{1}{2\omega} - \sqrt{\omega^2 + \frac{1}{4\omega^2} - (2M + 1)} \tag{34}$$

and the corresponding u field is

$$u_{\mathrm{Inc}} = \frac{i}{2}\exp\left[i(K_{M-}x - \omega t)\right]\left\{\frac{\sqrt{2(M+1)}\,\psi_{M+1}(y)}{\omega - K_{M-}} + \frac{\sqrt{2M}\,\psi_{M-1}(y)}{\omega + K_{M-}}\right\}. \tag{35}$$

The boundary condition to be satisfied at $x = 0$ is $u = 0$ there, so we must add a suitable superposition of outgoing waves (group velocity to the east) to accomplish this. The solution for the reflected waves is obtained as follows. The meridional velocity in the reflected wave field is

$$v_{\mathrm{Ref}} = \sum_{m=1}^{M} A_m \exp\left[i(K_{m+}x - \omega t)\right]\psi_m(y) + A_0 \exp\left\{i\left[\left(\omega - \frac{1}{\omega}\right)x - \omega t\right]\right\}\psi_0(y). \tag{36}$$

The corresponding zonal velocity field is

$$u_{\mathrm{Ref}} = \frac{i}{2}\sum_{m=1}^{M} A_m \exp\left[i(K_{m+}x - \omega t)\right]\left\{\frac{\sqrt{2(m+1)}\,\psi_{m+1}(y)}{\omega - K_{m+}} + \frac{\sqrt{2m}\,\psi_{m-1}(y)}{\omega + K_{m+}}\right\}$$

$$+ \frac{i\omega}{\sqrt{2}}A_0 \exp\left\{i\left[\left(\omega - \frac{1}{\omega}\right)x - \omega t\right]\right\}\psi_1(y) + A_{\mathrm{Kel}}e^{i\omega(x-t)}\psi_0(y). \tag{37}$$

Note that A_0 is the amplitude of a Yanai wave, and A_{Kel} the amplitude of a Kelvin wave. We now set

$$u_{\mathrm{Inc}} + u_{\mathrm{Ref}} = 0 \qquad \text{on } x = 0 \tag{38}$$

for all y. The resulting equations for the coefficients A_m are

$$A_M = -\frac{\omega - K_{M+}}{\omega - K_{M-}}, \tag{39a}$$

$$A_{M-1} = 0, \tag{39b}$$

$$A_{M-2} = -\sqrt{\frac{M}{M-1}}(\omega - K_{(M-2)+})\left[\frac{1}{\omega + K_{M-}} + \frac{A_M}{\omega + K_{M+}}\right], \tag{39c}$$

$$A_{m-2} = -\sqrt{\frac{m}{m-1}}\frac{\omega - K_{(m-2)+}}{\omega + K_{m+}} A_m \qquad \text{for } m = M-1 \text{ down to } m = 3, \tag{39d}$$

$$A_0 = \frac{\sqrt{2}\, A_2}{\omega(\omega + K_{2+})}, \tag{39e}$$

and

$$A_{\text{Kel}} = -\frac{iA_1}{\sqrt{2}\,(\omega + K_{1+})}. \tag{39f}$$

The important properties to note about this solution are that it is explicit, it involves a finite number of reflected waves, and that the Hermite functions $\psi_m(y)$ which are involved all have m equal to $M + 1$ or less. All the wave numbers K_m in the solution are real. This means that the latitude band over which the reflected wave is oscillatory in y is the same as the latitude band over which the incident wave is oscillatory in y. It is the possibility of the Yanai wave and equatorial Kelvin wave, both of which have group velocity to the east, that makes the reflection at a western boundary solvable in closed form. Since $A_{m-1} = 0$, we see that $A_{m-3} = 0$, $A_{m-5} = 0$, and so on. Only those waves with the same y symmetry as the incident wave enter the reflection problem.

The reflection of an incoming planetary wave at a western boundary works in exactly the same way, only in this case the incoming wave has wave number K_{m+}, and the reflected planetary waves have wave number K_{m-}. The Kelvin and Yanai waves come into the reflection exactly as before.

Reflections at Eastern Boundaries

The situation at an eastern boundary is quite different. If we consider an incoming wave, that is, one that has group velocity toward the east, with meridional velocity proportional to $\psi_M(y)$, and try a solution like the one in the preceding section for the reflected wave, we find that no finite series solution is possible. The reason is that the Kelvin and Yanai waves which have group velocity to the east are not available as part of the reflected solution off an eastern boundary since they are not "outgoing" from such a boundary.

Moore (1968) has demonstrated that the reflection at an eastern boundary can be represented by an infinite series solution, involving waves with meridional velocity proportional to $\psi_m(y)$, where m is equal to or greater than the M of the incoming wave. At any given frequency ω for the incident wave, there are a finite number of m's for which the wave numbers K_m are real, but for all m greater than a certain value the wave numbers K_m are complex. To solve the eastern boundary reflection problem, we use, in the reflected solution, those waves which have group velocity toward the west or (if K_m is complex) decrease exponentially to the west. The latitudinal extent of the response extends beyond the turning latitude for the incident wave. The contributions to the reflection which involve complex K_m, that is, the infinite series part of the solution, constitute a boundary layer near the eastern boundary. The principal result obtained by Moore (1968) was to prove that this boundary layer is in fact a meridional Kelvin wave propagating away from the equator along the eastern boundary. For y large and positive, the asymptotic form for this Kelvin wave near an eastern boundary at $x = l$ is

$$v \sim v_0 \sqrt{y} \exp\left[i\left(\omega t - \omega y + \frac{x}{2\omega} \right) - y(l - x) \right]\left[1 + O\left(\frac{1}{y}\right) \right], \qquad (40)$$

where v_0 is a constant. The identification of the infinite series Hermite function expansion with this asymptotic solution is too complicated to demonstrate here (see Moore, 1968). Since the Kelvin wave gets narrower (width $\sim 1/y$) as y increases, its amplitude increases ($\sim \sqrt{y}$) to conserve energy. Also, the constant phase lines for the wave are inclined with respect to the coast, owing to the β effect.

When an equatorially trapped wave is reflected at an eastern boundary, some of the energy is returned in equatorially trapped motions, and some escapes along the boundary in the meridional (boundary trapped) Kelvin wave.

Free Modes of a Closed Basin

It is also possible (Moore, 1968) to produce free modes in a closed basin (baroclinic normal modes) at discrete eigenfrequencies ω. These solutions consist of a finite number of equatorially trapped waves, plus an extra-equatorial boundary trapped Kelvin wave carrying energy away from the equator on the eastern boundary, and returning it to the equator along the western boundary.

It is doubtful that such normal modes play any role in the real ocean. Presumably, the effects of dissipation, irregular geometry, and mean flows would drastically alter the boundary Kelvin wave before it can get all the way around the basin. Both barotropic and baroclinic normal modes in a rectangular basin spanning the equator have been computed numerically by Mofjeld and Rattray (1975). Also, Longuet-Higgins and Pond (1970) have found the eigenfrequencies for the normal modes in a hemispherical basin bounded by a meridian, as a function of the equivalent depth. Both barotropic and baroclinic modes are included. The results of both these computations agree well with the predictions of Moore (1968).

B. Forced Motion

Vertical Structure

In trying to understand time-dependent equatorial motions forced by the surface wind stress, one is immediately faced with the problem of deciding how to model the input of momentum from the wind to the water. Since we are at the equator, classical Ekman layer theory is not valid. For analytic modeling, the alternative that is most commonly used is that of Lighthill (1969). The wind stress is treated as a body force distributed uniformly in depth over an upper well-mixed layer in the ocean. For linear problems, this works out very nicely, since the forcing can be projected directly onto the vertical normal modes of the system. This type of model for the effect of the wind stress is discussed in great detail in the appendix of Lighthill (1969).

If the depth of the well-mixed layer is h, we must expand the function $\mathscr{F}(z)$ given by

$$\mathscr{F}(z) = 1 \qquad 1 - h < z < 1$$
$$= 0 \qquad 0 < z < 1 - h \tag{41}$$

in normal modes to see how the surface stress projects onto each normal mode of the system. If $F_n(z)$ denotes the nth normal mode, what we need is an expansion like

$$\mathscr{F}(z) = \sum_{n=0}^{\infty} a_n F_n(z). \tag{42}$$

If the $F_n(z)$ are normalized, we simply have

$$a_n = \int_0^1 \mathscr{F}(z) F_n(z)\, dz = \int_{1-h}^1 F_n(z)\, dz. \tag{43}$$

In particular, since $F_n(z) = 1$, $a_0 = h$.

Also, since completeness implies

$$\int_0^1 \mathscr{F}^2\, dz = \sum_{n=0}^{\infty} a_n{}^2, \tag{44}$$

we have

$$\sum_{h=0}^{\infty} a_n{}^2 = h. \tag{45}$$

Since $a_0 = h$, this may also be written as

$$1 + \sum_{n=1}^{\infty} \left(\frac{a_n}{a_0}\right)^2 = \frac{1}{h}. \tag{46}$$

Clearly, the relative forcing amplitude of the baroclinic modes to the barotropic depends very sensitively on the assumed depth of the well-mixed layer. The actual sizes of the coefficients a_n for $n \geq 1$ depend on the detailed structure of $s(z)$, with equation 45 as an overall constraint. In the example Lighthill used, h was 0.05 (200 m out of a total depth of 4000 m), so

$$\sum_{n=1}^{\infty} \left(\frac{a_n}{a_0}\right)^2 = 19, \tag{47}$$

in his case. He worked out $(a_1/a_0)^2 = 13.9$, and claimed this was the only significant projection of the baroclinic modes. In fact, he somehow missed an eigenvalue, and the one he calls second baroclinic is actually the third. When the eigenfunction corresponding to the second eigenvalue is computed, it is found that $(a_2/a_0) = 4.97$, so the second baroclinic mode is also significantly forced. The important point here is that the relative forcing amplitudes (a_n/a_0) are very sensitive to the mixed layer depth h and the model chosen for the stratification $s(z)$. In any case, we expect a significant baroclinic response to the surface wind stress. Just as in the case of free waves, the barotropic response problem is a global one, and the equatorial response is not isolated from the rest of the ocean. Therefore, this discussion concentrates on the baroclinic response only.

Forced Motion in the Absence of Boundaries

We now consider the following problem. The ocean is initially at rest, and at time $t = 0$, we turn on a steady wind and study the equatorial response of an unbounded ocean. We consider only the baroclinic part of the response. The forcing is projected on to the baroclinic mode eigenfunctions which are described in the preceding section, and each baroclinic mode is studied separately.

The equations to be solved are

$$u_t - yv + p_x = F, \tag{48}$$

$$v_t + yu + p_y = G, \tag{49}$$

and

$$p_t + u_x + v_y = 0, \tag{50}$$

with initial conditions $u = v = p = 0$ at time $t = 0$. We discuss the two simplest cases separately, namely, $F = 1$, $G = 0$, and $G = 1$, $F = 0$. That is, we investigate the baroclinic response to a completely uniform wind stress turned on at $t = 0$. Since the stress is x independent and the ocean is unbounded, the response is also x independent. For a wind stress toward the east, we wish to solve

$$u_t - yv = 1, \tag{51}$$

$$v_t + yu + p_y = 0, \tag{52}$$

and

$$p_t + v_y = 0. \tag{53}$$

The resulting equation for v alone is

$$v_{tt} + y^2 v - v_{yy} = -y, \tag{54}$$

with $v = v_t = 0$ at $t = 0$ for all y. The solution may be written in the form

$$u = tu_1(y) + u_2(y, t), \tag{55}$$

$$v = v_1(y) + v_2(y, t), \tag{56}$$

and

$$p = tp_1(y) + p_2(y, t), \tag{57}$$

where the fields with subscript 2 are necessary since the steady $v = v_1(y)$ does not satisfy the initial condition $v = 0$ at $t = 0$. The solutions are easily expressed as series of Hermite functions in y (with time dependent coefficients for the subscript 2 fields). If we define

$$I_m = \frac{2^{1/2}\pi^{1/4}}{2^m m!} \sqrt{(2m)!}$$

which are the coefficients for the expansion of 1 in a series of even Hermite functions [i.e., $1 = \sum_{m=0}^{\infty} I_m \psi_{2m}(y)$], then the solutions are

$$\boldsymbol{u}_1(y) = -2 \sum_{m=0}^{\infty} \frac{I_m \psi_{2m}(y)}{(4m+3)(4m-1)}, \tag{58a}$$

$$\boldsymbol{v}_1(y) = -2 \sum_{m=0}^{\infty} \frac{\sqrt{m+1/2}}{4m+3} I_m \psi_{2m+1}(y), \tag{58b}$$

$$\boldsymbol{p}_1(y) = -\sum_{m=0}^{\infty} \frac{4m+1}{(4m+3)(4m-1)} I_m \psi_{2m}(y), \tag{58c}$$

$$\boldsymbol{u}_2(y,t) = \sum_{m=0}^{\infty} I_m \psi_{2m}(y) \left[\frac{2m+1}{(4m+3)^{3/2}} \sin\left(\sqrt{4m+3}\, t\right) + \frac{2m}{(4m-1)^{3/2}} \sin\left(\sqrt{4m-1}\, t\right) \right], \tag{59a}$$

$$\boldsymbol{v}_2(y,t) = \sum_{m=0}^{\infty} I_m \psi_{2m+1}(y) \frac{2\sqrt{m+1/2}}{4m+3} \cos\sqrt{4m+3}\, t, \tag{59b}$$

and

$$\boldsymbol{p}_2(y,t) = \sum_{m=0}^{\infty} I_m \psi_{2m}(y) \left[-\frac{2m+1}{(4m+3)^{3/2}} \sin\left(\sqrt{4m+3}\, t\right) \right.$$

$$\left. + \frac{2m}{(4m-1)^{3/2}} \sin\left(\sqrt{4m-1}\, t\right) \right]. \tag{59c}$$

The functions $\boldsymbol{u}_1$, $\boldsymbol{v}_1$, and $\boldsymbol{p}_1$ are shown in Figures 7a–7c. They were first described by Yoshida (1959). See the discussion in O'Brien and Hurlburt (1974). It is clear that the solution consists of an equatorially confined accelerating jet, and the extra-equatorial part of the solution simply corresponds to Ekman transport.

Perspective plots of $\boldsymbol{u}_2$, $\boldsymbol{v}_2$, and $\boldsymbol{p}_2$ as functions of y and t are shown in Figures 8a–8c. At any latitude y away from $y = 0$, these look like inertial oscillations, at least in the period $0 < t < y$. For $t > y$, it is clear that some equatorially generated disturbances have reached the latitude in question, and the inertial character of the oscillation is no longer evident.

Contour plots of the total $\boldsymbol{u}$, $\boldsymbol{v}$, and $\boldsymbol{p}$ fields as given by equations 55–57 are shown in Figures 9a–9c.

The analogous problem for a wind blowing from south to north involves solving the equations

$$\boldsymbol{u}_t - y\boldsymbol{v} = 0, \tag{60}$$

$$\boldsymbol{v}_t + y\boldsymbol{u} + \boldsymbol{p}_y = 1, \tag{61}$$

and

$$\boldsymbol{p}_t + \boldsymbol{v}_y = 0. \tag{62}$$

Since $\boldsymbol{u} = \boldsymbol{v} = \boldsymbol{p} = 0$ at $t = 0$, we see that $\boldsymbol{v}_t = 1$ at $t = 0$ for all y. The resulting equation for $\boldsymbol{v}$ is

$$\boldsymbol{v}_{tt} + y^2\boldsymbol{v} - \boldsymbol{v}_{yy} = 0, \tag{63}$$

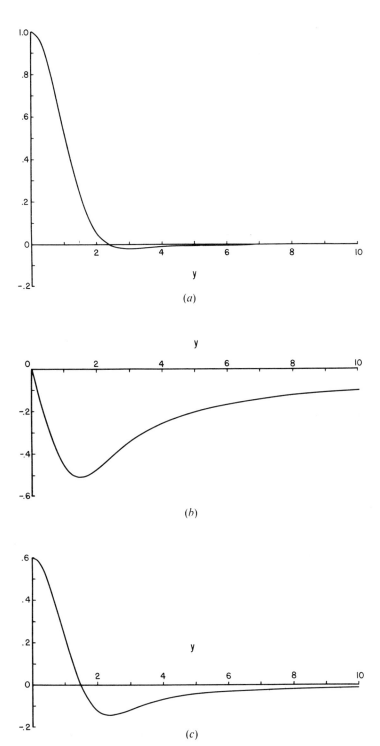

Fig. 7. (a) The zonal acceleration $u_t = u_1(y)$ for the Yoshida equatorial jet. (b) The corresponding meridional velocity. Note the asymptotic approach to Ekman transport ($v \sim -(1/y)$) for $y \geqslant 4$. (c) The corresponding pressure tendency $p_t = p_1(y)$.

339

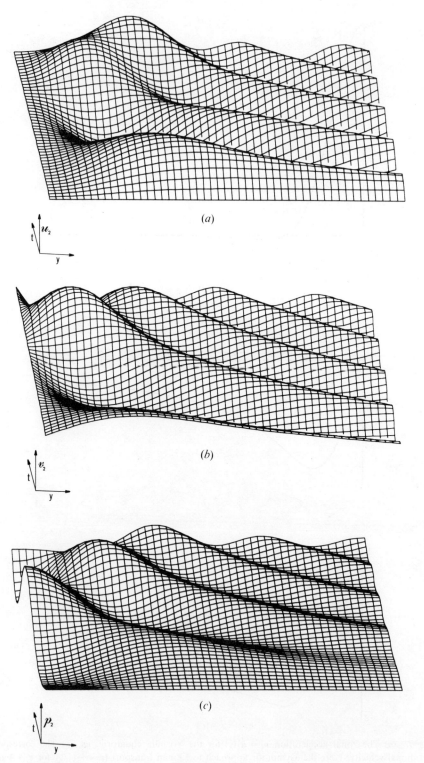

Fig. 8. (*a*) Perspective plot of $u_2(y, t)$. (*b*) Perspective plot of $v_2(y, t)$. (*c*) Perspective plot of $p_2(y, t)$.

340

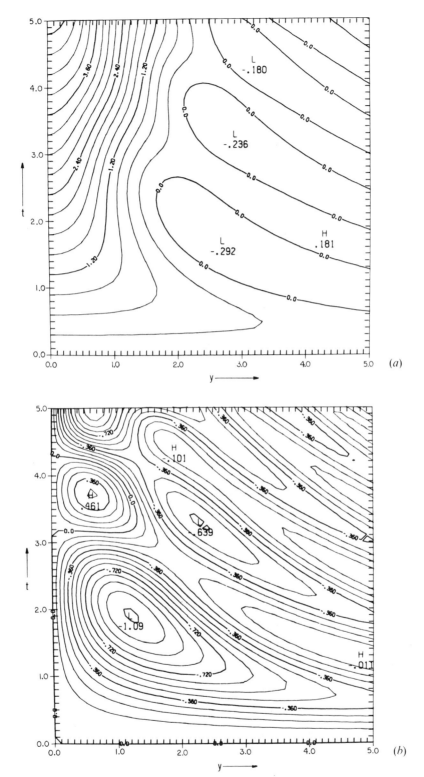

Fig. 9.   (a) Contour plot of the zonal velocity $u = tu_1(y) + u_2(y, t)$ as a function of y and t. (b) Contour plot of the meridional velocity $v = v_1(y) + v_2(y, t)$ as a function of y and t.

341

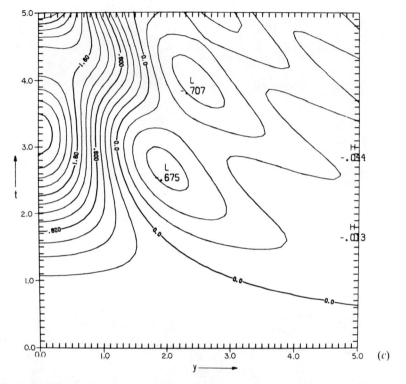

Fig. 9. (c) Contour plot of the pressure $p = tp_1(y) + p_2(y, t)$ as a function of y and t.

and the forcing enters only in the initial condition. The solutions for u, v, and p may be written in the form

$$u = u_3(y) + u_4(y, t), \tag{64}$$

$$v = v_4(y, t), \tag{65}$$

and

$$p = p_3(y) + p_4(y, t). \tag{66}$$

In contrast to the previous case, there is no accelerating jet in the solution. The explicit form of the solution is

$$u_3 = \sum_{n=0}^{\infty} I_n \psi_{2n+1}(y) \sqrt{n + \frac{1}{2}} \frac{4n + 3}{(4n + 1)(4n + 5)},$$

$$p_3 = \sum_{n=0}^{\infty} I_n \psi_{2n+1}(y) \sqrt{n + \frac{1}{2}} \frac{2}{(4n + 1)(4n + 5)},$$

$$u_4 = -\frac{1}{2} \sum_{n=0}^{\infty} I_n \psi_{2n+1}(y) \sqrt{n + \frac{1}{2}} \left[\frac{\cos(\sqrt{4n + 5}\,t)}{4n + 5} + \frac{\cos(\sqrt{4n + 1}\,t)}{4n + 1} \right],$$

$$v_4 = \sum_{n=0}^{\infty} I_n \psi_{2n}(y) \frac{\sin(\sqrt{4n + 1}\,t)}{\sqrt{4n + 1}},$$

and

$$p_4 = \frac{1}{2} \sum_{n=0}^{\infty} I_n \psi_{2n+1}(y) \sqrt{n + \frac{1}{2}} \left[\frac{\cos(\sqrt{4n+5}\,t)}{4n+5} - \frac{\cos(\sqrt{4n-1}\,t)}{4n+1} \right].$$

The solutions are shown in Figures 10–12. The extra-equatorial Ekman transport is clearly seen in the u_3 field. All the terms in v_4 are oscillatory in time, but a calculation of $\int_0^t v_4\, dt$ shows that what is really happening is a net northward displacement of the near equatorial surface waters, plus an oscillation about that net displacement. Of course, if we are looking at the second baroclinic mode, the subsurface water (at the undercurrent depth, say) is moving in the opposite direction from the surface water, and therefore would undergo a net displacement upwind. The steady models for the effect of cross-equatorial winds on the undercurrent also show a net upwind displacement, but it is not clear that these two models are related in any way (see Section 4.B).

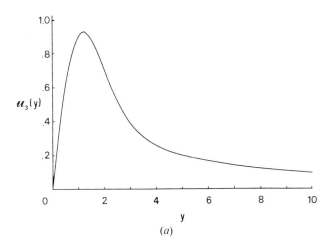

(a)

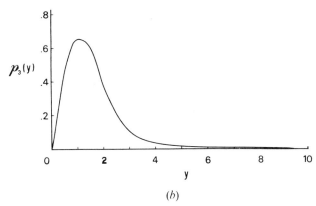

(b)

Fig. 10. (a) The function $u_3(y)$ describing the steady part of the zonal velocity generated by a uniform wind blowing toward the north. (b) The corresponding pressure field $p_3(y)$.

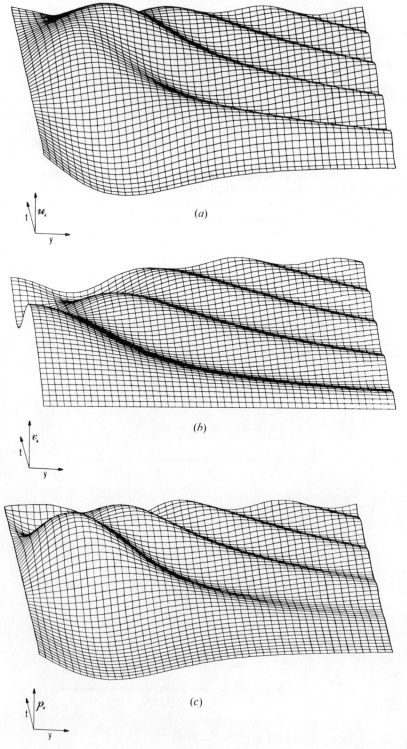

Fig. 11. (a) Perspective plot of $u_4(y, t)$. (b) Perspective plot of $v_4(y, t)$. (c) Perspective plot of $p_4(y, t)$.

344

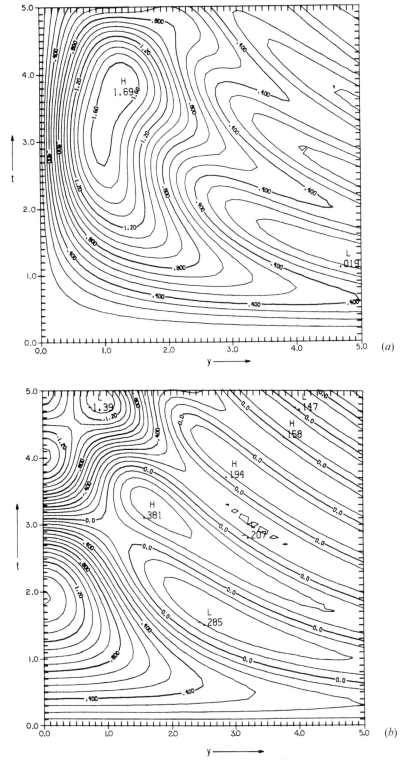

Fig. 12. (a) Contour plot of $u = u_3(y) + u_4(y, t)$ as a function of y and t. (b) Contour plot of $v_4(y, t)$ as a function of y and t.

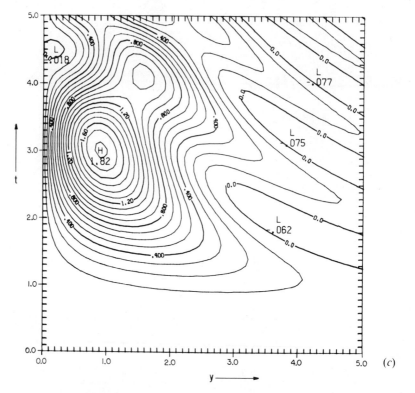

Fig. 12. (c) Contour plot of $p = p_3(y) + p_4(y, t)$ as a function of y and t.

The u_4, v_4, and p_4 solutions, at least in the extra-equatorial region, look like inertial oscillations for $t < y$. For $t > y$, it is again clear that near-equatorial disturbances have reached the latitude in question, and the inertial character of the oscillation is masked. Since $y = t$ is a characteristic curve of the equation for v, this is no surprise.

Effect of Meridional Boundaries

Forced models similar to those described above with boundaries present in the system have been explored both theoretically and numerically. See the works of Lighthill (1969), Gill (1975), O'Brien and Hurlburt (1974), Cane (1975), and Anderson and Rowlands (1976).

The effects of boundaries are complicated. The solutions show strong east–west asymmetry owing to the asymmetric properties of the free mode solutions, and to the boundary trapped Kelvin waves carrying energy away from the equator on eastern boundaries and toward the equator on western boundaries.

We discuss only one simple case, namely, the effect of a western boundary running straight north–south (i.e., along $x = 0$) on the accelerating Yoshida jet. We do not even discuss the complete solution to this problem, but only the grossest feature. It is clear from equation 58a for u_1 that the largest contribution is from the leading term, proportional to $\psi_0(y)$. The coefficient of the second term is equal to $1/7\sqrt{2}$ times the coefficient of the first. We discuss only the boundary response to that part of the u field proportional to $t\psi_0(y)$.

To make $u = 0$ on $x = 0$, we must add a free solution to cancel the forced solution there. Since $\psi_0(y)$ is the form for the y structure of an equatorial Kelvin wave, the free solution for u to be added is

$$(x - t)\psi_0(y)S(t - x)$$

where S denotes a step function. This carries with it a p field of the same form. This Kelvin wave extends only from $x = 0$ to $x = t$. Beyond $x = t$, boundary effects are not yet felt. In the region between $x = 0$ and $x = t$, the Kelvin wave p field provides a uniform pressure gradient p_x which replaces the acceleration u_t. The net result is that when the Kelvin wave reaches a given x (at time $t = x$), the u field stops accelerating and a uniform pressure gradient takes over.

The effect at an eastern boundary is quite different, since the Kelvin wave does not represent an outgoing solution there, and does not come into the boundary response.

An analytic-numerical model for the undercurrent in the Pacific has been explored by Gill (1975). He uses 25 modes in the vertical and takes the western boundary into account. Typical y–z sections show a pronounced undercurrent-like jet after a sufficient time. Both nonlinear and frictional effects have yet to be included in such models at this time (March 1975).

4. Steady-State Models

A. *Extra-Equatorial Currents*

One of the early triumphs of dynamical oceanography was Sverdrup's (1947) demonstration that the alternate bands of eastward and westward flowing geostrophic currents (see Fig. 1) in the tropics are a consequence of the curl of the wind stress (i.e., the vorticity transferred from the atmosphere to the ocean). Sverdrup assumed that the flow is linear and steady and that there is no motion below a certain depth. (It may alternatively be assumed that the ocean floor is flat and stressfree.) Under such conditions, the curl of the wind stress, τ, determines the meridional transport V:

$$V = \int_0^1 v\, dz = \hat{K} \cdot \operatorname{curl} \tau. \tag{67}$$

($\hat{K}$ is a unit vector in the positive z direction.) The vertically integrated continuity equation then yields the zonal transport except for a constant which is determined by imposing the condition that this transport vanishes at the eastern wall of the ocean basin.

Sarkisyan (1969) has pointed out that if the bathymetry of the ocean floor is taken into account, then the curl of the wind stress determines the transport normal to geostrophic contours, curves along which $L = $ constant:

$$\psi_x\left(\frac{y}{1-h}\right)_y - \psi_y\left(\frac{y}{1-h}\right)_x = -\hat{K} \cdot \operatorname{curl}\frac{\tau - \nabla\Phi}{1-h}, \tag{68}$$

where

$$\psi_y = \int_h^1 u\, dz; \qquad \psi_x = -\int_h^1 v\, dz; \qquad \Phi = \frac{1}{RoFr}\int_h^1 (1-z)\rho\, dz.$$

When $1 - h$, the depth of the ocean is constant, then equation 68 is identical to equation 67, and the transports are independent of the stratification of the ocean! Calculations for the North Atlantic show considerable differences between transports as calculated from equations 67 and 68 (Sarkisyan, 1969), but such computations are yet to be made for the tropics.

Kendall (1970) analyzed the additional data that have become available since Sverdrup (1947) compared his theory with measurements in the Pacific, and finds that some of Sverdrup's (1947) assumptions are invalid. The zonal velocity component of the North Equatorial Countercurrent in the Pacific is indeed in geostrophic balance, but the terms representing the nonlinear advection of momentum in the zonal momentum equation cannot be neglected. There are also indications from numerical models that Sverdrup's (1947) theory may be deficient. Cane (1975) finds that a purely meridional wind of constant intensity (and therefore zero curl) can generate eastward equatorial countercurrents. It is unclear which factors determine the location and intensity of these currents.

B. Equatorial Currents

A large number of theories have been proposed to explain the Equatorial Under-current since its rediscovery in 1952. The different theories each correspond to a different range of the nondimensional parameters (Ekman number, Rossby number, ...), that characterize the motion of a fluid between concentric rotating spheres. The discussion here is confined to the parameter range that corresponds to that of the observed equatorial currents. In particular, only parameter ranges for which the horizontal Coriolis terms are negligible are discussed. (The conditions necessary for these terms to be unimportant are given by Philander, 1973b.) The theoretical results are compared with measurements of the equatorial currents. For a review of these measurements, and for the appropriate references, the reader is referred to Philander (1973b).

Diffusive processes play an important role in all the steady-state models, so we start by describing measurements of turbulent diffusion near the equator. Williams and Gibson (1974) measured small-scale fluctuations of temperature in the core of the undercurrent at 0°N, 150°W, and toward its northern edge at 1°N, 150°W, at depths of 100 m. Their measurements indicate large vertical mixing even though the vertical stability of the fluid is high. They infer values of 25 and 27 cm^2/sec for the turbulent diffusion coefficients of momentum (v_v) and temperature (K_v), respectively, at the core of the undercurrent. At 1°N, 150°W, these coefficients have the values $v_v = 12$ and $K_v = 0.52$ cm^2/sec. Jones (1973) provides information about the variation of these values with depth. His measurements near 97°W on the equator enable him to calculate the Richardson number of the flow from the surface to 300 m. Because of the large shear below the core of the undercurrent, there is a region where the Richardson number is less than 1. Hence intense vertical mixing, owing to Kelvin–Helmholtz instabilities, is possible there. Jones (1973) infers that the values for the coefficients of turbulent diffusion vary from $v_v = 0.4$, $K_v = 2 \times 10^{-3}$ cm^2/sec at the depth of the core of the undercurrent (35 m), to values in excess of 100 cm^2/sec for v_v in the well-mixed layer below the core of the undercurrent. This mixed layer corresponds approximately with the shaded region in Fig. 2; it is also present in the Atlantic.

It is clear that models that parameterize turbulent diffusion should use a coefficient of eddy viscosity that not only varies with both depth and latitude, but that depends on the intensity of the undercurrent. All the models to be discussed next, except that of Robinson (1966), use constant values for v and K.

Constant-Density Models

The unusually deep mixed surface layer at the equator in the western Pacific is bounded below by a thermocline of high vertical stability (see Fig. 2). Because this thermocline inhibits the vertical transfer of momentum, a constant-density model in which the thermocline is replaced by a rigid surface should be appropriate for the study of motion in this mixed layer. The results of such models also elucidate the dynamics of equatorial currents in a region such as the eastern part of the Pacific, where there is no mixed surface layer, but the models cannot be expected to predict the detailed structure of these currents.

Consider a constant, zonal wind stress blowing over an ocean of constant density. Assume that the flow at extra-equatorial latitudes is linear and, except for a surface Ekman layer, inviscid. The pressure gradient created when the wind piles the water up against one wall of the ocean basin is then

$$P_x = E_v^{1/2} \tau^x. \tag{69}$$

This pressure force, which is independent of latitude, is opposite in direction to that in which the wind is blowing.

The depth of the surface Ekman layer, $(v_v/2\Omega \sin \theta)^{1/2}$, increases with decreasing latitude until, at a certain distance from the equator, friction is important at all depths. If we assume symmetry about the equator, then the linear zonal momentum equation at the equator is

$$-E_v u_{zz} + P_x = 0, \tag{70}$$

where the zonal pressure gradient is given by equation 69. Specification of the wind stress at the surface, and the condition that the flow must vanish at the floor of the (model) ocean, enables us to solve for the zonal velocity at the equator

$$u = \frac{\tau^x}{2E_v^{1/2}} z^2. \tag{71}$$

(The ocean floor cannot be assumed to be stressfree, for it is then impossible to have a steady-state circulation in a closed basin unless v varies with depth or the horizontal diffusion of momentum is introduced.) The zonal flow at the equator is in the direction of the wind stress even though the pressure force is in the opposite direction. We find that an eastward pressure force, generated by a westward wind, is necessary for the existence of the Equatorial Undercurrent. This example shows that such a pressure force is not sufficient.

In a linear model, a constant westward wind induces poleward flow in the surface layers where the westward flow is intense, and convergent equatorward flow at depth, where the westward flow is weak. The integral, over the depth of the ocean, of the northward flux of eastward momentum $\overline{uv}$ associated with this meridional circulation is shown in Fig. 13. When the direction of the wind stress is reversed, both u and v change sign, so that $\overline{uv}$ remains unchanged. Gill (1975) pointed out that, in the vicinity of the equator, $(\partial/\partial y)(\overline{uv}) < 0$, so that there is a convergence of eastward momentum. This explains why the nonlinear corrections to the linear zeroth-order flow, calculated by Robinson (1966), are eastward for both eastward and westward winds. Charney (1960) solved for the flow at the equator and demonstrated how, for a westward wind, the flow becomes progressively more eastward as the

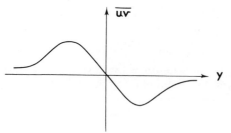

Fig. 13. Northward flux of eastward momentum near the equator (after Gill, 1975).

Rossby number is increased (Fig. 14). The flow studied by Charney (1960) is independent of longitude, except for a constant zonal pressure gradient, and is described by the equations

$$-E_v u_{zz} + Ro(vu_y + wu_z) - yv + E^{1/2}\tau_x = 0, \tag{72a}$$

$$-E_v v_{zz} + Ro(vv_y + wv_z) + yu + P_y = 0, \tag{72b}$$

$$P_z = 0, \tag{72c}$$

and

$$v_y + w_z = 0. \tag{72d}$$

By assuming symmetry about the equator, it is possible to derive equations that describe the flow in the equatorial plane, independently of the flow elsewhere. For very intense westward winds (large values of the Rossby number), the diffusion terms in equations 72 are important only in a thin surface boundary layer, while the deep inviscid flow has no vertical shear, and has its zonal component in geostrophic balance (Philander, 1971). This, however, is not the range of parameters that is of oceanographic interest. Charney (1960) obtains a profile that closely resembles the one observed at the equator in the well-mixed upper layer of the western Pacific

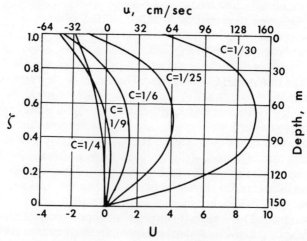

Fig. 14. The zonal flow at the equator as a function of depth, for different values of the parameter $C = v_v/(H^5\tau_0\,2\Omega/R)^{1/3}$ as predicted by a model described by equations (72) (after Charney, 1960).

when both the vertical diffusion and advection of momentum are important throughout the depth of this mixed layer. This is in accord with measurements that show that the Equatorial Undercurrent in the mixed layer of the western Pacific does not have its downstream flow in geostrophic balance. The appropriate scaling is

$$\text{Horizontal velocity} \sim \left(\frac{\tau_0^{\,4}R}{2\Omega v_v^{\,2}}\right)^{1/5} \sim 100 \text{ cm/sec,}$$

$$\text{Width of the current} \sim \left(\frac{\tau_0^{\,2}R^3}{8\Omega^3 v_v}\right)^{1/5} \sim 220 \text{ km,}$$

where the numerical values correspond to $v_v = 15$ cm^2/sec and $\tau_0 = 0.1$ cm^2/sec^2.

The model is found to be sensitive to the value assigned to v_v, the coefficient of eddy viscosity. A more sophisticated parameterization of processes that maintain the mixed surface layer in the western Pacific may improve this feature of the model.

If the meridional circulation is such as to transport westward momentum from the equator, then the flow at some distance from the equator will be more strongly westward. The model of Mikhailova et al. (1967), in which the intensity of the westward current adjacent to the Equatorial Undercurrent increases as the flow becomes more nonlinear, shows this effect. The westward flow observed next to the Equatorial Undercurrent must in part be due to this mechanism.

Whereas Charney's (1960) calculations are for the flow in the equatorial plane only, the computations of Charney and Spiegel (1971) give the meridional structure of the flow. They find that the fields develop cusps, discontinuous meridional derivatives, at the equator for sufficiently large Rossby numbers. The reason for this is that particles tend to conserve their vorticity as they move equatorward in each hemisphere, and arrive at the equator with a vorticity equal to that of the earth at their latitude of origin (Fofonoff and Montgomery, 1955). The cusps may be removed by the inclusion of terms representing the lateral diffusion of momentum in the model. The effect of lateral friction v_H is of secondary importance if v_v is 0(20 cm^2/sec), and provided v_H does not exceed 10^8 cm^2/sec. For larger values of v_H relative to the value of v_v, lateral friction modifies the flow considerably: the vertical diffusion of momentum is confined to a thin surface boundary at all latitudes, including the equator, so that the flow below the surface layers has no vertical shear. In a linear model (Gill, 1971), the vertically integrated zonal transport is zero at the equator, if the wind stress is constant. Westward flow near the surface is thus associated with an eastward flow at depth—an equatorial undercurrent. When the flow is nonlinear (McKee, 1973), this eastward current is in geostrophic balance. The Equatorial Undercurrent observed at the base of the mixed surface layer in the western Pacific is not in geostrophic balance, nor is the current without vertical shear. A model in which laterally diffusive processes are of primary importance thus seems inappropriate for a description of the flow in the upper mixed layer of the western Pacific.

We have thus far discussed the flow at the equator when the winds have a westward component. In the western Pacific, this is usually the case during the second half of the year when southeasterlies prevail. The winds there have a monsoonal character and reverse direction seasonally; during the first half of the year, northwesterlies usually prevail (Atkinson, 1971). There are no studies of the steady nonlinear circulation at the equator when the winds are eastward, but there are a few measurements that indicate that during periods of eastward winds, the surface flow is eastward while the flow at the base of the mixed layer is westward. (There are only a few such measurements, so that this may be a description of the flow during a period of transition from westward to eastward winds, rather than a description of the response of the ocean to steady eastward winds.) Currents with a similar vertical structure have been observed in the mixed surface layer of the Indian Ocean on an occasion when the winds were eastward (Taft and Knauss, 1967). The eastward winds generate a westward pressure force, but the Reynolds stresses, as pointed out earlier, transport eastward momentum equatorward irrespective of the direction of the zonal wind. This, however, is apparently not sufficient to maintain an eastward undercurrent.

To summarize, the constant-density model of Charney (1960) is reasonably successful in simulating the westward surface flow and eastward undercurrent in the mixed surface layer of the western Pacific when the winds are westward. The eastward pressure force created by this wind and the eastward momentum generated at the equator by Reynolds stresses are both crucial for the maintenance of the undercurrent. Constant-density models do not explain the permanent eastward current observed in the thermocline below this mixed layer. [Because of the two apparently independent eastward currents, one at the base of the mixed layer and one immediately below it in the thermocline, the subsurface eastward flow in the western Pacific sometimes has a double-celled structure; see Hisard et al. (1970).] Constant-density models also fail to explain the deep, westward current below the undercurrent in the equatorial thermocline. Stratification must be taken into account in a model that simulates these currents. The depth of the thermocline decreases in an eastward direction until, east of about 150°W, there is no mixed surface layer (see Fig. 2). The results described here enable us to understand the oceanic circulation in this region better, but constant-density models of course can not explain the details of the structure of the currents in the eastern part of the Pacific and in the Atlantic where there is usually no mixed surface layer.

Stratified Models

Westward winds generate an eastward pressure force, the presence of which appears to be a necessary condition for the existence of an equatorial undercurrent. Westward winds also cause divergent, poleward flow in the surface layers at the equator. In a constant-density model, one thus expects the Equatorial Undercurrent to be associated with vigorous equatorial upwelling. Yet measurements at the same longitudes in the eastern part of the Pacific at different times show that when the southeast trades are intense, usually in September and October, the undercurrent is weaker than it is during periods of weak equatorial winds such as occur in March and April. When the winds are light, there is often a marked absence of isotherms that dome near the equator—an indication of a lack of upwelling. The results of constant-density models are inconsistent with these measurements. Consider, however, the situation when stratification is taken into account. The westward winds will cause not only the sea surface, but also the thermocline, to slope in a zonal plane. The longitudinal density gradients shown in Fig. 2 have associated with them an eastward pressure force if pressure gradients at great depths are assumed small. If the winds abate temporarily, then this pressure force persists for a certain period of time because it is a consequence of the long-term effect of the winds. We inquire first into the nature of the thermohaline circulation when the ocean surface is temporarily stressfree, and later investigate the effect of "local" winds on this circulation.

The geostrophic flow associated with the eastward pressure force that is a consequence of the upward (from west to east) slope of surfaces of constant density in the upper ocean in the tropics has an equatorward component in both hemispheres. At the equator, where the Coriolis force vanishes, the convergent fluid flows down the pressure gradient. In the absence of surface winds, this eastward current extends from the surface through the thermocline. We should be able to determine the width and intensity of this current from a scale analysis, which must take into account that the zonal pressure gradient cannot be balanced by a Coriolis force at the equator. Measurements indicate that in the zonal momentum equation, the pressure term is comparable to the terms representing the nonlinear advection of momentum, whereas in the meridional momentum equation, a geostrophic balance is possible even close to the equator. The scale analysis that reflects this balance of terms yields (Robinson, 1960)

$$\text{Width} \sim \left(\frac{\Delta\rho g H R^2}{\rho_0 \Omega^2}\right)^{1/4} \sim 250 \text{ km,} \tag{73a}$$

$$\text{Zonal velocity} \sim \left(\frac{\Delta\rho g H}{\rho_0}\right)^{1/2} \sim 140 \text{ cm/sec,} \tag{73b}$$

and

$$\text{Meridional velocity} \sim \left(\frac{\Delta\rho^3 g^3 H^3 R^2}{\rho_0{}^2 \Omega^2 L^4}\right)^{1/4} \sim 35 \text{ cm/sec.} \tag{73c}$$

Here H denotes the depth scale of the Equatorial Undercurrent. Its numerical value has been taken to be 100 m, the depth of the extra-equatorial thermocline. This is a reasonable assumption, since it is equatorward flow in the thermocline that feeds the undercurrent. Note that the expressions in equations 73 are independent of coefficients of eddy viscosity or conductivity.

According to the scaling, the following terms are important in the equations of motion:

$$-E_v u_{zz} + Ro(uu_x + vu_y + wu_z) - yv + P_x = 0, \tag{74a}$$

$$-yu + P_y = 0, \tag{74b}$$

$$\frac{1}{RoFr}\rho + P_z = 0, \tag{74c}$$

$$u_x + v_y + w_z = 0, \tag{74d}$$

and

$$-\Gamma_1 \rho_{zz} + Ro(u\rho_x + v\rho_y + w\rho_z) = 0. \tag{74e}$$

Whether or not the diffusive terms can be neglected depends on the value of the parameter $[K_v{}^2 R^2/gH^5(\Delta\rho/\rho_0)]$.

To solve equations 74, boundary conditions must be specified. In particular, the motion in the equatorial region must merge smoothly with the flow in the extra-equatorial region where the dynamics is different. A description of the extra-equatorial flow, in the absence of surface winds, has been provided by Robinson and Welander (1963). In their model, the horizontal flow is in geostrophic balance; the upwelling of cold water balances the downward conduction of heat so as to give a thermocline near the ocean surface. The condition that this flow must match smoothly with that near the equator enables us to determine H (the depth scale of the Equatorial Undercurrent),

$$H \sim \left(\frac{k_v^2 R^2}{g(\Delta\rho/\rho_0)} \right)^{1/5}. \tag{75}$$

We now find that in equation 74e, the conduction of heat is as important as the advection of heat. Furthermore, if the Prandtl number is 0(1), then diffusive terms in equation 74a are as large as the advective terms.

The three-dimensionality of equations 74 poses serious mathematical difficulties, but the absence of an intrinsic zonal scale for the undercurrent permits the introduction of a similarity transformation that eliminates explicit x variations from equations 74 (Philander, 1973a). The transformation for the zonal velocity component, for example, is

$$u(x, y, z) = x^{-2\alpha} u(x^\alpha y, x^\beta z),$$

where α and β are constants. The surface boundary conditions, one of which specifies how the sea-surface temperature decreases from west to east, and the influx conditions at an extra-equatorial latitude enable us to determine α and β and to infer that the Equatorial Undercurrent becomes slower, narrower, and shallower in a downstream direction. Such variations are consistent with measurements over certain portions of the undercurrent (see, for example, Knauss, 1960). Since this current is maintained by an influx of fluid from the sides, its eastward transport can decrease downstream only if there is a loss of fluid through equatorial downwelling. Earlier we explained that, in the absence of surface winds, a thermal boundary layer near the surface, a thermocline, is possible when the downward conduction of heat is balanced by an upward advection of cold water. When heat is conducted and convected downward, as happens at the equator, then the horizontal advection of cold water becomes necessary. Because isotherms slope upward from west to east there is, at a given depth, a source of cold water to the east. Hence a westward current is necessary below the Equatorial Undercurrent if there is to be an equatorial thermocline near the surface. The numerical solutions to the similarity transformed equations 74 show the occurrence of equatorial downwelling and the presence of a deep westward equatorial current below the undercurrent (Philander, 1973a).

The results described here explain several features of the flow observed at the equator when the winds are light, a common occurrence in the eastern part of the Pacific and in the Atlantic in March and April. At these times, the surface flow at the equator is often eastward, a phenomenon referred to as the surfacing of the undercurrent. Though there is not much evidence of equatorial upwelling, the deeper isotherms are found to trough in a pronounced manner at the equator. This can in part be attributed to equatorial downwelling, but vertical mixing also plays a role. A deep westward current, named the Intermediate Equatorial Undercurrent by Hisard and Rual (1970), has been observed below the Equatorial Undercurrent at practically all longitudes.

The meridional circulation—equatorward flow near the surface where there is an eastward current, and poleward flow below the equatorial thermocline where there is a westward current—is such that the Reynolds stresses generate eastward momentum near the equator. As is the case in the constant-density model, this source of momentum, and the eastward pressure force associated with the sloping isopycnals, maintain the Equatorial Undercurrent. The pressure force appears to be necessary

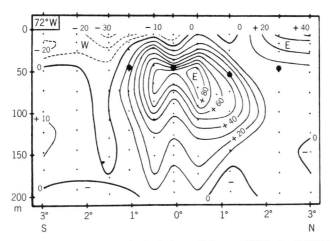

Fig. 15.   The zonal velocity component (cm/sec) relative to 215 m at 12°W in April 1964 (after Sturm and Voigt, 1966).

for the presence of the undercurrent, because this current disappears when the south-west monsoons over the Indian Ocean destroy the eastward pressure force by causing isopycnals to become horizontal. Information about the effect of an opposing (west-ward) pressure force on the Equatorial Undercurrent comes from measurements in the Gulf of Guinea where the southwesterly winds cause isopycnals to slope downward from west to east. (Southeasterly trades, which generate an eastward pressure force, prevail over the eastern part of the equatorial Atlantic.) The under-current appears to split symmetrically about the equator (see Fig. 15) when it enters the Gulf of Guinea. A similar phenomenon has been observed in the Pacific near 176°W where the undercurrent also encounters a counterpressure force. No attempt has yet been made to model these regions.

The discussion has thus far concerned the oceanic circulation due to density gradients during the temporary absence of surface winds. We next turn attention to the effect of "local" winds on this circulation. Westward winds cause a westward surface current so that the eastward flow is now an undercurrent. The more intense the winds are, the more intense is the westward surface flow, the deeper is the core of the undercurrent, and the lower is the maximum eastward velocity of the under-current. The westward surface flow is divergent and is accompanied by equatorial upwelling. The isotherms near the surface thus dome near the equator. This feature, together with the troughing of the deeper isotherms, results in a spreading of the equatorial thermocline. This phenomenon is so closely associated with the Equatorial Undercurrent that, when current measurements have been impossible, it has been taken as evidence of the presence of the Equatorial Undercurrent. This description of the effect of westward surface winds on the "thermally" driven Equatorial Under-current is based on numerical solutions of equations 74 with diffusive and nonlinear terms included in equation 74b so as to permit stress boundary conditions at the surface (Philander, 1973b). Figure 16 shows the computed zonal and meridional velocity components for different intensities of the westward winds.

The above results are consistent with measurements made at the equator. When the southeast trades at the equator are intense, usually in the autumn, the westward surface flow is stronger, and the eastward undercurrent weaker than when the winds

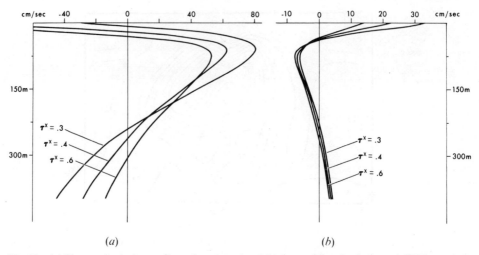

Fig. 16. (*a*) The zonal velocity profile at the equator, and (*b*) the meridional velocity at 1.5°N for various values of the wind stress (cm^2/sec^2) as predicted by the model described by equations (74), with $K_v = v_v = 1$ cm^2/sec^2.

are light. Near the equator, the upper isotherms ridge in a more pronounced manner, and the deeper isotherms trough in a less marked manner during periods of strong westward winds than during periods of light winds. Cross-equatorial winds are found to displace the core of the undercurrent in an upwind direction. Thus when the southeasterlies at the equator are intense, and after the onset of the southwest monsoons over the Indian Ocean, the core of the undercurrent is south of the equator. (Both stratified and constant-density models can simulate this effect. According to the models, zonal winds with a latitudinal shear also affect the position of the core of the undercurrent.) Though the measurements are indicative of this behavior of the equatorial currents under different wind conditions, the number of measurements is far too few to establish whether this behavior is systematic.

The results of the stratified model described here explain several features of the currents observed at the equator. The model, however, is subject to severe limitations because of the similarity transformation. It does not, for example, provide information about the origin and fate of the waters transported by these currents. To answer these questions, a fully three-dimensional model of the tropical circulation of a stratified ocean is required.

C. Circulation in a Closed Basin

According to Sverdrup's (1947) model of the tropical currents, the meridional transport at a given point depends on the local wind stress only. The transport across a circle of latitude running through the basin is therefore not necessarily zero. Stommel (1948) demonstrated that a western boundary current is necessary to close the circulation and to dissipate vorticity. He thus provided justification for Sverdrup's assumption that the zonal transport of the currents vanishes at the eastern wall of the ocean basin. This assumption implies that the zonal transport at a given point depends on conditions to the east of that point only. Charney's (1960) model for the Equatorial Undercurrent assumes that this current is a local phenomenon, determined

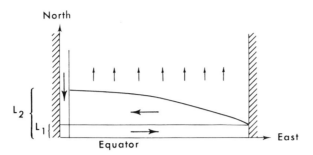

Fig. 17.   A schematic diagram of the wind-driven oceanic circulation in a basin that spans the equator.

by local conditions. How does the water transported eastward by this undercurrent return to the western side of the ocean basin? In an attempt to answer this question, Philander (1971) used a model similar to that of Stommel (1948) to study the oceanic circulation in an ocean basin that spans the equator. He showed that, for Charney's (1960) range of parameters, the flow within a distance $L_1 = (\tau_0^2 R^3/8\Omega^3 v_v)^{1/5}$ of the equator is indeed determined by local conditions. This, however, is not true of a more diffuse current of width $L_2 = (v_v RL^2/2\Omega H^2)^{1/3}$ in which the equatorial currents are embedded (see Fig. 17). The flow in this wider current, which returns the transport of the undercurrent to the western side of the basin, depends on conditions at the equator *and* at higher latitudes. By solving for flow patterns corresponding to different wind conditions, Philander (1971) showed that this current is necessary to close the circulation in nonequatorial latitudes, even when there is no Equatorial Undercurrent. A western boundary current provides the link between the flow near the equator and the flow at higher latitudes.

In Philander's (1971) model, the stress imposed on the ocean surface is transferred to the ocean floor. Gill (1975) has shown that in a model in which lateral friction is dominant, a boundary layer of width $(v_H LR/2\Omega)^{1/4}$ effects the transition from equatorial to extra-equatorial currents.

Though we have considerable information about individual currents in the tropics, little is known about the manner in which the circulation is closed. The results of the models described above are not inconsistent with the gross features of the observed circulation. Western boundary currents do provide a link between the currents at different latitudes and transfer water from the North and South Equatorial Currents to the eastward Equatorial Undercurrent and Equatorial Countercurrents. The North Equatorial Currents also provide water to the Gulf Stream (via the Gulf of Mexico) and Kuroshio Current. The eastward currents in the tropics lose water gradually as they proceed downstream, and have only small transports when they reach the eastern boundaries of the ocean basins. [The disappearance of the upwelling region off the coast of Peru is apparently correlated with unusually large transports of the North Equatorial Countercurrent (Wyrtki, 1973b; Namias, 1973), but it has not been established that this eastward transport of relatively warm water directly causes the advent of El Niño.] There is little information about the origin of the westward South Equatorial Current; perhaps we could identify that part of it which is within 3° latitude of the equator with the current of width L_2 in Fig. 17. The origin of the Intermediate Equatorial Current, and the role of the cold, northward flowing Benguela and von Humboldt Currents, also are poorly understood.

The circulation of a stratified ocean in a closed basin is yet to be investigated. Questions concerning the origin and fate of the equatorial currents cannot be answered with Philander's (1973a) model, because explicit zonal variations are eliminated by means of a similarity transformation. The extra-equatorial flow has to be specified so as to provide boundary conditions for this model. We have seen that in the case of a constant-density model, a special transition boundary layer has to be introduced near the equator in order to close the oceanic circulation. It would be of considerable interest to know how such a boundary layer affects the flow at the equator in a stratified model.

5. Instabilities of Tropical Currents

Descriptions of the Equatorial Undercurrent in the Pacific Ocean usually emphasize its symmetry about the equator in the spring and early summer, and its steadiness at all times of the year. In April 1971, however, Taft et al. (1974) evidenced an apparent northward movement of the core of the Equatorial Undercurrent at 150°W that could be interpreted as a meander. (They found that the zonal velocity at a depth of 102 m at 1°N, 150°W increased from 70 to 160 cm/sec over a period of 20 days; if it is assumed that this change is representative of changes in the zonal velocity profile at different latitudes, then the implied increase in transport is unacceptably large.)

Indications that the undercurrent in the Atlantic Ocean may meander are more numerous. Buchanan (1886), the discoverer of this current, noted that it flows in a southeastward direction. During the Equalant experiments in the 1960s, the high-salinity core at the equator, which is believed to be closely associated with the high-velocity core of the Equatorial Undercurrent, was found to have a meridional oscillation with a period between 2 and 3 weeks on several occasions (Rinkel, 1969). Preliminary results of GATE document a meander of the Equatorial Undercurrent with a similar period and a wavelength in excess of 2000 km.

It has recently been proposed (Philander, 1976) that these meanders of the Equatorial Undercurrent may be a consequence of instabilities associated with the large horizontal shear of the Undercurrent (in the Pacific Ocean), or of the surface currents (in the Atlantic Ocean). A model in which the ocean is assumed to consist of two layers of fluid of different density is used to study the influence of the β effect and equatorial divergence on the stability of the currents in the tropics. The importance of these effects is proportional to the ratio of the width of the currents L to the length scales $l_1 = \sqrt{U/\beta}$ and l_2 (the radius of deformation), respectively. It is found that both the β effect and divergence stabilize eastward currents, but both can destabilize westward currents.

The Equatorial Undercurrent has a (half) width L of approximately 150 km. The equatorial radius of deformation is 325 km and the length scale $\sqrt{U/\beta} = 250$ km. It follows that both the β effect and equatorial divergence stabilize the Equatorial Undercurrent so that instabilities of this current are unlikely in the Atlantic and Indian Oceans. Instabilities are possible in the central Pacific in March and April for the current is sufficiently intense there at that time. A stability analysis shows that for conditions similar to those measured by Taft et al. (1974), the undercurrent is unstable with a wavelength of about 900 km. Disturbances antisymmetric about the equator have the largest growth rate so that the waves cause the undercurrent to meander about the equator, and cause its core to oscillate in a meridional plane with a period of about 40 days.

The surface currents in the tropics are usually less intense, and they usually have smaller horizontal shears than the Equatorial Undercurrent. These currents, particularly the westward South Equatorial Current and the adjacent eastward North Equatorial Countercurrent, are nonetheless more likely to be unstable because both the β effect and equatorial divergence destabilize this current system. (The width of the region of maximum shear is approximately $L = 400$ km, the radius of deformation is 325 km, and the length scale $\sqrt{U/\beta} = 200$ km.) According to a stability analysis, westward propagating waves with a period between 2 and 3 weeks and a wavelength in excess of 2000 km are possible. The disturbances are antisymmetric about the equator and could cause the undercurrent (below the surface currents) to meander about the equator. These predictions are in reasonable agreement with the measurements in the Atlantic Ocean. The Atlantic, however, is barely 5000 km wide at the equator. Further studies are necessary to determine the effect of the coasts on the unstable waves. There are, as yet, no measurements that indicate that the surface currents in the Pacific are unstable. September and October would be the most likely months for the occurrence of such instabilities, for at that time the westward surface flow at the equator is most intense, and the neighboring eastward North Equatorial Countercurrent is also at its most intense (Wyrtki, 1974a; Tsuchiya, 1974). Further measurements are necessary to determine whether the horizontal shear is sufficiently large for the occurrence of unstable waves.

Acknowledgments

We wish to acknowledge the support of the National Science Foundation, the Atmospheric Sciences Section of the Division of Environmental Sciences (S.G.H.P.), and the Environmental Forecasting Program of the International Decade of Ocean Exploration (D. W. M.). We also wish to thank Ms. Marilyn Baker for her technical assistance in preparing the manuscript.

References

Anderson, D. L. T. and P. B. Rowlands, 1976. The Somali Current response to the South West Monsoon: The relative importance of local and remote forcing. Submitted to *J. Mar. Res.*

Atkinson, G. D., 1971. Forecaster's guide to tropical meteorology. *Tech. Rept. 240, Air Weather Serv. (MAC), U.S. Air Force.*

Buchanan, J. Y., 1886. On similarities in the physical geography of the great oceans. *Proc. Roy. Geogr. Soc. London*, **8**, 753–770.

Cane, M., 1975. A study of the wind-driven ocean circulation in an equatorial basin. Ph.D. thesis, Massachusetts Institute of Technology, Cambridge, Mass.

Charney, J. G., 1960. Non-linear theory of a wind-driven homogeneous layer near the equator. *Deep-Sea Res.*, **6**, 303–310.

Charney, J. G. and S. L. Spiegel, 1971. The structure of wind-driven equatorial currents in homogeneous oceans. *J. Phys. Oceanogr.*, **1**, 149–160.

Colin, C., C. Henin, P. Hisard, and C. Oudot, 1971. Le Courant de Cromwell dans le Pacifique central en février. *Cah. ORSTOM Ser. Oceanogr.*, **7**, 167–186.

Fofonoff, N. P. and R. B. Montgomery, 1955. The Equatorial Undercurrent in the light of the vorticity equation. *Tellus*, **7**, 518–521.

Gill, A. E., 1971. The Equatorial Current in a homogeneous ocean. *Deep-Sea Res.*, **18**, 421–431.

Gill, A. E., 1975. Models of Equatorial Currents. *Proc. Symp. Numer. Models Ocean Circ., Natl. Acad. Sci. Durham, Oct. 17–20, 1972.*

Hisard, P. and P. Rual, 1970. Courant Equatorial Intermediare de l'Ocean Pacifique et contre-courant adjacent. *Cah. ORSTOM Ser. Oceanogr.*, **8**, 21–45.

Hisard, P., J. Merle, and B. Voituriez, 1970. The Equatorial Undercurrent at 170°E in March and April 1967. *J. Mar. Res.*, **28**, 281–303.

Jones, J. H., 1973. Vertical mixing in the Equatorial Undercurrent. *J. Phys. Oceanogr.*, **3**, 286–296.

Kendall, T. R., 1970. *The Pacific Equatorial Countercurrent*. International Center for Environmental Research, Laguna Beach, Calif., 19 pp., 48 figs.

Knauss, J. A., 1960. Measurements of the Cromwell Current. *Deep-Sea Res.*, **6**, 265–286.

Knauss, J. A., 1963. The Equatorial Current Systems. In *The Sea*. Vol. 1. Wiley-Interscience, New York, pp. 235–252.

Knox, R. A., 1974. Reconnaissance of the Indian Ocean Equatorial Undercurrent near Addu Atoll. *Deep-Sea Res.*, **21**, 123–129.

Lighthill, M. J., 1969. Dynamic response of the Indian Ocean to the onset of the Southwest Monsoon. *Phil. Trans. Roy. Soc. London*, *A*, **265**, 45.

Longuet-Higgins, M. S., and G. S. Pond, 1970. The free oscillations of fluid on a hemisphere bounded by meridians of longitude. *Phil. Trans. Roy. Soc.*, **266** (1174), 193–223.

Matsuno, T., 1966. Quasi-geostrophic motions in the equatorial area. *J. Meteorol. Soc. Jap.*, **44**, 25–43.

McKee, W. D., 1973. The wind-driven equatorial circulation in a homogeneous ocean. *Deep-Sea Res.*, **20**, 889–899.

Mikhailova, E. N., A. I. Felzenbaum, and N. R. Shapiro, 1967. A contribution to the non-linear theory of currents at the equator. *Dokl. Akad. Nauk. SSSR*, **175**, 574–577.

Mofjeld, H. and M. Rattray, 1975. Barotropic Rossby waves in a zonal current: Effects of lateral viscosity. To be published in *J. Phys. Oceanogr.* **5** (3).

Moore, D., 1968. Planetary-gravity waves in an equatorial ocean. Ph.D. thesis, Harvard University, Cambridge, Mass.

Namias, J., 1973. Response of the Equatorial Countercurrent to the subtropical atmosphere. *Science*, **181**. 1244–1245.

O'Brien, J. J. and H. E. Hurlburt, 1974. Equatorial jet in the Indian Ocean: theory. *Science*, **184**, 1075–1077.

Philander, S. G. H., 1971. The equatorial dynamics of a shallow homogeneous ocean. *Geophys. Fluid Dyn.*, **2**, 219–245.

Philander, S. G. H., 1973a. The equatorial thermocline. *Deep-Sea Res.*, **20**, 69–86.

Philander, S. G. H., 1973b. Equatorial Undercurrent: Measurements and theories. *Rev. Geoph. Space Phys.*, **2** (3), 513–570.

Philander, S. G. H., 1976. Instabilities of Zonal Equatorial Currents. *J. Geophys. Res.*, **81** (21), 3725–3735.

Rinkel, M. O., 1969. Some features of relationships between the Atlantic Equatorial Undercurrent and its associated salinity core. In *Proc. Symp. Oceanogr. Fish. Resources Trop. Atl.* C. I. Abidjan, ed. UNESCO, Paris, pp. 193–212.

Robinson, A. R., 1960. The general thermal circulation in the equatorial regions. *Deep-Sea Res.*, **6**, 311–317.

Robinson, A. R., 1966. An investigation into the wind as the cause of the Equatorial Undercurrent. *J. Mar. Res.*, **24**, 179–204.

Robinson, A. R. and P. Welander, 1963. Thermal circulation on a rotating sphere, with application to the oceanic thermocline. *J. Mar. Res.*, **21**, 25–38.

Rossby, C. G. et al., 1939. Relation between variations in intensity of the zonal circulation of the atmosphere and displacements of the semi-permanent centers of actions. *J. Mar. Res.*, **2**, 38–55.

Sarkisyan, A. S., 1969. Deficiencies of barotropic models of oceanic circulation. *Izv. Acad. Sci. USSR, Atm. Oceanic Phys.*, **5**, 466–474.

Stommel, H., 1948. The westward intensification of wind-driven ocean currents. *EOS Trans. AGU*, **29**, 202–206.

Sturm, M. and K. Voigt, 1966. Observations of the structure of the Equatorial Undercurrent in the Gulf of Guinea in 1964. *J. Geophys. Res.*, **71**, 3105–3108.

Sverdrup, H. U., 1947. Wind-driven currents in a baroclinic ocean: With applications to the Equatorial currents of the Eastern Pacific. *Proc. Natl. Acad. Sci.*, **33**, 318–326.

Taft, B. A. and J. A. Knauss, 1967. The Equatorial Undercurrent of the Indian Ocean as observed by the Lusiad expedition. *Bull. Scripps Inst. Oceanogr.*, **9**.

Taft, B. A., B. M. Hickey, C. Wunsch, and D. Baker, 1974. Equatorial Undercurrent and deeper flows in the Central Pacific. *Deep-Sea Res.*, **21**, 403–430.

Tsuchiya, M., 1974. Variation of the surface geostrophic flow in the eastern Intertropical Pacific Ocean. *Fisheries Bull.*, **42** (4), 1075–1086.

Veronis, G., 1973. Large scale ocean circulation. In *Advances in Applied Mechanics*. Vol. 13. Academic Press, New York.

Williams, R. B. and C. H. Gibson, 1974. Direct measurements of turbulence in the Pacific Equatorial Undercurrent. *J. Phys. Oceanogr.*, **4** (1), 104–108.

Wyrtki, K., 1973a. Teleconnections in the Equatorial Pacific Ocean. *Science*, **180**, 66–68.

Wyrtki, K., 1973b. An equatorial jet in the Indian Ocean. *Science*, **181**, 262–264.

Wyrtki, K., 1974a. Sea level and the seasonal fluctuations of the Equatorial Currents in the Western Pacific Ocean. *J. Phys. Oceanogr.*, **4**, 91–103.

Wyrtki, K., 1974b. Equatorial Currents in the Pacific 1950 to 1970 and their relations to the Trade Winds. *J. Phys. Oceanogr.*, **4**, 372–380.

Yoshida, K., 1959. A theory of the Cromwell Current and of the equatorial upwelling. *J. Oceanogr. Soc. Jap.*, **15**, 154–170.

9. THE DIAGNOSTIC CALCULATIONS OF A LARGE-SCALE OCEANIC CIRCULATION

A. S. SARKISYAN

1. Initial equations, Their Simplifications and Transformations

A. Initial Equations and Boundary Conditions

The calculation of currents in the ocean requires an appropriate system of equations and boundary conditions. In studies of a large-scale circulation one must take into account sphericity of the earth; however, the Cartesian coordinates are simpler and more obvious. Therefore most equations and transformations are given in a Cartesian coordinate system and are applied to the northern hemisphere,[1] but some necessary equations are also given in a spherical system of coordinates. In Section 2.H we list the necessary equations and formulas for the southern hemisphere and in a spherical coordinate system. Thus we proceed from the following system of equations of thermo-hydrodynamics of the ocean. (The accepted designations are defined in the List of Symbols.) The equations of motion are as follows:

$$\frac{\partial u}{\partial t} + u\frac{\partial u}{\partial x} + v\frac{\partial u}{\partial y} + w\frac{\partial u}{\partial z} - lv = -\frac{1}{\rho_0}\frac{\partial p}{\partial x} + v\frac{\partial^2 u}{\partial z^2} + A_l\Delta u \tag{1}$$

$$\frac{\partial v}{\partial t} + u\frac{\partial v}{\partial x} + v\frac{\partial v}{\partial y} + w\frac{\partial v}{\partial z} + lu = -\frac{1}{\rho_0}\frac{\partial p}{\partial y} + v\frac{\partial^2 v}{\partial z^2} + A_l\Delta v \tag{2}$$

The equation of statics is

$$\frac{\partial p_1}{\partial z} = \rho_1 g \tag{3}$$

The equation of continuity of an incompressible fluid is

$$\frac{\partial u}{\partial x} + \frac{\partial v}{\partial y} + \frac{\partial w}{\partial z} = 0 \tag{4}$$

The equations of transport of heat and salt are

$$\frac{\partial T}{\partial t} + u\frac{\partial T}{\partial x} + v\frac{\partial T}{\partial y} + w\frac{\partial T}{\partial z} = \kappa_T\frac{\partial^2 T}{\partial z^2} + A_T\Delta T \tag{5}$$

$$\frac{\partial s}{\partial t} + u\frac{\partial s}{\partial x} + v\frac{\partial s}{\partial y} + w\frac{\partial s}{\partial z} = \kappa_S\frac{\partial^2 s}{\partial z^2} + A_s\Delta s \tag{6}$$

The equation of state in the Eckart form (Bryan, 1969b; Eckart, 1958) is

$$\rho_1 = \frac{p_1 + p_2}{1.000027[\chi + \alpha_0(p_1 + p_2)]} \tag{7}$$

The accepted designations are as follows: p_1 and ρ_1 are pressure and density; $p = p_1 - \rho_0\rho z$ and $\rho = \rho_1 - \rho_0$ are anomalies of pressure and density; $\alpha_0 = 0.698$;

$$\chi = 1779.5 + 11.25T - 0.0745T^2 - (3.8 + 0.01T)s \tag{8}$$

$$p_2 = 5890 + 38T - 0.375T^2 + 3s \tag{9}$$

363

Practically, we can substitute p_1 for the resting fluid column pressure in formula 7 (Bryan, 1969b). The system of equations 1–7 contains seven unknown functions: the components of current velocity u, v, and w, anomalies of pressure p, density ρ, and temperature and salinity T and s. For solving some problems of nonstationary circulation it is sufficient to have initial values for four of these functions, u, v, T, and s, from which one can determine the other three initial fields. The boundary conditions are as follows. At the ocean surface with $z = -\zeta_1(x, y, t)$

$$p = p_a \tag{10}$$

$$\rho_0 \nu \frac{\partial u}{\partial z} = -\tau_x, \qquad \rho_0 \nu \frac{\partial v}{\partial z} = -\tau_y \tag{11}$$

$$w = -\left(\frac{\partial \zeta_1}{\partial t} + u \frac{\partial \zeta_1}{\partial x} + v \frac{\partial \zeta_1}{\partial y}\right) \tag{12}$$

$$\frac{\partial T}{\partial z} = Q \tag{13}$$

or

$$T = T(x, y, t) \tag{14}$$

$$\frac{\partial s}{\partial z} = Q_1 \tag{15}$$

or

$$s = s(x, y, t) \tag{16}$$

At the ocean bottom with $z = H(x, y)$

$$u = v = w = 0 \tag{17}$$

or

$$\frac{\partial u}{\partial n} = \frac{\partial v}{\partial n} = 0 \tag{18}$$

and

$$w_H = u_H \frac{\partial H}{\partial x} + v_H \frac{\partial H}{\partial y} \tag{19}$$

$$\frac{\partial T}{\partial n} = \frac{\partial s}{\partial n} = 0 \tag{20}$$

or

$$T = T_H, \qquad s = s_H \tag{21}$$

where n is the normal to the bottom.

Now we speak of the boundary conditions over the horizontal. Generally u and v must be prescribed as functions of coordinates and time on the liquid part of lateral boundaries; on the solid part the conditions of sticking are accepted. However, most problems are solved without considering the lateral eddy viscosity; then it is sufficient to assign the velocity component normal to the boundary. In the studies of currents of large and middle scale the value of velocity averaged over the height can be prescribed without introducing large error. The boundary of the field is usually approximated by a broken line, each section of which is parallel to some axis of coordinates. Thus for u, v the boundary condition is, in fact, laid down in the form

$$\frac{1}{H}\int_0^H u\,dz = U_1, \qquad \frac{1}{H}\int_0^H v\,dz = V_1 \tag{22}$$

Sometimes we consider $U_1 = V_1 = 0$ on the liquid part of the boundary and on island contours. The more precise version of the boundary condition over the horizontal for u, v is discussed in Section 1.E. We have for temperature and salinity

$$\frac{\partial T}{\partial N} = Q', \qquad \frac{\partial s}{\partial N} = Q'_1 \tag{23}$$

or

$$T = T_b, \qquad s = s_b \tag{24}$$

where N is the normal to the lateral boundary. On the solid part of the lateral boundary $Q' = Q'_1 = 0$.

Now let us write some initial relations in a spherical coordinate system. The equations of motion are

$$\frac{\partial v_\lambda}{\partial t} + \frac{v_\theta}{a}\frac{\partial v_\lambda}{\partial \theta} + \frac{v_\lambda}{a\sin\theta}\frac{\partial v_\lambda}{\partial \lambda} + v_z\frac{\partial v_\lambda}{\partial z} + \frac{v_\theta v_\lambda}{a}\cot\theta$$
$$+ 2\omega\cos\theta v_\theta = -\frac{1}{\rho_0 a\sin\theta}\frac{\partial p}{\partial \lambda} + v\frac{\partial^2 v_\lambda}{\partial z^2} + \frac{A_l}{a^2}\Delta v_\lambda \tag{25}$$

$$\frac{\partial v_\theta}{\partial t} + \frac{v_\theta}{a}\frac{\partial v_\theta}{\partial \theta} + \frac{v_\lambda}{a\sin\theta}\frac{\partial v_\theta}{\partial \lambda} + v_z\frac{\partial v_\theta}{\partial z} - \frac{v_\lambda^2\cot\theta}{a}$$
$$- 2\omega\cos\theta v_\lambda = -\frac{1}{\rho a}\frac{\partial p}{\partial \theta} + v\frac{\partial^2 v_\theta}{\partial z^2} + \frac{A_l}{a^2}\Delta v_\theta \tag{26}$$

The equation of continuity of incompressible fluid is

$$\frac{\partial v_z}{\partial z} + \frac{1}{a\sin\theta}\left(\frac{\partial}{\partial \theta}\sin\theta v_\theta + \frac{\partial v_\lambda}{\partial \lambda}\right) = 0 \tag{27}$$

The equation of heat transfer is

$$\frac{\partial T}{\partial t} + \frac{v_\lambda}{a\sin\theta}\frac{\partial T}{\partial \lambda} + \frac{v_\theta}{a}\frac{\partial T}{\partial \theta} + v_z\frac{\partial T}{\partial z} = \kappa_T\frac{\partial^2 T}{\partial z^2} + \frac{A_T}{a^2}\Delta T \tag{28}$$

The equation for salt transfer is quite similar to the latter. The boundary condition on equation 19 for v_z in a spherical coordinate system is in the form

$$v_z = \frac{v_\theta^{(H)}}{a}\frac{\partial H}{\partial \theta} + \frac{v_\lambda^{(H)}}{a\sin\theta}\frac{\partial H}{\partial \lambda} \tag{29}$$

The Laplace operator is

$$\Delta = \frac{1}{\sin\theta}\left(\frac{\partial}{\partial \theta}\sin\theta\frac{\partial}{\partial \theta} + \frac{1}{\sin\theta}\frac{\partial^2}{\partial \lambda^2}\right) \tag{30}$$

B. Simplifications for Equations and Boundary Conditions for Large-Scale Stationary or Seasonal Currents

Let us transform the equation of statics (equation 3); we integrate from ζ_1 to z with due regard for boundary condition 10:

$$p_1 = p_a + g \int_{-\zeta_1}^{z} \rho_1 \, dz = p_a + g \int_{-\zeta_1}^{0} \rho \, dz + g \int_{0}^{z} \rho_1 \, dz$$

$$\approx p_a + g \int_{-\zeta}^{0} \rho_0 \, dz + g \int_{0}^{z} (\rho_0 + \rho) \, dz = p_a + \rho_0 g \zeta_q + \rho_0 g z + g \int_{0}^{z} \rho \, dz$$

Having excluded the pressure of a uniform fluid column from the discussion we have

$$p = p_a + \rho_0 g \zeta_1 + g \int_{0}^{z} \rho \, dz \tag{31}$$

It is more convenient to use a relative level ζ instead of a physical one ζ_1:

$$\zeta = \zeta_1 + \frac{p_a}{\rho_0 g} \tag{32}$$

Thus equation 3 and boundary condition 10 are substituted for a simple formula for the pressure anomaly:

$$p = \rho_0 g \zeta + g \int_{0}^{z} \rho \, dz \tag{33}$$

To estimate the orders of magnitudes of terms in some initial relations we turn to the dimensionless magnitudes. We should note that there are many cases of transition to dimensionless magnitudes described in the scientific literature and, it seems, we do not need to do this here. However, after we analyze dimensionless magnitudes we show below the sense and originality of the relations obtained. We have

$$x = L_0 \bar{x}, \qquad y = L_0 \bar{y}, \qquad z = h_0 \bar{z}, \qquad u = v_0 \bar{u}, \qquad v = v_0 \bar{u}, \qquad w = w_0 \bar{w}$$
$$t = t_0 \bar{t}, \qquad p = \mathscr{P}_0 \bar{p}, \qquad \rho = (\delta\rho)_0 \bar{\rho}, \qquad \zeta = \zeta_0 \bar{\zeta}, \qquad l = l_0 \bar{l}, \qquad \beta = \beta_0 \bar{\beta}$$

Firstly, let us treat large-scale motions in the baroclinic layer of a basin away from the equator. In this case, the geostrophic balance occurs; that is, the acceleration of the Coriolis force and pressure gradient are of the same order of magnitude:

$$l_0 v_0 = \frac{1}{\rho_0} \frac{\mathscr{P}_0}{L_0} \tag{34}$$

In addition, we use the following assumptions:

1. All terms in equation 4 are of the same order of magnitude:

$$\frac{v_0}{L_0} = \frac{w_0}{h_0} \tag{35}$$

2. All nonlinear terms in the equation of heat transfer are of the same order of magnitude; this is also true for the equation of salt transfer and for the equation of motion. These statements serve only to ascertain the validity of relation 35 and do not result in new formulas.

3. Terms with local derivatives by time coincide with nonlinear terms of the equations in order of magnitude:

$$\frac{1}{t_0} = \frac{v_0}{L_0} \tag{36}$$

4. Finally, it is quite natural to consider all the terms of formula 33 to be of the same order of magnitude:

$$\mathscr{P}_0 = \rho_0 g \zeta_0 = g(\delta\rho)_0 h_0 \tag{33a}$$

Relations 34–36 and 33a lead to the following formulas:

$$v_0 = \frac{g h_0 (\delta\rho)_0}{\rho_0 L_0 l_0}, \qquad w_0 = \frac{g(\delta\rho)_0}{\rho_0 l_0}\left(\frac{h_0}{L_0}\right)^2$$

$$t_0 = \frac{\rho_0 l_0 L_0{}^2}{g h_0 (\delta\rho)_0}, \qquad \mathscr{P}_0 = g h_0 (\delta\rho)_0, \qquad \zeta_0 = \frac{{}^p(\delta\rho)_0}{\rho_0} h_0 \tag{37}$$

In the above relations h_0 is the characteristic depth of the baroclinic layer of the ocean. Let us take $h_0 = 500$ m $= 5 \times 10^4$ cm. Having prescribed also the characteristic magnitudes $L_0 = 10^3$ km $= 10^8$ cm, $(\delta\rho)_0 = 10^{-3}$ g/cm³, $l_0 = 10^{-4}$/sec, $g = 10^3$ cm/sec² in formulas 37, we obtain (all magnitudes are in cgs units)

$$\mathscr{P}_0 = 5 \times 10^4, \qquad \zeta_0 = 50$$
$$v_0 = 5, \qquad w_0 = 2.5 \times 10^{-3}, \qquad t_0 = 2 \times 10^7 \tag{38}$$

If we pass on to dimensionless magnitudes and divide the equations of motion by $l_0 v_0$ and the equations of heat and salt transfer by $v_0/L_0 = w_0/h_0$, then small dimensionless multipliers appear before some terms. For instance, equation 1 takes the form

$$R_0\left(\frac{\partial\bar u}{\partial t} - \bar u\frac{\partial\bar u}{\partial\bar x} - \bar v\frac{\partial\bar u}{\partial\bar y} + \bar w\frac{\partial\bar u}{\partial\bar z}\right) - l\bar v = -\frac{\partial\bar p}{\partial\bar x} + E_v\frac{\partial^2\bar u}{\partial\bar z^2} + E_m\bar\Delta\bar u \tag{39}$$

where R_0 is the Rossby (Kibel) number:

$$R_0 = \frac{v_0}{l_0 L_0} \tag{40}$$

and E_v and E_m are the Ekman numbers for the vertical and horizontal turbulent viscosities, respectively:

$$E_v = \frac{v}{l_0 h_0{}^2}, \qquad E_m = \frac{A_l}{L_0{}^2 l_0} \tag{41}$$

The dimensionless equation of heat transfer is in the form

$$\frac{\partial\bar T}{\partial t} + \bar u\frac{\partial\bar T}{\partial\bar x} + \bar v\frac{\partial\bar T}{\partial\bar y} + \bar w\frac{\partial\bar T}{\partial\bar z} = \frac{1}{P_e}\frac{\partial^2\bar T}{\partial\bar z^2} + \frac{1}{\tilde P_e}\bar\Delta\bar T \tag{42}$$

where P_e and $\tilde P_e$ are, respectively, the Peclet numbers for the vertical and horizontal diffusion:

$$P_e = \frac{w_0 h_0}{\kappa_T}, \qquad \tilde P_e = \frac{v_0 L_0}{A_T} \tag{43}$$

The dimensionless analogues for equations 2 and 6 are identical with equations 39 and 42, so we do not list them here. The dimensionless multipliers appearing in these equations are small. For instance, at the values of characteristics given in equation 38 and coefficients of turbulence $v = 10^2$, $A_l = 10^7$, $\kappa_T = 1$, and $A_T = 10^6$, we have $R_0 = 5 \times 10^4$, $E_v = 5 \times 10^{-4}$, $E_m = 10^{-5}$, $P_e^{-1} = 10^{-2}$, and $\tilde{P}_e^{-1} = 2 \times 10^{-3}$. This means that the geostrophic balance is of dominant importance in equations of motion, and that in equations 5 and 6 the left sides are greater than the right sides. Turbulence plays an appreciable role only in the corresponding thin boundary layers. Let us note that even in intense currents like the Gulf Stream the Rossby number is not large. For instance, with $v_0 = 50$, $L_0 = 50$ km $= 5 \times 10^6$ cm; we have $R_0 = 0.1$; that is, even in the investigation of jet streams, the consideration of non-linear terms by the method of successive approximations is permissible in equations 1 and 2.

Let us discuss the characteristic value of w. Since the geostrophic balance dominates in the equations of motion, then, strictly speaking, we cannot consider all terms of the equation of continuity to be of the same order of magnitude. If we substitute u and v in equation 4 by

$$u_g = -\frac{1}{\rho_0 l}\frac{\partial p}{\partial y}, \qquad v_g = \frac{1}{\rho_0 l}\frac{\partial p}{\partial x} \tag{44}$$

then we obtain

$$\frac{\partial w}{\partial z} = -\frac{\beta}{l}v_g$$

That is, the characteristic value of w_0 in the baroclinic layer

$$w_0 = \frac{\beta_0 v_0 h_0}{l_0} = 10^{-3} \tag{45}$$

is several times less than the value obtained above ($\beta_0 = 2 \times 10^{-13}$). At great depths the characteristic value of the vertical velocity is determined from boundary condition 19. Taking the characteristic drop in the bottom relief $H_0 = 1$ km, the characteristic velocity $v_H = 1$ cm/sec, and characteristic horizontal scale $L_H = 500$ km, we obtain $w_H = v_H H_0/L_H = 0.5 \times 10^{-2}$ cm/sec, which on the contrary is several times greater than the value of w_0 given in equation 38. Thus the value of w_0 obtained when using formula 37 occupies the middle position between its value in the baroclinic and deep layers of the ocean. Taking into account that the relation 35 leads to a simple form of equations 39 and 42, and the change of w_0 several times does not change the qualitative conclusions, we keep formula 37 valid for the calculation of w_0. Finally, let us estimate the value of w determined from boundary condition 12. We substitute for v_0 the characteristic value of purely drift current velocity as large as $v^d = 25$ cm/sec. Despite this $w^d = v^d \zeta_c/L_0 = 10^{-5}$, which is less than the above-obtained characteristic values of the vertical velocity by at least two orders of magnitude. Therefore equation 12 can be substituted for the simple condition of the "rigid lid"; $w = 0$ at $z = 0$. Formulas 37 can also be used in the estimation of characteristic values of a large-scale deep-water circulation, substituting H for h_0, L_H for L_0, and reducing $(\delta\rho)_0$ by one order of magnitude. Then the characteristic values prove to be close to their values in the baroclinic layer of the ocean: v_0 somewhat decreases; w_0 increases; $\mathscr{P}_0$ decreases; etc.

In closing, let us describe the basic characteristic feature of formulas 37. To determine v_0, w_0, t_0, $\mathscr{P}_0$, and ζ_0 from these formulas, besides the external constants the internal characteristics of the baroclinic ocean $(\delta\rho)_0$ and h_0 must be given. The unknown values could be expressed completely by the external characteristics using the boundary conditions, but we consider it inexpedient. Boundary conditions of type 11 and 13 are applicable only to the estimation of characteristics in the upper boundary layer which is ≈ 10 m thick, but not for the baroclinic layer. On the other hand, the characteristic values $(\delta\rho)_0$ and h_0 are known with a degree of accuracy that is not worse than the accuracy of the values for tangential wind friction or inflows of heat and salts. In particular, we can determine h_0 using only an atlas of hydrological elements, for instance, by using the criterion that at the lower boundary of the baroclinic layer gradients of density are of magnitude less than their values at the ocean surface by one order. The use of formulas 37 underlines the circumstance that the baroclinity of seawater is a decisive factor in the dynamics of sea currents. Therefore all characteristic values of a large-scale current must be estimated not through the tangential wind friction but through the thermal factors, in the given case through $(\delta\rho)_0$. We support this statement below by concrete calculations. The estimates mentioned make it possible to introduce one simplification more. After substituting equation 3 for formula 31 or 33 and condition 12 for the condition of the "rigid lid" we can relate all boundary conditions 11 and 13–16 to the undisturbed surface of the ocean: $z = 0$.

We have seen that formulas 37 are valid for estimating the characteristic values not only for large-scale, but also for mesoscale intensive currents with a characteristic horizontal scale of 50 km. The equatorial currents are an exception. At the equator $l = 0$, and in the narrow band near the equator which is several hundred kilometers wide $\sin \varphi \ll 1$; therefore the geostrophic balance is not present. Here pressure gradients are basically compensated by nonlinear inertial terms. Equating the characteristic values of these terms to equation 2, we have

$$\frac{v_0}{t_0} = \frac{v_0^2}{L_0} = \frac{w_0 v_0}{h_0} = \frac{1}{\rho_0}\frac{\mathscr{P}_0}{L_0} \tag{46}$$

The characteristic values of $\mathscr{P}_0$ and ζ_0, as above, are calculated from equation 37. Thus we obtain for the equatorial zone the following formulas:

$$\mathscr{P}_0 = gh_0(\delta\rho)_0, \qquad \zeta_0 = \frac{(\delta\rho)_0}{\rho_0} h_0,$$

$$v_0 = \sqrt{\frac{gh_0(\delta\rho)_0}{\rho_0}}, \qquad w_0 = \frac{h_0}{L_0}\sqrt{\frac{gh_0(\delta\rho)_0}{\rho_0}}, \tag{47}$$

$$t_0 = \frac{L_0}{v_0} = L_0\sqrt{\frac{\rho_0}{gh_0(\delta\rho)_0}}$$

We take the characteristic values of L and h_0 here to be one order of magnitude less than those for large-scale currents away from the equator, taking into account that here we have a narrow jet of intense currents. This narrow jet is localized near the equator and stretches through the whole ocean; therefore it may seem that we must choose various characteristic scales for x and y. However, the jet meanders near the equator, deflecting now to the north, now to the south under the action of inertial forces (Kolesnikov et al., 1968; Monin, 1972, 1973); therefore we can keep the same characteristic scale L_0 for both directions. As is known from observations the equatorial currents abruptly change with depth, and their direction can reverse repeatedly (Kolesnikov et al., 1968; Kort et al., 1969). Therefore for h_0 we choose the value which is less by one order of magnitude than the depth of the baroclinic layer for mid-latitudes. Now $(\delta\rho)_0$ is also less than in mid-latitudes, but not by one order of magnitude. Hence we accept for the equatorial currents $L_0 = 10^7$, $h_0 = 5 \times 10^4$, and $(\delta\rho)_0 = 10^{-3}$; and from formulas 47 we obtain

$$\mathscr{P}_0 = 5 \times 10^3, \quad \zeta_0 = 5, \quad v_0 = 70, \quad w_0 = 3.5 \times 10^{-2}, \quad t_0 = 1.4 \times 10^5 \quad (48)$$

Comparing equations 48 with 38 we note the following. The anomalies of pressure and sea-surface topography are less than their corresponding values for mid-latitudes by one order of magnitude. However, the characteristic horizontal scale is also one order of magnitude less; therefore pressure gradients and slopes of the surface level are of the same order here as in mid-latitudes. Hence the same pressure gradient gives a velocity near the equator of one order of magnitude greater than in mid-latitudes. At these values of L_0 and v_0 it appears that t_0 is less by nearly two orders of magnitude than in mid-latitudes. It is possible that at the equator derivatives over time represent a small difference between large values, and therefore we cannot equate v_0/t_0 with other terms.

   Further, in equations 47, as in equations 37, the baroclinity of seawater is a decisive factor, but here the anomaly of density results in a greater value for the velocity of motion. At the equator it is necessary to take the inertial terms into account, for here they play a more important role than in jet currents in mid-latitudes.

   Now we consider the role of the turbulent viscosity at the equator. If we construct the analogue of equation 39 for the equator, then the term generated by the vertical turbulent viscosity possesses the multiplier π/P_e, where $\pi = v/\kappa$ is the Prandtl number, and P_e is the Peclet number. Taking into account the smallness of the characteristic vertical scale we let $v = 10$ cm^2/sec. This is not an underestimated but rather an overestimated value of v. Let us refer to the recent work (Bryan and Cox, 1967, 1968; Cox, 1970; Holland, 1973) in which even for the large-scale circulation v is taken equal to 10 or even to unity. At $v = 10$ and other characteristic values in equations 48 we obtain $\pi/P_e = 0.1$. Thus in the equatorial part the vertical turbulent viscosity is one order of magnitude less than inertial terms and pressure gradients, but the role of this factor is greater at the equator than in mid-latitudes. We can easily see in a similar way that the role of the lateral friction at the equator at $A_l \leqslant 10^7$ is one order of magnitude less than the role of the vertical viscosity.

C. Derivation of Equations for Auxiliary Functions

   In the preceding estimation of characteristic values we have not chosen the pressure gradients as a criterion by chance. In mid-latitudes the Coriolis force "withstands" the pressure, but at the equator the inertial force must withstand the pressure. For some reason or other (we discuss the reasons below) the pressure anomaly arises and causes motion; then as a result of this motion and other external and internal factors (the rotation of the earth, internal friction, etc.) forces arise that balance the field of pressure. Thus the pressure anomaly is the basic dynamic characteristic of the current. Only a purely drift current dictated by the direct drag action of wind in a boundless ocean can exist without the pressure anomaly. However, even in the boundless uniform ocean the nonuniformity of the wind field leads to generation of the slopes of the level. Therefore the determination of the field of pressure is a basic problem of ocean dynamics. We shall see below that the basic feature of the study of the dynamics of sea currents is a method of determining the pressure field. We can determine the pressure directly by constructing the appropriate differential equation for p, as in work by Marchuk (1967a, b). However, in the theory of sea currents another approach is more common: the problem is usually reduced to the determination of some auxiliary function. This section is devoted to two differential equations for such auxiliary functions. Now we rewrite equations 1 and 2 in the form

$$v \frac{\partial^2 u}{\partial z^2} + lv = \frac{1}{\rho_0} \frac{\partial p}{\partial x} + A \qquad (49)$$

$$v \frac{\partial^2 v}{\partial z^2} - lu = \frac{1}{\rho_0} \frac{\partial p}{\partial y} + B \qquad (50)$$

where A and B designate the sums of inertial terms and the effect of lateral exchange:

$$A = \frac{\partial u}{\partial t} + u \frac{\partial u}{\partial x} + v \frac{\partial u}{\partial y} + w \frac{\partial u}{\partial z} - A_l \Delta u \qquad (51)$$

$$B = \frac{\partial v}{\partial t} + u \frac{\partial v}{\partial x} + v \frac{\partial v}{\partial y} + w \frac{\partial v}{\partial z} - A_l \Delta v \qquad (52)$$

We satisfied ourselves above that expressions A and B are small for the large-scale circulation away from the equator. If we neglect these small values as a first approximation, then the simplified system 49–50, with due regard for boundary conditions 11 and 17, has the following approximate solution (Sarkisyan, 1969b, c):

$$u + iv = \frac{\tau_x + i\tau_y}{\rho_0 \, v\alpha(1 + i)} e^{-\alpha(1 + i)z} + \frac{1}{\rho_0 l} \left(i \frac{\partial p}{\partial x} - \frac{\partial p}{\partial y} \right) - \frac{\delta}{\rho_0 l} \left(i \frac{\partial p}{\partial x} - \frac{\partial p}{\partial y} \right)_{z = H} e^{(1 + i)\alpha(z - H)}$$

$$(53)$$

where $\alpha = (\omega \sin \varphi / v)^{1/2}$, $\delta = 1$, corresponding to boundary condition 17, and $\delta = 0$ corresponding to boundary condition 18. From 53 we obtain the following approximate expression for the components of the Ekman friction at the bottom (Sarkisyan, 1969c; Welander, 1959):

$$\rho_0 v \left(\frac{\partial u}{\partial z} \right)_{z = H} = \frac{\delta}{2\alpha} \left(\frac{\partial p}{\partial x} + \frac{\partial p}{\partial y} \right)_{z = H}$$

$$(54)$$

$$\rho_0 v \left(\frac{\partial v}{\partial z} \right)_{z = H} = \frac{\delta}{2\alpha} \left(\frac{\partial p}{\partial y} - \frac{\partial p}{\partial x} \right)_{z = H}$$

Formulas 54 can serve as the approximate boundary conditions instead of 17 or 18.

Now let us return to equations 49 and 50 and integrate them over the whole thickness of the ocean with due regard for boundary conditions 11 and 54. We have

$$lS_y = \frac{1}{\rho_0} \int_0^H \frac{\partial p}{\partial x} dz - \frac{\tau_x}{\rho_0} - \frac{\delta}{2\alpha\rho_0} \left(\frac{\partial p}{\partial x} + \frac{\partial p}{\partial y} \right)_{z = H} + \int_0^H A \, dz$$

$$(55)$$

$$lS_x = \frac{-1}{\rho_0} \int_0^H \frac{\partial p}{\partial y} dz + \frac{\tau_y}{\rho_0} + \frac{\delta}{2\alpha\rho_0} \left(\frac{\partial p}{\partial y} - \frac{\partial p}{\partial x} \right)_{z = H} - \int_0^H B \, dz$$

where

$$S_x = \int_0^H u \, dz, \qquad S_y = \int_0^H v \, dz \tag{56}$$

Further, equation 4, after integration over height with due regard for the above-mentioned boundary conditions for w (the condition of a "rigid lid" at the surface and condition 19 at the ocean bottom), leads to the relation

$$\frac{\partial S_x}{\partial x} + \frac{\partial S_y}{\partial y} = 0 \tag{57}$$

After cross-differentiation of equations 55 and 56, taking into account equation 57, we obtain

$$\beta S_y = -\frac{1}{\rho_0} J(H, p)_{z=H} + \frac{1}{\rho_0} \operatorname{rot} \tau - \frac{\delta}{2\alpha\rho_0} (\Delta p)_{z=H} + \frac{\partial}{\partial y} \int_0^H A \, dz - \frac{\partial}{\partial x} \int_0^H B \, dz \tag{58}$$

where J is the Jacobian operator, and rot τ is the vertical component of a vortex from tangential wind friction

$$J(H, p)_{z=H} = \left(\frac{\partial H}{\partial x} \frac{\partial p}{\partial y} - \frac{\partial H}{\partial y} \frac{\partial p}{\partial x} \right)_{z=H}$$

$$\operatorname{rot} \tau = \frac{\partial \tau_y}{\partial x} - \frac{\partial \tau_x}{\partial y} \tag{59}$$

While passing from equations 55 and 56 to 58 we have taken $\alpha = \text{const}$. Besides this simplification we repeatedly use the assumption $H = \text{const}$ in small terms, thus neglecting the secondary β effect and the secondary effect of bottom topography. As to the basic effects of sphericity of the earth and validity of the bottom topography, not only do we keep them but also we prove the importance of these factors. The basic β effect is the left side, and the effect of bottom topography is the first term of the right side of equation 58. The last two terms of the right side of equation 58 are simplified in the following manner. We neglect the terms containing w in expressions A and B; then we substitute u and v for their approximate values from 53 at $\delta = 0$ in the terms containing the derivatives over time. Finally, in all other terms we substitute u and v for their geostrophic approximations that is, for the second term of the right side of equation 53. As a result, neglecting also the secondary effects of variability of l and H, we obtain

$$\frac{\partial}{\partial y} \int_0^H A \, dz - \frac{\partial}{\partial x} \int_0^H B \, dz \approx \frac{1}{\rho_0 l} \frac{\partial}{\partial t} \operatorname{div} \tau - \frac{\beta}{l^2 \rho_0} \frac{\partial \tau_y}{\partial t}$$

$$- \int_0^H \left(\frac{\partial \Omega}{\partial t} + u_g \frac{\partial \Omega}{\partial x} + v_g \frac{\partial \Omega}{\partial y} \right) dz + A_l \int_0^H \Delta\Omega \, dz, \tag{60}$$

$$\operatorname{div} \tau = \frac{\partial \tau_x}{\partial x} + \frac{\partial \tau_y}{\partial y},$$

$$\Omega = \frac{\partial v_g}{\partial x} - \frac{\partial u_g}{\partial y} \approx + \frac{1}{\rho_0 l} \Delta p$$

Hence

$$\frac{\partial}{\partial y}\int_0^H A\,dz - \frac{\partial}{\partial x}\int_0^H B\,dz \approx \frac{1}{\rho_0 l}\frac{\partial}{\partial t}\operatorname{div}\boldsymbol{\tau} - \frac{1}{\rho_0 l}\int_0^H\left[\frac{\partial\Delta p}{\partial t} + \frac{1}{\rho_0 l}J(p,\Delta p)\right]dz$$

$$- \frac{\beta}{l^2\rho_0}\frac{\partial\tau_y}{\partial t} + \frac{A_l}{\rho_0 l}\int_0^H\Delta\Delta p\,dz \qquad (61)$$

Substituting equation 61 for 58 yields the following approximate expression for the vortex equation:

$$\beta S_y = -\frac{1}{\rho_0}J(H,p)_{z=H} + \frac{1}{\rho}\operatorname{rot}\boldsymbol{\tau} + \frac{1}{\rho_0 l}\frac{\partial}{\partial t}\operatorname{div}\boldsymbol{\tau} - \frac{\beta}{l^2\rho_0}\frac{\partial\tau_y}{\partial t} - \frac{\delta}{2\alpha\rho_0}(\Delta p)_{z=H}$$

$$- \frac{1}{\rho_0 l}\int_0^H\left[\frac{\partial\Delta p}{\partial t} + \frac{1}{\rho_0 l}J(p,\Delta p)\right]dz + \frac{A_l}{\rho_0 l_0}\int_0^H\Delta\Delta p\,dz \qquad (62)$$

Equation 62 serves as the initial relation from which one obtains two equations for the auxiliary functions. We choose the sea-surface topography and the mass transport stream function as auxiliary functions. The sea-surface topography is the pressure anomaly at the ocean surface $p_s = \rho_0 g\zeta$ with the accuracy of the constant multiplier and is related to the pressure anomaly through the simple equation 33. From this equation we also obtain relations between the sea-surface slope, density gradient, and pressure gradient. For instance

$$\left(\frac{\partial p}{\partial x}\right)_{z=H} = \rho_0 g\frac{\partial\zeta}{\partial x} + g\int_0^H\frac{\partial\rho}{\partial x}\,dz = \frac{\partial p_s}{\partial x} + g\int_0^H\frac{\partial\rho}{\partial x}\,dz \qquad (63)$$

$$\left(\frac{\partial p}{\partial x}\right)_{z=H} = \frac{\partial p}{\partial x}\bigg|_z - g\int_H^z\frac{\partial\rho}{\partial x}\,dz \qquad (64)$$

The relations between derivatives over y are written in a similar way. It is not difficult to obtain also the appropriate formula connecting the pressure anomaly with the function of the total transport current. Relations 57 make it possible to introduce a simple dependence between the function of the mass transport stream function and the components of the integral water flux

$$S_x = -\frac{\partial\Psi}{\partial y}, \qquad S_y = +\frac{\partial\Psi}{\partial x} \qquad (65)$$

and rewrite equations 55 and 56 in the form

$$\frac{\partial\Psi}{\partial x} = \frac{1}{\rho_0 l}\int_0^H\frac{\partial p}{\partial x}\,dz - \frac{\tau_x}{\rho_0 l} - \frac{\delta}{2\alpha\rho_0 l}\left(\frac{\partial p}{\partial x} + \frac{\partial p}{\partial y}\right)_{z=H} + \frac{1}{l}\int_0^H A\,dz \qquad (66)$$

$$\frac{\partial\Psi}{\partial y} = \frac{1}{\rho_0 l}\int_0^H\frac{\partial p}{\partial y}\,dz - \frac{\tau_y}{\rho_0 l} - \frac{\delta}{2\alpha\rho_0 l}\left(\frac{\partial p}{\partial y} - \frac{\partial p}{\partial x}\right)_{z=H} + \frac{1}{l}\int_0^H B\,dz \qquad (67)$$

On the right sides of these equations the first terms are basic; the rest of them can be neglected as a first approximation. For a rough estimation we neglect also the variability of l and H in the first terms and obtain the formula

$$\Psi \approx \frac{1}{\rho_0 l_1}\int_0^H p\,dz \qquad (68)$$

showing the physical sense of the relation between Ψ and p; the function of the total water flux accurate to a constant multiplier is an integral over height from the pressure anomaly. The equation for the sea-surface topography is obtained from equation 62 using relations like equations 63 and 64. We have

$$A_e \Delta\Delta\zeta + \frac{\partial}{\partial t}\Delta\zeta + \frac{g}{l}J(\zeta, \Delta\zeta) + \frac{l}{2\alpha H}\Delta\zeta + \frac{l}{H}J(H, \zeta) + \beta\frac{\partial\zeta}{\partial x}$$

$$\underbrace{}_{\text{VI}} \quad \underbrace{}_{\text{V}} \quad \underbrace{}_{\text{IV}} \quad \underbrace{}_{\text{III}} \quad \underbrace{}_{\text{II}} \quad \underbrace{}_{\text{I}}$$

$$= \frac{l}{\rho_0 gH}\operatorname{rot}\tau - \frac{1}{\rho_0 gH}\left[\beta\tau_x - \frac{\partial}{\partial t}\operatorname{div}\tau + \frac{\beta}{l}\frac{\partial\tau_y}{\partial t}\right] + f_1 \qquad (69)$$

$$\underbrace{}_{\text{VII}} \qquad \underbrace{}_{\text{VIII}} \quad \underbrace{}_{\text{IX}}$$

where

$$f_1 = \frac{A_l}{\rho_0 H}\int_0^H (H - z)\Delta\Delta\rho\, dz - \frac{1}{\rho_0 H}\int_0^H (H - z)\frac{\partial}{\partial t}\Delta\rho\, dz$$

$$\underbrace{}_{\text{VI}} \qquad \underbrace{}_{\text{V}}$$

$$- \frac{g}{\rho_0 lH}\int_0^H (H - z)J(\xi, \Delta\rho)\, dz + \frac{g}{\rho_0 lH}\int_0^H (H - z)J(\Delta\zeta, \rho)\, dz$$

$$\underbrace{}_{\text{IV}}$$

$$- \frac{g}{\rho_0{}^2 lH}\int_0^H J\left(\int_0^z \rho\, dz, \int_0^z \Delta\rho\, dz\right) dz - \frac{l}{2\alpha\rho_0 H}\int_0^H \Delta\rho\, dz$$

$$\underbrace{}_{\text{IV}} \qquad \underbrace{}_{\text{III}}$$

$$(70)$$

$$- \frac{l}{\rho_0 H}\int_0^H J(H, \rho)\, dz - \frac{\beta}{\rho_0 H}\int_0^H (H - z)\frac{\partial\rho}{\partial x}\, dz$$

$$\underbrace{}_{\text{II}} \qquad \underbrace{}_{\text{I}}$$

In order to obtain the equation for Ψ we return to the relations 66 and 67. On the right sides of these relations we retain only the two first terms and pass from gradients of p to slopes of the sea surface by formulas 33. As a result we obtain

$$\frac{\partial\zeta}{\partial x} = \frac{l}{gH}\frac{\partial\Psi}{\partial x} - \frac{1}{H\rho_0}\int_0^H (H - z)\frac{\partial\rho}{\partial x}\, dz + \frac{\tau_x}{\rho_0 gH} \qquad (71)$$

$$\frac{\partial\zeta}{\partial y} = \frac{l}{gH}\frac{\partial\Psi}{\partial y} - \frac{1}{H\rho_0}\int_0^H (H - z)\frac{\partial\rho}{\partial y}\, dz + \frac{\tau_y}{\rho_0 gH} \qquad (72)$$

The transition from equation 69 to the appropriate equation for Ψ is performed as follows. The first derivatives of ζ are substituted for the first derivatives of Ψ by using equations 71 and 72. To substitute the second derivatives we must differentiate equation 71 over x and equation 72 over y disregarding the variability of l and H (we neglect the secondary effects of variability of these values). We have

$$- A_l \Delta\Delta\Psi + \frac{\partial}{\partial t}\Delta\Psi + \frac{1}{H}J(\Psi, \Delta\Psi) + \frac{l}{2\alpha H}\Delta\Psi + \frac{l}{H}J(H, \Psi) + \beta\frac{\partial\Psi}{\partial x}$$

$$\underbrace{}_{\text{VI}} \quad \underbrace{}_{\text{V}} \quad \underbrace{}_{\text{IV}} \quad \underbrace{}_{\text{III}} \quad \underbrace{}_{\text{II}} \quad \underbrace{}_{\text{I}}$$

$$= \frac{1}{\rho_0}\operatorname{rot}\tau + \frac{1}{\rho_0}\frac{\partial\operatorname{div}\tau}{\partial t} - \frac{\beta}{\rho_0 l^2}\frac{\partial\tau_y}{\partial t} + \frac{\partial H}{\partial y}\frac{\tau_x}{\rho_0 H} - \frac{\partial H}{\partial x}\frac{\tau_y}{\rho_0 H} + \frac{\beta}{\rho_0 l}\tau_x + f_2 \quad (73)$$

$$\underbrace{}_{\text{VII}} \quad \underbrace{}_{\text{IX}} \qquad \underbrace{}_{\text{X}} \qquad \underbrace{}_{\text{VIII}}$$

where

$$f_2 = \frac{1}{H}\left(\frac{g}{\rho_0 l}\right)^2 \int_0^H (H - z)J(\rho, \Delta\rho)\, dz - \left(\frac{g}{\rho_0 l}\right)^2 \int_0^H J\left(\int_0^H \rho\, dz, \int_0^H \Delta\rho\, dz\right) dz$$

$$\underbrace{\qquad\qquad\qquad\qquad\qquad\qquad\qquad\qquad\qquad\qquad\qquad}_{\text{IV}}$$

$$-\frac{g}{2\rho_0\alpha H}\int_0^H z\Delta\rho\, dz - \frac{g}{\rho_0 H}\int_0^H zJ(H, \rho)\, dz \qquad (74)$$

$$\underbrace{\qquad\qquad}_{\text{III}}\qquad\underbrace{\qquad\qquad}_{\text{II}}$$

While deriving equation 73 we also neglected some small terms due to tangential wind friction. Despite a number of simplifications we obtained the rather bulky equation 73, from which we can derive, as a particular case, the basic relation of many well-known models in the theory of total mass transport.

D. Estimation of Orders of Magnitudes and Simplification of Equations for Auxiliary Functions

We begin with the estimation of the orders of magnitude of the terms in equation 69. The physical sense of terms on the left side of this equation is as follows: term VI is the effect of the lateral friction, V is the local derivative over time (from inertial terms), IV is the nonlinear inertial term, III is the effect of the bottom friction, II is the effect of bottom topography, and I is the β effect. On the right side term VII is the vortex from the tangential wind friction, VIII is the wind β effect, and IX is the effect of non-stationarity of wind. The other terms on the right side are dictated by the baroclinity of seawater; their numeration coincides with that of the corresponding terms on the left side which are of the same physical nature. Therefore term VI on the right side may be called the "baroclinic effect of lateral friction," I is the baroclinic β effect, etc.

The characteristic magnitudes of the corresponding values are designated by the same letters as in Section 1.B with one difference: in equation 69 H is the real depth of the ocean. Therefore, instead of the characteristic depth of the baroclinic layer h_0, we use the characteristic depth of the ocean itself, H_0. We choose the β effect as a criterion in equation 69. Equating the characteristic values of terms on the left and right sides of this equation enumerated by I yields

$$\zeta_0 = \frac{(\delta\rho)_0}{\rho_0} H_0 \qquad (75)$$

That is, for ζ_0 we obtain a formula similar to 47. Using formula 75 we can easily see that terms with the same number on the left and right sides of 69 have the same order of magnitude. Therefore, it is not necessary to determine the orders of magnitude of all terms of this equation.

Now let us accept that the inertial terms V and VI are of the same order of magnitude. This permits us to define the characteristic time scale

$$t_0 = \frac{l_0 L_0^2}{g\zeta_0} = \frac{L_0^2}{v_0} \qquad (76)$$

Then we turn to the dimensionless values in equation 69, divide the equation by the characteristic value of criterion $\beta_0 \zeta_0 / L_0$, and take into account formulas 75 and 76. As a result we obtain

$$-\varepsilon_6 \bar{\Delta}\bar{\Delta}\bar{\zeta} + \varepsilon_5 \frac{\partial}{\partial \bar{t}} \bar{\Delta}\bar{\zeta} + \frac{\varepsilon_5}{\bar{l}} J(\bar{\zeta}, \bar{\Delta}\bar{\zeta}) + \varepsilon_3 \frac{\bar{l}}{2\bar{\alpha}\bar{H}} \bar{\Delta}\bar{\zeta} + a_2 \frac{\bar{l}}{\bar{H}} J(\bar{H}, \bar{\zeta}) + \beta \frac{\partial \bar{\zeta}}{\partial \bar{x}}$$

$$= \varepsilon_7 \frac{\bar{l}}{\bar{H}} \overline{\text{rot }} \tau + \varepsilon_8 \frac{\bar{\beta}}{\bar{H}} \bar{\tau}_x + \varepsilon_9 \left(\frac{1}{\bar{H}} \frac{\partial}{\partial \bar{t}} \overline{\text{div }} \tau - \frac{\bar{\beta}}{\bar{l}} \frac{\partial \bar{\tau}_y}{\partial \bar{t}} \right) + \frac{L_0}{\beta_0 \zeta_0} f_1 \quad (77)$$

The subscript Arabic numerals of the small dimensionless parameters correspond to Roman numerals identifying terms. The parameter before term II is not small; therefore it is designated by a different letter. The dimensionless parameters have the following values:

$$\varepsilon_6 = \frac{A_l}{\beta_0 L_0{}^3}, \qquad \varepsilon_5 = \frac{1}{t_0 \beta_0 L_0}, \qquad \varepsilon_3 = \frac{\sqrt{l_0 \nu}}{H_0 L_0 \beta_0},$$

$$a_2 = \frac{l_0}{L_0 \beta_0}, \qquad \varepsilon_7 = \frac{l_0 \tau_0}{\rho_0 H_0 \beta_0 \zeta_0 g}, \qquad \varepsilon_8 = \frac{\tau_0 L_0}{\rho_0 g H_0 \zeta_0}, \qquad \varepsilon_9 = \frac{\tau_0}{t_0 \rho_0 g H_0 \beta_0 \zeta_0} \quad (78)$$

The prescribed characteristics have the same numerical values here as in Section 1.B. Namely,

$$l_0 = 10^{-4}, \quad \tau_0 = 1, \quad L_0 = 10^8, \quad A_l = 10^7, \quad \nu = 10^2, \quad \beta_0 = 2 \times 10^{-13} \quad (79)$$

The characteristic values $H_0 = 1$ km $= 10^5$ undergo little change, since this is not the thickness of the baroclinic layer any more but the characteristic depth of the ocean, and $(\delta\rho)_0 = 0.5 \times 10^{-3}$, since it is the characteristic density gradient average over height. Using formulas 75, 76, 33, and 34, we obtain the characteristic values

$$\zeta_0 = 50, \quad v_0 = 5, \quad \mathscr{P}_0 = 5 \times 10^4, \quad t_0 = 2 \times 10^7 \quad (80)$$

coinciding with those obtained in Section 1.B. With equations 79 and 80 we obtain the following values for the dimensionless parameters:

$$\varepsilon_6 = 0.5 \times 10^{-4}, \qquad \varepsilon_5 = 2.5 \times 10^{-3}, \qquad \varepsilon_3 = 3 \times 10^{-2},$$
$$a_2 = 5, \qquad \varepsilon_7 = 0.1, \qquad \varepsilon_8 = 2 \times 10^{-2}, \qquad \varepsilon_9 = 0.5 \times 10^{-4} \quad (81)$$

The estimates indicate the following. The effect of bottom topography is the only factor comparable with β effect on the left side of the equation. Further, ε_6 is five times greater than Ekman number E_m and ε_5 is as many times greater than the Rossby (Kibel) number as this (see Section 1.B). This means that the effect of inertial terms and lateral eddy viscosity in equation 69 is more important than the effect of these factors in equations of motion 1 and 2. Nevertheless, the smallness of parameters ε_6 and ε_5 indicates that we can neglect these factors as a first approximation even in the equation for ζ in the studies of large-scale stationary or seasonal currents. The β effect and the effect of the bottom topography are the basic factors on the left side of this equation.

If in expression 70 we pass on to dimensionless values and use formula 75, then we obtain the same dimensionless parameters before terms in each group as on the left side. Consequently, we may conclude that the baroclinic β effect and the joint effect of baroclinity and bottom topography (JEBAT) (groups I and II in expression 70) are the basic terms on the right side of equation 69. The direct effect of tangential wind friction, judging by value $\varepsilon_7 = 0.1$, plays a less important role than the density anomaly. The latter reflects the effect of heat and salt exchange and the indirect effect of wind field.

Passing on to the equation for Ψ we determine first of all the order of magnitude of this characteristic using relations 71, and 72. The left sides of these relations and the two first terms on the right sides are values of the same order of magnitude. We have

$$\frac{\zeta_0}{L_0} = \frac{l_0}{gH_0}\frac{\Psi_0}{L_0} = \frac{H_0(\delta\rho)_0}{\rho_0 L_0} \tag{82}$$

Hence

$$\Psi_0 = \frac{gH_0\zeta_0}{l_0} \quad \text{or} \quad \Psi = \frac{H_0{}^2 g(\delta\rho)_0}{l_0\rho_0}$$

Substituting the above-mentioned values of the characteristic values we obtain $\Psi_0 = 5 \times 10^{13}$ cm^3/sec or 50 sverdrups.

The transition to dimentionless values in equation 73 is quite similar to what we have already done with equation 69; all dimensionless parameters except for one parameter are the same. Because equations 69 and 73 are identical, in the transition to dimentionless values it is quite natural to choose the same criterion, the β effect. We have

$$-\varepsilon_6 \overline{\Delta}\overline{\Delta}\overline{\Psi} + \varepsilon_5 \frac{\partial}{\partial \bar{t}} \overline{\Delta}\overline{\Psi} + \varepsilon_5 \bar{J}(\overline{\Psi}, \overline{\Delta}\overline{\Psi}) + \varepsilon_3 \frac{\bar{l}}{2\bar{\alpha}\overline{H}} \overline{\Delta}\overline{\Psi} + a_2 \frac{\bar{l}}{\overline{H}} J(\overline{H}, \overline{\Psi}) + \beta \frac{\partial \overline{\Psi}}{\partial \bar{x}}$$

$$= \varepsilon_7 \operatorname{rot} \bar{\tau} - \varepsilon_9 \left[\frac{1}{\bar{l}} \frac{\partial}{\partial \bar{t}} \operatorname{div} \bar{\tau} - \frac{\bar{\beta}}{\bar{l}^2} \frac{\partial \bar{\tau}_y}{\partial \bar{t}} \right] + \varepsilon_{10} \left[\frac{\partial \overline{H}}{\partial \bar{y}} \frac{\bar{\tau}_x}{\overline{H}} - \frac{\partial \overline{H}}{\partial \bar{x}} \frac{\bar{\tau}_y}{\overline{H}} \right] + \varepsilon_8 \frac{\bar{\beta}\bar{\tau}_x}{\bar{l}} + \frac{L_0}{\beta_0 \Psi_0} f_2 \tag{83}$$

where $\varepsilon_{10} = \tau_0/\rho_0\beta_0\Psi_0 = 0.1$ and expressions and numerical values of a_2 and ε_3–ε_9 are given above. Owing to identity of the left sides of the equations 83 and 77 we discuss only the differences of their right sides.

On the right side of equation 83 we have an additional term, the joint effect of wind and bottom topography, which is at best of the same order of magnitude as the vortex from the tangential wind friction.

Unlike equation 77, on the right side of equation 83 only one *basic* term exists (JEBAT); this is group II in expression 74. We did not dwell on the estimation of this term because by using formula 82 we can easily see that the terms on the left and right sides of equation 73 with the same ordinal number are of the same order of magnitude. This analysis contracts with the traditional knowledge about rot τ as the basic factor forming Ψ field in the theory of total mass transport. Below we return repeatedly to this question regarding the basis of concrete calculations. Here we stress only that on the left side of equation 73, as in 69, the terms with the first derivatives prove to be the basic ones; that is, these are equations with small parameters at older derivatives.

E. Relations for Calculating Sea-Surface Topography at the Contour of the Basin Boundary

It follows from the preceding items that in order to solve a problem on the calculation of hydrological characteristics we must solve the corresponding differential equation for one of the auxiliary functions ζ and Ψ. In fact, we deal with the elliptic equation for Ψ and ζ.

But first we must define these functions on the contour of the basin boundary. When water flux is specified on the contour of the boundary we can easily determine Ψ; ζ must be defined on the contour by solving the corresponding differential equation.

Thus any problem can be solved using ζ or Ψ, and the latter can be easily determined on the contour. Consequently, it may seem that only Ψ need be chosen as an auxiliary function; many scientists hold such an opinion. Below we discuss other versions of the boundary condition, analyze the results of concrete calculations of Ψ and ζ, and state the reasons for which we choose mainly the function ζ. Meanwhile we construct equations for determining ζ on the basin contour. To do this we proceed from relations 63–67, making use of the fact that the integral water fluxes are specified on the boundary contour (on the solid part of the contour the fluxes are equal to zero.) We neglect inertial terms and lateral exchange and after transformations that are not complicated we obtain

$$\frac{\partial \zeta}{\partial x} = \frac{l}{gH} S_y + \frac{1}{\rho_0 H} \int_0^H \frac{\partial \rho}{\partial x} z\, dz - \frac{1}{\rho_0} \int_0^H \frac{\partial \rho}{\partial x} dz + \frac{\tau_x}{\rho_0 gH}$$

$$+ \frac{1}{2\alpha H}\left(\frac{\partial \zeta}{\partial x} - \frac{\partial \zeta}{\partial y}\right) + \frac{1}{2\alpha H \rho_0}\left(\int_0^H \frac{\partial \rho}{\partial x} dz + \int_0^H \frac{\partial \rho}{\partial y} dz\right) \tag{84}$$

$$\frac{\partial \zeta}{\partial y} = -\frac{l}{gH} S_x + \frac{1}{\rho_0 H} \int_0^H \frac{\partial \rho}{\partial y} z\, dz - \frac{1}{\rho_0} \int_0^H \frac{\partial \rho}{\partial y} dz + \frac{\tau_y}{\rho_0 gH}$$

$$+ \frac{1}{2\alpha H}\left(\frac{\partial \zeta}{\partial y} - \frac{\partial \zeta}{\partial x}\right) + \frac{1}{2\alpha H \rho_0}\left(\int_0^H \frac{\partial \rho}{\partial y} dz - \int_0^H \frac{\partial \rho}{\partial x} dz\right) \tag{85}$$

The physical sense of the terms of the right sides of equations 84 and 85 is as follows. The first terms are the integral fluxes of a fluid specified on the boundary contour. The second and third terms are the baroclinic part of the gradient current averaged over the height. The fourth terms are the effect of tangential wind friction on the contour of the basin boundary. The fifth terms are the effect of bottom topography on the contour of the basin boundary. The last terms are the baroclinic effect of bottom friction on the contour of the basin boundary.

If we pass to dimensionless values and estimate the orders of magnitudes of terms in equations 84 and 85, then following the procedure described in Section 1.B we can easily satisfy ourselves that the three first terms are basic and the last three are secondary. This fact is used below in the construction of simplified models for diagnostic calculations.

2. Numerical Models for Computing Currents from the Observed Density Field

A. Formulation of the Problem

At the beginning of this century Bjerknes (1900) and Sandström and Helland-Hansen (1903) elaborated the dynamical method of calculation of gradient currents from the observed density field. Though the authors intended the calculation only of relative currents, this method is practically used for the computation of ocean currents in the surface layer 1–1.5 km deep, supposing that the current velocity at this level is equal to zero. Numerous methods for determination of the depth of the reference level appeared after the well-known work of Defant (1941). Owing to its extraordinary simplicity the dynamic method won great recognition among oceano-graphers. However, this method has basic shortcomings. In real seas and oceans the vertical profile of gradient current is quite arbitrary. In some cases the current velocity at the definite depth reverses its sign with depth; in others it gradually decreases with depth, but has the same order of magnitude as that of the surface gradient current. In the third case the velocity does not practically decrease with depth. Hence there is no continuous zero surface (reference level), and this is the main shortcoming of the dynamic method. Further, the dynamic method does not allow for the direct effect of tangential wind friction, for the bottom topography, etc. Nor can we calculate a vertical velocity component by means of the dynamic method. Even without considering other shortcomings of the dynamic method, those mentioned above lead us to the necessity of working out another method. As we see in the following discussion, the method of solving the hydrodynamic problem of flow velocity calcula-tion from the observed density field is of theoretical and practical interest and, in particular, can serve as the alternative for the dynamic method.

The statement of the problem is as follows. Let there be the ocean or a sea basin of arbitrary coastline contours and arbitrary bottom topography; besides natural, solid boundaries the basin may also have sections of conditional boundary—liquid shorelines, through which the water flux is specified. The density field in the entire ocean is given; the tangential wind stress (or sea-level atmospheric pressure) is specified. We wish to solve the hydrodynamic problem of calculation of flow velocity in this basin from a known density and sea-level pressure fields (with due regard for boundary conditions). For solving this problem we should apply the set of equations and boundary conditions given in Section 1.A. It is not necessary to consider equations 5–7 and boundary conditions for temperature and salinity, since the density field is assumed to be given. Thus we have to consider only the hydrodynamic problem of flow velocity calculation in an inhomogeneous ocean. As we saw in the first section, we can essentially simplify the initial system of equations when studying the large-scale currents in a basin situated away from the equator. Using the simplifications based on the above estimates we describe some models for solving the formulated problem.

B. Simplest Model for Computing Sea-Surface Topography and Flow Velocity (D_1 Model)

The starting equations of motion for this model are 49 and 50, in which we set $A \equiv B \equiv 0$. The problem is reduced to solving a single equation for ζ obtained from equation 69 after neglecting the small terms. The result is

$$\underbrace{\frac{1}{2\alpha} \Delta \zeta}_{\text{III}} + \underbrace{J(H, \zeta)}_{\text{II}} + \underbrace{\frac{H\beta}{l} \frac{\partial \zeta}{\partial x}}_{\text{I}} = \underbrace{\frac{1}{\rho_0 g} \operatorname{rot} \tau}_{\text{VII}} - \underbrace{\frac{1}{2\alpha\rho_0} \int_0^H \Delta \rho \, dz}_{\text{III}}$$

$$- \underbrace{\frac{1}{\rho_0} \int_0^H J(H, \rho) \, dz}_{\text{II}} - \underbrace{\frac{\beta}{\rho_0 l} \int_0^H (H - z) \frac{\partial \rho}{\partial x} \, dz}_{\text{I}} \qquad (86)$$

It is seen from the above estimates that the main terms in equation 86 are those containing the first derivatives of ρ and ζ. Besides these terms we retain the effect of a wind stress and the effect of bottom friction. The latter allows us to solve the problem for an enclosed basin.

At the lateral boundaries of the basin the normal components of mass transport are specified. Usually, we approximate the boundary of a region by a broken line traced piecewise along one of the coordinate axes. At the sections of a broken coastline parallel to a meridian, S_x is given. We use for them a simplified version of relation 85. We have

$$\frac{\partial \zeta}{\partial y} = -\frac{l}{gH} S_x + \frac{1}{\rho_0 H} \int_0^{H} \frac{\partial \rho}{\partial y} z \, dz - \frac{1}{\rho_0} \int_0^{H} \frac{\partial \rho}{\partial y} \, dz = f(x, y) \qquad (87)$$

On zonal sections of the coastline we use a simplified variant of equation 84. The result is

$$\frac{\partial \zeta}{\partial x} = \frac{l}{gH} S_y + \frac{1}{\rho_0 H} \int_0^{H} \frac{\partial \rho}{\partial x} z \, dz - \frac{1}{\rho_0} \int_0^{H} \frac{\partial \rho}{\partial x} \, dz = f'(x, y) \qquad (88)$$

When passing from equations 84 and 85 to equations 87 and 88, we drop the small terms due to a wind stress and bottom friction at the coastline of the basin. The role of these factors is smaller than that of inertial terms, which we have also neglected, as we are considering the large-scale nonequatorial currents. The right sides of equations 87 and 88 are given. Having written these equations for all sections of the boundary, we obtain a closed system of equations for ζ at the coastline of a basin. Note that the problem of determination of ζ at the coastline s of a basin with the aid of equations 87 and 88 is inexactly set since the integral on a closed contour from the right-hand sides of these equations, generally speaking, differs from zero. We call the function $\zeta(x, y)$, minimizing an operator $((\partial \zeta/\partial s) - f')$, a quasi-solution of the problem. If $\|(\partial \zeta/\partial s) - f'\| < \varepsilon$, then a maximum of absolute value of difference between the calculated field of ζ and the solution of the correctly formulated problem is less than ε_1 (Ivanov, 1963). In Section 4 we show on the basis of numerous examples of calculation that the difference between ζ received from equations 87 and 88 and from the solution of correctly formulated problems 84 and 85 is very small.

Having computed the ocean surface topography we can easily find the pressure field and current velocity components. The pressure field is calculated from the simple formula 33. The horizontal flow velocity components can be found from formula 53, which can be easily simplified. First, when studying the large-scale currents, it is not necessary to consider their fine structure in the surface and bottom boundary layers. Besides, Ekman's simple spiral doesn't describe the currents' fine structure. Therefore we use this formula for calculating currents at levels the distance between which is more than Ekman's friction layer thickness (practically, more than 10 m for $v = 1$ cm/sec). Thus at the ocean surface the velocity components u and v are calculated from formulas

$$u = \frac{1}{2\rho_0 v\alpha} (\tau_x + \tau_y) - \frac{g}{l} \frac{\partial \zeta}{\partial y} \qquad (89)$$

$$v = \frac{1}{2\rho_0 v\alpha} (\tau_y - \tau_x) + \frac{g}{l} \frac{\partial \zeta}{\partial x} \qquad (90)$$

At the intermediate levels u and v can be determined from simple formulas of geostrophic current

$$u = -\frac{g}{l} \frac{\partial \zeta}{\partial y} - \frac{g}{\rho_0 l} \int_0^z \frac{\partial \rho}{\partial y} \, dz \qquad (91)$$

$$v = \frac{g}{l} \frac{\partial \zeta}{\partial x} + \frac{g}{\rho_0 l} \int_0^z \frac{\partial \rho}{\partial x} \, dz \qquad (92)$$

Deriving the formula for calculation of the vertical velocity component is also simple. From equation 4 and boundary condition $w = 0$ for $z = 0$ we have

$$w = -\int_0^z \left(\frac{\partial u}{\partial x} + \frac{\partial v}{\partial y}\right) dz \qquad (93)$$

Inserting the values of u and v from equation 53 into equation 93 and making use of equation 33, we obtain (out of boundary layers)

$$w = -\frac{1}{\rho_0 l} \operatorname{rot} \tau - \frac{\beta}{\rho_0 l^2} \tau_x + \frac{g\beta z}{l^2} \frac{\partial \zeta}{\partial x} + \frac{g\beta}{\rho_0 l^2} \int_0^z (z - \xi) \frac{\partial \rho}{\partial x} d\xi \qquad (94)$$

The components of wind stress are usually calculated by quadratic dependence between τ and wind velocity. However, the wind field is insufficiently studied and possesses great unsteadiness. Consequently, it is better to calculate τ by means of sea-level pressure p_a.

The relations

$$\tau_x = -\frac{1}{2\alpha'}\left(\frac{\partial p_a}{\partial x} + \frac{\partial p_a}{\partial y}\right), \qquad \tau_y = \frac{1}{2\alpha'}\left(\frac{\partial p_a}{\partial x} - \frac{\partial p_a}{\partial y}\right) \qquad (95)$$

which are readily obtained from Akerblom's model (Sarkisyan, 1954, 1956), can be regarded as the simplest version of such formulas. Here $\alpha' = (\omega \sin \varphi / v')^{1/2}$, where v' is the vertical kinematic turbulent mixing coefficient for air.

The described model is the most simple technique for solving the problem formulated. The model may be further generalized. However, its every simplification can lead us to contradictions. Let us show that the dynamic method is a particular case of this model. Supposing that we are dealing with the basin of minimal depth of 1 km, we can neglect the density anomaly at the bottom of the basin; that is, we assume that

$$\int_0^H \frac{\partial \rho}{\partial x} dz \approx \frac{\partial}{\partial x} \int_0^H \rho\, dz, \qquad \int_0^H \frac{\partial \rho}{\partial y} dz \approx \frac{\partial}{\partial y} \int_0^H \rho\, dz \qquad (96)$$

Further, remembering that the anomalies of density decrease with depth, we can assume

$$\left|\int \frac{\partial \rho}{\partial x} dz\right| > \frac{1}{H}\left|\int_0^H z \frac{\partial \rho}{\partial x} dz\right|$$

$$\int_0^H (H - z) \frac{\partial \rho}{\partial x} dz \approx H \int_0^H \frac{\partial \rho}{\partial x} dz \qquad (97)$$

Finally, let us drop the first term of the right side of equation 86. After these simplifications equation 86 takes the form

$$L(\zeta) = \frac{1}{2\alpha} \Delta\zeta + J(H, \zeta) + \frac{\beta H}{l} \frac{\partial \zeta}{\partial x} = -L\left(\int_0^H \rho\, dz\right) \qquad (98)$$

The solution of equation 98 is

$$\zeta_d = -\frac{1}{\rho_0} \int_0^H \rho \, dz \tag{99}$$

which represents the formula for computation of the sea-surface topography by the dynamic method (H is accepted as a reference level). This solution can be also applied to the boundary conditions. Indeed, suppose that the water flux across the lateral boundaries is given by the dynamic method; that is, S_x and S_y are determined at the contour from the formulas

$$S_x = \frac{g}{\rho_0 l} \int_0^H \frac{\partial \rho}{\partial y} z \, dz, \qquad S_y = -\frac{g}{\rho_0 l} \int_0^H \frac{\partial \rho}{\partial x} z \, dz \tag{100}$$

Then, remembering simplification 96, we can readily see that formula 99 satisfies boundary conditions 87 and 88. The approximate formula 99 is obtained even when instead of 86 we use the more general equation 69, and instead of equations 87 and 88 the more general conditions 84 and 85, that is, in the case of approximate allowance for nonlinear terms, effect of lateral exchange, and the bottom friction (Perederey and Sarkisyan, 1972; Perederey, 1974).

In Section D we show on the basis of specific calculations that the dynamic method is a reliable first approximation for solving equation 86. The D_1 model can be applied to the flow velocity calculation by the given density field (diagnostic calculations) away from the equator. In order to use the model for computation of the World Ocean large-scale currents we should construct a grid set in such a way that there are no grid points at the equator. For example, when calculating with a grid mesh of 5° the grid points nearest to the equator must be placed at the latitudes of 2.5°N and 2.5°S. Even at these latitudes we have exaggerated values of current velocities owing to the smallness of the Coriolis parameter. We can avoid this by replacing the value of sin 2.5° by sin 5° or sin 7.5° in formulas 89–92 and 94.

In order to study a multiconnected region (including large islands such as Antarctic and Australia) we may use the technique given in Bryan (1969) or Kamenkovich (1961). However, when making allowance for numerous small islands (such as New Zealand and Madagascar), this technique is unsuitable because it leads us to a great overload of the memory of the computer. However, we can take the small islands for the shallow regions of an ocean and consider that in these regions the water density is constant and a tangential wind friction at sea level is absent.

C. Simplest Model for Computing Mass Transport Stream Function and Flow Velocity (D'₁ Model)

The D_1 Model may be used for the computation of all hydrodynamic characteristics of quasi-stationary current. In particular, using relations 71 and 72 we can calculate a field of Ψ. However, another technique is more popular in the theory of ocean currents—the calculation of the field of Ψ and the other characteristics on the basis of the obtained Ψ field. In our opinion, however, this technique has led to conclusions requiring a critical approach. Besides, some peculiarities of current dynamics are easier to illustrate by computing the fields of ζ and Ψ independently of each other. We therefore consider the model for computing flow velocities with the aid of the auxiliary function Ψ. First we write equation 73 in a form convenient for the illustration of some models. The result is as follows:

$$
\underbrace{\frac{l}{H} J(H, \Psi)}_{\text{II}} + \underbrace{\frac{l}{2\alpha H} \Delta\Psi}_{\text{III}} + \underbrace{\beta \frac{\partial\Psi}{\partial x}}_{\text{I}} - \underbrace{A_l \Delta\Delta\Psi}_{\text{VI}}
$$

$$
+ \underbrace{\frac{\partial}{\partial t}\Delta\Psi}_{\text{V}} + \underbrace{\frac{1}{H} J(\Psi\Delta\Psi)}_{\text{IV}} = \underbrace{\frac{\operatorname{rot}\tau}{\rho_0}}_{\text{VII}} - \underbrace{\frac{g}{\rho_0 H} \int_0^H z J(H, \rho)\, dz + f_3}_{\text{II}}
$$

(101)

where

$$
f_3 = \underbrace{\frac{\beta}{\rho_0 l} \tau_x}_{\text{VIII}} - \underbrace{\frac{1}{\rho_0 H}\left(\frac{\partial H}{\partial x}\tau_y - \frac{\partial H}{\partial y}\tau_x\right)}_{\text{X}} + \underbrace{\frac{1}{\rho_0}\frac{\partial}{\partial t}\operatorname{div}\tau - \frac{\beta}{\rho_0 l^2}\frac{\partial \tau_y}{\partial t}}_{\text{IX}}
$$

$$
\underbrace{-\left(\frac{g}{\rho_0 l}\right)^2 \int_0^H J\left(\int_0^z \rho\,dz, \Delta\int_0^z \rho\,dz\right) dz - \left(\frac{g}{\rho_0 l}\right)^2 \int_0^H z J(\rho_0\Delta\rho)\,dz}_{\text{IV}}
$$

(102)

$$
\underbrace{-\frac{g}{2\alpha H\rho_0}\int_0^H z\Delta\rho\,dz}_{\text{III}}
$$

In equations 69, 73, and 101 the terms due to the same factor are marked by the same Roman numeral. The physical essence of these terms is explained in Section 1.D.

We have written equation 101 in a form from which we can obtain all the principal equations of many studies of the total mass transport theory. For example, setting $\rho \equiv 0$ in the right side of equation 101 we come to a basic equation of the total mass transport theory.

Arabic numerals on the dotted lines mean the following.

1. The first work of this theory of Shtokman (1946) provides the balance between the effect of lateral exchange and rot τ. Dotted line 1, connecting terms I and VII, illustrates the principal equation of Shtokman's theory.

2. A year later Sverdrup (1947) suggested a relation foreseeing the equilibrium between β effect and rot τ. This well-known work initiated an extensive range of

investigations connected with the integrated ocean circulation within the framework of total mass transport theory. Sverdrup's model is shown by dotted line 2.

3. In 1948 Stommel published a paper in which he showed that the main reason for the westward intensification of currents is the β effect. This model is illustrated by dotted line 3.

4. The main equation of Munk's work (1950) is denoted by dotted line 4.

5. Dotted line 5 is a principal relation of works of Neumann (1955, 1958) and Ivanov and Kamenkovich (1959).

6. Finally, dotted line 6 denotes the basic equation of Carrier and Robinson (1962).

Though the above list of principal works of total flow theory (1946–1962) is far from complete, it nevertheless is quite sufficient to be criticized. Historically, the theory of total mass transport emerged as a criticism of models of homogeneous ocean.

The founders of the total mass transport theory, Shtokman, Sverdrup, Munk, and others, considered that we were studying the integrated circulation of a baroclinic ocean. But there is no baroclinicity at all in the basic equation of this theory. All the versions of the total flow theory can be obtained from equation 101 only by neglecting the density anomalies, that is, setting $\rho \equiv 0$. There was no such assumption in the works of Shtokman, Sverdrup, and Munk, but there were other simplifications that led to the complete loss of baroclinity in the equation for Ψ. In deriving equation 69 we mentioned that many of the terms on both sides of the equation have the same physical essence and therefore are denoted by the same Roman numerals. The same refers to equations 73 and 101. In order to get equation 73 from 69 we made use of relations 71 and 72.

The left side of equation 101 is obtained from the first terms of the right sides of relations 71 and 72, and terms of the right side containing ρ are obtained from the second terms of the right sides of the same relations. We therefore denoted by the same Roman numerals the terms with the same physical essence on both sides of equation 101. For example, term III on the left side of this equation expresses the effect of bottom friction, and term III on the right side the baroclinic effect of the bottom friction; term II on the left side represents the effect of bottom topography, and term II on the right side the joint effect of baroclinity and bottom topography (JEBAT). If we neglect the effect of bottom topography, then not only does term II of the left side of equation 101 vanish but a basic term of the right side of this equation, JEBAT. Further, if in the derivation of equation 101 we simplify the nonlinear terms, considering that the integral from the product of two functions equals the product of the integrals from these functions (as is done in many works of the total mass transport theory), then term IV is retained on the left side of equation 101 but term IV on the right side vanishes. If, finally, we drop the bottom friction (or consider that $\partial u/\partial z = \partial v/\partial z = 0$ at the definite depth H_1), then term III on the left side and "the last of the Mohicans," the term on the right side with the same number, vanish together. For these three reasons we lost the effect of baroclinicity in the total mass transport theory, although some of the authors did not make the assumption $\rho \equiv 0$, considering that they were studying the integrated circulation of a baroclinic ocean.

In contrast to most work on the total mass transport theory, Neumann (1955, 1958) proves, proceeding from physical considerations, that allowance for JEBAT is essential. However, in concrete calculations he does not take this factor into account. It is true that in the works of Neumann we have the depth of baroclinicity H_1 instead of bottom topography, but this does not change the essence of the matter.

Term II on the right side of equation 101 is also in the work of Welander (1959).

But Welander considers that the effects of baroclinicity and bottom topography compensate each other in such a way that the height-averaged baroclinic ocean of variable depth behaves in the same way as a barotropic ocean of constant depth. For these reasons the works of Welander and Neumann refer to the total mass transport theory of a barotropic ocean. By means of scale analysis we have shown above that JEBAT is a principal term on the right side of equation 101. It is 5–10 times larger than rot τ. As we see from the scheme of equation 101, in the models of the total mass transport theory all arrows converge around rot τ. Some of the authors take into account one or two terms on the left side of this equation; others preserve the left side in more general form. A basic shortcoming of all these works, however, is the neglect of the main source of perturbation, JEBAT (at the same time the factor of less physical significance, rot τ, is preserved). The relation of Sverdrup does not apply in the baroclinic ocean of variable depth. This is shown by Sarkisyan (1969b) not only by means of qualitative estimates but also on the basis of diagnostic calculations.

Another problem discussed by Sarkisyan (1969b) is a comparison of the dynamic method and the method of total mass transport. As is known, the ocean surface topography, gradient currents in the surface layers, and, consequently, the integrated meridional transport in the baroclinic layer of the ocean are determined exclusively by the density field. In the theory of total mass transport the same quantity is calculated only by the wind field. Let us write the simple formulas of these two methods. According to Sverdrup's relation we have

$$S_y^{(Sv)} = \frac{1}{\rho_0 \beta} \,\text{rot}\, \tau \tag{103}$$

According to the dynamic method, as is seen from equation 100, we have

$$S_{yd} = \frac{g}{\rho_0 l_1} \int_0^{H_1} z \frac{\partial \rho}{\partial x} \, dz \tag{104}$$

where l_1 = const is some average value of the Coriolis parameter, and H_1 is the depth of a reference level. These two formulas are fundamentally different. When calculating the meridional transport from equation 103 we must assume that the wind field and Coriolis parameter are variable, but the seawater may be homogeneous. On the contrary, for the calculation of S_{yd} from equation 104 the seawater must be inhomogeneous, but the wind field and Coriolis parameter may be constant. Of course, the wind field is indirectly present in formula 104, to some extent, taking part in the redistribution of the density field. However, we do not have this indirect effect in formula 103; neither do we have the effect of heat and salt exchange processes. On the contrary, in formula 104 we have only the indirect effect of the wind, the effect of heat and salt exchange, and other factors due to the density field ρ.

From the very beginning of this century calculations have been carried out by the dynamic method. It is proved that the results of the calculations agree well with the observed data. Since 1947 the calculations have been performed with the aid of Sverdrup's relation and the results also seem to correspond to the observed data. Thus two contradictory relations repeatedly used in the theory of ocean currents on the one hand, and in the practice of oceanographic calculations on the other, have existed for more than a quarter of a century. In the generalized equations for auxiliary functions we have all the factors that have led us to both relations. We therefore have the opportunity to compare these relations and decide which of them better reflects the dynamics of stationary currents. For this purpose let us consider the initial relation 62. From the estimates given in Section 1.D it follows that in equation 62 we have only two principal factors: the left side and the first term on the right side. For analysis we must also retain the second term on the right side (rot τ), and therefore the simplified version of equation 62 takes the form

$$\beta S_y = -\frac{1}{\rho_0} J(H, p)_{z=H} + \frac{1}{\rho_0} \text{rot } \tau \tag{105}$$

Let us now consider two limiting cases.

1. $H = \text{const}$. In this case the Jacobian vanishes in equation 105 and we come to the relation of Sverdrup (equation 103).

2. $H \neq \text{const}$, $l = \text{const}$ ($\beta = 0$). Let us also neglect the second term on the right side of equation 105, which is small in comparison with the first term. We thus obtain

$$J(H, p)_{z=H} = 0 \tag{106}$$

Relation 106 is satisfied if we assume

$$\left(\frac{\partial p_H}{\partial x}\right)_{z=H} = \left(\frac{\partial p_H}{\partial y}\right)_{z=H} = 0 \tag{107}$$

We thus obtain the dynamic method in which the real bottom of the ocean serves as a reference level. In this case the sea-surface topography is determined from formula 99, but the total mass transport function from

$$\Psi_d = -\frac{g}{\rho_0 l_1} \int_0^H \rho z \, dz \tag{108}$$

Formula 108 is obtained from relations 71 and 72, if we take into account equation 99 and neglect the small terms due to a wind stress. Formula 104 easily follows from formula 108 if we remember the definition of the total mass transport function. Thus equations 103 and 104 are two limiting cases of relation 105. Because these two formulas do not agree physically, either both of them are wrong or only one of them is suitable. It is shown on the basis of diagnostic calculations (see Section 1.D) that S_{yd} is more or less close to the value of S_y, obtained from the general equation for ζ or Ψ, but $S_y^{(Sv)}$ is nearly one order of magnitude smaller.

Thus the qualitative analysis and the first calculations reveal the principal shortcomings of the total mass transport. But all the deficiencies of this theory will emerge only after performing a large number of computations.

For this purpose we proceed from equation 73. Remembering the estimates obtained in Section 1.D, we can further simplify this equation. The simplified form is

$$L(\Psi) = \underbrace{\frac{1}{2\alpha} \Delta\Psi}_{\text{III}} + \underbrace{J(H, \Psi)}_{\text{II}} + \underbrace{\frac{H\beta}{l}\frac{\partial\Psi}{\partial x}}_{\text{I}}$$

$$= \underbrace{\frac{H}{\rho_0 l} \text{rot } \tau}_{\text{VII}} - \underbrace{\frac{g}{2\rho_0\alpha l} \int_0^H z\Delta\rho \, dz}_{\text{III}} - \underbrace{\frac{g}{\rho_0 l} \int_0^H zJ(H, \rho) \, dz}_{\text{II}} \tag{109}$$

The differential operator of the left side of this equation is identical with the operator of equation 86. Both are derived on the basis of the same estimates of characteristic values. The only difference between them consists in the following: on the right side of equation 86 two dominant terms belong to groups I and II, whereas on the right-hand side of equation 109 only one principal term is of group II (JEBAT). This leads to considerable discrepancies in the behavior of functions ζ and Ψ. Sverdrup's relation is a particular case of equation 109. Consequently, we can verify it by specific diagnostic calculations performed on the basis of this equation. The boundary condition for solving equation 109 is obvious—Ψ is given at the contour of the ocean area being considered.

Having determined Ψ, one can easily find u, v, and w. The formulas for them readily follow from appropriate formulas of the D_1 model by means of relations 71 and 72.[1] The horizontal current velocity is calculated for the following formulas.

At the ocean surface:

$$u = -\frac{1}{\rho_0 l}\sqrt{\frac{v'}{v}}\frac{\partial p_a}{\partial y} - \frac{1}{H}\frac{\partial \Psi}{\partial y} + \frac{g}{\rho_0 lH}\int_0^H (H-z)\frac{\partial \rho}{\partial y}\,dz \tag{110}$$

$$v = \frac{1}{\rho_0 l}\sqrt{\frac{v'}{v}}\frac{\partial p_a}{\partial x} + \frac{1}{H}\frac{\partial \Psi}{\partial x} - \frac{g}{\rho_0 lH}\int_0^H (H-z)\frac{\partial \rho}{\partial x}\,dz \tag{111}$$

At the intermediate layers:

$$u = -\frac{1}{H}\frac{\partial \Psi}{\partial y} + \frac{g}{\rho_0 l}\int_z^H \frac{\partial \rho}{\partial y}\,dz - \frac{g}{\rho_0 lH}\int_0^H z\frac{\partial \rho}{\partial y}\,dz \tag{112}$$

$$v = \frac{1}{H}\frac{\partial \Psi}{\partial x} - \frac{g}{\rho l}\int_0^H \frac{\partial \rho}{\partial x}\,dz + \frac{g}{\rho_0 lH}\int_0^H z\frac{\partial \rho}{\partial x}\,dz \tag{113}$$

When passing from formulas 89 and 90 to 110 to 111, we replaced wind stress by P_a with the aid of formula 95.

The vertical component of flow velocity equals zero at the ocean surface, but at the other layers it is calculated from the formula

$$w = -\frac{1}{2\rho_0 l\alpha'}\Delta p_a + \frac{\beta}{2\alpha' l^2 \rho_0}\left(\frac{\partial p_a}{\partial x} + \frac{\partial p_a}{\partial y}\right) + \frac{\beta z}{lH}\frac{\partial \Psi}{\partial x}$$
$$- \frac{g\beta z}{H\rho_0 l^2}\int_0^H (H-z)\frac{\partial \rho}{\partial x}\,dz + \frac{g\beta}{\rho_0 l^2}\int_0^z (z-\zeta)\frac{\partial \rho}{\partial x}\,d\zeta \tag{114}$$

Equation 109 and formulas 110–114 constitute the D_1' model, that is, the simplest diagnostic model for computing currents from a known density field with the aid of the total mass transport function Ψ.

D. Allowance for Inertial Terms and Lateral Eddy Viscosity by Successive Approximations Method (D_2 Model)

There are large-scale intense currents (Gulf Stream, North Atlantic current, Kuroshio, etc.) in many regions of the World Ocean. In such regions allowance for nonlinear terms and for the effect of lateral exchange may yield a considerable correction (about 10–30%) to the calculated fields of current velocities even when we are studying only large-scale currents; on the other hand these factors are not so

---

[1] The last terms of these formulas are essential only in shallow regions of the basins.

large as to compete successfully with the other terms in equations of motion. We therefore need not construct a special numerical nonlinear model. In such cases the nonlinear terms can be taken into account by successive approximations with the aid of a corresponding generalization of model D_1. The derivation equations and formulas for the D_2 model is completely identical to that of D_1 model. The only difference consists in taking account of nonlinear terms and lateral eddy viscosity in the right sides of equations. We therefore write their final form.

The equation for the sea-surface topography is

$$L(\zeta) = \frac{l}{2\rho_0 gH\alpha} \Delta p_a - \frac{l}{2\alpha\rho_0 H} \int_0^H \Delta\rho \, dz - \frac{l}{\rho_0 H} \int_0^H J(H, \rho) \, dz$$

$$- \frac{\beta}{\rho_0 H} \int_0^H (H - z)\frac{\partial\rho}{\partial x} \, dz + \frac{l}{gH}\left[\frac{\partial}{\partial y}\int_0^H A' \, dz - \frac{\partial}{\partial x}\int_0^H B' \, dz\right] \quad (115)$$

where

$$A' = \frac{\partial u^2}{\partial x} + \frac{\partial uv}{\partial y} + \frac{\partial uw}{\partial z} - A_l\Delta u \quad (116)$$

$$B' = \frac{\partial uv}{\partial x} + \frac{\partial v^2}{\partial y} + \frac{\partial vw}{\partial z} - A_l\Delta v \quad (117)$$

The relations for flow velocity at the ocean surface are

$$u = -\frac{1}{\rho_0 l}\sqrt{\frac{v'}{v}}\frac{\partial p_a}{\partial y} - \frac{g}{l}\frac{\partial\zeta}{\partial y} + \left(A_l\Delta v - u\frac{\partial v}{\partial x} - v\frac{\partial v}{\partial y}\right)_{z=0} \quad (118)$$

$$v = -\frac{1}{\rho_0 l}\sqrt{\frac{v'}{v}}\frac{\partial p_a}{\partial x} + \frac{g}{l}\frac{\partial\zeta}{\partial x} - \left(A_l\Delta u - u\frac{\partial u}{\partial x} - v\frac{\partial u}{\partial y}\right)_{z=0} \quad (119)$$

and at the intermediate layers

$$u = -\frac{g}{l}\frac{\partial\zeta}{\partial y} - \frac{g}{\rho_0 l}\int_0^z \frac{\partial\rho}{\partial y} \, dz - \frac{B'}{l} \quad (120)$$

$$v = \frac{g}{l}\frac{\partial\zeta}{\partial x} + \frac{g}{\rho_0 l}\int_0^z \frac{\partial\rho}{\partial x} \, dz - \frac{A'}{l} \quad (121)$$

The computational flow sheet is as follows. First we calculate the first approximation of fields ζ, u, v, and w on the basis of the D_1 model. The only difference is that formula 93 is used for w instead of formula 94. Then we take into account inertial terms and lateral eddy viscosity in equation 115 and formulas 118–121 by successive approximations. At every $(n + 1)$th step approximation of w is determined by the $(n + 1)$th approximation of u and v with the aid of formula 93. Model D_2 differs from D_1 not only by inertial terms and the effect of lateral eddy viscosity on the right sides of relations 115 and 118–121, but also in a considerable discrepancy in boundary conditions. It is regarded that both components of flow velocity are given at the lateral boundaries as functions of three coordinates. In particular, at the solid sections of the boundary the condition of no motion is given.

We can easily use these boundary conditions for computing of the horizontal flow velocity. When replacing the derivatives of flow velocities on the right sides of formulas 118–121 by the finite-difference quotients at the grid points nearest to the boundary, the values of $u_{i,j,k}$ and $v_{i,j,k}$ are substituted for the given numbers at the boundary of the region.

In conclusion, a few words about the computation of ζ at the contour of a basin are in order. Let us derive formulas of type 84 and 85, but with due regard for A' and B'. Deducing these formulas offers no difficulty and therefore we write their final form. The result is

$$\left(1 - \frac{1}{2\alpha H}\right)\frac{\partial \zeta}{\partial x} - \frac{1}{2\alpha H}\frac{\partial \zeta}{\partial y} = \frac{l}{gH}S_y + \frac{1}{\rho_0 H}\int_0^H z\frac{\partial \rho}{\partial x}dz - \frac{1}{\rho_0}\int_0^H \frac{\partial \rho}{\partial x}dz + \frac{\tau_x}{\rho_0 gH}$$
$$+ \frac{1}{2\alpha H\rho_0}\left(\int_0^H \frac{\partial \rho}{\partial x}dz + \int_0^H \frac{\partial \rho}{\partial y}dz\right) - \frac{1}{gH}\int_0^H A'\,dz$$

(122)

$$\left(1 + \frac{1}{2\alpha H}\right)\frac{\partial \zeta}{\partial y} - \frac{1}{2\alpha H}\frac{\partial \zeta}{\partial x} = -\frac{l}{gH}S_x + \frac{1}{\rho_0 H}\int_0^H z\frac{\partial \rho}{\partial y}dz - \frac{1}{\rho_0}\int_0^H \frac{\partial \rho}{\partial y}dz + \frac{\tau_y}{\rho_0 gH}$$
$$+ \frac{1}{2\alpha H\rho_0}\left(\int_0^H \frac{\partial \rho}{\partial y}dz - \int_0^H \frac{\partial \rho}{\partial x}dz\right) + \frac{1}{gH}\int_0^H B'\,dz$$

(123)

In contrast to equations 84 and 85, we have taken into account the nonlinear terms and lateral eddy viscosity effect (expressions A' and B') in the relations 122 and 123. The formulas obtained are more complicated than those of 87 and 88. The main discrepancy between them is as follows. Formula 87 of the D_1 model is applicable only to the meridional sections of the ocean boundary, and formula 88 to the zonal. As to formulas 122 and 123, they remain valid for all sections of the boundary. This is due to the allowance for lateral eddy viscosity, and therefore both components of the water flux are assumed to be specified at every part of the boundary. Hence the relations 122 and 123 constitute a system of two equations for $\partial \zeta/\partial x$, $\partial \zeta/\partial y$. Let us denote the right sides of these equations by φ' and φ'_1, respectively, and solve this system of equations neglecting the terms multiplied by the square of small parameter $1/\alpha H$. The results are

$$\frac{\partial \zeta}{\partial x} = \frac{2\alpha H - 1}{2\alpha H - 2}\varphi' + \frac{1}{2\alpha H - 2}\varphi'_1$$

(124)

$$\frac{\partial \zeta}{\partial y} = \frac{2\alpha H}{2\alpha H - 1}\varphi'_1 - \frac{1}{2\alpha H - 2}\varphi'$$

(125)

The question of utilizing of these formulas for the determination of ζ at the boundary of a basin is discussed in Section 3.

E. Model for Computing Equatorial Currents (D_3 Model)

The preceding models are suitable for computing the World Ocean large-scale circulation without taking into account local peculiarities of equatorial currents. This can be obtained by a corresponding choice of the finite-difference grid. Namely, a grid is constructed in such a way that there are no grid points at the equator. The grid points nearest to the equator are situated at half-step distances north and south from it.

Another problem must be considered here. Supposing that at some distance to the north and south from the equator all the characteristics of the large-scale ocean circulation are known, we come to the problem of computing these quantities on the equator and in its environs. Then we form the difference grid of models D_1 and D_3 and calculate the World Ocean circulation, including the equator, by successive approximations.

In the D_1 model the simplifications are done with due regard for the smallness of inertial terms and supposition that $l \neq 0$. In that way we can set up a simple, economic scheme for the computation of the World Ocean currents by avoiding a narrow equatorial region of 4–5° width.

In the D_3 model we must include the inertial terms, but the region of model application is small. On the other hand we use the fact of Coriolis parameter smallness in the D_3 model.

Before we state the main points of model D_3 let us qualitatively analyze the peculiarities of currents in the equatorial zone on the basis of a linear model.

Setting $A \equiv B \equiv 0$ in equations 49 and 50 and solving them in regard to u and v with boundary conditions 11 and 17, we have (Sarkisyan, 1970)

$$u + iv = \frac{sh\,[(1 + i)\alpha(H - z)]}{(1 + i)\alpha v \rho_0\, ch\,(1 + i)\alpha H}\,(\tau_x + i\tau_y)$$

$$- \frac{1}{2(1 + i)\alpha}\left[\int_0^z e^{(1+i)\alpha(\xi - z)}F_1\,d\xi + \int_z^H e^{-(1+i)\alpha(\xi - z)}F\,d\xi\right]$$

$$- \frac{e^{-(1+i)\alpha H}}{2(1 + i)\alpha\, ch\,[(1 + i)\alpha H]}\int_0^H F_1\, ch\,[(1 + i)\alpha(\xi - z)]\,d\xi$$

$$- \frac{1}{2(1 + i)\alpha\, ch\,[(1 + i)\alpha H]}\int_0^H F_1\, sh\,[(1 + i)\alpha(H - \xi - z)]\alpha\xi \qquad (126)$$

where

$$\alpha = \sqrt{\pm\frac{l}{2v}}, \qquad F_1 = \frac{1}{\rho_0 v}\left(\frac{\partial p}{\partial x} + i\frac{\partial p}{\partial y}\right) \qquad (127)$$

The $\pm$ sign in equation 127 has the following meaning: the upper sign corresponds to the northern hemisphere and the lower sign to the southern hemisphere. If we simplify the right side of formula 126 for a nonequatorial region of the deep basin $(l \neq 0, H > 100\,\text{m})$, then we obtain the approximate formula 53. Here we are interested only in the equatorial zone in which everywhere $\sin \varphi \ll 1$.

In the equator environs all expressions like αH, $\alpha(\xi - z)$, etc., are infinitesimal quantities tending to zero when approaching the equator. We can therefore decompose all the exponentials on the right side of formula 126 into the degrees of corresponding arguments and even retain the small quantities of the first order, that is, consider $\exp\,[\pm(1 + i)\alpha H] \approx 1 \pm \alpha H(1 + i)$. After this substitution and some simple manipulations a part of expression 126 takes the final form, but some terms of it at the equator become indeterminate. Expanding them according to l'Hospital's rule and passing to the limit for $l \to 0$, we finally come to the following simple formula:

$$u + iv = \frac{H - z}{v\rho_0}(\tau_x + i\tau_y) + \int_0^z (z - \xi)F_1\,d\xi - \int_0^H (H - \xi)\,F_1 d\xi \qquad (128)$$

or

$$u = \frac{H-z}{\nu\rho_0}\tau_x - \frac{1}{\nu\rho_0}\int_0^z (z-\xi)\frac{\partial p}{\partial x}d\xi - \frac{1}{\rho_0\nu}\int_0^H (H-\xi)\frac{\partial p}{\partial x}d\xi \qquad (129)$$

$$v = \frac{H-z}{\nu\rho_0}\tau_y + \frac{1}{\nu\rho_0}\int_0^z (z-\xi)\frac{\partial p}{\partial y}d\xi - \frac{1}{\rho_0\nu}\int_0^H (H-\xi)\frac{\partial p}{\partial y}d\xi \qquad (130)$$

Formulas 129 and 130 can serve for the approximate calculations of u and v at the equator's environs. At the very equator they exactly satisfy equations 49 and 50 (for $A \equiv B \equiv 0$) and boundary conditions 11 and 17. From the equator poleward the exactness of these formulas takes a turn for the worse and at some critical latitude they become unsuitable, "giving way" to formula 53.

It is seen from formulas 129 and 130 that there are neither boundary layers nor an Ekman spiral at the equator. Here both pure drift currents (having the direction exactly with the wind) and the effect of bottom friction encompass the whole ocean thickness.

Let us investigate the direction of gradient currents on the equator. Letting $\partial p/\partial y = 0$, $\partial p/\partial x = (\partial p/\partial x)_0 e^{-\gamma z}$, $\gamma = \text{const} > 0$ for the sake of simplicity, from formulas 129 and 130 we obtain for the components of the gradient part of the flow velocity

$$u_g \approx -\frac{e^{-\gamma z}}{\nu\rho_0\gamma^2}\left(\frac{\partial p}{\partial x}\right)_0, \qquad v_g = 0 \qquad (131)$$

We assume that the regularity for the pressure gradient accepted takes place not only on the equator but also in the equatorial zone being considered. Then from the ordinary geostrophic relations we obtain the following.

For the northern hemisphere:

$$u_g^N = 0, \qquad v_g^N = \frac{e^{-\gamma z}}{2\omega \sin \varphi \rho_0}\left(\frac{\partial p_0}{\partial x}\right)_0 \qquad (132)$$

For the southern hemisphere:

$$u_g^S = 0, \qquad v_g^S = -\frac{e^{-\gamma z}}{2\omega \sin(-\varphi)}\frac{\partial p_0}{\partial x} = -\frac{e^{-\gamma z}}{2\omega \sin \varphi \rho_0}\frac{\partial p_0}{\partial x} \qquad (133)$$

Thus in the northern hemisphere a gradient current deviates to the right from the pressure gradient **G**, and in the southern hemisphere, to the left. At the equator gradient current takes the direction of the pressure gradient vector. This can be expressed schematically as in Fig. 1.

As is known from the practice of diagnostic calculations, in equatorial regions the sea level at the west coast is higher than that at the east, therefore a gradient current

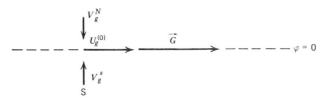

Fig. 1. Schematic picture of the gradient countercurrent at the equator.

is directed from the west to the east. On the sea surface the current is directed with the wind, that is, from east to west. With increasing depth the drift components of current fade more rapidly than do gradient components, and the Equatorial Under-current is formed.

Thus the simple linear model of the equatorial circulation is a possible qualitative explanation of the formation of the Equatorial Undercurrent (of the Cromwell and Lomonosov currents type). However, the equatorial currents have in fact a very complicated character, calling for three-dimensional analysis with due regard for nonlinear terms in the equations of motion.

For setting up just such a nonlinear model of equatorial currents, let us first deduce an equation for the determination of ζ. Let us integrate equations 1 and 2 with respect to z from O to H. Differentiating the first of them with respect to x, the second with respect to y and adding them, we obtain after simple transformations (with due regard for relations 33 and 57)

$$\underbrace{\Delta\zeta}_{\text{`II'}} + \underbrace{\frac{1}{H}\frac{\partial H}{\partial x}\frac{\partial \zeta}{\partial x} + \frac{1}{H}\frac{\partial H}{\partial y}\frac{\partial \zeta}{\partial y}}_{\text{I}} = \underbrace{-\frac{1}{\rho_0 H}\int (H-z)\Delta\rho\,dz}_{\text{II}}$$

$$\underbrace{-\frac{1}{\rho_0 H}\left[\frac{\partial H}{\partial x}\int_0^H \frac{\partial \rho}{\partial x}\,dz + \frac{\partial H}{\partial y}\int_0^H \frac{\partial \rho}{\partial y}\,dz\right]}_{\text{I}} \underbrace{-\frac{\beta}{gH}\int_0^H u\,dz}_{\text{III}}$$

$$\underbrace{-\frac{l}{gH}\int_0^H \left(\frac{\partial v}{\partial x} - \frac{\partial u}{\partial y}\right)dz}_{\text{V}} \underbrace{-\frac{1}{gH}\int_0^H \left(\frac{\partial A_1}{\partial x} + \frac{\partial B_1}{\partial y}\right)dz}_{\text{IV}}$$

$$\underbrace{+\frac{1}{gH\rho_0}\left(\frac{\partial \tau_x}{\partial x} + \frac{\partial \tau_y}{\partial y}\right)}_{\text{VI}} \underbrace{+\frac{1}{gH\rho_0}\left(\frac{\partial \tau_x{}^u}{\partial x} + \frac{\partial \tau_y{}^H}{\partial y}\right)}_{\text{VII}} \qquad (134)$$

where

$$\tau_x{}^H = \rho_0 v\left(\frac{\partial u}{\partial z}\right)_{z=H}, \qquad \tau_y{}^H = \rho_0 v\left(\frac{\partial v}{\partial z}\right)_{z=H} \qquad (135)$$

$$A_1 = \frac{\partial}{\partial x}u^2 + \frac{\partial}{\partial y}uv, \qquad B_1 = \frac{\partial}{\partial x}uv + \frac{\partial}{\partial y}v^2 \qquad (136)$$

In deriving equation 134 we have made one simplification: we have neglected the effect of bottom topography in lateral mixing terms. As a result, owing to relation 57, we do not have the terms due to horizontal mixing in equation 134. We have often neglected the secondary effect of bottom topography, retaining the main effect (in the given case, group I).

Because of relation 57 we also have no local time derivatives in equation 134. When constructing this equation we took into account the smallness of the parameter l in the environs of the equator. Therefore, the Coriolis force acceleration vorticity (group V) is on the right side of this equation and is taken into account by successive approximations. Some terms in equations 134 and 69 are formally alike. In fact, they differ considerably both in physical essence and in their orders of magnitude.

Therefore, the other order of numeration is accepted in equation 134. The terms on the right and left sides having the same essence are marked by the same number. The physical sense of the terms in group II is as follows: on the left side is the effect of the pressure gradient at the ocean surface, and on the right side is the integral from the purely baroclinic part of the pressure gradient. The sense of the other terms is obvious.

If we estimate the orders of magnitudes of terms in equation 134 with the aid of the technique cited in Section 1.D, we see that the terms of group II are dominant. So in contrast to equation 69 equation 134 has no small parameter before the highest-order derivative. If equation 69 is the vorticity equation, then equation 134 is the equation of divergence. Such a method of constructing the equation for the determination of ζ makes it possible to oppose the important terms of group II with nonlinear terms (group IV). We thus facilitate the problem of constructing a stable finite-difference scheme for the computation of ζ at the equator and its environs.

The set of equations for diagnostic calculations of currents at the equator and its environs (the D_3 model) is formed as follows. For the determination of ζ we proceed from equation 134. The components of the horizontal flow velocity are calculated from equations 1 and 2 without any simplifications. We determine the vertical velocity component from formula 93 and the pressure field from 33.

The method of computing ζ at the side boundaries is given in Section 2.G.

F. Models for Computing Stream-Type Currents (Models D_4 and D_5)

The D_2 model is simple enough, economical, and suitable for computing large-scale intense currents. Using this model, we can take into account the correction due to nonlinear terms and horizontal mixing in regions of intense currents by decreasing the horizontal grid step. It is supposed in this connection that these corrections are comparatively small (less than 20%). However, it may happen that the further decrease of grid steps causes an increase in the role of nonlinear terms, and therefore the numerical scheme is not steady. In other words, in the D_2 model the horizontal grid step is limited from below. We need another model for the investigation of smaller-scale currents.

In order to construct a new model, we refer to equations 49 and 50. In these equations we drop the terms containing w because of their smallness. After such simplifications the approximate solution of these equations with boundary conditions 11 and 17 takes the form (Sarkisyan, 1969b)

$$u + iv = \frac{1}{\rho_0 l}\left(i\frac{\partial p}{\partial x} - \frac{\partial p}{\partial y}\right) + \frac{1}{l}(iA_2 - B_2) + \frac{\tau_x + i\tau_y}{\alpha\rho_0 v(1+i)}e^{-\alpha z(1+i)}$$

$$-\left(\frac{1}{\rho_0 l}i\frac{\partial p}{\partial x} - \frac{\partial p}{\partial y}\right)_{z=n}e^{(1+i)\alpha(z-H)} - \frac{1}{l}(iA_2 - B_2)_{z=H}e^{(1+i)\alpha(z-H)} \qquad (137)$$

where

$$A_2 = \frac{\partial u}{\partial t} + u\frac{\partial u}{\partial x} + v\frac{\partial u}{\partial y} - A_l\Delta u$$

$$B_2 = \frac{\partial v}{\partial t} + u\frac{\partial v}{\partial x} + v\frac{\partial v}{\partial y} - A_l\Delta v \qquad (138)$$

If we exclude the thin bottom boundary layer (thickness of the order of 10 m), then the terms on the right side of equation 137 having the factor $\exp\left[(1 + i)\alpha(z - H)\right]$ can be neglected. Further, separating the real and imaginary parts and performing some simple transformations, we obtain at the ocean surface the following equations:

$$\frac{\partial u}{\partial t} + u\frac{\partial u}{\partial x} + v\frac{\partial u}{\partial y} - A_l\Delta v - lv = -\frac{1}{\rho_0}\frac{\partial p}{\partial x} - \frac{\alpha}{\rho_0}(\tau_y - \tau_x) \qquad (139)$$

$$\frac{\partial v}{\partial t} + u\frac{\partial v}{\partial x} + v\frac{\partial v}{\partial y} - A_l\Delta u + lu = -\frac{1}{\rho_0}\frac{\partial p}{\partial y} - \frac{\alpha}{\rho_0}(\tau_x + \tau_y) \qquad (140)$$

The equations for the intermediate layers may be obtained from equations 139 and 140 if we set $\tau_x \equiv \tau_y \equiv 0$. The essence of the simplifications is clear if we compare equations 139 and 140 with the corresponding equations 1 and 2: at the ocean surface we have approximately expressed the effect of vertical eddy viscosity in terms of the tangential wind-stress components; at the other levels we have dropped this effect. Moreover, we have neglected in the intermediate layers the terms containing w.

In order to derive the equation for ζ we refer to relation 58 in which we replace A and B by A_2 and B_2. Doing simple identical transformations (adding and subtracting terms such as $\partial u_g/\partial b$, $\partial v_g/\partial t$) and using formula 33, we obtain:

$$\frac{H}{l}\frac{\partial}{\partial t}\Delta\zeta - \frac{\beta H}{l^2}\frac{\partial}{\partial t}\frac{\partial\zeta}{\partial y} + \frac{1}{2\alpha}\Delta\zeta + J(H, \zeta) + \frac{H\beta}{l}\frac{\partial\zeta}{\partial x} = -\frac{H}{l\rho_0}\int_0^H\frac{\partial}{\partial t}\Delta\rho\,dz$$

$$-\frac{\beta}{\rho_0 l^2}\int_0^H(H - z)\frac{\partial}{\partial t}\frac{\partial\rho}{\partial y}\,dz + \frac{1}{\rho_0 g}\,\mathrm{rot}\,\tau - \frac{1}{2\alpha\rho_0}\int_0^H\Delta\rho\,dz - \frac{1}{\rho_0}\int_0^H J(H, \rho)\,dz$$

$$-\frac{\beta}{\rho_0 l}\int_0^H(H - z)\frac{\partial\rho}{\partial x}\,dz - \frac{\beta}{g}\int_0^H(v - v_g)\,dz + \frac{1}{g}\left[\frac{\partial}{\partial y}\int_0^H A_2'\,dz - \frac{\partial}{\partial x}\int_0^H B_2'\,dz\right] \quad (141)$$

where

$$A_2' = \frac{\partial(u - u_g)}{\partial t} + u\frac{\partial u}{\partial x} + v\frac{\partial u}{\partial y} - A_l\Delta u$$

$$B_2' = \frac{\partial(v - v_g)}{\partial t} + u\frac{\partial v}{\partial x} + v\frac{\partial v}{\partial y} - A_l\Delta v \qquad (142)$$

For carrying out the diagnostic calculations the first and second terms on the right side of this equation should be neglected. We retain them only in prognostic calculations.

The technique of computing ζ at the side boundaries is set forth in Section 2.G.

The iterative scheme of the D_4 model consists in the determination of u, v, and ζ from the three equations by time steps. In this case the pressure field in equations 139 and 140 is substituted for the fields of ζ and ρ from formula 33. In Section 3 we dwell on the difference method of solving this system. The vertical velocity component does not take part in iterations, but is easily calculated from the known u, v, and ζ with the aid of the following formula:

$$w = -\frac{1}{\rho_0 l}\,\mathrm{rot}\,\tau - \frac{\beta}{\rho_0 l^2}\tau_x + \frac{g\beta z}{l^2}\frac{\partial\zeta}{\partial x}$$

$$+ \frac{g\beta}{\rho_0 l^2}\int_0^z(z - \xi)\frac{\partial\rho}{\partial x}\,d\xi - \frac{1}{l}\int_0^z\left(\frac{\partial A_2}{\partial y} - \frac{\partial B_2}{\partial x}\right)dz \qquad (143)$$

We do not have the time derivatives on the right side of equation 143, since in diagnostic calculations the vertical velocity component is computed from the steady-state horizontal current velocity.

If we are calculating the mesoscale currents (requiring a small horizontal grid step), then scheme D_4 may be nonsteady. Moreover, in the regions of upwelling the vertical velocity component is large. In these cases the computations are carried out with the aid of model D_5, the essentials of which are as follows. Without any simplifications equations 1 and 2 are used for the computation of u and v. Formulas 33 and 93 are for P and w, respectively. For the calculation of ζ we use equation 141 in which A_2' and B_2' are supplemented with $w(\partial u/\partial z)$ and $w(\partial v/\partial z)$. The D_5 model is the most labor-consuming one for the computer. Here we have to solve the set of four equations for u, v, w, and ζ at every time step. It requires one or two orders of magnitude more machine time than do the other models. In addition, the D_5 model may be carried out only on computers having large, high-speed memory. Therefore, this model should be used only when the possibilities of the other models are exhausted. The difference methods of performance of this model are given in Section 3.

G. Determination of Sea-Surface Topography at the Solid Part of the Basin Contour

As we have seen above, when taking into account the inertial terms and the effect of lateral exchange we have to solve complicated relations similar to equations 122–125 for the determination of ζ at the boundary of a basin. This problem can be simplified; one such simplification was mentioned in Section 1. Namely, at the contour of the boundary, ζ can be approximately calculated from the simplest formula 99. We show on the basis of specific results of calculations the expediency of such a technique. Here we give the variant of the determination of ζ at the solid part of the contour of a basin.

When we make allowance for the lateral mixing effect, at the solid part of the contour we have the conditions of no motion:

$$u = v = \frac{\partial u}{\partial s} = \frac{\partial v}{\partial s} = \frac{\partial^2 u}{\partial s^2} = \frac{\partial^2 v}{\partial s^2} = 0$$

where s is a tangent to the contour of the basin. If the solid boundaries are assumed to be vertical walls, the derivatives for u and v also turn into zeros along the vertical. Thus we can significantly simplify equations 1 and 2 at the contour of the boundary. On latitudinal sections of the boundary equation 1 takes the form

$$\frac{1}{\rho_0} \frac{\partial p}{\partial x} = A_l \frac{\partial^2 u}{\partial y^2} \tag{144}$$

On the meridional sections of the boundary equation 2 becomes

$$\frac{1}{\rho_0} \frac{\partial p}{\partial y} = A_l \frac{\partial^2 v}{\partial x^2} \tag{145}$$

Therefore, for the computation of ζ at the latitudinal and meridional parts of the boundary we obtain:

$$\frac{\partial \zeta}{\partial x} = -\frac{1}{\rho_0 H} \int_0^H (H - z) \frac{\partial \rho}{\partial x} \, dz + \frac{A_l}{gH} \int_0^H \frac{\partial^2 u}{\partial y^2} \, dz \tag{146}$$

$$\frac{\partial \zeta}{\partial y} = -\frac{1}{\rho_0 H} \int_0^H (H - z) \frac{\partial \rho}{\partial y} \, dz + \frac{A_l}{gH} \int_0^H \frac{\partial^2 v}{\partial x^2} \, dz \tag{147}$$

In the next section we treat the numerical methods of solving equations 146 and 147. Here we consider the applications of the simple method given for the determination of ζ.

1. When calculating the currents in the large ocean basin it is sufficient to make allowance for inertial terms and lateral mixing in the small part of the basin where we have stream-type current. At the liquid section of the boundary the sea-surface topography ζ may be given as the solution to a problem of computing of a large-scale current in the entire basin, and at the solid parts, it is determined from equations 146 and 147.

2. When calculating equatorial currents, the boundaries of the zone being considered are at several degrees distance north and the south from the equator. We therefore can consider ζ at the liquid boundary to be known from the solution of linear problem.

At the solid parts of the boundary we must calculate ζ on the basis of relations 146 and 147.

3. When solving the problem of the World Ocean circulation, we have no liquid boundaries and therefore the ζ function at the coastlines can be calculated from the same simple formulas 146 and 147.

H. Basic Relations for the Southern Hemisphere and in Spherical Coordinates

All the preceding equations and formulas are written, for convenience sake, in rectangular coordinates for the northern hemisphere. However, many calculations of large-scale circulation are performed for the World Ocean with due regard for the sphericity of the earth. We give here the most important relations in spherical coordinates. Many formulas and terms are identical for both hemispheres. The double sign $\pm$ in front of some terms has the following meaning: the upper sign corresponds to the northern hemisphere and the lower sign to the southern one. Relations 84 and 85 for the determination of ζ at the contour of the basin have the following form in spherical coordinates:

$$\left(1 - \frac{1}{2\alpha H}\right)\frac{\partial \zeta}{\partial \theta} = -\frac{1}{\rho_0 H}\int_0^H (H-z)\frac{\partial \rho}{\partial \theta}\,dz$$

$$+ \frac{1}{2\alpha H \rho_0}\int_0^H \left(\frac{\partial \rho}{\partial \theta} \pm \frac{1}{\sin \theta}\frac{\partial \rho}{\partial \lambda}\right)dz \pm \frac{1}{2\alpha H \sin \theta}\frac{\partial \zeta}{\partial \lambda}$$

$$- \frac{1}{2\alpha' \rho_0 g H}\left(\frac{\partial p_a}{\partial \theta} \pm \frac{1}{\sin \theta}\frac{\partial p_a}{\partial \lambda}\right) + \frac{2\omega a \cos \theta}{g}v_\lambda \quad (148)$$

$$\left(1 - \frac{1}{2\alpha H}\right)\frac{\partial \zeta}{\partial \lambda} = -\frac{1}{\rho_0 H}\int_0^H (H-z)\frac{\partial \rho}{\partial \lambda}\,dz$$

$$+ \frac{1}{2\alpha H \rho_0}\int_0^H \left(\frac{\partial \rho}{\partial \lambda} \mp \sin \theta \frac{\partial \rho}{\partial \theta}\right)dz \mp \frac{\sin \theta}{2\alpha H}\frac{\partial \zeta}{\partial \theta}$$

$$- \frac{1}{2\alpha' \rho_0 g H}\left(\frac{\partial p_a}{\partial \lambda} \mp \sin \theta \frac{\partial p_a}{\partial \theta}\right) - \frac{2\omega a \sin \theta \cos \theta}{g}v_\theta \quad (149)$$

where

$$\alpha = \sqrt{\frac{\omega|\cos\theta|}{\nu}}, \qquad \alpha' = \sqrt{\frac{\omega|\cos\theta|}{\nu'}} \tag{150}$$

and v_λ and v_θ are the components of the current velocity, averaged over the height. Equations 86 and 109 for computing ζ and Ψ at interior points of the basin in spherical coordinates are as follows:

$$\frac{1}{2\alpha}\Delta\zeta \pm \frac{1}{\sin\theta}J(H,\zeta) + \frac{H}{|\cos\theta|}\frac{\partial\zeta}{\partial\lambda} = \frac{1}{2\alpha'\rho_0 g}\Delta p_a$$

$$-\frac{1}{2\alpha\rho_0}\int_0^H \Delta\rho\,dz \mp \frac{1}{\rho_0\sin\theta}\int_0^H \mathcal{J}(H,\rho)\,dz - \frac{1}{\rho_0|\cos\theta|}\int_0^H (H-z)\frac{\partial\rho}{\partial\lambda}\,dz \tag{151}$$

$$\frac{1}{2\alpha}\Delta\Psi \pm \frac{1}{\sin\theta}J(H,\Psi) + \frac{H}{|\cos\theta|}\frac{\partial\Psi}{\partial\lambda} = \frac{H}{4\omega\alpha'\rho_0\cos\theta}\Delta p_a$$

$$-\frac{g}{4\omega\alpha\rho_0\cos\theta}\int_0^H z\Delta\rho\,dz - \frac{g}{2\omega\rho_0\sin\theta|\cos\theta|}\int_0^H zJ(H,\rho)\,dz \tag{152}$$

Here,

$$J(b,c) = \frac{\partial b}{\partial\theta}\frac{\partial c}{\partial\lambda} - \frac{\partial b}{\partial\lambda}\frac{\partial c}{\partial\theta} \tag{153}$$

The expression of the Laplace operator in spherical coordinates is given in formula 30. The tangential wind-stress components are expressed in terms of the atmospheric pressure at sea level with the aid of the following formulas:

$$\tau_\lambda = \frac{1}{2a\alpha'}\left(\pm\frac{\partial p_a}{\partial\theta} - \frac{1}{\sin\theta}\frac{\partial p_a}{\partial\lambda}\right)$$

$$\tau_\theta = \frac{1}{2a\alpha'}\left(-\frac{\partial p_a}{\partial\theta} \mp \frac{1}{\sin\theta}\frac{\partial p_a}{\partial\lambda}\right) \tag{154}$$

which are analogous to formulas 95.

The formulas for the computation of the horizontal flow velocity components in terms of the total mass transport function are as follows:

$$v_\theta = v_\theta{}^d - \frac{1}{\alpha H\sin\theta}\frac{\partial\Psi}{\partial\lambda} + \frac{g}{2\omega\rho_0\alpha\sin\theta\cos\theta}\left[\int_z^H \frac{\partial\rho}{\partial\lambda}\,dz - \frac{1}{H}\int_0^H z\frac{\partial\rho}{\partial\lambda}\,dz\right]$$

$$v_\lambda = v_\lambda{}^d + \frac{1}{\alpha H}\frac{\partial\Psi}{\partial\theta} - \frac{g}{2\omega\alpha\rho_0\cos\theta}\left[\int_z^H \frac{\partial\rho}{\partial\theta}\,dz - \frac{1}{H}\int_0^H z\frac{\partial\rho}{\partial\theta}\,dz\right] \tag{155}$$

where $v_\theta{}^d$ and $v_\lambda{}^d$ are the purely drift current velocity components:

$$v_\theta{}^d = \frac{1}{2\omega a\rho_0\cos\theta}\sqrt{\frac{\nu'}{\nu}}\left(\frac{\partial p_a}{\partial\theta}\sin\alpha z \mp \frac{1}{\sin\theta}\frac{\partial p_a}{\partial\lambda}\cos\alpha z\right)e^{-\alpha z}$$

$$v_\lambda{}^d = \frac{1}{2\omega a\rho_0\cos\theta}\sqrt{\frac{\nu'}{\nu}}\left(\pm\frac{\partial p_a}{\partial\theta}\cos\alpha z + \frac{1}{\sin\theta}\frac{\partial p_a}{\partial\lambda}\sin\alpha z\right)e^{-\alpha z} \tag{156}$$

At the ocean surface the vertical velocity component is assumed to be zero, and at the other levels being considered it is determined from the formula

$$v_z = \mp \frac{1}{4\omega\rho_0 a^2 \cos\theta\, \alpha'}\Delta p_a - \frac{1}{4\alpha'\omega \cos^2\theta\, a^2\rho_0}\left(\frac{\partial p_a}{\partial\theta} + \frac{1}{\sin\theta}\frac{\partial p_a}{\partial\lambda}\right)$$
$$- \frac{z}{H\cos\theta\, a^2}\frac{\partial\Psi}{\partial\lambda} + \frac{gz}{2\omega\rho_0 H \cos^2\theta\, a^2}\int_0^H (H-z)\frac{\partial\rho}{\partial\lambda}\,dz$$
$$- \frac{g}{2\omega\rho_0 \cos^2\theta\, a^2}\int_0^z (z-\xi)\frac{\partial\rho}{\partial\lambda}\,d\xi \tag{157}$$

The formula for computing v_z with the aid of an auxiliary function ζ is

$$v_z = \mp \frac{1}{4\omega\rho_0 a^2\alpha' \cos\theta}\Delta p_a - \frac{\sin\theta}{4\alpha'\omega a^2\rho_0 \cos^2\theta}\left(\frac{\partial p_a}{\partial\theta} + \frac{1}{\sin\theta}\frac{\partial p_a}{\partial\lambda}\right)$$
$$+ \frac{gz}{2\omega a^2 \cos^2\theta}\frac{\partial\zeta}{\partial\lambda} + \frac{g}{2\omega a^2 \cos^2\theta\, \rho_0}\int_0^z (z-\xi)\frac{\partial\rho}{\partial\lambda}\,d\xi \tag{158}$$

Taking into account the inertial terms and lateral mixing, we obtain formulas for computing the horizontal flow velocity with the aid of ζ:

$$v_\theta = v_\theta{}^d - \frac{g}{2\omega a \sin\theta\cos\theta}\left[\frac{\partial\zeta}{\partial\lambda} + \frac{1}{\rho_0}\int_0^z \frac{\partial\rho}{\partial\lambda}\,dz\right] - \frac{A'}{2\omega a \cos\theta} \tag{159}$$

$$v_\lambda = v_\lambda{}^d + \frac{g}{2\omega a \cos\theta}\left[\frac{\partial\zeta}{\partial\theta} + \frac{1}{\rho_0}\int_0^z \frac{\partial\rho}{\partial\theta}\,dz\right] + \frac{B'}{2\omega a \cos\theta} \tag{160}$$

where

$$A' = v_\theta\frac{\partial v_\lambda}{\partial\theta} + \frac{v_\lambda}{\sin\theta}\frac{\partial\lambda_\lambda}{\partial\lambda} + a v_z\frac{\partial v_\lambda}{\partial z} + v_\theta v_\lambda \cot\theta - \frac{A_e}{a}\Delta v_\lambda \tag{161}$$

$$B' = v_\theta\frac{\partial v_\theta}{\partial\theta} + \frac{v_\lambda}{\sin\theta}\frac{\partial v_\theta}{\partial\lambda} + a v_z\frac{\partial v_\theta}{\partial z} - v_\lambda{}^2 \cot\theta - \frac{A_e}{a}\Delta v_\theta \tag{162}$$

In this case, at the contour of the basin ζ, the sea-surface topography is determined from the following relations:

$$\left(1 \mp \frac{1}{2\alpha H}\right)\frac{\partial\zeta}{\partial\lambda} = \frac{\sin\theta}{H\rho_0}\int_0^H (H-z)\left(\frac{1}{2\alpha H}\frac{\partial\rho}{\partial\theta} - \frac{1}{\sin\theta}\frac{\partial\rho}{\partial\lambda}\right)dz$$
$$- \frac{\sin\theta}{2\alpha H\rho_0}\int_0^H \left(\pm\frac{\partial\rho}{\partial\theta} - \frac{1}{\sin\theta}\frac{\partial\rho}{\partial\lambda}\right)dz - \frac{a\omega \sin 2\theta}{g}\left(v_\theta + \frac{1}{2\alpha H}v_\lambda\right)$$
$$- \frac{\sin\theta}{2\alpha H\rho_1 g}\sqrt{\frac{v'}{v}}\left(\mp\frac{\partial p_a}{\partial\theta} + \frac{1}{\sin\theta}\frac{\partial p_a}{\partial\lambda}\right) - \frac{\sin\theta}{gH}\int_0^H A'\,dz \tag{163}$$

$$\left(1 \mp \frac{\sin\theta}{2\alpha H}\right)\frac{\partial\zeta}{\partial\theta} = \frac{1}{H\rho_0}\int_0^H (H-z)\left[\frac{1}{2\alpha H \sin\theta}\frac{\partial\rho}{\partial\lambda} - \frac{\partial\rho}{\partial\theta}\right]dz$$
$$+ \frac{1}{2\alpha H\rho_0}\int_0^H \left(\frac{\partial\rho}{\partial\theta} \pm \frac{1}{\sin\theta}\frac{\partial\rho}{\partial\lambda}\right)dz + \frac{2\omega a \cos\theta}{g}\left(v_\lambda - \frac{1}{2\alpha H}v_\theta\right)$$
$$+ \frac{1}{2\alpha H\rho_0 g}\sqrt{\frac{v'}{v}}\left(\frac{\partial p_a}{\partial\theta} \pm \frac{1}{\sin\theta}\frac{\partial p_a}{\partial\lambda}\right) - \frac{1}{gH}\int_0^H B'\,\alpha z \tag{164}$$

Note that formulas 148 and 149 do not follow readily from 163 and 164 by neglecting the expressions A'_1 and B'_1. This was explained above, when deriving formulas 124 and 125.

For the computation of equatorial currents with the D_3 model we have the following equation for the sea-surface topography:

$$\Delta\zeta + \frac{1}{H}\left(\frac{\partial H}{\partial\theta}\frac{\partial\zeta}{\partial\theta} + \frac{1}{\sin^2\theta}\frac{\partial H}{\partial\lambda}\frac{\partial\zeta}{\partial\lambda}\right)$$

$$= -\frac{1}{H\rho_0}\int_0^H (H-z)\Delta\rho\,dz - \frac{1}{H\rho_0}\frac{\partial H}{\partial\theta}\int_0^H \frac{\partial\rho}{\partial\theta}\,dz$$

$$-\frac{1}{H\rho_0\sin^2\theta}\frac{\partial H}{\partial\lambda}\int_0^H \frac{\partial\rho}{\partial\lambda}\,dz + \frac{2\omega a}{gH}\cot\theta\left(\frac{\partial}{\partial\theta}\sin\theta S_\lambda - \frac{\partial S_\theta}{\partial\lambda}\right)$$

$$-\frac{2\omega a\sin\theta}{gH}S_\lambda - \frac{a}{gH\sin\theta}\int_0^H\left(\frac{\partial}{\partial\theta}\sin\theta B_1 + \frac{\partial A_1}{\partial\lambda}\right)dz$$

$$+\frac{a}{\rho_0 gH\sin\theta}\left(\frac{\partial\sin\theta\tau_\theta{}^H}{\partial\theta} + \frac{\partial\tau_\lambda{}^H}{\partial\lambda}\right) - \frac{1}{2\alpha'aH}\Delta p_a \qquad (165)$$

where

$$A_1 = \frac{1}{a}\frac{\partial v_\theta{}^2}{\partial\theta} + \frac{1}{a\sin\theta}\frac{\partial v_\theta v_\lambda}{\partial\lambda} + (v_\theta{}^2 - v_\lambda{}^2)\frac{\cot\theta}{a}$$

$$B_1 = \frac{1}{a}\frac{\partial v_\lambda v_\theta}{\partial\theta} + \frac{1}{a\sin\theta}\frac{\partial v_\lambda{}^2}{\partial\lambda} + 2v_\theta v_\lambda\frac{\cot\theta}{a} \qquad (166)$$

In conclusion we give the principal relations of the D_4 model in the spherical coordinate system. By analogy with equations 139 and 140 we have

$$\frac{\partial v_\lambda}{\partial t} + \frac{v_\theta}{a}\frac{\partial v_\lambda}{\partial\theta} + \frac{v_\lambda}{a\sin\theta}\frac{\partial v_\lambda}{\partial\lambda} + \frac{v_\theta v_\lambda}{a}\cot\theta - \frac{A_l}{a^2}\Delta v_\lambda + 2\omega\cos\theta v_\theta = -\frac{g}{a\sin\theta}\frac{\partial\zeta}{\partial\lambda}$$

$$-\frac{g}{\rho_0 a\sin\theta}\int_0^z\frac{\partial\rho}{\partial\lambda}\,dz - \frac{1}{\rho_0\sin\theta}\sqrt{\frac{v'}{v}}\frac{\partial p_a}{\partial\lambda} \qquad (167)$$

$$\frac{\partial v_\theta}{\partial t} + \frac{v_\theta}{a}\frac{\partial v_\theta}{\partial\theta} + \frac{v_\lambda}{a\sin\theta}\frac{\partial v_\theta}{\partial\lambda} - \frac{v_\lambda{}^2}{a}\cot\theta - \frac{A_l}{a^2}\Delta v_\theta - 2\omega\cos\theta v_\lambda = -\frac{g}{a}\frac{\partial\zeta}{\partial\theta}$$

$$-\frac{g}{\rho_0 a}\int_0^z\frac{\partial\rho}{\partial\theta}\,dz - \frac{1}{\rho_0}\sqrt{\frac{v'}{v}}\frac{\partial p_a}{\partial\theta} \qquad (168)$$

By analogy with equation 141 we obtain:

$$\frac{H}{2\omega\cos\theta}\frac{\partial}{\partial t}\Delta\zeta + \frac{aH\sin\theta}{2\omega\cos^2\theta}\frac{\partial}{\partial t}\frac{\partial\zeta}{\partial\theta} + \frac{1}{2\alpha}\Delta\zeta \pm \frac{1}{\sin\theta}J(H,\zeta) + \frac{H}{|\cos\theta|}\frac{\partial\zeta}{\partial\lambda}$$

$$= -\frac{H}{2\omega\rho_0\cos\theta}\int_0^H\frac{\partial}{\partial t}\Delta\rho\,dz - \frac{aH\sin\theta}{2\omega\cos^2\theta\rho_0}\int_0^H(H-z)\frac{\partial}{\partial t}\frac{\partial\rho}{\partial\theta}\,dz$$

$$+\frac{1}{2\alpha'\rho_0 g}\Delta p_a - \frac{1}{2\alpha\rho_0}\int_0^H\Delta\rho\,dz \pm \frac{1}{\rho_0\sin\theta}\int_0^H J(H\rho)\,dz$$

$$-\frac{1}{\rho_0|\cos\theta|}\int_0^H(H-z)\frac{\partial\rho}{\partial\lambda}\,dz - \frac{2\omega a\sin\theta}{g}\int_0^H(v_\theta - v_{\theta g})\,dz$$

$$-\frac{a}{g}\frac{\partial}{\partial\theta}\int_0^H A_2'\,dz - \frac{1}{\sin\theta}\frac{\partial}{\partial\lambda}\int_0^H B_2'\,dz \qquad (169)$$

where

$$A_2' = \frac{\partial(v_\lambda - v_{\lambda g})}{\partial t} + \frac{v_\theta}{a}\frac{\partial v_\lambda}{\partial \theta} + \frac{v_\lambda}{a \sin\theta}\frac{\partial v_\lambda}{\partial \lambda} - A_l \Delta v_\lambda$$

(170)

$$B_2' = \frac{\partial(v_\theta - v_{\theta g})}{\partial t} + \frac{v_\theta}{a}\frac{\partial v_\theta}{\partial \theta} + \frac{v_\lambda}{a \sin\theta}\frac{\partial \lambda}{\partial \lambda} - A_l \Delta v_\theta$$

In this case, to the formula for the vertical velocity calculated from equation 154 we must add the expression

$$v_z' = \frac{1}{2\omega \cos\theta a}\int_0^z \left(\frac{\partial A_2}{\partial \theta} - \frac{1}{\sin\theta}\cdot\frac{\partial B_2}{\partial \lambda}\right) dz$$

(171)

where

$$A_2 = A_2' + \frac{\partial v_{\lambda g}}{\partial t}, \qquad B_2 = B_2' + \frac{\partial v_{\theta g}}{\partial t}$$

(172)

3. Methods of Difference Approximations and Solution of Constructed Equations

A. Methods of Calculating Sea-Surface Level on the Contour of the Basin Boundary

As shown above, the determination of the surface level ζ and the mass transport stream function Ψ is an essential part of solving the problem for all the models of the diagnostic calculation of currents discussed. The corresponding equations were obtained to determine ζ on the contour of the basin boundary. Formally, these equations are the same for all models. In fact, there is the following significant difference between them. In the equations of motion of model D_1 we neglected all nonlinear terms and the effect of lateral exchange. We cannot satisfy the boundary conditions exactly here, but we can put the boundary condition for the current velocity average over the height. Thus the determination of function ζ from equations 124 and 125 is necessary to satisfy the condition of conservation of a fluid volume in the basin considered. As to the models D_2-D_5, here the components of current velocity themselves (independent of the field ζ) satisfy boundary conditions over the horizontal, being at each horizon. In these cases the boundary conditions for u and v are simply used once more for correct definition of ζ on the contour of the basin boundary.

Now let us consider the features common to equations for ζ on the basin boundary for the various diagnostic models. In model D_1 on meridional sections of the boundary $\partial\zeta/\partial y$ is given and on the zonal ones, $\partial\zeta/\partial x$. In other words, the derivative of the unknown function over the tangent to the contour is specified. In the other models, on each section of the boundary derivatives of ζ, both along the tangent and normal to the contour, are given. However, we do not need to use both derivatives. It is quite sufficient to use only the derivative over the tangent to the contour. In this case the method of determination of ζ on the contour of the basin boundary is identical for models D_1-D_5. Let us look at this universal version of the determination of ζ. For simplicity we give the scheme of the calculation of ζ on the contour of the closed, simple connected domain. Let us consider the part of the boundary shown in Fig. 2, and refer equations 87 and 88 to the middle parts of the corresponding links of this

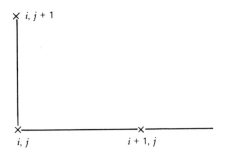

Fig. 2. Part of the boundary contour where each section is parallel to one of the coordinate axes.

boundary line. For the middle part of the segment $|i,j; \quad i+1,j|$ we use equation 88. Substituting the derivative of ζ for the central difference relation we have

$$\zeta_{i+1,j} - \zeta_{i,j} = \delta x \cdot f'_{i+1/2,j} \tag{173}$$

Using equation 87 for the middle part of the segment $|i,j; \quad i,j+1|$, we obtain in a similar way

$$\zeta_{i,j+1} - \zeta_{i,j} = \delta y \cdot f_{i,j+1/2} \tag{174}$$

From equations 173 and 174 we have

$$\zeta_{i+1,j} - 2\zeta_{i,j} + \zeta_{i,j+1} = \delta y \cdot f_{i,j+1/2} + \delta x \cdot f'_{i+1/2,j} \tag{175}$$

If we pass thus round the whole contour of the basin, we obtain the system of n equations of the form 175 with n unknown values $\zeta_{i,j}$. The system obtained can be solved, for instance, using the method of the trial run (Marchuk, 1973). The Gauss–Seidel procedure is a simple method for solution of this system. If one passes clockwise round the contour, for instance, then the iteration formula is of the form

$$\zeta_{i,j}^{n} = \tfrac{1}{2}(\zeta_{i+1,j}^{n} + \zeta_{i,j+1}^{n-1}) - \tfrac{1}{2}(\delta y \cdot f_{i,j+1/2} + \delta x \cdot f'_{i+1/2,j}) \tag{176}$$

The counting process goes on until the difference between two subsequent approximations becomes less than ε specified in advance. In fact, it is quite sufficient to take $\varepsilon = 10^{-2}$–10^{-3} cm. The counting process converges quickly, if one takes ζ calculated from formula 99 as the first approximation. It follows from equations 87 and 88 that ζ is defined within the accuracy of a constant. After the count is completed, it is convenient to subtract the average arithmetic (over the whole contour of the basin) value of ζ from each value of ζ_{ij}. The described procedure can be applied to systems 84 and 85 as well. In this case one must consider terms on the right sides of these equations containing ζ by means of successive approximations. Calculations of ζ on the contour and inside the basin are performed simultaneously in this case. The calculation of ζ with the aid of formulas 124 and 125 is a more reliable method for consideration of bottom friction. However, the concrete calculations below show us that it is insignificant, because while calculating currents of large or middle scale (grid step exceeding 50 km) we may neglect terms with small multiplier $1/2\alpha H$ in equations 84 and 85.

The calculation of ζ on the basis of equations 146 and 147 is an accurate account for boundary conditions in the calculation of ζ on the boundary contour in models D_2–D_5. The concrete calculations can be performed in accord with the scheme described above, taking into account the effect of the lateral exchange by successive approximations (from fields u and v determined at coastal points of the basin). Thus the determination of ζ reduces to the solution of the Dirichlet problem in all models.

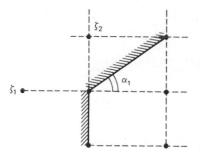

Fig. 3.   Part of the boundary contour where not all sections are parallel to the coordinate axes.

However, there exists a case when the complicated problem must be solved for the correct determination of ζ on the contour of the basin. With a more accurate approximation of the form of the coastline some parts of it will be situated at an acute angle to coordinates, as shown in Fig. 3.

In this case instead of equations of type 87 and 88, conditions of the following form apply on the boundary contour:

$$\frac{\partial \zeta}{\partial x} \cos \alpha_1 + \frac{\partial \zeta}{\partial y} \sin \alpha_1 = \Phi_{(0)}(x, y) \tag{177}$$

A simple and witty method for solving this problem is proposed by Marchuk (1969). The main point of Marchuk's method lies in writing the difference equation of form 86 not only for the internal points of the basin, but for the boundary ones as well. This, together with the equation of form 177, makes it possible to exclude fictitious points of type ζ_1 and ζ_2 (Fig. 3) and obtain an enclosed system of equations for the determination of values of ζ on the contour and inside the basin.

B. Methods of Approximations and Solution for Auxiliary Functions

Because the major numerical experiments are performed using the auxiliary function ζ, we therefore turn to equation 86. We denote the right side of this equation $\Phi(x, y)$ and rewrite it in the form

$$\frac{1}{2\alpha} \Delta \zeta + A^{(0)} \frac{\partial \zeta}{\partial x} + B^{(0)} \frac{\partial \zeta}{\partial y} = \Phi \tag{178}$$

where

$$A^{(0)} = \left(\frac{\beta H}{l} - \frac{\partial H}{\partial y} \right), \qquad B^{(0)} = \frac{\partial H}{\partial x} \tag{179}$$

In Section 1 we have already seen that equations for ζ and Ψ are the equations with small parameters at older derivatives. The development of steady schemes of the second order of accuracy is hampered for such equations. The shortcomings of such schemes are discussed, for instance, in Il'in (1969).

Method of Directional Differences

The method of directional differences is the simplest of the schemes of the first order of accuracy. Many works on the diagnostic investigation of currents have been performed by just this method. Therefore we begin the description of numerical methods with the method of directional differences. The essence of the method is quite simple: derivatives of the first order are substituted for the difference forward and backward depending on the signs of coefficients in such a way that diagonal terms possess maximum weights. For instance, in equation 178 we substitute the derivative with respect to x by the directional difference relation in the following way:

$$\delta x\left(\frac{\partial \zeta}{\partial x}\right)_{i,j} = \delta_1 \zeta_{i+1,j} + (1 - 2\delta_1)\zeta_{i,j} + (\delta_1 - 1)\zeta_{i-1,j} \tag{180}$$

where

$$\begin{aligned} \delta_1 &= 0 \quad \text{for } A_{i,j}^{(0)} < 0 \\ &= 1 \quad \text{for } A_{i,j}^{(0)} > 0 \end{aligned} \tag{181}$$

that is for $A_{i,j}^{(0)} > 0$

$$\delta x\left(\frac{\partial \zeta}{\partial x}\right)_{i,j} = \zeta_{i+1,j} - \zeta_{i,j} \tag{182}$$

and for $A_{i,j}^{(0)} < 0$

$$\delta x\left(\frac{\partial \zeta}{\partial x}\right)_{i,j} = \zeta_{i,j} - \zeta_{i-1,j}$$

Similarly,

$$\delta y\left(\frac{\partial \zeta}{\partial y}\right)_{i,j} = \delta_2 \zeta_{i,j+1} + (1 - 2\delta_2)\zeta_{i,j} + (\delta_2 - 1)\zeta_{i,j-1} \tag{183}$$

where

$$\begin{aligned} \delta_2 &= 0 \quad \text{for } B_{i,j}^{(0)} < 0 \\ &= 1 \quad \text{for } B_{i,j}^{(0)} > 0 \end{aligned} \tag{184}$$

If we write the finite-difference analogue of sum $A^{(0)}(\partial\zeta/\partial x) + B^{(0)}(\partial\zeta/\partial y)$, then $\zeta_{i,j}$ will have coefficient $[|A_{i,j}^{(0)}| + |B_{i,j}^{(0)}|]$ in this sum; that is, the diagonal predominance will be present in the system of algebraic equations obtained, independent of signs of coefficients $A^{(0)}$ and $B^{(0)}$. The Laplace operator is substituted in the common way for the central difference relation. After these transformations we obtain the following difference approximation of equation 178 on the grid G^k with steps δx, δy:

$$\begin{aligned} &\frac{1}{2\alpha_j}\frac{\zeta_{i-1,j} + \zeta_{i+1,j} - 2\zeta_{i,j}}{\delta x^2} + \frac{1}{2\alpha_j}\frac{\zeta_{i,j-1} + \zeta_{i,j+1} - 2\zeta_{i,j}}{\delta y^2} \\ &+ A_{i,j}^{(0)}\frac{\delta_1\zeta_{i+1,j} + (1 - 2\delta_1)\zeta_{i,j} + (\delta_1 - 1)\zeta_{i-1,j}}{\delta x} \\ &+ B_{i,j}^{(0)}\frac{\delta_2\zeta_{i,j+1} + (1 - 2\delta_2)\zeta_{i,j} + (\delta_2 - 1)\zeta_{i,j-1}}{\delta y} = \Phi_{i,j}. \end{aligned} \tag{185}$$

Finally, if we substitute $A^{(0)}$, $B^{(0)}$ for the central difference relations and solve the obtained equation relative to $\zeta_{i,j}$ then we derive the formula from which we can calculate $\zeta_{i,j}$ by the method of successive approximations. For simplicity, we present this formula for the case when the grid steps are equal in both directions ($\delta x = \delta y$):

$$\zeta_{i,j} = \frac{1}{c_{i,j}} \left\{ \left[\frac{1}{2\alpha_j} + A_{i,j}^{(0)} \delta_1 \right] \zeta_{i+1} + \left[\frac{1}{2\alpha_j} + (\delta_1 - 1)A^{(0)} \right] \zeta_{i-1,j} \right.$$

$$\left. + \left[\frac{1}{2\alpha_j} + B_{i,j}^{(0)} \delta_2 \right] \zeta_{i,j+1} + \left[\frac{1}{2\alpha_j} + (\delta_2 - 1)B_{i,j}^{(0)} \right] \zeta_{i,j-1} \right\} - \frac{(\delta x)^2}{c_{i,j}} \Phi_{i,j} \quad (186)$$

where

$$c_{i,j} = \frac{2}{\alpha_j} + |A_{i,j}^{(0)}| + |B_{i,j}^{(0)}|$$

$$\delta x \cdot A_{i,j}^{(0)} = \frac{\delta x \cdot \beta_j H_{i,j}}{l_j} - (H_{i,j} - H_{i,j-1}), \qquad \delta y \cdot B^{(0)} = H_{i+1,j} - H_{i-1,j} \quad (187)$$

The basic shortcoming of the method of directional differences is a significant calculation viscosity. We consider the extent to which it is reflected in the obtained equations, while discussing the results of calculations.

Il'in's Difference Approximation

Il'in proposed another procedure for difference approximation (1969). His aim was to develop a scheme of the second order of accuracy in such a way that accurate solutions of exponential type were the solutions of a difference equation in the boundary layer for the case of an equation with constant coefficients. For the case of a common differential equation with constant coefficients Il'in showed that the scheme really satisfies these requirements. The calculations showed that his method may be applied to equation 178 with variable coefficients $A^{(0)}$ and $B^{(0)}$ (Il'in et al., 1969). However, Il'in's scheme is, in fact, a scheme of the second order only at very small values of the horizontal step, or at small values of $A^{(0)}$ and $B^{(0)}$.

For values of $A^{(0)}$ and $B^{(0)}$ corresponding to real bottom topography and for reachable (for the large-scale circulation) values of the grid step, Il'in's approximation possesses a scheme viscosity comparable with that of the method of directional differences (Sarkisyan and Ivanov, 1972). Theoretically, being a scheme of the second order of accuracy, Il'in's scheme provides for the first order of accuracy. Nevertheless, in some cases Il'in's scheme has an advantage over that of directional differences; therefore a number of calculations have been performed using this scheme. Neglecting simple transformations we give the differential approximation of equation 178 by Il'in's method:

$$A_{i,j}^{(0)} \coth \left(\frac{\alpha_j A_{ij}^{(0)}}{2} \right) (\zeta_{i+1,j} + \zeta_{i-1,j} - 2\zeta_{i,j})$$

$$+ B_{i,j}^{(0)} \coth \left(\frac{\alpha_j B_{i,j}^{(0)}}{2} \right) (\zeta_{i,j+1} + \zeta_{i,j-1} - 2\zeta_{i,j})$$

$$+ A_{i,j}^{(0)}(\zeta_{i+1,j} - \zeta_{i-1,j}) + B_{i,j}^{(0)}(\zeta_{i,j+1} - \zeta_{i,j-1}) = 2\Phi_{ij} \quad (188)$$

In order to obtain an iterative formula from equation 188 we should solve this equation relative to $\zeta_{i,j}$. It should be borne in mind, however, that unlike the method of directional differences while using formula 188 we must distinguish, practically, the limiting cases when arguments of hyperbolic cotangents are too small ($<10^{-3}$) or too large (>10). In the former case we must multiply and divide terms with cotangents by $\alpha_j/2$ and apply the asymptotic value of $\lim_{x \to 0} x \coth x = 1$.

In the latter case we must take into account that

$$\lim_{x \to \pm \infty} \coth x = \pm 1$$

Now we speak briefly of a method for solving differential equations. Formula 186 is already the working formula for the calculation of $\zeta_{i,j}$ by successive approximations. The Gauss–Seidel procedure is very convenient for this purpose. For instance, in the case when the count begins with the upper right angle and ends at the lower left angle, the iteration formula has the form

$$\zeta_{i,j}^{(n+1)} = \frac{1}{c_{i,j}} \left\{ \left[\frac{1}{2\alpha_j} + A_{i,j}^{(0)} \delta_1 \right] \zeta_{i+1,j}^{(n+1)} + \left[\frac{1}{2\alpha} + (\delta_1 - 1) A_{i,j}^{(0)} \right] \zeta_{i-1,j}^{(n)} \right.$$

$$\left. + \left[\frac{1}{2\alpha_j} + B_{i,j}^{(0)} \delta_2 \right] \zeta_{i,j+1}^{(n+1)} + \left[\frac{1}{2\alpha} + (\delta_2 - 1) B_{i,j}^{(0)} \right] \zeta_{i,j-1}^{(n)} \right\} - \frac{(\delta x)^2}{c_{i,j}} \Phi_{i,j} \quad (189)$$

that is, as soon as the $(n+1)$th approximation of the unknown function is calculated at a given point, it is substituted in the computer memory in place of the nth approximation. On the boundary of the region the value of ζ is calculated in advance by means of the procedure described above. Below, after the analysis of results from concrete diagnostic calculations, we give additional methodical instructions on the selection of the most convenient auxiliary function, etc. After solving equation 188 relative to ζ, we obtain a formula of the form 186 on the basis of which the calculations are made using the Gauss–Seidel procedure.

Method of Increasing the Order of Approximation of Differential Schemes

Both methods described for the approximation of equation 178 have a scheme viscosity; the need arises to increase the order of approximation and minimize the scheme viscosity. A general method of increasing the order of approximation of difference schemes is presented by Marchuk (1973). In the work by Kochergin and Shcherbakov (1972) the application of this method to an equation like 178 is given. The method is quite simple and convenient to realize. It is based on the investigation of the solution of this equation by the method of directional differences on the sequence of grids.

Let the difference grid be G^h with steps δx and δy along the axes of x and y, respectively. Now we approximate equation 178 on the grid G^h by the method of directional differences. We denote the solution of the obtained differential equation 185 by ζ^h. Let us discuss now another difference grid G^{2h} with steps $2\delta x$ and $2\delta y$. Note that all calculated points coincide with those of the grid G^h. The solution of the problem on the grid G^{2h} is denoted by ζ^{2h}. Let us construct the corrector for the grid G^{2h}:

$$\bar{\zeta} = 2\zeta^h - \zeta^{2h} \quad (190)$$

It is easily seen that $\bar{\zeta}$ approximates equation 178 with the second order of accuracy. In order to show this we perform a Taylor series expansion $\zeta_{i,j+1}^h$, $\zeta_{j\pm1,j}^h$, $\zeta_{i,j\pm2}^{2h}$, and $\zeta_{i\pm2,j}^{2h}$ in the environs of the point $(2\delta x_i, 2\delta y_j)$. Now we substitute the Taylor series for $\zeta_{i,j\pm1}^h$ and $\zeta_{i\pm1,j}^h$ in equation 178 and perform a similar operation for the differential approximation constructed on the grid G^{2h}. Taking into account formula 190 for the corrector $\bar{\zeta}$ we obtain

$$\frac{1}{2\alpha}\Delta\bar{\zeta} + A^{(0)}\frac{\partial\bar{\zeta}}{\partial x} + B^{(0)}\frac{\partial\bar{\zeta}}{\partial y} - A^{(0)}\frac{\delta x^2}{3}\frac{\partial^3\zeta}{\partial x^3} - B^{(0)}\frac{\delta y^2}{3}\frac{\partial^3\zeta}{\partial y^3} + \cdots = \Phi \quad (191)$$

Thus on the basis of the method of directional differences with the aid of the simple corrector 190 we can obtain the solution of equation 178 with the second order of accuracy on the grid G^{2h}. After this, using a spline-polynomial (Alberg et al., 1967), the obtained solution interpolates on the grid G^h with the appropriate accuracy. In particular, the cubic spline-polynomial gives accuracy of the second order. Kochergin and Shcherbakov (1972) note that one can construct in a similar way the corrector of the third order of accuracy. However, it leads to only a little more precision in the solution. Namely, correction obtained on account of the corrector of the third order is four times less than that of the corrector of the second order. In Section 4 we see that the solution of a concrete problem in the ocean obtained by the method of directional differences with a step of 2.5° is made more precise by only 10% using a corrector of the second order (equation 190).

C. Finite-Difference Approximation and Solution of Equations of Models D_2–D_5

Although model D_2 is a considerable step forward in comparison with model D_1, we have no new equations here that must be solved. The equation for ζ and the method of calculation of this function at the contour of a basin, and formulas for the calculation of u, v, and w are analogous to those of the D_1 model. The inertial terms and the effect of lateral eddy viscosity are present on the right sides of the relations of the D_2 model and are taken into account by successive approximations. It means that in the D_2 model the horizontal grid step is limited from below. For a very small value of this grid step the inertial terms may turn out larger than the other terms on the right sides of relations 115–125. In this case the D_2 model becomes unsuitable. Thus the diagnostic calculations using the D_2 model may be carried out with the aid of methods stated for the D_1 model if we decrease the horizontal grid step to its minimum value. The magnitude of the minimum grid step depends on the maximum value of the density gradients in the region considered. In known diagnostic calculations with the D_2 model for the Gulf Stream region the minimum value of the grid step is equal to $\frac{5}{8}°$.

For constructing the finite-difference schemes of the D_3–D_5 models, let us do some preliminary transformations. The derivatives of the first and second orders over the height are present in the equations of these models. The direct substitution of these derivatives for the finite-difference quotients may lead to errors in the surface and bottom boundary layers. The object of the transformations made below is to lessen these errors.

Denote by u_0 the component of the flow velocity along the x axis at the surface of the ocean. The corresponding value of this function at the horizon nearest to the sea surface is denoted by u_1, and the distance between them, by h_1.

Performing the Taylor expansion of u_1 and limiting ourselves to three terms we obtain (with due regard for boundary condition 1):

$$u_1 = u_0 + h_1\left(\frac{\partial u}{\partial z}\right)_{z=0} + \frac{h_1^2}{2}\left(\frac{\partial^2 u}{\partial z^2}\right)_{z=0}$$

$$= u_0 + h_1\left(\frac{\partial u}{\partial z}\right)_{z=0} + \frac{h_1^2}{2v}\left(\frac{\partial}{\partial z}v\frac{\partial u}{\partial z}\right)_{z=0} - \frac{h_1^2}{2}\left(\frac{\partial u}{v\partial z}\frac{\partial v}{\partial z}\right)_{z=0}$$

Hence it follows that

$$\left(\frac{\partial}{\partial z} v \frac{\partial u}{\partial z}\right)_{z=0} = \frac{2v}{h_1^2}(u_1 - u_0) + \frac{2}{h_1 \rho_0} \tau_x - \frac{\tau_x}{\rho_0}\left(\frac{\partial v}{\partial z}\right)_{z=0} \qquad (192)$$

Here we consider only the case of a constant value of v; therefore, the last term on the right side of equation 192 is neglected in the subsequent transformations. Similarly, for the bottom boundary layer we have

$$vu_{k-1} = vu_k - h_k v\left(\frac{\partial u}{\partial z}\right)_{z=z_k} + vh_k^2\left(\frac{\partial^2 u}{\partial z^2}\right)_{z=z_k} \qquad (193)$$

The $z = z_k$ layer is the bottom of the ocean. Therefore, we can use the condition of no motion and from equation 1 it follows:

$$v\left(\frac{\partial^2 u}{\partial z^2}\right)_{z=H} = \frac{1}{\rho_0}\left(\frac{\partial p}{\partial x}\right)_{z=H}$$

As a result, from equation 193 we come to the formula

$$\tau_x{}^H = \rho_0 v\left(\frac{\partial u}{\partial z}\right)_{z=H} = \left(\frac{\partial p}{\partial x}\right)_{z=H} h_k - \frac{\rho_0 vu_{k-1}}{h_k} \qquad (194)$$

expressing the component of the bottom friction by the gradient of the bottom pressure and the value of the velocity at the last calculating layer.

Using formulas of equation 194 type for $\tau_x{}^H$ and $\tau_y{}^H$, we bring equation 134 of the D_3 model to the form:

$$\left(1 - \frac{h_k}{H}\right)\Delta\zeta + \frac{1}{H}\frac{\partial H}{\partial x}\frac{\partial \zeta}{\partial x} + \frac{1}{H}\frac{\partial H}{\partial y}\frac{\partial \zeta}{\partial y} = \frac{1}{\rho_0}\int_0^H \left(\frac{z}{H} + \frac{h_k}{H} - 1\right)\Delta\rho \, dz$$

$$- \frac{1}{\rho_0 H}\left[\frac{\partial H}{\partial x}\int_0^H \frac{\partial \rho}{\partial x} dz + \frac{\partial H}{\partial y}\int_0^H \frac{\partial \rho}{\partial y} dz\right] - \frac{\beta}{gH}\int_0^H u \, dz$$

$$+ \frac{1}{gH}\int_0^H \left(\frac{\partial v}{\partial x} - \frac{\partial u}{\partial y}\right)dz - \frac{1}{gH}\int_0^H \left(\frac{\partial A_1}{\partial x} + \frac{\partial B_1}{\partial y}\right)dz + \frac{1}{gH\rho_0}\left(\frac{\partial \tau_x}{\partial x} + \frac{\partial \tau_y}{\partial y}\right) \qquad (195)$$

Now let us reduce equation 1 to a form convenient for the calculation of equatorial currents. For that we substitute the time derivative for the directional finite-difference quotient and rewrite it in the form:

$$A_1\Delta u^{(n)} + v\frac{\partial^2 u^{(n)}}{\partial z^2} - u^{(n-1)}\frac{\partial u^{(n)}}{\partial x} - v^{(n-1)}\frac{\partial u^{(n)}}{\partial y}$$

$$- w^{(n-1)}\frac{\partial u^{(n)}}{\partial z} - \frac{u^{(n)}}{\delta t} + \frac{u^{(n-1)}}{\delta t} - lv^{(n-1)} - \frac{1}{\rho_0}\frac{\partial p}{\partial x} = F \qquad (196)$$

Equation 196 will serve for the determination of $u^{(n)}$, the flow velocity at the n time step. We refer the coefficients of this equation to the $(n + 1)$th time step and thus we linearize it. As a result it has the same form as equation 195 or 86. The discrepancy is that in equation 196 we have to deal with a function of the three independent variables.

The methods of solving equation 86 described in the preceding section may be generalized in the three-dimensional equation and applied to the solution of equation 196. The generalization of the method of the directional differences is trivial; therefore, we give only the finite-difference approximation of this equation given by the generalization of Il'in's method (Sarkisyan and Serebryakov, 1974). For this the derivatives of the first and second orders with respect to z are replaced by the central difference quotients of the formulas for nonequidistant grid points:

$$\frac{\partial u^{(n)}}{\partial z} = \frac{1}{\delta z_k + \delta z_{k+1}} \left[R_k(u^{(n)}_{i,j,k+1} - u^{(n)}_{i,j,k}) + \frac{1}{R_k}(u^{(n)}_{i,j,k} - u^{(n)}_{i,j,k-1}) \right] \qquad (197)$$

$$\frac{\partial^2 u^{(n)}}{\partial z^2} = \frac{2}{\delta z_k + \delta z_{k+1}} \left[\frac{1}{\delta z_{k+1}}(u^{(n)}_{i,j,k+1} - u^{(n)}_{i,j,k}) + \frac{1}{\delta z_k}(u^{(n)}_{i,j,k} - u^{(n)}_{i,j,k-1}) \right] \qquad (198)$$

where

$$\delta z_k = z_k - z_{k-1}, \qquad R_k = \frac{\delta z_k}{\delta z_{k+1}} \qquad (199)$$

The general formula for the calculation of $u^{(n)}_{i,j,k}$ has the form

$$u^{(n)}_{i,j,k} = \frac{1}{c_{i,j,k}} (u^{(n-1)}_{i,j,k} + c_{i+1,j,k} u^{(n)}_{i+1,j,k} + c_{i-1,j,k} u^{(n)}_{i-1,j,k} + c_{i,j+1,k} u^{(n)}_{i,j+1,k}$$

$$+ c_{i,j-1,k} u^{(n)}_{i,j-1,k} + c_{i,j,k+1} u^{(n)}_{i,j,k+1} + c_{i,j,k-1} u^{(n)}_{i,j,k-1} - \delta t F_{i,j,k}) \qquad (200)$$

where

$$c_{i+1,j,k} = \frac{\delta t}{\delta x} \frac{u^{n-1}_{i,j,k}}{2} \left(\coth \frac{u^{n-1}_{i,j,k} \delta x}{2A_l} - 1 \right),$$

$$c_{i-1,j,k} = \frac{\delta t}{\delta x} \frac{u^{n-1}_{i,j,k}}{2} \left(\coth \frac{u^{n-1}_{i,j,k} \delta x}{2A_l} + 1 \right),$$

$$c_{i,j+1,k} = \frac{\delta t}{\delta y} \frac{v^{n-1}_{i,j,k}}{2} \left(\coth \frac{v^{n-1}_{i,j,k} \delta y}{2A_l} - 1 \right), \qquad (201)$$

$$c_{i,j-1,k} = \frac{\delta t}{\delta y} \frac{v^{n-1}_{i,j,k}}{2} \left(\coth \frac{v^{n-1}_{i,j,k} \delta y}{2A_l} + 1 \right)$$

$$c_{i,j,k+1} = \frac{\delta t}{\delta z_{k+1}(\delta z_k + \delta z_{k+1})} (2\gamma_k - \overline{w}^{n-1}_{i,j,k} \cdot \delta z_k)$$

$$c_{i,j,k-1} = \frac{\delta t}{\delta z_k(\delta z_k + \delta z_{k+1})} (2\gamma_k + \overline{w}^{n-1}_{i,j,k} \cdot \delta z_{k+1})$$

$$c_{i,j,k} = 1 + \delta t \left[\frac{u^{n-1}_{i,j,k}}{\delta x} \coth \frac{u^{n-1}_{i,j,k} \delta x}{2A_l} + \frac{v^{n-1}_{i,j,k}}{\delta y} \coth \frac{v^{n-1}_{i,j,k} \delta y}{2A_l} \right. \qquad (202)$$

$$\left. + \frac{2\gamma_k}{\delta z_k \cdot \delta z_{k+1}} + \frac{(\delta z_{k+1} - \delta z_k)}{\delta z_k \cdot \delta z_{k+1}} \overline{w}^{n-1}_{i,j,k} \right]$$

$$\gamma_k = \frac{w^{n-1}_{i,j,k} \delta z_{k+1}}{2} \frac{R_k^2(e^{\eta \delta z_{k+1}} - 1) - (e^{-\eta \delta z_k} + 1)}{R_k(e^{\eta \delta z_{k+1}} - 1) + (e^{-\eta \delta z_k} + 1)} \qquad (203)$$

At the surface of an ocean $w = 0$. Moreover, instead of equation 198 we use formula 192 for the constant value of v here. As a result, the last term on the right side of a formula 200 and some of its coefficients for $k = 0$ take the form:

$$c_{i,j,1} = \frac{2v_0 \delta t}{z_1^2}, \qquad c_{i,j,-1} = 0,$$

$$c_{i,j,0} = 1 + \delta t \left[\frac{u_{i,j,0}^{n-1}}{\delta x} \coth \frac{u_{i,j,0}^{n-1} \delta x}{2A_l} + \frac{v_{i,j,0}^{n-1}}{\delta y} \coth \frac{v_{i,j,0}^{n-1} \delta y}{2A_l} + \frac{2v_0}{h_1} \right],$$

$$F_{i,j,0} = \frac{g(\zeta_{i+1,j}^n - \zeta_{i-1,j}^n)}{2\delta x} - l_j v_{i,j,0}^{n-1} - 2\frac{\tau_{x,i,j}}{h_1 \rho_0} \qquad (204)$$

The formulas for the other coefficients for $k = 0$ do not change.

Thus the working formula for u is set up. The corresponding formula for v has quite an analogous form. Equation 195 for ζ is solved by methods described in Section 3.B. The obtained system of algebraic equations is solved by means of the Gauss–Seidel technique. From Section 2.E we know that the D_3 model is, so to speak, an incrustation in the large-scale current model; that is, it is supposed that all the unknown large-scale characteristics have already been calculated avoiding the mesoscale peculiarities of the equatorial currents. Consequently, the boundary values of u, v, and ζ at the northern and southern boundaries of the equatorial zone are known. Inside this narrow belt (several degrees in width) the calculations are carried out by the following scheme. From equations 146 and 147, the first approximation of ζ is determined at the western and eastern boundaries of the equatorial zone with a small horizontal grid step. In doing this it is considered that in the first approximation $u = v = 0$. Then from equation 108 from the known values of ρ, τ_x, and τ_y inside the belt and from ζ at the boundary, we can determine the first approximation of ζ inside the belt (including the equator) for $u = v = 0$. The field of P is easily calculated from formula 33. Further, from the given values of ρ, τ_x, and τ_y and from relations like equation 200, we can determine the first approximation of u and v, setting $w = 0$. After that w is easily calculated from formula 93. Then in the same succession we determine the second approximation, etc. We continue the process of calculation of each function until the difference between the two successive approximations becomes less than the corresponding value of ε. The computing process must become steady since all the given magnitudes are stationary.

The time required to reach steady state is the time it takes to adapt the current field to the mass and wind fields.

In contrast to D_3, the D_4 model affords us the possibility of calculating stream-type currents in any region of the World Ocean. For example, this model may be used to calculate currents in the continental slope region of all continents. Consequently the regions of application and the volume of calculations can be large here. For decreasing the number of computations it is advisable to perform some simplifications in the D_4 model. In equations 139 and 140 we have neglected the terms containing w; moreover, we have obtained the approximate expression for the effect of the vertical eddy mixing on the surface currents. In the intermediate layers this factor is not taken into account. As a result, the flow velocity at each layer is calculated independently, facilitating the work with a short-access computer memory. The only connecting link between the layers is the pressure gradient (the sea surface topography and the density field). In equation 141 the nonlinear terms are written in a divergent form and therefore there is no need to assume that the terms containing w are small.

Let us set up an *explicit* finite-difference scheme to calculate u and v from equations 139 and 140. For that we substitute the time derivatives for one-directional differences and rewrite these equations in the following form:

$$u^{(n)} = u^{(n-1)} + \delta t\left[lv - \frac{1}{\rho_0}\frac{\partial p}{\partial x} + A_l \Delta u - u\frac{\partial u}{\partial x} v\frac{\partial u}{\partial y} - \frac{\alpha}{\rho_0}(\tau_y - \tau_x)\right]^{(n-1)} \quad (205)$$

$$v^{(n)} = v^{(n-1)} - \delta t\left[lu + \frac{1}{\rho_0}\frac{\partial p}{\partial y} + u\frac{\partial v}{\partial x} + v\frac{\partial v}{\partial y} - A_l \Delta u + \frac{\alpha}{\rho_0}(\tau_x + \tau_y)\right]^{(n-1)} \quad (206)$$

Further, denote the right side of equation 141 by Φ_1, replace the derivatives with respect to time by one-directional finite-difference quotients, and rewrite it in the form

$$\left(\frac{H}{l} + \frac{\delta t}{2\alpha}\right)\Delta\zeta^{(n)} + \left(\frac{\beta H}{l^2} + \delta t\frac{\partial H}{\partial x}\right)\frac{\partial \zeta^{(n)}}{\partial y} + \delta t\left(\frac{\beta H}{l} - \frac{\partial H}{\partial y}\right)\frac{\partial \zeta^{(n)}}{\partial x}$$
$$= \frac{H}{l}\Delta\zeta^{(n-1)} + \frac{\beta H}{l^2}\frac{\partial\zeta^{(n-1)}}{\partial y} + \Phi_1 \quad (207)$$

In the expression of Φ_1 we have the derivatives of ρ with respect to time. In diagnostic calculations these terms are equal to zero. Equation 207 is similar to equation 86 of the D_1 model for each moment of time. Therefore for solving it we can use the methods given in Section 3.B. A part of the contour of the basin boundary, for which the calculation by the D_1 model is carried out, may be liquid. On such a part of the boundary either the fluxes must be specified or ζ must be known from the solution of a nonlinear problem for the large-scale currents. On the solid part of the boundary ζ is determined by successive approximations with the aid of the technique suggested in Section 3.A.

The scheme of solving the problem by the D_4 model is as follows. From the known density field and with the aid of formula 99 we calculate the first approximation of ζ both on the contour and inside the basin. With the aid of formulas 91 and 92 the first approximation of a horizontal current velocity is calculated. Using these first approximations of unknown functions as the initial values and solving the system of equations 205–207 we can carry out computations to reach the steady state by time steps. For this on the right sides of formulas 205 and 206 the first derivatives of u and v are replaced by directional finite-difference quotients by the technique described in Section 3.B. The D_4 model is universal: it can be used for the computation of stream-type currents in any part of the World Ocean including the equator. The vertical component of the flow velocity is calculated using formula 94 after the computing process has become steady.

Finally, we turn to the D_5 model, the most complex of the models considered above. In this case the vertical component of the flow velocity is kept in the equations of motion. Therefore, it is necessary to solve the system of four equations of motion for u, v, w, and ζ. In this case equation 200 serves for the calculation of a horizontal current velocity u. The analogous formula we have for v. The sea surface topography is determined from equation 207. We use also formulas 114 and 33. Some methods of solving equations like 200 and 207 were considered above.

4. Review and Analysis of Diagnostic Calculations of Currents

A. Calculations of the Sea-Surface Level and Current Velocity

The statement of a diagnostic problem on the calculation of currents in a basin of arbitrary bottom topography by the specified fields of density, the deduction of an equation for ζ, the procedure of the calculation of ζ on the basin boundary, the method of directional differences for the determination of ζ inside the basin, simplifications of the equations of motion, formulas for the calculation of the three components of the current velocity, and the results of a series of calculations for five levels of the North Atlantic have been presented by Sarkisyan (1969c). The horizontal step was equal to 5°, the density field was taken from charts by Muromtsev (1963b), and the annual average field of the atmospheric pressure was also used.

Let us discuss briefly the basic results of the calculations and conclusions of this work. First of all, a series of numerical experiments was performed to elucidate the roles played by the various terms in the equation for ζ. Of course, many such numerical estimates and prognostic calculations of the density field are required to elucidate completely a physical cause of the formation of the fields in a baroclinic ocean. However, as we see below, the calculations are highly labor-consuming, and often it is quite sufficient to perform comparatively easy diagnostic calculations.

The first examples of calculation were concerned with the elucidation of the direct effect of the bottom topography on the reference level. The indirect effect of $H(x, y)$ is present already in the specified density field. The direct effect of this factor on the barotropic currents is substantial and is equally important to all the levels of the ocean. It is quite natural to expect in the baroclinic ocean that deep currents are much affected, currents of intermediate layers are affected to a lesser extent, and surface currents are in fact unaffected by H. However, these first calculations (Sarkisyan, 1969c) showed that taking into account the variability of H has a substantial effect even in the calculation of surface gradient currents of the baroclinic ocean. In the case of variable H, the ζ field bears a distinct azonal character illustrating two known centers, an anticyclonic one in the mid-latitudes and a cyclonic one in the region of Iceland; some small-scale vortices are also present. The assumption $H = \text{const} = 3$ km leads to a smoothed field of ζ. In order to make certain that the role of the bottom topography is important, the following experiment was carried out. Along a circle at 5°N a hypothetical ridge was constructed, rising from the bottom to the continental shelf level (200 m). As a result, a huge anticyclone, covering the area from 5 to 50°N, was formed in the ocean, and the intensity of the Iceland minimum decreased by four times. These changes are substantial if we take the following into account: (1) both in the basic version and in the version with an artificial ridge the fields of density and wind were the same; (2) in both cases everywhere on the boundary of the basin, including its liquid portion, the same boundary conditions were accepted, that is, the absence of water fluxes. We note that the numerical scheme then had a defect which caused errors of an order of 20–30%, but the basic qualitative conclusion based on these experiments remained invariable: in the investigation of the currents of the baroclinic ocean the consideration of the bottom topography is absolutely necessary. Smoothed bottom topography was certainly used in these calculations. During the process of calculation on the electronic computer it was additionally smoothed automatically due to too great a horizontal grid step; however, the basic specific features of the relief remained, since the ocean depth varied from 200 m on lateral boundaries to 6 km in the area of deep-water plateaus. We can be convinced from the study described that the presence of an upright (but not vertical) continental slope is a basic peculiarity of the accepted bottom topography. We return to the problem of topography more than once and list the corresponding maps, including those for the North Atlantic.

The β effect was another object of the numerical analysis by Sarkisyan (1969c). After the publication of the well-known work by Stommel (1948) the importance of this factor does not provoke any doubt. Equation 86, however, differs principally from Stommel's equation of the sea-surface level by Stommel. The difference lies in the fact that in the left side of equation 86 additional terms due to the bottom topography occur, and in the right side, JEBAT. The importance of these factors is shown by simple estimates and by the numerical experiments just described. However, that is not the point. Besides Stommel's well-known β effect, the right side of equation 86 contains another β effect, which we call the baroclinic β effect. Therefore it was necessary to investigate the role played by this factor in the more general equation 86. The version that ignores the β effect leads to the following result. In the mid-latitudes the Azores maximum shifted eastward, the intensity of currents of the western coast decreased, and the intensity of the Canary Current increased. These changes are quite natural and correspond to the role of the β effect discovered by Stommel. But there are other results; the Labrabor current did not change at all as a result of neglecting the β effect. It is necessary to take into account the following: in the version with $H = $ const, there are no surface gradient currents resembling the Gulf Stream or the Labrador current in spite of the fact that the β effect is kept. Finally, the numerical analysis has shown that the effect of bottom topography is greater than the β effect in the regions of the continental slope and slopes of underwater ridges. As a result of the analysis of these calculations in (Sarkisyan, 1969c) it was concluded that the β effect is not a decisive factor in the formation of intense western boundary currents. The factors include, in addition to the β effect, bottom topography, configuration of coastline, position of atmospheric active centers in relation to the edge of the coastline of the ocean (for instance, the Azores maximum of the atmospheric pressure is situated far from the western coast and the Iceland minimum is next to Greenland) and even the zonal field of a tangential wind friction. With regard to the latter factor, the unknown and known characteristics in the work by Sarkisyan (1969c) were given as a sum of zonal components (independent of longitude) and nonzonal deviations. Later this procedure was rejected because it makes calculations more bulky. However, for analysis it is sometimes convenient to apply this technique. In particular, it has been shown (Kochergin et al., 1972, 1973; Sarkisyan et al., 1974) that the zonal fields of density and wind cause a zonal current far from coasts, which on meeting the meridional coasts of continents generates intense coastal currents.

The third series of numerical experiments refer to the elucidation of the problem of which of two factors is more important, the direct effect of a tangential wind stress or inhomogeneity of the water density. The given field of density contains, of course, the indirect effect of the wind, but the direct effect of this factor on the gradient currents is pronounced in equation 86, as terms containing τ_x and τ_y. Some words about the necessity of performing such an analysis are in order. If $\rho \equiv 0$ is accepted in equation 86, then a model for a homogeneous ocean is obtained. The development of this problem was begun by Ekman (1905, 1906, 1923); we call it simply Ekman's model. Gradient currents in this model are determined solely from the direct effect of a tangential wind stress. Another extreme is the dynamic method, in which the field of pressure is defined exclusively from the field of ρ without regard for the direct effect of the wind. Equation 86 contains both these factors, and the relative role of each of them can be elucidated. First a version of the homogeneous ocean was calculated. The sea surface inclination proved to be very small (of the order of 10 cm for the whole ocean); such an inclination produces gradient currents of the order of 1 cm/sec and less. Another version, when $\tau_x \equiv \tau_y \equiv 0$ and ρ is specified, leads to a field of ζ very close to the basic version (with a sea surface inclination greater than 1.5 m for the whole area). A general conclusion on the basis of these two versions is clear; the model of a homogeneous ocean is too strong an idealization; in calculating ocean currents, it is absolutely necessary to account for the baroclinity of seawater.

Calculations of currents resulted in the following. At the ocean surface gradient currents (the calculation of a purely drift current is simple and the result is evident) show two well-known gyres: an anticyclonic one in the area of 5–50°N, and a cyclonic one off Iceland. The velocity of the gradient current is underestimated and is equal to 5–20 cm/sec. Along the western coast at a depth of $z/H = 0.1$ a countercurrent arises which was predicted by Stommel and measured by Swallow and Worthington (1961). The velocity of the countercurrent reaches 20 cm/sec. A countercurrent exists under the northern trade current and also under the Canary Current but at a greater depth now, $z/H = 0.5$. One can see the influence of the Mid-Atlantic Ridge upon the field of current at the middle of the basin. As a whole, deep currents are of the same order of magnitude as the surface ones; they are reciprocally connected with each other. Based on these calculations a theory of the baroclinic layer by Lineikin (1955, 1957) can be criticized with regard to a supposition of the attenuation of pressure gradients with depth. A similar supposition forms the foundation for the dynamic method.

Now we turn briefly to the results of the calculation of the vertical component of velocity. First, the validity of a "rigid lid" was verified by calculating w from the known values of u, v, and ζ. The vertical component of velocity obtained from formula 12 proved to be really small (10^{-6}–10^{-5} cm/sec). It is of the same order of magnitude for the lower boundary of the Ekman friction layer, that is, w obtained due to rot τ. Its value increases with depth by two orders of magnitude, to 10^{-3} at great depths. Fields of the vertical velocity correlate with the bottom topography rather than with the wind field.

In conclusion we speak of two further results of the work under discussion (Sarkisyan, 1969c). In addition to terms of equation 86, $\partial\zeta/\partial t$ was also retained; it was obtained from equation 12. This made it possible to solve the Koshi problem, using a quite unreal initial condition of an absence of currents in the presence of a stationary density field. Nevertheless, the process quickly becomes steady (in 1–2 days). Thus because the establishment of the density field is very slow and the current velocity adjusts very quickly to the density field, while solving a prognostic problem we can neglect derivatives with respect to time in the equations of motion and retain these derivatives only in equations of diffusion of density (heat, salts). The work had a number of defects, which were eliminated by subsequent calculations, to be described below. On the whole, the results of this work (Sarkisyan, 1969c) were encouraging; it could be seen that an alternative to the dynamic method had been found and that such diagnostic calculations should be continued.

Results of subsequent series of experiments performed for the same region are given by Sarkisyan et al. (1967). Smoothing of the bottom topography on the contour and inside the basin was carried out without change in the density field as follows: the Mid-Atlantic Ridge was "removed," the continental slope was stretched by 1000–1500 km, and the bottom topography on the basin contour was smoothed partially or completely. Because the field of density remains the same each time, as calculations have shown, the smoothing of the topography if the maximum depth is 6 km little affects ζ. A change in the bottom topography at levels of 1–2 km or more basically affects deep currents. The ζ field is sensitive, for the most part, to changes in the topography in the upper layer of the ocean, and in particular to changes in the topography of the basin contour (on a solid part of the basin the depth of a vertical wall is accepted as equal to 200 m). Another experiment (Sarkisyan et al., 1967) investigates seasonal variations of the currents. All the above-described results refer to the annual mean fields of density and atmospheric pressure. The annual mean field of atmospheric pressure at sea level for January and July given by Sarkisyan et al. (1967) indicates the presence of a great difference between the summer and winter wind fields. However, the sea surface level for the summer and winter are very close to each other. This can be attributed to the fact that the direct effect of the wind plays a secondary role in the formation of the ζ field. The density fields play a basic role. In Muromtsev's charts density fields were smoothed heavily: in upper layers their winter values differ little from the summer ones; below the "active" layer they are naturally identical. Thus a seasonal run of gradient currents in the upper ocean layer can first of all be obtained on account of a clear seasonal run of density field in the same layer.

Bolgurtsev (1966, 1968) calculated of the sea surface topography and currents of the Antarctic sector of the Pacific Ocean using the density fields of Muromtsev (1963a) and the charts of the atmospheric pressure at sea level. The field of density was specified for nine levels from the surface to a depth of 4 km, with a horizontal grid step of 5°. Bolgurtsev drew the same conclusions that Sarkisyan (1969c) did. In particular, it was found that the inhomogeneity of the density of seawater is a basic factor in the formation of the sea level slope and in the deep circulation of the Antarctic waters. The drift component of the velocity is 40–45 % of the surface current but does not exceed 8 % in the mass transport of the Antarctic gyre. Assuming that $H = \text{const}$ results in a zonal field of ζ. Nonzonal gradient currents and the vertical component of the velocity are due to the bottom topography. We should note the importance of these results; before Bolgurtsev's calculations one could suppose that a purely drift transport in the Antarctic latitudes was basic, since the density anomalies were comparatively small here. It follows from Bolgurtsev's calculations that here the thermohaline factors also affect, in a decisive way, the mass transport in the entire thickness of the ocean. Bolgurtsev, Kozlov, and Molchanova (1968, 1969) investigated the currents in this region. Unlike the work discussed above, the authors specified density not in the whole ocean water column, but at its surface only. At other levels the density was calculated from the hyperbolic model of Kozlov (1968):

$$\rho_1 = \rho_H + \frac{\rho_{(0)}}{1 + (z/h)} \tag{208}$$

where ρ_1 is water density, ρ_H is the density at the ocean bottom, $\rho_{(0)}$ is the difference between densities at the ocean surface and bottom, and parameter h is found by solving the corresponding equation. On the whole, the authors came to the same conclusions as did Sarkisyan et al. (1967) and Sarkisyan (1969c); they noted, however, that modeling leads to the smoothing of the ζ field and values of the current velocity are underestimated by 2–2.5 times.

The sea-surface topography and the flow velocity in the Caribbean Sea were calculated by Sarkisyan and Knysh (1969). In spite of the fact that the Caribbean Sea is a part of the system of the general circulation of the Atlantic Ocean, a special numerical analysis is necessary here since a large grid step, applicable to the whole ocean, cannot describe the circulation of this sea. This concerns the Gulf of Mexico as well.

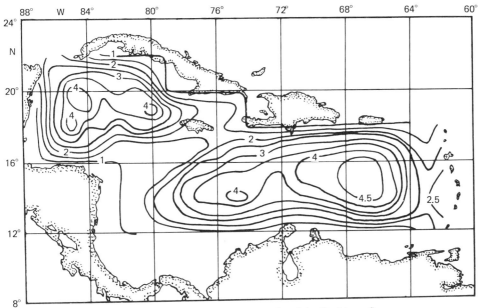

Fig. 4   The smoothed bottom topography of the Caribbean Sea.

Smoothed bottom topography used in the calculations is shown in Fig. 4. Vertical walls of a depth of 1 km are accepted on the side boundaries; the minimum depth inside the basin is the same. If H is greater than 1 km in the straits, then a real depth is taken. The density field used was constructed by V. V. Rossov on the basis of scarce data of observations for a winter season at six levels, from the surface down to 1 km. Calculations were performed for five levels with a horizontal grid step of 1°. The field

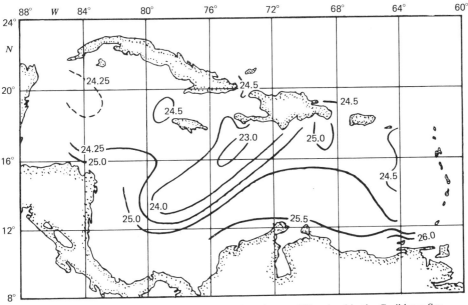

Fig. 5.   The field of the conventional density (σ_t) at the $z = 100$ m level in the Caribbean Sea.

TABLE I

Levels (m)	0	100	200	300	500	1000
$\delta\sigma_t$	2	2	1.1	0.5	0.5	0.1

of conventional density (σ_t) for a level of $z = 100$ m is given in Fig. 5. The difference between maximum and minimum values of σ_t for all levels is shown in Table I.

Thus the horizontal gradients of density were small and decayed rapidly with depth. In spite of this, the baroclinity of seawater is a basic factor in forming the field of ζ and gradient currents. Figure 6 presents the calculated field of ζ. We should pay attention to the point denoted by a heavy arrow with coordinates 12°N, 75°W. Here σ_t was overestimated for unity by mistake. Firstly, this caused a considerable change in the ζ field at this point; secondly, as a repeated calculation with the correction showed, the effect of the error is localized in the direct vicinity of the given point. These properties are typical of the diagnostic calculations. One can judge the directions and values of velocities of surface gradient currents from the ζ field. The velocity of total (purely drift + gradient) surface currents is shown in Fig. 7. The velocity current ranges from 10 to 80 cm/sec; the maximum value of the velocity at the surface was equal to 85 cm/sec. Figure 7 indicates the presence of an inflow of the Atlantic Water to the Caribbean Sea through the Mona Strait. Instrument measurements made from on board the research vessel "Akademik Vernadsky," after these calculations, evidenced this. All the same, it is necessary to repeat these calculations on the basis of the future charts of density, constructed from the more abundant observational data. There is no need to describe in detail the results of calculations of currents in the Caribbean Sea; they supported the conclusions drawn for the North Atlantic by Sarkisyan (1969c). We note only a single new result. As is known, a huge

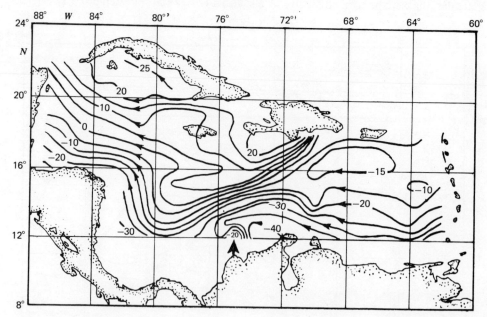

Fig. 6.   The calculated sea-surface topography of the Caribbean Sea. Arrows on isolines show the direction of a surface gradient current.

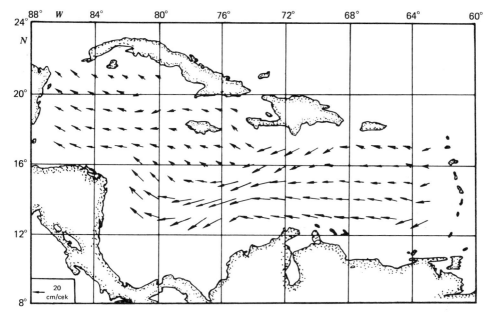

Fig. 7.   The chart of calculated total resultant surface currents of the Caribbean Sea.

water mass passes through this area which, flowing through the Gulf of Mexico, generates the Florida current. We could ask if it is possible to obtain the basic picture of the circulation in the Caribbean Sea only from water fluxes and the wind field. To answer this question the following experiment has been made. The inflow of the Atlantic water of 30 sverdrups (1 sv = 10^6 m^3/sec) was specified through the Lesser Antilles Straits and the outflow through the Yucatan Strait. The tangential wind stress was also taken into account, but $\rho \equiv 0$ was accepted for the whole Caribbean Sea. It has been proved that comparatively great velocities of currents appear only very near the straits (with specified water fluxes) and in shallow regions. In deep-water regions forming the basic part of the Caribbean Sea, a slow current of the order of 1 cm/sec is uniform in the entire water column of the sea. Thus even in such a small basin where the circulation might be determined by the inflow and outflow, the specification of these quantities does not permit the construction of the picture of the currents. For the same water fluxes the picture of currents can be quite different, depending on the density variation of the seawater. Let us note that the obtained currents are in good agreement with the results of Ichye (1972).

The next step in the diagnostic investigation of currents of the North Atlantic is the work by Sarkisyan and Pastuchov (1970). As compared with Sarkisyan (1969c) and Sarkisyan et al. (1967) the following generalizations are made here:

1. Nonlinear inertial terms and the effect of lateral mixing in the equation of motion and in the equation for the sea-surface topography are taken into account using the method of successive approximations (the D_2 model).

2. A grid step over the horizontal is made two times less in the whole ocean, and four times in the Gulf Stream area.

3. The number of calculated levels is increased: all 10 available density charts by Muromtsev are taken into account and fields of current velocities at eight levels are calculated.

4. Water fluxes have been specified on a number of sections of the liquid boundary.

5. The effect of bottom friction at the side boundaries and a number of other factors from the results of diagnostic calculations are investigated.

An increase in the number of calculated levels and a decrease in the horizontal step (to 2.5° in the whole ocean and to 1.25° in the Gulf Stream area) proved to be the most effective of these generalizations. If in this region velocity of a gradient current was of the order of 15–20 cm/sec (Sarkisyan, 1969c), then it increased to 20–30 cm/sec. At the same time it was elucidated that a further decrease of the step is inexpedient since the field of density by Muromtsev is too smooth. Another consequence of finer horizontal and vertical grid spacing is the decrease of the direct effect of the bottom topography on surface gradient currents. The sea-surface topography for $H = $ const of course differs substantially from the case where $H \neq$ const, but there is not such a striking difference as in Sarkisyan (1969c). In Sarkisyan and Pastuchov (1970) a new simplified procedure of the determination of ζ on the contour of the basin was applied; however, in subsequent calculations procedure was discarded, because it requires strong smoothing of the basin depth on the boundary (and consequently near the boundary).

The calculations showed that consideration of inertial terms slightly smooths the ζ field and changes its amplitude by only 6%. The effect of taking into account the lateral eddy viscosity (even with an overestimated value of $A_e = 10^8$ cm²/sec) has proved to be even less—a decrease in the amplitude of ζ by 3–4%. Both these factors complicate the calculations and produce insignificant effects. Therefore, we may neglect them while performing a diagnostic analysis of large-scale currents. The allowance for equations 122 and 123, which have the multiplier $1/2\alpha H$, is also negligible; the amplitude of ζ on the basin contour (despite the overestimated value $v = 10^2$ cm²/sec) changes by only 7% and this change, in fact, does not affect the field of ζ inside the basin. Taking into account the mass fluxes through a liquid part of the boundary proved to be more significant. Specifying the balanced water fluxes at only 13 points of the liquid part changed the amplitude of ζ on the basin contour by 20%. Now we list the results of some of the other numerical experiments of Sarkisyan and Pastuchov (1970). At $\rho = 0$ the sea-surface topography is unrealistically smooth, with an amplitude of only 15 cm, whereas taking ρ into account leads to a field of ζ with an amplitude of 135 cm. The assumption that $\rho = 0$ at only four points of the Gulf Stream area resulted in the change of the level of the given region to 23 cm. Finally, it was noted that the effect of the bottom topography is important, first of all, because of the fact that the right side of equation 115 has a product of bottom topography and gradients of integrals of ρ (JEBAT). Thus the density field is a basic indicator of stationary sea currents.

Let us sum up the preliminary results from the diagnostic calculations of velocity of gradient currents by means of the sea-surface topography.

1. Above we noted the contradictions between two approaches in the theory and calculation of stationary or seasonal currents: models of a homogeneous ocean consider only the direct effect of the tangential wind stress in the determination of ζ, and the dynamic method takes into account only the field of ρ. Diagnostic calculations, taking into account both these factors, indicated that models of a homogeneous ocean cannot serve as a foundation for any estimates of the velocity for stationary ocean currents. The field of ρ is a basic indicator which reflects the process of heat and salt exchange, as well as the indirect effect of the wind field.

2. In the investigation of a large-scale quasi-stationary circulation with a step over the horizontal of $1.25°$ or more, we may neglect nonlinear terms and the effect of lateral eddy viscosity. In the calculation of the sea-surface topography the bottom topography is the factor of second importance, after the density anomaly. The effect of the bottom topography on gradient currents was investigated by Ekman. However, the main point is not that $H(x, y)$ changes the picture of wind currents; another circumstance is more important: the joint effect of baroclinity and the bottom topography (JEBAT), which is present in the right side of the equation for ζ, is 5–10 times greater than the direct effect of the wind. The variability of the topography just in the upper baroclinic layer of the ocean affects, basically, currents in the upper ocean layer. The change in H at great depths slightly affects ζ. The assumption that $H =$ const leads to the loss of one of the two basic factors reflecting the baroclinity of seawater.

3. Apart from the β effect, the importance of which was shown by Stommel (1948, 1958), the equation for ζ contains a more important factor, the baroclinic β effect. If the β effect redistributes only the currents induced by the wind, then if baroclinic β effect is taken into account in the right side of equation 72, a term is kept which is 5–10 times greater than the wind force.

4. Beginning with the well-known work by Stommel (1948), the β effect was accepted to be a dominating factor in the intensification of currents off an eastern coast. This was concluded only because the baroclinity of the seawater was not taken into account and because of some other simplifications. The matter is much more complicated. Continental slope, heat advection by a pure drift current (Bolgurtsev and Kozlov, 1969; Bryan and Cox, 1967), etc. take part in the intensification of currents near boundaries of the ocean (and intense currents also exist near the eastern walls).

5. The vertical component of velocity is small in the surface layer and increases with depth by two orders of magnitude. It is engendered, first of all, by the baroclinity of seawater and the bottom topography. The vortex from the tangential wind stress plays a secondary role in the formation of the field of v_z at great depths.

B. Calculation of the Sea-Surface Level, Mass Transport Stream Function, and Flow Velocity

Diagnostic calculations of currents, using the mass transport stream function with due regard for JEBAT, started after the publication of the work by Sarkisyan (1969b). Some qualitative conclusions from this work have been given in Section 2.C. Figures 8 and 9 present the comparison between the meridional transport obtained with due regard for JEBAT and that obtained from Sverdrup's relation. These values are given on the same scale in Fig. 8a, whereas the characteristics obtained from Sverdrup's relation are increased by one order of magnitude in Figs. 8b, 9a, and 9b. Subsequent calculations (Sarkisyan and Ivanov, 1971, 1972; Sarkisyan and Keondzhyan, 1972) indicated that values of S_y and v shown in Figs. 8 and 9 are overestimated owing to the inaccuracy of density charts by Muromtsev (1963b). However, a qualitative conclusion drawn on the basis of these calculations is reliable; in the calculation of S_y it is necessary to take JEBAT into account. Based on the work by Sarkisyan (1969b) a number of diagnostic calculations of fields of Ψ, ζ, u, v, and w have been made. We dwell briefly on the analysis of some of these calculations.

Sarkisyan and Ivanov (1971) presented for the first time a solution of equation 109 with due regard for all its terms. This made it possible to estimate the relative roles of the baroclinity and rot τ in the formation of the mass transport stream function.

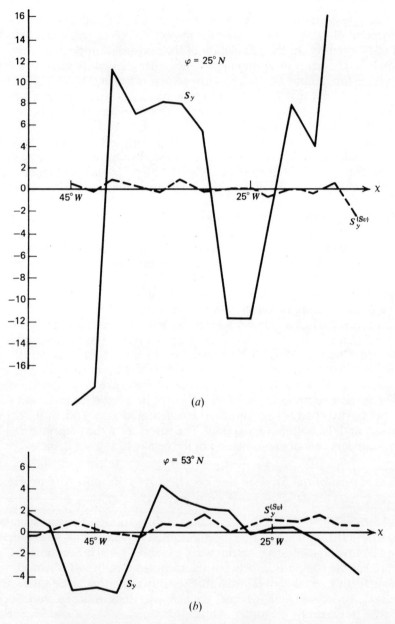

Fig. 8. (a) The meridional mass transport through the latitude circle $\varphi = 53°N$ (the North Atlantic). Solid line, $10^{-5} S_y$ (cm^2/sec), obtained with due regard for JEBAT; dashed line, $10^{-5} S_y^{(Sv)}$, Sverdrup's mass transport in the same scale. (b) The meridional mass transport through the latitude circle $\varphi = 25°N$. Solid line, $10^{-5} S_y$ (cm^2/sec), obtained with due regard for JEBAT; dashed line, $10^{-4} S_y^{(Sv)}$, Sverdrup's mass transport with the scale increased by one order of magnitude.

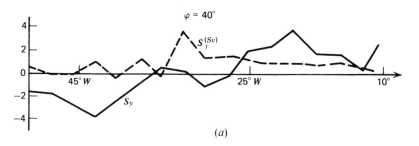

(a)

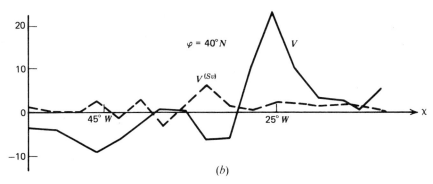

(b)

Fig. 9.   (a) The meridional mass transport through the latitude circle $\varphi = 40°$N. Solid line, $10^{-6} S_y$, obtained with due regard for JEBAT; dashed line, $10^{-5} S_y^{(Sv)}$, Sverdrup's mass transport with the scale increased by one order of magnitude. (b) The height-averaged magnitude of the meridional component of the flow velocity at the latitude circle $\varphi = 40°$N. Solid line $V = -(1/H) - S_y$ (cm/sec), obtained with due regard for JEBAT; dashed line $10V^{(Sv)} = -(10/H) - S_y^{(Sv)}$ (cm/sec), Sverdrup's velocity with the scale increased by one order of magnitude.

Calculations have been performed for the area of 5–65°N in the North Atlantic with a grid step of $\delta\theta = \delta\lambda = 2.5°$. The atmospheric pressure at sea level, density charts by Muromtsev (1963b) for nine levels (0, 25, 50, 100, 200, 300, 500, 1000, and 1500 m), and the smoothed bottom topography (Fig. 10) served as initial data. Let us discuss some of the versions of these calculations. First, the field of Ψ was obtained for the barotropic ocean with a constant depth, that is, the traditional solution of Stommel's problem obtained on the basis of Sverdrup's relation with the western intensification due to the β effect. As a result two well-known gyres (a subtropical anticyclone and a subpolar cyclone) were obtained with the western intensification of the integral mass transport. If the baroclinic effect of bottom friction (group III) is taken into account on the right side of this equation, the picture does not change qualitatively. Variations of Ψ is a barotropic ocean due to bottom topography are more significant. The bottom topography used in these calculations was of the following form. The minimum depth inside the basin was taken to be equal to 1500 m. Upright walls of the same depth served as lateral boundaries. At other points the bottom topography was of the form shown in Fig. 10. Taking into account the bottom topography significantly transformed the picture of the field of Ψ without changing the amplitude of this function. The basic variations of Ψ were obtained due to the allowance for JEBAT (group II). In doing so, two methods of taking this factor into account were tested.

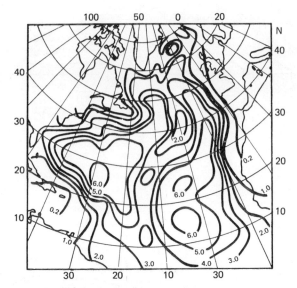

Fig. 10.   The smoothed bottom topography of the North Atlantic.

(1) Values of density anomalies were extrapolated from 1–1.5 km to the ocean bottom using the formula:

$$\rho_z = \rho_{1500} + 0.1 \frac{\rho_{1500} - \rho_{1000}}{500} (z - 1500), \qquad 1500 \leqslant z \leqslant H \qquad (209)$$

(2) Below 1.5 km the seawater was accepted to be homogeneous. In the first case, one huge cyclonic gyre with water fluxes equal to 100 sverdrups was obtained; water fluxes for the western and eastern coasts were equally intense. The second version of the calculations listed in Fig. 11 differs completely from the first one. Neither version is like the classic picture of the total mass transport obtained repeatedly for a barotropic ocean of constant depth. It was found that: (a) the mass transport stream function is basically formed by the density field of the deep layers; (b) baroclinity of oceanic water persists for the most part, due to taking into account JEBAT; and (c) rot τ plays a less significant role in the formation of Ψ. In the cited work calculations of the flow velocity were also performed for all nine levels. The analysis of calculations showed that the surface gradient current field differs considerably from the integral transport. There are regions of intense surface currents in which the integral transport is not large. In a number of regions (in particular, in the Gulf Stream area) deep currents are opposite to surface ones. On the other hand, the deep-water circulation is in quite good qualitative agreement with the integral picture obtained with due regard for the baroclinity of deep layers. Thus the integral circulation does not correlate with surface currents.

Finally, calculations showed that intense currents exist not only near the western but also near the eastern coast of the ocean. The latter cannot be obtained in terms of the classical theories of Sverdrup, Stommel, and Munk. First of all to obtain a realistic picture of currents for various levels the accurate allowance for JEBAT is required. We should note that despite a great difference between the two versions of calculation of Ψ, the surface currents calculated from them are close to each other and are realistic. They depict two well-known gyres and the western intensification of

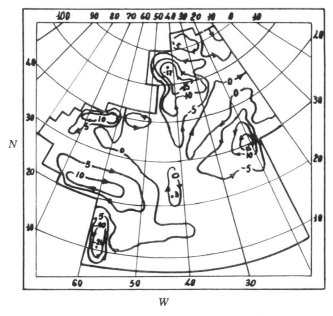

Fig. 11.   The mass transport stream function. $\psi \times 10^{-6}\,\mathrm{m}^3/\mathrm{sec}$ when the ocean is taken to be homogeneous below 1.5 km.

currents. Only deep currents differ much. Thus inaccuracies of the density anomaly of the deep layers produce considerable errors in the field of deep currents and strongly transform the Ψ field, but have little affect on the surface currents.

The direct continuation of the described calculations has been published by Sarkisyan and Ivanov (1972). Another version is discussed here for the extrapolation of the density anomalies at great depths:

$$\rho_z = \rho_{1500} - \frac{\rho_{1500}(z - 1500)}{H - 1500} \tag{210}$$

This corresponds to a linear attenuation of the density anomaly from the value at a depth of 1500 m to zero at the ocean bottom. A comparison is given between variations of the sea-surface topography ζ and the mass transport stream function, depending on the allowance for the baroclinity of deep layers, bottom topography, etc. The ζ and Ψ curves at two sections are shown in Fig. 12 to illustrate the effect of the bottom topography. The response of these two functions to the baroclinity of deep layers is illustrated in Fig. 13. It follows from Fig. 13 that the second version of extrapolation is "the golden mean" between the first version and absence of taking into account the baroclinity of deep layers. In a new version of extrapolation, the amplitude of Ψ proved to be two to three times less than that in the first case, but the field as before had nothing in common with the classic field obtained for $\rho \equiv 0$ and $H = \mathrm{const}$. Thus the second version of extrapolation proved to be more realistic, but probably it is better to use the physically more grounded extrapolation

$$\frac{1}{\sigma_t}\frac{\partial \sigma_t}{\partial z} = \frac{w_1^2}{gz^2} \qquad (w_1 = \mathrm{const}) \tag{211}$$

suggested in work Monin et al., (1970).

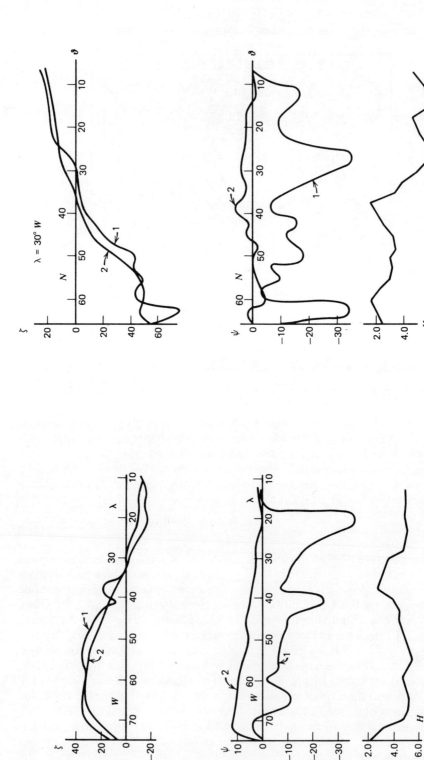

Fig. 12. The sea-surface topography ζ (cm) and the mass transport stream function $\psi \times 10^{-6}$ m^3/sec at the section along (a) the 35° N latitude and (b) the 30° W meridian. 1, The case of the real bottom topography; 2, for $H = \text{const} = 1.5$ km. The section of the bottom topography (km) is shown below.

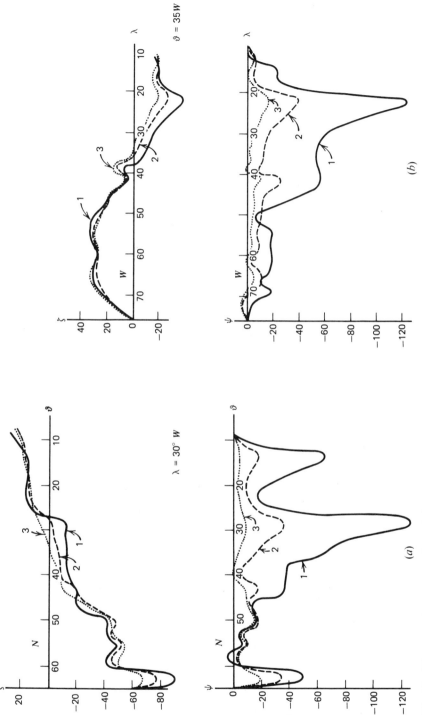

Fig. 13. The sea-surface topography and the mass transport stream function (*a*) along the 30°W meridian and (*b*) along the 35°N latitude by the different methods of extrapolation of density anomalies at great depths (Sarkisyan and Ivanov, 1972). 1, The first method of extrapolation; 2, the second variant of the extrapolation, and 3, the case when the ocean is taken to be homogeneous below 1.5 km.

The figures show that the field of ζ and, consequently, surface currents react little to the variations of the bottom topography and the baroclinicity of deep layers, whereas the field of Ψ is very sensitive to these changes. The matter is as follows. JEBAT is a decisive factor in the right side of equation 109, whereas the right side of equation 86 also contains the group I, the baroclinic β effect, which is of the same order of magnitude as JEBAT. The baroclinic β effect is, in general, determined by the baroclinicity of upper layers and depends little upon the variability of the bottom topography.

Simultaneously with the calculations of Sarkisyan and Ivanov (1971) and independently of them calculations of currents in the Pacific Ocean were performed by Kozlov (1971). Besides the fact that Kozlov performed calculations for a different ocean, there is another significant difference: the observed density anomaly is used only at two levels; at others it is determined using a simple interpolation formula. An example of such parametric specification of the density anomaly has already been given (see formula 208).

Despite such a difference, conclusions drawn by Kozlov are in good agreement with those of Sarkisyan and Ivanov (1971) and Sarkisyan (1969a). In particular, Kozlov notes the following. (1) Taking the bottom topography into account transforms the classical field of Ψ, obtained from using the Sverdrup–Stommel model considerably. (2) Even more striking variations of Ψ occur when JEBAT is taken into account. (3) The field of Ψ does not reflect the surface circulation but does reflect the currents at 1–2 km depths. (4) The wind field has a comparatively lower effect on the integral circulation in a baroclinic ocean with uneven bottom. To illustrate this we give two figures from the work by Kozlov. Figure 14a presents a classical picture of the integral circulation in a homogeneous ocean of the constant depth. Figure 14b shows the field of Ψ for a baroclinic ocean with variable depth. These two figures differ from each other approximately in the same way as the corresponding figures of Sarkisyan and Ivanov (1971). Thus calculations performed simultaneously by different authors using different methods and for different oceans resulted in the same conclusion supporting that of Sarkisyan (1969b), that JEBAT is a basic factor on the right side of equation 109.

The work by Ivanov (1971) is devoted to the calculation of the vertical component of the flow velocity. The author has calculated v_z for the North Atlantic based on calculated fields of Ψ and ζ (Sarkisyan and Ivanov, 1971, 1972) using formulas like 94 and 114. It was noted that both versions of the calculation coincide only in the upper ocean layer. The difference is significant for the lower layers. We believe that this difference is due to the density field for lower layers constructed by rough extrapolation. Ivanov shows that though gradients of the bottom topography are not present in an explicit form in formulas 94 and 114, the ocean depth variability substantially affects v_z through fields of Ψ and ζ. Ivanov also supported the result of Sarkisyan (1969b) concerning rapid increase of v_z with depth from the surface to the bottom friction layer.

Sarkisyan and Perederey (1972) present a numerical solution of equation 86 for the South Pacific Ocean. The field of ζ calculated by the dynamic method is also given for comparison. The dynamic method proved to be a good first approximation for solving equation 86, whereas the model for a homogeneous ocean leads to the field of ζ, the amplitude of which is less by one order of magnitude. This work contains a table of latitudinal average of absolute values of individual groups of terms of this equation which we list in Table II.

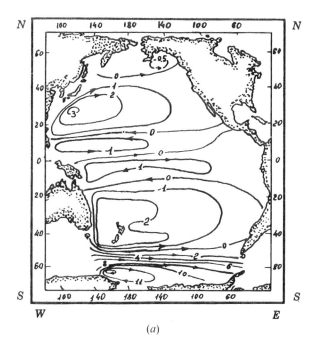

(a)

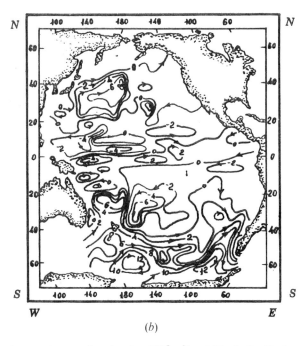

(b)

Fig. 14.   The mass transport stream function $\psi \times 10^{-7}$ m^3/sec (a) in the Pacific Ocean by the model of a homogeneous ocean of constant depth, and (b) in the baroclinic ocean of variable depth (Kozlov, 1971).

TABLE II

| Group | \multicolumn{17}{c}{Latitude (°)} |||||||||||||||||
	4	6	8	10	12	14	16	18	20	22	24	26	28	30	32	34	36
I	163	114	98	76	59	50	48	46	32	18	16	14	14	24	11	11	7
II	41	47	49	52	50	58	72	113	99	96	100	97	89	112	149	111	67
III	4	3	2	2	3	3	6	3	4	2	1	1	2	2	2	1	1
VII	8	6	4	6	4	4	4	2	2	2	2	2	2	2	1	2	3

It is seen from Table II that in the real baroclinic ocean of variable depth, the β effect and JEBAT are quantities of the same order of magnitude, and rot τ and the friction with the bottom are one order of magnitude less.

The idea that the dynamic method is the reliable first approximation in diagnostic calculations of horizontal currents has received further development by Perederey (1972, 1974), Perederey and Sarkisyan (1972) and Kozlov (1973). In particular, it has been shown that by some simplifications and transformations of the right sides of equations 69 and 73 their solutions can be obtained in the form of formulas 99 and 108.

In Perederey (1972, 1974) and Sarkisyan and Perederey (1972) the results of calculations of horizontal currents at 21 levels are stated. But because of the big horizontal grid step ($\delta\theta = \delta\lambda = 5°$) the velocity of a gradient current in an open ocean is not large (of the order of 3–5 cm/sec); the boundary currents are about 10–15 cm/sec.

Among the works published in the USSR some of the most detailed results have been obtained by Sarkisyan and Keondzhyan (1972), Sarkisyan and Keondzhyan (1975),[1] and Keondzhyan (1972). The calculations were made for the North Atlantic (from 7 to 60°N) with a horizontal grid step of 1.25° using the density field restored by Suchovey (1971). In contrast to the charts of Muromtsev, the density field of Suchovey is restored on the basis of data statistical analysis at points on the small grid (1.25°). The detailed density field and a small step of the finite-difference approximation have led to interesting results. In the Gulf Stream region the flow velocity increased to 70–80 cm/sec (in the preceding calculations it was 10–20 cm/sec). The field of the mass transport stream function suffered great changes; for example, the Gulf Stream, Labrador, and other intense currents assumed more distinct forms. The calculations have confirmed all the qualitative conclusions made in previous investigations. Since it is clear that in the formation of the ζ and Ψ fields the terms due to the baroclinity of seawater play the main role, the following question arises: what is the minimum thickness of the upper baroclinic layer of the ocean, the effect of which is comparable to that of rot τ?

It has been shown by Sarkisyan and Perederey (1972) and Sarkisyan and Keondzhyan (1972) that the baroclinicity of an upper layer of 50 m thickness already contains a driving, causing the same change in the sea-surface level as rot τ. As for

---

[1] The authors are very sorry for the errors of this paper. For the purpose of shortening it, some figures were withdrawn in the process of editing. As a result, unfortunately, there is nonagreement between the figures and their descriptions. Figures that are not included in the paper are described in the text as Figs. 8 and 9. Charts 8 and 9 correspond to Figs. 10 and 11 in the text. Finally, charts 10 and 11 present variants of Ψ calculation that are not described in the text.

the integral transport, here we need the baroclinity of a layer of 500 m thickness in order to compete with the effect of rot τ. Consequently, when calculating ζ and surface currents, we can even neglect in equation 86 (in the first approximation) the effect of rot τ. The same cannot be done in equation 109. The calculations made by Keondzhyan (1972) confirmed that at least in the Atlantic Ocean the main part of seasonal changes of the sea-surface topography is due to the baroclinity of the sea-water. This conclusion coincides with that of Sarkisyan et al. (1967). Shown in Fig. 15 is the stationary surface current field of the North Atlantic obtained by Keondzhyan. This chart corresponds both qualitatively and quantitatively to the modern concepts of the large-scale circulation in the North Atlantic. Next Fig. 16 shows the current field at the $z = 500$ m horizon. It is seen that at this depth the jet of the Gulf Stream is narrow and the velocities have decreased. Although at the surface we obtained velocities equal to 50–100 cm/sec, here they are only 15–50 cm/sec. The influence of the bottom topography on the jet of the Gulf Stream (Warren, 1962) is also seen. With increasing depth the further decrease of velocities in the Gulf Stream region takes place; at the 600–1200 m horizons the countercurrent appears.

This is demonstrated by the chart of currents at the $z = 3000$ m, given in Fig. 17. The countercurrent under the Gulf Stream forecasted by Stommel and measured by Swallow and Worthington (1961) is clearly seen.

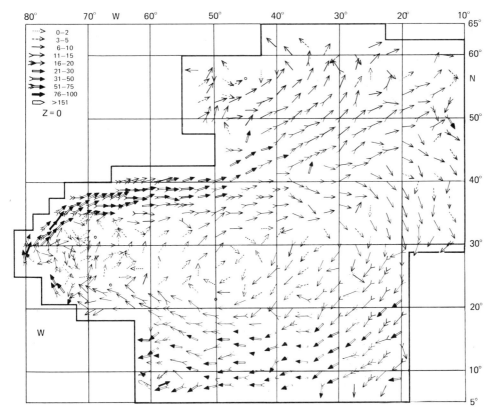

Fig. 15.   The field of the annual mean surface currents in the North Atlantic.

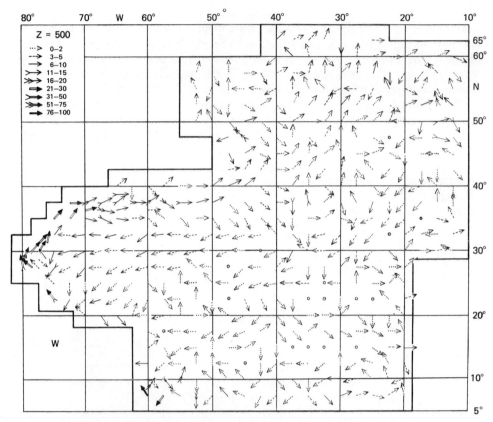

Fig. 16.   The field of the current at the level $z = 500$ m of the North Atlantic.

It is shown by Sarkisyan and Keondzhyan (1975) that the integral transport of the upper layer of 800 m thickness and that of the lower layer from 800 to 2000 m have contrary directions in many regions of the North Atlantic and especially in the Gulf Stream region.

Diagnostic calculations were carried out in the Pacific Ocean also by Makarevich and Fel'zenbaum (1972) by a modified variant of the D_1' model. Unfortunately, the authors adduce only the results obtained when taking into account the baroclinity of an upper layer of 200–1000 m thickness (the depth of the isoline $\rho_0 = 1.0274 \, \mathrm{g/cm^3}$). As we know from the preceding calculations, the Ψ field is formed mainly by the baroclinity of lower layers. It is natural, therefore, that Makarevich and Fel'zenbaum (1972) underestimated the role of this factor.

C.   Diagnostic Calculations by the Method of Bryan

In 1969 Bryan suggested a numerical method for a prognostic calculation of thermohydrodynamical characteristics of an ocean of variable depth. This method has been widely adopted for prognostic calculations. Some results of calculations by this method are adduced by Holland in Chapter 1 of this volume. If we assume the density field to be specified, then a particular case of Bryan's method can serve for the diagnostic method of the calculation of Ψ and the current velocity field. Bryan uses an

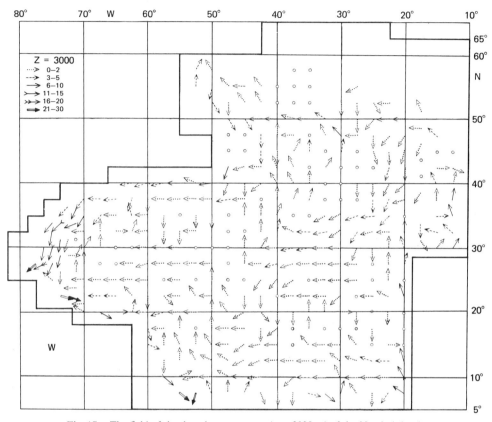

Fig. 17.   The field of the deep-layer currents ($z = 3000$ m) of the North Atlantic.

equation of the D_5 model, that is, the initial equations 1–4 without subsequent simplifications. Let us dwell on two diagnostic calculations carried out by Bryan's method.

Holland and Hirschman (1972) carried out an interesting diagnostic investigation of the Atlantic Ocean currents from 10°S to 50°N with a grid step of 1° and maximum number of calculated horizons equal to 14. Note that these parameters are close to those of Sarkisyan and Keondzhyan (1972) and Keondzhyan (1972) mentioned above (the region from 7 to 65°N, a grid step of 1.25°, and 16 horizons). We state now some results of the work of Holland and Hirschman and compare them with the other diagnostic calculations, especially those of Sarkisyan and Keondzhyan (1972) and Keondzhyan (1972). In spite of the difference between the model of Bryan and the above-mentioned models, the comparison is rightful for the following reasons. As was mentioned by Holland and Hirschman, the nonlinear terms (just as the direct effect of a tangential wind stress) were essential only in an equatorial zone of 3–4° width when calculating with a grid step of 1°. Below we show that even with a grid step of 0.25° the contribution of nonlinear terms is not large in the Gulf Stream region. All the aforementioned diagnostic calculations refer to the nonequatorial region. Holland and Hirschman carried out the following main numerical calculations.

The first case is the model of a homogeneous ocean of variable depth. The known anticyclonic eddy of the integral transport with westward intensification was obtained.

The maximum integral water flux is in good agreement with the results of Sarkisyan and Ivanov (1971), but the form of isolines of Ψ differs essentially. This means that the fields of the tangential wind stress of these two studies are close to each other, but the bottom topographies differ significantly.

As is known, in the model of a homogeneous ocean for the same wind field, the form of isolines of Ψ is on the whole defined by the sign of the derivative $(\partial/\partial y)(l/H)$. The parameter l, of course, is the same in all the studies; only the bottom topography has a different form. Depending on H, we can obtain, for example, either westward or eastward intensification of the Ψ field (Kochergin and Klimok, 1971). For obtaining the westward intensification we must assume $H = \text{const}$, or have the greatly smoothed bottom topography (Sarkisyan and Ivanov, 1971; Bryan and Cox, 1972).

The second case is the baroclinic ocean of constant depth. In this case the isolines of Ψ are smoother, and the maximum water flux has increased from 12 to 28, but no qualitative changes are observed: the region of the North Atlantic considered is covered by one huge anticyclonic eddy.

The third case is the baroclinic ocean of variable depth. The greatest changes due to JEBAT are observed only in this case. This conclusion agrees also qualitatively with that of Sarkisyan and Ivanov (1971). However, considerable quantitative discrepancies are noted; they are due to the differences in the density fields. The density fields used by Sarkisyan and Ivanov (1971) were greatly smoothed. The case is somewhat different when comparing the results of Holland and Hirschman with the calculations of Sarkisyan and Keondzhyan (1972) and Keondzhyan (1972). The last calculations have been carried out by the same model as in Sarkisyan and Ivanov (1971), but on the basis of a good-quality density field. As a result, good agreement is obtained between the calculations of Sarkisyan and Keondzhyan (1972) and Keondzhyan (1972) and the results of Holland and Hirschman.

Shown in Fig. 18 is the Ψ field obtained with due regard for JEBAT by Holland and Hirschman. We see that, as in Sarkisyan and Ivanov (1971), we have here a lot of

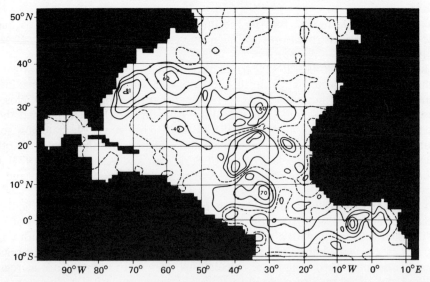

Fig. 18.   The mass transport stream function for the third case, contoured at intervals of 20 sverdrups (after Holland and Hirschman 1972). The dashed lines indicate the $\psi = 0$ contours, which separate regions of clockwise and counterclockwise flow in the interior.

chaotic cyclonic and anticyclonic eddies instead of the huge classical subtropical anticyclone. If the water flux in the Gulf Stream region, reaching 60–80 sverdrups, is the usual, then the water flux in the region of 32–38°W, 22–26°N (deep-water plateau to the east of the Mid-Atlantic Ridge) exceeding 100 sverdrups cannot be considered to be a normal phenomenon. This fact was explained above: the established tradition of comparing the integral transport with the surface currents is wrong. Moreover, we cannot even regard that the Ψ field reflects the integral circulation of the upper 1 km of the ocean. There may be good correlation between them in some region and none in the others. The Ψ field depends strongly on the bottom topography; it is formed mainly as a result of the density anomalies at great depths (containing considerable errors) and, first of all, reflects the deep-layer circulation. That is why in spite of the complicated system of cycles of Fig. 18, Holland and Hirschman have obtained a very realistic and logical chart of surface currents.

Figure 19 shows another result obtained by Holland and Hirschman. As is seen from the chart, the water flux of the Gulf Stream in the third case differs essentially from that of the other variants. We have supplemented this chart with one more curve, the value of water flux, obtained by Sarkisyan and Keondzhyan (1972) with due regard for JEBAT. The good agreement between these results and those of Holland and Hirschman is obvious. However, there is not such good agreement between the calculations for the rest of the ocean. Probably the density anomalies used in these works differ, especially at great depths. However, the charts of the sea-surface topography (see Fig. 6 of Sarkisyan and Keondzhyan, 1972, and Fig. 9 of Holland and Hirschman, 1972) agree very well. The ζ field is due mainly to the density anomalies of the upper layer of 1.5 km thickness. Here the accuracy of ρ is quite satisfactory and and therefore good agreement between the results of these two works has been obtained.

Cox (1972) made the first diagnostic investigation of the global circulation of the World Ocean. Cox has calculated the integral water flux and currents by the method of Bryan with horizontal grid steps equal to $2 \times 2°$; the maximum number of horizons is nine. Cox carried out two experiments.

The first experiment was on currents in a homogeneous ocean of variable depth. These results of Bryan and Cox agree well with the aforementioned calculations of Sarkisyan and Ivanov (1971), Kozlov (1971), and Holland and Hirschman (1972) for the northern hemisphere. In particular, the water flux of the Gulf Stream in Sarkisyan and Ivanov (1971), Cox (1972), and Holland and Hirschman (1972) changes within 12–15 sverdrups. We have a somewhat complicated picture in the southern hemisphere, which is due to the effect of the bottom topography. The main peculiarity of the mass transport stream function in the southern hemisphere is the distinctly expressed nonzonality of the Antarctic Circumpolar Current.

The second experiment was on the diagnostic calculation of the integral transport in the World Ocean. This experiment is more interesting; the results are given in Fig. 20. We see the complicated system of cycles of the circulation typical for all the diagnostic calculations of the Ψ field. The water fluxes of the Gulf Stream and Kuroshio have increased to four times those in a homogeneous ocean. In general, Cox remarks, all the known gyres have become two to three times more intensive, and the water flux of the Antarctic Circumpolar Current has increased 10 times. The integral water fluxes do not agree often with the known surface currents, either in direction or in intensity of the currents. For example, in the region between the Brazilian and Falkland Currents (about 40°S) a powerful anticyclonic gyre is obtained, with the water flux 2.5 times that of the Gulf Stream water flux. Cox divides the ocean

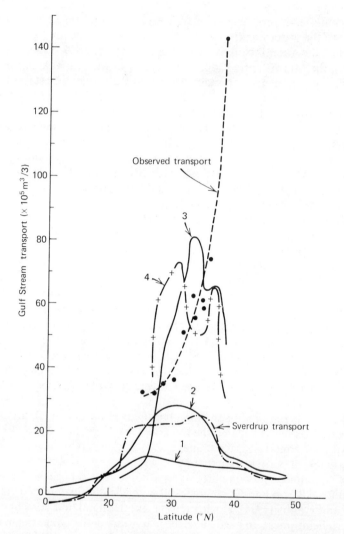

Fig. 19. The mass transport of the Gulf Stream as a function of latitude by theoretical and experimental data according to Holland and Hirschman (1972). 1, The case of a homogeneous ocean of variable depth; 2, the baroclinic ocean of constant depth; 3, the baroclinic ocean of variable depth; 4, the baroclinic ocean of variable depth according to Sarkisyan and Keondzhyan (1972).

into a number of layers, calculates the integral water fluxes at these layers, and shows that in many regions of the World Ocean the water flux several times reverses its sign. Finally, one more result of the calculations of Cox should be mentioned. In the case of a homogeneous ocean, the streamlines everywhere, with the exception of boundary layers, follow the H/l isolines. In the case of a baroclinic ocean, the streamlines, as a rule, cross underwater mountains.

All these peculiarities obtained by Cox are in good agreement with the aforementioned diagnostic calculations, carried out by other authors for different regions of the World Ocean.

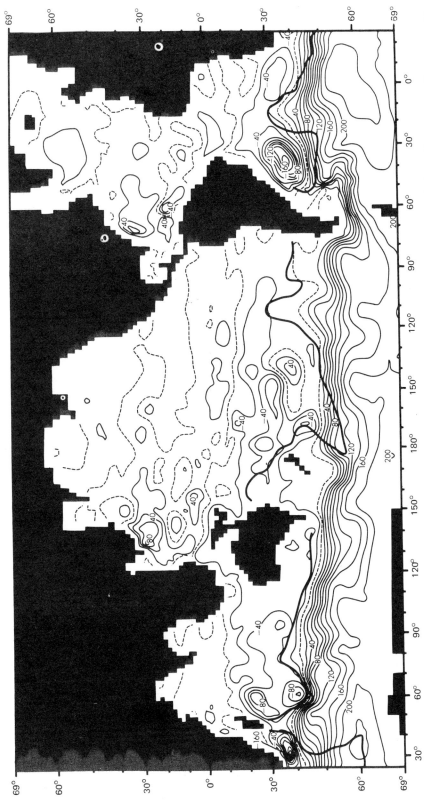

Fig. 20. The mass transport stream function after experiment II of Cox (1972). Transport through Drake Passage = 184 sverdrups. The heavy line represents H cosec φ = 6 km.

D. New Diagnostic Calculations of Currents

Here we include work carried out in 1973–1974. Bulatov, Demin, and Poyarkov calculated ζ and currents by the D_1 model for all 24 standard levels of the Atlantic Ocean with a horizontal grid step of 5°. At the liquid part of the boundary ζ was determined by the dynamical method. Shown in Fig. 21 is the sea-surface topography obtained by these authors. The chart is in very good agreement with the modern

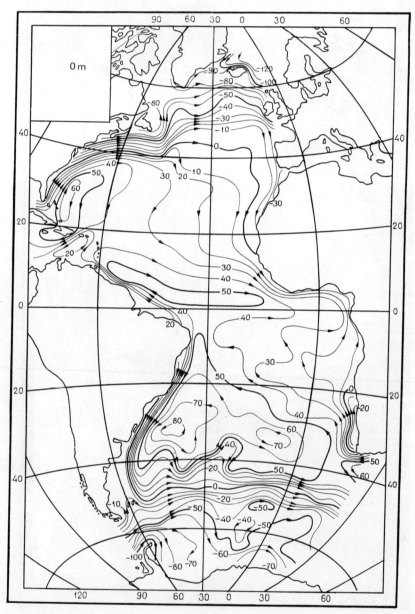

Fig. 21.   The sea-surface topography of the Atlantic Ocean in cm. The arrows indicate the direction of surface gradient currents.

concepts of the general surface circulation of the Atlantic Ocean. Because of the large horizontal grid step the magnitudes of the gradient currents (the maximum values are 10–20 cm/sec) are somewhat underestimated. In contrast to Sarkisyan and Pastuchov (1970), the authors show that the density anomaly at large depths introduces a considerable contribution in the surface gradient currents (about 30%). Figure 22 illustrates the field of the vertical velocity component at the $z = 10$ m level obtained by the same authors. The regions of upwelling are shaded. Comparison of Fig. 21 with Fig. 22 reveals that the regions of the anticyclonic circulation correspond

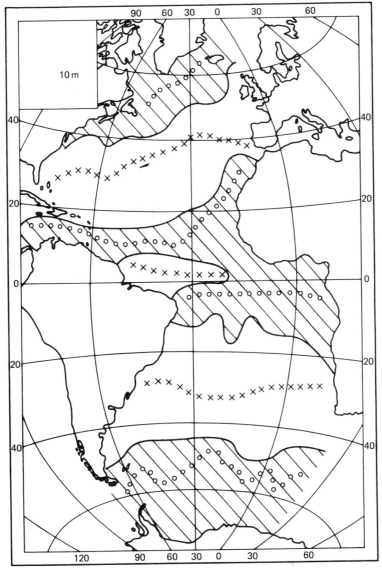

Fig. 22. The qualitative picture of the field of the vertical circulation at the $z = 10$ m level of the Atlantic Ocean. Zones of upwelling are shaded. Circles and crosses denote the regions of maximum descending and ascending currents, respectively.

to the zones of descending currents, and the regions of the cyclonic circulation to the zones of ascending currents. The magnitude of the vertical velocity at this level is of the order of 10^{-5}–10^{-4} cm/sec. With increasing depth the magnitude of w increases up to 10^{-3} cm/sec. At deep layers w correlates well with the bottom topography.

The calculations of ζ and Ψ fields and currents by the D_1 and D'_1 models were carried out for the North Atlantic (between 7 and 65°N) by Stashkevich. Two main questions about (1) a choice of the optimum horizontal grid step and (2) boundary conditions along the horizontal were cleared up. The calculations were carried out with the grid steps 2.5° and 1.25° for 16 levels with the density fields of Suchovey (1971) and Rzheplinski (1975). It was shown that in the main part of an ocean a horizontal grid step of 2.5° is quite sufficient to obtain the characteristics of the large-scale motion. Here the amplitudes calculated with the step of 2.5° are only 3 cm smaller than that with the step of 1.25°. In the Gulf Stream region the difference between amplitudes reaches 15%. Of course, in regions of such intense currents the step of 1.25° is also too large. The problem of boundary conditions is of great importance. We have the vertical walls on the solid parts of the side boundaries in all the calculations of the large-scale currents. The depth of these walls is considerably larger than that of a continental shelf (often more than 500–1000 m). Consequently, the boundary is at a considerable distance from the coast. If we assume that the component of velocity normal to such a "coast" equals zero, then this condition does not correspond to the density anomaly there. The calculations were carried out with difference versions of the boundary conditions. The minimum difference between the current fields calculated in terms of ζ and Ψ near the boundary (far from the boundary these two fields naturally coincide) served as a criterion for a choice of the best version. The version of the calculation of ζ and Ψ by the dynamic method on the side boundaries turned out the best one.

The diagnostic calculations of fields ζ, Ψ, and currents by the D_1 and D'_1 models were carried out by Rzheplinski (1975). The region considered was doubly connected. The grid steps along the latitude and longitude were equal to 1 and 2°, respectively. In the environs of the island the grid step was halved. It was shown that the integral transport in a homogeneous model was three to five times less than that in a baroclinic one. The distribution of the surface currents obtained is more or less realistic. For obtaining more exact results in the future we should construct more reliable density fields of the region considered.

The diagnostic investigation of the circulation of the Tropic Atlantic waters was carried out by Koshlyakov and Enikeev (1974), Koshlyakov (1961), and Enikeev and Koshlyakov (1973). The main merits of these investigations are, in my opinion, an accurate statistical analysis of the initial observed data and the construction of the equally smoothed density fields and the bottom topography. The authors proceed from a special case of equation 109; namely, they do not make allowance for friction with the bottom. Simplified in such a way, an equation of the first order is solved by the method of characteristics; the main shortcoming of this approach is the essential limitation on the bottom topography (see Koshlyakov and Enikeev, 1974, p. 57). In concrete calculations a minimum depth of the ocean was taken as equal to 3.6 km, the grid step about 2°, and the number of layers, 14. The obtained fields of the flow velocity are underestimated, but qualitatively realistic.

Drozdov (1974) carried out the diagnostic calculations of the currents in the north-west part of the Pacific Ocean by the D_1 model. The step along the horizontal was 1°, the number of layers, 24. Figure 23 illustrates the sea-surface topography obtained by Drozdov. The meandering jet of the Kuroshio is distinctly seen. There are anticyclonic eddies to the right of the jet, and cyclonic ones to the left. The velocity of the gradient current in the core of the Kuroshio reaches 1 m/sec. The figure also shows that the baroclinity at great depths (from 1.5 km to the bottom) does not essentially effect on the surface gradient currents. This differs from the results of Bulatov, Demin, and Poyarkov for the following reasons:

1. the density fields are different in these two works, and it is not known which of them is more reliable:
2. different scale phenomena are studied in these investigations (Drozdov takes the grid step to be five times less than that of Bulatov, Demin, and Poyarkov):
3. there are also differences in the technique of the calculations.

The problem needs further consideration, but it is quite obvious that the principal part of the ζ field is formed at the expense of the baroclinicity of the upper 1.5 km layer. Using the data of the Atlantic polygon-70, Rzheplinski (1974) carried out the diagnostic calculations by the D_1 and D_1' models for 19 layers with a step along the horizontal equal to $0.5°$. Below we see that calculations with such a small grid step lead, even within the framework of the linear model, to reliable results. In Rzheplinski's experiment, the density field was measured only in the upper layer of 1.5 km thickness.

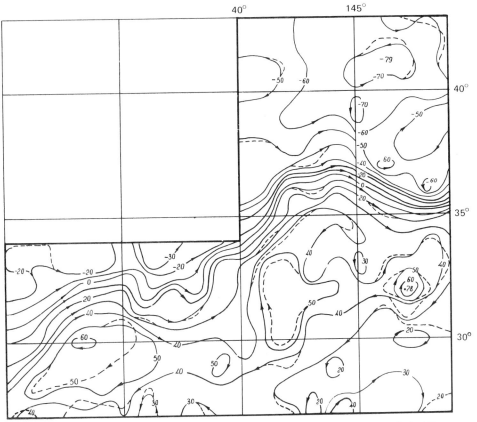

Fig. 23.   The sea-surface topography in the northwest part of the Pacific Ocean. The dashed lines correspond to the case when the ocean is considered to be homogeneous below the $z = 1500$ m level.

The polygon region is not large; there are only 4 × 4 calculated points. It is natural to clear up the question whether the values of Ψ and ζ are defined by the boundary conditions or by the right sides of equations 86 and 109. The boundary of the basin is everywhere liquid; the values of Ψ and ζ at the boundary are determined by the dynamic method. The solution of the complete linear problem is represented in the form of the sum of two solutions. The first of them is the solution of these equations with zero right sides and specified values $ζ_d$ and $Ψ_d$ at the boundary, the second with zero boundary conditions and specified right sides of these equations. It turned out that even in such a small region the solution is defined, on the whole, by the right sides of the equations. The amplitude of Ψ in the second solution is equal to that of the full solution, but we have a small change in the character of the Ψ field. The amplitude of ζ is less by 2 cm (13 instead of 15). The boundary values of $Ψ_d$ and $ζ_d$ give only a general background of these fields.

The diagnostic investigation of currents in the Gulf Stream region was carried out by Sarkisyan and Seidov (1975) and Seidov (1974, 1975). The nonlinear terms and the effect horizontal mixing were taken into account here (the D_4 model). The grid step along the horizontal was 0.25°, the number of layers, 16. The calculations were made with density fields taken from Suchovey (1971). These fields are interpolated linearly from the values at points of the 1.25° grid. The basin being considered is a part of the North Atlantic, for which the calculations have been carried out before with the same density fields, but using the D_1 model (Sarkisyan and Keondzhyan, 1972; Keondzhyan, 1972). The ζ fields at the boundary of the basin and flow velocities were taken from the results of calculations given in those studies. The result of the calculation of the ζ field is presented in Fig. 24. It is seen that the effect of nonlinear terms and lateral mixing ($A_l = 10^6$ cm²/sec) is small, even with the grid step of 0.25°. The flow velocity of a linear variant differs from that of a nonlinear variant by 5–15%. Shown

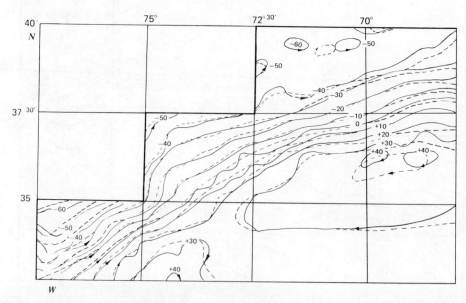

Fig. 24.   The sea-surface topography in the Gulf Stream region, obtained with due regard for nonlinear terms and horizontal mixing (D_4 model). The dashed lines indicate the ζ field obtained without making allowance for these factors (D_1 model).

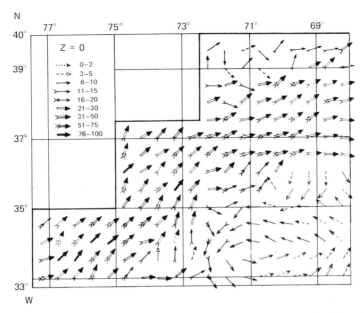

Fig. 25.   The field of surface gradient currents in the Gulf Stream region, obtained by the nonlinear model with the horizontal grid step of 0.25°.

in Fig. 25 is the field of surface gradient currents. The lessening of a horizontal grid step and partly making allowance for nonlinearity have led to the distinct jet current of the Gulf Stream region. On the basis of these detailed calculations Seidov (1975) managed to set up interesting trajectories of water particles of the region considered. The effect of values of A_l on the current fields was studied. Figure 26 presents the average for the whole volume course of the kinetic energy depending on A_l. Apparently, the 10^5-10^6 cm^2/sec values are quite acceptable whereas a value of 10^7 leads to an undesirable smoothing of the results.

Most of the abovementioned work refers to the nonequatorial region of the World Ocean. In these regions the effect of nonlinear terms and lateral eddy viscosity was very small. Holland and Hirschman (1972) note that these factors are essential only

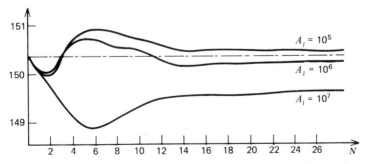

Fig. 26.   The variation of the average kinetic energy depending on values of A_l. The numbers of iterations are shown along the abscissa. The broken line indicates the value of the average kinetic energy of the linear variant.

in the equatorial zone of 3–4° width. This is distinctly seen also from Fig. 27, taken from Demin and Sarkisyan (1974). Here the calculations were carried out by the modified version of the D_2 model. Figure 27 shows that the corrections due to the nonlinearity rapidly increase approaching the equator, and consequently, in the zone of 2–3° width this nonequatorial model becomes unsuitable. In this work as in Sarkisyan and Pastuchov (1970), it is shown that the effect of the bottom friction at the contour of a basin can be neglected when the large-scale circulation is studied.

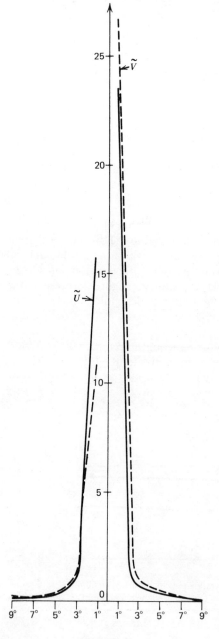

Fig. 27.   The increase of average latitudinal nonlinear corrections to the flow velocity when approaching the equator (Demin and Sarkisyan, 1974).

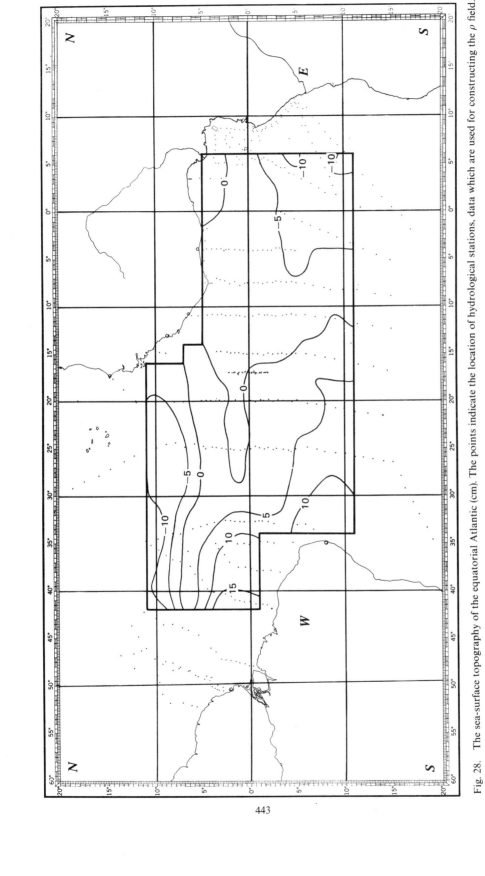

Fig. 28. The sea-surface topography of the equatorial Atlantic (cm). The points indicate the location of hydrological stations, data which are used for constructing the ρ field.

443

We do not review here the work on the calculations of equatorial currents, but refer the reader to the literature (Crommell, Montgomery, and Stroup, 1954; Gill, 1972; Kolesnikov et al., 1968; Kort, Burkov, and Chetokillo, 1969; Metcalf, Voorhis, and Stalcup, 1962; Sarkisyan, 1969; Voigt et al., 1969; Yoshida, 1967). Also, an interesting investigation was carried out recently by O'Brien and Hurlburt (1974). We concentrate on the new work by Sarkisyan and Serebryakov (1974) which is a necessary supplement to the many diagnostic calculations mentioned above. In this work the calculations are carried out in the equatorial Atlantic (from 2°S to 2°N) by the simplified variant of the D_3 model, on the basis of the data observed by the international investigations of EQUALANT-II. Away from the narrow equatorial zone of 4° width the calculations were made using the D_2 model. In the equatorial belt seven levels (0, 75, 100, 150, 200, 400, and 600 m) were considered. The step along the longitude was taken as 2°, and along the latitude, 1°. The depth H of the ocean was taken equal to 800 m. Shown in Fig. 28 is the sea-surface topography of the equatorial Atlantic. This chart is in good qualitative agreement with that of Holland and Hirschman (1972). Note that the sea-surface topography here, as in the other regions of the ocean, is formed mainly by the baroclinity of the seawater. The ζ is very smooth. At the equator the sea-level elevation increases when moving from west to east; the amplitude is equal to 25 cm.

At the beginning of Section 2.E we showed that at the equator such a longitudinal gradient of the ocean surface gives rise to a current from west to east. At the ocean surface a purely drift current driven by the trade winds overwhelms a gradient current, but with increasing depth the velocity of a purely drift current decreases and the velocity of a gradient current becomes the main part of the current. As a result, the compensating equatorial undercurrent is formed. As for the main east–west gradient trade-wind current, in my opinion it is most likely caused by thermohaline factors, just as the trade winds in the atmosphere are caused by irregular heating. To some extent the wind field takes part indirectly in the formation of the gradient trade-wind transfer in the ocean by affecting the distribution of the density anomalies. This oceanic trade current leads to the west–east gradient of the sea surface. All the conclusions made on the basis of diagnostic calculations need the confirmation of prognostic calculations.

It is important to investigate the role of nonlinear terms in the dynamics of equatorial currents. Figures 29a and 29b show the vertical section of the zonal velocity component along the three meridians. First of all we note that at the surface layer of 50–100 m thickness, the trade-wind current directed to the west dominates. Only on the Greenwich meridian is the surface current directed to the east, but it is probably a branch of the Guinea Current. The Equatorial Undercurrent is under this layer. For the two given sections the results of calculations by linear and nonlinear models are close to each other. As for the section along the Greenwich meridian, the role of nonlinear terms turns out to be very important. This is quite enough to conclude that the nonlinear terms play an important role in the dynamics of equatorial currents. The same results were obtained by Holland and Hirschman (1972).

Let us give one more example from the calculations of Sarkisyan and Serebryakov (1974). Figure 30 shows the influence of the Coriolis force on equatorial currents. It is seen that at the equator itself the angle between two arrows is not large, but it increases when moving away from the equator. The Coriolis force deviates the current to the right in the northern hemisphere, and to the left in the southern hemisphere. In spite of the fact that the β effect is indirectly incorporated in the given density field, the direct influence of this factor is distinctly and realistically noted in the equatorial region.

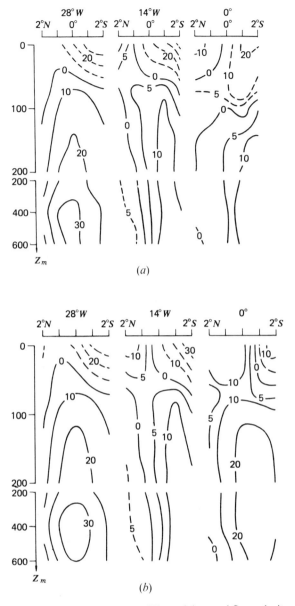

Fig. 29. (a) Meridional section along the three meridians of the zonal flow velocity component. Isotachs of currents, directed to the west, are indicated by the dotted lines. (b) The same, but without making allowance for nonlinear terms in the equations of motion.

The work of Sarkisyan and Serebryakov (1974) has fundamental shortcomings: the constructed density field contains errors and the horizontal grid step is large. Therefore, the Atlantic Equatorial Undercurrent (the Lomonosov Current) is not seen in all the sections; the velocity of the undercurrent is underestimated. But even these calculations have shown that combining the D_1 model with the D_3 model allows the diagnostic investigation of the World Ocean currents. Besides, a number of qualitative conclusions were made on the basis of diagnostic calculations.

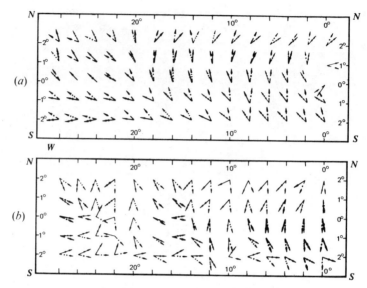

Fig. 30. (a) The velocity of the surface current with due regard for (solid lines) and without taking into account (dotted lines) the Coriolis force. (b) The same for the $z = 600$ m level.

E. Discussion and Conclusions Drawn from Diagnostic Calculations

We proceed from the main difficulty of any diagnostic method of calculation, namely, from the problem of field adjustment. The fields of wind, density, and the bottom topography are specified independently of each other. The calculations have shown that the stationary or seasonal surface gradient currents due to the direct effect of the wind are about one order of magnitude less than the effect of the density anomaly. On the average, the baroclinic currents do not decay very rapidly with depth, so even at great depths the horizontal inhomogeneity of the density remains the main factor in the dynamics of the large-scale ocean currents.

The purely drift currents calculated independently of gradient currents are the main contribution of the direct influence of the wind. The errors in the calculation of the purely drift currents are present in any theoretical models. We do not dwell on this question here.

The case is somewhat different with the bottom topography. The bottom topography has a considerable effect on the gradient currents, especially the deep-sea currents. Thus the adjustment of fields of ρ, and H is more important than that of fields of ρ and τ. One of the methods of such adjustment is the smoothing of fields of ρ, τ, and H with the same smoothing scale (Koshlyakov and Enikeev, 1974). It is not enough, however. The statistical agreement of these fields on the basis of the mutual correlation function is necessary (Belyaev et al., 1970). In the future, fields of temperature, salinity, and density can give more exact information about current fields if they agree statistically with the H field. However, this condition is not observed in most of the density fields by which the calculations of currents have been and will be carried out. Our task, therefore, is to clear up to what extent this affects the results of the calculations.

We have shown above that ζ_d, calculated from formula 99, is the main part of the unknown function and the solution of the simplified equation 98. Since equation 86 is linear, one can represent its solution as the sum of two functions. The first of them is ζ_d.

The second, ζ', is the solution of the equation

$$\frac{1}{2\alpha} \Delta\zeta' + J(H, \zeta') + \frac{H\beta}{l}\frac{\partial\zeta'}{\partial x} = \frac{1}{\rho_0 g} \text{ rot } \tau + \frac{\beta}{\rho_0 l}\frac{\partial}{\partial x} \int_0^H z\rho' \, dz \qquad (212)$$

In relations 99 and 212 $\rho' = \rho_1 - \rho(z)$ is the density anomaly. The standard density $\rho(z)$ is subtracted in order to exclude the dependence of the results of the calculation on the choice of a constant ρ_0 (Perederey, 1974).

Let us call ζ_d the full dynamical sea level, and ζ' the nondynamical correction of the level. Indeed, ζ_d is a surface level defined by the dynamical method. The bottom topography serves as the reference level. Usually, ζ_d accounts for 60–80% of the value of ζ. The value of ζ' is not large; it makes up 20–40% of the level anomalies ζ. However, without this correction the velocity would fade monotonously with increasing depth at any point of the ocean. The nondynamic correction represents the pressure anomaly at the ocean bottom (Kozlov, 1973).

The main influence of the bottom topography on ζ is due to JEBAT. Equation 212, in contrast to 86 has no JEBAT. The influence of this factor is in formula 212. Suppose that side boundaries of the ocean are upright walls of H_0 depth. If we neglect the baroclinity of the water in layers situated deeper than H_0, then there is no influence of the bottom topography in equation 212. Thus the main influence of the bottom topography on ζ is that of underwater elevations and slopes of depths less than 1–1.5 km. Of course, in the operator of the left side of equation 212 we have H, but ζ' itself is small. Thus the errors arising due to the nonagreement of fields of ρ and H, have comparatively smaller effects on the ζ field and, consequently, the surface gradient currents. This property of ζ was proved by numerous calculations and is shown in Fig. 12. The considerable dependence of ζ on changes in H arises only when these changes take place in the upper, essentially baroclinic, layer of the ocean; where we have the basic baroclinity as well as variable bottom topography, JEBAT becomes the important factor in the formation of the ζ field. Thus as for the calculation of ζ, we should speak first of all not how much the fields of ρ and H agree, but about the minimum ocean depth which accurs in any calculation. We take up this question below.

Now let us turn to the Ψ field. It is seen from Fig. 12 that the Ψ field is, in contrast to ζ, very "capricious"; it reacts strongly to changes in the bottom topography and the seawater baroclinity independently of where these changes take place, that is, whether H changes in the upper baroclinic layer or in the lower layers. Moreover, Ψ is more sensitive to changes in these characteristics in the deep layers, that is, where errors in observations of the ρ field are close to the real changes of the field. Thus the inaccuracy of our knowledge about the baroclinity of the deep layers and nonagreement of the ρ and H fields strongly affects the Ψ field and, consequently, the deep currents. This is not a result of mere chance: the sea-surface topography obtained by Holland and Hirschman (1972) is in good agreement with an analogous field obtained by Sarkisyan and Ivanov (1971). As for the Ψ fields, there is nothing in common between the results of these studies.

Thus for obtaining a reliable Ψ field, it is necessary to keep two conditions:

1. the density fields of the deep layers must be sufficiently exact;
2. these fields must agree with the H field.

Unfortunately, for the present there is no possibility of implementing the first (and main) condition; therefore, the Ψ fields obtained by different authors differ fundamentally.

Now let us consider the minimum depth of the ocean and boundary conditions along the horizontal accepted in works of current dynamics generally, and in diagnostic calculations in particular. The assumption that there are upright walls at the solid side boundaries of the basin makes it possible not only to avoid some difficulties of hydrodynamical character but also to smooth the bottom topography. In the diagnostic calculations of the Atlantic Ocean the minimum depth of upright walls accepted by different authors varies within inadmissibly wide limits—from 100 to 3600 m. We consider that the numerical method must not impose rigid demands on the depth of upright walls: it must cope with the solution of the problem at any depth of upright walls, including variable ones. It will be possible then to choose this minimum depth from physical considerations. The majority of aforementioned works satisfy these demands. However some authors choose this depth at will and do not discuss the reasons for their choice. Choosing H at the solid part of the contour of the order of 100 m, and often 1–1.5 km, we essentially step back from the real shores inside the basin and deal with liquid boundaries. The ρ field is specified at these boundaries and, generally speaking, does not satisfy the condition of zero mass flux. Accepting this condition we deliberately assume the contradiction between the boundary values of ζ and Ψ on the one hand, and the density field at the contour on the other hand. The numerical experiments have shown that the best way is the determination of ζ at the contour of the basin by formula 99. In other words, when solving equation 212 we assume $\zeta' = 0$ at the contour of the basin. It refers to the liquid boundaries too, since the mass fluxes are nowhere known exactly. Calculating the auxiliary function Ψ is worse. In this case the solution also exists, Ψ_d, defined by formula 108. However, this solution is obtained at the cost of neglecting the β effect. Therefore, it is reliable only in small basins such as the Baltic Sea (Kowalik et al., 1974) and possesses basic errors when calculating the large-scale oceanic currents. And yet Ψ_d is closer to the total Ψ than the mass transport stream function obtained from the simple relation of Sverdrup. The latter defines only the lesser part of the total mass transport (approximately 10–40%). Thus when we are trying to specify more realistic mass fluxes at the contour of the basin, the question of specifying Ψ at this contour becomes complicated. Specifying $\Psi = 0$ at the contour of the boundary seems to be the solution, but it is suitable only for models of a homogeneous basin, and for baroclinic ones with non-deep upright walls (about 50–100 m).

The marked strong dependence of Ψ on the configuration of the bottom topography and on the baroclinity of deep layers, uncertainty in boundary conditions, etc., impel the choice of ζ as the more convenient auxiliary function. Of course, the question from which to proceed, Ψ or ζ, is not so important. However, we should at least take into account that there will often be significant discrepancies between the Ψ fields obtained by different authors for the same basin. It is more reliable to compare the ζ fields or currents in the surface layer, since currents at great depths and the Ψ field have, for the present, fundamental errors. In models D_1-D_5 we deliberately avoid the boundary layer. The accuracy of observations of ρ hardly allows us to obtain the fine peculiarities of currents in these layers. Besides, we must apply here the relations that more exactly reflect turbulent motion. We consider, however, that the large-scale gradient current depends little on the fine structure of boundary layers. We have also paid little attention to deep currents, considering them insufficiently exact. The analysis of calculations has shown that when we are calculating the sea-surface topography, the homogeneous model is quite unsuitable. Even in the Baltic Sea, with average depths of the order of several tens of meters, JEBAT proves to be essential (Kowalik et al., 1974). In deep seas and oceans the direct effect of a tangential wind stress forms about 10% of a gradient current; that is, its effect is within the limits of errors of the method. Comparing the Ekman model to the dynamic method, we undoubtedly choose the latter. In integral water fluxes rot τ plays a more essential, but not the main role. Therefore, the relation of Sverdrup is not the basic relation when calculating S_y. The dynamic method is another extreme; that is, the calculation of Ψ_d from formula 108 leads to results that are closer to the total S_y than $S_y^{(sv)}$. But the discrepancies between Ψ and Ψ_d are nevertheless basic. Thus equation 109 has, unlike equation 86, no simple reliable solution. Despite their shortcomings, the diagnostic calculations provide reliable information about current field in surface layers, and qualitatively realistic information at great depths is provided by the density field. The charts in Figs. 15, 21, and 25 speak for themselves. All the attempts of improvement of the simplest linear model D_1 on the basis of existing density fields have not led to considerable results. Even in the Gulf Stream region, when calculating with a grid step of 0.25°, the nonlinear terms and lateral eddy viscosity have not led to fundamental corrections. The density field is of the main importance. With a given density field, the simple D_1 model is quite enough to calculate currents in all the regions of the World Ocean, except the equatorial zone of 3–4° width.

In the equatorial region as well as in the other regions, the ρ field is the main indicator of currents. At the equator, in contrast to the other regions, we must take into account nonlinear terms in conjunction with higher demands on the given density field.

Calculations carried out with different steps along the horizontal coordinates have shown that in the main part of the ocean the step of 2–2.5° may serve as an optimum one along the horizontal. In the region of intense currents the grid step must be one order of magnitude less. Generally, the choice of a grid step depends on the scale of smoothing of the ρ field, but if δx and δy are larger than 2–2.5°, then it leads to the smoothed ζ field and underestimated velocities. For example, with a step of 5° in the Gulf Stream region we obtain gradient current velocities of an order of 10–20 cm/sec, and at steps 0.5–1.25°, of 70–80 cm/sec. Steps along the vertical depend on the initial information. We should choose the variable step in the vertical in order to use all the available information about the density field (for example, all 24 standard depths). Generally, we have no difficulties whatever in diagnostic calculations with small steps. Computational mathematics and techniques have exceeded the possibilities of observed data. Increasing the accuracy of diagnostic calculations depends, first of all, on the accuracy of the density field.

Owing to the presence of solution 99, the results of diagnostic calculations weakly depend on boundary conditions. Even in the small region consisting of 4 × 4 calculated points, the solution is basically determined by density anomalies at interior points. The ζ field depend on the coefficient of the vertical turbulent mixing even less. Varying of this parameter from 1 to 100 cm²/sec lead to changes in the amplitude of ζ of about 3–5%. The result obtained for different values of A_l are very interesting. Sarkisyan and Seidov (1975) show that the average kinetic energy changes weakly at values of A_l within the limits of 10^5–10^6 cm²/sec. If $A_l \geqslant 10^7$ cm²/sec, it leads to the essential decrease of the average kinetic energy, that is, to the underestimation of the flow velocity. This means that the computational viscosity in the method of directed differences proves to be less than the physical one. Consequently, this method does not engender the essential viscosity when calculating the nonlinear terms. It will be recalled that by increasing the accuracy of approximation (when calculating ζ with a grid step of 2.5°), Kochergin and Scherbakov only managed to increase the precision of the method of directed differences by 10%.

Conclusions

1. The density field is the main indicator of stationary and seasonal large-scale currents in seas and oceans. Diagnostic calculations are not satisfactory if the density field is greatly smoothed or contains errors. Consequently, the very important need in oceanography is the accumulation of new hydrological data in order to construct more reliable atlases of density fields.

2. When processing the observed data of density field one should implement the statistical agreement of density fields and the bottom topography. Existing density fields, which do not statistically agree with H, lead to fundamental errors when calculating deep currents and, especially, the Ψ field. Errors of density charts at great depths and lack of agreement between the fields of H and ρ affect surface gradient currents weakly.

3. Diagnostic calculations are the generalization of the dynamic method of calculating currents and serve as an alternative to the dynamic method. The simplest version of the diagnostic calculation is as follows. ζ_d is determined by the simple formula 99, that is, by the full dynamic method; the bottom topography serves as the reference level. The correction to ζ is determined by solving equation 212 with the boundary condition $\zeta' = 0$ at the contour of the boundary of the basin; u, v, and w are calculated by simple formulas 89–94 of the D_1 model. This simplest method is quite enough for the calculation of currents using a grid step greater than 0.5° and in regions situated more than 2° from the equator. At the equator and its environs one must take into account nonlinear terms (see the model of Bryan, models D_3, D_5, etc.).

4. In diagnostic calculations of currents one can consider even a small region in which the number of calculated points along the horizontal is more than 5×5. The boundary of the region may even be completely liquid. In essence, the solid part of the region is considered to be liquid too, since the depth of the upright wall is more than 100 m. Inaccuracy in setting boundary values (by the dynamic method) affect only the points nearest to the boundary. In each interior point of the area the ρ field at given and neighboring points is the main factor forming the fields of ζ, Ψ, and currents.

5. When the density field is prepared, then the choice of a horizontal step is of next priority.

The horizontal grid step must be sufficiently small lest the calculations lead to underestimated values of velocities. In the main part of the World Ocean the maximum admissible grid step is 2–2.5°; in regions of intense currents and especially in regions of coastal upwelling (O'Brien, 1972) the grid step must be one order of magnitude less.

6. The tangential wind-stress vorticity (rot τ) plays a secondary role in the formation of gradient currents in the surface layer. Here purely drift currents are the main contribution of the wind.

7. At great depths we have few observations; here the horizontal gradients of density are so small that they are close to the errors in observations, interpolations, etc. Moreover, the ρ fields agree poorly with the H field. Therefore, the currents at great depths are inexactly calculated. We have especially great inaccuracies in the

calculations of Ψ. Therefore, the values of Ψ obtained by different authors differ fundamentally. However, the fields of sea-surface gradient currents, calculated by such different fields of Ψ, are very close to each other, if the fields of ρ are close in the upper layer of 1–1.5 km thickness.

8. Historically established notions of the convergence of isolines of Ψ near the western coast, representing the well-known intense coastal currents (Gulf Stream, Kuroshio, etc.), require critical reconsideration. In many regions of the World Ocean the integral mass transport does not coincide with the mass transport of surface layers, and is even contrary to it. In order to compare some average mass transports with currents, we should find, at the minimum, the height-averaged currents of the upper (0.5–1 km thickness) and the lower layers, as was done by Sarkisyan and Keondzhyan (1972). Still better is determination of the mass transports within several layers, as was done by Cox (1972). Finally, the best method of all is to investigate currents at each level. At present the diagnostic calculations of currents have been carried out at 14–24 levels. Characteristics averaged over the whole ocean thickness reflect, on the whole, the mean circulation at great depths and contain the depth-integrated errors of currents at these layers.

9. In the deep barotropic ocean there are not the large gradient current velocities (5–10 cm/sec and more) observed in the real ocean. Generally speaking, the barotropic ocean does not exist in nature; this idealization, devised by theorists, has become old-fashioned.

Intense, steady, gradient, stationary, or seasonal currents exist only where significant horizontal density gradients, covering a substantial thickness of the ocean, are observed. If there is no western intensification in isolines of density fields (independent of whether they are obtained, on the basis of observed data or by means of prognostic calculations), then there is no western intensification in the calculated fields of currents.

10. Even now the diagnostic calculations can partially answer the basic question of the dynamics of sea currents; what are the main factors responsible for the large-scale circulation of the World Ocean? It is clear that these factors have not a mechanical but a thermohydrodynamic character. The tangential wind stress causes, on the whole, the purely drift mass transport to play a secondary role in the total mass transport. The wind gradient mass transport is also small, at least in the upper layer (0.5–1 km thickness). The processes of heat and, probably, salt exchange are more important. Thus the question of the role of one factor or another (including rot τ) in the dynamics of upper-layer currents is reduced to the question of its role in the redistribution of the density field.

11. The diagnostic calculations are of great practical interest. After proper processing of observed data it is possible to construct atlases of density fields in seas and oceans. From these Atlases and with the aid of diagnostic calculations we can construct atlases of the horizontal and vertical circulation in the seas and oceans. There will be no necessity for diagnostic calculations at all, if we manage to obtain more reliable and exact charts of density fields by prognostic calculations than are available in the observed data.

List of Symbols

u, v, w	Flow velocity components along the axes x, y, z of the Cartesian coordinate system; the x axis points eastward, the y axis points northward, and the z axis is directed vertically down; t is time

ρ_0	Constant value of the density
ρ_1, p_1	Density and pressure, respectively
ρ, p	Density and pressure anomalies, respectively
A_l, v	Horizontal and vertical kinematic turbulent mixing coefficients
T, s	Temperature and salinity anomalies, respectively
p_a	Atmospheric pressure at sea level
τ_x, τ_y	Tangential wind-stress components
A_T, κ_T	Horizontal and vertical temperature diffusion coefficients
A_s, κ_s	Horizontal and vertical salinity coefficients
Q, Q_1	Heat and salt fluxes (with accuracy to a constant) through the ocean surface
Q^1, Q'_1	Heat and salt fluxes through side boundaries of the basin
V_1, U_1	Depth-averaged horizontal components of the flow velocity
T_b, s_b	Temperature and salinity values at the side boundaries of the basin
v_λ, v_θ, v_z	Flow velocity components along the axes λ, θ, and z of the spherical coordinate system; λ is the longitude (the λ axis pointing eastward); $\theta = 90° - \varphi$, the latitude complement; φ-Latitude
n, N	Normals to the bottom and side boundary of the ocean, respectively
H	Ocean depth at the given point
a	Mean radius of the earth
ω	Angular velocity of the earth
g	Acceleration of gravity
ζ_1	Elevation of the open ocean surface above the initial undisturbed state
$\zeta = \zeta_1 + \dfrac{p_a}{\rho_0 g}$	Modified sea-surface topography; positive values of ζ_1 and ζ correspond to the elevation of the sea-surface topography
l	Coriolis parameter
$\beta = \dfrac{dl}{dy}$	Variation of the Coriolis parameter with latitude
L_0	Characteristic horizontal scale
h_0	Characteristic vertical scale (characteristic depth of baroclinity)
$\mathscr{P}_0, (\delta\rho)_0$	Characteristic values of pressure and density anomalies, respectively. A subscript zero denotes the characteristic value of the given quantity
u_g, v_g	Geostrophic current velocity components
u^d, v^d, w^d	Purely drift current velocity components
$u_H, v_H, v_\theta^{(H)}, v_\lambda^{(H)}$	Subscript or superscript H corresponds to the values of given characteristics at the bottom of the ocean
$R_0 = \dfrac{v_0}{l_0 L_0}$	Rossby (Kibel) number
$E_v = \dfrac{v}{l_0 h_2{}^2}, \; E_m = \dfrac{A_l}{l_0 L_0{}^2}$	Ekman numbers corresponding to the vertical and horizontal turbulent viscosity

$P_e = \dfrac{w_0 h_0}{\kappa_T}, \tilde{P}_e = \dfrac{v_0 L_0}{A_T}$ Peclet numbers for the vertical and horizontal diffusions

$\pi = \dfrac{v}{\kappa}$ Prandtl number

$\alpha = \sqrt{\cos \varphi / v}, \delta = 1 \text{ or } 0$ Conditions for sticking and slipping at the bottom, respectively.

$S_x = \displaystyle\int_0^H u \, dz, S_y = \int_0^H v \, dz$ Total mass transport components

Δ Laplace operator

J Jacobian operator

$\bar{u}, \bar{v}, \bar{w}$ Nondimensional flow velocity components along the axes x, y, z. The overbar denotes the nondimensional value of the given characteristic

L_H Characteristic horizontal scale at the ocean bottom layer

$\text{rot } \tau = \dfrac{\partial \tau_y}{\partial x} - \dfrac{\partial \tau_x}{\partial y}$ Vertical component of the tangential wind stress vortex

p_s Pressure anomaly at ocean surface

JEBAT   Joint effect of baroclinicity and bottom topography

1 sverdrup $10^{12} \text{ cm}^3/\text{sec}$

v' Vertical turbulent viscosity coefficient for air

H_1 Depth of the "reference level" in the dynamic method and depth of baroclinity in the theory of the ocean baroclinic layer

$S_y^{(Sv)}$ Integral meridional mass transport after Sverdrup (equation 103)

S_{yd} Integral meridional mass transport in the ocean baroclinic layer by the dynamical method (equation 104)

Ψ_d Mass transport stream function with "reference level" at the ocean bottom (equation 108)

Model D_1 Linear diagnostic model of ζ and flow velocity calculation. For the main relations of this model see equations 86–94

Model D_1' Linear diagnostic model of Ψ and flow velocity calculation. For the main relations of this see equations 109–114

Model D_2 Diagnostic model with due regard for lateral mixing and inertial terms. See relations 115–125

Model D_3 Diagnostic model for the calculation of equatorial currents

Model D_4 Nonlinear diagnostic model. For the main relations of this model see equations 139–143

Model D_5 Most general diagnostic model. For the description of this model see Section 2.F

G   Pressure vector gradient

$A = \dfrac{\partial u}{\partial t} + \dfrac{\partial^2 u}{\partial x} + \dfrac{\partial uv}{\partial y} + \dfrac{\partial uw}{\partial z} - A_l \Delta u$

$B = \dfrac{\partial v}{\partial t} + \dfrac{\partial uv}{\partial x} + \dfrac{\partial v^2}{\partial y} + \dfrac{\partial vw}{\partial z} - A_l \Delta v$

$$A' = \frac{\partial u^2}{\partial x} + \frac{\partial uv}{\partial y} + \frac{\partial uw}{\partial z} - A_l \Delta u \left.\right\}$$

$$B' = \frac{\partial uv}{\partial x} + \frac{\partial v^2}{\partial y} + \frac{\partial vw}{\partial z} - A_l \Delta v$$

$$A_1 = \frac{\partial u^2}{\partial x} + \frac{\partial uv}{\partial y} \left.\right\}$$

$$B_1 = \frac{\partial uv}{\partial x} + \frac{\partial v^2}{\partial y}$$

$$A_2 = \frac{\partial u}{\partial t} + u\frac{\partial u}{\partial x} + v\frac{\partial u}{\partial y} - A_l \Delta u \left.\right\}$$

$$B_2 = \frac{\partial v}{\partial t} + u\frac{\partial v}{\partial x} + v\frac{\partial v}{\partial y} - A_l \Delta v$$

$$A_2' = \frac{\partial (u - u_g)}{\partial t} + u\frac{\partial u}{\partial x} + v\frac{\partial u}{\partial y} - A_l \Delta u$$

$$B_2' = \frac{\partial (v - v_g)}{\partial t} + u\frac{\partial v}{\partial x} + v\frac{\partial v}{\partial y} - A_l \Delta v$$

$$A^{(0)} = \frac{\beta H}{l} - \frac{\partial H}{\partial y} \left.\right\}$$

$$B^{(0)} = \frac{\partial H}{\partial x}$$

Acknowledgment

The author has discussed many aspects of the diagnostic calculations with academician G. I. Marchuk, Drs. Kirk Bryan, William R. Holland, and Vadim F. Kozlow, and many other scientists. These discussions helped to improve the contents of the chapter. Most of the manuscript has been translated into English by young oceanographer Dr. Tengiz A. Dzhioev. The author is very grateful to all of his colleagues and to interpreter Tamara I. Stroganora.

References

Alberg, Y. H., E. H. Nilson, and J. L. Walsh, 1967. *The Theory of Splines and their Applications.* Academic Press, New York.

Belyaev, V. G., I. E. Timchenko, and V. D. Yarin, 1970. A statistical adjustment of hydrophysical fields. *Morsk. Gidrofiz. Issled.* **1** (47).

Bjerknes, V., 1900. Das dynamischen Prinzip der Zirkulations-bewegung in der Atmosphäre. *Het. Leit.*

Bolgurtsev, B. N., 1966. Calculation of currents in the Pacific Ocean region of the Antarctic. *Izv. Akad. Nauk SSSR, Ser. Fiz. Atmos. Okeana*, **2** (11).

Bolgurtsev, B. N., 1968. A circulation of surface and deep layer waters in Atlantic region of the Pacific. *Izv. Akad. Nauk SSSR, Ser. Fiz. Atmos. Okeana*, **4** (10).

Bolgurtsev, B. N. and V. F. Kozlov, 1969. An approximate method of calculation of density field and currents in a baroclinic ocean. *Izv. Akad. Nauk SSSR, Ser. Fiz. Atmos. Okeana*, **5** (7).

Bolgurtsev, B. N., V. F. Kozlov, and L. A. Molchanova, 1968. The results of currents calculation in the Pacific Ocean region of the Antarctic. *Izv. Akad. Nauk SSSR, Ser. Fiz. Atmos. Okeana*, **4**, (6), 622–632.

Bryan, K., 1969a. A numerical method for the study of the circulation of the world ocean. *J. Comput. Phys.*, **4** (3).

Bryan, K., 1969b. Climate and the ocean circulation, Part III. The ocean model. *Mon. Weather Rev.*, **97** (11).

Bryan, K. and M. D. Cox, 1967. A numerical investigation of the oceanic general circulation. *Tellus*, **19** (1).

Bryan, K. and M. D. Cox, 1968. A nonlinear model of an ocean driven by wind and differential heating, Parts I and II. *J. Atmos. Sci.*, **25** (6), 945–978.

Bryan, K. and M. Cox, 1972. The circulation of the world ocean: a numerical study, Part I. A homogeneous model. *J. Phys. Oceanogr.*, **2** (4), 319–335.

Carrier, G. F. and A. R. Robinson, 1962. On the theory of the wind-driven ocean circulation. *J. Fluid Mech.*, **12**, Part I, 49–80.

Cox, M. D., 1970. A mathematical model of the Indian Ocean. *Deep-Sea Res.*, **17** (1), 47–75.

Cox, M., 1975. A baroclinic numerical model of the world ocean: preliminary results. *Numerical Models of Ocean Circulation. Proceedings Symposium Held in Durham, N.H., Oct. 17–20, 1972*, National Academy of Sciences, Washington, D.C., 1975.

Crommell, T., R. B. Montgomery and E. D. Stroup, 1954. Equatorial undercurrent in Pacific Ocean revealed by new methods. *Science*, **119**, 648–649.

Defant, A., 1941. Die absolute Topographie des physikalishen Meeresniveaus und der Drüflächen, sowie die Wasserbewegungen im Atlantischen Ozean. Deutsche Atlantishe Exped. "Meteor" 1925–1927, *Wiss. Ergeb.*, **6** (2).

Demin, Yu. L. and A. S. Sarkisyan, 1974. On calculations of currents in an oceanic basin involving the equator. *Izv. Akad. Nauk SSSR, Fiz. Atmos. Okeana*, **10** (11).

Drozdov, V. N., 1974. Diagnostic calculations of the stationary currents for Kuroshio region. *Ekspress-informatsiya VNIIGMI MTSD*, **12** (5).

Eckart, C., 1958. Properties of water II. The equation of state of water and sea water at low temperatures and pressures. *Am. J. Sci.*, **256** (4).

Ekman, V. W., 1905. On the influence of the earth rotation on ocean currents. *Ark. Mat. Astron. Fys.*, **2** (11), 1–52.

Ekman, V. W., 1906. Beiträge zur Theorie der Meeresströmungen. *Ann. Hydrogr. Marit. Meteorol.*, **34** (9).

Ekman, V. W., 1923. Über Horizontalzirkulation bei Winderzeugten Meeresströmungen. *Ark. Mat. Astron. Phys.*, **17** (26).

Enikeev, V. Kh. and M. N. Koshlyakov, 1973. Geostrophic currents in the tropical Atlantic. *Okeanologiya*, **13** (6).

Gill, A. E., 1975. Models of equatorial currents. *Numerical Models of Ocean Circulation. Proceedings Symposium Held in Durham, N.H., Oct. 17–20, 1972*, National Academy of Sciences, Washington, D.C., 1975.

Holland, W. R., 1973. Baroclinic and topographic influences on the transports in western boundary currents. *Geophys. Fluid Dyn.*, **4**, 187–210.

Holland, W. B. and A. D. Hirschman, 1972. A numerical calculation of the circulation in the north Atlantic ocean. *J. Phys. Oceanogr.*, **2** (4), 336–354.

Ichye, T., 1972. Experimental circulation modelling within the gulf and the Caribbean. *Contrib. Phys. Oceanogr. Gulf of Mexico*, **2**.

Il'in, A. M., 1969. The difference scheme for the differential equation with a small parameter at the senior derivative. *Math. Notes*, **6** (2).

Il'in, A. M., V. M. Kamenkovich, T. G. Zhgurina, and M. N. Silgina, 1969. On the calculation of total mass transport in the world ocean. *Izv. Akad. Nauk SSSR, Ser. Fiz. Atmos. Okeana*, **5** (11).

Ivanov, V. F., 1971. Calculation of vertical velocity in the North Atlantic by total mass transport stream-function and sea-surface topography. *Morsk. Gidrofiz. Issled.*, **6** (56).

Ivanov, V. K., 1963. On the noncorrectly formulated problems *Math. Sb.*, **61** (2).

Ivanov, Yu. A. and V. M. Kamenkovich, 1959. A bottom topography as a main factor, forming the nonzonality of the Antarctic circumpolar current. *Dokl. Akad. Nauk SSSR*, **128** (6).

Kamenkovich, V. M., 1961. On the integration of the sea currents' theory equations in multicollected regions. *Dokl. Akad. Nauk SSSR*, **138** (5).

Keondzhyan, V. P., 1972. The diagnostic calculations of flow velocities at the 16 horizons of the North Atlantic. *Izv. Akad. Nauk SSSR*, **8** (12).

Kochergin, V. P., and V. I. Klimok, 1971. On the effect of the bottom relief on oceanic circulation. *Izv. Akad. Nauk SSSR, Ser. Fiz. Atmos. Okeana*, **7** (8).

Kochergin, V. P. and A. V. Shcherbakov, 1972. An investigation of difference scheme with a small parameter at the senior derivative. *Numerical Models of Ocean Circulation*. USSR Academy of Sciences, Siberian Branch, Novosibirsk.

Kochergin, V. P., V. I. Klimok, and A. C. Sarkisyan, 1973. Diagnostic and prognostic calculations with high accuracy. Preprint. *Novosibirskoe otdelenie AN SSR, Vichislitel'ni Tsentr*.

Kochergin, V. P., A. S. Sarkisyan, and V. I. Klimok, 1972. Numerical experiments on the density field calculations in the North Atlantic. *Meteorol. Gidrol.*, **8**, 54–61.

Kolesnikov, A. G. et al., 1968. On discovery, experimental investigation and elaboration of theory of the Lomonosov current. *Izd. MGI AN USSR*, Sevastopol.

Kort, V. G., V. A. Burkov, and K. A. Chekotillo, 1969. New data on the equatorial currents in West Pacific. *Dokl. Akad. Nauk SSSR*, **172** (2).

Koshlyakov, M. N., 1961. A calculation of the ocean's deep layer circulation. *Okeanologiya*, **1** (6).

Koshlyakov, M. N. and V. Kh. Enikeev, 1974. An integral circulation of the tropical Atlantic waters. *Izv. Akad. Nauk SSSR, Ser. Fiz. Atm. Okeana*, **10** (1).

Kowalik, Z., A. S. Sarkisyan, and A. Staskewicz, 1974. On the model of steady density-driven circulation in the Baltic Sea. *Proc. ICES*.

Kozlov, V. F., 1968. An application of one-parameter models of the density to the investigation of thermokaline circulation in a finite depth ocean. *Izv. Akad. Nauk SSSR, Ser. Fiz. Atmos. Okeana*, **4** (6), 622–632.

Kozlov, V. F., 1971. Some results of approximate calculation of the Pacific Ocean circulation. *Izv. Akad. Nauk SSSR, Ser. Fiz. Atmos. Okeana*, **7** (4), 421–430.

Kozlov, V. F., 1973. On the methodics of ocean currents' calculation by the given density field. *Meteorol. Gidrol.*, **1**.

Lineikin, P. S., 1955. On determination of the thickness of the baroclinic sea layer. *Dokl. Akad. Nauk SSSR*, **101** (3).

Lineikin, P. S., 1957. Basic problems of the dynamic theory of the baroclinic sea layer. *K. Gidrometeoizdat*.

Makarevich, V. A. and A. I. Fel'zenbaum, 1972. A calculation of currents in the Pacific Ocean (a numerical experiment). *Morsk. Gidrofiz. Issled.*, Sevastopol Izd. MGI AN UrSSR, **2** (58), 40–45.

Marchuk, G. I., 1967a. On the equations of the baroclinic ocean dynamics. *Dokl. Akad. Nauk SSSR*, **173** (6), 1317–1320.

Marchuk, G. I., 1967b. On the nonlinear problems of the ocean circulation. *Dokl. Akad. Nauk SSSR*, **176**, (1), 80–83.

Marchuk, G. I., 1969. On numerical solution of Puankare problem for the oceanic circulation. *Dokl. Akad. Nauk SSSR*, **185** (5).

Marchuk, G. I., 1973. *Methods of Numerical Mathematics*. Publishing House "Nauka," Siberian branch, Novosibirsk.

Metcalf, W. G., A. D. Voorhis, and M. C. Stalcup, 1962. The Atlantic Equatorial Undercurrent. *J. Geophys. Res.*, **67** (6).

Monin, A. S., 1972. On the inertional movement on a rotating sphere. *Izv. Akad. Nauk SSSR, Ser. Fiz. Atmos. Okeana*, **8** (10).

Monin, A. S., 1973. On the meandering of equatorial currents. *Izv. Akad. Nauk SSSR, Mechan. Zhidk. Gaza*, **3**.

Monin, A. S., V. G. Neumann, and V. N. Filushkin, 1970. On the density stratification in an ocean. *Dokl. Akad. Nauk SSSR*, **191** (6).

Munk, W. H., 1950. On the wind-driven ocean circulation. *J. Meteorol.*, **7** (2), 79–93.

Muromtsev, A. M., 1963a. Atlas of the temperature, salinity and density of the Pacific Ocean. *Izv. Akad. Nauk SSSR, Moskva*.

Muromtsev, A. M., 1963b. The main hydrological features of the Atlantic Ocean. Appendix 2.

Neumann, G., 1955. On the dynamics of wind-driven ocean currents. *Meteorol. Papers*, **2** (4) (New York Univ. Press).

Neumann, G., 1958. On the mass transport of wind-driven currents in a baroclinic ocean with applications to the North Atlantic. *Z. Meteorol.*, **12** (4–6), 138–147.

O'Brien, J. J., 1975. Models of coastal upwelling. *Numerical Models of Ocean Circulation. Proceedings*

Symposium Held in Durham, N.H., Oct. 17–20, 1972, National Academy of Sciences, Washington, D.C., 1975.

O'Brien, J. J. and H. E. Hurlburt, 1974. Equatorial jet in the Indian Ocean: theory. *Science*, **184**, 1075–1077.

Perederey, A. I., 1972. A calculation of the sea surface and deep layer currents in the South Pacific. *Morsk. Gidrofiz. Issled.*, Sevastopol, Izd. MGI AN UrSSR, **1**.

Perederey, A. I., 1974. Effect of the bottom topography and deep layer baroclinicity on the sea surface level. *Izv. Akad. Nauk SSSR, Ser. Fiz. Atmos. Okeana*, **9** (8).

Perederey, A. I. and A. S. Sarkisyan, 1972. The exact solutions of several transformed equations of the sea currents' dynamics. *Izv. Akad. Nauk SSSR, Ser. Fiz. Atmos. Okeana*, **8** (10).

Rzheplinsky, D. G., 1974. The diagnostic calculations of mezoscale currents on the geophysical polygon in Atlantic. *Okeanologiya*, **3**.

Rzheplinsky, D. G., 1975. Calculations and investigation of the ocean circulation in the Iceland region. *Izv. Akad. Nauk SSSR, Ser. Fiz. Atmos. Okeana*, **1**.

Sandström, I. W. and B. Helland-Hansen, 1903. Über die Berechnung von Meeresströmungen. *Res. Norw. Fish Mar. Inst.*

Sarkisyan, A. S., 1954. Calculation of the stationary wind-driven currents in an ocean. *Izv. Akad. Nauk SSSR, Ser. Geofiz.* **6**.

Sarkisyan, A. S., 1956. A generalization of the Ekman's theory. *Izv. Akad. Nauk SSSR, Ser. Geofiz.*, **6**, 669–672.

Sarkisyan, A. S., 1961. On the role of pure wind advection of density in dynamics of wind-driven currents of a baroclinic ocean. *Izv. Akad. Nauk SSSR, Ser. Geofiz.* **9**, 1396–1407.

Sarkisyan, A. S., 1969a. On Dynamics of currents in the Equatorial Atlantic. *Proc. Symp. Oceanogr. Fish. Res. Trop. Atl. Abidjan, Ivory Coast.* UNESCO, Paris.

Sarkisyan, A. S., 1969b. Deficiencies of barotropic models of ocean circulation. *Izv. Akad. Nauk SSR, Ser. Fiz. Atmos. Okeana*, **5**, 8.

Sarkisyan, A. S., 1969c. *Theory and Computation of Ocean Currents.* Translation available from U.S. Department of Commerce and the National Science Foundation, Washington, D.C.

Sarkisyan, A. S., 1970. A theoretical model of flow velocity calculation in an oceanic basin including equator. *Morsk. Gidrofiz. Issled.*, Izd. MGI AN UkSSR, **2** (48).

Sarkisyan, A. S. and V. F. Ivanov, 1971. Joint effect of baroclinicity and bottom topography as an important factor in sea currents' dynamics. *Izv. Akad. Nauk SSSR, Ser. Fiz. Atmos. Okeana*, **6**, (2).

Sarkisyan, A. S. and V. F. Ivanov, 1972. A comparison of different methods of a baroclinic ocean currents' calculation. *Izv. Akad. Nauk SSSR, Fiz. Atmos. Okeana*, **8** (4), 403–418.

Sarkisyan, A. S. and V. P. Keondzhyan, 1972. Calculation of the sea surface topography and the total mass transport stream-function for the North Atlantic. *Izv. Akad. Nauk SSSR, Ser. Fiz. Atmos. Okeana*, **8** (11).

Sarkisyan, A. S. and V. P. Keondzhyan, 1975. Review of numerical ocean circulation models using the observed density field. *Numerical Models of Ocean Circulation. Proceedings of The Symposium Held in Durham, N.H., Oct. 17–20, 1972.* National Academy of Sciences, Washington, D.C., 1975.

Sarkisyan, A. S. and V. V. Knysh, 1969. An experiment of calculation of Caribbean sea currents and the sea surface topography. *Meteorol. Gidrol.*, **3**, 83–93.

Sarkisyan, A. S. and A. V. Pastuchov, 1970. The density field as a main indicator of the stationary sea currents. *Izv. Akad. Nauk SSSR, Ser. Fiz. Atmos. Okeana*, **6** (1), 64–75.

Sarkisyan, A. S. and A. I. Perederey, 1972. The dynamical method as a first approximation when calculating a sea surface topography of a baroclinic ocean. *Meteorol. Gidrol.*, **4**.

Sarkisyan, A. S. and D. G. Seidov, 1975. A diagnostic calculations of jet-like currents with a non-linear model. *Okeanologiya*, **15** (1).

Sarkisyan, A. S. and A. A. Serebryakov, 1974. On some results of diagnostic calculations of the equatorial currents. *Morsk. Gidrofiz. Issled.*, Izd. MGI AN UrSSR, Sevastopol, **3** (66).

Sarkisyan, A. S., V. P. Kochergin, and V. I. Klimok, 1974. Numerical models and calculations of thermohydrodynamic properties of large-scale circulation of the ocean. *Proc. Symp. Difference and Spectral Methods for Atmosphere and Ocean Dynamics Problems, 17–22 Sept. 1974, Novosibirsk.*

Sarkisyan, A. S., V. V. Knysh, K. U. Vasil'eva, and N. M. Kireeva, 1967. The examples of the calculation of the flow velocities three components by the given density field. *Collected papers: Hydrophysical and Hydrochemical Investigations in the Atlantic and Black Sea.* Naukova Dymka, pp. 24–35.

Seidov, D. G., 1974. Numerical experiments for calculation jetlike currents. *Ekspress-informatsiya VNIIGMI MTSD*, **12** (5) (Obninsk).

Seidov, D. G., 1975. A calculation of hydrodynamic characteristics by given density in the Gulf Stream region. *Izv. Akad. Nauk SSSR, Ser. Fiz. Atmos. Okeana*, **1**.

Shtokman, V. B., 1946. The equations of the wind-driven total mass transport in an homogeneous sea. *Dokl. Akad. Nauk SSSR*, **54** (5).

Stommel, H. 1948. The westward intensification of wind-driven ocean currents. *Trans. Am. Geophys. Union*, **29** (2), 202–206.

Stommel, H., 1958. *The Gulf Stream*. University of California and Cambridge University Presses, 202 pp.

Suchovey, V. F., 1971. The restorement of the hydrological elements' field on the basis of measurement data. *Morsk. Gidrofiz. Issled.*, **3** (53).

Sverdrup, H. U., 1947. Wind-driven currents in a baroclinic ocean with application to the Equatorial Currents of the Eastern Pacific. *Contrib. Scripps Inst. Oceanogr.*, July 1947, 318–326.

Swallow, J. C. and L. V. Worthington, 1961. An observation of a deep countercurrent in the Western North Atlantic. *Deep-Sea Res.*, **8** (1), 1–19.

Voigt, K., M. Sturm, Moeckel, and E. Bengelsdorff, 1969. Salinity–temperature–velocity profiles in the equatorial waters of the gulf of Guinea areas. *Proc. Symp. Oceanogr. Fish. Res. Trop. Atl., Abijan I. E.*, UNESCO, Paris, 1969.

Warren, B. A., 1962. Topographic influences on the path of the Gulf-Stream. *Tellus*, **15** (2), 167–183.

Welander, P., 1959. On the vertically integrated mass transport in the oceans. In *The Atmosphere and the Sea in Motion*. B. Bolin, ed. New York, pp. 95–101.

Yoshida, K., 1967. Circulation in the eastern tropical oceans with special references to upwelling and undercurrents. *Jap. J. Geophys.*, **4**, (2).

10. TRAPPED COASTAL WAVES AND THEIR ROLE IN SHELF DYNAMICS

P. H. LeBlond and L. A. Mysak

The hydrodynamics of continental shelf areas has been receiving increased attention. A pressing need for practical answers to problems concerned with fisheries and pollution as well as a deeper realization of the importance of boundary processes in understanding ocean dynamics have provoked a bloom of observational, theoretical, and numerical investigations. Whenever a need for prompt answers to economically relevant problems is felt, it is often assumed that a numerical model of the situation, providing a fully controllable simulacrum of the real world, can be used as a self-sufficient tool to solve such problems. This is an assumption that is commonly encountered in the field of ecology, for example. The construction of a reliable predictive model of large-scale hydrodynamic phenomena, however, is far from a simple matter; the difficulties involved are well known to anyone involved in numerical weather forecasting. A useful prognostic model must be based on intimate knowledge of the physical processes underlying the observed (and computed) phenomena. Failure to understand the parts and their interrelations leaves one with a computed whole that is just as incomprehensible as the reality that it models. It is in this light, in an attempt to summarize what is known about one broad area of time-dependent coastal phenomena, that this review of edge waves and continental shelf waves is presented.

Our attention concentrates on wave motions that are topographically trapped by depth gradients; consequently we do not discuss the well-known Kelvin and Poincaré wave solutions which propagate along a vertical boundary, and for which we refer the reader to a recent review by Platzman (1971).

1. Basic Formulation of the Problem

We review the properties of trapped waves propagating along a rectilinear coast bordered by a region of variable depth. Axes are chosen as shown in Fig. 1, with the coast lying at $x = 0$, along the y axis, and the depth $H = H(x)$ only. In typical oceanic situations, the presence of vertical density stratification allows the simultaneous existence of internal and surface wave motions. Any real propagating disturbance is likely to consist of a superposition of many vertical modes. In a fluid with a flat bottom, the vertical and horizontal dependences of the oscillations can be separated into a pair of coupled eigenvalue problems (Kamenkovich, 1973; Gill and Clarke, 1974) and each mode may be studied separately. When the depth is variable, this separation technique is no longer applicable. We do not venture into the complexities of the nonseparable problem and concentrate our attention on barotropic trapped waves in a nonstratified fluid; the stratified problem is considered only briefly in connection with deep-sea boundary currents and upwelling problems in Section 11 and edge waves in Section 12.

Most observed trapped waves have wavelengths that are large compared to the depth of the water, at least in the shallow areas where most of their amplitude is concentrated. The theory of trapped coastal waves has thus focused on the analysis of shallow-water waves. We follow suit and base our analysis on the so-called long-wave

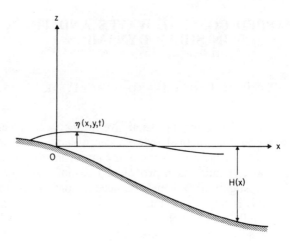

Fig. 1.   The coordinate system. The y axis is directed into the plane at 0 and is along the mean coastline.

or shallow-water equations. A few solutions have of course been developed for deep-water waves, starting with Stokes's (1846) original paper on fundamental-mode edge waves on a linear bottom profile. It took more than a century before Stokes's work was extended by Ursell (1952) to include the whole spectrum of possible modes. Ursell's solution was generalized to include the Coriolis force due to the vertical component of rotation by Saint-Guily (1968). The influence of both components of the Coriolis force on fundamental-mode edge waves was investigated by Johns (1965). The great bulk of the theory, however, is based on the shallow-water equations.

Denoting the displacement of the free surface from its equilibrium level by η, and the velocity components in the x and y directions by u and v, the unforced linearized shallow-water equations for $H = H(x)$ take the familiar form

$$u_t - fv = -g\eta_x \tag{1}$$

$$v_t + fu = -g\eta_y \tag{2}$$

$$(uH)_x + v_y H = -\eta_t \tag{3}$$

with the subscripts indicating differentiation. In equations 1 and 2, g is the acceleration of gravity, and $f = 2\Omega \sin\theta$, where Ω is the angular rotation rate of the earth and θ the latitude. Implicit in the use of the shallow-water equations is the assumption that the bottom slope is small enough for the boundary condition on the sloping bottom to be replaced by that appropriate for a flat bottom: w (the vertical velocity) $\simeq 0$ on $z = -H(x)$. It is nevertheless also assumed that the bottom slope is sufficiently large that its influence on the vorticity balance greatly exceeds that of the β effect. These two assumptions are by no means contradictory: all that is needed for the latter to hold is that $H_x \gg H/R$, where R is the radius of the earth (Rhines, 1969). This certainly holds since, typically, shelf/slope regions have $H_x = 0(5 \times 10^{-3})$ and $H/R = 0(10^{-4})$; hence the variation of f with latitude is safely neglected in most coastal areas where the shallow-water equations are applicable.

The velocity components may be expressed in terms of the derivatives of η:

$$Lu = -g(\eta_{xt} + f\eta_y) \tag{4}$$

$$Lv = -g(\eta_{yt} - f\eta_x) \tag{5}$$

with $L = \partial_{tt} + f^2$.

Substitution of equations 4 and 5 into 3 yields an equation for η alone:

$$H\Delta\eta_t + H_x\eta_{xt} + fH_x\eta_y - \frac{L}{g}\eta_t = 0 \tag{6}$$

where $\Delta = \partial_{xx} + \partial_{yy}$. For waves traveling parallel to the coast we let

$$\eta = F(x)\exp\left[i(ky - \omega t)\right] \tag{7}$$

in which, without loss of generality, we take $k > 0$; ω can have either sign. Inserting equation 7 into 6, we obtain the following equation for F:

$$(HF')' + \left(\frac{\omega^2 - f^2}{g} - k^2 H - \frac{fk}{\omega}H'\right)F = 0, \qquad 0 < x < \infty \tag{8}$$

Derivatives are now indicated by primes. The appropriate boundary conditions on F stem from (1) the necessity that the energy density of trapped modes tend to zero as $x \to \infty$, that is, that $HF^2 \to 0$ as $x \to \infty$, and (2) from the condition that there be no mass flux through the coastal boundary, that is, that $Hu = 0$ at $x = 0$. For all realistic topographies, H is bounded as $x \to \infty$; even in model topographies where H is not bounded, such as in the semi-infinite sloping beach model examined in Section 3, $H(x)$ increases less rapidly than exponentially as $x \to \infty$. As F will be seen to decay exponentially at large x for trapped waves, the first condition reduces to

$$F \to 0 \qquad \text{as } x \to \infty \tag{9}$$

Using equations 4 and 7, the second condition is written in terms of F as

$$H(fkF - \omega F') = 0 \qquad \text{at } x = 0 \tag{10}$$

If $H(0) \neq 0$, this condition reduces to

$$fkF(0) - \omega F'(0) = 0 \tag{10a}$$

whereas if $H(0) = 0$ (as in the case of a sloping beach), equation 10 holds provided F is differentiable at $x = 0$;

$$|F'(0)| < M \text{ (a constant)} \tag{10b}$$

Since differentiability implies continuity, equation 10b guarantees that the sea surface as well as its slope (and thus also the velocity components u and v) are well-behaved at the coastline.

We note that since equation 6 is of third order in the time derivative (through L), or equivalently that the coefficient of F in equation 8 is a cubic in ω, it is conceivable that for a given k there exist three distinct roots for ω, corresponding to three different waves. What kinds of solutions can we expect to find for various classes of topographies $H(x)$? A partial answer to this question was first given by Stokes (1846), who considered the case of a sloping beach of constant inclination β:

$$H(x) = x \tan \beta = \alpha x, \qquad 0 < x < \infty \tag{11}$$

The solutions found by Stokes in a nonrotating fluid behave like

$$F \propto e^{-kx \cos \beta} \tag{12}$$

and have frequencies

$$\omega^2 = gk \sin \beta \tag{13}$$

The classical Stokes edge wave described by this dispersion relation can travel in either direction along the coast, at a speed $\omega/k = (g \sin \beta/k)^{1/2}$ which, for small angles of inclination ($\beta \ll 1$), is much smaller than the deep-water gravity wave speed $(g/k)^{1/2}$. In view of equation 12, the energy of the Stokes edge wave is effectively confined to within a single wavelength of the shoreline. It was not until over a century later that Eckart (1951) established, on the basis of long-wave theory, that the Stokes edge wave is only the gravest of an infinity of modes whose energy is trapped against the coast. Eckart obtained the following dispersion relation for the nth mode:

$$\omega_n^2 = gk(2n + 1) \tan \beta, \qquad n = 0, 1, 2, \ldots \qquad (14)$$

which for $n = 0$, agrees with equation 13 for very gentle bottom slopes ($\beta \ll 1$). For the nth mode wave, there are n nodal lines parallel to the shore; this oscillatory behavior is superimposed on the same kind of exponential decay (equation 12) as for the fundamental mode ($n = 0$) described by Stokes.

Ursell (1952) derived and experimentally verified the exact theory of edge waves without recourse to the long-wave approximations; in particular, he found that

$$\omega_n^2 = gk \sin [(2n + 1)\beta], \qquad n = 1, 2, \ldots \qquad (15)$$

provided

$$(2n + 1)\beta \leqslant \frac{\pi}{2} \qquad (16)$$

The condition 16 arises in Ursell's analysis through the necessity of keeping the particle velocities finite everywhere within the fluid. This condition also implies that for any given β only a finite number of edge wave modes are physically permissible, a restriction not found in the shallow-water theory. However, we note that for small slopes and low mode numbers, that is, for $(2n + 1)\beta \ll 1$, equation 15 reduces to Eckart's shallow-water wave result 14. The restriction on the possible number of modes is of little practical relevance: for $\beta = 10^{-2}$, solutions are possible up to $n = 78$. Nobody has ever observed that many modes! Reid (1958) studied the effect of the Coriolis force on shallow-water edge waves and showed that rotation gives rise to slightly different phase speeds for right-bounded edge waves (for which the phase propagates with the coast on its right) and left-bounded edge waves. Reid also found that for any given mode n there exists a "quasi-geostrophic" wave which, for a given sign of f, propagates in one direction only. This wave is a low-frequency second-class oscillation, whereas the two edge waves are first-class modes: there are three possible waves for a given k, arising from the cubic in ω mentioned above. Reid's unified solution is presented in detail in Section 3.

The question as to whether trapping occurs in more realistic topographies, where $H \to H_0$, a constant, as $x \to \infty$ for example, may be resolved by direct analysis of equation 8 with different forms for $H(x)$, as is done in Sections 6–9. However, a great amount of useful information as to the behavior of waves over any type of topography may also be obtained from a study of the trapping mechanisms from the point of view of ray theory.

2. Topographic Wave Trapping: Ray Theory

A straightforward understanding of the mechanism of topographic wave trapping may be obtained from the methods of ray theory. Although ray theory is strictly applicable only for waves that are short compared to a typical length scale over which the medium varies, a restriction to which we have no intention of binding ourselves, its geometrical simplicity makes it a precious tool in apprehending the manner in which gravity and topographic planetary waves are trapped by coastal topography.

Let us first consider gravity waves. Since their group velocity c_g increases with depth for all frequencies, it is sufficient to consider long waves, for which the phase velocity c and c_g are equal:

$$c = c_g = (gH)^{1/2} \tag{17}$$

Any results derived for long waves in a nonrotating fluid will be qualitatively similar for shorter waves, in deep water, or for rotationally influenced long waves. Consider a situation where $H = H(x)$ varies only normally to a rectilinear coast (Fig. 1). From simple ray theory, the frequency ω and the longshore wavenumber $k = |\mathbf{k}| \sin \theta$ are invariant along rays and Snell's law holds in the form

$$\frac{\sin \theta}{c} = \frac{\sin \theta_0}{c_0} \tag{18}$$

where θ is the angle between the direction of a ray and the x axis (see Fig. 2), and θ_0 and c_0 are the values of θ and c at the shore, where it is assumed that H does not vanish (a restriction that is not necessary but is convenient in this context). A ray

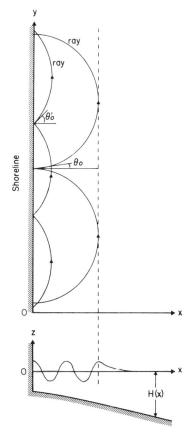

Fig. 2. Gravity wave refraction on a sloping bottom. Top: plan view showing two rays, making angles of incidence θ_0 and θ'_0 with the coast. The position of the caustic line for one of the rays is shown by a dashed line. Bottom: side view, showing the oscillatory behavior inshore of the caustic and the exponential decay offshore.

emanating from the shore at an angle θ_0 is refracted back toward the shore when $\sin \theta = 1$, that is, at that value of x given by

$$\frac{c(x)}{c_0} = \operatorname{cosec} \theta_0 \tag{19}$$

provided of course that the bottom does not level out before a sufficiently high value of c has been reached. Given a maximum depth H_m, to which corresponds a maximum speed c_m, all rays satisfying

$$\operatorname{cosec} \theta_0 \leqslant \frac{c_m}{c_0} \tag{20}$$

are refracted back onshore, that is, are trapped by the topography.

For a coastal angle θ_0 there is thus a caustic line at a position given by equation 19. The x dependence of the oscillations is in the form of a standing wave between the shore and the caustic and exponentially decaying outwards from the caustic (see Fig. 2). It is clear from equation 20 that a caustic will not be found for all longshore wave numbers, as characterized by the angle θ_0, and that at least for nonzero frequencies, trapping will occur only above a minimum longshore wave number. The exact position of the caustic, if it exists, will depend on the form of $H(x)$, and will in general have to be found by numerical techniques, to which ray theory lends itself quite naturally.

Taking the existence of a caustic paralleling the coast as a condition for wave trapping, we readily extend the above picture to curvilinear boundaries. The caustic obviously need not be at a uniform distance from shore. Wherever the caustic is interrupted, such as at a sharp corner in the coastline, energy may leak out of the gap; the trapped modes become leaky.

A recent review of the ray theory of gravity wave propagation over variable depth has been given by Shen (1975). A more specific application to trapping by rectilinear

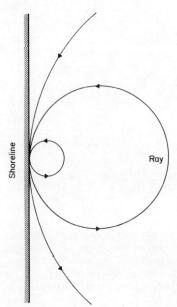

Fig. 3.   Ray paths for short topographic Rossby waves near a coast at which the depth vanishes (from Smith, 1971).

and curvilinear geometries is found in Shen et al. (1968). A general treatment of high-frequency trapped wave has also been published by Smith (1970).

Short Rossby waves propagate at a speed proportional to the magnitude of the geostrophic vector

$$\mathbf{G} = H\nabla\left(\frac{f}{H}\right) \tag{21}$$

Smith (1971) showed that near a simple pole of G, such as at a shoreline where H vanishes, the ray paths are circles tangent to the shore (Fig. 3). Topographic Rossby waves are thus trapped in coastal regions.

Wave trapping in a rotating fluid with arbitrary bathymetry has also been examined by Odulo (1975), on the basis of a direct, but qualitative, analysis of equation 8. It is readily seen that equation 8 has oscillatory solutions (in the x direction) only when the coefficient of the F term is positive. An examination of the condition under which this criterion is satisfied yields the regions in frequency—wave number space in which edge waves and topographic Rossby waves can exist.

3. Trapped Waves on a Sloping Beach with $f \neq 0$

The linear beach profile was the first one investigated and the wave solutions found trapped against it illustrate most of the fundamental properties of trapped coastal waves. For the semi-infinite sloping beach (equation 11), equation 8 for the amplitude $F(x)$ takes the form

$$(xF')' + (\mu k - k^2 x)F(x) = 0, \qquad 0 < x < \infty \tag{22}$$

where

$$\mu = \frac{\omega^2 - f^2}{\alpha g k} - \frac{f}{\omega} \tag{23}$$

The appropriate boundary conditions for this topography are equations 9 and 10b. For a given $k > 0$, equation 22 and the boundary conditions constitute an eigenvalue problem for the eigenvalue μ and the corresponding eigenfunction $F(x) \equiv F_\mu(x)$. Once these are determined, then equation 23 is used to find the dispersion relation $\omega = \omega(k; \mu)$ corresponding to each eigensolution.

One standard way of solving this problem is by the method of power series; however, this is somewhat of a tedious calculation because of the presence of the regular singular point at $x = 0$. The solution of equation 22 and the boundary condition 10b is proportional to a Laguerre function. In order to satisfy the boundary condition at infinity, μ must be a positive odd integer, in which case the solution reduces to a decaying exponential function multiplied by a Laguerre polynomial. Thus one of the remarkable mathematical features of edge wave theory is that even though the domain is semi-infinite, the spectrum (i.e., μ) is discrete, a fact embellished upon by Ursell (1952).

An alternative and elegant method of solving equation 22 and the boundary conditions 9 and 10b is by means of the Laplace transform. Since the coefficients of equation 22 are linear in x, this equation transforms into a first-order separable differential equation for the Laplace transform of F which can be readily integrated. The inverse Laplace transform can then be evaluated by elementary contour integration techniques to yield the eigenfunctions $F_\mu(x)$ and the eigenvalues μ.

Let us first introduce new variables as follows:

$$\chi = kx, \qquad G(\chi) = F\left(\frac{\chi}{k}\right) \tag{24}$$

Then equations 22, and 10b become, upon recalling that we chose $k > 0$,

$$(\chi G')' + (\mu - \chi)G(\chi) = 0, \qquad 0 < \chi < \infty \tag{25}$$

$$|G'(0)| < M_0; \qquad G \to 0 \quad \text{as } \chi \to \infty \tag{26}$$

We define the Laplace transform of $G(\chi)$ by

$$\bar{G}(\chi) = \int_0^\infty e^{-\chi s} G(\chi)\, d\chi \tag{27}$$

Then, provided G is of exponential order and the first condition of equation 26 holds, the application of equation 27 to equation 25 gives

$$(s^2 - 1)\bar{G}' + (s - \mu)\bar{G}(s) = 0$$

whose solution, apart from an arbitrary multiplicative constant, is

$$\bar{G}(s) = \frac{(s - 1)^{(\mu - 1)/2}}{(s + 1)^{(\mu + 1)/2}} \tag{28}$$

Hence the inverse of the Laplace transform of equation 28 gives

$$G(\chi) = \frac{1}{2\pi i} \int_{-i\infty + \gamma}^{i\infty + \gamma} e^{s\chi} \frac{(s - 1)^{(\mu - 1)/2}}{(s + 1)^{(\mu + 1)/2}}\, ds \tag{29}$$

where $\gamma > 1$ so that the path of integration passes to the right of the singularities at $s = \pm 1$. For arbitrary μ these singularities are branch points, and it can be shown that as $\chi \to \infty$, the singularity at $s = 1$ gives rise to a contribution to $G(\chi)$ which goes as e^χ. This clearly violates the second condition in equation 26. However, we can get around this difficulty by choosing the following values for μ:

$$\frac{\mu - 1}{2} = n = 0, 1, 2, \ldots$$

or equivalently,

$$\left. \begin{array}{l} \\ \\ \\ \end{array} \right\} \tag{30}$$

$$\mu = 2n + 1 = 1, 3, 5, \ldots$$

which eliminates the singularity at $s = 1$! Then for choice 30, $(\mu + 1)/2 = n + 1$ and the only singularity in the integrand of equation 29 is the pole of order $n + 1$ at $s = -1$. Thus without loss of generality, we can take $\gamma = 0$ and

$$G(\chi) \equiv G_n(\chi) = \frac{1}{2\pi i} \int_{-i\infty}^{i\infty} e^{s\chi} \frac{(s - 1)^n}{(s + 1)^{n+1}}\, ds \tag{31}$$

To evaluate equation 31 for $\chi > 0$, we close the contour to the left and apply Cauchy's integral formula; this gives

$$G_n(\chi) = \frac{1}{n!} \left\{ \frac{d^n}{ds^n} e^{s\chi}(s - 1)^n \right\}_{s=-1}, \qquad n = 1, 2, \ldots$$

from which we obtain the eigenfunctions $F_n(x)$:

$$
\left.\begin{aligned}
F_0 &= e^{-kx} \\
F_1 &= e^{-kx}(1 - 2kx) \\
F_2 &= e^{-kx}(1 - 4kx + 2k^2x^2) \\
&\;\;\vdots \\
F_n &= e^{-kx}L_n(2kx)
\end{aligned}\right\}
\tag{32}
$$

where $L_n(z)$ is the nth-degree Laguerre polynomial. Hence the surface elevation $\eta_n(x, y, t)$ corresponding to the nth mode has n nodal lines parallel to the coast and a simultaneous decaying behavior with e-folding distance k^{-1}.

Combining equations 23 and 30 we obtain the following implicit dispersion relation for each mode:

$$
\omega_n{}^3 - [f^2 + (2n + 1)g\alpha k]\omega_n - fg\alpha k = 0, \qquad n = 0, 1, 2, \ldots
\tag{33}
$$

Thus for any given $k\ (>0)$ and mode n there are three roots ω_n implied by equation 33. We denote these frequencies by $\omega_{jn}(k)$, $j = 1, 2, 3$. It can be shown that the roots $\omega_{jn}(k)$ are real and have the following properties:

$$
\left.\begin{aligned}
\sum_{j=1}^{3} \omega_{jn} &= 0 \\
\prod_{j=1}^{3} \omega_{jn} &= fg\alpha k
\end{aligned}\right\}
\tag{34}
$$

Equation 34 implies that when $f > 0$, two of the roots are negative and the third is positive. Thus in the northern hemisphere the positive root, which we denote by $j = 1$, is a *left-bounded wave* (as defined in Section 1) moving in the positive y direction. We denote the two negative roots by $j = 2, 3$; these are *right-bounded waves* (see equation 7). Further, when $f = 0$, equation 33 reduces to

$$
[\omega_n{}^2 - (2n + 1)g\alpha k]\omega_n = 0
$$

which we rewrite as

$$
\omega_{1n} = [(2n + 1)g\alpha k]^{1/2}, \qquad \omega_{2n} = -[(2n + 1)g\alpha k]^{1/2}
$$

in agreement with Eckart's solution 14, and $\omega_{3n} = 0$, corresponding to steady currents (no waves). Thus the roots ω_{jn}, $j = 1, 2$ are identified with the first-class edge wave modes as modified by rotation, whereas the root ω_{3n} is recognized as a second-class wave mode, trapped by rotation. The three dispersion relations implied by equation 33 for $n = 0, 1, 2, 3$ are plotted in Fig. 4. For the lowest or gravest mode $n = 0$, equation 33 can be factored:

$$
\left\{\frac{\omega_{10}}{f} - \left[\frac{1}{2} + \left(\frac{1}{4} + \frac{g\alpha k}{f^2}\right)^{1/2}\right]\right\}\left\{\frac{\omega_{20}}{f} - \left[\frac{1}{2} - \left(\frac{1}{4} + \frac{g\alpha k}{f^2}\right)^{1/2}\right]\right\}\left\{\frac{\omega_{30}}{f} + 1\right\} = 0
\tag{35}
$$

The edge wave roots clearly exhibit the phenomenon of rotational splitting of the frequencies: the left-bounded wave ($j = 1$) moves faster than the right-bounded wave ($j = 2$) by an amount f/k. However, the group velocity, though modified by rotation, has the same magnitude for both waves. As $k \to 0$, $\omega_{10} \to f$ whereas $\omega_{20} \to 0$.

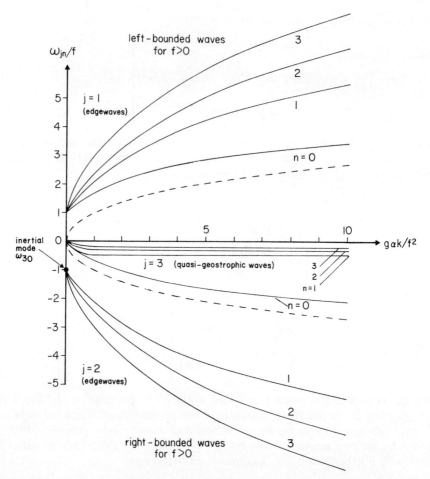

Fig. 4.   The dispersion relations for edge and quasi-geostrophic waves on a semi-infinite sloping beach of slope α. The dashed curves correspond to the lowest-mode ($n = 0$) Stokes edge wave solution in the absence of rotation ($\omega_{jn}^2 = g\alpha k, j = 1, 2$), and are shown for comparison. The dispersion curves are labeled by mode number n. Note the existence of the ω_{30} inertial wave solution at $k = 0$, $\omega = -f$ (adapted from Reid, 1958).

When $f = 0$, on the other hand, both ω_{10} and $\omega_{20} \to 0$ as $k \to 0$. The third root of equation 35 corresponds to an inertial oscillation of infinite wavelength ($k = 0$), which may be obtained from an analysis of the primitive equations 1–3. For the higher modes ($n \geqslant 1$), the edge wave frequency curves are still slightly asymmetric about $\omega = 0$; however, they all have the property that $\omega_{1n} \to f$ and $\omega_{2n} \to -f$ as $k \to 0$. The higher-mode second-class waves, on the other hand, have the property that $\omega_{3n} \to 0$ as $k \to 0$ and that $|\omega_{3n}/f| \ll 1$ for all $k > 0$. It is because of this latter inequality that Reid (1958) introduced the term quasi-geostrophic to describe these waves. They are effectively topographic planetary waves, trapped against the coast. Consequently, they have a relatively large vorticity compared with that of the rotationally modified first-class edge waves. Also, in contrast with edge waves, for a fixed k the speed of propagation of the quasi-geostrophic waves decreases with increasing mode number n.

4. Observations of Edge Waves

For a long time edge waves were regarded as a mere curiosity of hydrodynamics. For instance in his discussion of the gravest-mode solution of Stokes, Lamb (1945, p. 447) states that "it does not appear that the type of motion here referred to is very important." However, there is now a growing body of evidence which suggests that edge waves are common in occurrence and of practical importance. We defer the discussion on the observations of quasi-geostrophic waves until later and consider first the evidence for the existence of first-class edge waves.

Munk et al. (1956) examined sea-level records taken on the eastern coast of the United States during the passage of several hurricanes and squalls during 1954 and found that these storms excited gravest-mode ($n = 0$) edge waves, with typical amplitudes of 1 m, periods of about 6 hr, and wavelengths of several hundred kilometers. Detailed calculations of the initial value problem leading to the generation of edge waves by atmospheric pressure systems have been presented by Greenspan (1956) for the case $f = 0$ and by Kajiura (1958) for $f \neq 0$. Observations from the California coast show a different situation: a continuum of edge waves noise, in the period range of 10–30 min, always seems to be present. Munk et al. (1956) suggested

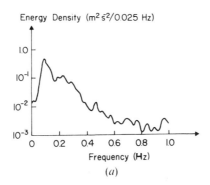

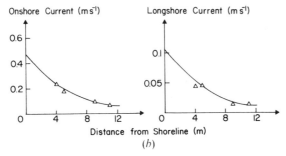

Fig. 5. (a) Edge wave measurements taken from Slapton Beach, South Devon, England, at an offshore distance of 9.0 m, in August 1972, a typical onshore current spectrum. The broad peak at 0.2 Hz corresponds to the observed breaker frequency in the surf zone; the subsequent decay at higher frequencies is typical of wind–wave spectrum. The additional sharp peak at half the breaker frequency (0.1 Hz) is suggestive of a subharmonic edge wave generated by the breakers. (b) Measurements (triangles) of onshore and longshore current amplitudes at 0.1 Hz, which exhibit the pure decaying behavior of a gravest-mode edge wave. The solid lines show the theoretical decay for the gravest-mode edge wave with a frequency of 0.1 Hz and a corresponding wavelength of 32 m, as determined by the dispersion relation for edge waves on the profile (6.1) with $H_0 = 7.05$ m and $a = 3.4 \times 10^{-2}$/m (from Huntley and Bowen, 1973).

that this wave continuum might be generated by atmospheric internal gravity waves, traveling just above the sea surface. Very high-frequency edge waves of the gravest mode have also been observed by Huntley and Bowen (1973) off the south Devon coast (Fig. 5). It is also well established by now that earthquake-generated long waves (tsunamis) incident from the open ocean upon the continental shelf/slope region can also excite edge waves (see, e.g., Munk et al., 1956; Aida, 1967; Kajiura, 1972; Fuller, 1975). Their presence has also been indirectly detected through their role in exciting seiche action in coastal bays (Lemon, 1975). Laboratory experiments on edge waves have been reported by Galvin (1965).

Finally, edge waves have been shown to be of fundamental importance in the dynamics and the sedimentology of the nearshore zone through their interaction with ocean swell and surf to produce rip current patterns. This topic deserves a section of its own.

5. Edge Waves, Rip Currents, and Beach Cusps

Rip currents are known and feared by many ocean-beach bathers; they appear in seaside folklore as the dreaded "undertow." Though it has long been obvious that the rip currents must form the return flow for water thrown ashore by breaking waves, a satisfactory explanation of their dynamics and of their spacing is quite recent and is to be found in an interaction process between edge waves and the in-coming swell, as described by Bowen (1969) and Bowen and Inman (1969).

An explanation of rip currents has also led to a better understanding of the forma-tion of the striking cusp patterns seen along many ocean beaches (Fig. 6) and of their relation to edge waves. Equally impressive crescentic bars (Fig. 7) occurring offshore between headlands have also been related to edge waves (Bowen and Inman, 1971).

Fig. 6.   Beach cusps from Musquodoboit Harbour, N.S., Canada (courtesy of E. M. Owens).

Fig. 7.   Crescentic bars from the coast of Algiers (from Bowen and Inman, 1971).

We now outline the basic features of the interaction between the incoming swell and edge waves which lead to the formation of rip currents, cusps, and crescentic bars.

A breaking wave loses much more of its energy than of its momentum, and continues shoreward as an identifiable, if rather distorted, wave form. Let us assume, rather heuristically, that the energy of the wave may still be written as

$$E = \tfrac{1}{2}\rho g a^2 \tag{36}$$

after breaking, with a the wave amplitude. Similarly, we also take for the radiation stress of the surf the expression derived by Longuet-Higgins and Stewart (1964) for small-amplitude shallow-water waves; in diagonal form, it is given by

$$S_{ij} = E \begin{pmatrix} \tfrac{3}{2}, & 0 \\ 0, & \tfrac{1}{2} \end{pmatrix} \tag{37}$$

Let us consider plane, normally incident waves, in the coordinate system of Fig. 1. Because the wave amplitude decreases shoreward, the shoreward flux of x momentum (S_{xx}) also decreases; a gradient of radiation stress is set up which pushes water onto the beach. Under equilibrium conditions, a pressure gradient arises which exactly cancels the radiation stress gradient. The resulting wave setup has been discussed by Longuet-Higgins and Stewart (1963). Now let us suppose that the incident wave amplitude is not exactly uniform along its crest. Regions of higher and lower wave energy density follow each other along the shore and lateral gradients of radiation stress arise. These lateral gradients push water into the region of lower energy density and set up a cellular pattern of broad onshore flows compensated by concentrated offshore rip currents.

Averaged over many wave periods, the shallow-water equations in the nearshore area take the form

$$\bar{u}_j \frac{\partial \bar{u}_i}{\partial x_j} = - \frac{1}{\rho(\bar{\eta} + H)} \frac{\partial S_{ij}}{\partial x_j} - g \frac{\partial \bar{\eta}}{\partial x_i} - R_i \tag{38}$$

where $\bar{u}_i$ and $\bar{\eta}$ are the mean current and surface displacement induced by the radiation stress divergence $\partial S_{ij}/\partial x_j$, H is the equilibrium water depth as measured in the absence of waves, and R_i is a frictional force including bottom and lateral friction:

$$R_i = - \frac{\partial}{\partial x_j}\left(A_H \frac{\partial \bar{u}_i}{\partial x_j}\right) + \frac{c\bar{u}_i}{\bar{\eta} + H} \tag{39}$$

The parameter A_H is a lateral eddy viscosity and c a bottom friction coefficient. Solutions of equation 38 arising out of a longshore breaker amplitude a_b varying as

$$a_b = a_0(1 + \varepsilon \cos \lambda y) \tag{40}$$

have been computed by Bowen (1969). Streamline patterns for the solutions with $c = 0$ and with two values of the Reynolds number are shown in Fig. 8. The rip currents are clearly concentrated by nonlinear effects, a phenomenon that is readily explained in terms of conservation of potential vorticity $\mathbf{P} = (\nabla \times \bar{\mathbf{u}})/(\bar{\eta} + H)$. As the rip current flows outward in regions of larger H its vorticity must also increase,

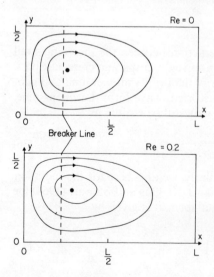

Fig. 8. Streamline patterns for rip currents generated by an edge wave interaction; only one half of a longshore cycle is shown in each case. Top: a linear solution. Bottom: a nonlinear solution, with Reynolds number $= UL/A_H = 0.2$ (from Bowen, 1969).

and with it the gradient $\partial \bar{u}/\partial y$: the rip current becomes narrower as it progresses outward until the streamlines diverge again to feed the onshore flow.

How is all this related to edge waves? The breaker modulation (equation 40) that had to be assumed to produce cellular current patterns results from the superposition of the incoming surf and of a standing edge wave, the characteristics of which are determined by the beach slope and by the separation of the headlands which bound the beach. Figure 9 shows how the amplitude of a uniform swell and that of a standing edge wave add up to modulate the total amplitude of the incoming surf. Notice that there is a minimum in net elevation at each alternate antinode of the standing edge wave so that the spacing of the rip currents is equal to the wavelength of the edge wave. Observations reported by Bowen and Inman (1969) have confirmed this conclusion, and the experiments they performed in a wave tank of uniform slope shown that edge waves are rapidly excited by a second-order resonant interaction with the incoming swell. On beaches bounded by headlands, rip currents are commonly found to occur at the center of beach cusps (Shepard et al., 1941). Given the flow pattern illustrated in Fig. 8, erosion would tend to occur in the convergent accelerating flow at the head of the rip. The position of the cusps is then consistent with that of the rip current pattern. The formation of crescentic bars (Fig. 7) occurring shoreward of sandy bays in regions of small tidal range has also been successfully explained by Bowen and Inman (1971) in terms of the sediment transport produced by rip current patterns.

The reader should not, however, be left with the impression that the situation is always so clear-cut and simply explained. Complications arise when the incoming swell does not fall normally onto the beach and sets up longshore currents which

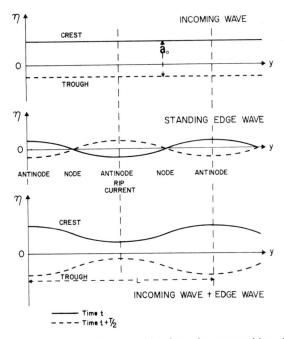

Fig. 9. The longshore wave crest modulation resulting from the superposition of a uniform incident wave and of a standing edge wave, each of period T.

may completely smudge the neat picture painted above. On the subject of longshore currents, the reader should consult a recent review article by Longuet-Higgins (1972), which clearly summarizes the work to date. Further, even for normally incident swell, it is difficult to involve an interaction with standing edge waves in determining the spacing of rip currents and cusps on extremely long beaches, not bounded by clearly defined headlands. Direct interaction mechanisms between the incoming swell and the rip currents themselves have been invoked, with a certain degree of success, by LeBlond and Tang (1974) and by Hino (1974) to explain rip current spacing on unbounded coastlines. The whole field of study of the interactions between swell and edge waves and their nonlinear coupling with longshore currents remains a fertile ground for further investigation. In this respect, we should mention the recent work of James (1974a, b) on nonlinear wave and longshore currents, which should form a solid basis for future studies. Other aspects of edge wave theory have been studied by Kenyon, such as their modification by a shear current (Kenyon, 1972), and the Stokes drift that they produce (Kenyon, 1969).

6. Trapped Waves on Other Topographies

From the discussion of tsunami-generated edge waves in Section 4 it is clear that the semi-infinite sloping-beach model is not adequate to describe phenomena that involve trapped waves that are generated by or connected to deep-sea motions for which a finite depth far from the coast is important. Also, in practice a typical shelf width can be comparable to or even less than the wavelength, in which case the refraction mechanism for edge waves discussed in Section 2 is no longer effective. Finally, there also arises the question of the validity of the shallow-water theory when the depth tends to infinity, as for a semi-infinite sloping beach. Thus for these (and other) reasons, there has been a recent proliferation of studies of trapped waves over more realistic shelf/slope topographies.

Generally speaking all the shelf/slope models studied in the last decade or so fall into two categories: in the interval $0 < x < L \leqslant \infty$ they are either (a) concave upward or (b) concave downward. However, in each case the depth far from the coast

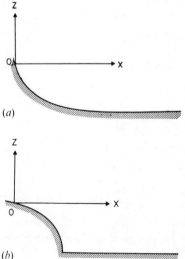

Fig. 10. The two basic shelf/slope topographies treated in the literature: (a) profiles with upward concavity; (b) profiles concave downward.

is constant (see Fig. 10). Since in practice most continental shelf/slope topographies are of type b, the bulk of the literature deals with topographies of this type. Accordingly, we derive explicit solutions for three type b topographies in Sections 7–9. However, an interesting solution for a type a topography was found by Ball (1967). He investigated the behavior of edge waves and quasi-geostrophic waves over an exponential depth profile of the form

$$H(x) = H_0(1 - e^{-ax}), \qquad 0 < x < \infty \tag{41}$$

Huntley and Bowen (1973) found that equation 41 is a reasonably accurate model of Slapton Beach, South Devon, where, as was mentioned in Section 4, they detected the lowest-mode edge wave. The profile 41 leads to a hypergeometric equation for $F(x)$ whose appropriate solutions for trapped modes are hypergeometric (Jacobi) polynomials. There are two attractive limiting cases of this profile, and the corresponding solutions:

1. For large a, equation 41 models a profile of constant depth with an abrupt edge arbitrarily close to the vertical; in this case the solutions degenerate into the barotropic Kelvin wave mode.

2. For large H_0 and small a such that the slope $\alpha = aH_0$ is kept constant, equation 41 models a uniform slope for arbitrarily large distances from the coast; in this case the solutions degenerate into Reid's solutions for a semi-infinite beach (Section 3).

It is interesting to note that in the short-wavelength limit, the same asymptotic frequency $(\omega_n = f/(2n + 1))$ is found by Ball for the second-class modes as was obtained by Reid for a linear depth profile. Such short topographic waves are clearly so strongly trapped near the shore that they are not influenced by the shape of the deep-water bathymetry. Finally, Ball also noted that for long wavelengths, such that $ka \ll 1$, *trapped* edge waves cannot exist; this of course is due to the breakdown of the trapping mechanisms discussed in Section 2. This feature of long edge waves has also been explored further by Clarke (1974), for a variety of both type a and type b shelves. Grimshaw (1974), on the other hand, derived upper and lower bounds for the gravest-mode edge wave on an arbitrary topography of type a. However, rotation is ignored in his work so that the results are valid only for high frequencies.

7. Trapped Waves on a Sloping Shelf of Finite Width

A type b topography which readily illustrates the effect of a sloping shelf of finite width is given by

$$
\begin{aligned}
H(x) &= \frac{dx}{l}, \qquad 0 < x < l \\
&= D \text{ (constant)}, \qquad x > l
\end{aligned}
\tag{42}
$$

This model was introduced by Robinson (1964) to study very low-frequency and very long trapped quasi-geostrophic waves. Equation 42 represents a shelf region $(0 < x < l)$ of uniform slope $\alpha = d/l$ which sharply drops off to a deep-sea region $(x > l)$ of constant depth $D > d$. For the topography 42, the appropriate solution of equation 8 which satisfies conditions 9 and 10b is

$$
\begin{aligned}
F(x) &= Ae^{-kx}L_\nu(2kx), \qquad 0 < x < l \\
&= Be^{-Kx}, \qquad x > l
\end{aligned}
\tag{43}
$$

where $v = (-1 + \mu)/2$, with μ as given by equation 23 and

$$K = \left[k^2 + \frac{(f^2 - \omega^2)}{gD} \right]^{1/2} \tag{44}$$

which must be real and positive for trapped waves. $L_v(z)$ is the Laguerre function, which has the series representation

$$L_v(z) = 1 - vz - \frac{v(-v + 1)z^2}{(2!)^2} - \cdots$$

and which is related to the confluent hypergeometric function by $L_v(z) = {}_1F_1(-v; 1; z)$. For $v = n = 0, 1, 2, \ldots, L_v$ reduces to a Laguerre polynomial of degree n. An extensive discussion of the properties of the Laguerre function is contained in Pinney (1946).

Examination of equations 44 and 43 clearly shows that for the high-frequency edge waves ($\omega^2 \gg f^2$), the waves cannot be trapped when k is small. When $K^2 < 0$ the spectrum is continuous and these very long waves are the so-called leaky edge wave modes first discussed by Snodgrass et al. (1962) for a flat shelf. They are also known as topographically-modified Poincaré waves (Munk et al., 1970). For quasi-geostrophic waves ($\omega^2 < f^2$), $K > 0$ for all $k > 0$ and trapping occurs at all wavelengths.

At the edge of the shelf ($x = l$), we require that η and Hu be continuous. This leads to two homogeneous equations for A and B. For a nontrivial solution the determinant of coefficients must vanish, which implies the following implicit dispersion relation:

$$L_v(2\kappa)\left\{ \left[1 + \frac{\delta\Delta(1 - \sigma^2)}{\kappa^2} \right]^{1/2} + \frac{1}{\sigma} - \Delta\left(1 + \frac{1}{\sigma} \right) \right\} + 2\Delta L'_v(2\kappa) = 0 \tag{45}$$

where $\sigma = \omega/f$, $\kappa = kl$, $\delta = f^2l^2/gd$, and $\Delta = d/D$. The square root term inside the curly brackets is proportional to K (see equation 44) and the quantity v is related to σ, κ, and δ by

$$\sigma^3 - \left[1 + \frac{(2v + 1)\kappa}{\delta} \right]\sigma - \frac{\kappa}{\delta} = 0 \tag{46}$$

which is equivalent to the cubic for ω given by equation 33.

The relationship 45 was first derived and analyzed by Mysak (1968a). For small Δ, corresponding to a large drop from the shelf to the deep-sea region, the waves on the shelf are weakly coupled to those in the deep-sea region and hence a first approximation for the implicit dispersion relation is obtained by setting

$$L_v(2\kappa) = 0 \tag{47}$$

Physically, this approximation means that the waves on the shelf have a node at the edge of the shelf; since typically $\Delta = 4 \times 10^{-2}$ ($d = 200$ m, $D = 5000$ m), this approximation introduces an error of only a few percent. For a given κ (the nondimensional wave number), equation 47 is satisfied for a countably infinite number of discrete values of $v = v_0, v_1, v_2, \ldots$. The first three of these have been tabulated as a function of κ by Mysak (1968a). Then, corresponding to these values of v_n, the cubic 46 can be solved for the frequency functions $\sigma = \sigma(j, v_n, \kappa)$. For each mode v_n, the roots $j = 1$ and 2 correspond to the edge wave dispersion relations whereas the root $j = 3$ corresponds to the quasi-geostrophic wave frequency function. In the limit of a wide shelf, or equivalently, as $\kappa \to \infty$, the numbers $v_n \to n = 0, 1, 2, \ldots$, the values for a semi-infinite shelf. In computing the values of v_n, however, the only values of κ

which are allowed are those for which $K > 0$ (trapped wave condition). It was pointed out by Munk et al. (1970) that in using the approximation 47, Mysak omitted (except for the case $\omega = -f$) the very long Kelvin wave mode implicit in equation 45. A careful analysis of equation 45 reveals that the gravest-mode edge wave for $j = 2$, which is a right-bounded wave for $f > 0$, continuously merges into the barotropic Kelvin wave mode as $\kappa \to 0$ (see Fig. 11). It is this curve that separates the right-bounded edge waves from the low-frequency quasi-geostrophic waves, for which Robinson (1964) coined the term "continental shelf waves." However, Robinson considered only the very low-frequency end of the spectrum ($\sigma \ll 1$, $\kappa \ll 1$) under the additional approximation $\delta \ll 1$, in which case the waves are nondispersive and non-divergent. Mysak (1967a) also analyzed this end of the frequency spectrum for continental shelf waves traveling around a large circular continent with the same shelf/slope topography.

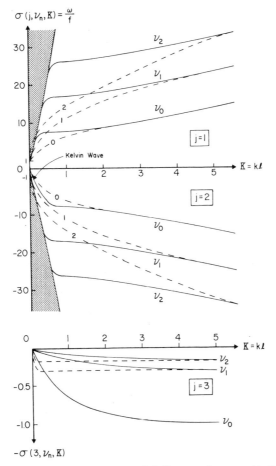

Fig. 11. Plots of edge ($j = 1$ and 2) and continental shelf or quasi-geostrophic ($j = 3$) wave dispersion relations for a sloping shelf of finite width (solid lines) and semi-infinite width (dashed lines) with $\delta = f^2 l^2/gd = 0.027$ [$f = 0.73 \times 10^{-4}$ rad/sec, $d = 200$ m, $l = 10^5$ m (or $\alpha = 2 \times 10^{-3}$)]. The different modes ν_n or n are indicated on the curves. There is no gravest-mode quasi-geostrophic dispersion relation shown since it is an inertial oscillation with $k = 0$. The shaded region corresponds to the continuous spectrum of topographically modified Poincaré waves; it is bounded by the hyperbolas $\omega/f = \pm(1 + gDk^2)^{1/2}$ (with $D = 5 \times 10^3$ m) corresponding to $K = 0$ (adapted from Mysak, 1968a).

In Fig. 11 the frequency functions $\sigma(j, v_n, \kappa)$ are plotted for the first three modes v_0, v_1, v_2 and are compared with the semi-infinite shelf solutions of Reid discussed in Section 3. Note that for $\kappa \lesssim 1$ the curves for both edge waves and shelf waves are numerically quite distinct from those due to Reid, thus revealing the importance of a finite-depth deep-sea region. The shaded region in the edge wave diagram corresponds to the frequency–wave number region defined by $K^2 \leqslant 0$ and contains the continuous spectrum of topographically modified Poincaré waves.

8. Trapped Waves on a Flat Shelf

Snodgrass et al. (1972) introduced the single flat step model as defined by equation 48 for a continental shelf/slope region in order to investigate nonrotational ($f = 0$) edge waves on the California borderland. To our knowledge they were the first to make the distinction between the continuous spectrum of leaky modes and the classical discrete spectrum of trapped wave motions. The existence of these two types of waves is also well-known in other contexts, in layered elastic media (Brekhovskikh, 1960) and for electromagnetic wave propagation (Wait, 1962), for example. An analysis of data from bottom pressure gauges at La Jolla and at San Clemente Island, about 100 km seaward, showed that in the period range of 5 min to 5 hr there was a fairly even partition between the leaky and trapped modes, with rms amplitudes of about 0.6×10^{-2} m! In a later experiment, however (Munk et al., 1964), it was found that most of the energy in this period range was contained in the first few trapped modes and that it was fairly evenly split between northward and southward propagating waves. Buchwald and de Szoeke (1973) also used the flat shelf model to study the generation of edge waves off the east Australian coast. The flat shelf model and more complicated multistepped models have been used extensively in the Japanese literature on edge waves (see, for example, Aida, 1967; Aida et al., 1968). The low-frequency quasi-geostrophic waves and the Kelvin wave propagating along a flat shelf were first studied by Larsen (1969). Although he derived a general dispersion relation for all frequencies, Larsen did not discuss the rotationally modified edge wave modes. In their lengthy study of tidal propagation along the coast of California, Munk et al. (1970) performed a thorough analysis of all types of waves that travel along a flat shelf profile. The dispersion curves they obtained are very similar to those found for a sloping shelf of finite width (Fig. 11). The only fundamental difference between the results for the two profiles is that only the lowest-mode shelf wave appears for the flat shelf; all the higher modes disappear and the shelf wave spectrum is reduced to a single wave associated with the single step change in topography.

In the interest of simplicity, we shall analyze the two classes of waves separately, first considering edge waves on a flat shelf, with $f = 0$ (thereby removing the Kelvin and shelf wave modes) and then shelf waves in a rotating ocean with a rigid lid (thereby removing the Kelvin and edge wave modes).

A. Edge Waves

For the flat shelf topography given by

$$
\begin{aligned}
H(x) &= H_1, \qquad 0 < x < l \\
&= H_2, \qquad x > l
\end{aligned}
\tag{48}
$$

with $H_2 > H_1$, and $f = 0$, equation 8 for the x dependence of the wave amplitude becomes

$$F_i'' + \left(\frac{\omega^2}{gH_i} - k^2\right)F_i = 0, \qquad i = 1, 2 \tag{49}$$

where the subscripts $i = 1, 2$ refer to the shelf and to the deep-sea regions, respectively. Following Buchwald and de Szoeke (1973) we now introduce the nondimensional quantities

$$\chi = \frac{x}{l}, \qquad \kappa = kl, \qquad \Omega = \frac{\omega l}{(gH_1)^{1/2}} \tag{50}$$

and

$$\gamma = \left(\frac{H_2}{H_1}\right)^{1/2} > 1 \tag{51}$$

In terms of these quantities, the amplitude variations for the two regions are governed by

$$F_1''(\chi) + (\Omega^2 - \kappa^2)F_1(\chi) = 0, \qquad 0 < \chi < 1 \tag{52}$$

$$F_2''(\chi) + \left(\frac{\Omega^2}{\gamma^2 - \kappa^2}\right)F_2(\chi) = 0, \qquad \chi > 1 \tag{53}$$

The long-wave propagation speeds in the deep and shallow parts of the basin are $(gH_2)^{1/2}$ and $(gH_1)^{1/2}$, respectively, or in nondimensional form, γ and 1. The nondimensional wave speed for the coupled system will be denoted by $c = \Omega/\kappa$. Depending on whether (a) $c > \gamma$, (b) $1 < c < \gamma$, or (c) $1 > c$, we find (a) leaky modes, for which the energy may propagate away from the coast, (b) trapped modes, for which wave propagation occurs on the shelf only and the energy decays away exponentially in the deep-sea region, or (c) virtual modes, for which no wave propagation is possible.

The appropriate boundary conditions on the boundaries of the domain $0 < \chi < \infty$ are

$$F_1'(0) = 0 \tag{54}$$

and

$$F_2 \begin{cases} \text{bounded for leaky modes} \\ \to 0 \qquad \text{for trapped modes} \end{cases} \text{as } \chi \to \infty \tag{55}$$

At the depth discontinuity ($\chi = 1$) continuity of surface elevation and of normal mass flux imposes the matching conditions

$$F_1(1) = F_2(1) \tag{56}$$

$$F_1'(1) = \gamma^2 F_2'(1) \tag{57}$$

Leaky Modes: $c > \gamma$

The solutions of equations 52 and 53, subject to the conditions 54–57 may be written

$$F_1(\chi) = A \cos \varepsilon \cos \mu_1 \chi \tag{58}$$

$$F_2(\chi) = A \cos \mu_1 \cos [\mu_2(\chi - 1) + \varepsilon] \tag{59}$$

where $\mu_1 = \Omega \cos \theta_1 = \kappa \cot \theta_1, \mu_2 = \Omega \gamma^{-1} \cos \theta_2 = \kappa \cot \theta_2, \tan \varepsilon = (\mu_1/\gamma^2 \mu_2) \tan \mu_1$, and A is an arbitrary amplitude constant.

Combining equations 58 and 59 with the wave dependence, namely, $\exp[i(\kappa Y - \Omega \tau)]$ where $Y = y/l$ and $\tau = t(gH_1)^{1/2}/l$, we see that these solutions represent long waves approaching the shelf at an angle θ_2 to the normal, being reflected and refracted at $\chi = 1$ and reflected from the coast at $\chi = 0$, with $\theta_1(<\theta_2)$ the angle between the wave number vector and the normal to the coast on the shelf. The conservation of phase at the edge of the shelf implies that Snell's law holds:

$$\sin \theta_2 = \gamma \sin \theta_1 \tag{60}$$

Trapped Modes: $1 < c < \gamma$

Total internal reflection now occurs at $\chi = 1$, and $\theta_2 > \sin^{-1}(\gamma)$. The solutions are now expressed as

$$F_1(\chi) = A \cos \mu_3 \chi \tag{61}$$

$$F_2(\chi) = A \cos \mu_3 \exp[-\rho(\chi - 1)] \tag{62}$$

where

$$\mu_3{}^2 = \Omega^2 - \kappa^2, \qquad \rho^2 = \kappa^2 - \frac{\Omega^2}{\gamma^2} \qquad (\rho > 0) \tag{63}$$

and

$$\mu_3 \tan \mu_3 = \rho \gamma^2 \tag{64}$$

It is theoretically possible to eliminate μ_3 between equations 63 and 64 to arrive at an implicit form for the dispersion relation $\Omega = \Omega(\kappa)$.

A detailed discussion of this relation is given by Snodgrass et al. (1962) and by Buchwald and de Szoeke (1973). We mention here only that for each mode n the

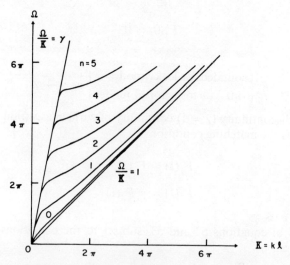

Fig. 12.   The dispersion relations of the first six trapped edge wave modes ($n = 0, 1, \ldots, 5$) on a flat shelf with $\gamma^2 = H_2/H_1 = 16$ and no rotation ($f = 0$). Because $f = 0$, the mirror image of these curves reflected across the kl axis also exists but has not been plotted.

waves are similar to the edge waves whose dispersion relations are shown in Fig. 11, except for the asymmetry due to the presence of rotation, which does not arise here (see Fig. 12), since we have taken $f = 0$. In particular, we note that all the modes are contained in the wedge bounded by the lines $\Omega = \kappa$ and $\Omega = \gamma\kappa$; the trapped waves travel at a velocity intermediate between that of waves in depth H_1 and that of waves in depth H_2 $(1 < c < \gamma)$.

Virtual Modes: $c < 1$

In this case, the χ dependence is exponential everywhere; there are no solutions corresponding to real wave numbers and hence no wave propagation.

B. Shelf Waves

The amplitudes of observed shelf waves usually do not exceed a few centimeters (Fig. 15) and hence their analysis may be performed under the restrictions of the rigid-lid hypothesis (Rhines, 1969). If

$$\frac{f^2 L^2}{gH} \ll 1 \tag{65}$$

where L is a typical horizontal length scale of the motion, variations in the level of the free surface play a negligible role in the vorticity balance. For a typical mid-latitude shelf, $f \simeq 10^{-4}$ rad/sec, $L \simeq 5 \times 10^4$ m (an average shelf width), $H \simeq 10^2$ m and $f^2 L^2/gH \simeq 2.5 \times 10^{-2}$, and equation 65 holds; in the deep-sea region, with $H \simeq 5 \times 10^3$ m equation 65 remains satisfied for $L < 10^6$ m.

Eliminating η from equations 1 and 2, we find the vorticity equation

$$(v_x - u_y)_t + f(u_x + v_y) = 0 \tag{66}$$

The rigid-lid form of the continuity equation 3 is

$$(Hu)_x + (Hv)_y = 0 \tag{67}$$

We introduce the transport stream function $\psi(x, y, t)$ defined by

$$\psi_x = Hv, \qquad \psi_y = -Hu \tag{68}$$

and which satisfies equation 67 identically. Substituting for ψ in equation 66, and recalling that we consider only $H = H(x)$ and uniform f, we find an equation for ψ:

$$\left(\frac{\psi_x}{H}\right)_{xt} + \frac{1}{H}\psi_{yyt} - \psi_y\left(\frac{f}{H}\right)_x = 0 \tag{69}$$

Assuming wave solutions of the form

$$\psi(x, y, t) = \phi(x)\exp\left[i(ky - \omega t)\right], \qquad k > 0 \tag{70}$$

we obtain for $\phi(x)$,

$$\left(\frac{1}{H}\phi'\right)' + \left[\frac{fk}{\omega}\left(\frac{1}{H}\right)' - \frac{k^2}{H}\right]\phi = 0, \qquad 0 < x < \infty \tag{71}$$

with primes indicating differentiation.

For the step topography of equation 48, $H' = 0$ except at the depth discontinuity, and equation 71 reduces to

$$\phi_i'' - k^2\phi_i = 0, \qquad i = 1, 2 \tag{72}$$

in each region ($i = 1, 2$ refer to the shelf and to the deep-sea regions, respectively). As for the boundary and matching conditions, at the coast, $Hu = 0$ for all y; hence

$$\phi_1(0) = 0 \tag{73}$$

In the deep-sea region, the amplitude of the trapped waves decays:

$$\phi_2 \to 0 \quad \text{as} \quad x \to \infty \tag{74}$$

For continuity of the normal transport Hu at $x = l$, it is necessary that

$$\phi_1(l) = \phi_2(l) \tag{75}$$

Finally, continuity of pressure across the discontinuity implies (see equation 2) continuity of $v_t + fu$ and hence

$$\frac{1}{H}\left(\phi' + \frac{fk}{\omega}\phi\right) \text{ continuous across } x = l \tag{76}$$

The solutions of equation 72 which satisfy the four conditions 73–76 are

$$\phi_1 = A_1 \sinh kx, \qquad 0 < x < l \tag{77}$$

$$\phi_2 = (A_1 \sinh kl)e^{-k(x-l)}, \qquad x > l \tag{78}$$

subject to the dispersion relation

$$\frac{\omega}{f} = -\frac{(\gamma^2 - 1)}{1 + \gamma^2 \coth kl} \tag{79}$$

in which $\gamma > 1$ is defined as earlier by equation 51. This dispersion relation is shown graphically by the dashed line in Fig. 20. For $f > 0$, equation 79 shows that $\omega < 0$; the phase speed is negative and the wave is right-bounded. The topographic nature of the wave is made evident by changing the relative magnitudes of H_1 and H_2: for propagation along a trench, $\gamma < 1$, and the wave reverses its direction of propagation. The dispersion relation 79 was first obtained by Larsen (1969) as a limiting case of his divergent solution for trapped quasi-geostrophic waves; it is also a special case of the dispersion relation obtained by Niiler and Mysak (1971) for shelf waves in the presence of a laterally sheared coastal current. Finally, in the long-wavelength limit $kl \ll 1$, equation 79 reduces to a Robinson-type shelf wave dispersion relation:

$$\frac{\omega}{f} = -(1 - \gamma^{-2})kl$$

In the short-wavelength limit $kl \gg 1$ (or equivalently for a wide shelf) on the other hand, we obtain the frequency for a trapped escarpment oscillation discovered by Rhines (1969):

$$\frac{\omega}{f} = \frac{1 - \gamma^2}{1 + \gamma^2}$$

9. Shelf Waves on an Exponential Shelf

Equation 71 for nondivergent shelf waves has constant coefficients for an exponential shelf profile. This fact was exploited by Buchwald and Adams (1968) in their study of shelf waves off the east Austrialian coast. They considered a depth profile of the form

$$H(x) = H_1 e^{bx}, \qquad 0 < x < l$$
$$\qquad = H_2, \qquad\qquad x > l \tag{80}$$

where $H_2 = H_1 e^{bl}$, so that H is continuous at $x = l$. Appropriate values for the shelf near Sydney, Australia are $H_1 = 67$ m, $H_2 = 5 \times 10^3$ m, $l = 8 \times 10^4$ m, and $b = 3.4 \times 10^{-5}$ m^{-1}. The actual depth profile and the fitted exponential are compared in Fig. 13. The solution of equation 71 for the profile 80 and subject to the conditions 72–76 is straightforward; the resulting dispersion relation is given implicitly by the pair of equations

$$m^2 + k^2 + b^2 + \frac{2bfk}{\omega} = 0 \tag{81}$$

$$\tan ml = \frac{-m}{(b + k)} \tag{82}$$

As in the case of edge waves on the flat shelf, it is possible to eliminate m from equations 81 and 82 to obtain a single relation between ω and k in the form

$$\frac{\omega}{f} = G(kl; bl) \tag{83}$$

The dispersion curves are shown graphically in Fig. 14 for the first five modes with $bl = 2.7$. We note that for $f > 0$, each mode is right-bounded ($\omega < 0$, as may be inferred directly from equation 81), as in all the previous shelf models discussed. However, in contrast to all the other shelf wave dispersion relations seen so far, we notice that for short waves ($kl \gg 1$), the group velocity is in a direction opposite to the phase velocity. The group velocity changes sign by passing through zero at an intermediate wave number. The occurrence of an extremum in group velocity is usually associated with an Airy phase, for which the decay rate is slower than at other wave numbers. In this case, the Airy phase would not propagate, and one would expect enhanced persistence of locally generated phenomena at those wave numbers where the group velocity is equal to zero. Observations by Cutchin and Smith (1973)

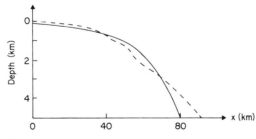

Fig. 13. Comparison of the shelf/slope topography near Sydney (dashed line) with the exponential model given by equation 80 (from Buchwald and Adams, 1968).

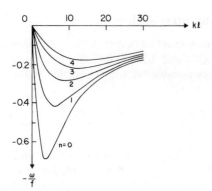

Fig. 14.   The dispersion relations of the first five modes of nondivergent shelf waves on an exponential shelf with decay parameter $bl = 2.7$ and $f > 0$ (from Buchwald and Adams, 1968).

of low-frequency motions on the Oregon shelf, where a depth profile very similar to that of equation 80 is found, have shown peaks of the coherence spectrum which nearly coincide with the frequencies at which the group velocity vanishes. The situation is far from clear-cut, but it is tempting to think, with Cutchin and Smith, that this coincidence is not accidental.

The depth profile given by equation 80 differs from all the profiles considered earlier in that H'/H remains bounded for all x. It has recently been found by Odulo (1975) and Huthnance (1975) in the framework of general studies of coastal trapped waves that whenever H'/H is bounded for all x, the group velocity of shelf waves is always negative for some value of k and the dispersion curves have the general form shown in Figure 14.

Finally, because of the nature of the dispersion relation, a long shelf wave traveling on an exponential shelf could backscatter energy in the form of shorter shelf waves, upon encountering topographical irregularities. Thus shelf wave scattering provides a mechanism through which energy could be transferred between different scales of motion and in different directions.

The effect of a free surface on shelf waves on an exponential shelf has been examined by Buchwald (1973). Though not important in oceanic situations, free surface divergence is relevant to the interpretation of laboratory experiments, for which the parameter $f^2 L^2/gH \sim 0(1)$ (see Caldwell et al., 1972).

10. Observations of Shelf Waves

Continental shelf waves were first observed along the Australian coast by Hamon (1962, 1963, 1966). In his first paper on this topic in 1962, Hamon presented the spectra of daily mean sea level and atmospheric pressure fluctuations at Sydney and Coffs' Harbour (situated 500 km to the north of Sydney) on the east Australian coast, and at Lord Howe Island, situated about 800 km eastward of Sydney. Surprisingly, he found that even at very low frequencies (periods greater than a few days) the daily mean[1] sea level on the shelf did not respond as an inverse barometer. That is, corresponding to an increase of 1 mbar (10^2 Pa) in the atmospheric pressure, the sea surface did not decrease by 1 cm (0.01 m), as would be expected for static deformations. In particular, Hamon found that at Sydney and Coff's Harbour the sea level was depressed only about half the expected amount. He also found that the

---

[1] Henceforth in this discussion on shelf wave observations we omit the phrase "daily mean" for convenience.

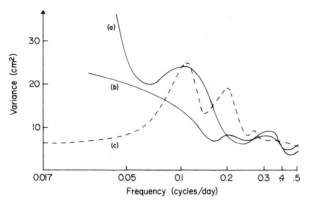

Fig. 15. The spectra of adjusted sea levels at Sydney, Australia, during (*a*) "winter" (April–Sept. 1958) and (*b*) "summer" (Oct. 1957–March 1958); and (*c*) the atmospheric pressure for the period July 1957–Dec. 1958 (adapted from Hamon, 1962).

spectra of the adjusted sea level (defined as the sea level minus the negative of the atmospheric pressure measured in centimeters of water) were peaked at 9 and 5 days, corresponding to the winter and summer peaks in the atmospheric pressure spectrum (see Fig. 15). At Lord Howe Island, on the other hand, the sea surface did respond as an inverse barometer. Hamon (1962, 1963) also performed a coherence and a lag analysis between the two coastal stations and found that for periods longer than 3 days, the adjusted sea level at Sydney led that at Coff's Harbour by about 1 day (see Figs. 16 and 17). This result suggests the presence of a low-frequency, nondispersive left-bounded wave which travels northward along the continental shelf. To test this hypothesis further, Hamon (1966) performed a more extensive lag analysis of the

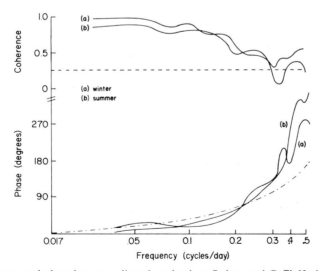

Fig. 16. Coherence and phase between adjusted sea levels at Sydney and Coff's Harbour. The dashed line indicates the 95% confidence limit for coherence. The dot-dash line stands for a phase difference corresponding to Sydney leading Coff's Harbour by 1 day at all frequencies (see also Fig. 17) (from Hamon, 1962, 1963).

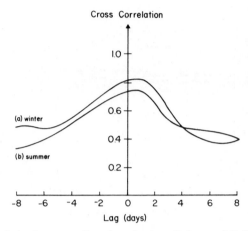

Fig. 17. The cross correlation between adjusted sea levels at Sydney and Coff's Harbour; positive lag means Sydney leads Coff's Harbour (from Hamon, 1962 and 1963).

adjusted sea level between Eden (37°S) and five stations to the north; these data suggested that such a northward-traveling wave propagates along the whole east Australian coast (Fig. 18). The slope of the dashed line in Fig. 18 gives an average speed of 4 m/sec for this wave.

Robinson (1964) showed that the observed nonbarometric sea level behavior on the east Australian continental shelf could be due to a resonant response of the adjusted sea level over a shelf topography of the form of equation 42 to large-scale moving weather systems. Mysak (1967b) further substantiated this explanation on the basis of Fourier transform and time series methods. Mysak showed that provided linear bottom friction is included with a friction coefficient of $0(10^{-8}/\text{sec})$, the theoretical response spectrum of the adjusted sea level that is forced by a numerical model for the observed atmospheric pressure spectrum compares quite favorably with the

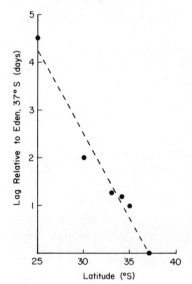

Fig. 18. Cumulative lags in adjusted sea levels as a function of latitude, for the east Australian coast (from Hamon, 1966).

dominant features of the observed adjusted sea-level spectrum. Robinson (1964) also attributed the observed lead to a low-frequency nondispersive continental shelf wave. In the limiting case $\kappa^2 = k^2 l^2 \ll 1$, $\sigma^2 \ll 1(\omega^2 \ll f^2)$, and $\delta = f^2 l^2/gd \ll 1$, equation 47 reduces to

$$J_0(2\sqrt{q}) = 0 \qquad \left(q = \frac{-fkl}{\omega}\right)$$

From the first zero (lowest-mode solution) of this equation, Robinson calculated a theoretical phase speed of 2.5 m/sec, a value considerably lower than the observed speed. Upon addition of a continental slope region of finite constant slope to produce a topography closer to type b (Fig. 10), Mysak (1967a) found that the gravest-mode speed was increased by about 30%. A comparable speed was also obtained by Buchwald and Adams (1968) using the exponential shelf model (equation 80): they obtained a speed of 2.8 m/sec for the lowest-mode phase speed in the (nondispersive) limit $kl \ll 1$. In another paper on shelf wave generation, Adams and Buchwald (1969) proposed that the observed nonbarometric behavior might be due to the low-frequency response of the sea surface to localizing moving wind stresses, rather than to large-scale moving weather systems.

However, the topographic modifications discussed above do not totally eliminate the discrepancy between theory and observations. This is perhaps not too surprising since no account was taken of the main oceanic features off the east Australian coast: the east Australian current and its associated stratification (Hamon, 1965). These offshore features were crudely modeled by Mysak (1967a), who complemented the purely topographic model of equation 42 with a two-layer deep-sea stratification and a basic horizontal deep-sea current of constant magnitude flowing southward in the upper layer alongside the edge of the shelf (see Fig. 7 in Mysak, 1967a). Mysak found that for a relative density difference of 2.5×10^{-3} between the two layers and a current speed of 1 m/sec, the northward-propagating lowest-mode shelf wave for low frequencies had a speed of 4 m/sec, in excellent agreement with Hamon's observations.

Shelf waves have subsequently been observed in many other parts of the world. They have been detected on the west Australian coast (Hamon, 1966), the Oregon coast (Mooers and Smith, 1968; Cutchin and Smith, 1973), the North Carolina coast (Mysak and Hamon, 1969), the west coast of Scotland (Cartwright, 1969), the north Mediterranean coast (Saint-Guily and Rouault, 1971) and the west Florida coast (P. P. Niiler, 1976). Also, in the more recent studies, current meter records rather than sea-level fluctuations have been analyzed to confirm the existence of shelf waves and to investigate their properties more closely.

11. The Role of Shelf Waves in Coastal and Deep-Ocean Dynamics

Over the past few years it has become increasingly evident from both theory and observations that continental shelf waves are intimately connected with many dynamic and hydrographic features of coastal and deep-ocean regions. For example, they appear to interact (either passively or actively) with western boundary currents such as the East Australian Current (as mentioned briefly in Section 10) and the Gulf Stream. Furthermore, they probably play an important role in the phenomenon of coastal upwelling. In this section we discuss briefly these aspects of shelf wave theory.

In attempting to model the effect of the East Australian Current on the propagation properties of shelf waves, Mysak (1967a) found that at very low frequencies, there arises, in the phase speed of the lowest-mode shelf wave, a strong Doppler shift due to a constant surface current in the deep-sea region. The relationship between the phase speed and the current speed V_0 is nearly linear, and for typical values of the various parameters involved the phase speed decreases by nearly 1 m/sec when the current speed increases from 0 to 2 m/sec (see Fig. 19). It was thus suggested by Mysak that this passive coupling (involving no energy exchange) between shelf waves and a deep-sea current might be used to detect seasonal changes in the speed of a deep-sea current by measuring seasonal changes in the phase speed of the lowest-mode shelf wave. This application of shelf wave theory has been partially justified by Mysak and Hamon (1969) on the basis of an analysis of sea level and Gulf Stream current data taken near the North Carolina coast (see Figs. 8 and 9 in Mysak and Hamon, 1969).

One of the mechanisms that has been suggested to explain the origin of the familiar Gulf Stream meanders is a hydrodynamic instability involving trapped coastal waves. It has been hypothesized that the meanders may be due to unstable baroclinic planetary waves traveling parallel to the coast as time-dependent perturbations of a basic baroclinic jet (Orlanski, 1969; Hansen, 1970). Orlanski used as a basic state a two-layer model for the stratification along with a laterally sheared horizontal current confined to the upper layer; a bottom topography of type b (Fig. 10) was used in the model. Orlanski showed the mean potential energy due to the stratification to be

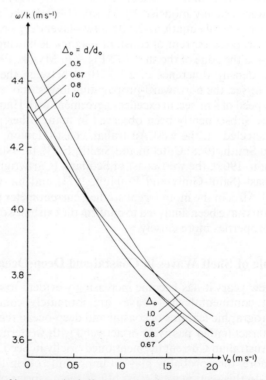

Fig. 19. Phase speed of lowest-mode shelf wave versus deep-sea current speed for various values of Δ_0 and for parameters applicable to the East Australian Current. The relative density difference in the deep-sea region is 2.5×10^{-3}, the shelf width is 5×10^4 m, $f = -0.73 \times 10^{-4}$ rad/sec; $d = $ maximum shelf depth and $d_0 = $ depth of the upper layer in the deep-sea region (from Mysak, 1967a).

the main source of energy for the unstable waves, and on the basis of numerical solutions, demonstrated that the most unstable waves have wavelengths and phase speeds which agree quite well with the observations. However, the theoretical growth rate calculated for the most unstable wave was much higher than that observed (Hansen, 1970). Niiler and Mysak (1971), on the other hand, showed that for a small range of wave numbers, the lowest-mode shelf wave becomes unstable when interacting with a laterally sheared, basic barotropic current of the form

$$V(x) = \frac{V_0 x}{l} \qquad\qquad 0 < x < l$$

$$= V_0\left(2 - \frac{x}{l}\right) \qquad l < x < 2l$$

$$= 0 \qquad\qquad x > 2l$$

flowing along the step topography 48. In this model, the wave extracts energy from the kinetic energy of the shear flow. For parameters applicable to the Gulf Stream and with the topography of the Blake Plateau region, the dispersion relation for Re(ω) takes the form shown in Fig. 20. We note that for $0 < kl < 2.7$, the shear flow effectively "splits" the dispersion relation 79 and stable shelf waves can travel in both

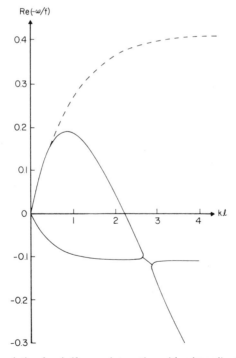

Fig. 20.   The dispersion relation for shelf waves interacting with a laterally sheared horizontal flow that models the Gulf Stream near the Blake Plateau. The Rossby number for the current is 0.22 and the flat shelf topography is characterized by $H_2/H_1 = \gamma^2 = 1.8$. The dashed line corresponds to the dispersion relation 79, applicable when no current is present (adapted from Niiler and Mysak, 1971).

directions along the shelf for most of this wave number range. For $kl > 2.9$, there are also two distinct stable waves but they are both left-bounded. Finally, for $2.7 < kl < 2.9$, the two real roots for ω coalesce and become complex conjugates of each other, one root being stable (decaying exponentially with time) and the other unstable. Further, both waves travel with the flow (i.e., are left-bounded). The most unstable wave (i.e., that wave with a wave number corresponding to the most rapid growth rate) has an e-folding time of 13 days, which is about half Orlanski's value. Its phase speed of 11 km/day, and wavelength of 180 km agree reasonably well with the range of observed values due to Hansen, namely, 4–9 km/day and 200–400 km. Lee (1975) considered the forced problem for this model and showed that an impulsively applied and slowly moving wind stress is very efficient in generating unstable shelf waves. Lee also found that the amplitudes of the associated current patterns have maximum values at the outer edge of the flow ($x = 2l$).

One of the most active areas of research in coastal oceanography today is the phenomenon of coastal upwelling on the eastern side of mid-latitude ocean basins. It has recently been suggested that shelf waves play an important role in the overall dynamical picture of this phenomenon (Gill and Clarke, 1974). Observations indicate that coastal upwelling at mid-latitudes is induced primarily by the equatorward component of wind stress. Off the Oregon coast, for example, upwelling events are characterized in part by the rising of the main thermocline very close to the shoreline (Smith, 1974); off the west coast of Africa it is at the edge of the continental shelf that the main thermocline rises in response to strong equatorward winds (J. S. Allen, personal communication). Finally, in the northern hemisphere many observations indicate that upwelling motions are usually accompanied by a poleward jet in the lower layer and an equatorward flow in the upper layer.

In contrast to much of the earlier theoretical work (for example, see O'Brien and Hurlburt, (1972), recent time-dependent models of coastal upwelling are three-dimensional (i.e., include longshore variations). This extension to three dimensions is based on a recognition of the possible importance of trapped coastal waves. Movements of the thermocline at any one point of the coast (or at the edge of the shelf) are influenced by the wind stress at earlier times at other positions along the coast because information can be transmitted along in the longshore direction by shelf waves and internal Kelvin waves.

Suginohara (1974) has recently completed an interesting numerical study which exhibits the importance of internal Kelvin waves and shelf waves in upwelling very near the coast, as would apply off Oregon. He introduced a two-layer model for the stratification, extending right to the coast, and assumed that the barotropic mode is horizontally nondivergent (as in Section 9). Suginohara then examined the response of this model to an equatorward wind stress of finite extent in the longshore direction. For a flat bottom ocean (no shelf) he found that the upwelling adjacent to the coast occurs within a narrow band with a width of the order r, the internal Rossby radius of deformation. The results showed that the upwelling process can be interpreted as the generation and propagation of internal Kelvin waves. Associated with the upwelling, Suginohara found an equatorward flow in the upper layer. Similar results were also obtained by Gill and Clarke (1974) for the more general case of a continuous stratification. When topography of type b is included, with the stratification extending over the shelf, Suginohara found that the poleward undercurrent is delayed for a few days due to the generation of shelf waves. However, after the forcing has stopped, right-bounded shelf waves propagate northward away from the upwelling region and the poleward undercurrent develops fully. Gill and Clarke (1974) also studied upwelling in a model including a topography and stratification essentially similar to the form taken by Mysak (1967a). They showed that movements of the thermocline [of $0(1$ m)] just below the edge of the shelf are correlated with small [$0(0.01$ m)] sea-level changes at the coast, and suggested on that basis that information about coastal currents and upwelling could be obtained by monitoring tide gauge records at the coast. Gill and Clarke also demonstrated that the response of this model to an equatorward wind stress is dominated by the lowest-mode shelf wave.

The importance of both shelf waves and internal Kelvin waves in upwelling phenomena motivated Allen (1975) and Wang (1975) to investigate the coupling between these two types of waves in a two-layer ocean with stratification extending over the shelf. Allen calculated that for an exponential shelf (Section 9) and the rigid-lid approximation (Section 8), the strength of the coupling is controlled by the magnitude of the parameter $\lambda = r/b$, where $b = H/H'(x)$ is a characteristic horizontal length scale of the topography. Off the Oregon coast, for example, this parameter is generally less than unity, which allowed Allen to analyze the coupling problem by means of an ordinary perturbation expansion in λ. He found that for $kl \ll 1$ (where l is the total shelf/slope width), an $0(1)$ Kelvin wave is accompanied by an $0(\lambda)$ barotropic motion which extends across the shelf and that an $0(1)$ shelf wave is accompanied by a weaker $0(\lambda^2)$ baroclinic motion. For intermediate scales in both the longshore and offshore directions [i.e., effectively for $kl = 0(1)$], shelf waves are strongly coupled to baroclinic effects. Finally, for very short scales ($kl \gg 1$), the shelf wave motions are primarily confined to the bottom layer and the waves are "bottom trapped" (see also Section 12). Wang (1975) confirmed many of the above results when the free surface is included and also studied the case of the flat shelf model (equation 48) as did also Kajiura (1974). In particular, Wang found that when the shelf and internal Kelvin wave speeds are comparable, resonant coupling occurs. In such a case the relative amplitude distributions of the resonant modes are very sensitive to the value of the internal Rossby radius, r.

In the above upwelling studies we note that many of the phenomena occur on a scale or order r, the internal radius of deformation. For a continuously stratified ocean r is equal to NH/f (where N = Brunt–Väisälä frequency); typically, $r = 0(10$ km) for most coastal upwelling regions. It is thus apparent that coastal and topographic irregularities, which are generally also of this scale or slightly smaller, are likely to have a significant influence on the propagation properties of shelf and internal Kelvin waves and hence affect some of the results on upwelling. This influence has been studied numerically for the case of isolated topographic features (a headland and a large canyon) that were incorporated into upwelling models (Hurlburt, 1974; Kishi and Suginohara, 1975; and Peffley and O'Brien, 1975). However, typical coastlines and bottom reliefs are marked by extensive irregularities, both in the offshore and longshore directions. Such extensive irregularities affect trapped coastal waves in two basic ways. The irregularities scatter energy from one coherent mode into other modes and cause a particular coherent mode to be attenuated as it propagates along the coast. Based on a stochastic model for an infinitely long irregular coastline, Mysak and Tang (1974) and Fuller (1975) have examined respectively the behavior of barotropic Kelvin waves and edge waves in the presence of such irregularities. A similar study for smaller-scale shelf waves and internal Kelvin waves has not as yet been performed, however; such a calculation would certainly advance our understanding of coastal upwelling in the real ocean. It would definitely be advisable to carry out such a study before further costly large-scale upwelling observational programs are undertaken.

12. Trapped Waves in a Stratified Fluid

In a stratified fluid of nonuniform depth, it is not generally possible to separate the vertical and horizontal dependences of the wave variables and to split the wave motion into a series of uncoupled (in a linear theory) modes, according to the method used, for example, by Gill and Clarke (1974). Some special cases, however, are more accessible to analysis, and a number of problems where the stratification and the bottom slope occur in different regions have also been considered.

Edge waves in a stratified fluid have been only briefly touched upon in the literature. This is perhaps not surprising, since edge waves are generally closely confined to the shoreline and are thus unlikely to be affected by the deep-sea stratification. Mysak (1968b) showed that edge waves traveling along an abrupt topography of the form of equation 42, bordering a two-layer deep-sea region, are only weakly coupled to long internal waves propagating in the deep ocean.

Greenspan (1970), on the other hand, found solutions for trapped surface and internal waves in a continuously and uniformly stratified fluid lying over the semi-infinite sloping beach profile given by equation 11. In his paper, the rotation of the earth is not included ($f = 0$), but the full vertical dynamics are preserved: no hydrostatic approximation is made. Greenspan found a lowest-mode solution which is almost identical to the fundamental mode of Stokes's solution. The dispersion relation and the horizontal currents are unaffected by the stratification for that mode, but perturbations of the pressure and density fields arise from the presence of the mean density gradient. The higher-mode solutions are much more complicated and may be examined in the original paper. More recently, Odulo (1974) has studied this lowest-mode edge wave in a rotating stratified fluid over a constant slope.

The solutions of Mysak (1967a, 1968b) and of Gill and Clarke (1974) for trapped wave propagation on a nonstratified sloping shelf adjacent to a stratified deep-sea region of uniform depth do not describe trapped wave propagation in a stratified fluid, but merely the coupling of trapped wave solutions in a homogeneous fluid to offshore internal waves. A wave solution which is intimately dependent for its existence on both stratification and bottom slope was discovered by Rhines (1970), as the short-wavelength limit of topographic planetary waves in a stratified fluid away from coastal boundaries. For waves that are appreciably larger than the internal Rossby radius of deformation ($\lambda \gg NH_0/f$, where λ is the wavelength, N is the Brunt–Väisälä frequency, and H_0 is a reference depth), bottom slopes have the same kind of dynamic influence as f variations, but for short scales ($\lambda \ll NH_0/f$), the waves degenerate into simple buoyancy oscillations of frequency

$$\omega = N \sin \beta \sin \varepsilon \qquad (84)$$

where β is the bottom slope angle and ε is the angle between the wave crests and the bottom contours. These buoyancy oscillations consist of a sloshing back and forth across the bottom contours and are trapped near the bottom by the Coriolis force. They have also been discussed by Needler and LeBlond (1973) and by Allen (1975). Although some recent deep-sea current measurements reported by Swallow (1971) have been interpreted as bottom-trapped waves by Rhines (1970), the evidence for their existence is by no means conclusive. Nevertheless Bell (1974) has proposed that the interaction of such waves with small topographic features (0(1 km)) is responsible for the generation of internal waves in the deep ocean which in turn induce vertical mixing in the bottom nepheloid layer.

Acknowledgments

This work was partly supported by the National Research Council of Canada through operating grants A7490 (P.H.L.) and A5201 (L.A.M.). We wish to thank A. J. Bowen for providing us with the photographs shown in Figs. 6 and 7.

References

Adams, J. K. and V. T. Buchwald, 1969. The generation of continental shelf waves. *J. Fluid Mech.*, **35**, 815–826.

Aida, I., 1967. Water level oscillations on the continental shelf in the vicinity of Miyagi–Enoshima. *Bull. Earthquake Res. Inst.*, **45**, 61–78.

Aida, I., T. Hatori, M. Koyama, and K. Kajiura, 1968. A model experiment of long-period waves travelling along a continental shelf. *Bull. Earthquake Res. Inst.*, **46**, 707–739 (in Japanese).

Allen, J. S., 1975. Coastal trapped waves in a stratified ocean. *J. Phys. Oceanogr.*, **5**, 300–325.

Ball, F. K., 1967. Edge waves in an ocean of finite depth. *Deep-Sea Res.*, **14**, 79–88.

Bell, T. H., Jr., 1974. Vertical mixing in the deep ocean. *Nature*, **251**, 43–44.

Bowen, A. J., 1969. Rip currents, 1. Theoretical investigations. *J. Geophys. Res.*, **74**, 5467–5478.

Bowen, A. J. and D. L. Inman, 1969. Rip currents. 2. Laboratory and field observations. *J. Geophys. Res.*, **74**, 5479–5490.

Bowen, A. J. and D. L. Inman, 1971. Edge waves and crescentic bars. *J. Geophys. Res.*, **76**, 8662–8671.

Brekhovskikh, L. M., 1960. *Waves in Layered Media*. Academic Press, New York, 561 pp.

Buchwald, V. T., 1973. On divergent shelf waves. *J. Mar. Res.*, **31**, 105–115.

Buchwald, V. T. and J. K. Adams, 1968. The propagation of continental shelf waves. *Proc. Roy. Soc. A*, **305**, 235–250.

Buchwald, V. T. and R. A. de Szoeke, 1973. The response of a continental shelf to travelling pressure disturbances. *Aust. J. Mar. Freshwater Res.*, **24**, 143–158.

Caldwell, D. R., D. L. Cutchin, and M. S. Longuet-Higgins, 1972. Some model experiments on continental shelf waves. *J. Mar. Res.*, **30**, 39–55.

Cartwright, D., 1969. Extraordinary tidal currents near St. Kilda. *Nature*, **223**, 928–932.

Clarke, D. J., 1974. Long edge waves over a continental shelf. *Dtsch. Hydrogr. Z.*, **27**, 1–8.

Cutchin, D. L. and R. L. Smith, 1973. Continental shelf waves: Low frequency variations in sea level and currents over the Oregon continental shelf. *J. Phys. Oceanogr.*, **3**, 73–82.

Eckart, C., 1951. Surface waves in water of variable depth. *Marine Physical Laboratory of the Scripps Institute of Oceanography, Wave Rept. No. 100*, S.I.O. Ref. 51-12 (unpublished manuscript), 99 pp.

Fuller, J. D., 1975. Edge waves in the presence of an irregular coastline. M.Sc. thesis, University of British Columbia, 64 pp.

Galvin, C. J., 1965. Resonant edge waves on laboratory beaches (abstract). *Trans. Am. Geophys. Union*, **46**, 112.

Gill, A. E. and A. J. Clarke, 1974. Wind-induced upwelling, coastal currents and sea-level changes. *Deep-Sea Res.*, **21**, 325–345.

Gill, A. E. and E. H. Schumann, 1974. The generation of long shelf waves by the wind. *J. Phys. Oceanogr.*, **4**, 83–90.

Greenspan, H. P., 1956. The generation of edge waves by moving pressure distributions. *J. Fluid Mech.*, **1**, 574–592.

Greenspan, H. P., 1970. A note on edge waves in a stratified fluid. *Stud. Appl. Math.*, **49**, 381–388.

Grimshaw, R., 1974. Edge waves: a long-wave theory for oceans of finite depth. *J. Fluid Mech.*, **62**, 775–791.

Hamon, B. V., 1962. The spectrums of mean sea level at Sydney, Coff's Harbour, and Lord Howe Island. *J. Geophys. Res.*, **67**, 5147–5155.

Hamon, B. V., 1965. The east Australian current, 1960–1964. *Deep-Sea Res.*, **12**, 899–922.

Hamon, B. V., 1966. Continental shelf waves and the effects of atmospheric pressure and wind stress on sea level. *J. Geophys. Res.*, **71**, 2883–2893.

Hansen, D. V., 1970. Gulf Stream meanders between Cape Hatteras and the Grand Banks. *Deep-Sea Res.*, **17**, 495–511.

Hino, M., 1974. Theory on formation of rip-current and cuspidal coast. *Proc. 14th Int. Conf. Coastal Eng.*, Copenhagen.

Huntley, D. A., and A. J. Bowen, 1973. Field observations of edge waves. *Nature*, **243**, 160–162.

Hurlburt, H. E., 1974. The influence of coastline geometry and bottom topography on the eastern ocean circulation. Ph.D. thesis, Florida State University, 103 pp.

Huthnance, J. M., 1975. On trapped waves over a continental shelf. *J. Fluid Mech.*, **67**, 689–704.

James, I. D., 1974a. Non-linear waves in the nearshore region: shoaling and set-up. *Estuarine Coastal Mar. Sci.*, **2**, 207–234.

James, I. D., 1974b. A non-linear theory of longshore currents. *Estuarine Coastal Mar. Sci.*, **2**, 235–249.

Johns, B., 1965. Fundamental mode edge waves over a steeply sloping shelf. *J. Mar. Res.*, **23**, 200–206.

Kajiura, K., 1958. Effect of Coriolis force on edge waves. (II) Specific examples of free and forced waves. *J. Mar. Res.*, **16**, 145–157.

Kajiura, K., 1972. The directivity of energy radiation of the tsunami generated in the vicinity of a continental shelf. *J. Oceanogr. Soc. Japan*, **28**, 260–277.

Kajiura, K., 1974. Effect of stratification on long period trapped waves on the shelf. *J. Oceanogr. Soc. Japan*, **30**, 271–281.

Kamenkovich, V. M., 1973. *Osnoví Dinamiki Okeana*. Gidromet, Leningrad, 240 pp.

Kenyon, K. E., 1969. Note on Stokes' drift velocity for edge waves. *J. Geophys. Res.*, **74**, 5533–5535.

Kenyon, K. E., 1972. Edge waves with current shear. *J. Geophys. Res.*, **77**, 6599–6603.

Kishi, M. J. and N. Suginohara, 1975. Effects of longshore variation of coastline geometry and bottom topography on coastal upwelling in a two-layer model. *J. Oceanogr. Soc. Japan*, **31**, 48–50.

Lamb, H., 1945. *Hydrodynamics*. 6th ed. Dover, New York, 738 pp.

Larsen, J. C., 1969. Long waves along a single-step topography in a semiinfinite uniformly rotating ocean. *J. Mar. Res.*, **27**, 1–6.

LeBlond, P. H. and C. L. Tang, 1974. On energy coupling between waves and rip currents. *J. Geophys. Res.*, **79**, 811–816.

Lee, C. A., 1975. The generation of unstable waves and the generation of transverse upwelling: two problems in geophysical fluid dynamics. Ph.D. thesis, University of British Columbia, 157 pp.

Lemon, D. D., 1975. Observations and theory of seiche motions in San Juan Harbour, B.C., M.Sc. thesis, University of British Columbia, 81 pp.

Longuet-Higgins, M. S., 1972. Recent progress in the study of longshore currents. In *Waves on Beaches, and Resulting Sediment Transport*. R. E. Mayer, ed. Academic Press, New York, pp. 203–248.

Longuet-Higgins, M. S. and R. W. Stewart, 1963. A note on wave set-up. *J. Mar. Res.*, **21**, 4–10.

Longuet-Higgins, M. S. and R. W. Stewart, 1964. Radiation stress in water waves; a physical discussion with applications. *Deep-Sea Res.*, **11**, 529–562.

Mooers, C. N. K., and R. L. Smith, 1968. Continental shelf waves off Oregon. *J. Geophys. Res.*, **73**, 549–557.

Munk, W., F. Snodgrass, and G. Carrier, 1956. Edge waves on the continental shelf. *Science*, **123**, 127–132.

Munk, W., F. Snodgrass, and M. Wimbush, 1970. Tides off-shore: Transition from California coastal to deep-sea waters. *Geophys. Fluid Dyn.*, **1**, 161–235.

Munk, W. H., F. E. Snodgrass, and F. J. Gilbert, 1964. Long waves on the continental shelf: an experiment to separate trapped and leaky modes. *J. Fluid Mech.*, **20**, 529–554.

Mysak, L. A., 1967a. On the theory of continental shelf waves. *J. Mar. Res.*, **25**, 205–227.

Mysak, L. A., 1967b. On the very low frequency spectrum of the sea level on a continental shelf. *J. Geophys. Res.*, **72**, 3043–3047.

Mysak, L. A., 1968a. Edgewaves on a gently sloping continental shelf of finite width. *J. Mar. Res.*, **26**, 24–33.

Mysak, L. A., 1968b. Effects of deep-sea stratification and current on edgewaves. *J. Mar. Res.*, **26**, 34–43.

Mysak, L. A., and B. V. Hamon, 1969. Low frequency sea level behaviour and continental shelf waves. *J. Geophys. Res.*, **74**, 1397–1405.

Mysak, L. A., and C. L. Tang, 1974. Kelvin wave propagation along an irregular coastline. *J. Fluid Mech.*, **64**, 241–261.

Needler, G. T., and P. H. LeBlond, 1973. On the influence of the horizontal component of the Earth's rotation on long period waves. *Geophys. Fluid Dyn.*, **5**, 23–46.

Niiler, P. P. and L. A. Mysak, 1971. Barotropic waves along an eastern continental shelf. *Geophys. Fluid Dyn.*, **2**, 273–288.

Niiler, P. P., 1976. Observations of low-frequency motions on the western Florida continental shelf. *J. Phys. Oceanogr.* In press.

O'Brien, J. J. and H. E. Hurlburt, 1972. A numerical model of coastal upwelling. *J. Phys. Oceanogr.*, **2**, 14–20.

Odulo, A. B., 1974. Edge waves in a rotating stratified fluid at an inclined shore. *Izv. Akad. Nauk SSSR, Atmos. Oceanic phys.*, **10**, 188–189.

Odulo, A. B., 1975. Rasprostranenie dlinnikh voln v okeane peremennoy glubini. *Okeanologiya*, **15** (1).

Orlanski, I., 1969. The influence of bottom topography on the stability of jets in a baroclinic fluid. *J. Atmos. Sci.*, **26**, 1216–1232.

Peffley, M. B. and J. J. O'Brien, 1975. A three-dimensional simulation of coastal upwelling off Oregon. Submitted to *J. Phys. Oceanogr.*

Pinney, E., 1946. Laguerre functions in the mathematical foundations of the electromagnetic theory of the paraboloidal reflector. *J. Math. Phys.*, **25**, 49–79.

Platzman, G., 1971. Ocean tides and related waves. In *Mathematical Problems in the Geophysical Sciences.* Vol. 14. W. H. Reid, ed. American Mathematical Society, Providence, R.I., pp. 239–291.

Reid, R. O., 1958. Effects of Coriolis force on edge waves. (I) Investigation of the normal modes. *J. Mar. Res.*, **16**, 109–144.

Rhines, P., 1971. A comment on the Aries observations. *Phil. Trans. Roy. Soc. London A*, **270**, 461–463.

Rhines, P. B., 1969. Slow oscillations in an ocean of varying depth. Part I. Abrupt topography. *J. Fluid Mech.*, **37**, 161–189.

Rhines, P. B., 1970. Edge-, bottom-, and Rossby waves in a stratified fluid. *Geophys. Fluid Dyn.*, **1**, 273–302.

Robinson, A. R., 1964. Continental shelf waves and the response of sea level to weather systems. *J. Geophys. Res.*, **69**, 367–368.

Saint-Guily, B., 1968. Ondes de frontière dans un bassin tournant dont le fond est incliné. *Compt. Rend. Acad. Sci. Paris, Ser. A*, **266**, 1291–1293.

Saint-Guily, B. and C. Rouault, 1971. Sur la présence d'ondes de seconde class dans le golfe du Lion. *Compt. Rend. Acad. Sci. Paris*, **272**, 2661–2663.

Shephard, F. P., K. O. Emery, and E. C. Lafond, 1941. Rip currents: a process of geological importance. *J. Geol.*, **49**, 337–369.

Shen, M. C., 1975. Ray method for surface waves on fluid of variable depth. *SIAM Rev.*, **17**, 38–56.

Shen, M. C., R. E. Meyer, and J. B. Keller, 1968. Spectra of water waves in channels and around islands. *Phys. Fluids*, **11**, 2289–2304.

Smith, R., 1970. Asymptotic solutions for high-frequency trapped wave propagation. *Phil. Trans. Roy. Soc. London, A*, **268**, 289–324.

Smith, R., 1971. The ray paths of topographic Rossby waves. *Deep-Sea Res.*, **18**, 477–483.

Smith, R. L., 1974. A description of current, wind, and sea-level variations during coastal upwelling off the Oregon coast, July–August, 1972. *J. Geophys. Res.*, **79**, 435–443.

Snodgrass, F. E., W. H. Munk, and G. R. Miller, 1962. Long-period waves over California's continental borderland. Part I. Background spectra. *J. Mar. Res.*, **20**, 3–30.

Stokes, G. G., 1846. Report on recent researches in hydrodynamics. *Rept. 16th Meet. Brit. Assoc. Adv. Sci. Southampton, 1846.* John Murray, London, pp. 1–20; *Math. Phys. Papers*, **1**, 167.

Suginohara, H., 1974. Onset of coastal upwelling in a two-layer ocean by wind stress with longshore variation. *J. Oceanogr. Soc. Japan*, **30**, 23–33.

Swallow, J. C., 1971. The Aries current measurements in the western North Atlantic. *Phil. Trans. Roy. Soc. London, A*, **270**, 451–460.

Ursell, F., 1952. Edge waves on a sloping beach. *Proc. Roy. Soc., A*, **214**, 79–97.

Wait, J. R., 1962. *Electromagnetic Waves in Stratified Media.* Macmillan, New York, 372 pp.

Wang, Dong-Ping, 1975. Coastal trapped waves in a baroclinic ocean. *J. Phys. Oceanogr.*, **5**, 326–333.

II. GEOLOGY

11. MODELING OF SAND TRANSPORT ON BEACHES AND THE RESULTING SHORELINE EVOLUTION

PAUL D. KOMAR

1. Introduction

A standard technique of physical oceanographers and engineers is the development of numerical models which simulate on a computer the flow of a fluid. Continuity and momentum flux equations are utilized to govern the fluid flow. The continuity equation keeps track of the total mass of water and its distribution within the system; the momentum equation evaluates the forces acting on individual water elements, the forces that cause their motion. In a finite-difference computer model the total volume of fluid is divided into small fixed cubic volumes and the flow from one cube to the next is followed through time, the time being likewise incremented in small intervals. By such a scheme the physical oceanographers and engineers have successfully modeled the motions of ocean waves, tides, a range of open ocean and shallow-water currents, as well as the flow in estuaries, rivers, and pipes.

Computer simulation of sediment motion, in contrast, is more in an infant stage of development. Only very simple schemes have been developed and in only very restricted areas. One of the problems, of course, is that we are not as certain as to the equations that govern sediment transport as we are of the momentum flux equation that controls fluid flow. For example, there is no general agreement among scientists as to the evaluation of sand transport in rivers or on continental shelves.

At this stage, models of sediment movement along beaches have probably received the greatest attention. Part of the reason for this is that we understand better the evaluation of sediment transport along beaches under wave action. Since I am most interested in this area, this chapter centers mainly on such nearshore models. They serve to illustrate the development of such models, as the techniques are much the same no matter what the application. The principal purpose of this review is to introduce the techniques when applied specifically to sediment movements. Harbaugh and Bonham-Carter (1970) give a thorough presentation on the general philosophy of simulation modeling and the techniques employed. Their book should be consulted for additional methods.

In order to develop computer models of beach sedimentation, we must first review the background on the evaluation of water and sediment motions in the nearshore region. The sand transport equation is analogous to the momentum flux equation, our attention having shifted from the motions of the water to the movement, erosion, and deposition of beach sand. We also must formulate an equation for the continuity of sand on the beach, in obvious parallel to the continuity of water. Next the beach system is divided into a series of cells, the finite-difference elements. Finally, examples of the results of such applications to the configuration of the shoreline are presented.

2. Evaluation of the Sediment Transport

The development of such models of course requires evaluation of the sediment transport rate along the beach. There is a considerable literature on this which can be only briefly reviewed here, the two or three equations needed in our models being presented.

When waves break at an angle to the shoreline their energy and momentum are expended, producing longshore currents and sand transport. Attempts at evaluation of the sand transport rate have relied mainly on empirical correlations with

$$P_l = (ECn)_b \sin \alpha_b \cos \alpha_b \tag{1}$$

where $(ECn)_b$ is the energy flux of the waves evaluated at the breaker zone, and α_b is the breaker angle, the angle the wave crests make with a parallel to the shoreline. Inman and Bagnold (1963) pointed out that the littoral transport rate should be expressed as an immersed weight transport rate, I_l, rather than as a volume transport rate, S_l. The two are related by

$$I_l = (\rho_s - \rho)ga'S_l \tag{2}$$

where ρ_s and ρ are the sand and water densities, respectively, and a' is the correction factor for the pore space of the beach sand (approximately 0.6 for most beach sands). The cgs units of S_l are cubic centimeters per second and of I_l are dynes per second. I_l is related to P_l through

$$I_l = 0.77(ECn)_b \sin \alpha_b \cos \alpha_b \tag{3}$$

The empirical 0.77 factor is dimensionless and is based on a fit to the available field data shown in Fig. 1.

The advantages of using the immersed weight transport rate I_l rather than the volume transport rate S_l are that I_l takes into consideration grains of differing densities, and I_l and P_l have the same units so that equation 3 is dimensionally correct. However, in the simulation models we need to evaluate the volume of sand moving from one shoreline cell to the next under the wave action. For this purpose we require S_l rather than I_l. For a quartz sand beach ($\rho_s = 2.65$ g/cm^3), equation 3 then becomes

$$S_l = (6.85 \times 10^{-5})(ECn)_b \sin \alpha_b \cos \alpha_b \tag{4}$$

A units change factor has also been included such that if $(ECn)_b$ is given in units of ergs per centimeter per second, the value of S_l obtained is in units of cubic meters per day, the form most suitable for the models.

Equations 3 and 4 apply only if the sand transport is caused by longshore currents due to waves breaking obliquely at the shoreline. A more fundamental examination of sand transport under combined waves and currents was undertaken by Bagnold (1963) and applied to the littoral zone by Inman and Bagnold (1963). The transport is related to the longshore current velocity $\bar{v}_l$ through

$$I_l = 0.28(ECn)_b \frac{\bar{v}_l}{u_m} \tag{5}$$

where u_m is the maximum orbital velocity evaluated at the breaker zone, related to the energy of the breaking waves, E_b, and the water depth at breaking, h_b, by

$$u_m = \left[\frac{2E_b}{\rho h_b}\right]^{1/2} \tag{6}$$

The 0.28 coefficient of equation 5 is based on the field data of Komar and Inman (1970).

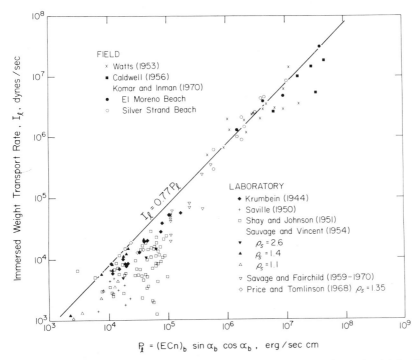

Fig. 1. The immersed weight sand transport rate on beaches, I_l, versus P_l of equation 1 (after Komar and Inman, 1970).

When the longshore current $\bar{v}_l$ is generated by waves breaking at an angle to the shoreline, the current is given by

$$\bar{v}_l = 2.7 u_m \sin \alpha_b \cos \alpha_b \qquad (7)$$

(Komar and Inman, 1970). Equation 7 shows better agreement with the available field and laboratory measurements of longshore currents than any of the other equations that have been proposed for evaluating $\bar{v}_l$. In addition, Longuet–Higgins (1970) provides a derivation of a relationship basically the same as equation 7 through application of the radiation stress, the momentum flux of the waves. Note that when $\bar{v}_l$ of equation 7 is substituted in equation 5, equation 3 results. This demonstrates the fact that equations 3 and 4 apply only when the longshore current and sand transport result from an oblique wave approach of the waves to the shoreline. If the current $\bar{v}_l$ is due to causes other than an oblique wave approach, then equation 5 must be used directly to evaluate the sand transport rate. An example is where the currents are associated with a nearshore cell circulation, the longshore currents feeding rip currents. For application in models, equation 5 can also be converted to yield the volume sand transport rate S_l by using equation 2.

3. The Continuity Equation

As diagrammed in Fig. 2, the shoreline is divided into a series of cells of uniform width Δx and with individual lengths $y_1, \ldots, y_{i-1}, y_i, y_{i+1}, \ldots, y_n$ beyond some base line. The narrower the cells (the smaller Δx), the more nearly the series of cells approximates the true shoreline. Each individual cell is like that depicted in Fig. 3.

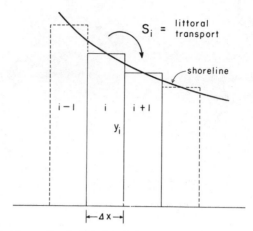

Fig. 2. Shoreline divided into cells of width Δx. Changes in the shoreline are produced by the littoral drift, S_l, from one cell to the next (from Komar, 1973).

Subsequent changes in the shoreline are brought about by littoral drift S_l (m³/day), which shifts sand from one cell to the next. Simple considerations of continuity of sand movement show that ΔV_i, the net change in volume of sand in the ith cell, is given by

$$\Delta V_i = (S_{i-1} - S_i)\Delta t \qquad (8)$$

where S_i is the rate of littoral drift from cell i to cell $i + 1$, and S_{i-1} is the littoral drift into cell i from cell $i - 1$; Δt is an increment of time (days). Note that ΔV_i is positive or negative depending on the relative rates at which sand is transported into and out of the cell. $\Delta V_i = 0$ if the rate into the cell equals the rate out.

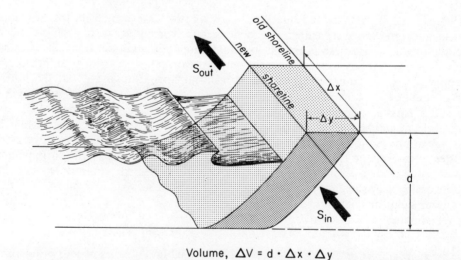

Volume, $\Delta V = d \cdot \Delta x \cdot \Delta y$

Fig. 3. One shoreline cell demonstrating how a change in sand volume ΔV within the cell is produced by the littoral drift in and out of the cell and how this is reflected in a change in the shoreline position Δy (after Komar, 1973).

The parameter ΔV_i must be reflected in a change in the position of the shoreline, that is, a change in the value of y_i. If Δy_i is the change in y_i in the increment of time Δt, then

$$\Delta V_i = d\Delta y_i \Delta x \tag{9}$$

the beach deposition or erosion being depicted as a wedge shown in Fig. 3. The depth d is chosen such that $d \cdot \Delta y_i$ equals the cross-sectional area of littoral sand deposition or erosion, and is approximately equal to the water depth at wave breaking.

Combining equations 8 and 9 gives

$$\Delta y_i = (S_{i-1} - S_i)\frac{\Delta t}{d\Delta x} \tag{10}$$

relating the shoreline position change to the sand transport rate into and out of the individual cell. The parameters Δt, Δx, and d are set within a given model so it remains only to determine values of Δy_i for each cell from evaluation of the littoral drift. Note that in equation 10, Δy_i is positive when $S_{i-1} > S_i$, indicating net deposition, and erosion occurs (Δy_i is negative) when $S_i > S_{i-1}$.

If the cell has other sources of sand supply or losses besides littoral drift, these can easily be included. For example, if the cell is at the mouth of a river supplying sand at the rate S_r (m³/day), then equation 10 can be modified to

$$\Delta y_i = (S_r + S_{i-1} - S_i)\frac{\Delta t}{d\Delta x} \tag{11}$$

Equations 10 and 11 are forms of a continuity equation for beach sand. Decreasing the size of the finite elements to their limits leads to

$$\frac{dy}{dt} = -\frac{1}{d}\frac{dS}{dx} \tag{12}$$

which has the more familiar form of a continuity equation.

In the models it is important that the Δy_i values remain relatively small so that there are no sudden "jumps" in the shoreline configuration. This usually requires that the time increment Δt be kept small. If Δx is decreased, it is necessary that Δt be correspondingly decreased.

In the models, the angle α_i which the shoreline makes with a parallel to the x axis, between the i and $i + 1$ cells, is given by

$$\tan \alpha_i = \frac{y_i - y_{i+1}}{\Delta x} \tag{13}$$

If incoming waves in deep water make an angle α_0 with the x axis direction, then the breaker angle at the shore is $\alpha_b = \alpha_i \pm \alpha_0$, which is most easily obtained with the trigonometric identity

$$\tan \alpha_b = \tan (\alpha_i \pm \alpha_0) = \frac{\tan \alpha_i \pm \tan \alpha_0}{1 + \tan \alpha_i \tan \alpha_0} \tag{14}$$

A certain amount of care must be taken, and sign conventions established for the angles, to obtain the proper breaker angles and transport directions.

4. Example Model: Jetty Blockage of Littoral Drift

Any simulation model of shoreline changes simply involves the following:

1. defining an initial shoreline configuration,
2. establishing the sources of sand to the beach such as rivers and possible losses of sand from the beach,
3. giving offshore wave parameters (height, period, energy flux, approach angle),
4. indicating how littoral transport of sand along the beach is to be governed by the wave parameters; and
5. determining how the shoreline is altered from its initial configuration under these conditions at increments of time Δt for some total span of time.

As an example of this procedure, this section develops a simple model of the blockage of a littoral sand drift under an oblique wave approach by a jetty built transverse to the drift. It examines only the updrift side of the jetty where sand accumulates, not downdrift of the jetty, where erosion occurs.

In the model an initially straight shoreline segment 5 km in length is divided into 200 cells, each 25 m wide ($\Delta x = 25$ m). At one end of the segment is a jetty which blocks the drift so that one boundary condition in the model is that the transport is zero at that end. This means that for cell 200 there is sand transport into the cell but none out. In the models the wave energy flux is set at $(ECn)_b = 1.0 \times 10^8$ erg/cm · sec with a deep-water angle of approach $\alpha_0 = 15°$. This energy flux can of course correspond to an unlimited number of combinations of wave heights and periods; a few within the expected range of periods are shown in Table I. It is seen that this energy flux generally represents relatively small waves.

The oblique wave approach causes a littoral drift into the 5-km segment of beach at a rate of 1712.5 m³/day (from equation 4), and this becomes a second boundary condition. With these two boundary conditions and an initially straight shoreline, the simulated drift blockage by the jetty is run on a time increment of $\Delta t = 0.1$ day for a total time of 365 days. The littoral drift between adjacent cells is computed with equation 4, the breaker angles being obtained from equation 14. The computer

TABLE I

Combinations of Wave
Periods and Heights
that Yield the Energy
Flux $ECn = 1.0 \times 10^8$
erg/cm · sec

Period (sec)	Height (cm)
2	72
4	51
6	42
8	36
10	32
12	30

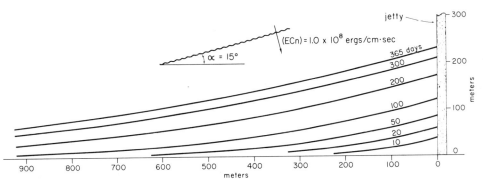

Fig. 4. Simulation model for the blockage of the littoral drift by a jetty. $\Delta x = 25$ m and $\Delta t = 0.1$ day.

program routine prints out the shoreline configuration after each 5-day accumulation so that changes through time can be studied.

Figure 4 shows the resulting simulated accumulation of sand behind the jetty for 10–365 days. The smooth shoreline curves are drawn through the midpoints of the individual cells as illustrated in Fig. 5. This figure also illustrates results using $\Delta x = 50$ m versus $\Delta x = 25$ m. It is apparent that $\Delta x = 25$ m is already much closer to the

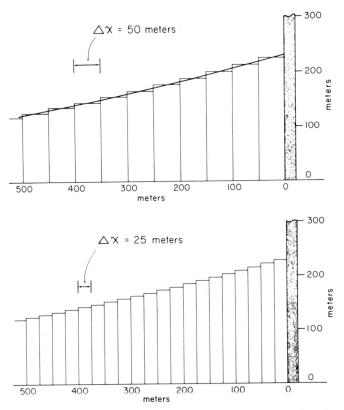

Fig. 5. Differences in the model of Fig. 4 using $\Delta x = 50$ m rather than 25 m. Although $\Delta x = 25$ m gives a closer approximation to the smooth shoreline, in this case little difference is noted. In other models small shoreline irregularities may be important, requiring the smaller Δx value.

smooth-curve version but actually the differences are very small in this example and $\Delta x = 50$ m would have been satisfactory. The advantage of using $\Delta x = 50$ m is that it halves the number of computations from that when $\Delta x = 25$ m, at a corresponding saving in computer time. Smaller Δx values are required in more complicated models, especially where there will be smaller irregularities in the shoreline configuration rather than the simple smooth shorelines seen in Fig. 5.

Figure 6 gives variations in the wave breaker angle along the shoreline with increasing distance from the jetty at 25, 100, and 365 days. It is seen that in each case the breaker angle approaches zero at the jetty position itself; this is because the sand transport must decrease to zero at the jetty and since the energy flux $(ECn)_b$ does not become zero, α_b must become zero in equation 4. With increasing distance from the jetty the littoral drift progressively increases (approaching 1712.5 m³/day far from the jetty). Looking at it in the other direction, the breaker angle α_b decreases as the jetty is approached since sand is deposited along the way and progressively less and less sand remains to be transported. One advantage of models of this sort is that they provide the ability to investigate such systematic changes in the wave breaker angle and littoral drift as the jetty is approached.

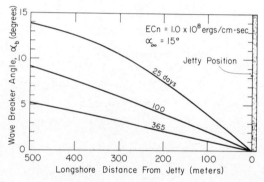

Fig. 6.   Variations in the wave breaker angle along the shoreline with increasing distance from the jetty, obtained from the model of Fig. 4 at 25, 100, and 365 days. Simulation models enable the study of variations of this type.

In simulation models one can also investigate time variations in the shoreline development. For example, one could perhaps answer such questions as how long after jetty construction would the beach noticeably begin to build outward at a point, say, 3 km updrift of the jetty? Answers such as this might have real application to problems involved in jetty construction, although in actual case studies one would have to model time variations in the wave energy flux and direction.

Price, Tomlinson, and Willis (1973) examined changes in a beach brought about by the construction of a long groin or jetty blocking the longshore drift. They developed a numerical model just like that above except that it was on a scale of a wave basin rather than on a prototype scale. Of particular interest is that they then compared the results of their computer model with actual tests within a wave basin. The results are shown in Fig. 7, and it is seen that the computer simulated model compares favorably with the actual shoreline changes experienced in a laboratory wave basin. This indicates both that the numerical approach is valid and that equation 3, which they used to evaluate the littoral drift, is correct.

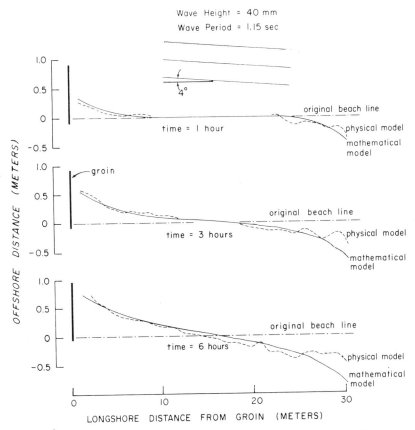

Fig. 7.   Computer simulation model of sand blockage by a groin, compared with a physical model in a laboratory wave basin (after Price, Tomlinson, and Willis, 1973).

5. Example with a Sediment Source: the River Delta

In a previous paper (Komar, 1973), I studied the equilibrium configurations of deltas through application of computer models. Most of the examples presented in this section are derived from that paper.

When a stream or river enters a large body of water the sediments it is transporting tend to be deposited in the form of a delta. The delta growth is enhanced by the quantity of sand carried by the river; opposing the delta growth is the wave action and resulting littoral drift which acts to redistribute the sediment along the coast. Simulation models enable us to study the equilibrium delta shape wherein its curvature is adjusted in just such a way that the waves impinging on the shoreline provide precisely the energy and breaker angle required to transport the load of sand supplied to the beach by the river. With such models we can vary the wave energy or river sediment input and see how the equilibrium delta configuration responds.

Figure 8 shows the growth of a simulated delta with a river supplying sand at the rate $S_r = 2 \times 10^4$ m^3/day. The increments used were $\Delta x = 100$ m and $\Delta t = 0.1$ day. Initially there is a rapid build-out of sand near the river mouth but this build-out decreases as the shoreline begins to make an appreciable angle to the incoming waves

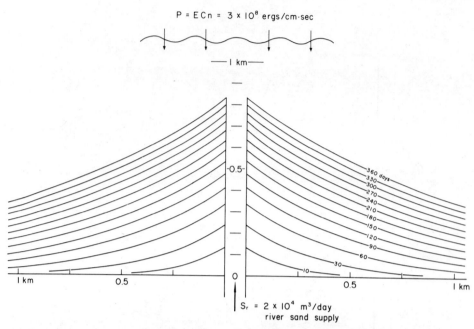

Fig. 8.   Growth of a river delta to equilibrium as simulated on a computer (from Komar, 1973).

so that the waves are better able to remove the sand from near the river mouth. It is seen that the spacings between the successive 30-day shorelines progressively decrease in the first 150–180 days and apparently approach a constant growth rate. This steady-state configuration must be the equilibrium configuration for the delta for the given wave conditions and sediment supply. The breaker angle decreases progressively away from the river mouth. At the river mouth itself $\alpha_b = 38.3°$ which, according to equation 4, is precisely the angle needed for the waves with energy flux $ECn = 3 \times 10^8$ erg/cm · sec to transport all the sand supplied by the river. Deposition occurs along the length of the delta so that the greater the distance from the river, the smaller the quantity of river sand remaining to be transported and the smaller the required breaker angle.

Figure 9 demonstrates the effects of varying wave energy flux. Similar responses could be expected from changing river sediment supply, keeping the waves the same. As expected, the greater the wave energy flux, the flatter the resulting equilibrium shoreline. This result is due to smaller values of the breaker angles required for redistributing the river sediment load.

In my complete paper on delta growth (Komar, 1973), I also considered effects of waves arriving at an angle to the initially straight shoreline producing asymmetric deltas, and the effects of the delta growing out over a bottom sloping seaward. The results of those tests can be found in that paper.

With no sediment sources or losses within an embayment the waves redistribute the sand until $\alpha_b = 0$ everywhere and transport ceases. The shoreline configuration then takes on the shape of the refracted wave crests. This can be shown in simulation models (Komar, 1976) and many actual examples in nature can be found. If the wave approach varies, the beach within the embayment wobbles, attempting to remain with its shoreline parallel to the wave crests. With a river source of sand within

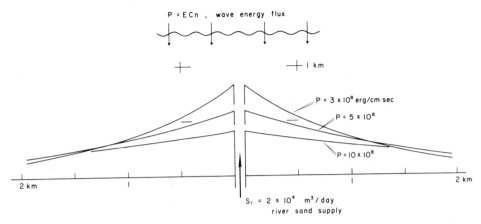

Fig. 9.   Computer simulation models of delta shape for 365 days of growth as influenced by different levels of the wave energy flux, maintaining the same sediment supply from the river. The higher the wave energy flux, the flatter the resulting delta (from Komar, 1973).

the embayment a delta is constructed and the waves now break at angles to the shoreline, necessary for the redistribution of sand throughout the embayment. Figure 10 illustrates with a computer simulation model the growth of a delta within an embayment.

Figure 11 demonstrates one possible type of application of computer models of shoreline changes. In this example a river first enters the ocean with its mouth at the 0 km position, but after 10 yr of delta growth the mouth suddenly shifts to the flank of the original delta. The resulting growth lines are indicated in Fig. 11, and these growth lines can be imagined to represent series of old beach ridges. It is seen that a complex relationship of beach ridges results from this simple shift of the river mouth. The shoreline erodes at the first position of the river mouth after its shift because the waves continue to transport sand alongshore, but the river source has moved. Ridges are truncated while at the same time the delta builds rapidly at the new mouth.

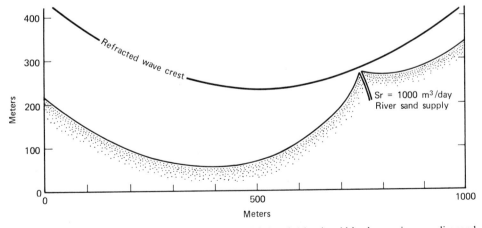

Fig. 10.   Equilibrium shoreline configuration of a pocket beach 1 km in width where a river supplies sand as shown. In this example $ECn = 1.5 \times 10^7$ erg/cm sec (from Komar, 1976).

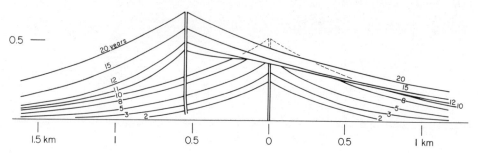

Fig. 11.   Computer simulation of the growth of a delta with the river mouth at 0 km for 10 yr and then a shift in the mouth to the flank of the delta. The growth lines can be viewed as series of beach ridges. The dashed shorelines give the extent of the original delta after 10 yr which was then eroded back when the mouth of the river shifted position.

Much more sand moves alongshore to the left of the new mouth than to the right, owing to the mass of sand around the old river mouth causing reduced breaker angles in that direction. The ridges on the left are therefore more widely spaced than on the right. It is apparent that such computer models can be used to test hypotheses on the sequence of events in the growth of real systems of beach ridges. Not only might the sediment source shift position as in this example, but the wave direction and energy flux could change and alter the patterns of ridges. One intriguing possibility is that the model of Fig. 11 can be run backward as well as forward through time. One could start with the configuration of ridges as shown in Fig. 11 and compute the sand transport along the beach, but have it move backward, that is, the reverse of the actual transport direction. Running time in reverse, the sand would systematically be removed from the delta and beach ridges, and be pushed back into the river. This approach would provide a still more powerful technique for unraveling the history of development of beach ridge groups.

6. Applications of Shoreline Simulation Models

Although the models presented above are relatively simple, they demonstrate the application of computer simulation models to investigations of shoreline configuration under steady input conditions. However, one of the principal advantages of such models is that they allow the parameters to be varied through space and time. For example, rather than using one set of wave conditions as in the previous examples, we could attempt to model the wave climate, having realistic distributions of wave periods, heights, and directions, the model selecting the given set of wave conditions from those distributions. This would combine a probabilistic aspect with our otherwise deterministic model, the result being closer to nature. In our delta models we could equally well vary the river discharge through time and therefore the rate at which sand is added to the beach by the river.

The next major stage in the development of simulation models of shoreline processes would be to couple the models to one of the computer routines (for example, Wilson, 1966) which compute wave refraction patterns. This would involve a two-dimensional array to account for offshore topography and hence would considerably increase the overall complexity of the models. However, such models would be more realistic and onshore–offshore shifts of sand could be included as well as longshore transport, once the processes responsible for these transports are better understood. It is this level of simulation model that would be most applicable to examination of real case studies of shoreline changes, both natural and man-made.

The first application of this sort is that of Motyka and Willis (1975), who modeled the effects of offshore dredging on the wave refraction and in turn the alteration of the shoreline. The results have application to offshore mining of sand and shingle and answer such questions as how deep must the water be and how much material can be safely removed. Figure 12 shows one example of their results, for a 4-m deep hole in 7 m of water, 500 m offshore. The root-mean-square wave height was 0.4 m and periods of 5 and 8 sec were used. Directions were selected to give a net longshore sediment transport of 30,000 m³/yr. The model performed the following sequence of operations:

1. Calculate breaking wave conditions from deep-water wave conditions by refraction over the inshore seabed.
2. Calculate rates of longshore sediment transport on the beach from the breaking wave conditions.
3. Calculate changes in beach plan shape.
4. Distribute accretion and erosion over the inshore seabed.
5. Recalculate the wave refraction and return to 2.

Realistic profiles of the offshore seabed were utilized with a dredged hole inserted. The results in Fig. 12 show that the dredge hole causes erosion of the shoreline in its lee and deposition (shoreline advance) to either side. The pattern is asymmetric owing to the superimposed littoral drift of 30,000 m³/yr and overall oblique wave

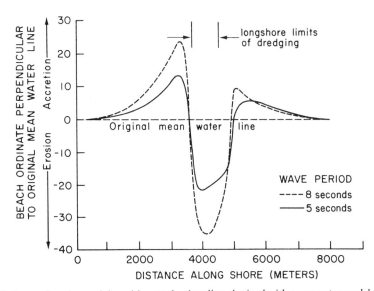

Fig. 12.   Patterns of erosion and deposition on the shoreline obtained with a computer model to simulate the effects of offshore dredging on the wave refraction pattern which in turn causes the beach changes. This represents the first model where wave refraction computer routines are coupled with models of shoreline changes (after Motyka and Willis, 1975).

approach. With results such as those of Fig. 12, Motyka and Willis investigated the importance of dredge-hole geometry (wall steepness, hole relief, diameter) on the resulting beach erosion, the water depth at which the dredging is done, and what wave conditions have the greatest effects.

Other related practical applications might be to investigate the effects of construction of offshore islands (for nuclear power plants or deep-water ship ports), and of a variety of nearshore breakwaters, jetties, or groins. In each case the obstacle affects the wave refraction pattern and creates a wave diffraction zone in its lee. This alteration of the wave patterns may in turn cause changes in the long-term shoreline configuration. With computer models it may be possible to predict what these shoreline changes will be and take measures to prevent property erosion prior to construction. An example of this application is shown in Fig. 13, where shoreline changes result from the construction of jetties on the Siuslaw River, Oregon. The 1889 and 1974 shorelines are real positions, whereas the others have been obtained with a computer simulation model with $\Delta x = 50$ m and $\Delta t = 2$ days. The 1889 shoreline is used in the model as the initial shoreline prior to jetty construction, the starting configuration of the model. The waves are made to approach parallel to the shoreline to the far right of the jetties beyond the effects of the river mouth. Closer to the jetties the waves reach the shoreline with oblique angles and therefore cause a sand transport toward the jetties, filling the "embayment" caused by the jetty construction. Some finite-element cells are eliminated close to the jetty as they fill out to the jetty and can accept no additional sand. The shoreline advance shown in Fig. 13, obtained in the model, agrees quite closely with the actual observed shoreline changes that occurred following jetty construction. The main disagreement results from the jetty construction initially not having kept pace with the shoreline advance so that some sand bypassed the jetty and entered the estuary. Note that the model predicts that not only will there be deposition near the jetty, but there will be erosion of the beach further from the jetty to supply sand to the accreting area. Such a prediction would have been useful at the time the jetties were actually constructed, especially if homes were located in this erosion area. The model makes it possible to predict actual quantities of beach erosion and deposition at various positions along the coast. There is considerable potential for application to future considerations of jetty construction.

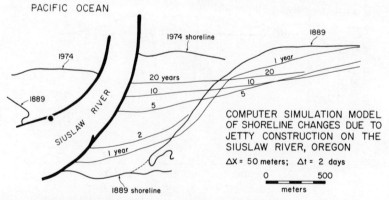

Fig. 13.   Computer simulation model of the shoreline changes that result from the construction of jetties on the Siuslaw River, Oregon. The 1889 shoreline, which is an actual shoreline, was used as the starting configuration for the model. The shoreline positions shown agree reasonably well with the actual changes that occurred due to the jetty construction. In this model $\Delta x = 50$ m and $\Delta t = 2$ days.

7. Summary

Computer simulation models of sediment transport along the beach and the resulting shoreline changes are in an infant stage of development. They have been previously used to examine geomorphological problems concerning the equilibrium shoreline configuration; for example, the equilibrium shape of a delta where the waves are in balance with the sediment supplied by the river and so are able to redistribute the sediment alongshore while maintaining the delta shape. Such examinations are oversimplified when compared with nature, necessarily so if we are to study piecemeal the several factors affecting the shoreline configuration.

When applied to actual case studies, the simulation models must be more realistic and include wave refraction, wave climate (variability in wave height, period, and direction), and such things as changes in the river discharge and sediment load through time. Only the study of Motyka and Willis (1975), applied to the effects of offshore dredging, attempts such a realistic model. However, the potential for such models is considerable and many examples can be expected in the near future.

References

Bagnold, R. A., 1963. Mechanics of marine sedimentation. In *The Sea*. Vol. 3. M. N. Hill, ed. Wiley-Interscience, New York, pp. 507–528.

Harbaugh, J. W. and G. Bonham-Carter, 1970. *Computer Simulation in Geology*. Wiley-Interscience, New York, 575 pp.

Inman, D. L., and R. A. Bagnold, 1963. Littoral processes. In *The Sea*. Vol. 3. M. N. Hill, ed. Wiley-Interscience, New York, pp. 529–553.

Komar, P. D., 1973. Computer models of delta growth due to sediment input from rivers and longshore transport. *Geol. Soc. Am. Bull.*, **84**, 2217–2226.

Komar, P. D., 1976. *Beach Processes and Sedimentation*. Prentice-Hall, Englewood Cliffs, N.J., 429 pp.

Komar, P. D. and D. L. Inman, 1970. Longshore sand transport on beaches. *J. Geophys. Res.* **75** (30), 5914–5927.

Longuet-Higgins, M. S., 1970. Longshore currents generated by obliquely incident sea waves, 1. *J. Geophys. Res.*, **75** (33), 6778–6789.

Motyka, J. M. and D. H. Willis, 1975. The effect of refraction over dredged holes. In *Proc. 14th Conf. Coastal Eng.*, pp. 615–625.

Price, W. A., K. W. Tomlinson, and D. H. Willis, 1973. Predicting the changes in the plan shape of beaches. In *Proc. 13th Conf. Coastal Eng.*, pp. 1321–1329.

Wilson, W. S., 1966. A method for calculating and plotting surface wave rays. *U.S. Army Corps Eng. Coastal Eng. Res. Cent. Tech. Memo. 17*, 57 pp.

12. PROBLEMS IN THE MODELING OF TRANSPORT, EROSION, AND DEPOSITION OF COHESIVE SEDIMENTS

M. W. OWEN

1. Introduction

Most of the estuaries and major tidal rivers of the world contain considerable quantities of cohesive silts and muds, which form the seabed and are transported to and fro by the tidal currents. These estuaries very often form important trade highways to the sea, and it is desirable to be able to predict the depths of water for the navigation of the trading vessels, and to forecast any changes that may occur as a result of schemes to develop the estuary, such as the dredging of new shipping channels. The various docks, harbors, and riverside berths necessary for the transfer of cargoes and passengers are generally designed to have comparatively slack water, and thus experience deposition of cohesive sediments. It is necessary to be able to predict the rates of accretion and consequent maintenance dredging in these areas, and to suggest ways in which conditions might be improved. The estuaries are also often the centers of industry and commerce, which use the tidal waters as a convenient sink for waste materials of all sorts. Some of these pollutants, particularly organics and heavy metals, may react with the cohesive silts and muds, which serve as transporting, storing, and catalytic agents that may either expand or reduce the severity of the pollution.

A proportion of the cohesive sediment, together with any organisms, organic materials, or pollutants which have become attached, passes out of the estuary system into the ocean, gradually moving toward hollows in the topography of the seabed, where it accumulates, consolidates, and undergoes various chemical and organic changes. Eventually it becomes the sedimentary rock which later marine geologists will examine with interest to try to determine the nature of the environment at the time of its deposition.

From the foregoing it can be seen that it is very important to be able to model the behavior of these cohesive sediments, both as a tool for the better management of our present-day estuaries, and to give an understanding of the factors governing the formation of sedimentary rocks on the ocean floor. After examining the various processes involved in the behavior of these cohesive sediments, this chapter describes some of the problems involved in modeling, and goes on to discuss recent attempts to develop working mathematical models of the transport of cohesive sediments in estuaries.

2. Basic Properties of Cohesive Sediments

Although silts and muds can have a median particle diameter of up to about 40 μm, their cohesion depends primarily on the presence of a significant proportion of very fine particles, with sizes typically less than about 2 μm. These cohesive particles are formed of the clay minerals, usually kaolinite, illite, chlorite, and montmorillonite. Detailed discussion of the structure and behavior of these clay minerals can be found elsewhere (e.g., Grim, 1959); a simplified description is sufficient here. Each of the clay minerals has a layered structure, and forms flaky, platelike crystals which carry negative charges around their edges. As the particles become smaller, the perimeter of the crystals becomes proportionally greater, so that the charge per unit mass of clay increases. In a solution containing a low concentration of cations, the clay particles repel each other; with high cation concentrations the charges on the clay particles are neutralized, and if the particles are sufficiently close together they are held by the electronic bonding forces between the various atoms in each crystal. This simplified system is complicated by the fact that some of the clay minerals, particularly montmorillonite, can absorb some of the ions from solution into the basic layered structure of the clay, causing the clay crystal to adjust its size and surface charge. The cohesion of a clay thus depends on the type of clay mineral present, on the particle sizes, and on the quantity and type of cations present in solution.

Two different types of cohesion of fine particles occur in the presence of organic material or of marine organisms. Particles of organic material, themselves electrically charged, can act as nuclei, and attract crystals of clay minerals, forming clay–organic–clay particles. In addition certain marine organisms can bind fine particles together with the mucus they secrete. Neither of these processes can be readily quantified, however, and they are generally ignored in attempts at modeling the behavior of cohesive sediments.

The distribution of the clay minerals varies considerably from estuary to estuary, depending largely on the type of rock and the amount of rainfall in the catchment area of the river that enters at the head of that estuary (Ehlmann, 1968). Within a given estuary, however, there is very little variation in the relative proportions of the clay minerals present (Biddle and Miles, 1972).

In order to model changes in cohesion in a particular estuary, therefore, the clay mineralogy need not be considered, and attention is given instead to predicting changes in the concentration of cations in solution, which in estuaries generally means predicting changes in salinity.

3. Movement Under Tidal Currents

The movement of cohesive sediments under tidal currents can generally be considered as a cycle of four processes: erosion from the bed by high velocities, transport by the tidal flow, deposition to the bed about slack water, and consolidation of the deposited bed. Each of these processes needs to be understood before accurate modeling can commence. However, despite the efforts of several research workers, a considerable amount of research is still needed before a thorough knowledge of the factors governing each process is obtained.

A. Erosion

Cohesive sediments are eroded from the bed whenever the shear stress, τ, exerted on the bed by the flowing water exceeds a certain value, known as the critical shear stress for erosion, τ_e, which is sufficient to break the cohesive bonding of the particles. However, there is no way as yet of predicting the cohesive strength of the bed surface. It seems reasonable to assume that it depends on the basic cohesion of the sediment, in other words on the clay mineralogy and the cations in solution. But does it also depend on the state of the sediment, such as its compacted density, or the arrangement of the sediment particles within the bed? For recently deposited sediments, such as those found in estuaries, Partheniades (1965) concluded that the critical shear stress for erosion and the rate at which erosion occurred were independent of the overall density of the sediment. However, for more compacted sediments, such as those found in riverbanks, Carlson and Enger (1962) found a statistically significant correlation between the critical shear stress for erosion and the density of the sediment, expressed as a percentage of the maximum Proctor density. The results of Krone (1962) indicated that the cohesive strength of the surface layer of recently deposited sediments depends on the structure of the deposited bed. In particular he found that for San Francisco Bay muds, there were five distinct structures, each with its own value of critical shear stress. These structures existed in layers within the bed, with the weakest at the surface, but each layer was comparatively thin, and the basic or primary structure was only a few centimeters below the bed surface. The cohesive strength of this primary structure was very similar in value to that measured by Partheniades (1965).

B. Transport

Once eroded from the bed, the cohesive sediments are transported along the estuary by the tidal currents. The distribution of the sediment within the flow depends on the relative magnitudes of the turbulence of the flow and of the settling velocity of the sediment (Camp, 1943). With proportionally high settling velocities the sediment tends to travel along near the bed, whereas with lower settling velocities the sediment is mixed in suspension throughout the depth of flow. However, the settling velocity of a cohesive sediment is not a constant parameter, since single particles very rarely exist in the flow. Because of the cohesive forces between clay particles, any particles that collide in the flow adhere to each other to form flocs of widely differing size and density. Since settling velocity is a function of both size and density of the flocs, the settling velocity itself changes as the process of flocculation proceeds. The cohesive forces binding the particles together were discussed earlier. The collisions necessary to bring two or more particles close enough for these forces to act can be the result of Brownian motion, differential shear of the flow, or differential settling of the particles in suspension. The frequencies of collisions due to Brownian motion and differential settling depend on the temperature of the suspension, the suspended concentration, and the particle density. The frequency of collision due to the internal shearing of the flow depends on the particle concentration, the particle volume, and the local velocity gradient. In estuarine systems the velocity gradients are usually considerable, and flocculation occurs primarily as a result of this internal shearing. In oceanic systems, however, flocculation due to Brownian motion and differential settling is more significant. Flocs cannot grow in size indefinitely, being limited by the maximum fluid shear stress they can withstand. Krone (1963) gives the relationship between maximum floc size and shear stress in laminar flow as

$$R_{max} = \frac{\tau_{max} \Delta R}{du/dz}$$

where R_{max} = maximum floc radius
ΔR = surface roughness of the floc
τ_{max} = maximum bonding shear strength
$\dfrac{du}{dz}$ = local velocity gradient

Although the main factors governing the flocculation process are fairly well known and understood, it is still not possible to predict the size, density, and hence settling velocity of the resulting flocs. For this reason, attempts have been made to measure settling velocities directly in the laboratory (Krone, 1962; McLaughlin, 1959; Pierce and Williams, 1966; Migniot, 1968). However, each of these studies, although able to reproduce salinity, temperature, and suspended concentration, could not adequately represent the structure of turbulence found in estuarine and deep-sea situations, where velocity gradients have been shown to be an important factor in the flocculation process, in terms of causing interparticle collisions and of limiting the maximum size of the resulting flocs. For this reason Owen (1971) developed a method of measuring in the field the settling velocities of the flocs in their actual state as they exist in estuaries, and thus including the effects not only of the natural values of concentration, salinity, and temperature, but also of the turbulent structure of the flow. The results showed that settling velocity is very strongly dependent on the suspended concentration and turbulence, and less dependent on variations in salinity. Settling velocities can typically vary from about 1 μm/sec at very low concentrations to about 10 mm/sec at higher concentrations compared with the value of about 8 mm/sec found for fine sand.

C. Deposition

The deposition of cohesive sediments at exactly slack water is comparatively easy to predict, since the rate of deposition is simply the product of the suspended concentration and settling velocity of the sediment flocs. Deposition in slowly moving water is more difficult to determine, however. The cohesive sediment continues to settle toward the bed owing to its settling velocity, but can adhere to the bed only if the shear stress exerted by the flow is lower than the initial bonding strength of the flocs to the bed surface. The shear stress below which deposition can proceed is known as the limiting shear stress for deposition, τ_d. Einstein and Krone (1962) showed by laboratory experiments that once shear stresses are below this value, the rate of deposition is proportional to the ratio $(1 - \tau/\tau_d)$ where τ is the actual shear stress at the bed.

As is the case for the critical shear stress for erosion, there is no known way of predicting the value of the limiting shear stress for deposition of cohesive sediments, and measurements have to be carried out for each particular silt or mud. In almost all cases studied so far the value of τ_d is significantly smaller than τ_e, so that theoretically there exists a range of shear stress $\tau_d < \tau < \tau_e$ when neither erosion nor deposition of cohesive sediment can occur. However, there is considerable debate as to whether or not this situation actually exists. Krone (1962) deduced from laboratory experiments that simultaneous deposition and erosion can occur, and argued that this was inevitable given the statistical distribution of the bonding strengths of clay particles. To some extent the work of Partheniades and Kennedy (1966) supports the simultaneous existence of deposition and erosion. Their results showed that deposition does not continue indefinitely, but ceases at some equilibrium concentration which, for a given bed shear stress, is a fixed proportion of the initial concentration. It has not yet been possible, however, to find a convincing reason for this to occur.

D. Consolidation

During the formation of a bed of cohesive sediment flocs are continually being deposited on the surface of the bed while the buried flocs are consolidating, so that there exists a profile of density and strength within the bed. The cohesive sediment continues to consolidate for a considerable length of time before achieving its ultimate density. Various authors (Krone, 1962; Migniot, 1968; Owen, 1970) have shown that consolidation occurs in fairly distinct phases, during each of which the mean bed density is proportional to the logarithm of time. In the initial phase, lasting perhaps for about 10 hr or so, densities increase rapidly as the water of the bed escapes through the interstices of the deposited flocs. During the second phase consolidation proceeds more slowly as the water escapes by percolation through the bed or by drainage wells. This second phase lasts perhaps 500 hr, and is followed by very slow consolidation during the third phase. During each of these phases the profile of density and strength within the bed undergoes slow changes, with wide variations in density from the surface to the bed of the deposited sediment at the beginning of each phase, and near-uniform densities and strengths throughout the depth towards the end of each phase. The existence of these profiles, particularly those of shear strength, can be particularly important in a situation where rapid erosion is occurring, since the erosion can proceed only down to a level in the bed where the cohesive strength of the sediment can withstand the applied shear stress of the flow.

During the formation and initial stages of the consolidation of a bed of cohesive sediment the shear strength of the sediment is sufficiently low for the mass of sediment to flow as a plastic fluid along quite shallow slopes. In estuaries with high suspended concentrations of cohesive sediment, the rate of deposition at near slack water is considerably higher than the rate at which consolidation can proceed, and layers up to several meters thick of this low-density mud are formed, variously known as fluid mud, sling mud, liquid mud, and other local names. These fluid mud layers cannot be transported by tidal currents, since because of the profile of shear strength within the layer a higher shear stress would be needed at the surface to cause movement than would be needed to cause erosion of the bed: erosion would thus occur before the bulk movement of the layer. However, these layers can flow under gravity in a fashion rather similar to density currents, although for each mud there is a critical slope necessary for movement, of the order of 1 : 500 to 1 : 100, depending on the density of the fluid mud. The movement of this fluid mud is thus more of a function of bed topography than of the actual tidal currents.

4. Modeling of the Movements of Sediments

In the preceding section dealing with the various processes and the behavior of cohesive sediments in tidal situations, it was seen that for a given sediment the erosion and deposition depend on the shear stress applied by the flow and on the salinity of the water; the transport in suspension depends on the flow velocity and turbulence and on the sediment settling velocity; and the sediment settling velocity is itself a function of turbulence, suspended matter concentration, and salinity. In order to model the transport of cohesive sediments, therefore, it is necessary to reproduce not only the actual processes of erosion, transport, deposition, and consolidation, but also changes in the various factors that govern them. In an estuary, for instance, it is necessary to reproduce the tidal propagation and the movement of the dissolved salt in some detail. However, the motion of the water, salt, and sediment are interactive: for instance, not only does the salinity affect the critical shear stress for erosion, it also affects the distribution of velocities in the flow, leading in extreme cases to complete stratification. Similarly high velocities cause erosion of the bed which deepens the channel and changes the tidal velocities. Odd and Owen (1972) presented a rather complicated diagram (Fig. 1), which shows all the known possible interactions between the motion of the water, dissolved salts, and suspended sediment in a muddy estuary. In this diagram the external forces governing the tidal system are enclosed by triangles: the main factors causing the periodic motion and changes in the various

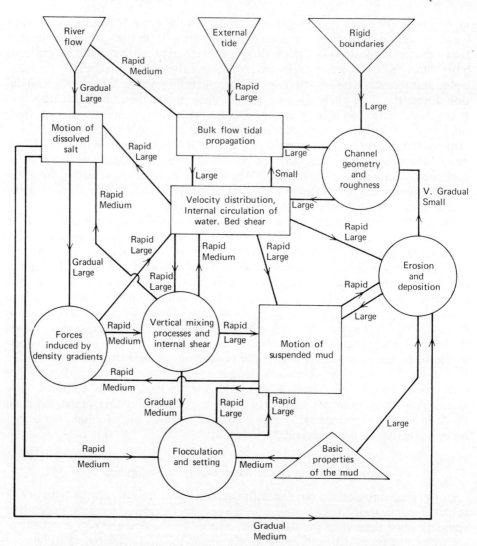

Fig. 1. Interaction of tidal processes in an estuary with a muddy environment.

quantities are the ocean tide at the mouth of the estuary and the river flow at its head. The major processes of the motion of the water, the dissolved salt, and the suspended mud are represented by the rectangular boxes: the motion of the water is divided into two parts, (1) the bulk or mean properties of the tidal propagation, such as tidal elevations or average velocities, together with (2) the detailed structure of the flow, such as velocity and shear stress gradients, internal circulations, etc. The factors enclosed in circles are various secondary parameters and processes that occur simultaneously with and contribute to the major processes.

The interactions between the various tidal processes and other parameters included in the diagram are not of equal importance, nor do they all act within the same time scale. For instance, the river flow at the head of an estuary can have a large effect on the motion of the dissolved salt, but changes in river flow would take several tides to significantly alter the salinity distribution. The interaction between river flow and motion of dissolved salt is therefore large but gradual. The effect of river flow on bulk flow tidal propagation is medium but rapid: tide levels and mean velocities are changed only by a fairly small amount, but the effect is felt almost immediately.

Figure 1 is drawn for an estuarine situation; for a deep-sea or coastal situation most of the processes and parameters still apply but the scale and response time of the interactions are altered. The discharge of the various rivers feeding the ocean would have virtually no effect, for instance, on the salinity or the tidal flow propagation. Also the geometry and roughness of the ocean flow would have only a medium effect on the tidal propagation.

Altogether there are about 25 different interactions shown in Fig. 1: those directly affecting the movement of the cohesive sediments have been referred to in the preceding section, and it would be out of place to discuss each of the others at present. However, it will be appreciated that the ideal model of the movement of the cohesive sediment would reproduce accurately each process mentioned in this diagram, and all the various interactions between these processes. Needless to say the modeling of cohesive sediments is still a long way from perfection!

A. Physical Modeling

Prior to the introduction of modern, fast computers the only possible way of attempting to model the movement of cohesive sediments was by means of physical scale models. Physical models can be designed and operated in such a way that good similarity is obtained between the model and nature in terms of bulk flow tidal propagation, velocity distribution within the flow, shear stresses at the bed, and motion of dissolved salt, although mixing processes and internal shear stresses are not so well reproduced. In addition, attempts to model cohesive sediments were encouraged by the fact that reasonable success had been achieved on many occasions by the use of scale models of rivers and estuaries containing noncohesive sediments, principally fine to medium sands. However, the processes of movement, and hence the problems of modeling the movement of noncohesive sediments, are considerably different from those of cohesive sediments. Sands begin to move at a certain critical shear stress, which is mainly a function of their particle size and density (Shields, 1936) but movement is initially as bed load, with particles rolling or bouncing along the bed. As the shear stress increases ripples are formed, and only at considerably higher velocities does a significant proportion of the total transport occur owing to sand particles in suspension. Deposition of sand onto the bed is mainly caused by the cessation of movement, rather than a separate process. In very many situations, therefore, in order to model the movement of sands in an estuary it is only necessary to reproduce to scale the correct critical shear stress: all other parameters are of secondary importance, serving mainly to determine the time scale of the sand movement rather than the basic patterns or changes occurring in the estuary.

However, as we have seen, the processes of movement of cohesive sediments are considerably more complicated. The critical shear stress for erosion is significantly greater than the limiting shear stress below which deposition occurs. The sediment is transported by the tidal currents almost entirely in suspension, so that the settling velocity of the flocs must be adequately reproduced, bearing in mind that the settling velocity can itself vary. At each position the balance between erosion on one part of the tidal cycle and deposition on another must be approximately correct. Given that our knowledge of the details of the various processes involved is still very scanty, it seems an impossible task to correctly scale simultaneously each of the important parameters. Nevertheless attempts have been made to do so, notably by Orgeron (1968), who reproduced the behavior of cohesive sediments in a scale model of the Vilaine Estuary.

It is not exactly clear how the correct model sediment was chosen, but it appears that emphasis was placed mainly on obtaining the correct settling velocity. The physical scales of the model were 1 : 750 horizontally and 1 : 100 vertically, so that the settling velocity scale works out as 1 : 1.33. To this end the settling velocity of Vilaine mud at a typical suspended concentration and salinity was measured in the laboratory, and another natural mud, from an unnamed source, was chosen which could give the correctly scaled equivalent velocity. To achieve this settling velocity, however, the natural mud was placed in the model at the very high suspended concentration of 60,000 mg/l, and the model water was dosed with a flocculating agent, sodium pyrophosphate, at a rate of up to 1000 mg/l. Evidently the correct settling velocities were obtained, but although it was stated that the model sediment should also be able to reproduce the deposition, consolidation, and erosion of the estuary sediment, no indication is given as to whether or not this was actually achieved. Nevertheless the model gave reasonable agreement in terms of patterns of sediment movement with those actually found in the Vilaine estuary.

Other investigators have carried out studies, not so well documented, using non-cohesive model sediments to reproduce cohesive estuary materials. In certain circumstances this approach can be reasonably successful, particularly when the movement of the sediment is dominated by one process. In a situation where erosion predominates, for example, such as scour downstream of a tidal structure, then modeling can proceed reasonably satisfactorily on the basis of choosing a cohesionless model sediment having a critical shear stress correctly scaled to the critical shear stress for erosion of the natural silt or mud. Similarly for a deposition dominated situation, such as siltation in a harbor, the model should attempt to reproduce the limiting shear stress for deposition, and if possible the typical settling velocities also.

Physical models have the advantage, therefore, that they can be designed to reproduce the tidal flow and salinity variations reasonably accurately. In certain circumstances, particularly in situations where a single process is dominant, they can give fairly satisfactory and useful results of the movement of cohesive sediments. Where all processes of movement are equally important, however, the difficulties of correctly modeling the cohesive sediment are immense.

B. Mathematical Modeling

The use of mathematical or numerical modeling techniques has made a significant contribution to the advancement of knowledge in almost all research areas in the past few years. Even where the basic theory was well-known, the calculations necessary to predict a given process were frequently extremely time-consuming, often to the extent that they were never attempted. With the development of high-speed computers this situation has changed dramatically. This is no less true for hydraulic modeling where numerical models are now applied to rivers, estuaries, coasts, and oceans, to study tidal propagation, flood routing, thermal pollution, siltation, littoral drift, harbor design, wave refraction, etc., in some cases merely replacing at less cost techniques already in existence, such as physical modeling of tidal propagation, and in other instances opening up entirely new approaches to these problems.

As mentioned in the preceding section, attempts to represent the movement of cohesive sediment in physical models for other than simple situations have achieved only limited success, and it is doubtful if much more progress could be made in this direction, because of the difficulties of producing a suitable model sediment. With mathematical models, however, virtually any process can be reproduced, provided that the basic factors governing that process are known, and that equations describing it can be formulated: a mathematical or numerical model can only be as accurate as the basic knowledge of the process being reproduced. As we have seen, there is still a great deal of uncertainty surrounding our knowledge of the erosion of cohesive sediments, of their settling velocities, of the deposition process, and of the consolidation of the deposited bed. The mathematical modeling of these processes must therefore proceed in the knowledge that our understanding of cohesive sediments is far from complete. In addition, the advent of mathematical techniques in hydraulic modeling has highlighted gaps in our present knowledge of the details of the flow of water. In physical models one could reasonably assume that, although our understanding of vertical mixing and factors governing velocity gradients was not complete, a correct choice of scales would almost always lead to satisfactory agreement between model and natural conditions: the model in effect obeys the same laws as the natural flow, whatever those laws might be. In a mathematical model, however, it is necessary to know the details of all processes, and to formulate laws governing them. A substantial amount of research is therefore still needed to fully understand each of the various processes shown in Fig. 1, particularly on the mixing processes for momentum, salinity, and suspended concentrations in tidal flows, the processes of cohesive sediment movement, and also the interactions between all these processes. In the meantime mathematical modeling has to proceed on the basis of the best information available, and the success or failure of the model will depend almost entirely on the assumptions made in the descriptions of those processes that are not fully understood, and the relevance of those assumptions to the particular situation being studied. The fact that very few mathematical models of cohesive sediment movement have been reported in the literature is perhaps an indication of the difficulties of arriving at suitable methods of reproducing little understood processes.

Estuary Models

The earliest reference to a fairly comprehensive and reasonably successful mathematical model of cohesive sediment movement in a tidal situation was a paper by Owen and Odd (1970) describing a model of a well mixed estuary having a rectangular cross section, which they applied to the Thames Estuary, England.

Flow velocities and suspended sediment concentrations in the Thames, as in most estuaries, change significantly with depth, with the most rapid variation near the bed. Since these changes play an important role in the movement and longitudinal distribution of suspended sediment, it was necessary to try to simulate these variations. However, because the mixing processes and the turbulent stresses within the flow are not yet fully understood, a major simplification was made, and the flow of water was divided into two unequal horizontal layers, within each of which it was assumed that all the properties were uniformly distributed with depth at their respective depth-averaged values. The lower layer adjacent to the bed has a constant thickness and was considered as a boundary region; in well-mixed estuaries the velocity in this layer increased from zero at the bed to almost its mainstream value, and the concentration of suspended sediment decreased from a very high value near the bed to its average value. The remaining depth of flow—the upper layer—varied with the tide. The model was essentially divided into three submodels. Firstly the bulk flow tidal propagation was calculated; then the flow of water in the lower of the two horizontal layers was calculated, and the flow in the upper layer deduced from the bulk flow; finally the sediment movement was calculated using the velocities and shear stresses obtained from the two-layer flow calculations. The motion of the dissolved salt was not modeled per se, although a mean density gradient due to longitudinal salinity variations was included in the calculations of the distribution of flow between the two layers. By dividing the numerical model into these three compartments and by omitting the tidal motion of the dissolved salt, several of the interactions shown in Fig. 1 were absent. For instance, the model effectively assumed no effect of internal circulation on water bulk flow tidal propagation, which is probably not radically in error for well mixed estuaries. Secondly it was assumed that changes in bed level due to erosion or deposition had no effect on channel geometry and thus no effect on the tidal propagation.

Because of the comparatively simple geometry of the Thames, in all the calculations the assumption was made of a rectangular channel of varying width and depth. The bulk flow tidal propagation was calculated from the classical equations of motion and continuity, which for well-mixed flow in a rectangular channel become

$$\frac{1}{g}\frac{\partial U}{\partial t} + \frac{\partial h}{\partial x} + \frac{d}{2\rho}\frac{\partial \rho}{\partial x} + \frac{f|U|U}{8gd} = 0 \tag{1}$$

and

$$\frac{\partial}{\partial x}[b(x)\,dU] + b(x)\frac{\partial h}{\partial t} = 0 \tag{2}$$

where U = depth-averaged flow velocity
h = water surface elevation
d = total depth of water
ρ = water density
$\dfrac{\partial \rho}{\partial x}$ = mean longitudinal density gradient throughout the depth
f = Darcy–Weisbach friction factor
$b(x)$ = channel width, a function of distance along the estuary

The derivation of the equations of motion for flow in the lower of the two horizontal layers is given in detail in a later reference to the same model (Odd and Owen, 1972). Even with the simplification of dividing the flow into two layers several assumptions were necessary, principally that the instantaneous velocity profile approaches the logarithmic profile of steady flow, so that the internal shear stress increased linearly with depth. Again this is probably a reasonable assumption for well-mixed estuaries, except near slack water when friction terms in the equation of motion are comparatively small anyway. With these assumptions the equations of motion and continuity were as follows:

$$\frac{1}{g}\frac{\partial U_L}{\partial t} + \frac{\partial h}{\partial x} + \frac{d}{\rho}\left(\frac{\partial \rho}{\partial x}\right)_{\text{bed}} + \frac{f_L|U_L|U_L}{8gd} = 0 \tag{3}$$

$$d_L\frac{\partial}{\partial x}[U_L b(x)] + b(x)V = 0 \tag{4}$$

where suffix L denotes the depth-mean value in the lower layer, and V is the vertical velocity between the upper and lower layers.

The friction factors are a function of the flow Reynolds number and were calculated from the Colebrooke–White equation. From equations 1 and 2, the values of $\partial h/\partial x$ and d were calculated and used in equations 3 and 4 to calculate U_L and V. The mean velocity in the upper layer, U_U, was then calculated to satisfy the continuity of the bulk flow.

The movement of the cohesive sediment in suspension must satisfy the mass balance equation. In other words, in the bulk flow situation the difference between the total quantity of sediment entering and leaving a particular length of channel must equal the change in the quantity stored in that length plus or minus sediment which has been deposited onto or eroded from the bed. Where the flow is considered as two layers there is an additional term to consider: the transport of suspended mud between these hypothetical layers. For the two-layer situation, therefore, the equations of mass balance for a rectangular channel are given as follows, for the lower and upper layers, respectively:

$$\frac{\partial}{\partial t}(bd_L C_L) + \alpha_L \frac{\partial}{\partial x}(U_L b_L C_L) = b\left[\left(\frac{dm}{dt}\right)_e - \left(\frac{dm}{dt}\right)_d - \left(\frac{dm}{dt}\right)_i\right] \tag{5}$$

$$\frac{\partial}{\partial t}(bd_U C_U) + \alpha_U \frac{\partial}{\partial x}(U_U b_U C_U) = b\left(\frac{dm}{dt}\right)_i \tag{6}$$

C_L and C_U are the depth-averaged concentrations in the lower and upper layers, respectively, and the factors α_U and α_L are coefficients to correct for the effects of using the product of the mean values of velocity and concentration instead of the mean of the products. It should be noted that these equations contain the additional assumption that the sediment is transported only in suspension, and that it travels along with and at the same velocity as the water.

The terms on the right side of equation 5 represent respectively the rate of erosion (per unit area) from the bed, the rate of deposition onto the bed, and the rate of mass transfer across the hypothetical interface between the two layers. The evaluation of these three terms obviously plays a considerable part in the correct reproduction of the movement of the sediment, but none of the processes involved is yet completely understood, and several different expressions could presumably have been adopted to describe them. In this particular model, Odd and Owen used simple expressions based on the work of Partheniades (1965) and Krone (1962). Erosion occurred whenever the applied shear stress τ was greater than the critical shear stress τ_e, and was given by the expression

$$\left(\frac{dm}{dt}\right)_e = M\left(\frac{\tau}{\tau_e} - 1\right) \tag{7}$$

where M is a fixed constant. Similarly deposition was allowed to occur only when $\tau < \tau_d$, the limiting shear stress for deposition, and was given by

$$\left(\frac{dm}{dt}\right)_d = C_L V_S\left(1 - \frac{\tau}{\tau_d}\right) \tag{8}$$

where V_S is the settling velocity of the suspended sediment evaluated at the relevant concentration, salinity, and turbulence level.

The use of these two expressions denies the simultaneous existence of deposition and erosion at the bed. Since $\tau_d < \tau_e$, it also gives rise to periods during the tide, and $\tau_d < \tau < \tau_e$ when the bed plays no active part in the sediment movement.

The general expression governing the rate of mass transfer across a horizontal plane in the flow is given by

$$\left(\frac{dm}{dt}\right)_i = V_i C_i - V_S C_i + \varepsilon_i \left(\frac{\partial C}{\partial Z}\right)_i \qquad (9)$$

where V_i is the vertical exchange of water at the interface, C_i the local concentration, V_S the local settling velocity, ε_i the local coefficient of vertical mixing, and $(\partial C/\partial Z)_i$ the local vertical concentration gradient. In the case of mass transfer between two hypothetical well-mixed layers, various methods of evaluating this generalized expression could be used, and Owen and Odd suggested the equation

$$\left(\frac{dm}{dt}\right)_i = V[H\{-V\}C_U + H\{V\}C_L] - V_S C_U + \frac{2\varepsilon}{d}(C_L - C_U) \qquad (10)$$

where the Heaviside function $H\{V\}$ has the values

$$H\{V\} = 1 \text{ when } V \text{ is positive (flow upward)}$$
$$H\{V\} = 0 \text{ when } V \text{ is negative (flow downward)}$$

and V_S is evaluated at concentration C_U.

The coefficient of vertical mixing in fact varies throughout the depth, and in homogeneous flow the theoretical mean depth value is $U_* d/15$, where U_* is the shear velocity at the bed $[U_* = (\tau/\rho)^{1/2}]$. However, the presence of even small degrees of stratification or vertical density gradients can significantly reduce the vertical mixing, and the value used in this model was given as $EU_* d$, where E was a constant determined by trial and error during the initial operation of the model.

The mass balance equations 5 and 6 were combined with the equations of continuity of water in each layer,

$$\frac{\partial}{\partial x}(U_L b d_L) = -bV \qquad (11)$$

$$\frac{\partial}{\partial x}(U_U b d_U) = -b\left(\frac{\partial d_U}{\partial t} - V\right) \qquad (12)$$

to give the following simplified equations governing the movement of sediment in the model:

$$d_L \frac{\partial C_L}{\partial t} + \alpha_L U_L d_L \frac{\partial C_L}{\partial x} = \alpha_L D_L V + \left(\frac{dm}{dt}\right)_e - \left(\frac{dm}{dt}\right)_d - \left(\frac{dm}{dt}\right)_i \qquad (13)$$

$$d_U \frac{\partial C_U}{\partial t} + \alpha_U U_U d_U \frac{\partial C_U}{\partial x} = -\alpha_U C_U V + \left(\frac{dm}{dt}\right)_i + (\alpha_U - 1)\frac{\partial d_U}{\partial t} \qquad (14)$$

The operation of the mathematical model thus reduces to the solution of three pairs of simultaneous equations, each solved at frequent intervals of time during the tidal cycle and space along the estuary. The solutions of equations 1 and 2 give the bulk flow tidal propagation: the results together with the solution of equations 3 and 4 give the flow tidal propagation in the two horizontal layers: the flow in the two layers and the solution of equations 13 and 14 give the variations of suspended concentrations and the movement of the sediment throughout the tide and along the whole length of the estuary. The solution of equations 1 and 2, and of 3 and 4, was based on the Lax–Wendroff explicit finite-difference method, and the method of characteristics used to solve equations 13 and 14.

This model was then applied to the Thames Estuary, England. The main features of this estuary are its relatively uniform depth of about 7.6 m at mean tide, and the approximately exponential variation of cross-sectional area and channel width along its length. The estuary is approximately 100 km long, being 7000 m and 85 m wide at its seaward and landward limits, respectively. The trumpet shape of the lower estuary causes the local tidal range to rise from a mean value of about 4.3 m at its seaward limit to a maximum mean value of 5.6 m some 60 km upstream. Further inland the effects of frictional resistance predominate and reduce the mean range to about 3.5 m at the lock at the head of the estuary. Saline water intrudes about 64 km upstream under normal conditions, and the flow is relatively well mixed in depth, with a variation from bed to surface of about 1–2 g/l. The longitudinal density gradient caused by the intrusion of saline water sets up a net landward movement of water near the bed in the lower reaches, and the fluvial flow (averaging 70 m^3/sec) causes a net seaward movement in the upper reaches. The greater part of the estuary has a hard bed of gravel, clay, and chalk, with one major exception between 45 and 53 km upstream where there are extensive deposits of mud, and where dredging has been necessary in the past to provide adequate navigational depths. The null point, or point of zero residual current near the bed, is located in this area, and it is believed that the siltation is due mainly to mud being transported into the region both from the upstream and the downstream directions. The purpose of the model of Owen and Odd was to reproduce the movement of the cohesive mud in the estuary, and to predict changes as the result of the construction of a tidal barrier about half way along the estuary.

From equations 7, 8, and 9 it will be seen that to apply the model to a particular estuary, it is necessary to know the values of the various parameters in the erosion, deposition, and mixing processes, namely, the values of M, τ_e, τ_d, and V_S, and of the vertical mixing coefficient. In this case the erosion parameters M and τ_e were obtained from laboratory measurements on samples of Thames mud, using a recirculating flume and increasing the flow of water until erosion commenced, and equating the rate of erosion to the rate of increase of suspended concentration. The deposition parameter τ_d was measured in the same laboratory flume by gradually reducing the flow until deposition commenced.

Settling velocities were initially measured in laboratory settling columns, at various salinities and concentrations. Later, measurements were taken of settling velocities in the actual Thames Estuary, using the new sampling/settling tube developed by Owen (1971). The in situ settling velocities were found to be considerably higher than those measured in the laboratory, reflecting the importance of turbulence in flocculation. The effect of salinities in the range 0.7–20 g/l was found to be insignificant, and the settling velocity was finally expressed in the form

$$V_S = KC^n \tag{15}$$

where $n = 1$ during spring tides, and $n = 2$ during neap tides. At a suspended concentration of 100 mg/l the settling velocities on spring and neap tides were equal, at 2.00 mm/sec. The vertical diffusion coefficient was initially assumed to equal its theoretical mean value for homogeneous flows, but with this value mixing was too great, resulting in virtually equal suspended concentrations in the two layers, which is not the case in the natural estuary. The final adopted value was $U_*d/150$, one-tenth of the theoretical value. The values of α_L and α_U in equations 5 and 6 were taken to be 1.0; in other words, this particular effect of adopting a two-layer approach was ignored.

   The results obtained from this model were compared with extensive measurements taken in the actual estuary. The computed longitudinal distributions of time-averaged concentrations varied smoothly along the estuary, and agreed fairly well with the field observations, as illustrated by Fig. 2. Although the absolute values were not in close agreement, both distributions had their maxima in the region known as the Mud Reaches, and both were skewed in the seaward direction. The redistribution of the bed by erosion and deposition during the tides also resulted in most of the silt being eroded away throughout the estuary, and deposited in a short reach centered on Barking (Fig. 3), thereby reflecting the known distribution of silt. The computed variations of suspended concentrations in each reach during a tide were also compared with field observations, and Fig. 4 shows such a comparison for Halfway Reach during a spring tide. Such a comparison is a very severe test of a highly idealized model, and the computed values differ considerably from the observations. However, the peak and low concentrations in each layer generally have the correct order of magnitude, and occur at the correct phases of the tide.

   After showing that the results agreed fairly well with field observations, Odd and Owen went on to apply the model to predict the effects of a half-tide barrier at various

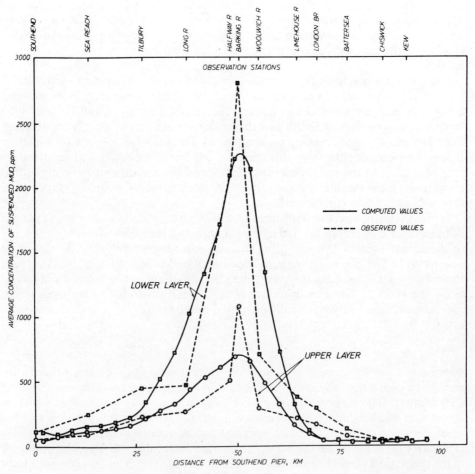

Fig. 2.   Longitudinal distribution of time-averaged concentrations, spring tides.

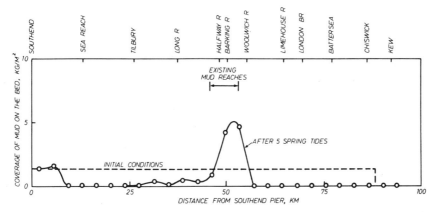

Fig. 3. The effect of repeating spring tides on the distribution of mud on the bed.

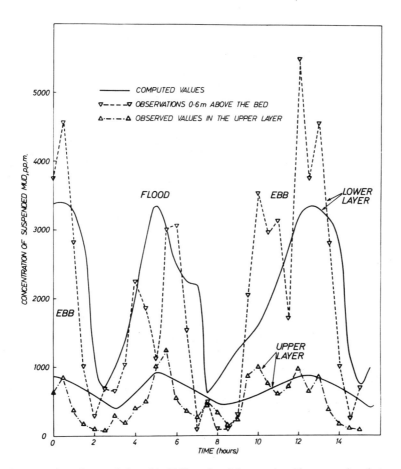

Fig. 4. Concentration of suspended mud in Halfway Reach for a spring tide; comparison between model and nature.

sites on the distribution of mud in the estuary, both in suspension and also on the channel bed.

Two more mathematical models of the transport of cohesive sediments in estuaries have recently been formulated at the Hydraulics Research Station, Wallingford, England, which are very similar in their basic concepts to the two-layer Thames model developed by Odd and Owen, but with one or two interesting differences. The actual estuaries involved are also significantly different from the Thames. The first model was of the Great Ouse (Hydraulics Research Station, 1974), a tidal river system discharging into the Wash, on the east coast of England. The Great Ouse and its various tidal tributaries total some 200 km in length, with a typical width of 150 m and depth of 6 m. The mean tidal range at its mouth is about 4.8 m. At certain times during the year, the river system is relatively well mixed in depth, with salinity intrusion extending some 20 km upstream. At other times, however, the freshwater flow in the incoming rivers is sufficiently high to give noticeable stratification of the flow, with a saline wedge extending about 10 km upstream. There are essentially three distinct sizes of sediment being transported to and fro by the tidal currents in the estuary, cohesive mud, very fine sand, and fine sand, all of which are transported mainly in suspension.

In addition to all the processes described by Fig. 1, therefore, the Great Ouse model, developed by Odd, must include the parallel processes of erosion, transport, and deposition for the sand fractions, as well as the possible additional interactions between the movement of the sand and of the cohesive mud. Once again the model was based on a two-layer system, but in this case both layers could vary in thickness during the tide. The lower layer was defined essentially as the saltwater layer, with the hypothetical interface being defined by the position of the maximum vertical salinity gradient, known as the halocline. As before, the concentrations, velocities, and salinities were assumed to be uniformly distributed in each layer. In order to model the stratification of the flow, the tidal motion of the dissolved salts was reproduced by considering the equation of continuity of salt, and allowing for mixing across the hypothetical interface due to vertical turbulent Reynolds flux, and for longitudinal dispersion within each layer. The transport and deposition of the cohesive mud was treated in the same way as the Thames model, using settling velocity measurements obtained in the actual estuary with the Owen settling/sampling tube. The erosion process was refined, however, by the inclusion of what were termed slack water deposits. From field measurements it was noticed that at slack water a thin layer of fluffy mud settled on the bed for a short while, until the velocities increased again and the current took it off the bed, apparently all at once. This mechanism was allowed for in the model by assuming that the deposited mud did not consolidate sufficiently to be considered as part of the bed until an overburden of 10 kg dry weight of sediment per square meter was achieved. When the velocities increased, the top 10 kg/m^2 was allowed to go into suspension instantaneously as soon as the critical shear stress was exceeded while the erosion of the underlying mud proceeded according to an expression of the type given by equation 7. The concentrations of suspended mud in the Great Ouse system are generally significantly lower than those of very fine and fine sand, so that attention has naturally been focused on obtaining good agreement between the model and the estuary in terms of sand transport. However, the reproduction of mud transport also appears to be reasonably successful.

The second Wallingford model, based broadly on the Thames two-layer model, is being developed for the River Avon, a comparatively short tidal estuary discharging into the Severn Estuary, England. The tidal river is about 16 km long, with a typical width of about 130 m. A significant feature of the estuary is the comparatively steep bed slope of 1 : 2000, so that bed levels vary between 7.5 m below mean sea level at the seaward limit to 0.5 m above mean sea level at the weir which forms the upriver tidal limit. This steep bed slope, coupled with a tidal range in the Severn Estuary of about 12 m during mean spring tides, gives high velocities on the late ebb tide and also indicates that parts of the estuary are virtually dry, apart from a limited freshwater flow, for a large part of the tide. In addition, salinities are virtually unchanged either along the estuary or through the depth of flow. The high tidal range in the Severn Estuary gives rise to high concentrations of cohesive mud in suspension, rising to 8000 mg/l over a wide area. The mud at these concentrations is transported into the River Avon during flood tides, traveling fairly well mixed throughout the depth of flow, and is then deposited at slack high water, giving significant depths of mud on the bed. During the ebb tide this mud is eroded, at first relatively gradually, giving fairly low suspended concentrations; then during the late ebb tide the very high velocities cause rapid erosion, and extremely high suspended concentrations, so that the estuary is transformed into what is effectively a river of fluid mud. In order to model these processes the flow was divided into two layers, as before, but in addition the deposited mud was treated as a stationary layer of varying thickness. In other words the interaction between the deposited mud and the channel geometry, and hence tidal velocities, was reproduced. A further departure from the Thames model was that the thickness of the lower flow layer was determined by the suspended concentrations and was essentially a layer of fluid mud. Within this lower layer also the velocity and suspended concentration were not taken as uniformly distributed, but were assumed to vary linearly with depth. The velocity increased from zero at the base of this layer, to the mainstream velocity at the interface of the upper layer. The suspended concentration at the base of this lower flow layer was the concentration at which fluid mud would cease to flow, which was taken as 75,000 mg/l, whereas the concentration at the top of this layer was the value at which the suspension begins to behave more like fluid mud than a free suspension, which was taken as 10,000 mg/l.

In the upper flow layer, the velocities and suspended concentration were taken as uniformly distributed with depth. Figure 5 shows in diagrammatic form the differences in the basic two-layer model schemes used for the Thames, Great Ouse, and River Avon models.

In the Avon model the expressions governing erosion, deposition, and mixing across the interface were virtually identical to those used in the Thames model, equations 7–9, although some difficulty was experienced in adopting a suitable term for the vertical concentration gradient at the interface of the lower and upper flow layer interface.

The results of this mathematical model of the Avon have not yet been published, but the initial indications are that the results agree reasonably satisfactorily with measurements of suspended concentrations, velocities, and mud bed depths measured over spring and neap tides at five positions in the actual estuary.

The two-layer model of the transport of cohesive sediments originally developed by Owen and Odd has thus been used with reasonable success on three different estuaries. Each of these estuaries is comparatively narrow, so that they may all be treated as essentially one-dimensional in plan. However, they differ in the scale of the transport of the cohesive sediment. The Thames is a relatively well-mixed estuary, with moderately high suspended concentrations, but with very little fluid mud present, so that the lower of the two horizontal flow layers does not suffer from being of fixed thickness. The Great Ouse can be highly stratified at certain times, and contains a mixture of cohesive mud with very fine and fine sand; the mud is present only in

Thames and Great Ouse

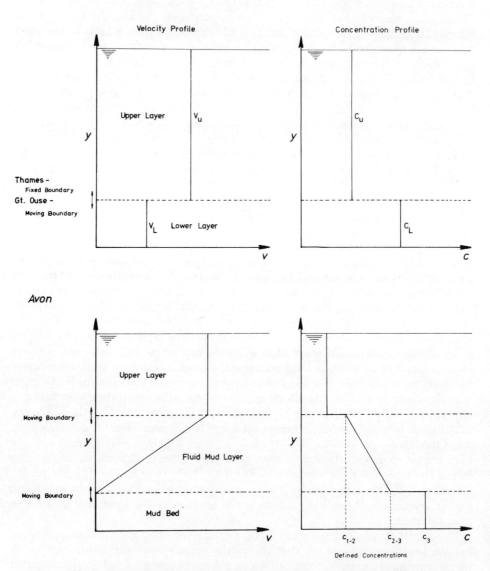

Fig. 5. Schematization of the velocity and suspended concentration profiles.

fairly low concentrations. Because of the stratification, the lower layer in this case was defined essentially as the saltwater layer. The Avon is another well-mixed estuary, but with very high suspended concentrations, and with a substantial amount of fluid mud present. Because of this, the lower layer was defined essentially as the fluid mud layer. All three models used basically the same expressions for the erosion, transportation, and deposition of the cohesive sediment, with the addition for the Great Ouse of a slack water deposit eroded instantly at high velocities. None of the models considered the consolidation of the deposited mud bed and the effect that this has on the critical shear stress for erosion.

The basic idea of treating the estuary as two horizontal layers is obviously a gross simplification of the actual velocity, suspended sediment concentration, and salinity vertical profiles, and it is perhaps somewhat surprising that reasonably successful results can be obtained from a model formulated on this basis. Greater accuracy of modeling could be achieved by adopting three, four, or even multilayer models, although several assumptions would be necessary to predict the vertical mixing between the various layers, since very little is known of the turbulent mixing processes in estuaries, especially where vertical density gradients occur, either due to vertical salinity or suspended material concentration gradients. Carried to its conclusion, a multilayer model becomes very similar to a two-dimensional model, with grid points in both the longitudinal and depth directions. However, in a multilayer model the layers would probably be allowed to expand and contract as the tide rose and fell, whereas a two-dimensional model would probably have a vertical grid fixed in space, so that the number of "layers" would increase and decrease as the tide rose and fell. Shubinski and Krone (1970) proposed a model of cohesive sediment transport in a stratified estuary which was apparently a combination of a multilayer and two-dimensional model. The vertical dimension was divided by a triangular grid, but the number of grid points in the full depth of water was constant all along the estuary; in other words, the triangular elements were larger at the seaward end of the model than at the landward end. The erosion process in this model was based on Krone's identification of a layered structure within the bed, with each layer having its own critical shear stress. For modeling purposes this critical shear stress was assumed to be linearly proportional to the maximum depth of overburden experienced by that particular layer of the bed. During periods of high velocity the mud bed was assumed to erode instantly down to the level at which the shear strength of the bed just equals the shear stress applied by the flow. No details were given of the proposed processes for transportation of the sediment, in terms of settling velocity variations or of vertical mixing of sediment. The deposition process was based on the work of Einstein and Krone (1962) for low suspended concentrations (less than 300 mg/l), giving the expression

$$\frac{C}{C_0} = \exp\left[-\frac{k'V_s}{d}\left(1 - \frac{\tau}{\tau_d}\right)t\right] \qquad (16)$$

where C is the concentration at time t, C_0 is the concentration at $t = 0$, d is the depth of water, and k' is a constant. This expression for deposition is in fact very similar to that used by Odd and Owen (1972) also based on Einstein and Krone's work, being mainly an integration with respect to time of equation 8 for a fixed volume of water in which the sediment is suspended. No further references to Shubinski and Krone's proposed model have been published to date, and it is not clear whether the model was ever used or seen to be successful. The basic expressions used for deposition and erosion appear to be quite justifiable, and it is likely that the biggest difficulties would arise in reproducing the vertical mixing and the varying settling velocities throughout the depth of flow. In the same paper, Shubinski and Krone also referred to a model of San Francisco Bay, which would be two-dimensional in plan, but well mixed throughout the depth of flow. This model would use the same basic expressions for deposition and erosion. Again no further reference to this model has appeared in the literature, but it probably forms the basis of a very recent publication from the University of California by Ariathurai (1974). Although I have not yet seen this thesis, it is known to be a report on a two-dimensional model developed under Krone to predict siltation or erosion rates caused by sudden changes in channel cross section, such as a jetty protruding from the banks of an estuary or a harbor constructed alongside the main channel, and is reported to give very good agreement with observed siltation patterns.

Marine Models

Most of the discussion so far has been based on attempts to model the transport of cohesive sediments in estuaries. In many ways the modeling of sediment transport in marine situations is simpler than for estuaries, since suspended concentrations and salinities vary comparatively little during the tidal cycle, and the tidal range is very small compared with the total depth of water. In some situations also the ocean currents are significantly greater than the tidal currents, and the latter can be ignored in the calculation of the total sediment transport and of the deposition or erosion at the sea floor. Based on the expression used for erosion and deposition in the Thames two-layer model, Harrison and Owen (1971) arrived at a relatively simple expression for the siltation or erosion at the seabed, making the following assumptions for the flow properties:

1. The tidal range is insignificant compared with the depth of flow.
2. The suspended material concentration and salinity are constant in time and space throughout the layer near the bed from which the deposited material is derived.
3. The shear velocity, U_*, varies sinusoidally during the tidal cycle, having the equation

$$U_* = U_{*0} \sin \frac{2\pi t}{T} \tag{17}$$

where U_{*0} is a constant, equal to the maximum value of shear velocity, and T is the tidal period.
4. The properties of the silt bed are constant throughout the thickness of material removed in one erosion period.
5. The deposited silt remaining on the bed after the subsequent erosion period is completed has a constant density, ρ_b.

By using equations 7 and 8 for erosion and deposition, replacing τ/τ_e by $(U_*/U_{*e})^2$ and τ/τ_d by $(U_*/U_{*d})^2$, and integrating over a complete tidal cycle, Harrison and Owen arrived at the siltation or erosion at the sea bed.

With the particular assumptions made, this integral has an analytic solution, of the form

$$S\rho_b = \frac{2}{\pi}(CV_S D - ME) \tag{18}$$

where

$$D = f\left(\frac{U_{*0}}{U_{*d}}\right)$$

$$E = f\left(\frac{U_{*0}}{U_{*e}}\right)$$

and S is the siltation rate in terms of change in bed level per unit time, being positive when net accretion occurs, and negative when net erosion occurs over a tidal cycle. Equation 18 can also be expressed in the form

$$\frac{S\rho_b}{CV_S} = f\left(\frac{U_{*d}}{U_{*e}}, \frac{U_{*0}}{U_{*e}}, \frac{M}{CV_S}\right) \tag{19}$$

and Harrison and Owen produced a diagram plotting $S\rho_b/CV_S$ against U_{*0}/U_{*e}, giving families of lines for varying U_{*d}/U_{*e} and M/CV_S.

This relatively simple expression of siltation was used with some success in the calculation of siltation rates in dredged channels in the estuary of the River Plate, South America, where the assumptions were closely matched by the actual conditions. In other situations these particular assumptions may not be justified, but alternative expressions could be substituted for the variation of shear velocity with time, and also to allow the variation of concentration with time. The integration might well have to be carried out numerically, but would still be a very simple form of mathematical model. If concentrations within the layer near the bed from which the deposited material is derived vary significantly with depth, or with distance on the scale of the area being studied, then the mathematical modeling would become very similar to that used in estuaries.

5. Evaluation of Physical Parameters

All the models of cohesive sediment transport that have been developed so far necessitate the evaluation of the following physical parameters of the particular sediment:

Erosion   Critical shear stress, τ_e, which may
                be a function of overburden
                Constant of rate of erosion, M
Deposition   Limiting shear stress, τ_d
                Settling velocity, V_S

No one has yet been able to secure a good correlation between any of these parameters and the more commonly measured soil mechanics properties, for example, vane shear strength, bulk density, plasticity index, and size grading. In the meantime, therefore, special measurements have to be made directly of these parameters, or of other closely related properties. For the shear stresses τ_e and τ_d and for the erosion constant M this generally means that quantities of the sediment have to be transported to the laboratory, and tests of erosion and deposition carried out in experimental channels or flumes (Krone, 1962; Partheniades, 1965), measuring directly the values of these parameters. Recently Migniot (1968) has found a good correlation between critical shear stress and a yield strength of the form

$$\tau_e = 0.25\tau_y \qquad \text{for } \tau_y > 2 \text{ N/m}^2$$
$$\tau_e = 0.32\tau_y^{1/2} \qquad \text{for } \tau_y < 1 \text{ N/m}^2$$

Although this correlation reduces the need for flume tests to determine τ_e, the yield stress still has to be determined experimentally in the laboratory, and in addition flume tests are still necessary to determine the erosion constant M.

The difficulties of evaluating the settling velocity, V_S, have been referred to earlier. The most reliable method to date appears to be the use of the Owen sampling/settling tube to measure the settling velocities in the actual estuary or sea, and covering as wide a range as possible of the suspended matter concentration, salinity, and turbulence to obtain the best correlation possible with each of these factors.

The possible extension of mathematical modeling of cohesive sediments into multi-layer or two-dimensional models (length and depth) depends heavily on the evaluation of vertical mixing processes in estuaries and seas, particularly where vertical density gradients occur. Much elaborate fieldwork is necessary to measure turbulent Reynolds fluxes at as wide a range of conditions as possible to establish methods of predicting vertical mixing, if only for the actual situation for which a model is being attempted. Fully three-dimensional models will present few problems once the two-dimensional (length and depth) approach has been solved.

6. Conclusion

Mathematical models of the transport, deposition, and erosion of cohesive sediments are still in their infancy, owing to our incomplete understanding of the basic processes of erosion, flocculation, deposition, consolidation, and vertical mixing in the flow. Nevertheless, surprisingly good agreement between model and natural conditions can be obtained with very simple models, even in estuaries where flow depths, velocities, suspended concentrations, and salinities all change significantly during the tidal cycle. Considerable additional research is necessary before mathematical models can be applied with absolute confidence to every situation where transport of cohesive sediments occurs.

References

Ariathurai, C. R., 1974. A finite element model of cohesive sediment transportation. Ph.D. thesis, University of California, Davis.

Biddle, P. and J. H. Miles, 1972. The nature of contemporary silts in British estuaries. *Sediment. Geol.*, 7, 23–33.

Camp., T. R., 1943. The effect of turbulence on retarding settling. *Proc. 2nd Hyd. Conf., Univ. Iowa Stud. Eng. Bull. 27*, pp. 307–317.

Carlson, E. J. and P. F. Enger, 1962. Studies of tractive forces of cohesive soils in earth canals. *U.S. Bur. Reclam. Hydrol. Branch Rept., 504.*

Ehlmann, A. J., 1968. Clay mineralogy of weathered products and of river sediments, Puerto Rico. *J. Sediment. Petrol.*, 38 (2), 885–894.

Einstein, H. A. and R. B. Krone, 1962. Experiments to determine modes of cohesive sediment transport in salt water. *J. Geophys. Res.*, 67, 1451–1461.

Grim, R. E., 1959. Clay minerals. Symposium on physico-chemical properties of soils. *Proc. Am. Soc. Civ. Eng.*, 85, (SM2).

Harrison, A. J. M. and M. W. Owen, 1971. Siltation of fine sediments in estuaries. Paper D1, *Proc. 14th Congr., Int. Assoc. Hydr. Res., Paris.*

Hydraulics Research Station, 1974. Wash water storage scheme, numerical model studies of the Great Ouse estuary; concepts, schematic representation, theory and numerical methods. *Hydraul. Res. Stn. Wallingford, England, Rept. DE 13.*

Krone, R. B., 1962. Flume studies of the transport of sediment in estuarial shoaling processes. University of California Hydraulic Engineering Laboratory and Sanitation Research Laboratory, Berkeley, Calif.

Krone, R. B., 1963. A study of the rheologic properties of estuarial sediments. University of California Hydraulic Engineering Laboratory and Sanitation Research Laboratory, Berkeley, Calif.

McLaughlin, R. T., 1959. The settling properties of suspensions. *Proc. Am. Soc. Civ. Eng.*, 85, (HY 12), 9–41.

Migniot, C., 1968. Étude des proprietes physiques de différents sédiments très fins et de leur comportement sous des actions hydrodynamiques. *La Houille Blanche*, 7, 591–620.

Odd, N. V. M. and M. W. Owen, 1972. A two-layer model of mud transport in the Thames Estuary. *Proc. Inst. Civ. Eng. Suppl.*, 9, 175.

Orgeron, C., 1968. Études sur modèle reduit de la sédimentation dans l'estuaire de la Vilaine après construction du barrage d'Arzal. *La Houille Blanche,* **7,** 621–630.

Owen, M. W., 1970. Properties of a consolidating mud. *Hydraul. Res. Stn., Wallingford, England, Rept. INT 83.*

Owen, M. W., 1971. The effect of turbulence on floc settling velocities. Paper D4, *Proc. 14th Congr., Int. Assoc. Hydraul. Res., Paris 1971.*

Owen, M. W. and N. V. M. Odd, 1970. A mathematical model of the effect of a tidal barrier on siltation in an estuary. *Int. Conf. Tidal Power, Halifax, Nova Scotia, May 1970.*

Partheniades, E., 1965. Erosion and deposition of cohesive soils. *Proc. Am. Soc. Civ. Eng.,* **91** (HY1), 105–139.

Partheniades, E. and J. F. Kennedy, 1966. The depositional behaviour of fine sediment suspensions in a turbulent fluid motion. Paper No. 4.8, *Tenth Conf. Coastal Eng., Tokyo, Japan, Sept. 1966.*

Pierce, T. J. and D. J. Williams, 1966. Experiments on certain aspects of sedimentation of estuarine muds. *Proc. Inst. Civ. Eng.,* **34,** 391–402.

Shields, A., 1936. Anwendung der Aehnlichkeitsmechanik und der Turbulenzforschung auf die Geschiebebewegung. *Mitt. Preuss. Versuchsanst. Wasserbau Schiffbau,* Berlin, 1936.

Shubinski, R. P. and R. B. Krone, 1970. A proposed model of sediment processes in estuaries. *ASCE Conf. Water Resources in the Seventies, Memphis, Tenn., Jan. 1970.*

13. MODELING OF SEDIMENT TRANSPORT ON CONTINENTAL SHELVES

J. Dungan Smith

1. Introduction

In recent years the importance of continental shelf sediment transport to marine geological and ocean engineering projects has been widely recognized; yet because of experimental and theoretical difficulties, many fundamental questions still are unanswered. At this time, it is useful to take stock of what is known and what must be accomplished to improve the general capability for modeling erosion and sediment transport processes in the marine environment. This is done herein by presenting the components of a continental shelf sediment transport model in which both theoretical and empirical expressions are used to provide estimates of the essential quantities. Some parameters are determined fairly accurately and others are merely constrained estimates. In several instances untested theory is relied on, and in many cases simple mathematical methods have been chosen over more general but more complicated ones because the more difficult approach is not warranted by available experimental data. By one means or another an attempt is made to surmount each barrier with the aim of focusing attention both on what can be calculated by present methods and on the areas that require substantial research effort.

A. Sediment Transport Models in Geology

At present most sediment transport models are quite elementary in character, and suffer substantially from a lack of proper basic fluid mechanical input; however, their construction is an important part of any marine geologic investigation. Those based upon a solid foundation in mechanics serve to rationalize and extend experimental results in addition to providing a predictive capability. Moreover, fairly general sediment transport models are essential for stratigraphic purposes. One rarely is lucky enough to measure the extreme event destined to leave an imprint in the geologic record, so the marine geologist is forced to rely on the most pertinent, best available sediment transport model in order to couple the physical characteristics of marine strata with the processes that produce them. Even more dependent upon sediment transport calculations is the historical geologist concerned with reconstruction of past environments. He cannot hope to make a flow measurement and must rely totally on fluid mechanical and sediment transport models, whether cast loosely in the form of general principles or in the precise language of mathematics.

In the past, most scientists concerned with the geologic record have found the mathematical approach to be too costly in effort relative to the benefits that could be derived from it. Also, marine and historical geologists appear to have been confused by the variety of approaches to sediment transport theory and the apparently conflicting statements made by authors in this field. Some of the muddled situation has been due to a lack of appreciation for the wide variety of methods necessary to solve the basic problems being faced by a diverse group of scientists and engineers, but much of it has arisen because of a desire on the part of many workers to generalize answers to particular problems without properly understanding their implications or considering the fundamental constraints imposed upon the subject by basic fluid mechanical principles. Fortunately in recent years several attempts have been made to put the field of sediment transport on a sound theoretical basis and in the process the differences between various approaches have become better understood. At the same time a number of important advances have been made in the neighboring fields of physical oceanography and boundary layer mechanics providing a sound basis for future sediment transport modeling efforts. Increasing interest in the coastal zone and large lakes now promises to produce the theoretical and experimental information necessary for accurate modeling of broad continental shelves, shallow seas, and enclosed basins over the next decade, and development of these techniques will be of great benefit not only to the marine geologist and ocean engineer, but also to the historical geologist. It appears that the time is nearly at hand when the stratigraphic benefits of mathematical models will outweigh their cost.

B. A Means of Approaching Sediment Transport Models

In order to make progress in the solution of complicated natural problems it is necessary to identify potential simplifications. In many cases certain subcomponents can be partially or totally decoupled from others, reducing the extent of a particular problem that must be addressed in an individual investigation. The continental shelf sediment transport problem can be divided into three such parts:

1. regional physical oceanography,
2. boundary layer mechanics,
3. specific sediment transport processes.

Each of these subcomponents can be investigated to a reasonable degree independently of the others; moreover, the most efficient means of studying each one is to some extent different.

The first area lies within the realm of the physical oceanographer; however, it is poor practice to assume that the type of results necessary for a sediment transport model will be produced automatically either from a physical oceanographic field measurement program or theoretical investigation. The goals of the physical oceanographer are substantially different from those of the marine geologist. Often the thrust of a continental shelf circulation model is an accurate prediction of flow in the surface layers, rather than near the seabed, and bottom topography usually is considered as an irritating complication to be treated only when absolutely essential and even then to be approximated in the simplest possible manner. In contrast, for sediment transport purposes accurate computation of current speed and direction near the actual seabed including all important topographic effects is essential. Experience has shown that the best approach to modeling current fields for sediment transport purposes is a joint effort in which the oceanographer and the geologist make a concerted attempt to physically understand each others scientific problems.

Sediment transport models begin with the physical oceanographic input. If the near-bottom velocity field is not known with sufficient accuracy, there is no point in proceeding further, and at present physical oceanographic knowledge is marginal in this respect. Fortunately, this situation can be expected to change immensely in the next few years and as the physical processes active on the shelf become better understood, numerical models for this region will become more general and more useful.

Nevertheless, for some time to come the most accurate results are likely to be derived from those models constrained by measured current and density data obtained throughout the water column at the shelf edge and at judiciously selected sites near the seabed.

Although the specific details of bottom current modeling lie within the domain of the basic physical oceanographic problem, proper treatment of the benthic boundary layer is an essential part of the marine geologic investigation. The most efficient physical oceanographic models attempt to parameterize the boundary layer, then permit slip at the seabed. This is desirable because turbulent flow adjacent to the sea floor yields a logarithmic velocity profile and requires a logarithmic grid; whereas in the interior a linear grid is necessary. In a quasi-steady model the simplest, most economical approach is to decouple the two flows at the top of the logarithmic layer and evaluate the boundary shear stress with a drag coefficient. More complicated approaches are required in unsteady models. From an experimental point of view, most standard current meters are not constructed to be deployed a few tens of centimeters from the seabed, and the real reason for examining this part of the flow in detail lies in its interaction with the sediments and organisms of the sea floor. Therefore, it is best investigated by the scientists working with this interaction. Marine boundary layer computations generally are based upon a combination of theoretical and experimental results, the latter from laboratory as well as field situations. Of particular interest to the marine geologist and benthic ecologist is computation of the instantaneous shear stress on the actual boundary as opposed to that averaged over a large section of the seabed, thereby including form drag as well as skin friction. Because of this, proper treatment of both unsteady and nonuniform effects as they affect the boundary shear stress are the most pressing boundary layer problems that must be handled in marine geologic models.

Sediment transport is a small-scale phenomenon; in most instances it is best studied in the laboratory. Once the physical, chemical, and biological conditions prevailing at the seabed have been elucidated, they usually can be reproduced in flumes and wave tanks more easily than careful sediment transport measurements can be made in the field. Sediment is transported either as bed load or as suspended load. In the former case, the material is confined to a thin layer near the seabed and travels with a velocity somewhat less than that of the fluid; whereas in the latter situation the sediment grains are dispersed throughout a thick layer and, being remote from the boundary, travel with the mean downstream component of the flow velocity. Bed load transport rates can be computed from one of several moderately well tested empirical and semi-empirical equations. Although none of these even approximates a final stage of development, the greatest source of error lies more with computing the correct value of boundary shear stress to insert in them than with the values predicted by the equations, at least when used within their range of applicability. In the suspended sediment case concentration profiles can be determined reasonably accurately, but there still is the important question of how to calculate the reference concentration at the top of the bed load layer. This stands out as one of the most fundamental problems in marine sediment transport.

The core of this chapter concentrates on the boundary layer and sediment transport aspects of modeling, the physical oceanographic facet of the problem being treated in Part I of this volume. The aim is to provide procedures that can be used as they stand for crude estimates on any continental shelf outside the nearshore region and that can be modified with additional research to improve substantially the accuracy of sediment transport modeling.

2. Computation of Boundary Shear Stress in Steady Flow

In order to carry out accurate sediment transport computations, a means of determining boundary shear stress averaged over an area a few tens of grain diameters in scale is required. From a dynamic point of view, the volume flux of sediment transported as bed load is related to the shear velocity times the excess shear stress at high transport rates and to the shear velocity times the square of the excess shear stress at low transport rates. In addition, the volume transport of suspended sediment is dynamically related to the shear velocity to a power greater than unity times the excess shear stress. One of the fundamental problems in sediment transport arises from this extreme sensitivity of transport rates to what is called here the local skin friction, a parameter that is very difficult either to measure or to calculate accurately in flows over the type of boundaries that is typically found on continental shelves. The source of this difficulty is the general lack of horizontal uniformity and, therefore, the production of a series of upward diffusing momentum defects. These, in effect, cause the velocity profiles to curve, matching a logarithmic region at some distance

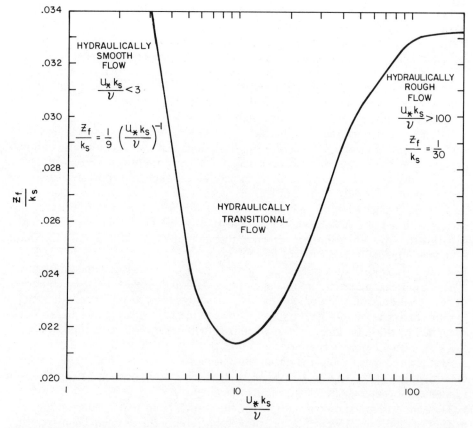

Fig. 1.   Relationship between z_f/k_s and $u_* k_s/v$ for hydraulically transitional flow according to the data of Nikuradse (as presented by Schlichting, 1960, p. 523). For $u_* k_s/v < 3$, the flow is called hydraulically smooth and $z_f = 0.11 v/u_*$; whereas for $u_* k_s/v > 100$, the flow is called hydraulically rough and $z_f = 0.033 k_s$. In many, if not most, interesting sediment transport situations, the flow is hydraulically transitional and a figure such as this one must be used to compute the roughness parameter z_f in the logarithmic velocity profile equation.

from the seabed with a high u_* and z_0 to one near the bed with a lower u_* and a z_0 related to the grain roughness or viscous sublayer thickness. In the upper region, the flow is horizontally uniform and the topographic features of the seabed act as roughness elements. Form drag on them is transmitted to the upper layer through a spatially variable pressure gradient and appears there as a Reynolds shear stress. Extrapolating the velocity profile from the upper region to the seabed yields a roughness parameter many times that due to the grain roughness, and the u_* for this upper layer is commensurately higher. In contrast, if the flow near the boundary is averaged along lines of constant distance from the seabed, then the spatial variation is removed and in the vicinity of the boundary a new logarithmic profile is produced. In this case, the bed roughness is that associated with the sand grains and in nonsediment transporting situations z_0 can be computed from the diagram redrawn as Fig. 1 from the data of Nikuradse (Schlichting, 1960, p. 523). The slope of the velocity profile in this lower region is related only to skin friction, defined here as the boundary shear stress exerted on a region of the scale of a few tens of grain diameters. Of course, in actuality this shear stress is caused predominantly by form drag on the sand grains just as the apparent outer or total boundary shear stress is caused primarily by form drag on bed irregularities.

Figure 2, based on data from a manuscript on spatially averaged flow over wavy

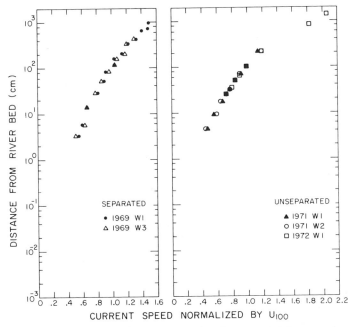

Fig. 2. Spatially averaged velocity profiles in separated and unseparated flow over large sand waves. Each data point was obtained from the average of between 12 and 60 velocity measurements procured a fixed distance from the river bed but at varying downstream positions. The features over which these flow measurements were made ranged from 1.5 to 3 m in height and 70 to 96 m in length and were in a 14 to 17 m deep reach. The influence of these large sand waves plus a set of smaller ones on the velocity profile is displayed through their general convex shape. Moreover, close examination reveals that the profiles are comprised of three straight line segments smoothly merging into each other. Starting from the boundary these are associated with skin friction on the seabed, the skin friction on the seabed plus the form drag due to the smaller dunes, and the skin friction on the seabed plus the form drag due to both the smaller and larger dunes.

boundaries by Smith and McLean (in press), shows this effect very clearly. The data presented in this diagram are mean current speeds for various levels over five separate Columbia River sand waves of substantially different shapes. In the 1969 experiments the flow separated, but in the 1971 and 1972 investigations it did not. This came about because higher mean speeds during the latter two years caused sediment to go into suspension at the wave crests eliminating the usual sharp discontinuities at the upper sides of the lee slopes, thus causing a considerable decrease in the slopes. During the separated flow experiments current speeds 100 cm from the seabed ranged from 43 to 54 cm/sec, whereas during the unseparated flow studies, current speeds ranged from 81 to 85 cm/sec. The data presented in Fig. 2 are normalized by the values at this level. In each of the two cases agreement between experiments is well within the experimental error; the greater scatter associated with the 1969 data is due to the use of a less accurate set of current meters in the earlier investigations. The ones employed in later years are described by Smith (1974).

Owen (1964) has shown that intense bed load transport produces high values of z_0 under aeolian conditions. Smith and McLean (in press) find a similar effect and suggest that Owen's method of calculating the roughness length can be generalized from $z_0 = 1.033 \times 10^{-2} u_*^2/g$, which appears to be applicable only when quartz sand is transported by air, to

$$z_0 = \frac{26.4(\tau_b - \tau_c)}{(\rho_s - \rho)g} + z_n \tag{1}$$

where z_n is the Nikuradse z_0. The coefficient in the latter expression was obtained from the 1971–1972 Columbia River data rather than being derived from Owen's paper, so that it is directly applicable to the problems of current concern. Nevertheless, the two values differ by less than 50% over a wide range of conditions.

Figure 2 shows the substantial effect that boundary topography can have on the near bed flow and, in particular, on the shear velocity. In the unseparated flow case these results yield a ratio of total boundary shear stress (τ_T) to skin friction (τ_f) of 5.7, whereas in the separated flow case this ratio is 4.1. Unfortunately these values are applicable only to the specific cases that were examined in the Columbia River, and a general means of computing this ratio is required. This can be provided by dividing the flow into layers and matching the velocity at the boundaries of each zone.

In the upper layer the velocity profile is

$$u = \frac{u_T}{k} \ln \frac{z}{z_T} \qquad z \geqslant z_* \tag{2a}$$

and in the lower layer it is

$$u = \frac{u_f}{k} \ln \frac{z}{z_f} \qquad z \leqslant z_* \tag{2b}$$

Equating these expressions at $z = z_*$, then rearranging and squaring the result gives

$$\frac{\tau_f}{\tau_T} = \frac{u_f^2}{u_T^2} = \left[1 - \frac{\ln (z_T/z_f)}{\ln (z_*/z_f)} \right]^2 \tag{3}$$

but to be of any value a relationship for the matching level z_* in terms of known parameters is required. It seems reasonable to assume that z_* is proportional to the mean thickness of the momentum defect zone and to use an expression for this internal boundary layer thickness (δ) in separated flow derived by Smith (1969). Averaging δ over one wavelength yields a mean thickness of $\delta_{av} = \sigma_0 z_0(\lambda/z_0)^\nu$ where $\sigma_0 = 0.219$

and $v = 0.834$ for $10^3 \leqslant \lambda/z_0 \leqslant 10^5$, but $\sigma_0 = 0.127$ and $v = 0.881$ for $10^5 \leqslant \lambda/z_0 \leqslant 10^7$. Thus $z_* = \sigma_1 z_0 (\lambda/z_0)^v$. Using values of z_* obtained only from the very similar and most accurate 1971 and 1972 Columbia River sand wave experiments yields $\sigma_1 = 0.425\sigma_0$. The value of this coefficient is close to unity and probably differs from this number because some of the weak nonlinearities of the actual problem are neglected in the boundary layer model. Nevertheless, in the case of a periodic boundary roughness this expression for z_* in terms of λ and z_0 should provide a satisfactory approximation to the necessary relationship.

In order to apply these results to the continental shelf an additional relationship is required because z_T is not always known. In an investigation of the form drag exerted on pressure ridges (i.e., on sea ice by Arctic winds) Arya (1975) suggests employing a drag coefficient for this purpose. Using this suggestion and defining

$$\langle u \rangle = \frac{1}{2H} \int_{z_f}^{2H} u \, dz \cong \frac{1}{2H} \int_{z_T}^{2H} u \, dz \cong \frac{u_T}{k}\left(\ln \frac{2H}{z_T} - 1\right) \tag{4}$$

where H is the height of the obstruction, noting that the stress on the seabed due to form drag is $\tau_T - \tau_f$, and defining a drag coefficient for the obstacle yield

$$\frac{\tau_T - \tau_f}{\tau_T} = \left(\frac{C_D H}{2k^2 \lambda_*}\right)\left(\ln \frac{2H}{z_T} - 1\right)^2 \tag{5a}$$

or

$$\frac{\tau_f}{\tau_T} = 1 - \left(\frac{C_D H}{2k^2 \lambda_*}\right)\left(\ln \frac{H}{z_f} - \ln \frac{z_T}{z_f} - 1\right)^2 \tag{5b}$$

where λ_* is the mean spacing of the bed forms measured parallel to a streamline. From equation 3,

$$\ln \frac{z_T}{z_f} = \ln \frac{z_*}{z_f}\left[1 - \left(\frac{\tau_f}{\tau_T}\right)^{1/2}\right] \tag{6}$$

so

$$\left(\frac{\tau_f}{\tau_T}\right) = 1 - \left(\frac{C_D H}{2k^2 \lambda_*}\right)\left\{\left(\ln \frac{2H}{z_f} - 1\right) - \ln \frac{z_*}{z_f}\left[1 - \left(\frac{\tau_f}{\tau_T}\right)^{1/2}\right]\right\}^2 \tag{7}$$

and finally

$$\begin{aligned}
\left(\frac{\tau_f}{\tau_T}\right)^{1/2} = &- \left(\frac{(C_D H(2k^2 \lambda_*)[\ln(2H/z_*) - 1][\ln(z_*/z_f)]}{(C_D H/2k^2 \lambda_*)[\ln(z_*/z_f)]^2 + 1}\right) \\
&\pm \left[\left(\frac{(C_D H/2k^2 \lambda_*)[\ln(zH/z_*) - 1][\ln(z_*/z_f)]}{(C_D H/2k^2 \lambda_*)[\ln(z_*/z_f)]^2 + 1}\right)^2 \right. \\
&\left. + \left(\frac{1 - (C_D H/2k^2 \lambda_*)[\ln(zH/z_*) - 1]^2}{(C_D H/2k^2 \lambda_*)[\ln(z_*/z_f)]^2 + 1}\right)\right]^{1/2}
\end{aligned} \tag{8}$$

In order to employ this equation the geometric properties of the seabed and a drag coefficient for the bed forms must be known. Using the Columbia River sand wave data and solving equation 8 for the drag coefficient gives $C_D = 0.15$ for separated flow and 0.70 for unseparated flow. Clearly more research in this area is required, but even when used with an estimated drag coefficient equation 8 produces reasonable results and provides the essential expression for relating τ_T and τ_f.

In a continental shelf sediment transport model the appropriate bed geometry and roughness parameters are chosen; then $(\tau_f/\tau_T)^{1/2}$ is computed from equation 8. This ratio is constant for a given sediment type and ripple field geometry; however, the latter may vary to some extent with flow conditions. The required value of τ_T is obtained from measurements made in the outer logarithmic layer or from a suitable physical oceanographic theory. Here the velocity measured at a known distance from the boundary or that calculated at the outer edge of the logarithmic layer is multiplied by $k(\ln z_R/z_f - \ln z_T/z_f)^{-1}$ to give u_T. The first term in this factor is computed directly from the reference elevation z_R and an estimate of z_f, whereas the second is obtained from equation 6. Of course iteration is required to find the exact value of z_f.

In principle the seabed geometry could be left to adjust or partially adjust to flow conditions, but in practice ripple and dune formation calculations have not been developed to the state where usable results can be obtained for this purpose. Therefore, for an accurate boundary shear stress decomposition, good seabed photography under a variety of flow conditions and a substantial amount of experience in relating high wave number seabed morphology to sediment type and bottom currents are required. In the absence of such information, specification of seabed topography in the light of general experimental information on ripple and dune properties is the only possible route. Further experimental and theoretical work on the bed form problem is essential to the improvement of sediment transport computations; however, this aspect of the subject lies in the domain of Chapter 14.

In using the shear stress decomposition just described, it should be recalled that the accuracy of the procedure depends in part upon the relationship between z_* and λ_*. The values used in this paper are based on a theory for boundary layer growth in separated flow as it relates to the wavelength of large sand waves. At this scale it seems to be quite accurate, but when extrapolated to small ripples this approach may lead to an error of unknown magnitude. Careful laboratory experiments in this area would be of considerable value and might be performed in conjunction with a set of ripple drag coefficient measurements.

3. Computation of Boundary Shear Stress in Uniform, Unsteady Flow

Away from the seabed, current and wind–wave velocity fields can be added vectorally with reasonable accuracy, but the effect that this superposition produces on the benthic boundary layer and, more particularly, on the boundary shear stress must be examined. To begin this discussion the boundary layer caused by a simple oscillatory flow must be considered, that due to a steady current having been described in the preceding section. In a progressive gravity wave, the ratio of temporal to convective accelerations scales as the ratio of particle to phase velocity; in the situation at hand this ratio is very small, permitting the boundary layer under normal wind waves to be treated as uniform in the horizontal direction. If this layer is viscous, then its thickness is given by $(v/\omega)^{1/2}$ which is the order of 2 mm for the wind waves that interact with the seabed on the continental shelf.

According to Collins (1963) the critical Reynolds number for an oscillatory flow of this type is $[U_0(v/\omega)^{1/2}/v]_c = [U_0/(\omega v)^{1/2}]_c \cong 113$, and Jonsson (1967) uses 250; therefore, for velocities in excess of a few tens of centimeters per second the boundary layer will be turbulent and the eddy viscosity will be equal to $k(u_*)_m\delta_\omega$. Writing $\delta_\omega \sim$

$(K/\omega)^{1/2}$, where K is an eddy viscosity, yields a boundary layer thickness of $\delta_\omega = (u_*)_m/\omega$. As is shown below, the logarithmic layer has a thickness on the order of $\delta_l = 0.02\delta_\omega$. Although $(u_*)_m/U_0$ varies with $\omega z_0/U_0$ the dependence is weak and for wind waves flowing over a geometrically smooth sand bed, this ratio is on the order of 0.03. Therefore $\delta_\omega \approx 0.03 U_0/\omega$ and $\delta_l \approx 6 \times 10^{-4} U_0/\omega$, this ratio indicating that turbulent boundary layers on the order of a few centimeters in thickness with logarithmic regions on the order of 1 mm high are to be expected under storm conditions on continental shelves. Jonsson (1967) has examined this problem and provides some evidence for the existence of such a logarithmic layer. More recently, Komar and Miller (1975) have used the friction factors computed by Jonsson to show that the experimental data on initial motion of sediment under wind waves can be plotted on a Shield's diagram with reasonable accuracy. These important results of Komar's confirm the basic validity of Jonsson's arguments and suggest that the approach to be taken here in computing instantaneous boundary shear stress cannot be too far in error.

In the case where both wind waves and currents are present near the seabed the problem is somewhat more complicated. In essence, a thick boundary layer is caused by the currents and a thin one by the wind waves. However, each motion is affected by turbulence produced from the other. The motions are coupled through the momentum diffusion coefficient or eddy viscosity, and a solution for the total boundary shear stress can be obtained only if a reasonably general means of computing this parameter is devised. Unfortunately not enough is yet known about the mechanics of turbulent boundary layers to provide much guidance in this endeavor, so the approach to be taken in this paper relies on similarity theory.

The equation governing uniform flow in the vicinity of the seabed due to a combination of wind waves and quasi-steady currents is

$$\frac{\partial u}{\partial t} = -\frac{1}{\rho}\frac{\partial p}{\partial x} + \frac{\partial}{\partial z} K(z)\frac{\partial u}{\partial z} \tag{9}$$

The appropriate boundary conditions are $u \to u_\infty(t)$ as $z \to \infty$ and $u \to 0$ as $z \to z_0$. The momentum equation is linear so it can be separated into the two following parts:

$$0 = \frac{\partial}{\partial z} K(z)\frac{\partial u_1}{\partial z} \tag{10a}$$

$$\frac{\partial u_2}{\partial t} = \frac{\partial U_\infty}{\partial t} + \frac{\partial}{\partial z} K(z)\frac{\partial u_2}{\partial z} \tag{10b}$$

where $u = u_1 + u_2$ and where $U_\infty(t)$ is the oscillatory velocity outside the boundary layer. Solution of the former expression for an eddy viscosity of the form ku_*z gives the familiar logarithmic profile, $u_1 = (u_*/k)\ln z/z_0$, and a solution to the latter equation for a sinusoidally varying interior velocity field is obtained by postulating $U_\infty = U_0\mathcal{R}(e^{i\omega t})$ and $u_2 = U_0\mathcal{R}\{e^{i\omega t}[1 - Z(z)]\}$ where $\mathcal{R}$ denotes the real part of a complex function. Substituting into equation 10b yields

$$i\omega Z = \frac{\partial K}{\partial z}\frac{\partial Z}{\partial z} + K\frac{\partial^2 Z}{\partial z^2} \tag{11}$$

and for $K = K_0 z$, as is the case in the immediate vicinity of the seabed, this reduces to

$$\frac{-i\omega Z}{K_0} + Z' + zZ'' = 0 \qquad (12)$$

The resulting solution is in terms of Kelvin functions (Bessel functions of complex argument). After the boundary conditions are satisfied, the temporal and spatial components of the velocity field are recombined, and the real part of the expression is found, then

$$u_2 = U_0\bigg(\cos \omega t - \frac{(\text{ker } \xi)(\text{ker } \xi_0) + (\text{kei } \xi)(\text{kei } \xi_0)}{\text{ker}^2 \xi_0 + \text{kei}^2 \xi_0} \cos \omega t$$

$$- \frac{(\text{ker } \xi)(\text{kei } \xi_0) - (\text{kei } \xi)(\text{ker } \xi_0)}{\text{ker}^2 \xi_0 + \text{kei}^2 \xi_0} \sin \omega t\bigg) \qquad (13)$$

where $\xi = 2(\omega z/K_0)^{1/2}$ and $\xi_0 = 2(\omega z_0/K_0)^{1/2}$. The zero-order Kelvin functions ker and kei are tabulated by Abramowitz and Stegun (1964, p. 431) and a brief description of their properties, including a graph of each, is given also (following p. 379). Equation 13 is plotted for $\xi_0 = 0.0338$ in a semilogarithmic manner in Fig. 3,

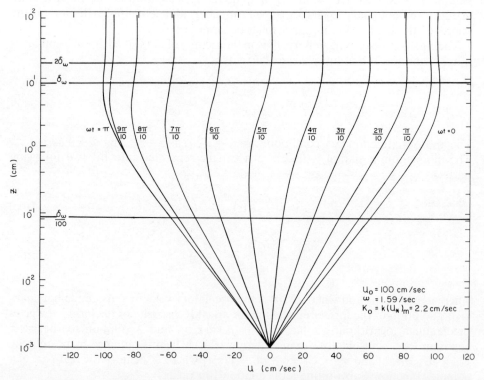

Fig. 3.   Typical velocity field for the oscillatory turbulent boundary layer under a wind wave. The scale thickness for the boundary layer, given as $\delta_\omega = (u_*)_m/\omega$, is shown on the left side of the diagram. Note that by $2\delta_\omega$ the velocity field is essentially that of the inviscid interior flow and that below $\delta_\omega/100$ the velocity profiles are logarithmic; moreover, at the times of maximum flow they are logarithmic to $\delta_\omega/10$.

emphasizing the logarithmic velocity profile near the seabed. In this similarity region the local shear stress is approximately equal to the boundary shear stress so the shear velocity is obtained by writing

$$u_2 = U_0[\cos \omega t - F_1(\xi, \xi_0) \cos \omega t + F_2(\xi, \xi_0) \sin \omega t] \qquad (14)$$

and differentiating with respect to z to get

$$\left. \frac{\partial u_2}{\partial z} \right|_{z_0} = \frac{u_*^2}{K_0 z_0} = \frac{U_0 \xi_0}{2 z_0} [-F_1'(\xi_0) \cos \omega t + F_2'(\xi_0) \sin \omega t]$$

$$= \frac{U_0 \xi_0}{2 z_0} \{[F_1'(\xi_0)]^2 + [F_2'(\xi_0)]^2\}^{1/2} \cos \left[\omega t + \arctan \frac{F_2'(\xi_0)}{F_1'(\xi_0)} \right] \qquad (15)$$

Here the prime denotes differentiation with respect to ξ. The boundary shear stress is

$$\tau_b = \rho \frac{U_0 K_0 \xi_0}{2} \{[F_1'(\xi_0)]^2 + [F_2'(\xi_0)]^2\}^{1/2} \cos \left[\omega t + \arctan \frac{F_2'(\xi_0)}{F_1'(\xi_0)} \right] \qquad (16a)$$

and the maximum value of u_* is given by

$$(u_*)_m = \left\{ \frac{U_0 K_0 \xi_0}{2} [(F_1'(\xi_0)^2 + (F_2'(\xi_0)^2]^{1/2} \right\}^{1/2} \qquad (16b)$$

where

$$F_1'(\xi_0) = \frac{(\mathrm{ker}_1 \, \xi_0 + \mathrm{kei}_1 \, \xi_0) \, \mathrm{ker} \, \xi_0 + (\mathrm{kei}_1 \, \xi_0 - \mathrm{ker}_1 \, \xi_0) \, \mathrm{kei} \, \xi_0}{\sqrt{2} \, (\mathrm{ker}^2 \, \xi_0 + \mathrm{kei}^2 \, \xi_0)} \qquad (17a)$$

$$F_2'(\xi_0) = \frac{-(\mathrm{ker}_1 \, \xi_0 + \mathrm{kei}_1 \, \xi_0) \, \mathrm{kei} \, \xi_0 + (\mathrm{kei}_1 \, \xi_0 - \mathrm{ker}_1 \, \xi_0) \, \mathrm{ker} \, \xi_0}{\sqrt{2} \, (\mathrm{ker}^2 \, \xi_0 + \mathrm{kei}^2 \, \xi_0)} \qquad (17b)$$

In this expression ker_1 and kei_1 are first-order Kelvin functions, the tables for which are presented by Abramowitz and Stegun (1964). In order to simplify computation of the shear velocity under wind waves the quantities

$$\{[F_1'(\xi_0)]^2 + [F_2'(\xi_0)]^2\}^{1/4} \left(\frac{\xi_0}{2} \right)^{1/2}$$

and $\arctan F_2'(\xi_0)/F_1'(\xi_0)$ are plotted in Fig. 4. When $\xi_0 = 0.02$ $\tau_b = 0.123(\rho K_0 U_0)$ according to the figures and if $K_0 = k(u_*)_m$ this gives $\tau_b = 2.40 \times 10^{-3} U_0^2$. As ξ_0 gets smaller the numerical values both decrease and become less sensitive to it. For a boundary layer due only to a single wave train the assumption that $K_0 = k(u_*)_m$ is valid and yields

$$\frac{(u_*)_m}{U_0} = \frac{k \xi_0}{2} \{[F_1'(\xi_0)]^2 + [F_2'(\xi_0)]^2\}^{1/2} \qquad (18)$$

However,

$$\xi_0 = 2 \left(\frac{\omega z_0}{K_0} \right)^{1/2} = 2 \left(\frac{\omega z_0}{U_0} \right)^{1/2} \left(\frac{U_0}{K_0} \right)^{1/2} \qquad (19)$$

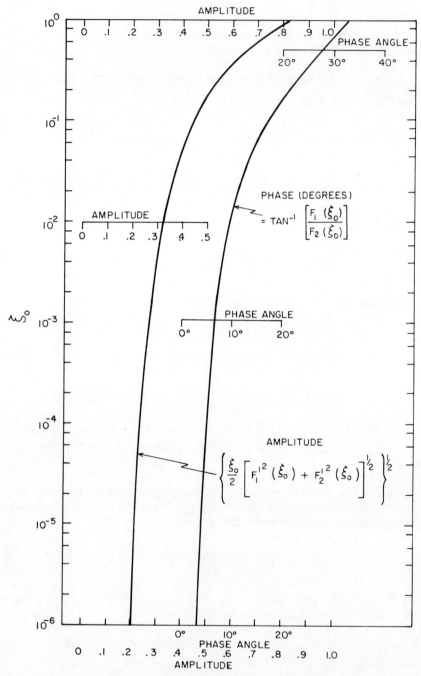

Fig. 4. Amplitude and phase angle of the nondimensional boundary shear stress $u_*/(K_0 U_0)^{1/2}$ as a function of $\xi_0 = 2(\omega z_0/K_0)^{1/2}$. Note that the amplitude varies only between 0.32 and 0.20 over the four decades below $\xi_0 = 10^{-2}$ and that the phase angle varies only from 10 to 4° over this range. The theory requires z_0 to be very small relative to $\omega/(u_*)_m$ so results for $\xi_0 > 10^{-1}$ cannot be relied on.

so for a fixed von Karman's constant, $K_0/U_0 = F_3(\omega z_0/U_0)$, a function that is plotted in Fig. 5. With the assistance of this graph, the boundary shear stress due to a purely oscillatory flow over a geometrically smooth seabed with roughness properties characterized by z_f can be calculated. As is the case in steady uniform flow, z_f must be computed from equation 1 or the roughness Reynolds number $R_* = u_* k_s / v$ by iteration. The graph of z_f/k_s in terms of R_*, presented in Fig. 1, is required for the latter purpose. Fortunately the variation of $(u_*)_m$ with U_0 as shown in Fig. 5 is weak so this task is not a difficult one.

At any point on the continental shelf, the velocity just above the seabed due to wind waves or swell with known deep-water properties can be computed employing small-amplitude wave theory. From the specified wave frequency, this computed value of U_0, and an estimate of z_0, $(u_*)_m/U_0$ can be found from Fig. 5. If only skin friction is considered, the best first guess for z_0 usually is 1.0×10^{-3} cm; however, form drag due to ripples or z_0 adjustments due to bedload transport may be included in an approximate manner using the techniques described previously. Once an estimate of $(u_*)_m$ is obtained, new values of R_*, hence z_0, hence $\omega z_0/U_0$, and hence $(u_*)_m/U_0$ can be computed in the nonsediment transporting situation. The procedure is repeated until it converges to the desired accuracy; however, this usually takes only a few iterations. Owing to its dependence on R_*, which in turn varies with u_*, strictly speaking z_0 is a function of time. However, in many problems the boundary layer is in transition between hydraulically rough and hydraulically smooth flow, so z_0 does not vary too drastically over the high shear velocity part of the cycle. Moreover, the nonlinear nature of the sediment transport process reduces the requirement for great accuracy except near the time of maximum boundary shear stress, and use of a constant z_0 computed from $(u_*)_m$ is not likely to be a serious source of error.[1]

Up to this point emphasis has been placed upon computation of boundary shear stress in a flow devoid of mean currents. This is of concern in the nearshore zone and in some areas with particularly weak tidal and wind-driven flows. However, the most interesting condition in regard to continental shelf sediment transport occurs when both currents and wind waves are superimposed. Were the tidal flow on the continental shelf of a reversing type, then δ_ω would be on the order of 100 m and the logarithmic layer would be on the order of a few meters. Therefore, the tidal currents would appear to be steady on the wind–wave time scale and the turbulence generated by the two oscillatory motions would affect different regions of the boundary layer. Rotary tidal currents and unsteady motions of lower frequency produce the same results. The eddy viscosity contributed by each independent component of the flow increases linearly with distance from the boundary in the region immediately above the seabed. Moreover, an eddy viscosity based on the sum of the two shear velocities is equivalent to that which would be produced by the shear of the combined motions, suggesting that a suitable solution to the general problem might be obtained if such an eddy viscosity profile were used. In this case the predicted boundary shear stress is not likely to be far from the correct value and there is no point in over complicating the problem in light of the number of secondary effects such as might be associated with boundary topography that have not been accounted for. In the layer just above the seabed adding diffusion coefficients, that is, adding velocity scales, gives

---

[1] It should be noted here that much of what has been said about the oscillatory boundary layer under wind waves also is applicable to the unsteady boundary layer driven by reversing tidal currents. Under tidal situations where phenomena such as net sediment transport are of concern, a good first estimate often can be obtained by computing z_0 as described here and holding it constant with respect to time.

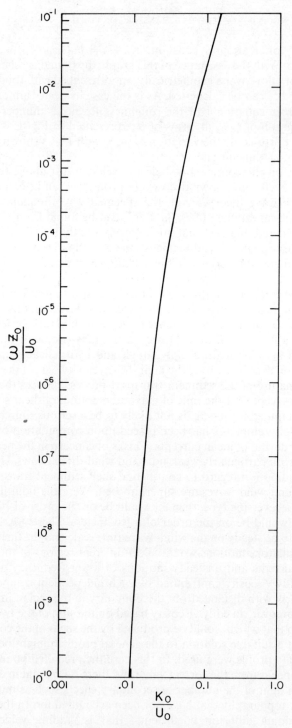

Fig. 5. Graph of (K_0/U_0) vs. $(\omega z_0/U_0)$. This diagram is provided so that $K_0 = k(u_*)_m$ can be computed in terms of U_0, ω, and z_0 for the situation where the oscillatory motion is the only one present. The roughness parameter z_0 is obtained as in the steady flow situation and iteration may be required. The graph has sufficient range to be used both for wind-wave and tidal frequencies.

$K = k[(u_*)_1 + (u_*)_m]z$. This is the region of greatest concern as the ultimate goal of the calculation is to determine u_* for the combined flow.

The solution to equation 10a in the lower layer, but with a combined eddy viscosity, is

$$u = \frac{(u_*)_1}{k}\left[1 + \frac{(u_*)_m}{(u_*)_1}\right]^{-1} \ln\frac{z}{z_0} \tag{20}$$

Thus the presence of wind waves has the apparent effect of increasing von Karman's constant and lowering the ratio of u/u_* for a given z/z_0. That is, by increasing the eddy viscosity in this layer, downward diffusion of momentum is increased; thus for a given velocity at a fixed level from the seabed and a given roughness. a higher mean boundary shear stress results. Moreover, if $(u_*)_m/(u_*)_1 = \frac{1}{4}$ the boundary shear stress is increased by 56 %, indicating the importance of this effect even for small wind waves. For $z > \delta_\omega$ the solution to equation 10a with $K = ku_{*_1}\{[z + \delta_\omega(u_*)_m/(u_*)_1]\}$ is

$$u = \frac{(u_*)_1}{k}\ln\left(\frac{z + \delta_\omega(u_*)_m/(u_*)_1}{z_1}\right) \tag{21a}$$

where

$$z_0 = (1 + \gamma)\delta_\omega\left(\frac{z_0}{\delta_\omega}\right)^{1/(1+\gamma)} \tag{21b}$$

and $\gamma = (u_*)_m/(u_*)_1$. The solution to equation 10b in the inner layer is as given by equation 13 where $K_0 = k[(u_*)_1 + (u_*)_m]$ is used instead of $k(u_*)_m$. For the outer layer the unsteady velocity field is uniform with distance from the boundary and is independent of the character of the eddy viscosity as long as this coefficient is less than $k[(u_*)_1 + (u_*)_m]z$ at every level. In the neighborhood of the seabed, the total velocity field is obtained by adding equations 20 and 13.

As far as bed load transport of sediment is concerned, the oscillatory component averages to zero, so the major effect is through the increased boundary shear stress in the current-driven part of the flow, owing in turn to the increased near-bed momentum diffusion coefficient. Further enhancement of sediment transport by waves occurs only when the local boundary shear stress rises to the point where a substantial amount of material is put into suspension. In either the bed load or suspended load case, sediment transport ceases after the boundary shear stress drops below its critical value within a time on the order of δ_s/w_s. Here δ_s is the scale height of the layer containing sediment. In the bed load case δ_s is of the order D, and this time scale is small compared to the period of the wind waves. On the other hand, in the suspended load situation, this time scale can be much longer than a wave period. To account for the latter effect, the shear velocity is computed as described here, then the time-dependent sediment concentration field is computed as described below. These are multiplied together and averaged in time.

In actuality, waves of a variety of frequencies impinge upon the continental shelf, but only a subset of these motions contributes measurably to the oscillatory velocity field near the bottom. From the incident wave spectrum, shelf topography, and small-amplitude wave theory, velocity component spectra can be computed for each point on the continental shelf. Furthermore, these can be used in conjunction with the methods outlined previously to compute shear velocity spectra. It probably is reasonable to assume that the maximum shear velocity for each wave can be used as the scale for the diffusion coefficient over that period. Although the turbulence field generated by a single wave lasts for longer than one period, it is decaying, and properly accounting for this effect would complicate the problem immensely and un- necessarily. Moreover, to do so with reasonable accuracy is beyond the current state of knowledge in regard to the production and decay of turbulence in unsteady boundary layers over rough terrain.

With this simplification, each wave acts independently and the average bed load flux over a long time interval is equal to the average of the bed load fluxes over each cycle weighted by the fraction of time the seabed is subjected to waves of that period. This is a vector average, and the waves and currents need not be traveling in the same direction. The wind waves are independent of each other because the problem essentially begins again with a new eddy viscosity after each cycle, and they are independent of the current direction because a relationship identical to that in equation 10a, but with v instead of u as a variable, can be written; then this expression can be added vectorally to the preceding equation and solved in the direction of the actual flow. Unfortunately, in the case where the excess shear stress is high enough to put a substantial amount of sediment into suspension during a wave cycle, this simple decomposition of the sediment transport problem is less accurate because the net flux depends to a greater extent upon the exact sequence of events.

4. Bed Load Transport of Sediment

The system of equations governing erosion and transportation of sediment in a turbulent flow is extremely complicated and a number of assumptions must be made before results of any practical value can be obtained. Much of what is known about this subject is based on the equations for conservation of mass, both for the sediment and the transporting fluid, and on two very powerful postulates. The first of these is the bed load assumption; it states that when material moves in a thin layer near the seabed the only flow variable on which the transport process depends is the local skin friction. The second is the suspended load hypothesis which states that outside the bed load layer the sediment velocity is equal to the water velocity minus the still-water settling velocity for the sediment particles. In each of these cases the sediment can be treated either as a single component or as many, each representing a size or specific gravity class of the overall sample.

Currently two separate types of bed load transport models are popular. The first, favored by civil engineers, focuses on the interaction between individual sediment particles and the fluid. Collisions between grains are neglected and momentum exchange is treated through the equation for forces exerted on approximately spherical particles by a fluid. The other approach, favored by many geologists and oceanographers, is based on the assumption that sediment particles are held in suspension by the dilitation that must occur when a fluid containing a high concentra- tion of sediment is sheared. In this model momentum transfer is primarily by particle– particle interactions. Although both groups try to apply their equations over the entire range of concentrations, it is clear from the assumptions on which they are based that this cannot be valid. The latter approach, developed by Bagnold (1956, 1966, 1973) and extended to the nearshore region by Inman and Bagnold (1963) and Komar and Inman (1970), generally is produced from an energetic argument, whereas the discrete particle approach has been developed from a specific mechanistic model.

In the low particle concentration model it is assumed that bed load transport occurs when the particle settling velocity is large relative to the shear velocity. Under these conditions, when the shear is increased across a sediment grain, the pressure drops at its top owing to the Bernoulli effect creating a force in the direction of the shear. This lift force can exceed the grain weight if the instantaneous boundary shear stress is high enough and if the grain protrudes above the mean bed level. Until the particle is removed from the bed, the drag force on it is transmitted to the grains beneath. Once lifted from the bed, the shear across the particle decreases drastically and the vertical component of the pressure gradient across it all but disappears. The drag force is no longer balanced and the particle accelerates in the downstream direction. As long as the settling velocity of the grain is large relative to the mean shear velocity, the particle returns to the boundary after one hop, extracting horizontal momentum as it moves through the fluid but losing it in an inelastic collision when it hits the bed. If the particle lands in an area of high local shear it may be lifted off the bed again immediately, and if not, it may remain at rest for a substantial length of time before making another hop. For true bed load these hops are on the order of a few grain diameters in height and 100 grain diameters in length. However, all bed load equations also are used outside the range of applicability for the physical model, this being made possible by the inclusion of empirically adjusted coefficients.

The equation that must be solved to determine the exact particle trajectory for a given initial force is fairly complicated, and the model has not yet been developed properly from a mathematical point of view. Nevertheless, three important attempts to approximate a solution have been made. The earliest, by Einstein, culminated in his bed load function paper (Einstein, 1950). He did not try to solve the particle equations of motion, but rather assumed the trajectories to be similar and scaled by particle diameter. To close the problem Einstein related the probability of the instantaneous lift exceeding grain weight to a Gaussian distribution of pressure fluctuations centered about the mean lift as calculated using a logarithmic velocity profile. In this approach the threshold of sediment motion is expressed implicitly in a probabilistic manner. In contrast Yalin (1963) postulated the existence of a critical mean boundary shear stress as given by the Shields diagram and attempted to solve the particle equations of motion; but in order to do so, he had to simplify them considerably. From his solution, Yalin was able to find the sediment particle velocity averaged over the bed load layer, but rather than computing the concentration of sediment in this layer, he postulated that it was proportional to the normalized excess shear stress. In the most complete work available to date Owen (1964) attempted to solve the particle equations of motion, and by approximating them he was able to find both the sediment velocity and concentration fields in the bed load layer. Although Owen's results are for the transportation of sediment by wind rather than water, they are of interest because they most closely approximate a proper solution to the bed load transportation problem under low concentration conditions.

It has been argued that the Bagnold approach is more fundamental because it is based on conservation of energy. However, all the above-mentioned bed load equations can be cast in that form and their intricacies attributed to the specific models required to determine the so-called efficiency factor. In essence, the difference between the two main approaches to bed load transport is one of concentration range. In many marine and fluvial situations only weak sediment transport occurs even during extreme events, and bed load equations of the Einstein and Yalin type are preferable. On the other hand, if one is concerned with beach processes, accurate computation of bed load transport under flood conditions in dry climates or slurries in flumes, then the Bagnold approach may be preferable. It is particularly important to be aware of the fundamental premises upon which these equations are based and of how to choose among them. Unfortunately marine geologists working on continental shelf sediment transport must be familiar with all the approaches and must be able to use each where it provides the best results.

In most natural systems boundary shear stress varies from place to place and the bed load assumption tightly ties the local sediment transport to this parameter. Given that the constants for the various bed load equations have been evaluated in a field situation or in a flume with a specific boundary geometry, the equations can be expected to be accurate only under similar conditions. For example, the boundary shear stress measured by Sternberg and Kachel (1971) was evaluated using the logarithmic profile method over rippled beds; thus the "efficiency factor" correction presented by Sternberg (1972) is valid only when the equations are applied to analogously rippled beds. In contrast, in many of the experiments upon which the Yalin equation is based, care was taken to ensure that the bed was flat. In the former case one is faced with the problem of deciding whether the rippled bed in the area with which he is concerned has the same ratio of skin friction to form drag as those over which Sternberg and Kachel worked, and in the latter case he is faced with the problem of computing the flat bed boundary shear stress. Although no conclusive statement can be made on the subject, I have found it very difficult to guarantee a fixed ratio of skin friction to total boundary shear stress in many marine sediment transport problems and prefer to use the Yalin equation in conjunction with the technique outlined in Section 2.

The Yalin bed load equation can be written in terms of the volume flux of sediment per unit flow width as

$$Q_s = a_1 u_f SD\left[1 - \frac{1}{a_2 S} \ln\left(1 + a_2 S\right)\right] \tag{22}$$

where

$$a_1 = 0.635$$

$$a_2 = 2.45\left(\frac{\rho}{\rho_s}\right)^{0.4}\left[\frac{\tau_c}{(\rho_s - \rho)gD}\right]^{1/2}$$

Here u_f is the local shear velocity, τ_f is the flat bed boundary shear stress, τ_c is the critical shear stress for the initiation of sediment motion given in terms of grain and fluid parameters in Fig. 6, ρ is the fluid density, ρ_s is the sediment density, g is the acceleration due to gravity, and D is the nominal grain diameter. The parameter $S = \tau_f/\tau_c - 1$ is the normalized excess shear stress. When $a_2 S \ll 1$ equation 22 reduces to

$$Q_s \cong \frac{a_1 a_2 D u_f S^2}{2} \tag{23}$$

Moreover, for quartz sand in seawater a_2 ranges from 0.30 to 0.42, so this asymptotic expression can be used for values of skin friction less than 30% above the critical boundary shear stress.

The bed load equations described here can be applied directly to the continental shelf without further consideration only when the near-bottom velocity field due to wind waves is negligible. In all bed load equations root-mean-square turbulent effects are assumed to be similar and scaled by the mean boundary shear stress; therefore, they are accounted for through the empirical coefficients. This is a suitable approach for unidirectional flow, but it breaks down when a large oscillatory velocity field is imposed on the problem. In the more general situation, use of the instantaneous boundary shear stress in an equation that could be time averaged to produce the

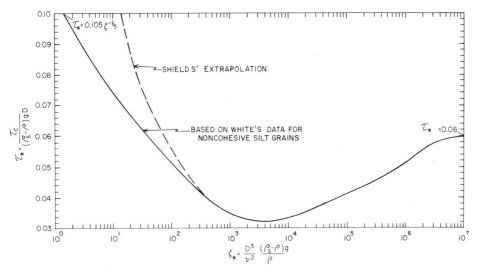

Fig. 6. Nondimensional boundary shear stress required to initiate sediment transport as a function of the nondimensional number representing fluid and sediment properties. The dashed curve and the solid one to the right side of the branch are from the Shields' diagram as modified by Vanoni (1964) and usually plotted in terms of $(u_*)_c D/\nu = (\zeta_* \tau_*)^{1/2}$. The solid line to the left of the branch is based on data of White (1970) for noncohesive material of silt size. For large ζ_*, $\tau_* = 0.06$, and for small ζ_*, White's expression yields an asymptotic value of $\tau_* = 0.105\zeta_*^{-1/5}$. Strictly speaking, Shields' diagram and curves derived from it such as the one presented here apply only to geometrically smooth beds of well-sorted sedimentary material for which $D = k_s$.

proper rectification effect would be preferable, but the experimental data necessary to construct such an equation are not presently available and a less rigorous approach must be taken. In the case where the wave-induced boundary shear stress is on the order of or larger than the mean boundary shear stress due to tidal and wind-driven currents, inclusion of the former as part of the mean shear stress in bed load calculation leads to the least error. Here time-dependent turbulent fluctuations in the current field with periods comparable to those of the wind waves must be excluded from consideration as part of the "mean" flow, as their effects have been accounted for through the empirical coefficients of the bed load equation.

In general the mean boundary shear stresses due to wind waves and lower frequency currents can be added, but this approach produces a possible error in estimating "instantaneous" bed load transport comparable to the additional sediment flux caused by any turbulent fluctuations in the mean current with periods on the order of, or longer than, the wave period. In percentage terms this error is small as long as the boundary shear stress due to the wind waves is large, and if the bed load flux is averaged over a long period of time the error disappears.

5. Suspended Sediment Transport

A. Basic Concepts

Although computation of boundary shear stress with reasonable accuracy is prerequisite to all sediment transport models, determination of concentration fields for each of the important size and specific gravity classes of the bed material also is an essential part of the problem. As a transport process, the bed load mechanism is relatively inefficient, so in most continental shelf models accurate treatment of suspended load is of primary concern. In order to compute normalized concentration profiles in uniform flow, only the spatial dependence of the momentum diffusion coefficient for the fluid is required; however, to determine sediment concentration at a reference level near the seabed, accurate information on local skin friction also is necessary. Furthermore, to compute sediment transport rates accurate velocity profiles are required. Usually the diffusion coefficient in the sediment-bearing flow is approximated by that for the analogous clear water situation. This is satisfactory when the sediment concentration is low; however, the presence of suspended sediment stratifies the boundary layer, inhibiting vertical velocities in the turbulence field, thus decreasing the amount of sediment put into suspension especially in regard to the coarsest size fractions. Generally this effect is treated in channel flow by adjusting von Karman's constant according to an empirically derived relationship, rather than computing the Richardson number profile, as is typical in studies of the atmospheric boundary layer under stable conditions.[2] Fortunately in the continental shelf sediment transport situation, concentrations often are low. If not, the methods of Smith and McLean (in press) can be used.

The concentration field of suspended sediment is derived from equations expressing conservation of mass for each size and specific gravity class that the sediment sample is divided into, plus an additional equation for the transporting fluid. Mass conservation is a strong constraint in all sediment transport problems, and fortunately, few additional assumptions are required in order to obtain sediment concentration profiles of reasonable accuracy. In order to permit use of a continuum mechanical approximation, sediment velocities and concentrations must be defined at all points. The former is the value that a sediment particle would have were it present at the reference location, and the latter is the probability of a sediment particle being found at that location. The use of these definitions puts investigation of multicomponent or multiphase flow on a sound mathematical basis. If all the sediment present in the flow were derived from the boundary, then some time after the shear velocity dropped below its critical value the concentration would go to zero; however, the sediment velocity would not because if a particle were introduced into the fluid it would have a finite velocity according to the formalism.

From a practical point of view sediment concentration can be thought of as the volume of sedimentary material in a specified region divided by the volume of that region as long as the number of particles in any given size class remains sufficiently large. This means that as the concentration decreases, the volume of fluid over which the measurement must be made increases, and ultimately in a flow with gradients of any sort, measurements become extremely inaccurate. In fact, proper field measurement of suspended sediment concentration is extremely difficult and for that reason empirical confirmation of suspended sediment concentration equations often is lacking. These experimental difficulties mean that theoretical and empirical techniques must be combined in order to provide answers to the pressing continental shelf sediment transport problems, and that the computations carried out here are necessary not only to provide theoretical estimates for continental shelf sediment transport models, but also to guide and assist the marine geologist and ocean engineer concerned with making field measurements of these quantities. Although the following discussion is fairly mathematical and some functions unfamiliar to many marine

[2] See Businger and Arya (1974) for a model of the stable atmospheric boundary layer [and Smith and McLean (in press) for a discussion of sediment-induced stratification corrections.]

geologists may be used, neither the general concepts nor application of the results to sediment transport problems are particularly difficult. Some of the theory presented here builds on the careful studies of Dobbins (1944) and Hunt (1954); these works, plus the paper by Smith and Hopkins (1972), serve as a source of background information on the general suspended sediment problem.

In a particular experimental situation, empirical coefficients can be added to theoretically derived equations in order to improve their agreement with data and these almost always can be heuristically justified, but in the case of a continental shelf sediment transport model, such coefficients are of little value because no accurate means of computing their magnitude under a wide variety of circumstances is available. Therefore, the approach taken here is to use as few adjustable coefficients as possible and to provide a numerical value or a means of calculating each one.

If the suspended sediment equations are to be applied to a seabed roughened with bed forms, then a two-layer velocity and eddy diffusion coefficient profile should be used; however, in most cases the majority of sediment is in the outer layer. Under these circumstances the proper shear velocity to use in calculating the eddy viscosity and flow velocity is the one for the outer region, whereas skin friction is the proper boundary shear stress to use in computing the concentration at the base of the suspended sediment layer. For the time-dependent flows of interest on the continental shelf it is unlikely that these shear velocities vary in proportion to one another, so a method whereby they can be treated separately is required. The theory presented subsequently in this section uses only a single logarithmic region, but can be generalized if necessary. The reference-level concentration computation is carried out separately, employing an excess shear stress based on skin friction.

Conservation of mass yields N equations of the form

$$\frac{\partial \varepsilon_n}{\partial t} + \nabla \cdot (\mathbf{u}_n \varepsilon_n) = 0 \tag{24}$$

where ε_n is the instantaneous volume concentration of sediment in class n and $\mathbf{u}_n$ is the instantaneous velocity of this material. The additional equation conserving mass for the fluid is analogous. Volume concentration is defined such that

$$\varepsilon_s = \sum_{n=1}^{N} \varepsilon_n \quad \text{and} \quad \varepsilon_s + \varepsilon_w = 1 \tag{25}$$

the extra component being the transporting medium. Averaging each of these equations over a period long compared to that of the turbulent fluctuations in the boundary layer but short compared to the time scale of tidal and lower-frequency wind-driven currents on continental shelves, approximating the turbulent mass fluxes by gradient type diffusion, and assuming the flow to be horizontally uniform yield N equations of the form

$$\frac{\partial \bar{\varepsilon}_n}{\partial t} = \frac{\partial}{\partial z}\left(-\bar{w}_n \bar{\varepsilon}_n + K_n \frac{\partial \bar{\varepsilon}_n}{\partial z}\right) \tag{26}$$

and one equation of the form

$$\frac{\partial \bar{\varepsilon}_s}{\partial t} = \frac{\partial}{\partial z}\left[\bar{w}_w\left(1 - \bar{\varepsilon}_s\right) + K_w \frac{\partial \bar{\varepsilon}_s}{\partial z}\right] \tag{27}$$

where $\bar{\varepsilon}_s$ is the total sediment concentration. The expression $\bar{\varepsilon}_w = 1 - \bar{\varepsilon}_s$ has been used for the concentration of the transporting fluid in the latter equation. For a brief but somewhat more detailed derivation of these equations the reader is referred to the discussion presented by Smith and Hopkins (1972).

In suspended sediment problems it is usually accurate to set the sediment velocity equal to the fluid velocity minus the settling velocity of the sediment,

$$\mathbf{u}_n = \mathbf{u}_w - \mathbf{w}_n$$

Making this assumption, taking $\bar{\varepsilon}_n \ll 1$ so that $w_n = -w_n$, and noting that all the diffusion coefficients should be the same as that for momentum, yield a set of equations of the form

$$\frac{\partial \bar{\varepsilon}_n}{\partial t} = \frac{\partial}{\partial z}\left(w_n \bar{\varepsilon}_n + K(z)\frac{\partial \bar{\varepsilon}_n}{\partial z}\right) \tag{28}$$

Taking $K = K_0 z$ as required in the "law of the wall" region and writing $dv = K_0\, dt$ and $p_n = w_n/K_0$ yields

$$\frac{\partial \bar{\varepsilon}_n}{\partial v} = \frac{\partial}{\partial z}\left(p_n \bar{\varepsilon}_n + z\frac{\partial \bar{\varepsilon}_n}{\partial z}\right) \tag{29}$$

These equations are valid through the boundary layer except in a thin region near the bed in which $\bar{\varepsilon}_n$ is of order one. In this zone the concentration of sediment is too high to neglect the equation expressing conservation of mass for the fluid so

$$\frac{\partial \bar{\varepsilon}_s}{\partial v} = \frac{\partial}{\partial z}\left(\frac{-\bar{w}_s}{ku_*}\bar{\varepsilon}_s + z\frac{\partial \bar{\varepsilon}_s}{\partial z}\right) \tag{30}$$

must be combined with

$$\frac{\partial \bar{\varepsilon}_s}{\partial v} = \frac{\partial}{\partial z}\left(\frac{\bar{w}_w}{ku_*}(1 - \bar{\varepsilon}_s) + z\frac{\partial \bar{\varepsilon}_s}{\partial z}\right) \tag{31}$$

This gives $\bar{w}_w = w_s \bar{\varepsilon}_s + \text{const}$; however, in uniform flow problems with fixed boundaries the vertical velocity component is zero when there is no suspended sediment. Thus $\bar{w}_w = 0$ when $\bar{\varepsilon}_s = 0$ so const $= 0$, and the second equation becomes

$$\frac{\partial \bar{\varepsilon}_s}{\partial v} = \frac{\partial}{\partial z}\left(p_s \bar{\varepsilon}_s(1 - \bar{\varepsilon}_s) + z\frac{\partial \bar{\varepsilon}_s}{\partial z}\right) \tag{32}$$

Here $p_s = w_s/K_0$ and w_s is the concentration weighted settling velocity discussed by Hunt (1954).

In this thin layer near the bed, changes in concentration with respect to time are small even when the flow is unsteady, owing to its lack of capacity to store additional suspended sediment, so the left side of equation 32 can be set equal to zero and the right side integrated to give

$$p_s \alpha_0 + p_s \bar{\varepsilon}_s(1 - \bar{\varepsilon}_s) + z\frac{\partial \bar{\varepsilon}_s}{\partial z} = 0 \tag{33}$$

where $p\alpha_0$ is a constant of integration. Rearranging this expression and integrating a second time with respect to z gives

$$\ln\left(\frac{\bar{\varepsilon}_s + \alpha_1}{1 - \bar{\varepsilon}_s + \alpha_1}\right) = -p_s(2\alpha_1 + 1)\ln z$$

or

$$\frac{\bar{\varepsilon}_s + \alpha_1}{1 - \bar{\varepsilon}_s + \alpha_1} = \frac{\bar{\varepsilon}_a + \alpha_1}{1 - \bar{\varepsilon}_a + \alpha_1} \left(\frac{z_a}{z}\right)^{p_s(2\alpha_1 + 1)} \tag{34}$$

where $\alpha_1 = \frac{1}{2}[(1 + 4\alpha_0)^{1/2} - 1]$ and $\bar{\varepsilon}_a$ is the concentration at z_a. Finally, writing $q_s = p_s(2\alpha_1 + 1)$ and $A_\alpha = (\bar{\varepsilon}_a + \alpha_1)/(1 - \bar{\varepsilon}_a + \alpha_1)$ and rearranging gives the steady suspended sediment concentration profile for the case of an arbitrary sediment flux at the top of the layer:

$$\bar{\varepsilon}_s = \frac{(1 + \alpha_1)A_\alpha(z_a/z)^{q_s} - \alpha_1}{1 + A_\alpha(z_a/z)^{q_s}} \tag{35}$$

Although any reference level in the zone where $\mathbf{u}_s = \mathbf{u}_w - \mathbf{w}_s$ can be chosen, it is best to take z_a as the base of this region. Moreover, this level can be thought of as an incremental distance above the height to which saltating particles would rise were the material being transported as bed load. According to the observations of Einstein (1950) the bed load transport zone is only a few grain diameters thick. Although a rigorous definition of z_a is required in order to compute the sediment concentration at this level, the results are not sensitive to the exact height that is chosen, except under slurry conditions which are best treated by other means. Therefore, the Einstein estimate cannot be very much in error in regard to giving the level below which the suspended sediment assumption breaks down, and his value of $z_a = 2D$ is used here.

B. Suspended Sediment in Steady, Uniform Flow

In a steady, uniform situation if the upward diffusive and downward advective fluxes of sediment just balance at some level, as is the case at the upper boundary of the flows of concern here, then α_0 and α_1 are zero and equation 33 shows that these fluxes balance everywhere in the interior as long as the concentration fields are continuous. Moreover, this result is independent of the type of diffusion coefficient that is used and the nature of the layers into which the problem may be divided for computational ease. In this important situation

$$\frac{\bar{\varepsilon}_s}{1 - \bar{\varepsilon}_s} = \frac{\bar{\varepsilon}_a}{1 - \bar{\varepsilon}_a} \left(\frac{z_a}{z}\right)^{p_s} \tag{36}$$

and

$$\bar{\varepsilon}_s = \frac{[\bar{\varepsilon}_a/(1 - \bar{\varepsilon}_a)](z_a/z)^{p_s}}{1 + [\bar{\varepsilon}_a/(1 - \bar{\varepsilon}_a)(z_a/z)^{p_s}]} \tag{37}$$

for $z_a \leqslant z \leqslant z_b$. Above the "law of the wall" zone the momentum diffusion coefficient no longer increases linearly with z, but it can be approximated in many problems by $K = ku_* h/\beta_1$ where h is a characteristic length and β_1 is a coefficient. For a channel flow the inverse amplitude coefficient β_1 is 6.24 and the characteristic length h is the depth of the fluid. In this case, $z_b = 0.2h$ and equation 28 yields

$$\frac{\bar{\varepsilon}_s}{1 - \bar{\varepsilon}_s} = \frac{\bar{\varepsilon}_b}{1 - \bar{\varepsilon}_b} \exp\left(-\frac{p_s \beta_1(z - z_b)}{h}\right) = \frac{\bar{\varepsilon}_a}{1 - \bar{\varepsilon}_a}\left(\frac{z_a}{z_b}\right)^{p_s} \exp\left(-\frac{p_s \beta_1(z - z_b)}{h}\right) \tag{38}$$

for $z_b \leqslant z \leqslant h$. Here $\bar{\varepsilon}_b$ is the concentration at the top of the inner layer. Moreover, $\bar{\varepsilon}_s \ll 1$ for $z \geqslant z_b$ and the concentration field in the outer region is given sufficiently accurately by

$$\bar{\varepsilon}_s = \frac{\bar{\varepsilon}_a}{1 - \bar{\varepsilon}_a} \left(\frac{z_a}{z_b}\right)^{p_s} \exp\left(-\frac{p_s \beta_1 (z - z_b)}{h}\right) \tag{39}$$

where $(z_a/z_b)^{p_s} = (10D/h)^{p_s}$. Of course in the continental shelf sediment transport problem a different interior solution must be employed, but it is useful to examine the simplest problems first in order to obtain some physical insight into the systems with which one is dealing. Also it is necessary to use this result in order to explain some fluvial results that bear on the marine geologic problem.

In the steady continental shelf situation there is a benthic Ekman layer with a thickness on the order of u_*/f. Although this region of frictional influence is modified by any upwelling or downwelling forced on the system by the presence of the coast, the turbulent structure of this layer probably does not differ substantially from that found in analogous horizontally uniform situations. Even in the more general uniform Ekman layer situation there is a dearth of information on the momentum diffusion coefficient profile; however, the data reported by Smith (1974) and McPhee (1974) taken under sea ice in the Beaufort Sea yield a maximum eddy viscosity of $K_e \cong 0.012u_*^2/f$. Letting $h_e = 0.45 \, (u_*/f)$ as suggested by the experiment gives $K_e = 0.0667ku_* h_e$ and $\beta_1 = 15$.

These Arctic Ekman layer measurements were made during a three-day storm. The mixed layer was very nearly neutrally stable, and being devoid of wind–wave mixing, provides a good model for a steady benthic boundary layer. Moreover, this experiment is the only one known to me in which complete profiles of density, velocity, and Reynolds stress have been procured simultaneously throughout an entire oceanic boundary layer. On the second day of the three-day storm current speeds between 20 and 25 cm/sec produced an Ekman spiral with a 23° turning angle. The "law of the wall" region extended about 3 m under the ice, whereas the Ekman depth was $h_e \cong 0.45u_*/f \cong 35$ m. The eddy viscosity computed by McPhee (1974) appears to increase linearly through the inner region and then more or less follow the profile calculated by Deardorf (1972) for a neutral atmospheric boundary layer. The fact that the turbulent fields presented by Smith (1974) and McPhee (1974) are in general agreement with theoretical studies of the atmospheric boundary layer and differ primarily because of the effects of under-ice topography in the experimental situation, suggests that flow in benthic Ekman layers probably can be computed with reasonable accuracy using the approaches currently under development for both steady and unsteady atmospheric boundary layers. In any case, $0.45u_*/f$ appears to provide a suitable estimate of the Ekman depth and $0.03u_*/f$ appears to provide a reasonable measure of the thickness of the logarithmic region. These results imply that in temperate latitudes under typical storm conditions, the Ekman layer can develop fully only on the outer continental shelf. The turning angle is reduced on the inner shelf and velocity fields for sediment transport computations in the outer layer must come from field measurements or from an accurate physical oceanographic model in this region.

Of particular interest in many sediment transport problems is the total volume of sediment per unit area of seabed stored in suspension by the turbulence field. In horizontally uniform but unsteady flow the rate of erosion is equal to the rate of change of this parameter (denoted by V_s) divided by the volume concentration of sediment in the bed ε_0, and from a geologic point of view the total amount of sediment that can be put into suspension by a particular storm, $(V_s)_{max}$, is directly proportional to the depth of erosion that results from the storm and the thickness of the layer that is deposited immediately after it. In fact V_s is the tie that binds stratigraphy to physical oceanography. Also, dividing V_s by the depth of the flow yields the vertically averaged concentration—a parameter often used in studies of fluvial sediment transport. By definition

$$V_s = \int_{z_a}^{z_s} \bar{\varepsilon}_s \, dz = \int_{z_a}^{z_b} \bar{\varepsilon}_s \, dz + \int_{z_b}^{z_s} \bar{\varepsilon}_s \, dz \tag{40}$$

where z_s denotes the free surface. Substituting for $\bar{\varepsilon}_s$ from equations 36 and 39, assuming $\bar{\varepsilon}_s \ll 1$, and integrating give

$$
\begin{aligned}
V_s &= \frac{z_a}{1 - p_s}\left(\frac{\bar{\varepsilon}_a}{1 - \bar{\varepsilon}_a}\right)\left[\left(\frac{z_b}{z_a}\right)^{1 - p_s} - 1\right] \\
&\quad + \left(\frac{\bar{\varepsilon}_a}{1 - \bar{\varepsilon}_a}\right)\left(\frac{h}{p_s\beta_1}\right)\left(\frac{z_a}{z_b}\right)^{p_s}\left\{1 - \exp\left[-p_s\beta_1\left(1 - \frac{z_b}{h}\right)\right]\right\} \\
&= 2D\left(\frac{\bar{\varepsilon}_a}{1 - \bar{\varepsilon}_a}\right)\left[\left(\frac{h}{10D}\right)^{1 - p_s}\left(\frac{1}{1 - p_s} + \frac{5}{\beta_1 p_s}[1 - \exp(-0.8p_s\beta_1)]\right) - \frac{1}{1 - p_s}\right]
\end{aligned}
\tag{41}
$$

for $p_s \neq 1$. The only variable in this equation that remains unknown is $\bar{\varepsilon}_a$. As $p_s \to 0$, $V_s \to \bar{\varepsilon}_a h$ as expected, whereas for $p_s \to \infty$, $V_s \to 1/(1 - p_s) \to 0$. In the region between these asymptotes V_s varies with p_s as shown in Fig. 7.

In the derivation of his bed load equation Yalin assumes $\bar{\varepsilon}_a = \gamma_Y S$; however, in order to obtain the correct relationship as the boundary shear stress goes to infinity, it is necessary to rewrite this as

$$\bar{\varepsilon}_a = \frac{\varepsilon_0 \gamma_s S}{1 + \gamma_s S} \tag{42}$$

In the latter expression the concentration at level z_a can never exceed that which is permissible, even though in most situations $\gamma_s S \ll 1$ and this correction essentially is a formality. Substituting this expression into equation 41 gives an equation of the form

$$V_s = \frac{2\gamma_s \varepsilon_0 DS}{1 + \gamma_s S} \, f_1\left(p_s, \frac{D}{h}; \beta_1\right) \tag{43}$$

where p_s and D/h are variables and β_1 is a parameter. The Laursen (1958) total load formula used by Smith and Hopkins (1972) in order to obtain a crude estimate of V_s for the Washington Continental Shelf has the form

$$V_s = DS\left(\frac{D}{h}\right)^{1/6} f_2\left(\frac{u_*}{w_s}\right) \tag{44}$$

With the exceptions that $(D/h)^{1/6}$ has been factored out of the function f_2 and $1 + \gamma_s S$ has been excluded from the denominator, equation 44 is of the same general form as equation 43. Moreover, it is clear from both expressions that the D/h dependence is weak and the fact that Bogardi (1958) tried to revise the Laursen equation with regard to its grain diameter dependence indicates that $(D/h)^{1/6}$ was not in good agreement with his experimental data. Although neither the Laursen nor the Bogardi expression is particularly reliable over a wide range of conditions in rivers, and neither is directly applicable to continental shelves, their structure provides confirmation of

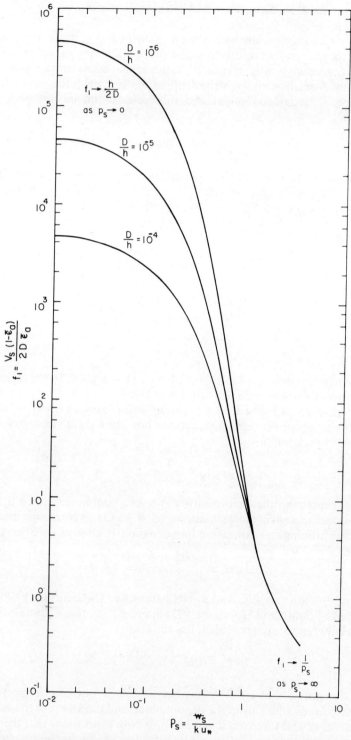

Fig. 7. Variation of the total volume of suspended per unit area of seabed with p_s. For small values of p_s the function goes to $h/(2D)$ and $V_s \to \bar{\varepsilon}_a h$, whereas for large values of p_s this function goes to 0 as $1/p_s$. Therefore for large p_s the only sediment remaining in the fluid is that being transported in the bed load layer.

564

the general form of the expression given in equation 43, including the nearly linear dependence on excess shear stress.

Smith and McLean (in press) find $\gamma_s = 0.0112$ for medium sands when defined as in equation 42. In contrast Einstein (1950) postulates that $\bar{\varepsilon}_a$ is equal to the concentration of sediment in the bed load layer. Then using his bed load function to evaluate Q_s, he then finds $\bar{\varepsilon}_a$ through the empirical relationship

$$\bar{\varepsilon}_a \cong \frac{\alpha_2 Q_s}{2u_* D} \tag{45}$$

where $\alpha_2^{-1} \cong 11.6$. Hopefully, experimental data capable of providing an accurate test of this expression and equation 42 will become available in the next few years, but until they do, both the form and numerical values employed in these expressions must be considered gross estimates, especially when used for silt sized materials. The present inability to compute $\bar{\varepsilon}_a$ accurately is one of the most fundamental difficulties in marine sediment transport.

The volume flux of suspended sediment is defined as

$$Q_s = \int_{z_0}^{h} \bar{u}_s \bar{\varepsilon}_s \, dz = \int_{z_0}^{h} \bar{u}_w \bar{\varepsilon}_s \, dz \tag{46}$$

where $\bar{\varepsilon}_s$ is summed over all classes. Therefore, once the suspended sediment concentration field is known for each sedimentary component, it can be added to all the others, multiplied times the velocity, and integrated. In a continental shelf situation, horizontal diffusion of suspended sediment is negligible, and horizontal advection probably does not change the profiles substantially. Therefore, as a first approximation, one might use a one-dimensional suspended sediment theory in conjunction with the actual velocity field computed from physical oceanographic considerations and attribute the convergences and divergences to deposition and erosion. It should be recognized that the neglect of advection in the sediment diffusion equation eliminates the possibility of properly treating unusual erosion and deposition patterns induced by specific topographic features; however, such patterns should be studied as special problems rather than be included in an overall continental shelf sediment transport model.

In many problems of geophysical interest $A_\varepsilon = \bar{\varepsilon}_a/(1 - \bar{\varepsilon}_a) < 1$ but $\bar{\varepsilon}_a$ is not small enough to permit writing $A_\varepsilon \cong \bar{\varepsilon}_a$. Under these circumstances V_s can be found by expanding the denominator of the integrand with the binomial theorem as follows:

$$
\begin{aligned}
V_s &= \int_{z_a}^{\infty} \frac{A_\varepsilon (z_a/z)^{p_s}}{1 + A_\varepsilon (z_a/z)^{p_s}} \, dz = -z_a \int_{1}^{0} \left(\frac{z_a}{z}\right)^{-2} \frac{A_\varepsilon (z_a/z)^{p_s}}{1 + A_\varepsilon (z_a/z)^{p_s}} \, d\left(\frac{z_a}{z}\right) \\
&= -z_a \int_{1}^{0} \left[A_\varepsilon \left(\frac{z_a}{z}\right)^{p_s - 2} - A_\varepsilon^2 \left(\frac{z_a}{z}\right)^{2p_s - 2} + A_\varepsilon^3 \left(\frac{z_a}{z}\right)^{3p_s - 2} \right. \\
&\qquad \left. - A_\varepsilon^4 \left(\frac{z_a}{z}\right)^{4p_s - 2} + \cdots \right] d\left(\frac{z_a}{z}\right) \\
&= -z_a \int_{1}^{0} \left[\sum_{m=1}^{\infty} (-1)^{m+1} A_\varepsilon^m \left(\frac{z_a}{z}\right)^{m p_s - 2} \right] d\left(\frac{z_a}{z}\right) \\
&= -z_a \sum_{m=1}^{\infty} \frac{(-1)^{m+1} A_\varepsilon^m (z_a/z)^{m p_s - 1}}{m p_s - 1} \Bigg|_{1}^{0} = z_a \sum_{m=1}^{\infty} \frac{(-1)^{m+1} A_\varepsilon^m}{m p_s - 1} \tag{47}
\end{aligned}
$$

for $p_s \neq 1/m$. The restriction on p_s is not a severe one and can be removed by breaking the series into two parts then independently integrating the offending term. However, in most cases it is easier and just as accurate to avoid the singularities and extrapolate across them.

The volume flux of suspended sediment in the nonlinear layer is given by

$$
\begin{aligned}
Q_s &= \int_{z_a}^{\infty} \left(\frac{u_*}{k} \ln \frac{z}{z_0} \right) \frac{A_\varepsilon (z_a/z)^{p_s}}{1 + A_\varepsilon (z_a/z)^{p_s}} \, dz \\
&= \frac{z_a u_*}{k} \int_1^0 \ln \left(\frac{z_0}{z} \right) \left(\frac{z_a}{z} \right)^{-2} \frac{A_\varepsilon (z_a/z)^{p_s}}{1 + A_\varepsilon (z_a/z)^{p_s}} \, d\left(\frac{z_a}{z} \right) \\
&= \frac{z_a u_*}{k} \int_1^0 \ln \left(\frac{z_a}{z} \frac{z_0}{z_a} \right) \sum_{m=1}^{\infty} (-1)^{m+1} A_\varepsilon^{\,m} \left(\frac{z_a}{z} \right)^{mp_s - 2} d\left(\frac{z_a}{z} \right) \\
&= \frac{z_a u_*}{k} \ln \left(\frac{z_a}{z} \frac{z_0}{z_a} \right) \sum_{m=1}^{\infty} \frac{(-1)^{m+1} A_\varepsilon^{\,m} (z_a/z)^{mp_s - 1}}{mp_s - 1} \Big|_1^0 \\
&\quad + \frac{-z_a u_*}{k} \sum_{m=1}^{\infty} \frac{(-1)^{m+1} A_\varepsilon^{\,m} (z_a/z)^{mp_s - 1}}{(mp_s - 1)^2} \Big|_1^0 \\
&= \frac{u_* z_a}{k} \left[\ln \frac{z_a}{z_0} \left(\sum_{m=1}^{\infty} \frac{(-1)^{m+1} A_\varepsilon^{\,m}}{mp_s - 1} \right) + \left(\sum_{m=1}^{\infty} \frac{(-1)^{m+1} A_\varepsilon^{\,m}}{(mp_s - 1)^2} \right) \right]
\end{aligned}
\tag{48}
$$

Again there is a removable singularity at $p_s = 1/m$. In this case as in the preceding one the resulting series can be calculated easily on a digital computer or a programmable calculator for the particular p_s and A_ε of concern.

If the level to which the suspended sediment rises extends significantly outside the linear eddy viscosity layer then integrals representing V_s and Q_s can be broken into two parts, one corresponding to each of the layers. Carrying out these integrations presents no additional complications and for the situation where $\bar{\varepsilon}_s \ll 1$ in the upper layer

$$
\begin{aligned}
V_s = z_a A_\varepsilon \Bigg(& \sum_{m=1}^{\infty} \frac{(-1)^{m+1} A_\varepsilon^{m-1}}{mp_s - 1} \left[1 - \left(\frac{z_a}{z_b} \right)^{mp_s - 1} \right] \\
& - \left(\frac{z_a}{z_b} \right)^{p_s - 1} \frac{h}{p_s z_b \beta_1} \left\{ 1 - \exp \left[-p_s \beta_1 \left(1 - \frac{z_b}{h} \right) \right] \right\} \Bigg)
\end{aligned}
\tag{49}
$$

Similarly, the volume flux of suspended sediment above z_a is given by

$$
\begin{aligned}
Q_s = z_a A_\varepsilon \frac{u_*}{k} \Bigg\{ & \sum_{m=1}^{\infty} \frac{(-1)^{m+1} A_\varepsilon^{m-1}}{mp_s - 1} \left[\ln \left(\frac{z_a}{z_0} \right) - \left(\frac{z_a}{z_b} \right)^{mp_s - 1} \ln \left(\frac{z_b}{z_0} \right) \right] \\
& + \sum_{m=1}^{\infty} \frac{(-1)^{m+1} A_\varepsilon^{m-1}}{(mp_s - 1)^2} \left[1 - \left(\frac{z_a}{z_b} \right)^{mp_s - 1} \right] \Bigg\} \\
& + \frac{-\beta_1}{(p_s \beta_1)^3} \left(\frac{z_a}{z_b} \right)^{p_s - 1} \frac{h}{z_b} \left(\left(\frac{p_s^2 \beta_1^2}{2} (1 + 2B_1) - 1 \right) \exp \left[-p_s \beta_1 \left(1 - \frac{z_b}{h} \right) \right] \right. \\
& \left. - \left\{ p_s^2 \beta^2 \left[-\frac{1}{2} \left(\frac{z_b}{h} \right)^2 + \frac{z_b}{h} + B_1 \right] + p_s \beta_1 \left(1 - \frac{z_b}{h} \right) - 1 \right\} \right)
\end{aligned}
\tag{50}
$$

In both cases the singularities at $p_s = 1/m$ are still present. The interior velocity profile used in the latter expression is

$$u_w = \frac{\beta_1(u_*/k)(2z/h - z^2/h^2 + 2B_1)}{2} \tag{51}$$

where $B_1 = \beta_1^{-1} \ln z_b/z_0 - (1 - z_b/2h)(z_b/h)$ and $z_b/h = \frac{1}{2} \pm (\beta_1 - 4)^{1/2}/(4\beta_1)^{1/2}$. For $\beta_1 = 6.24$, $B_1 = \beta_1^{-1} \ln(0.065h/z_0)$ and $z_b/h = 0.20$. As $A_\varepsilon < 1$ and $\bar{\varepsilon}_s \ll 1$ for $z > z_b$ in most practical problems, the assumption that $\mathbf{u}_s = \mathbf{u}_w - \mathbf{w}_s$ is quite accurate for $z > z_a$, and as the basic equation merely conserves mass, use of equations 50 and 51 cannot lead to much error as long as a good estimate of $\bar{\varepsilon}_a$ is employed in them.

C. Suspended Sediment in Uniform, Unsteady Flow

Although steady-state suspended sediment concentration theory provides an upper limit for V_s and Q_s computations in most continental shelf sediment transport models, time dependence must be included for accurate estimates of these fields. The nonlinear variation of both bed load and suspended load with u_* results in greatly enhanced sediment transport during periods of high currents and negligible sediment transport during periods of low flow. Taking an average boundary shear stress substantially underestimates the net sediment transport. Moreover, computations made by Smith and Hopkins (1972) for the Washington Continental Shelf clearly indicate that most sediment transport takes place under a few extreme events. This finding suggests a simplification, namely, that the most reasonable approach to continental shelf sediment transport problems is to identify the extreme events and to make computations with the greatest accuracy for these periods. They typically are the hurricanes, typhoons, and severe winter storms that produce high currents and large waves.

In horizontally uniform but unsteady flow where $\bar{\varepsilon}_n \ll 1$ the suspended sediment profile is given by the solution to equation 29 near the seabed and equation 28 outside the "law of the wall" region. Under the circumstances where (1) there is a step change in $\bar{\varepsilon}_{an}$ from zero to some finite value, (2) $\bar{\varepsilon}_n$ goes to zero as z goes to infinity because not enough time has elapsed for sediment to diffuse to the sea surface, and (3) p_n is independent of time, a similarity solution to the former equation is possible. Assuming $\bar{\varepsilon}_n = \bar{\varepsilon}_n(\eta)$ where $\eta = z/\upsilon$ reduces equation 29 to

$$\eta\bar{\varepsilon}_n''(\eta) + [(1 + p_n) + \eta]\bar{\varepsilon}_n'(\eta) = 0 \tag{52}$$

Furthermore, postulating $\bar{\varepsilon}_n' = A_\eta \exp f_2(\eta)$ where A_η is a constant with respect to η yields $\eta f'_2(\eta) + (1 + p_n) + \eta = 0$ or $f_2(\eta) = -(1 + p_n)(\ln \eta) - \eta$ and

$$\bar{\varepsilon}_n' = A_\eta \left(\frac{1}{\eta}\right)^{1+p_n} e^{-\eta} \tag{53}$$

Therefore,

$$\bar{\varepsilon}_n = A_\eta \int_\infty^\eta \left(\frac{1}{\eta}\right)^{1+p_n} e^{-\eta} \, d\eta \tag{54}$$

For $p_n \ll 1$ the variation of $\eta^{-1}e^{-\eta}$ with η is much more rapid than the variation of η^{-p_n} so equation 55 can be approximated by bringing η^{-p_n} outside the integral and noting that the remaining part of the original integrand yields an exponential integral (Abramowitz and Stegun, 1964, pp. 228ff.):

$$\bar{\varepsilon}_n \cong A_\eta \left(\frac{1}{\eta}\right)^{p_n} \int_\infty^\eta \frac{1}{\eta} e^{-\eta}\, d\eta = -A_\eta \left(\frac{1}{\eta}\right)^{p_n} E_1(\eta) \tag{55}$$

The exact expression for equation 54 can be obtained in series form as

$$\bar{\varepsilon}_n = A_\eta \int_\infty^\eta \sum_{m=0}^\infty \frac{(-1)^m}{m!}\, \eta^{m-p_n-1}\, d\eta = A_\eta \left(\sum_{m=0}^\infty \frac{(-1)^m \eta^{m-p_n}}{m!(m-p_n)} - \chi_1\right) \tag{56}$$

where

$$\chi_1 = \lim_{q\to\infty} \sum_{m=0}^\infty \frac{(-1)^m q^{m-p_n}}{m!(m-p_n)} \tag{57}$$

and $p_n \neq m$. The singularity is similar to those described above and can be removed if necessary. The profile given by equation 56 when divided through by A_η is shown in Fig. 8 for $p_n = 0.1, 0.3, 0.5, 0.7,$ and 0.9. As η gets large this curve falls off exponentially. When $\eta \to 0$, $\bar{\varepsilon}_n/A_\eta$ gets large; however, this ratio cannot exceed a value of order 10^{-1} because in this region the previously neglected nonlinear terms become important. Close examination of Fig. 8 suggests that the large η asymptote might be extrapolated all the way to the boundary in order to obtain a first approximation to the full solution. A correction to this zero-order profile then could be effected by perturbation methods if a more accurate solution were required.

If the nonlinear layer is below η_α, then at any time the volume of sediment per unit area of bed in suspension above $z = z_\alpha$ is given by

$$V_n = \int_{z_\alpha}^\infty \bar{\varepsilon}_n\, dz = v \int_{\eta_\alpha}^\infty \bar{\varepsilon}_n\, d\eta = vA_\eta \sum_{m=0}^\infty \frac{(-1)^{m+1}\eta_\alpha^{m-p_n+1}}{m!(m-p_n)(m-p_n+1)} - \chi_1\eta_\alpha - \chi_2 \tag{58}$$

where

$$\chi_2 = \lim_{q\to\infty} \sum_{m=0}^\infty \frac{(-1)^{m+1}q^{m-p_n+1}}{m!(m-p_n)(m-p_n+1)} = p_n\chi_1 \tag{59}$$

and $p_n \neq m, m+1$.

The steady-state and similarity solutions represent two asymptotic situations in regard to the general suspended sediment problem in uniform flow. These are particularly valuable for the insight that they give in simple situations; however, they cannot be relied on to provide answers over the wide range of conditions faced in most sediment transport problems. For this purpose a more general technique is required. One such approach is suggested, but not developed, by Smith and Hopkins (1972). These authors recommend looking for a separable solution to equation 29 where the time dependence is sinusoidal, then expanding the results for a general temporal variation by Fourier's method. The simplest means of approaching this problem is to look for a solution of the form $\bar{\varepsilon}_n = \bar{\varepsilon}_\omega \mathscr{R}[e^{i\omega t}Z(z)]$ where $\mathscr{R}$ denotes the real part of the complex function. Making this substitution gives $-i\omega/K_0 Z + (p+1)Z' + zZ'' = 0$ which has a solution in terms of Bessel functions with a complex argument. This result is analogous to the one obtained previously for the unsteady flow problem and can be written in terms of Kelvin functions of order

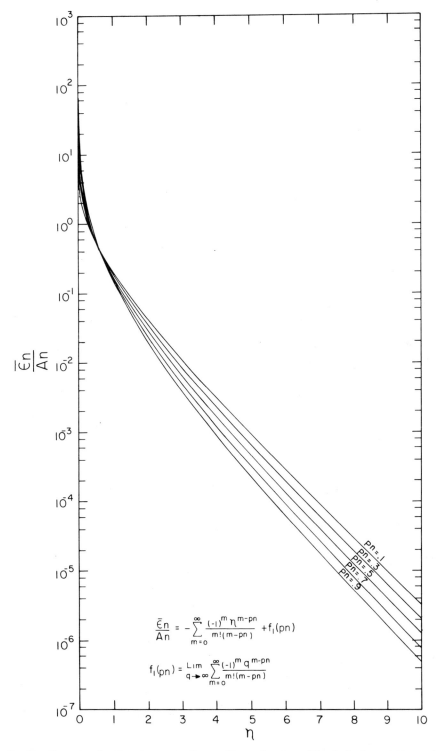

Fig. 8.   Nondimensional sediment concentration profile as a function of the similarity variable η and the parameter p_n for the case of a step increase in ε_a from zero to some finite value.

569

p_n. When the sediment concentration at the reference level $z = z_a$ oscillates sinusoidally with time around a mean value and the boundary layer is of infinite depth, then the concentration field is

$$\bar{\varepsilon}_n = \bar{\varepsilon}_{cn}\left(\frac{\xi_a}{\xi}\right)^{2p_n} + \bar{\varepsilon}_{\omega n}\left(\frac{\xi_a}{\xi}\right)^{p_n}\left(\frac{(\ker_{p_n}\xi)(\ker_{p_n}\xi_a) + (\kei_{p_n}\xi)(\kei_{p_n}\xi_a)}{\ker^2_{p_n}\xi_a + \kei^2_{p_n}\xi_a}\cos\omega t\right.$$

$$\left. + \frac{(\ker_{p_n}\xi)(\kei_{p_n}\xi_a) - (\kei_{p_n}\xi)(\ker_{p_n}\xi_a)}{\ker^2_{p_n}\xi_a + \kei^2_{p_n}\xi_a}\sin\omega t\right)$$

$$= \bar{\varepsilon}_{cn}\left(\frac{z_a}{z}\right)^{p_n} + \bar{\varepsilon}_{\omega n}\left(\frac{z_a}{z}\right)^{p_n/2}[F_1(\xi;\xi_a, p_n)\cos\omega t + F_2(\xi;\xi_a, p_n)\sin\omega t] \qquad (60)$$

where $\xi = 2(\omega z/K_0)^{1/2}$ and $\xi_a = 2(\omega z_a/K_0)^{1/2}$. Here $\bar{\varepsilon}_{cn}$ is the mean value of concentration at $z = z_a$, $\bar{\varepsilon}_{\omega n}$ is the amplitude of the sinusoidal variation, ω is the frequency of this temporal oscillation, K_0 is von Karman's constant times the velocity scale for the diffusion coefficient, and $\ker_{p_n}$, $\kei_{p_n}$ are Kelvin functions of order p_n. At $z = z_a$, $F_1(\xi_a, p_n) = 1$ and $F_2(\xi_a, p_n) = 0$ so $\bar{\varepsilon}_n(z_a, t) = \bar{\varepsilon}_{cn} + \bar{\varepsilon}_{\omega n}\cos\omega t$; however, $\bar{\varepsilon}_n$ cannot be negative so $\bar{\varepsilon}_{\omega n} \leqslant \bar{\varepsilon}_{cn}$ is required.

This result can be generalized to give the suspended sediment concentration field for any size and specific gravity class not interacting with any other above $z = z_a$ as long as p_n is constant with respect to time. However, to accomplish this task the concentration of sediment in class n at $z = z_a$ must be expressible in terms of a Fourier series. Generalizing the concentration field given in equation 60 to

$$\bar{\varepsilon}_n(z, t) = \left(\frac{z_a}{z}\right)^{p_n}A_0 + \left(\frac{z_a}{z}\right)^{p_n/2}\left\{\sum_{m=1}^{\infty}A_m[F_1(\xi;\xi_a, p_n)_m\cos\omega_m t + F_2(\xi;\xi_a, p_n)_m\sin\omega_m t]\right.$$

$$\left. + \sum_{m=1}^{\infty}B_m[F_1(\xi;\xi_a, p_n)_m\sin\omega_m t + F_2(\xi;\xi_a, p_n)_m\cos\omega_m t]\right\} \qquad (61a)$$

yields

$$\bar{\varepsilon}_n(z_a, t) = A_0 + \sum_{m=1}^{\infty}(A_m\cos\omega_m t + B_m\sin\omega_m t) \qquad (61b)$$

at $z = z_a$ as required. A means of finding $\bar{\varepsilon}_n(z_a, t)$ in terms of $(\tau_b/\tau_c - 1)$ is given in the preceding section and methods for computing τ_b are described earlier. Therefore equation 61b reduces to

$$\gamma_n\left(\frac{\tau_b}{\tau_c} - 1\right) = A_0 + \sum_{m=1}^{\infty}(A_m\cos\omega_m t + B_m\sin\omega_m t) \qquad (62)$$

from which the coefficients A_0, A_m, and B_m can be computed in the usual manner using a Fourier transform.

A separate set of coefficients must be computed for each size and specific gravity class used in this expression, because τ_c depends upon the properties of the sedimentary material and γ_n is equal to a constant times the fraction of material in class n available at the seabed for transport. For a sediment transport model all the expressions necessary to calculate $\bar{\varepsilon}_n(z, t)$ can be programmed easily employing generally available subroutines for computing Bessel functions and Fourier transforms. Therefore, $\bar{\varepsilon}_n$ can be determined to the accuracy of the empirical expressions for $\bar{\varepsilon}_n(z_a, t)$ in terms of $(\tau_b/\tau_c - 1)$ and for p_n in terms of u_*. It should be noted that $\bar{\varepsilon}_n$ is slowly varying with $p_n(t)$ relative to $\cos\omega t$ so rather than holding p_n absolutely constant it is acceptable to let it change slowly with time in a particular problem if necessary. Strictly speaking, this procedure should be justified mathematically in each problem for the particular time dependence chosen for p_n.

A typical one harmonic, time-dependent sediment concentration field obtained from equation 60, for $\xi_a = 2[2\omega D/(ku_*)_m]^{1/2} = 0.03$ and $p_n = 0.1$, is shown in Fig. 9. Transforming the origin of the oscillatory part to -1.0 and multiplying by the curve labeled steady solution gives the actual concentration field as a function of ωt when $\bar{\varepsilon}_{cn} = \bar{\varepsilon}_{\omega n}$. This figure shows that upward diffusion of sediment occurs over one half-cycle and downward diffusion over the other. The temporal disturbance penetrates a distance on the order of $\delta_\omega = (u_*)_m/\omega$ into the fluid with oscillations in the concentration profile being heavily damped with distance from the boundary. Taking the thickness of the suspended sediment layer to be $\delta_\omega = (u_*)_m/\omega < h$, where h is the flow depth, gives $T < 2\pi h/(u_*)_m$, so for $h = 100$ m and $(u_*)_m = 2.0$ cm/sec, the period of the oscillation T must be less than 9 hr for the infinite depth solution to be applicable. However, this computation is based on the assumption that the eddy viscosity increases linearly to the free surface; in fact it drops off drastically outside the "law

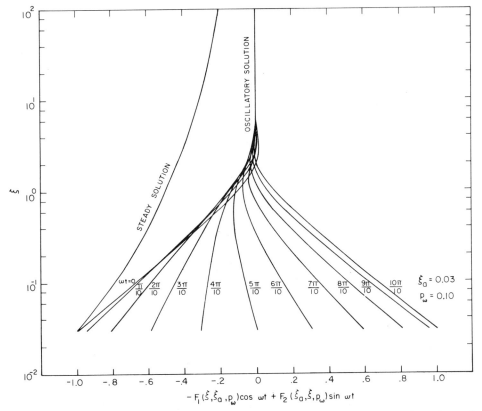

Fig. 9. A typical suspended sediment concentration field normalized by $\bar{\varepsilon}_a$ for a time dependence of $\cos \omega t$. The actual concentration field is given by the product of the curves labeled "oscillatory solution" and the curve labeled "steady solution." The net sediment concentration must be zero or positive. Shifting the origin of the oscillatory solution to -1.0 and multiplying by the absolute value of the steady solution yields the normalized concentration field that would result from a normalized reference concentration varying sinusoidally from 0 to 2.

of the wall" region, suggesting that a matched solution similar to the one used in the steady case is required. In the interior region the eddy viscosity is more or less constant and the proper length scale is $\delta_I = (K/\omega)^{1/2} = [(u_*)_m h/\beta_1 \omega]^{1/2}$. Because the constant eddy viscosity region includes most of the flow, the infinite depth interior solution is satisfactory when $\delta_I < h$ or $T < 2\pi\beta_1 h/(u_*)_m$ and the period of permissible oscillations is increased by $\beta_1 = 0(10)$. If temporal variations of longer period are required in a particular problem then the finite-depth solutions are easily obtained by employing the two additional Kelvin functions $\mathrm{ber}_{p_n} \xi$ and $\mathrm{bei}_{p_n} \xi$. The proper boundary condition at the free surface is that there is no net sediment flux through it.

In the oscillating boundary layer considered previously the velocity profile goes asymptotically to that given by the steady flow equation as the seabed is approached. A similar relationship might be expected in the suspended sediment problem, and if one can be found, inclusion of a thin nonlinear layer near the boundary may be possible. To look for an asymptotic expression, the Kelvin functions are best written as

$$\mathrm{ker}_p \xi + i\,\mathrm{kei}_p \xi = \frac{\pi}{2} i J_p(\xi e^{3\pi i/4}) - \frac{\pi}{2} Y_p(\xi e^{3\pi i/4}) \tag{63}$$

where J_p is a Bessel function of the first kind of order p and Y_p is a Bessel function of the second kind of order p. Thus

$$\mathrm{ker}_p \xi = -\frac{\pi}{2} \frac{J_p(\xi e^{3\pi i/4}) \cos p\pi - J_{-p}(\xi e^{3\pi i/4})}{\sin p\pi} \tag{64a}$$

$$\mathrm{kei}_p \xi = \frac{\pi}{2} J_p(\xi e^{3\pi i/4}) \tag{64b}$$

where $e^{3\pi i/4} = (-i)^{1/2}$ and $\xi = 2(\omega z/K_0)^{1/2}$. The series expansion for $J_p(\zeta)$ is

$$J_p(\zeta) = \left(\frac{\zeta}{2}\right)^p \sum_{k=0}^{\infty} \frac{(-\zeta^2/4)^k}{k!\,\Gamma(p+k+1)}$$

where Γ denotes the gamma function (Abramowitz and Stegun, 1964, p. 253). In the limit as $\zeta \to 0$, $J_p(\zeta) \to (\zeta/2)^p/\Gamma(p+1) \to 0$ and $J_{-p}(\zeta) \to (\zeta/2)^{-p}/\Gamma(-p+1) \to \infty$ as long as p is not a negative integer. Use of these expressions in conjunction with equations 64a and 64b shows that as $\xi \to 0$

$$\mathrm{ker}_p \xi \to \frac{\pi(\xi/2)^{-p}(e^{3\pi i/4})^{-p}}{2\Gamma(-p+1)\sin(\pi p)} \tag{65a}$$

$$\mathrm{kei}_p \xi \to \frac{\pi(\xi/2)^p(e^{3\pi i/4})^p}{2\Gamma(p+1)} \to 0 \tag{65b}$$

Therefore, as $\xi \to \xi_a$

$$F_1(\xi; \xi_a, p_s) \to \left(\frac{\xi_a}{\xi}\right)^{p_s} \tag{66a}$$

$$F_2(\xi; \xi_a, p_s) \to \left(\frac{\xi_a}{2}\right)^{2p_s} \ll 1 \tag{66b}$$

and

$$\bar{\varepsilon}_s = \bar{\varepsilon}_a\left(\frac{z_a}{z}\right)^{p_s} + \bar{\varepsilon}_\omega\left(\frac{\xi_a}{\xi}\right)^{2p_s} = \bar{\varepsilon}_a\left(\frac{z_a}{z}\right)^{p_s} + \bar{\varepsilon}_\omega\left(\frac{z_a}{z}\right)^{p_s} \tag{67}$$

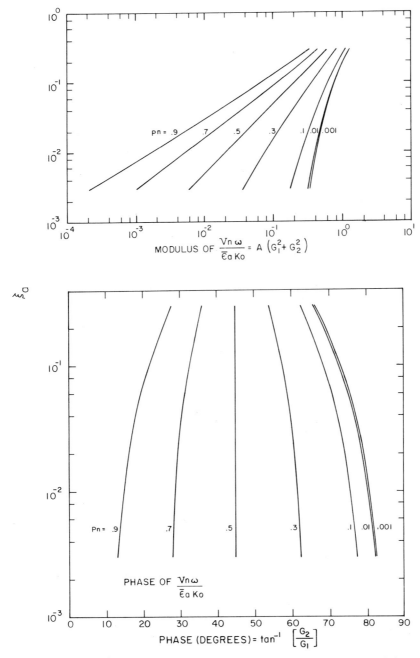

Fig. 10. Amplitude and phase angle of the nondimensional volume of material suspended above a point on the seabed per unit area of bed as a function of ε_a and p_n for a sinusoidally varying reference concentration ε_a.

In the steady flow problem for $\bar{\varepsilon}_s \ll 1$

$$\frac{\bar{\varepsilon}_s + \alpha_1}{1 + \alpha_1} = \frac{\bar{\varepsilon}_a + \alpha_1}{1 - \bar{\varepsilon}_a + \alpha_1} \left(\frac{z_a}{z}\right)^{(1 + 4\alpha_0)^{1/2} p_s} \tag{68}$$

from equation 34 so $(1 + 4\alpha_0)^{1/2} = 1$ and $\alpha_0 = \alpha_1 = 0$. As long as $\bar{\varepsilon}_s \ll 1$ in the upper part of the region where this asymptotic expression is valid then the nonlinear solution can be employed in the lower layer. Moreover, one can check on the validity of this weak assumption by examining the sediment concentration computed with $\alpha_0 = 0$ from equation 37 at the level where the concentration profile is no longer approximated by the asymptotic expression. This fortunate result means that in cases where the bed material acts as a single component the entire suspended sediment profile can be determined with accuracy far greater than $\bar{\varepsilon}_a$ can be computed. In cases where the suspended sediment field must be broken into separate size and specific gravity classes the nonlinear region must be treated as described for the steady uniform flow problem by Hunt (1954).

As in the steady flow situation, variation of the total volume of suspended sediment with p_n and ξ_a is of considerable interest and is given by

$$\begin{aligned} V_n = \bar{\varepsilon}_a \frac{\xi_a K_0 \sqrt{2}}{\omega(\ker_{p_n}^2 \xi_a + \kei_{p_n}^2 \xi_a)} & [(\ker_{p_n-1} \xi_a - \kei_{p_n-1} \xi_a) \ker_{p_n} \xi_a \\ & + (\kei_{p_n-1} \xi_a + \ker_{p_n-1} \xi_a) \kei_{p_n} \xi_a] \cos \omega t \\ & + [(\ker_{p_n-1} \xi_a - \kei_{p_n-1} \xi_a) \kei_{p_n} \xi_a \\ & - (\kei_{p_n-1} \xi_a + \ker_{p_n-1} \xi_a) \ker_{p_n} \xi_a] \sin \omega t \end{aligned} \tag{69}$$

where the actual volume of suspended sediment at any time is the difference between the steady solution and V_n. The variation of this parameter with p_n is shown in Fig. 10.

An expression generally called the *erosion equation* is derived by defining the sea-bed to be the top of the uppermost layer of nonmoving grains and then integrating equation 24 from below this boundary to above the free surface of the fluid. The two discontinuities must be crossed and this is accomplished by conserving mass at each one. The equation that results relates the rate of deposition to V_s and Q_s in the form

$$\frac{\partial \eta}{\partial t} = \frac{-1}{\varepsilon_0} \left(\frac{\partial V_s}{\partial t} + \nabla \cdot \mathbf{Q}_s\right) \tag{70}$$

Using this relationship in a continental shelf sediment transport model, once the V_s and Q_s fields have been computed, permits areas of erosion and deposition to be identified.

6. Geologic Implications

Up to this point the emphasis has been placed on procedures for carrying out sediment transport computations on continental shelves. These are designed to be used with a comprehensive physical oceanographic model in order to predict sediment fluxes and erosion rates. Nevertheless, certain general conclusions of geological importance can be derived directly from the nature of the sediment transport equations. The highly nonlinear character of these equations means that the most severe storms have a substantially greater effect than subsequent less intense ones. Erosion to a depth of many centimeters under typical shelf conditions is possible, and the equations derived in this chapter show that these sediments are redeposited in a graded layer. A homogeneous mixture of the material transported as bed load is found first. This is followed by a layer comprised of the sediments that were transported in suspension; the degree of sorting improves as the size decreases. The general grading results from the fact that the finer particles diffuse to higher levels in the flow during the extreme conditions and then settle back to the seabed more slowly at the end of the event. The sorting decreases with distance from the top of the layer because the finer materials are found at all levels, and some are deposited with the coarse sediment.

The type of deposition that occurs in an area where the sediment transport is of bed material, that is, where the sedimentary material entering the area during a particular event is derived from the surrounding sea floor, is given by the frequency of events that erode the sea floor sufficiently deeply to leave a mark on the stratigraphic record times the thickness of the layer that they erode divided by the long-term mean sedimentation rate. If this nondimensional number is large, then only the coarsest material is added to the strata and layering is almost nonexistent, whereas if this number is small, the geological deposit is graded with the bed load and coarsest suspended load material at the base of layer. The ratio of layer thickness to erosion depth increases toward unity as this parameter gets smaller. Typically the upper part of each graded layer is eroded by subsequent storms, and only a small fraction of the fine material entering the area gets deposited in a layer that will be preserved. Therefore, the material at the sediment–water interface is likely to be transient, and investigations of surface sediments have to be interpreted very carefully if stratigraphic implications are to be derived from them. Once deposited, organisms may substantially rework the strata but this is not always the case; even when it is, a clear understanding of the physical processes operating in a particular depositional environment is essential if a full interpretation of the geologic record is to be obtained. Sediments homogenized owing to the action of benthic animals probably can be distinguished from the heavy homogeneous deposits that result when the above-mentioned nondimensional number gets large if a full suite of modern microsedimentological techniques is employed.

7. Conclusions

An attempt has been made in this chapter to provide a means for computing sediment transport in a complicated natural system, in particular on the continental shelf. Emphasis has been focused on boundary layer mechanics and on the physics of sediment transport. Unfortunately much of the theory presented here has not been confirmed by actual field measurements, and some may exceed our current experimental capability. Although critical components of the theory are consistent with what I have found to be the best available experimental data, further tests are required and these undoubtedly will bring about substantial improvements.

The approach taken in this chapter is founded upon the assumption that particle–fluid interactions are more important than particle–particle interactions. This certainly is true in situations where sediment concentrations are low, but remains open to question in the region near the bed under intense bed load and suspended load transport. This fundamental question must be addressed carefully and answered. In addition, the accuracy of boundary shear stress computations must be examined carefully under these circumstances, for only when nonuniform effects are properly treated and residual errors accurately estimated will one be able to distinguish among available bed load and suspended load equations and make suggestions for their improvement.

Probably the weakest link in the procedure outlined herein is computation of the sediment concentration at a reference level. The approaches that have been used are based neither on comprehensive theoretical calculations nor on careful experiments addressed specifically to this problem. Research in both these areas is needed critically; moreover, accurate measurements of time-dependent sediment concentration fields made simultaneously with precision measurements of current velocity and seabed topography are required to determine the accuracy with which computations can be carried out following the procedures outlined in this chapter.

Once these gaps are filled and reasonably good physical oceanographic models of open continental shelves become available, the exciting prospect of carrying out sediment transport computations for recent and paleo-oceanographic situations, and thereby predicting observed stratigraphy, may be realized. At present, continental shelf sediment transport modeling is in an embryonic stage awaiting breakthroughs in sediment mechanics and physical oceanography.

Acknowledgments

Many of the ideas expressed in this chapter were generated during boundary layer and sediment transport investigations supported by NSF Grant GA14178 and AEC Contract AT(45-1)-2225-TA-25. The substantial assistance provided by C. E. Long in computing and plotting the functions presented in many of the figures as well as a number of fruitful discussions with C. E. Long and K. F. Jones about the ideas presented here are gratefully acknowledged, as are the long hours of difficult typing by Rise Mercier.

References

Abramowitz, M. and I. A. Stegun, 1964. *Handbook of Mathematical Functions with Formulas, Graphs, and Mathematical Tables.* Dover, New York, pp. 228ff., p. 253.

Arya, S. P., 1975 (in press). A drag partition theory for determining the large scale roughness parameter and wind stress on Arctic pack ice. *AIDJEX Bull.*, **28**.

Bagnold, R. A., 1956. The flow of cohesionless grains in fluids. *Phil. Trans. Roy. Soc. London*, A, **249**, 235–297.

Bagnold, R. A., 1966. An approach to the sediment transport problem from general physics. *U.S. Geol. Surv. Prof. Paper 422I*, 37 pp.

Bagnold, R. A., 1973. The nature of saltation and "bed-load" transport in water. *Proc. Roy. Soc. London*, A, **332**, 473–504.

Bogardi, J. L., 1958. The total sediment load of streams: a discussion. *Proc. Am. Soc. Civ. Eng. J. Hydraul. Div.*, **84** (HY6), 74–79.

Businger, J. A. and S. P. S. Arya, 1974. The height of the mixed layer in the stably stratified planetary boundary layer. *Adv. Geophys.*, **18A**, 74–92.

Collins, J. I., 1963. Inception of turbulence at the bed under periodic gravity waves. *J. Geophys. Res.*, **68**, 6007–6014.

Deardorf, J. W., 1972. Numerical investigation of neutral and unstable planetary boundary layers. *J. Atmos. Sci.*, **29**, 91–115.

Dobbins, W. E., 1944. Effect of turbulence on sedimentation. *Trans. Am. Soc. Civ. Eng.*, **109**, 629–678.

Einstein, H. A., 1950. The bed-load function for sediment transportation in open channel flows. *U.S. Dept. Agr. Soil Conserv. Serv., Tech. Bull. 1026*, 71 pp.

Hunt, J. N. (1954). The turbulent transport of suspended sediment in open channels. *Proc. Roy. Soc. London*, A, **224**, 322–335.

Inman, D. L. and R. A. Bagnold, 1963. Littoral processes. In *The Sea.* Vol. 3. M. N. Hill, ed. Wiley-Interscience, New York, pp. 529–553.

Jonsson, I. G., 1967. Wave boundary layer and friction factor. *Proc. 10th Conf. Coastal Eng.*, pp. 127–148.

Komar, P. D. and D. L. Inman, 1970. Longshore sand transport on beaches. *J. Geophys. Res.*, **75**, 5914–5927.

Komar, P. D. and M. C. Miller, 1975. Sediment threshold under oscillatory waves. *Proc. 14th Conf. Coastal Eng.*, pp. 756–775.

Laursen, E. M., 1958. The total sediment load of streams. *Proc. Am. Soc. Civ. Eng., J. Hydraul. Div.*, **84** (HYI), 1530-1–1530-36.

McPhee, M. G., 1974. An experimental investigation of the boundary layer under pack ice. Ph.D. thesis, University of Washington.

Owen, P. R., 1964. Saltation of uniform grains in air. *J. Fluid Mech.*, **20**, 225–242.

Schlichting, H., 1960. *Boundary layer theory.* 4th ed., McGraw-Hill, New York.

Smith, J. D., 1969. Studies of non-uniform boundary layer flows. In *Investigations of Turbulent Boundary Layer and Sediment Transport Phenomena as Related to Shallow Marine Environments*, Part 2, U.S. At. Energy Comm. Contract AT (45-1)-1752. Ref: A69-7. Department of Oceanography, University of Washington.

Smith, J. D., 1974. Turbulent structure of the surface boundary layer in an ice-covered ocean. In *Proceedings of a Symposium on The Physical Processes Responsible for the Dispersal of Pollutants in the Sea Particularly in the Nearshore Zone.* Vol. 167. Rapports et Proces-Verbaux, Int. Council Exploration of the Sea, pp. 53–65.

Smith, J. D. and T. S. Hopkins, 1972. Sediment transport on the continental shelf off of Washington and Oregon in light of recent current measurements. In *Shelf Sediment Transport.* D. J. P. Swift, D. B. Duane, and O. H. Pilkey, eds. Dowden, Hutchison & Ross, Stroudsburg, Pa., pp. 143–180.

Smith, J. D. and S. R. McLean, (in press) Spatially averaged flow over wavy boundaries. *J. Geophys. Res.*

Sternberg, R. W., 1972. Predicting initial motion and bed load transport of sediment particles in the shallow marine environment. In *Shelf Sediment Transport.* D. J. P. Swift, D. B. Duane, and O. H. Pilkey, eds. Dowden, Hutchinson & Ross, Stroudsburg, Pa., pp. 61–82.

Sternberg, R. W. and N. B. Kachel, 1971. Transport of bed load as ripples during an ebb current. *Mar. Geol.*, **10**, 229–244.

Vanoni, V. A., 1964. Measurements of critical shear stress for entraining fine sediments in a boundary layer. *Calif. Inst. Technol., Keck Hydraul. Lab. Rept. KH-R-7*, 47 pp.

White, S. J., 1970. Plane bed thresholds of fine grained sediments. *Nature*, **228**, 152–153.

Yalin, M. S., 1963. An expression for bed load transportation. *J. Hydraul. Div. Proc. Am. Soc. Civ. Eng.*, **89** (HY3), 221–250.

14. THEORETICAL MODELS OF FLOW NEAR THE BED AND THEIR IMPLICATIONS FOR SEDIMENT TRANSPORT

P. A. TAYLOR AND K. R. DYER

1. Introduction

Flow above a noncohesive granular bed frequently gives rise to transport of sediment both as suspended matter and bed load. This movement can deform an initially plane bed and produce a range of bed forms. Such features are common in the sea.

Measurements of sand transport in flumes have shown that transverse bed forms occur if fluid velocities are sustained above the threshold of movement (Gilbert, 1914; Liu, 1957; Simons et al., 1961). These wavelike features start as fairly low-amplitude ripples which have avalanche lee slopes with angles equal to the angle of repose of the grains (about 30°). In steady uniform flow these features propagate their form down-flow (Southard and Dingler, 1971). With increasing flow, the ripples transform into larger features, dunes, which have dimensions related to the flow depth. The geometric properties of these dune features have been examined by Yalin (1964), who shows that there is a critical dune height to water depth ratio of $\frac{1}{6}$, above which two-dimensional dunes should not occur, and that the wavelength of the dunes is approximately five times the water depth. The critical steepness (height to wavelength ratio, $2a/L$) is $\frac{1}{30}$. The dunes are distinguished from the ripples whose size is a function of grain diameter, but the two features can exist together. The critical height to depth ratio is supported by the spectral analysis of Hino (1968).

Simons et al. (1961) found in flume experiments, with water depths up to 30 cm and velocities up to 90 cm/sec, on a 0.45-mm diameter sand, that dunes were restricted to Froude numbers between 0.38 and 0.60. Their steepness was the same as ripples produced at lower Froude numbers, but the ratio of dune height to flow depth was $\frac{1}{3}$ to $\frac{1}{5}$. The dunes progressed downstream at 0.015 to 0.36 cm/sec, the higher rates being associated with steeper slopes and shallower water. They also found that a plot of u_*/w_s against $u_* k_s/v$ (where w_s is the fall velocity of grains of diameter k_s, u_* is the friction velocity, and v is the kinematic viscosity), for a variety of grain sizes, gave a series of curves as limits to the formation of particular bed forms. These curves suggested that ripples may no longer form when $k_s > 2.5$ mm and dunes may not exist when $k_s > 5$ mm.

Analysis of wavelike bed forms has been carried out in terms of continuous spectra by Nordin and Algert (1966), Hino (1968), and Jain and Kennedy (1974).

In the sea features similar to dunes occur. They are generally known as sand waves and can reach large sizes with heights of 18 m and wavelengths of 900 m (Stride, 1963). Their maximum height is $\frac{1}{10}$ to $\frac{1}{5}$ of the water depth. Associated with these features are smaller ones, typically 1 m in height, known as megaripples or dunes, which are the marine analogues of the ripples seen on dunes in flumes and whose size does not appear to be related to the water depth.

Sand waves are common on continental shelves and form extensive fields in some areas. The field off the coast of Holland has an area of about 15000 km^2 (McCave, 1971). Sand waves are associated with banks (Kirby and Kelland, 1972; Langhorne, 1973) and the continental shelf edge (Cartwright, 1959); they occur in hollows (Harvey, 1966), and as intermediate stages on a bed load transport path (Stride, 1963;

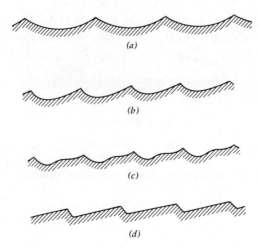

Fig. 1. Forms of marine sand waves, after Van Veen (1935). Vertical scale exaggerated. (a) trochoidal, (b) asymmetric trochoidal, (c) cat-back, (d) progressive.

Belderson and Stride, 1966). Sand waves have a variety of cross-sectional and plan forms. Van Veen (1935) distinguished four forms of sand wave (Fig. 1); trochoidal, asymmetric trochoidal, cat-back, and progressive. These were related to the tidal currents, the symmetrical forms occurring in areas of equal ebb and flood currents and the asymmetric forms in areas dominated by one tidal stage. Variation in form of sand waves between flood and ebb stages has been reported by several authors. Cat-back forms may also be indicative of crest bifurcation.

The variation in plan form has been detailed by Kenyon and Stride (1968). They distinguish long crested, sinuous, and barchan-shaped sand waves, and suggest a sequence of form with sediment supply and with the shape of the tidal current ellipse. For sinuous sand waves it is suggested that there is a plentiful supply of sediment and an elongated tidal current ellipse. For straight crested sand waves there is a more restricted sediment supply. Barchan-shaped features are formed where the current ellipse is almost circular and the sediment supply is restricted.

Unlike features in flumes, however, sand waves occur in areas where the Froude number is generally less than 0.1, and though they are normally found in areas of sand size material, gravel waves in material of mean size 25 mm occur (Dyer, 1971). Also, the slopes of the faces of the sand waves are different from dunes in flumes. Even in the asymmetric form the angle of the steeper, lee slope seldom exceeds 10° and that of the stoss, or upstream, slope is generally half this value (Langhorne, 1973). Ludwick (1972) finds lee slopes averaging 1.5°, only occasionally reaching 6°. Terwindt (1971) obtains values ranging up to 12°. In spite of the lower lee slopes the steepness of the sand waves is similar to that of dunes in unidirectional flow, typically being about $\frac{1}{30}$. This means that the stoss slopes must be steeper for sand waves. Symmetrical sand waves often show greater steepness than asymmetric ones.

The movement of sand waves has been studied by many people. Jones et al. (1965) on a sand bank south of the Isle of Man measured rates of advance of the sand wave crests of 50–100 cm/day. Both Langeraar (1966) and Terwindt (1971) found the movements of sand waves in the southern North Sea over several years to be less than the errors of navigation. Kirby and Oele (1975), however, measured movements of up to 40 m in 3 yr on the Sandettie Bank. Langhorne (1973) reports intermittent movements near the Thames Estuary of up to 25 m/yr, but which were more a flexing of the crest than a steady widespread advance. Salsman et al. (1966) working in a tidal bay in Florida measured rates of advance averaging 1.35 cm/day and Ludwick (1972) obtained movements of 35–150 m/yr at the mouth of Chesapeake Bay. Ludwick also found that the migration of symmetrical sand waves was insignificant.

In spite of the large amount of field description of marine sand waves, it is still not clear to what extent movement of these features reflects the sediment transport rate. Sediment moving up to the crest may be carried off into suspension. Also the presence of dunes with avalanche faces superimposed on the sand waves may indicate sediment traveling as a carpet over the form of the sand wave. This could produce higher transport rates than would be calculated from the movement of the sand wave itself. McCave (1971) uses the argument of an increasing proportion of suspended load to account for the decrease in sand wave height to the north off the Dutch coast.

In the absence of good field measurements of sediment transport rates over a sand wave and their variation over a tidal cycle, quantitative predictions are difficult to make. Understanding of these processes may be assisted by modeling the water flow and applying existing sediment transport theory. In general much of this theory relates to horizontal averages over any bed form that might be present, whereas we require relationships that are valid locally. Some theories that might be adapted to meet this requirement are considered in the next section.

2. Turbulent Flow and Sediment Movement above Plane Beds

In order to keep things as simple as possible we take the constant stress layer near a rough wall as the underlying theme for our review. We can then assume that we are dealing with a region of flow near the bottom of a deep turbulent boundary layer which is driven by a stress (τ) applied at the top of the layer and within which we may neglect body forces and externally applied pressure gradients. Clearly this is far from true in many important sediment transport situations and differs somewhat from the open channel flow favored in engineering studies. It does, however, considerably simplify the situation although it precludes, for the present, effects associated with finite water depth and the presence of a free surface. The flow near the bed is assumed to be hydraulically rough (see Schlichting, 1968) so that the flow is independent of viscosity. This implies that the drag on the wall is essentially comprised of form drag on the individual roughness elements rather than the viscous shear stress on a smooth wall and needs modification in cases where the sediment is very fine. Under these circumstances it is well established, experimentally and from dimensional arguments (e.g., Monin and Yaglom, 1971, p. 274), that, in the absence of any density stratification, the mean velocity profile may be represented by

$$\overline{U}(z) = \frac{u_*}{\kappa} \ln \frac{z}{z_0} \tag{1}$$

where z is the distance from the bed, u_* ($= \tau/\rho$) is the friction velocity, ρ being the water density, and κ is von Karman's constant, which we take as 0.4 in spite of recent suggestions (Businger et al., 1971) that the value may be somewhat lower in geophysical situations. The roughness of the underlying surface is characterized by the roughness length, z_0, the logarithm of which is the value of the intercept of the velocity profile when extrapolated to $\overline{U} = 0$ as in Fig. 2. For mathematical modeling purposes

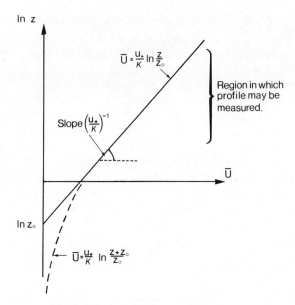

Fig. 2.   The logarithmic velocity profile.

it is convenient to modify the equation above and assume, as Rossby has suggested (see, e.g., Sutton, 1953, p. 83), that the profile is of the form

$$\overline{U}(z) = \frac{u_*}{\kappa} \ln \frac{z + z_0}{z_0} \tag{2}$$

so that $\overline{U} = 0$ on $z = 0$. For boundaries composed of uniform sand z_0 is found, empirically, to be approximately related to the sand grain diameter, k_s, by

$$z_0 \simeq \frac{k_s}{30}$$

(see Schlichting, 1968). We should not expect equations 1 or 2 to be valid very close to the wall (say, $z < 10\, k_s$) but they do give a convenient theoretical framework within which to work. In terms of mixing length (l) and eddy viscosity (K_M) models of turbulent boundary layers we would have, corresponding to the profile of equation 2:

$$l(z) = \kappa(z + z_0)$$

and

$$K_M(z) = u_* l(z) = \kappa u_*(z + z_0) \tag{3}$$

Alternatively we may make use of the turbulent energy as a flow variable and set

$$K_M(z) = \lambda^{1/2} E^{1/2} l(z) \tag{4}$$

where λ is a constant equal to the equilibrium value of the ratio u_*^2/E (0.16–0.32 being typical values) and $E[= \frac{1}{2}(\overline{u'^2} + \overline{v'^2} + \overline{w'^2})]$ is the turbulent energy/ρ. This approach appears to have originated in the work of Kolmogorov, Prandtl, and Rotta (see Monin and Yaglom, 1971, p. 378) and has recently been applied to several atmospheric boundary layer problems.

The foregoing view of unidirectional flow above a rough surface will clearly not always be applicable. If the surface were not sufficiently rough, or the flow not at a sufficiently high Reynolds number, then a slightly different approach is needed. This may be found, for example, in Yalin (1972) or Schlichting (1968). We could also modify the model quite easily to consider the open channel flow situation.

We now ask whether or not a given flow above a boundary consisting of loose, cohesionless, uniform solid particles will induce any movement of bed material. Such sediment movement is usually divided into the two categories of suspended load and bed load but the division is not entirely clear-cut. We consider them separately.

A. Suspended Load

The modeling of suspended load can be undertaken with turbulent diffusion keeping sediment in suspension against gravitationally produced settling velocities. The basis for these models is discussed at length in the recent texts by Raudkivi (1967), Graf (1971), and Yalin (1972). Implicit in the treatment of suspended sediment movement via a diffusion equation is that the horizontal velocity of the sediment is identical with that of the fluid, whereas the vertical velocities differ by the terminal settling velocities of the solid particles of sediment. With mean velocity components of the fluid as $\overline{U}$, $\overline{W}$ (in a two-dimensional situation with the z axis vertical) and with $\overline{C}$ representing the mean concentration of sediment (mass fraction, volume fraction, or grams per cubic centimeter) we can write an equation for conservation of C, for small concentrations of a uniform grain size sediment, as

$$\frac{\partial \overline{C}}{\partial t} + \overline{U}\frac{\partial \overline{C}}{\partial x} + (\overline{W} - w_s)\frac{\partial \overline{C}}{\partial z} = \frac{\partial}{\partial x}\left(K_s\frac{\partial \overline{C}}{\partial x}\right) + \frac{\partial}{\partial z}\left(K_s\frac{\partial \overline{C}}{\partial z}\right) \qquad (5)$$

We assume that turbulent diffusion can be represented by an isotropic, but not necessarily constant, diffusion coefficient K_s. If we have steady horizontally homogeneous flow above an infinite horizontal plane surface then equation 5 reduces to

$$-w_s\frac{\partial \overline{C}}{\partial z} = \frac{\partial}{\partial z}\left(K_s\frac{\partial \overline{C}}{\partial z}\right)$$

which integrates to give

$$w_s\overline{C} = -K_s\frac{\partial \overline{C}}{\partial z} \qquad (6)$$

as the equilibrium situation with settling balanced by upward diffusion (for an alternative derivation see Monin and Yaglom, 1971, p. 412). Hunt (1954) derives a modified form of this equation for use when the volume occupied by the sediment is no longer small. With $K_s = K_M$ (and defined by equation 3 or 4) and also assuming (which may be rather suspect) that z_0 is the same for momentum and sediment concentration, we may integrate equation 6 to give an equilibrium profile

$$\overline{C} = C_0\left(\frac{z + z_0}{z_0}\right)^{-w_s/\kappa u_*} \qquad (7)$$

where C_0 is a surface concentration of sediment. The latter is a very necessary concept if we wish to consider nonequilibrium situations but it is presumably impossible to measure except by extrapolation. We might reasonably expect C_0 to be dependent on bed shear stress in addition to the physical nature of the bed.

In practice equation 7 is usually used (see Graf, 1971, p. 173) in a form (due to Rouse, 1937) relating to flow in an open channel and comparing $\bar{C}$ to the concentration at a reference level. In this context there appears to be quite good agreement with laboratory and field data in the form of equation 7 but not necessarily in the value of the exponent, $-w_s/\kappa u_*$ (Chien, 1956; Graf, 1971, p. 174). These differences could be due to the nonequality of K_s and K_M and may also be associated with the way in which the settling velocities were determined. One factor, which we have so far neglected, is the stable density gradient induced by decreasing sediment concentrations as we move further away from the bed. As long as concentrations and density changes remain very small the eddy diffusivity may be expected to remain similar to that given by equation 3 or 4 but if the suspended sediment has a significant stabilizing effect on the fluid there will be a tendency to inhibit turbulent fluctuations and to reduce the effective eddy diffusivity. This effect will normall be more pronounced the further one moves from the boundary. Observations in open channel flows indicate that observed values of the exponent in equation 7 are sometimes less than the predicted value of $-w_s/\kappa u_*$ ($\kappa = 0.4$) and that this discrepancy increases with increasing sediment concentration. This observation is consistent with a stratification effect, as are some of the velocity profiles observed in such flow situations (Vanoni and Nomicos, (1960). These are often interpreted in terms of a decrease in κ. It appears to us to be appropriate, at this stage, to introduce some ideas based on studies of the atmospheric boundary layer under stable stratification.

Most modern accounts of the surface or constant flux layer of the atmospheric boundary layer (the lowest 10–30 m) use Monin–Obukhov (1954) similarity as a basis. In terms of eddy viscosity or diffusivity we can, in the framework of mixing-length theory, express K_M in the form

$$K_M = \frac{\kappa u_*(z + z_0)}{\Phi_M} \tag{8}$$

where the "nondimensional wind shear" Φ_M is assumed to be a function of $z/\mathscr{L}$. The Monin–Obukhov length, $\mathscr{L}$, can be defined by

$$\mathscr{L} = \frac{\bar{\rho}u_*^3}{\kappa g \rho'w'}$$

If we set

$$\overline{\rho'w'} = -K_s\frac{\partial\bar{\rho}}{\partial z}$$

and

$$\rho = \rho_0(1 + \gamma C) \tag{9}$$

where ρ_0 is the clear fluid density, and assume $K_s = K_M$, this gives

$$\mathscr{L} = -\frac{u_*^2 \Phi_M}{\kappa^2 g\gamma(z + z_0)(\partial\bar{C}/\partial z)} \tag{10}$$

Recent observations in stably stratified, constant flux atmospheric boundary layers suggest that Φ_M is adequately represented by

$$\Phi_M = 1 + \beta\frac{(z + z_0)}{\mathscr{L}} \tag{11}$$

(the $+z_0$ term is very small and included for mathematical convenience). The constant β has a value of about 5.2 (Webb, 1970; Businger et al., 1971; Dyer, 1974), which is used later in this chapter. Equation 11 corresponds to the log-linear velocity profile

$$\bar{U} = \frac{u_*}{\kappa}\left(\ln\frac{z + z_0}{z_0} + \beta\frac{z}{\mathscr{L}}\right) \tag{12}$$

in situations where both u_*, the buoyancy flux $(-g\overline{\rho'w'}/\rho_0)$, and hence $\mathscr{L}$ are constant throughout the layer. In our situation the buoyancy flux will not be constant with height and we need to apply equation 11 in a situation slightly different from that in which it has been established. For simplicity the preceding analysis has been given in terms of mixing-length theory, rather than in terms of models using the turbulent energy equation. Adaptation to such models could be undertaken with a few extra assumptions without altering the results that follow.

With these reservations we can write equation 6 as

$$w_s\bar{C} = \frac{-\kappa u_*(z + z_0)(\partial\bar{C}/\partial z)}{1 + \beta(z + z_0)/\mathscr{L}} \tag{13}$$

If we assume as before that the lower boundary condition is $\bar{C} = C_0$ on $z = 0$, and if we let $\zeta = \ln[(z + z_0)/z_0]$ and $X = \ln(\bar{C}/C_0)$, we can rewrite the equation above, after some manipulation, as

$$\frac{dX}{d\zeta} = -B(1 + Ae^Xe^\zeta) \tag{14}$$

where

$$A = \frac{\beta w_s \kappa g\gamma C_0 z_0}{u_*^3} \quad \text{and} \quad B = \frac{w_s}{\kappa u_*}$$

This will integrate to give equilibrium profiles defined by

$$X = -B\zeta - \ln\left[1 + \frac{AB}{(1 - B)}(e^{(1 - B)\zeta} - 1)\right] \tag{15}$$

The corresponding velocity profiles take the similar form

$$U = \frac{u_*}{\kappa}\left[\zeta + \frac{1}{B}\ln\left\{1 + \frac{AB}{(1 - B)}(e^{(1 - B)\zeta} - 1)\right\}\right] \tag{16}$$

and are essentially mirror images of the X profiles.

The results given here for velocity and sediment concentration profiles have been previously presented by Barenblatt (1953, 1955) (see Monin and Yaglom, 1971, p. 416 for a brief summary) using Kolmogorov's ideas for closure of the turbulent energy equation. This work seems to have passed unnoticed in the recent Western literature on the subject.

We illustrate the form of the profiles with some examples in Fig. 3. In choosing typical values it is convenient not to be tied by a relationship between z_0 and suspended sediment particle diameter, and so in most of the cases we assume that the material in suspension may be much finer than that causing the drag; that is, we have flow above a sandy gravel.

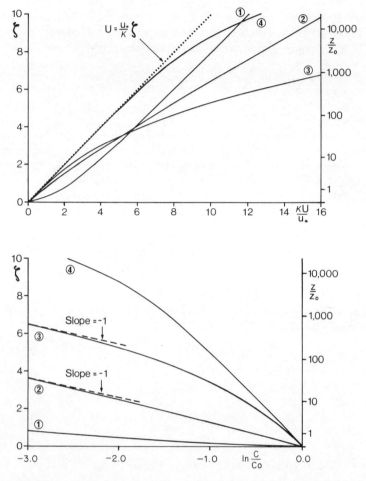

Fig. 3.   Velocity and sediment concentration profiles for the four cases discussed in the text.

Case 1: $B = 1.4$, $A = 5.0$

These values, for moderate flow above quite coarse sandy gravel, are attained if we set $u_* = 4$ cm/sec, $w_s = 2.24$ cm/sec (particle diameter $\simeq 0.02$ cm), $z_0 = 1.0$ cm, and $\gamma C_0 = 0.07$. In this case, with $B > 1$, the sediment concentration falls off rapidly with z and the influence on the velocity profile is confined to the lower layers. At higher elevations the velocity profile retains the same slope as in the sediment free situation.

Case 2: $B = 0.56$, $A = 0.32$

These values arise if we set $u_* = 10$ cm/sec, corresponding to a quite strong current, and keep all other parameters as in Case 1. In this case the sediment concentration still falls off quite rapidly away from the bed but as a result of the increasing effectiveness of stratification as we move away from the bed the velocity profile is modified for all z. In particular, for large ζ we have, provided $B < 1$,

$$U \simeq \frac{u_*}{B\kappa}\left[\zeta + \ln\frac{AB}{1-B}\right] \tag{17}$$

which could be interpreted as a logarithmic velocity profile with modified values of z_0 and the von Karman constant κ, this value being simply w_s/u_*. Indeed in the open channel flow situation and in the sea (McCave, 1973) this interpretation is often given to observations of the velocity profile (see Graf, 1971, p. 179 for details). It may be cautiously applied to some flume observations of Vanoni and Nomicos (1960). For set III of their observations, for which the bed was essentially flat and $B < 1$ they give $w_s/u_* \simeq 0.25$, whereas the effective von Karman constant computed from the velocity profile is approximately 0.23.

Case 3: $B = 0.2$, $A = 0.1$

Here we have taken $u_* = 4$ cm/sec, $w_s = 0.32$ cm/sec [$d \sim 0.006$ cm], $\gamma C_0 = 0.1$, and $z_0 = 0.1$ cm. Again this is sandy gravel but with a much finer sand than before. This gives a lower settling velocity and in consequence the sediment is able to diffuse through a deeper layer. The effect on the velocity profile is quite pronounced. It is perhaps worth noting that the velocity profiles are plotted after scaling with u_* and if we instead scaled with respect to $\bar{U}$ at 10 m, say, we would see that increasing sediment concentrations have the effect of reducing drag coefficients.

Case 4: $B = 0.2$, $A = 0.001$

This is as Case 3 but with $z_0 = 0.001$ cm, which is not very far from the value appropriate to the (hydraulically smooth) flow above this rather fine sand. The stratification effects appear less pronounced near the wall but if we take account of the different z_0 values in the vertical scale the results are not dissimilar from Case 3.

The foregoing discussion of models of suspended sediment transport in horizontally homogeneous conditions above plane beds should indicate that there is still room for some improvement. In particular, an extension and reevaluation of Barenblatt's ideas might be undertaken.

Recently Mei (1969), Apmann and Rumer (1970), and Hjelmfelt and Lenau (1970) have considered a leading edge problem where an initially clear fluid encounters an abrupt change in bottom conditions from a solid to a sediment surface. By using equation 5 with $\partial/\partial t = 0$ and neglecting horizontal diffusion they solve the equation for suspended sediment concentration with the approximation that $\bar{U}$ is constant and with either $K_s = $ const or (Hjelmfelt and Lenau) $K_s = \kappa u_* z[1 - (z/h)]$, where h is the depth of channel considered. The effects of density stratification are not included and they assume $B \ll 1$. Again we may draw analogies with the meteorological situation where the modeling of similar "internal boundary layers" has been undertaken. The techniques used there could perhaps form the basis for a study of the leading edge problem for sediment transport.

B. Bed Load

The movement of bed load on essentially plane beds is somewhat more straightforward to deal with and is usually regarded as being governed by the nondimensional Shields parameter:

$$\theta = \frac{\tau}{(\sigma - \rho)gk_s} \tag{18}$$

where σ is the grain density. At the threshold of movement there is a critical value of this parameter and $\theta_{cr} \sim 0.06$ for steady, uniform flow over an hydraulically rough bed. Dunes form at about $\theta = 0.10$ and are eventually washed out and the bed becomes planar again at about $\theta = 0.4$ (Bagnold, 1956). The movement of grains over dunes has been described by many authors (e.g., Raudkivi, 1963). If the bed slopes significantly, as may be the case locally when sand or gravel waves are present, θ becomes modified by the component of the grain weight acting downslope. This has been considered by Bagnold (1956) and by Fredsøe (1974), who replaces θ by θ^*, defined as

$$\theta^* = \theta + \mu \tan \beta \qquad (19)$$

where $\tan \beta$ is the slope of the bed, measured positive if the flow is up the slope, and μ is a dynamic friction coefficient. He expects its value to be about 0.1. This value corresponds to the critical threshold $\theta_{cr}^* = 0$ on a slope of 22°.

Several bed load transport equations have been developed for planar beds which contain modifications that make them applicable over sloping ones. The work of Bagnold (1956, 1966) has been quite widely used in the oceanographic literature (e.g., Sternberg, 1972). Bagnold (1966) considers that the rate of work done in pushing the bed load along the bed against a frictional resistance is $i_b (\tan \alpha + \tan \beta) \cos \beta$, where i_b is the bed load transport rate and α is the angle of frictional resistance. The power in the flow available to move the sediment is $\tau \overline{U}$, where $\overline{U}$ is the mean stream velocity. Thus

$$i_b = \frac{\tau \overline{U} e_b}{(\tan \alpha + \tan \beta) \cos \beta} \qquad (20)$$

where e_b is an efficiency factor which has a value of the order 0.1, varying slightly with flow velocity and grain size (Bagnold, 1966). The angle of frictional resistance α can have values in the range $\tan \alpha = 0.32$–0.75 depending on the moving grain size, concentration, and rate of shearing. However, it is often considered to equal the angle of repose of the material. Equation 20 shows that the bed load transport rate on an avalanche slope is notionally infinite. In fact such high grain concentrations would be produced that the dispersive pressure would effectively increase $\tan \alpha$ above the static frictional value. Equation 20 is not supposed to be valid for $\theta < 0.4$ and is essentially the same formula as that developed by Bagnold (1956), except that the threshold shear stress is considered negligible compared with the applied shear stress. In the earlier paper the available fluid power is considered as $(\tau - \tau_{cr})\tau^{1/2}$ or $(\theta - \theta_{cr})\theta^{1/2}$. This formula fits measured transport values over a wider range of flow conditions. For lower flow stages the efficiency factor has been shown by Kachel and Sternberg (1971) from measurement of ripple migration rates to be a function of the excess shear above the threshold.

Other useful bed load transport equations include those by Meyer-Peter and Müller (1948) and Einstein (1950). The former differs from the Bagnold (1956) formula essentially by considering the fluid power as $(\theta - \theta_{cr})^{3/2}$, but at high flow stages the results are similar. Both these approaches, however, require separation of the shear stress on the bed into two components, the friction component available to move sediment and a form drag component. This requires that the formulas be applied for average transports over areas larger than an individual sand wave.

3. Flow over Sand Waves

Measurements of flow over ripples and dunes in flumes have shown that the flow pattern has several characteristic features. On the upstream face of the dune the rising elevation of the bed causes an acceleration of the near-bed flow and there is a decrease in surface pressure. The velocity gradient near the bed and also the bed shear stress consequently increase toward the crest, to a maximum value about equal to that for the same flow over a flat bed (Raudkivi, 1963). Upstream of the crest where the slope starts to become convex rather than concave, the velocity gradients and the bed shear stress diminish slightly. At the crest, where the steep avalanche slope commences and pressure is at a minimum, flow separation often occurs. The zone of high current shear leaves the boundary and rejoins it about six ripple heights downstream of the crest (Raudkivi, 1963). Beneath this zone of high shear is the lee-eddy or ground-roller, in which the flow is a weakly rotating vortex giving flow up the steep slope. At the reattachment point the turbulent intensities are greatest and surface pressures maximum (Raudkivi, 1966). Rifai and Smith (1971) show turbulent intensities, normalized by the local mean velocity, reaching 45% near the reattachment point with minimum intensities of 15% occurring near the crest. At a level about one wave height above the crest the turbulent intensities were almost uniform over crest and trough. At this level the mean flow streamlines are approximately sinusoidal, though out of phase with the actual sand boundary. Rifai and Smith (1971) also show that the macroscale of the turbulence is related to the height of the features.

The shear stress distribution is discontinuous, being virtually zero in the sheltered zone and rising to a maximum near the crest. The total drag includes the form drag as well as the surface drag or shear stress. The form drag is the integral of the longitudinal component of the pressure distribution over the length of the wave and the total drag over the wave can be more than twice that on a flat bed (Khanna, 1970).

Separation, however, does not necessarily occur over all features. If the pressure increase on the lee side is not too abrupt then the flow need not separate, though there will be deceleration of the flow near the boundary. Flume and wind tunnel experiments have also been carried out on symmetrical, generally sinusoidal, wavelike forms. Motzfeld (1937) investigated flow over a sequence of three sinusoidal waves of varying steepness. He failed to obtain separation even with steepnesses of $\frac{1}{10}$. Smith (1969) over two waves of steepness $\frac{1}{15}$ also did not obtain separation. However, Hsu and Kennedy (1971) consider separation to occur at steepnesses greater than $\frac{1}{17.5}$. Benjamin (1959) has reviewed the evidence for the occurrence of separation and concludes that it occurs with increasing fetch and at steepnesses greater than about $\frac{1}{20}$. Motzfeld (1937) obtained separation on a trochoidal wave profile of the same steepness as one which with a sinusoidal profile did not separate. As one might expect separation generally does occur with a sharp crest and a lee slope of avalanche steepness. On the other hand, it is not necessary that separation occur over marine sand waves because of the lower slopes involved.

Nonseparating flow over wavy surfaces has been examined by several workers. Hsu and Kennedy (1971) report results in wavy tubes for two wave steepnesses and show that the maximum bed shear stress occurs 26° and 18° upstream of the crest, the larger value being for the steeper wave. The maximum shear stress was also 25% larger for the steeper wave, though the trough values remained about the same. The velocity profiles were matched with a power-law distribution with an exponent varying over the wave form. Comparison of the turbulence intensities with those for a straight tube showed somewhat greater intensities in the longitudinal component, but values less than half those in a straight tube for the component normal to the wall. On the other hand, Kendall (1970) obtained values for shear stress that had a maximum 70° upwind of the crest. This is considerably at variance with the results of other authors.

Khanna (1970) gives a series of illustrative velocity profiles over a steep $\frac{1}{15}$, asymmetrical, but rounded, profile and found that the intensity of turbulence reached a maximum of 15% in the troughs and was 6% near the crest, somewhat lower than values for separated flows. The use of turbulent intensities scaled with respect to local velocities in this context is possibly misleading. In the case of nonseparating flows we might expect the highest surface values of the turbulent kinetic energy to occur at or just upstream of the crest.

Measurement of the flow over sand waves in the sea has been carried out by Dyer (1970). The velocity profiles, which are similar to those in nonseparating flows in the laboratory, are described. A positive shear stress of some magnitude was found in the lee, and no separation was evident. The maximum shear stress at the crest was about four times that in the trough. The velocity profiles could also be matched to a power law with exponent varying with position over a sand wave. Velocity profiles have also been obtained by Ludwick (1974), and Terwindt (1971) reports velocities of 0.8 m/sec, 0.5 m above a sand wave crest and 0.5 m/sec the same height above the trough.

Several groups are at present actively engaged in measuring turbulent parameters in the boundary layer near the seabed that might affect sand wave generation (Gordon, 1974; Heathershaw, 1974) and over sand waves themselves (J. D. Smith, personal communication). The presence of dunes superimposed on sand waves not exhibiting separation suggests that the dune features contain most of the bed load which is moving over a more stationary sand wave form rather like traffic on a hilly road.

4. Theoretical Models of Flow above Sand Waves and of Bed Form Mechanics

Much of the interest in the flow above sand waves arises in relation to the study of the formation of the waves themselves, often referred to as bed form mechanics. Perhaps because the flow is then of secondary interest, there has been a tendency to treat it in a rather approximate way. In his classic work on the topic Exner (1925) (see Graf, 1971; or Raudkivi, 1967) used a one-dimensional steady-state model and, in effect, assumed $\overline{U}$ to be uniform throughout a flow cross section, in a channel of finite depth. Reynolds (1965) and Raudkivi (1966) have developed similar one-dimensional models while Gradowczyk (1968) using time-dependent equations for the flow, links the problem of bed form development to surface wave propagation.

Many models assume the flow to be inviscid, irrotational, and governed by potential flow theory. The basic solution for such a flow over a simple sinusoidal wavy bottom is given by Milne-Thompson (1960). Several authors, for example, Anderson (1953), Kennedy (1963, 1969), Hayashi (1970), and Jain and Kennedy (1974) have linked this to equations for sediment transport and studied the evolution of bed forms as a stability problem. Potential flow theory has also been used by Mercer (1971) and Mercer and Haque (1973) who, in a study concentrating on the flow, allow for different bed forms and for the existence of a separated eddy zone in the lee of the sand wave crest. Mercer and Haque also discuss the use of a model of rotational, but still frictionless, flow to match observed velocity and pressure data.

Smith (1970), Engelund (1970), and Fredsøe (1974) have added internal friction to the fluid flow equations via an eddy viscosity coefficient. This coefficient (K_M) is assumed constant but the basic velocity profiles used by Engelund and Fredsøe also make use of a "slip velocity," U_{b0}, at the bed. The profiles used by Engelund are of the form

$$\overline{U}(z) = U_{b0} + \frac{du_*^2}{K_M}\left(\frac{z}{d} - \frac{1}{2}\left(\frac{z}{d}\right)^2\right) \tag{21}$$

for uniform flow in a channel of depth d, whereas Smith's form has $U_{b0} = 0$. For flow over a wavy boundary small perturbations from this form are considered. Fredsøe, with a slightly different basic profile, extends the analysis to second order and considers the causes of the asymmetric development of dunes. These are also discussed by Smith, as is the situation in which there is separation behind the crests. A model of flow downstream of the point of separation based on boundary-layer concepts is proposed by Smith (1969).

Closely related to the study of flow above sand waves is the investigation of wind–wave generation. The main differences are that the lower boundary is moving and that more attention is given to the normal stresses, which appear to be responsible for most of the energy transfer from wind to wave, than to the shear stresses. The theories of Miles (1957), Benjamin (1959), and others, for flow above monochromatic small-amplitude water waves have recently been extended by Long (1971), Davis (1972), and Townsend (1972), who include the effects of changes in shear stress in their models. In these cases this is achieved by adding the turbulent energy equation to the system to be solved and making appropriate closure hypotheses at that level. In particular they follow Bradshaw et al. (1967) in assuming that τ is proportional to E, which allows them to close the set of equations. Townsend gives some results for a fixed wave at the bottom of a deep turbulent boundary layer with $\ln(kz_0) = -8$, where k is the wave number $(2\pi/L)$ of the underlying sinusoidal bed. This could, for example, correspond to a wavelength of about 19 m if we take $z_0 = 0.1$ cm. As a result of the linearization all quantities have a sinusoidal variation with x, the horizontal coordinate perpendicular to the wave crests. The surface shear stress perturbation (from its average over the bed form) predicted for this case has amplitude $4.6\ ak\rho u_0^2$ where ρu_0^2 is the average horizontal stress on the bed and a is the amplitude of the underlying wave. The shear stress maximum is predicted to occur at a phase of $21.6°$ $(0.06L)$ upstream of the crest, independent of wave amplitude. The assumption of linearity may be a little too restrictive if we are interested in flow over fully developed sand waves. Townsend argues that the linear approximation fails for $ak > 0.1$. This value, of maximum wave slope, corresponds to a "steepness" of about $\frac{1}{30}$ for a sinusoidal bed form, about the middle of the range of values in which we are interested.

5. A Numerical Model of Flow above "Gentle Topography"

Recently Taylor et al. (1976) and Taylor and Gent (1974) have started to develop a series of numerical models of flow above "gentle topography" by which they imply that there is no separation of the mean flow. Particular situations considered so far include the flow above an essentially sinusoidal, periodic lower boundary and flow above an isolated two-dimensional hill. They find it convenient to conformally map the flow region considered prior to attempting a numerical solution of their governing equations so that the lower boundary coincides with one of the coordinate axes. As an example Taylor et al. (1976) use the transformation

$$z = \zeta^* + ia\ \exp(ik\zeta^*) \tag{22}$$

where $z = x + iy$ and $\zeta^* = \xi + i\eta$, between real (x, y) space and the transformed (ξ, η) region. In this case the lower boundary (in real space) is given, parametrically, by

$$\begin{aligned} y_b &= a\cos k\xi \\ x &= \xi - a\sin k\xi \end{aligned} \tag{23}$$

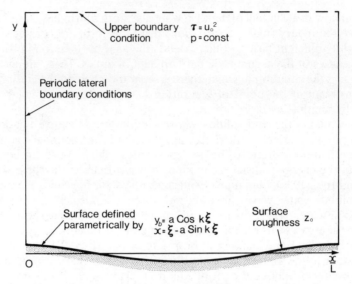

Fig. 4.   Boundary conditions for model of flow over sand waves.

which is quite close to a cosine wave with wave number k and amplitude a as shown in Fig. 4. Note that y is now the vertical coordinate. For any transformation of essentially this type the governing mean momentum equations, in (ξ, η) space, may be written, for steady mean flow and excluding gravity and hydrostatic pressures, as,

$$\bar{U} \frac{\partial \bar{U}}{\partial \xi} + \bar{V} \frac{\partial \bar{U}}{\partial \eta} - \frac{1}{2J} (J_\eta \bar{U}\bar{V} - J_\xi \bar{V}^2) = - \frac{\partial \bar{p}^*}{\partial \xi} - \frac{\partial}{\partial \xi} (\overline{u'^2} - \overline{v'^2})$$

$$- \frac{\partial}{\partial \eta} (\overline{u'v'}) + \frac{1}{2J} (J_\xi (\overline{u'^2} - \overline{v'^2}) + 2J_\eta \overline{u'v'})$$

(24)

and

$$\bar{U} \frac{\partial \bar{V}}{\partial \xi} + \bar{V} \frac{\partial \bar{V}}{\partial \eta} - \frac{1}{2J} (J_\xi \bar{U}\bar{V} - J_\eta \bar{U}^2) = - \frac{\partial \bar{p}^*}{\partial \eta} - \frac{\partial}{\partial \xi} (\overline{u'v'})$$

$$+ \frac{1}{2J} (J_\eta (\overline{v'^2} - \overline{u'^2}) + 2J_\xi \overline{u'v'}) \qquad (25)$$

where $\bar{U}$, $\bar{V}$ and u', v' are mean and fluctuating velocity components parallel to the ξ, η coordinate system and $\bar{p}^* = \bar{p} + \overline{v'^2}$. $J = |dz/d\zeta^*|^{-2}$ is the Jacobian of the transformation. In addition the mean continuity equation can be written as

$$\frac{\partial}{\partial \xi} (J^{-1/2}\bar{U}) + \frac{\partial}{\partial \eta} (J^{-1/2}\bar{V}) = 0 \qquad (26)$$

There are now various possibilities for the closure of this system of equations. The choice made by Taylor et al. was to add the turbulent energy equation to the system above and, in addition to closure assumptions therein, to assume by analogy with equation 4 that

$$-\overline{u'v'} = \lambda^{1/2} l E^{1/2} \frac{\partial}{\partial \eta} (J^{1/2} \overline{U}) \tag{27}$$

The mixing length was specified as

$$l = \kappa(\eta + z_0)(1 - e^{-k\eta}) + \kappa(s + z_0)e^{-k\eta} \tag{28}$$

so as to be proportional to $\kappa(s + z_0)$ near the bed and $\kappa(\eta + z_0)$ well away from it, s being the distance, in (x, y) space, from the lower boundary along the lines $\xi = $ const. A value of 0.25 was used for λ in the results presented here. In addition they assumed a fixed partitioning of the turbulent energy equation with

$$\overline{u'^2} = 1.0E; \qquad \overline{v'^2} = 0.35E; \qquad \overline{w'^2} = 0.65E$$

They apply periodic lateral boundary conditions and set $U = V = \partial E/\partial \eta = 0$ on the lower boundary, $\eta = 0$; and $\bar{p}^* = 0$, $E = u_0^2/\lambda$, and $\tau = u_0^2$ on the upper boundary, taken at about $\eta = 2.5L$.

The basic case considered is for a sand or gravel wave with wavelength 10 m and amplitude 0.25 m giving $L/z_0 = 10{,}000$ with $z_0 = 0.1$ cm and $ak = 0.157 (\simeq$ maximum wave slope). Results are obtained for the mean velocity and pressure fields, shear stress, and turbulent energy distribution and for the stresses on the bed. In the context of sediment transport these are perhaps the most interesting and are shown in Fig. 5. The variation in shear stress with position on the wave is much as we might expect but is larger than we would anticipate on the basis of irrotational flow theory. The total horizontal stress on the bed now has a form drag contribution which in this case amounts to 22% of the total; thus the average shear stress on the bed is only $0.78u_0^2$. The results obtained show that the pressure field slowly decays with distance from the bed, keeping more or less the same phase, but that the turbulent shear stress and

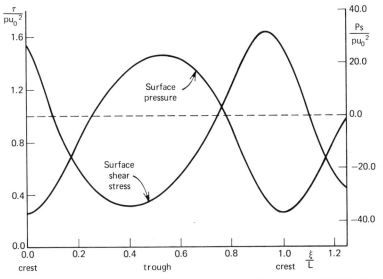

Fig. 5. Surface shear stress and pressure over the wave form of Fig. 4. $L/z_0 = 10{,}000$, $ak = 0.157$.

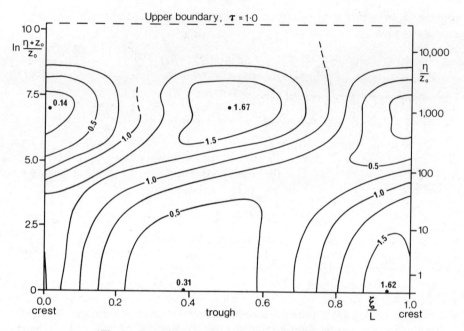

Fig. 6. Shear stress contours; $L/z_0 = 10,000$, $ak = 0.157$.

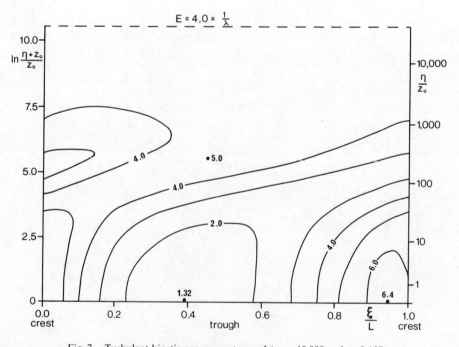

Fig. 7. Turbulent kinetic energy contours; $L/z_0 = 10,000$, $ak = 0.157$.

turbulent energy fields have phases (in ξ) which advance with η. In the case of the stress field we find elevated maxima and minima at a height of about $0.1L$ above the surface and positioned about $180°$ out of phase with the surface features. Thus there is a shear stress maximum above the trough and a minimum above the crest as shown in Fig. 6. It should be noted that the vertical scale in this figure is logarithmic and that actual depths or thicknesses associated with the elevated extrema are much greater than for the surface features. The turbulent energy field, shown in Fig. 7, displays a similar pattern but with less pronounced elevated extrema. Taylor et al. obtain good agreement with Townsend's (1972) linear theory for small-amplitude waves and compare their results with Kendall's (1970) and other data.

It should be stressed that their model assumes a deep turbulent boundary layer with no free surface effects and no mean flow separation and requires that the flow has attained true periodicity. In practice this necessitates the flow to be observed after passing over a series of uniform waves—probably at least 10—if quantitative agreement is to be expected. The model can be adapted to deal with asymmetric waves by changing the conformal mapping (equation 22). Some results for this case are given in the next section.

6. Implications of the Model for Sediment Transport

A. Symmetrical Sand Wave

The model discussed in Section 5 was used to calculate the shear stress distribution over a symmetrical wave. The wave form was almost sinusoidal, but had slightly sharpened crests and consequently could approximate a symmetrical trochoidal sand wave. Its steepness was $\frac{1}{20}$, which gave maximum slopes of about $9°$ (Fig. 8). The shear stress distribution is shown in Fig. 8a. The maximum shear stress occurs $19°$ upstream of the crest, which is somewhat less than that predicted by Benjamin (1959), but compares with that obtained on a wave of steepness $\frac{1}{45}$ by Hsu and Kennedy (1971). The minimum stress, which is about $\frac{1}{5}$ of the maximum value, occurs about $45°$ upstream of the trough. Flow in the opposite direction across the sand wave produces a mirror image distribution about the crest.

Let us consider a tidal cycle of equal magnitude currents in each direction with a step change between. The bed load transport can be approximated using equation 20 with the fluid power calculated as $(\tau/\rho u_0^2)^{3/2}$, and considering for the moment the efficiency factor e_b as a constant. The resulting curves for the variation in relative bed load transport rate across the sand wave are shown in Fig. 8b. The relationship of the transport rate with bed shear stress and the effect of bed slope produces a maximum transport rate $27°$ upstream of the crest and a minimum $54°$ upstream of the trough. Continuity requires that $\partial y_b/\partial t \propto \partial i_b/\partial x$. Consequently deposition occurs when i_b is decreasing and erosion when i_b is increasing respectively with x. Figure 8c shows diagrammatically the differential of the bed load transport curves. Maximum erosion occurs halfway between trough and crest on the upstream side of the sand wave and maximum deposition about $25°$ downstream of the crest. This would lead to a downstream progression of the bed form for unidirectional flow. The net effect over a tidal cycle (Fig. 8d) would be to produce deposition within about $65°$ either side of the crest, with a maximum at the crest, and maximum erosion in the trough. This would cause the sand wave to assume a more trochoidal form with a sharper crest and a heightening leading to increased steepness. With a more trochoidal form it is possible that erosion on one face during one stage of the tide would balance the deposition on it

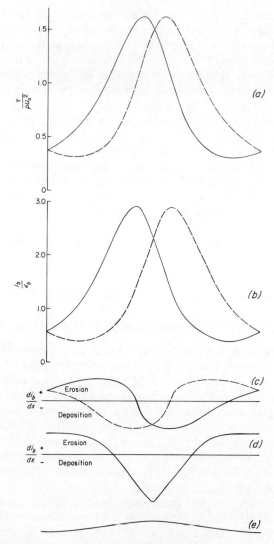

Fig. 8. Shear stress and bedload transport for flow in opposite directions over a symmetrical sand wave. Solid line, flow left to right; broken line, flow right to left. (a) Normalized shear stress distribution; (b) bed load transport rates; (c) erosion and deposition; (d) net erosion and deposition; (e) sand wave shape.

when the flow is in the opposite direction, leading to a stable wave form. If e_b were a variable and a function of τ, the width of the zones of erosion and deposition would be altered, and the rate of heightening of the sand wave either increased or decreased depending on whether e_b decreased or increased with τ, respectively.

B. Asymmetric Sand Wave

The shear stress distribution was also calculated for flow both ways over an asymmetric bed form. The steepness was again $\frac{1}{20}$, but the lee slope had a maximum angle of 14° and the stoss slope about $6\frac{1}{2}°$ (Fig. 9e). In form it approximated an asymmetric

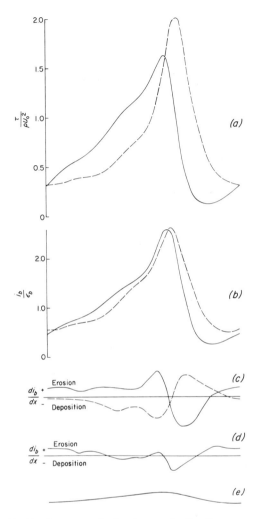

Fig. 9. Shear stress and bed load transport for flow in opposite directions over an asymmetric sand wave. Solid line, flow left to right; broken line, flow right to left. (*a*)–(*d*) as in Fig. 8.

trochoidal sand wave. In this case, however, it was simplest to consider equal stresses at the top of the constant stress layer rather than equal currents for the opposing flow directions. This is because of the different values of form drag for flow with or against the asymmetry. In the tidal situation it can be thought of as equal water surface slopes rather than equal velocities.

Flow with the asymmetry is almost separating (Fig. 9*a*), the shear stress falling almost to zero about halfway down the lee face. On the stoss side the stress increases almost linearly to a maximum at the crest. For flow against the asymmetry the maximum stress occurs some way down the steeper slope and is about 25% larger than the maximum for the other flow direction. Also the flow is obviously less near to separating. The relative bed load transport rates were calculated as before. Because of the surface slopes the inequalities in the shear stress distributions (Fig. 9*a*) are less apparent in the bed load transport curves for the two directions (Fig. 9*b*), which are almost in phase and of similar amplitude. However, the erosion rate of the stoss slope during flow with the asymmetry is slightly greater than the deposition during the opposite flow, especially in the vicinity of the trough (Fig. 9c). On the steep lee slope deposition is greater than erosion just in the lee of the crest. The net effect over a tidal cycle (Fig. 9d) would be a steepening of the lee slope and a slight movement of the crest to the right. This would lead to separation and the feature would develop into a progressive type sand wave. The process would be enhanced if there were an asymmetry in the tidal flows giving a residual current in the direction of the bed form asymmetry, as is often observed. A mechanism by which an asymmetric wave could be maintained as nonseparating, yet move in the direction of the steeper facing slope, may be related to the possible functional relationship between e_b and τ. If the two curves in Fig. 9b can be made coincident and have maximum and minimum values slightly downstream of the crest and trough, respectively, the whole wave form is shifted without modification. Attempts to match the curves by making $e_b = f(\tau)$ have been unsuccessful, but have led to the conclusion that the excess shear rather than the efficiency factor e_b may have a different relationship to the shear stress depending on whether the flow is accelerating or decelerating.

7. Conclusions

Semi-empirical models of turbulent boundary layers can be applied to flow near mobile beds and used to predict the transport of sediment. All such models are somewhat unsatisfactory in that empirical closure hypotheses have to be made at some level. We would favor the use of a system based on closure in the turbulent energy equation, but the so-called "higher-order closure" techniques proposed by Lumley and Khajeh-Nouri (1974) and now being applied in several atmospheric boundary layer situations could in the future possibly be applied to sediment transport problems.

It should be stressed that the models presented of flow and bed load transport over sand waves are rather tentative. Development of these models, for example, to allow for finite water depth and the inclusion of a threshold bed stress is desirable. The inclusion of suspended sediment, possibly using Barenblatt's approach, is a further possibility but it is important that the models be carefully tested against experimental data, obtained both in flumes and in the field.

Acknowledgments

Some of the flow modeling work was undertaken by Mr. J. M. Keen while employed as a research assistant to P. A. Taylor. Funds for this were provided under a Natural Environment Research Council Grant (GR3/1932).

References

Anderson, A. G., 1953. The characteristics of sediment waves formed by flow in open channels. *Proc. 3rd Midwest. Conf. Fluid Mech., Minneapolis*, pp. 379–395.

Apmann, R. P. and R. R. Rumer, 1970. Diffusion of sediment in developing flow. *Proc. Am. Soc. Civ. Eng.*, **96** (HY1), 109–123.

Bagnold, R. A., 1956. Flow of cohesionless grains in fluids. *Proc. Roy. Soc. London, A*, **249**, 235–297.

Bagnold, R. A., 1966, An approach to the sediment transport problem from general physics. *U.S. Geol. Surv. Prof. Paper 422I.*

Barenblatt, G. F., 1953. On the motion of suspended particles in a turbulent stream. *Prikl. Matem. Mekh.*, **17**, 261–274 (Engl. trans.).

Barenblatt, G. F., 1955. Concerning the motion of suspended particles in turbulent flow occupying a half-space, or flat open channel of finite depth. *Prikl. Matem. Mekh.*, **19**, 61–88 (English translation by National Lending Library Russian Translating Programme).

Belderson, E. H. and A. H. Stride, 1966. Tidal current fashioning of a basal bed. *Mar. Geol.*, **4**, 237–257.

Benjamin, T. B., 1959. Shearing flow over a wavy boundary, *J. Fluid Mech.*, **6**, 161–205.

Bradshaw, P., P. H. Ferris, and N. P. Atwell, 1967. Calculation of boundary-layer development using the turbulent energy equation. *J. Fluid Mech.*, **28**, 593–616.

Businger, J. A., J. C. Wyngaard, I. Izumi, and E. F. Bradley, 1971. Flux-profile relationships in the atmospheric surface layer. *J. Atmos. Sci.*, **28**, 181–189.

Cartwright, D. E., 1959. On submarine sand waves and tidal lee waves. *Proc. Roy. Soc. London, A*, **253**, 218–241.

Chien, N., 1956. The present status of research on sediment transport. *Trans. Am. Soc. Civ. Eng.*, **125**, 833–884.

Davis, R. E., 1972. On the prediction of the turbulent flow over a wavy boundary, *J. Fluid Mech.*, **52**, 287–306.

Dyer, A. J., 1974. A review of flux-profile relationships. *Boundary-Layer Meteorol.*, **7**, 363–372.

Dyer, K. R., 1970. Current velocity profiles in a tidal channel. *Geophys. J.*, **22**, 153–161.

Dyer, K. R., 1971. The distribution and movement of sediment in the Solent, Southern England. *Mar. Geol.*, **11**, 175–187.

Einstein, H. A., 1950. The bed load function for sediment transportation in open channel flows. *U.S. Dept. Agric. Soil Conserv. Ser. Tech. Bull. 1026.*

Engelund, F., 1970. Instability of erodible beds. *J. Fluid Mech.*, **42**, 225–244.

Exner, F. M., 1925. Über die Wechselwirkung zwischen Wasser und Geschiebe in Flüssen. *Sitzber. Akad. Wiss (Wien)*, **3–4**, 165–180.

Fredsøe, J., 1974. On the development of dunes in erodible channels. *J. Fluid Mech.*, **64**, 1–16.

Gilbert, K. G., 1914. The transportation of debris by running water. *U.S. Geol. Surv. Prof. Paper 86.*

Gordon, C. M., 1974. Intermittent momentum transport in a geophysical boundary-layer. *Nature*, **248**, 392–394.

Gradowczyk, M. H., 1968. Wave propagation and boundary instability in erodible bed channels. *J. Fluid Mech.*, **33**, 93–112.

Graf, W. H., 1971. *Hydraulics of Sediment Transport.* McGraw-Hill, New York, 513 pp.

Harvey, J. G., 1966. Large sand waves in the Irish Sea. *Mar. Geol.*, **4**, 49–55.

Hayashi, T., 1970. Formation of dunes and anti-dunes in open channels. *Proc. Am. Soc. Civ. Eng.*, **96** (HY2), 357–366.

Heathershaw, A. D., 1974. "Bursting" phenomena in the sea. *Nature*, **248**, 394–395.

Hino, M., 1968. Equilibrium-range spectra of sand waves formed by flowing water. *J. Fluid Mech.*, **34**, 565–573.

Hjelmfelt, A. T. and C. W. Lenau, 1970. Non-equilibrium transport of suspended sediment. *Proc. Am. Soc. Civ. Eng.*, **96** (HY7), 1567–1586.

Hsu, S. T. and J. F. Kennedy, 1971. Turbulent flow in wavy pipes. *J. Fluid Mech.*, **47**, 481–502.

Hunt, J. N., 1954. The turbulent transport of suspended sediment in open channels. *Proc. Roy. Soc. London, A*, **224**, 322–335.

Jain, S. C. and J. F. Kennedy, 1974. The spectral evolution of sedimentary bedforms. *J. Fluid Mech.*, **63**, 301–314.

Jones, N. S., J. M. Kain, and A. H. Stride, 1965. The movement of sand waves on Warts Bank, Isle of Man. *Mar. Geol.*, **3**, 329–336.

Kachel, N. B. and R. W. Sternberg, 1971. Transport of bedload as ripples during an ebb current. *Mar. Geol.*, **10**, 229–244.

Kendall, J. M., 1970. The turbulent boundary-layer over a wall with progressive surface waves. *J. Fluid Mech.*, **41**, 259–281.

Kennedy, J. F., 1963. The mechanics of dunes and anti-dunes in erodible bed channels. *J. Fluid Mech.*, **16**, 521–544.

Kennedy, J. F., 1969. The formation of sediment ripples, dunes and anti-dunes. *Ann. Rev. Fluid Mech.*, **1**, 147–168.

Kenyon, N. H. and A. H. Stride, 1968. The crest length and sinuosity of some marine sand waves. *J. Sediment. Petrol*, **38**, 255–259.

Khanna, S. D., 1970. Experimental investigation of form of bed roughness. *Proc. Am. Soc. Civ. Eng.*, **96** (HY10), 2029–2040.

Kirby, R. and N. C. Kelland, 1972. Adjacent stable and apparently mobile linear sediment ridges in the Southern North Sea. *Nature*, **238**, 111–112.

Kirby, R. and Oele, E., 1975. The geological history of the Sandettie–Fairy Bank area, southern North Sea. *Phil. Trans. Roy. Soc. Ser. A*, **279**, 257–267.

Langeraar, W., 1966. Sand waves in the North Sea. *Hydrogr. Newsletter*, **1**, 243–246.

Langhorne, D. N., 1973. A sandwave field in the Outer Thames Estuary, Great Britain. *Mar. Geol.*, **14**, 129–143.

Liu, H. K., 1957. Mechanics of sediment-ripple formation. *Proc. Am. Soc. Civ. Eng.*, **83** (HY2), paper 1197.

Long, R. B., 1971. On generation of ocean waves by a turbulent wind. Ph.D. thesis, University of Miami, Florida.

Ludwick, J. C., 1972. Migration of tidal sand waves in Chesapeake Bay entrance. In *Shelf Sediment Transport: Process and Pattern*. D. J. P. Swift, D. B. Duane, and O. H. Pilkey, eds. Dowden, Hutchinson & Ross, Stroudsburg, Pa., pp. 377–410.

Ludwick, J. C., 1974. Tidal currents and zig-zag sand shoals in a wide estuary entrance. *Geol. Soc. Am. Bull.*, **85**, 717–726.

Lumley, J. L. and B. Khajeh-Nouri, 1974. Computational modelling of turbulent transport. *Adv. Geophys.*, **18A**, 169–192.

Mei, C. C., 1969. Non uniform diffusion of suspended sediment. *Proc. Am. Soc. Civ. Eng.*, **95** (HY1), 581–584.

Mercer, A. G., 1971. Analytically determined bedform shape. *Proc. Am. Soc. Civ. Eng.*, **97** (EM1), 175–180.

Mercer, A. G. and M. I. Haque, 1973. Ripple profiles modelled mathematically. *Proc. Am. Soc. Civ. Eng.*, **99** (HY3), 441–459.

Meyer-Peter, E. and R. Müller, 1948. Formulae for bed-load transport. *Proc. 2nd Cong. I.A.H.R., Stockholm, June 1948*.

Miles, J. W., 1957. On the generation of surface waves by shear flows. *J. Fluid Mech.*, **3**, 185–204.

Milne-Thompson, L., 1960. *Theoretical Hydrodynamics*. 4th ed. Macmillan, New York.

Monin, A. S. and A. M. Obukhov, 1954. Basic turbulent mixing layers in the surface layer of the atmosphere. *Akad. Nauk. SSSR Trud. Geofiz. Inst.*, **24** (151), 163–187.

Monin, A. S. and A. M. Yaglom, 1971. *Statistical Fluid Dynamics*, (Engl. transl.). Vol. 1. M.I.T. Press, Cambridge, Mass., 769 pp.

Motzfeld, H., 1937. Die turbulente Strömung an Welligen Wänden. *Z. Angew. Math. Mech.*, **17**, 193–212.

McCave, I. N., 1971. Sand Waves in the North Sea off the coast of Holland. *Mar. Geol.*, **10**, 199–225.

McCave, I. N., 1973. Some boundary-layer characteristics of tidal currents bearing sand in suspension. *Mem. Soc. Roy. Sci. Liege 6^e Ser.*, 187–206.

Nordin, C. F. and J. H. Algert, 1966. Spectral analysis of sand waves. *Proc. Am. Soc. Civ. Eng.*, **92** (HY5), 95–114.

Raudkivi, A. J., 1963. Study of sediment ripple formation. *Proc. Am. Soc. Civ. Eng.*, **89** (HY6), 15–33.

Raudkivi, A. J., 1966. Bed forms in alluvial channels. *J. Fluid Mech.*, **26**, 507–514.

Raudkivi, A. J., 1967. *Loose Boundary Hydraulics*. Pergamon Press, Oxford, 331 pp.

Reynolds, A. J., 1965. Waves on the erodible bed of an open channel. *J. Fluid Mech.*, **22**, 113–133.

Rifai, M. F. and K. V. H. Smith, 1971. Flow over triangular elements simulating dunes. *Proc. Am. Soc. Civ. Eng.*, **97**, (HY7), 963–976.

Rouse, H., 1937. Modern conceptions of the mechanics of turbulence. *Trans. Am. Soc. Civ. Eng.*, **102**, 436–505.

Salsman, G. G., W. H. Tolbert, and R. G. Villars, 1966. Sand-ridge migration in St. Andrews Bay, Florida. *Mar. Geol.*, **4**, 11–19.

Schlichting, H., 1968. *Boundary-Layer Theory*. 6th ed. McGraw-Hill, New York, 747 pp.

Simons, D. B., E. V. Richardson, and M. L. Albertson, 1961. Flume studies using a medium sand (0.45 mm). *U.S. Geol. Surv. Water Supply Paper 1498A*.

Smith, J. D., 1969. Investigations of turbulent boundary layers and sediment-transport phenomena as related to shallow marine environments. Part 2: Studies of non-uniform boundary layer flows. *Report A69-7*, Dept. Oceanography, University of Washington.

Smith, J. D., 1970. Stability of a sand bed, subjected to a shear flow of low Froude number. *J. Geophys. Res.*, **75**, 5928–5940.

Southard, J. B. and J. R. Dingler, 1971. Flume study of ripple propagation behind mounds on flat sandy beds. *Sedimentology*, **16**, 251–263.

Sternberg, R. W., 1972. Predicting initial motion and bedload transport of sediment particles in the shallow marine environment. In *Shelf Sediment Transport*. D. J. P. Swift, D. B. Duane, and O. H. Pilkey, eds. Dowden, Hutchinson & Ross, Stroudsburg, Pa., pp. 61–81.

Stride, A. H., 1963. Current-swept sea floors near the southern half of Great Britain. *Quart. J. Geol. Soc. London*, **119**, 175–199.

Sutton, O. G., 1953. *Micrometeorology*. McGraw-Hill, New York, 333 pp.

Taylor, P. A. and P. R. Gent, 1974. A model of atmospheric boundary-layer flow above an isolated two dimensional "hill," an example of flow above "gentle topography." *Boundary-Layer Meteorol.*, **7**, 349–362.

Taylor, P. A., P. R. Gent, and J. M. Keen, 1976. Some numerical solutions for turbulent boundary-layer flow above rough wavy surfaces. *Geophys. J. R. Astron. Soc.*, **44**, 177–201.

Terwindt, J. H. J., 1971. Sand waves in the southern bight of the North Sea. *Mar. Geol.*, **10**, 51–67.

Townsend, A. A., 1972. Flow in a deep turbulent boundary-layer over a surface distorted by water waves. *J. Fluid Mech.*, **55**, 719–735.

Vanoni, V. A. and G. N. Nomicos, 1960. Resistance properties of sediment laden streams. *Trans. Am. Soc. Civ. Eng.*, **125**, 1140–1175.

Van Veen, J., 1935. Sand waves in the southern North Sea. *Natl. Hydrogr. Rev.*, **12**, 21–29.

Webb, E. K., 1970. Profile relationships: the log-linear range and extension to strong stability. *Quart. J. Roy. Meteorol. Soc.*, **96**, 67–90.

Yalin, M. S., 1964. Geometrical properties of sand waves. *Proc. Am. Soc. Civ. Eng.*, **90** (HY5), 105–119.

Yalin, M. S., 1972. *Mechanics of Sediment Transport*. Pergamon Press, Oxford, 290 pp.

15. COMPUTER SIMULATION OF TURBIDITY CURRENT FLOW AND THE STUDY OF DEEP-SEA CHANNELS AND FAN SEDIMENTATION

PAUL D. KOMAR

1. Introduction

Computer simulation models of turbidity currents have been employed principally in the study of the morphology of deep-sea channels. For example, many channels show progressive increases in width and relief along their lengths. The models allow the investigation of how these changes might be related to the varying hydraulic parameters of the flows that occupy the channels. Only very limited attempts have been made to utilize simulation models to understand patterns of sediment deposition resulting from individual flows.

Numerical models employ the equations that govern the flow of turbidity currents. As in other fluid flow problems these are a momentum equation and continuity equations. Since turbidity currents are a two-phase system, two continuity equations are required, one for the sediment comprising the flow, and a second for the water. This chapter summarizes these equations and discusses how they are employed in the development of models. Some discussion is also required as to what we presently believe are the expected ranges of flow velocities, densities, thicknesses, and so on. This information is of course required in any model. It should be recognized at the outset that a deep-sea turbidity current has never been seen or measured. All we have is indirect information on their properties, and there is no general agreement on this. Since we lack direct measurements of the flow parameters, the models must obviously be very subjective. On the other hand, the models become especially important owing to the fact that the flows cannot be observed directly.

The chapter concludes with a review of the various models of turbidity currents that have been employed. These are almost entirely flows that are channelized and therefore can be dealt with using two-dimensional models. Three-dimensional models, such as of a spreading sheet–flow turbidity current, have not been developed, although there is obvious application of such models.

2. Equations Governing the Flow of Turbidity Currents

Several experimental studies have shown that density and turbidity currents consist of a head with a characteristic shape, followed immediately by a thinner flow (Schmidt, 1910, 1911; Ippen and Harleman, 1952; Keulegan, 1957, 1958; Middleton, 1966a). The head, neck, and body of a turbidity current are illustrated in Fig. 1 along with the notation to be used.

The experiments conducted by Middleton (1966a) demonstrate that the velocity of the head may be predicted from the simple formula

$$v = C\left[\frac{\rho_t - \rho}{\rho_t} gh_1\right]^{1/2} \tag{1}$$

where v is the velocity, ρ_t is the flow density, ρ is the density of water, g is the acceleration of gravity, and h_1 is the thickness of the head (Fig. 1). C is a dimensionless coefficient for which Middleton's (1966a) and Keulegan's (1957, 1958) experiments

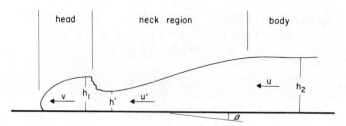

Fig. 1   The head, neck, and body of a turbidity current traveling down a slope of angle β.

yield the approximate value 0.75. It is of special interest to note that this equation shows no dependence on the bottom slope. Middleton also shows experimentally that the value of C is essentially independent of the slope. It was shown by Benjamin (1968) that this lack of dependence on the slope is due to the head not being driven by a gravity component acting down the slope, but rather by the larger piezometric pressure existing inside the current than in the lighter water ahead. Balancing this hydrostatic driving force against the resisting horizontal momentum that is imparted to the overlying fluid, Benjamin (1968) derived an equation for the motion of the head that is essentially the same as equation 1. On theoretical grounds, Benjamin deduced a value $C = 0.74$ for the coefficient of equation 1, which is amazingly close to the value determined experimentally.

When the flow becomes essentially uniform, that is, when the velocity, density, and thickness become nearly constant in the direction of flow, the motion of the body is governed by a Chezy-type equation of the form

$$u = \left[\frac{\rho_t - \rho}{\rho_t} \, gh_2 \, \frac{\sin \beta}{(1 + \alpha)c_f} \right]^{1/2} \tag{2}$$

where $\sin \beta$ is the bottom slope, h_2 is the thickness of the body (Fig. 1), c_f is a drag coefficient, and α is the ratio of the drag on the flow at the upper interface to the drag on the bottom. Like the Chezy equation for river flow, equation 2 can be derived by balancing the gravity force component on the flow down the slope against the retarding frictional drag force (Komar, 1971, p. 1481). Therefore, the physical forces important in the flow of the body are different from those that are significant in the motion of the head. Several experimental studies have demonstrated the general validity of equation 2 when applied to density currents (see review in Middleton, 1966b, p. 628). Middleton has demonstrated that the value of the parameter α is dependent upon the densimetric Froude number:

$$Fr = \frac{u}{\{[(\rho_t - \rho)/\rho_t]gh\}^{1/2}} \tag{3}$$

α increasing with increasing Fr.

Dividing equation 1 by equation 2 yields

$$\frac{v}{u} = 0.75 \left(\frac{h_1}{h_2} \right)^{1/2} \frac{1}{Fr} \tag{4}$$

for the ratio of the velocity of the head to the velocity of the body of the turbidity current. Equation 4 is plotted in Fig. 2 together with the data of Middleton (1966a). It is seen that at high values of the Froude number the body will move faster than the

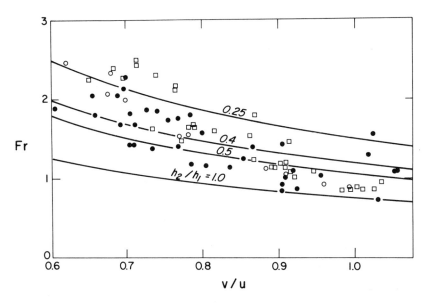

Fig. 2.   Curves of the ratio of the head to the body flow velocities, v/u, versus the Froude number, Fr, for a series of h_2/h_1 values as given by equation 4. The data is from Middleton (1966a, Fig. 16), the symbols representing experiments in three different flumes (from Komar, 1972).

head $(v/u < 1)$, and the head will be considerably thicker than the body of the flow $(h_2/h_1 < 1)$. The extra flux of sediment brought forward to the head from the body will be lifted upward within the head and returned back toward the body by means of interface waves breaking directly behind the head (Middleton, 1966a, p. 541; Benjamin, 1968). Such flow characteristics might be expected to be found in flows as they pass through submarine canyons which have steep slopes and therefore Froude numbers greater than critical (Komar, 1971).

It is apparent from equation 4 that as Fr decreases, v/u increases; the head tends to surge forward ahead of the body. This was actually observed in the experiments of Middleton (1966a) and can be seen in the data points of Fig. 2 extending into the range of $v/u > 1$. Middleton correctly explained this as resulting from the inability of the flow to reach a steady-state equilibrium within the 5 m length of the experimental flume. As discussed in Komar (1972), the turbidity current cannot be stable if the head is traveling faster than the body. Instead, there must be some readjustment of the thickness ratio h_2/h_1 in equation 4 to offset the decrease in Fr in order to maintain the head and body at the same speed. It was concluded that in the deep-sea conditions of low Froude number we are interested in the case where the head and body maintain the same velocity $(v/u = 1)$. Under such conditions equation 4 yields a relationship between h_2/h_1 and Fr which is shown in Fig. 3. It is seen that as Fr is reduced (the bottom slope decreases), the ratio h_2/h_1 increases since the body thickness h_2 must correspondingly increase over the thickness of the head h_1 if the body is to keep pace with the head. The relative thickness of head and body has important consequences in that it determines which is spilling most from the confines of a deep-sea channel. This becomes important later in the development of the models and how they relate to the relief of the channels.

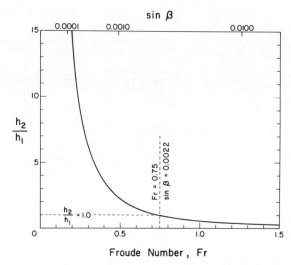

Fig. 3.   The ratio of the body to head thickness, h_2/h_1, as a function of the Froude number or bottom slope for a flow in which the head and body are traveling at the same rate ($v/u = 1$) (from Komar, 1972).

As indicated, equation 2 applies to steady, uniform flow. This relationship is satisfactory for the analysis of most flow conditions as changes in velocity, density, thickness, and channel slope are generally sufficiently small that equation 2 is a satisfactory approximation. However, when the changes are rapid, as in a hydraulic jump, then the full momentum flux equation must be utilized. This equation is an expression of Newton's second law of motion: any increase or decrease in the momentum flux must be brought about by forces acting on the fluid. The momentum flux equation balances the change in momentum against these forces. Application is made to a control volume as shown in Fig. 4, two cross sections of the channel separated by a distance l. If $\bar{u}_1$, h_1, and ρ_{t1} are the average velocity, thickness, and density, respectively, of the flow entering the control volume on its up-channel end, then the momentum flux entering the control volume (per unit width) is $(\rho_{t1}\bar{u}_1 h_1)\bar{u}_1$. Similarly the momentum flux out of the control volume is $(\rho_{t2}\bar{u}_2 h_2)\bar{u}_2$, where the subscript 2 refers to the lower channel cross section. Changes in this momentum flux must be balanced by forces such as the gravity component acting along the

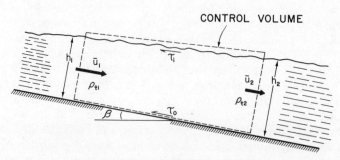

Fig. 4.   Control volume formed by two cross sections of the channel separated by a distance l, where τ_0 is the frictional drag on the bottom and τ_i the interface drag.

channel slope, pressure-gradient forces, and the forces of frictional drag on the flow. Making such a balance yields

$$\rho_{t2}\bar{u}_2{}^2 h_2 - \rho_{t1}\bar{u}_1{}^2 h_1 = -(\tau_0 + \tau_i)l + [(\bar{\rho}_t - \rho)g\bar{h}\sin\beta]l$$

$$- \frac{g}{2}[(\rho_{t2} - \rho)h_2{}^2 - (\rho_{t1} - \rho)h_1{}^2] \tag{5}$$

where τ_0 and τ_i are respectively the drag on the bottom and upper interface of the flow so that $\tau_0 + \tau_i$ is the total drag. Actually, there should be a third drag term, that due to the internal friction arising from the suspended sediment. This cannot be properly evaluated yet, and attempts at including its effects involve increasing the drag coefficients above the values expected for water flow alone. Placing $\tau_i = \alpha\tau_0$, the drag stress sum becomes

$$\tau_0 + \tau_i = (1 + \alpha)\tau_0 = (1 + \alpha)c_f\rho_t\bar{u}^2 \tag{6}$$

having used the well-known quadratic-stress law to evaluate the bottom stress. In practice in the models, the stress is evaluated at both channel cross sections of the control volume, the average of the two being taken to represent the average frictional drag within the control volume. As has already been discussed, the parameter α is dependent upon the Froude number, and values can be found in Middleton (1966b). There is considerable scatter in the data on α, so that at present its value is highly uncertain. The evaluation of the drag coefficient c_f is likewise very uncertain. It has generally been taken to be on the order of 0.0035–0.0050 (Johnson, 1964; Komar, 1969, 1971). These values are somewhat higher than would be expected for the flow of water in a channel, increased to include the internal friction of the sediment comprising the flow. This range of drag coefficients also yields acceptable velocities for expected flow densities and thicknesses.

The gravity component force term of equation 5 is also an average for the entire control volume. Therefore the average density $\bar{\rho}_t$ and average thickness $\bar{h}$ are used. Again, in application these are taken as the averages of the values of these parameters at the two channel cross sections, the ends of the control volume.

When the flow is uniform, $\rho_{t1} = \rho_{t2}$, $h_1 = h_2$, and $\bar{u}_1 = \bar{u}_2$, and equation 5 reduces to

$$\tau_0 + \tau_i = (\rho_t - \rho)gh\sin\beta \tag{7}$$

a balance between the gravity component tending to accelerate the flow, and the frictional drag retarding the flow. Making the same evaluations of the frictional drag given in equation 6, equation 7 becomes the Chezy-type equation 2. Thus equation 2 is in effect the momentum flux equation for uniform flow.

By considering the energetics of the system, Bagnold (1962) has developed a model for turbidity currents in which autosuspension of the sediments provides the power to drive the flow. Balancing this source of available power against the loss of power owing to bottom and water-turbidity current interface drag and the power utilized to maintain the suspension of sediments, Bagnold (1962) obtained the relationship

$$(\rho_s - \rho)gN\bar{h}(\bar{u}\sin\beta - w) = (\tau_0 + \tau_i)\bar{u} \tag{8}$$

in which ρ_s is the density of the sediment grains forming the current, N is the mean volume concentration, and w is the settling velocity of the grains. Utilizing equation 6 for the frictional drag, and

$$N = \frac{\rho_t - \rho}{\rho_s - \rho} \tag{9}$$

equation 8 becomes

$$(\rho_t - \rho)gh(\bar{u}\sin\beta - w) = (1 + \alpha)c_f\rho_t\bar{u}^3 \tag{10}$$

For sand-size sediments and finer, the settling velocity w is sufficiently small that the energy required to maintain the suspension is much smaller than the other terms in equation 10. Neglecting this term, equation 10 reduces to equation 2. Since equation 2 is simpler and gives satisfactory results, it is generally employed rather than equation 10. However, equation 10 is important in that it gives an expression for the auto-suspension limit. Note that equation 10 limits acceptable values of the velocity $\bar{u}$ to

$$\bar{u} \geq \frac{w}{\sin\beta} \tag{11}$$

Otherwise the available power of the flow is insufficient to maintain the sediment in autosuspension. This expression also places an upward limit on the grain size of sediment that can be carried in autosuspension within a given flow. This has important consequences to deposition from the turbidity current which will be examined later.

For a given flow density, there may be two possible solutions for $\bar{u}$ according to equation 10. However, the lower $\bar{u}$ solution corresponds to the relationship between $\bar{u}$ and ρ_t where the velocity decreases with increasing density. This is physically un-reasonable so it is apparent that we are interested in the larger $\bar{u}$ solution where $\bar{u}$ increases with increasing density. The limit between the two sets of solutions can be obtained by differentiating equation 10 and setting $d\rho_t/d\bar{u} = 0$ (Komar et al., 1974). This gives

$$\bar{u} = \frac{3}{2}\frac{w}{\sin\beta} \tag{12}$$

which is precisely 1.5 times greater than the normal autosuspension limit and places a higher value on the lower limit of the possible range of flow velocities.

In the development of models, equations are also needed for the continuity of the transporting materials. In river flow the total amount of water is conserved so that the discharge of water into a control volume must equal the discharge out. The situation is somewhat more complex in a turbidity current in that there are two components, the solid sediment grains and the intergranular water. Each must be conserved and separate continuity equations may be written for each material.

Consider first the sediment within the flow. If the flow has an average volume concentration N of sediment, then Nh is the total volume of suspended sediment over a unit area of channel and the discharge or flux of sediment per unit channel width becomes

$$\text{sediment flux} = Nh\bar{u}$$

The continuity equation compares the sediment flux through two sections of channel and can be stated

$$N_2h_2\bar{u}_2 = N_1h_1\bar{u}_1 + q_sl \tag{13}$$

where the subscripts again refer to the two ends of the control volume of Fig. 4, being separated by a channel distance l. q_s denotes the average rate of sediment loss (q_s negative) per unit area of channel floor due to deposition, channel overspill, and so on, or the rate of sediment gain (q_s positive) due to erosion within that length of channel. In sedimentation models where deposition and erosion are the important aspects of the model, a separate continuity equation may be written for each grain size present in the flow.

Similarly, the continuity equation for the water portion of the turbidity current becomes

$$(1 - N_2)h_2\bar{u}_2 = (1 - N_1)h_1\bar{u}_1 + q_w l \tag{14}$$

where q_w is the rate of water entrainment (q_w positive) or loss (q_w negative) through the interface of the flow. When $N_1 = N_2 = 0$ equation 14 reduces to the normal discharge equation for rivers.

Both equations 13 and 14 are on a unit width basis. If changes in the channel width are expected, then this variable must be included as well. The resulting equations can be found in Komar (1973).

It should be recognized that equations 5, 10, 13, and 14 are strictly valid only if the sediment concentration N or flow density ρ_t is constant vertically through the flow thickness. If there are appreciable variations in these parameters, then, for example, the sediment flux through a unit width of channel section would be

$$\text{sediment flux} = (\overline{Nu})h = \int_0^h Nu\,dz$$

where z is the vertical coordinate axis. Average values of N and u are not separable from $\overline{Nu}$, which is the integrated average of the product of N and u. This would be true for the integrated momentum flux as well. However, if the flow is composed of autosuspended sediments, then there should not be appreciable vertical changes in N and ρ_t. The greatest variations in these parameters would be at the upper interface of the flow due to dilution by the overlying water. Variations at the bottom of the flow could be produced if the current is transporting a bed and suspended load as well as an autosuspended load. This would generally be important only where the current is waning and depositing its sediment load. Considerations of vertical variations in sediment concentration and therefore flow density would improve the models but greatly add to their complexity. Owing to uncertainties in how to deal theoretically with the causes of such vertical variations, and because we have no measurements from prototype flows, such considerations may be unwarranted presently. However, if deposition from turbidity currents is to be better understood, such considerations will become necessary.

3. Flow Parameters of Deep-Sea Turbidity Currents

With the exception of estimates of the flow velocities of turbidity currents obtained from sequences of cable breaks, no direct measurement of the flow characteristics has been possible. Thus unlike other models of fluid flow, it is not presently possible to check the accuracy of the simulation models against the prototype. This section briefly reviews what information is available on the flow parameters, or what are believed to be reasonably expected ranges of values. These are needed in the models to follow.

A. Thickness

From the appearance of levees along submarine channels, like those in Fig. 5, Menard (1964, p. 203) deduced that the flows are sufficiently thick to spill out of the channels and spread laterally for some distance. Deep-sea channels commonly

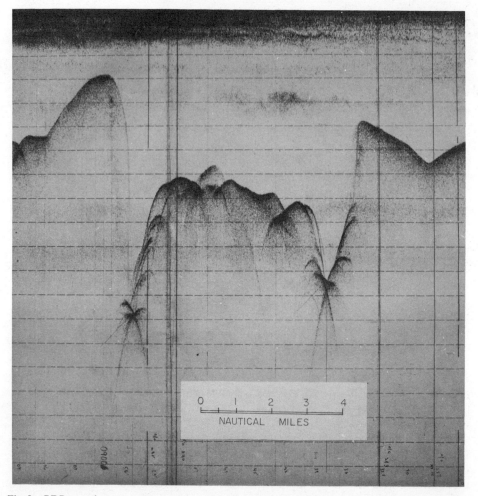

Fig. 5.   PDR record on a course approximately east–west through a meander in the Monterey deep-sea channel, showing pronounced levee development on the outside of the meander. As viewed, the flow comes out of the page through the right crossing and reenters through the left crossing. Horizontal distance between the two outside levees is approximately 12 km (from Komar, 1969).

have reliefs of 100–200 m, sometimes as much as 300 m, so it appears that turbidity currents in the deep sea can attain appreciable thicknesses. The flows that spill from the channels are probably exceptionally large turbidity currents. Most others would remain entirely confined to the channel for most of its length, spilling only near the terminal end of the channel where the relief is generally observed to decrease [the supra-fan area of Normark (1970)]. Griggs and Kulm (1970) were able to trace individual flows through Cascadia Channel off the northwest United States by their content of Mazama ash. Even these Holocene flows, which are probably appreciably smaller than Pleistocene flows, reached thicknesses of about 117 m. In sheet flow outside channels the flows are of course generally thinner, being reduced in thickness as they spread. It is difficult to estimate flow thickness there. Menard (1964, p. 204) obtained some idea by examining the height above the abyssal plain to which isolated abyssal hills are covered with turbidities and also the height of low ridges which deflect the currents and produce lee basins. On this basis he determined that in flowing across the gently sloping abyssal plains turbidity currents have thicknesses limited to a few tens of meters.

B. Velocity

Sequences of breakage of submarine cables give us our best estimate of turbidity current velocity; the most famous of these is the Grand Banks earthquake and flow (Heezen and Ewing, 1952). Menard (1964, pp. 207–213) reinterpreted the path of the flow and estimated velocities ranging from 20 m/sec down to 10 m/sec, decreasing in the direction of flow as the bottom slope was reduced from 5 m/km to 0.5 m/km. Although a variety of interpretations are possible, the flow velocity estimates all lie approximately within this range. Similarly, Krause et al. (1970) obtained average velocities of 8–14 m/sec (30–50 km/hr) from cable breaks in the western New Britain Trench, the Solomon Sea.

Other estimates of the velocities of turbidity currents are indirect. Utilizing the cross-channel slope of the flow deduced from differences in levee heights on the two sides, such as those of Fig. 5, Komar (1969) calculated velocities on the order of 8–22 m/sec, the value depending on the density of the flow. Taking acceptable values for the flow density and thickness, and slopes from deep-sea channels, equation 2 was shown to yield the same approximate range of velocities. Finally, considerations of the stresses required to transport gravels and cobbles, found both in ancient turbidite deposits and the present deep-sea fans, indicate that the velocities of flows exceeded about 7 m/sec (Komar, 1970). Thus there is agreement from the several lines of evidence as to the expected range of flow velocities of deep-sea turbidity currents.

C. Densities

The most uncertainty and disagreement concerns the expected range of turbidity current densities. Values ranging from about 1.05 to 2.0 g/cm^3 have been suggested. Kuenen (1950) was the main proponent of "high-density" currents, generally taken to mean the range 1.5–2.0 g/cm^3. The available evidence indicates that these densities are too high for turbidity currents and would be more typical of nonturbulent mud flows.

Increasing concentration of sand-size sediment in the flows will act to increase the internal friction. As discussed by Bagnold (1962), at a grain concentration of about 9% the free distance between the grains is equal to the grain diameter and collisions will become important. Increases in concentration above about 9% ($\rho_t = 1.15$ g/cm^3) within a flow would thus lead to a rapid increase in the internal friction and thus a dissipation of the momentum and energy of the flow.

This collision limit does not apply to the fines in the flow so it is uncertain as to how much could be added to raise the density above the 1.15 g/cm^3 value. An indication of the limit of concentration of fines in the flow is given by the study of Dangeard et al. (1965), who generated a series of flows in the laboratory, composed of industrial kaolin with a median diameter of 4.5 μ. At concentrations lower than a critical value of 375 g/l turbidity currents were formed whose velocity and distance of travel increased as the concentration increased. With concentrations higher than 375 g/l mud flows were produced in which a further increase in concentration reduced the velocity and travel distance. This critical 375 g/l corresponds to a flow density of about 1.4 g/cm^3. Flows consisting of Aiguillon clay yielded a critical concentration of about 250 g/l $[\rho_t = 1.26$ g/cm$^3]$ so that apparently this transition depends on the sediment material. In addition, these values cannot be expected to apply to large-scale flows in the ocean, but can be taken as only suggestive.

Increasing concentration of either fines or sand-size sediment tends to damp out the turbulence of the flows (Bagnold, 1954). The suspended material has the greatest effect on the small eddies, the larger eddies which are more important to the maintenance of the suspension being less influenced. Damping of eddy motions is less important in the large deep-sea turbidity currents than in the small laboratory models, where the Reynolds numbers are small and the scale of eddies small. Little is known on the effects of various sediment concentrations on the turbulence. Lane (1940) quotes Eliassen as indicating that when the sediment concentration in a river reaches about 5% by weight (density about 1.06 g/cm^3) the eddies become damped and "water flows in a straight line motion one never sees when the water is clear." Undoubtedly, large-scale eddies were still present at such concentrations. Lane (1940) summarized data on exceptional sediment concentrations found in rivers. Values reached as high as 50% by weight which corresponds to a density of 1.45 g/cm^3. At somewhat higher concentrations the clay–water mixture would begin to behave as a non-Newtonian plastic.

In conclusion, turbidity currents composed principally of sand-size material probably have densities ranging somewhere between about 1.05 and 1.30 g/cm^3. Flows composed mostly of fine-grained sediment could have rather wider ranges of possible densities. Fortunately, nearly all the deep-sea turbidity currents appear to be composed chiefly of sand.

4. Examples of Application of Models

Most of the computer simulation models that have been developed of deep-sea turbidity currents have as their aim the investigation of the relationship between the hydraulics of the flows and the morphology of the submarine channels through which the flows travel. The reason for this is that, although we cannot see or measure the turbidity currents, we can detect their channels and many studies have been centered on the investigation of such channels.

Turbidity currents generated within submarine canyons remain channelized when they reach the deep-ocean floor and cross submarine fans. Each canyon has a sediment-constructed fan and channel as its natural extension. Menard (1964, p. 204) noted that many channels have the same general characteristics in which the relief is relatively small at the canyon mouth, gradually increases along the channel length in the direction of flow, finally reaching a maximum of some 200–300 m, and then decreases with further advance along the channel. One example is shown in Fig. 6, Cascadia Channel off the northwest coast of the United States. This channel was extensively studied by Griggs and Kulm (1970) and their data utilized in models of turbidity current flow along the channel by Komar (1973).

Menard (1964, p. 215) has suggested that the progressive deepening observed in many channels in the direction of flow reflects the increase in the flow thickness brought about by a hydraulic jump. Komar (1971) has investigated hydraulic jumps in turbidity currents utilizing the equations presented above. The control volume utilized is shown in Fig. 7. Since there are rapid changes in the flow thickness, velocity, and density, the full momentum flux relationship, equation 5, must be employed. This is solved together with the two continuity relationships, equations 13 and 14. It is found that for the average submarine canyon–channel system the hydraulic

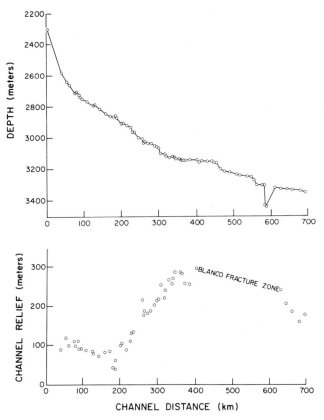

Fig. 6. Longitudinal axis gradient and the profile relief of Cascadia Channel off the northwest coast of the United States (after Griggs and Kulm, 1970).

jump would occur at the mouth of the canyon where the slope suddenly decreases, causing the flow to change from supercritical to subcritical. For the average canyon–channel system, the flow thickness more than doubles during the jump, and its velocity is halved. The entrainment of water through the interface of the flow during the jump is significant in reducing the density of the flow. However, the reduction is not sufficient to reduce the density of a slide (1.50 g/cm^3) to a level more reasonable for a turbidity

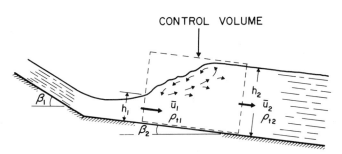

Fig. 7. Control volume for the analysis of a hydraulic jump in a turbidity current where the flow passes from supercritical ($Fr > 1$) to subcritical ($Fr < 1$) (from Komar, 1971).

current. The density prior to the jump must already have been reduced to approximately 1.30 g/cm³ for the jump to convert the flow to a low-density turbidity current. The hydraulic jump completes itself within on the order of 1 km channel length. Clearly, then, the channel relief increase such as that shown in Fig. 6 cannot be due to a hydraulic jump as it continues for hundreds of kilometers.

Komar (1973) suggested that the progressive channel increase seen in Fig. 6 and in other channels might reflect the continuity of turbidity current flow. When the turbidity current decreases in velocity due to a decrease in bottom slope, the thickness of the flow must correspondingly increase to maintain continuity of discharge according to equations 13 and 14. A series of simulation models were developed to test this hypothesis, the models used to examine the changes in the hydraulics of the flows along the channel length to determine how these relate to the morphology of the channels. The changes were sufficiently slow for equation 2 to provide a satisfactory evaluation of the velocity. The continuity relationships, equations 13 and 14, were solved together with equation 2 for control volumes like that of Fig. 4. The length of the control volume was set at $l = 1$ km and the exit velocity, thickness, and density from one control volume became the entrance velocity, thickness, and density for the next control volume. Thus the calculations for the turbidity current were obtained on 1-km increments for the entire length of the channel. The thickness of the head also had to be considered because at steeper slopes the head would be thicker than the body of the flow and the channel relief might therefore be controled by spill from the head.

Figure 8 shows the results for a hypothetical channel where the bottom slope is related to the channel distance D (in kilometers) by

$$\sin \beta = 0.005 e^{-0.01D}$$

This relationship yields an exponential decrease in channel slope from an initial value of 0.005 at $D = 0$ to 0.00025 at $D = 300$ km. These model slopes encompass the range of slopes found in real channels. In the model of Fig. 8 the density was held

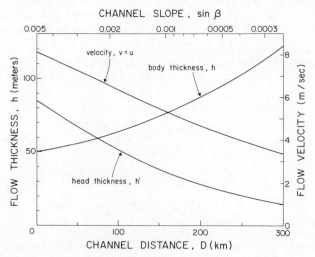

Fig. 8. Flow variations of a computer model turbidity current whose density is constant at $\rho_t = 1.20$ g/cm³, flowing along a hypothetical deep-sea channel (from Komar, 1973).

constant at 1.20 g/cm³; that is, no erosion or deposition of sediment was included and no water entrainment considered. The initial body thickness was placed at 50 m, but at $D = 0$ the head would then have been 84 m thick owing to the relationship of equation 4. As the bottom slope decreased in the direction of flow, the velocity progressively decreased from 8.1 m/sec at $D = 0$ to 3.3 m/sec after 300 m of travel along the channel. At the same time the body thickness increased steadily from its initial value of 50 to 124 m; the body more than doubled in thickness. This increase was without any entrainment of water by the flow as it traveled the length of the channel. Models were also developed with varying amounts of water entrainment and density reduction included. These models showed that reasonable amounts of dilution by water entrainment could produce considerable increases in flow thickness, sufficient to account for the channel relief changes observed in real channels. However, the real channels also increase in width along their lengths, and additional calculations indicated that simple dilution could not account entirely for the change in cross-sectional area of Cascadia Channel along its length. Much of the increase could be explained by progressive dilution of the flows passing through the channel, but subsequent channel erosion (deepening and widening) had to be called upon to account for the entire increase in cross-sectional area. Ness and Kulm (1973) applied the same approach to the analysis of Surveyor Channel in the Gulf of Alaska and found that changes in the channel relief could be explained through continuity considerations.

Van Andel and Komar (1969) developed a simple computer model of turbidity current flow within an enclosed basin, related to turbidite deposits found in ponded sediments on the flank of the Mid-Atlantic Ridge. The turbidite sequences show peculiarities that can be explained by assuming that the currents rebounded repeatedly from the pond walls and made several transits across the pond before coming to rest. Estimates of the turbidity current densities and thicknesses were determined from the pore water content of the turbidites by the approach of Kuenen (1966). The sediment comprising the flows was bimodal, consisting of calcareous foraminifera (mean settling velocity $w = 0.6$ cm/sec) and very fine lutum (settling velocity $w = 0.0015$ cm/sec). The computer model was developed to test the hypothesis of a repeatedly rebounding turbidity current. The deposition from the flow was given by

$$q_s = -N_a w_a - N_b w_b$$

where the subscripts a and b refer to the two modes comprising the flow, N and w being the volume concentrations and settling velocities. As shown in Fig. 9, it was hypothesized that the flow would first pass through a hydraulic jump as it reached the flat floor of the pond, having been generated on the steep walls. Analyses of the hydraulic jump were first performed. Once the flow moved across the flat floor it would progressively decrease in velocity as drag retarded its motion. The thickness would progressively decrease as the sediment comprising the flow settled. The coarse mode would settle faster than the lutum, so that during the first passage of the flow across the basin the deposit would be relatively richer in calcareous foraminifera. As the turbidity current reaches the opposite wall, the flow should climb the wall until its kinetic energy is converted to potential energy, whereupon it will flow back down. The turbidity current has some length, so that the return flow probably collides with its own tail which is still flowing toward the wall. The effect would be to induce considerable turbulence with the result that a substantial amount of energy of the flow is lost. The details of this reflection could not be modeled so this energy loss was arbitrarily set at 33% of the kinetic energy of the flow just prior to meeting the wall;

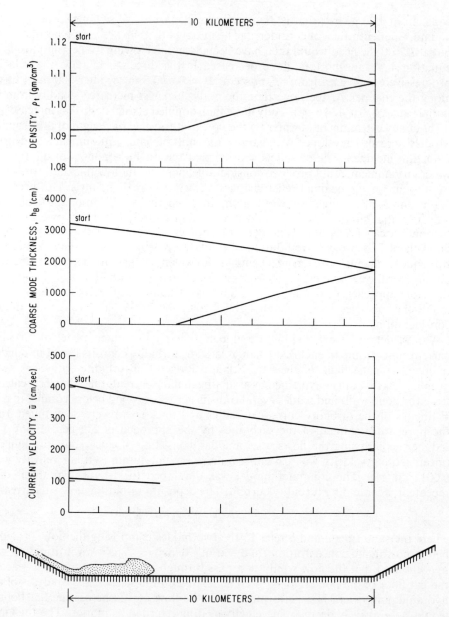

Fig. 9.   Progressive changes in average velocity, thickness of the coarse-mode distribution within the flow, and average density of a model turbidity current starting with an initial velocity of 407 cm/sec, thickness of 32 m, and density of 1.12 g/cm³ (from van Andel and Komar, 1969).

this energy loss is somewhat less than the loss due to a full hydraulic jump. The model of Fig. 9 utilized an initial thickness of 32 m and density of 1.12 g/cm³ obtained from the pore water analyses. The initial velocity of 407 cm/sec was obtained from the hydraulic jump considerations. The progress of the turbidity current was followed in steps of $l = 100$ m back and forth across the basin. According to the model the last of the coarse mode would be deposited after 6.4 km of the second passage has been completed. This distance would depend on the initial velocity of the flow and whether the turbulence generated during the reflection would help resuspend this coarse mode. In the process of losing its coarse mode, the density of the model turbidity current decreases from the initial value of 1.120 g/cm^3 to 1.092 g/cm^3 when the flow is composed entirely of lutum. These density changes agree very well with those determined from the pore water. The model flow shows a steady decrease in velocity from its initial value of 407 cm/sec to 253 cm/sec at the end of the first passage and then to 206 cm/sec because of the loss of energy at the wall. Because of the lower velocity during the second passage, more sediment would be deposited than during the first passage, a result also found in cores from the basin. Although greatly simplified, models like this can be tested against the patterns of sediment accumulation. In this case the model did much to confirm the hypothesis that the sequences of ponded sediments resulted from turbidity currents that rebounded from the basin walls, making repeated passages across the flat bottom of the basin.

The sedimentation model of van Andel and Komar (1969) is for a waning current over a flat bottom, the sediments simply settling to the bottom as the flow loses its momentum. For flows over a sloping bottom consideration must be given to the relationship between the flow hydraulics and sediment grain sizes that can remain in the flow without being deposited. Johnson (1966) gives a good discussion of this problem and applies the analysis to turbidites from the abyssal plain at the western end of the Madeira–Cape Verde Basin, studied by Belderson and Laughton (1966). Johnson points out that bed load transport could move sediment for only a very small fraction of the total travel distance of a turbidity current. Sediments found in turbidites far from the generation site must primarily have been transported as an autosuspension load. Once a given grain size is no longer in autosuspension, it will soon deposit and be found in the sediment accumulation. As discussed by Johnson (pp. 121–122) there will be a lag in the deposition such that the grain size material that is deposited will be controlled by the bottom slope for some 25 km up-current from the site, not the slope at the site itself. Deposition from the turbidity current model can, therefore, be including by considering the autosuspension limit given by equation 11. Both the flow velocity $\bar{u}$ and bottom slope $\tan \beta$ generally decrease in the direction of flow as the turbidity currents travel out onto the abyssal plains. Therefore, according to equation 11, finer and finer material will be deposited. Johnson (1966) used equation 11 to estimate the flow velocity of the turbidity current, knowing the settling velocity w of the deposited material and $\tan \beta$ from the ocean bottom surveys. With his revised form of the autosuspension limit, for the same w and $\tan \beta$, Panicker (1972) obtained velocities 10 times as large as those calculated by Johnson (1966).

Three-dimensional flows, such as a spreading sheet–flow turbidity current, are especially important in the depositional phase. This occurs when the flow spills out of the confines of a channel or becomes less confined in the supra-fan (Normark, 1970) region of a submarine fan. Turbidity currents generated from slumps where there are no channels would also spread as sheet flow. Johnson (1962) gives some consideration to spreading flows. Fietz and Wood (1967) performed laboratory experiments of three-dimensional density currents. Computer simulation models of three-dimensional flows have not been attempted.

The applications considered above all employ the integral form of the momentum flux and continuity relationships, the equations given in this chapter. With the integral form one considers only average velocities, densities, and sediment concentrations. This is entirely adequate for modeling deep-sea turbidity currents in that so little is known about the actual flows or about the distribution of sediment concentration that could be theoretically expected in a flow. At our present state of knowledge more detailed analyses are not warranted.

The differential form of the momentum flux equations, also known as the Navier–Stokes equations, could provide more detail on velocity distributions within the flows and the overlying water. Simulation models utilizing the Navier–Stokes equations would presently be satisfactory for flows composed of very fine-grained material where vertical density gradients through the flow thickness will be less of a problem. Their best application would be to density currents consisting of a dense liquid such as saltwater flowing under another less dense liquid.

Daly and Pracht (1968) have developed computer simulation models utilizing the Navier–Stokes equations in the study of density current surges. An example of the product of their study is shown in Fig. 10. A rectangular grid of Eulerian calculation cells is employed, 60 cells wide and 20 cells high. In Fig. 10 the right region is initially occupied by a heavy fluid of density 1.2 g/cm^3, indicated by the smaller dots. The lighter liquid on the left (heavier dots) has a density of 1.0 g/cm^3. The computer-produced marker particle plots serve to demonstrate the surge motion. It is seen in this example that as the heavy liquid drops, its surface is lowered and at the same time the surface of the lighter liquid is raised. This sets up a wave that travels to and fro, reflecting from the end of the grid and complicating the motion of the density surge.

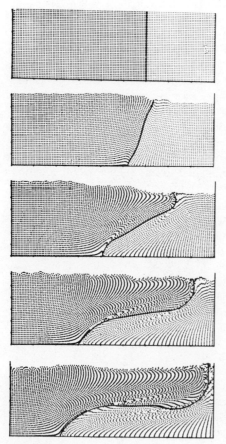

Fig. 10. Computer simulation model of a density surge with a free surface, obtained by solution of the Navier–Stokes equations. The denser fluid is on the right, depicted by the smaller dots (from Daly and Pracht, 1968).

TABLE I

Values of the Coefficient C of Equation 1 for the Motion of the Head of a Density Flow, Obtained with Numerical Models of Density Flows Where $\rho_t = 1.2$ g/cm^3 (Daly and Pracht, 1968)

Slope	Velocity, v	C
0	0.195	0.70
0.04	0.196	0.71
0.08	0.199	0.72
0.16	0.206	0.74

A model comparable but with an upper boundary prevented the development of such a wave and hence the motion of the density surge was somewhat simpler. In this way Daly and Pracht were able to investigate the effects of slope, viscosity, and surface tension on density surges. The models were compared with one another, with available experimental observations, and with analytical equations. Table I shows the results for the velocity of the front of the surge, which gives a model-determined evaluation of the coefficient C in equation 1 for the head of a turbidity current. It is seen that the values agree closely with the 0.75 experimental value of Keulegan (1957, 1958) and Middleton (1966a), and with the theoretical 0.74 value of Benjamin (1968). The model results also show little effect of the bottom slope, again agreeing with the observations and theory. Daly and Pracht also duplicated in a simulation model a laboratory experiment performed by Middleton (1966a). For these calculations they used a grid which was 15 cells high and 60 cells long, each cell being 2 cm on a side. A density current was modeled which corresponded to Middleton's experiment number

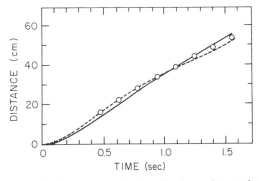

Fig. 11.   Travel-time curve for the front of a density flow. The data points are for an experimental flow of Middleton (1966b); the curves are based on numerical models of the same flow by Daly and Pracht (1968). The solid and the dashed curves represent different approximations in the models (from Daly and Pracht, 1968).

7. A comparison of the front positions as functions of time for the model and experiment are shown in Fig. 11. The two model curves result from different techniques of making the computations. It is seen that there is excellent agreement between the models and experimental results. The study of Daly and Pracht demonstrates that numerical models for the study of density currents provide a useful supplement to analytic and experimental investigations.

References

Bagnold, R. A., 1954. Experiments on a gravity-free dispersion of large solid spheres in a Newtonian fluid under shear. *Proc. Roy. Soc., London Ser. A*, **225**, 49–63.

Bagnold, R. A., 1962. Auto-suspension of transported sediment; turbidity currents. *Proc. Roy. Soc., London, Ser. A*, **265** (1322), 315–319.

Belderson, R. H. and A. S. Laughton, 1966. Correlation of some Atlantic turbidities. *Sedimentology*, **7**, 103–116.

Benjamin, T. Brooke, 1968. Gravity currents and related phenomena. *J. Fluid Mech.*, **31** (Pt 2), 209–248.

Daly, B. J. and W. E. Pracht, 1968. Numerical study of density-current surges. *Phys. Fluids*, **11**, 15–30.

Daneard, L., C. Larsonneur, and C. Migniot, 1965. Les courants de turbidité les coulées boueuses et les glissements: résultats d'expériences. *Compt. Rend. Acad. Sci., Paris*, **216** (9), 2123–2126.

Fietz, T. R. and I. R. Wood, 1967. Three-dimensional density current. *J. Hydraul. Div., Proc. Am. Soc. Soc. Civ. Eng.* (HY6), 1–23.

Griggs, G. B. and L. D. Kulm, 1970. Sedimentation in Cascadia Deep-Sea Channel. *Geol. Soc. Am. Bull.*, **81**, 1361–1384.

Heezen, B. C. and M. Ewing, 1952. Turbidity currents and submarine slumps and the 1929 Grand Banks earthquake. *Am. J. Sci.*, **250**, 849–873.

Ippen, A. T. and D. R. F. Harleman, 1952. Steady-state characteristic of subsurface flow. *Natl. Bur. Stand. Circ. 521*, pp. 79–93.

Johnson, M. A., 1962. Physical oceanography: turbidity currents. *Sci. Prog., London*, **50** (193), 257–273.

Johnson, M. A., 1964. Turbidity currents. *Oceanogr. Mar. Biol. Ann. Rev.*, **2**, 31–43.

Johnson, M. A., 1966. Application of theory to an Atlantic turbidity-current path. *Sedimentology*, **7**, 117–129.

Keulegan, G. H., 1957. Thirteenth progress report on model laws for density currents. An experimental study of the motion of saline water from locks into fresh water channels. *U.S. Natl. Bur. Stand. Rept. 5168*.

Keulegan, G. H., 1958. Twelfth progress report on model laws for density currents. The motion of saline fronts in still water. *U.S. Natl. Bur. Stand. Rept. 5831*.

Komar, P. D., 1969. The channelized flow of turbidity currents with application to Monterey Deep-Sea Fan Channel. *J. Geophys. Res.*, **74** (18), 4544–4558.

Komar, P. D., 1970. The competence of turbidity current flow. *Geol. Soc. Am. Bull.*, **81**, 1555–1562.

Komar, P. D., 1971. Hydraulic jumps in turbidity currents. *Geol. Soc. Am. Bull.*, **82**, 1477–1488.

Komar, P. D., 1972. Relative significance of head and body spill from a channelized turbidity current. *Geol. Soc. Am. Bull.*, **83**, 1151–1156.

Komar, P. D., 1973. Continuity of turbidity current flow and systematic variations in deep-sea channel morphology. *Geol. Soc. Am. Bull.*, **84**, 3329–3338.

Komar, P. D., L. D. Kulm, and J. C. Hartlett, 1974. Observations and analysis of bottom turbid layers on the Oregon continental shelf. *J. Geol.*, **82**, 104–111.

Krause, D. C., W. C. White, D. J. W. Piper, and B. C. Heezen, 1970. Turbidity currents and cable breaks in the western New Britain Trench. *Geol. Soc. Am. Bull.*, **81**, 2153–2160.

Kuenen, Ph. H., 1950. Turbidity currents of high density. *Int. Geol. Congr. Rept. 18th Sess., Great Britain*, pt. 8, pp. 44–52.

Kuenen, Ph. H., 1966. Matrix of turbidities: experimental approach. *Sedimentology*, **7**, 267–297.

Lane, E. W., 1940. Notes on limit of sediment concentration. *J. Sediment. Petrol.*, **10**, 95–96.

Menard, H. W., 1964. *Marine Geology of the Pacific*. McGraw-Hill, New York, 271 pp.

Middleton, G. V., 1966a. Experiments on density and turbidity currents. I. Motion of the head. *Can. J. Earth Sci.*, **3**, 523–546.

Middleton, G. V., 1966b. Experiments on density and turbidity currents. II. Uniform flow of density currents. *Can. J. Earth Sci.*, **3**, 627–637.

Ness, G. E. and L. D. Kulm, 1973. Origin and development of surveyor deep-sea channel, *Geol. Soc. Am. Bull.*, **84**, 3339–3354.

Normark, W. R., 1970. Growth patterns of deep-sea fans. *Am. Assoc. Petrol. Geol. Bull.*, **54**, 2170–2195.

Panicker, N. N., 1972. Prediction of bottom current velocities from sediment deposit on the sea bed. In *Sedimentation, Symposium to Honor Professor H. A. Einstein.* H. W. Shen, ed. and publ., Ft. Collins, Colorado, pp. 19–1 to 19–18.

Schmidt, W., 1910. Gewitter und Böen, rasche Druckanstiege. *Sitzber. Akad. Wiss. Wien, Math. Naturw. Kl. Part IIa*, **119**, 1101.

Schmidt, W., 1911. Zur Mechanik der Böen. *Z. Meteorol.*, **28**, 355.

van Andel, Tj. H. and P. D. Komar, 1969. Ponded sediments of the mid-Atlantic ridge between 22° and 23° North latitude. *Geol. Soc. Am. Bull.*, **80**, 1163–1190.

16. COMPUTER SIMULATION OF CONTINENTAL MARGIN SEDIMENTATION

JOHN W. HARBAUGH AND GRAEME BONHAM-CARTER

1. Introduction

Until recent years, to "simulate" meant to feign or imitate, or to assume the appearance of something without representing it in reality. In the 1940s, however, simulation took on a new sense beginning with the work of John von Neumann, who applied "Monte Carlo analysis" in dealing with problems related to the shielding of nuclear reactors. These problems could not be solved by physical experimentation because of costs and physical hazards involved, and their solution by conventional mathematical methods was too complicated. As an alternative, Monte Carlo methods were employed, in which random processes having known statistical probability distributions were simulated mathematically.

When high-speed digital computers appeared in the early 1950s, the meaning of simulation began to change, for simulation then came to signify the use of mathematical models that could be manipulated by computers to perform kinds of experiments. Almost for the first time, social scientists and business theorists found that they could perform controlled experiments by using computers to carry out the arithmetic and logic operation embodied in simulation models. Simulation, viewed in this manner, is thus a class of techniques that involve setting up a model of a real system and then performing experiments on the model.

We may define a system as a set of dynamically interrelated components. If we change any part of a system, repercussions are felt throughout the system. Interdependence of components thus is generally necessary in the definition of a system. Establishing the boundaries of a system is also part of its definition. It could be argued, for example, that the entire universe is a system. But within such a universal system, it is expedient to define smaller systems, as for example, the solar system, the earth's atmosphere, hydrosphere, and lithosphere, a river drainage system, and the biological components of the soil. Likewise, the gross economy of the world is a system, but a factory and a supermarket are also systems. Thus each system must be arbitrarily defined by specifying its boundaries and its components. At this point it is obvious that we are invariably faced with a hierarchy of systems, lesser systems being nested within larger systems. A hierarchy of systems defined for a seacoast provides an example (Fig. 1). The gross system might be considered to include the sea, beaches, streams, deltas, and cliffs. Within this large system, however, one might wish to isolate a delta as a system in itself, and on a still smaller scale, a single distributary within the delta could be considered as a system, and so on, ad infinitum.

In defining a system, the manner of defining the boundaries of the system is critical. We can distinguish physical boundaries and abstract boundaries. The physical boundaries of a sedimentary basin, for example, might be defined by the areal dimensions of the body of water enclosing the basin. An abstract boundary, on the other hand, might be defined by the inclusion of certain geological variables, and omission of others. Furthermore, boundaries may be arbitrarily defined so that the system is either "open" or "closed." Closed systems, as their name implies, are isolated in the sense that they neither receive input from nor provide output to the "outer world" or to the larger systems that surround them. Though models of closed systems may

623

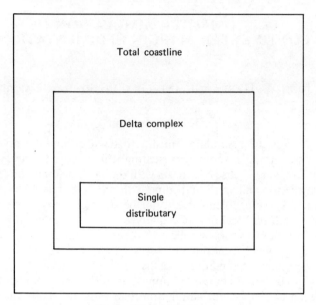

Fig. 1. Nesting of systems.

be simpler to create, it is obvious that most closed systems are unrealistic. Real world systems are almost invariably open, being continually affected by external factors, and in a state of dynamic equilibrium. Factories have inputs (raw materials) and outputs (finished products); their performance adjusts to the market need for products and the rate of inflow of materials. Similarly a coastal system has inputs (streams, longshore currents) and outputs (turbidity currents); its performance, embodied as beaches, deltas, lagoons, and so on, is continually being adjusted to "external" changes such as rates of sediment input, eustatic sea-level fluctuations, and storms.

Because a system consists of an assemblage of complexly interrelated parts, it may be exceedingly hard to predict the effect of altering the state of a particular variable, or of changing the system structure. Man has first to simplify the system conceptually and then to represent it with a model. Models are artificial systems that attempt to portray the characteristics of real systems. There are many types of models, such as physical, conceptual, and graphic models, but in many cases the most powerful and flexible way of representing a system is with mathematical models. Mathematical models may seek complete and precise solutions to certain aspects of systems, or they may be used for computer simulation, in which case "solutions" are obtained by observing the model's performance on a computer. Dynamic simulation is the operation of the model of a system in such a way that the behavior of the real system is reproduced to some degree as the model moves through time.

Implicit in any dynamic system is the flow of time. All the processes are time dependent; transport of materials and transformation of energy are meaningless without flow of time. In representing a dynamic system, it is convenient to freeze time and look initially at a static "snapshot" of the system. The condition of the system during a "snapshot" is given by the "state" of each variable. If we deal with digital models, time can then be moved forward in a series of discrete steps, the system state altering by an increment at each step. In this way a dynamic system can be built up from a series of static snapshots.

Simulation models can be classified according to whether they are static or dynamic, and whether they are deterministic or probabilistic (or hybrid in that they combine both deterministic and probabilistic components). A deterministic model is one in which there is no element of chance. In a dynamic-deterministic model the state of the model at any point in time is completely predetermined. In a probabilistic model, however, there is a degree of uncertainty. The state of a dynamic-probabilistic model at any subsequent moment in time cannot be precisely predicted because of one or more components of uncertainty (stochastic components).

Many deterministic models may be regarded as special cases of stochastic models. For example, the process by which heat is conducted from a warm object to a cooler object is, in essence, a stochastic process which depends upon the random motion of molecules. On the other hand, the number of molecules in any real situation is so large that there is virtually absolute certainty that heat will flow from warm to cold. Thus for practical purposes, we can assume that the process of heat conduction is deterministic, and can effectively represent it with a deterministic model.

Dynamic-probabilistic models are of particular importance in geology because most geologic processes can be regarded as possessing random components. Most dynamic-probabilistic models also contain deterministic components, and are therefore "hybrid" in that respect. An example is provided by the carbonate ecology model developed by Harbaugh (1966), described subsequently, which contains a number of interdependent deterministic and probabilistic components.

2. Representation of Space

Many geologic problems involve spatial relationships. Thus representation of space is of prime importance in a geologic simulation model. The geologic model builder has a number of choices concerning representation of space. He must consider the number of dimensions to be represented (one, two, or three) and whether space is to be represented in continuous or discrete form. Figure 2a provides an example of the representation of spatial relationships with a continuous curving line. Although the line represents a topographic profile that occupies two dimensions, the line itself can be thought of as a function of the single independent variable y. If we can write a function that represents the line, we can represent the line in continuous form, graphing it by supplying different values of the independent variable y, and in turn obtaining values of the dependent variable z.

An alternative method is to approximate a line in discrete form. For example, the same topographic profile can be divided into a series of columns of equal width along the y axis, with the height of each column approximating the line at that location (Fig. 2b). The height of columns can be represented by a sequence of numbers. If the shape of the profile is complex, it might be difficult to find a function that describes it, but it is simple to represent the profile with a set of numbers, no matter how complicated its shape. Furthermore, computing languages such as FORTRAN readily provide for representation of sequences of numbers in array form.

Representation of surfaces, in contrast to lines, involves an additional dimension. The choices for representation are similar, however. Surfaces can be represented by continuous functions, such as polynomials or harmonic functions involving two independent variables, or they may be approximated in discrete form by a series of rectangular cross-sectioned prisms of different height (Fig. 3). The height of the prisms can be represented as a series of numbers stored in a two-dimensional

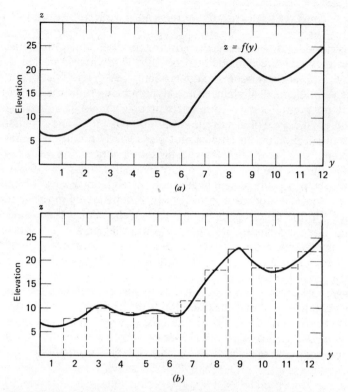

Fig. 2. Topographic profile represented by (*a*) the line in which *z* is a function of *y* and (*b*) by a series of discrete columns. The height of each column can be specified by numerical values in an array.

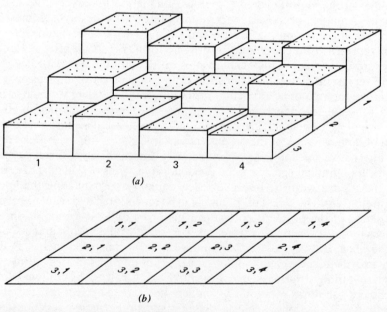

Fig. 3. (*a*) Meshwork of square cross-sectioned prisms whose heights provide approximation to irregular surface. (*b*) Two-dimensional array whose elements represents heights of prisms. Column and row indexes of array also serve a geographic coordinates. Additional numbers (not shown) are needed to represent heights.

626

array, the indexes of the elements in the array specifying the geographic location of each prism.

Although continuous functions are widely used to describe spatial relationships in many computing applications, they have limited usefulness in geologic simulation applications for representation of space. There are two reasons for this. First the spatial relationships of many geologic features are so complex that it would be difficult or practically impossible to represent them by mathematical functions. Secondly, many geologic relationships are concerned with qualities, such as rock types or presence of different kinds of fluids. Qualities are readily represented in a discrete system, whereas they are difficult to represent with conventional functional relationships. Finally, most dynamic geologic simulation models must incorporate materials accounting systems. The nature of accounting, which involves crediting and debiting of discrete quantities of materials, inherently requires discrete representation of space. Figure 4, which deals with the representation of three-dimensional spatial relationships, also portrays the manner in which qualities can be represented. When qualities are represented by numbers, they can be stored and manipulated in computer arrays. We should note, however, that a two-dimensional array can store information about bodies that exist in three-dimensional space. The two-dimensional array of Fig. 3b stores information that can be thought of as representing either a surface (the tops of the prisms) or a solid body (the aggregation of prisms themselves), both of which occupy three-dimensional space. The simulation examples described here employ discrete representation of space. However, there are many inherent advantages in continuous functions if they can be employed, and simulation model builders should consider their use if feasible.

As we have pointed out, arrays provide effective means of storing information about the properties of materials. A simulation program representing a marine sedimentation model, for example, might contain quantitative information on porosity, permeability, sorting, and proportions of sand, silt, and mud. In addition

TABLE I

Example of Representation of Sediment
Properties in Three-Dimensional Space
in FORTRAN Programs

Property Represented	FORTRAN Array[a]
Porosity (%)	PORE (I, J, K)
Permeability (millidarcys)	PERM (I, J, K)
Sorting	SORT (I, J, K)
Sand (%)	SAND (I, J, K)
Silt (%)	SILT (I, J, K)
Mud (%)	MUD (I, J, K)
Color (red, gray, green, black)	COLOR (I, J, K)
Foraminifera (presence or absence)	FORAM (I, J, K)

[a] I, J, and K are identifiers representing array indexes in the three dimensions.

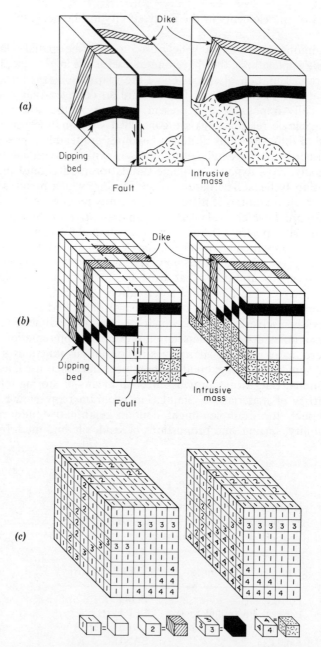

Fig. 4. Method of representation of qualitative relationships in three-dimensional space. (a) Block diagram (split apart to reveal details) employing conventional graphic symbols. (b) Block diagram in which space has been compartmented into rectangular cells. Conventional graphic symbols are still employed. (c) Substitution of numbers, which form three-dimensional array, for conventional graphic symbols (from Harbaugh, 1966).

628

it might contain qualitative information on sediment color, and on the presence or absence of foraminifera. This information could be stored in various ways. A simple and direct way would be to create an array for each property (Table I). The various arrays could then be thought of as coinciding in space, each occupying the same area or volume as the others.

3. Materials Accounting

Dynamic simulation models that purport to be realistic should adhere to the provision that they neither symbolically "create" nor "destroy" matter. This requirement can be met by employing accounting methods that contain information on all materials that enter, leave, or remain within the boundaries of the system represented by the model. The net difference between materials entering and materials leaving a system during a simulation run should be equal to the overall gain or loss of materials. In other words,

$$accumulation = input - output$$

This concept is extremely simple and is widely applied, commonly being termed *mass balance* or *materials balance*. Furthermore, the requirements are such that not only must the gross amounts of materials moving into and through the system be accounted for, but each different type of material must be subject to the same bookkeeping rigor. Horn and Adams (1966), Bowen and Inman (1966), and Neumann and Land (1968) provide examples of geologic applications of materials accounting methods. Applications in other fields include economic input–output analysis (Leontief, 1966), which provides an approach that might be adapted for accounting of energy relationships in geology.

Although the techniques used in materials accounting are simple in principle, they tend to make large demands on computer storage requirements. Consider, for example, the problem of storing information on sedimentary materials in a modest sedimentation model that employs a geographic grid containing 50×50 cells. If there are five different types of sediment to be accounted for over an interval of 10 time increments, the total number of items to be stored in the accounting system after the 10 time increments have elapsed is $50 \times 50 \times 5 \times 10 = 125,000$. This is a staggering volume of information for a small simulation model. Furthermore, this figure is merely the amount of information to be stored at the end of the run. It does not include the large number of arithmetic and logic operations involved in the accounting operations themselves.

The delta depositional model of Bonham-Carter and Sutherland (1968) provides a good example of an accounting system. Their model employs a geographic "accounting grid" to receive sediment deposited at the mouth of the river. The model provides for sediment of various specified grain sizes. As the river pours out into the sea, the settling trajectories of sediment particles are calculated (Fig. 5). The particles are assigned to particular cells in the accounting grid, depending on where they come to rest. During each time increment, the amount of sediment of each grain size that enters each cell is tabulated.

The delta model employs a four-dimensional FORTRAN array SED(I, J, L, N) for accounting purposes. The indexes I and J pertain to the rows and columns in the horizontal plane of the geographic grid. Index L denotes the particular sediment size fraction and N is an index of the time increment. Array SED is thus capable of containing all essential information about the sediment that has been deposited.

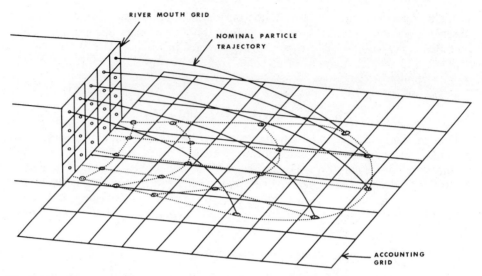

Fig. 5. Use of accounting grid in delta sedimentation model. River flowing in rectangular channel at left pours into sea (upper edge of river mouth grid coincides with sea level). Distribution of sediment size fractions in river channel at its mouth is accounted for by river mouth accounting grid. During each time increment, amount of each sediment size fraction leaving river mouth is equal to amount deposited on sea floor (from Bonham–Carter and Sutherland, 1968).

The numerical values of elements in the array describe the amount of sediment of each size deposited in thickness units per cell. The array is indexed according to location, type of sediment, and time increment. The time increment also serves as a stratigraphic index, because the sediment can be regarded as deposited in consecutive layers, each layer corresponding to a time increment.

4. Control

Control is an essential aspect of all dynamic systems. Control is almost synonymous with negative feedback. Without negative feedback control, a system would have a very short life; it would "blow up," so to speak. Negative feedback is inherent in many aspects of systems, whether simple or complex. Furthermore, negative feedback is present in many aspects of mathematical models of systems. Because of the circuitous or continuously closed nature of loops in models of such systems, most of the components of the model may be regarded as parts of the control system. Given such complexity, it is difficult to isolate parts of the system for analysis without reference to other parts. Consider, for example, the conceptual model of shallow-water marine sedimentation shown in Fig. 6. Accumulation of sediment affects isostatic adjustment, which in turn affects basin configuration, affecting accumulation of sediment. The components of this loop tend to regulate each other. In addition, there are other loops which, in total, provide a complex network of interdependent, overlapping control systems.

Before proceeding further let us examine the nature of inputs to a system model. Endogenous components are linked with feedback loops and are thus "inside" a system, while *exogenous* components feed into a system, but do not receive feedback from the system and therefore are not controlled by the system. We can distinguish

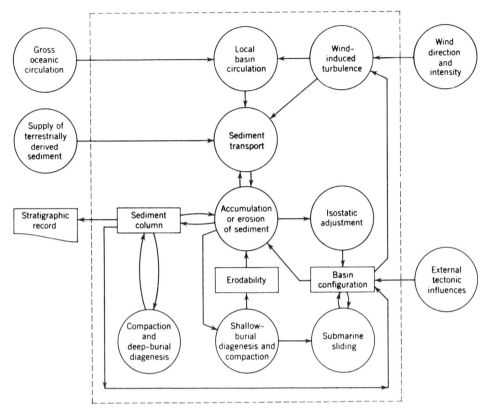

Fig. 6. Theoretical system diagram showing principal processes or classes of processes that affect shallow-water marine sedimentation. Dashed line separates exogenous processes that are outside the system as defined, in that they supply inputs to the system but do not in turn receive feedback from the system. Endogenous processes are inside the box outlined by the dashed line.

three principal ways in which exogenous inputs to the system influence the system (Fig. 7): the simplest case is one in which a single exogenous input is simply fed into and incorporated within the system (Fig. 7a). A slightly more complex situation occurs when there are multiple exogenous inputs that interact with each other, but there is no feedback (Fig. 7b). A still more complex, though more realistic, situation is one in which there are multiple exogenous inputs and internal feedback (Fig. 7c).

Some control systems concepts within a simple dynamic, mathematical model of a sedimentary basin, which incorporates ideas advanced by Sloss in 1962, are shown in a flow diagram in Fig. 8. The system as defined provides for two exogenous inputs,

1. sediment from outside the system
2. depth to base level or wave base.

Inside the artifically bounded system, the sediment is transported and the sediment sizes are fractionated according to depth, and in turn the water depth modified.

The Sloss model deals principally with deposition of clastic sediments on continental shelves. Sediments are delivered to the edge of the sea by streams flowing

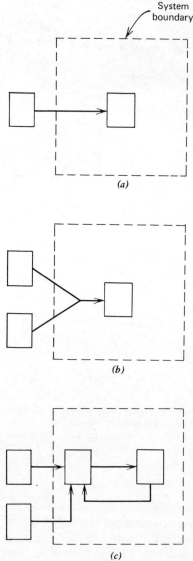

Fig. 7. Diagram to illustrate effect of (a) single exogenous input without feedback within the system; (b) multiple exogenous inputs which interact but in which there is no feedback; and (c) internal feedback in addition to exogenous inputs.

across a coastal plain. When brought to the sea, each sediment particle is transported until it finds a position of rest and becomes available for incorporation in the sedimentary sequence. The position of rest is a function of kinetic energy (waves and currents), material (composition and particle size), and boundary conditions (bottom slope and roughness). The interaction of these factors produces an equilibrium surface or base level, above which a particle cannot come to rest and below which deposition and burial are possible. At any instant in time, given an adequate supply of sediment, the interface between water and sediment tends to coincide with base level. Successive interfaces representing successive instants of time can be interpreted as a record of the relative rate of subsidence of the depositional basin.

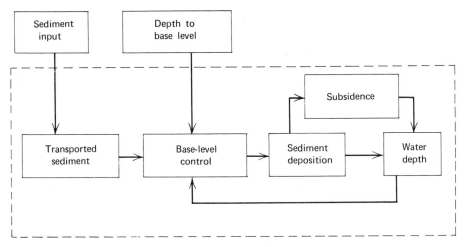

Fig. 8.   Flow chart of simple basin model in which crustal subsidence is provided for. Subsidence is provided for by an endogenous loop. In addition, an external or exogenous influence on subsidence could be provided.

The gross geometry of shape S of a body of sedimentary rocks varies as a function of the quantity Q of material supplied to the depositional site, the rate of subsidence R at the site, the rate of dispersal D, and the nature of the materials supplied M:

$$S = f(Q, R, D, M)$$

In this expression, R is a measure of the rate of subsidence expressed as a receptor value, defined as the available volume below base level created per unit time by subsidence. The proportion of different particle sizes of sedimentary material M is assumed to be constant. This assumption implies that weathering and erosion in a heterogeneous source area yield coarse, medium, and fine clastic particles to the depositional area in unchanging proportions.

With these assumptions, Sloss developed the process elements and the resulting stratigraphic responses in a series of diagrams. Figure 9 shows hypothetical responses when the quantity Q of material supplied and receptor value R are varied. In Fig. 9a more material is supplied than can be handled by dispersal and subsidence ($Q > R$), resulting in deposition of a regressive sequence. The results where $Q = R$ and $Q < R$ are illustrated in Figs. 9b and 9c, respectively.

A. Computer Model

A computerized extension of the Sloss model incorporates the following factors:

1. quantity of material supplied, which may include from one to five different sediment size fractions,
2. initial geometry of the sedimentary basin, expressed as water depth,
3. tectonic warping (subsidence) through time and from place to place in the basin,
4. position of base level or equilibrium surface defined with respect to sea level for each particle-size class.

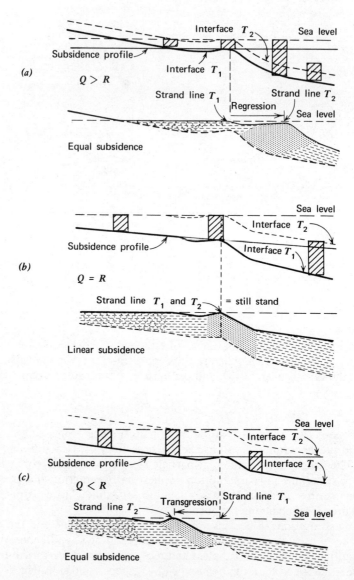

Fig. 9. Sequence of cross sections which illustrate hypothetical responses of Sloss (1962) model under three different conditions: upper part of each diagram shows interfaces at time T_1 and T_2. Diagonally ruled bars represent amount of sediment supplied during interval between T_1 and T_2, whereas receptor value is denoted by lowering of interface with respect to sea level between T_1 and T_2, as shown by subsidence profile. Lower part of each diagram shows facies that will result. (a) When quantity Q of sediment supplied is greater than receptor R value ($Q > R$), coupled with uniform subsidence, the model responds by deposition of regressive facies sequence. (b) When $Q = R$, coupled with subsidence rate that increases linearly seaward, the model responds by producing facies which have vertical boundaries. (c) When $Q < R$, coupled with uniform subsidence, the model yields transgressive sequence.

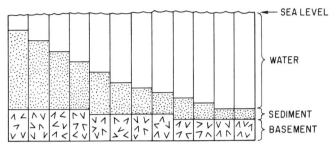

Fig. 10. Subdivision of two-dimensional sedimentary basin into series of discrete vertical columns representing water, sediment, and basement (from Bonham–Carter and Harbaugh, 1970).

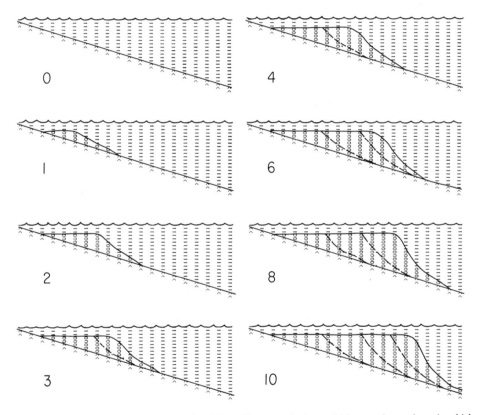

Fig. 11. Sequence of cross sections produced by sedimentary basin model in experimental run in which crustal subsidence does not occur and which involves sediment of a single size class. Time increments 0, 1, 2, 3, 4, 6, 8, and 10 are shown. Water is symbolized by horizontal lines and sand by 0's. Model responds by producing a regressive sequence of deltaic deposits. Slope of deposits is affected by initial slope of sea floor and by "decay constant" k, which has been set at 0.5. Sediment–water interface is plotted every third time increment, thus forming series of stratigraphic time lines (from Bonham–Carter and Harbaugh, 1970).

The model treats only two spatial dimensions, representing a vertical section through a sedimentary basin. The other horizontal dimension is not represented, but could be incorporated if a more advanced version were to be developed. As in the rock mass model, time is divided into discrete steps. Space is represented by a sequence of columns, each of unit width, which represent water depth and thickness increments of various sediment types (Fig. 10). Subsidence (or uplift) is represented by sliding the columns up or down relative to sea level. FORTRAN arrays are used to store water depths and sediment thicknesses for each vertical column for each time increment, similar to the manner described earlier. Each column of sediment and water is displayed as a row of symbols on the computer printer, and the resulting sequence of rows forms a cross section through the sedimentary basin (Fig. 11).

Sediment Transport and Deposition

Transport and deposition of sediment is treated heuristically in the computer model. It is assumed that during any time interval, a certain increment of sediment (the sediment "load") enters the basin from a source area on one side of the basin, and is then transported from column to column. Deposition may take place in each column, the amount deposited being debited from the sediment load, and the remainder passed on to the next column, where the process is repeated. The sequence proceeds from the sediment-source side of the basin toward the seaward side. The amount of sediment deposited in a particular column depends on (a) the amount of sediment available for deposition and (b) the water depth in that column relative to base level. Mass balance is observed by accounting for all sedimentary materials as they move through the system.

The rules governing transportation and deposition of sediment are extremely simple and are outlined below. Part of the sediment load reaching a particular column is deposited if the water depth is greater than depth to base level. In columns that contain water sufficiently deep so that base level exerts no control, the proportion of sediment deposited for each particle-size class is represented by a curve that declines exponentially toward the seaward side of the basin (Figs. 12a and 12b). The accounting system arithmetic involved in this process can be envisioned as follows. Let the sediment load entering the basin be L, and the proportion of this load that is deposited in the first column is k. Thus the amount deposited is kL, and the remaining load that is shunted on to the next column in the sequence is $L - kL$ or $L(1 - k)$. The quantities deposited in successive columns are illustrated in Table II.

The amount deposited in the second column is found by multiplying the remaining load by k, giving $kL(1 - k)$. In column 3, the remaining load is $L(1 - k)^2$ and the

TABLE II

Column	Sediment Load	Sediment Deposited
1	L	kL
2	$L(1 - k)$	$kL(1 - k)$
3	$L(1 - k)^2$	$kL(1 - k)^2$
4	$L(1 - k)^3$	$kL(1 - k)^3$
5	$L(1 - k)^4$	$kL(1 - k)^4$
$\vdots$	$\vdots$	$\vdots$
n	$L(1 - k)^{n-1}$	$kL(1 - k)^{n-1}$

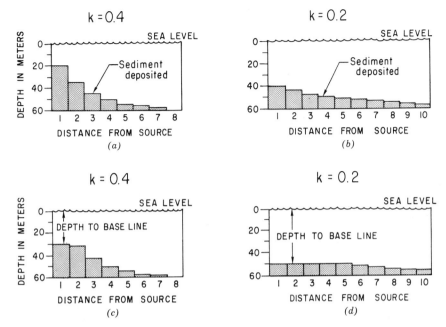

Fig. 12. Four diagrams representing vertical sections through sedimentary basin into which uniform volume of sediment is supplied from source at left. Initial water depth is uniform. Diagrams (a) and (b) assume that base level does not exert influence and illustrate effects of varying the "decay constant" k. Diagrams (c) and (d) illustrate the effects of two different base levels, coupled with values of k equivalent to those of diagrams (a) and (b), respectively. Volume of sediment introduced in all four diagrams is equivalent to a column 100 m high (from Bonham-Carter and Harbaugh, 1970).

amount deposited is $kL(1 - k)^2$. Subsequent columns are treated similarly. Thus we may generalize for the nth column where the sediment load remaining is $L(1 - k)^{n-1}$ and the amount deposited is $kL(1 - k)^{n-1}$. This relationship is an adaptation of the familiar law of growth and decay in which k is the decay constant.

The significance of different values of k is open to interpretation. It is clear from Figs. 12a and 12b that k pertains to slope of deposits. If slope were solely a function of grain size, larger values of k might correspond to coarse sediment capable of reposing on steeper slopes and being less mobile than fine sediment. Maximum slope angle and mobility, however, do not bear a simple relationship to grain size. Thus it is difficult to interpret k in terms of a simple physical relationship; instead it should be regarded as a parameter used in conjunction with a relationship that is largely heuristic.

If the model dealt only with the transport of sediment according to the growth and decay law, the calculations would be simple and could be readily carried out by hand. If we introduce constraints, however, the solutions are not simple. If we introduce base level, above which sediment of a particular particle size cannot come to rest, we must continually check the elevation of the sediment–water interface in each column to ensure that it is not above base level. Figures 12c and 12d illustrate the effect of base level as a constraint, employing specified values of k and holding other factors constant.

The model incorporates various decision rules that govern sediment deposition. We can distinguish three situations:

1. Where the sediment–water interface is above or equal to base level (as specified for a particular particle-size class), deposition is not possible, and all the load is shunted on to the next column.
2. Where sediment–water interface is slightly below base level, only part of the quantity of sediment that would otherwise be deposited is accommodated. Sufficient sediment is deposited to bring the column to base level, and the remainder is passed on to the next column.
3. Where sediment–water interface is sufficiently far below base level, all the sediment available for deposition is accommodated.

These relationships can be expressed algebraically. If water depth in a column is D, and depth to base level is B, then the amount deposited S is given by one of the following relationships. If water depth is less than, or equal to depth to base level, no sediment is deposited:

$$S = 0 \qquad D \leq B$$

If water depth minus the quantity kL is less than or equal to depth to base level, the amount deposited is equal to depth minus base level:

$$S = D - B \qquad (D - kL) \leq B$$

Otherwise, the amount deposited equals the quantity kL, and base level has no influence on sedimentation; thus

$$S = kL$$

Figures 12c and 12d illustrate these algebraic base level control relationships. In Fig. 12c, depth to base level is 30 m. Without base level control of the total of 100 m of sediment supplied, 40 m would have been deposited in the first column, reducing its water depth to 20 m. But 20 m is shallower than base level. Thus deposition of $D - B = 60 - 30 = 30$ m of sediment takes place in column 1, and the remaining load of 70 m of sediment passes on to column 2. The sediment load reaching column 2 is equivalent to $100 - 30 = 70$ m of sediment. The new value of kL is thus $0.4 \times 70 = 28$. This time, if all 28 m of sediment are deposited, water depth is given by $D - kL = 60 - 28 = 32$ which is greater than depth to base level. Thus 28 m is deposited in column 2. Similar calculations are carried out for columns 3–8. In Fig. 12d, depth to base level is 50 m and $k = 0.2$. Base level imposes a depth limit for sediment in the first five columns. In columns 6–10, the amount deposited in each column dies away exponentially.

The treatment of more than one grain size further complicates the calculations. Each grain size is assigned

1. a specific value of k,
2. an initial sediment load to be released from the source for each time increment,
3. a depth to base level.

In an experimental run, many columns receive sediment of more than one grain size during a time increment. Mixtures of sediment of varying size in a particular

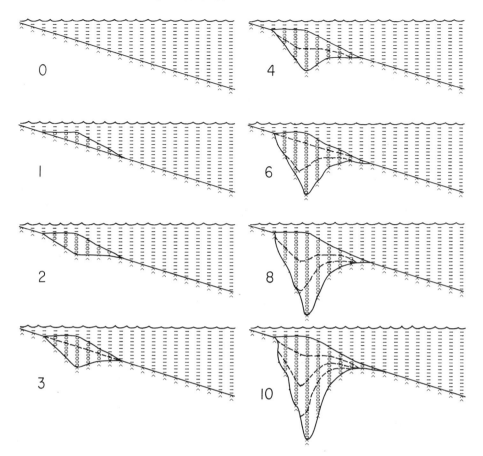

Fig. 13.   Response of model when controlling parameters are similar to those in experiment shown in Fig. 11, except that subsidence takes place in each column in proportion to volume of sediment deposited in that column. Each increment of subsidence lags behind deposition by one time increment. Note that configuration of sediment–water interface is unchanged after first time increment. Greatest amount of subsidence occurs where maximum quantity of sediment is deposited (from Bonham–Carter and Harbaugh, 1970).

column are graphically represented by using a symbol that represents the particle-size class that forms the largest proportion of the volume deposited during the particular time increment.

Crustal Subsidence

The effect of subsidence, or conversely, eustatic changes in sea level, can be represented by adding or subtracting values to the water depth in each column. Increasing water depth in a single column by some number of meters has the effect of depressing the entire column downward by the same number of meters. In this model there are two ways in which water depth values can be changed. First, provision is made in the program for adding a specific increment to water depth in each column during each time increment. The increment to water depth may vary from column to column, but is constant for each time interval. This simulates uniform subsidence with time and also may be used to represent changes in sea level. Alternatively, subsidence

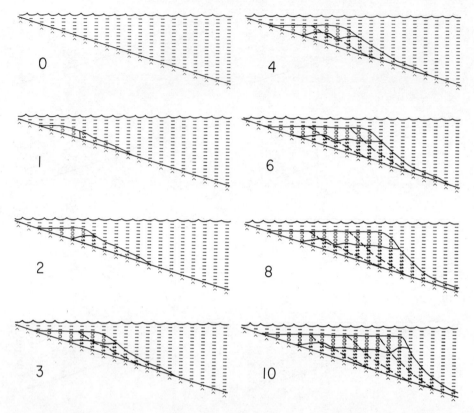

Fig. 14. Computer-printed cross sections through sedimentary basin which does not subside. Numbers identify time increments. Water is symbolized by horizontal lines, sand by 0's, and silt by $'s. Sea floor is presumed to have sloped uniformly to right at start of simulation run. Five "characters worth" of sand and five of silt were supplied from left to right during each time increment. Time lines, which are added every third time increment, are represented by dashed lines. Cross sections, being purely hypothetical, lack scale.

can be related to deposition by a simple proportionality constant. Each column can be depressed by an amount equal to thickness of sediment deposited multiplied by a proportionality constant F. At one extreme, if F equals one, the amount of subsidence equals the amount of deposition. At the other extreme, subsidence does not occur if F equals zero.

Subsidence need not be instantaneous with deposition. As an alternative, subsidence may lag behind deposition by some whole number of time increments. For example, if the lag length is three time increments, subsidence will occur at the end of every third time increment, and the amount of subsidence in a particular column is obtained by multiplying F by the total quantity of sediment deposited in that cell during the previous three time increments. The lag cannot be made shorter than one time increment because of the division of time into discrete steps in the model. Unit lag causes virtually instant response. It could be argued that there may be an appreciable lag between the loading of the actual crust and its subsequent subsidence. The model makes it convenient to experiment with the effects of different lag lengths.

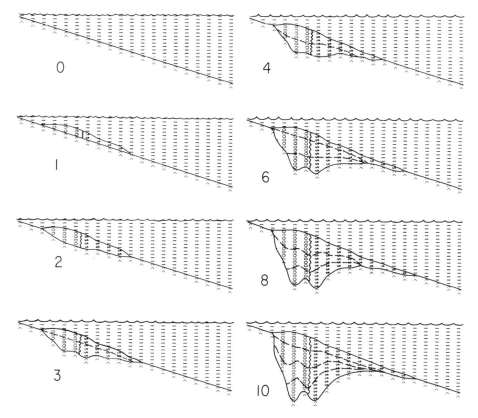

Fig. 15.   Results in which subsidence, following initial deposition, keeps pace with deposition of sediment.

B. Experiments with Model

The response of the model in a no-subsidence situation is shown in Fig. 11. As the cross sections reveal, without subsidence and with only a single grain size, the model produces a "deltaic" deposit that grows progressively out into deep water, forming a series of "foreset" beds that dip progressively more steeply as the delta builds outward.

A similar experiment, in which crustal subsidence occurs, is shown in Fig. 13. The sediment–water interface remains at a constant elevation after the first time increment, each column subsiding by an amount equal to the thickness of sediment deposited immediately before. Under these conditions, the response of the crust is to subside the most where the maximum amount of sediment has been deposited. The overall form of the deposits is that of a lens, reminiscent of the lens-shaped mass of Cenozoic sediments of the Gulf Coast.

Three other experiments with the model are illustrated. In Fig. 14, a regressive sequence has been produced by introducing two grain sizes and making subsidence equal to zero. At time 1, sand is deposited close to shore, and silt is deposited farther from shore. Beginning at time 2, and continuing during subsequent time increments, sand begins to build out over silt; the effect of base level (set to three depth units for sand) is to cause the zone of maximum deposition to migrate progressively

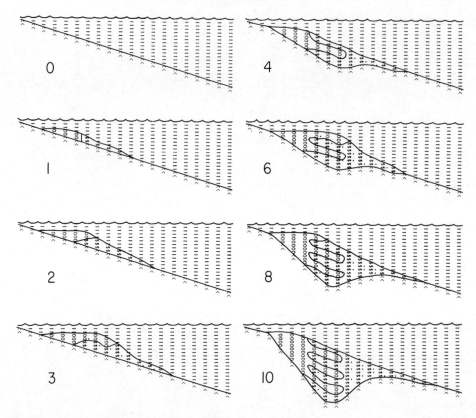

Fig. 16. Results in which crustal subsidence lags behind deposition. Subsidence compensating for sediment load takes place every third time increment.

seaward. Note that time lines intersect the facies boundary. Irregularities in the facies boundary are produced by numerical rounding associated with the use of whole numbers of graphic symbols on the computer printout.

In the experiment shown in Fig. 15, the same controlling parameters are employed as in Fig. 14, except that rate of subsidence is made equal to rate of sedimentation, with a lag of one time increment. As in Fig. 13, the configuration of the sediment–water interface remains constant through time, and the site of maximum deposition remains in one place. Because there are two facies, however, we note that a vertical boundary separates the sand and silt facies.

The effect of lag is introduced in the experiment shown in Fig. 16. Note the interfingering of facies. It seems possible that many naturally occurring examples of interfingering are due to lag in a feedback loop. If this is true, explanations for cyclic sedimentation may not require elaborate hypotheses of worldwide changes in sea level, but could be alternatively explained by lag.

5. Probabilistic Depositional Ecology Model

The last model to be discussed involves a number of probabilistic components, in contrast to the models described previously, which are strictly deterministic. In principle, the probabilistic depositional model could be used for simulating behavior

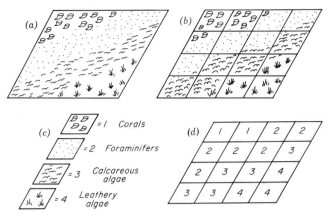

Fig. 17. Means of representing geographic distribution of organism communities: (a) sea floor is popu-lated by different communities that are continuous and are portrayed graphically: (b) sea floor has been divided into square cells: (c) graphic symbols are assigned numerical equivalents: (d) two-dimensional array of integers contains essentially same information as graphic symbols in (a) except for some loss of detail owing to relative coarseness of discrete cells (from Harbaugh, 1966).

of organism communities under a variety of conditions, ranging from bottom-dwelling marine organisms to aggregations of land plants. In practice, the model has been applied to shallow-water, carbonate-secreting marine organism communities. The components of the model that involve interaction between organism commu-nities are probabilistic, and make extensive use of methods that involve Markov chains. The probabilistic mechanisms in the model are allied with third-order, single-step Markov chains with nonstationary transition probabilities. The transition probabilities are recomputed at each step by methods which are simple in principle, although complicated in practice.

Representation of organism communities within the model is simple. The area which the organism communities occupy is divided into a grid with square cells of arbitrary size. Each cell contains only one organism community at a time. A com-munity can be defined in the model as consisting of one or more species, but each community type is regarded as the basic unit, rather than species or individual organisms. In the computer programs representing the model (Harbaugh, 1966; Harbaugh and Wahlstedt, 1967), each organism community occupying a specific cell is symbolized by an integer (Fig. 17). In turn, the integers are stored in three-dimensional arrays. Two of the array dimensions pertain to geographic dimensions, and the third represents successive time planes (or successive strata, depending on the intended meaning).

A. Migration and Succession of Organism Communities

Migration and succession of organism communities in the model involve a method in which the different organism community types are multiplied by weighting factors and then one is selected at random. The process might be likened to the process of placing balls with numbers painted on them in an urn and withdrawing one of the balls at random. The probability of drawing a ball with a particular number is propor-tional to the number of balls possessing that number with respect to the total number of balls in the urn. For example, if there are a total of 1000 balls in the urn, and 150 balls have number 5 painted on them, the probability of drawing a ball with 5 painted on it is 0.150. The program representing the model, in effect, puts a succession of balls representing organism communities into an urn and then draws one at random for each cell during each time increment.

The probability pertaining to the selection of the next community to occupy a particular cell is influenced by preceding events and is therefore Markovian. The model assumes that three previous events influence the selection, each separated by a single time increment. Thus in some respects the selection process is very similar to a third-order Markov chain (Harbaugh and Bonham-Carter, 1970, chapter 4). There are, however, important differences in that not only do previous occupants of the particular cell in question influence the selection process, but their geographic neighbors also exert influence. Probabilistic lateral migration incorporates the effect of neighboring cells through their contribution to the transition probability values. The model is constructed so that the influence of neighboring cells decreases with distance from the cell under consideration.

The influence of neighboring cells applies to each particular time increment. In the selection of a new occupant for a particular cell, only the occupants of previous time increments are considered, not the neighbors for which the new occupants are being chosen. The reason for this is inherent in the construction of digital dynamic models. Because of the necessity of taking discrete steps in carrying out computing operations, simultaneous events cannot be directly represented in the model. In other words, we cannot carry out calculations that represent the behavior of the occupant of a particular cell simultaneously with those in other cells. Instead, we must perform all the computing operations one by one that pertain to a particular time increment before we pass to the next time increment.

In selecting the occupant of each cell, the model compares the occupants of the cell under consideration for the three preceding time increments and, in turn, supplies an additional or secondary weighting factor which is used to further modify the contribution (as the imaginary numbered balls supplied to the urn). A secondary weighting factor is calculated for each of the three preceding time increments, and is multiplied times the weighted contribution based on the neighbor-cell effect (or primary weighting factor).

The secondary weighting factor serves three purposes:

1. It permits the degree of persistence of an organism community to exert influence on succeeding events. For example, if a given type of organism community has occupied a given cell for three successive time increments, the probability of the same community occupying the same cell in the forthcoming time increment is greater than it would be if different communities had occupied the cell during the three preceding time increments. The effect of this may be likened to inertia in that long-established communities may tend to resist subsequent change much more than communities whose occupancy has been brief.

2. Secondly, it permits physical environmental factors to exert influence on succession. For example, to represent the influence of a favorable environmental factor, a weighting factor greater than one is used, whereas an unfavorable environmental factor is expressed by use of a weighting factor between 1.0 and 0.0. A weighting factor of zero would eliminate any contribution from the particular organism community for the particular time increment, and therefore represents the effect of an environmental factor that is totally intolerable.

3. Thirdly, the secondary weighting factor permits the degree to which a succession of occupants of a particular cell approaches an ideal ecologic succession to exert influence. The degree of "closeness" or "farness" in an ecologic succession can be expressed numerically. For example, if there are 12 communities symbolized by numbers such that community 1 is a pioneer community and community 12 is the climax community (for a given set of environmental conditions), given sufficient time increments, the pioneer community (1) should gradually be replaced by communities symbolized by higher and higher numbers until the climax community (12) is reached. Thus although there is a tendency for the succession to be unidirectional (i.e., toward the climax), momentary reversals can occur as a result of random fluctuations and major reversals can be produced by major changes in environmental factors (including catastrophic events).

B. Representation of Physical Factors

In dealing with physical factors in the model, we may distinguish between representation of the physical factors themselves, and representation of their effects on organism communities. Representation of some physical factors is simple, whereas others are more complex. For example, depth of water (or height above sea level) is represented by an array of numbers, each of which signifies elevation with respect to sea level for the cell that it represents. Changes from one time increment to the next can be represented by increments or decrements in elevation which represent tectonic movements and sedimentation. Each cell is treated independently for accounting purposes, and may be likened to a prism, square in cross section, which is free to move up and down.

The deposition of clastic sediment in the model is represented with methods that are roughly similar to those used in the two-dimensional sedimentary basin model described earlier. Deposition of carbonate sediment is treated differently, however, in that carbonate sediment is assumed to be formed solely through secretion by organism communities. The production of carbonate sediment is treated as a function of the type of community and the relative vitality of the community. For example, a phylloid algae community could be assumed to produce more carbonate sediment per cell per time increment than a brachiopod community. The relative vitality of the community under the prevailing environmental conditions also has large influence. For example, a phylloid algae community on a shallowly submerged marine bank would have high carbonate productivity, whereas the same community in deeper water might produce very little carbonate sediment.

C. Simulation of Carbonate Marine Banks

The model was used in a series of experiments in which the goal was to simulate the environmental conditions in which certain Upper Pennsylvanian marine deposits were formed in southeastern Kansas. These deposits include locally thickened, lenslike masses of limestone, termed marine banks, that are composed principally of leaflike (phylloid) calcareous algae. One of the aspects of the problem of origin of the banks is understanding the factors that cause them to be localized. They appear to have been formed adjacent to an ancient shoreline, where the lobate, deltalike deposits formed as a result of debouching streams strongly influenced the position of the banks (Fig. 18).

The environmental responses of a Pennsylvanian phylloid algae community are clearly difficult to ascertain, and can only be inferred. In simulation experiments

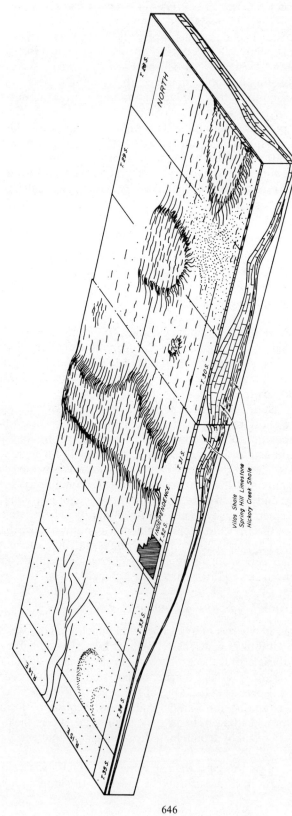

Fig. 18. Block diagram showing interpretation of geography and geology during deposition of marine-bank deposits in Late Pennsylvanian time in southern Kansas. Block is about 40 mi long in north–south direction and 15 mi wide in east–west direction (from Harbaugh, 1960).

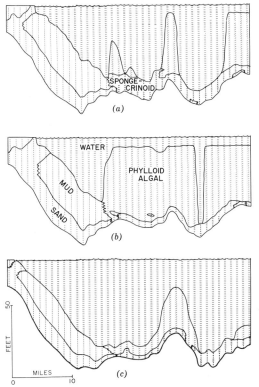

Fig. 19. Series of geologic cross sections showing different responses of phylloid algae community to conditions assumed in experiments. (a) Response of algae is too weak; (b) too strong; (c) appropriate. North is toward right. Except for lines and large identifying letters, symbols have been printed directly by computer's line printer. Symbol representing sediment type of greatest volumetric importance per time increment is printed.

conducted with the model, the problem was approached heuristically, with a great deal of trial-and-error probing. An example of trial-and-error probing is shown in Fig. 19. This figure contains three 40-mi-long, north–south geologic section through an area in southeastern Kansas in which deposition of Pennsylvanian sediments has been simulated. The three sections show the response of the model to different "settings" which affect vitality and sediment-contributing ability of phylloid algae. In Fig. 19a the response of the algae is too weak, in Fig. 19b, too strong, and in Fig. 19c response of the algae is similar to that inferred from observations of the actual deposits (Harbaugh, 1960). Thus the vitality factor and sediment-contributing ability of the algae has been adjusted in the model until the algae "perform properly" under the assumed environmental conditions.

6. Conclusion

In conclusion, we should point out that the simulation models described here are empirical in that they tend not to involve simple physical principles. Other simulation models, for example, may involve the solution of differential equations in the representation of the processes whose parameters and rates are capable of being represented by differential equations. Many geologic processes, however, cannot presently be represented by a series of differential equations, and therefore can be represented only by models that involve a variety of heuristic and arbitrary assumptions. In spite of the fact that the representation of many geologic processes in simulation models is necessarily heuristic, these models may still be very valuable in demonstrating interrelationships. For example, the simple sedimentation model presented in Figs. 8–16 is revealing in that it demonstrates the interdependent effects of sediment supply and subsidence on the geometric configuration of the resulting facies. On the other hand, the sediment-transporting process and the subsidence process are entirely empirical in the way that they are represented. In fact, most geologic "models," whether in mathematical simulation form or simply as conceptual models, involve a blending of components or assumptions that have varying degrees of validity. One major advantage of constructing a mathematical simulation model is that the assumptions, however theoretical versus empirical they may be, must be explicitly expressed if they are to be incorporated in the simulation model. The mere necessity of an explicit representation requires careful thought as to the mode of the processes, and of the manner in which they react in their interdependent roles.

References

Bonham-Carter, G. F. and J. W. Harbaugh, 1968. Simulation of geologic systems: an overview. *Computer Contribution 22*. Kansas Geological Survey, pp. 3–10.

Bonham-Carter, G. F. and J. W. Harbaugh, 1970. Stratigraphic simulation models. In *Advances in Data Processing for Biology and Geology, Systematics Association Symposium held at Cambridge England, September 24–26, 1969*.

Bonham-Carter, G. F. and Alex J. Sutherland, 1967. Diffusion and settling of sediments at river mouths: a computer simulation model. *Trans. Gulf Coast Assoc. Geol. Socs.*, **17**, 326–338.

Bonham-Carter, G. F. and A. J. Sutherland, 1968. Mathematical model and FORTRAN IV program for computer simulation of deltaic sedimentation. *Computer Contribution 24*. Kansas Geological Survey, 56 pp.

Bowen, A. J. and D. L. Inman, 1966. Budget of littoral sands in the vicinity of Point Arguello, Calif. *U.S. Army Coastal Eng. Res. Cent. Tech. Memo. 19*, 41 pp.

Harbaugh, J. W., 1960. Petrology of marine bank limestones of Lansing Group (Pennsylvanian), southeast Kansas. *Kansas Geol. Sur. Bull. 142*, pt. 5, pp. 189–234.

Harbaugh, J. W., 1966. Mathematical simulation of marine sedimentation with IBM 7090/7094 computers. *Computer Contribution 1*. Kansas Geological Survey, 52 pp.

Harbaugh, J. W., 1967. Computer simulation as an experimental tool in geology and paleontology. In *Essays in Paleontology and Stratigraphy*, Raymond C. Moore Commemorative Volume. *Univ. Kansas Dept. Geol. Spec. Publ. 2*, pp. 368–389.

Harbaugh, J. W. and G. F. Bonham-Carter, 1970. *Computer Simulation in Geology*. Wiley, New York, 575 pp.

Harbaugh, J. W., G. F. Bonham-Carter, and W. M. Merrill, 1971. Programs for computer simulation in geology. *Final Tech. Rept., Off. Nav. Res. Geogr. Branch, Contract N00014-67-0112-004*. Department of Geology, Stanford University, Palo Alto, Calif., 337 pp.

Harbaugh, J. W. and D. F. Merriam, 1968. *Computer Applications in Stratigraphic Analysis*. Wiley, New York, 282 pp.

Harbaugh, J. W. and W. J. Wahlstedt, 1967. FORTRAN IV program for mathematical simulation of marine sedimentation with IBM 7040 or 7094 computers. *Computer Contribution 9*. Kansas Geological Survey, 40 pp.

Horn, M. K. and J. A. S. Adams, 1966. Computer-derived geochemical balances and element abundances. *Geochim. Cosmochim. Acta*, **30**, 279–297.

Krumbein, W. C., 1964. A geological process-response model for analysis of beach phenomena: *Northwestern Univ. Geol. Dept., Tech. Rept. 8*, pp. 1–15.

Krumbein, W. C., 1968. FORTRAN IV computer program for simulation of transgression and regression with continuous-time Markov models. *Computer Contribution 26*. Kansas Geological Survey, 38 pp.

Leontif, W., 1966. *Input–Output Economics*. Oxford University Press, New York, 257 pp.

Neumann, C. A. and L. S. Land, 1968. Algal production and lime mud deposition in the Bight of Abaco; a budget. Abstract *Geological Society of America, Annual Meeting, Mexico City, November 1968*.

Schwarzacher, W., 1967. Some experiments to simulate the Pennsylvanian rock sequence of Kansas. *Computer Contribution 18*. Kansas Geological Survey, pp. 5–14.

Sloss, L. L., 1962. Stratigraphic models in exploration, *J. Sediment. Petrol.*, **32**, 415–422.

III. CHEMISTRY

17. THERMODYNAMIC MODELS FOR THE STATE OF METAL IONS IN SEAWATER

FRANK J. MILLERO

1. Introduction

In recent years there has been increasing concern for the geochemical fate of trace metals in the marine environment. Many so-called "base line" studies have been made to determine the total concentration of these metals in the solution phase, the gas phase, and the solid phase (i.e., on and in living and nonliving material). Although these studies have been useful in indicating the fate of trace metals in the marine environment, little progress has been made in determining the mechanisms involved (physical, chemical, and biological) in the transfer of metals between phases. Part of the reason for this lack of understanding of trace metals in the marine environment comes from the difficulties in detecting experimentally the true activity (both thermodynamic and biochemical) of the trace metals. Owing to their low concentrations conventional electrode methods of determining thermodynamic activity are difficult to apply. Although it is possible to detect by polarography the activity of some heavy metals (Baric and Branica, 1967; Branica et al., 1969; Malijkovic and Branica, 1971; Zirino and Healy, 1970, 1971, 1972), we are at present forced to use models to estimate the thermodynamic activity of most trace metals in order to make comparisons with biological studies.

The importance of knowing the true biological activity (rather than total concentration) of metal ions in seawater has been suggested by a number of workers (Barber and Ryther, 1969; Barber et al., 1971; Lewis et al., 1971; Davey et al., 1970, 1973; Erickson, 1972; Steeman-Nielson and Wium-Anderson, 1970). For example, Erickson (1972) has found that the growth depression by copper of the diatom *Thalassiosira pseudonana* varies with season and location of collection. These variations in copper toxicity are probably due to the presence of differing concentrations of free copper arising from the formation of complexes with natural ligands in seawater. More recently, Davey et al. (1973) have demonstrated by using EDTA in artificial seawater that this diatom can be used as a bioassay of available or active copper in natural waters. They also showed that the toxicity concentration of copper determined by the bioassay was similar to that determined by a copper-specific ion electrode (the diatom died shortly before free copper was seen by the electrode). Although these workers did not directly show that the reduced copper toxicity was due to naturally occurring organic ligands, their results in artificial seawater with EDTA do suggest this role.

Since considerable attention has recently been directed toward the use of synthetic ligands (like NTA, nitriloacetic acid), the effect of organic ligands on the marine environment must be given serious consideration. Three consequences of the addition of such a ligand are possible (Davey et al., 1973):

    1. metals may become solubilized and transported from regions of high concentration to uncontaminated regions;
    2. owing to over-complexation, necessary metals may not be available for the growth of marine organisms;
    3. by reducing the natural metal toxicity increasedly pathogenic organisms could occur.

    The important role that free or complexed metal ions play in productivity has also been pointed out by a number of workers. For example, Barber and Ryther (1969; Barber et al., 1971) reported that phytoplankton blooms in newly upwelled waters off the Peru coast are conditioned by the organics in seawater. They suggested that the organic compounds are necessary to make the trace metals available for growth. However, Steeman-Nielson and Wium-Anderson (1970) suggested that the organic conditioning results from the reduction in copper toxicity (owing to complexation). Although at present it is not possible to state with certainty which of the suggested roles of organic ligands (or both) is true, we can state that it is the form of the metal (not the total amount) that is important in determining its biological and geochemical activity.

    In this chapter the thermodynamic models that can be used to define the state or structure of metal ions in seawater are discussed. Whether the thermodynamic activities can be directly related to biological activities remains to be seen. It must also be kept in mind that the most probable thermodynamic state may not be the state found in the marine environment, owing to the slow kinetics of the formation and degradation of metal complexes.

    To understand the state and structure of an ion in seawater there are a number of steps one must take. Millero (1971c, 1974a, b) prefers to examine the state of an ion in seawater by considering two processes: (1) the ion–water interactions that occur when an ion is transferred from the ideal gas state to an infinitely large reservoir of water (i.e., where ions cannot interact with one another)

$$M^+(\text{ideal gas}) \longrightarrow M^+(\text{infinite dilute solution}) \qquad (1)$$

and (2) the ion–ion interactions that occur when an ion is transferred from infinite dilution to seawater, where interactions of all the ions in the mixture affect the state (i.e., plus–plus, plus–minus, and minus–minus interactions)

$$M^+(\text{infinite dilute solution}) \longrightarrow M^+(\text{seawater}) \qquad (2)$$

These two processes are discussed in greater detail in the next sections.

2. Ion–Water Interaction Models

    To understand the behavior of ions in seawater it is important to be able to understand the interactions of ions with water molecules. To study these ion–water interactions one must study the thermodynamic and transport properties of electrolytes at infinite dilution. In practice it is not possible to make direct measurements at infinite dilution; thus the infinite dilution thermodynamic properties are extrapolated from experimental results at finite low concentrations (with the aid of the Debye–Hückel equations for long-range ion–ion interactions). Since one normally studies

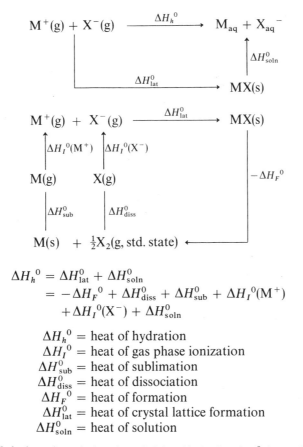

$$\Delta H_h{}^0 = \Delta H_{lat}^0 + \Delta H_{soln}^0$$
$$= -\Delta H_F{}^0 + \Delta H_{diss}^0 + \Delta H_{sub}^0 + \Delta H_I{}^0(M^+)$$
$$+ \Delta H_I{}^0(X^-) + \Delta H_{soln}^0$$

where

$\Delta H_h{}^0$ = heat of hydration
$\Delta H_I{}^0$ = heat of gas phase ionization
ΔH_{sub}^0 = heat of sublimation
ΔH_{diss}^0 = heat of dissociation
$\Delta H_F{}^0$ = heat of formation
ΔH_{lat}^0 = heat of crystal lattice formation
ΔH_{soln}^0 = heat of solution

Fig. 1. Methods used to calculate the enthalpies of hydration, $\Delta H_h{}^0$, for an electrolyte.

the ion–water interactions in solution where no ion–ion interactions occur, it is necessary to select an initial state (i.e., the gas phase) devoid of ion–ion interactions. The initial state normally selected is that of ions in a vacuum at an infinitely low pressure (i.e., the ideal gas). One considers then the changes in properties such as free energy $\Delta G_h{}^0$, enthalpy $\Delta H_h{}^0$, and entropy $\Delta S_h{}^0$ for the process denoted by equation 1 (called, respectively, the hydration free energies, enthalpies, and entropies). The methods used to calculate these thermodynamic hydration functions are discussed elsewhere. (Robinson and Stokes, 1959; Bockris and Reddy, 1973; Latimer, 1952; Rosseinsky, 1965). The methods used to calculate the $\Delta H_h{}^0$ are shown in Fig. 1.

Since we are interested in the transfer of ions (M^+) rather than electrolytes (MX), it is necessary to make some nonthermodynamic assumptions concerning the differences between the properties of cations and anions. The details of such methods are discussed elsewhere (Robinson and Stokes, 1959; Bockris and Reddy, 1973; Latimer, 1952; Rosseinsky, 1965; Noyes, 1964; Gurney, 1953; Millero, 1971b, 1972). Once the selection is made for the absolute thermodynamic quantity of one ion (usually the proton), the values for the other ions can be easily determined by the additivity principle:

$$\Delta H_h{}^0(MX) = \Delta H_h{}^0(M^+) + \Delta H_h{}^0(X^-) \qquad (3)$$

Values of ΔG_h^0, ΔH_h^0, and ΔS_h^0 for some metal ions are given in Table I (Rosseinsky, 1965).

To truly understand ion–water interactions one must know the structure of water. Since the structure of water is very complex, one must use simple models for the interaction between ions and water molecules. These models serve as mental pictures and produce approximately what occurs in the real system. The better they are able to predict the experimental properties of the real system, the better they serve as aids to understanding the real system. Figure 2 shows some sketches of the models frequently used to discuss ion–water interactions.

The continuum model is an example of a crude model that can serve as an approximation for the real system. Drude and Nernst (1884) first used this model to explain the decrease in volume that occurs when an electrolyte is dissolved in water. Born (1920) popularized the model and his name is normally attached to its use. In the model an ion is pictured as a solid sphere of radius r bearing a charge Ze (where Z is

TABLE I
The Thermodynamics of Hydration of Ions at 25°C[a]

Ion	r (Å)	$-\Delta G_h^0$ (kcal/mol)	$-\Delta H_h^0$ (kcal/mol)	$-\Delta S_h^0$ (kcal/mol · deg)
H^+	—	260.5	269.8	31.3
Li^+	0.60	122.1	132.1	33.7
Na^+	0.95	98.2	106.0	26.2
Ag^+	1.26	114.5	122.7	27.6
K^+	1.33	80.6	85.8	17.7
Tl^+	1.40	82.0	87.0	16.7
Rb^+	1.48	75.5	79.8	14.8
NH_4^+	1.60	—	84.8	—
Cs^+	1.69	67.8	72.0	14.1
Cu^+	0.96	136.2	151.1	50.0
Be^{2+}	0.31	—	594.6	—
Mg^{2+}	0.65	455.5	477.6	74.3
Ni^{2+}	0.72	494.2	518.8	82.4
Co^{2+}	0.74	479.5	503.3	80.0
Zn^{2+}	0.74	484.6	506.8	74.5
Fe^{2+}	0.76	456.4	480.2	79.8
Mn^{2+}	0.80	437.8	459.2	72.1
Cu^{2+}	0.96	498.7	519.7	73.9
Cd^{2+}	0.97	430.5	449.8	65.2
Ca^{2+}	0.99	380.8	398.8	60.8
Hg^{2+}	1.10	436.3	—	—
Sr^{2+}	1.13	345.9	363.5	59.2
Pb^{2+}	1.20	357.8	371.9	47.4
Ba^{2+}	1.35	315.1	329.5	48.5
Al^{3+}	0.50	1,103.3	1,141.0	126.6
Fe^{3+}	0.64	1,035.5	1,073.4	127.5
Cr^{3+}	0.69	—	1,079.4	—
Y^{3+}	0.93	859.5	891.5	107.6
Sc^{3+}	0.81	929.3	962.7	112.5
La^{3+}	1.15	—	811.9	—

[a] Taken from the work of Rosseinsky (1965).

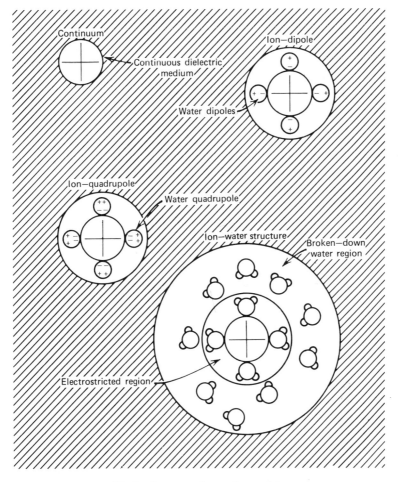

Fig. 2.   Ion–water interaction models.

the valence and e is the electrostatic charge) and the solvent is a structureless continuous dielectric medium. If electrostatics is used, $\Delta G_h{}^0$ (kcal/mol) is given by (at
25°C)

$$\Delta G_h{}^0 = -\left(\frac{Ne^2Z^2}{2r}\right)\left(1 - \frac{1}{D}\right) = -163.89\frac{Z^2}{r} \tag{4}$$

where N is Avogadro's number, r is the radius (Pauling, 1940) in angstrom units
(1 Å = 1 × 10^{-8} cm), and D is the dielectric constant of water (78.36 at 25°C) (Owen
et al., 1961). By appropriate differentiation of equation 4 with respect to temperature
(T), it is possible to determine the other thermodynamic hydration functions $\Delta S_h{}^0$
(cal/deg · mol) and $\Delta H_h{}^0$ (kcal/mol)

$$\Delta S_h{}^0 = \left(\frac{Ne^2Z^2}{2r}\right)\left(\frac{\partial \ln D}{\partial T}\right)_P = -9.649\frac{Z^2}{r} \tag{5}$$

$$\Delta H_h{}^0 = \left(-\frac{Ne^2Z^2}{2r}\right)\left[1 - \frac{1}{D} - \left(\frac{T}{D}\right)\left(\frac{\partial \ln D}{\partial T}\right)_P\right] = -166.78\frac{Z^2}{r} \tag{6}$$

A comparison of the experimental values of ΔG_h^0, ΔH_h^0, and ΔS_h^0 plotted versus Z^2/r is shown in Figs. 3–5. As is quite apparent from these figures, the Born model offers a reasonable first approximation to the magnitude, radius, and charge dependence of ΔG_h^0, ΔH_h^0, and ΔS_h^0. A close examination of the data shows a number of significant deviations. For example, values of ΔH_h^0 for the transition metals (Ca to Zn) given in Fig. 6 do not increase in magnitude with increasing atomic number (decreasing radius). This is because the $3d$ orbitals are not spherically symmetrical (i.e., the hydrated water molecules do not have the same energy) (Bockris and Reddy, 1973).

By further differentiation of equations 4–6 it is possible to obtain information about the size and structure of the hydration sphere. From the pressure dependence of $\bar{G}^0$ (elect), one obtains the volume change (called electrostriction)

$$\bar{V}^0(\text{elect}) = \left(\frac{-NZ^2e^2}{2Dr}\right)\left(\frac{\partial \ln D}{\partial P}\right)_T = -4.175\frac{Z^2}{r} \qquad (7)$$

Further differentiation of the solution component of equation 4 ($\bar{G}^0(\text{elect}) = NZ^2e^2/2Dr$) with respect to T and P gives the electrostriction partial molal expan-

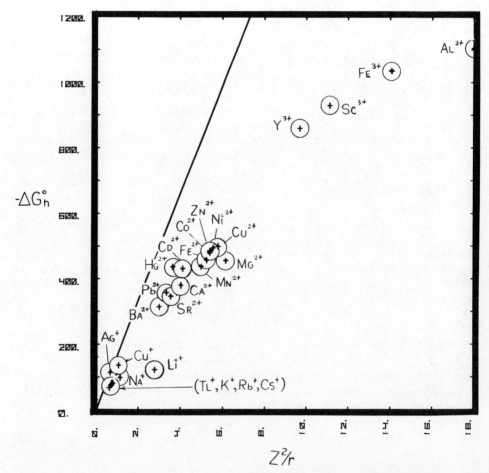

Fig. 3.   The free energies of hydration, ΔG_h^0, for various cations plotted versus Z^2/r at 25°C. The straight line is from the continuum model.

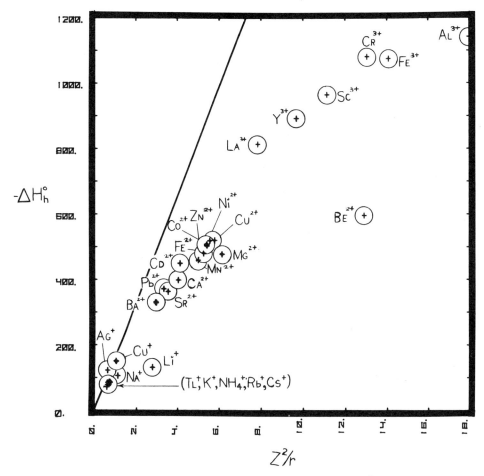

Fig. 4. The enthalpies of hydration, ΔH_h^0, for various cations plotted versus Z^2/r at 25°C. The straight line is from the continuum model.

sibility $(\bar{E}^0 = \partial \bar{V}^0/\partial T)$ and compressibility $(\bar{K}^0 = -\partial \bar{V}^0/\partial P)$:

$$\bar{E}^0(\text{elect}) = \left(\frac{-NZ^2e^2}{2Dr}\right)\left[\frac{\partial^2 \ln D}{\partial T \partial P} - \left(\frac{\partial \ln D}{\partial T}\right)_P\left(\frac{\partial \ln D}{\partial P}\right)_T\right] = -2.74 \times 10^{-2}\frac{Z^2}{r} \qquad (8)$$

$$\bar{K}^0(\text{elect}) = \left(\frac{NZ^2e^2}{2Dr}\right)\left[\left(\frac{\partial^2 \ln D}{\partial P^2}\right)_T - \left(\frac{\partial \ln D}{\partial P}\right)_T^2\right] = -8.31 \times 10^{-4}\frac{Z^2}{r} \qquad (9)$$

Similar differentiation of the solution component of equation 4 with respect to temperature yields the electrostatic partial molal entropy $(\partial \bar{G}^0/\partial T = -\bar{S}^0)$ and heat capacity $[\partial(\bar{S}^0/T)/\partial T = \partial \bar{H}^0/\partial T = \bar{C}_P^0]$

$$\bar{S}^0(\text{elect}) = \left(\frac{NZ^2e^2}{2Dr}\right)\left(\frac{\partial \ln D}{\partial T}\right)_P = -9.65\frac{Z^2}{r} \qquad (10)$$

$$\bar{C}_P^0(\text{elect}) = \left(\frac{NZ^2eT^2}{2Dr}\right)\left[\left(\frac{\partial^2 \ln D}{\partial T^2}\right)_P - \left(\frac{\partial \ln D}{\partial T}\right)_P^2\right] = -12.96\frac{Z^2}{r} \qquad (11)$$

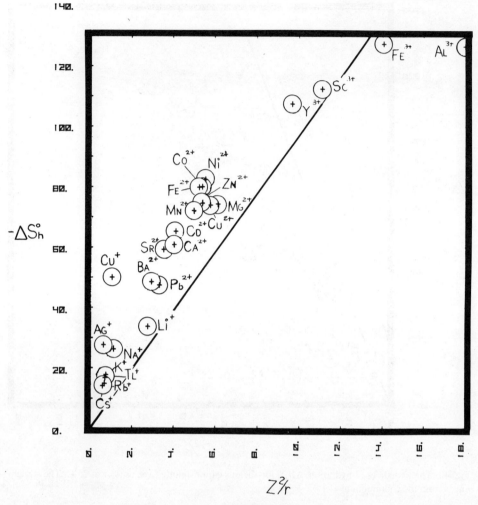

Fig. 5. The entropies of hydration, $\Delta S_h{}^0$, for various cations plotted versus Z^2/r at 25°C. The straight line is from the continuum model.

The partial molal properties of ions in solution contain a minimum of two terms, an intrinsic contribution and an electrical contribution. For example, for the partial molal volume of an ion, we have

$$\bar{V}^0(\text{ion}) = \bar{V}^0(\text{int}) + \bar{V}^0(\text{elect}) \tag{12}$$

where the intrinsic partial molal volume $\bar{V}^0(\text{int})$ is equal to the size of the ion $[\bar{V}^0(\text{cryst}) = (4\pi N/3)r^3 = 2.52r^3$, when r is expressed in angstrom units] plus the packing effects, and the electrical or electrostriction partial molal volume $\bar{V}^0(\text{elect})$ is the decrease in volume owing to ion–water interactions. Thus to plot the various partial molal properties versus Z^2/r one must estimate the intrinsic term. For $\bar{V}^0(\text{int})$ one can use the semiempirical values from (Millero, 1972; Hepler, 1957; Mukerjee, 1961)

$$\bar{V}^0(\text{int}) = 4.48r^3 \tag{13}$$

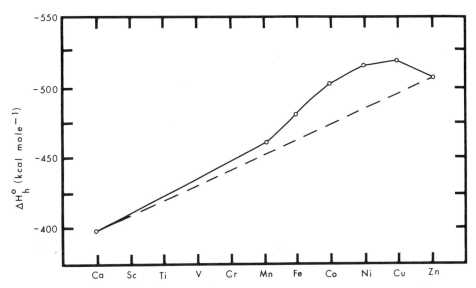

Fig. 6.   The enthalpy of hydration, ΔH_h^0, from Ca to Zn showing the transition metal effect.

and for $\bar{S}^0(\text{int})$ one can use (Laidler, 1956; Latimer et al., 1939; Powell and Latimer, 1951)

$$\bar{S}^0(\text{int}) = \tfrac{3}{2} \ln [\text{A.W.}] \qquad (14)$$

where [A.W.] is the atomic weight. A plot of $\bar{V}^0(\text{elect})$ and $\bar{S}^0(\text{elect})$, calculated by using equations 13 and 14, versus Z^2/r is shown in Figs. 7 and 8. Although the general features of these figures agree with the Born model, a look at the fine structure shows some discrepancies. For example, the $\bar{V}^0(\text{elect})$ of many divalent and trivalent ions appear (Mukerjee, 1961) to be nearly independent of r and the charge dependence of $\bar{V}^0(\text{elect})$ does not appear (Millero, 1971b, 1972) to be directly related to Z^2.

A number of workers (Noyes, 1964; Laidler, 1956, Latimer et al., 1939, Powell and Latimer, 1951; Laidler and Pegis, 1957; Glueckauf, 1964; Padova, 1963; Desnoyers et al., 1965) have attempted to extend the Born model to account for these discrepancies by adjusting the size of the radii [e.g., Latimer et al., (1939); Powell and Latimer, (1951) have added 0.85 Å to cations and 0.1 Å to anions to obtain a linear plot of $\bar{S}^0(\text{elect})$] and adjusting the dielectric constant of the solvent as one approaches an ion [e.g., Laidler and Pegis, (1957) have suggested the effective dielectric constant is ~ 2 near an ion]. These methods, however, do not consider the structure of the water molecule. In recent years workers have used structural hydration models to explain ion–water interactions (Millero, 1971b, c, 1972; Bockris and Reddy, 1973; Frank and Wen, 1957; Desnoyers, in press). The structural models consider the interactions between an ion and a water molecule in terms of ion–dipole interactions, ion–quadrupole interactions, and effects related to the structure of water. Although a full discussion of these models is beyond the scope of this chapter, they can be summarized as follows: by considering the interaction of an ion with a water dipole one can account for the molecular structure of the water molecule, and by considering the interaction of an ion with a water quadrupole, one can account for the difference in the thermodynamic properties of cations and anions of the same size. The water structure effects lead to a region between the oriented dipoles in the electrostricted

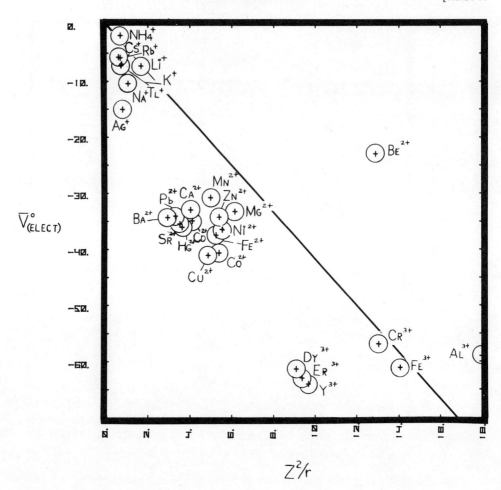

Fig. 7. The $\overline{V}^0$ (elect) for various cations plotted versus Z^2/r at 25°C. The straight line represents the continuum model.

region and the bulk water; the water molecules are partially oriented by the ion and also affected by the bulk water structure. Many workers in recent years (Millero, 1971c; Frank and Wen, 1957) have divided ions into two classes:

1. structure makers, which have a net effect of making more structure around the ion,
2. structure breakers, which have a net effect of breaking down the structure of water.

In general, the use of these terms is ambiguous because we know little about the structure being made or broken. By confining our arguments to hydration effects [$\overline{V}^0$(elect)], it is possible to discuss ion–water interactions using equation 12 or its equivalent for other thermodynamic properties.

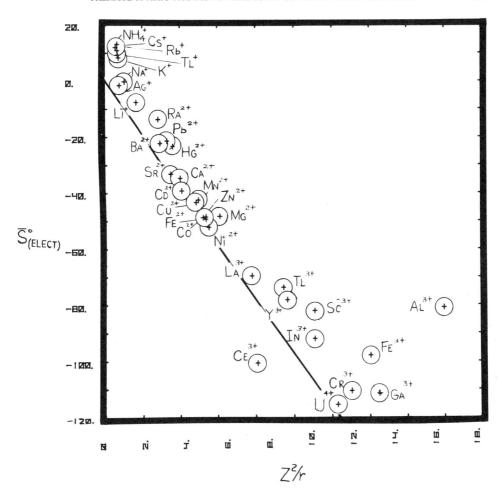

Fig. 8. The $\bar{S}^0$ (elect) for various cations plotted versus Z^2/r at 25°C. The straight line represents the continuum model.

Recently we have examined (Millero and Masterson, 1974; Millero et al., 1974) the hydration of the major cations and anions of seawater; we now briefly discuss these results. If we use a hydration model for ion–water interactions, the $\bar{V}^0$(elect) can be related to the number of water molecules affected by the ion (i.e., the hydration number, h) (Padova, 1963; Desnoyers, in press; Millero and Masterson, 1974; Millero et al., 1974)

$$\bar{V}^0(\text{elect}) = \bar{V}^0(\text{ion}) - \bar{V}^0(\text{int}) = h(V_E^0 - V_B^0) \qquad (15)$$

where V_E^0 is the molal volume of water in the electrostricted region and V_B^0 is the molal volume of water in the bulk phase (18.0 cm³/mol). Owing to difficulties of determining $\bar{V}^0$(int) it is not possible to solve this equation. By differentiating equation 12 with respect to pressure we have the partial molal compressibility

$$\bar{K}^0(\text{ion}) = \bar{K}^0(\text{int}) + \bar{K}^0(\text{elect}) \qquad (16)$$

If we assume $\bar{K}^0(\text{int}) \simeq 0$, we can combine this equation with the differential of equation 15 (assuming h and V_E^0 are not functions of pressure)

$$\bar{K}^0(\text{ion}) = \bar{K}^0(\text{elect}) = -\frac{\partial \bar{V}^0(\text{elect})}{\partial P} = h\left(\frac{\partial V_B^0}{\partial P}\right) = -hV_B^0\beta_B^0 \qquad (17)$$

where $\beta_B^0 = -(1/V_B^0)(\partial V_B^0/\partial P)$ is the compressibility of bulk water (45.25×10^{-6} bar at 25°C). By rearrangement of equation 17 we have for the hydration numbers

$$h = -\frac{\bar{K}^0(\text{ion})}{V_B^0\beta_B^0} \qquad (18)$$

Hydration numbers for some cations and anions calculated from equation 18 are given in Table II. By examining $\bar{V}^0(\text{elect})$ and $\bar{K}^0(\text{ion}) \simeq \bar{K}^0(\text{elect})$ for various ions it is possible to calculate $(\bar{V}_E^0 - \bar{V}_B^0)$. Combining equations 7 and 9 or 15 and 17, we have

$$\bar{V}^0(\text{elect}) = -\left[\frac{(V_E^0 - V_B^0)}{V_B^0 \beta_B^0}\right]\bar{K}^0(\text{elect}) = k\bar{K}^0(\text{elect}) \qquad (19)$$

A plot of $\bar{V}^0(\text{elect})$ versus $\bar{K}^0(\text{ion})$ is shown in Fig. 9. From Fig. 9, we obtain $k = 4800$ bars, which can be compared to

$$k = \frac{-(\partial \ln D/\partial P)}{\partial^2 \ln D/\partial P - (\partial \ln D/\partial P)^2} = 5000 \text{ bars} \qquad (20)$$

TABLE II

Hydration Numbers of Some Solute at 25°C Determined from Compressibility Data[a]

Ion	h	Ion	h
Cations		*Anions*	
Li^+	2.8	F^-	5.6
Na^+	3.7	Cl^-	2.0
K^+	2.9	Br^-	1.2
Rb^+	2.9	I^-	0.1
Cs^+	2.5	OH^-	6.4
Ag^+	3.4	SO_4^{2-}	8.6
NH_4^+	0.4	CO_3^{2-}	12.8
Mg^{2+}	7.8		
Zn^{2+}	5.9		
Cu^{2+}	7.6	*Ion pairs*	
Cd^{2+}	8.5	$MgSO_4^0$	15.3
Ca^{2+}	6.5	$MnSO_4^0$	14.2
Ba^{2+}	9.2	$LaSO_4^+$	17.3
La^{3+}	14.7	$LaFeCN_6^0$	18.3

[a] Calculated from the partial molal compressibility data tabulated by Millero et al. (1974) using equation 18, where $\beta_B^0 = 45.25 \times 10^{-6}$ bar and $V_B^0 = 18.0$ cm³/mol.

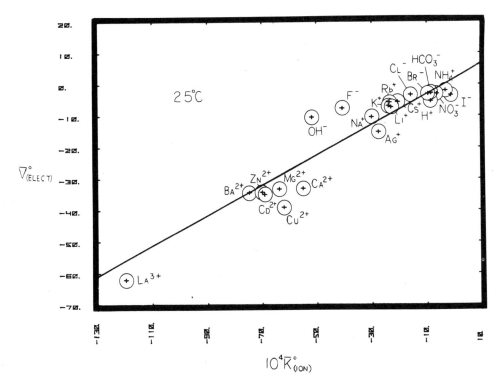

Fig. 9. The $\bar{V}^0$ (elect) plotted versus $\bar{K}^0$ (ion) for various ions at 25°C.

from the Born model. Using the value of $k = 4800$ bars, we obtain $(V_E^0 - V_B^0) = -3.9$ cm³/mol. Combining this value with $V_B^0 = 18.0$ cm³/mol, we find $V_E^0 = 14.1$ cm³/mol, which is a lot larger than the crystal molal volume of water $\bar{V}^0$(cryst) $= 2.52 \times (1.38)^3 = 6.6$ cm³/mol or the value corrected for packing effects $\bar{V}^0$(int) $= 4.48 \times (1.38)^3 = 11.8$ cm³/mol. Thus the water molecules in the electrostricted region are not as tightly packed as one might expect; part of this difference, however, may be due to the water molecules in the so-called broken-down region. The solution properties of the other partial molal properties can also be treated by using the hydration model (Desnoyers, in press). Further work is necessary to determine if $\bar{K}^0$(int) is actually zero or must be considered when calculating hydration numbers.

3. Ion–Ion Interaction Models

Now that we have a reasonable understanding of the structure of an ion in solution at infinite dilution, we consider what happens as the concentration is increased. To obtain an understanding of these ion–ion interactions experimentally, the activity coefficient as well as its pressure ($\bar{V} - \bar{V}^0$) and temperature ($\bar{H} - \bar{H}^0$) dependence are studied. By studying the thermodynamic properties of two ion systems (M^+, X^-), one can obtain an understanding of plus–minus interactions and by studying three ion systems (M^+, N^+, X^- or M^+, X^-, Y^-) one can study plus–plus and minus–minus interactions. As in the case of ion–water interactions, it is useful to use models (see Fig. 10) to examine these interactions. A thorough discussion of the models used

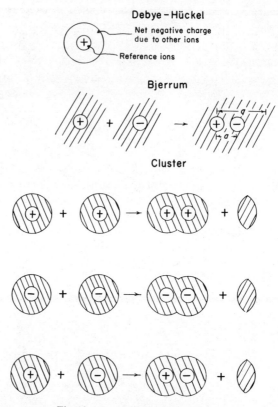

Fig. 10.   Ion–ion interaction models.

to treat ion–ion interactions is given elsewhere (Millero, 1971c, 1974a, b). Some of the models include the continuum models [the Debye–Hückel theory (1923) and Bjerrum (1926) ion pairing theory]. These models assume that the nonideal behavior of an electrolyte is due entirely to electrical effects. The structural models attempt to account for hydration effects as well as specific interactions. The more recent cluster theories of Friedman (1962) make no attempt to separate the electrical and non-electrical interactions (except in the limit). They also consider the importance of all the possible interactions in solution (plus–plus, plus–minus, and minus–minus).

The starting point for all the discussions of ion–ion interactions is the Debye–Hückel theory. The theory predicts that the mean activity coefficient ($f_\pm$) of an electrolyte is given by

$$-\ln f_\pm = \frac{S_f I_V^{1/2}}{1 + A_f \mathring{a} I_V^{1/2}} \tag{21}$$

$$S_f = \frac{1}{v} \sum v_i Z_i^2 (DT)^{-3/2} 1.8243 \times 10^6 \tag{22}$$

$$A_f = 50.29(DT)^{-1/2} \tag{23}$$

where S_f and A_f are constants related to absolute temperature (T) and the dielectric constant (D) of water (for a 1–1 electrolyte, $S_f = 0.5116$ and $A_f = 0.3292$ at 25°C),

$I_V = \frac{1}{2}\sum v_i Z_i^2 c_i$ is the molar ionic strength (v_i is the number, Z_i is the charge, and c_i is the molarity of ionic species i), and $\mathring{a}$ is the ion size parameter in angstrom units. This equation serves as a limit in dilute solutions; however, it fails at the high ionic strength of seawater because of

1. defects in some of the basic assumptions (e.g., treating ions as point charges in a continuous dielectric medium)
2. deviations that occur due to noncoulombic effects such as hydration (the Debye–Hückel theory considers only electrical effects).

Equations for other thermodynamic properties such as the relative molal volume $\bar{V} - \bar{V}^0 = RT(\partial \ln f_\pm/\partial P)$ and the relative molal enthalpy $\bar{L} = \bar{H} - \bar{H}^0 = RT^2 (\partial \ln f_\pm/\partial T)$ can be derived by appropriate differentiation of equation 21:

$$\bar{V} - \bar{V}^0 = \frac{\frac{3}{2}S_V I_V^{1/2}}{1 + A_f \mathring{a} I_V^{1/2}} + \frac{W_V I_V}{1 + A_f \mathring{a} I_V^{1/2}} \tag{24}$$

$$\bar{H} - \bar{H}^0 = \frac{\frac{3}{2}S_H I_V^{1/2}}{1 + A_f \mathring{a} I_V^{1/2}} + \frac{W_H I_V}{1 + A_f \mathring{a} I_V^{1/2}} \tag{25}$$

where the theoretical limiting slopes are given by

$$S_V = 2.303 v R T S_f \left(\frac{\partial \ln D}{\partial P} - \frac{\beta}{3} \right) \tag{26}$$

$$S_H = -2.303 v R T S_f \left(\frac{\partial \ln D}{\partial T} + \frac{1}{T} + \frac{\alpha}{3} \right) \tag{27}$$

(derived by differentiation of the terms S_f and $I_V^{1/2}$ which are both functions of pressure and temperature). The symbol $\alpha = (1/V)(\partial V/\partial T)$ is the expansibility and the symbol $\beta = -(1/V)(\partial V/\partial T)$ is the compressibility (V is the specific volume of water). At $25°C$ $\frac{3}{2}S_V = 2.802$ cm$^3 \cdot$ l$^{1/2}$/mol$^{3/2}$ and $\frac{3}{2}S_H = 707.1$ cal $\cdot$ l$^{1/2}$/mol$^{3/2}$ for a 1–1 electrolyte. The symbols for the coefficients W_V and W_H result from the differentiation of the term $(1 + A_f \mathring{a} I_V^{1/2})$ and cannot normally be evaluated since the effects of temperature and pressure on the ion size parameter ($\partial \ln \mathring{a}/\partial T$ and $\partial \ln \mathring{a}/\partial P$) are not known.

The classical method of examining the deviations from the Debye–Hückel theory in concentrated solutions is to use various extended forms involving one or more arbitrary constants (Robinson and Stokes, 1959). The difference between this form and the experimental data is attributed to noncoulombic effects. For example, Guggenheim (1935) used the equation

$$-\log \gamma_\pm (MX) = \frac{A Z_M Z_X I_m^{1/2}}{1 + I_m^{1/2}} + v B_{MX} m$$

$$= \log \gamma_\pm (\text{elect}) + B I_m^{1/2} \tag{28}$$

where at $25°C$ $A = 0.5108$, $v = 2 v_M v_X/(v_M + v_X)$, I_m is the molal ionic strength, m is the molality, and $B = 2 v B_{MX}/(v_M Z_M^2 + v_X Z_X^2)$. By differentiating equation 28 with respect to temperature and pressure it is possible to examine the specific interaction model as a function of temperature and pressure (i.e., $\partial B_{MX}/\partial T$ and $\partial B_{MX}/\partial P$).

Robinson and Stokes (1959) used similar methods; however, they examined the differences between log $\gamma_\pm$ and log $\gamma_\pm$(elect) by using a hydration model.

$$\ln \gamma_\pm = \ln \gamma_\pm(\text{elect}) - \frac{h}{v} \ln a_w + \ln \left[\frac{1 - (h - v)m}{55.51} \right] \tag{29}$$

where a_w is the activity of water ($a_w = p/p_0$, p and p_0 are the vapor pressure of solution and water) and m is the molality.

The most popular method of treating the deviations from the Debye–Hückel theory in concentrated solutions is the ion pairing method of Bjerrum (1926). This method assumes that short-range interactions can be represented by the formation of ion pairs

$$M^+ + A^- \longrightarrow MA^0 \tag{30}$$

A characteristic association constant is assigned to this formation

$$K_A = \frac{a_{MA^0}}{a_M a_A} = \left(\frac{[MA^0]}{[M^+][A^-]} \right) \left(\frac{\gamma_{AM^0}}{\gamma_M \gamma_A} \right) \tag{31}$$

where a_i, $[i]$, and γ_i are, respectively, the activity, molal concentration, and activity coefficient of species i. There are four classes of ion pairs:

1. complexes—when the ions are held in contact by covalent bonds;
2. contact ion pairs—when the ions are in contact and linked electrostatically (with no covalent bonding);
3. solvent-shared ion pairs—pairs of ions linked electrostatically, separated by a single water molecule;
4. solvent-separated ion pairs—pairs of ions linked electrostatically but separated by more than one water molecule.

Bjerrum (1926) defined the distance between oppositely charged ions which can be classified as being associated by $q = Z_+ Z_- e^2/2DkT$, where Z_i is the charge on the ion i; e is the electrostatic charge; D is the dielectric constant; k is the Boltzmann constant; and T is the absolute temperature. In this treatment, two ions of opposite charge are considered to form an ion pair when they are between $\mathring{a}$, the ion size parameter, and q. This can include ion pairs of classes 2, 3, and 4.

The association constant of the Bjerrium method is given by

$$K_A = \left(\frac{4\pi N}{1000} \right) \left(\frac{Z_+ Z_- e^2}{DkT} \right)^3 Q(b) \tag{32}$$

where $b = |Z_+ Z_-| e^2/\mathring{a}DkT$ and the function of $Q(b)$ is given by Robinson and Stokes (1959). The theory predicts greater ion pair formation, the higher the valencies and the smaller the dielectric constant of the solvent, which is in agreement with experimental results.

Many workers have criticized the theory because of the arbitrary cutoff distance. It has now been superseded by other theories (Davies, 1962; Fuoss, 1958; Monk, 1961; Nancollas, 1966). For example, the model of Fuoss considers only anions on the surface of a cation in volume, $v = 2.52\mathring{a}^3$ to be ion pairs. Fuoss obtained

$$K_A = \frac{4\pi N \mathring{a}^3}{3000} \exp\left[\frac{Z_+ Z_- e^2}{D\mathring{a}kT} \right] \tag{33}$$

where the first term is the excluded volume around the cation. Others have made further elaborations on these methods and discussed the shortcomings of the model (see Nancollas, 1966).

By differentiating equations 32 and 33 with respect to pressure and temperature, one can determine the theoretical volume, enthalpy, and entropy change for the ion pairing process:

$$-\Delta \bar{V}^0 = RT\left(\frac{\partial \ln K_A}{\partial P}\right) \tag{34}$$

$$\Delta \bar{H}^0 = RT^2\left(\frac{\partial \ln K_A}{\partial T}\right) \tag{35}$$

$$\Delta \bar{S}^0 = RT\left(\frac{\partial \ln K_A}{\partial T}\right) \tag{36}$$

Hemmes (1972) has recently estimated the $\Delta \bar{V}^0$ for ion pair formation by using the Fuoss and Bjerrum constants and their pressure dependence. He obtained from the Fuoss equation

$$\Delta \bar{V}^0 = \left(\frac{Z_+ Z_- e^2 N}{\mathring{a} D}\right)\left(\frac{\partial \ln D}{\partial P}\right) - RT\beta \tag{37}$$

where the second term involving the compressibility of water (β) is needed because both the Fuoss and Bjerrum association constants are based on the molarity scale. From the Bjerrum equation he obtained

$$\Delta \bar{V}^0 = RT\left[3 + \left(\frac{\exp(b)}{Q(b)b^3}\right)\left(\frac{\partial \ln D}{\partial P}\right) - \beta\right] \tag{38}$$

For large values of b, the compressibility term is negligible ($RT\beta = 1.1$ cm³/mol at 25°C) compared to the first term in both equations. Since $(\partial \ln D/\partial P)_T$ is positive for water (and other solvents), $\Delta \bar{V}^0$ is positive for the association of an ion pair in agreement with experimental results. For $\mathring{a} = 7.36$ Å, $\Delta \bar{V}^0 = 8.98$ cm³/mol from equation 37, and for $\mathring{a} = 4.19$ Å, $\Delta \bar{V}^0 = 6.89$ from equation 38 (Hemmes, 1972). These equations can be used only for outer-sphere complexes. The ion size parameter needed in the calculation can be back-calculated from equations 32 and 33 by using the measured K_A. For example, Hemmes calculated for the formation of $LaFe(CN)_6{}^0$, $\mathring{a} = 7.36$ Å, and $\Delta \bar{V}^0 = 8.98$ cm³/mol from the Fuoss equations, and $\mathring{a} = 4.19$ Å and $\Delta \bar{V}^0 = 6.89$ cm³/mol from the Bjerrum equation, which is in reasonable agreement with the measured value of 8.0 cm³/mol (Hamman, 1964). For $MgSO_4{}^0$, he obtained $\Delta \bar{V}^0 = 7.42$ and 4.86 cm³/mol, respectively, from the Fuoss and Bjerrum equations, compared with the measured values of 7.3 cm³/mol (Fisher, 1962). For $MnSO_4{}^0$, he calculated $\Delta \bar{V}^0 = 8.3$ and 5.0, respectively, from the Fuoss and Bjerrum equations, compared with the experimental value of 7.4 cm³/mol (Fisher and Davis, 1965).

For ion pairs that form inner-sphere complexes like $LaSO_4{}^+$ (Fisher and Davis, 1967) and $EuSO_4{}^+$ (Hale and Spedding, 1972), the predicted $\Delta \bar{V}$'s are much smaller than the directly measured values. It is thus possible to use the magnitude of the experimental $\Delta \bar{V}$'s to infer the structure of the ion pair.

The effect of temperature on ion pair formation calculated from the theory of Bjerrum and Fuoss has recently been examined by Prue (1969). For the Fuoss equation he obtained

$$\Delta H^0 = -RTb\left[1 + \left(\frac{\partial \ln D}{\partial \ln T}\right)\right] \tag{39}$$

$$\Delta S^0 = R \ln\left(\frac{4\pi N \mathring{a}^3}{3000}\right) - Rb\left(\frac{\partial \ln D}{\partial \ln T}\right) \tag{40}$$

Because $(\partial \ln D/\partial \ln T)$ is negative for pure water, one would expect ΔH^0 to be positive (endothermic) or unfavorable to association. Negative values of ΔH^0 can be accounted for by adding a covalent contribution to K_A.

Because the second term of equation 40 is normally larger than the first term, the Fuoss theory predicts that ΔS^0 should be positive for ion pair formation. In general, the ΔH^0, ΔS^0, and ΔV^0 for ion pair formation are more sensitive to changes of solvent than is ΔG^0 [similar to the behavior for the dissociation of acids and bases (Millero et al., 1969)]. In recent years, a number of workers have determined ΔH^0 by calorimetry (Nancollas, 1966, 1970). Various correlations between ΔH^0 and ΔS^0 have also been discussed (Nancollas, 1966).

More recent studies of electrolyte solutions have been made by using the cluster expansion method (Friedman, 1962; Wood and Reilly, 1970). This method in simple terms considers all the interactions in a solution and makes no attempt to separate coulombic and noncoulombic terms. For example, for the major sea salts (NaCl + $MgSO_4$) there are a number of possible interactions to consider:

Interactions	Possible Types
$\oplus-\oplus$	Na–Na, Mg–Mg, Na–Mg
$\ominus-\ominus$	Cl–Cl, SO_4–SO_4, Cl–SO_4
$\oplus-\ominus$	Na–Cl, Mg–SO_4, Mg–Cl, Na–SO_4

These interactions can be represented by the following cross-square diagram:

By studying the mixtures along the side of this diagram one can obtain some information about $\oplus-\oplus$ and $\ominus-\ominus$ interactions; by studying individual salts and the sum around the sides ($MgSO_4 = MgCl_2 + Na_2SO_4 - 2NaCl$) one can study $\oplus-\ominus$ interactions. The cross terms represent the mixtures (or simple seawater). Since the $\oplus-\oplus$ and $\ominus-\ominus$ terms are small, the total activity coefficients can be estimated from

$$\log \gamma_T(MX) = \log \gamma_\pm^0(MX) + \sum \oplus-\oplus \text{ terms} + \sum \ominus-\ominus \text{ terms} \tag{41}$$

where $\log \gamma_\pm^0(MX)$ is the value for MX itself at the ionic strength of the mixture and the other terms are related to excess mixing terms. For example, for NaCl in seawater the

$$\sum \oplus\text{-}\oplus = (\text{Na–Mg}) + (\text{Na–K}) + (\text{Na–Ca}) + \cdots \tag{42}$$

$$\sum \ominus\text{-}\ominus = (\text{Cl–SO}_4) + (\text{Cl–HCO}_3) + (\text{Cl–Br}) + \cdots \tag{43}$$

where the terms in parentheses are weighted according to the composition of the mixture.

4. State of Metal Ion in Seawater

Before we examine the use of the methods described in the last section to determine the activity of a metal in seawater

$$a_M = [\text{M}]_T \gamma_T(\text{M}) \tag{44}$$

by estimating the total stoichiometric activity coefficient, we examine some of the factors that control the state of an ion in seawater. There are four major factors that control the state of a dissolved ion in seawater:

1. The Eh.
2. The pH.
3. The inorganic ligands.
4. The organic ligands.

The Eh of seawater may control the oxidation state of a metal ion. For a metal ion that can exist in two oxidation states we have

$$\text{Ox} + ne^- = \text{Red} \tag{45}$$

where Ox is the oxidized form, Red is the reduced form, and n is the number of electrons (e^-) transferred. The equilibrium constant is given by

$$\log K = \log a_{\text{Red}} - \log a_{\text{Ox}} + n\,pE \tag{46}$$

where a_i is the activity of i and $pE = -\log a_e$, the log of the activity of an electron. Rearranging this equation we have

$$pE = pE^0 + \left(\frac{1}{n}\right) \log\left(\frac{a_{\text{Ox}}}{a_{\text{Red}}}\right) \tag{47}$$

where $pE^0 = (1/n) \log K$. Since $pE = Eh/(2.303RT/F)$, we have the more familiar form

$$Eh = Eh^0 + \left(\frac{2.303RT}{nF}\right) \log\left(\frac{a_{\text{Ox}}}{a_{\text{Red}}}\right) \tag{48}$$

where $Eh^0 = (2.303RT/nF) \log K = (0.0591/n) \log K$ at 25°C.

The upper theoretical limit (Stumm and Morgan, 1970; Whitfield, 1969; Breck, 1972, 1974) of the pE or Eh of oxygenated water and perhaps seawater is controlled by the reaction

$$\tfrac{1}{2}O_2(g) + 2H^+ + 2e = H_2O(l) \tag{49}$$

Using $\log K = 41.6$, $\log a_{H_2O} = -0.01$, and pH = 8.1 at 25°C, the pE is given by

$$pE = \frac{\log P_{O_2} + 50.7}{4} \tag{50}$$

at $\log P_{O_2} = -0.69$, $pE = 12.5$ or $Eh = 0.73$ V. The experimentally measured values (about 0.5–0.6 V) (Whitfield, 1969; Breck, 1972, 1974; Liss et al., 1973) for the Eh of open seawater are lower than this theoretical value. By using the reaction

$$O_2 + 2H^+ + 2e^- = H_2O_2 \tag{51}$$

one obtains (Breck, 1974) a lower theoretical $pE = 8.5$ or $Eh = 0.5$ V (taking $[H_2O_2] = 10^{-11}$), which is closer to the experimentally determined values.

For anoxic conditions the negative pE or Eh are thought to be controlled by the reactions (Breck, 1974)

$$SO_4^{2-} + 9H^+ + 8e^- = HS^- + 4H_2O \tag{52}$$

$$SO_4^{2-} + 8H^+ + 6e^- = S^0(s) + 4H_2O \tag{53}$$

Using $\log K = 34.0$ and 36.6, respectively, for reactions 52 and 53, the pE is given by (Breck, 1974)

$$pE = \frac{-\log(HS^-) - 41.4}{8} \tag{54}$$

$$pE = \frac{-\log(S^0) - 30.4}{6} \tag{55}$$

at $(HS^-) = 10^{-3}$–10^{-6}, $pE = -4.8$ to -4.4 or $Eh = -0.28$ to -0.26.

Recently Kester et al., (1976) have examined the pH and Eh environments that one encounters in marine waters. In Fig. 11 these environments are shown for various pH's and Eh's. The upper and lower limits are determined by the properties of water. The cross-textured and dotted bands represent the stability band for oxygen (0.1–6.0 ml/l) and sulfide (10^{-3}–10^{-6} M) concentrations as given by reactions 49 and 52. Most ocean waters have pH values between 7.6 and 8.3 and Eh values greater than 0.2 V. A sample calculation of the oxidation state of a metal in seawater is made for the iron system recently examined by Kester and Byrne (1972). The two oxidation states of iron are related by

$$Fe^{3+} + e^- = Fe^{2+} \tag{56}$$

Using equation 48 the oxidation state can be determined from

$$Eh = Eh^0 + \left(\frac{2.3RT}{nF}\right)\log\left(\frac{a_{Fe^{3+}}}{a_{Fe^{2+}}}\right) \tag{57}$$

Using $Eh^0 = 0.771$ V at 25°C (Latimer, 1952), we obtain $a_{Fe^{3+}}/a_{Fe^{2+}} = 2.0 \times 10^{-1}$–$1.8 \times 10^{-4}$ when $Eh = 0.73$–0.55 V in oxygenated seawater. For anoxic seawater, we obtain $a_{Fe^{3+}}/a_{Fe^{2+}} = 2.43 \times 10^{-18}$ when $Eh = -0.27$ V. It should be pointed out that these ratios are at infinite dilution and do not include the effects of complex formation (which usually stabilizes a system against reduction). To determine the concentration ratios at higher ionic strengths, one must estimate the stoichiometric activity coefficients of the ions

$$\frac{[Fe^{3+}]_T}{[Fe^{2+}]_T} = \left[\frac{a_{Fe^{3+}}}{a_{Fe^{2+}}}\right]\left[\frac{\gamma_T(Fe^{2+})}{\gamma_T(Fe^{3+})}\right] \tag{58}$$

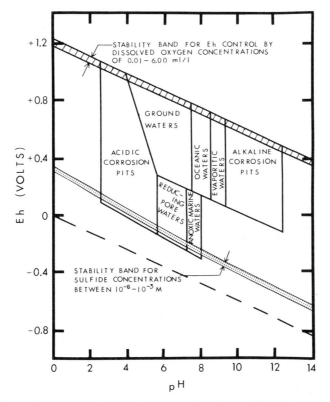

Fig. 11.   The *Eh*–pH diagram for various marine systems (from Kester, 1974). The upper and lower limits are governed by the reactions $2H_2O \rightarrow O_2 + 4H^+ + 4e^-$ and $2H^+ + 2e^- \rightarrow H_2$, respectively.

Kester and Byrne (1972) have also examined the ion pairing of Fe^{3+} and Fe^{2+} in seawater which can be used to estimate the activity coefficient ratio. Although redox calculations are quite simple, some of the problems inherent in these calculations are as follows:

1. Equilibrium is not reached (i.e., the process is kinetically controlled).
2. Biological activity may change the oxidation state.
3. Photochemical processes may control the state.
4. Other important species may be neglected (e.g., organic complexes).
5. Unreliable analytical and thermodynamic data are available for the actual system.

The pH can affect the metals directly by hydrolysis equilibrium

$$M^{2+} + H_2O = M(OH)^+ + H^+ \tag{59}$$

and by affecting the form of the ligands

$$HCO_3^- = H^+ + CO_3^{2-} \tag{60}$$

The effect of pH on the state of iron in marine waters has been examined by Kester et al. and Kester and Byrne (1972). These results are shown in Fig. 12. The major form of iron at the pH and *Eh* of seawater is $Fe(OH)_3^0$. Since the OH^- concentration

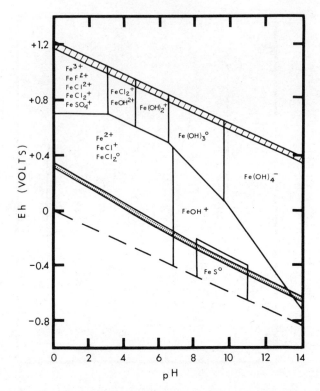

Fig. 12.   The *Eh*–pH diagram for the iron system (from Kester et al. 1976).

is a function of pH, the relative forms of Fe^{3+} are quite pH dependent. Near the pH
limits of 7 and 10, there is a transition between $Fe(OH)_2{}^+ \rightleftarrows Fe(OH)_3{}^0$ and $Fe(OH)_3{}^0$
$\rightleftarrows Fe(OH)_4{}^-$. These results suggest that iron can alternate between a cationic and
anionic species, which would strongly influence the ion-exchange characteristics of
colloidal iron as well as its transport properties.

The inorganic ligands affecting metals include the major anionic components of
seawater (Cl^-, $SO_4{}^{2-}$, $HCO_3{}^-$, $CO_3{}^{2-}$, Br^-, $B(OH)_4{}^-$, and F^-) and some of the
minor anionic components (OH^-, $H_2PO_4{}^-$, and $NO_3{}^-$ to mention a few). It is
difficult to classify the types of organic ligands one must consider, since little is known
about the composition of the organics in seawater. EDTA is frequently used as a
model; however, the more important organic ligands may be made up of humic
and fulvic acids and their derivatives. For the present, we are not concerned with
these organic ligands, although we realize they may be of great importance. It should
be pointed out that all the model calculations (Kester and Byrne, 1972; Morel and
Morgan, 1972; Dyrssen and Wedborg, 1974), made by using various organic ligands,
indicate that organic ligands have little effect on the speciation of trace metals (i.e.,
compared to the major inorganic ligands).

We now examine the use of the ion pairing model, the specific interaction model,
and the cluster model to estimate the total activity coefficient of the major ions of
seawater. Owing to the large amount of thermodynamic data available for the major
sea salts it is possible to apply all these methods. As we see later, for the minor or trace
metals, we can use at present only the ion pairing methods.

In applying the ion pairing method one makes the assumption that

$$a_M = [M^+]_F \gamma_F(M^+) \tag{61}$$

where $[M^+]_F$ is the free metal concentration and $\gamma_F(M^+)$ is the activity coefficient of the free ion. Combining this relationship with equation 44, we have

$$\gamma_T(M^+) = \frac{[M^+]_F}{[M^+]_T} \gamma_F(M^+) \tag{62}$$

The fraction of free ions, $[M^+]_F/[M^+]_T$, is determined from

$$\frac{[M^+]_F}{[M^+]_T} = \frac{1}{1 + \sum K_A^*(i)[L(i)]_F} \tag{63}$$

where K_A^* is the stoichiometric association constant for

$$M^a + L_i^b \longrightarrow ML_i^{a+b} \tag{64}$$

given by

$$K_A^*(i) = K_A(i) \frac{\gamma_F(M^a)\gamma_F(L_i^b)}{\gamma_F(ML_i^{a+b})} \tag{65}$$

where $K_A(i)$ is the thermodynamic association constant, L_i is a ligand, and $\gamma_F(i)$ is the activity coefficient of the free ions. The K_A^* can thus be determined in various ionic media (Pytkowicz and Kester, 1969) [i.e., by neglecting $\gamma_F(i)$] or by using infinite dilution K_A's (Garrels and Thompson, 1962; Atkinson et al., 1973); both methods should agree, providing reasonable values can be determined for $\gamma_F(i)$ and the ionic media are nonreacting. In Tables III and IV a comparison is made of the experimental γ_T values with those estimated for the free ions γ_F obtained by using the specific

TABLE III

Comparison of Measured and Calculated Activity Coefficients for Sea Salts at 25°C and $I = 0.7^a$

Salt	Measured	Ion Pairing	Specific Interaction	Cluster
HCl	—	—	—	0.696
NaCl	0.668	0.66–0.69	0.668	0.668
KCl	—	0.62–0.64	0.647	0.641
MgCl$_2$	—	0.46–0.50	0.466	0.463
CaCl$_2$	—	0.45–0.47	0.455	0.448
H$_2$SO$_4$	—	—	—	0.386
Na$_2$SO$_4$	0.378	0.32–0.35	0.370	0.374
K$_2$SO$_4$	—	0.30–0.33	0.357	0.355
MgSO$_4$	—	0.13–0.15	0.157	0.160
CaSO$_4$	—	0.13–0.14	0.157	0.159
CaCO$_3$	—	—	0.069	—
NaHCO$_3$	—	—	0.612	—

[a] Taken (Millero, 1974a, b) from the work of Leyendekkers (1972), Whitfield (1973), and Robinson and Wood (1972).

TABLE IV

Comparison of the Calculated and Measured Total Activity Coefficients (γ_T) for the Major Ions in Seawater at 25°C and $I = 0.7^a$

Ion	Measured	Ionic Strength	Ion Pairing	Specific Interaction Model
H^+	0.74	0.85	0.74	0.74
Na^+	0.68	0.71	0.70	0.68
Mg^{2+}	0.23	0.29	0.25	0.23
Ca^{2+}	0.21	0.26	0.22	0.21
K^+	0.64	0.63	0.62	0.63
Sr^{2+}	—	0.25	0.22	—
Cl^-	0.68	0.63	0.63	0.66
SO_4^{2-}	0.11	0.22	0.10	0.11
HCO_3^-	0.55	0.68	0.43	0.59
CO_3^{2-}	0.02	0.21	0.02	0.03
F^-	—	0.68	0.31	—
OH^-	—	0.65	0.11	0.56
$B(OH)_4^-$	0.26	0.68	0.38	—

^{a} Taken from the compilation of Millero (1974a, b).

interaction model, the cluster model, and the ion pairing model. As is quite apparent from Table IV, the ion pairing estimates for γ_T are in better agreement than those estimated for the free ions by using the ionic strength principle or some simple extension of the Debye–Hückel theory (e.g., equation 28 with $B = 0.2$, the Davies equation).

To apply the specific interaction model (Leyendekkers, 1972; Whitfield, 1973) to estimate total activity coefficients we must use the equations

$$\log \gamma_T(MX) = \log \gamma(\text{elect}) + \left(\frac{\nu_M}{\nu}\right)\sum_X B_{MX}[X]_T + \left(\frac{\nu_X}{\nu}\right)\sum_M B_{MX}[M]_T \quad (66)$$

$$\log \gamma_T(M) = \left(\frac{Z_M}{Z_X}\right)\log \gamma(\text{elect}) + \sum_M B_{MX}[X]_T \quad (67)$$

$$\log \gamma_T(X) = \left(\frac{Z_X}{Z_M}\right)\log \gamma(\text{elect}) + \sum_X B_{MX}[M]_T \quad (68)$$

The total activity coefficients calculated from equations (66–68) for the major sea salts and ions of seawater are given in Tables III and IV. The results are in excellent agreement with the measured values. Calculations using the cluster theory have been made by Robinson and Wood (1972). The equations they used are similar to equation 66; however, terms are included to consider the excess free energies related to three ion interactions (M^+ and N^+, X^- and M^+, X^- and Y^-) (equation 41). The results are in excellent agreement with the estimates from the methods of Guggenheim (Millero, 1974a; Leyendekkers, 1972; Whitfield, 1973). This is not surprising since the $\oplus$–$\oplus$ and $\ominus$–$\ominus$ interaction terms are small for free energies (however, for enthalpies they are very important) (Millero, 1974a).

In conclusion, all the methods yield comparable results for γ_T for the major ionic components of seawater. These facts lead one to believe that the success of the popular ion pairing methods may be fortuitous, and the existence of such species in seawater requires further studies. Before we examine the methods used in my laboratory to study the structure and existence of ion pairs, it is useful to review the use of the ion pairing methods in studying trace metals. It should be pointed out that owing to the lack of reliable thermodynamic data, we are normally forced to use the ion pairing methods to estimate γ_T for trace metals in seawater. Although it is difficult to experimentally determine γ_T for salts in seawater, the effect of temperature and pressure on γ_T (related to relative molal enthalpies $\bar{L}_2$ and molal volumes $\bar{V}_2$) can be determined or estimated with reasonable accuracy.

Recently, there have been two studies on the use of ion pairing methods in determining the speciation of the heavy metals (Cu, Zn, Cd, Pb) (Dyrssen and Wedborg, 1974; Zirino and Yamamoto, 1972). Zirino and Yamamoto in their study examine the effect of pH as well as the formation of Cl^-, HCO_3^-, CO_3^{2-}, OH^-, and SO_4^{2-} ion pairs. Their results are shown in Figs. 13–16. Since the complexes on Zn^{2+}, Pb^{2+}, and Cu^{2+} are mainly with OH^- and CO_3^{2-}, the speciation of these metals shows a large pH dependence (see Figs. 17–20). Dyrssen and Wedborg (1974) considered the same complexes as well as the Br^-, F^-, glycine, and mixed $OH–Cl$ complexes; their results are also given in Figs. 13–16. Dyrssen and Wedborg also made calculations on complexes of Hg^{2+}. They found that the major species include Hg^{2+} (0.1%), $HgCl_2^0$ (2.9%), $HgCl_3^-$ (11.4%), $HgCl_4^{2-}$ (62.8%), $HgBr_2^0$ (0.1%), $HgClBr^0$ (2.1%), $HgCl_2Br^-$ (7.6%), $HgCl_3Br^{2-}$ (11.7%), $HgClBr_2^-$ (0.5%), $HgCl_2Br_2^{2-}$ (0.6%), and $HgOHCl^0$ (0.2%).

It is interesting to compare the results of these two studies because they point out the problems involved in determining the speciation of trace metals. The percent free ions determined by both studies are in reasonable agreement: Cu^{2+} (0.7–1.0%), Zn^{2+} (16.1–17%), Cd^{2+} (1.7–2.5%), and Pb^{2+} (2–4.5%). The forms for the complexes, however, show wide deviations. Both studies indicate that neutral hydroxyl complexes are very important, which may be an indication that these species are associated with colloidal material $[Fe(OH)_3^0]$ in seawater. The pH dependence of the complexes are in good agreement with polarographic studies (Zirino and Yamamoto, 1972). The pH dependence of the complexes may be important when the particles sink and are incorporated into the sediments (the lower pH may return trace metals to the free form). The full significance of these models must await further studies.

The difficulties in using the ion pairing models are clearly pointed out by this brief examination of two independent studies. The existence or nonexistence of various ion paired forms requires confirmation by other types of measurement. For example, the effect of pressure on the thermodynamic association constant yields the volume change

$$\frac{\partial \ln K_A}{\partial P} = \frac{-\Delta \bar{V}_A^0}{RT} \tag{69}$$

and the temperature effects yield the enthalpy change

$$\frac{\partial \ln K_A}{\partial T} = \frac{\Delta \bar{H}_A^0}{RT^2} \tag{70}$$

The $\Delta \bar{V}_A^0$ and $\Delta \bar{H}_A^0$ are given by

$$\Delta \bar{V}_A^0 = \sum \bar{V}^0 (\text{Products}) - \sum \bar{V}^0 (\text{Reactants}) \tag{71}$$

$$\Delta \bar{H}_A^0 = \sum \bar{H}^0 (\text{Products}) - \sum \bar{H}^0 (\text{Reactants}) \tag{72}$$

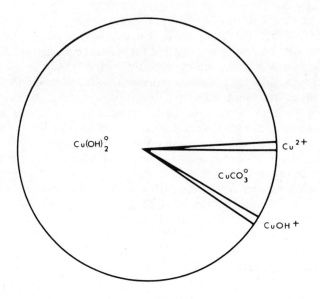

Zirino & Yamamoto

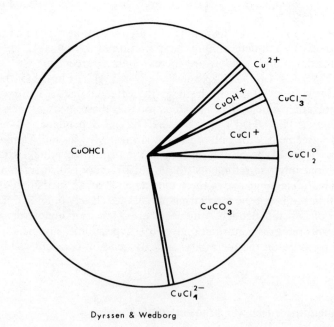

Dyrssen & Wedborg

Fig. 13. The speciation of Cu^{2+} in seawater at 25°C and pH = 8.1 obtained by Zirino and Yamamoto (1972) and Dyrssen and Wedborg (1974).

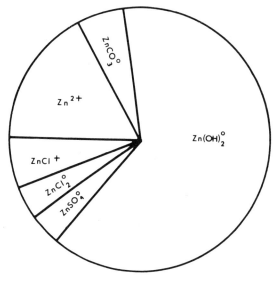

Zirino & Yamamoto

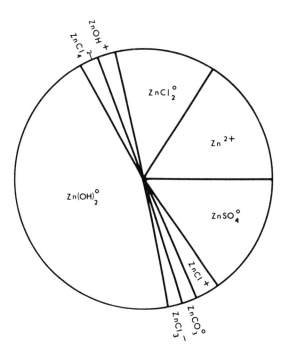

Dyrssen & Wedborg

Fig. 14. The speciation of Zn^{2+} in seawater at $25°C$ and $pH = 8.1$ obtained by Zirino and Yamamoto (1972) and Dyrssen and Wedborg (1974).

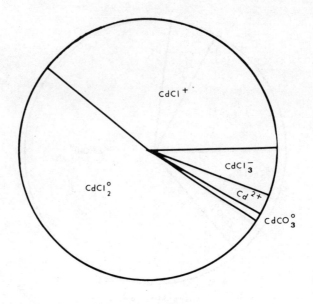

Zirino & Yamamoto

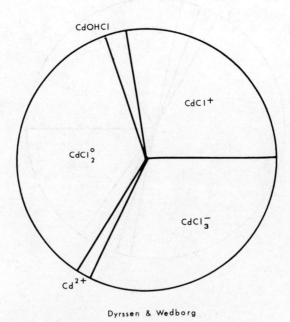

Dyrssen & Wedborg

Fig. 15. The speciation of Cd^{2+} in seawater at $25°C$ and pH = 8.1 obtained by Zirino and Yamamoto (1972) and Dyrssen and Wedborg (1974).

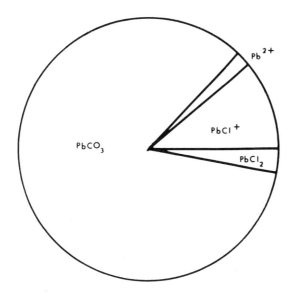

Zirino & Yamamoto

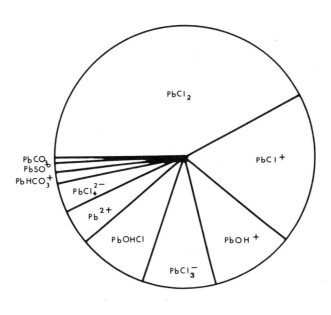

Dyrssen & Wedborg

Fig. 16.   The speciation of Pb^{2+} in seawater at 25°C and pH = 81. obtained by Zirino and Yamamoto (1972) and Dyrssen and Wedborg (1974).

681

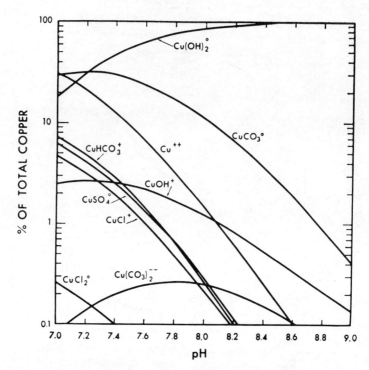

Fig. 17.   The effect of pH on the speciation Cu^{2+} in seawater at 25°C (from Zirino and Yamamoto, 1972).

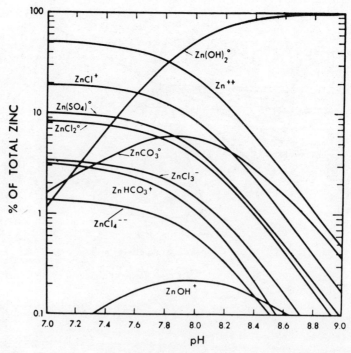

Fig. 18.   The effect of pH on the speciation Zn^{2+} in seawater at 25°C (from Zirino and Yamamoto, 1972).

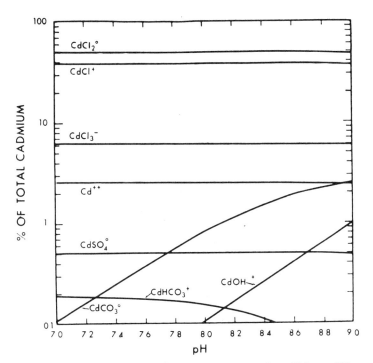

Fig. 19.   The effect of pH on the speciation Cd^{2+} in seawater at 25°C (from Zirino and Yamamoto, 1972).

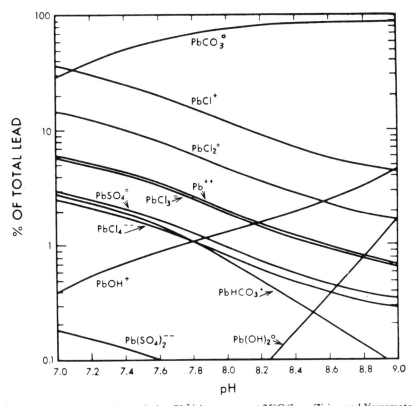

Fig. 20.   The effect of pH on the speciation Pb^{2+} in seawater at 25°C (from Zirino and Yamamoto, 1972).

where $\bar{V}^0$ and $\bar{H}^0$ are the infinite dilution partial molal volumes and enthalpies. The enthalpies and volumes can lead to more information concerning the structure of the ion pair.

In my laboratory, we have initiated studies (Leung and Millero, 1975a, b; Millero, 1969, 1970, 1971a; Millero and Berner, 1972; Millero et al., 1972; Perron et al., 1975a, b; Ward and Millero, 1974) on the effect of pressure and temperature (i.e., volume and heat measurements) on ion pairing systems. We now briefly summarize these methods and discuss our results for the formation of $MgSO_4{}^0$ ion pair. The observed apparent molal [$\phi(\text{obs})$] property (volume, expansibility, compressibility, enthalpy, and heat capacity) of an electrolyte (MA) thought to form an ion pair is divided into two components (Leung and Millero, 1975b; Millero, 1970; Millero and Masterson, 1974; Millero et al., 1974)

$$\phi(\text{obs}) = \alpha\phi(M^+, A^-) + (1 - \alpha)\phi(MA^0) \tag{73}$$

where α is the fraction of free ions and $\phi(M^+, A^-)$ and $\phi(MA^0)$ are the apparent molal property for the free ions and the ion pair. By rearranging this equation, we have

$$\phi(MA^0) = \frac{\phi(\text{obs}) - \alpha\phi(M^+, A^-)}{1 - \alpha} \tag{74}$$

$$\Delta\phi(MA^0) = \frac{\phi(\text{obs}) - \phi(M^+, A^-)}{1 - \alpha} \tag{75}$$

where the property change $\Delta\phi(MA^0)$ is given by

$$\Delta\phi(MA^0) = \phi(MA^0) - \phi(M^+, A^-) \tag{76}$$

At infinite dilution $\Delta\phi(MA^0)$ is equal to the partial molal property. From equations 73, 74, or 75, it is possible to determine $\Delta\phi$ or ϕ for an ion pair providing one can determine α and $\phi(M^+, A^-)$ in some independent manner. Since the K_A is normally known, α can be determined from

$$K_A = \frac{1 - \alpha}{\alpha^2 f_\pm{}^2 m_T} \tag{77}$$

where $f_\pm$ is the mean activity coefficient of the free ions (f of MA^0 is assumed to be 1.0), and m_T is the total molality. The value of $f_\pm$ can be estimated in dilute solutions from equation 21 or 28 and in more concentrated solutions (Robinson and Stokes, 1959) from the stoichiometric mean activity coefficients ($\gamma_\pm$)

$$f_\pm = \frac{\gamma_\pm}{\alpha} \tag{78}$$

Combining equations 77 and 78, we have

$$\alpha = 1 - K_A(\gamma_\pm)^2 m_T \tag{79}$$

Values of ϕ for the free ions (M^+, A^-) can be estimated by three methods (Millero, 1971a): (1) from an extended Debye–Hückel equation

$$\phi(M^+, A^-) = \phi^0 + \frac{S\sqrt{\alpha m}}{1 + B\mathring{a}\sqrt{\alpha m}} \tag{80}$$

where $\mathring{a}$ is an adjustable parameter; (2) from the additivity principle (Millero, 1971a; Millero and Masterson, 1974; Millero et al., 1974)

$$\phi(M^+, A^-) = \phi(M^+, Cl^-) + \phi(N^+, A^-) - \phi(N^+, Cl^-) \qquad (81)$$

where N^+ is the cation Li^+, Na^+, or K^+; and (3) from a semiempirical relationship (Millero, 1969)

$$\phi(M^+, A^-) = \phi^0 + A\left(\frac{Z_+^2}{r_+} + \frac{Z_-^2}{r_-}\right) + B \qquad (82)$$

where $A = 0.372$ and $B = 1.666$ for the estimation of the $\bar{V}$ of free ions in seawater at $35\%_0$ S and $25°C$. A comparison of the observed ϕ_V and ϕ_K for $MgSO_4$ solutions with free ions estimated from equations 80 and 81 are shown in Figs. 21 and 22. As is quite apparent from these figures, the ϕ_V and ϕ_K for the free ions are lower than the observed values for all the various estimates. As discussed elsewhere (Millero and Masterson, 1974; Millero et al., 1974), we feel that the free ion values of ϕ_V and ϕ_K determined from the additivity methods yield the most consistent and reliable values of $\Delta\phi_V$ and $\Delta\phi_K$. A plot of the calculated values of $\Delta\phi_V$ and $\Delta\phi_K$ are shown in Figs. 23 and 24. The infinite dilution values for $\Delta\bar{V}^0$ and $\Delta\bar{K}^0$ obtained at various temperatures are given in Table V; also given in this table are the values of $\bar{V}^0$ and $\bar{K}^0$ for the ion pair $MgSO_4^0$. The $25°C$ value for $\Delta\bar{V}^0$ is in good agreement with the value of 7.2 cm^3/mol determined by Fisher (1962) from high-pressure conductance work. These results can be used to obtain a better understanding of the structure of the ion pair.

It should be pointed out that the $\Delta\bar{V}^0$ and $\Delta\bar{K}^0$ determined in these studies are for the total ion paired species (i.e., both inner- and outer-sphere types). Eigen and

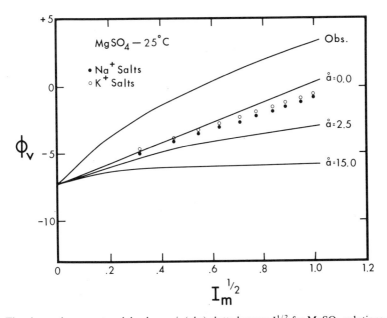

Fig. 21.   The observed apparent molal volume, ϕ_V(obs) plotted versus $I_m^{1/2}$ for $MgSO_4$ solutions compared to various estimates for the free ions.

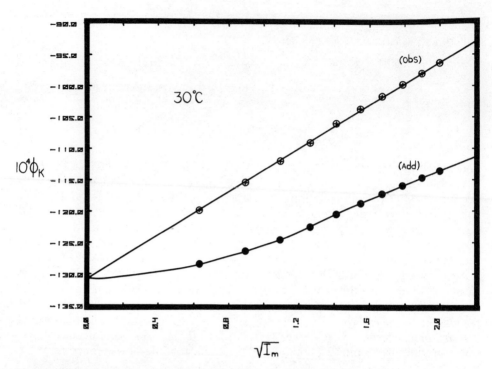

Fig. 22. The observed apparent molal compressibility, ϕ_K(obs) plotted versus $I_m^{1/2}$ for MgSO$_4$ solutions compared to various estimates for the free ions.

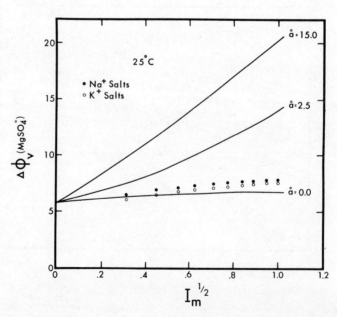

Fig. 23. The apparent molal volume change $\Delta\phi_V$, for the formation of MgSO$_4^0$.

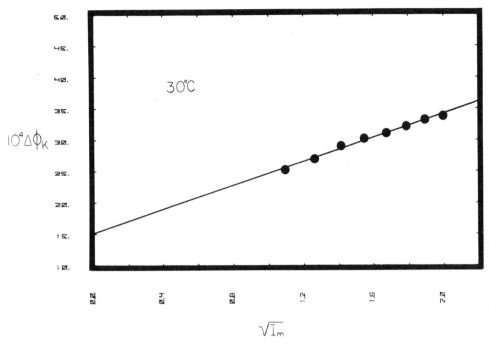

Fig. 24. The apparent molal compressiblity change, $\Delta\phi_K$, for the formation of $MgSO_4^0$.

Tamm (1962) have proposed a four-state (three-step) model for the formation of $MgSO_4^0$. The model is given by

$$Mg^{2+}(aq) + SO_4^{2-}(aq) \overset{K_I}{\rightleftharpoons} Mg^{2+}(H_2O)_2SO_4^{2-} \overset{K_{II}}{\rightleftharpoons} Mg^{2+}(H_2O)SO_4^{2-}$$

$$\qquad m_1 \qquad\qquad\qquad\qquad m_2 \qquad\qquad\qquad m_3$$

$$\overset{K_{III}}{\rightleftharpoons} Mg^{2+}SO_4^{2-} \qquad\qquad\qquad (83)$$

$$m_4$$

Table V

The Partial Molal Volume and Compressibility for the Formation of $MgSO_4^0$ at Various Temperatures[a]

Temp. (°C)	$\Delta\bar{V}^0$	$10^4\Delta\bar{K}^0$	$\bar{V}^0$	$10^4\bar{K}^0$
0	—	23.3	-3.5 ± 0.4	146.3
15	—	11.7	—	134.1
25	7.8 ± 0.6	—	0.6 ± 0.6	—
30	—	$10.2 \pm$	—	120.3
45	—	10.4	—	115.2
50	6.0 ± 0.8	—	-2.1 ± 0.8	—

[a] Taken from the work of Millero et al. (1974; Millero and Masterson, 1974). The volumes are in units of cm^3/mol and the compressibilities are in units of $cm^3/bar \cdot mol$.

The conventional K_A is given by

$$K_A = K_I K_{II} K_{III} + K_I K_{II} + K_I \tag{84}$$

Values for K_I, K_{II}, and K_{III} equal to 50, 1.96, and 0.17, respectively, have been proposed by Atkinson and Petrucci (1966). Using this model the total ϕ_V or ϕ_K is given by

$$\phi_V(MgSO_4{}^0) = \left(\frac{m_2}{m_P}\right)\phi_V[Mg^{2+}(H_2O)_2SO_4{}^{2-}] + \left(\frac{m_3}{m_P}\right)\phi_V[Mg^{2+}(H_2O)SO_4{}^{2-}]$$

$$+ \left(\frac{m_4}{m_P}\right)\phi_V[Mg^{2+}SO_4{}^{2-}] \tag{85}$$

where the total paired species $m_P = (1 - \alpha)m_T = m_2 + m_3 + m_4$. By differentiating equation 84 with respect to pressure the total $\Delta \bar{V}_A{}^0$ is given by

$$\Delta \bar{V}_A{}^0 = \Delta \bar{V}_I{}^0 + \left(\frac{m_4}{m_P}\right)\left[\Delta \bar{V}_{III}{}^0 + \Delta \bar{V}_{II}{}^0 + \frac{\Delta \bar{V}_{II}{}^0}{K_{III}}\right] \tag{86}$$

Eigen and Tamm (1962) have estimated $\Delta \bar{V}_I{}^0 = 0$, $\Delta \bar{V}_{II}{}^0 = 14\text{–}18$, and $\Delta \bar{V}_{III}{}^0 = 3$ cm³/mol. As discussed elsewhere (Millero and Masterson, 1974) these values are not self-consistent with the values of $m_2/m_P = 0.3$, $m_3/m_P = 0.60$, and $m_4/m_P = 0.10$ determined from Atkinson and Petrucci's work and our value for $\phi_V(MgSO_4{}^0)$. By making $\Delta \bar{V}_{II}{}^0 = 10$ one obtains a set of self-consistent results. By examining (Millero and Masterson, 1974; Millero et al., 1974) the $\bar{V}^0$ and $\bar{K}^0$ for the total ion pair it is quite clear that the electrostriction is quite large [~ 46 cm³/mol for $\bar{V}^0$(elect)] and that a large number of water molecules are associated with the total ion pair (15 water molecules). Let us now examine the use of the Fuoss model (or the continuum model) and the hydration model discussed earlier.

If we differentiate the theoretically derived K_A from the Fuoss (1958) methods with respect to pressure we obtain equation 37. Upon further differentiation with respect to pressure, we obtain

$$\Delta \bar{K}_A{}^0 = \left(\frac{Z_+ Z_- N e^2}{\mathring{a}}\right)\left[\frac{1}{D}\left(\frac{\partial^2 \ln D}{\partial P^2}\right) - \frac{1}{D}\left(\frac{\partial \ln D}{\partial P}\right)^2\right] + RT\frac{\partial \beta}{\partial P} \tag{87}$$

These equations yield $\Delta \bar{V}_A{}^0 = 6.15$, 7.42, and 8.94 cm³/mol; $10^4\Delta\bar{K}_A{}^0 = 7.2$, 13.8, and 19.5 cm³/mol·bar, respectively, at 0, 25, and 50°C. The 25°C results are in reasonable agreement with the measured values; however, the temperature dependence is completely wrong. These results are similar to what one finds for the dissociation of weak acids and bases in aqueous solutions (Millero, 1971b).

Another way one can examine these results is to use the hydration methods described earlier. Using these methods the $\Delta \bar{V}_A{}^0$ and $\Delta \bar{K}_A{}^0$ are given by

$$\Delta \bar{V}_A{}^0 = -\Delta h(V_E{}^0 - V_B{}^0) \tag{88}$$

and

$$\Delta \bar{K}_A{}^0 = -\Delta h(V_B{}^0 \beta_B{}^0) \tag{89}$$

Using $V_E - V_B = -3.9$, we obtain $\Delta h = 2.0$ at 25°C from equation 88 compared to $\Delta h = 1.3$ obtained from equation 89. These small values for Δh are what one would expect for the formation of an outer-sphere complex (Eigen and Tamm, 1962). For the formation of an inner-sphere complex one might expect $\Delta h = h(Mg^{2+}, SO_4{}^{2-}) = 16.4$ at 25°C. Using this value the percent of contact ion pairs would be given by

(1.3–2.0)/16.4 or 8–12%, which is in good agreement with the values determined from ultrasonic measurements (Atkinson and Petrucci, 1966) and Raman spectra data (Davis and Oliver, 1965).

Both the continuum model and the hydration model predict the $\Delta \bar{V}_A^0$ should be proportional to $\Delta \bar{K}_A^0$. A plot of $\Delta \bar{V}_A^0$ versus $\Delta \bar{K}_A^0$ for the formation of ion pairs determined from direct measurements of K_A at various pressures are shown in Fig. 25. The $\Delta \bar{V}_A^0$'s and $\Delta \bar{K}_A^0$'s have been determined from the directly measured K_A's by using the equation (Lown et al., 1968)

$$RT \ln \left(\frac{K_A^P}{K_A^0} \right) = -\Delta \bar{V}_A^0 P + 0.5 \Delta \bar{K}_A^0 P^2 \tag{90}$$

where P is the applied pressure. As is quite apparent from Fig. 25, $\Delta \bar{V}_A^0$ is linearly related to $\Delta \bar{K}_A^0$ for ion pair formation. This is similar to what is found for the dissociation of acids and bases (Lown et al., 1968). The slope is equal to 3700 bars, which is smaller than the value of 4700 bars determined for the ionization of acids and bases. The ion pairing value is also smaller than the value of 4800 bars found earlier for the free ion plot of $\bar{V}^0$(elect) versus $\bar{K}^0$(ion). It thus appears that the value of $V_E^0 - V_B^0$ determined from the hydration model

$$V_E^0 - V_B^0 = -k V_B^0 \beta_B^0 \tag{91}$$

is -3.9 for free ions and -3.0 for ion pairs.

We recently (Leung and Millero, 1975b; Perron et al., 1975a) examined the ΔH and ΔC_P data for $MgSO_4$ solutions by the same techniques outlined above. In

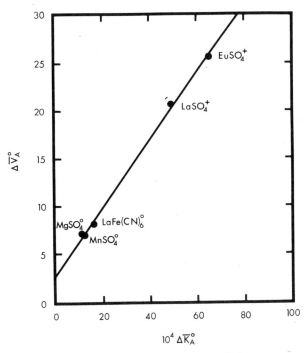

Fig. 25. The partial molal volume change $\Delta \bar{V}_A^0$ plotted versus $\Delta \bar{K}_A^0$ for the formation of ion pairs.

future work we plan to investigate the ion pairs of transition metal inorganic complexes by the same methods. Hopefully this work will yield valuable information about the structure of ion pairs thought to exist in the oceans.

One must keep in mind that all the thermodynamic models discussed in this chapter deal with the static structure of the ion or ion pair. This structure may not be identical to the kinetic species. For example, thermodynamically an ion pair may be thought to exist; however, its lifetime may be short compared to the time it takes to move through the solution (under a concentration or electrical potential). In this case the static speciation is different from the dynamic speciation. A similar argument holds for the hydration of an ion or ion pair. For most simple inorganic ions and ion pairs the lifetimes are very short; thus one might not expect the kinetic behavior to be strongly influenced by an inorganic speciation.

Acknowledgment

I wish to acknowledge the support of the Office of Naval Research (N00014-67-A-0201-0013) and the Oceanographic Section of the National Science Foundation (GA-40532) for this study. A large part of this chapter is based on a paper presented at the Third Symposium for the Chemistry of the Mediterranean Sea in Rovinj, Yugoslavia. I also wish to thank Dr. Wayne C. Duer for his helpful comments and constructive criticism.

References

Atkinson, G., M. O. Dayhoff, and D. W. Ebdon, 1973. Computer modeling of inorganic equilibria in seawater. In *Marine Electrochemistry*. J. B. Berkowitz et al., eds. The Electrochemical Society, Inc., Princeton, N.J., pp. 124–138.

Atkinson, G. and S. Petrucci, 1966. Ion association of magnesium sulfate in water at 25°C. *J. Phys. Chem.*, **70**, 3122–3128.

Barber, R. T., R. C. Dugdale, J. J. MacIsaac, and R. L. Smith, 1971. Variations in phytoplankton growth associated with the source and conditioning of upwelling water. *Invest. Pesq.*, **35**, 171–193.

Barber, R. T., and J. H. Ryther, 1969. Organic chelators: factors affecting primary production in the Cromwell Current upwelling. *J. Exptl. Mar. Biol. Ecol.*, **3**, 191–199.

Baric, A. and M. Branica, 1967. Polarography of seawater. I, Ionic state of cadmium and zinc in seawater. *J. Polarogr. Soc.*, **13**, 4–8.

Bjerrum, N., 1926. Ionic association I. Influence of ionic association on the activity of ions at a moderate degree of association. *Kgl. Danske Vidensk. Selsk. Mat.-Fys. Medd.*, **7**, 1–48.

Bockris, J. O'M. and A. K. N. Reddy, 1973. *Modern Electrochemistry*. Vol. 1. Plenum/Rosetta, New York, pp. 45–286.

Born, M., 1920. Volumes and heats of hydration of ions. *Z. Physik*, **1**, 45.

Branica, M., M. Tekek, A. Baric, and L. Jeftic, 1969. Polarographic characterization of some trace elements in seawater. *Rapp. P-V Reun. Comm. Int. Explor. Sci. Med. Mediterr.*, **19**, 929–933.

Breck, W. G., 1972. Redox potentials by equilibration. *J. Mar. Chem.*, **30**, 121–139.

Breck, W. G., 1974. Redox levels in the sea. In *The Sea*. Vol. 5. E. Goldberg, ed. Wiley, New York, pp. 153–179.

Davey, E. W., J. H. Gentils, S. J. Erickson, and R. Betzer, 1970. Removal of trace metals from marine culture media, *Limnol. Oceanogr.*, **15**, 586–488.

Davey, E. W., M. J. Morgan, and S. J. Erickson, 1973. A biological measurement of the copper complexation capacity of seawater. *Limnol. Oceanogr.*, **18**, 993–997.

Davies, C. W., 1962. *Ion Association*. Butterworths, London.

Davis, A. R. and B. G. Oliver, 1965. Raman spectroscopic evidence for contact ion pairing in aqueous magnesium sulfate solutions. *J. Phys. Chem.*, **69**, 2595–2598.

Debye, P. and E. Hückel, 1923. Electrolyte solutions. *Phys. Z.*, **24**, 185.

Desnoyers, J. E., in press. Ionic solute hydration. *XV Consiel International de Chemie Solvay, Brussels, June 1972, Proceedings*.

Desnoyers, J. E., R. E. Verrall, and B. E: Conway, 1965. Electrostriction in aqueous solutions of electro-lytes. *J. Chem. Phys.*, **43**, 243–250.

Drude, P. and W. Nernst, 1884. Electrostriction of free ions. *Z. Phys. Chem. (Frankfurt)*, **15**, 79.

Dyrssen, D. and M. Wedborg, 1974. Equilibrium calculations of the speciation of elements in seawater. In *The Sea*. Vol. 5. E. Goldberg, ed. Wiley, New York, pp. 181–195.

Eigen, M. and K. Tamm, 1962. Sound absorption in electrolytes as a consequence of chemical relaxation I. Relaxation theory of stepwise dissociation. *Z. Elektrochem.*, **66**, 93–107.

Erickson, S. J., 1972. Toxicity of copper to a marine diatom in unenriched inshore seawater. *J. Phycol.*, **8**, 318–323.

Fisher, F. H., 1962. The effect of pressure on the equilibrium of magnesium sulfate. *J. Phys. Chem.*, **66**, 1607–1611.

Fisher, F. H. and D. F. Davis, 1965. The effect of pressure on the dissociation of magnesium sulfate ion pairs in water. *J. Phys. Chem.*, **69**, 2595–2598.

Fisher, F. H. and D. F. Davis, 1967. Effect of pressure on the dissociation of the ($LaSO_4$) complex ion. *J. Phys. Chem.*, **71**, 819–822.

Frank, H. and W-Y Wen, 1957. Structural aspects of ion–solvent interactions in aqueous solutions. *Discuss. Faraday Soc.*, **24**, 113–140.

Friedman, H. L., 1962. *Ionic Solution Theory*. Wiley-Interscience, New York.

Fuoss, R. M., 1958. Ionic association III. The equilibrium between ion pairs and free ions. *J. Am. Chem. Soc.*, **80**, 5059–5061.

Garrels, R. M. and M. E. Thompson, 1962. A chemical model for seawater at 25°C and one atmosphere. *Am. J. Sci.*, **260**, 57–66.

Glueckauf, E., 1964. Heats and entropies of ions in aqueous solution. *Trans. Faraday Soc.*, **60**, 572–577.

Guggenheim, E. A., 1935. Thermodynamic properties of aqueous solutions of strong electrolytes. *Phil. Mag.*, **19**, 588–643.

Gurney, R. W., 1953. *Ionic Processes in Solution*. Dover, New York.

Hale, C. F. and F. H. Spedding, 1972. Effect of high pressure on the formation of aqueous $EuSO_4^+$ at 25°C. *J. Phys. Chem.*, **76**, 2925–2929.

Hamman, S. D., P. J. Pearce, and W. Strauss, 1964. The effect of pressure on the dissociation of lanthanum ferricyanide ion pairs in water. *J. Phys. Chem.*, **68**, 375–380.

Hemmes, P., 1972. The volume changes of ionic association reactions. *J. Phys. Chem.*, **76**, 895–900.

Hepler, L. G., 1957. Partial molal volume of aqueous ions. *J. Phys. Chem.*, **61**, 1426–1428.

Kester, D. R. and R. H. Byrne, Jr., 1972. Chemical forms of iron in seawater. In *Ferromangonese Deposits on the Ocean Floor*. D. R. Horn, ed. Lamont, New York, pp. 107–116.

Kester, D. R., T. P. O'Connor, and R. H. Byrne, Jr. (1976).Solution chemistry. Solubility and absorption of equilibria of iron, cobalt and copper in marine systems. Thallassia, Yugoslavia. In press.

Laidler, K. J., 1956. The entropies of ions in aqueous solution. I. Dependent on charge and radius, *Can. J. Chem.*, **34**, 1107–1113.

Laidler, K. J. and C. Pegis, 1957. The influence of dielectric saturation on the thermodynamic properties of aqueous ions. *Proc. Roy. Soc. (London)*, **241**, 80–92.

Latimer, W. M., 1952. *Oxidation Potentials*. 2nd ed. Prentice-Hall, Englewood Cliffs, N.J.

Latimer, W. M., K. S. Pitzer, and C. M. Slansky, 1939. The free energy of hydration of gaseous ions, and absolute potential of the normal calomel electrode. *J. Chem. Phys.*, **7**, 108–111.

Leung, W. and F. J. Millero, 1975a. The enthalpy of dilution of some 1–1 and 2–1 electrolytes in aqueous solution. *J. Chem. Thermodyn.*, **7**, 1067–1078.

Leung, W. and F. J. Millero, 1975b. The heat of formation of magnesium sulfate ion pairs. *J. Soln. Chem.*, **4**, 145–159.

Lewis, A. G., A Ramnarine, and M. S. Evans, 1971. Natural chelators—an indication of activity with the calanoid capepod, Euchaeta japonica. *Mar. Biol.*, **11**, 1–4.

Leyendekkers, J. V., 1972. The chemical potentials of seawater components. *Mar. Chem.*, **1**, 75–88.

Liss, P. S., J. R. Herring, and E. D. Goldberg, 1973. The iodide/iodate system in seawater as a possible measure of redox potential. *Nature*, **242**, 108–109.

Lown, D. A., H. R. Thirsk, and Lord Wynne-Jones, 1968. Effect of pressure on ionization equilibria in water at 25°C. *Trans. Faraday Soc.*, **64**, 2073–2080.

Malijkovic, D. and M. Branica, 1971. Polarography of seawater II. Complex formation of cadmium with EDTA. *Limnol. Oceanogr.*, **16**, 779–785.

Millero, F. J., 1969. The partial molal volume of ions in seawater. *Limnol. Oceanogr.*, **14**, 376–385.

Millero, F. J., 1970. The apparent and partial molal volume of aqueous sodium chloride solutions at various temperatures. *J. Phys. Chem.*, **74**, 356–362.

Millero, F. J., 1971a. Effect of pressure on sulfate ion association in seawater. *Geochim. Cosmochim. Acta*, **35**, 1089–1098.

Millero, F. J., 1971b. The molal volumes of electrolytes. *Chem. Rev.*, **71**, 147–176.

Millero, F. J. 1971c. The physical chemistry of multicomponent salt solutions. In *Biophysical Properties of Skin*. H. R. Elden, ed. Wiley, New York, pp. 329–376.

Millero, F. J., 1972. The partial molal volume of electrolytes in aqueous solutions. In *Water and Aqueous Solutions*. R. A. Horne, ed. Wiley, New York, pp. 519–595.

Millero, F. J., 1974a. Seawater as a multicomponent electrolyte solution. In *The Sea*. E. D. Goldberg, ed. Wiley-Interscience, New York, pp. 3–80.

Millero, F. J., 1974b. The physical chemistry of seawater. *Ann. Rev. Earth Planet. Sci.*, **2**, 101–150.

Millero, F. J. and R. A. Berner, 1972. Effect of pressure on carbonate equilibria in seawater. *Geochim. Cosmochim. Acta*, **36**, 92–98.

Millero, F. J., E. V. Hoff, and L. Kahn, 1972. The effect of pressure on the ionization of water at various temperatures from molal volume data. *J. Soln. Chem.*, **1**, 309–327.

Millero, F. J. and W. L. Masterton, 1974. The volume change for the formation of magnesium sulfate ion pairs at various temperatures. *J. Phys. Chem.*, **79**, 1287–1294.

Millero, F. J., G. K. Ward, F. K. Lepple, and E. V. Hoff, 1974. The isothermal compressibility of aqueous NaCl, MgCl$_2$, Na$_2$SO$_4$ and MgSO$_4$ solutions from 0 to 45°C at one atmosphere. *J. Phys. Chem.*, **78**, 1636–1643.

Millero, F. J., C. H. Wu, and L. G. Hepler, 1969. Thermodynamics of ionization of acetic and chloroacetic acid in ethanol–water mixtures. *J. Phys. Chem.*, **73**, 2453–2455.

Monk, C. B., 1961. *Electrolytic Dissociation*. Academic Press, London.

Morel, F. and J. Morgan, 1972. A numerical method for computing equilibria in aqueous chemical systems. *Environ. Sci. Technol.*, **6**, 58–67.

Mukerjee, P., 1961. On ion-solvent interactions. Part I. Partial molal volumes of ions in aqueous solutions. *J. Phys. Chem.*, **65**, 740–744.

Nancollas, G. H., 1966. *Interactions in Electrolyte Solutions*. Elsevier, New York.

Nancollas, G. H., 1970. The thermodynamics of metal–complex and ion–pair formation. *Coord. Chem. Rev.*, **5**, 379–415.

Noyes, R. M., 1964. Assignment of individual ionic contributions to properties of aqueous ions. *J. Am. Chem. Soc.*, **86**, 971–979.

Owen, B. B., R. C. Miller, C. E. Milner, and H. L. Cogan, 1961. The dielectric constant of water as a function of temperature and pressure. *J. Phys. Chem.*, **65**, 2065–2070.

Padova, J., 1963. Ion solvent interaction. II. partial molal volume and electrostriction: a thermodynamic approach. *J. Chem. Phys.*, **39**, 1152–1557.

Pauling, L., 1940. *The Nature of the Chemical Bond*. 3rd ed. Cornell University Press, Ithaca, N.Y.

Perron, G., J. E. Desnoyers, and F. J. Millero, 1975a. Apparent molal volumes and heat capacities of alkaline earth chlorides in water at 25°C. Submitted to *Can. J. Chem.*

Perron, G., J. E. Desnoyers, and F. J. Millero, 1975b. Apparent molal volumes and heat capacities of some sulfates and carbonates in water at 25°C. Submitted to *Can. J. Chem.*

Powell, R. E. and W. M. Latimer, 1951. The entropy of aqueous solutes. *J. Chem. Phys.*, **19**, 1139–1141.

Prue, J. E., 1969. Ion pairs and complexes: free energies, enthalpies and entropies, *J. Chem. Educ.*, **46**, 12–16.

Pytkowicz, R. M., and D. R. Kester, 1969. Harned's rule behavior of NaCl–Na$_2$SO$_4$ solutions explained by an association model. *Am. J. Sci.*, **267**, 217–229.

Robinson, R. A. and R. H. Stokes, 1959. *Electrolyte Solutions*. Butterworths, London.

Robinson, R. A. and R. H. Wood, 1972. Calculations of the osmotic and activity coefficients of seawater at 25°C. *J. Soln. Chem.*, **1**, 481–488.

Rosseinsky, D. R., 1965. Electrode potentials and hydration energies. Theories and correlations. *Chem. Rev.*, **65**, 467–490.

Steeman-Nielson, E. and S. Wium-Anderson, 1970. Copper ions as poison in the sea and fresh water. *Mar. Biol.*, **6**, 93–97.

Stumm, W. and J. J. Morgan, 1970. *Aquatic Chemistry*. Wiley, New York, p. 583.

Ward, G. K. and F. J. Millero, 1974. The effect of pressure on the ionization of boric acid in aqueous solutions from molal volume data. *J. Soln. Chem.*, **3**, 417–430.

Ward, G. K. and F. J. Millero, 1975. The effect of pressure on the ionization of boric acid in sodium chloride and seawater. *Geochim. Cosmochim. Acta.*, **39**, 1595–1604.

Whitfield, M., 1969. Eh as an operational parameter in estuarine studies. *Limnol Oceanogr.*, **14**, 547–558.

Whitfield, M., 1973. A chemical model for the major electrolyte component of seawater based on the Bronsted–Guggenheim hypothesis. *Mar. Chem.*, **1**, 251–266.

Wood, R. H. and P. J. Reilly, 1970. Electrolytes. *Ann. Rev. Phys. Chem.*, **21**, 287–406.

Zirino, A. and M. L. Healy, 1970. Inorganic zinc complexes in seawater. *Limnol. Oceanogr.*, **15**, 956–958.

Zirino, A. and M. L. Healy, 1971. Voltammetric measurement of zinc in the northeastern tropical Pacific Ocean. *Limnol. Oceanogr.*, **16**, 773–778.

Zirino, A. and M. L. Healy, 1972. pH controlled differential voltammetry of certain trace transition elements in nature waters. *Environ. Sci. Technol.*, **6**, 243–249.

Zirino, A. and S. Yamamoto, 1972. A pH dependent model for the chemical speciation of copper, zinc, cadmium, and lead in seawater. *Limnol. Oceanogr.*, **17**, 661–671.

18. MIGRATIONAL PROCESSES AND CHEMICAL REACTIONS IN INTERSTITIAL WATERS*

ABRAHAM LERMAN

1. Introduction

Geochemical reactions taking place in oceanic sediments bear on two broad groups of processes outside the mineral-pore water system. One group includes the processes of the present and recent past, centering primarily on the role of sediments as suppliers or scavengers of chemical species in ocean water. Recycling of the nutrients, man-produced stable and radioactive substances, and recovery of other chemical species from settling material and young sediments to solution: these are the main processes relevant to the current state of ocean water.

The other group includes processes of historical significance. The physical and chemical diagenesis of oceanic sediments must be coped with in any interpretation of the history of the ocean and the Earth's crust. Physical and mineralogical changes observable in oceanic sediments supply much of the information needed in the studies of older marine sediments preserved on continents. Formation of new minerals, selective dissolution of shells of marine organisms, cementation, and more subtle changes in the chemical composition of the solids in contact with pore waters— these are some of the more important processes that bear on the interpretation of the sedimentary record.

Figure 1 is a schematic diagram of the geochemical processes taking place in sediments, and of their roots. Rivers and the atmosphere add to the ocean the products of crustal weathering, extraterrestrial matter, and gaseous products of volcanic and photochemical origin. Input to the ocean consists of particulate and dissolved materials; biological activity produces new particles from dissolved materials in the ocean. Inorganic detritus and biogenic particles reaching the ocean floor and ocean water trapped in the interparticle pore space are the main avenues of transport to the sediment (dashed-line arrows in Fig. 1 indicate dissolved material). Some dissolution of terrigenous detritus is likely to take place in transit through ocean water, and biogenic solids can partly dissolve on the way from the photic zone to oceanic depths (Mackenzie and Garrels, 1965; Berger, 1967; Lerman et al., 1974). Continuing dissolution of some of the solid phases in sediments contributes to the concentrations of dissolved species in pore waters. Dissolution of calcite, biogenic silica, and decomposition of organic matter are reflected in the concentrations of alkalinity, calcium, silica, and phosphorus and nitrogen species in pore waters, as has been demonstrated by different investigators. Return of dissolved species to the ocean by means of molecular diffusional flux upward across the sediment–water interface has been discussed in the literature for the tracer and major constituents of ocean water (for example, Goldberg and Koide, 1963; Mangelsdorf et al., 1969; Li et al., 1969; Fanning and Pilson, 1974; Berner, 1974; Goldhaber and Kaplan, 1974; Manheim and Sayles, 1974; Lerman, 1975).

* This work was supported by the Oceanography Section, National Science Foundation, through NSF Grant No. OCE 75-13844. I also thank F. T. Mackenzie and R. M. Garrels for discussions of the various topics treated in this chapter, and *American Journal of Science* for permission to reproduce material in Figs. 8 and 9.

695

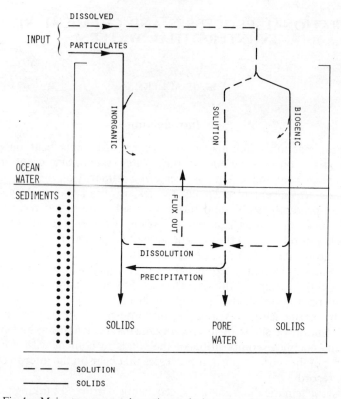

Fig. 1.   Major transport and reaction paths in ocean water and sediments.

Acceptance of the picture of a steady-state ocean implies that input of dissolved species must be balanced by their removal. For stable nonvolatile species, the path of removal is from ocean water to sediments. If input and removal in a steady-state ocean consisted solely of dissolved load, then the relative proportions of the chemical species in sediments would have been the same as in the input load. However, for some of the seven stable constituents of ocean water and sediments (Si, Al, Fe, Ca, Na, K, and Mg, in decreasing order of mean crustal abundance), the relative abundances in oceanic sediments differ considerably from the relative abundances in the dissolved input. Taking the abundance of Ca as 100 in dissolved input and in oceanic sediments, the relative abundances of the seven elements are as follows (from data in Garrels and Mackenzie, 1971, tables 4.9 and 8.4):

	Ca	Si	Na	Mg	K	Fe	Al
Dissolved input to ocean	100	41.1	27.5	25.4	12.9	4.66	0.064
Mean oceanic sediments	100	931.0	32.4	41.4	54.2	148.0	242.0

Tabulation of the relative abundances shows that Si, Fe, and Al are in particular more abundant in sediments than in the dissolved load received by the ocean. Even if the Fe and Al analyses of river waters, on which the estimates of input to the ocean are based, include some particulates and are therefore too high (F. T. Mackenzie, personal communication), the contrast of the relative abundances in solution and oceanic sediments is still great. The higher relative abundance of some of the elements in sediments emphasizes the importance of particulate flux and its major share in the makeup of oceanic sediments.

Global changes in tectonics, climates, and evolution of marine biota in the past left imprints on the mineralogical composition and textures of oceanic sediments. Literature on the mineralogy and stratigraphy of oceanic sediment cores abounds in detailed records of such changes which took place at various times during the last 10^8 years of the ocean history. Variations in mineralogical composition and texture of a sediment column add, in general, to the complexity of the chemical reactions between solids and constituents of pore waters.

2. Migration and Reaction Models

A. General Approach

Sedimentation is one of the mechanisms of entry of dissolved and particulate matter into sediments, for this is an advective process involving entrapment of pore water and solid particles, and their movement away from the sediment–ocean water interface.

Chemical reactions in sediments set up concentration (and chemical potential) gradients that are responsible for the other transport mechanism—molecular diffusion. Bioturbation as a mechanism of transport affecting sediments and pore waters near the top of the sediment column has been likened to a diffusional process (Schink et al., 1975), although its effectiveness may vary from one locality to the next.

The homogeneity of oceanic sediments over large areas of the ocean floor dictates, as an approximation, a one-dimensional approach: migration of chemical species is viewed as taking place only in the vertical dimension, and the nature and availability of data leave in many cases little choice for anything but one-dimensional models. A one-dimensional approach would not apply to those areas where compositional changes in horizontal directions are comparable to the vertical gradients. Such situations are likely to exist in nearshore areas, and at those localities where ground water inflow is significant (cf. Ristvet et al., 1973; Manheim and Sayles, 1974).

Because the main goal of modeling is to provide a quantitative picture of geochemical processes and their evolution through time, the question of whether oceanic sediments and pore waters are a steady- or transient-state system is important (Lerman, 1971).

Information on the distribution of chemical species in sediments is by itself usually insufficient to determine whether the system is in a transient or steady state. The chemical and transport rate parameters are commonly not known for a given sediment, although they can be estimated if assumptions are made that a particular migration-reaction model applies to the processes controlling the observed distribution. If the rate parameters computed from the model show that characteristic times within the system (for example, reaction half-lives) are significantly shorter than the age of the system, then the steady-state assumption may be acceptable. For transient states, the mathematical structure of the models is as a rule more involved. This, however, is not the main obstacle to making direct comparison tests between steady-state and transient models: the latter require information about the initial conditions in the system, as well as stipulation of the starting point on the time scale. The knowledge of initial conditions in transient models is particularly important for those geochemical processes the time scales of which are short by comparison with the ages of sediment sections. The emphasis of this chapter is on steady-state models. Discussions of how good an approximation a steady-state model is to a particular process are included in the sections dealing with specific processes in sediment.

B. General Theory

The behavior of a chemical species present in pore water and solid particles can be described by two equations, one for the pore water and the other for the solids.

For pore water, the flux of a dissolved species (F, in units of mass per unit time per unit area of sediment surface) can be written as

$$F = -\phi\left(D\frac{\partial C}{\partial z} - UC\right) \tag{1}$$

where ϕ is the porosity (volume fraction of sediment occupied by water) in the plane perpendicular to the flux direction, D is diffusion coefficient in pore water, C is concentration (mass per unit volume of pore water), and U is the velocity of pore water movement relative to the sediment (Fig. 2). The vertical distance coordinate z is taken as $z = 0$ at the sediment–water interface, increasing downward. The term $-D(\partial C/\partial z)$ is the diffusional flux as defined by Fick's first law of diffusion. The advective velocity U is positive if pore water moves with time away from the sediment–water interface at $z = 0$ (see Section 3.B).

For solid particles containing the same chemical species, the flux across any horizontal plane in sediment is

$$F_s = -\phi\left(D_s\frac{\partial C_s}{\partial z} - U_sC_s\right) \tag{2}$$

where subscript s denotes quantities associated with the solids. C_s, concentration of the species in solids, is in units of mass per unit volume of pore water (the same as C in equation 1). The sedimentation rate of particles, U_s, may or may not be equal to the advective term for pore water, U in equation 1. The diffusion coefficient D_s can be visualized as diffusion on the particle surfaces or in the adsorbed layer. The possible

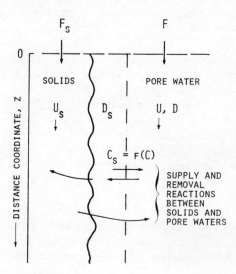

Fig. 2. Fluxes in sediment. Refer to equations 1 and 2. Dashed line denotes schematically the adsorbed layer adjacent to the solid particles' surface. Advection of sediment particles (U_s) and pore water (U) is relative to the sediment–water interface at $z = 0$. D_s is molecular diffusion coefficient in the adsorbed layer or on solid surface. Exchange reactions are represented by $C_s = f(C)$, as given in equation 5.

magnitude of D_s is discussed in Section 3.A, whereas at this stage it is sufficient to point out its physical significance and the functional similarity between equations 1 and 2. From the general mass conservation condition (for example, Jost and Oel, 1964) in each of the phases, Fick's second-law relationships can be written for the species in pore water and solids. In pore water

$$\frac{\partial(\phi C)}{\partial t} = -\frac{\partial F}{\partial z} - \phi\left(\begin{array}{c}\text{Removal} \\ \text{rate terms}\end{array}\right) + \phi\left(\begin{array}{c}\text{Supply} \\ \text{rate terms}\end{array}\right) \quad (3)$$

and in solids

$$\frac{\partial(\phi C_s)}{\partial t} = -\frac{\partial F_s}{\partial z} - \phi\left(\begin{array}{c}\text{Removal} \\ \text{rate terms}\end{array}\right)_s + \phi\left(\begin{array}{c}\text{Supply} \\ \text{rate terms}\end{array}\right)_s \quad (4)$$

The sum of equations 3 and 4 gives the rate of change in total concentration, $C_{total} = C + C_s$, in terms of the diffusional and advective fluxes, and the chemical removal and supply processes. Equations 3 and 4 as written above define a general model of migration and reactions in sediments, where reacting species are associated with the solids and pore water. To make use of the models, the removal and supply rate terms in the equations must be known, and the flux derivatives $\partial F/\partial z$ and $\partial F_s/\partial z$ should be substituted for explicitly from equations 1 and 2.

Concentrations of a species in pore water and solids may be related to one another by fast equilibrium exchange or adsorption processes, such that concentration on the solids is always and everywhere some function (f) of the concentration in pore water,

$$C_s = f(C) \quad (5)$$

For adsorption, $f(C)$ can be given by any of the known adsorption models. One of the more commonly used is the Freundlich adsorption equilibrium

$$C_s = KC^n \quad (6)$$

where K and n are empirically determined constants $(0.1 < n < 1)$. If $n = 1$, equation 6 describes an exchange equilibrium in a dilute solution,

$$C_s = KC \quad (7)$$

With the latter relationship, mathematical treatment of the migration-reaction models derivable from equations 3 and 4 is much simpler, whereas the power-law relationship in equation 6 makes the differential equations nonlinear. Some particular cases of diffusion of nonreactive species and Freundlich adsorption have been discussed by Crank (1956); additional references to work by others on adsorption and exchange in sediments are given in Section 3.A.

If a chemical species exchanges between pore water and solids and a relationship between the concentrations in the two phases is known, such as equation 5, then the behavior of the species in sediments or pore water can be expressed in terms of either C_s or C alone. The relationships for F (equation 1), F_s (equation 2), and C_s (equation 7) can be used in equations 3 and 4, and addition of the latter two equations gives a more general relationship for the rate of concentration change in pore water:

$$\frac{\partial C}{\partial t} = \frac{1}{\phi(K + 1)}\frac{\partial}{\partial z}\left[\phi(D + KD_s)\frac{\partial C}{\partial z}\right] - \frac{1}{\phi(K + 1)}\frac{\partial}{\partial z}\left[\phi(U + KU_s)C\right]$$

$$- \frac{C}{\phi}\frac{\partial\phi}{\partial t} - \frac{R_i + R_{i,s}}{K + 1} + \frac{S_i + S_{i,s}}{K + 1} \quad (8)$$

where R_i and S_i stand for the removal and supply rates in pore water, and $R_{i,s}$ and $S_{i,s}$ are similar processes on solids.

All the migration-reaction models discussed subsequently are derivable from equation 8. For example, a simple case of diffusion of a nonreactive species in pore water is obtainable from equation 8 by taking the sediment porosity (ϕ) and diffusion coefficient (D) as constant and independent of t and z, and setting the advection, removal, supply, exchange, and D_s terms equal to zero. Then equation 8 reduces to

$$\frac{\partial C}{\partial t} = D \frac{\partial^2 C}{\partial z^2}$$

which is the conventional form of Fick's second law of diffusion.

3. Chemical and Physical Contributions to Migration of Reacting Species

Equation 8 relates the rate of change in concentration in pore water to diffusional (D, D_s) and advective (U, U_s) transport, sediment compaction [$\phi(t, z)$], chemical exchange (K), removal (R_i), and supply (S_i) processes. The significance of the individual processes and their interrelationships suitable for obtaining specific solutions of equation 8 are discussed in this section.

A. Diffusion in Pore Water and on Solids

Electrical conductance of adsorbed ionic layers accounts for some fraction of the conductance by the pore water solution (van Olphen, 1957, 1963). Studies of surface conductance and ionic diffusion in clay–water mixtures (van Schaik and Kemper, 1966; Ellis et al., 1970), as well as theoretical work on the behavior of adsorbed molecules on solid surfaces (review in Meares, 1968), support the possibility of diffusional migration of aqueous species within adsorbed layers of molecular dimensions, adjacent to the solid particle surfaces. The two modes of diffusion—in pore water and in the adsorbed layer on solids—are represented by the diffusion coefficients D and D_s in the flux equations 1 and 2. The relationships between the two fluxes can be demonstrated if one considers an exchange equilibrium as given by equation 7, and diffusion of a conservative species in pore water and in the adsorbed layer (Fig. 2),

$$\frac{\partial C}{\partial t} = D \frac{\partial^2 C}{\partial z^2} \tag{9}$$

$$\frac{\partial C_s}{\partial t} = D_s \frac{\partial^2 C_s}{\partial z^2} \tag{10}$$

where the diffusion coefficients (D and D_s) are assumed constant and independent of the distance coordinate z. Using $KC = C_s$ (equation 7), summation of equations 9 and 10 gives a relationship for diffusion in pore water

$$\frac{\partial C}{\partial t} = \frac{D + KD_s}{K + 1} \cdot \frac{\partial^2 C}{\partial z^2} \tag{11}$$

The constant coefficient $(D + KD_s)/(K + 1)$ is an effective diffusion coefficient of the species in pore water, in the presence of instantaneous exchange with the solid and diffusion on the solid surface,

$$D' \equiv \frac{D + KD_s}{K + 1} \tag{12}$$

If there is no diffusion in the adsorbed layer ($D_s = 0$), the effective diffusion coefficient D' reduces to the form $D/(K + 1)$, as has been derived for diffusion with adsorption in solution (Crank, 1956) and used in studies of diffusion in sediments (Duursma and Eisma, 1973; Aston and Duursma, 1973).

Diffusion in the adsorbed layer increases the value of the effective diffusion coefficient in pore water and enhances the diffusional flux. A simple numerical example, based on the values of D and K comparable to those reported for oceanic and lake sediments, demonstrates this. If for a particular sediment and chemical species D and K are known to be $D \simeq 6 \times 10^{-6}$ cm^2/sec and $K \simeq 3 \times 10^2$, and the effective diffusion coefficient (measured or computed from a model) is $D' = 4 \times 10^{-8}$ cm^2/sec, then D_s can be estimated from equation 12 as

$$D_s \simeq D' - \frac{D}{K} = 4 \times 10^{-8} - \frac{6 \times 10^{-6}}{3 \times 10^2} = 2 \times 10^{-8} \text{ cm}^2/\text{sec}$$

That D_s should be smaller than D (in general, $0 \leqslant D_s < D$) is supported by the fact that surface electrical conductance is appreciably lower than the bulk solution conductance.

Geometric hindrance of diffusion in solution by solid particles in a porous medium results in a value of the diffusion coefficient D that is smaller than the value for the same chemical species in a free solution. Geometric hindrance is caused by tortuosity of the diffusional free path and it is commonly referred to as tortuosity (cf. Li and Gregory, 1974). Relationships between the electrical conductivity of water-filled sediments (and other materials), diffusion coefficients, and porosity can be combined in the so-called structure or formation factor (Klinkenberg, 1951; Barrer, 1968; Manheim and Waterman, 1974) that is strictly valid for those porous media in which the solids are not conducting. Alternative relationships between diffusion coefficients and porosity are based on theoretical derivations and empirical work which indicate that D can be represented as a power series function of ϕ (Helfferich, 1962, p. 302; Youngquist, 1970). A fairly successful relationship for porous media is

$$D = D_0 \phi^2 \tag{13}$$

where D_0 is the diffusion coefficient in a free solution ($\phi = 1$). It has also been noted that diffusion coefficients in porous sediments, obtained by a number of investigators, can be approximated by equation 13 (Manheim, 1970; Berner, 1975).

B. Pore Water Advection and Sediment Porosity

If a sediment column grows by deposition and a plane of reference is always taken as the sediment–water interface, then pore water and sediment particles move away from the reference plane in the downward direction.

1. If compaction of sediments is insignificant, porosity (ϕ) remains constant and, consequently, $\partial\phi/\partial t = 0$ and $\partial\phi/\partial z = 0$ in equation 8. In this case the rates of pore water (U) and sediment (U_s) movement relative to the sediment–water interface are equal and constant with depth. However, porosity of oceanic sediments decreases approximately exponentially with depth (for example, Emery and Rittenberg, 1952; von Huene and Piper, 1973; Keller and Bennett, 1973) such that only at greater depths ϕ becomes nearly constant and therefore $\partial\phi/\partial z = 0$. Also, when sedimentation rate is steady, there is no change in porosity with time in the deeper layers of sediments and $\partial\phi/\partial t = 0$. When the sediment porosity remains constant (that is, $\partial\phi/\partial t = 0$ and $\partial\phi/\partial z = 0$), equations 3 and 4 can in many cases be solved explicitly.

2. If compaction of sediment is significant, porosity (ϕ) changes with depth and $\partial\phi/\partial z \neq 0$. For the chemical processes taking place in the upper part of the sediment column, the region below the sediment–water interface may in some cases be treated as a semi-infinite column with its origin at the sediment–water interface ($z = 0$). Then in the presence of constant rates of deposition and compaction, the porosity profile preserves its shape and porosity at any depth relative to $z = 0$ remains constant. Under this condition $\partial\phi/\partial t = 0$ but, as explained above, $\partial\phi/\partial z \neq 0$.

The relationships between the advection rate of pore water (U) and depth-dependent porosity have been discussed by a number of authors for particular (Anikouchine, 1967; Tzur, 1971) and general cases (Berner, 1971, 1975). In principle, conservation of the fluxes of pore water and sediment particles downward from the sediment–water interface requires the following two conditions to be satisfied:

$$\frac{\partial(\phi U)}{\partial z} = 0 \quad \text{and} \quad \frac{\partial[(1 - \phi)U_s]}{\partial z} = 0 \tag{14}$$

If compaction attains a steady value deeper in the sediment ($z \to \infty$) and there is no flow of pore water relative to the sediment particles, then $U_\infty = U_{s,\infty}$, and integration of equation 14 gives

$$\phi U = \phi_\infty U_{s,\infty} \tag{15}$$

and

$$(1 - \phi)U_s = (1 - \phi_\infty)U_{s,\infty} \tag{16}$$

Thus when the rate of sedimentation in the region of constant porosity (ϕ_∞, $U_{s,\infty}$) and the porosity profile are known, the rates of pore water (U) and sediment particles advection (U_s) at any depth can be obtained from equations 15 and 16.

3. In the presence of other types of pore water flow, such as artesian flow or horizontal advection, U may be independent of U_s and no simple relationships between the two may exist.

4. In a sediment layer growing continuously, the top and bottom boundaries recede with time one from the other. If migrational and chemical processes take place within the entire thickness of such a layer, from the sediment–water interface down to the bottom boundary, then strictly the sediment porosity is not constant in time nor with depth (that is, $\partial\phi/\partial t \neq 0$ and $\partial\phi/\partial z \neq 0$). Explicit relationships for porosity as a function of time and depth in sediments can be derived using empirical information on the rates of sedimentation and rate of compaction as a function of load and sediment type (Athy, 1930; Hamilton, 1959, 1971). Explicit forms of $\phi(t, z)$ can then be used in equations 3, 4, and 8, although no simple solutions of the equations may exist and numerical methods may have to be used. Such cases of $\phi(t, z)$ are not treated in this chapter.

Application of steady-state models to continuously growing sediment layers is justified as an approximation if the rates of chemical reactions and migration within a layer are faster than the change in the layer dimensions. For example, at the rate of sedimentation of 1 cm/1000 yr, a layer will increase in thickness 100 m in 10^7 yr. Diffusional transport characterized by the diffusion coefficient value of $D = 3 \times 10^{-6}$ cm^2/sec $\simeq 10^2$ cm^2/yr will be effective within a 100-m thick layer on a shorter time scale of

$$t \simeq \frac{(10^4 \text{ cm})^2}{10^2 \text{ cm}^2/\text{yr}} = 10^6 \text{ yr}$$

Similarly, the computed rates of chemical reactions in pore waters of deep ocean sediments, discussed in subsequent sections of this paper, are characterized by half-lives of the order of 10^6 yr, and these time scales are comparable to the diffusion-distance time scale.

5. In a sediment layer of constant thickness, points 1–3 apply. However, vertical advection of pore water through a layer may be $U = 0$ if the position of the layer remains constant relative to the bottom of the sediment column (while the sediment–water interface moves upward and away from it). This may be the case of lithologically distinct layers at some depth in the sediment.

If the layer moves upward with the sediment–water interface, or at a slower rate, then pore water advection may have to be taken into account, by analogy with equations 15 and 16. For a chemically distinct layer to move within the sediment column, chemical reactions responsible for the layer boundaries must keep pace with the sedimentation.

C. Chemical Supply and Removal Rates

The removal and supply rate terms, in equations 3, 4, and 8, denoting the rates in units of mass per unit time per unit volume of pore water, can stand for any combination of such processes as supply by settling particles, addition of dissolved constituents owing to biological productivity, dissolution, radioactive decay, biodegradation, precipitation, and exchange between solids and pore waters.

A chemical species in pore water may be supplied by dissolution of one mineral phase and it may be removed by precipitation of another phase. If dissolution and precipitation refer to two solid phases only, and if they can be treated as first-order chemical reactions, then the supply rate term in equation 3 is

$$k_1(C_{s,1} - C) \qquad \text{for } C_{s,1} > C \tag{17}$$

and the removal rate term is

$$-k_2(C - C_{s,2}) \qquad \text{for } C > C_{s,2} \tag{18}$$

where $C_{s,i}$ are the solubilities of the dissolving and precipitating phases, C is concentration in pore water, and k_i is reaction rate constant (time^{-1}). Here the dependence of the reaction rate constant on the solid surface area (Stöber, 1967; Lerman et al., 1973) is included in k_i. The production and removal rates in equations 17 and 18 can be added, giving

$$-(k_1 + k_2)C + (k_1C_{s,1} + k_2C_{s,2}) \equiv -kC + J \tag{19}$$

where the combined reaction rate constant $k = k_1 + k_2$ (time^{-1}) and the combined zero-order term $J = k_1C_{s,1} + k_2C_{s,2}$ (mass/time · volume) can be treated as net removal and addition rates (Lerman, 1975). An obvious advantage of this approach is that only two parameters (k and J) are involved, rather than four. In a system where dissolution is balanced by precipitation only (that is, no diffusion, advection, or other chemical processes), a steady-state concentration C_{ss} is derivable from the equality of the two rates given by equations 17 and 18 as

$$C_{ss} = \frac{J}{k} \tag{20}$$

In a pore water solution that is far from equilibrium with the dissolving and precipitating phases, the rates of the two reactions become, respectively, $J \simeq k_1C_{s,1}$ and $-k_2 C$.

4. Factors Controlling Concentration Profile Shape

The shape of a concentration against depth profile in pore water is the main source of information from which the values of one or more of the parameters D, U, K, k, and J can be derived. Because in situ measurements of these and related parameters in sediments are difficult, and independent information on diffusion coefficients, sedimentation, and reaction rates may be inadequate, this section examines the effects that the choice of a model has on the interpretation of a concentration profile, as well as on the estimates of chemical kinetic parameters derivable from such a model.

A. Sediment–Water Interface Region

Advection and Reaction Rates

A concentration profile increasing or decreasing exponentially with depth and attaining a steady value (cf. Fig. 3) is controlled by dissolution and precipitation reactions resulting in the flux out or into the sediment. Profiles of such shape, in which steady concentration values are attained at depths ranging from less than 1 m to several meters, have been reported for silica, nitrogen species, and phosphate in marine sediments (Anikouchine, 1967; Berner, 1974; Fanning and Pilson, 1974;

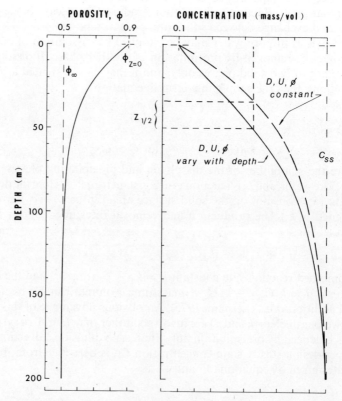

Fig. 3.   Variable porosity and concentration profiles of a reacting species in pore waters. Porosity profile given by equation 34 with $\alpha = 0.035$/m. Concentration profile computed using equation 36 with $k = 4.87 \times 10^{-6}$/yr, $U_\infty = 2$ cm/1000 yr, $C_{ss} = 1.0$, and $C_{z=0} = 0.1$. Concentration profile for the case of constant porosity (dashed curve) computed using equation 22 with the same parameters. $z_{1/2}$ is the depth of mid-concentration range.

Lerman, 1975; Schink et al., 1975). Interpretation of such profiles as steady-state distributions of dissolved species, controlled by dissolution, precipitation, diffusion, and advection in sediments of constant porosity leads to the following differential equation derivable from equation 8,[1]

$$D \frac{\partial^2 C}{\partial z^2} - U \frac{\partial C}{\partial z} - kC + J = 0 \tag{21}$$

The solution of equation 21 is

$$C = \frac{J}{k} + \left(C_{z=0} - \frac{J}{k} \right) \exp \left[\frac{zU}{2D} - z \left(\frac{U^2}{4D^2} + \frac{k}{D} \right)^{1/2} \right] \tag{22}$$

and if there is no advection ($U = 0$) equation 22 reduces to

$$C = \frac{J}{k} + \left(C_{z=0} - \frac{J}{k} \right) \exp \left[-z \left(\frac{k}{D} \right)^{1/2} \right] \tag{23}$$

where $C_{z=0}$ is concentration at sediment–water interface, and $J/k = C_{ss}$ is the concentration at "infinite" depth, as defined in equation 20.

The depth z at which the midpoint of the concentration range occurs,

$$\frac{C - J/k}{C_{z=0} - J/k} = \frac{1}{2}$$

makes it possible to determine the power exponent of e in equations 22 and 23. With additional information available on D and U, the rate constant k and, subsequently, J can be computed. If advection is neglected and k is estimated by using equation 23 from an observed concentration profile, then the estimated value of $k = k'$ is smaller than the correct value of k in equation 22. For a given concentration profile, the values of k' ($=k$ in equation 23) and k in equation 22 are interrelated as follows:

$$k = k' + U \left(\frac{k'}{D} \right)^{1/2} \tag{24}$$

This shows that for a given profile the computed value of the reaction rate constant k is higher if the rate of sedimentation or pore water advection is taken into account.

The rate of sedimentation or water advection can generally be neglected if in equation 22

$$\frac{U^2}{4D} \ll k \tag{25}$$

In deep-water oceanic sediments where the rates of sedimentation are commonly between 0.1 and 2 cm/1000 yr, and molecular diffusion coefficients in pore waters are of the order of 10^{-6} cm²/sec (Li and Gregory, 1974), an order of magnitude value of the $U^2/4D$ quotient is

$$\frac{U^2}{4D} = \frac{(1 \times 10^{-4} \text{ to } 2 \times 10^{-3} \text{ cm/yr})^2}{4 \times (30 \text{ to } 150 \text{ cm}^2/\text{yr})} = 10^{-10} \text{ to } 10^{-8}/\text{yr}$$

[1] At steady state, C is a function of z only and the equations can be written in terms of total (d/dz) rather than partial derivatives. The latter are used in this chapter for notational simplicity.

Thus for reactions characterized by first-order rate constants $k > 10^{-8}$/yr, or half-lives of 10^7 yr or shorter, the advective term in the migration-reaction equations for pore water can be neglected.

Advection may be important if there is strong adsorption of a dissolved species by solid particles in the sediment. When adsorption is strong (large K), the main mode of transport is by sedimentation, and the role of diffusion in pore water diminishes as uptake by solids becomes more effective (Lerman and Lietzke, 1975). If in addition to dissolution (J) and precipitation ($-kC$) there is ion exchange between pore water and some solid phases in the sediment, then the system can be described by the two equations

$$\frac{\partial C}{\partial t} = D \frac{\partial^2 C}{\partial z^2} - U \frac{\partial C}{\partial z} - kC + J \tag{26}$$

$$\frac{\partial C_s}{\partial t} = -U \frac{\partial C_s}{\partial z} \tag{27}$$

Substitution of KC for C_s in equation 27, and addition of the two equations with the left-hand side $(K + 1) \partial C / \partial t = 0$ for the steady-state case, gives the result

$$D \frac{\partial^2 C}{\partial z^2} - U(K + 1) \frac{\partial C}{\partial z} - kC + J = 0 \tag{28}$$

Equation 28 differs from 21 only in the value of the constant coefficient of the advective term. Therefore, the solution of equation 28 is identical with that of equation 22, except for $U(K + 1)$ replacing U. By reasoning similar to that given for relationship 25, pore water advection rate U in the presence of adsorption taking place may be neglected if the following relationship holds:

$$\frac{U^2(K + 1)^2}{4D} \ll k \tag{29}$$

Adsorption of some trace metals, as measured by radiotracer methods in laboratories using seawater and marine sediments (Duursma and Eisma, 1973; see Section 5.B), is characterized by values of K as high as 10^2–10^3. Accordingly, the value of the quotient $U^2/4D \leqslant 10^{-8}$/yr given previously can be 4–6 orders of magnitude higher in the presence of strong adsorption—10^{-4} to 10^{-2}/yr. Under these conditions, advection can be neglected only if reaction rates are very fast on the geological time scale and their half-lives are shorter than 10^2 yr. Such relatively fast reaction rates, with k of the order of 10^0–10^1/yr, or half-lives shorter than 1 yr, characterize dissolution of opal in oceanic sediments (Hurd, 1973) and dissolution of clay minerals and zeolites in seawater under laboratory conditions (Lerman, Mackenzie, and Bricker, 1975).

Radionuclides diffusing in pore water and undergoing exchange with solid phases represent a somewhat different case, because the nuclides decay both in pore water and in the adsorbed state on solids. For a diffusing radionuclide, the migration-reaction equations are

$$\frac{\partial C}{\partial t} = D \frac{\partial^2 C}{\partial z^2} - U \frac{\partial C}{\partial z} - \lambda C \tag{30}$$

$$\frac{\partial C_s}{\partial t} = -U \frac{\partial C_s}{\partial z} - \lambda C_s \tag{31}$$

where λ is the decay rate constant (yr^{-1}). At steady state, using KC for C_s and adding equations 30 and 31, the equation for the radionuclide concentration in pore water is

$$\frac{D}{K+1}\frac{\partial^2 C}{\partial z^2} - U\frac{\partial C}{\partial z} - \lambda C = 0 \tag{32}$$

If the radionuclide concentration tends to zero with depth owing to decay, then the solution of equation 32 is identical to that of equation 22 with $J = 0$ and $D/(K+1)$ replacing D. In this case the role of advection in controlling the concentration profile in sediments is given by

$$\frac{U^2(K+1)}{4D} \ll \lambda \tag{33}$$

For K between 10^2 and 5×10^3, the quotient in equation 33 is $< 10^{-4}$/yr. Thus for those radionuclides the decay half-lives of which are shorter than 10^3–10^4 yr, diffusional transport in pore water from a source at the sediment–water interface is not significantly affected by sedimentation and, consequently the advective term in the migration-reaction equation 32 and in its solution may be neglected. Among the naturally occurring radionuclides, ^{226}Ra of half-life 1.6×10^3 yr is probably a borderline case for sedimentation affecting transport in deep-ocean sediments.

In shallower sections of the ocean where the rates of sedimentation are higher, transport in sediments due to continuous deposition becomes progressively more important for those chemical species and radionuclides the behavior of which in deep-ocean sediments is controlled primarily by their reaction and decay rates.

Porosity and Reaction Rates

The rate of change in porosity with depth is usually greatest below the sediment–water interface (for example, compilation of data in Berner, 1971, p. 90). A generalized equation describing a decrease in porosity with depth in oceanic sediments (Fig. 3) can be written as

$$\phi = \phi_\infty + (\phi_0 - \phi_\infty)e^{-\alpha z} \tag{34}$$

where ϕ_∞ is constant porosity at depth, ϕ_0 is the porosity value at the sediment water interface ($z = 0$), and α is a constant. Typical values of ϕ_0 are between 0.7 and 0.9, and ϕ_∞ between 0.4 and 0.6. If an overall decrease in porosity from the sediment surface downward is by a factor of 1.5–2, some effect on concentration profile shapes is to be expected owing to the ϕ dependence of diffusion and pore water advection (Sections 3.A and 3.B), as well as to the changing proportions of solids and pore water. When porosity varies with depth, a concentration profile in Fig. 3 can be described by the following equation (compare equations 8, 21, and 26):

$$\frac{\partial C}{\partial t} = \frac{1}{\phi}\frac{\partial}{\partial z}\left[\phi\left(D\frac{\partial C}{\partial z} - UC\right)\right] - kC + J \tag{35}$$

At steady state, using $D = D_0\phi^2$ (equation 13) and $U = \phi_\infty U_{s,\infty}/\phi$ (equation 15), equation 35 becomes

$$\frac{\partial^2 C}{\partial z^2} + \left(\frac{3}{\phi}\frac{\partial\phi}{\partial z} - \frac{\phi_\infty U_\infty}{D_0\phi^3}\right)\frac{\partial C}{\partial z} - \frac{kC - J}{D_0\phi^2} = 0 \tag{36}$$

Using $\partial\phi/\partial z$ from equation 34 in equation 36, solution of equation 36 gives C as a function of depth in sediment, as shown by the solid-curve profile in Fig. 3. The equation was solved numerically, using an integration procedure developed for one-boundary value problems (Stein, 1972).

The concentration profile computed for the case of depth-dependent porosity should be compared with the profile computed for a constant-porosity sediment, using equation 22. As a whole, the differences between the two concentration profiles are relatively small on the scale of a tenfold concentration change from 0.1 at the sediment–water interface to 1.0 at depth.

A relevant question to ask is, what would be the effects on the computed values of k and J if they were estimated from an observed profile neglecting the porosity variation in the sediment? The rate constant k can be estimated from equation 22 using the depth $z = z_{1/2}$ at which the midpoint of the concentration range occurs (Fig. 3):

$$k = \left(\frac{U}{z_{1/2}} + \frac{D \ln 2}{z_{1/2}^2} \right) \ln 2 \tag{37}$$

For different species, $z_{1/2}$ ranges from 10^{-1} to 10^{1} m below the sediment–water interface. In deep ocean where the rates of sedimentation are slow, the $z_{1/2}^2$ term is more important and therefore the shape of the profile has a relatively significant effect on the computed value of k. Referring to the concentration profiles in Fig. 3, for the constant-porosity case $z_{1/2} = 35$ m, whereas for the variable porosity case $z_{1/2} = 50$ m. The rate constants k, computed for the two profiles using equation 37 with $U = 2 \times 10^{-3}$ cm/yr and $D = 100$ cm^2/yr, differ by about a factor of two: $k = 4.3 \times 10^{-6}$/yr from the constant-porosity profile, and $k = 2.2 \times 10^{-6}$/yr from the variable porosity profile, if the porosity change is ignored. Variation by a factor of two in the computed value of k may in many cases be admissible, particularly if it is viewed within the scope of other uncertainties, such as in the values of U, D, and analytical errors associated with pore water concentration data.

B. Sediment Layers

Diagenetic behavior of chemical species in pore waters of distinct sediment layers can be controlled by the processes within the layer as well as its boundaries. A distinct layer within a sedimentary sequence may be recognizable by its mineralogical composition, as well as by the presence of mineralogically different beds at its upper and lower boundaries. Migration-reaction models aimed at describing steady-state processes in pore waters of sediment layers commonly stipulate constant concentrations at the layer boundaries as the boundary conditions for solution of the diagenetic equations. Information on the nature and rates of processes within the layer is derivable from application of a particular model to the observed data, in a manner similar to other pore water scenarios dealt with in the preceding sections.

Because the shape of a concentration against depth profile in pore water has much to do with the model that may be applied to the data, this section deals with the more important physical and chemical processes that can be responsible for the observed profile shapes.

In a sediment layer shown schematically in Fig. 4, the vertical dimension coordinate z increases from $z = 0$ at the upper boundary to $z = h$ at the lower boundary. Steady-state concentrations at the upper and lower boundaries are C_1 and C_2.

The following five cases of physical and chemical controls of the concentration profile shape are considered below.

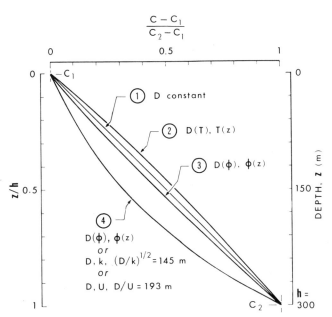

Fig. 4. Effects of temperature gradient, porosity gradient, precipitation reaction, and pore water advection on the shape of a concentration profile at steady state, controlled by constant concentrations C_1 and C_2 at the sediment layer boundaries.

Diffusivity (D) and Porosity (ϕ) Constant with Depth

If diffusion alone is responsible for the steady-state distribution of a chemical species in pore water, and its concentrations at the layer boundaries are C_1 and C_2, then the solution of the diffusion equation

$$\frac{\partial}{\partial z}\left(\phi D \frac{\partial C}{\partial z}\right) = 0 \tag{38}$$

is

$$\frac{C - C_1}{C_2 - C_1} = \frac{z}{h} \tag{39}$$

According to equation 39, concentration in pore water (C) varies linearly with depth from C_1 to C_2. The value of the diffusion coefficient D cannot be determined from such a profile unless the flux is independently measured. The profile is shown in Fig. 4, labeled 1.

Temperature Effect: Temperature Gradient, D Temperature Dependent, ϕ Constant with Depth

In the presence of a temperature gradient (dT/dz) within the sediment layer, the diffusion equation 38 can be written as

$$D\frac{\partial^2 C}{\partial z^2} + \frac{\partial C}{\partial z} \cdot \frac{dD}{dT} \cdot \frac{dT}{dz} = 0 \tag{40}$$

The relationship between D and T for the temperature range from 0 to 25°C in seawater, derivable from the data of Li and Gregory (1974), is

$$D = D_0(1 + aT) \tag{41}$$

where $0° \leqslant T \leqslant 25°C$, $a \simeq 0.04/\text{deg}$, and D_0 is the value of D at $0°C$. A linear increase in temperature with depth can be taken as

$$T = T_0 + bz \tag{42}$$

where $b \simeq 3 \times 10^{-4}$ deg/cm is the mean geothermal gradient, and T_0 is the temperature at $z = 0$. Using equations 41 and 42 in 40, the solution is

$$\frac{C - C_1}{C_2 - C_1} = \frac{\ln\left[1 + abz/(1 + aT_0)\right]}{\ln\left[1 + abh/(1 + aT_0)\right]} \tag{43}$$

If a and b are small, the right side of equation 43 approaches the ratio z/h, which is the solution for the case of depth-constant diffusivity given in equation 39. The concentration profile in the presence of the thermal gradient, computed using equation 43 with $T_0 = 3°C$, is shown in curve 2 in Fig. 4. The effect of the geothermal gradient on the concentration profile shape is small, and departures from the straight line amount only to a few percent on the concentration scale.

Advection Effect: Pore Water Flow Rate U, D, and ϕ Constant with Depth

In the presence of pore water advection (constant rate U), the equation for the layer

$$D\frac{\partial^2 C}{\partial z^2} - U\frac{\partial C}{\partial z} = 0 \tag{44}$$

has the solution

$$\frac{C - C_1}{C_2 - C_1} = \frac{e^{zU/D} - 1}{e^{hU/D} - 1} \tag{45}$$

For small U, the right side of equation 45 approaches the solution for the case of constant diffusivity, z/h in equation 39. For the scale distance (D/U) value of $D/U = 193$ m, the profile computed using equation 45 is shown in curve 4 in Fig. 4. The effect of advection on the concentration profile shape is fairly well pronounced and it amounts to departures from the straight line of the order of 10% of the entire concentration range.

For the diffusion coefficient $D \simeq 3 \times 10^{-6}$ cm²/sec, the advection rate is $U \simeq 5$ cm/1000 yr. Positive U implies movement of pore water downward relative to the layer boundaries; upward flow would be represented by a negative U.

Reaction Effect: Precipitation or Decay in Pore Water (k), D and ϕ Constant with Depth

A steady-state distribution of a dissolved species controlled by the boundary concentrations, diffusion, and first-order removal reaction or radioactive decay is given by

$$D\frac{\partial^2 C}{\partial z^2} - kC = 0 \tag{46}$$

the solution of which is

$$C = \frac{C_1 \sinh(h - z)(k/D)^{1/2} + C_2 \sinh z(k/D)^{1/2}}{\sinh h(k/D)^{1/2}} \tag{47}$$

For the diffusion-reaction scale distance of $(D/k)^{1/2} = 145$ m, the concentration profile is curve 4 in Fig. 4, and it is identical with the profile computed from the diffusion-advection model with the scale distance of $D/U = 193$ m. Taking the diffusion coefficient $D = 3 \times 10^{-6}$ cm^2/sec, the value of the precipitation reaction rate constant is $k = 4.5 \times 10^{-7}$/yr. Rate constants of this order of magnitude are characteristic of some of the concentration profiles of Ca, Mg, and SiO$_2$ in deep-ocean pore waters (Lerman, 1975).

Porosity Effect: ϕ Varies with Depth, D is ϕ Dependent

If porosity varies with depth within the layer, differentiation of equation 38 gives

$$\phi D \frac{\partial^2 C}{\partial z^2} + \left(\phi \frac{\partial D}{\partial z} + D \frac{\partial \phi}{\partial z} \right) \frac{\partial C}{\partial z} = 0 \tag{48}$$

To obtain an explicit solution of equation 48, the relationship $D = D_0 \phi^2$ given in equation 13 and a linear porosity gradient within the layer can be used:

$$\phi = \phi_0 \left(1 - \frac{zP}{h} \right) \tag{49}$$

where ϕ_0 and ϕ_h are the porosity values at the top and bottom boundaries of the layer, and the constant coefficient P is

$$P = 1 - \frac{\phi_h}{\phi_0} \tag{50}$$

A linear porosity gradient within the layer (equation 49) was taken for simplicity, although an exponentially decreasing porosity profile, as in Fig. 3 and equation 34, could also be used to demonstrate the porosity effect on concentration at steady state.

The solution of equation 48 is

$$\frac{C - C_1}{C_2 - C_1} = \frac{1/(1 - zP/h)^2 - 1}{1/(1 - P)^2 - 1} \tag{51}$$

If porosity changes little with depth then P is small and the right side of equation 51 approaches the ratio z/h, which is the solution for a linear concentration gradient controlled only by diffusion (equation 39). Equation 51 shows that departures of the concentration profile from a straight line are controlled by the magnitude of porosity change with depth, as reflected in the value of the quotient P. A 6% change in porosity within the layer ($P = 0.06$) gives a concentration profile that deviates only slightly from the linear profile (curve 3, Fig. 4). The effect is comparable in magnitude to the effect of the thermal gradient on concentration profile but in opposite direction. A greater change in porosity within the layer, of the order of 40% ($P = 0.4$), results in a more pronouncedly curved profile (curve 4, Fig. 4). The latter profile is identical with the profiles computed from the diffusion-advection and diffusion-reaction models discussed in the preceding examples.

Summary of Effects

The effects on a steady-state concentration profile of a dissolved species in a sediment layer were discussed for constant diffusivity, temperature gradient, advection of pore water, precipitation reaction, and porosity gradient. A linear gradient, characteristic of D constant with depth, can also result from a combination of a slight temperature increase and porosity decrease with depth (curves 2 and 3, Fig. 4), owing to the fact that the two effects may mutually cancel. Although the physical mechanisms in the two cases that control the linear shape of the concentration profile are very different, no gross error in the interpretation of results would arise if the weak temperature and porosity gradients were neglected, and D were considered constant within the pore water layer.

The fact, however, that a curvilinear profile can be interpreted in the simplest case as the result of diffusion with advection, or diffusion through a porosity gradient, or diffusion with chemical reaction, indicates that a profile shape alone does not provide unique information on the physical and chemical processes responsible for the observed shape. Therefore, any model developed on the basis of the profile shape only may be underdetermined. In practice, the choice of a model should be governed by such information as the presence or absence of reactive mineral phases in the sediment, information on sediment porosity, and, more significantly, information on the continuity of a concentration profile within the sediment. The latter is difficult to obtain owing to the fact that samples of pore waters in long sediment cores are often taken several meters or tens of meters apart, such that any possible occurrence of mineralogically different layers acting as sources or sinks between adjacent samples can remain undetected.

5. Case Studies: Sediment–Water Interface Region and Below

Within the upper few meters of the sediment, the migrational behavior of several chemical species has been analyzed by a number of investigators in terms of diffusion-advection-reaction models. The studies include silica, phosphate, nitrogen species, manganese, and radioactive decay products of the uranium and thorium isotopes in sediments (Anikouchine, 1967; Berner, 1974; Fanning and Pilson, 1974; Goldberg and Koide, 1963; Hurd, 1973; Lerman, 1975; Schink et al., 1975). Most of the models used to describe the behavior of chemical species near the sediment–water interface are based on steady-state equations similar in principle to equation 21. The region below the sediment–water interface dealt with in such models is of the order of 10^{-1}–10^0 m.

Some recently reported data on NH_4^+ concentrations in long sections of deep-ocean cores (Gieskes, 1974) make it possible to analyze the diagenetic behavior of this species from the sediment–water interface down, over distances of hundreds of meters.

A. Ammonia in Deep-Ocean Sediments

The NH_4^+ concentrations in pore waters of two cores from the Indian Ocean (Figs. 5, 6) increase from the sediment–water interface (first samples taken at 4 and 5 m depth), go through a maximum, and decline. In both cases, the NH_4^+ concentration maxima correlate with SO_4^{2-} minima, a feature that has been attributed to biogenic sulfate reduction and organic matter decomposition reactions in the sediments (Gieskes, 1974). In Fig. 5, the approximately constant concentration trend of NH_4^+ with depth below 500 m also coincides with a similar trend of SO_4^{2-}.

At site 241 (Fig. 5) the higher NH_4^+ concentrations above 200 m can be interpreted as owing to decomposition of organic matter in sediments. If the pore water column is continuous, NH_4^+ diffuses upward and downward from the depth of concentration maximum. Then the constant concentration at greater depths implies that the NH_4^+ flux from above and production in the sediment are balanced by removal of NH_4^+ from pore water. There is no evidence of NH_4^+ oxidation nor assimilation by microorganisms in deep sediments. [Oxidation products of NH_4^+ have been looked for

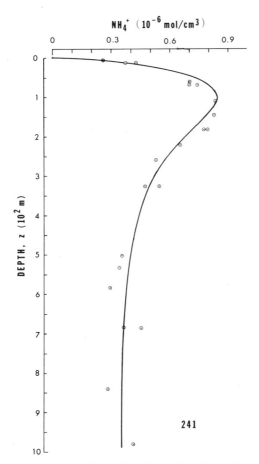

Fig. 5. Ammonia in pore waters at site 241, Indian Ocean (data from Gieskes, 1974). Concentration profile computed using equation 54 with the values of the chemical and migrational parameters given in the text. (Two different concentrations at one depth indicate analysis of pore waters squeezed at 5 and 23°C.)

but not detected near the sediment–water interface in the sediments of Santa Barbara Basin (Sholkovitz, 1973).] A plausible explanation for the NH_4^+ sink in deep sediments is its incorporation in clay minerals (Stevenson and Cheng, 1972), either as a trace cation or by exchange with K^+.

For a sink to exist, the mechanism of NH_4^+ migration into solids must be a one-way process analogous to an irreversible removal. Such a sink may be visualized as a first-order reaction, analogous to $-k(C - C_s)$ given in equation 18. Because the organic matter content of sediments commonly decreases with depth, the source of NH_4^+ can be represented by a production rate term in which the production rate decreases exponentially with depth, as has been demonstrated (Berner, 1974) for anoxic sediments near the sediment–water interface ($J_* e^{-\beta z}$ mol/cm^3 · yr, where J_* and β are constants).

An alternative interpretation can be a production rate that decreases exponentially with depth and tends to a constant value, such as $J(1 + Ae^{-\beta z})$, where A and β are constants.

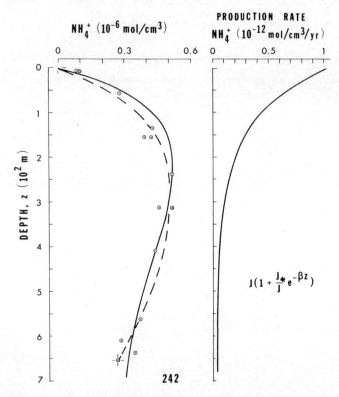

Fig. 6. Ammonia in pore waters at site 242, Indian Ocean (data from Gieskes, 1974). Solid-line concentration profile computed using equation 54 and the values of migrational and chemical parameters given in the text. The NH_4^+ production rate for this profile is given in the graph on the right. Dashed-line concentration profile computed for the one-layer model, equation 56, using the parameter values given in the text.

Mathematically, the two models give identical results, as will be shown below. A steady-state NH_4^+ profile will be described by the following equation:

$$D \frac{\partial^2 C}{\partial z^2} - U \frac{\partial C}{\partial z} - k(C - C_s) + J_* e^{-\beta z} = 0 \qquad (52)$$

From equation 52, denoting $J \equiv kC_s$ and $A \equiv J_*/J$, the following equation is obtained

$$D \frac{\partial^2 C}{\partial z^2} - U \frac{\partial C}{\partial z} - kC + J(1 + A e^{-\beta z}) = 0 \qquad (53)$$

which shows that the two models of ammonia production and reaction are identical.
   With the boundary conditions

$$z = 0: \quad C = C_{z=0} \quad \text{and} \quad z \to \infty: \quad C = C_{ss} \equiv \frac{J}{k}$$

the solution of equation 52 is

$$C = \frac{J}{k} - \frac{J_* e^{-\beta z}}{D\beta^2 + \beta U - k}$$

$$+ \left(C_{z=0} - \frac{J}{k} + \frac{J_*}{D\beta^2 + \beta U - k} \right) \exp \left[\frac{zU}{2D} - z \left(\frac{U^2}{4D^2} + \frac{k}{D} \right)^{1/2} \right] \qquad (54)$$

Equation 54 was used to compute the concentration against depth profiles of NH_4^+ shown in Figs. 5 and 6. For both sites the rate of pore water advection (U) was taken as equal to the rates of sedimentation reported for the younger sediments as ~ 3 cm/1000 yr. (Values of U in the range from ~ 0 to 6 cm/1000 yr affect the computed concentrations only within a few percent.) The diffusion coefficient of NH_4^+ in pore water was taken as $D = 3 \times 10^{-6}$ cm^2/sec for the two profiles, a value close to $D = 3.5 \times 10^{-6}$ cm^2/sec computed for NH_4^+ in some anoxic sediments (Berner, 1974). The production and reaction rate parameters used in computation of the profiles shown in Figs. 5 and 6 are the following:

Site 241	Site 242
$C_{z=0} = 0$	$C_{z=0} = 0$
$C_{ss} = \dfrac{J}{k} = 0.35 \times 10^{-6}$ mol/cm^3	$C_{ss} = \dfrac{J}{k} = 0.20 \times 10^{-6}$ mol/cm^3
$J_* = 7.5 \times 10^{-12}$ mol/cm$^3 \cdot$ yr	$J_* = 0.95 \times 10^{-12}$ mol/cm$^3 \cdot$ yr
$k = 7.0 \times 10^{-7}$/yr	$k = 2.0 \times 10^{-7}$/yr
$\beta = 0.021$/m	$\beta = 0.0087$/m

The NH_4^+ concentration peak in pore waters at site 242 (Fig. 6) is not as well pronounced as the peak at site 241 (Fig. 5), and this feature is reflected in the lower value of the production rate term J_* as well as the slower decay with depth in the production rate at site 242. The decay in the production rate with depth is controlled by β: at site 241 the NH_4^+ production rate decreases to $1/e$ of its sediment–water interface value at depth 50 m, whereas a similar decrease in the production rate at site 242 occurs at depth 115 m.

The net production rate of NH_4^+ at site 242 given by the sum $J + J_* e^{-\beta z}$ is shown as a function of depth in Fig. 6.

The values of NH_4^+ production rate $(J_* e^{-\beta z})$ in the sediments of sites 241 and 242 in the Indian Ocean are orders of magnitude lower than the production rates in shallower-water anoxic sediments. From the data in Berner (1974), the rate of NH_4^+ production in the sediments of Somes Sound, Maine, is about 3.5×10^{-6} mol/cm$^3 \cdot$ yr, and in Santa Barbara Basin, off California, the rate is about 100 times lower, 0.035×10^{-6} mol/cm$^3 \cdot$ yr. The much slower rates in deep-ocean sediments, as obtained for sites 241 and 242, may be owing to the same factors that may be responsible for the slower NH_4^+ production rate in the Santa Barbara Basin (Berner, 1974): organic matter preserved in older sediments is more resilient to decomposition, and the rates of its decomposition are slower.

The trend of the NH_4^+ concentrations in pore waters at site 242 (Fig. 6) does not display a clearly defined region of zero concentration gradient in the deeper part of the sediment column. Therefore, the concentration profile can be considered as occurring within a sediment layer, between depths 0 and 650 m, and its interpretation by means of a one-layer model should be tested in addition to the semi-infinite column model discussed above (equation 54).

For a one-layer system the boundary conditions are

$$z = 0: \quad C = 0 \quad \text{and} \quad z = h: \quad C = C_h$$

Removal of NH_4^+ from pore waters by a chemical reaction will not be considered ($k = 0$ in equation 52) and only the contribution of the production term $J_* e^{-\beta z}$ will be included in the model. Then, using the above boundary conditions, the solution of equation 52 is

$$C = \frac{J_*(1 - e^{-\beta z})}{D\beta^2 + \beta U} + \left[C_h - \frac{J_*(1 - e^{-\beta h})}{D\beta^2 + \beta U} \right] \frac{e^{zU/D} - 1}{e^{hU/D} - 1} \tag{56}$$

The dashed-line profile in Fig. 6 was computed from equation 56 with the values of $D = 3 \times 10^{-6}$ cm²/sec and $U = 3 \times 10^{-3}$ cm/yr as used previously, and the following values of the other parameters for site 242, one-layer:

$$C_h = 0.27 \times 10^{-6} \text{ mol/cm}^3 \qquad J_* = 0.65 \times 10^{-12} \text{ mol/cm}^3 \cdot \text{yr}$$
$$h = 650 \text{ m} \qquad \beta = 0.009/\text{m}$$

The values of the NH_4^+ production rate (J_*) and its decay with depth (β) are comparable to those derived from the semi-infinite model. Although the two concentration profiles shown in Fig. 6 are fairly close to the data and to each other, the fundamental difference between the two models is in the presence or absence of NH_4^+ removal reactions.

A ballpark estimate of the rate of NH_4^+ removal by clays in the deeper part of the sediment column can be obtained using the reaction rate parameters derived from equation 54 and listed for sites 241 and 242. The rate of NH_4^+ migration to clays in the deeper sediment is $kC \simeq 3 \times 10^{-7}$ mol/cm³ $\times 5 \times 10^{-7}$/yr $= 1.5 \times 10^{-13}$ mol/cm³ · yr. Because the potassium content of 1 cm³ of an "average" argillaceous sediment is of the order of 10^{-4} mol/cm³, the migration rate of NH_4^+ into solids is too slow to cause a substantial replacement of K^+ in clays. Removal of K^+ from pore waters, rather than its addition, at sites 241 and 242 is indicated by the K^+ concentrations decreasing continuously with depth (Gieskes, 1974). At the removal rate of NH_4^+ computed above, during a time period of 8×10^7 yr only about 1×10^{-5} mol NH_4^+ would be incorporated in the clays contained in 1 cm³ of sediment (about 0.5 cm³ solids), and this figure would be less than 10% of K^+ present in the clays.

B. ^{226}Ra: Diffusion and Exchange

Migration of ^{226}Ra has been reported to take place in deep-ocean sediments (Goldberg and Koide, 1963). The parent nuclide ^{230}Th enters the sediment on solid particles, whereas the daughter ^{226}Ra exchanges with pore water and diffuses out of the sediment. According to the diffusion-decay model of Goldberg and Koide (1963), the apparent diffusion coefficient of ^{226}Ra in sediment is approximately 1×10^{-9} cm²/sec.

From equations 3 and 4, the model of ^{226}Ra behavior in sediments can be written in the following form:

$$\frac{\partial C}{\partial t} = D \frac{\partial^2 C}{\partial z^2} - U \frac{\partial C}{\partial z} - \lambda C \tag{57}$$

$$\frac{\partial C_s}{\partial t} = -U \frac{\partial C_s}{\partial z} - \lambda C_s + \lambda_{Th} C_{s, Th} \tag{58}$$

Equation 57 describes migration of ^{226}Ra (concentration C) in pore water by molecular diffusion and advection. Equation 58 describes the rate of concentration change of ^{226}Ra on the solid phase where it forms by decay of ^{230}Th (the ^{226}Ra production rate is $\lambda_{Th} C_{s, Th}$), decays, and migrates due to advection of sediment. In equations 57 and 58 the sediment porosity is taken as constant with depth, such that the rates of pore water and solid sediment advection are the same (U cm/yr). Ion-exchange equilibrium for ^{226}Ra is assumed according to equation 7, $C_s = KC$. Using this relationship for C_s in equation 58, the two equations can be combined into one that defines a steady state ^{226}Ra profile in pore water,

$$\frac{D}{K+1} \frac{\partial^2 C}{\partial z^2} - U \frac{\partial C}{\partial z} - \lambda C + \frac{\lambda_{Th} C_{s, Th}}{K+1} = 0 \tag{59}$$

Concentrations in bulk sediment (that is, solids plus pore water) are given by $C(K + 1)$, where C is the solution of equation 59 with appropriate boundary conditions. The effective diffusion coefficient of $\sim 1 \times 10^{-9}$ cm^2/sec reported by Goldberg and Koide (1963) is, in terms of equation 59, the ratio $D/(K + 1)$. Taking the molecular diffusion coefficient of ^{226}Ra in ocean pore water as $D \simeq 2 \times 10^{-6}$ cm^2/sec (cf. Li and Gregory, 1974), the exchange constant K for ^{226}Ra can be estimated as

$$K \simeq \frac{2 \times 10^{-6}}{1 \times 10^{-9}} = 2000$$

(Remember that K is a dimensionless number by definition, as C and C_s are both concentrations per unit volume of pore water). Most of the experimental results on exchange and adsorption of radionuclides report distribution factors analogous to K but given in units of a ratio (concentration per unit volume of solution)/(concentration per unit mass of solids). This difference in units makes direct comparisons with literature data on uptake of other radionuclides difficult. A comparison can be made with K values computed from concentration profiles of ^{90}Sr and ^{137}Cs in freshwater Great Lakes sediments: for ^{90}Sr, $K \simeq 120$, and for ^{137}Cs, K is between 3×10^3 and 10×10^3 (Lerman and Lietzke, 1975).

6. Case Studies: Reactions and Migration in Sediment Layers

Chemical and migrational processes taking place within a sediment layer may differ from those taking place above and below it owing to such features as differences in the mineralogy or processes controlling concentrations at the layer boundaries. More or less abrupt changes in concentration gradients in pore waters are usually taken as indications of a different chemical and physical regime. At steady state, interpretation of the chemical and migrational processes in pore waters of sediment layers is to a large extent based on the shape of concentration profiles, as was discussed in more detail in Section 4.B. In this section, some concentration profiles of Ca, Mg, and SiO$_2$ in pore waters are analyzed in terms of migration-reaction models aimed at one-, two-, and three-layer systems.

A. One-Layer Models

A reciprocal relationship between the Ca and Mg concentrations in pore waters has been reported in many deep-ocean sediment cores: Ca concentration increases with depth, whereas Mg decreases. The relationships between Ca and Mg have been inferred to arise from such reactions as dolomitization, dissolution of calcite, incorporation of Mg in clays, and reactions of pore waters with basalt leading to release of Ca from feldspars and uptake of Mg by clays (Drever, 1971, 1974; Sayles et al., 1973; Gieskes, 1973). Irrespective of whether Ca and Mg in pore water come only from ocean water or from additional mineral sources as well, concentration profiles of the two species provide information on the nature and rates of chemical reactions taking place in sediments. Pore water data and gross mineralogical composition of sediments are in principle not sufficient to indicate the reacting solid phases, insofar as the amounts of reacting phases may be a small fraction of the total amount of solids. Thus the model approach to sediment–pore water reactions can tell in what directions the reactions go and how fast.

The importance of chemical reactions in contributing to concentration profiles of Ca and Mg (at least in some sediments) is illustrated in Fig. 7. The curvature of the profiles indicates removal of Ca from pore water and addition of Mg that may be interpreted, respectively, as precipitation and dissolution. The porosity of sediments changes from $\phi_0 \simeq 0.65$ near the sediment–water interface to $\phi_h \simeq 0.5$ at 200 m depth (Hayes et al., 1972). This decrease in porosity is insufficient to account for the curvature of the two profiles, as shown by dashed lines in Fig. 7. A concentration profile based on diffusion and reaction (in a sediment column of constant porosity), fits the reported data better. For Ca, the scale distance of $(D/k)^{1/2} = 113$ m corresponds to the first-order precipitation rate constant $k = 7.4 \times 10^{-7}$/yr, if the diffusion coefficient is taken as $D = 3 \times 10^{-6}$ cm^2/sec.

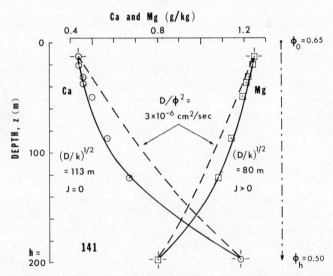

Fig. 7. Calcium and magnesium in pore waters at site 141, Eastern Atlantic (data from Waterman et al., 1972). Solid curves are concentration profiles computed using the diffusion-reaction model, equation 60. Scale distances $(D/k)^{1/2}$ for Ca and Mg are shown in the figure. Dashed curves are concentration profiles computed from equation 51 describing diffusion of a nonreactive species in a sediment layer of variable porosity. Porosity gradient taken as linear (data from Hayes et al., 1972).

A number of Ca and Mg profiles from other deep ocean drilling sites, and the computed concentrations fitting the reported data are shown in Figs. 8 and 9. The computed curves were obtained from the solution of equation 21, solved with the following boundary conditions:

$$z = 0: \quad C = C_1 \quad \text{and} \quad z = h: \quad C = C_2$$

where $z = 0$ is the upper boundary of the layer (not necessarily the sediment–water interface) and $z = h$ is the lower boundary. The solution of equation 21 is

$$C = \frac{J}{K} + \frac{(C_1 - J/k)e^{Uz/2D} \sinh R(h - z) + (C_2 - J/k)e^{U(z - h)/2D} \sinh Rz}{\sinh Rh} \quad (60)$$

where

$$R = +\left(\frac{U^2}{4D^2} + \frac{k}{D}\right)^{1/2}$$

The solution given in equation 60 is valid for any combination of advection and dissolution: $U = 0$, $U \neq 0$, $J = 0$, or $J \neq 0$. When D and U are known, the reaction rate constant k can be evaluated using the boundary concentrations C_1 and C_2, and an estimated concentration ($C_{h/2}$) at midpoint of the layer at $z = h/2$. Derivation of the relationships for computation of k and J by this procedure is given in Lerman (1975).

Analysis of the published data on Ca and Mg in pore waters of deep ocean sediments indicates that precipitation or removal reactions ($k > 0$) are always revealed by the model (except for the linear concentration gradients). With regard to the addition or dissolution rate term (J), three classes of profiles are recognizable.

1. $J = 0$ for Ca and Mg. This type of profiles (Fig. 8) indicates removal of the two species from pore waters. For example, at site 148 from the Aves Ridge, west of the Lesser Antilles, both Ca and Mg data indicate removal from pore water at the rates

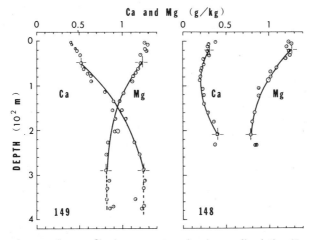

Fig. 8.   Calcium and magnesium profiles in pore waters showing no dissolution ($J = 0$), site 148, Aves Ridge, west of Lesser Antilles. Dissolution of a Ca-containing phase ($J > 0$ for Ca, $J = 0$ for Mg) is shown by the profiles at site 149, central Venezuelan Basin (from Lerman, 1975). Corroborating evidence for dissolution and precipitation reactions discussed in the text.

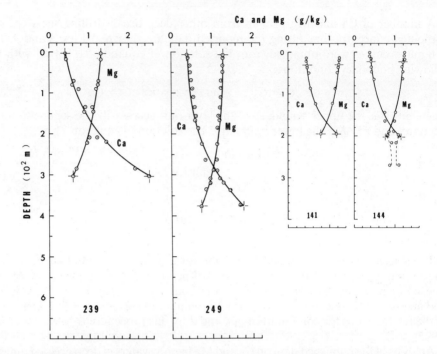

Fig. 9.   Calcium and magnesium concentration profiles showing removal of Ca from pore waters and addition of magnesium to pore waters ($J = 0$ for Ca, $J > 0$ for Mg). Data for the eastern (site 141) and western Atlantic (site 144) from Waterman et al. (1972); data for the western Indian Ocean (sites 239 and 249) from Gieskes (1974). Parameters for computed profiles given in Table I (from Lerman, 1975).

between 7×10^{-12} and 12×10^{-12} mol/cm$^3 \cdot$ yr. The similarity of the rates suggests (but is not a proof) that dolomite may be forming in the sediments. Perhaps significantly, determinations of the pH, Ca, and alkalinity (water samples squeezed from cores in the laboratory) have shown that the pore waters at site 148 are supersaturated with respect to calcite and dolomite (Hammond, 1973; Takahashi et al., 1973).

2. $J > 0$ for Ca, $J = 0$ for Mg (Fig. 8). The values of J indicate addition of Ca to pore water. In one case, pore waters at site 149 in the central Venezuelan Basin have been reported as undersaturated with respect to calcite (Hammond, 1973). The addition rate of Ca, $J = 1.75 \times 10^{-11}$ mol/cm$^3 \cdot$ yr, derived from the model, is compatible with the measured undersaturation, as well as the abundance of calcite in the sediments (Edgar et al., 1973, p. 101).

3. $J = 0$ for Ca, $J > 0$ for Mg. Profiles of this type are common, and some are shown in Fig. 9. In eight sediment sections, the values of J for Mg are between 3×10^{-11} and 14×10^{-11} mol/cm$^3 \cdot$ yr, although the nature of the Mg-releasing phases are not known.

The kinetic parameters and other relevant data for the Ca and Mg profiles are summarized in Table I.

Another one-layer model describing the occurrence of NH$_4^+$ in pore waters was discussed in Section 5.A, equation 56.

TABLE I

Parameters Used in Computation of Ca and Mg Concentration Profiles in Pore Waters at Four Deep Sea Drilling Project Sites[a]

D.S.D.P. Site No.		h (m)	Concentration (10^{-3} g/cm^3)			U (10^{-3} cm/yr)	k $(10^{-7}/\text{yr})$	J $(10^{-10} \text{ g/cm}^3 \cdot \text{yr})$
			C_1	C_2	$C_{h/2}$			
141	Ca	184	0.44	1.18	0.60	0	7.41	0
						0.3	7.23	
	Mg		1.25	0.80	1.12	0	14.7	18.3
						0.3	14.0	17.5
144	Ca	178	0.42	1.12	0.56	0	8.46	0
						0.3	7.71	
	Mg		1.21	0.80	1.09	0	15.2	18.4
						0.3	14.8	17.9
239	Ca	298	0.42	2.49	0.95	0	4.18	0
						0.8	3.68	
	Mg		1.31	0.62	1.11	0	5.55	7.3
						0.8	4.80	6.3
249	Ca	360	0.41	1.72	0.63	0	3.64	0
						0.8	3.27	
	Mg		1.28	0.77	1.18	0	7.37	9.4
						0.8	6.60	8.4

[a] References to sedimentation rates ($U > 0$) and description of the method for computation of k and J given in Lerman (1975). Diffusion coefficients of Ca and Mg in pore water taken as $D = 3 \times 10^{-6} \text{ cm}^2/\text{sec}$ (equation 60).

B. Two-Layer Model

Diagenetic reactions and migration of dissolved species in pore waters can extend beyond the boundaries of a mineralogically distinct sediment layer, and in such situations the sediment–pore water system may have to be treated as made of two- or three-layers (compare Figs. 10 and 11). Information on the chemical reaction parameters and transport characteristics can be derived from two-layer models in a manner similar to that outlined for a one-layer model in Section 6.A. Because for multilayer systems the algebraic form of the migration-reaction equations becomes much more complex and less convenient for estimation of the kinetic parameters, it is a considerable advantage if a multilayer sediment sequence can be subdivided into several two-layer sections. An example of this approach is shown in Fig. 10 which illustrates a layer containing more soluble SiO_2 phases than the sediments above and below it. The layer with the higher SiO_2 concentration in pore water is a source supplying SiO_2 to the younger and older sediments, where other precipitation and dissolution reactions control its concentrations in pore waters. The depth of the SiO_2 concentration maximum can be taken as a boundary dividing a three-layer system into two two-layer sections, one above and the other below the maximum. This procedure is permissible insofar as there is no diffusional flux across the plane of concentration maximum or minimum, where the concentration gradient is zero.

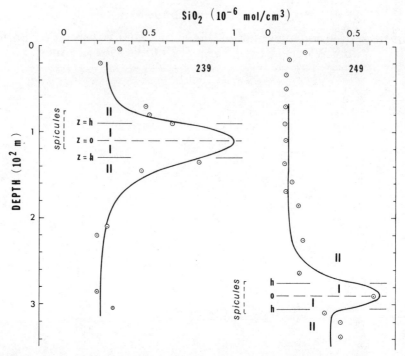

Fig. 10.   Two-layer model of silica in pore waters. Data for sites 239 and 249, Indian Ocean, from Gieskes (1974). Kinetic parameters derived from the model are given in Table II.

(The condition of no flux applies if there is no pore water advection, $U = 0$, across the plane of concentration maximum).

The migration-reaction model for a two-layer system is based on diffusion and first-order dissolution and precipitation reactions at steady state. The equations for the source layer (subscript I) and outer layer (subscript II) are as follows:

$$D_I \frac{\partial^2 C_I}{\partial z^2} - k_I C_I + J_I = 0 \tag{61}$$

$$D_{II} \frac{\partial^2 C_{II}}{\partial z^2} - k_{II} C_{II} + J_{II} = 0 \tag{62}$$

where the vertical distance coordinate z is positive and it increases from $z = 0$ at the depth of the concentration maximum (Fig. 10) to $z = h$ at the boundary of the source layer. Note that this coordinate system applies to migration both up and down from the depth of maximum. The model applies to each of the two-layer sections above and below $z = 0$. Equations 61 and 62 can be solved with the following boundary conditions:

$$z = 0: \quad \frac{dC_I}{dz} = 0 \tag{63}$$

$$z = h: \quad C_I = C_{II} \quad \text{and} \quad D_I\left(\frac{dC_I}{dz}\right) = D_{II}\left(\frac{dC_{II}}{dz}\right) \tag{64}$$

$$z \to \infty: \quad C_{II} = C_{II,s} \equiv \frac{J_{II}}{k_{II}} \tag{65}$$

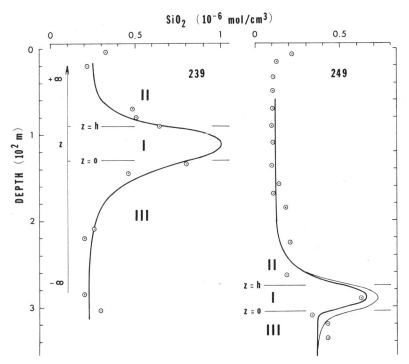

Fig. 11. Three-layer model of silica in pore waters. Same data as in Fig. 10. Parameters used in computation of continuous concentration profiles are given in Table III. At site 249, the thinner line concentration peak corresponds to a slower rate of SiO_2 precipitation in layer III; a faster rate produces a smaller concentration peak and a more uniform concentration in layer III (heavy line).

Boundary condition 63 indicates that there is no flux across the plane designated $z = 0$. Condition 64 states that the concentrations and fluxes in pore water are continuous at the boundary between the two layers. Condition 65 stipulates that in the outer layer concentration tends to a steady-state value $C_{II,s}$. With the above boundary conditions, the solutions of 61 and 62 are

$$C_I = C_{I,s} + \frac{(C_{II,s} - C_{I,s}) \cosh z(k_I/D_I)^{1/2}}{\cosh h(k_I/D_I)^{1/2} + (k_I D_I)^{1/2}(k_{II} D_{II})^{-1/2} \sinh h(k_I/D_I)^{1/2}} \qquad (66)$$

for $0 \leqslant z \leqslant h$, and

$$C_{II} = C_{II,s} + \frac{(C_{II,s} - C_{I,s}) \exp \{(h - z)(k_{II}/D_{II})^{1/2}\} \sinh h(k_I/D_I)^{1/2}}{(k_{II} D_{II})^{1/2}(k_I D_I)^{-1/2} \cosh h(k_I/D_I)^{1/2} + \sinh h(k_I/D_I)^{1/2}} \qquad (67)$$

for $z \geqslant h$.

Equations 66 and 67 define the concentrations in pore water of each layer in terms of the solubilities in each layer ($C_{I,s}$ and $C_{II,s}$), reaction rate constants (k_I and k_{II}), and diffusion coefficients (D_I and D_{II}). If the diffusion coefficients and solubilities are known, the rate constants can be evaluated from the concentration data. The concentration values at the boundary $z = 0$ ($C_{I,z=0}$) and at the boundary between the two layers at $z = h$ ($C_{I,z=h}$) can be substituted consecutively for C_I in equation 66 giving the relationships for k_I and k_{II}:

$$k_I = D_I \frac{(\cosh^{-1} X)^2}{h^2} \qquad (68)$$

TABLE II

Two-Layer Model of SiO_2 in Pore Waters at Sites 239 and 249, Indian Ocean[a]

Site	h (m)	SiO_2 Concentration (10^{-6} mol/cm³)				Rate Constant (10^{-5}/yr)	
		$C_{I,s}$	$C_{II,s}$	$C_{I,z=0}$	$C_{I,z=h}$	k_I	k_{II}
239							
Up from $z = 0$	20	1.5	0.25	1.0	0.7	2.59	5.00
Down from $z = 0$	20	1.5	0.22	1.0	0.8	1.78	1.27
249							
Up from $z = 0$	15	1.5	0.12	0.66	0.50	1.55	3.17
Down from $z = 0$	15	1.5	0.37	0.66	0.45	2.02	125

[a] Rate constants k_I and k_{II} computed using equations 68–70. $D_I = D_{II} = 3 \times 10^{-6}$ cm²/sec taken in the model.

where X is a dimensionless quotient of the three concentration values,

$$X = \frac{C_{I,z=h} - C_{I,s}}{C_{I,z=0} - C_{I,s}} \tag{69}$$

and

$$k_{II} = k_I \left[\frac{(C_{I,z=h} - C_{I,s}) \tanh h(k_I/D_I)^{1/2}}{C_{II,s} - C_{I,z=h}} \right]^2 \tag{70}$$

Note that concentration at the peak ($z = 0$), $C_{I,z=0}$, is lower than the solubility value $C_{I,s}$, owing to diffusion of the species out of the source layer and precipitation.

For the SiO_2 profiles in the pore waters shown in Fig. 10, k_I and k_{II} were computed using the concentration and diffusion coefficient values listed in Table II. The solubility of the silicate phase in the source layer was taken as $C_{I,s} = 1.5$ mmol/l, which is approximately midpoint of the range of amorphous SiO_2 solubilities between 0 and 25°C (cf. Siever, 1962; Hurd, 1973); the temperatures at depths 100–300 m in the sediment can be between 5 and 12°C.

The lower part of hole 239 ($C_{II,s} = 0.2$ mmol/l) and the upper part of hole 249 ($C_{II,s} = 0.12$ mmol/l) are clayey sediments containing montmorillonite, some illite and mica, and palygorskite. Experimental results on dissolution of different clay minerals in seawater at room temperature, reported in Lerman et al. (1975), indicate that muscovite and montmorillonite tend to steady-state solubility values between 0.12 and 0.2 mmol SiO_2/l. The similarity of the computed rate constants for the profiles in holes 239 and 249 should be noted.

C. Three-Layer Model

If information is available on the diffusivities, reaction rate constants, and solubilities of individual phases in three consecutive sediment layers, a model described in this section can be used to compute a continuous concentration profile and to determine the position of the concentration minimum or maximum in the source layer. (In the two-layer model of the preceding section, the source layer was subdivided into two parts, not necessarily of equal thickness, at the plane of the concentration maximum.)

A three-layer section of sediment (Fig. 11) may be characterized by chemical reactions, sediment porosities, and diffusion coefficients that differ from one layer to another. If pore water advection is significant, then the flow rates through each layer would be different if porosity changes from layer to layer. To preserve generality, the three-layer model given below allows for the diffusion coefficients, advection rates, and chemical kinetic parameters that are characteristic of each layer but constant within that layer:

$$D_I \frac{\partial^2 C_I}{\partial z^2} - U_I \frac{\partial C_I}{\partial z} - k_I C_I + J_I = 0 \tag{71}$$

$$D_{II} \frac{\partial^2 C_{II}}{\partial z^2} - U_{II} \frac{\partial C_{II}}{\partial z} - k_{II} C_{II} + J_{II} = 0 \tag{72}$$

$$D_{III} \frac{\partial^2 C_{III}}{\partial z^2} - U_{III} \frac{\partial C_{III}}{\partial z} - k_{III} C_{III} + J_{III} = 0 \tag{73}$$

where subscripts I, II, and III refer to the three layers. The distance coordinate z increases from negative values in the lower layer (III) to positive values in layer I and II. The boundary conditions needed to solve the system of three simultaneous differential equations are as follows:

$$z \to -\infty: \quad C_{III} = C_{III,s} \equiv \frac{J_{III}}{k_{III}} \tag{74}$$

$$z = 0: \quad C_I = C_{III} \quad \text{and} \quad \phi_I\left(D \frac{dC}{dz} - UC\right)_I = \phi_{III}\left(D \frac{dC}{dz} - UC\right)_{III} \tag{75}$$

$$z = h: \quad C_I = C_{II} \quad \text{and} \quad \phi_I\left(D \frac{dC}{dz} - UC\right)_I = \phi_{II}\left(D \frac{dC}{dz} - UC\right)_{II} \tag{76}$$

$$z \to \infty: \quad C_{II} = C_{II,s} \equiv \frac{J_{II}}{k_{II}} \tag{77}$$

The solutions of equations 71–73 are as follows:

$$C_I = \frac{J_I}{k_I} + K_{1,I} \exp(r_{1,I}z) + K_{2,I} \exp(r_{2,I}z) \tag{78}$$

$$C_{II} = \frac{J_{II}}{k_{II}} + K_{2,II} \exp(r_{2,II}z) \tag{79}$$

$$C_{III} = \frac{J_{III}}{k_{III}} + K_{1,III} \exp(r_{1,III}z) \tag{80}$$

The constants K and r in the solutions are defined in terms of the constant coefficients D, U, k, and J, and the solubilities of the reacting phases in the three layers ($C_{I,s}$, $C_{II,s}$, and $C_{III,s}$). The relationships for the four constants K and four r are listed below, with additional notations introduced for brevity.

Reintroducing from equation 60

$$R = +\left(\frac{U^2}{4D^2} + \frac{k}{D}\right)^{1/2} \tag{81}$$

the exponential constants r are given by the following relationships:

$$r_{1,\mathrm{I}} = \left(\frac{U}{2D} + R\right)_{\mathrm{I}} \tag{82}$$

$$r_{2,\mathrm{I}} = \left(\frac{U}{2D} - R\right)_{\mathrm{I}} \tag{83}$$

$$r_{1,\mathrm{III}} = \left(\frac{U}{2D} + R\right)_{\mathrm{III}} \tag{84}$$

$$r_{2,\mathrm{II}} = \left(\frac{U}{2D} - R\right)_{\mathrm{II}} \tag{85}$$

For the constants $\mathbf{K}$, the following intermediate notations will be introduced:

$$A_1 = 1 - \frac{\phi_{\mathrm{I}}(Dr_1 - U)_{\mathrm{I}}}{\phi_{\mathrm{III}}(Dr_1 - U)_{\mathrm{III}}}, \qquad B_1 = 1 - \frac{\phi_{\mathrm{I}}(Dr_2 - U)_{\mathrm{I}}}{\phi_{\mathrm{III}}(Dr_1 - U)_{\mathrm{III}}} \tag{86}$$

$$A_2 = 1 - \frac{\phi_{\mathrm{I}}(Dr_1 - U)_{\mathrm{I}}}{\phi_{\mathrm{II}}(Dr_2 - U)_{\mathrm{II}}}, \qquad B_2 = 1 - \frac{\phi_{\mathrm{I}}(Dr_2 - U)_{\mathrm{I}}}{\phi_{\mathrm{II}}(Dr_2 - U)_{\mathrm{II}}} \tag{87}$$

$$L_{13} = C_{\mathrm{I},s} - C_{\mathrm{III},s} + \frac{(C_s \phi U)_{\mathrm{I}} - (C_s \phi U)_{\mathrm{III}}}{\phi_{\mathrm{III}}(Dr_1 - U)_{\mathrm{III}}} \tag{88}$$

$$L_{12} = C_{\mathrm{I},s} - C_{\mathrm{II},s} + \frac{(C_s \phi U)_{\mathrm{I}} - (C_s \phi U)_{\mathrm{II}}}{\phi_{\mathrm{II}}(Dr_2 - U)_{\mathrm{II}}} \tag{89}$$

Using the notation from equations 81–89, the four constants $\mathbf{K}$ are as follows:

$$\mathbf{K}_{1,\mathrm{I}} = -\frac{L_{13}}{A_1} - \frac{L_{13}/A_1 - L_{12}\exp\left(-r_{1,\mathrm{I}}h\right)/A_2}{A_1 B_2 \exp\left[(r_{2,\mathrm{I}} - r_{1,\mathrm{I}})h\right]/(A_2 B_1) - 1} \tag{90}$$

$$\mathbf{K}_{2,\mathrm{I}} = \frac{L_{13}/A_1 - L_{12}\exp\left(-r_{1,\mathrm{I}}h\right)/A_2}{B_2 \exp\left[(r_{2,\mathrm{I}} - r_{1,\mathrm{I}})h\right] - B_1/A_1} \tag{91}$$

$$\mathbf{K}_{1,\mathrm{III}} = C_{\mathrm{I},s} - C_{\mathrm{III},s} + \mathbf{K}_{1,\mathrm{I}} + \mathbf{K}_{2,\mathrm{I}} \tag{92}$$

$$\mathbf{K}_{2,\mathrm{II}} = [C_{\mathrm{I},s} - C_{\mathrm{II},s} + \mathbf{K}_{1,\mathrm{I}}\exp\left(r_{1,\mathrm{I}}h\right) + \mathbf{K}_{2,\mathrm{I}}\exp\left(r_{2,\mathrm{I}}h\right)]\exp\left(-r_{2,\mathrm{II}}h\right) \tag{93}$$

The arithmetic involved in computation of the eight constants $\mathbf{K}$ and r is much simplified if there is no pore water advection ($U = 0$) and diffusion coefficients in the individual layers are equal ($D_{\mathrm{I}} = D_{\mathrm{II}} = D_{\mathrm{III}}$). For this particular case, model profiles were computed as shown in Fig. 11. The rate constants and solubilities used in computation (listed in Table III) are similar to those derived from the two-layer models, and this accounts for the agreement with the reported SiO_2 data in holes 239 and 249.

The location of a concentration maximum or minimum in the middle layer (I) is controlled by the diffusivity, advection and reaction rates, and the solubilities of the reacting phases in the three layers. The location of the concentration maximum ($z_{\max}$) can be computed by differentiating equation 78 and equating with zero, $dC_{\mathrm{I}}/dz = 0$. Then

$$z_{\max} = \frac{1}{r_{1,\mathrm{I}} - r_{2,\mathrm{I}}} \ln\left(-\frac{\mathbf{K}_{2,\mathrm{I}}r_{2,\mathrm{I}}}{\mathbf{K}_{1,\mathrm{I}}r_{1,\mathrm{I}}}\right) \tag{94}$$

TABLE III

Three-Layer Model of SiO$_2$ in Pore Waters at Sites 239 and 249, Indian Ocean[a]

Site No.	h (m)	SiO$_2$ Concentration (10^{-6} mol/cm^3)			Rate constants (10^{-5}/yr)		
		$C_{I,s}$	$C_{II,s}$	$C_{III,s}$	k_I	k_{II}	k_{III}
239	40	1.5	0.25	0.22	2.18	5.00	1.27
249	30	1.5	0.12	0.37	1.78	3.17	125.0
							12.5

[a] See equations 78–80. k_I is the mean of two estimates for layer I given in Table II. Two values of k_{III} refer to the two SiO$_2$ peaks shown at site 249: 125×10^{-5} for the heavy line, 12.5×10^{-5} for the light line. $D_I = D_{II} = D_{III} = 3 \times 10^{-6}$ cm^2/sec and $U = 0$ taken in the model.

The difference $(r_{1,1} - r_{2,1})$ is always positive, as follows from equations 82 and 83, and the quotient in parentheses must be positive for the logarithm to exist.

7. Fluxes Across the Sediment–Water Interface

Persistence of concentration gradients in pore waters over large sections of the ocean floor is the main indication that sediments contribute to ocean water. Perturbation of the ionic mass balance of seawater by the bacterial reduction of sulfate in sediments may lead to development of very weak concentration gradients of such cations as Na$^+$, K$^+$, and Mg^{2+} near the sediment–water interface with no diffusional fluxes at steady state (Ben-Yaakov, 1972). Such gradients, however, are negligibly small by comparison with the pronounced gradients of NH$_4^+$, Ca^{2+}, and Mg^{2+}, shown in Figs. 5–9, and they do not affect the fluxes computed from the concentration profiles. The fluxes of NH$_4^+$, Ca^{2+}, and Mg^{2+} across the sediment–water interface of deep-ocean sediments will be examined in this section and compared with some other estimates made from other oceanic environments.

A. Fluxes of Ammonia

The flux of NH$_4^+$ across the sediment–water interface is, from equation 1,

$$F = -\phi D \left.\frac{\partial C}{\partial z}\right|_{z=0} \tag{95}$$

The advective term (UC) is not included in equation 95 because the NH$_4^+$ concentration at the sediment–water interface is near zero, $C_{z=0} \simeq 0$. Concentration gradient dC/dz, obtainable from equation 54, gives the following relationship for the flux:

$$F_{z=0} = \frac{-\phi D J_*}{D\beta + U - k/\beta} - \phi D \left(\frac{J_*}{D\beta^2 + \beta U - k} - C_{ss}\right)\left(\frac{U}{2D} - R\right) \tag{96}$$

where $C_{ss} = J/k$ (equation 20) and R was defined under equation 60 and in equation 81. In terms of the distance coordinate z as used in this paper, the flux down from the sediment–water interface is characterized by $F > 0$, whereas the flux from sediments

TABLE IV

Ammonia Fluxes out of Sediments in Different Oceanic Environments
(equation 96)

Locality	NH_4^+ Flux[a] $(mol/cm^2 \cdot yr)$	References
Indian Ocean, site 242	-4.8×10^{-9}	This chapter
Indian Ocean, site 241	-1.8×10^{-8}	This chapter
Santa Barbara Basin	-5.0×10^{-6}	Berner, 1974
Kaneohe Bay, Oahu, Hawaii	-4.9×10^{-6}	Thorstenson and Mackenzie, in preparation, 1976
Somes Sound, Maine	-2.6×10^{-5}	Berner, 1974
Devil's Hole, Harrington Sound, Bermuda	-1.3×10^{-3} (range -3×10^{-4} to -4×10^{-3})	Thorstenson and Mackenzie, 1974
Saanich Inlet fjord, Vancouver Is., British Columbia	-1.8×10^{-3}	Thorstenson and Mackenzie, in preparation, 1976

[a] Negative sign indicates flux upward across the sediment–water interface.

to overlying water is given by $F < 0$. Using the values of the parameters given in Section 5.A, the fluxes of NH_4^+ across the sediment–water interface at sites 241 and 242 in the Indian Ocean were computed and entered in Table IV ($\phi_{z=0} = 0.7$ was taken for the two profiles).

If all the ammonia migrating from pore waters is oxidized in the ocean to nitrite and nitrate, then the NH_4^+ flux provides a measure of the combined NO_2^-–NO_3^- supply to ocean water from sediments. The NH_4^+ fluxes from deep ocean sediments given in Table IV are considerably smaller than the NH_4^+ fluxes reported from shallower water environments. The flux from the Santa Barbara Basin sediments, where the sediment–water interface is approximately at 550 m water depth, is the next value higher than the deep ocean fluxes. The organic matter-rich sediments of Somes Sound, Maine, and Saanich Inlet fjord, British Columbia, are characterized by NH_4^+ fluxes that are 10^4–10^5 times greater than the deep-ocean fluxes. The strong NH_4^+ flux from the recent carbonate sediments in Devil's Hole, Harrington Sound, Bermuda, may be accounted for by the shallowness of the hole (25 m) and presumably rapid burial of organic matter associated with biogenic carbonates. (The range of the NH_4^+ fluxes in Devil's Hole is based on several NH_4^+ concentration profiles measured at different times in the sediments at Devil's Hole, Harrington Sound (Thorstenson and Mackenzie, 1974).)

It should be reminded that the NH_4^+ flux out of the sediments is, in terms of the model used to describe the concentration profiles in Figs. 5 and 6, the net difference between the NH_4^+ production from decomposing organic matter and uptake of NH_4^+ by clays.

B. Fluxes of Calcium and Magnesium

Concentrations of Ca and Mg in pore waters of the uppermost several meters of the sediment column (Figs. 7–9) are within a few percent of mean ocean water concentrations. Therefore, the fluxes at the upper boundaries of the concentration

TABLE V

Calcium and Magnesium Fluxes in Pore Waters
Near the Sediment–Water Interface of Some
Deep-Ocean Sediments[a]

	Flux (10^{-6} g/cm$^2 \cdot$ yr)			
	Site 141	Site 144	Site 239	Site 249
Ca				
$U > 0$	0.038	0.045	-0.95	0.36
$U = 0$	-0.039	0.12	-1.16	0.19
Mg				
$U > 0$	1.02	0.98	1.43	0.96
$U = 0$	0.76	0.73	0.73	0.25

[a] Flux values computed taking into account pore water advection ($U > 0$) and neglecting it ($U = 0$); equations 97 and 1, with data from Table I (negative sign indicates flux upward at $z = 0$; positive, flux downward into the sediment).

profiles ($z = 0$ of the model) can be considered as representative of the sediment–water interface region, provided no other sources or sinks of the chemical species exist between the sediment–water interface and the point of concentration C_1 ($z = 0$) used in the model.

Concentration gradients of Mg (Figs. 7–9) near the sediment–water interface are negative, indicating that the fluxes are in the downward direction. Typical values are of the order of 10^{-6} g/cm$^2 \cdot$ yr, as computed from equations 97 and 1 and summarized in Table V. It has been estimated (Drever, 1974) that the Mg flux of the order of 2.5×10^{-5} g/cm$^2 \cdot$ yr over the entire ocean floor would match the river input of Mg to the ocean. As far as the fluxes in deep-ocean sediments may be considered representative of the entire ocean floor, the values computed from the data for the Atlantic and Indian Ocean sediments (Table V) account for less than 10% of the river input to the ocean.

The flux equation 1 applies to migration of Ca and Mg in pore waters. Although the Ca concentration gradients shown in Figs. 7 and 9 are positive, this does not necessarily indicate that there is flux of Ca *out* of the sediment: the algebraic sign and direction of the flux depends on the relative magnitudes of the diffusional and advective fluxes in the difference $[D(dC/dz) - UC]$ in equation 1. The diffusional flux term is obtainable from equation 60 and it applies to any combination of production and advection rate values ($J = 0$, $J \neq 0$, $U = 0$, or $U \neq 0$):

$$D\frac{\partial C}{\partial z}\bigg|_{z=0} = \frac{D(C_1 - J/k)[(U/2D)\sinh Rh - R\cosh Rh] + DR(C_2 - J/k)e^{-hU/2D}}{\sinh Rh}$$

(97)

where R and other parameters are as defined under equation 60. The diffusional fluxes computed from equation 97 for the Ca profiles at deep-ocean sites 141, 144, and 239 (Fig. 9) are in the range

$$D\left(\frac{\partial C}{\partial z}\right)_{z=0} = 6 \times 10^{-8} \text{ to } 2 \times 10^{-6} \text{ g/cm}^2 \cdot \text{yr}$$

The advective flux term $(-UC)_{z=0}$ for the same profiles $(C_{z=0} \simeq 4 \times 10^{-4} \text{ g/cm}^3)$ is characterized by values in the range

$$\begin{aligned}(-UC)_{z=0} &= -(2 \times 10^{-4} \text{ to } 1 \times 10^{-3} \text{ cm/yr}) \times 4 \times 10^{-4} \text{ g/cm}^3 \\ &= -8 \times 10^{-8} \text{ to } -4 \times 10^{-7} \text{ g/cm}^2 \cdot \text{yr}\end{aligned}$$

The fivefold variation in U used in the above computation includes the effects of sediment compaction on the pore water advection rates (see equation 15). The values of the diffusional and advective flux terms computed above show that the algebraic sum of the two can give either a positive or negative term that would indicate, respectively, the flux $F = -\phi[D(\partial C/\partial z) - UC]$ upward or downward across the plane $z = 0$. Thus the pore water advection rate U is important to the determination of Ca fluxes in pore waters and it may result in the fluxes of direction opposite to that suggested by the concentration gradients. Note that the values of U of the order of 10^{-4} cm/yr have virtually no effect (Table I) on the estimates of the reaction rate constant k in the steady-state model of equation 60, but they are, as shown above, important to the flux computations.

The fluxes of Ca at the upper boundary of the concentration profiles at deep ocean sites 141, 144, 239, and 249 were calculated for the case of no pore water advection $(U = 0)$ and the case of pore water advection at $z = 0$ taken equal to the mean sedimentation rate $(U \simeq U_s)$. The results for the two cases are given in Table V and they show the importance of the advective contribution to the flux direction $(\phi_{z=0} = 0.7$ was taken for all the profiles).

8. Cementation and Fluxes in Sediments

Cementation of oceanic sediments resulting in a reduction of porosity from about 50% to a lower value and achieved from in situ sources requires an equivalent reduction in volume within the sediment column, at the expense of the dissolving source. Because cementation of relatively thick sedimentary units by precipitated carbonates and silica is commonly not associated with an evidence of extensive dissolution of adjacent sediments, the sources of cement in many cases are presumably located outside the area of a lithified section. For the early stages of lithification of oceanic sediments owing to dissolution and precipitation reactions and migration of dissolved species, limiting estimates of the time scales involved are discussed with reference to the removal of Ca and SiO_2 from pore waters, as well as the diagenetic reactions evidenced by the SiO_2 migration from source layers (Figs. 7–10).

A. Calcium

The removal of Ca from pore water, interpreted as precipitation of a Ca-containing phase, can be used to estimate the amount of solids formed during the lifetime of the 200–300 m thickness of oceanic sediments, taken as between 30×10^6 and 80×10^6 yr. The rate of Ca removal is given by the term $-kC$ in equation 21, and this can be viewed as an upper-limit estimate of the precipitation rate: the rates given by the relationship $-k(C - C_s)$ would be lower. In the computation it will be assumed that

Ca and other dissolved species characterized by similar removal rates (Lerman, 1975) form solids of mean gram-formula weight of 100 g/mol. Then the volume of solids formed in 1 cm^3 of sediment during time period Δt is

$$kC\phi\Delta t \times \frac{100}{\rho} = \frac{(5 \times 10^{-7}/\text{yr})(1 \times 10^{-3}/40 \text{ mol/cm}^3)(0.5)(5 \times 10^7 \text{ yr})(100 \text{ g/mol})}{2.5 \text{ g/cm}^3}$$

$$= 1.25 \times 10^{-2} \text{ cm}^3 \text{ of solids in 1 cm}^3 \text{ of sediment}$$

This amount is only about 2.5% of the volume of solids originally present in the sediment of 50% porosity. The values of the individual parameters used in the preceding computation can vary within a factor of 2 as shown, for example, by the Ca concentrations in pore waters in Figs. 7 and 9, where concentrations between 0.5×10^{-3} and 1.5×10^{-3} g/cm^3 are common. As a whole, allowing for the higher values of concentration and time of precipitation, reduction of porosity by 10%, such as from 50 to 40%, may be considered an upper limit when vertical transport alone is responsible for redistribution of dissolved species.

B. Silica Diagenesis and Chert

Petrographic evidence of the formation of chert in oceanic sediments indicates that amorphous silica (opal) transforms first to disordered crystobalite and tridymite microcrystalline aggregates, and finally to microcrystalline quartz (Heath and Moberly, 1971; von Rad and Rösch, 1972; Berger and von Rad, 1972). There are apparently no chert layers younger than 25×10^6 yr, and the more common occurrences are in sediments of age $> 70 \times 10^6$ yr. The evidence for the formation of chert either by replacement of opal in the solid state or by dissolution and precipitation is not conclusive (Heath and Moberly, 1971), although pore water data alone cannot resolve this question.

The rates of dissolution and precipitation of SiO$_2$ in deep-ocean sediments, derived in Section 6.B, can be used to estimate the rates of chert formation. For a sediment layer containing silicate phases of higher solubility, the material mass balance of SiO$_2$ within the layer can be formulated as

$$\text{Mass precipitated} = \text{mass dissolved} - \text{fluxes out} \tag{98}$$

A mathematical formulation of equation 98, with the notation of Fig. 11 for layer I, is

$$\Delta t \int_0^h \phi k_1 C_1 \, dz = \phi J_1 h \Delta t - (F\uparrow + F\downarrow)\Delta t \tag{99}$$

where Δt is the length of time the steady-state processes in the layer have been going on, and the arrows denote the fluxes across the upper and lower boundaries, respectively. (In the three-layer model the layer boundaries are at $z = h$ and $z = 0$, whereas in the two-layer model the boundaries are alternatively at $z = h$.)

Because the rate of SiO$_2$ dissolution in the two- and three-layer models is taken as constant ($J_1 = k_1 C_{1,s}$), the rate of precipitation is given by the term $-k_1 C_1$ which is an acceptable approximation as long as the concentrations in pore water (C_1) are far from an equilibrium with the precipitation phase. If the solubility of the latter is similar to the solubility of quartz, ≤ 0.1 μmol SiO$_2$/cm^3, the approximation does not introduce gross errors for such SiO$_2$ concentrations as those in the source layers of sites 239 and 249, shown in Fig. 10.

In terms of the two-layer model (Fig. 10), the flux across each boundary of the layer is obtainable from equation 66:

$$F_{z=h} = -\phi D_{\rm I}\left(\frac{dC_{\rm I}}{dz}\right)_{z=h} = \frac{\phi(C_{\rm I,s} - C_{\rm II,s})}{(k_{\rm I}D_{\rm I})^{-1/2}\coth h(k_{\rm I}/D_{\rm I})^{1/2} + (k_{\rm II}D_{\rm II})^{-1/2}} \tag{100}$$

To use equation 100 for computation of the flux at the upper bounbary, the values of $k_{\rm II}$ and $D_{\rm II}$ should be those given for the upper sediment layer (Table III, Fig. 11). For the flux at the lower boundary, the values of $k_{\rm III}$ and $D_{\rm III}$ $(=D_{\rm I})$ should be used from Table III for $k_{\rm II}$ and $D_{\rm II}$. By this procedure, the total fluxes $F = F + F$ out of the SiO_2-rich layers, taking for the sediment porosity $\phi = 0.5$, are

Site 239: $F = 2.53 \times 10^{-8}$ mol/cm² · yr
Site 249: $F = 2.20 \times 10^{-8}$ mol/cm² · yr

The rates of dissolution of SiO_2 solids, given by the term $\phi J_{\rm I}h$ in equation 99, are (from Table III)

Site 239: $\phi J_{\rm I}h = 6.56 \times 10^{-8}$ mol/cm² · yr
Site 249: $\phi J_{\rm I}h = 4.02 \times 10^{-8}$ mol/cm² · yr

Thus the steady-state fluxes from the SiO_2-rich layers remove between 40 and 50% of dissolving SiO_2, while the remainder precipitates within the layers. The volume of precipitated solid is a measure of replacement of the more soluble original material, such as biogenic opal or volcanic glass. A mean volume $(\overline{V}_p)$ of SiO_2 precipitated in 1 cm³ of bulk sediment within the source layer is, from equation 99,

$$\overline{V}_p = \frac{\Delta t}{\rho h}\int_0^h \phi k_{\rm I}C_{\rm I}\, dz \tag{101}$$

Using the fluxes and dissolution rates given earlier in this section, and the times $\Delta t = 30 \times 10^6$ yr for the age of the source layer at site 239 and $\Delta t = 80 \times 10^6$ yr for the source layer at site 249 (Middle Miocene and Middle Cretaceous ages, respectively, cited by Gieskes, 1974), the volumes of SiO_2 precipitated within the source layers are, from equations 99 and 101,

Site 239: $\overline{V}_p = \dfrac{(6.56 - 2.53) \times 10^{-8} \times 60 \times 3 \times 10^7}{2.5 \times 4 \times 10^3}$

= 0.007 cm³ SiO_2/cm³ sediment

Site 249: $\overline{V}_p = \dfrac{(4.02 - 2.20) \times 10^{-8} \times 60 \times 8 \times 10^7}{2.5 \times 3 \times 10^3}$

= 0.012 cm³ SiO_2/cm³ sediment

Precipitation of SiO_2 as derived from the model for the sediments at sites 239 and 249 replaces only small fractions—on the average, 1–2%—of the solid phases within the source layers. However, the fluxes of SiO_2 out of the two source layers as given in this section, can result in silicification of thin—20–40 cm—sediment layers above and below the SiO_2 source layer. For this to be accomplished in times ranging from 30×10^6 to 80×10^6 yr, SiO_2 carried by the fluxes across the upper and lower boundaries of the source layer (Fig. 10) must precipitate within 20- to 40-cm-thick layers, reducing in the process their porosity from 50% to almost 0. Cherts in oceanic sediments commonly occur as thin (centimeters) layers, and the preceding computation indicates that such layers can form by precipitation of SiO_2 transported by diffusional fluxes from outside, within time periods shorter than 30×10^6 yr.

The slowness of replacement of amorphous SiO_2 by a less soluble phase within the SiO_2-source layers as estimated for sites 239 and 249 requires an explanation. The computed rate of precipitation depends on the rate constant k_I. The latter, derived from equations 69 and 68, strongly depends on the thickness of the source layer (h) owing to the term h^2 in equation 68. (A reaction rate constant does not depend on the physical dimensions of the system, but the shape of a concentration against depth profile does. If sediment layers of different thickness contain SiO_2 concentration peaks of the same magnitude, this suggests that different solid phases are responsible for the SiO_2 concentrations in pore waters.) The thicknesses of the source layers at sites 239 and 249 (Fig. 10) were estimated visually as 40 and 30 m. Such thicknesses cannot be considered a rule in oceanic sediments, and much thinner layers of biogenic and volcanogenic SiO_2 phases can occur. If the SiO_2 concentration peak such as the one at site 249 (Fig. 10) occurred within a layer 2 m thick, the rate constant k_I (Table III) would have been $k_I = 1.8 \times 10^{-5} \times (30/2)^2 = 4.05 \times 10^{-3}$/yr. This value approaches the rate constants of 10^{-2} to 10^0/yr characteristic of SiO_2 diagenesis in sediments on the continental and insular shelves and near the sediment–water interface (Lerman, 1975). The higher the precipitation rate within the source layer, the lower are the diffusional fluxes of SiO_2 out of the layer (see equation 100 for the dependence of F on small h and large k_I). Using the higher value of k_I computed above, the mean volume of SiO_2 precipitated within a 2-m-thick source layer in 20×10^6 yr, is approximately

$$\overline{V}_p \simeq \frac{0.5 \times 4.05 \times 10^{-3} \times 5 \times 10^{-7} \times 60 \times 2 \times 10^7}{2.5}$$

$$= 0.49 \text{ cm}^3 \text{ } SiO_2/\text{cm}^3 \text{ sediment}$$

Thus nearly the entire volume of solid phases initially present in the sediment (0.5 cm³) can be replaced by dissolution and precipitation processes. This order of magnitude calculation illustrates in principle that the rates of formation of chert are compatible with the dissolution and precipitation rates of SiO_2 in oceanic sediments, derivable from steady-state models.

To complete this section, one refinement to the model of silica diagenesis will be introduced. The concentration value of $C_{I,s}$ in equation 66 was taken as the solubility of amorphous silica and interpreted, according to the model, as $C_{I,s} = J_I/k_I = 1.5 \times 10^{-6}$ mol/cm³. If, however, $C_{I,s}$ is interpreted not as the amorphous silica (opal) solubility but as some lower concentration characteristic of a closed system where opal dissolves and another SiO_2 phase precipitates at the same time, then higher values of the precipitation rate constant can be derived from the model.

Using the definitions of J and k given in equation 19, the following relationships apply to dissolution and precipitation of silica in layer I (Figs. 10 and 11):

$$J_I \equiv k_I C_{I,s} = k_1 C_{s,1} + k_2 C_{s,2} \tag{102}$$

$$k_I = k_1 + k_2 \tag{103}$$

where $C_{s,1}$ is the solubility of the dissolving phase, $C_{s,2}$ is the solubility of the pre-cipitating phase, and k_1 and k_2 are the corresponding first-order reaction rate constants. From the two equations, the precipitation rate constant k_2 is given by

$$k_2 = \frac{k_I(C_{I,s} - C_{s,1})}{C_{s,2} - C_{s,1}} \tag{104}$$

For k_2 to be positive, the difference terms in equation 104 must be of the same sign. The difference $(C_{s,2} - C_{s,1}) < 0$ because the solubility of the precipitating phase is lower than that of the dissolving phase. Also, the steady-state concentration $C_{I,s}$ characteristic of a system where dissolution and precipitation take place simulta-neously cannot be greater than the equilibrium solubility of the more soluble phase or, in other words, $(C_{I,s} - C_{s,1}) < 0$.

If $C_{I,s}$ is known then k_1 can be determined from the model; also, if the solubilities of the dissolving and precipitating phases, $C_{s,1}$ and $C_{s,2}$, are known, the precipitation rate constant sought (k_2) can be computed from equation 104. With reference to the more complete SiO_2 concentration profile at the Indian Ocean site 249 (Fig. 10), a sample computation is given as follows.

Assuming the value of $C_{I,s} = 0.7 \times 10^{-6}$ mol/cm^3 and taking the values of other parameters for SiO_2 at site 249 given in Table II, k_1 from equation 68 is

$$k_I = 2.21 \times 10^{-4}/\text{yr}$$

which is one order of magnitude higher than the value based on $C_{I,s} = 1.5 \times 10^{-6}$ mol/cm^3 and used previously. Although there is no evidence for the value of $C_{I,s} = 0.7 \times 10^{-6}$ the closer it is to the concentration peak, the higher is the computed value of the parameter k_1 (see equation 69). Taking for the solubility of opal $C_{s,1} = 1.5 \times 10^{-6}$ mol/cm^3 and for the solubility of the precipitating phase $C_{s,2} \simeq 0.1 \times 10^{-6}$ mol/cm^3 (approximating the solubility of quartz in pore waters), the precipitation rate constant is

$$k_2 = \frac{2.21 \times 10^{-4}(0.7-1.5) \times 10^{-6}}{(0.1-1.5) \times 10^{-6}} = 1.26 \times 10^{-4}/\text{yr}$$

The difference between the solubilities of quartz and opal is extreme, and a higher solubility $C_{s,2}$ of some other precipitating phase would give a higher value of the rate constant k_2, by a factor of less than 10.

According to the dissolution and precipitation model, with opal and quartz as the starting and end members, the volume of SiO_2 precipitated during a period of 80×10^6 yr, is

$$\frac{\phi k_2(C - C_{s,2})\Delta t}{\rho} = \frac{0.5 \times 1.3 \times 10^{-4}(0.6-0.1) \times 10^{-6} \times 60 \times 8 \times 10^7}{2.5}$$

$$= 0.06 \text{ cm}^3 \text{ SiO}_2/\text{cm}^3 \text{ sediment} \tag{105}$$

This estimate indicates replacement of the order of 10% of opal in the source layer. Much more extensive replacement can take place in thinner layers, as was explained earlier in this section with reference to the source layer thickness at site 249 and its effect on the estimates of k_1 using a steady-state model.

9. Summary and Conclusions

Migration and reactions of chemical species in sediments can be modeled in terms of two fluxes, the flux on deposited particles and flux in interstitial water. The models of behavior of chemical species in sediments are based on the processes of (1) chemical interactions between solids and pore waters, (2) diffusional transport in pore water and on solid surfaces, and (3) advective transport by the moving sediment particles and pore waters. Analysis of the relative importance of the individual processes in sediments shows that temperature and porosity gradients have only small effects on the distribution of chemical species in interstitial waters; advective transport can be of some importance when sedimentation rates are higher than about 3 cm/1000 yr; molecular diffusion in interstitial water and the chemical reactions of dissolution, precipitation, and adsorption are the main processes controlling such species as calcium, magnesium, silica, and ammonia.

Steady-state models can be used to analyze the behavior of chemical species in oceanic sediments if the time scales of the migrational and chemical reaction processes are short by comparison with the rates of change in the thickness of the sediment column owing to its growth by deposition. Steady-state is an acceptable approximation to distributions of many reactive species within the upper 10^2 m of the sediment column.

The fluxes in interstitial waters near and across the sediment–water interface were derived from the migration-reaction models and computed for Mg, Ca, and NH_4^+ at a number of deep-ocean sites. Within the upper few meters of the sediment column, the fluxes of Mg are downward. The computed fluxes are lower than the flux of Mg from river input to the ocean and this suggests, within the limits of available data, that higher fluxes should exist elsewhere on the ocean floor to balance the Mg input.

The direction of the Ca fluxes near the sediment–water interface varies from up to down among the several concentration profiles analyzed. Advection of pore water in the sediment–water interface region has a strong effect on the direction of the Ca flux.

For NH_4^+, the fluxes across the sediment–water interface were computed for two sites in the western Indian Ocean. The fluxes are 10^5–10^6 times lower than the values characteristic of shallower water sediments, richer in organic matter. The variation of the NH_4^+ flux from the ocean floor with location and depth bears on the estimates of nitrogen return from sediments to ocean water.

Sediment layers containing more soluble silicate phases can act as sources of SiO_2 that migrates to the younger and older sediments. For such systems, characterized by SiO_2 concentration peaks in pore water, the chemical reaction rate parameters can be derived from two- and three-layer models based on diffusional (and advective) transport in pore water and dissolution and precipitation reactions. The rates of SiO_2 precipitation reactions derived for near-surface sediments are in general higher than the reaction rates computed for the deeper sediments, containing mixtures of biogenic and terrigenous silicates. However, the SiO_2 concentration data for deep-ocean sediments are often represented by widely spaced samples, and application of the model to thick layers of sediment leads to average values of reaction rates that are characteristic of mineralogically heterogeneous units. A finer detail can give a better picture of the thickness of SiO_2 source layers within which the rates of dissolution and precipitation can be comparable to those of the biogenic and pyroclastic sediments sampled closer to the sediment–water interface.

The fluxes of SiO_2 out of source layers, of the order of 10^{-8} mol $SiO_2/cm^2 \cdot yr$, are sufficient to precipitate SiO_2 and form chert within tens of centimeter-thick layers on the time scale of 30×10^6 yr. Another mode of chert formation by replacement of opal within a sediment layer requires, according to the model, the rates of dissolution and precipitation comparable to those of the sediment–water interface region. An overall picture of the formation of chert by replacement of opal and precipitation of less soluble SiO_2 phases can be accounted for by steady-state models based on silicate reaction rates derived for pore waters and time scales of the order of 10^7 yr.

References

Anikouchine, W. A., 1967. Dissolved chemical substances in compacting marine sediments. *J. Geophys. Res.*, **72**, 505–509.

Aston, S. R. and E. K. Duursma, 1973. Concentration effects on ^{137}Cs, ^{65}Zn, ^{60}Co, and ^{106}Ru sorption by marine sediments, with geochemical implications. *Neth. J. Sea Res.*, **6**, 225–240.

Athy, L. F., 1930. Density, porosity and compaction of sedimentary rocks. *Am. Assoc. Petrol. Geol. Bull.*, **14**, 1–24.

Barrer, R. M., 1968. Diffusion and permeation in heterogeneous media. In *Diffusion in Polymers*, J. Crank and G. S. Park, eds. Academic Press, New York, pp. 165–217.

Ben-Yaakov, S., 1972. Diffusion of sea water ions.—I. Diffusion of sea water into a dilute solution. *Geochim. Cosmochim. Acta*, **36**, 1395–1406.

Berger, W. H., 1967. Foraminiferal ooze: solution at depths. *Science*, **156**, 383–385.

Berger, W. H. and U. von Rad, 1972. Cretaceous and Cenozoic sediments from the Atlantic Ocean. In *Initial Reports of the Deep Sea Drilling Project*. Vol. XIV. U.S. Government Printing Office, Washington, D.C., pp. 787–954.

Berner, R. A., 1971. *Principles of Chemical Sedimentology*. McGraw-Hill, New York, 240 pp.

Berner, R. A., 1974. Kinetic models for the early diagenesis of nitrogen, sulfur, phosphorus, and silicon in anoxic marine sediments. In *The Sea*. Vol. 5. E. D. Goldberg, ed. Wiley, New York, pp. 427–450.

Berner, R. A., 1975. Diagenetic models of dissolved species in the interstitial waters of compacting sediments. *Am. J. Sci.*, **275**, 88–96.

Crank, J., 1956. *The Mathematics of Diffusion*. Oxford. 347 pp.

Drever, J. I., 1971. Magnesium-ion replacement in clay minerals in anoxic marine sediments. *Science*, **172**, 1334–1336.

Drever, J. I., 1974. The magnesium problem. In *The Sea*. Vol. 5, E. D. Goldberg, ed. Wiley, New York, pp. 337–357.

Duursma, E. K. and D. Eisma, 1973. Theoretical, experimental and field studies concerning reactions of radioisotopes with sediments and suspended particles of the sea. Part C: Applications to field studies. *Neth. J. Sea Res.*, **6**, 265–324.

Edgar, N. T., J. B. Saunders, et al., 1973. Site 146/149. In *Initial Reports of the Deep Sea Drilling Project*. Vol. XV. U.S. Government Printing Office, Washington, D.C., pp. 17–168

Emery, K. O. and S. C. Rittenberg, 1952. Early diagenesis of California basin sediments in relation to origin of oil. *Am. Assoc. Petrol. Geol. Bull.*, **36**, 735–806.

Ellis, J. H., R. I. Barnhisel, and R. E. Phillips, 1970. The diffusion of copper, manganese, and zinc as affected by concentration, clay mineralogy, and associated anions. *Soil Sci. Soc. Am. Proc.*, **34**, 866–870.

Fanning, K. A. and M. E. Q. Pilson, 1974. The diffusion of dissolved silica out of deep-sea sediments. *J. Geophys. Res.*, **79**, 1293–1297.

Garrels, R. M. and F. T. Mackenzie, 1971. *Evolution of Sedimentary Rocks*. Norton, New York, 397 pp.

Gieskes, J. M., 1973. Interstitial water studies, Leg 15—alkalinity, pH, Mg, Ca, Si, PO_4, and NH_4. In *Initial Reports of the Deep Sea Drilling Project*. Vol. XX. U.S. Government Printing Office, Washington, D.C., pp. 813–829.

Gieskes, J. M., 1974. Interstitial water studies, Leg 25. In *Initial Reports of the Deep Sea Drilling Project*. Vol. XXV. U.S. Government Printing Office, Washington, D.C., pp. 361–394.

Goldberg, E. D. and M. Koide, 1963. Rates of sediment accumulation in the Indian Ocean. In *Earth Science and Meteoritics*. J. Geiss and E. D. Goldberg, eds. North-Holland, Amsterdam, pp. 90–102.

Goldhaber, M. B. and I. R. Kaplan, 1974. The sulfur cycle. In *The Sea*. Vol. 5. E. D. Goldberg, ed. Wiley, New York, pp. 569–655.

Hamilton, E. L., 1959. Thickness and consolidation of deep-sea sediments. *Geol. Soc. Am. Bull.*, **70**, 1399–1424.

Hamilton, E. L., 1971. Elastic properties of marine sediments. *J. Geophys. Res.*, **76**, 579–604.

Hammond, D. C., 1973. Interstitial water studies, Leg 15, a comparison of the major element and carbonate chemistry data from Sites 147, 148, and 149. In *Initial Reports of the Deep Sea Drilling Project*. Vol. XX. U.S. Government Printing Office, Washington, D.C., pp. 831–850.

Hayes, D. E., A. C. Pimm, and others, 1972. Site 141. In *Initial Reports of the Deep Sea Drilling Project*. Vol. XIV. U.S. Government Printing Office, Washington, D.C., pp. 217–247.

Heath, G. R. and R. Moberly, Jr., 1971. Cherts from the Western Pacific, Leg 7, Deep Sea Drilling Project. In *Initial Reports of the Deep Sea Drilling Project*. Vol. VII. U.S. Government Printing Office, Washington, D.C., pp. 991–1007.

Helfferich, F., 1962. *Ion Exchange*. McGraw-Hill, New York, 624 pp.

Hurd, D. C., 1973. Interactions of biogenic opal, sediment and seawater in the Central Equatorial Pacific. *Geochim. Cocmochim. Acta*, **37**, 2257–2282.

Jost, W. and H. J. Oel, 1964. Diffusion. *Encyclopedia Britannica*, **7**, 421–427.

Keller, G. H. and R. H. Bennett, 1973. Sediment mass physical properties—Panama Basin and Northeastern Equatorial Pacific. In *Initial Reports of the Deep Sea Drilling Project*. Vol. XVI. U.S. Government Printing Office, Washington, D.C., pp. 499–512.

Klinkenberg, L. J., 1951. Analogy between diffusion and electrical conductivity in porous rocks. *Geol. Soc. Am. Bull.*, **62**, 559–563.

Lerman, A., 1971. Time to chemical steady-states in lakes and ocean. *Adv. Chem. Ser.*, **106**, 30–76.

Lerman, A., 1975. Maintenance of steady state in oceanic sediments. *Am. J. Sci.*, **275** (in press).

Lerman, A., D. Lal, and M. F. Dacey, 1974. Stokes settling and chemical reactivity of suspended particles in natural waters. In *Suspended Solids in Water*. R. J. Gibbs, ed. Plenum Press, New York, pp. 17–47.

Lerman, A. and T. A. Lietzke, 1975. Uptake and migration of tracers in lake sediments. *Limnol. Oceanogr.*, **20**, 497–510.

Lerman, A., F. T. Mackenzie, and O. P. Bricker, 1975. Rates of dissolution of aluminosilicates in sea water. *Earth Planet. Sci. Lett.*, **25**, 82–88.

Lerman, A., F. T. Mackenzie, and L. N. Plummer, 1973. Mineral dissolution and precipitation: S-shaped kinetics. *Am. Geophys. Union Trans.*, **54**, 341.

Lerman, A. and H. Taniguchi, 1972. Strontium-90—diffusional transport in sediments of the Great Lakes. *J. Geophys. Res.*, **77**, 474–481.

Li, Y. H., J. L. Bischoff, and G. Mathieu, 1969. The migration of manganese in the Arctic Basin sediment. *Earth Planet. Sci. Lett.* **7**, 265–270.

Li, Y. H. and S. Gregory, 1974. Diffusion of ions in sea water and in deep-sea sediments. *Geochim. Cosmochim. Acta*, **38**, 703–714.

Mackenzie, F. T. and R. M. Garrels, 1965. Silicates: reactivity with sea water. *Science*, **150**, 57.

Mangelsdorf, P. C., Jr., T. R. S. Wilson, and E. Daniell, 1969. Potassium enrichments in interstitial waters of recent marine sediments. *Science*, **165**, 171–172.

Manheim, F. T., 1970. The diffusion of ions in unconsolidated sediments. *Earth Planet. Sci. Lett.*, **9**, 307–309.

Manheim, F. T. and F. L. Sayles, 1974. Composition and origin of interstitial waters of marine sediments, based on deep sea drill cores. In *The Sea*. Vol. 5, E. D. Goldberg, ed. Wiley, New York, pp. 527–568.

Manheim, F. T. and L. S. Waterman, 1974. Diffusimetry (diffusion constant estimation) on sediment cores by resistivity probe. In *Initial Reports of the Deep Sea Drilling Project*. Vol. XXII. U.S. Government Printing Office, Washington, D.C., pp. 663–670.

Meares, P., 1968. Transport in ion-exchange polymers. In *Diffusion in Polymers*, J. Crank and G. S. Park, eds. Academic Press, New York, pp. 373–428.

Ristvet, B. L., F. T. Mackenzie, D. C. Thorstenson, and R. H. Leeper, 1973. Pore-water chemistry and early diagenesis of nearshore marine sediments. *Am. Assoc. Petrol. Geol. Bull.*, **57**, 801–802.

Sayles, F. L., F. T. Manheim, and L. S. Waterman, 1973. Interstitial water studies on small core samples, Leg 15. In *Initial Reports of the Deep Sea Drilling Project*. Vol. XX. U.S. Government Printing Office, Washington, D.C., pp. 783–804.

Schink, D. R., N. L. Guinasso, Jr., and K. A. Fanning, 1975. Processes affecting the concentration of silica at the sediment-water interface of the Atlantic Ocean, *J. Geophys. Res.*, **80**, 3013–3031.

Sholkovitz, E., 1973. Interstitial water chemistry of the Santa Barbara Basin sediments. *Geochim. Cosmochim. Acta*, **37**, 2043–2073.

Siever, R., 1962. Silica solubility, 0°–200°C, and the diagenesis of siliceous sediments. *J. Geol.*, **70**, 127–150.

Stein, J., 1972. *Computer Program BSSODE*. Northwestern University, Vogelback Computing Center, Program Library No. NUCC227, Evanston, Ill., 6 pp., mimeogr.

Stevenson, F. J. and C. N. Cheng, 1972. Organic geochemistry of the Argentine Basin sediments: carbon-nitrogen relationships and Quaternary correlations. *Geochim. Cosmochim. Acta*, **36**, 653–671.

Stöber, W., 1967. Formation of silicic acid in aqueous suspensions of different silica modifications. *Adv. Chem. Ser.*, **67**, 161–182.

Takahashi, T., L. A. Prince, and L. J. Felice, 1973. Interstitial water studies, Leg 15, dissolved carbon dioxide concentrations. In *Initial Reports of the Deep Sea Drilling Project*. Vol. XX. U.S. Government Printing Office, Washington, D.C., pp. 885–876.

Thorstenson, D. C. and F. T. Mackenzie, 1974. Time variability of pore water chemistry in recent carbonate sediments, Devil's Hole, Harrington Sound, Bermuda. *Geochim. Cosmochim. Acta*, **38**, 1–19.

Tzur, Y., 1971. Interstitial diffusion and advection of solute in accumulating sediments. *J. Geophys. Res.*, **76**, 4208–4211.

van Olphen, H., 1957. Surface conductance of various ion forms of bentonite in water and the electrical double layer. *J. Phys. Chem.*, **61**, 1276–1280.

van Olphen, H., 1963. *Clay Colloid Chemistry*. Wiley-Interscience, New York, 301 pp.

van Schaik, J. C. and W. K. Kemper, 1966. Chloride diffusion in clay-water systems. *Soil Sci. Soc. Am. Proc.*, **30**, 22–25.

von Huene, R. and D. J. W. Piper, 1973. Measurements of porosity in sediments of the lower continental margin, deep-sea fans, the Aleutian Trench, and Alaskan Abyssal Plain. In *Initial Reports of the Deep Sea Drilling Project*. Vol. XVIII. U.S. Government Printing Office, Washington, D.C., pp. 889–895.

von Rad, U. and H. Rösch, 1972. Mineralogy and origin of clay minerals, silica and authigenic silicates in Leg 14 sediments. In *Initial Reports of the Deep Sea Drilling Project*. Vol. XIV. U.S. Government Printing Office, Washington, D.C., pp. 727–751.

Waterman, L. S., F. L. Sayles, and F. T. Manheim, 1972. Interstitial water studies on small core samples, Leg 14. In *Initial Reports of the Deep Sea Drilling Project*. Vol. XIV. U.S. Government Printing Office, Washington, D.C., pp. 753–762.

Youngquist, G. R., 1970. Diffusion and flow of gases in porous solids. In *Flow Through Porous Media*, American Chemical Society, Washington, D.C., pp. 57–69.

19. SEDIMENTARY CYCLING
MODELS OF GLOBAL PROCESSES

FRED T. MACKENZIE AND ROLAND WOLLAST

1. Introduction

Geochemical models ("global dispersion models," "element cycling models") describing the cycles of elements at the earth's surface represent the most frequent approach to study of the dispersion of materials, natural or pollutant, on a global basis. These models permit estimation of the importance of modification of natural chemical cycles owing to perturbations introduced by man's activities. Geochemical models generally describe transport paths and fluxes among a limited number of physically well-defined portions of the earth referred to as spheres or reservoirs. The majority of these global models is concerned with the description of transfer of material from the continents to the atmosphere or oceans and vice versa. The main transport paths and their distribution for major elements are rather well known, and the natural fluxes among the three reservoirs have been estimated successfully. The quality of these models is controlled by the mass balance for each reservoir. If a steady state is assumed for the model, which is usually the case for natural element cycles on a geological scale, there is no accumulation or removal in the reservoir and the mass balance is equal to zero.

In the case of the minor elements, like the heavy metals, their distributions in all the compartments necessary for the model are commonly less well known than for the major elements. Data on fundamental chemical and biological characteristics of the elements are generally insufficient to predict their behavior in each reservoir. Also, the input of some of these minor elements into the natural system owing to man's activities is comparable to the natural input, thus invalidating an assumption of steady state. Global dispersion models of minor elements are limited at present, and their quality poor, even in the case of a three-reservoir model.

To describe or predict the consequences of man's input of pollutants into the environment, a three-box model is insufficient. Therefore, it is necessary to subdivide the system into more reservoirs such as stratosphere, troposphere, oceanic mixed layer, oceanic deep layer, biosphere, etc. This subdivision greatly complicates description of the system because it is necessary to predict and calculate a greater number of transport paths and fluxes among reservoirs. Also, for the environmentalist, it is necessary to predict the concentrations of pollutants within "compartments" and rates of transfer among compartments, such as mercury concentrations in water, plankton, and sediments, and mercury transfers among these compartments.

Because of the many global models available for the dispersion of substances, we do not attempt to describe the mathematics or results of each model. We do, however, consider the factors involved in construction of global dispersion models:

1. Definition of boundaries of system and subsystem (reservoirs).
2. Prediction and evaluation of transport paths.
3. Problems of evaluating fluxes.
4. Mathematics of modeling of global systems.

Then we consider models of the sedimentary cycle of major elements and of the minor elements mercury and manganese as examples of global modeling.

739

2. General Considerations

In development of a global model, it is necessary to separate the system of interest from its natural surroundings. In general, the boundaries of a natural system are defined by the scale of the phenomena of interest and by *previous knowledge* of possible interactions between the system and its surroundings. Global dispersion models involve consideration of phenomena on a worldwide scale, resulting in subdivision of the earth into a number of physically well-defined spheres—boxes or reservoirs. Thus the term "box model" is commonly applied to models of this type.

The number of reservoirs considered in modeling of the global dispersion of a substance depends on *previous knowledge* of the way in which the substance of interest is distributed about the earth's surface; that is, on the number and direction of transport paths for the substance. A transport path is a directional property of the system, providing the route by which a substance is transported from one box to another.

On a global basis for a simple three-box model involving the reservoirs land, atmosphere, and ocean, the major agents responsible for transport of materials from one reservoir to another are, for example, streams, groundwater flow, rain, ice, and wind. If, however, the system of interest is subdivided into more boxes, such as subdivision of the atmosphere into stratosphere and troposphere and subdivision of the ocean into mixed layer and deep layer, then advective, diffusional, or biological processes become important agents of material transfer. For example, upwelling currents and settling of biological debris in the ocean are important means of material transfer between the oceanic mixed and deep layers.

Fundamentally, the manner in which a substance is transported from one box to another depends on its physicochemical characteristics. A substance with a high water solubility would be expected to move about the earth's surface via agents involving water movement. On the other hand, a substance with a high solubility in fatty tissue,

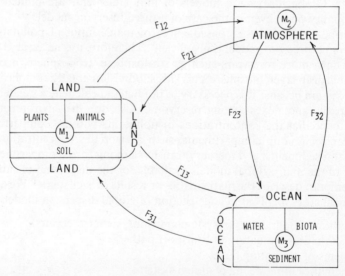

Fig. 1. Schematic diagram of a three-box global cycling model. F_{ij} represents flux of a substance A between reservoirs, where $i, j = 1, 2, 3$ and $i \neq j$. M_i is mass of A in reservoir, where $i = 1, 2, 3$. See text for discussion.

for example, DDT, would be expected to be involved in biological processes and to exhibit significant material transport by organism movement.

If it can be demonstrated that there is a means of transporting a substance from one earth sphere to another, then the spheres should be considered boxes or reservoirs in a global model. If, on the other hand, there are several sources for transport of materials from one box to another, but no known intersource transport paths, then the individual sources within a box are considered to be compartments in a global model. Figure 1 demonstrates this distinction between reservoir and compartment for a three-box, land–atmosphere–ocean global model of an hypothetical substance A.

In Fig. 1, three reservoirs of substance A are shown with interreservoir transport paths. The ocean reservoir is subdivided into compartments of water, biota, and sediment, and the land reservoir into compartments of plants, animals, and soil. Masses of A in these compartments are unknown, and although intercompartmental transport paths are suspected between some compartments, no material flux values are known, except for those between reservoirs. The flux of a substance is the mass of the substance transferred along a transport path between reservoirs and usually is defined in terms of mass per unit time (e.g., grams/year). Material transport for substance A in the system of Fig. 1 involves (1) gaseous transport from the sea and land surfaces and return in rainfall, (2) river transport in the dissolved and suspended load, and (3) return to the land via uplift of sedimentary rocks. Fluxes of A are known for these transport paths. As is shown later, this hypothetical model resembles in some respects the tentative model for the global cycle of mercury.

3. Global Transport Paths

Transport paths represent the major agents responsible for distribution of materials about the earth's surface. In global models of a substance's dispersion, these agents must describe to a significant degree the transport paths among reservoirs. Their nature may have various origins and may be physical, chemical, or biological processes. The quantitative evaluation of their contribution is probably the most critical point of model building and necessitates a good knowledge of all the interactions possible between the reservoirs and of the dynamics of the physical, chemical, and biological processes involved. The complexity of the task increases rapidly when subsystems are considered, where advective and diffusive processes, as well as biological processes, within the reservoirs define transport paths. Although it may be easy to estimate the transport of a dissolved substance from the continents to the ocean by streams, it is more difficult to estimate the substance's dispersion within the ocean by advective and diffusive processes involving biotic uptake and decomposition.

We now discuss first transport paths in the three-box model, and then main transport paths between subsystems are presented briefly. The last section is devoted to transport paths of pollutants.

A. Transport Paths Between Continents, Oceans, and Atmosphere

Table I shows the major agents of material transport to the oceans. Erosion and weathering of rocks by rainwater constitute the most efficient mobilization process at the earth's surface, and streams transport each year nearly 90 % of the total material entering the oceans. Transport related to the cycle of water (streams, icebergs,

TABLE I

Agents of Material Transport to the Oceans[a]

Agent	Mass (10^{14} g/yr)	Percent of Total Transport	Remarks
Streams	39 (dissolved) } 183 (suspended) }	90	Present dissolved load 17%; suspended 73% of total; more nearly equal in geologic past
Ice	17	7	Carries ground-up, fine-grained rock debris as well as boulder-sized material; composition similar to average sedimentary rocks. Chiefly from Greenland and Antarctica and distributed by icebergs
Ground-water	4	2	Poor estimate; major area of ignorance with respect to possible contamination
Shore erosion	2	1	Sands, silts, and muds derived from erosion of shorelines by waves, tides, and currents. Composition like stream-suspended load
Dust	0.5	0.2	Dust delivered to ocean related to deserts and wind patterns; Sahara major source for Atlantic Ocean. Composition like average sedimentary rock
Volcanic	0.4	0.2(?)	Poor estimate; lava and gases from earth's interior. Explosive volcanism producing dust may be important in climatic control; also, submarine volcanism and alteration is a source of some materials, particularly in the deep sea, e.g., Mn
Total	~240		

[a] After Garrels, Mackenzie, and Hunt, 1973; Garrels and Mackenzie, 1971.

groundwater) represents 98% of total transport. Material transport from the continent to the ocean via the atmosphere involves two principal paths:

1. Wind stress on the land surface results in advection of solid matter (dust) into the atmosphere; the dust later falls on the sea surface as particulate matter in rain or as dry fallout. Approximately $0.5–5 \times 10^{14}$ g of continental dust is ejected in this manner into the atmosphere each year (Garrels and Mackenzie, 1971; Goldberg, 1971).

2. Substances generated as volatile materials at the land surface escape to the atmosphere; these gases may later dissolve directly in seawater, or are carried to the sea surface dissolved in rain. Of particular interest are the gases generated by

decomposition of organic matter in soils, e.g., CO_2, CH_4, NH_3, N_2O, H_2S, and SO_2. The reduced species are generally oxidized in the atmosphere.

The great bulk of solids transported to the oceans in the suspended load of streams or as dust via the atmosphere is deposited as sediment without reacting significantly with seawater.

If the oceans do not change composition during time, however, then the river-derived dissolved materials entering the oceans must be removed. Major investigations currently are going on to determine the reactions and sinks that remove dissolved materials from ocean water (e.g., see Goldberg, 1974). Balances have been obtained for a number of elements; however, the only unequivocal balance, perhaps, is that of calcium carbonate. It can be demonstrated that the $CaCO_3$ input into the oceans, as dissolved Ca^{2+} and HCO_3^-, is balanced by deposition of $CaCO_3$, principally in skeletal organisms.

Once materials enter sediments, they may be remobilized and returned to seawater. An excellent example of this remobilization involves the bacterial oxidation of organic matter in the interstitial waters of sediments. Anaerobic decomposition results in the release of dissolved species of carbon, nitrogen, and phosphorus to the pore waters. These dissolved constituents may diffuse out of sediments into overlying water. Metals incorporated in organic matter also may be released to the pore water during organic decomposition and precipitate in another solid phase or diffuse upward into waters overlying the sediment.

Burial of sediments followed by uplift supplies fresh sediment for the processes of soil formation.

The sea surface itself is a source of materials for the atmosphere, and in the case of some substances, for the continents. The most obvious example is water. It evaporates from the sea surface or is ejected mechanically by breaking waves or in aerosol generated by the impact of rain on the sea surface. About 9% of the water entering the atmosphere from the oceans falls on land to enter the stream and groundwater systems; the rest rains back onto the sea surface.

Gases generated in seawater may leave the sea surface to enter the atmosphere. This transport path is particularly important for gases produced by decomposition of organic matter and respiration in the ocean, that is, CO_2, CO, NH_3, N_2O, and SO_2. Aerosol ejected into the atmosphere by bursting bubbles results in transport of not only water but also sea salts to the atmosphere. Aerosol containing chloride and other sea salts is transported by winds over the continents where precipitation as rain or dry fallout occurs. Indeed, about 50% or more of the chloride carried by streams to the oceans annually comes from the sea.

B. Transport Paths Between Subreservoirs

Transport of material between the marine and terrestrial biospheres and the oceanic mixed layer and the troposphere, respectively, is of major ecological importance. Exchange between these subreservoirs for substances involved in biological processes is strongly affected by the photosynthetic and respiratory activity of organisms. For example, land plants consume CO_2 from the troposphere during photosynthesis and release O_2, but during plant decomposition in soils, O_2 is consumed and CO_2 released. Nitrogen is fixed by nitrogen-fixing bacteria in soils and transferred to plants as nutrients; a small amount is released to the atmosphere as NH_3 and N_2O, or to the rivers as NO_3^-, after degradation of organic matter.

In the ocean, CO_2 from the atmosphere enters the mixed layer where it is removed during photosynthesis. Decomposition of the phytoplankton or animals that have consumed the plants results in return of CO_2 to the atmosphere. Dissolved substances transported to the oceanic mixed layer by rivers, such as calcium phosphate and dissolved silica, are consumed by organisms to form their skeletons and removed from the mixed layer by sinking to the deep layer after death of the organisms. Many heavy metals necessary to life as trace elements are probably submitted to a similar process. The dead biological material passing through the mixed layer and entering the oceanic deep layer continues to undergo oxidative decomposition or simply inorganic dissolution.

Return of the dissolved substances resulting from these processes from the deep layer to the mixed layer can be by upwelling of water at open ocean divergences and along margins of oceans. Also, concentration differences between the upper and lower layers in the oceans may create a diffusive transport of material between the two layers.

Substances, such as CO_2, that remain in the atmosphere for a long time are well mixed and consequently their proportions are constant with elevation. Substances generated at the earth's surface, like carbon monoxide and water vapor, show decreasing proportions in dry air with increasing elevation. Ozone, on the other hand, which is formed in the stratosphere, mixes downward into the troposphere where it is destroyed. Its proportion in dry air thus increases with increasing elevation up to the ozone maximum in the stratosphere.

Dust in the atmosphere exhibits a nearly exponential decrease in concentration with increasing elevation to the base of the stratosphere where it increases sharply. The maximum at 18–25 km is thought to be due to volcanic ejections, and more recently, explosion of atomic bombs.

Transport between the troposphere and the stratosphere occurs principally at low and high latitudes where advective and diffusive processes are important.

C. Transport Paths of Pollutants

It is necessary to emphasize that materials have been circulating through the reservoirs and by transport paths similar to those shown in Fig. 1 for hundreds of millions of years. In a sense, the earth has a natural metabolism. The sedimentary, or exogenic, cycle of materials describes the circulation of substances moving from the weathering and erosion of rocks to the atmosphere, to the oceans and streams, from the atmosphere to the biota and back again, and return to the continents via uplift. A number of pollutants can be considered simply as increased injections of materials into these natural cycles—addition to the atmosphere of gases, carbon dioxide, carbon monoxide, nitrogen and sulfur compounds, and mercury and its compounds; additions to streams and groundwaters of toxic minor elements such as zinc, arsenic, cadmium, selenium, and lead; and additions to oceans and lakes of nitrogen- and phosphorus-containing nutrients, animal wastes, petroleum, and radioactive nuclides.

Other pollutants are compounds that have been synthesized by man for agricultural, industrial, etc., purposes. These materials are new to the earth's surface; however, they enter the general scheme of material transfer and storage. For example, the halogenated hydrocarbons and PCB's are found distributed throughout the earth's surface environment. To understand the dispersal of these materials, it is necessary to appreciate the fact that they also enter cycles that natural substances have been circulating in for much of geologic time.

4. Problems of Evaluating Fluxes

It is appropriate at this stage to consider the problems associated with evaluation of fluxes and reservoir masses in global dispersion models. These problems fall under two general headings: those associated with basic data, and those derived from lack of understanding of processes operating in the system of concern.

We use a variety of examples involving the agents of transport of materials to the oceans (Table I) to illustrate the problems involved.

One of the major inputs into the ocean arises from the transport by rivers of dissolved and suspended matter. The yearly input by rivers of major dissolved constituents (Ca, Mg, Na, K, HCO_3, Cl, SO_4) to the ocean is reasonably well known (e.g., Livingstone, 1963), but only recently, Gibbs (1972) revised downward the average salinity of the world's rivers by about 5%. Data for dissolved organic species, including nutrients, and trace element concentrations of the world's rivers, are scarce. For example, virtually no data for these species are available for the Congo and Amazon Rivers, systems which deliver about 20% of the water discharge to the ocean yearly, and at this time are little affected by pollution. Data for the chemical composition of the suspended load of rivers are even more scarce than those for the dissolved load. Nearly 80% of the yearly input of suspended matter to the sea is carried by rivers draining Southeast Asia (Holeman, 1968). To date, no systematic study of the major element composition of this material appears in the literature. Needless to say, no systematic data are available for the suspended trace elements or organic concentrations of these rivers.

It is interesting to note that estimates of total "pollutant sulfur" in the world's rivers range from less than 5 to about 30%. It is true, however, that there are few sulfur isotopic studies of rivers, fundamental data that could be used to estimate man's contribution to the natural cycle of sulfur.

Aside from lack of data involving fluxes of river-borne trace elements to the ocean, there are a number of unsolved problems dealing with the processes affecting trace elements transported to the ocean by streams. First, the analyses of some "dissolved" trace elements in river and coastal systems are suspect. The dissolved fraction is usually defined as that which on filtration passes through a 0.45-μ filter. However, concentrations of trace elements in rivers obtained in the above manner are usually several to many orders of magnitude greater than the maximum concentrations allowed by equilibrium values. This result has led many investigators to conclude that the elements are actually present in disequilibrium concentrations. Another interpretation is that filtration through a 0.45-μ filter does not completely remove suspended colloidal material, and analyses of the concentrations of "dissolved" trace elements in river water include suspended material (Kennedy and Zellweger, 1974; Jones et al., 1974). There is need for sequential filtration of river waters through several filter sizes to obtain accurate values of dissolved trace element concentrations.

Some trace element concentrations in the suspended load of sediments vary with particle size; in general, the smaller the particle size, the higher the concentration (Martin et al., 1973). Thus actual inputs of suspended trace element masses into coastal regions via rivers are fraught with difficulties related to the above fractionation. This fractionation problem is of particular importance to the trace element economy of the deep ocean.

Second, from a process viewpoint, we have scant information on the chemical precipitation mechanisms, and their variance with environmental conditions, of trace element removal from river plumes entering the sea. Indeed, as mentioned previously, such information is lacking even for major elements. It is known that some minor elements are abstracted from seawater by biological processes and concentrated in the organic or skeletal portions of marine organisms. However, concentration factors, rates of overturn, and sedimentation rates of minor elements in organisms are poorly known. The first *detailed* account of minor element concentrations in marine plankton was published in 1973 (Martin and Knauer).

Once an element is transported by sedimentation of organic or inorganic debris to the sediment–water interface, it may undergo remobilization during subsequent burial. The chemical reactions resulting in remobilization and the rates of release of minor elements from sediments to overlying water are in their infancy of study.

A second major input of material to the ocean involves wind transport of dust from the continents. This source of material is particularly important to deep-sea budgets of elements because the dust-derived material sedimented in the deep sea may amount to as much as 50% of the sedimentation rate (Goldberg, 1971). It was not until the last decade that this source of material for the ocean received major attention. The total flux from continent to ocean of 5×10^{14} g/yr has been estimated by indirect means involving estimates of atmospheric dust loads and frequency of atmospheric washout (Goldberg, 1971). The estimate of Garrels and Mackenzie (1971) is an order of magnitude less. The composition of the dust is thought to resemble average crustal rock (average sedimentary rock probably provides a better estimate); however, actual measurements of the elemental composition of the dust have been accomplished only during the last decade. Data on the concentrations of minor elements in dust are just beginning to appear in the literature (e.g., Hoffman et al., 1972; Duce et al., 1974; Chester and Stoner, 1974). It is not known whether the dust as it sinks through the water column after sedimenting on the sea surface releases elements to the seawater or removes them.

It is interesting to note that for some elements, those termed "volatile" such as Hg, As, and Se, concentrations in dust particles seem much too low when compared with concentrations in average sedimentary rock (see Section 6.B). This observation suggests that these elements may be transported through the atmosphere in the gaseous, as well as particulate, state. Concentrations of these, and other trace elements, in the gaseous state in the atmosphere are poorly known, as are rates of transfer (residence times) in the atmosphere. It is shown in a subsequent section that gaseous transport of Hg between the earth's surface and the atmosphere and return is a very important pathway of Hg transport on a global basis.

Although we know only meager details of material transport to the oceans via streams, our knowledge concerning material transport to the ocean by ice, ground water, and volcanic emanations is even less. Estimates of total fluxes of materials for these agents have been made and their proportions are given in Table I. We have some crude estimates of supply of major elements to the world's oceans by these agents (Garrels and Mackenzie, 1971); however, little is known concerning their significance as agents of minor element transfer on a global basis. Groundwater is of particular concern because of possible contamination by man's activities and eventual release of contaminants to seawater. Indeed, the actual volume of groundwater reaching the sea and its composition are virtually unknown.

5. Mathematical Formulation

A. General Aspects

Box models involve only simple mathematical formulations, but require the critical judgment of an experienced practitioner. The basic relations used express the mass balance for each box representing a system and for all the systems together (the global mass balance).

The interactions of the system with the outside provide the inputs and outputs to the system. The physical, chemical, or biological transformations of the substances in the system provide the evolution terms which represent the production or the consumption of the substances. The mass balance for a system requires that the change of mass of any substance in the system, dM_s/dt, be equal to the input flux M_i plus the production term P minus the output flux M_o and the consumption term C. The general differential equation is

$$\frac{dM_s}{dt} = M_i + P - M_o - C. \tag{1}$$

Generally, the various terms of the mass balance equation are expressed with respect to the mean concentration of the substance in the various phases considered. Each of the terms in equation 1 is generally composite because there are different possible transport agents controlling the inputs and the outputs in the model, and various processes of removal or production in the system. Also, the substance may occur in different phases (dissolved chemical substances, suspended particles, living organisms) and as various chemical species.

An initial simplification is to consider the total concentration of the substance in a number of limited but representative phases. In the case of mercury, for instance, we need not consider each of the numerous mercury compounds, but simply the total mercury concentration in each phase. These phases may be, for example, in the ocean, the aqueous phase, the mineral suspension, and the plankton. For each of these phases a mass balance equation is written describing the evolution in a particular phase plus a mass balance equation for the whole system. If various species of mercury are considered, a mass balance equation can be written for each of the species in the various phases considered.

Input and output terms are selected for the known transport agents between reservoirs. For advective processes like river discharge, sedimentation, and upwelling, these terms may be expressed as the product of the total flux of the transport agent times the mean concentration of the substance considered in that flux. For diffusive processes like exchange owing to eddy diffusion between the lower and upper layers of the atmosphere or the ocean, or molecular diffusion from sediment pore waters to overlying seawater, the flux of the substance is equal to the product of the diffusion coefficient times the concentration gradient of the substance. The production and consumption terms represent kinetic equations describing the evolution of the concentration of the substance with time. These equations contain multiple inter-action terms that generally are little known by chemists and biologists. Thus it is necessary to obtain a simplified description in terms of control parameters (Nihoul, 1974). These parameters arise from the initial demarcation of the system, and the necessity of restricting the variables used and of formulating the laws of their evolution in a simple and tractable way. They are in most cases determined by separate models, experimental data, or theoretical reflections. For instance, if we want to describe the uptake of a substance in the sea by phytoplankton, total primary pro-ductivity can be used as the control parameter. For conservative chemical species, those that are not consumed or produced (this is the case also when one considers the total concentration of the elements), the terms P and C are equal to zero and equation 1 contains only the input and output fluxes, M_i and M_o, respectively.

If we consider in a model m species and n phases, the number of differential equations that needs to be solved simultaneously is equal to $m \times n$. This number is reduced to n when only the total concentration of a substance is taken into account. The resolution of the system is considerably simplified if the assumption is made that the system is at steady state. Steady-state models are usually satisfactory if concern is with the average properties of a system over long periods of time. Most models of worldwide material dispersal are steady state, based on the fact that, from geological evidence, the compositions of the atmosphere, ocean, and earth's crust have remained constant for at least a few hundreds of million years. This result stems from the fact that the various fluxes between the reservoirs and the production and consumption processes were constant during time. Equation 1 is then reduced to

$$M_i + P - M_o - C = 0 \tag{2}$$

and for a conservative chemical species

$$M_i = M_o \tag{3}$$

where all the terms are independent of time. The assumption of steady state is very useful because it is then possible to estimate one of the unknown terms of equations 2 and 3. This assumption has been used extensively in computations of steady-state masses and rate constants in systems of simultaneous chemical reactions and in studies of mass balances in "box models" as used in oceanography, geochemistry, meteorology, and other fields (e.g., Craig, 1957; Broecker, 1963; Lal and Peters, 1967; Eriksson, 1971; Lerman, Mackenzie, and Garrels, 1974; Wollast, 1974).

Attempts have also been made to construct time-dependent concentration models, for example, of CO_2 in the atmosphere (Machta, 1972), of DDT in the surface environment (Woodwell, Craig, and Johnson, 1971), and of phosphorus in the surface environment (Lerman, Mackenzie, and Garrels, 1974). In these models, fluxes for transfer of materials between reservoirs may be considered as a function of time; additional time-dependent inputs to reservoirs from outside the system may be evaluated.

With this brief introduction to modeling of global systems, we now look at some mathematical aspects of steady-state models.

B. Steady-State Modeling

In this section, mathematical formalism is presented for a simple three-box model. It is shown that subdivision of a box into subreservoirs increases the number of mass balance equations necessary for solution of the model and necessitates knowledge, in most cases, of advective and diffusive processes in the system of interest.

Three-Box Model

Figure 2 illustrates an example of the mathematical formulation of a three-reservoir model (land–ocean–atmosphere) for a conservative substance having a gaseous state. Carbon dioxide is an example. Conservatism implies that there is neither net production or consumption of the substance in the system. The fluxes considered are as follows:

    1. River discharge to oceans; total flux of the substance equals total concentration in rivers, C_r, times total discharge, V_r. $C_{r, \text{diss}}$ and $C_{r, \text{sus}}$ are, respectively, concentrations of the substance in the dissolved and suspended load of rivers.

    2. Sedimentation in the oceans; total flux is rate of oceanic sedimentation, V_s, times concentration, C_s, of substance in marine sediments.

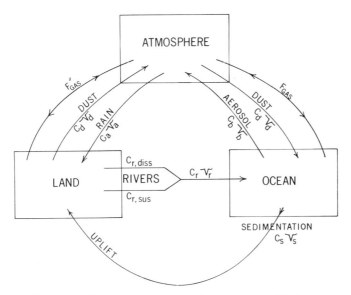

Fig. 2. Schematic diagram of a three-box, steady-state, global cycling model of a substance. See text for discussion.

3. Continental dust; total flux of substance to the ocean is flux of dust from continents to ocean, V_d, times concentration, C_d, of substance in dust.

4. Rainfall over continents; total flux of substance in rain is total rainfall over continents, V_a, times concentration, C_a, of substance in rain.

5. Sea aerosol; total flux of substance to atmosphere from ocean is total aerosol flux, V_b, times concentration, C_b, of substance in aerosol produced by bursting bubbles at sea surface.

6. Gaseous exchange; difference in fluxes between atmosphere and land or atmosphere and sea surface and return, F'_{gas} and F_{gas}, respectively.

The mass balance equations for the boxes considered and for the whole system, assuming steady state, and that there is net gaseous transport from ocean to land via the atmosphere, are as follows:

1. Land

$$F'_{gas} + C_a V_a + C_s V_s = C_r V_r + C_d V_d$$

2. Ocean

$$C_r V_r + C_d V_d = C_s V_s + C_b V_b + F_{gas}$$

3. Atmosphere

$$F_{gas} + C_b V_b + C_d V_d = C_a V_a + F'_{gas} + C_d V_d$$

Dust flux terms usually cancel out because dust does not undergo chemical modification in the atmosphere.

4. Total system: steady-state condition that there is no net flux in or out of the system.

This steady (or stationary) state model contains seven unknown fluxes (V_r, V_s, V_d, V_b, V_a, F_{gas}, F'_{gas}) and five unknown concentrations (C_d, C_a, C_s, C_b, C_r). There are, however, two independent relations; thus the model is solvable if only two variables are unknown.

It is possible to simplify these equations by imposing conditions. For example, for a minor element with low volatility, the terms F_{gas} and F'_{gas} drop out. If aerosol transport of the element is negligible, the array of equations reduce to the following:

1. Land

$$C_s V_s = C_r V_r + C_d V_d$$

2. Ocean

$$C_d V_d + C_r V_r = C_s V_s$$

3. Atmosphere

$$C_d V_d = C_d V_d$$

In this case, one of the relations is trivial, and only one equation can be used if values for all the variables are unknown.

Expanded Ocean Reservoir

The boxes, or reservoirs, of Fig. 2 can be expanded into a number of subreservoirs. For example, the ocean may be subdivided into upper mixed layer, lower deep layer, and sediments (Fig. 3). This subdivision commonly necessitates inclusion of advective and diffusive terms in the mass balance equations for the subreservoirs. These terms represent processes of upwelling and turbulent diffusion in ocean waters and molecular

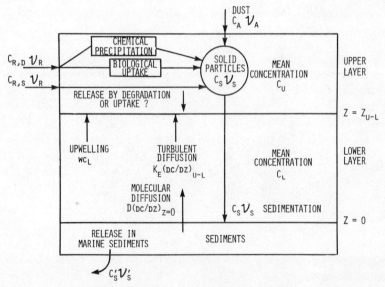

Fig. 3.   Schematic diagram of an example of an "expanded" ocean reservoir; ocean has been subdivided into subreservoirs of upper layer, lower layer, and sediments. See text for discussion.

diffusion in sediments. The array of mass balance equations for an "expanded" ocean model where a substance is transported to the ocean by streams and in airborne dust is as follows:

1. Mass balance for the system "ocean"

$$C_a V_a + (C_{r,\,\text{diss}} + C_{r,\,\text{sus}}) V_r = C_s V_s$$

where C is concentration, V is flux, and subscripts are as in preceding section.

2. Mass balance for upper layer

$$C_a V_a + (C_{r,\,\text{diss}} + C_{r,\,\text{sus}}) V_r + W C_1 + K_e \left(\frac{dc}{dz}\right)_{u-1} = C'_s V'_s$$

where the upwelling term, WC, is composed of the velocity, W (cm/yr), at which water upwells times the concentration, C, of the substance in the water. The turbulent diffusion term consists of the eddy diffusion coefficient, K_e (cm^2/sec), times the concentration gradient, dc/dz, of the substance between layers. The term $C'_s V'_s$ is explained below.

3. Mass balance for lower layer

$$D\left(\frac{dc}{dz}\right)_{z=0} = W C_1 + K_e \left(\frac{dc}{dz}\right)_{u-1} + C'_s V'_s$$

where $D(dc/dz)$ is the molecular diffusion term representing transport of the substance out of marine sediments into overlying seawater. D is the molecular diffusion coefficient (cm^2/sec) and dc/dz is the concentration gradient of the substance in the pore water of sediments. The net sedimentation term $C'_s V'_s$ is not necessarily equal to $C_s V_s$ because the substance may be remobilized in sediments after sedimentation and released to pore waters and overlying seawater.

4. Mass balance for marine sediments

$$C_s V_s = C'_s V'_s + D\left(\frac{dc}{dz}\right)_{z=0}.$$

Thus in the simple three-box model of Fig. 2, if an "expanded" ocean reservoir is used, there are five mass balance equations representing the reservoirs and subreservoirs atmosphere, land, oceanic upper layer, oceanic lower layer, and sediments, plus the condition that the system is steady state. Also, diffusion and advective terms are introduced into the mass balance equations for each subreservoir of the ocean.

6. Modeling of Sedimentary Cycles of Elements: Examples

A. Steady-State Sedimentary System

General Aspects

Perhaps one of the most significant advances in our understanding of the ocean in the last decade or two has been the recognition that its composition cannot be viewed as a separate entity but must be considered as part of the major sedimentary cycle of the elements (exogenic cycle). This cycle involves the reservoirs of atmosphere, biosphere, ocean, and sedimentary lithosphere, with the land being an ephemeral source of the products of weathering. The residence time of an element in a particular

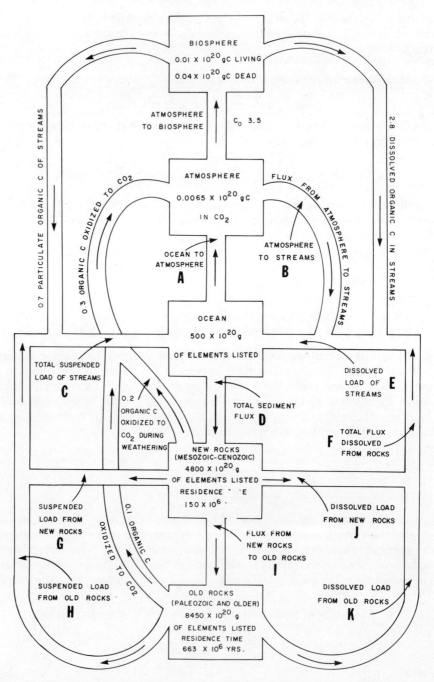

Fig. 4. Steady-state model for circulation of elements in the sedimentary cycle. Fluxes are in units of 10^{14} g/yr. C_o denotes organic carbon; C_i inorganic carbon (Garrels and Mackenzie, 1972).

	A	B
Si	0.00	0.00
Al	0.00	0.00
Fe	0.00	0.00
Mg	0.30	0.30
Ca	0.90	0.90
Na	1.68	1.68
K	0.18	0.18
Ti	0.00	0.00
S	0.84	0.84
Cl	2.10	2.10
C_i	2.50	2.50
C_o	3.20	0.00
	11.70	8.50

	C	D	E	F
Si	11.16	13.68	2.52	2.52
Al	3.01	3.03	0.00	0.00
Fe	2.00	2.00	0.00	0.00
Mg	0.28	1.60	1.62	1.32
Ca	0.61	5.66	5.94	5.04
Na	0.36	1.20	2.52	0.84
K	0.58	1.30	0.90	0.72
Ti	0.13	0.13	0.00	0.00
S	0.05	0.65	1.44	0.60
Cl	0.00	0.96	3.06	0.96
C_i	0:20	1.60	3.90	1.40
C_o	0.70	0.30	2.80	0.00
	19.08	32.11	24.70	13.40

	G	H	I	J	K
Si	5.58	5.58	6.21	1.89	0.63
Al	1.39	1.62	1.62	0.00	0.00
Fe	1.00	1.00	1.00	0.00	0.00
Mg	0.14	0.14	0.56	0.90	0.42
Ca	0.43	0.18	1.72	3.49	1.55
Na	0.23	0.14	0.25	0.72	0.12
K	0.20	0.38	0.47	0.63	0.09
Ti	0.06	0.07	0.07	0.00	0.00
S	0.05	0.00	0.09	0.51	0.09
Cl	0.00	0.00	0.10	0.85	0.11
C_i	0.13	0.07	0.56	0.91	0.49
C_o	0.00	0.00	0.10	0.00	0.00
	9.21	9.18	12.75	9.90	3.50

753

reservoir is dictated by the element mass in the reservoir and the input or output rate of the element; at steady state, input rate equals output. If the individual reservoirs are in steady state with respect to an element, then the system as a whole is in steady state for that element. In a system of this nature, the ocean is viewed as a chemical reactor, or reflux condenser, in which materials are added, react, and leave. This view contrasts sharply with that of the early part of this century in which the ocean was viewed as a storage bin or accumulator of elements transported to it principally by streams.

The concept of the ocean as a reservoir in the sedimentary rock cycle provides a means for assessing the fate of the elements after entering the ocean. Recently, Garrels and Mackenzie (1972) developed a steady-state quantitative model for the sedimentary cycle. This model is shown in Fig. 4 and is used to discuss the sources, rates of additions, and reactions involving elements in seawater. The model was developed on the basis of several assumptions outlined by Garrels and Mackenzie (1972). The most important assumption here is that the material balance in the model was accomplished assuming that seawater composition does not change through time. The internal consistency of the model with a number of known variables, such as the chemical composition of the dissolved and suspended load of rivers, suggests that this assumption is applicable at least to Phanerozoic time, and perhaps as far back as

TABLE II

Element Reservoir Sizes and Residence Times

Element	Reservoir Size $(10^{20}$ g$)$	Reservoir Size $(10^{20}$ mol$)$	Residence Time T^a $(10^6$ yr$)$	Reservoir Size $(10^{20}$ g$)$	Reservoir Size $(10^{20}$ mol$)$	Residence Time T^c $(10^6$ yr$)$
	Ocean			Lithosphere		
Na	162.38	7.06	193	345	15.00	288
K	5.85	0.15	8.1	507.13	12.97	390
Mg	19.68	0.82	14.9	610.08	25.42	381
Ca	6.00	0.15	1.2	1986.40	49.66	351
Si	0.0392	0.0014	0.016	6160.84	220.03	50
Al	—	—	—	1527.66	56.58	504
Fe	—	—	—	961.52	17.17	431
Ti	—	—	—	66.95	1.39	515
S	13.44	0.42	22.4	157.12	4.91	242
Cl	288.40	8.24	300	209.45	5.90	218
C_{inorg}	0.396	0.033	0.1^b	610.08	50.84	381
C_{org}	0.007	0.00058	0.0025	125.04	10.42	417
	Atmosphere			Biosphere		
C_{inorg}	0.00648	0.00054	0.0026	—	—	—
C_{org}	—	—	—	0.0504	0.0042	0.0144

^{a} Corrected for atmosphere cycling.
^{b} Includes C input from atmospheric CO_2 used in weathering.
^{c} Includes input flux of both dissolved and suspended stream loads.
Source: Garrels and Mackenzie, 1972.

1.5×10^9 yr—the period of time for which we have sedimentary rock types similar to today. The rest of the discussion is keyed to Fig. 4 and Table II.

The model is constructed with four reservoirs: biosphere, atmosphere, ocean, and sedimentary lithosphere. The sediments are divided into New Rocks (Cenozoic and Mesozoic) and Old Rocks (Paleozoic and Precambrian). Weathering and diagenetic changes result in the conversion of New Rocks to Old and transfer of much dissolved material from the land to the ocean. The admixture of metastable phases of detrital minerals, biochemical precipitates, and authigenic minerals is slowly converted to stable mineral assemblages of Old Rocks; that is, shales develop a monotonous mineralogy of chlorite, illite, quartz, and lose $CaCO_3$, whereas limestones become high in magnesium. The mineralogical changes that muddy sediments undergo during stabilization are shown diagrammatically in Fig. 5. The triangle connects the three-phase assemblage quartz–illite–chlorite; this assemblage tends to be the final product of postdepositional change of initial multiphase assemblages of minerals in muddy sediments.

The ratio of dissolved load derived from New Rocks to that from Old Rocks is high, showing that much of the stream load entering today's ocean is dominated by materials from New Rocks. Bicarbonate, Ca, SiO_2, SO_4, Mg, Na, and Cl are the principal dissolved species in rivers, and these are the ones lost from New Rocks as they are converted to Old.

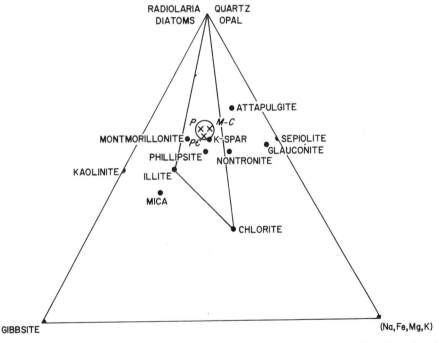

Fig. 5. Schematic diagram illustrating variety of phases found in muddy sediments plotted as a function of atomic percents of Al, Si, and Na + Fe + Mg + K. Circled area encloses composition of average Precambrian (PC), Paleozoic (P), and Mesozoic–Cenozoic (M–C) shales. The triangle connects phases thought to be the final phase assemblage in the post depositional alteration of muddy sediments to shales (Garrels and Mackenzie, 1974).

The most recently uplifted rocks are being destroyed by mechanical erosion and selectively leached by chemical erosion to form a new sedimentary mass very similar to their original compositions. In the course of time the remaining New Rocks should take on the characteristics of today's Old Rocks, with the formation of a new sedimentary mass with a chemical and mineralogical composition like present New Rocks. The rocks of this new "Era," to judge from the chemistry of the dissolved load of present-day streams, will be enriched in their proportions of limestone and evaporites over the original New Rock deposits. If this picture is accurate in its broad long-term relationships, no matter how much it is modified by shorter-term period-to-period relationships, the kinds of detrital and dissolved materials being brought to the oceans have not changed greatly since the Late Precambrian, and a mean seawater composition *approximately* in equilibrium with calcite, K-feldspar, illite–montmorillonite solid solution, and chlorite is implied (Helgeson and Mackenzie, 1970).

Element Cycles

Silicon

To begin the discussion of element cycles, it is convenient to start with silicon. During the past decade silicon has been the focus of considerable debate on the controls of seawater composition (cf. Sillén, 1961; Holland, 1965; Mackenzie and Garrels, 1966a; Harris, 1966; Pytkowicz, 1967; Calvert, 1968; Gregor, 1968; Siever, 1968a, b; Helgeson and Mackenzie, 1970; Liss and Spencer, 1970; Broecker, 1971; Lisitzin, 1972; Heath, 1974; Wollast, 1974). This debate was initiated by L. G. Sillén in his address at the first International Oceanographic Congress in 1959 at which he proposed an equilibrium model for the ocean. In his model, silicate minerals played a major role in determining the pH and cation concentrations of seawater. Prior to this idea, it was generally thought that biological activity alone controlled the distribution of dissolved silica in the ocean, although previous attention had been paid to silicate interactions with seawater (e.g., Grim et al., 1949).

Following Sillén's lead, Mackenzie and Garrels (1966a, b) attempted to draw up a mass balance between river waters and seawater, in a manner similar to Conway (1942, 1943) and Heck (1964). To accomplish the balance, we found it necessary to synthesize new clay minerals from reactions involving the suspended sediment load of streams and dissolved silica and cations as they entered the ocean. At that time, it was felt that the silicate synthesis reactions occurred very early, perhaps in the water column itself. These reactions also entailed the return of CO_2 used during weathering of rock minerals to the atmosphere, a requirement necessary to prevent CO_2 drain on the atmosphere. A schematic "reverse weathering" reaction is as follows:

$$\text{degraded silicate} + H_4SiO_4 + HCO_3^- + \text{cations} \longrightarrow$$
$$\text{new clay mineral} + CO_2 + H_2O \quad (4)$$

Experimental data on reaction of silicates with seawater (Mackenzie and Garrels, 1965; Mackenzie et al., 1967; Siever, 1968a) show that silicate minerals can release or take up dissolved silica from silica-deficient and silica-spiked seawater, respectively. Experimental and field observations, however, have not supported the magnitude of very early silicate synthesis originally suggested by us. However, Ristvet et al., (1973) showed that apparently "reverse weathering" reactions do obtain during the early stages of burial diagenesis of sediments in Kaneohe Bay, Hawaii.

The problem can be further developed by reference to Fig. 4. Silicon delivered to the oceans by streams is principally found in silicate minerals and quartz of the suspended load, but about 20% is carried as H_4SiO_4 in the dissolved load. The smaller residence time of Si than those of Fe, Al, Ti in the sedimentary rock reservoir (Table II) reflects transport of Si in the dissolved load. The dissolved silica is derived from weathering reactions of the general type (Garrels and Mackenzie, 1971):

$$\text{silicate} + CO_2 + H_2O \longrightarrow \text{clay mineral} + HCO_3^- + 2H_4SiO_4 + \text{cations} \tag{5}$$

At today's rate of dissolved silica transfer to the ocean, silicate weathering would require 3.5×10^{12} mol CO_2/yr. At this rate, it would take only 15,000 yr to remove all the 5.4×10^{16} mol of CO_2 presently in the atmosphere by silicate weathering alone. The implication is that there must be reactions that return CO_2 to the atmosphere. Indeed, it is likely that these reactions instead of occurring in the water column obtain during diagenesis and metamorphism and involve silica (Wollast and De Broeu, 1971; Siever, 1968b; Wollast, 1974). Mackenzie and Garrels (1966b) pointed out that if all the silica entering the ocean today via streams accumulated as amorphous silica or its diagenetic products, chalcedony, and quartz, the sedimentary rock mass would contain a greater proportion of silicon than observed stored in these minerals.

Recent interstitial water analyses of modern deep-sea and shallow-water marine sediments (cf. Presley and Kaplan, 1970; Bischoff and Ku, 1970, 1971; Drever, 1971a; Ristvet et al., 1973; Nissenbaum et al., 1972; Manheim and Sayles, 1974) shed some light on the site of silicate regeneration. In interstitial waters that are chemically different for one or more elements from normal seawater, Mg^{2+} is generally depleted from the interstitial waters. Both K^+ and Ca^{2+} enrichments and depletions have been observed, whereas Na^+ either decreases or does not change significantly. In general, HCO_3^- is enriched and SO_4^{2-} depleted with respect to normal seawater, particularly in waters extracted from cores of anoxic sediments. Dissolved silica is enriched in interstitial waters; Heath (1974) obtained an average interstitial SiO_2 concentration for 800 determinations of 0.4 mol/l (24 ppm), a value less than that at saturation with amorphous silica at in situ sea bottom temperatures.

It is true, however, that not all interstitial waters have been extracted at in situ temperatures. Because dissolved silica and K^+ concentrations in interstitial waters appear to be dependent on the temperature of extraction of the waters (Fanning and Pilson, 1971; Mangelsdorf et al., 1969), some of the data for these species are particularly suspect.

Nevertheless, various reactions have been proposed to account for pore water chemical changes. Drever (1971a, b) suggested that interstitial water depletions of Mg^{2+} and SO_4^{2-} in anoxic sediments may be due to Mg replacement of Fe in clay minerals and precipitation of the Fe as iron sulfide, thus depleting the pore waters in both SO_4^{2-} and Mg^{2+}. K^+ and Mg^{2+} removal have been attributed to formation of illite and chlorite or sepiolite, respectively (i.e., Presley and Kaplan, 1970; Manheim and Sayles, 1974; Wollast, 1974).

Wollast (1974) recently proposed a reaction that has direct bearing on the silicate regeneration problem. From analysis of the species concentration profiles of Hole 149 of the Deep Sea Drilling Project, he suggested the following overall reaction to account for the Mg^{2+} decrease and Ca^{2+} and dissolved silica increase observed in pore waters extracted from core samples at this site:

$$6SiO_{2(am)} + 4CaCO_3 + 4Mg^{2+} + 7H_2O = 2Mg_2Si_3O_6(OH)_4 \cdot 1.5H_2O$$
$$\text{(sepiolite)}$$
$$+ 4CO_2 + 4Ca^{2+} \tag{6}$$

Here we have a diagenetic reaction that results in the consumption of SiO_2 and return of CO_2 to the atmosphere. Indeed, Wollast (1974) calculates that 3.5×10^{14} g/yr of silica react during diagenesis to make new silicate minerals; if all the silica reacted as in reaction 6, a flux of CO_2 out of the ocean would obtain nearly equivalent to that necessary to compensate for the removal rate by silicate weathering reactions.

Iron, titanium, and aluminum

Iron, titanium, and aluminum are transported to the ocean principally in solids of the suspended load of rivers. Their similar residence times in the sedimentary mass (Table II) show that the composition of today's suspended load is not greatly different from that of the past (Garrels and Mackenzie, 1972).

Iron deserves some special attention. Iron is transported in the stream-suspended load as discrete particles of ferric oxide, as ferric oxide coatings on detrital grains, and in the structures of silicate minerals. Iron is remobilized after deposition in anaerobic sediments where the ferric iron is reduced to ferrous and precipitated as the ferrous monosulfides greigite and mackinawite, later to be transformed to pyrite. The oxidation of buried organic matter with oxygen derived from the bacterial reduction of $SO_4{}^{2-}$ provides the electrons for the reduction of ferric iron. The mass of ferrous sulfide formed in modern anaerobic environments appears to be principally controlled by the presence of organic matter that can be metabolized by sulfate-reducing bacteria, although the availability of iron and sedimentation rate are also important factors (Berner, 1973). As mentioned previously, ferrous iron for sulfide mineral formation may also be derived from the structure of clay minerals. There is some suggestion that the ferric:ferrous ratio of shales decreases with increasing age, pointing to continuous reduction of ferric iron by carbonaceous material even during the later stages in the postdepositional alteration of muddy sediments.

Magnesium and potassium

The residence times of magnesium and potassium in the sedimentary rock mass are less than those of Fe, Ti, Al, and Si, but greater than those of Ca or Na, reflecting Mg and K's tendency to travel to the ocean in the dissolved, as well as suspended, load of streams. In the steady-state system, approximately 20% of the potassium and magnesium entering the ocean in dissolved form recycles through the atmosphere back to the continents.

Dissolved magnesium is derived from the weathering of carbonate minerals, (e.g., dolomite, magnesian calcites), and a variety of silicates (e.g., olivine, biotite). The schematic reactions are as follows:

$$MgCO_3 + CO_2 + H_2O = Mg^{2+} + 2HCO_3{}^- \tag{7}$$

and

$$MgSiO_3 + 2CO_2 + 3H_2O = Mg^{2+} + 2HCO_3{}^- + H_4SiO_4 \tag{8}$$

Notice that in both reactions CO_2 is consumed; in a steady-state system, these reactions would need to be reversed during deposition and burial for the ocean–atmosphere system to remain constant with respect to CO_2. It is known, however, that a large percentage of the dissolved magnesium in streams is derived from weathering of carbonate minerals but only a very small portion of this magnesium is being removed from the ocean in magnesian calcites and dolomite. Thus reaction 7 is not reversed significantly in the ocean or during early burial of sediments.

In the model presented here the magnesium balance is maintained by transfer of magnesium from shales to carbonates during advanced diagenesis. The higher

Mg : Ca ratio of Old Rocks reflects the tendencies for New Rocks to lose their calcium by selective leaching and for Mg to leave the shales of New Rocks during diagenesis and be incorporated in the carbonate rocks of Old Rocks.

Drever (1974) reviewed the magnesium balance in detail and concluded that present removal mechanisms of Mg from the ocean could account for only 50% of the river flux. These removal mechanisms involve carbonate mineral formation, ion exchange, glauconite formation, Mg–Fe exchange during burial of sediments, and burial of interstitial water. It is interesting to note, however, that complete removal of the stream flux of magnesium in interstitial waters would require a negative concentration gradient in pore waters of only 0.2 mg/l · cm over the entire sea floor. Gradients of this magnitude are not unusual for interstitial waters of sediments. Also, if Wollast's (1974) value of 3.5×10^{14} g/yr of silica removal by the neoformation of clay minerals is correct, and if only one-third of this silica enters minerals like sepiolite, then all the Mg entering the ocean could be removed by reaction 6. Drever pointed out, however, that if all the Mg is entering new mineral phases, the MgO content of marine sediments should be 1–2% greater than that of river sediments. Although data are sparse, the magnesium content of river sediments is not significantly lower than that of marine sediments; thus at present there is no satisfactory explanation for removal of all the stream-derived Mg flux from the ocean.

Sepiolite as a sink of magnesium may react with kaolinite to form chlorite during later diagenesis according to the reaction:

$$5Mg_2Si_3O_6(OH)_4 + 2Al_2Si_2O_5(OH)_4 + 20H_2O$$

$$\text{sepiolite} \qquad \text{kaolinite}$$

$$= 2Mg_5Al_2Si_3O_{10}(OH)_8 + 13H_4SiO_4 \quad (9)$$

$$\text{chlorite}$$

The silica generated may be lost from the shale via fluid movement and account in part for the decrease of SiO_2 in shales with increasing geologic age.

Potassium is derived principally from the weathering of potassium silicates. The weathering process involves CO_2 drain on the atmosphere. Mechanisms of removal of potassium from the ocean at present, such as zeolite formation or clay mineral exchange, are not of sufficient magnitude to account for all the potassium coming in via streams. Recently, Hart (1970, 1973) has suggested that a significant fraction of the stream-derived potassium may be removed by the formation of K-rich smectite during low-temperature weathering of basalts in the deep sea.

Potassium, like magnesium, tends to be retained in sediments; however, with increasing time the potassium present in the micas and feldspars in shales of New Rocks migrates into the interlayers of mixed-layer illite/montmorillonite. Also, it appears that the potassium-to-aluminum ratio of shaly rocks increases with time, suggesting that potassium in shales is enriched by additions of potassium from carbonate rocks and sandstones via subsurface waters, a reversal of the pathway suggested for magnesium.

Sodium and chlorine

The mineral halite dissolves to release Na^+ and Cl^- to streams. Sodium is also derived from the weathering of feldspars and other silicates. The residence time of sodium in rocks is about 70×10^6 yr longer than that of chlorine, reflecting the fact that part of the sodium is transported in the suspended load of streams. About two-thirds of the sodium in the sedimentary cycle circulates with chlorine. In a steady-state system, the sodium in streams balanced by chlorine would precipitate as halite in evaporite basins or be recycled from the sea back to the atmosphere. On a global basis, the values for the fluxes of sodium and chlorine from the sea to the continents are poorly known.

The remaining one-third of the sodium circulates through the sedimentary cycle associated with silicate minerals in shales. This sodium is found in the suspended load of streams, particularly as plagioclase feldspar, and in the dissolved load derived from the weathering of feldspars and balanced by bicarbonate and sulfate. Weathering of Na-silicates promotes a drain on the CO_2 of the atmosphere; maintenance of a CO_2 balance necessitates deposition of silicates accompanied by release of CO_2 equal in amount to that used during weathering. Silicate reactions involving sodium in the ocean today are precipitation of zeolites, and clay-mineral ion exchange. Hart (1973) suggested that the total flux of stream-derived sodium to the ocean can be accounted for by greenschist-grade metamorphism of basalts at ridge crests. It is true, however, that most of the sodium in the sedimentary cycle circulates with chloride, and halite deposits are distributed sporadically throughout the Phanerozoic rock column. Thus in times of very minor evaporite deposition, as today, it is likely that sodium and chlorine could accumulate in the ocean.

Sulfur

The cycling of sulfur through the sedimentary system is in reality much more complex than shown by the model. Estimates of total sulfur in sedimentary rocks have increased markedly, as major evaporite sequences have been discovered in unsuspected locations. For example, the Joint Oceanographic Institute's Deep Earth Sampling (JOIDES) program has drilled into evaporites in the deeps of the Gulf of Mexico and in the Mediterranean Sea.

As shown in Fig. 4, no sulfur moves directly from land into the atmosphere; all such movement is lumped into the transfer to the oceans via the dissolved load of streams. There is, in fact, a significant flux of H_2S and SO_2 from land directly to the atmosphere and thence to the ocean. It is derived from the partial oxidation of pyrite and of the sulfur in dead terrestrial organisms. In the model, this flux is put from rocks into streams, before being passed into the ocean.

The overall mobility of sulfur is emphasized, however, by its short residence time of 245×10^6 yr. Evidence is accumulating for a correlation between the mobilization of sulfur and magnesium in the sedimentary cycle. There has been a tendency to assume that sulfur in streams that comes from rocks is mostly balanced by calcium, derived from the minerals gypsum and anhydrite. It now appears that magnesium has as much influence on the movement of sulfur as does calcium. It was mentioned that Drever (1971b) tied fixation of magnesium in clay minerals to sulfate reduction and pyrite crystallization. The reverse of this process, oxidation of pyrite to sulfuric acid and iron oxide, with release of magnesium and other cations from shales by subsequent acid attack on the clay minerals, may account for one important tie between magnesium and sulfur in streams.

Whereas the total sulfur in sedimentary rocks may well be nearly constant with time, the ratio of oxidized to reduced sulfur has changed. The marked temporal variations in $\delta^{34}S$ of evaporite deposits are direct evidence for this conclusion. Most investigators (cf. Li, 1972) calculate that of the total sulfur in the ocean and in sedimentary rocks, sulfide and sulfate sulfur are about equal in mass today. McKenzie (1972) estimates that the ratio of reduced to oxidized sulfur deposited from the oceans may have ranged as much as $4:1$ to $1:4$. It is likely, but not yet proven, that times of major evaporite deposition in the Phanerozoic correlate with low $\delta^{34}S$ values of evaporite deposits and with low sulfate concentrations in seawater.

Garrels and Perry (1974) have recognized a major feedback loop in the sedimentary cycle involving sulfur-bearing minerals and atmospheric O_2 and CO_2. Sulfide weathering involves an O_2 drain on the atmosphere, resulting in the production of both ferric iron and sulfate. When these oxidized constituents are deposited in the sea, they react with organic matter originally derived from photosynthesis to make iron sulfide again. Carbon dioxide is released and through photosynthesis can react to restore O_2 to the atmosphere.

Today sulfur is either accumulating in the ocean or being removed chiefly as reduced sulfur by bacterial reduction on the sea floor (Berner, 1972). The rate of formation of oxidized sulfur minerals is negligible. On the other hand, because $\delta^{34}S$ in evaporite deposits has not changed markedly since the Cretaceous (Holser and Kaplan, 1966), the current situation probably has not endured more than a few million years.

Calcium and carbon

Calcium and carbon are intimately related. Calcium cycles through the system dominantly in the solution flux balanced by bicarbonate ion, but about 20% of the calcium circulates in the suspended load of streams as calcite or dolomite. Calcium is separated from magnesium when it is precipitated from the oceans; the skeletal materials of carbonate-secreting organisms are made up of nearly pure calcite, aragonite, or magnesian calcites that range up to 20 wt % magnesium carbonate. The material that eventually is buried in new sediments contains approximately 6 wt % magnesium carbonate (Chave, 1954), so as mentioned previously, a large part of the magnesium flux into new sediments must find mineral sinks other than carbonates.

The carbon cycle has been discussed in detail by many authors (cf. Bolin, 1970; Pytkowicz, 1967, 1972, 1973; Johnson, 1970). In the model of Fig. 4, attention is focused on the oceanic, atmospheric, and biospheric net fluxes necessary to maintain the inorganic–organic carbon balances between streams and rocks. The concentrations of dissolved organic carbon and of particulate organic carbon in streams are not well known (Stumm and Morgan, 1970), but the fluxes given are probably minimal. They are certainly much larger than the organic carbon flux into sediments from the ocean, giving rise to the requirement of a significant net of oxidation over photosynthesis in the oceans, and a reverse requirement for the land. On the other hand, the flux of organic carbon into the atmosphere from the oxidation of old carbon in the weathering of rocks must almost exactly balance the flux of organic carbon into New Rocks, for the organic content of rocks is nearly independent of rock age (Broecker, 1970). The average overall land and sea net of photosynthesis over oxidation is only 0.1% or so of the total mass of C photosynthesized, as determined from the average organic content of present-day sediments and the mass of sediments deposited each year. The restoration of this tiny deficiency through oxidation of old carbon that has passed through the sedimentary cycle and is currently being weathered shows that over geologic time intervals, the huge amounts of carbon fixed and oxidized may have differed by even less than 0.1%.

In the balance shown among streams, ocean, and atmosphere, there is an interesting compensation between inorganic and organic carbon that is required for steady state. About 80% of the inorganic carbon flux into streams is required for the weathering of carbonate minerals, and about 20% for the weathering of silicates (Garrels and Mackenzie, 1971; Li, 1972). Furthermore, the average carbonate mineral weathered contains several times as much magnesium as that deposited as carbonates in the ocean. Therefore, only the CO_2 abstracted from the atmosphere in the weathering of the calcium component of carbonates is restored to the atmosphere when carbonates are precipitated from the ocean, leaving more than 20% of the CO_2 converted to bicarbonate in streams (that used for silicate minerals plus that required for $MgCO_3$) to accumulate in the oceans. If there were no compensatory mechanism, carbon dioxide would be drained out of the atmosphere in a few tens of thousand years. In the model the drain is balanced by CO_2 restored to the atmosphere from part of the net of oxidation over photosynthesis in the oceans. This relation defines an important site of interchange between the organic and inorganic carbon cycles. The total flux of organic carbon out of the sea as carbon dioxide, 3.2×10^{14} g/yr, is of the same order of magnitude as that calculated by Kroopnick (1971) from his studies of oxidation of carbon in the deep sea, about 2.5×10^{14} g/yr.

The residence time of inorganic carbon in rocks, 381×10^6 yr, is expected from the ready solubility of carbonate minerals, which leads to a high flux of inorganic carbon out of New Rocks, as a consequence of both diagenesis and weathering (Perry and Hower, 1970).

B. Trace and Minor Elements

Some General Comments

Trace elements, that is, those elements that usually occur in low concentrations in rocks, soils, waters, atmosphere, and biota, have received considerable attention during the past decade because some are known to be toxic at low concentrations in the human body. Examples of such elements are mercury, lead, arsenic, cadmium, selenium, and vanadium.

Because of the lack of numerical data, few global models of trace element cycles have been developed; therefore, their environmental impact on a global basis is poorly known. However, because of man's activities, local concentrations of some of these elements have been multiplied many times, particularly in waters, so that they are toxic to man. On a global basis an estimate of man's contribution of metals to their natural exogenic cycles can be made by comparison of mining production, emission rates to atmosphere owing to man's activities, worldwide atmospheric rainout, and total river load of the element (Goldberg, 1971; Table III).

Calculation of the values for the rainout of metals from the atmosphere to the earth's surface deserves some explanation. The calculation is based on estimates of the concentrations of metals in atmospheric particulates recovered from "clean air." These concentration values in units of nanograms per cubic meter (ng/m^3) were then multiplied by the atmospheric volume up to an elevation of 5000 m. The standard assumption was then made that this volume of air is completely washed free of particulates 40 times each year; this is equivalent to a rain about every 10 days. For nickel, for example, the calculation is as follows:

$$\text{(concentration in air} = 1.2 \text{ ng/m}^3)\text{(volume of atmosphere to 5000 m} =$$
$$510 \times 10^{16} \text{ cm}^2 \times 5 \times 10^5 \text{ cm} = 2.55 \times 10^{22} \text{ cm}^3 = 2.55 \times 10^{18} \text{ m}^3)$$
$$\text{(rainouts per year, 40)} = 0.12 \times 10^{12} \text{ g Ni/yr}$$

Table III shows that mining production for many metals approaches the total stream load; indeed for the metals Pb, Cu, Cr, Hg, and Zn, mining production apparently equals or exceeds the rate at which the metal is transported to the oceans in the dissolved and particulate load of streams. The emission rate of Pb and Hg to

TABLE III

Comparison of Mining Production of Some Metals, Metal Emission Rates to Atmosphere Owing to Man's Activities, Worldwide Atmospheric Rainout, and Total Stream Load (Units of 10^{12} g/yr)[a]

Metal	Mining	Emission	Atmospheric Rainout	Stream Load	Interference Index (Atmosphere) (Emission/Rainout) × 100(%)
Pb	3	0.40	0.31	0.42	129
Cu	6	0.21	0.19	0.82	111
V	0.02	0.09	0.02	2.4	450
Ni	0.48	0.05	0.12	1.2	42
Cr	2	0.05	0.07	1.7	71
Sn	0.2	0.04	—	0.27	—
Cd	0.014	0.004	—	0.04	—
As	0.06	0.05	0.19	0.3	26
Hg	0.009	0.01	0.008	0.005	125
Zn	5	0.73	1.04	1.8	70
Se	0.002	0.009	0.03	0.02	30
Ag	0.01	0.003	—	0.03	—
Sb	0.07	0.03	0.03	0.09	100

[a] Mining production from *Statistical Abstract*, 1973; emission rates calculated from average of United States and European urban atmosphere metal concentrations (cf. Harrison et al., 1971; Dams, 1974), average urban atmosphere particulate load of 100 $\mu g/m^3$, and total 1968 worldwide particulate production of 193.9×10^{12} g; estimates of worldwide atmospheric rainout from "clean air" metal concentrations (cf. Dams, 1974; Hoffman et al., 1972; Chester and Stoner, 1974), atmospheric volume to 5000 m of 2.55×10^{18} m^3, and 40 "rains"/y; stream load calculated from worldwide runoff (0.32×10^{20} g/yr), average metal concentrations in streams (Turekian, 1969), and total worldwide river suspended load of 183×10^{14} g/yr (Holeman, 1968) and average metal concentrations of shale (cf. Krauskopf, 1957) to represent river-suspended load.

the atmosphere from man's activities already equals or exceeds the natural stream flux, and for the metals Cu, Sn, Cd, As, Zn, Se, Ag, and Sb, emission rates are within an order of magnitude of the stream fluxes—values well within the calculation errors (owing to poor data for emission rates and for concentrations of metals in the dissolved and particulate loads of streams) inherent in Table III.

Comparison of emission rates to the atmosphere with atmospheric rainout is particularly informative. Regardless of the errors inherent in these calculations, it is evident from the interference indices in Table III that emission rates for all the metals shown are within an order of magnitude of rates of particulate fallout over the earth's surface, suggesting that metal input to the atmosphere owing to industrial activities, combustion of fossil fuels, etc., rivals natural inputs.

It is also interesting to note that some of those elements, such as Hg, As, Se, and Pb, that are considered as potentially detrimental to human health are selectively emitted into the earth's atmosphere by natural processes. Sedimentary rocks are a major source of atmospheric particulates. Figure 6 illustrates that those trace elements with relatively low oxide boiling points (a measure of volatility) are more concentrated in the atmosphere with respect to their content in sedimentary rocks than elements with high boiling points. In other words, elements like Se, As, and Cd have high volatilities; elements like Pb, Zn, and Ni have moderate volatilities; and Al, Ti, and Sc are present in the atmosphere at concentrations similar to their crustal source. Figure 6 also shows that significant concentrations of some high volatility elements are present in analyses of atmospheric particulates at lower concentrations than would be expected if oxide boiling points are a reasonable index of potential volatility. These elements are probably present in the atmosphere in a state other than particulate—possibly gaseous. Recently, Johnson and Brannan (1974) showed that mercury in the atmosphere over Tampa Bay, Florida, is primarily in "volatile" form, present as gaseous mercury(II)-type compounds, methylmercury(II)-type compounds, and metallic mercury vapor. Mercury in particulates composed *less* than 10% of total mercury in the atmosphere. Mercury, as well as As and Se, in organic compounds in soils and waters can be volatilized by bacterial processes (cf. Wood, 1974; Wollast, Billen, and Mackenzie, 1975), and selectively emitted to the atmosphere. Zoller et al. (1974) point out that high element enrichment factors in atmospheric particulates suggest an initial vapor phase for the element owing to either high- or low-temperature processes. Obvious natural processes are volcanism, biological mobilization, and fractionation at the sea surface during production of atmospheric sea salt particles.

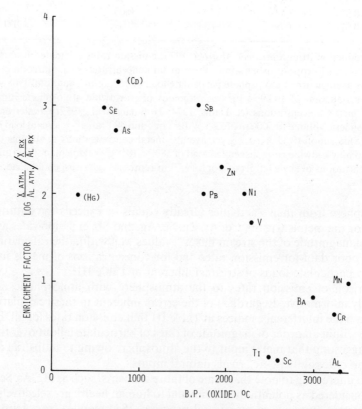

Fig. 6.   Enrichment factors for elements in atmospheric particulars plotted as a function of boiling points of the oxides of the elements. Oxide boiling point is considered to be a measure of the "volatility" of an element.

We further suggest that those elements with high enrichment factors fall into two groups: (1) elements, like Hg, As, and Se, that are emitted to the atmosphere in vapor form and later removed as dissolved gases in rain; and (2) elements, like Pb, Zn, and V, which also in part may be released to the atmosphere in vapor form but condense in the atmosphere and are removed principally as solid particles in rain or dry fallout.

The important point is that some of the trace elements that are toxic at low concentrations are selectively released to the atmosphere from crustal materials. It is known that industrial activities involving fuel combustion and production of stack dust and gases also release these elements to the atmosphere. Indeed, it is difficult to determine whether an atom of Hg in the atmosphere is derived from natural processes or from man's activities (see Section 6.B).

Mercury

The details inherent in the construction of trace metal global cycling models are discussed here. Mercury is used as an example to provide an appreciation for the types of data and approximations used in construction of these models.

An estimate of the present-day cycle of mercury in the global environment is shown in Fig. 7. Tables IV and V give sources of data used to calculate mercury reservoir contents and transfers between reservoirs of Fig. 7. Details of the calculations are given in the reference and remarks column of the tables. To facilitate discussion, the global model of Fig. 7 was reduced to the four-box model of Fig. 8, and a pre-man global model of mercury cycling was constructed (Fig. 9). The basis for construction of simplified pre-man and present-day models of mercury is given below.

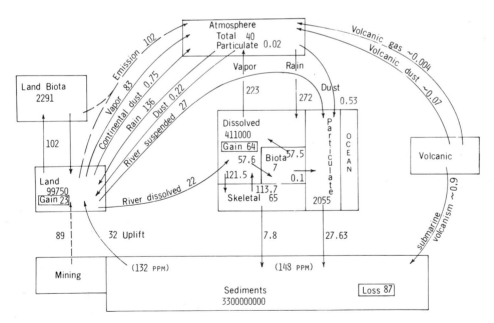

Fig. 7.   Present-day global cycle of mercury. Fluxes are in units of 10^8 g/yr; reservoir masses in units of 10^8 g. Sources of data and calculations are given in Tables IV and V.

TABLE IV

Mercury Content of Geochemical Reservoirs

Reservoir	Hg Content $(10^8$ g)	References and Remarks
Sediments	3,300,000,000	Computed from average values of Hg in shales (150 ppb), sandstones (75 ppb), and carbonates (75 ppb) as obtained from literature search and evaluation (cf. Saukov, 1946; Stock and Cucuel, 1934; Turekian and Wedepohl, 1961; Cameron and Jonasson, 1972; USGS PP 713, 1970); proportions of shale, carbonate, and sandstone in sedimentary mass of 0.75, 0.14, and 0.11, respectively; and total sedimentary mass of 25000×10^{20} g (Garrels and Mackenzie, 1971) $$(0.75 \times 150 \times 10^{-9} + 0.14 \times 75 \times 10^{-9} + 0.11 \times 75 \times 10^{-9})(25000 \times 10^{20})$$ $$= 3,3000,000,000 \times 10^8 \text{ g}$$
Land	99,750	Computed from land area (total land less ice, 133×10^{16} cm^2), assumed soil thickness (60 cm), density (2.5 g/cm^3) and mean Hg content of soil (50 ppb, see USGS PP 713, 1970) $$133 \times 10^{16} \times 60 \times 2.5 \times 50 \times 10^{-9}$$ $$= 99,750 \times 10^8 \text{ g}$$
Land biota	2,291	Computed from carbon content of land biota (450×10^{15} g C; Bolin, 1970) and mean Hg:C weight ratio in land plants (5.09×10^{-7}; Deevey, 1970; Shacklette, 1970) $$450 \times 10^{15} \times 5.09 \times 10^{-7} = 2291 \times 10^8 \text{ g}$$
Oceanic biota (organic fraction)	7	Computed from carbon content of oceanic biota (5×10^{15} g C; Bolin, 1970) and mean Hg:C weight ratio in oceanic biota (1.44×10^{-7}; Redfield et al., 1963; Martin and Knauer, 1973) $$5 \times 10^{15} \times 1.44 \times 10^{-7} = 7 \times 10^8 \text{ g}$$
Dissolved (oceanic)	411,000	Computed from total volume of ocean (1.37×10^{21} l) and mean concentration of dissolved Hg (0.030 μg/l; see USGS PP 713; Leatherland et al., 1971) $$1.37 \times 10^{21} \times 30 \times 10^{-9} = 411,000 \times 10^8 \text{ g}$$

TABLE IV (*Continued*)

Reservoir	Hg Content $(10^8 g)$	References and Remarks
Skeletal particulate (oceanic)	65	Computed from mean suspended particulate concentration of world's ocean (1 mg/l), mean percentage of skeletal particulate in oceanic surface layer (15%, Lisitzin, 1972), oceanic surface layer depth (300 m), oceanic surface area $(360 \times 10^{12} \text{ m})$, and mean Hg concentration of skeletal particulate (0.40 ppm; Martin and Knauer, 1973)

$$360 \times 10^{16} \times 3 \times 10^4 \times 1 \times 10^{-6} \times 0.15$$
$$\times 0.40 \times 10^{-6} = 65 \times 10^8 \text{ g}$$

Terrigenous particulate (oceanic)	2055	Computed from mean suspended particulate concentration of world's oceans of 1 mg/l; mean concentration of Hg in shales (150 ppb, see first item), and oceanic volume (1.37 $\times 10^{21}$ l). Includes organic and nonskeletal mineral particulate

$$1.37 \times 10^{21} \times 1 \times 10^{-3} \times 150 \times 10^{-9}$$
$$= 2055 \times 10^8 \text{ g}$$

Atmosphere (total Hg)	40	Computed from mass of atmosphere $(5.1 \times 10^{21} \text{ g})$, density of atmosphere of $1.3 \times 10^3 \text{ g/m}^3$, and mean atmospheric Hg content (1 ng/m³; Weiss, et al., 1971)

$$5.1 \times 10^{21} \div 1.3 \times 10^{-3} \times 1 \times 10^{-9}$$
$$= 40 \times 10^8 \text{ g}$$

Atmosphere (particulate Hg)	0.02	Computed from mean dust content of atmosphere $(5 \times 10^{-6} \text{ g/m}^3$, Goldberg, 1971), assumed height of significant dust concentration in atmosphere (5000 m), surface area of earth $(510 \times 10^{12} \text{ m}^2)$, and mean concentration of Hg in shales (see first item)

$$510 \times 10^{12} \times 5 \times 10^3 \times 5 \times 10^{-6} \times 150 \times 10^{-9}$$
$$= 0.02 \times 10^8 \text{ g}$$

767

TABLE V

Fluxes of Mercury Between Geochemical Reservoirs[a]

Direction of Flux	Flux (10^8 g/yr)	References and Remarks
Land to:		
Land biota	102	Computed from Bolin's (1970) value for total amount of carbon fixed annually by land plants (2×10^{16} g C/yr); and mean Hg:C weight ratio in land plants (5.09×10^{-7}; Deevey, 1970; Shacklette, 1970)
		$2 \times 10^{16} \times 5.09 \times 10^{-7} = 102 \times 10^8$ g/yr
Atmosphere (vapor)	83	Computed from total pre-man mercury flux to atmosphere of 250×10^8 g/yr (Weiss et al., 1971) and assumption that at steady state about one-third of this flux is derived from land and returned to land (approximately proportional to land area)
		$0.33 \times 250 \times 10^8 = 83 \times 10^8$ g/yr
Atmosphere (continental dust)	0.75	Computed from total continental dust flux (5×10^{14} g/yr; Goldberg, 1971), and mean concentration of Hg in shales (150 ppb) as an estimate of the Hg concentration in continental dust
		$5 \times 10^{14} \times 150 \times 10^{-9} = 0.75 \times 10^8$ g/yr
Ocean (river suspended)	27	Computed from total suspended load of world's rivers (183×10^{14} g/yr; Garrels and Mackenzie, 1971) and mean concentration of Hg in shales (150 ppb) as an estimate of the Hg concentration in river-suspended material
		$183 \times 10^{14} \times 150 \times 10^{-9} = 27 \times 10^8$ g/yr
Ocean (river dissolved)	22	Computed from total yearly runoff (3.2×10^{16} l/yr; Garrels and Mackenzie, 1971) and mean concentration of Hg in rivers (0.07 ppb; Turekian, 1969)
		$3.2 \times 10^{16} \times 70 \times 10^{-9} = 22 \times 10^8$ g/yr

[a] Reservoir gains or losses computed on basis of difference between inputs and outputs to reservoir. Sediments deposited today contain an average mercury concentration of 148 ppb, whereas those of the past contain 132 ppb.

TABLE V (*Continued*)

Direction of Flux	Flux (10^8 g/yr)	References and Remarks
Atmosphere (emission)	102	Computed on the basis of estimates for the following sources of mercury vapor to the atmosphere:

chlor–alkali production 30.0×10^8 g/yr
coal–lignite combustion 27.4
oil–gas combustion 22.9
cement manufacturing 1.2
roasting sulfide ores 20.0
$\overline{}$
$101.5 \simeq 102$

Atmosphere to: Land (rain)	136	Computed from total present-day mercury flux to atmosphere of 408×10^8 g/yr (see text for discussion) and assumption that about one-third of this flux is returned to land from the atmosphere (approximately proportional to land area)

$$0.33 \times 408 \times 10^8 = 136 \times 10^8 \text{ g/yr}$$

Land (dust)	0.22	Computed on the assumption that the continental dust flux (0.75×10^8 g/yr) will fall out over the ocean and land in proportion to their areas (30% land)

$$0.30 \times 0.75 \times 10^8 = 0.22 \times 10^8 \text{ g/yr}$$

Ocean (rain)	272	Computed from total present-day mercury flux to atmosphere of 408×10^8 g/yr (see text for discussion) and assumption that about two-thirds of this flux is returned to the ocean from atmosphere (approximately proportional to ocean area)

$$0.67 \times 408 \times 10^8 = 272 \times 10^8 \text{ g/yr}$$

Ocean (dust)	0.53	Computed on the assumption that the total continental dust flux will fall out over the ocean and land in proportion to their areas (70% ocean)

$$0.70 \times 0.75 \times 10^8 = 0.53 \times 10^8 \text{ g/yr}$$

TABLE V (*Continued*)

Direction of Flux	Flux (10^8 g/yr)	References and Remarks
Ocean to atmosphere	223	Computed from difference between total present-day mercury flux to atmosphere (408×10^8 g/yr) and fluxes of land-derived emission (102×10^8 g/yr) and vapor (83×10^8 g/yr) to atmosphere
		$$408 \times 10^8 - (102 \times 10^8 + 83 \times 10^8)$$ $$= 223 \times 10^8 \text{ g/yr}$$
Ocean Skeletal to sediments	7.8	Computed on basis of total mass of SiO_2 and $CaCO_3$ skeletons deposited per year (7.5×10^{14} g/yr, 1.2×10^{15} g/yr, respectively; Wollast, 1974; Garrels and Mackenzie, 1971) and mean skeletal Hg concentration of 0.4 ppm (Martin and Knauer, 1973)
		$$1.95 \times 10^{15} \times 0.4 \times 10^{-6} = 7.8 \times 10^8 \text{ g/yr}$$
Dissolved to biota	57.6	Computed from rate of carbon fixation by oceanic biota (4×10^{16} g C/yr; Bolin, 1970) and mean Hg:C ratio of organic fraction of oceanic biota (1.44×10^{-7}; Redfield et al., 1963; Martin and Knauer, 1973)
		$$4 \times 10^{16} \times 1.44 \times 10^{-7} = 57.6 \times 10^8 \text{ g/yr}$$
Biota to particulate	0.1	Computed to provide a maximum estimate of flux of Hg to particulate reservoir from newly dead oceanic biota. Based on sedimentation rate of organic carbon (73.2×10^{12} g/yr; Garrels et al., 1974) and mean Hg:C weight ratio of oceanic organic matter (1.44×10^{-7}; Redfield et al., 1963; Martin and Knauer, 1973)
		$$73.2 \times 10^{12} \times 1.44 \times 10^{-7} = 0.1 \times 10^8 \text{ g/yr}$$
Biota to dissolved	57.5	Computed on basis of rate of fixation of Hg by oceanic biota (57.6×10^8 g/yr) less that sedimented in organic matter (0.1×10^8 g/yr)
		$$57.6 \times 10^8 - 0.1 \times 10^8 = 57.5 \times 10^8 \text{ g/yr}$$
Dissolved to skeletal	121.5	Computed on basis of SiO_2 uptake in photic zone (250×10^{14} g/yr; Wollast, 1974), $CaCO_3$ uptake in photic zone (53.7×10^{14} g/yr; Pytkowicz, 1973) and mean Hg concentration of skeletal material (0.4 ppm; Martin and Knauer, 1973)
		$$303.7 \times 10^{14} \times 0.4 \times 10^{-6} = 121.5 \times 10^8 \text{ g/yr}$$

TABLE V (*Continued*)

Direction of Flux	Flux (10^8 g/yr)	References and Remarks
Skeletal to dissolved	113.7	Computed on basis of dissolved to skeletal flux (121.5×10^8 g/yr) minus skeletal to sediment flux (see above, 7.8×10^8 g/yr) $121.5 \times 10^8 - 7.8 \times 10^8 = 113.7 \times 10^8$ g/yr
Particulate to sediments	35.43	Computed by summation of all particulate inputs into sediment; skeletal (7.8×10^8 g/yr), particulate organic (0.1×10^8 g/yr), dust (0.53×10^8 g/yr), and river suspended (27×10^8 g/yr) $7.8 \times 10^8 + 0.1 \times 10^8 + 0.53 \times 10^8 + 27 \times 10^8$ $= 35.43 \times 10^8$ g/yr
Land biota to land	102	Estimate balanced by flux from land to land biota
Volcanic to: Atmosphere (gas)	0.0044^b	Computed from estimate of steady state release of "excess volatiles" (4.4×10^{14} g/yr; Rubey, 1951; Garrels and Mackenzie, 1971) to earth's surface during geologic time, and assumption that mercury concentration of condensed volatiles is similar to that of volcanic fumaroles and hot spring waters (average about 1 ppb; USGS PP 713, 1970) $4.4 \times 10^{14} \times 1 \times 10^{-9} = 0.0044 \times 10^8$ g/yr
Atmosphere (dust)	0.07^b	Computed from estimate of volcanic dust flux (1.5×10^{14} g/yr; Goldberg, 1971) and mean concentration of Hg in igneous rocks (47 ppb) $1.5 \times 10^{14} \times 47 \times 10^{-9} = 0.07 \times 10^8$ g/yr
Sediments	0.9^b	Computed from estimate of total Mn released to sediments yearly by submarine volcanism ($\sim 300 \times 10^{10}$ g/yr; see Corliss, 1970; Hart, 1970, for data) and Hg:Mn weight ratio of basalts (3.13×10^{-5}) $300 \times 10^{10} \times 3.13 \times 10^{-5} = 0.9 \times 10^8$ g/yr
Sediments to land	32	Computed on basis of present-day denudation rate ($=$depositional rate; 240×10^{14} g/yr; Garrels and Mackenzie, 1971) and mean Hg content of sedimentary rocks (132 ppb) $240 \times 10^{14} \times 132 \times 10^{-9} = 32 \times 10^8$ g/yr
Mining	89	Estimate from *Statistical Abstract*, 1973

^{b} Flux not included in balance.

PRESENT   CYCLE

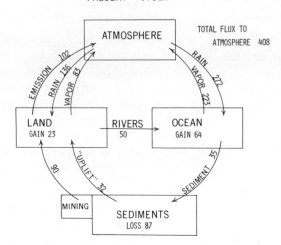

Fig. 8.   Simplified global model of present-day cycle of mercury. Fluxes are in units of 10^8 g/yr; reservoir masses in units of 10^8 g. Total flux to atmosphere from earth's surface is 408×10^8 g/yr.

PRE·MAN   CYCLE

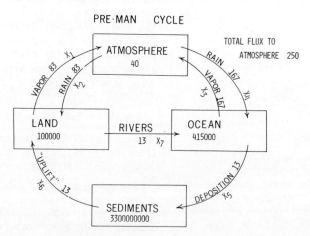

Fig. 9.   Simplified global model of pre-man cycle of mercury. Fluxes are in units of 10^8 g/yr; reservoir masses in units of 10^8 g. Total flux to atmosphere is 250×10^8 g/yr. X_i, where $i = 1, 2, 3, 4, 5, 6, 7$, represents notation used in text for fluxes.

Basis for Construction of mercury cycles

The calculation procedure is based on the general principles outlined in Section 5.

Reservoirs and masses. Four major reservoirs and masses were considered, as follows.

ATMOSPHERE: Hg mass computed from mass of atmosphere (5.2×10^{21} g), density of atmosphere (1.3×10^3 ng/m^3), and mean atmospheric mercury content (1 ng/m^3).

$$5.2 \times 10^{21} \div 1.3 \times 10^3 \times 1 \times 10^{-9} = 40 \times 10^8 g$$

OCEANS: Hg mass computed from total oceanic volume (1.37×10^{21} l) and mean concentration of dissolved Hg in seawater (0.03 μg/l).

$$1.37 \times 10^{21} \times 30 \times 10^{-9} = 411{,}000 \times 10^8 \text{ g}$$

Mercury in suspended matter in the ocean is about 2100×10^8 g; this value added to the dissolved mass gives a total oceanic Hg burden of about $415{,}000 \times 10^8$ g.

SEDIMENTS: Hg mass computed from average values of Hg in shales (150 ppb), sandstones (75 ppb), and carbonates (75 ppb); relative proportions of shale, carbonate, and sandstone in sediments (75:14:11, respectively); and total sedimentary mass ($25{,}000 \times 10^{20}$ g).

$$(0.75 \times 150 \times 10^{-9} + 0.14 \times 75 \times 10^{-9} + 0.11 \times 75 \times 10^{-9})$$
$$\times (25{,}000 \times 10^{20}) = 3{,}300{,}000{,}000 \times 10^8 \text{ g}$$

LAND: Hg mass computed from land area less that covered by ice (133×10^{16} cm²), average worldwide soil thickness (60 cm), soil density (2.5 g/cm³), and mean Hg content of soil (50 ppb).

$$133 \times 10^{16} \times 60 \times 2.5 \times 50 \times 10^{-9} = 100{,}000 \times 10^8 \text{ g}$$

Pre-man cycle. For each reservoir, assuming the pre-man cycle of mercury was in steady state, we may write the following equations representing fluxes into and out of a reservoir:

$$x_1 + x_3 = x_2 + x_4 \text{ (atmosphere)}$$
$$x_4 + x_7 = x_3 + x_5 \text{ (oceans)}$$
$$x_5 = x_6 \text{ (sediments)}$$
$$x_2 + x_6 = x_1 + x_7 \text{ (land)}$$

where x represents material flux in units of grams per year (see Fig. 9). Thus there are seven unknowns but only four equations. The following three conditions enable the solution of the array of simultaneous linear algebraic equations:

1. $x_2 = x_4/0.5$; Hg fallout on land and sea surfaces is proportional to their surface areas.
2. $x_5 = x_6 = 13 \times 10^8$ g/yr; steady-state assumption computed by using the pre-man denudation rate of the continents (100×10^{14} g/yr) and the average Hg content of sediments (132 ppb).

$$100 \times 10^{14} \times 132 \times 10^{-9} = 13 \times 10^8 \text{ g/yr}$$

Major assumption is that the rate of supply of "new" land surface via uplift is equal to rate of denudation of continents.
3. $x_1 + x_3 = 250 \times 10^8$ g/yr $= x_2 + x_4$; steady-state assumption computed by using the average Hg content of Greenland ice from 800 BC to 1952 (0.06 ppb) and assuming that the concentration in precipitation averages the same worldwide. Total global precipitation is 4.2×10^{20} g/yr, so total mercury flux to earth's surface is 60×10^{-12} g Hg/g precipitation $\times 4.2 \times 10^{20}$ g precipitation $= 250 \times 10^8$ g/yr (Weiss et al., 1971).

Present cycle. The present cycle of mercury is not in steady state because the land and ocean reservoirs are gaining Hg, whereas the sedimentary rock reservoir is being depleted of mercury. The atmosphere is a steady-state reservoir. To derive the present cycle of Hg, the following calculations and conditions were satisfied:

1. Flux from oceans to sediments computed on basis of (a) mass of siliceous and calcareous skeletons deposited annually (7.5×10^{14} g/yr and 1.2×10^{15} g/yr, respectively) and average skeletal Hg concentration (0.4 ppm),

$$1.95 \times 10^{15} \times 0.4 \times 10^{-6} = 7.8 \times 10^8 \text{ g/yr}$$

(b) steady-state assumption that river-suspended load, calculated from total suspended load of world's rivers (183×10^{14} g/yr) and average Hg concentration of shales (150 ppb) ($183 \times 10^{14} \times 150 \times 10^{-9} = 27 \times 10^8$ g/yr), passes through the ocean without reaction; and (c) steady-state assumption that airborne dust falling on ocean surface, calculated from continental dust flux (5×10^{14} g/yr), average Hg concentration of shales (150 ppb), and fallout of 70% of the dust on the sea surface ($5 \times 10^{14} \times 150 \times 10^{-9} \times 0.70 = 0.53 \times 10^8$ g/yr), passes through the ocean without reaction. Summation gives:

$$7.8 \times 10^8 + 27 \times 10^8 + 0.53 \times 10^8 = 35 \times 10^8 \text{ g/yr}$$

2. Flux from sediments to land computed on basis of present denudation rate (240×10^{14} g/yr) and average Hg content of sediments (132 ppb):

$$240 \times 10^{14} \times 132 \times 10^{-9} = 32 \times 10^8 \text{ g/yr}$$

3. Flux from land to oceans computed by summation of dissolved Hg load of world's rivers, calculated from total yearly runoff (0.32×10^{20} g/yr) and average Hg concentration of world's rivers (0.07 ppb) ($0.32 \times 10^{20} \times 0.07 \times 10^{-9} = 22 \times 10^8$ g/yr), and particulate Hg load of world's rivers, calculated from total suspended load (183×10^{14} g/yr) and average concentration of Hg in shales (150 ppb) ($183 \times 10^{14} \times 150 \times 10^{-9} = 27 \times 10^8$ g/yr). Summation gives $22 \times 10^8 + 27 \times 10^8 = 50 \times 10^8$ g/yr).

4. Mining flux of 90×10^8 g/yr taken from *Statistical Abstract*, 1973.

5. Total emission flux of 102×10^8 g Hg/yr obtained from summation of atmospheric Hg emissions owing to man's activities:

Source	Emission Rate (10^8 g/yr)
Chlor–alkali production	30.0
Coal–lignite combustion	27.4
Oil–gas combustion	22.9
Cement manufacturing	1.2
Sulfide ore roasting	20.0
	101.5 = 102

6. Condition that the ocean to atmosphere flux increased by one-third of the pre-man flux to 223×10^8 g Hg/yr. Using this condition and the above fluxes, the remaining fluxes of Fig. 8 can be obtained and the changes in reservoir masses calculated. Justification for this condition is that the present total flux of Hg to the atmosphere from the sea and land surfaces is about 410×10^8 g/yr, a value in accord with an increase of Hg in glacial ice of about twice pre-man concentrations. Weiss et al. (1971) observed that precipitation on the Greenland ice sheet prior to 1952 contained

0.06 $\pm$ 0.02 ppb Hg, whereas those waters deposited as ice from 1952 to 1965 contained 0.13 $\pm$ 0.05 ppb Hg, an observation in good agreement with the model prediction.

Discussion

Major transfer in the pre-man sedimentary cycle was between the atmosphere and the earth's surface. The total flux of 250×10^8 g/yr is 20 times greater than that involved in land–stream–ocean–sediment transfer. The underlying cause is high volatility of metallic mercury. This volatility is strikingly demonstrated by the fact that the ratio Hg/Al in atmospheric particulates is 200 times greater than the Hg/Al ratio of crustal rocks, the major source of particulates. Organomercuric substances in soil and sediments are known to degrade, releasing mercury vapor to the atmosphere. This process is bacterially mediated and probably also accounts for release of mercury from surface waters.

Weiss et al.'s (1971) maximum estimate of the degassing flux to the atmosphere of 1.5×10^{11} g Hg/yr probably represents an extreme value because they assume Hg is washed out of the *whole* atmosphere 40 times a year. It is likely that the concentration of Hg in the atmosphere shows an exponential decrease with increasing elevation similar to other gases originating at the earth's surface (CO for example). Thus significant mass transfer to the earth's surface is limited to the lower portion of the atmosphere. Also, they assume the washout rate represents only the land degassing rate; however, as we show in our models, a significant portion of the total flux to the atmosphere comes from the sea surface.

Indeed, the total flux of mercury to the atmosphere (2.5×10^{10} g/yr) prior to man's activities recorded in the Greenland snow by Weiss et al. may not be explained by considering only degassing in the soil zone. At the present rate of denudation (6.4×10^{-3} cm/yr) and for a mean difference between the concentration of Hg in shales and soils (100 ppb), total volatilization would represent only 21×10^8 g/yr. However, rain contains rather large amounts of mercury, probably as Hg^0, which may be easily recycled from soils to the atmosphere. During the methylation–demethylation cycle, volatile Hg^0 species represents an important intermediary product and may account for release of mercury to the atmosphere from soils, sediments, and surface waters.

In the case of seawater, for example, the occurrence of 0.0001 μgHg/l as Hg^0 is enough to assure a transfer of 167×10^8 g Hg/yr from the oceans to the atmosphere, assuming an exchange coefficient across the air–water interface D/Z equal to 7×10^4 cm/yr (Broecker, 1963).

The vapor is washed out of the atmosphere to the land and sea surfaces during rains. There is uncertainty about the details of the process; presumably mercury vapor dissolves in water droplets of clouds as elemental mercury. Figure 9 shows that vapor transport from the land and sea surfaces to the atmosphere before man was balanced by equal transfer of mercury in rain back to these surfaces.

The residence times of mercury in the four reservoirs were as follows:

Atmosphere	60 days
Land	1000 yr
Ocean	3200 yr
Sediments	2.5×10^8 yr

The relatively short residence time of mercury in the atmosphere reemphasizes the high vapor pressure of metallic mercury.

The pre-man mercury cycle serves as background for the present-day cycle (Fig. 8). The most obvious differences between the pre-man and present cycles are the higher fluxes between reservoirs of the present cycle. These increased fluxes are principally due to the increased rate of mercury input into the land reservoir from mining, from emissions of mercury to the atmosphere during chlor–alkali production, combustion of fossil fuels, cement manufacturing, and roasting of sulfide ores, and by many other processes in which metallic mercury is involved.

The estimated total vapor flux of mercury to the atmosphere in the model of the present cycle is 408×10^8 g/yr, an increase of about 60% over the pre-man cycle. The input to the atmosphere is assumed to be removed in rain over the land and sea in proportion to their surface areas. The residence time of 36 days for mercury in the atmosphere justifies this assumption. Rain falling on the ocean results in a net input of mercury into the ocean surface of 49×10^8 g/yr; this value plus the addition of Hg to oceans via streams (50×10^8 g/yr) minus the sedimentation rate (35×10^8 g/yr) gives an accumulation of mercury in the ocean of 64×10^8 g/yr. Although some of this mercury enters living biota, most of it is dissolved in seawater.

Several factors justify this conclusion. The total mass of mercury in living oceanic biota today is only 7×10^8 g, and the mercury concentrations in the organic and skeletal phases of the biota are only 0.12 and 0.40 ppm, respectively. The uptake of mercury in the photic zone by the biota is about 58×10^8 g Hg/yr. It would be expected that if a large part of the oceanic gain of mercury each year entered the biota the total mass of mercury and its concentration in the biota would be greater than observed. The calculated gain in the ocean exceeds the biotic uptake rate. A simple calculation further justifies the conclusion: assuming an ocean surface area of 360×10^{16} cm^2 and a mixed layer depth of the ocean of 200 m, the rate of addition of mercury to this layer would result in an increase of concentration of about 0.0001 g Hg/l · yr. This increase in total mercury is large enough to increase elemental mercury (Hg^0) concentration sufficiently in seawater to account for the increased rate of evasion of mercury vapor from the sea surface today as compared with pre-man time (223×10^8 versus 167×10^8 g/yr).

It is interesting to compare further the present total flux of mercury vapor to the atmosphere with the pre-man flux. The present rate is about 1.6 times the pre-man rate. Thus we would predict that concentrations of mercury in glacial ice formed during recent time would be about twice as great as those of the past. Weiss et al. (1971) observed that waters deposited in the Greenland ice sheet prior to 1952 contained 0.06 ± 0.02 ppb Hg, whereas those waters deposited as ice from 1952 to 1965 contained 0.13 ± 0.05 ppb Hg, an observation in harmony with the conclusions of the preceding discussion.

The net result of mining and utilization of mercury on land and the increased mercury content of rain over land is a calculated increase of about 0.02% in the mercury content of the land surface (soil). Such an increase suggests that the degassing rate of the land surface during recent times may be greater than that prior to man's interference with the mercury cycle.

Mining, mercury utilization by man, and emissions have increased the total mercury content of rivers about four times. Today's rivers carry about equal masses of mercury in dissolved and solid form to the oceans. This relation is difficult to understand because on the average suspended sediment in rivers contains three to four times more mercury than the water. The problem may be that analyses of "dissolved" mercury in waters include very fine, suspended matter and mercury–organic complexes.

In Fig. 10, an attempt has been made to assess changes in ocean water mercury

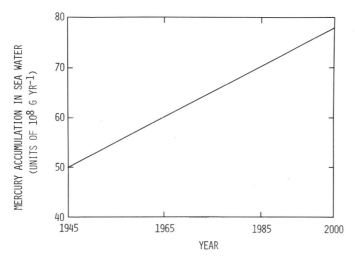

Fig. 10. Example of a calculation assessing changes in accumulation of mercury in upper 100 m of ocean owing to man's activities. As a limiting case, linear increases in mercury emission to the atmosphere owing to man's activities and in the total mercury flux to the atmosphere from the earth's surface were assumed for the period 1945–2000.

concentration owing to man's activities. The model assumes as a limiting case a linear increase of 1.7×10^8 g/yr in mercury emissions to the atmosphere owing to man's activities, and a linear increase of 3.6×10^8 g/yr in the total mercury flux to the atmosphere during the period 1945–2000. These assumptions are based on evaluation of pre-man and present-day global cycles of mercury and the fact that mercury utilization by man has been increasing at a linear rate with time.

The integrated accumulation during the period 1945–2000 of about 3500×10^8 g could result in an increase in the concentration of mercury in the oceanic surface layer (upper 100 m) of 30 %. This change could be less or more than the estimate depending on the future rate at which mercury is released to the atmosphere by man's activities.

In summary, man's contributions to the mercury cycle rival the natural fluxes. The major pathways affected are from land to atmosphere and from land to ocean. Mercury occurs in oceanic biota, but it is not likely that on a global basis there has been a significant increase in their mercury content.

Manganese

General

Pre-man and present-day global cycles of manganese are shown in Figs. 11 and 12. Evaluation of these cycles is based on the same type of considerations outlined in detail for mercury. Because of space, details are not presented here.

There are several striking differences between the cycles of mercury, a volatile, trace element, and manganese, a nonvolatile minor element. First, the major exchange of manganese between the atmosphere and the pre-man earth's surface is due to continental dust being swept into the atmosphere by winds and then falling back onto the earth's surface. Today this dust flux is augmented by manganese emitted to the atmosphere in particulate form owing to industrial activities. Bacterial processes

PRE-MAN CYCLE OF MANGANESE

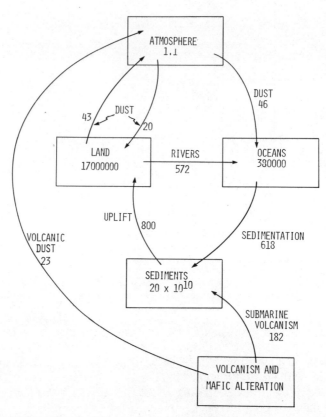

Fig. 11.   Pre-man global cycle of manganese. Fluxes are in units of 10^{10} g/yr; reservoir masses in units of 10^{10} g.

or high-temperature release and condensation processes do not appear to be important for manganese; that is, no volatile manganese species is important in the element's global cycle.

Second, subsea volcanism and mafic rock alteration appear to be significant sources of manganese to the exogenic cycle, as suggested by several authors (e.g., Goldberg, 1965; Krishnaswamy and Lal, 1972); the subsea flux is about 20% of the river flux of manganese today. This value of about 300×10^{10} g/yr, based on the material balance of today's manganese cycle (Fig. 12), agrees remarkably well with a recent estimate made by Wolery and Sleep (1975) of $100 - 300 \times 10^{10}$ g/yr based on estimates of the flux of water through oceanic ridges and its manganese concentration.

Third, the total river flux of manganese to the ocean today is nearly three times the pre-man flux. This increase represents principally an increase in the denudation rate of the land's surface from about 100×10^{14} g/yr pre-man to today's rate of about 225×10^{14} g/yr. Because this increase in denudation reflects principally an increase in the suspended load of rivers owing to man's deforestation and agricultural activities

PRESENT DAY CYCLE OF MANGANESE

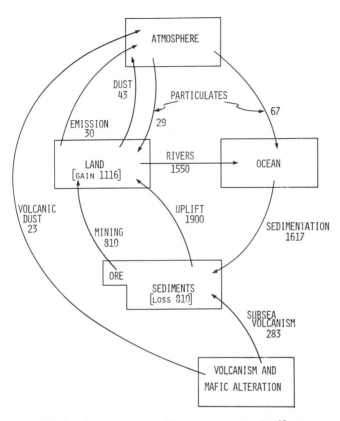

Fig. 12.   Present-day global cycle of manganese. Fluxes are in units of 10^{10} g/yr; reservoir masses in units of 10^{10} g.

(Judson, 1968) and manganese is concentrated in the ferric oxide coatings on suspended material and in the suspended particles (Gibbs, 1973), the land to ocean manganese flux is higher today than in the past.

Fourth, man's activities have changed the global cycle of manganese. Manganese in particulate emissions owing to man's industrial activities rivals the natural input of continental dust to the atmosphere (30×10^{10} g versus 43×10^{10} g). The mining of manganese ore has resulted in a net gain for the land reservoir and a net loss from the sediment reservoir. There is no evidence for change in the oceanic reservoir of dissolved Mn.

Manganese cycle and calcium carbonate

It has been noted by several authors (e.g., Pettijohn, 1957; Garrels and Mackenzie, 1971) that the percentage of carbonate rocks preserved in the total sedimentary rock mass today is greater than that calculated by geochemical calculations involving the conversion of average igneous rock to average sedimentary rock. Garrels and Mackenzie (1971) pointed out that this discrepancy could be resolved if a significant

mass of calcium were derived from the submarine alteration and leaching of volcanogenic mafic rocks and sediments. The overall chemical reaction proposed for this
alteration process is

basaltic rock + $2CO_2$ + H_2O = Na-feldspar + clay minerals

$$+ (Ca^{2+})carbonate + 2HCO_3^- + Fe\ oxides + SiO_2$$

Interestingly, this overall reaction is very similar to reactions proposed for the
alteration of basalts and basaltic sediments in the deep sea today (Hart, 1973). The
principal mobile components appear to be calcium and bicarbonate which after
release from the basalt may go elsewhere to be deposited inorganically or in the
skeletons of calcareous plankton. Today, most of the $CaCO_3$ is deposited in the deep
sea in biogenic deposits. It is possible that a large mass of these deposits is removed
from the exogenic cycle by subduction of sea floor at continental margins. Prior to the
Cretaceous and evolution of calcareous foraminifera, $CaCO_3$ was deposited principally in shallow-water environments, and consequently was available for cycling
through the exogenic cycle.

There are about 60×10^{20} mol of calcium and carbon locked up in limestones
today; of this amount, Garrels and Mackenzie (1971) computed that about half the
calcium, or 30×10^{20} mol, came from the submarine alteration of basaltic materials.
This value represents an average depositional rate of $CaCO_3$ over the past 3.5×10^9 yr
of 0.9×10^{12} mol/yr. We may now inquire whether or not present-day submarine
basaltic alteration rates are compatible with this average depositional rate.

From the present-day Mn flux owing to subsea volcanism and alteration (Fig. 12),
we can compute a calcium flux. If as a first approximation we assume that the manganese is quantitatively removed during alteration of basalt, for a 0.2% MnO
concentration in basalt, the mass of altered basalt necessary to supply the 300×10^{10}
g/yr of volcanically derived manganese is

$$\text{mass of basalt} = \frac{10^2 \times 300 \times 10^{10}}{0.15} = 20 \times 10^{14} \text{ g/yr}$$

At today's rate of formation of new sea floor of 3 km²/yr (Chase, 1972), this mass
represents alteration of a layer of basalt only 0.26 km thick, an estimate similar to
that of Wolery and Sleep (1975) of 0.19 km based on the independent considerations
mentioned above. Thus the mass of basalt altered in the deep sea amounts to $20 \times
10^{14}$ g/yr/240×10^{14} g/yr, or only about 8% of the continental denudation rate
today.

For a basalt containing 10.5% CaO, this mass of altered basalt could release
$(10.5) (20 \times 10^{14})/10^2 = 2.1 \times 10^{14}$ g/yr CaO, or 3.8×10^{12} mol Ca^{2+}/yr, to seawater, assuming all the calcium is quantitatively removed from the basalt by alteration
and leaching processes; a rate that is about 31% of the annual rate at which streams
carry calcium to the oceans. If all the calcium precipitated as $CaCO_3$ according to the
reaction

$$Ca^{2+} + CO_2 + H_2O = CaCO_3 + 2H^+$$

then 3.8×10^{12} mol CO_2/yr would be required for precipitation. This value is about
23% of the annual CO_2 requirement for weathering of continental rocks.

From the above calculations, it can be seen that the submarine alteration and
leaching of basaltic materials may be an important source of calcium to the exogenic
cycle. Indeed, the calculations further suggest that less than 25% of the calcium

originally in basaltic materials need be removed during submarine alteration and leaching processes to account for the additional $CaCO_3$ in the total sedimentary rock mass, a percentage not inconsistent with observations of calcium loss from basaltic materials in the deep sea today (Hart, 1973).

7. Summary and Conclusions

Sedimentary cycling models describing global processes of element circulation at and near the earth's surface are in their infancy today. A steady-state model for the major elements was presented, and shown to be descriptive of element circulation in the exogenic cycle for the past 600×10^6 yr. This model is successful on a long-term basis in predicting sources and sinks of the major elements in the exogenic cycle and steady-state fluxes of elements among reservoirs.

For minor elements, however, like the heavy metals, reservoir contents, transport paths, and fluxes between reservoirs are poorly known. An assumption of steady state today for all minor elements probably is not apropos because inputs of some of these elements into the natural system owing to man's activities are comparable to, or exceed, natural inputs.

Global modeling of element cycles involves knowledge of the physicochemical properties of the element and its compounds, of major reservoirs and subreservoirs in which the element is found and quantities in these boxes, and of major transport paths and fluxes of substances between reservoirs. For many elements, particularly minor and trace elements, data are insufficient to predict their behavior in a particular reservoir, or to evaluate their transport paths or fluxes between reservoirs.

Sedimentary cycling models can provide a basis for evaluating man's impact on the natural cycle of an element. For mercury, a volatile trace element, it was shown that man's industrial activities have resulted in an increase of mercury vapor emission from the land surface to the atmosphere of about 2.2 times the pre-man flux. Based on the present-day global model for mercury and on a linear increase in mercury utilization until the year 2000, it was predicted that mercury concentration in the upper 100 m of the ocean could increase by 30%.

For manganese, a nonvolatile minor element, the major impact of man on its global cycle involves mining and emission of manganese from the land surface to the atmosphere in particulates produced in industrial activities. To construct the present-day cycle of manganese, it was necessary to derive a significant flux of manganese to the exogenic cycle from submarine alteration of basaltic materials. Further calculations showed that such processes could account for the "excess" calcium carbonate in the sedimentary lithosphere.

Acknowledgment

We thank our colleague, Robert M. Garrels, for many hours of pleasant discussion of the ideas and concepts in this paper.

References

Berner, R. A., 1972. Sulfate reduction, pyrite formation, and the oceanic sulfur budget. In *The Changing Chemistry of the Oceans*. D. Dyrssen and D. Jagner, eds. Wiley-Interscience, New York, pp. 347–362.
Berner, R. A., 1973. Pyrite formation in the ocean. In *Proc. Sym. Hydrogeochem. Biogeochem.* Vol. 1, *Hydrogeochemistry*. The Clarke Company, Washington, D.C., pp. 402–417.

Bischoff, J. L. and T. L. Ku, 1970. Pore fluids of recent marine sediments: I. Oxidizing sediments òf 20° N, continental rise to mid-Atlantic ridge. *J. Sediment. Petrol.*, **40**, 960–972.

Bischoff, J. L. and T. L. Ku, 1971. Pore fluids of modern marine sediments: II. Anoxic sediments of 35° to 45° N, Gibralter to mid-Atlantic ridge. *J. Sediment. Petrol.*, **4**, 1008–1017.

Bolin, B., 1970. The carbon cycle. *Sci. Am.*, **223**, 125–132.

Broecker, W., 1963. Radioisotopes and large-scale oceanic mixing. In *The Sea*. Vol. 2. M. N. Hill, ed. Wiley-Interscience, New York, pp. 88–108.

Broecker, W. S., 1970. A boundary condition on the evolution of atmospheric oxygen. *J. Geophys. Res.*, **75**, 3353–3357.

Broecker, W. S., 1971. A kinetic model for the composition of sea water. *Quant. Res.*, **1**, 188–207.

Calvert, G. E., 1968. Silica balance in the ocean and diagenesis. *Nature*, **219**, 919–920.

Cameron, E. M. and I. R. Jonasson, 1972. Mercury in Precambrian shales of the Canadian Shield. *Geochim. Cosmochim. Acta*, **36**, 985–1006.

Chave, K. E., 1954. Aspects of the biogeochemistry of magnesium. 2. Calcareous sediments and rocks. *J. Geol.*, **62**, 587–599.

Chester, R. and J. H. Stoner, 1974. The distribution of Mn, Fe, Cu, Ni, Co, Gd, Cr, V, Ba, Sr, Sn, Zn, and Pb in same soil-sized particulates from the lower troposphere over the world ocean. *Mar. Chem.*, **2**, 157–188.

Conway, E. J., 1942. Mean geochemical data in relation to ocean evolution. *Roy. Irish Acad. Proc.*, **48**, 119–159.

Conway, E. J., 1943. The chemical evolution of the ocean. *Roy. Irish Acad. Proc.*, **48**, 161–212.

Corliss, J. B., 1970. Mid-ocean ridge basalts: I. The origin of sub-marine hydrothermal solutions. II. Regional diversity along the mid-Atlantic ridge. Ph.D. Thesis, University of California at San Diego.

Craig, H., 1957. The natural distribution of radiocarbon and the exchange time of carbon dioxide between atmosphere and sea. *Tellus*, **9**, 1–17.

Dams, J. A., 1974. Personal communication.

Deevey, E. S., Jr., 1970. Mineral cycles. *Sci. Am.*, **223**, 149–158.

Drever, J. I., 1971a. Early diagenesis of clay minerals, Rio Ameca Basin, Mexico. *J. Sediment. Petrol.*, **41**, 892–894.

Drever, J. I., 1971b. Magnesium iron replacements in clay minerals in anoxic marine sediments. *Science*, **172**, 1334–1336.

Drever, J. I., 1974. The magnesium problem. In *The Sea*. Vol. 5. E. D. Goldberg, ed. Wiley-Interscience, New York, pp. 337–355.

Duce, R. D., P. L. Parker, and C. S. Giam, 1974. *Pollutant Transfer to the Marine Environment*, Kingston, R.I., 55 pp.

Erikkson, E., 1971. Compartment models and reservoir theory. *Ann. Rev. Ecol. Syst.*, **2**, 67–84.

Fanning, K. A. and M. E. Q. Pilson, 1971. Interstitial silica and pH in marine sediments: Some effects of sampling procedures. *Science*, **173**, 1228–1231.

Garrels, R. M. and F. T. Mackenzie, 1971. *Evolution of Sedimentary Rocks*. W. W. Norton, New York, 397 pp.

Garrels, R. M. and F. T. Mackenzie, 1972. A quantitative model for the sedimentary rock cycle. *Mar. Chem.*, **1**, 27–41.

Garrels, R. M., F. T. Mackenzie, and C. Hunt, 1974. *Man's Contributions to Natural Chemical Cycles*. Authors' copyright, 220 pp.

Garrels, R. M. and F. T. Mackenzie, 1974. Chemical history of the oceans deduced from post-depositional changes in sedimentary rocks. In *Studies in Paleo-Oceanography, Soc. Econ. Paleontol. Mineral., Spec. Publ.*, **20**. W. W. Hay, ed. Tulsa, Okla. pp. 193–204.

Garrels, R. M. and E. A. Perry, Jr., 1974. Cycling of carbon, sulfur, and oxygen through geologic time. In *The Sea*. Vol. 5. E. D. Goldberg, ed. Wiley-Interscience, New York, 303–336.

Gibbs, R. M., 1972. Water chemistry of the Amazon River. *Geochim. Cosmochim. Acta*, **36**, 1061–1066.

Gibbs, R., 1973. Mechanisms of trace metal transport in rivers. *Science*, **150**, 71–73.

Goldberg, E. D., 1965. Minor elements in sea water. In *Chemical Oceanography*. Vol. 1. J. P. Riley and G. Skirrow, eds., Academic Press, New York.

Goldberg, E. D., 1971. Atmospheric dust, the sedimentary cycle, and man. *Geophysics*, **1**, 117–132.

Goldberg, E. D., ed., 1974. *The Sea*. Vol. 5. Wiley-Interscience, New York, 895 pp.

Gregor, C. B., 1968. Silica balance of the ocean. *Nature*, **219**, 360–361.

Grim, R. E., R. S. Dietz, and W. F. Bradley, 1949. Clay mineral composition of some sediments from the Pacific Ocean off the California coast and Gulf of California. *Geol. Soc. Am. Bull.*, **60**, 1785–1808.

Harris, R. C., 1966. Biological buffering of oceanic silica. *Nature*, **212**, 275–276.

Harrison, P. R., K. A. Rahn, R. Dams, J. A. Robbins, J. W. Winchester, S. S. Brar, and D. M. Nelson, 1971. Areawide trace metal concentrations measured by multielement neutron activation analysis. *APCA J.*, **21**, 563–570.

Hart, R., 1970. Chemical exchange between sea water and deep ocean basalts. *Earth Planet. Sci. Lett.*, **9**, 269–279.

Hart, R., 1973. A model for chemical exchange in the basalt-seawater system of oceanic layer II. *Can. J. Earth Sci.*, **10**, 799–816.

Heath, G. R., 1974. Dissolved silica and deep-sea sediments. In *Studies in Paleo-Oceanography*, *Soc. Econ. Paleontol. Mineral. Spec. Publ.*, **20**. W. W. Hay, ed. Tulsa, Okla., pp. 77–93.

Heck, E. T., 1964. Ocean salt. *W. Va. Geol. Econ. Surv. Bull.*, **28**, 1–42.

Helgeson, H. C. and F. T. Mackenzie, 1970. Silicate-sea water equilibria in the ocean system. *Deep-Sea Res.*, **17**, 877–892.

Hoffman, G. L., R. A. Duce, and E. J. Hoffman, 1972. Trace metals in the Hawaiian marine atmosphere. *J. Geophys. Res.*, **77**, 5322–5329.

Holeman, J. N., 1968. The sediment yield of major rivers of the world. *Water Res. Res.*, **4**, 737–747.

Holland, H. D., 1965. The history of ocean water and its effect on the chemistry of the atmosphere. *Proc. Natl. Acad. Sci.*, **53**, 1173–1183.

Holser, W. T., and I. R. Kaplan, 1966. Isotope geochemistry of sedimentary sulfates. *Chem. Geol.*, **1**, 93–135.

Johnson, E. D. and R. S. Brannan, 1974. Distribution of atmospheric mercury species near ground. *Environ. Sci. Technol.*, **12**, 1003–1008.

Johnson, F. J., 1970. The oxygen and carbon dioxide balance in the earth's atmosphere. In *Global Effects of Environmental Pollution*. G. F. Singer, ed. Springer, New York, pp. 4–11.

Jones, B. F., V. C. Kennedy, and G. W. Zellweger, 1974. Comparison of observed and calculated concentrations of dissolved Al and Fe in stream water. *Water Res. Res.*, **10**, 791–793.

Judson, S., 1968. Erosion of the land. *Am. Scientist*, **56**, 356–374.

Kennedy, V. C. and G. W. Zellweger, 1974. Filter pore-size effects on the analysis of Al, Fe, Mn, and Ti in water. *Water Res. Res.*, **10**, 585–590.

Krauskopf, K. B., 1957. *Introduction to Geochemistry*. McGraw-Hill, New York, 721 pp.

Krishnaswamy, S. and D. Lal., 1972. Manganese nodules and budget of trace solubles in oceans. In *The Changing Chemistry of the Oceans*. D. Dyrssen and D. Jagner, eds. Wiley-Interscience, New York, pp. 306–327.

Kroopnick, P., 1971. Oxygen and carbon in the oceans and atmosphere; stable isotopes as tracers—for consumption, production, and circulation models. Ph.D. thesis, Scripps Institute of Oceanography, La Jolla, Calif., 230 pp.

Lal, D. and B. Peters, 1967. Cosmic ray-produced radioactivity on the Earth. In *Handbuch der Physik*. Vol. 46, No. 2. S. Flügge, ed. Springer-Verlag, Berlin, pp. 551–612.

Leatherland, T. M. and J. D. Burton, 1971. Mercury in a coastal environment. *Nature*, **231**, 440–442.

Lerman, A., F. T. Mackenzie, and R. M. Garrels, 1974. Modeling of geochemical cycles: problems of phosphorus as an example. *Geol. Soc. Am. Mem. 142*. E. H. T. Whitten, ed. In press.

Li, Y. H., 1972. Geochemical mass balance among lithosphere, hydrosphere, and atmosphere. *Am. J. Sci.*, **272**, 119–137.

Lisitzin, A. P., 1972. Sedimentation in the world's oceans. *Soc. Econ. Paleontol. Mineral., Spec. Publ.*, **17**, 218 pp.

Liss, P. S. and C. P. Spencer, 1970. Abiological processes in the removal of silica from seawater. *Geochim. Cosmochim. Acta*, **34**, 1073–1088.

Livingstone, D. A., 1963. Chemical composition of rivers and lakes. In *Data of Geochemistry* 6th ed. M. Fleischer, ed., *U.S. Geol. Surv. Prof. Paper 440-G*.

Machta, L., 1972. The role of the oceans and biosphere in the carbon dioxide cycle. In *The Changing Chemistry of the Oceans*. D. Dyrssen and D. Jagner, eds. Wiley, New York, pp. 121–145.

Mackenzie, F. T. and R. M. Garrels, 1965. Silicates—reactivity in seawater. *Science*, **150**, 57–58.

Mackenzie, F. T. and R. M. Garrels, 1966a. Chemical mass balance between rivers and oceans. *Am. J. Sci.*, **264**, 507–525.

Mackenzie, F. T. and R. M. Garrels, 1966b. Silica-bicarbonate balance in the ocean and early diagenesis. *J. Sediment. Petrol.*, **36**, 1075–1084.

Mackenzie, F. T., R. M. Garrels, O. P. Bricker, and F. Bickley, 1967. Silica in sea water: Control by silicate minerals. *Science*, **155**, 1404–1405.

Mangelsdorf, P., T. R. S. Wilson, and E. Daniell, 1969. Potassium enrichments in interstitial waters of marine sediments. *Science*, **165**, 171–174.

Manheim, F. T. and F. L. Sayles, 1974. Composition and origin of interstitial waters of marine sediments, based on deep sea drill cores. In *The Sea*. Vol. 5. E. D. Goldberg, ed. Wiley-Interscience, New York, pp. 527–568.

Martin, J. H. and G. A. Knauer, 1973. The elemental composition of plankton. *Geochim. Cosmochim. Acta*, **37**, 1639–1654.

Martin, J.-M., G. Kulbicki, and A. J. De Groot, 1973. Terrigenous supply of radioactive and stable elements to the ocean. In *Proc. Symp. Hydrogeochem. Biogeochem., Tokyo, Japan, Sept. 7–9, 1970*. The Clarke Company, Washington, D.C., pp. 463–483.

McKenzie, J. A., 1972. A mathematical model for the isotopic balance of sulfur in the oceans. In preparation.

Nihoul, J. C. J., 1974. Mathematical model for the study of marine pollution. In *Model Mathematique, Rapport de Synthese*. Vol. 1. Liege, Belgium, pp. 3–27.

Nissenbaum, A., B. M. Presley, and I. R. Kaplan, 1972. Early diagenesis in reducing fjord, Saanich Inlet, British Columbia—I. Chemical and isotopic changes in major components of interstitial water. *Geochim. Cosmochim. Acta*, **36**, 1007–1027.

Perry, E. A., Jr., and J. Hower, 1970. Burial diagenesis in Gulf Coast pelitic sediments. *Clays Clay Min.*, **18**, 165–177.

Pettijohn, F. J., 1957. *Sedimentary Rocks*. 2nd ed. Harper and Row, New York, 718 pp.

Presley, B. J. and I. R. Kaplan, 1970. Interstitial water chemistry. Deep Sea Drilling Project, Leg 4. In *Initial Reports of the Deep Sea Drilling Project*. Vol. 4. U.S. Government Printing Office, Washington, D.C., 415.

Pytkowicz, R. M., 1967. Carbonate cycle and the buffer mechanism of recent oceans. *Geochim. Cosmochim. Acta*, **31**, 63–73.

Pytkowicz, R. M., 1972. The chemical stability of the oceans and CO_2 system. In *The Changing Chemistry of the Oceans*. D. Dyrssen and D. Jagner, eds. Wiley-Interscience, New York, pp. 147–152.

Pytkowicz, R. M., 1973. The carbon dioxide system in the oceans. *Swiss J. Hydrol.*, **35**, 8–28.

Redfield, A. C., B. H. Ketchum, and F. A. Richards, 1963. The influence of organisms on the composition of sea water. In *The Sea*. Vol. 2. M. N. Hill, ed. Wiley-Interscience, New York, pp. 26–77.

Ristvet, B. L., F. T. Mackenzie, D. C. Thorstenson, and R. H. Leeper, 1973. Pore water chemistry and early diagenesis of nearshore marine sediments. *Bull. Am. Assoc. Petrol. Geol.*, **57**, 801–802.

Rubey, W. W., 1951. Geologic history of sea water, an attempt to state the problem. *Geol. Soc. Am. Bull.*, **62**, 1111–1147.

Saukov, A. A., 1946. Geochemistry of mercury (Geokhim. Rtuti). *Akad. Nauk. SSSR, Inst. Geol. Nauk.*, **78** (Mineralogo-Geokhim., Ser. No. 17).

Shacklette, H. T., 1970. Mercury content of plants. In *Mercury in the Environment, U.S. Geol. Surv. Prof. Paper 713*, pp. 35–36.

Siever, R., 1968a. Establishment of equilibrium between clays and sea water. *Earth Planet. Sci. Lett.*, **5**, 106–110.

Siever, R., 1968b. Sedimentological consequences of a steady-state ocean-atmosphere. *Sedimentology*, **11**, 5–29.

Sillén, L. G., 1961. The physical chemistry of sea water. In *Oceanography*. M. Sears, ed. *Publ. Am. Assoc. Adv. Sci.*, **67**, 549–581.

Statistical Abstract of the United States, 1973. U.S. Department of Commerce, Washington, D.C., 1014 pp.

Statistical Yearbook, 1972. UNESCO, Statistical Division, Paris.

Stock, A. and F. Cucuel, 1934. Die verbreitung des quicksilbers. *Naturwiss.*, **22**, 390–393.

Stumm, W. and J. J. Morgan, 1970. *Aquatic Chemistry*. Wiley-Interscience, New York, 583 pp.

Turekian, K. K. and K. H. Wedepohl, 1961. Distribution of the elements in some major units of the earth's crust. *Geol. Soc. Am. Bull.*, **72**, 175.

Turekian, K. K., 1969. The oceans, streams, and atmosphere. In *Handbook of Geochemistry*. Vol. 1. K. H. Wedepohl, ed. Springer-Verlag, New York, pp. 297–323.

U.S. Geol. Surv. Prof. Paper 713, 1970. Mercury in the environment. U.S. Government Printing Office, Washington, D.C., 67 pp.

Weiss, H. V., M. Koide, and E. D. Goldberg, 1971. Mercury in the Greenland ice sheet: Evidence of recent input by man. *Science*, **174**, 692–694.

Wolery, T. J. and N. H. Sleep, 1975. Hydrothermal circulation and geochemical flux at mid-ocean ridges. *J. Geol.* (in press).

Wollast, R. and F. De Broeu, 1971. Study of the behaviour of dissolved silica in the estuary of the Scheldt. *Geochim. Cosmochim. Acta*, **35**, 613–620.

Wollast, R., 1974. The silica problem. In *The Sea*. Vol. 5. E. D. Goldberg, ed. Wiley-Interscience, New York, pp. 359–392.

Wollast, R., G. Billen, and F. T. Mackenzie, 1975. Behavior of mercury in natural systems and its global cycle. In *Proc. NATO Sci. Comm. Mtg. on Toxicology of Heavy Metals and Organohalogens*. McIntyre and C. F. Mills, eds. In press.

Wood, J. M., 1974. Biological cycles for toxic elements in the environment. *Science*, **183**, 1049–1052.

Woodwell, G. M., P. P. Craig, and H. A. Johnson, 1971. DDT in the biosphere: Where does it go? *Science*, **174**, 1101–1107.

Zoller, W. H., E. S. Gladney, and R. A. Duce, 1974. Atmospheric concentrations and sources of trace metals at the South Pole. *Science*, **183**, 198–200.

IV. BIOLOGY

20. MODELING

R. C. DUGDALE

1. Introduction

The central problem in modeling nutrient uptake by marine phytoplankton is to understand the reaction of the phytoplankton to characteristic nutrient regimes. These reactions or adaptations result in modification of simple nutrient uptake kinetics largely through rearrangement of cell structure and physiology. These processes are moderately well known as a result of the advanced state of modern cell biology and of extensive work with nutrient-limited cultures in the laboratory. Field measurements using tracer nitrogen, phosphorus, and silicon compounds provide the link in understanding between laboratory and natural population processes. The approach used in this chapter is to discuss first nutrient uptake kinetics, then the adaptations that take place. Combining an understanding of these two processes should result in accurate nutrient-based growth models.

It is appropriate to note here that the classical paper by Ketchum (1939) on uptake of nitrate and phosphate by *Phaeodactylum tricornutum* set the standards for the modern study of nutrient uptake kinetics. Although these data originally were not interpreted in terms of biochemical kinetics, the quality is such that some of the most reliable kinetic constants have been computed from them (Dugdale, 1976a).

2. Nutrient Uptake

A. Michaelis–Menten Kinetics

Monod (1942), working with bacteria in chemostat culture limited by carbon source, found that the growth rate versus substrate concentration curve followed the expression:

$$\mu = \mu_{max} \frac{S}{K_s + S} \tag{1}$$

where μ = specific growth rate, μ_{max} = maximum specific growth rate of the population under the prevailing environmental conditions, and K_s = concentration of limiting nutrient in the medium at which $\mu = \mu_{max/2}$. The chemostat is a simple device in which the growth rate of the culture organism is set by the dilution rate, D, the rate at which medium is added divided by the volume of the reactor or culture vessel:

$$D = \frac{F}{W} \tag{2}$$

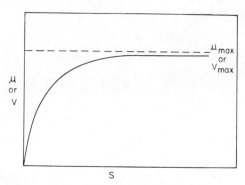

Fig. 1.   Growth rate or uptake of limiting nutrient as a function of limiting nutrient concentration according to equations 1 and 3.

where F = flow rate and W = volume of the reactor. The proportion of the nutrients in the medium is such that one becomes limiting. Under these conditions the uptake of limiting nutrient and, at steady state, the growth rate of the culture are controlled on the rectangular hyperbola described by equation 1 (Fig. 1). The Michaelis–Menten expression for enzyme kinetics is virtually identical to equation 1 and has been adapted to describe the kinetics of nutrient uptake:

$$V = V_{\max} \frac{S}{K_s + S} \tag{3}$$

where V = specific uptake rate of nutrient normalized to the nutrient content of the cells, that is, $V = (1/S)(dS/dt)$, and the hyperbola is the same as in Fig. 1 with the axes relabeled as shown. This expression is assumed to describe the transport of nutrient from the outside to the inside of the cell and thus the growth rate at steady state, since in balanced growth the specific uptake rates of all required nutrients are equal to each other and to the specific growth rate:

$$\frac{1}{x} \frac{dx}{dt} = \frac{1}{S_1} \frac{dS_1}{dt} = \frac{1}{S_2} \frac{dS_2}{dt} \ldots$$

where x = number of cells per unit volume and $S_1, S_2, \ldots$ is the cellular concentration of various nutrients. A full description of the chemostat and its operation can be found in Herbert et al. (1956). The applicability of this simple nutrient-controlled growth model to the growth of phytoplankton in the sea was suggested by Dugdale (1967). This suggestion was followed quickly by evidence from the laboratory that the Michaelis–Menten hyperbolic relationship could be applied to the uptake of nitrogen (Eppley et al., 1968; Caperon, 1968).

Evidence that the uptake of nitrate and ammonium for natural populations in eutrophic regions follows Michaelis–Menten kinetics was obtained with [15]N (MacIsaac and Dugdale, 1969). P. J. Williams (1973) showed that departure from Michaelis–Menten kinetics in mixed populations would be slight and probably undetectable under most conditions. The case for the use of Michaelis–Menten kinetics to describe the uptake of silicate, phosphate, nitrate, ammonium, urea, and other compounds in marine phytoplankton now is overwhelming (Dugdale, 1976a). Many models already have incorporated expression 3 to describe limiting nutrient uptake with considerable success (Dugdale, 1975). The models are, however, accurate only under conditions of temporary low nutrient concentration for reasons that will become apparent.

3. Growth

Although the Michaelis–Menten expression adequately describes the curve for nutrient concentration versus uptake rate, the applicability of the Monod expression (equation 1) in which growth rate is substituted for uptake rate has proved to be inadequate when applied to the chemostat culture of marine phytoplankton grown on limiting nutrients other than carbon. Droop (1968), Caperon and Meyer (1972), Eppley and Renger (1974), Fuhs et al. (1972), Harrison (1974), Davis (1973), and Conway (1974), variously working with vitamin B_{12}, nitrate, silicate, and phosphate, all reported that the dilution rate versus limiting substrate curve did not agree with

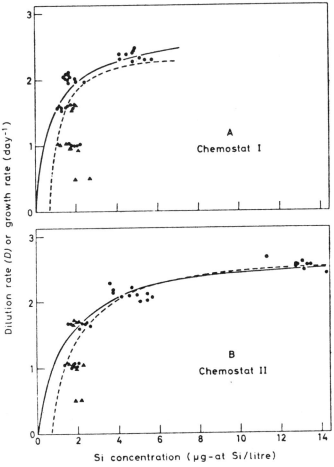

Fig. 2. Relationship between dilution rate and silicate concentration in chemostat reactors containing *Thalassiosira pseudonana* (*Cyclotella nana*). Solid lines represent hyperbolas fitted to high dilution rates (>1.25/day) with the curve intercepting the origin. Dashed lines represent hyperbolas fitted to low dilution rates (<0.75/day) and assuming an x axis intercept of 0.70 μg-at Si/l (from Paasche, 1973).

equation 1. The disagreement lies in the observation that changes in dilution rate are not accompanied by the expected changes in limiting nutrient concentration in the reactor. The reason for this disparity is the decreased amount of limiting nutrient contained in each cell under nutrient-limited conditions, a phenomenon that does not seem to occur when a carbon source limits the growth of bacteria in chemostat culture.

Paasche (1973), working with *Thalassiosira pseudonana* in chemostat culture under silicate limitation, found the simple Monod model fit his data quite well at high dilution rates (Fig. 2). Departures from a simple hyperbola at low dilution rates are apparent in the figure, and Paasche suggested that there was a finite value of silicate at which growth rate was reduced to zero. A plot of silicate uptake against silicate concentration also showed a hyperbolic pattern with a suggestion of a positive intercept on the abscissa. Paasche's results provide considerable justification for the use of the simple Monod model at relatively high growth rates. The appearance of an intercept on the abscissa is due to distortion of the kinetics at low dilution rates, a point to be discussed in detail later. In Caperon and Meyer's experiments with nitrogen-limited cultures of marine phytoplankton, a finite value of S was observed when uptake had ceased. This value, S_0, was small, 0.05–0.07 $\mu g \cdot at/l$, at the limit of detectability. Caperon and Meyer (1972) prefer to include the term in a modified expression:

$$\rho = \frac{\rho_{max}(S - S_0)}{K_s + (S - S_0)} \tag{4}$$

where ρ = absolute rate of uptake or transport rate, usually expressed on a per-cell basis. The necessity for retaining this complication is unclear at present. In any event it is more likely that there is complex curvature in the low-concentration region and that both variables go to zero simultaneously.

The distinction between V, nutrient specific uptake rate with units identical to those of specific growth rate, t^{-1}, and transport rate, ρ, with units of mass taken up per unit population per unit time, is important. These symbols and associated definitions are consistent with those used in previous publications (e.g., Dugdale, 1967).

4. Coupling of Growth and Nutrient Uptake

A. Steady-State Conditions

The uptake of nutrient and growth rate are related to each other at steady state by a proportionality constant, Q, the amount of limiting nutrient per cell:

$$V = \frac{\rho}{Q} \tag{5}$$

At steady state $V = \mu$ as shown above. The mechanisms that bring about the balance between nutrient uptake and growth are complex, but may be conceptualized as a chain of enzymatic reactions. It is not necessary to be concerned with all of them. For our purposes, two steps suffice: the first is the one already discussed, nutrient uptake, and the second is the first linked enzymatic step within the cell. The functions of the permeases in the cell membrane are to control the amount of nutrient entering the cell in the event of a surplus within the cell and to pump up the nutrient concentration within the cell to levels where the following reactions can take place efficiently. As a general rule, the substrate concentrations in a linked set of enzyme reactions float around the K_m value. (K_m is used when describing enzyme kinetics and is analogous to K_s). Usually cytoplasmic enzymes have K_m values on the order of millimoles, whereas in marine phytoplankton the K_s values for the uptake of nitrate, ammonium, phosphate, and silicate are all about 1 μmol. Packard and Blasco (1974) have measured the K_m for nitrate reductase in natural populations and report values of about 100 μmol. The K_m of the enzyme is relatively constant in ordinary enzymes; that is, the affinity of the enzyme for the substrate is not easily changed. However, $\rho_{\max}$ is a function of the concentration of enzyme and this is one of the means by which an organism adjusts its machinery to match the available resource for growth. We return to this subject later; for the time being the assumption is made that the concentration and thus the $V_{\max}$ of the first cyctoplasmic enzyme remain constant over a range of growth rates.

Droop (1968), working with vitamin B_{12}-limited growth, and Caperon (1968), working with nitrogen-limited growth in chemostat culture of marine forms, found a hyperbolic relationship between growth rate and the cell quota, Q. The data for vitamin B_{12} are shown in Fig. 3. The same relationship holds for silicate (Paasche, 1973). Caperon and Meyer (1972) fit these curves to a hyperbola of the form:

$$\mu = \frac{\mu_m(Q - Q_0)}{K_Q + (Q - Q_0)} \tag{6}$$

and use equations 4–6 to solve for the steady-state relationship between growth rate and reactor substrate concentration in the chemostat. They consider $\rho_{\max}$ to be a variable and add an expression to the equation for $\rho_{\max}$ as a function of the pre-existing growth rate. Caperon's and Meyer's variable $\rho_{\max}$ is a result of normalizing uptake to cell carbon, however, and from their Table 1 no apparent pattern of $\rho_{\max}$ normalized to cell number as a function of dilution rate can be seen. For the time being, $\rho_{\max}$ is treated as if it remained unchanged with dilution rate.

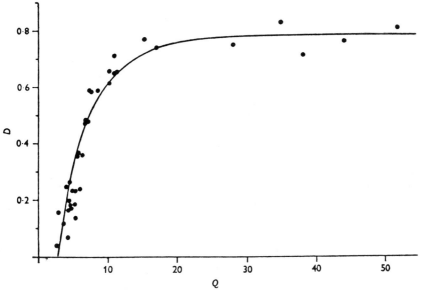

Fig. 3. Relationship between dilution rate and cell quota in vitamin B_{12}-limited chemostat culture of *Monochrysis lutheri* (from Droop, 1968).

Droop (1968) proposed a simpler hyperbolic expression:

$$\mu = \mu'_m \frac{(Q - K_Q)}{Q} \tag{7}$$

using μ'_m to make a distinction between this parameter and the μ_{max} of equation 1. Caperon and Meyer (1972) point out that equation 6 is equivalent to equation 7 if $K_Q = Q_0$, and that for some of their results, the assumption is approximately correct and results in small errors. In equation 7, Q_0 in Droop's original formulation has been replaced by K_Q. This convention is followed throughout this chapter. Equation 7 when rearranged gives

$$\mu Q = \mu'_m(Q - K_Q) \tag{8}$$

Plots of μQ (uptake) against Q invariably give a linear relationship according to Droop (1973), who gives examples from both his own data and that of others. For example, uptake of vitamin B_{12} by *Skeletonema costatum* is plotted against cell quota in Fig. 4. Droop refers to the parameter obtained from the slope of the line as μ'_m to distinguish it from the true cellular μ_{max}. The variability of μ'_m is treated below. An estimate of K_Q is obtained from the x axis intercept.

We now develop a simple model relating growth and uptake, primarily to show the underlying processes responsible for experimental observations. For a growth equation, Droop's hyperbolic expression 8 is used; for uptake, the simple Michaelis–Menten hyperbola, intercepting the origin, is used:

$$\rho = \rho_m \frac{S}{K_s + S} \tag{9}$$

It is useful to arrange equation 9 in terms of specific uptake rates since these are equivalent to growth rates at steady state. Rearranging equation 5,

$$\rho = VQ \tag{10}$$

and at steady state

$$\rho = \mu Q \tag{11}$$

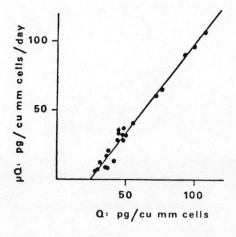

Fig. 4. Data from vitamin B_{12}-limited chemostat cultures of *Monochrysis lutheri* fitted to equation 8 (from Droop, 1973).

Substituting equation 11 into the left side of equation 9,

$$\mu Q = \rho_m \frac{S}{K_s + S} \qquad (12)$$

and

$$\mu = \frac{\rho_m}{Q} \frac{S}{K_s + S} \qquad (13)$$

The term ρ_m/Q in equation 13 is a nutrient-specific maximum uptake rate we call V'_m to indicate that it is a variable:

$$V'_m = \frac{\rho_m}{Q} \qquad (14)$$

and substituting into equation 13,

$$\mu = V'_m \frac{S}{K_s + S} \qquad (15)$$

Deviations from Monod kinetics (equation 1) are largely described by the variation in V'_m. If Q is constant, Monod kinetics is followed. When Q is reduced as a result of chronic nutrient scarcity, V'_m increases according to equation 14, with the result that the cell's ability to grow at low nutrient supply rates is enhanced simply because less limiting nutrient is used in producing the next cell. This is a somewhat tricky point that must be understood in employing nutrient specific uptake rates. For example, Eppley and Renger (1974), working with *Thalassiosira pseudonana*, show that the nutrient specific V_{max} increases with increasing nutrient deficiency, that is, with decreasing Q as predicted by equation 14. A plot of Q against V'_m should yield a hyperbola of the form $y = 1/x$ if ρ_m is a constant. Eppley's and Renger's (1974) data for *Thalassiosira pseudonana* fall near the line for $\rho = 0.017$ pg-at/cell/hr as shown in Fig. 5. The values for ρ_m for ammonium and nitrate for the lowest dilution rate are greater than the values for the higher dilution rates, hence the actual V'_m curve rises more steeply than the simple hyperbola plotted.

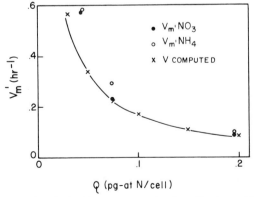

Fig. 5. Relationship between Q and V'_m with constant ρ_m; data from Eppley and Renger (1974); x = points computed according to equation 14.

To couple growth and uptake we need only obtain an expression for Q from equation 8:

$$Q = \frac{K_Q}{1 - (\mu/\mu'_m)}$$ (16)

substituting equation 16 into 14

$$V'_m = \frac{\rho_m((1 - \mu)/\mu'_m)}{K_Q}$$ (17)

Equation 17 shows that V'_m varies directly with ρ_m or inversely as K_Q. For a given population or species the exact value of V'_m depends also upon the ratio of μ to μ'_m. The limits are ρ_m/K_Q at vanishingly low growth rates and zero at $\mu = \mu'_m$. The latter result is a consequence of the lack of provision for a maximum Q in Droop's hyperbola (equation 8). The importance of equation 17 is that its value may be of overriding importance in the competition for nutrients under chronically low growth and nutrient supply rates. From equation 15 it can be seen that to compete, cells must have a high V'_m or a low K_s. Eppley and Renger (1974) were unable to detect differences in K_s in *Thalassiosira pseudonana* grown at different dilution rates and the variation in K_s for nitrate and ammonium measured for different species and different nutrients up to now is surprisingly small. However, Guillard et al. (1973) found differences in the K_s for silicate in two clones of *Thalassiosira pseudonana*, one isolated from an estuarine region and the other from the open sea, the latter having the lower value.

Although the value of ρ_m appears in some cases to be constant for a species grown under varying nutrient limitation conditions, K_Q and μ'_m vary in *Skeletonema costatum* grown at different dilution rates under either nitrogen or silicon limitation (Harrison et al., 1976). Droop plots are shown for their silicate data in Fig. 6. The data appear to fall along two distinct lines: the points obtained at dilution rates lower than 0.05/hr fall on the line with the lower slope, and the higher-slope line is comprised of points obtained at dilution rates greater than 0.08/hr. The sets of parameters obtained from the slopes and intercepts are as follows:

$D > 0.08$/hr	$D < 0.05$/hr
$\mu'_m = 0.120$/hr	$\mu'_m = 0.052$/hr
$K_Q = 0.024$ pg-at/cell	$K_Q = 0.006$ pg-at/cell

Virtually identical results were found for ammonium-limited growth of *Skeletonema costatum* (Harrison et al., 1976):

$\mu'_m = 0.14$/hr	$\mu'_m = 0.05$/hr
$K_Q = 0.065$ pg-at/cell	$K_Q = 0.005$ pg-at/cell

It appears that the populations observed at low dilution rates are "shifted down," a term in use for some time in the microbial growth literature (Schaechter, 1968). As shown in Fig. 7, a reduced maximal growth rate is the end of a chain of effects resulting from the reduction of nutrient availability. We return to the intermediate effects below, because these concern the reactions of populations to transients in nutrient concentration. In *Skeletonema costatum* shift down in μ'_m is accompanied by an even greater reduction in K_Q. The net effect is a large increase in V'_m and a reduction in the range of μ over which V'_m retains positive values, as can be seen in μ versus V'_m curves for the two population states in Fig. 8. There is a crossover point

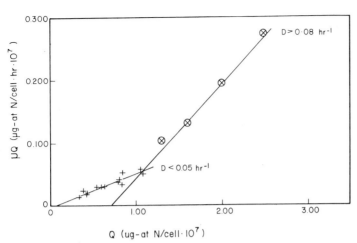

Fig. 6.   Data from nitrogen-limited cultures of *Skeletonema costatum* showing change in slope and intercepts according to equation 8 at dilution rates of about 0.05/hr.

near $\mu = 0.044$/hr. Below that dilution or growth rate, the shifted-down population successfully competes with the shifted-up population for nutrient, assuming the K_s values do not differ. Above $D = 0.044$/hr, shifted-up, high-μ'_m cells take on the advantage, and at slightly higher dilution rates, the low-μ'_m population must wash out because $V'_m < D$.

The steady-state model described above is similar in most details to those proposed by Caperon and Meyers (1972), Droop (1973), and Fuhs et al. (1972), although the last chose to use exponential functions to describe the saturation curves. Each model features a more rapid rise in growth rate with increasing external substrate than would be predicted by Michaelis–Menten kinetics.

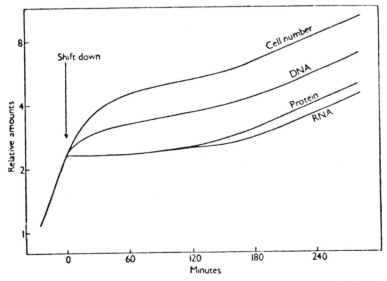

Fig. 7.   Schematic representation of the sequence of shift down in cultures of bacteria (from Schaechter, 1968).

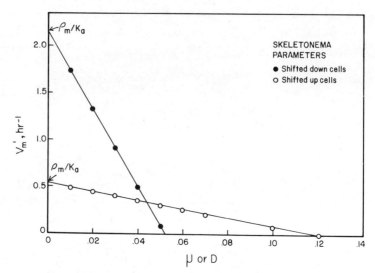

Fig. 8. V'_m computed as a function of dilution rate for two population states of *Skeletonema costatum* indicated in Fig. 6, $\rho_m = 0.01286$ pg-at N/cell for both curves; shifted down $\mu'_m = 0.052/\text{hr}$, $K_Q = 0.006$ pg-at N/cell; shifted up $\mu'_m = 0.120/\text{hr}$, $K_Q = 0.024$ pg-at N/cell.

B. *Transient Uptake of Limiting Nutrient*

Phytoplankton cells usually grow in the sea at some rate less than the $\mu_{\max}$ characteristic for the species. Many factors must interact in an optimum fashion to allow growth to proceed at the highest rates observed in batch culture, that is, at $\mu_{\max}$. At growth rates less than $\mu_{\max}$, cells must reduce the amount of nutrient entering to match their requirements to reduce the energy expended in transporting nutrients against the outside to inside gradient. Although control of nutrient uptake might be accomplished by reducing the number of active sites, that is, the concentration of the permease, it appears from the work of Conway et al. (1976) on chemostat culture of *Skeletonema costatum* that the permease capacity remains constant and that its activity is controlled on a time scale of minutes (Dugdale, 1976b). The result is that Michaelis–Menten kinetics is not obeyed over some upper portion of the nutrient uptake hyperbola, resulting in a truncated uptake hyperbola as shown in Fig. 9. The truncated uptake rate is ρ_i, that is, internally controlled uptake rate, the value decreasing with more and more unfavorable conditions such as reduced average light levels and decreasing nutrient supply rates.

The value of ρ_i can be found by analysis of the time series response of a phytoplankton population to the sudden addition of limiting nutrient. As shown in Fig. 10, the initial high uptake ρ_s is followed by a period of constant uptake, ρ_i, that gives way to external, nutrient-controlled uptake, ρ_e, when the concentration of limiting nutrient has decreased to a value corresponding to the truncation point T_s on the uptake hyperbola. The value of ρ_s may provide an estimate of ρ_m; however, the sampling interval must be made very small to obtain such an estimate as control is applied to the permease very soon after the addition of the nutrient. The value of ρ_i cannot be predicted accurately now, in part because the existence of hyperbola truncation has been recognized so recently (Davis, 1973) in culture work, although the phenomenon was suspected previously from field studies using ^{15}N (MacIsaac and

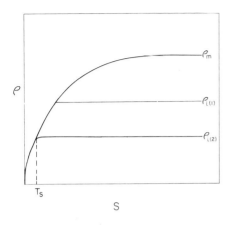

Fig. 9. Truncation of uptake hyperbola brought about by shift down.

Dugdale, 1972). Evidence for the control of ρ_i for nitrate and ammonium uptake in natural eutrophic populations by average ambient light levels is accumulating (MacIsaac, personal communication) and data obtained by Davis (1973) suggests control of ρ_i by light in chemostat populations of *Skeletonema*. Shift down of ρ_i by nitrogen limitation is evident in data obtained by Conway (1974), who operated three ammonium-limited reactors in series at a dilution rate of 0.04/hr. The *Skeletonema costatum* population in the first reactor had a ρ_i of 0.006 pg-at/cell/hr, and ρ_i was reduced in succeeding reactors to 0.0037 and 0.0033 pg-at/cell/hr. Thus under many suboptimal growth conditions we can expect that $\rho_i < \rho_{max}$ and that only at nutrient concentrations below the truncation point, T_s, is Michaelis–Menten kinetics obeyed. To assess the effect on the growth of the cell or phytoplankton population, the state of shift up must be related to nutrient specific uptake rate. The relationship of transport rate to growth is exactly analogous to that given in equation 14 for ρ_m and V'_m:

$$V_i = \frac{\rho_i}{Q} \tag{18}$$

The cost to the organism of shifting down is that its maximum immediately realizable growth rate, $\mu_i = V_i$, is reduced directly with ρ_i, but can be increased by further reduction in Q. The sequence of shift-up events and the consequences for μ_i are shown

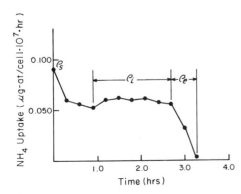

Fig. 10. Transient response of ammonium-limited *Skeletonema costatum* to addition of ammonium at $T = 0$ (from Conway, 1974).

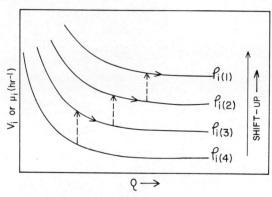

Fig. 11. Q versus V_i hyperbola for increasing values of ρ_i. Dashed lines and arrows represent the possible kinetic changes occurring during a series of increases of dilution rate in a chemostat culture of marine diatoms showing shift up in ρ_i.

in Fig. 11. The interesting and intuitively unexpected feature shown is that with increasingly favorable nutrient conditions resulting in increased Q, the value of V_i and hence the maximum immediately realizable growth rate, μ_i, is reduced rather than increased. In the chemostat, washout with increasing dilution rate and increased Q can be avoided only through a shift up, that is, by increasing ρ_i. The manner in which shift up and nutrient sparing interact in natural phytoplankton populations should be one of the major research areas in phytoplankton research in the next decade if we are to be able to model nutrient uptake over the full range of oceanic environmental conditions.

3. Interaction Between Limiting and Nonlimiting Nutrients

Conway (1974) and Davis (1973) observed during their perturbation experiments in chemostats that the initial high uptake period following addition of limiting nutrient was accompanied by the suppression of uptake of nonlimiting nutrients. One such experiment is shown in Fig. 12. Perturbation experiments were carried out on shipboard during the JOINT-I cruise off northwest Africa in 1974 using natural

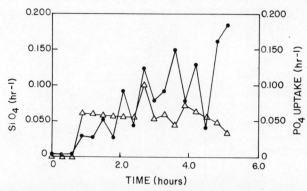

Fig. 12.   Uptake of nonlimiting nutrients by ammonium-limited culture of *Skeletonema costatum* following addition of ammonium at $T = 0$ (from Conway, 1974). Triangles and circles designate phosphate and silicate uptake, respectively.

populations concentrated by a reverse-flow technique. The results were similar to those obtained in the chemostat populations (Harrison and Davis, 1976), and the values of V_i obtained corresponded very closely to the saturated uptake rate obtained for nitrate and ammonium by the ^{15}N technique. The implication to be drawn from these results is that with limited energy resources in the cell, priority is given to the uptake of the nutrient in shortest supply.

A. Effect of Temperature

The μ_{max} of marine phytoplankton appears to be affected in a straightforward manner, showing a Q_{10} of about 1.8–1.9 (Dugdale, 1976a). Harrison (1974) in his studies with *Skeletonema costatum* in chemostat culture, noted that a decrease in temperature from 18 to 12°C resulted in a transition of the population from a condition of clear silicon limitation to one showing little or no morphological evidence of such limitation. This observation is predicted by equation 16 where it can be seen that a decrease in μ_m results in an increase in Q, the cell quota. Equation 16 is imperfect in the sense that no upper limit is set on the value that Q may take. However, within limits the expression may be modified to incorporate a term modifying the value of μ'_m. For example, the expression given by Packard et al. (1971) may be used:

$$\mu'_m = \mu_{max} \frac{Ea}{R} \left(\frac{1}{T_{max}} - \frac{1}{T_t} \right)$$
(19)

where T_{max} is the optimum temperature for growth of a species, T_t is the ambient temperature, R is the universal gas constant, and Ea is the slope of the line obtained by plotting μ against $1/T$. V'_m is affected also according to equation 17, where it can be seen that temperature-induced decrease of μ'_m results in a linear decrease in V'_m.

B. Effect of Light

Coupling between light and ammonium and nitrate uptake is extremely tight, as indicated by the fact that in [15]N K_{LT} experiments with natural populations good hyperbolic curves are virtually always obtained (MacIsaac and Dugdale, 1972). Falkowski (1975) has provided the physiological basis for these observations by showing in *Skeletonema costatum* that a nitrate-activated ATPase located in the cell membrane is responsible for the uptake of nitrate and probably a specific ATPase exists for ammonium as well. Further, he has found that the Photosystem I generates most of the ATP energizing these uptake enzymes or permeases. In preliminary field experiments Falkowski and Stone (1975) have obtained data that suggest that nutrient uptake has priority for ATP over the Calvin cycle. Shift-up can be fitted into these findings by suggesting that the process begins with the uptake of nitrogen or other nutrients, followed eventually by an increase in the photosynthetic mechanism and increased carbon fixation. Under saturating nutrient conditions, the shift-up level, ρ_i, of the population probably is set by the average light levels experienced over the previous day or days. At present no data exist with which to test this hypothesis. However, Davis (1976) has studied the interaction between silicate limitation and light limitation in *Skeletonema costatum* grown in chemostat culture. In these experiments he placed neutral density screens around reactors at steady state with $D = 0.04/hr$ and followed the course of population events, either attaining a new steady state at light levels above 0.043 ly/min or washing out below that light level. Analysis of his experiments shows that no change in ρ_i occurred during adaptation, and the mean value was 0.00234 pg-at/cell/hr, in agreement with the results of Harrison et al. (1976). The major change observed was an increase in Q for silicon, resulting in a decrease in V_i (equation 18). The observed increase in Q is analogous to that occurring with decrease in temperature and it seems likely that μ'_m is a function of light level also and varies directly with it. Equation 16 predicts that with μ held constant, as in the chemostat, decreasing μ'_m results in increasing Q. Eppley and Dyer (1965) suggest a hyperbolic expression for the relation between growth rate and light level:

$$\mu = \mu_{max} \frac{\bar{I}}{K_I + \bar{I}} \tag{20}$$

where $\bar{I}$ is the mean light intensity. The expression was designed for nutrient-saturated growth conditions and we may replace μ with μ'_m for the nutrient-limited case:

$$\mu'_m = \mu_{max} \frac{\bar{I}}{K_I + \bar{I}} \tag{21}$$

If μ_{max} and K_I are known for the species or for a population, Q can be computed from equation 16 after first computing μ'_m from equation 21. If the shift-up state of the population, ρ_i, is known, the V_i versus Q operating line is defined by equation 18.

Substituting ρ_i for ρ_m in equation 17 gives

$$V_i = \rho_i \frac{1 - (\mu/\mu'_m)}{K_Q} \tag{22}$$

and substituting μ'_m from equation 21 into 22,

$$V_i = \rho_i \frac{1 - \{\mu/[\mu_{max}(\bar{I}/(K_I + \bar{I}))]\}}{K_Q} \tag{23}$$

Using Davis' (1976) value for $K_I = 0.07$ ly/min, $\mu_{max} = 0.12$/hr, and $\rho_i = 0.00233$ pg-at/cell (silicon), along with an appropriate value for K_Q from Harrison et al. (1976), the curve of V_i versus $\bar{I}$ shown in Fig. 13 was obtained. The curve predicts washout

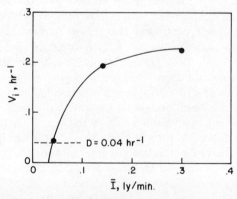

Fig. 13.   Computed values of V_i as a function of incident radiation. $\bar{I}$ based on data on *Skeletonema costatum* from Davis, 1973; parameters are $K_I = 0.07$ ly/min, $\mu_{max} = 0.12$/hr, $\mu = D = 0.04$/hr, $\rho_i = 0.0023$, $K_Q = 0.006$ pg-at Si/cell.

at light intensities below 0.042 ly/min; the dilution rate in these experiments was 0.04/hr. Davis (1973) found that at 0.042 ly/min his cultures barely were able to maintain steady state, and at all lower levels washout occurred. The analysis cannot be carried further since values of K_Q were not obtained in these experiments. Other expressions relating growth and light intensity also could be substituted into equation 22 in the manner illustrated here.

The short-term effect of reduced light intensity on nitrate and ammonium uptake has been studied in natural populations using ^{15}N (MacIsaac and Dugdale, 1972). Hyperbolic curves fitting the Michaelis–Menten expression usually are obtained with the K_I values in the range of 1–10% of surface radiation. These results agree well with Falkowski's finding that light-produced ATP drives the nitrate and ammonium uptake systems. A model designed to predict the uptake of nitrate in the water column was successful when compared to ^{15}N uptake data at some stations off the Peru coast (Dugdale and MacIsaac, 1971). Specific uptake rates were calculated from the light hyperbola and nitrate hyperbola expressions and the lower of the two values selected. The $V_{\max}$ for nitrate was reduced depending upon the amount of ammonium present in the water.

C. Multiple-Nutrient Limitation

In a recent series of experiments, Droop (1974) tested the following hypotheses:

1. That the effect of nutrients was multiplicative (Droop, 1973):

$$\frac{\mu}{\mu'_m} = \left(1 - \frac{K_{QA}}{Q_A}\right)\left(1 - \frac{K_{QB}}{Q_B}\right)(\cdots) \text{ etc.} \tag{24}$$

2. That only one nutrient controlled growth rate independently of the concentration of other potentially limiting nutrients in the medium, in his terminology, the "threshold" hypothesis. Using *Monochrysis lutheri* and working with phosphorus and vitamin B_{12} he was able to show that the threshold hypothesis was substantiated and the multiplicative was repudiated by the data. The correct formulation then is

$$\frac{\mu}{\mu'_m} = 1 - \frac{K_{QA}}{Q_A} \qquad \text{or} \qquad \frac{\mu}{\mu'_m} = 1 - \frac{K_{QB}}{Q_B} \tag{25}$$

There is a complication in that the value of K_Q increases above its true level when that nutrient is no longer limiting. Droop refers to nonlimiting values of K_Q as K'_Q and assumes that all internal nutrient processes can be described by hyperbolas:

$$\frac{\mu}{\mu'_m} = 1 - \frac{K'_{QA}}{Q_A} = 1 - \frac{K'_{QB}}{Q_B} \tag{26}$$

When nutrient A is limiting, equation 26 becomes

$$\frac{\mu}{\mu'_m} = 1 - \frac{K_{QA}}{Q_A} = 1 - \frac{K'_{QB}}{Q_B} \tag{27}$$

and

$$\frac{K_{QA}}{Q_A} = \frac{K'_{QB}}{Q_B} \tag{28}$$

Limitation by a specific nutrient is recognized in this scheme by the stability of K_{Q_A} with increasing concentration of nonlimiting nutrient, B. The value of K'_{Q_B} varies to maintain the equality indicated in equation 28 and is larger than its value when B is the limiting nutrient; that is, $K_{Q_B} < K'_{Q_B}$.

4. Internal Pool Models

Considerable insight into the control of growth and cell composition by nutrients has been gained from the chemostat research discussed above. Droop (1974) hopes that these cell quota models will alleviate the necessity to consider the biochemical details of internal cell processes involved in the incorporation of nutrients into new cells. He points out that the kinetics of nitrogen, silicon, phosphorus, and vitamin B_{12} can all be described by the cell quota technique and assumes that the proportion of active nutrient in the cell must be proportional to the total amount of the nutrient in the cell. There seems little doubt that the cell quota models will be very useful. However, they suffer from a major drawback, namely, that these models apply only to steady-state conditions. Although they have been enormously useful in the analysis of chemostat results, their application to nature is limited.

Internal pool models, on the other hand, are able to handle the problem of transient, non-steady states and also should give the same steady-state solutions as the cell quota models. These models were proposed by F. M. Williams (1965) and provide for an internal pool into which nutrient from the exterior flows under control of cell membrane processes and from which nutrient flows into the biochemical pathways that eventually result in growth. Grenney et al. (1973) have constructed an internal pool model for nitrogen-limited growth that worked well in predicting the results of their experiments. The internal pool model of Davis et al. (1975) appears to predict silicon-limited growth of *Skeletonema costatum* in chemostat culture.

5. Prospects for Modeling Nutrient Uptake and Growth

We can anticipate that reasonably soon, good cell quota models for specific marine species will be available. For example, the data available on *Monochrysis lutheri, Isochrysis galbana, Thalassiosira pseudonana*, and *Skeletonema costatum* are substantial and growing rapidly. These models are enormously useful in understanding growth of these organisms in continuous culture and give insight into nutrient-limited growth in the sea.

The internal pool models need to be developed further so that predictions under transient conditions can be made, since, for example, the outcome of competition between species of phytoplankton under rapidly changing conditions probably depends strongly on the time response scales as well as on steady-state parameters such as μ_{max}, K_s, etc. Internal pool models easily accommodate biochemical reality. For example, it appears that in natural populations of phytoplankton in upwelling regions, ambient ammonium acts to reduce the uptake of nitrate but does not affect the concentration or activity of nitrate reductase within the cell (Blasco, personal communication). Rhee's (1973) findings on control of phosphate uptake are based on biochemical kinetic theory and can be incorporated easily into nutrient pool models.

Incorporation of shift-up into nutrient uptake models is important for prediction of washout points in chemostats and for response of natural populations to perturbations, especially of changes in light and temperature. Unfortunately, shift-up has been recognized in marine phytoplankton only recently and the manner in which ρ_i, the internal nutrient uptake control level, or uptake hyperbola truncation point T_s, varies is not known quantitatively. However, this information is likely to be available soon and we can look forward to a period of consolidation and synthesis of the knowledge of nutrient uptake and its relationship to phytoplankton growth.

References

Caperon, J., 1968. Population growth response of *Isochrysis galbana* to nitrate variation at limiting concentrations. *Ecology*, **49**, 866–872.

Caperon, J., and J. Meyer, 1972. Nitrogen-limited growth of marine phytoplankton. II. Uptake kinetics and their role in nutrient limited growth of phytoplankton. *Deep-Sea Res.*, **19**, 619–632.

Conway, H. L., 1974. The uptake and assimilation of inorganic nitrogen by *Skeletonema costatum* (Grev.) Cleve. Ph.D. thesis, University of Washington, 126 pp.

Conway, H. L., P. J. Harrison, and C. O. Davis, 1976. Marine diatoms grown in chemostats under silicate or ammonium limitation. II. Transient response of *Skeletonema costatum* to a single addition of the limiting nutrient. *Mar. Biol.* In press.

Davis, C. O., 1973. Effects of changes in light intensity and photoperiod on the silicate-limited continuous culture of the marine diatom *Skeletonema costatum* (Grev.) Cleve. Ph.D. thesis, University of Washington, 123 pp.

Davis, C. O., 1976. Continuous culture of marine diatoms under silicate limitation. II. Effect of light intensity on growth and nutrient uptake of *Skeletonema costatum*. *J. Phycol.* In press.

Davis, C. O., N. F. Breitner, and P. J. Harrison, 1976. Continuous culture of marine diatoms under silicate limitation. III. A model of silicate-limited diatom growth. In preparation.

Droop, M. R., 1968. Vitamin B_{12} and marine ecology. IV. The kinetics of uptake, growth and inhibition in *Monochrysis lutheri*. *J. Mar. Biol. Assoc. U.K.*, **48**, 689–733.

Droop, M. R., 1973. Some thoughts on nutrient limitation in algae. *J. Phycol.*, **9**, 264–272.

Droop, M. R., 1974. The nutrient status of algal cells in continuous culture. *J. Mar. Biol. Assoc. U.K.*, **54**, 825–855.

Dugdale, R. C., 1967. Nutrient limitation in the sea: Dynamics, identification and significance. *Limnol. Oceanogr.*, **12**, 685–695.

Dugdale, R. C. and J. J. MacIsaac, 1971. A computational model for the uptake of nitrate in the Peru upwelling region. *Invest. Pesq.*, **35** (1), 299–308.

Dugdale, R. C., 1975. Biological Modelling I. In *Modelling of Marine Ecosystems*. J. Nihoul, ed. Elsevier, Amsterdam, pp. 187–205.

Dugdale, R. C., 1976a. Nutrient cycling and primary production. In *Ecology of the Sea*. D. Cushing and J. Walsh, eds. Blackwell, London. In press.

Dugdale, R. C., 1976b. Marine diatoms grown in chemostats under silicate or ammonium limitation. III. Consequences of shift-up phenomena in modelling of phytoplankton production. In preparation.

Eppley, R. W., and D. L. Dyer, 1965. Predicting production in light-limited continuous cultures of algae. *Appl. Microbiol.*, **3**, 833–837.

Eppley, R. W., J. N. Rogers, and J. J. McCarthy, 1968. Half saturation constants for uptake of nitrate and ammonium by marine phytoplankton. *Limnol. Oceanogr.*, **14**, 912–920.

Eppley, R. W. and E. M. Renger, 1974. Nitrogen assimilation of an oceanic diatom in nitrogen-limited continuous culture. *J. Phycol.*, **10**, 15–23.

Falkowski, P. G., 1975. Nitrate uptake in marine phytoplankton: (nitrate, chloride) activated adenosine triphosphatase from *Skeletonema costatum* (Bacillariophycaea). *J. Phycol.*, **11**, 323–326.

Falkowski, P. G., and D. P. Stone, 1975. Nitrate uptake in marine phytoplankton: Energy sources and the interaction with carbon fixation. *Mar. Biol.*, **32**, 77–84.

Fuhs, G. W., S. D. Demmerle, E. Canelli, and M. Chen, 1972. Characterization of phosphorus-limited plankton algae (with reflections on the limiting-nutrient concept). *Limnol. Oceanogr. Spec. Symp.*, **1**, 113–133.

Grenney, W. J., D. A. Bella, and H. C. Curl, Jr., 1973. A mathematical model of the nutrient dynamics of phytoplankton in a nitrate-limited environment. *Biotech. Bioeng.*, **15**, 331–358.

Guillard, R. L., P. Kilham, and T. A. Jackson, 1973. Kinetics of silicon-limited growth in the marine diatom *Thalassiosira pseudonana* Hasle and Heimdal (= *Cyclotella nana* Hustedt). *J. Phycol.*, **9**, 233–237.

Harrison, P. J., 1974. Continuous culture of the marine diatom *Skeletonema costatum* (Grev.) Cleve under silicate limitation. Ph.D. thesis, University of Washington, 141 pp.

Harrison, P. J., H. L. Conway, and R. C. Dugdale, 1976. Marine diatoms grown in chemostats under silicate or ammonium limitation. I. Cellular chemical composition and steady state growth kinetics of *Skeletonema costatum*. *J. Phycol.* In press.

Harrison, P. J., and C. O. Davis, 1976. Use of the perturbation technique to measure nutrient uptake rates for natural phytoplankton populations. *Deep-Sea Res.* In press.

Herbert, D., R. Elsworth, and R. C. Telling, 1956. The continuous culture of bacteria: a theoretical and experimental study. *J. Gen. Microbiol.*, **14**, 601–622.

Ketchum, B. H., 1939. The absorption of phosphate and nitrate by illuminated cultures of *Nitzschia closterium*. *Am. J. Bot.*, **26**, 399–407.

MacIsaac, J. J., and R. C. Dugdale, 1969. The kinetics of nitrate and ammonia uptake by natural populations of marine phytoplankton. *Deep-Sea Res.*, **16**, 45–57.

MacIsaac, J. J., and R. C. Dugdale, 1972. Interactions of light and inorganic nitrogen in controlling nitrogen uptake in the sea. *Deep-Sea Res.*, **19**, 209–232.

Monod, J., 1942. *Recherches sur la Croissance des Cultures Bacteriennes*. 2nd ed. Hermann and Cie, Paris, 210 pp.

Paasche, E., 1973. Silicon and the ecology of marine plankton diatoms. I. *Thalassiosira pseudonana* (*Cyclotella nana*) grown in a chemostat with silicate as limiting nutrient. *Mar. Biol.*, **19**, 117–126.

Packard, T. T., M. L. Healy, and F. A. Richards, 1971. Vertical distribution of the respiratory electron transport system in marine plankton. *Limnol. Oceanogr.*, **16**, 60–70.

Packard, T., and D. Blasco, 1974. Nitrate reductase activity in upwelling regions. II. Ammonia and light dependence. *Tethys*, **6**, 269–280.

Rhee, G-Yull, 1973. A continuous culture study of phosphate uptake, growth rate and polyphosphate in *Scenedesmus sp. J. Phycol.*, **9**, 495–506.

Schaechter, M., 1968. Growth: cells and populations. In *Biochemistry of Bacterial Cell Growth*. J. Mandelstam and K. McQuillen, eds. Wiley, New York, pp. 136–162.

Williams, F. M., 1965. Population growth and regulation in continuously cultured algae. Ph.D. thesis, Yale University, New Haven.

Williams, P. J. LeB., 1973. The validity of the application of simple kinetic analysis to heterogenous microbial populations. *Limnol. Oceanogr.*, **18**, 159–165.

21. MODELING THE PRODUCTIVITY OF PHYTOPLANKTON

Trevor Platt, Kenneth L. Denman, and Alan D. Jassby

1. Preamble

The study of the modeling of phytoplankton productivity should begin with the central process of photosynthesis; we treat in some detail the mathematical representation of empirical data on production and light, and the integration of these expressions over depth and time. We then consider the modifications necessary when the effects of other environmental variables are taken into account, and the manner in which these modified expressions for production are used in models of phytoplankton biomass. Following is a discussion of the significance of nonlinearities in the production models, the characterization and implication of spatial heterogeneity in the biomass distribution, and the fundamental limitations on predictability in the plankton ecosystem. Finally we indicate some perspectives, both experimental and theoretical, for future work in this field.

Before we embark on our discussion of production models we mention the various ways in which the models may be classified and state the limitations which apply to each class; we refer to this classification frequently throughout the chapter.

The results generated by a particular model become of value only when we reach a solid opinion regarding their purpose, precision, accuracy, and robustness; we should not expect too much from our calculations and we should tread very carefully when we use a model in a context for which it was not designed. The following remarks have a certain general applicability since they apply to models describing any mechanism, but we illustrate the discussion with examples drawn from phytoplankton production.

One might first classify models according to the motivation behind their construction. Thus we recognize various ambitions (by no means mutually exclusive) in the mind of the modeler when he approaches the study of a particular system:

1. To find a convenient and economical way to summarize a collection of data, that is, to determine the parameters of a simple quantitative relationship that is *descriptive* of a large number of observations on the system.

2. To be able to make quantitative predictions about future states of the system: a *predictive* representation.

3. To discover the mechanisms by which the system operates and to express this knowledge in the form of quantitative relationships: an *explanatory* model.

We agree with Lucas (1964) that whatever the original motivation of the modeler, the model will eventually be used, by him or by somebody else, for making predictions (or even management decisions) about the system. This being the case, it is wise to proceed from the outset, whether we seek a model that is primarily descriptive or primarily explanatory, on the assumption that the final goal of our work is prediction. This is particularly important in fields, including phytoplankton productivity, that have an economic component.

One might also classify models, following the terminology of Lucas (1964), into rational and empirical models according to their internal structure. *Empirical models* are constructed with little regard to the internal mechanism of the system. Their purpose is to describe, with more or less precision, the essential characteristics of the available data relating two or more variables (e.g., the dependence of phytoplankton production on environmental factors). The construction of empirical models is therefore the exercise of curve fitting. Empirical models may be used for predictive purposes and the precision associated with the prediction may be increased almost indefinitely (within the range of the independent variable for which data are available) by increasing the number of arbitrary parameters to be fitted. However, this violates the principle of parsimony, and it is a characteristic of these models that the fitted parameters are usually uninterpretable; they cannot be identified with any tangible properties of the real system. The most extreme example of the empirical model is seen when the available data are described in terms of generalized functions such as the polynomials of arbitrary order or the spherical harmonics. The greatest failing of the empirical model, particularly if we bear in mind our ultimate goal of predictability, is that, because of the arbitrariness and the absence of structure in the model, predictions made outside the range of the independent variable(s) for which we have prior data are quite unreliable. This is a serious weakness because predictions connected with resource management or environmental quality often involve extrapolation outside the range of prior experience.

In contrast to empirical models, *rational models* are more highly structured and admit a minimum of arbitrariness. Rational models are based on experimental or observational data and on accepted knowledge about the way various components of the real system operate. From these, and with reasonable assumptions concerning less well-known features of system behavior, models are derived containing parameters that are interpretable in the real system. These models are usually robust in the sense that they can often be applied with reliable results to predict values of the independent variable(s) outside the range of the experimental or observational data and also in that they are more likely to be generalizable to other comparable systems. Of course, this classification is not a mutually exclusive one either, and we usually find that empiricism creeps in when rationality is exhausted.

Rational models may be *mechanistic* (*reductionist*) or *holistic* according to whether the modeler explains the organization he finds at one level of system description in terms of the organization at a lower or higher level of system description. Reductionist models often become lost in a wealth of detail and therefore too complicated to be useful. On the other hand, for phytoplankton production, it is doubtful whether the appropriate macrovariables have been identified for the construction of holistic models. In this class we should also mention the nonlinear statistical mechanical model, which stands between holism and reductionism, and which has been proposed as a unified approach to the theory of living systems in general (Yates et al., 1972) and ecosystems in particular (Platt and Denham, 1975b).

A final distinction should be made also between deterministic and probabilistic models. By *deterministic* we mean a rational model that is so structured that the value of the dependent variable is predicted exactly (but not necessarily accurately, depending on the quality of the model) from exact values of the independent variable(s). By contrast, *stochastic* (probabilistic) models are used to calculate the probability that a future value of the dependent variable will lie between certain specified limits, using values of the independent variable(s) which could be either exact or expressed in the form of a probability. The central objective of the probabilistic approach is predictability. It should be emphasized that the adoption of a stochastic formulation does not imply a conviction that the system is truly random; it is often used to conceal (or admit) our ignorance of the deterministic details. The development of these two classes of model requires different kinds of data. Generally stochastic models are more easily constructed from data in the form of time series, and it is perhaps for this reason that stochastic formulations have been largely unexploited in the field of phytoplankton production.

Let us agree that the long-term objective of our modeling (even explanatory modeling) is to improve our capability for prediction, and that the ultimate test of our models will be their influence on our predictive power. By this criterion, the most successful and most generally applicable models will be those with the largest component of rationality.

In this connection, several points are worth bearing in mind. First, there is the problem of scale. The larger the spatial and temporal scales of the model, the lower the precision of the model's prediction. This is discussed in some detail below, but for the moment let us recognize that an imprecise prediction is not necessarily a bad one; it might be inevitable. And then there is the problem of universality. There is a fundamental difference between a model which purports to describe, say, the primary production as a function of light intensity over the world's oceans and one which has been derived for the same relationships in a specific location at a specific time of the year. In the former case, the model would have uniform, but low, precision throughout its domain of application; in the latter case we can usually take advantage of more specific information in the parameter values to improve the precision over the much restricted range of applicability.

This is related in a way to the problem of scale, since an increase in the size of domain of interest usually implies an increase in the range of the independent variable(s). Sometimes this can lead to results which are little more than truisms and are of limited utility. For example, it is possible to make a predictive equation relating the average yield of fisheries to the average phytoplankton production for all the fisheries of the world. But the equation would not be very useful since for any specific fishery we are not very interested in knowing what its mean yield should be, given its mean productivity. It is much more important to know what the actual yield is (i.e., to explain the residual between the observed yield and the regression estimate), and to know how sensitive the future yield is to annual variations in phytoplankton productivity.

This brings us to the final point which we want to raise in this section: the problem of knowing what to model (how to choose the independent variable). Often, as hinted above, prediction of variances may be at least as important as prediction of means (Platt and Denman, 1975b). The independent variable should be chosen to be observable (for verification) and to answer the questions we want to ask. An excellent example of the simplification that can be achieved through a proper choice concerns the relationship between phytoplankton and zooplankton in the sea. Much effort had been expended over many years without clarification of the relative abundance of these two groups at given locations in the sea. Once it was realized by Steele (1974a) that what was required was a model that predicted the joint-probability distribution of phytoplankton and zooplankton (i.e., a phase-plane representation), the problem became immediately more understandable and the models open to verification by relatively unsophisticated sampling methods.

2. Photosynthesis—Light Models

In this section we discuss the construction of models describing the core process in primary production, namely, photosynthesis–light models. All models of production in the plankton community, however broad in scope and however detailed in structure, must contain one of these core models and it is therefore worthwhile to consider their derivation at some length.

A. Preliminary Remarks

It is possible to define gross primary productivity (P_g) in a number of different ways. For example, P_g may refer to the rate at which radiant energy is used in the production of ATP and reduced NADP or it may refer to the rate at which inorganic carbon is incorporated into organic carbon in the cell. Note that these two definitions are not necessarily exactly equivalent. For example, some of the ATP formed in photo-phosphorylation may be exchanged across the chloroplast and would be unavailable as reducing energy in the dark fixation of CO_2 (Ried, 1970). Since we are more interested here in the ecological rather than the biophysical aspects of photosynthesis, and since our models have a strong empirical flavor we adopt the carbon definition, which corresponds better with the measurement techniques used in oceanography.

We deal only with the production of phototrophic organisms. The treatment of chemoautotrophy and heterotrophy involves real problems; a different set of measurement procedures is involved, and there have been very few attempts to model this kind of production. Although it is generally accepted that phototrophy is quantitatively the most important form of primary production in the sea, there is no general agreement on how large the difference might be between phototrophy and chemotrophy under various circumstances.

We define respiration (R) also in terms of carbon as the rate at which inorganic carbon is being released from cellular organic carbon. Net productivity (P_n) is simply $P_g - R$. It is useful to distinguish also net particulate production (P_p) which differs from P_n by the rate of excretion (E) of organic carbon from the cell: $P_p = P_n - E$. For herbivores, P_p is the quantity of interest rather than P_n (although E of course may reappear in particulate form if assimilated by bacteria or adsorbed to inanimate particles).

Additional complications arise in models where the respiration R is included explicitly (various examples are given in Patten, 1968) since it is normally assumed in aquatic productivity models that R is independent of light intensity. For higher plants, this assumption was shown to be false 20 years ago with the discovery of photo-respiration (Decker, 1955). A similar process has been observed for many algae (Tolbert, 1974) where respiration in the light may be several times higher than that in the dark. Preliminary attempts to quantify the complications introduced by photo-respiration into phytoplankton models have already been made (Winter et al., 1975) and one might hope that it would not be too long before realistic and explicit formulations of R in terms of carbon balance can be made.

For the present it seems preferable to concentrate our attention on models in which the dependent variable is P_p. This is the only carbon flow for which we are able to make routine measurements with any accuracy; the standard ^{14}C method seems to yield P_p in practice (Eppley and Sloan, 1965; Ryther and Menzel, 1965), and it can be shown theoretically that in any but the most unrealistic compartment model for the flux of carbon between the cell and its environment, the rate of influx of ^{14}C (even for short incubation times) is not proportional to P_g.

Given the facts that P_p is the quantity that is transferred up the food chain, that P_p is the only carbon flow which can be estimated accurately, and that our goal is to construct predictive (and thus verifiable) models, we can conclude that models in which P_p is the dependent variable are to be preferred. It has to be admitted, however, that most of the models which treat "primary production" of phytoplankton purport to be models of P_g. Thus in what follows we use the unqualified term "primary production" and the unsubscripted symbol P (mg C/m^3/hr) to denote the dependent variable with the understanding that although the intent is usually to model P_g, other kinds of data have often been used for fitting and verification.

Among the factors that influence photosynthesis in the algal cell we may distinguish external factors, such as light, temperature, nutrients, and water motion relative to the cell; internal factors, such as pigments, nutrient pools, and enzyme concentrations; and cellular architecture, such as external shape and position and size of organelles.

In practice, however, there are as yet no models that treat all (or even a few) of these factors simultaneously, and it is found that most of the variability in the photosynthesis of a given organism can be attributed to variations in light. The simplest possible models for the primary productivity of a unit element of water volume thus contain as independent variables only light and the abundance of the cells. For these highly simplified models, a particularly suitable index of the relative productive capacity of the photosynthesizing cells is the concentration of chlorophyll a. For example, Platt and Subba Rao (1970a), studying the spring bloom in an inlet off the coast of Nova Scotia, found that 64% of the variance of P not accounted for by light could be explained by changes in the concentration of chlorophyll a. The suitability of chlorophyll a depends not only on its central role in the photosynthetic process. As argued in Platt (1969) and Platt and Subba Rao (1970a) the quantity of chlorophyll a per cell changes in response to environmental factors such as nutrient concentration, light, and temperature and therefore, in a sense, contains some information about the recent physiological history of a kind that we are not yet able to model explicitly.

In what follows it will often be convenient to choose as dependent variable the primary productivity per unit of biomass $P^B \equiv P/B$, where B is the biomass with dimensions [mass × length^{-3}]. Unless specified otherwise, the units for B will be mg Chl a/m^3 and for P^B, mg C/mg Chl a/hr.

Various difficulties accompany the choice of chlorophyll a as the index of B. In principle, by normalizing P relative to chlorophyll a, the graph of P^B against light should become independent of chlorophyll a. However, a problem arises because of the clumped distribution of the chlorophyll molecules. The photosynthetic pigments may have a profound effect on the penetration of light through individual cells even when the concentration of cells in the medium is so low that pigments make a negligible contribution to the macroscopic extinction coefficient. If the number of cells in an elemental volume doubles while the pigment concentration per cell stays constant, the P^B versus light curve should be unaffected even though the pigment concentration per unit volume has doubled. If, however, the pigment concentration per cell doubles while the number of cells stays constant, the pigment concentration per unit volume will increase by the same factor of two, but the P^B versus light curve will now be modified by self-shading of chlorophyll within individual chloroplasts. It might be possible to recognize this effect by including, say, the chlorophyll a:cell volume ratio in our models. Other things being equal, P^B should be a decreasing function of this ratio. In the field, however, the difficulty of making routine measurements of living phytoplankton cell volume remains formidable.

An additional problem associated with chlorophyll concerns the coupling of photosynthesis–light models to larger models which include processes such as sinking, grazing, and transport in an attempt to predict biomass changes in phytoplankton. Measurement of the other processes of these models is often made in terms of carbon, so that the carbon:chlorophyll ratio is needed to reduce all terms to common units. In practice this ratio varies over an order of magnitude (from ~ 20 to 200; Steele and Baird, 1962) and depends on light, temperature, and nutrient regimes (Eppley, 1972). A universal predictive equation for the carbon:chlorophyll ratio does not exist. In particular cases the difficulty may be tackled head on by attempting to measure the ratio directly, as was done in the budget experiment of Platt and Conover (1971).

B. Semiempirical Models

The models discussed in this section are semiempirical in that they represent attempts to summarize large numbers of observations but are relatively devoid of explanations of the observations. The models are rational to the extent that the parameters in them are interpretable. A schematic form of the observed relationship which we are trying to model is shown in Fig. 1.

The simplest quantitative expression of the relationship between photosynthesis and light is that implied by Blackman (1905) in his principle of limiting factors. Expanding on the work of Liebig (see Section 3.B), he hypothesized that the rate of any physiological process (e.g., photosynthesis) is proportional to some single rate limiting factor. No one factor remains rate limiting indefinitely as its availability is increased, and above a certain value the physiological process becomes independent of it and another factor becomes rate limiting. For the process of photosynthesis and the limiting-factor light intensity I (W/m^2) the hypothesis translates as follows:

$$P^B = P_m^B \frac{I}{I_k^{(1)}}, \qquad I \leqslant I_k^{(1)}$$

$$= P_m^B, \qquad I > I_k^{(1)} \tag{1}$$

where P_m^B is the maximum productivity per unit biomass that can be achieved at the prevailing levels of the other independent variables and $I_k^{(1)}$ is the light intensity at which this maximum is achieved, that is, at which photosynthesis is just saturated.

For real data it is found that equation 1 overestimates P^B at suboptimal light levels. It is these intermediate light intensities that are of particular interest and the application of equation 1 is correspondingly limited. However, ease of manipulation and other considerations may favor the use of this expression for the light response in certain circumstances, and it is still employed in oceanographic models, for example, that of Winter et al. (1975), which describes the productivity of Puget Sound (a model where biomass is in units of carbon).

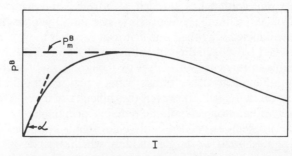

Fig. 1. Graph illustrating a typical relationship between gross primary productivity per unit biomass (P^B) and irradiance (I). The interpretations of the assimilation number (P_m^B) and initial slope (α) are indicated.

For real populations of algal cells, the curve of P^B versus I is much smoother than the Blackman equation would indicate: the saturation rate P_m^B is approached asymptotically. Ryther and Yentsch (1957) described a graphical analysis to take into account this curvature. For various depths the dimensionless productivity P^B/P_m^B was plotted against surface light intensity. On the assumption that this normalized curve was sufficiently general to apply to a wide variety of water types, a measurement of P_m^B would be all that was required to determine the depth profile of photosynthesis in any particular area.

A simple analytic description of curvature in the light response takes the form of a rectangular hyperbola, translated and rotated:

$$P^B = \frac{P_m^B(I/I_k^{(2)})}{1 + I/I_k^{(2)}} \tag{2}$$

Here P_m^B is the light-saturated photosynthetic rate as before and $I_k^{(2)}$ is the light intensity for which $P^B \equiv P_m^B/2$. Equation 2, first suggested by Baly in 1935, is the classical formula for the photosynthesis–light curve in higher plants and has since been applied to algal cultures, e.g., Tamiya, 1951. An advantage of using this formulation for the light response is that it is the same mathematical form as the equation popularly used to describe the dependence of photosynthesis on inorganic nutrients. This point has been exploited by Platt and Subba Rao (1975) to obtain an expression for the relative importance of light and nutrients in controlling primary productivity.

It is found that the initial slope of equation 2 falls off too fast compared to measured curves. Correspondence with data can be improved (without any gain in understanding) by rewriting equation 2 as follows:

$$P^B = P_m^B \frac{I/I_k^{(3)}}{[1 + (I/I_k^{(3)})^2]^{1/2}} \tag{3}$$

where $I_k^{(3)}$ is the light intensity for which $P^B = P_m^B/\sqrt{2}$. This formulation, first suggested for photosynthesis in 1936 by Smith, was used for phytoplankton by Winokur (1948) and later applied by Talling (1957a, b) to the problem of estimating the total productivity beneath a unit area of water surface.

The models discussed so far ignore the possibility of photoinhibition, the decrease in photosynthesis which accompanies light intensities large compared to that required to saturate photosynthesis (Fig. 1). Steele (1962) proposed an expression which reproduces the main characteristics of this phenomenon:

$$P^B = P_m^B \frac{I}{I_k^{(4)}} \exp\left(1 - \frac{I}{I_k^{(4)}}\right) \tag{4}$$

Here $I_k^{(4)}$ is the value of I for which P^B is maximal. Note that in this formulation the slopes on either side of the maximum are not independent. B is in carbon units (mg C/m^3) so that P^B and P_m^B have the dimension [time^{-1}] in Steele's model.

The fitting of the parameters in this model requires some care. P_m^B was obtained directly from data collected by Myers (1953) but the remaining parameter $I_k^{(4)}$ was determined in a more indirect fashion. It was assumed that at low light intensities the slope of the curve of P versus I depends only on pigment concentration, that is, the initial slope of the (unnormalized) curve $P(I)$ is proportional to concentration of chlorophyll a, with proportionality constant equal to, say, α:

$$\left(\frac{\partial P}{\partial I}\right)_{I \to 0} = \alpha \cdot [\text{Chl } a] \tag{5}$$

Dividing each side by B (in terms of carbon), letting β denote the ratio of carbon to chlorophyll a, and substituting the right side of equation 4 for $P^B = P/B$ leads to the following expression for $I_k^{(4)}$ (where $e = 2.7183$):

$$I_k^{(4)} = \frac{e\beta P_m^B}{\alpha} \tag{6}$$

In practice, Steele used a fixed value for α suggested by the work of Duysens (1956) whereas β was a parameter capable of wide variation depending on local conditions and the physiological history of the cells.

This approach requires knowledge of the three parameters α, β, and P_m^B as opposed to two parameters in the models discussed previously, but it is based on a better understanding of the fundamental process. Whereas the initial slope depends on the light reactions, the maximum (unnormalized) P_m is a function of the enzymes associated with the dark reactions and is not directly related to pigment concentrations, a state of affairs with which Steele's model is fully consistent. For purely empirical models, however, it may be expedient to express P_m relative to chlorophyll concentration and so avoid the necessity of finding β. Although there may be no direct causal relationship between P_m and pigments, the dark enzymes as well as the pigments are contained in the chloroplasts, and would thus tend to be correlated with the total number of chloroplasts and with each other.

Normalizing P_m to chlorophyll rather than carbon may be justified by comparing the relative variability of P_m^B standardized in this way with that of the initial slope standardized in the same way. This initial slope is easily identified with α of equation 5. The necessary data are collected from a variety of sources by Parsons and Takahashi (1973a, Tables 16 and 17). The standard errors for the initial slope α were 18% of the mean in algal cultures (number of cases $N = 9$) and 18% in natural populations ($N = 10$). For P_m^B (usually defined to be the assimulation number) the variability was 17% for cultures ($N = 11$) and 19% for natural populations ($N = 20$). It is clear then that there is no more variability involved in normalizing P_m to chlorophyll than there is in normalizing initial slope to chlorophyll, a procedure justified on physiological grounds.

In a different approach to the problem of light inhibition, Vollenweider (1965) proposed a modification of Smith's function (equation 3):

$$P^B = P_m^{B\prime} \frac{I/I_k^{(5)}}{[1 + (I/I_k^{(5)})^2]^{1/2}[1 + (aI/I_k^{(5)})^2]^{n/2}} \tag{7}$$

where $P_m^{B\prime}$, $I_k^{(5)}$, a, and n are constants. In general, $P_m^{B\prime}$ is not the optimum photosynthesis as in the other models, but can be readily reparameterized in terms of it: $P_m^B = P_m^{B\prime}/\delta$, where $\delta(a, n)$ is a function of a and n (Fee, 1969). $I_k^{(5)}$ is then the light intensity for which $P^B = P_m^{B\prime}[2(1 + a^2)^n]^{-1/2}$ and its exact specification depends on the choices for the two "additional" parameters a and n.

Equation 7 is capable of fitting a large variety of P^B versus I curves exhibiting photoinhibition, and is the most successful produced to date for describing experimental data. This is not surprising in view of the number (four) of free parameters which are available for fitting. Several drawbacks accompany this flexibility. The necessity to select four parameters simultaneously involves elaborate nonlinear fitting procedures. The parameters of previous models, on the other hand, can be estimated by simple measurements on the observed curves. Furthermore, the two additional parameters a and n have no fundamental interpretation of interest to either ecologists or physiologists, and the characteristic light intensity $I_k^{(5)}$ cannot be specified independently of a and n. Equation 7 is also more cumbersome to integrate over depth and time. Since the equation does not advance our understanding, one might equally have chosen a fourth-degree polynomial with advantages of simpler parameter estimation and analytic integrability and probably without loss of fidelity in describing experimental data.

A case can be made, however, that more than two parameters are needed to represent P^B versus I curves where the range of I is wide enough to include photoinhibition. If only two parameters are used, the slopes of the curve on either side of its maximum cannot be independent, even though we have no reason to suppose that the mechanisms responsible for photoinhibition are other than independent of the mechanisms which control the initial slope. A two-parameter model, therefore, cannot be expected to give a good fit to all parts of a P^B versus I curve simultaneously and in fact Steele's model has sometimes been inadequate to simulate data showing photoinhibition (Section 2.D).

A four-parameter model for the relation between photosynthesis and light has also been proposed by Parker (1974):

$$P^B = P_m^B \left[\frac{I}{I_k^{(6)}} \left(\frac{I_k^{(7)} - I}{I_k^{(7)} - I_k^{(6)}} \right)^{(I_k^{(7)} - I_k^{(6)})/I_k^{(6)}} \right]^b \tag{8}$$

where $I_k^{(6)}$ is the irradiance for which P^B is maximum (P_m^B), $I_k^{(7)}$ is the irradiance for which photoinhibition causes P^B to drop to zero, and b is some fitted constant. Equation 8 has the advantage over equation 7 that three of the four parameters $[P_m^B, I_k^{(6)},$ and $I_k^{(7)}]$ can be easily interpreted. It retains the disadvantage of requiring cumbersome fitting procedures. An additional drawback of equation 8 is the feature that $(\partial P^B / \partial I)_{I \to 0} = 0$, which is difficult to interpret in the context of what is known about the quantum yield of photosynthesis at low light levels. It is not considered further here.

An alternative is to fit different functional forms to the data on either side of P_m^B. One formulation of this approach which is particularly simple to integrate consists of a second-degree polynomial from the origin to P_m^B and a linear decay to describe the photoinhibition over the light intensities of interest:

$$P^B = P_m^B \left[\frac{I}{I_k^{(8)}} - \frac{1}{4} \left(\frac{I}{I_k^{(8)}} \right)^2 \right], \qquad I \leqslant 2I_k^{(8)}$$

$$= -cI + (P_m^B + 2cI_k^{(8)}), \qquad 2I_k^{(8)} < I \leqslant 2I_k^{(8)} + \frac{P_m^B}{c} \tag{9}$$

$$= 0, \qquad\qquad\qquad\qquad\qquad \text{otherwise}$$

where $2I_k^{(8)}$ is the light intensity at which P_m^B occurs and c is the slope of the photoinhibition.

C. Reparameterization of the Semiempirical Models

The models given in equations 1–4, 7, and 9 contain two parameters P_m^B and $I_k^{(i)}$, $i = 1, 2, 3, 4, 5, 8$, which (except in the case of Vollenweider's model and equation 9 when additional parameters must be specified) specify completely the light response of the phytoplankton. The assimilation number P_m^B is common to all the models but the magnitude and interpretation of the characteristic light intensity $I_k^{(i)}$ differs between them. In this section we attempt to unify the previous work by replacing the $I_k^{(i)}$ with a single parameter.

According to Ross (1970) a well-chosen parameter should have both stability (in the sense of small variance and lack of correlation with the other parameters) and interpretability (in the sense of an intrinsic interest which transcends its role in a particular model).

The initial slope α and the assimilation number P_m^B meet these requirements. As discussed above both have relatively small standard errors $\sim 20\%$ of the mean (although individual values may differ by an order of magnitude). Both parameters are recognized and accepted in phytoplankton ecology; the assimilation number has been studied extensively (see tabulation in Platt and Subba Rao, 1975) and α is equivalent to the maximum photosynthetic efficiency k_b (Platt, 1969) per unit of chlorophyll a. P_m^B and α are uncoupled physiologically, the one being controlled in the dark reactions and the other in the light reactions of photosynthesis, although they may be correlated if they adjust in similar ways to seasonal trends.

Making the formal solution of $(\partial P^B/\partial I)_{I \to 0} = \alpha$ to eliminate $I_k^{(i)}$ from equations 1–4 and 7 yields

$$P^B = \alpha I, \qquad I \leqslant \frac{P_m^B}{\alpha}$$

$$\qquad\quad = P_m^B, \qquad I > \frac{P_m^B}{\alpha} \tag{10}$$

$$P^B = \frac{P_m^B \alpha I}{P_m^B + \alpha I} \tag{11}$$

$$P^B = \frac{P_m^B \alpha I}{[(P_m^B)^2 + (\alpha I)^2]^{1/2}} \tag{12}$$

$$P^B = \alpha I \exp\left[-\left(\frac{\alpha I}{P_m^B e}\right)\right] \tag{13}$$

$$P^B = \frac{\alpha I (\delta P_m^B)^{n+1}}{[(\delta P_m^B)^2 + (\alpha I)^2]^{1/2}[(\delta P_m^B)^2 + (a\alpha I)^2]^{n/2}} \tag{14}$$

In making the transformation for Steele's model, we have assumed that P^B and P_m^B are normalized relative to chlorophyll, obviating the use of β. All the models are then specified entirely by P_m^B and α (for equation 14 we assume that a and n are fixed beforehand), and the $I_k^{(i)}$, which are peculiar to each model, have been eliminated.

If the range of I is sufficiently wide that light inhibition has to be taken into account, a third (and maybe fourth) *independent* parameter has to be chosen if the representation is to be adequate. Since there are as yet no "inhibition parameters" firmly entrenched in the literature, the choice is open. In Vollenweider's expression, even when the parameter a is fixed, the resulting three-parameter model suffers from having no interesting interpretation for n. It seems preferable to formulate any model containing more than two parameters so that it can be reparameterized in terms of α, P_m^B, and some photoinhibition parameter(s). For example, one possible choice for the third parameter in equation 9 is the ratio $\gamma \equiv P^B(I)/P_m^B$, when I is twice the intensity at which P^B is maximum. With this choice, $c = \alpha(1 - \gamma)/2$, and equation 6 can be rewritten as

$$P^B = \alpha I - \frac{\alpha^2}{4P^B_m} I^2, \qquad I \leqslant \frac{2P^B_m}{\alpha}$$

$$= \alpha \frac{(\gamma - 1)}{2} I + (2 - \gamma)P^B_m, \qquad \frac{2P^B_m}{\alpha} < I \leqslant \frac{2P^B_m(2 - \gamma)}{\alpha(1 - \gamma)}$$

(15)

In some cases a plateau is maintained at the level of P^B_m over a large range of light intensities before those at which photoinhibition occurs. Two linear portions may be appropriate in these cases, one representing the plateau and one the region in which photoinhibition occurs. An additional (fourth) parameter is then required, namely, the light intensity which marks the separation between these two linear portions.

D. Refining the Semiempirical Models

The photosynthesis–light curve for phytoplankton is not constant in space–time, but is influenced by short-term changes in water chemistry and temperature, the species composition of the phytoplankton, diurnal rhythms in cellular metabolism, and adaptation to longer-period fluctuations in environmental factors (including light). Thus the parameters that describe the curve are not constant, and an application of any of the models in a particular situation requires that the parameter values be chosen to suit the local conditions.

The reparameterization of the models in terms of P^B_m and α suggests a unified approach to this problem. Since any P^B versus I curve may be characterized completely by these two parameters (and photoinhibition parameters where appropriate) it seems unprofitable to search for additional representations of the photosynthesis–light relationship; it would be more worthwhile to devote our future attention to the investigation of the relationship between these parameters and environmental factors.

Parsons and Takahashi (1973a) have summarized measurements of the assimilation number P^B_m and the initial slope α; a comprehensive tabulation of assimilation numbers measured on natural populations is given in Platt and Subba Rao (1975) and an indication of the temporal variation in the initial slope for populations living in a particular coastal inlet is given in Platt (1969, 1975). The assimilation number varies from about 0.1 to 20 mg C/mg Chl a/hr and the initial slope from 0.01 to 0.2 mg C/mg Chl a/hr/(W/m^2). Clearly the range of variation in both parameters is too great for a single equation, based on their mean values, to be successful for routine use in prediction. One of the goals of our research should be to seek ways to restrict the possible range of parameter variation for particular problems, and in the remainder of this section we point out a few directions that might be helpful in this regard.

A number of workers have drawn attention to the observation that the distribution of photosynthesis with depth in widely differing water types is remarkably similar (e.g., Vollenweider, 1970); the curve usually goes through a maximum at a depth where the available light is 30–50% of its surface value. In this model, Steele (1962) incorporated this observation in the formal assumption that the maximum P^B_m would occur always at the depth where the light available was half that at the surface:

$$I_k^{(4)} = \frac{\bar{I}_0}{2}$$

(16)

where $I_k^{(4)}$ is as in equation 4 and $\bar{I}_0$ is the mean surface light intensity. Combining with equation 6 gives the following expression for P_m^B:

$$P_m^B = \frac{\alpha \bar{I}_0}{2e\beta} \tag{17}$$

where, as already pointed out, P_m^B in Steele's model is the photosynthesis normalized to carbon biomass so that P_m^B has the unit hr^{-1}. If P_m^B is normalized to chlorophyll a, that is, if P_m^B is expressed in units of mg C/mg Chl a/hr, it is easily seen that equation 17 becomes

$$P_m^B = \frac{\alpha \bar{I}_0}{2e} \tag{18}$$

Steele's assumption thus is equivalent to the hypothesis that the assimilation number P_m^B is proportional to the mean daily surface light intensity.

Although this is doubtless an oversimplification, even aside from the improbability noted by Steele that the cells would be able to adapt in this way to indefinitely low light intensities, it does illustrate one approach to the representation of the response of P^B versus I curves to changes in the daily mean of available light. Since the time constant for this adaptation appears to be several days (Steemann Nielsen and Park, 1964), we could, for example, let P_m^B be an empirical linear function of the mean light intensity during the preceding few days. This is suggested merely as a conceptual basis for further field work, rather than a tried-and-true procedure.

Other researchers have taken a theoretical rather than empirical approach to the same problem. Tooming (1970), for example, studied the variational problem of determining what values of P_m^B would maximize daily net productivity and arrived at a conclusion similar to Steele's. Laisk (1970) and Chartier (1970) developed expressions for P_m^B based on cell physiological models. Although these other workers were concerned primarily with higher plants, there is no reason why their analyses cannot be extended to phytoplankton populations.

The assimilation number is known to vary with environmental factors. Eppley (1972) has provided a most useful summary of the effect of temperature on P_m^B including the construction of curves illustrating the nature of the relationship. A power-law dependence seems to be indicated although a linear relationship between P_m^B and temperature may be an adequate approximation in practice (Takahashi et al., 1973; Stadelma et al., 1974). The assimilation number changes also with nutrient concentration (Ichimura, 1967; Takahashi et al., 1973) but no quantitative expression has yet been tested adequately to express this dependence.

Little organized information is available on environmental control of α, though its measured variance is as great as that of P_m^B. Its range, however, is substantially smaller than that of P_m^B (Parsons and Takahashi, 1973a). Unpublished data from this laboratory show that there is a regular seasonal cycle of initial slope. The belief that α should be less variable than P_m^B ignores that the initial slope of a photosynthesis–light curve is determined by the ability of the cell to trap incident light as well as its quantum yield (i.e., its ability to convert this absorbed light into chemical energy). For example, if accessory pigments increase without a corresponding increase in chlorophyll a, then the initial slope of P_m^B versus I will probably increase. Even if α did represent only quantum yield, the variable demands made on photosynthetically produced ATP by processes other than those of the reductive pentose cycle could be expected to result in a variable quantum yield in nature. The accumulating evidence therefore indicates that for natural populations of phytoplankton, α will prove to be not so (relatively) invariable as had been hoped by some authors (e.g., Dunstan, 1973; Bannister, 1974).

E. Integration of the Models

The calculation of the total production beneath unit area of the sea surface, that is, the integration over *depth* of $P^B(z)$, requires knowledge, in addition to the P^B versus I equation, of the distribution $I(z)$ of available light with depth z. The normal procedure is to use the expression $I(z) = I'_0 e^{-\kappa z}$ where I'_0 is the energy immediately below the sea surface and κ is an extinction coefficient calculated either as the average extinction coefficient for photosynthetically active radiation (PhAR) or as the extinction of the most penetrating visible wavelength. Neither of these methods is really adequate. Talling (1957b) suggested that a single coefficient 1.33 times as large as that of the most penetrating wavelength might reduce errors to a minimum, but this is no more than a rule of thumb.

It is worthwhile to calculate $I(z)$ as accurately as possible. Jassby and Powell (1975) have suggested a more detailed model for light penetration which approximates the PhAR portion of the spectrum with three wave bands. It is particularly important to try to take into account the relatively high attenuation in the red band, which leads to a drastic decrease in available light in the first few meters below the surface (Ivanoff, 1975).

All models with extinction coefficients independent of depth assume that the chlorophyll is uniformly distributed throughout the portion of the water column under study. When this assumption does not hold, and if the distribution of chlorophyll a with depth is known, then $I(z)$ can be approximated by:

$$I(z) = I'_0 \exp\left[-\left(\kappa_w z + \kappa_s \int_0^z B(\zeta)\, d\zeta\right)\right] \qquad (19)$$

where $B(\zeta)$ is the concentration of chlorophyll a at depth ζ (mg Chl a/m^3), κ_s is the spectral extinction coefficient per unit of chlorophyll a (m^2/mg Chl a), and κ_w is the extinction coefficient of the water in the absence of any chlorophyll (m^{-1}). The term $\int_0^z B(\zeta)\, d\zeta$, which can be referred to as "cumulative phytoplankton cover," is analogous to the concept of cumulative leaf area index commonly used in models of radiation penetrating higher plant stands (Talling, 1970).

Where possible it is best to measure $I(z)$ directly in energy units, or to estimate the depth average of the extinction coefficient for visible energy, using an instrument capable of giving submarine radiation in energy units independent of wavelength (e.g., Platt et al., 1970b). Since photosynthesis is a quantum process, the desirability of expressing PhAR in terms of quanta rather than energy has sometimes been emphasized. Morel and Smith (1974), however, have shown that the total quanta can be estimated within $\pm 5\%$ from total energy, or vice versa, in the ocean. The difference between expressing irradiance in terms of quanta or the more easily measured energy value thus is insignificant.

It must be remembered that representation of the irradiance profile solely in terms of macroscopic extinction coefficients may be only a gross approximation to the actual solar radiation incident on individual cells. In particular, the microstructure of phytoplankton distribution may be responsible for significant local modifications of the incident solar energy. For example, colonial forms or cells of other species in clumped distributions may absorb less light than random distributions of single cells, since in the former there is a higher probability of being shaded by another cell. This problem has been considered in its simplest form by Tamiya et al. (1953).

Additional complications are related to the movement of cells within the water column. For example, phytoplankton moving up and down within the mixed layer will experience a variable light regime, the effects of which are difficult to assess even qualitatively. By ignoring this phenomenon, we are tacitly assuming that the phytoplankton is photosynthesizing as if it were at a depth which averages the fluctuations experienced by vertical movement. Because of the nonlinear relation of P with depth z, this assumption will not apply in general. Complicated light regimes may also result from disturbances of the water surface (Dera and Gordon, 1968).

The transformation of P^B versus I curves into P^B versus z curves requires that we know also the variation with depth of α, P_m^B, and B. We may then determine $\int_z P$, the total primary production of the water column (mg C/m^2/hr). Generally only numerical integration is possible, but for theoretical explorations, analytic integration with α, P_m^B, and B assumed independent of depth can be useful. This assumption may not be too unrealistic in the case of a well-stirred mixed layer. It can be convenient sometimes to set the upper limit of integration as infinity since the primary production usually becomes insignificant at some depth above the bottom or even just below the mixed layer; that is, $\int_z P = \int_0^\infty P^B(\zeta)B(\zeta)\,d\zeta$.

All the models given may be integrated analytically for constant α, B, and P_m^B; equation 7 can be integrated only for special cases of a and n and we take the case $a = n = 1$ as an illustrative example. Vollenweider (1965) pointed out that for the models given in equations 3, 4, and 7 we can express the depth integrals in the general form

$$\int_z P = \int_0^\infty P(\zeta)\,d\zeta = \left(\frac{P_m}{\kappa}\right) f_i\left(\frac{I_0}{I_k^{(i)}}\right), \qquad i = 1, 3, \text{ or } 7$$

where f_i is a simple function. It is not difficult to show that the integrals of equations 1, 2, and 9 can be cast in the same general form. In particular, if we work with the re-parameterized equations 10–15 and calculate

$$\int_z P = B \int_0^\infty P^B(\zeta)\,d\zeta = \frac{P_m}{\kappa}\, f_i\left(\frac{\alpha I_0'}{P_m^B}\right)$$

we find

$$
\begin{aligned}
f_1(x) &= x, & x &\leqslant 1 \\
&= 1 + \ln x, & x &> 1
\end{aligned}
\tag{20}
$$

$$f_2(x) = \ln(1 + x) \tag{21}$$

$$f_3(x) = \ln\left[(1 + x^2)^{1/2} + x\right] \tag{22}$$

$$f_4(x) = e\left[1 - \exp\left(-\frac{x}{e}\right)\right] \tag{23}$$

$$f_5(x) = 2\tan^{-1}\left(\frac{x}{2}\right) \tag{24}$$

$$
\begin{aligned}
f_6(x) &= x - \frac{1}{8}x^2, & x &\leqslant 2 \\
&= \left(\frac{5}{2} - \gamma\right) + \left(\frac{\gamma - 1}{2}\right)x + (2 - \gamma)\ln\left(\frac{x}{2}\right), & 2 &< x \leqslant \frac{2(2 - \gamma)}{(1 - \gamma)}
\end{aligned}
\tag{25}
$$

These functions $f_i(x)$, which are plotted in Fig. 2, afford a direct intercomparison of the different models. In making the calculations we have taken P_m^B as 3.4 mg C/mg Chl a/hr and α as 0.1 mg C/mg Chl a/hr/(W/m²) (Parsons and Takahashi, 1973a, using the conversion 1 klux $\simeq 4$ W/m²; Strickland, 1958). With these parameter values the (dimensionless) argument x of $f_i(x)$ takes the value $x = 1$ when the available light (PhAR) just below the sea surface is ~ 34 W/m². The daily average light intensities at mid-latitudes for December and June (14 W/m² and 105 W/m² PhAR; Strickland, 1958) correspond to $x = 0.4$ and $x = 3$, respectively. The average daily *maximum* light intensities for December and June correspond to $x = 1$ and $x = 8$. The maximum instantaneous light intensity liable to be encountered under any natural conditions is equivalent to $x \simeq 15$. The parameter γ of equation 15 has been taken as 0.7 (Ryther, 1956; fig. 2).

Inspection of Fig. 2 shows that, with the chosen parameter values, equations 22–25 are not very different for $x < 1$, to the point where it is immaterial which one is used to describe the data. Even up to $x = 8$, none of these four functions lies more than 8% on either side of their mean. In any given situation, for which α and P_m^B are determined specifically to provide the best fit to the available data, one model may well do a noticeably better job than the others.

Equations 20 and 21 lie respectively well above and below the other curves in Fig. 2, particularly for large x; they have not been widely used for actual estimation of integral primary productivity in aquatic systems. The model corresponding to $f_1(x) = x$ in equation 20, however, has figured in many theoretical studies (e.g., Riley, 1946); it is particularly convenient (and justifiable) for the investigation of photosynthesis at discrete depths with low light intensities.

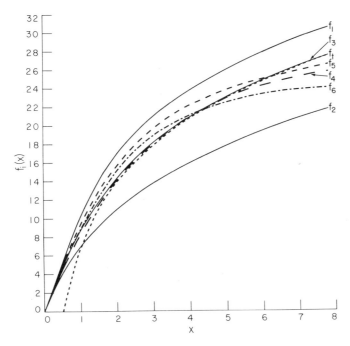

Fig. 2. Graphs of the functions $f_i(x)$, $i = 1, \ldots, 6$, given in equations 17–22. The function $f_i(x) = \ln (2x)$ is shown as well. All values of $f_i(x)$ have been multiplied by 10.

Talling (1957b) first applied equation 3 to the estimation of $\int_z P$. He did not attempt an analytic integration, but determined planimetrically that, for large values of I'_0, $\int_z P \simeq P^B_m \cdot z'$, where z' is the depth at which $I = I^{(3)}_k/2$. This may be expressed equivalently in our generalized form as $\int_z P = (P_m/\kappa) f_t(\alpha I'_0/P^B_m)$ where $f_t(x) = \ln(2x)$. As was first pointed out by Vollenweider (1965) the exact integral is given by our equation 22, and it may be verified without difficulty that Talling's empirical solution approaches the exact solution for large x. There are serious problems, however, for small x (Fig. 2), and for $x < \frac{1}{2}$ the function $f_t(x)$ is negative. Talling (1961) proposed the ad hoc function $f_t(x) = \ln(1 + x)$ to be used at low light intensities. In our rationalized scheme it can be seen that this function is none other than equation 21. Despite the fact that it contains no explicit mention of photoinhibition, Talling's prescription seems to work reasonably well as a rule of thumb (Rodhe, 1965); subsequent work shows, however, that it offers nothing for routine modeling that cannot be obtained more systematically, and with more precision, from the analytic solution given above.

Steele's model, which does account for photoinhibition, has been used with variable success in integrated form (equation 23). Parsons and Anderson (1970) found it applicable to primary production measurements in the North Pacific, but Taguchi (1972) noted very poor correspondence between measurements from the Bering Sea and predictions based on this model. In both studies, the parameter values suggested by Steele (1962) were found to be inappropriate for the particular water masses under investigation and were adjusted accordingly. Vollenweider (1965) suggested that Steele's model could be improved by the substitution of a third parameter $I^{(4)'}_k$ in place of the $I^{(4)}_k$ in the exponent of equation 4, that is, by making the rate of decline in the photoinhibited phase independent of the initial rate of increase in the light-limited phase. The proposed equation was

$$P^B = P^{B'}_m \left(\frac{I}{I^{(4)}_k}\right) \exp\left(1 - \frac{I}{I^{(4)'}_k}\right) \tag{26}$$

The symbol $P^{B'}_m$ is used because, in the modified formulation with $I^{(4)'}_k \neq I^{(4)}_k$, $P^{B'}_m$ is no longer the physiological maximum rate P^B_m which appears in the original equation 4, a point which may be verified by setting the derivative with respect to I of the modified equation equal to zero. In fact, it may be shown that $P^{B'}_m/I^{(4)}_k = P^B_m/I^{(4)'}_k$, so that the modified equation is identical in all respects to the original equation. Parker (1973) too has suggested a modification of Steele's equation 4 by raising the right side to some fitted constant. Unfortunately this modification would introduce a new parameter with no interesting interpretation, an undesirable practice as we have discussed earlier.

Integration over depth of Vollenweider's four-parameter model (equation 7) can be done numerically only if no restrictions are put on a and n. Fee (1973a, b) has made use of numerical integration of this model to calculate $\int_z P$, including cases where the depth profile of chlorophyll concentration is nonuniform.

In certain not uncommon circumstances we can still make analytic integrations over depth when the condition that chlorophyll concentration be independent of depth is relaxed. For example, a chlorophyll maximum near the thermocline may be described by a polynomial of low degree. For $\alpha I'_0/P^B_m < 2$ (i.e., at subsaturating light intensities) integration of equation 25 is a simple matter, involving terms like $z^j e^{-\kappa z}$ and $z^j e^{-2\kappa z}$, where j is an integer. This procedure could be particularly applicable to stations where the shape of the chlorophyll profile was approximately constant, at least for periods of days.

Integration over *time* of the productivity models has proved to be a more difficult task than integration over depth except when unrealistic representations of $I_0'(t)$ were used. Talling (1957b), for example, assumed that incident light was constant over the day and equal to the mean intensity. Steele (1962) suggested the use of a triangular wave for approximating daily irradiance, that is, a linear increase in incident solar energy during the first half of the day and a linear decrease during the second half. Both representations permit the analytical integration over time of equations 20–25. A more realistic description of the time variation of surface light intensity is the "standard light day" proposed by Vollenweider (1965):

$$I_0'(t) = \frac{I_0^{m'}}{2}\left(1 + \cos\frac{2\pi t}{\lambda}\right) \tag{27}$$

where $I_0^{m'}$ is the daily maximum intensity (W/m^2), λ is a tabulated day-length factor (hr), and the time t (hr) is measured with respect to an origin at local noon. Ikushima (1967) suggested a similar expression for $I_0'(t)$:

$$I_0'(t) = I_0^{m'} \sin^3\frac{\pi t}{\lambda} \tag{28}$$

where t is now measured with respect to an origin at the beginning of the day.

With either of these formulations of the time dependence of surface light intensity and with the depth-integrated form of one of the photosynthesis–light models, we would normally be obliged to resort to numerical integration to find $\int_t \int_z P$, the total production in the water column for the day. This is no particular handicap if we are interested only in computational models for a specific time and place, but it can be a serious obstacle to the progress of theoretical studies which may require a general analytic expression for $\int_t \int_z P$. On days with the complex light regimes that can be produced by scattered cloudiness, there is, of course, no alternative to numerical integration.

In at least two cases, where the maximum daily light intensity is below that required for saturation, analytic integration is possible over the standard light day. For the case of equation 20 when $\alpha I_0^{m'}/P_m^B \leqslant 1$, the integration is trivial. Equation 25 may also be integrated over time when $\alpha I_0^{m'}/P_m^{B'} \leqslant 2$. It is desirable to rewrite equation 27 in the form $I_0'(t) = I_0^{m'} \cos^2(\pi t/\lambda)$; then

$$\int_t \int_z P = \frac{P_m}{\kappa} x_m \int_{-\lambda/2}^{+\lambda/2}\left(\cos^2\frac{\pi t}{\lambda} - \frac{1}{8} x_m \cos^4\frac{\pi t}{\lambda}\right) d\tau = \frac{P_m}{\kappa} x_m\left(\frac{\lambda}{2} - \frac{3 x_m \lambda}{64}\right) \tag{29}$$

where $x_m = \alpha I_0^{m'}/P_m^B \leqslant 2$. This equation will give a fairly good approximation to daily production in the water column under certain conditions. For temperate latitudes, it should be applicable at least during the winter season. For $I_0^{m'}$ much in excess of $2P_m^B/\alpha$, photoinhibition will cause equation 29 to overestimate the realized primary production.

Two limitations of this discussion of the time dependence of phytoplankton production should be emphasized. The first is that no account has been taken of intrinsic physiological rhythms in the physiological parameters. The diurnal rhythm has been much studied in the phytoplankton (and has been recently reviewed by Sournia, 1974) but the conclusions obtained so far do not seem to have been consistent enough for incorporation into theoretical models. The second point, discussed at some length by

Patten (1968), is that the assumption of uniform distribution of biomass with depth, if acceptable at time zero, becomes progressively less acceptable as time increases and production goes on in a light field which decreases exponentially with depth. Winter et al. (1975) have made a similar point about the averaging properties of the quantity $P^B B$.

F. Mechanistic Models

The characterization of daily primary productivity in the water column and the conclusions drawn therefrom rest ultimately on the use of empirical P^B versus I relationships. There are at least two reasons for extending the search for mechanistic models to replace the empirical ones. First, as we have indicated earlier, mechanistic models are, in the long run, going to give us a better predictive capability than empirical descriptions. Second, for the most efficient prediction of primary production, we want to be able to deduce the values of the parameters α, P_m^B, and γ for a given time and place without the necessity to carry out lengthy and detailed experiments. It was suggested above that α and P_m^B might be simple functions of recent light, temperature, and nutrient levels. It is possible that these parameters might also be predictable from measurements of a few subcellular quantities. For example, there is evidence accumulating from studies of higher plants that P_m^B is controlled largely by the levels of ribulose diphosphate carboxylase, a possibility which certainly merits further investigation in the phytoplankton (cf. Hellebust and Terborgh, 1967; Morris and Farrell, 1971).

Elucidation of the relationships between productivity and enzyme levels based on chemical kinetics should certainly lead to results of interest to phytoplankton ecologists. An example of this approach is contained in the work of Zlobin (1973). Most of the chemical–kinetic models of photosynthesis have not been derived specifically for phytoplankton; they may nevertheless be of considerable interest to our purpose. An excellent example is the investigation of periodic phenomena in photosynthesis by Chernavskaya and Chernavskii (1961). These authors have suggested that photosynthetic rhythms result in part from oscillations in the dark reactions and have derived explicit equations for these oscillations in terms of the biochemical kinetics.

G. Respiration and Vertical Structure

The P^B versus I models presented so far were all intended as descriptions of *gross* primary productivity per unit pigment; this is implicit in the feature that $P^B = 0$ when $I = 0$ for all of these models. Extension to models of net productivity requires a mathematical formulation for R (or the normalized $R^B = R/B$). A number of researchers have made the simple assumption that R is proportional to biomass (various examples in review by Patten, 1968). Tooming (1967) has suggested a slightly different formulation, assuming that R is proportional to the maximum photosynthetic rate:

$$R = rP_m \tag{30}$$

where r is some (dimensionless) proportionality constant. The rationale behind equation 30 is the belief that P_m closely reflects the standing crop and hence the respiratory demands of the plants.

Many models of respiration in higher plants incorporate explicitly the effects of altered respiration in the light. There is no reason why these same formulations cannot be applied to phytoplankton. Tooming (1970), for example, extended equation 30 to account for light respiration by assuming that the additional carbon loss in the light was proportional to gross primary productivity:

$$R = r_1 P_m + r_2 P_g \tag{31}$$

where r_1 and r_2 are (dimensionless) proportionality constants. McCree (1970) made a similar modification of the assumption that R is directly proportional to biomass:

$$R = r_3 B + r_4 P_g \tag{32}$$

where r_3 (mg C/mg Chl a/hr when B is in mg Chl a/m^3) and r_4 (dimensionless) are constant. As a final example, Peisker's (1974) analysis of the initial steps in photosynthetic carboxylation led to the approximation:

$$R_L \propto \frac{[O_2]}{[CO_2]} P_g \tag{33}$$

where R_L is the additional respiration observed in the light.

A number of authors have combined expressions for respiration with models for gross productivity to produce a formulation for net productivity ($P_n = P_g - R$). This analytical expression for P_n permits us to estimate the depths of certain points in the water column which have been identified for one reason or another by researchers in production ecology. Perhaps the first of these to be recognized was the *average daily compensation depth* (Sverdrup et al., 1942) which can be defined as that depth z_c where the daily average of P_g and R are equal. Since $P_n < 0$ below z_c, it is easily seen that z_c must be the depth which maximizes the daily integral net primary productivity above that depth, that is, the depth that maximizes $\int_0^{24} \int_0^{z_c} P_n \, d\zeta \, d\tau$. An explicit expression for z_c is obtained by solving

$$\frac{\partial}{\partial z_c} \left(\int_0^{24} \int_0^{z_c} P_n \, d\zeta \, d\tau \right) = 0 \tag{34}$$

for z_c. For example, in the extremely simplified case when $P_g^B = \alpha I$, R^B and B are constants, and $I(z) = I_0' e^{-\kappa z}$, then

$$\int_0^{24} \int_0^{z_c} P_n \, d\zeta \, d\tau = \left\{ \frac{\alpha \bar{I}_0'}{\kappa} [1 - \exp(-\kappa z_c)] B - R^B z_c B \right\} 24 \tag{35}$$

where $\bar{I}_0'$ is the daily mean I_0'. The solution to equation 34 is then

$$z_c = \frac{1}{\kappa} \ln \left(\frac{\alpha \bar{I}_0'}{R^B} \right) \tag{36}$$

as Vollenweider (1970) has deduced from a different viewpoint. Substitution of typical values for the parameters in the right side of equation 36 leads to the provocative conclusion that z_c in nature is of the same order as the thermocline depth.

The *critical depth* (Sverdrup, 1953), sometimes referred to as the *column compensation depth* (Talling, 1957), is simply that depth z_{cr} where

$$\int_0^{24} \int_0^{z_{cr}} P_n \, d\zeta \, d\tau = 0 \tag{37}$$

Making the same assumptions that lead to equation 35, Sverdrup demonstrated that the critical depth satisfies the relationship

$$\frac{z_{cr}}{1 - \exp(-\kappa z_{cr})} = \frac{1}{\kappa}\left(\frac{\alpha \bar{I}_0}{R^B}\right) \tag{38}$$

Equation 38 expresses the approximate relationship between the mean daily light intensity and the depth below which no net growth can occur. Cushing (1962) refined the analysis by permitting a more realistic P_g versus I relationship and found an acceptable agreement with empirical observations.

Patten (1965) has taken issue with these analyses, primarily because they do not account for the possibility of adaptation in such parameters as R^B. He suggested that changes in R^B were, in fact, of a magnitude to allow $P_n > 0$, regardless of the position of the thermocline. In the course of his argument, Patten introduced the *cancellation depth* z_{ca} which is defined by

$$\int_0^{24}\int_0^{z_{ca}} P_g \, d\zeta \, d\tau = \int_0^{24}\int_0^{z_b} R \, d\zeta \, d\tau \tag{39}$$

where z_b is the depth of the bottom. Vollenweider (1970) has compared Sverdrup's and Patten's arguments in some detail and concluded that the validity of one or the other depends on the time scale being considered. He also pointed out, however, that some respiration must always be present and that the minimum possible respiration established a limiting critical depth which might be of biological significance in less transparent waters.

With expressions for P_n at our disposal, explanations of observations on phytoplankton variables can be attempted from a holistic extension of these expressions rather than a mechanistic one. An example will make the distinction more clear. If we are interested in predicting phytoplankton biomass distributions, one alternative is to include with P_n expressions for other processes such as sinking, grazing, and turbulent transport in an equation for $\partial B/\partial t$ (Section 3.C). Such an approach attempts a mechanistic or reductionist explanation for observations of phytoplankton biomass.

The other alternative is to combine the expression for P_n with some accessory hypothesis concerning the macroscopic behavior of phytoplankton communities; for example, we might postulate that phytoplankton biomass tends toward a value which maximizes the productivity of the mixed layer. Solution for B of the extremum problem

$$\frac{\partial}{\partial B}\left(\int_0^{24}\int_0^{z_t} P_n \, d\zeta \, d\tau\right) = 0 \tag{40}$$

where z_t is the thermocline depth, would then constitute a holistic recipe for the problem of accounting for observed biomass. Substitution of B back into equation 35 then provides an estimate of daily net production in the mixed layer for a system maximizing this quantity with respect to biomass. Steele and Menzel (1962) have applied this procedure to Steele's (1962) model, using an empirical relationship developed by Riley for the dependence of κ on B and including a term for the influence of nitrate on P_n. Agreement with the observed chlorophyll a concentrations during three winters in the Sargasso Sea was good, and the results thus were consistent with the concept of maximized primary productivity in the mixed layer.

These analyses of net productivity in the mixed layer suffer from several obvious deficiencies. For example, either light saturation and inhibition are ignored (Sverdrup,

1953) or, if they are included (Steele, 1962), a realistic integration over time is not used. In addition, respiration is considered to be constant with depth and time, an assumption we have emphasized previously to be unrealistic. The use of these models, regardless of their simplicity and their inability to predict in detail, should be encouraged, for they sometimes lead to satisfactory explanations of gross behavior and often indicate clearly which relationships merit more detailed empirical investigations.

3. Comprehensive Models

In the preceding section we gave an elementary treatment of primary production models which depend only on the relationship between photosynthesis and light. In this section we consider more comprehensive or global models which attempt to include explicitly most or all of the variables which might influence primary production. Apart from illustrating the general structure of these wider models, the survey will serve to justify the considerable space we have devoted to photosynthesis–light models and will help to account for the relatively high degree of success associated with the common practice of ignoring all variables other than light in the construction of P^B models.

A. Regression Models

To exploit the possibilities offered by regression models, which consitute an extreme class of empirical models, one tries to collect data as frequently as possible on as many as possible of the variables which might influence the dependent variable of interest, in our case the primary production. The total data series is then examined for the presence of systematic trends or patterns; to this extent the procedure might be described as exploratory.

The statistical procedures involved are well known. The simplest is to calculate partial correlation coefficients between different pairs of variables. These variables may be either measured quantities or derived from measured quantities (if one is trying to identify new state variables). The purpose of *partial correlation analysis* is twofold: first, a high correlation coefficient between an independent variable and the dependent variable may indicate a causal relationship or at least a correlation useful for prediction; second, high correlations between independent variables may indicate redundant information in the data series and may help to provide the basis for a rational reduction in the number of parameters measured in future studies.

A more sophisticated procedure for identifying and grouping highly correlated parameters is a *cluster- or association-analysis* of partial covariances (Pielou, 1969; Williams, 1971). In such an analysis, one attempts to group together variables which are correlated and to rank them according to their degree of association; this procedure is also useful to identify redundancy in the data. There are two basic types of cluster analysis, divisive and agglomerative. In the *divisive* method, all the parameters are first grouped under the parameter that has the highest degree of association with all other parameters. This large group is split successively into lesser groups as other variables are identified which have progressively smaller degrees of internal association with the other members of their groups. The *agglomerative* method proceeds by progressive fusion of individual variables into larger and larger groups until all possible groupings have been made. This second method has two disadvantages. The first is computational: all the calculations must be carried out before the parameters with the highest degree of association can be identified. The second is that a single fortuitous or random high correlation between any two parameters can result in a spurious pairing of the parameters which will be carried through the whole procedure.

Where a single parameter has been identified clearly as the dependent variable of interest, stepwise multiple regression represents a more quantitative procedure for identifying the most important independent parameters. First, a linear regression is carried out between the dependent variable and each of the other parameters in turn, and one of these is identified as explaining the greatest fraction of the variance in the dependent variable. Next, a series of two-parameter multiple linear regressions is carried out to find the next most important independent variable, that is, the variable that reduces the residual variance by the greatest amount. The procedure then is repeated for three-parameter regressions, and so on. By calculating in this way how much the residual variance is reduced by including each additional parameter, one can judge post facto how many parameters were necessary to attain the desired precision of prediction in the most efficient way.

Several investigators have carried out primary production studies of sufficient magnitude to allow the use of these correlation and regression techniques. Goldman et al. (1968) measured 28 different variables at nine depths in Lake Maggiore, Italy on 14 different days at roughly weekly intervals. They calculated partial correlations, an agglomerative cluster analysis, and a stepwise multiple regression with primary production as the dependent variable.

In the multiple regression, light alone accounted for 64 % of the variation in primary production. The addition of temperature accounted for a further 8 %, and the inclusion of oxygen accounted for an additional 7 %. Diversity, silicates, and biomass, the next three most important variables, together accounted for only 3 % of the remaining unexplained variance. Surprisingly, phytoplankton productivity, as measured by the ^{14}C method, was inversely related to total phytoplankton biomass (cell volume). However, the phytoplankton productivity was also inversely related to temperature. Goldman et al. explained these observations by noting that during the warmest month, high nighttime respiration may have reduced considerably the primary productivity as measured by the ^{14}C technique.

Two companion correlation studies of primary production were made by Platt and Subba Rao (1970a) and Platt et al. (1973). Spring-bloom periods were sampled in two coastal inlets, St. Margaret's Bay and Bedford Basin, the one exposed and the other semi-enclosed. In each case 12 parameters were sampled at six depths about 12 times at roughly twice-weekly intervals. The dependent variable was $k_b = P(z)/I(z)$, which is proportional to the initial slope α of the photosynthesis–light curve at low light intensities (Platt, 1969); the dependence of primary production on light was therefore implicit.

From the partial correlation analyses highly significant correlations were found in both inlets between chlorophyll and k_b, and between PO_4 and NO_3. The derived quantities $k_b/Chl\ a$ and k_b were both correlated with species diversity; Goldman et al. also found a similar parameter of production efficiency to be correlated with species diversity. In St. Margaret's Bay, chlorophyll a explained 64 % of the variance in k_b; including depth explained an additional 9 %, cell numbers 7 %, NO_3 4 %, and PO_4 1 %. In Bedford Basin, chlorophyll a explained 82 % of the variance, and no other parameters made a significant further reduction in the residual variance. Temperature was not included as a regression variable.

Although Platt and Subba Rao (1970a) and Platt et al. (1973) did not attempt cluster analyses, we present here divisive cluster analyses (Figs. 3 and 4) of their covariance tables to illustrate the method as an alternative to the agglomerative procedure used by Goldman et al. (1968). The method we have used is essentially that outlined by Williams et al. (1966) and by Pielou (1969) for quantitative correlation measures. All partial correlation coefficients less than the value expected at the 5 % probability level have been set equal to zero. The normalized covariance r_{ij}^2 for each variable i were summed over all other j variables and the variable with the largest $\sum_{j \neq i} r_{ij}^2$ was placed at the top of the cluster at a level equal to its $\sum_{j \neq i} r_{ij}^2$. The second highest was identified and branched at a level equal to its $\sum_{i \neq i} r_{ij}^2$. The next highest variable was identified and branched at the level of its $\sum_{j \neq i} r_{ij}^2$ from the previously identified variable with which it had the highest single r_{ij}, and so on. Thus variables located near the top of the diagram have the highest degree of association with the other variables. Variables placed near the bottom show little correlation and are assumed to be relatively unimportant in describing phytoplankton production.

In Fig. 3, the analysis for St. Margaret's Bay, chlorophyll a biomass is the most important variable and the primary production efficiency, k_b, is associated with it. On the same branch is cell number, which was the most important variable after chlorophyll a in determining k_b. In Bedford Basin (Fig. 4) k_b was again grouped with chlorophyll a, but in this case, NO_3 was as highly associated with other variables as was chlorophyll a. A rather interesting feature is that in both cases, the group con-

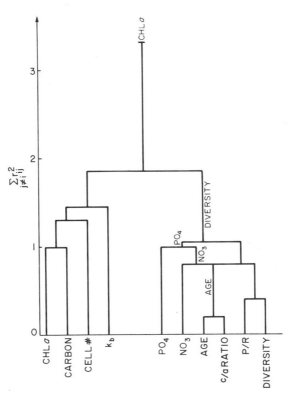

Fig. 3. Divisive cluster analysis of the primary production and related parameters for St. Margaret's Bay, Nova Scotia, calculated from correlation tables presented in Platt and Subba Rao (1970a).

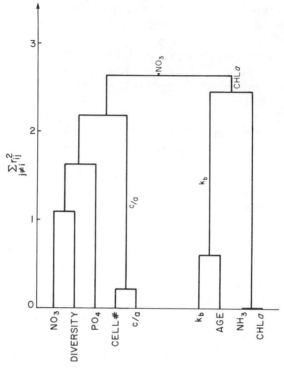

Fig. 4.   Divisive cluster analysis of the primary production and related parameters for Bedford Basin, Nova Scotia, calculated from correlation tables presented in Platt et al., 1973.

taining k_b and chlorophyll a was distinct from the group containing nutrients, species diversity and the Chl c/Chl a pigment ratio.

A correlation and regression analysis of a more global nature (in the geographical sense) was done by Brylinsky and Mann (1973), who analyzed data collected from 55 lakes and reservoirs during the International Biological Programme. They found that the latitude, and other variables related to light, were most important in predicting primary production. Their best predictive regression was on the two independent variables, latitude and chlorophyll a. Lake morphology was not important and nutrient concentrations were important only when lakes at similar latitudes were compared.

From these three studies one can conclude that for the empirical prediction of primary production, light and chlorophyll a concentration are the most useful variables to measure. For most natural situations, other variables, such as nutrients, may be regarded as making a strictly second-order contribution to the primary production and its prediction. However, in certain perturbed aquatic systems, a knowledge of nutrient concentrations would be indispensable for adequate prediction of primary production. One is thus led to an important conclusion regarding the use of regression models for prediction: no single regression equation is universally applicable and calibration of the regression model must be performed for each system of interest.

In addition to prediction, regression and correlation analyses do provide limited insight into the behavior of the system under study. They can help to quantify the loss in predictability we incur by limiting our system description to a few key parameters. However, there are drawbacks. The analyses are most often strictly linear, so that important nonlinear processes may be overlooked. Furthermore, this approach can tell little or nothing about processes operating in the system at frequencies higher than the sampling frequency. Typically, primary productivity investigations based on regression models have been limited in scope to the study of features with characteristic periods of weeks or more.

Probably the best use for these techniques is in the identification of key variables which might be studied efficiently and profitably in a further detailed observational program aimed at the construction of a more refined predictive model.

B. The Photosynthesis–Light Model In Its Environmental Context

In this section we consider how the core models of primary production and light may be modified to account for variations in other environmental variables such as inorganic nutrients and temperature. We do not go into details of the specific mechanisms of nutrient uptake, since they are covered in Chapter 20; we seek only to place the basic model in its proper environmental setting.

We have suggested in the last section that a rational way to introduce the effects of other environmental variables into the core model of phytosynthesis would be through their influence on the values of the physiological parameters α, P_m^B, and possibly one or more photoinhibition parameters. For the moment this suggestion must be considered as a program for future research rather than an actual possibility, since the systematic investigations required have not yet been carried out.

Instead, we look at what can be done within the present state of knowledge. Two general approaches are in use. One consists in formulating an expression for the productivity P^B in which each of the relevant environmental variables appears as a multiplicative factor (e.g., Parsons and Takahashi, 1973b); the other consists in deciding for the given situation which environmental variable is likely to have the most control over P^B, then invoking a "minimum principle" to infer that this variable must be *the* limiting factor and modeling in detail the relationship between P^B and this factor (e.g., Dugdale and MacIsaac, 1971).

In our opinion, the use of minimum principles has led to some confusion in phytoplankton ecology. What is usually called Liebig's law of the minimum states "that the total *crop* [our emphasis] of any organism will be determined by the abundance of the substance that, in relation to the needs of the organism, is least abundant in the environment" (Hutchinson, 1973). Note that the law refers to the limitation of crop or biomass and not to the limitation of the rates of physiological processes. The validity of Liebig's law is intuitively obvious for single species, although in multispecific communities the total crop can be determined simultaneously by the abundances of more than one substance (see discussion by Hutchinson, 1973). Note that the relative time scales and phase relationships of growth and nutrient recycling are quite different in planktonic systems compared to agricultural systems, so that the relevance of Liebig's law to our problem is not as obvious as to that of predicting field crops.

Occasionally a principle similar to (but distinct from) Liebig's law has been proposed, namely, that the rate of any physiological process (such as photosynthesis) is determined at any one time by a single environmental factor, often called *the* limiting factor. This hypothesis, known as the principle of limiting factors, was advanced by Blackman (1905), although it is often erroneously identified with Liebig's law. The concept that process *rates* are dependent on a single limiting factor is, of course, not generally true and insistence on its validity could lead to spurious results. Part of the confusion could be eliminated if the term "limited" (used in the sense of nutrient-limited or light-limited conditions for phytoplankton growth) were replaced by "suboptimal." Again, confusion arises if terminology commonly used in connection with culture experiments is carried over too readily to the discussion of measurements made under natural conditions; in culture studies it is a simple matter to arrange for a given variable to be truly limiting to growth, but such a simplified situation pertains probably only rarely in the photic layer of the sea. In nature it would be unreasonable to suppose that all but one environmental variable take optimal values and we antici-pate that primary production will be regulated by several environmental variables simultaneously. In measurements made on natural populations of the response of photosynthesis to a given environmental variable, the parameters of the model describing the response will be functions of the other environmental variables which are present at suboptimal values. Thus in a given natural situation, the approach of structuring a predictive model and choosing the parameters "according to the Liebig principle that only one factor (light or nitrate) can limit the rate of reaction at a given instant in time" (Dugdale and MacIsaac, 1971) cannot be recommended; in the field, there is no a priori justification for supposing that primary production could not be controlled by the two (or more) variables simultaneously.

If we cannot, in general, identify one single factor limiting growth in a given natural situation, we can compare environmental variables to see which is exerting the greater control over production at a given time (Platt and Subba Rao, 1975). As a simple illustration, we suppose that the response of the phytoplankton to each of the en-vironmental factors (not temperature) can be described by a rectangular hyperbola (usually called a Michaelis–Menten equation when the environmental factors are nutrients); this is not too unreal an assumption within certain limitations to be dis-cussed below. Let μ be the specific growth rate of phytoplankton (dimensions of time^{-1}) and let the response function to the environmental variable S be a hyperbola with parameters μ_m and K_s. The simple ratio $\mu \equiv \mu_m/K_s$ cannot be used for com-parison purposes because its units would depend on the units of S so that it would not be possible to compare, say, the effect of nitrate with that of light.

Instead we define a new dimensionless number $S_* \equiv S/K_s$ indicative of the mag-nitude of the controlling variable relative to its half-saturation constant. Clearly, when $S = K_s$, $S_* = 1$ and the normalized Michaelis–Menten equation may be written:

$$\mu(S_*) = \frac{\mu_m S_*}{1 + S_*} \tag{41}$$

A measure of the dependence of μ on S_* at any given value of S_* is

$$\frac{d\mu}{dS_*} = \frac{\mu_m}{(1 + S_*)^2} \tag{42}$$

The relative strength of dependence of μ on two environmental variables i and j may therefore be expressed by the ratio $\xi(i, j)$ where

$$\xi(i, j) \equiv \frac{\mu_{mi}}{\mu_{mj}} \left[\frac{1 + S_*^j}{1 + S_*^i} \right]^2 \tag{43}$$

For particular values of S_*^i and S_*^j, $\xi(i, j)$ is a single number indicative of the relative dependence of specific growth rate on variable i compared to variable j. For any given species or combination of species, the maximum specific growth rate under optimal (nonlimiting) conditions should be the same for data describing the response of μ to both variables i and j in the steady state. In this case

$$\xi(i, j) = \frac{(1 + S_*^j)^2}{(1 + S_*^i)^2} \tag{44}$$

While we are discussing the application of the Michaelis–Menten formalism, several points are worth emphasizing. The first concerns the estimation of the parameters μ_m and K_s from experimental data. Most of the information is contained in the points for which $S < K_s$, and the experimental design should be arranged such that most of the observations are made in this region; most often one sees experiments designed exactly the opposite way, that is, with most observations made in the saturation or near-saturation range of the substrate concentration. Again, it is of the utmost importance to use a statistically acceptable method of fitting the data to the model equation. The most popular technique is a linearization method in which the data are fitted to a straight line in either the variables μ^{-1} and S^{-1}, or S/μ and S;

$$\frac{1}{S} = \frac{1}{\mu}\left(\frac{\mu_m}{K_s}\right) - \frac{1}{K_s} \quad \text{or} \quad \frac{S}{\mu} = \frac{S}{\mu_m} + \frac{K_s}{\mu_m} \tag{45}$$

Statistically, this method is simply unacceptable and K_s values calculated by it are quite unreliable (especially in the first equation). The method can be salvaged if the data for low substrate concentrations are weighted appropriately (Wilkinson, 1961; Dowd and Riggs, 1965), but it is preferable to use one of the generalized non-linear fitting algorithms such as those of Glass (1967) or Conway et al. (1970). Even using one of these techniques it will be found difficult to reduce the coefficient of variation of K_s much below 20%.

The second caution about following too blindly the Michaelis–Menten formalism concerns the extrapolation of parameters derived from culture experiments to prediction about natural populations. One difficulty is that whereas laboratory measurements are made on monospecific populations (or more exactly, clones), natural phytoplankton communities are invariably polyspecific, even during blooms (e.g., Platt and Subba Rao, 1970b). As pointed out by Williams (1973), although the μ_m's for the individual species add linearly to give an average μ_m for the community, no such simple relationship exists for the K_s. Again, if there is any rationality in using the Michaelis–Menten expression to describe nutrient uptake, the experiments should be made in the steady state. But with the odd exception, such as Caperon (1968) and Caperon and Meyer (1972a, b), the experiments are usually made on batch cultures. Except during periods of extreme environmental perturbation, the pelagic zone can probably be described fairly well by a steady-state representation.

A final difficulty with the Michaelis–Menten formalism is the frequently observed lack of agreement between K_s derived from nutrient uptake measurements and K_s for cell growth, that is, between experiments designed to measure $\mu(S) = S^{-1} \, dS/dt$ and those designed to measure $\mu(S) = B^{-1} \, dB/dt$ (e.g., MacIsaac and Dugdale, 1972).

These remarks on the Michaelis–Menten expression apply equally to its use in the other type of extended production model which we consider in this section, namely, that in which the core expression for photosynthesis is modified by a series of multiplicative factors. An example of this type of calculation is given by Parsons and

Takahashi (1973b), who applied it to the discussion of the environmental control of cell size in various phytoplankters. Formally, we can write

$$P(I, S_*^1, S_*^2, \ldots, S_*^n, \theta) = P^B(I)g_1(S_*^1)g_2(S_*^2)\cdots g_n(S_*^n)h(\theta) \qquad (46)$$

where $g_n(S_*^n)$ is the normalized Michaelis–Menten function for the nth micronutrient, and $h(\theta)$ is a function of the temperature θ.

It has been shown in culture experiments (Caperon and Meyer, 1972a) that cell growth depends less on nutrient concentration at the time of measurement than on the history of nutrient uptake by the cells prior to the measurement. This kind of information is included in the model of Winter et al. (1975) by writing the Michaelis–Menten expressions as functions of the lagged time $(t - t_L)$:

$$\mu(t) = \frac{\mu_m(t)S(t - t_L)}{K_s(t) + S(t - t_L)} \qquad (47)$$

In their calculations for Puget Sound, the lag time t_L was ~ 3 days.

The effect of temperature on phytoplankton growth still awaits clarification. Eppley (1972) has provided an excellent review of the available data, in which it appears that we could express the dependence of maximum growth on θ as a simple power function; Goldman and Carpenter (1974) have suggested the use of the Arrhenius equation. But there still remains a great deal of systematic experimentation to be done.

C. Extended Models for the Biomass

There exists a large class of models designed to predict the biomass $B(X, Y, Z, T)$ at specific times and places given the initial biomass condition, a more or less complex expression for P^B, and a series of terms describing the coupling of the phytoplankton to the physical environment and to other biological communities, such as the herbivores. Patten (1968) has given a comprehensive review of such models.

We do not propose to enter into the detailed structure of such models, but to give instead a general discussion of the problems associated with them. A typical model of this type would include first some representation of the core model describing the photosynthesis–light response (suitably modified for the influence of inorganic nutrients), the sinking of cells through and out of the photic zone, the grazing by zooplankton, the effect of turbulent diffusion in three dimensions, and horizontal and vertical advection. Perhaps the best developed and most careful example of these extended models is the one published by Winter et al. (1975) for the Puget Sound. In addition to the factors just listed, the model includes the effect of runoff, winds, and tides; a time-lagged Michaelis–Menten function for the effect of nutrients; and a complex function to describe the respiration of the phytoplankton which attempts to account for the phenomenon of photorespiration.

If the construction of such a model is contemplated, various problems should be resolved at the outset; these are discussed in Platt and Conover (1971). Perhaps the most important among them include the selection of the fundamental time and space scales in the model; the decision on whether the model will be free running or whether it will be corrected periodically (how often?) by checking the initial conditions against direct measurement; and the decision on what will constitute a statistically adequate set of data for verification of the model.

Selection of the space scales and specification of the verification data must be done in the context of what is known about the spatial structure of phytoplankton populations; this is taken up in a later section. Selection of the time scales should take into account the maximum time rates of change of the parameters. The biggest problem, however, is how long the model can be allowed to run without being readjusted to match the real world. Because the model is not perfect, it is bound to accumulate errors and the standard error of the estimate will grow with time. The modeler should have a clear idea of how far into the future his model can reasonably be expected to give reliable forecasts. This question is taken up in Section 6, where it is shown that there are fundamental limitations on prediction time for these extended models. These limitations do not arise through inadequacies of the functional relationships chosen to represent the real world; they would exist even if our equations were exact. They are due rather to our inability to specify without error the initial biomass condition (the same *kind* of problem as is involved in specifying suitable verification data).

Platt and Conover (1971) addressed these questions for the production of phytoplankton in a semi-enclosed marine basin of small (~ 5 km) physical dimension. It was considered that an essential prerequisite to the construction of a quasi-continuous biomass model was the demonstration that the continuity equation for phytoplankton biomass could be balanced in the short term (say, 24 hr) by direct measurement and calculation. This was done and the budget equation was balanced within 10%; analysis of the possible sources of error showed that this agreement was probably fortuitous. No attempt was made to construct a longer running model. Jassby and Goldman (1974a) have tackled the problem of balancing budget equations for phytoplankton biomass over seasonal time scales; they concluded that either present measurement techniques are inadequate or certain processes not usually taken into account in biomass models are, in fact, significant.

Although these extended models are of limited value for prediction of biomass at future times greater than a few days, they can be useful for showing the relative influence of the various physical and biological factors in controlling the distribution of biomass. To make this comparison the equation should be written in terms of dimensionless parameters (Platt and Denman, 1975a). For the sake of illustration consider the following differential equation for the biomass $B(X, Y, Z, T)$:

$$\frac{\partial B}{\partial T} + \mathbf{U} \cdot \nabla_H B + W \frac{\partial B}{\partial Z} = D_H \nabla_H{}^2 B + D_V \frac{\partial^2 B}{\partial Z^2} + P_m^B \cdot B \left(\frac{\alpha I_z}{P_m^B + \alpha I_z} \right)$$

$$- R_m \{ 1 - \exp \left[-\rho (B - B_0) \right] \} \tag{48}$$

Here $\mathbf{U}$ and W are the mean horizontal and vertical water velocities; D_H and D_V are the coefficients of turbulent diffusion in the horizontal and vertical; and the last term is a modified Ivlev expression (Parsons et al., 1967) describing the grazing by zooplankton in terms of three parameters: a rate constant ρ, a maximum ration R_m, and a threshold biomass B_0. We introduce the dimensionless variables $x, z, t, b, \mathbf{u}, w, i$ defined by

$$B = \phi b$$
$$X = \chi x$$
$$Z = \psi z$$
$$T = (P_m^B)^{-1} t$$
$$W = (v - \upsilon) w$$
$$\mathbf{U} = v \chi \psi^{-1} \mathbf{u}$$
$$I_z = P_m^B \alpha^{-1} i$$

where ϕ is a characteristic biomass; χ and ψ are horizontal and vertical length-scales; v is a characteristic vertical velocity and υ is a mean sinking rate for phytoplankton. The scaled form of the equation can now be written as (Platt and Denman, 1975a):

$$P_m^B \frac{\partial b}{\partial t} + \frac{v}{\psi}\left(\mathbf{u}\cdot\nabla_H b + w\frac{\partial b}{\partial z}\right) - \frac{\upsilon}{\psi}w\frac{\partial b}{\partial z} = \frac{D_H}{\chi^2}\nabla_H^2 b + \frac{D_V}{\psi^2}\frac{\partial^2 b}{\partial z^2}$$

$$+ \left[P_m^B\left(\frac{i}{1+i}\right) - \rho R_m\right]b + R_m \rho b_0 + R_m \sum_{n=2}^{\infty}(-1)^n\frac{\rho^n}{n!}\phi^{n-1}(b-b_0)^n \qquad (49)$$

where the exponential in the last term has been expanded in a Taylor series and the differential operators ∇_H and ∇_H^2 are now nondimensional.

The relative magnitudes of the coefficients in this general equation may now be calculated (Table I), using the literature data collected in Platt and Denman (1975a).

TABLE I

Relative Sizes of Coefficients in the General Equation
for Biomass[a]

Quantity	Range (sec^{-1})	Process
P_m^B	$(0.5\text{–}3.0)\times 10^{-5}$	Phytoplankton turnover
$\dfrac{v}{\psi}$	$10^{-9}\text{–}10^{-3}$	Upwelling
$\dfrac{\upsilon}{\psi}$	$10^{-8}\text{–}10^{-4}$	Sinking
$\dfrac{D_H}{\chi^2}$ [b]	$10^{-8}\text{–}10^{-3}$	Horizontal diffusion
$\dfrac{D_V}{\psi^2}$	$10^{-7}\text{–}4\times 10^{-3}$	Vertical diffusion
$P_m^B\,\dfrac{i}{1+i}$	$0\text{–}3.0\times 10^{-5}$	Growth at suboptimal light intensity (depth)
ρR_m	$10^{-12}\text{–}10^{-5}$	Grazing
$\dfrac{\phi_0}{\phi}$ [c]	$0\text{–}1$ (dimensionless)	Proximity to grazing threshold

[a] These coefficients each have the dimensions of a frequency, and they may be compared with the fundamental frequency which is the phytoplankton turnover rate. The reciprocals of the coefficients give the characteristic time scales of the processes corresponding to the terms which they multiply.
[b] The possible range of $D_H\chi^{-2}$ is limited since the diffusion coefficient is not independent of the length scale (e.g., see Bowden et al., 1974).
[c] Note that in the case $\phi_0\phi^{-1} \geq 1$, grazing is zero, by definition.

Rather than divide the equation through by P_m^B to make the coefficients dimensionless, we have chosen to retain the dimensions of frequency. The reciprocals of these coefficients then represent the characteristic time scales of the processes represented by the terms which they multiply. These time scales may then be compared with the fundamental time scale for the problem, the turnover time of the phytoplankton.

The tabulated ranges are rather wide, extending over several orders of magnitude for each of the time scales involved, reflecting the wide variety of situations studied in oceanography from estuarine to deep-ocean conditions. The table shows that, given the appropriate conditions it is possible for any one of the processes listed (phytoplankton turnover, upwelling, sinking, diffusion and grazing) to dominate the general equation describing $b(x, y, z, t)$. In any given situation (such as an upwelling system, an estuarine system, a continental shelf system, or a deep-ocean system) the parameters will be specified much more closely and a direct comparison will be possible between the various terms involved.

4. Nonlinear Effects in Phytoplankton Models

Much recent work in theoretical biology has emphasized the importance of non-linearities in complex living systems (Prigogine and Nicolis, 1971), some authors going so far as to propose a nonlinear statistical mechanical approach as a new paradigm of theoretical biology (Yates et al., 1972). A fundamental characteristic of nonlinear systems is their tendency toward periodic behavior, even for aperiodic boundary conditions; the solutions to the system differential equations are in general cyclic in the spatiotemporal domain. New structures may arise in the system as a direct consequence of the nonlinear processes acting on random fluctuations (Wiener, 1964; Nicolis and Auchmuty, 1974). These structures may be present when the system is far from thermodynamic equilibrium and can be maintained only at the cost of a steady energy supply; they are commonly called *dissipative structures*. Since the incidence of fluctuations increases with the number of degrees of freedom, this mechanism of structure generation should be particularly important in ecosystems; the ecological evidence has been reviewed recently by Platt and Denman (1975b).

In the particular case of phytoplankton ecology, we find that nonlinear effects are important, first in that they control the spatial heterogeneity of biomass; second, in that they provide a mechanism by which variance at a particular geometrical scale can be transferred to other scales; and third, in that they govern how fluctuations in the forcing variables (e.g., light intensity) are transmitted to the dependent variables (e.g., primary production).

In a differential equation, a nonlinear term is a term in which the dependent variable (either alone or in combination with the independent variables) or its derivatives occur raised to a power greater than one. Consider two dependent variables $B(x, t)$ and $H(x, t)$ which are functions of the time and of one space dimension. Simple non-linear differential equations describing their rate of change with time would be, for example,

$$\frac{\partial B}{\partial t} = aB + bB^2 + cBH$$

$$\frac{\partial H}{\partial t} = dH + eH^2 + fBH$$

$$(50)$$

where a, b, c, d, e, and f are arbitrary constants (if b and e are both equal zero, these equations are just the Lotka–Volterra equations). The terms aB and dH are linear because B and H occur in the first power only. On the other hand, the terms bB^2, eH^2, cBH, and fBH are nonlinear in that they are second-order terms. In general, the presence of nonlinear terms in a differential equation makes solution of the equation difficult if not impossible: the principle of superposition of solutions cannot be applied (in fact, this may be taken as a defining criterion for nonlinearity). Nonlinear equations have several interesting properties which are illustrated below in the context of phytoplankton ecology.

A. Wave Number Representation

The representation of system equations in wave number space was first applied to plankton ecology by Steele (1974b), whose treatment we follow partially in this section. In equation 50, suppose B represents biomass of phytoplankton and H represents herbivore biomass. The nonlinear terms B^2 and H^2 might represent density-dependent effects and the BH terms grazing interactions. If $(0, L)$ is the interval of x over which B and H can vary, we may express B and H as infinite Fourier series (including, for simplicity, only the cosine terms):

$$B = \sum_{n=0}^{\infty} B_n \cos nkx$$

$$H = \sum_{n=0}^{\infty} H_n \cos nkx \tag{51}$$

where $k = 2\pi/L$ and the time dependence is now contained in $B_n(t)$ and $H_n(t)$. Then the first member of equation 50 becomes

$$\frac{\partial}{\partial t}\left(\sum_{n=0}^{\infty} B_n \cos nkx\right) = a \sum_{n=0}^{\infty} B_n \cos nkx + b \sum_{p=0}^{\infty} \sum_{q=0}^{\infty} B_p B_q \cos pkx \cos qkx$$

$$+ c \sum_{r=0}^{\infty} \sum_{s=0}^{\infty} B_r H_s \cos rkx \cos skx \tag{52}$$

The last two terms may be rewritten to give

$$\frac{\partial}{\partial t}\left(\sum_{n=0}^{\infty} B_n \cos nkx\right) = a \sum_{n=0}^{\infty} B_n \cos nkx$$

$$+ \tfrac{1}{2}b \sum_{p=0}^{\infty} \sum_{q=0}^{\infty} B_p B_q \left[\cos(p+q)kn + \cos(p-q)kn\right]$$

$$+ \tfrac{1}{2}c \sum_{r=0}^{\infty} \sum_{s=0}^{\infty} B_r H_s [\cos(r+s)kx + \cos(r-s)kn] \tag{52a}$$

The cosine functions are independent of the time; they should cancel for each wave number n separately; this leads to an equation of the following form for each wave number:

$$\frac{dB_n}{dt} = aB_n + \tfrac{1}{2}b\left(\sum_{n=p+q} B_p B_q + \sum_{n=|p-q|} B_p B_q\right) + \tfrac{1}{2}c\left(\sum_{n=r+s} B_r H_s + \sum_{n=|r-s|} B_r H_s\right) \tag{53}$$

where the summations are carried over all wave number pairs whose sum or difference is equal to wave number n. Thus for a system described by the pair of equations 50, fluctuations at two wave numbers p and q, say, can contribute by interference to the fluctuations at the two other wave numbers $(p + q)$ and $|p - q|$. Conversely, the time variation of the nth wave number contains, in general, the effects of fluctuations at all other pairs of wave numbers whose sum or difference happens to equal n. Linear systems cannot transfer energy between wave numbers in this way.

In a nonlinear system in which one or more of the independent variables is strongly cyclic with a characteristic wave number (i.e., "periodic forcing"), the nonlinearities in the system may cause the periodicity to shift to a totally different wave number. It is this property, more than any other, which renders our intuition inadequate to predict even the qualitative behavior of nonlinear systems.

Furthermore, as emphasized by Steele (1974b) spatial fluctuations in the dependent variables can lead to a shift in the mean state of the system. Thus calculating the mean of B as the zeroth wave number from equation 53 with $n = 0$:

$$\frac{dB_0}{dt} = aB_0 + bB_0 B_0 + cB_0 H_0 + \frac{1}{2} \sum_{r=1}^{\infty} (bB_r H_r + cB_r H_r) \qquad (54)$$

The last term on the right represents the shift in the mean value B_0 caused by inhomogeneities in the spatial domain. A similar shift may be calculated for H_0. Thus the nonlinearities act on the spatial variability to give mean values which are different from those which the same system would have were it uniform in space. This phenomenon has important implications in the context of the "patchiness" of the plankton which is treated in Section 5.

B. Nonlinearities in the Photosynthesis–Light Equation

In this section we calculate the effect of fluctuations in the available light on the nonlinear photosynthesis–light equation. All the models we presented on photosynthetic response to light were nonlinear; here, for purposes of illustration, we choose equation 15, $P^B = \alpha I - \frac{1}{4}(\alpha I)^2/P_m^B$; $I \leqslant 2P_m^B/\alpha$. What happens if $I(t)$ is allowed to fluctuate with time? Let us consider a definite interval of time $(0, T)$. Apart from the annual cycle, the strongest cycles we might expect in I would have periods of 1 day and 3–5 days (characteristic period for the passage of weather systems in mid-latitudes). We might also expect smaller variations in the light incident on phytoplankton cells resulting from fluctuations about their mean depth excited by the characteristic periodic processes of the particular physical oceanographic environment (e.g., internal waves, surface waves, tides, and so on). Suppose then that the interval $(0, T)$ represents 1 month. Our treatment is oversimplified to the extent that we suppose the parameter values of the P^B versus I model stay constant during this period; in reality we know that the physiological parameters can change in response to environmental conditions, but we have only sketchy information on the response times.

As before we expand $I(t)$ and $P^B(t)$ in Fourier series:

$$I = \sum_{n=0}^{\infty} I_n \cos\left(\frac{2\pi nt}{T}\right); \qquad P^B = \sum_{n=0}^{\infty} P_n^B \cos\left(\frac{2\pi nt}{T}\right)$$

Following the method of the preceding section, we arrive at equations for the Fourier coefficients of the type

$$P_n^B = \alpha \left[I_n - \frac{\alpha}{8P_m^B} \left(\sum_{r+s=n} I_r I_s + \sum_{|p-q|=n} I_p I_q \right) \right] \qquad (55)$$

We can find the mean production for the month (T) by calculating the zeroth Fourier coefficient:

$$P_0^B = \alpha\left[I_0 - \frac{\alpha}{P_m^B}I_0^2 - \frac{\alpha}{P_m^B}\sum_{r=1}^{\infty}I_rI_r\right] \tag{56}$$

Hence for an *average* light intensity I_0 the mean production P_m^B is reduced (by an amount equal to the last term on the right) from the mean production which would be attained with a *constant* intensity I_0.

To estimate the magnitude of this effect, let us calculate the coefficient of variation of production, σ_{PB}/P^B, from equation 15. To second order $\sigma_{PB}^2 = \sigma_I^2(\partial P^B/\partial I)^2$; then

$$\frac{\sigma_P}{P^B} = \frac{\sigma_I}{I}\left(1 - \frac{\alpha}{2P_m^B}I\right)\left(1 - \frac{\alpha}{4P_m^B}I\right)^{-1} \tag{57}$$

The denominator may be expanded by the binomial theorem since $I \leqslant P_m^B/\alpha$. This gives:

$$\frac{\sigma_{PB}}{P^B} \simeq \frac{\sigma_I}{I}\left[1 - \frac{\alpha I}{4P_m^B} - \left(\frac{\alpha I}{4P_m^B}\right)^2 - \left(\frac{\alpha I}{4P_m^B}\right)^3 - \cdots - \left(\frac{\alpha I}{4P_m^B}\right)^n - \cdots\right] \tag{57a}$$

The coefficient of variation of P^B is thus smaller than that of I; the nonlinear response of photosynthesis to light therefore acts as a stabilizing or buffering mechanism which protects the primary production against fluctuations in available light. The corollary, however, is that the total production is less in fluctuating light than it would be for the same total quantity of light supplied at a constant rate.

That the environmental fluctuations (in this case, fluctuations in the available light) are damped out in the phytoplankton production is entirely a result of the nonlinear response of the primary production to incident light. If the response were linear, the coefficients of variation would be equal, that is, $\sigma_{PB}/P^B = \sigma_I/I$. However, the damping or stabilizing effect of the nonlinearity is quite fortuitous: if the P^B versus I curve in Fig. 1 had upward or positive curvature rather than downward or negative curvature, the effect of the nonlinearity would then be to magnify or enhance the environmental fluctuations.[1] Thus the coefficient of variation technique presented here can give a quantitative measure of the effectiveness of a particular nonlinearity in acting as a stabilizing or destablizing factor in a system subjected to environmental fluctuations.

5. Spatial Heterogeneity in Phytoplankton Biomass and its Turnover

In the sea, phytoplankton populations are highly structured in three spatial dimensions and there is an intimate relationship between this organization and the dynamic structure of the physical environment. Vertical structuring results from the existence in the sea of vertical density gradients and from the exponential decrease with depth of the available light. These factors impose special constraints on the dynamics of phytoplankton populations which are altogether different from those giving rise to structure in the horizontal. It is the inhomogeneity in the horizontal, the so-called patchiness (Fig. 5), that we wish to elaborate on in this section.

The principal aims of the study of spatial heterogeneity in the plankton are (1) to be able to characterize it, in particular to discover its characteristic scales under

---

[1] Vollenweider (1970) pointed out the stabilizing effect of the nonlinear dependency on light, but with respect to production integrated over depth; in Fig. 2 the curves that include photoinhibition are the ones with the greatest negative curvature.

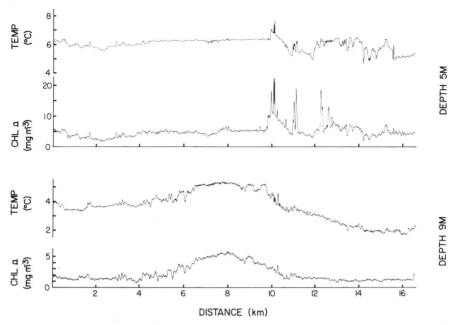

Fig. 5.   Series of chlorophyll α and temperature, obtained simultaneously at depths of 5 and 9 m along a 16.5-km transect in the lower St. Lawrence River estuary on June 23, 1973 (from Denman and Platt, 1975).

different conditions; (2) to account for its origin and persistence, especially to determine the strength of the coupling between the physical and biological components of the ecosystem; and (3) to assess its significance for the behavior and stability of the ecosystem. In the context of modeling the production of phytoplankton, such knowledge is important because it allows us to estimate the reliability of our sampling of the real world and therefore to estimate the probable magnitude of the observational error on the initial biomass condition used in the calculation of predictability time of extended biomass models (see next section); it helps in the selection of the fundamental space scales and in the parameterization of spatial fluctuations for the construction of grid models for biomass; also in grid models it indicates the effects on prediction at the largest scales through aliasing errors from fluctuations at scales smaller than the mesh size; it tells us about the magnitude and distribution of fluctuations, important for the calculation of the effects of thresholds and nonlinearities (preceding section) in the system equations (Platt and Denman, 1975b); it relates to the crucial question of stability in the plankton ecosystem (Steele, 1974b); and finally it is important for generalized trophodynamic models of the marine ecosystem because for many organisms (including commercial fish species) with a planktonic larval stage, it is thought that survival of the larvae is critically dependent on their finding a higher-than-average food concentration at some key phase of their life history (Conover, 1968; Jones, 1973).

A. Characteristic Scales

The investigation of spatial inhomogeneity in the phytoplankton is still in its infancy. From the work that has been done, however, we can begin to set some limitations on the scales of patchiness which are known to occur, but always with the caveat that it is impossible to dissociate completely the results obtained from the sampling methods used.

We discuss only those studies that treat fluctuations of the phytoplankton biomass as a whole, usually indexed by the concentration of chlorophyll *a* which can be measured *in situ* automatically; a few investigations (e.g. Richerson et al., in press) have looked at individual species separately, but because we are interested in predicting the production of the entire phytoplankton population, we mention them no further. Platt et al. (1970a) examined the variance of chlorophyll measurements made at 10 stations dispersed over areas of different sizes in an exposed bay. The variance between stations tended toward a constant value for sampling areas $\gtrsim 2.5$ km^2. We might suppose that this corresponds to a characteristic length scale lying between the square root of the minimum area and the square root of one tenth of the area, 0.5–1.6 km. In a study of spatial variations in phytoplankton turnover time in a semi-enclosed bay (Platt and Filion, 1973), the area occupied by six sampling stations which showed considerable variance 60% of the time and negligible variance 40% of the time corresponded to a length scale, calculated in the same way, of 0.8–1.9 km. Primitive attempts at spectral analysis of point-sampled data on chlorophyll concentration in the nearshore zone (Platt et al., 1970a) suggested a scale of ~ 1–2 km for the plankton patches. Variance spectra calculated for chlorophyll and temperature fluctuations in the surface layer of the Gulf of St. Lawrence (Denman and Platt, 1975; Platt and Denman, 1975b) indicate a characteristic scale for the phytoplankton $\lesssim 5$ km. Comparison of spatial spectra of chlorophyll and the spectra of velocity fluctuations (thought to be free from contamination by internal waves) in Lake Tahoe (Powell et al., 1975) suggests a scale ~ 100 m for the phytoplankton (Fig. 6).

In addition to estimates of characteristic scales from field measurements, one can also attempt theoretical calculation, but progress here has been minimal also. The earliest approach was that of Kierstead and Slobodkin (1953) who saw the critical length scale as that for which the opposing tendencies of turbulent diffusion and phytoplankton growth were just balanced. Allowing for the dependence of the diffusion coefficient on length scale (Okubo, 1971), it may be shown (Platt and Denman, 1975a) that this assumption is equivalent to the following equation for the characteristic phytoplankton scale ξ_c (cm):

$$\xi_c \approx (10^2 P_m^B)^{-1.15} \tag{58}$$

where P_m^B (normalized to carbon) is in sec^{-1}. This would give a characteristic length scale for phytoplankton distribution of between 50 and 100 m. Okubo (1974) calculates a range 10^2–10^3 m for ξ_c using a somewhat different method.

There are at least two reasons why such a simplistic treatment would tend to *underestimate* the length scale for phytoplankton. First, the boundary conditions on the original Kierstead and Slobodkin formulation precluded the possibility of growth outside a domain of size ξ_c. It would be more realistic to postulate growth outside the patch, but at a different rate from that inside. Second, the effect of grazing by herbivorous zooplankton has been neglected. This effect may be introduced as in equation 48, the extended biomass equation. The appropriate length scale ξ_c is then (Platt and Denman, 1975a).

$$\xi_c \approx (10^2 [P_m^B - R_m \rho])^{-1.15} \tag{59}$$

This expression is valid for situations where the phytoplankton abundance is well above the grazing threshold: it gives values of $\xi_c \sim 1$ km where phytoplankton growth and zooplankton grazing are almost equal. One difficulty that probably invalidates

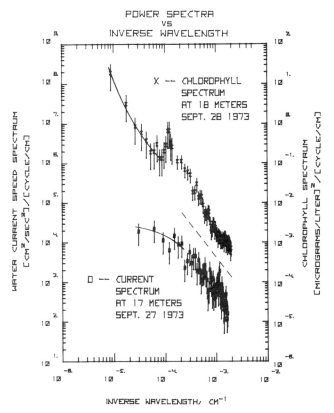

Fig. 6.   Variance spectra for chlorophyll and current speed from Lake Tahoe, California. Error bars are 80% confidence limits. The broken line has the slope $-\frac{5}{3}$ (from Powell et al., 1975).

the whole concept of the Kierstead and Slobodkin calculation of a critical scale is that it considers effects at one wavelength only. In the real world there is a continuous cascade of variance from larger to smaller scales, as can be seen from the variance spectra of Platt (1972), Fig. 7; Denman and Platt (1975a), Fig. 8; and Powell et al. (1975), Fig. 6; which tends to concentrate variance near the particular wavelength for which the Kierstead and Slobodkin method attempts to calculate the erosion of variance by turbulent diffusion. Because of the energy cascade, even "patches" at the largest scales are not stable or of a constant configuration, as the concept of a critical length scale would suggest.

The horizontal length scales relevant to this discussion are thus in the range 10^2–10^4 m. The corresponding range of time scales would be from 10^4 sec (minimum generation time of phytoplankton) to 10^6 sec (lifetime of the largest zooplankton patch ever studied systematically; Cushing, 1963). This range of length scales is probably the least understood in physical oceanography (Platt, 1974).

B. Significance of Spatial Heterogeneity

Interpretation of the variance spectra of chlorophyll concentration (Figs. 6–8) suggest that for length scales below about 100 m the distribution of phytoplankton is controlled primarily by turbulence; for scales greater than 100 m it is thought that

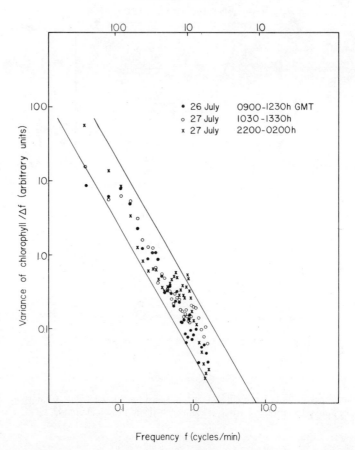

Fig. 7.   Variance spectrum of chlorophyll concentration from a fixed station at a depth of 8 m for a sampling interval of 0.33 min. The upper scale of wavelengths was derived by assuming that the turbulent field was advected past the station at a speed of 0.2 m/sec (from Platt, 1972).

spatial variability in such parameters as P_m^B and community structure begins to play a role in shaping the chlorophyll distribution. The requisite spatial inhomogeneities have already been observed for phytoplankton community structure in lakes (Richerson et al., 1970) and for phytoplankton turnover time in the sea (Platt and Filion, 1973). In the zooplankton, Frontier (1973) has found a similar dependence of variability in spatial scale which he attributes to the influence of turbulence. Parallel results have been found by Fasham et al. (1974).

Because the horizontal variability in phytoplankton biomass and turnover time is ephemeral and not fixed in space, its effects tend to be diminished when measurements at a fixed station are averaged over time. This is illustrated by Platt (1975), who analyzed the importance of spatial and temporal variability in the estimation of annual production by phytoplankton in the Bedford Basin.

The significance of spatial heterogeneity has only just begun to be explored in theoretical models. One of the earliest ecological examples was provided by Segel

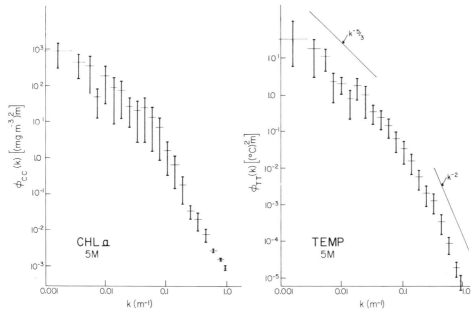

Fig. 8.   Variance spectra for the upper two series of Fig. 5. The lines with slopes $-\frac{5}{3}$ and -2 are not fitted but are included for reference (from Denman and Platt, 1975).

and Jackson (1972), who showed that in the presence of diffusion nonlinear inter-actions between populations in adjacent trophic levels could bring about the break-down of a steady state and the development of spatial inhomogeneities at a certain critical wavelength. Steele (1974b) studied similar equations for plankton and found equivalent results, showing also how the variance could be transferred between length scales (see preceding section). Okubo (1974) has calculated the effect of diffusion-induced instability on populations of interacting species.

Richerson et al. (1970), extending the work of Margalef (1967), developed the con-cept of the pelagic environment as a mosaic of diverse microhabitats in which the different species of the phytoplankton community have different growth character-istics; the structure is considered to be sufficiently stable in time to maintain the observed spatial variability in the phytoplankton but not stable to the point that the population of each microhabitat becomes monospecific. This concept of "contem-poraneous disequilibrium" was found to be a satisfying explanation for the results of Platt and Filion (1973) on spatial variations in P_m^B. Levin and Paine (1974), in an investigation of the importance of spatial heterogeneity in general ecological models, arrived at an essentially similar concept of the natural community "as a spatial and temporal mosaic of small-scale systems, recognizing that the individual component islands or patches cannot be considered closed. Rather they are part of an integrated patchwork, with individual patches constantly exchanging materials." Environmental perturbations disturb natural successional sequences within the patches and prevent them from ever reaching equilibrium; it is appropriate then to disregard the properties of the individual patch in favor of the macroscopic properties of their ensemble. Levin and Paine consider the spatial heterogeneity of such profound importance for com-munity dynamics that they postulate the patch as the fundamental unit of community structure. For the phytoplankton, the implications are still to be explored; but we do know that the fundamental time scale of environmental disturbance would be the typical period within which weather systems pass through an area, that is, ~ 12 hr associated with cumulus convection in the tropics, or 3–5 days associated with frontal disturbances in mid-latitude regions.

6. Fundamental Limitations on Predictability

A. Characteristics of the Marine Ecosystem

The thermodynamic characteristics of the marine ecosystem imply that there will be serious difficulties in the way of an adequate theoretical analysis: it is a nonlinear, nonconservative (dissipative), open thermodynamic system. Making predictions about particular aspects of the marine ecosystem (about the production of phytoplankton, say) involves many problems in common with those of making predictions about the weather: both the atmospheric circulation and the phytoplankton community are inherently nonlinear systems with, as we have seen above, the capacity to transfer energy or variance between the many temporal and spatial scales which characterize them; both systems have virtually infinite numbers of degrees of freedom and are intrinsically noisy, that is, they exhibit a high degree of variability which, as we have also shown above, can result in a shift of the mean or equilibrium state (if one exists), implying that the capability to predict variance is for practical purposes at least as important as the ability to predict means; and finally, neither system can be described in complete detail either observationally or through a closed set of state equations.

Many scientists have held (implicitly if not explicitly) the view that, at those scales for which quantum effects do not apply, nature will turn out in the long run to be essentially deterministic; the problems associated with environmental variability and incomplete system determination would eventually become resolved as our observational and analytic capabilities become more refined and our dissection of the system becomes ever finer in space and time. But we shall summarize a body of work in meteorology which, collectively, makes a strong case that, apart from logistic difficulties, there exist fundamental limitations on the degree of predictability that can be attained for a noisy, nonlinear, dissipative system.

B. Predictability in the Atmosphere

In the late 1960s, the Global Atmospheric Research Program was organized with the intention of collecting a complete set of data, worldwide in scope, for testing numerical forecasting models for large-scale weather features. So far, only two (very costly) pilot studies have been carried out: the Barbados Oceanographic and Meteorological Experiment (in 1969) and the GARP Atlantic Tropical Experiment (in 1974). It is already clear that the desired improvement in predictive capability (at time periods > 2 weeks for the largest-scale motions) may not be forthcoming, not for operational reasons but because of fundamental theoretical obstacles. Recent studies which bear on this problem for the atmosphere include Robinson (1967, 1971), Lorenz (1969), Leith (1971), and Leith and Kraichnan (1972). The principal tools employed are scale analysis, statistical theory, and numerical simulation; the papers most relevant to the problem of predictive modeling of primary production in the sea are those of Lorenz (1969) and Robinson (1971).

Lorenz represents the atmosphere by a set of simple deterministic equations which allow for motion on many spatial scales described by a characteristic wave number spectrum of kinetic energy. It is commonly believed in meteorology that the fundamental limitation on predictability is uncertainty about the initial state of the system because of observational errors and of the spatial coarseness of the observing grid. Lorenz finds, however, that two initial states differing by a small "observational error" can evolve, within a finite time interval, into two states which are as dissimilar as would be two states chosen randomly from all possible states of the system. The length of the time interval required for this to occur is one measure of predictability time; it cannot be increased significantly by reducing the size of the initial error. This limitation on predictability is attributed to the nonlinear transfer of error variance from the smallest scales via intermediate scales to the largest scales of motion. Suppose that at the smallest scales, an initial difference (uncertainty) between two states is doubled after an elapsed time of ~ 5 min; suppose also that at the largest scales uncertainties double after ~ 5 days. Reducing the initial error at the smallest scale by a factor of two would (roughly) double the time needed for it to reach the second smallest scale. Thus halving the initial uncertainty implies that ~ 10 min rather than 5 min are required for it to magnify to the next larger scale, but thereafter the uncertainty is propagated through the system to larger scales *exactly as before*, and the predictability time remains essentially unchanged at ~ 5 days.

The slope of the kinetic energy spectrum is indicative of the predictability of the system; if the spectrum falls off faster than k^{-3} the system is predictable in the sense that reducing the initial error by a given amount results in a linear increase in predictability time; if, as in most turbulent systems, the spectrum falls off more slowly than k^{-3}, the system may be indistinguishable from an indeterministic system in the sense that there will be an inherent (finite) range of predictability which cannot be increased significantly by reducing the uncertainty of the initial state to any value different from zero. In the atmosphere Lorenz estimates that this predictability time is less than 6 days for the largest scales of motion ($\sim 10^4$ km).

These conclusions would remain valid even if we had absolute certainty about the equations which govern the motion of the system; but as Lorenz emphasizes, this is far from being the case, and if we cannot find the equations of motion for the atmosphere, what hope do we have for prediction in the pelagic ecosystem of the sea?

Robinson (1971) reaches conclusions similar to those of Lorenz (1969) by a more direct approach. He uses the primitive equations of fluid motion to formulate simple estimates of error growth in an atmosphere with a specified spectrum of kinetic energy. One set of these estimates agrees remarkably well with those of Lorenz (Table II); that measure of predictability is the time taken for the average kinetic energy of a feature of the flow on a given scale of motion to be drained completely by dissipation. He concludes that there is a fundamental limit to the predictive power of a model of any dissipative system which is solved by forward integration of the equations of state and an initial value condition. For any given scale of interest, the dissipation of energy (variance) to smaller scales can be parameterized only approximately.

Another body of evidence, at first sight contradictory to the view that the upper limit to prediction of the largest atmospheric scales is less than 10 days, deals with the *persistence* of anomalies in the atmospheric motions for periods of weeks, months, years, and even centuries in the case of secular climatological cycles (Kutzbach and Bryson, 1974; Lamb, 1974). Atmospheric anomalies have been documented by Namias (1969, 1970a, b, 1973) which persist for periods of several months to several years, and which can be related to features in the oceanic environment. These features

TABLE II

Various Estimates of Atmospheric Predictability Time for Different Scales of Motion

Scale	Lorenz	Robinson-1 (Time of Removal by Dissipation of Average Kinetic Energy from a Wave Band)	Robinson-2 (Time for 50% dilution of an "Eddy" by Diffusion)
100 m	3 min		
1 km	13 min		
10 km	1 hr	1 hr	1.5 hr
100 km	4 hr	4 hr	6 hr
1000 km	1 day	1 day	1.5 day
10000 km	5.5 days	5.5 days	9 days

are believed to result from the nonlinear coupling of the atmospheric system with that of the ocean whose characteristic time scales are considerably larger (Manabe and Bryan, 1969). One might speculate that the phytoplankton system, whose characteristic time scales vary from ~ 1 day (typical doubling time) through 2–4 weeks (typical duration of bloom) to 6–12 months (typical successional period), will exhibit similar persistent anomalies as it responds to more slowly varying oceanic and atmospheric forcing, and to interaction with biological elements with longer characteristic times.

C. Predictability in the Ocean

The predictability estimates of Lorenz and of Robinson for the atmosphere cannot be carried over directly to oceanography because of differences in the shape of the kinetic energy spectrum and in the rate of dissipation of kinetic energy. Also, the Lorenz and Robinson calculations usually require an estimate of the spectrum of kinetic energy per unit wave number, but energy spectra in the ocean are usually measured in terms of frequency rather than wave number (from current meters at a fixed point) and the application of Taylor's frozen-field hypothesis to convert frequencies to wave numbers often cannot be justified.

Robinson (1971) did have, however, another measure of predictability derived from the time taken for 50% dilution of matter contained in a circle of radius $\frac{1}{2}l$ by diffusion through its circumference; thus

$$\Delta t = \frac{l^2 D_H^{-1}(2^{1/2} - 1)^2}{8} \tag{60}$$

where D_H is the horizontal eddy coefficient of turbulent diffusion. These estimates (Table II) are about 50% larger than the more rigorous measures discussed above. The reason is that they consider the effects of diffusion only at scales smaller than the scale of interest; the dissipative transfer of kinetic energy from large scales through the scale of interest to smaller scales is ignored.

For the upper layer of the ocean, we can exploit equation 60 to get similar estimates of predictability time by using an empirical formula relating the eddy coefficient D_H to linear scale l (Okubo, 1971):

$$D_H \simeq 0.0103 \, l^{1.15} \tag{61}$$

TABLE III

Estimates of Predictabil-
ity Times in the Ocean[a]

Scale	Predictability Time
10 m	10 min
100 m	1.5 hr
1 km	10 hr
10 km	3 days
100 km	3 weeks

[a] Using the Robinson-2 formula
with Okubo's (1971) values for
the horizontal diffusion coefficient.

Substitution of equation 61 into 60 leads to the estimates of predictability time for physical structures in the sea given in Table III.

To estimate the characteristic lifetimes of horizontal anomalies in phytoplankton abundance we must take into account additionally that the cells are reproducing and therefore tending to prolong the existence of the anomalies. Using Okubo's (1971) estimates of diffusion, Platt and Denman (1975a) calculate that structures in phytoplankton abundances larger than about 100 m are stable against diffusion; that is, their prediction time tends to infinity. If phytoplankton growth and zooplankton grazing are just balanced, the critical dimension is considerably larger. These calculations, however, are inconsistent with the real world, first in that deformation of the structures by horizontal current shear present in the larger scale eddies is ignored, and second in that the turnover time (P^B) of the cells is treated as constant, whereas experience shows that decline of phytoplankton bloom is associated with diminished physiological vigor.

The arguments presented in this section lead to the conclusion that attempts to predict biomass of phytoplankton at a particular location in the sea by continuous integration of a comprehensive production model over extended time periods should not be expected to yield reliable results. Instead, it might prove more rewarding to seek ways to characterize mathematically the demonstrable persistence of the pelagic ecosystem.

7. Perspectives for Modeling Phytoplankton Production

It is clear that almost nowhere in the modeling of the productivity of marine phytoplankton are we much beyond the semiempirical stage of development. The models presently available for P^B versus I curves describe the observations quite satisfactorily and it seems fruitless to search for ever-more-complex formulations of the rectangular hyperbola. What we need now are experimental and theoretical investigations on the adaptation times of the basic parameters P_m^B and α, and on the effects of environmental variables other than light on the magnitudes of these parameters. Since most of the temporal variance in primary production can be accounted for by variations in light, it seems best to base our predictive equations on the P^B versus I curve and introduce the second-order environmental variables through variation of the parameters. Pigment concentration and light comprise the minimum data set which can be used for the prediction of primary production and there may be considerable redundancy in trying to introduce all relevant environmental variables as multiplicative factors, since changes in these variables are probably accompanied by changes in the chlorophyll content per cell and by adaptations in the parameters of the P^B versus I curve.

The effect of second-order factors in simplistic models is manifested in the amount of residual variance unexplained by the model, and this error variance or noise often contains a large quantity of potential information. The information will be largely inaccessible, however, unless proper care is taken to fit models by statistically sound methods and to ensure that adequate statistical control is exercised over the measurements. The latter condition is particularly difficult to meet for natural populations, and it is correspondingly difficult to make objective judgments about the significance of a given environmental factor or to choose between alternative models. Economic priorities usually prohibit adequate replication of measurements at deep-sea stations; this is not necessarily the case for inshore stations, particularly those readily accessible from a field laboratory.

There is a need for more effort to be directed to the construction of explanatory models based on a better understanding of the enzyme chemistry of photosynthesis. There is a need also to pay attention to the quality of what is being produced as well as its quantity. It is known that the relative amount of protein synthesized depends on the wavelength of the available light (Kowallik, 1970) and that the underwater light field is often very sharply peaked in coastal waters. The qualitative aspect of primary production is particularly relevant to predictions for use in aquiculture. It should be pointed out too that the account we have given is totally nontaxonomic. We are simply not at the stage of being able to model the subtleties of interaction between particular species, although this may be feasible in special cases. It does seem that macroscopic generalizations about species composition and its rate of change are sometimes possible (Jassby and Goldman, 1974b).

In view of the rather limited success which we have had in predicting primary production using the methods available until now, it would be worth constructing purely stochastic models relating the system input (solar radiation) and the output of interest, say, the daily production for the water column $\int_t \int_z P$. We have only begun to understand the significance of spatial heterogeneity in the pelagic ecosystem. It seems to be an essential element of system stability and to provide the raw material for the emergence of new system structure through the action of nonlinearities in the state equations. At the very least it places a fundamental limitation on our ability to make useful predictions from extended biomass models.

List of Symbols

b	Dimensionless biomass (also an arbitrary constant in several equations)
b_0	Dimensionless threshold biomass for grazing
B	Biomass (mg Chl a/m^3; mg C/m^3)
B_0	Threshold biomass for grazing (mg Chl a/m^3; mg C/m^3)
D_H, D_V	Eddy coefficients of horizontal and vertical diffusion (m^2/hr)
E	Excretion (mg C/m^3/hr)
$I_z, I(z)$	Downwelling irradiance at depth z (W/m^2)
I_0	Daily mean downwelling irradiance at sea surface (W/m^2)

$I_0', \bar{I}_0'$	Downwelling irradiance just below the sea surface and its daily mean (W/m^2)
$I_0^{m'}$	Daily maximum downwelling irradiance just below sea surface (W/m^2)
k_b	Photosynthetic "efficiency" (m^{-1})
k	Wave number (m^{-1})
L, l	Wavelength, length scale (m)
P, P_g	Gross primary productivity $(\text{mg C/m}^3/\text{hr})$
P_m	Maximum gross primary productivity $(\text{mg C/m}^3/\text{hr})$
$P(z)$	Gross primary productivity at depth z (W/m^3)
P_n	Net primary productivity $(\text{mg C/m}^3/\text{hr})$
P_p	Net particulate primary productivity $(\text{mg C/m}^3/\text{hr})$
P^B	Gross primary productivity per unit biomass $(\text{mg C/mg Chl } a/\text{hr}; \text{hr}^{-1})$
P_m^B	Maximum gross primary productivity per unit biomass $(\text{mg C/mg Chl } a/\text{hr}; \text{hr}^{-1})$
$\int_z P$	Gross primary productivity of water column $(\text{mg C/m}^2/\text{hr})$
$\int_t \int_z P$	Daily gross primary production of water column (mg C/m^2)
R	Respiration $(\text{mg C/m}^3/\text{hr})$
R_L	Light respiration $(\text{mg C/m}^3/\text{hr})$
R^B	Respiration per unit biomass $(\text{mg C/mg Chl } a/\text{hr}; \text{hr}^{-1})$
R_m	Maximum loss due to grazing $(\text{mg Chl } a/\text{hr})$
S, K_s	Intensity or concentration of environmental factor in Michaelis–Menten expression. $S = K_s$ when $\mu = \mu_m/2$ (arbitrary units)
S_*	Normalized intensity or concentration $(= S/K_s)$
T, t, t_L	Time (hr)
z	Depth below water surface (m)
z_c	Average daily compensation depth (m)
z_{cr}	Critical depth (m)
z_{ca}	Cancellation depth (m)
z_t	Thermocline depth (m)
α	Initial slope of P^B versus I curves $[\text{mg C/mg Chl } a/\text{hr}/(\text{W/m}^2)]$
β	Ratio of C to Chl a in phytoplankton $(\text{mg C/mg Chl } a)$
γ	$P^B(I)/P_m^B$, when I is twice the intensity at which P_m^B occurs
θ	Temperature $(^\circ\text{K})$
κ	Extinction coefficient of water (m^{-1})
κ_s	Extinction coefficient per unit Chl a $(\text{m}^2/\text{mg Chl } a)$
κ_w	Extinction coefficient of water containing no chlorophyll a (m^{-1})
λ	Day length (hr)
μ, μ_m	Specific growth rate of phytoplankton, and its maximum (hr^{-1})
ρ	Grazing rate constant $[(\text{mg Chl } a/\text{m}^3)^{-1}]$

References

Baly, E. C. C., 1935. The kinetics of photosynthesis. *Proc. Roy. Soc., London*, **117B**, 218–239.

Bannister, T. T., 1974. Production equations in terms of chlorophyll concentration, quantum yield, and upper limit to production. *Limnol. Oceanogr.*, **19**, 1–12.

Blackman, F. F., 1905. Optima and limiting factors. *Ann. Bot. Lond.*, **19**, 281–295.

Bowden, K. R., D. P. Krauel, and R. E. Lewis, 1974. Some features of turbulent diffusion from a continuous source at sea. *Adv. Geophys.*, **18**, 315–329.

Brylinsky, M. and K. H. Mann, 1973. An analysis of factors governing productivity in lakes and reservoirs. *Limnol. Oceanogr.*, **18**, 1–14.

Caperon, J., 1968. Population growth response of *Isochrysis galbana* to nitrate variation at limiting concentrations. *Ecol.*, **49**, 866–872.

Caperon, J. and J. Meyer, 1972a. Nitrogen-limited growth of marine phytoplankton I. Changes in population characteristics with steady-state growth rate. *Deep-Sea Res.*, **19** (9), 601–618.

Caperon, J. and J. Meyer, 1972b. Nitrogen-limited growth of marine phytoplankton II. Uptake kinetics and their role in nutrient limited growth of phytoplankton. *Deep-Sea Res.*, **19** (9), 619–632.

Chartier, Ph., 1970. A model of CO_2 assimilation in the leaf. In *Prediction and Measurement of Photosynthetic Productivity*. I. Setlik, ed. Pudoc, Wageningen, pp. 307–315.

Chernavskaya, N. M. and D. S. Chernavskii, 1961. Periodic phenomena in photosynthesis. *Sov. Phys. Usp.*, **3**, 850–865.

Conover, R. J., 1968. Zooplankton—life in a nutritionally dilute environment. *Am. Zool.*, **8**, 107–118.

Conway, C. R., N. R. Glass, and J. C. Wilcox, 1970. Fitting nonlinear models to biological data by Marquardt's algorithm. *Ecol.*, **51**, 503–507.

Cushing, D. H., 1962. An alternative method of estimating the critical depth. *J. Cons. Perm. Int. Explor. Mer.*, **27**, 131–140.

Cushing, D. H., 1963. Studies on a *Calanus* patch. 2. The estimation of algal productive rates. *J. Mar. Biol. Assoc. U.K.*, **43** (2), 339–347.

Decker, J. P., 1955. A rapid post-illumination deceleration of respiration in green leaves. *Plant Physiol.*, **30**, 82–84.

Denman, K. L. and T. Platt, 1975. Coherences in the horizontal distributions of phytoplankton and temperature in the upper ocean. *Mém. Soc. r. Sci. Liège*, **7**, 19–30.

Dera, J. and H. R. Gordon, 1968. Light-field fluctuations in the photic zone. *Limnol. Oceanogr.*, **13**, 697–699.

Dowd, J. E. and D. S. Riggs, 1965. A comparison of estimates of Michaelis-Menten kinetic constants from various linear transformations. *J. Biol. Chem.*, **240**, 863–869.

Dugdale, R. C. and J. J. MacIsaac, 1971. A computation model for the uptake of nitrate in the Peru upwelling region. *Invest. Pesq.*, **35** (1), 299–308.

Dunstan, W. M., 1973. A comparison of the photosynthesis-light intensity relationship in phylogenetically different marine microalgae. *J. Exp. Mar. Biol. Ecol.*, **13**, 181–187.

Duysens, L. N. M., 1956. Energy transformations in photosynthesis. *Ann. Rev. Plant Physiol.*, **7**, 25–50.

Eppley, R. W., 1972. Temperature and phytoplankton growth in the sea. *Fish. Bull.*, **70**, 1063–1085.

Eppley, R. W. and P. R. Sloan, 1965. Carbon balance experiments with marine phytoplankton. *J. Fish. Res. Bd. Can.*, **22**, 1083–1097.

Fasham, M. J. R., M. V. Angel, and H. S. J. Roe, 1974. An investigation of the spatial pattern of zooplankton using the Longhurst–Hardy plankton recorder. *J. Exp. Mar. Biol. Ecol.*, **16**, 93–112.

Fee, E. J., 1969. A numerical model for the estimation of photosynthetic production, integrated over time and depth, in natural waters. *Limnol. Oceanogr.*, **14**, 906–911.

Fee, E. J., 1973a. A numerical model for determining integral primary production and its application to Lake Michigan. *J. Fish. Res. Bd. Can.*, **30**, 1447–1468.

Fee, E. J., 1973b. Modelling primary production in water bodies: a numerical approach that allows vertical inhomogeneities. *J. Fish. Res. Bd. Can.*, **30**, 1469–1473.

Frontier, S., 1973. Étude statistique de la dispersion du zooplankton. *J. Exp. Mar. Biol. Ecol.*, **12**, 229–262.

Glass, N. R., 1967. A technique for fitting nonlinear models to biological data. *Ecol.*, **48**, 1010–1013.

Goldman, C. R., M. Gerletti, P. Javornicky, U. Melchiorri-Santolini, and E. de Amezaga, 1968. Primary productivity, bacteria, phyto- and zooplankton in Lake Maggiore: correlations and relationships with ecological factors. *Mem. Ist. Ital. Idrobiol.*, **23**, 49–127.

Goldman, J. C. and E. J. Carpenter, 1974. A kinetic approach to the effect of temperature on algal growth. *Limnol. Oceanogr.*, **19** (5), 756–766.

Hellebust, J. A. and J. Terborgh, 1967. Effects of environmental conditions on the rates of photosynthesis and some photosynthetic enzymes in *Dunaliella tertiolecta* Butcher. *Limnol. Oceanogr.*, **12**, 559–567.

Hutchinson, G. E., 1973. Eutrophication. *Am. Sci.*, **61**, 269–279.

Ichimura, S., 1967. Environmental gradient and its relation to primary productivity in Tokyo Bay. *Rec. Oceanogr. Works Jap.*, **9**, 115–128.

Ikushima, I., 1967. Ecological studies on the productivity of aquatic plant communities. III. Effect of depth on daily photosynthesis in submerged macrophytes. *Bot. Mag., Tokyo*, **80**, 57–67.

Ivanoff, A., 1975. *Propriétés Physiques et Chimiques des Eaux de Mer*. Vol. 2. Librairíe Vuibert, Paris.

Jassby, A. D. and C. R. Goldman, 1974a. Loss rates from a lake phytoplankton community. *Limnol. Oceanogr.*, **19** (4), 618–627.

Jassby, A. D. and C. R. Goldman, 1974b. A quantitative measure of succession rate and its application to the phytoplankton of lakes. *Am. Nat.*, **108** (963), 688–693.

Jassby, A. D. and T. Powell, 1975. Vertical patterns of eddy diffusion during stratification in Castle Lake, California. *Limnol. Oceanogr.*, **20** (4), 530–543.

Jones, R., 1973. Density dependent regulation of the numbers of cod and haddock. *Rapp.-V. Reun. Cons. Perm. Int. Explor. Mer.*, **164**, 156–173.

Kierstead, H. and L. B. Slobodkin, 1953. The size of water masses containing plankton blooms. *J. Mar. Res.*, **12** (1), 141–147.

Kowallik, W., 1970. Light effects on carbohydrate and protein metabolism in algae. In *Photobiology of Micro-organisms*. P. Halldal, ed. Wiley-Interscience, New York, pp. 333–379.

Kutzbach, J. E. and R. A. Bryson, 1974. Variance spectrum of Holocene climatic fluctuations in the North Atlantic sector. *J. Atmos. Sci.*, **31**, 1958–1963.

Laisk, A., 1970. A model of leaf photosynthesis and photorespiration. In *Prediction and Measurement of Photosynthetic Productivity*. I. Setlik, ed. Pudoc, Wageningen, pp. 295–306.

Lamb, H. H., 1974. Fluctuations in climate. *Nature*, **251**, 568.

Leith, C. E., 1971. Atmospheric predictability and two-dimensional turbulence. *J. Atmos. Sci.*, **28** (2), 145–161.

Leith, C. E. and R. H. Kraichnan, 1972. Predictability of turbulent flows. *J. Atmos. Sci.*, **29**, 1041–1058.

Levin, S. A. and R. T. Paine, 1974. Disturbance, patch formation and community structure. *Proc. Nat. Acad. Sci. U.S.*, **71**, 2744–2747.

Lorenz, E. N., 1969. The predictability of a flow which possesses many scales of motion. *Tellus*, **21** (3), 289–307.

Lucas, H. L., 1964. Stochastic elements in biological models; their sources and significance. In *Stochastic Models in Medicine and Biology*, J. Gurland, ed. Publication No. 10 of the Mathematics Research Center, U.S. Army, The University of Wisconsin, Madison, pp. 355–383.

McCree, K. J., 1970. An equation for the rate of respiration of white clover plants grown under controlled conditions. In *Prediction and Measurement of Photosynthetic Productivity*. I. Setlik, ed. Pudoc, Wageningen, pp. 221–229.

MacIsaac, J. J. and R. C. Dugdale, 1972. Interactions of light and inorganic nitrogen in controlling nitrogen uptake in the sea. *Deep-Sea Res.*, **19** (3), 209–232.

Manabe, S. and K. Bryan, 1969. Climate and the ocean circulation. *Mon. Weather Rev.*, **97** (11), 739–827.

Margalef, R., 1967. Some concepts relative to the organization of plankton. *Oceanogr. Mar. Biol. Ann. Rev.*, **5**, 257–289.

Morel, A. and R. C. Smith, 1974. Relation between total quanta and total energy for aquatic photosynthesis. *Limnol. Oceanogr.*, **19**, 591–600.

Morris, I. and K. Farrell, 1971. Photosynthetic rates, gross patterns of carbon dioxide assimilation and activities of ribulose diphosphate carboxylase in marine algae grown at different temperatures. *Physiol. Plant.*, **25**, 372–377.

Myers, J., 1953. Growth characteristics of algae in relation to the problem of mass culture. In *Algal Culture from Laboratory to Pilot Plant*. J. S. Burlew, ed. Carnegie Institution of Washington Publication 600, Washington, D.C., pp. 37–54.

Namias, J., 1969. Autumnal variations in the North Pacific and North Atlantic anticyclones as manifestations of air-sea interactions. *Deep-Sea Res.*, **16**, Suppl., 153–164.

Namias, J., 1970a. Climatic anomaly over the United States during the 1960's. *Science*, **170**, 741–743.

Namias, J., 1970b. Macroscale variations in sea-surface temperatures in the North Pacific. *J. Geophys. Res.*, **75** (3), 565–582.

Namias, J., 1973. Thermal communication between the sea surface and the lower troposphere. *J. Phys. Oceanogr.*, **3** (4), 373–378.

Nicolis, G. and J. F. G. Auchmuty, 1974. Dissipative structures, castastrophes and pattern formation: a bifurcation analysis. *Proc. Nat. Acad. Sci. U.S.*, **71**, 2748–2751.

Okubo, A., 1971. Oceanic diffusion diagrams. *Deep-Sea Res.*, **18** (8), 789–802.

Okubo, A., 1974. *Diffusion-induced instability in model ecosystems: another possible explanation of patchiness.* Chesapeake Bay Institute, Tech. Rep. 67, 17 pp.

Parker, R. A., 1973. Some problems associated with computer simulation of an ecological system. In *The Mathematical Theory of the Dynamics of Biological Populations*. M. S. Bartlett and R. W. Hiorns, eds. Academic Press, London, pp. 269–288.

Parker, R. A., 1974. Empirical functions relating metabolic processes in aquatic systems to environmental variables. *J. Fish. Res. Bd. Can.*, **31**, 1550–1552.

Parsons, T. R. and G. C. Anderson, 1970. Large scale studies of primary production in the North Pacific Ocean. *Deep-Sea Res.*, **17**, 756–776.

Parsons, T. R., R. J. LeBrasseur, and J. D. Fulton, 1967. Some observations on the dependence of zooplankton grazing on the cell size and concentration of phytoplankton blooms. *J. Oceanogr. Soc. Jap.*, **23**, 10–17.

Parsons, T. R. and M. Takahashi, 1973a. *Biological Oceanographic Processes.* Pergamon Press, Oxford, 186 pp.

Parsons, T. R. and M. Takahashi, 1973b. Environmental control of phytoplankton cell size. *Limnol. Oceanogr.*, **18** (4), 511–515.

Patten, B. C., 1965. Community organization and energy relationships in plankton. *Oak Ridge Natl. Lab. Rep.*, **3634**, 1–60.

Patten, B. C., 1968. Mathematical models of plankton production. *Int. Rev. Ges. Hydrobiol.*, **53**, 357–408.

Peisker, M., 1974. A model describing the influence of oxygen on photosynthetic carboxylation. *Photosynthetica*, **8**, 47–50.

Pielou, E. C., 1969. *An Introduction to Mathematical Ecology.* Wiley, New York, 286 pp.

Platt, T., 1969. The concept of energy efficiency in primary production. *Limnol. Oceanogr.*, **14**, 653–659.

Platt, T., 1972. Local phytoplankton abundance and turbulence. *Deep-Sea Res.*, **19** (3), 183–188.

Platt, T., 1974. Spatial inhomogeneity in the oceans. In *Modelling of Marine Systems.* J. C. J. Nihoul, ed. Elsevier Oceanography Series No. 10, Amsterdam.

Platt, T., 1975. Analysis of the importance of spatial and temporal heterogeneity in the estimation of annual production by phytoplankton in a small, enriched, marine basin. *J. Exp. Mar. Biol. Ecol.*, **18**, 99–109.

Platt, T. and R. J. Conover, 1971. Variability and its effect on the 24 h chlorophyll budget of a small marine basin. *Mar. Biol.*, **10**, 52–65.

Platt, T. and K. L. Denman, 1975a. A general equation for the mesoscale distribution of phytoplankton in the sea. *Mém. Soc. r. Sci. Liège*, **7**, 31–42.

Platt, T. and K. L. Denman, 1975b. Spectral analysis in ecology. *Ann. Rev. Ecol. Syst.*, **6**, 189–210.

Platt, T., L. M. Dickie, and R. W. Trites, 1970a. Spatial heterogeneity of phytoplankton in a near-shore environment. *J. Fish. Res. Bd. Can.*, **27**, 1453–1473.

Platt, T. and C. Filion, 1973. Spatial variability of the productivity: biomass ratio for phytoplankton in a small marine basin. *Limnol. Oceanogr.*, **18** (5), 743–749.

Platt, T., B. Irwin, and D. V. Subba Rao, 1973. Primary productivity and nutrient measurements on the spring phytoplankton bloom in Bedford Basin 1971. *Fish. Res. Bd. Tech. Rep.*, **423**, 46 pp.

Platt, T., E. Larsen, and R. Vine, 1970b. Integrating radiometer: A self-contained device for measurement of submarine light energy in absolute units. *J. Fish. Res. Bd. Can.*, **27**, 181–191.

Platt, T. and D. V. Subba Rao, 1970a. Energy flow and species diversity in a marine phytoplankton bloom. *Nature*, **227**, 1059–1060.

Platt, T. and D. V. Subba Rao, 1970b. Primary production measurements on a natural plankton bloom. *J. Fish. Res. Bd. Can.*, **27**, 887–899.

Platt, T. and D. V. Subba Rao, 1975. Primary production of marine microphytes, In *Photosynthesis and Productivity in Different Environments*, International Biological Program, Vol. 3, Cambridge University Press. pp. 249–280.

Powell, T. M., P. J. Richerson, T. M. Dillon, B. A. Agee, B. J. Dozier, D. A. Godden, and L. O. Myrup, 1975. Spatial scales of current speed and phytoplankton biomass fluctuations in Lake Tahoe. *Science*. **189**, 1088–1090.

Prigogine, I. and G. Nicolis, 1971. Biological order, structure and instabilities. *Quart. Rev. Biophy.*, **4**, 107–148.

Richerson, P., R. Armstrong, and C. R. Goldman, 1970. Contemporaneous disequilibrium, a new hypothesis to explain the "paradox of the plankton." *Proc. Natl. Acad. Sci., U.S.*, **67** (4), 1710–1714.

Richerson, P. J., B. J. Dozier, and B. R. Maeda, in press. The structure of phytoplankton associations in Lake Tahoe (California–Nevada). *Verh. Internat. Verein, Limnol.*, **19** (to be published).

Ried, A., 1970. Energetic aspects of the interaction between photosynthesis and respiration. In *Prediction and Measurement of Photosynthetic Productivity.* I. Setlik, ed. Pudoc, Wageningen, pp. 231–246.

Riley, G. A., 1946. Factors controlling phytoplankton populations on Georges Bank. *J. Mar. Res.*, **6**, 54–73.

Robinson, G. D., 1967. Some current projects for global meteorological observation and experiment. *Quart. J. Roy. Meteorol. Soc.*, **97** (398), 409–418.

Robinson, G. D., 1971. The predictability of a dissipative flow. *Quart. J. Roy. Meteorol. Soc.*, **97**, 300–312.

Rodhe, W., 1965. Standard correlations between pelagic photosynthesis and light. In *Primary Productivity in Aquatic Environments*. C. R. Goldman, ed. University of California Press, Berkeley, pp. 365–381.

Ross, G. J. S., 1970. The efficient use of function minimization in nonlinear maximum likelihood estimation. *Appl. Statist.*, **19**, 205–221.

Ryther, J. H., 1956. Photosynthesis in the ocean as a function of light intensity. *Limnol. Oceanogr.*, **1**, 61–70.

Ryther, J. H. and D. W. Menzel, 1965. Comparison of the ^{14}C-technique with direct measurement of photosynthetic carbon fixation. *Limnol. Oceanogr.*, **10**, 490–492.

Ryther, J. H. and C. S. Yentsch, 1957. The estimation of phytoplankton production in the ocean from chlorophyll and light data. *Limnol. Oceanogr.*, **2**, 281–286.

Segel, L. A. and J. L. Jackson, 1972. Dissipative structure: An explanation and an ecological example. *J. Theor. Biol.*, **37**, 545–559.

Smith, E. L., 1936. Photosynthesis in relation to light and carbon dioxide. *Proc. Nat. Acad. Sci., Washington*, **22**, 504.

Sournia, A., 1974. Circadian periodicities in natural populations of marine phytoplankton: A review. In *Advances in Marine Biology*. Vol. 12. F. S. Russell and M. Yonge, eds. Academic Press, New York, 325–389.

Stadelma, P., J. E. Moore, and E. Pickett, 1974. Primary production in relation to temperature structure, biomass concentration, and light conditions at an inshore and offshore station in Lake Ontario. *J. Fish. Res. Bd. Can.*, **31**, 1215–1232.

Steele, J. H., 1962. Environmental control of photosynthesis in the sea. *Limnol. Oceanogr.*, **7**, 137–150.

Steele, J. H., 1974a. *The Structure of Marine Ecosystems*. Harvard University Press, Cambridge, Mass., 128 pp.

Steele, J. H., 1974b. Spatial heterogeneity and population stability. *Nature*, **248**, 83.

Steele, J. H. and I. E. Baird, 1962. Carbon-chlorophyll relations in cultures. *Limnol. Oceanogr.*, **7**, 101–102.

Steele, J. H. and D. W. Menzel, 1962. Condition for maximum primary production in the mixed layer. *Deep-Sea Res.*, **9**, 39–49.

Steemann Nielsen, E. and T. S. Park, 1964. On the time course in adapting to low light intensities in marine phytoplankton. *J. Cons. Int. L'Expl. Mer*, **29**, 19–24.

Strickland, J. D. H., 1958. Solar radiation penetrating the ocean. A review of requirements, data, and methods of measurement, with particular reference to photosynthetic productivity. *J. Fish. Res. Bd. Can.*, **15**, 453–493.

Sverdrup, H. U., 1953. On conditions for the vernal blooming of phytoplankton. *J. Cons. Perm. Int. Explor. Mer*, **18**, 287–295.

Sverdrup, H. U., M. W. Johnson, and R. H. Fleming, 1942. *The Oceans*. Prentice-Hall, Englewood Cliffs, N.J., 1087 pp.

Taguchi, S., 1972. Mathematical analysis of primary production in the Bering Sea in summer. In *Biological Oceanography of the Northern North Pacific Ocean*. A. Y. Takenouti, ed. Idemitsu Shoten, Tokyo, pp. 253–262.

Takahashi, M., K. Fujii, and T. R. Parsons, 1973. Simulation study of phytoplankton photosynthesis and growth in the Fraser River Estuary. *Mar. Biol.*, **19**, 102–116.

Talling, J. F., 1957a. Photosynthetic characteristics of some fresh-water plankton diatoms in relation to underwater radiation. *New Phytol.*, **56**, 29–50.

Talling, J. F., 1957b. The phytoplankton population as a compound photosynthetic system. *New Phytol.*, **56**, 133–149.

Talling, J. F., 1961. Photosynthesis under natural conditions. *Ann. Rev. Plant Physiol.*, **12**, 133–154.

Talling, J. F., 1970. Generalized and specialized features of photoplankton as a form of photosynthetic cover. In *Prediction and Measurement of Photosynthetic Productivity*. I. Setlick, ed. Pudoc, Wageningen, pp. 431–445.

Tamiya, H., 1951. Some theoretical notes on the kinetics of algal growth. *Bot. Mag., Tokyo*, **64**, 167–173.

Tamiya, H., E. Hase, K. Shibata, A. Mituya, T. Iwamura, T. Nihei, and T. Sasa, 1953. Kinetics of growth of *Chlorella*, with special reference to its dependence on quantity of available light and temperature. In *Algal Culture from Laboratory to Pilot Plant*. J. S. Burlew, ed. Carnegie Institution of Washington Publication 600, Washington, D.C., pp. 204–232.

Tolbert, N. E., 1974. Photorespiration. In *Algal Physiology and Biochemistry*. W. D. P. Stewart, ed. University of California Press, Berkeley, pp. 474–504.

Tooming, H., 1967. Mathematical model of plant photosynthesis considering adaptation. *Photosynthetica*, **1**, 233–240.

Tooming, H., 1970. Mathematical description of net photosynthesis and adaptation processes in the photosynthetic apparatus of plant communities. In *Prediction and Measurement of Photosynthetic Productivity*. I. Setlik, ed. Pudoc, Wageningen, pp. 103–113.

Vollenweider, R. A., 1965. Calculation models of photosynthesis-depth curves and some implications regarding day rate estimates in primary production measurement. In *Primary Productivity in Aquatic Environments*. C. R. Goldman, ed. University of California Press, Berkeley, pp. 425–457.

Vollenweider, R. A., 1970. Models for calculating integral photosynthesis and some implications regarding structural properties of the community metabolism of aquatic systems. In *Prediction and Measurement of Photosynthetic Productivity*. I. Setlik, ed. Pudoc, Wageningen, pp. 455–472.

Wiener, N., 1964. On the oscillations of nonlinear systems. In *Stochastic Models in Medicine and Biology*. J. Gurland, ed. Mathematics Research Center Publications No. 10, University of Wisconsin Press, Madison, pp. 167–174.

Wilkinson, G. N., 1961. Statistical estimations in enzyme kinetics. *Biochem. J.*, **80**, 324–332.

Williams, P. J. LeB., 1973. The validity of the application of simple kinetic analysis to heterogeneous microbial populations. *Limnol. Oceanogr.*, **18** (1), 159–165.

Williams, W. T., 1971. Principles of clustering. *Ann. Rev. Ecol. Syst.*, **2** (4028), 303–326.

Williams, W. T., J. M. Lambert, and G. N. Lance, 1966. Multivariate methods in plant ecology: V. Similarity analyses and information-analysis. *J. Ecol.*, **54**, 427–445.

Winter, D. F., K. Banse, and G. C. Anderson, 1975. The dynamics of phytoplankton blooms in Puget Sound, a fjord in the northwestern United States. *Mar. Biol.*, **29**, 139–176.

Winokur, M., 1948. Photosynthesis relationship of *Chlorella* species. *Am. J. Bot.*, **35**, 207–214.

Yates, F. E., D. J. Marsh, and A. S. Iberall, 1972. Integration of the whole organism—A foundation for a theoretical biology. In *Challenging Biological Problems: Directions towards their solutions*. J. A. Behnke, ed. 25th Anniversary Volume, The American Institute of Biological Sciences, Oxford University Press, New York, pp. 110–132.

Zlobin, V. S., 1973. Osnovi prognozirovaniya pervichnoi produktivnosti foticheskovo sloya okeana. Murmansk, 515 pp. (in Russian).

22. ZOOPLANKTON DYNAMICS

JOHN H. STEELE AND MICHAEL M. MULLIN

1. Introduction

Modeling of marine ecosystems has tended to concentrate on two trophic levels, the phytoplankton and the fish. In the former case this has been possible because the plant material could be considered conceptually and measured analytically as a single unit. In the latter case the major commercial fish species were treated one by one, mainly in relation to the effects of fishing effort on stock abundance, since the fishing industry supplied the necessary data. Thus the intermediate steps in the food web have been the poor relations, lacking a simplifying assumption; a simple, analytical technique for distinguishing trophic levels; and an economic pressure. Usually the herbivores or benthic detritus feeders entered a simulation model only as a means of supplying or disposing of biological matter required by those parts of the ecosystem treated as the central components. For benthos there is, so far, no evidence of a change in this attitude, but the zooplankters are now being considered as animals rather than as sources or sinks. This recognition requires that they be treated not as "biomass" but as organisms having definite patterns of growth, reproduction, and mortality. Thus the parameters of population dynamics—fecundity, age structure, age-specific birth and death rates—are more important in determining the behavior of an ecosystem than the simpler concepts of flow of organic matter.

In the past, where zooplankton has been introduced into a model, factors such as filtering, respiration, and excretion rates have often been taken as fixed proportions of the hypothetical "biomass" rather than being related to more detailed information on behavior and metabolism. In the literature there are now considerable experimental data on these aspects for several species of zooplankton. This information can be used to provide some idea of the functional relations which could be used in a simulation of a zooplankter's response to variations in its environment. The development of such theoretical descriptions is critical to the inclusion of these animals, as animals, in more general simulations of ecosystems. Thus although some examples are given of zooplankton as "biomass" and mention is made of some preliminary work with zooplankton as "animals," the emphasis of this chapter is the functional relations used to describe the various "building blocks." It is these relations that are more likely to be generally applicable than any particular model utilizing them.

There are still, however, certain limitations which must be imposed here. Zooplankters are a very heterogeneous group, defined more by how they are caught than by their position in the food web. Any net haul, and particularly a series of hauls with different sizes of mesh, is likely to contain animals that feed on bacteria, or are carnivores, as well as those feeding on the phytoplankton. Yet nearly all models incorporating zooplankton consider the entire catch as the herbivores feeding in the upper layers of the sea. There are good reasons for this; herbivorous copepods are the largest single group in the zooplankton; effectively all the primary production must literally pass through them. In turn they, or their feces and excreta, are the predominant source of food for the rest of the system. There are also less satisfactory reasons—lack of information on the quantities of other groups within the plankton such as the microzooplankton, the carnivores, and the deep plankton communities of the open ocean. Especially, we have few experimental data on the feeding behavior or metabolism of these populations. Because of these factors, the emphasis here is on the pelagic herbivores as part of a food chain from nutrients and phytoplankton to some, usually unspecified, predator. The inherent simplification in this approach must be borne in mind where it is not referred to explicitly.

In one sense, the attempt to deal with herbivores as animals is merely transferring the problem a further step up the trophic ladder, since the same problems exist for the carnivores. To treat predation as a general mortality effect on the herbivores is to ignore the same problems that arise with the herbivores themselves. It is possible that a large, rare predator may exert an effect on the species composition of a community, and therefore on the flow of organic matter, which is disproportionate to the fraction of organic matter actually passing to its population. Such a species has been termed a "keystone predator" (Paine, 1969) and acts by feeding preferentially on a species of prey that would dominate the community by competition if not held in check by predation. Such concepts indicate the open-ended nature of any ecosystem and emphasize the necessary artificialities in a simulation which must close off these upper ends in some simplistic manner.

Lastly, the effects of the physical environment, particularly light and water movements, usually enter models through their direct effect on nutrient distributions and photosynthesis, yet these also affect the animal populations. The distributions of animals in three dimensions depend on their vertical migrations and their horizontal transport by water movements, which can vary with depth. This component of any ecosystem model is also considered here, particularly in relation to the problem of the response of the system to perturbations. Such perturbations may be damped out by the functional responses of the plankton to such factors as food concentration, or the damping may depend on the effects of processes in the physical environment. The relative importance of these two mechanisms is a central problem in plankton dynamics and examples of the use of modeling will relate to this problem.

2. Zooplankton as Biomass

It is simplest to begin with the general interactions between nutrients (N), phytoplankton (P), and herbivorous zooplankton (H). A schematic representation gives

$$\frac{dN}{dt} = -\text{phytop. uptake } (N, P) + \text{zoop. excretion } (H) \tag{1}$$

$$\frac{dP}{dt} = \text{phytop. growth } (N, P) - \text{zoop. grazing } (P, H) \tag{2}$$

$$\frac{dH}{dt} = \text{zoop. growth } (P, H) - \text{predation } (H) \tag{3}$$

It can be seen that the zooplankton has an effect on all three trophic levels. The terms containing H are not simple functions, just as the other terms are not, as is known from previous work. All require knowledge of the kinetics of metabolism at each trophic level. Yet often this is not available, particularly for the zooplankton and particularly when only data concerning biomass (i.e., wet weights, dry weights, or organic matter in net hauls) are available.

One approach, not strictly modeling, is to assume that

$$\frac{dN}{dt} = \frac{dP}{dt} = \frac{dH}{dt} = 0$$

and use field data for regression analyses which may provide some indication of the structure of equations 1–3 at quasi-steady state. As an example, Blackburn (1973) used data on chlorophyll, zooplankton, and micronekton from the eastern tropical Pacific as indices of phytoplankton, herbivores, and carnivores. There were significant regressions between pairs of parameters. In particular,

$$\text{zoop.} \propto (\text{Chl})^{0.7} \tag{4}$$

For the micronekton the regressions

$$\text{micronekton} \propto (\text{zoop.})^k \tag{5}$$

had values of k between 0.2 and 1.0, depending on the group of micronekton. Also, the most significant correlations occurred with micronekton data lagging 4 months behind the zooplankton. These results are from an area where seasonal cycles are expected to be small compared with those in higher latitudes. The conclusion that as primary production increases there are corresponding increases in other parts of the food chain is not unexpected. Blackburn suggests that relations 4 and 5 indicate that efficiency of transfer of material up the trophic pyramid is probably greater in oligotrophic than eutrophic situations.

This analysis is given here as an indication of the uses and the limitations of a purely statistical approach to trophic interactions. It poses the challenge of whether simulation modeling can do any better.

As a first step, consider the implication of these relations in terms of equations 2 and 3. Take as a particular form

$$\frac{dP}{dt} = aP - bP^{\alpha}H \tag{6}$$

$$\frac{dH}{dt} = cP^{\alpha}H - dH^{\beta} \tag{7}$$

If equation 4 is accepted for the steady state then from equation 6, $\alpha = 0.3$ and from equation 7, $\beta = 1.4$. There are certain immediate conclusions. Firstly, the simplest prey–predator relations (Lotka–Volterra), for which $\alpha = \beta = 1$, would not be a satisfactory formulation. In particular, it would not be true that predation rate on the herbivores can be defined by a single constant d times the herbivorous biomass. The biomass of predators, and so the predation rate, will in turn depend on the herbivorous biomass, as indicated by equation 5. Thus any rounding off of the trophic equations by a constant predation term is likely to be inadequate. The second conclusion, that the grazing rate of H on P is of the form P^{α}, $\alpha < 1$, is an acceptable *rough* approximation to the experimental data which is considered later. Thirdly, the efficiency of H feeding on P defined as growth/grazing $= c/b$ is independent of the values given to α and β, and this implies that no conclusions about efficiency can be drawn from these regressions.

One method of taking account of variable predation is to use actual data on predators. For example, Riley (1947) used an equation of the form

$$\frac{dH}{dt} = H(cP - R_H - C_H) \tag{8}$$

where R_H is the respiratory rate of the herbivores determined from laboratory experiments. The value of c was determined from a phytoplankton equation but had an added arbitrary upper limit set as $cP \leqslant 0.08/\text{day}$. Finally, the mortality,

$C_H = K_1 \times$ (number of *Sagitta*) $+ K_2$, the constants K_1 and K_2 being determined statistically to give the best fit of the model to data from Georges Bank. Obviously, the relation of theory to observation is not a test of a prediction from a model but rather a test of the extent to which certain relatively simple assumptions about the nature of interactions between different trophic levels can explain the variations in a particular set of data. Riley's success in doing this has encouraged considerable use of models basically the same as his, for a variety of situations (see Di Toro et al., 1975; Riley, 1963). However, the need for an arbitrary upper limit to assimilation is an indication of the problems inherent in the simplification of zooplankton to one parameter H, and a linear grazing relation with P.

One improvement over the use of a single value for H has been to divide the zooplankton into major groups. Thus Parker (1973), dealing with Kootenay Lake, considered copepods and cladocerans separately, and Vinberg and Anisimov (1969) used large and small herbivores with different feeding habits. Also, nonlinear grazing relations were used. Parker took grazing on P as

$$\alpha P e^{-\beta P}$$

where α and β are constants. These and other coefficients were fitted by "subjective evaluation of simulation patterns" and reasonable agreement with data was found over 3 yr. However, a dominant feature of this lake system is the very marked seasonal variation in advection of water through the lake; this may determine the main patterns in the biological variables.

Vinberg and Anisimov considered a self-contained system and used grazing relations of the form

$$\alpha\{1 - \exp[-\beta(P - P_0)]\}$$

with grazing zero for $P \leqslant P_0$. The authors state that it was necessary to introduce P_0 to avoid complete extermination *in the model* of a trophic level. This theoretical use of thresholds to avoid extermination in such simple self-contained systems is known (Holling, 1965; Steele, 1974b) but the empirical conclusion from Parker's work is that seasonal cycles in physical factors can override the difficulties arising from purely biological interactions. The relevance of these questions to marine plankton communities is discussed more fully at a later stage.

These examples also show the development of models based on simple food *webs* rather than food *chains*. Any detailed study of feeding habits (e.g., Hardy, 1924) usually shows many pathways. The problem is how to handle very large numbers of links between organisms. Following Isaacs (1973) we take, as an extreme example, an "unstructured food web" where everything eats everything else, including itself. Assume that all the animals have the same rates of growth (k_1), metabolism (k_2), and defecation of particulate food material (k_3) defined as fractions of food intake, so that

$$k_1 + k_2 + k_3 = 1$$

Unlike Isaac's web, plants are not included since it is not easy to imagine them as having the same k_r as animals. They can be considered as the source of energy. If M is the total biomass of the animal ecosystem then the energy flow would be as follows:

where the input balances the total metabolic losses. The fractions $k_1 M$ and $k_3 M$ represent particulate matter that is recycled within the system. The loss of biomass by metabolism, $k_2 M$, must be replaced to maintain a steady state, and therefore the input also has this value. Since all the animals have the same k_r, their growth rates are proportional to their biomasses. In terms of functions, if not of actual animals, M can be divided into herbivores, carnivores, and detritivores. Thus the biomass would have the following proportions:

$$\frac{\text{Herbivores}}{k_2} = \frac{\text{Carnivores}}{k_1} = \frac{\text{Detritivores}}{k_3}$$

If, for a very simple example, $k_1 = k_2 = k_2 = \frac{1}{3}$, then the biomasses in these groups would be equal. This is very different from a straight food chain without recycling where on the same metabolic basis:

$$\frac{\text{Herbivores}}{3} = \frac{\text{Carnivores}}{1} = \frac{\text{Detritivores}}{1}$$

This example emphasizes the consequences of ignoring structure within the food web. These arise from taking the contents of benthic grab collections or large mesh nekton net hauls as "detritivores" or "carnivores" and forgetting the recycling of material which goes on within these communities.

These considerations of herbivores as biomass show that useful deductions can be made. Especially for studies of populations of phytoplankton it may be adequate to use a single parameter for grazing, and the general concepts from this point of view have been reviewed before (Riley, 1963). Such studies of phytoplankton usually stress the effects of physical variables in changing the phytoplankton populations. These factors are certainly important, but Cushing (1959) and others have pointed out that this may be overstressed by the excessive simplicity of the portrayal of the herbivores. Thus it is necessary to look at the probable intricacies that can arise from a fuller consideration of grazing, growth, metabolism, reproduction, and mortality of copepods.

3. Zooplankton as Animals

Equation 3 can be expanded to

$$\frac{dH}{dt} = \text{ingestion} - \text{defecation} - \text{metabolism} - \text{predation} \qquad (9)$$

Assume H is composed of i cohorts of copepods with weights W_i and numbers Z_i; then

$$H = \sum W_i Z_i$$

$$\frac{dH}{dt} = \sum \left(W_i \frac{dZ_i}{dt} + Z_i \frac{dW_i}{dt} \right)$$

by comparison with equation 9,

$$\frac{dW_i}{dt} = \frac{1}{Z_i} (\text{ingestion} - \text{defecation} - \text{metabolism}) \qquad (10)$$

$$- \frac{dZ_i}{dt} = \frac{1}{W_i} (\text{predation}) \qquad (11)$$

Equation 10 determines the change in weight of an individual copepod as the sum of its individual gains and losses of energy; equation 11 represents the effects of predation on a particular cohort as a function of numbers in that cohort, assuming that all death is due to predation.

If WO is the weight of the naupliar stage at which feeding starts and WN is the weight of the adult, then for each cohort relations of the form

$$ZO = F\left(P, ZN, \frac{WN}{WO}\right) \qquad (12)$$

indicate the requirements for some function defining recruitment (ZO) in terms of food available, adult numbers (ZN), and the ratio of adult to naupliar weight. This function includes not merely reproductive capacity but also any mortality before the feeding naupliar stage is reached.

Equations 10–12 form the basis for a portrayal of zooplankton as animals and each of the five components—ingestion, defecation, metabolism, reproduction, and predation—is considered in turn.

A. Ingestion

To begin, one must stress again that zooplankton includes a variety of feeding types from predators, such as chaetognaths which pursue mobile prey, to species such as salps which unselectively filter small particles on a mucous net. Here we are concerned with ingestion by those species—mainly crustaceans—that remove relatively small, immobile particles by capture on a meshwork of coarse setae. Mathematical formulation for these species may not be appropriate for other types of feeding mechanism.

The terms "filtering" or "grazing" have been applied to the rate at which a zooplankter collects suspended particles, assuming that zooplankton feeds by relatively constant, mechanical filtration. What is usually computed is the volume of water containing the collected particles; the actual volume passing through the meshwork of setae exceeds this computed volume if the filtration process is not perfectly efficient. This simplification, possibly usable in some simple biomass models, is not supported by detailed experimental work. Thus we use "ingestion rate," I, defined as the rate of intake of food per unit time per animal. This is a function of

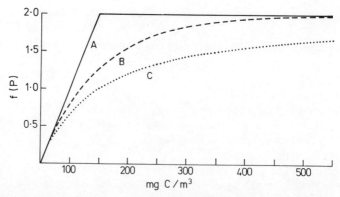

Fig. 1. Three functional relations for ingestion, I, as a function of food concentration, P. A, B, and C correspond to equations 13, 14, and 15 with $I_m = 2$, $\delta = 100$, $P_0 = 50$.

both the food concentration and the weight of the animal. A first step is to assume
that these variables can be separated in the form

$$I = f(P) \cdot W^{\alpha}$$

The value of α has not been exactly determined but Gauld (1951) suggested $\alpha = \frac{2}{3}$
and Paffenhöfer's (1971) data are in general agreement with this. In these, and nearly
all other experiments, food was supplied as unialgal cultures so that as they grew the
copepods fed on the same size of particle throughout development. Natural phytoplankton can provide a wide selection of possible sizes of food items but, as in the
experiments, a first approximation is to assume that all sizes of copepods are feeding
on one total population, P, defined in biomass units such as mg C/m^3. There are
three functions used to define $f(P)$ (Fig. 1):

$$f(P) = I_m \frac{P - P_0}{\delta}, \qquad P_0 \leqslant P < P_0 + \delta$$

$$= I_m, \qquad\qquad P \geqslant P_0 + \delta \tag{13}$$

$$f(P) = I_m \left\{ 1 - \exp\left[\frac{-(P - P_0)}{\delta} \right] \right\} \tag{14}$$

$$f(P) = I_m \frac{P - P_0}{\delta + P - P_0} \tag{15}$$

In all three, $f(P) = 0$ when $P \leqslant P_0$. Each relation depends on three constants and
satisfies the same three conditions:

1. $f(P) = 0$ when $P = P_0$
2. $f(P) = I_m \dfrac{P - P_0}{\delta}$ for P slightly greater than P_0
3. $f(P) \to I_m$ as $P \to \infty$

These correspond to three facets of experimental studies of ingestion:

1. There may or may not be a threshold (P_0) below which the animals do not feed.
2. When feeding starts the ingestion rate increases in proportion to the increase in
food concentration.
3. As food concentration goes to high values the ingestion rate tends to become
constant.

The best choice between these relations is in doubt (Frost, 1974; Mullin et al.,
1975). One reason for this is that there are still many simplifications implicit in these
formulas. For example, the maximal ingestion rate, I_m, is not independent of the
feeding of the animal prior to the measurement (Frost, 1972; McAllister, 1970).
More controversial is the existence of $P_0 > 0$. Feeding experiments with unialgal
cultures in the laboratory usually give zero (e.g., Frost, 1972) or very low values
(e.g., Frost, 1975). On the other hand, studies in which a natural assemblage of
particulate matter is the source of food (e.g., Adams and Steele, 1966; Parsons et al.,
1969) indicate that feeding ceases at a threshold concentration, P_0, significantly
different from zero. This latter finding is teleologically attractive because it provides
the phytoplankton with a refuge in low density so that they cannot be grazed to

extinction. It is a technical necessity in many models of phytoplankton–herbivore interactions (e.g., Vinberg and Anisimov, 1969; Steele, 1974a; Walsh, 1975; and the next section).

An apparently large P_0 in experimental determinations of the relations 13–15 may result from the simplifications in assuming that all particle sizes in a natural assemblage are equally available to all sizes of zooplankton. Further, as has been pointed out before (Steele, 1974a), the use of a "mathematical" threshold may arise from a range of biological activities other than simple food intake; alternative mechanisms are discussed in relation to vertical migration.

B. Selection of Food

All sizes of particles are not equally accessible to a zooplankter as food. Thus δ in equations 13–15 is probably related directly to the relative sizes of the zooplankter and the particles (Frost, 1974; Ambler and Frost, 1974). The morphology of the food-collecting organs sets upper and lower limits to the size of particles which can be captured and ingested. For zooplankters entrapping particles in a mucous net, these morphological constraints seem to be of primary importance in determining which particles will be ingested and which will not (e.g., Gilmer, 1974, concerning pteropods; Madin, 1974, concerning salps). Morphological constraints are also important for setous feeders such as copepods; it is often the case that an increase in bodily size, with corresponding increase in feeding appendages, is correlated with an increase in the maximum size of particle which can be eaten, but does not necessarily reduce the ability to feed on small particles (Wilson, 1973).

Within these morphological constraints, there is increasing evidence of selectivity [ingestion of larger particles in a proportion greater than their relative concentration in the water (Harvey, 1937; Mullin, 1963)] and of switching [changes in selectivity in response to changes in relative concentrations (Richman and Rogers, 1969)]. The quantitative description of selectivity and switching has been discussed extensively in the general ecological literature (Ivlev, 1955; Jacobs, 1974) but often these indices have little predictive value since they are not based on a description of a mechanism by which selectivity occurs. Another approach (Holling, 1965; Ware, 1973) is to use information on the behavior and sensory physiology of the predator and prey to predict patterns of selective feeding.

If we accept that copepods generally feed preferentially on larger particles (Poulet, 1973, 1974), there are also indications (Wilson, 1973; Berman and Richman, 1974) that, within this, they "select" the largest *abundant* particle and feed unselectively on the smaller ones. The question is whether this is true "selection" of individual cells, or whether it is the result of changes in the filtering appendages. This provides a point of departure, temporarily, from the real world of data to illustrate the modeling process by considering ways in which presence of large cells might affect ingestion of small cells. Greatly simplified, the filter can be considered a fan consisting of a fixed number of setae radiating from a central point. The spacing between the setae can be increased or decreased (discussed by Wilson, 1973). If we assume a constant rate of flow of water through the setal fan and, for simplicity, $I \ll I_m$, $P_0 = 0$, then it is obvious that, for a given cell diameter, d, the maximum retention by the filter will occur when the angle between setae is such that the spacing between setae at the tip of the fan just equals this diameter. At larger angles, cell retention should increase in proportion to cell diameter (see Frost, 1972). Assume that for large diameter, d_L, and small diameter, d_s, cells, the angles of the fan for maximum intake are θ_L and θ_s, respectively. If there is a combined food supply with biomasses per unit volume of B_L and B_s, and if the filtering rate per unit area of the fan is f, then

$$I = (B_L R^2 + B_s r^2)\theta \cdot \frac{f}{2} \qquad \theta_s < \theta < \theta_L$$

where $R =$ radius of fan, $r = R \cdot \theta_s/\theta$; thus

$$I = \left(B_L \theta + \frac{B_s \theta_s^2}{\theta} \right) R^2 \frac{f}{2}$$

This has a minimum at $\theta_{min} = \sqrt{B_s/B_L}\,\theta_s$, if $\theta_s < \theta_{min} < \theta_L$. Thus there are two possible maxima (Fig. 2) at

$$I_L = \left(B_L \theta_L + \frac{B_s \theta_s}{\theta_L} \right) R^2 \frac{f}{2}$$

$$I_s = (B_L \theta_s + B_s \theta_s) R^2 \frac{f}{2}$$

and

$$I_L = I_s \qquad \text{for} \quad \frac{B_L}{B_s} = \frac{\theta_s}{\theta_L} = \frac{d_s}{d_L}$$

This suggests that for two cell sizes presented simultaneously there is a maximal food intake found by feeding selectively on the larger or smaller cells. The switchover point would occur when the ratio of biomasses is near the ratio of the inverse of the diameters. Further, in a switch in preference from small to large cells, the ingestion rate on the former would decrease. This has been observed in certain experiments (Petipa, 1959; Anraku and Omori, 1963; Mullin, 1966). However, in an experiment with varying proportions of large (paired) and smaller (single) cells of *Ditylum brightwellii*, Richman and Rogers (1969) found that increasing the proportion of the larger cells

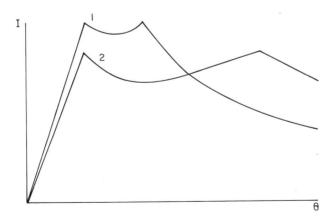

Fig. 2.   Two examples of total ingestion as a function of spread, θ, of the setous fan in the presence of mixed populations where (1) the diameter, d_L, of the large cells is twice the diameter, d_s, of the small cells and (2) the diameter of the large cells is four times that of small cells.

caused an increase in the filtering rate for them without any decrease in the rate for the smaller cells. The difference in volume between cells was relatively small (1.5:1) yet the increase in filtering was by a factor of four. These cells were relatively large and the authors suggest that the copepods were actively hunting the bigger cells.

Figure 2 shows two examples of mixed populations of two kinds of cells where $I_L = I_s$. For case 2, in which $d_L = 4 d_s$, the presence of a small fraction of the biomass (20%) as large cells (and even smaller numerically, 0.4%) provides a relatively uniform ingestion rate for a wide range of θ. Thus although switching is theoretically an optimal procedure, it is likely that selection to ensure capture of the scarcer, larger cells would be a normal strategy.

Although some methods of selection may be relatively mechanical and could be introduced into models, there are others which are too elusive at present. Yet these must be assessed before any full description of the feeding process is made. Furthermore, until this is done it is not possible to simulate the details of competitive interactions between different species of zooplankton (Steele, 1974a). This is one reason why only the simpler food chain with one type of copepod is amenable to numerical analysis at present.

C. Assimilation

The actual gain to a feeding zooplankter is the organic matter that is assimilated from the gut, rather than that ingested. The measurement of efficiency of assimilation (assimilation/ingestion) is surprisingly difficult (Edmondson and Winberg, 1971) and the results quite variable (see Conover, 1964, for review). The apparently simple, direct measurement of assimilation using radioisotopic tracers can be complex to interpret (e.g., Conover and Francis, 1973), and the quantitative recovery of feces is difficult. Even if discrete fecal pellets are formed, there may be other losses of ingested but unassimilated material (Johannes, 1964), and some of the food captured and killed may be lost prior to actual ingestion (Dagg, 1974).

Some investigators have held that zooplankters engage in "superfluous feeding" in dense concentrations of phytoplankton, resulting in extensive production of rapidly sinking fecal material from which rather little organic matter has been assimilated, but experiments by Conover (1966) and Corner, Head, and Kilvington (1972) have failed to confirm this. A variation of the "superfluous feeding" hypothesis is the suggestion of "spoilation," based on observations of feeding copepods, that at high concentrations of phytoplankton, cells (especially those of large size) may be broken open but only partly ingested (in the extreme case, ingesting only organelles such as the chloroplasts). This suggestion has merit for carnivores (Dagg, 1974), but has not been supported by Corner, Head, and Kilvington (1972) for herbivorous copepods.

Lacking better information, it would seem that the rate of assimilation, A, could be computed either as a constant fraction of the rate of ingestion, I (e.g., Steele, 1974a, who used $A = 0.7I$), or as a fraction of I which decreases as I increases. For example (Isaacs, personal communication):

$$A = 0.3I\left(3.0 - \frac{I}{I_m}\right)$$

This equation establishes an efficiency which approaches 90% for low values of I and decreases to 60% near the maximal ingestion, I_m.

D. *Respiration*

The major metabolic loss of organic matter from a population is undoubtedly through respiration, and for purposes of modeling the zooplankton, respiration and excretion can probably be considered to be the same process. The total rate of metabolic loss, M, can be split into three components with different relations (Fig. 3) to rate of food intake I. There is assumed to be a resting or standard metabolism, M_1, independent of food supply. The respiratory costs of foraging for and capturing food, M_2, should decrease as food concentration, and so $f(P)$, increases. Lastly, there is a cost of assimilating and biochemically transforming the food (specific dynamic action, M_3), proportional to A. Where separation of these components has been attempted, M_1 is usually derived indirectly by extrapolating regressions of M on independently measured activity back to zero activity. An estimate of M_3 can be obtained by comparing the respiratory rate of an animal without food ($M_1 + M_2$) with the actual amount of assimilated energy necessary to hold the animal's weight constant ($M_1 + M_2 + M_3$). For example, Conover and Lalli (1974) compared M as determined by respirometry, by weight loss during starvation, and by the difference between assimilation and growth of a planktonic carnivore.

In most direct measurements, animals are restrained in a fairly small container. Techniques for estimating metabolic losses in nature, such as measuring the rate of loss of a radionuclide after recapturing an individual which had been labeled and then released for a known period, or measuring a rate related to respiration, such as heartbeat, by telemetry of an unconfined animal (see Petrusewicz and Macfadyen, 1970, for review) seem inapplicable to zooplankton. Studies of energy flow in natural fish populations indicate that total respiration can be reasonably approximated by twice the standard respiratory rate, that is, $M = 2M_1$ (see Mann, 1969, for review), but the applicability of this generalization to zooplankton, or indeed to any unstudied population, is unknown.

In nature, many adult zooplankters undergo diurnal vertical migrations of tens or even hundreds of meters. By comparing the sizes of the oil storage organs in stage V copepodite *Calanus helgolandicus* captured near the surface in early morning, presumably after a night of feeding, with those of animals taken in late morning in deep water, presumably after a downward migration, Petipa (1966) calculated that the respiratory cost of migrating might be as much as 35 times the standard rate, M_1. The assumption that all the animals caught at depth migrated down from the surface that morning is open to question, of course. In contrast, a hydrodynamical analysis of the power expenditure of swimming by another copepod led Vlyman (1970) to conclude that, at known speeds of vertical migration, this activity results in an additional respiratory burden which is trivial. Another hydrodynamical analysis by Klyashtorin

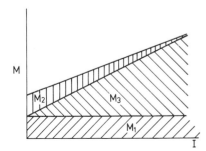

Fig. 3.   Metabolic losses in relation to ingestion (see text).

and Yarzhombek (1973) indicates that the cost associated with vertical migration is 20–40 % of M_1.

It is known that respiratory rate varies with availability of food (reviewed by Marshall, 1973; Steele, 1974a). In experiments with a wide range of concentration of food supply Mullin and Brooks (1970b) found that the ratio growth/food intake was relatively constant. These results suggest that M_3 is dominant. In his model of a simple phytoplankton–zooplankton ecosystem, Steele (1974a) found it important for the stability of the system that respiration by the zooplankton be proportional to ingestion, in effect making M_2 and M_3 the major components of M.

There remains the relation of metabolism to weight of the animal. If we take this as W^β, then if $\alpha < \beta$ (where α is the analogous effect of weight on ingestion), gross efficiency of growth would decrease with increasing weight, and vice versa. Any difference between α and β may not be very marked; Mullin and Brooks (1970a) found no significant difference in gross growth efficiency during all stages of growth of two species of copepods, except in the adult when a difference is to be expected.

In summary, metabolism can probably be expressed as

$$M = (M_1 + M_3 \cdot I)W^\beta \tag{16}$$

with $\beta = \alpha$ and $M_2 = 0$ as a first approximation.

E. Growth

In the formalism used here, the rate of increase in weight, dW/dt, is derived from other processes. There are several functional relations describing weight increase in terms of weight only, such as the von Bertelanffy curve derived from

$$\frac{dW}{dt} = gW^{2/3} - 3rW$$

which has an asymmetric sigmoid shape. The original derivation was based on the argument that assimilation was proportional to $W^{2/3}$ whereas loss by respiration was proportional to W. This appears inadequate in relation to the preceding section, but, more important, by ignoring the effects of concentration of food on g and r it eliminates the interactions with lower trophic levels. If, however, the herbivores were being used as the base of a food chain then a relation of this kind might be used as an approximation to the whole life span (as in fish dynamics; Beverton and Holt, 1957).

F. Reproduction

We must define the number of juveniles that recruit to the feeding population as a result of both reproduction and mortality during the egg and earlier naupliar stages before feeding starts.

Depending on the biology of the species under consideration (and perhaps the purposes of a particular model), eggs are released by the adult female as a single brood, a series of clutches, or continuously throughout some time span J. For example, for *Calanus finmarchicus* in the Clyde (Marshall and Orr, 1955), a few days after fertilization, eggs are laid for a period of several weeks. For several species of copepods, laboratory experiments has shown that the presence of abundant food acts as a stimulus to egg laying (e.g., Marshall and Orr, 1955; Corkett and McLaren, 1969; Nassogne, 1970; Comita and Comita, 1966). The rate of production of eggs has a similar, quasi-hyperbolic relation to that expected for growth. Although these may not be the same, the simplest assumption is that the female, instead of converting food assimilated to growth, transfers it to egg production.

The first problem is the role of the males in the population. Again the simplest assumption is that they feed at the same rate as females with no egg production. Then if WN is the adult weight for the species and ZN the number at time t, egg production is

$$S = X \int_J ZN \left(\frac{dW}{dt}\right)_{W=WN} dt \tag{17}$$

The efficiency term, X, is the conversion of increase in biomass by the adult population into eggs, including the "wasted" growth in the males.

Thereafter there are losses of energy as the eggs develop and, especially, mortality of eggs and nauplii. These losses, again on the simplest assumption, can be included in the constant, X. The importance of this factor can be illustrated by considering the potential production. Given a high concentration of food, a female *Calanus* can produce about 400 eggs; when these recruit the nauplius weighs about 0.2 μg C compared with an adult weight of 100 μg C. The rates of ingestion of nauplii and adults are in the ratio $(WO/WN)^{0.7} = 0.013$. Thus the potential food requirements of the offspring are about five times those of the female parent. If the plankton consisted of a single cohort of mainly adult females which produced the maximum possible recruits, these could put an excessive demand on the available food. To avoid these Malthusian consequences, yet retain adequate recruitment at low population densities, it may be necessary to have some density dependent factor in the recruitment relation, for example,

$$\text{recruitment} = \frac{S}{1 + X'S} \tag{18}$$

The difficulty is that, if control is in terms of the *total* number of recruits then, biologically, this function cannot be performed by the individual adult females. The most likely predators are other large copepods since the egg diameters 100–150 μ are within the range of particle sizes which they can take.

G. Predation

Mortality of the herbivorous copepods can be attributed to a great variety of predators, vertebrate and invertebrate. There is a very much smaller amount of experimental work on this aspect than on the growth of copepods. It has been shown that there is size selection by invertebrate predators (Reeve and Walter, 1972). Thus predation is unlikely to be on the biomass of herbivores as a single unit but may depend on the biomass in particular size categories or cohorts. Since the younger stages, although numerous, may have a small total biomass, predation as a function of numbers only could also be too great a simplification. There is a need to explore the kinds of population distributions observed in the sea; the following analysis indicates the data requirements and the nature of the results that could be obtained (see Mullin and Brooks, 1970c, for an example).

Copepods increase in size by shedding an exoskeleton and forming a new and larger one. The developmental stages are identifiable, and a plankton sample can provide data on numbers and average weights in each stage. Take N_j to be the number in stage j with average weight S_j. Take the average time in the stage as T_j and predation rate as p_j. Assume there is a stable population structure—a rather dangerous assumption but often the only one possible. Then if n_j is the rate at which copepods leave the $(j-1)$th stage and enter the jth stage,

$$N_{j-1} = n_j \frac{\exp(p_{j-1}T_{j-1}) - 1}{p_{j-1}}$$

$$N_j = n_j \frac{1 - \exp(-p_j T_j)}{p_j}$$

Taking $p_j = p_{j-1}$ gives

$$\frac{N_{j-1}}{N_j} = \frac{\exp(p_j T_{j-1}) - 1}{1 - \exp(-p_j T_j)}$$

This shows that it is essential to have not only numbers in each stage but also growth rates expressed as time spent in each stage.

[It must be stressed that this is a different representation of the population from that used in equations 10 and 11 where a particular cohort is being followed as its individuals grow in weight. Here the animals are considered as passing through distinct developmental stages.]

It is known that duration of each stage, T_j, increases as the animal grows, but detailed values for T_j are usually not available. A comparison of bodily and metabolic processes may help to provide relative values so that only the total life span is needed. The increase in weight from one stage to the next is, very approximately, by a factor of two (Paffenhöfer, 1971; Mullin and Brooks, 1970a) so that $S_{j+1} = 2S_j$.

If as discussed earlier, ingestion and metabolism are proportional to (weight)$^{2/3}$, then in this context rate of growth is proportional to $S_j^{2/3}$. Thus, as a possible approximation,

$$\frac{dS_j}{dt} = \frac{(\text{weight at end of interval}) - (\text{weight at start of interval})}{T_j}$$

$$= \frac{S_{j+1} - S_{j-1}}{2T_j} = \frac{3S_j}{4T_j}$$

If this is proportional to $S_j^{2/3}$, then $S_j^{1/3} = \alpha T_j$, where α is some constant. That is, T_j will increase in rough proportion to the length of the copepod. Also $T_j/T_{j-1} = 2^{1/3}$. Using these relations

$$\alpha = \frac{S_k^{1/3} \sum_{j=0}^{k-1} 2^{-j/3}}{T^*}$$

where S_k is the weight of the adult stage, k is the number of stages, and T^* is the time from first feeding juvenile to start of the adult stage.

These relations for p_j and α would allow the rate of mortality to be calculated for each stage from a knowledge of the abundance and average bodily weight of that stage, plus an estimate of the total duration of feeding juvenile stages. As an illustration, assume $p_j = p$ for all j. Table I gives an example for, again, a hypothetical copepod, assuming each adult produces 200 eggs at the end of its adult stage and then dies. A fixed mortality ($p = .15$ in this case) keeps numbers in the earlier, naupliar stages roughly constant and numbers per stage only decrease later, giving a fallacious impression of greater predation on the later stages. As a result, biomass is concentrated in the older animals. If the intensity of predation depends on biomass in individual

Table I

Distribution of a Copepod Population by Stages under Constant Predation

Stage (j)	NIII	NIV	NV	NVI	CI	CII	CIII	CIV	CV	CVI
Weight (S_j)	0.15	0.30	0.60	1.20	2.4	4.8	9.6	19	38	76
Days (T_j)	1.0	1.3	1.6	2.0	2.5	3.2	4.0	5.0	6.3	8.0
Nos. (N_j)	100	111	111	107	95	80	58	37	20	9
Biomass	15	33	67	128	228	384	557	703	760	684

stages, then it should be relatively high in the older animals. A detailed study of a *Calanus* patch (Cushing and Tungate, 1963) suggested (Steele, 1974a) that mortality within the copepodite stages was mainly at the fifth stage. This could also be due, behaviorally, to selective predation by fish on the larger copepodites. In contrast to the constant mortality used to calculate the distributions in Table I, nauplii in nature probably have much higher rates of mortality than do copepodites (Parsons et al., 1969; Mullin and Brooks, 1970c).

H. Temperature Effects

Feeding, respiration, and other processes are dependent upon temperature. If, as has been done here, we assume a general relation to weight of any metabolic rate, R,

$$R = aW^b \qquad 0.67 \leqslant b \leqslant 1.0$$

Then it is usually assumed that temperature affects the value of a but not b, although some studies (e.g. Ikeda, 1970; Champalbert and Gaudy, 1972) have indicated that b may also be a function of temperature. Further, a can sometimes be expressed as

$$a = cQ_{10}^{(0.1t-1)}$$

where t = temperature, c = metabolic rate at 10°C for an animal of unit weight, and $Q_{10} = R_{t^*+10}/R_{t^*}$ is the ratio of metabolic rates for a difference of 10°C from some temperature t^*.

Normally animals do not respond so simply, and it is necessary to determine the relation experimentally over the required range of temperatures. McLaren (1963) reviewed several formulations for the effect of temperature on metabolic rates and strongly urged the use of Bělehrádek's equation, of which the classical "normal curve" of Krogh is a special case, in preference to the Q_{10}. This equation has the form $R = g(t + \alpha)^h$, where g and h are constants governing mean slope and degree of curvature of the relationship, and α is a "biological zero" which in effect shifts the temperature scale. The parameters of this equation would still have to be determined for several situations, since α reflects state of acclimatization.

In using either formulation determined for, say, respiration, one would assume that temperature affects all metabolic processes similarly so that the ratio of two metabolic rates would be independent of temperature. Thus in theoretical work, the rates would be speeded up or slowed down but the general pattern and interrelations would not be affected. If, however, feeding had a different Q_{10} from respiration then the gross efficiency of growth (= growth/ingestion) would vary with temperature (compare Mullin and Brooks, 1970a; Reeve, 1963; Reeve, 1970, for contradictory results concerning this efficiency in zooplankton).

I. Vertical Migration

The simple model which treated herbivores as numbers and weights did not have any spatial components. Yet one of the most distinctive features of the behavior of some species of zooplankton is their vertical migration. A variety of evolutionary explanations might be given in terms of avoidance of predators, metabolic balance in relation to temperature gradients, or searching for food. The main consequence, however, is that some herbivores do not live continuously in the area of maximum food concentration but, daily, may enter regions where food is scarce or effectively non-existent. The effect of this on ingestion of food may be less than expected since starved animals (Frost, 1972) or nocturnally feeding ones (McAllister, 1970) may have higher rates of ingestion than those feeding continuously. Further, as McAllister (1970) showed by a simulation model, it is possible that larger populations of both phyto-plankton and zooplankton may result from a nocturnal grazing rhythm. Thus the energy flow through the ecosystem may not be decreased by this behavior pattern.

There is a further consideration, concerning the stability of the populations. When describing the formulas for ingestion, the advantages of a "threshold" were described in terms of refuge in low numbers. Formally (Steele, 1974b) this is achieved by any sigmoid relation which, at low food concentrations, has ingestion, as a function of food, increasing more rapidly than the growth rate of the food (Fig. 4). For a food supply defined by a single concentration P, both effects refer to this concentration. In a vertically structured environment, with vertical migration, this no longer holds. As a simple example (Fig. 5) consider the euphotic zone as a mixed layer of depth Z_m, with concentration P_m, and below this a decrease in P which has a gradient α, that is,

$$P = P_m \qquad\qquad 0 \leqslant Z \leqslant Z_m$$

$$P = P_m - \alpha\left(\frac{Z}{Z_m} - 1\right) \qquad Z_m \leqslant Z \leqslant Z^* = Z_m\left(1 + \frac{P_m}{\alpha}\right)$$

$$P = 0 \qquad\qquad Z^* \leqslant Z \leqslant Z_b$$

Suppose any zooplankter spends equal times at all depths between surface and bottom, Z_b, and has an ingestion rate

$$I(P) = I_m \frac{P}{\delta + P}, \qquad \text{i.e., equation 15 with } P_0 = 0$$

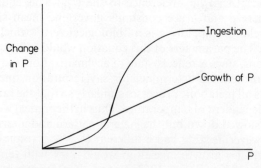

Fig. 4. Rates of increase and decrease in P due to growth and ingestion. The sigmoid shape for ingestion tends to stabilize P at the point of intersection.

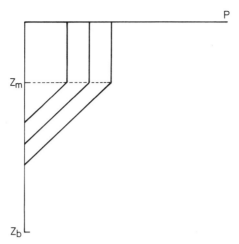

Fig. 5. A very simple representation of distributions of P with depth in an upper mixed layer $(0 - Z_m)$ and a deeper thermocline $(> Z_m)$.

This curve is such that $d^2I/dP^2 < 0$ for all P. Consider the average ingestion rate as the population of zooplankton migrates vertically

$$I = \frac{1}{Z_b} \int_0^{z_b} I(P)\, dz = \frac{Z_m}{Z_b} I(P_m) + \frac{I_m Z_m}{\alpha Z_b} \int_0^{P_m} \frac{P\, dP}{P + \delta}$$

$$= \frac{I_m Z_m}{Z_b} \left[\frac{P_m}{P_m + \delta} + \frac{P_m}{\alpha} - \frac{\delta}{\alpha} \log \left(1 + \frac{P_m}{\delta} \right) \right] \qquad \text{for } \frac{P_m}{\delta} \ll 1$$

$$I = \frac{I_m Z_m}{Z_b} \left(\frac{P_m}{\delta} + \frac{P_m^2 (\delta/2\alpha - 1)}{\delta^2} \right)$$

and

$$\frac{d^2 I}{dP_m^2} > 0 \qquad \text{for } \delta > 2\alpha$$

Thus if growth at low concentrations is linearly related to P_m, under certain conditions grazing *as a function of P_m* can have a sigmoid curve. This contrived example illustrates the possible way in which spatial variations can convert a dynamically unstable relation into a potentially stable one. In the next section this effect is considered in more detail in a simulation model.

J. Horizontal Variations

The same general question arises with horizontal as with vertical heterogeneity— can horizontal dispersion affect the phytoplankton–herbivore relations? In particular, can lateral eddy diffusion overcome the effects of instabilities that arise in a purely temporal model with no grazing threshold? We know that there can be a critical "patch" size, L_c, for phytoplankton such that any perturbations at larger scales would tend to increase in amplitude (see Platt et al., Chapter 22 of this volume). Further, when we consider the grazing interactions of zooplankton with the phytoplankton, it is to be expected that perturbations at small scales may be transferred to larger scales through the nonlinear interactions inherent in the phytoplankton–zooplankton equations (Steele, 1974b, c; this is also described in Chapter 22). However, because of the complexities in such nonlinear systems, it is not possible to get fully quantitative answers to the added effect of nonlinear ingestion rates. For an indication of the added effects of grazing, it is necessary to use small perturbation methods. Reverting to herbivore biomass, H, the simplest form for the prey–predator relations incorporating the grazing relation 15 without a threshold and with diffusion in one dimension, is

$$\frac{\partial P}{\partial t} = aP - \frac{bPH}{1 + cP} + \frac{\kappa \partial^2 P}{\partial x^2} \tag{19}$$

$$\frac{\partial H}{\partial t} = \frac{b^1 PH}{1 + cP} - dH + \frac{\kappa \partial^2 H}{\partial x^2} \tag{20}$$

($1 + cP$ is used, rather than $\delta + P$, so that conditions as $c \to 0$ can be indicated.)

The linearized equations for small perturbations, p, h about the steady state P^*, H^* are

$$\frac{\partial p}{\partial t} = \frac{acP^*p}{1 + cP^*} - \frac{aP^*h}{H^*} + \frac{\kappa \partial^2 p}{\partial x^2} \tag{21}$$

$$\frac{\partial h}{\partial t} = \frac{dH^*p}{P^*(1 + cP^*)} + \frac{\kappa \partial^2 h}{\partial x^2}$$

For an interval $(0, L)$, assuming there is no flux of P or H through the boundaries, the relation of k to the critical length can be indicated most simply by finding the conditions on L for $\partial p/\partial t = \partial h/\partial t = 0$ when a patch is taken as $p = A \cos 2\pi x/L$, $h = B \cos 2\pi x/L$. Putting these in equation 21 and eliminating A, B gives

$$\frac{8\kappa\pi^2}{L^2} = \frac{acP^*}{1 + cP^*} \left(1 \pm \sqrt{1 - \frac{4b^1}{ac^2P^*}} \right)$$

Using the maximum value of the right side gives

$$L \geqslant 2\pi \sqrt{\frac{\kappa(1 + cP^*)}{acP^*}}$$

From these equations, as $c \to \infty$,

$$L \to L_c = 2\pi \sqrt{\frac{\kappa}{a}}$$

which is the result for phytoplankton only. As $c \to 0$, $L \to \infty$, which indicates that the simple Lotka–Volterra equations plus diffusion are stable to small perturbations at all wavelengths. Between these limits there still exists a critical wavelength below which perturbations may be damped out but above which small perturbations could increase. Because of the nonlinearities in this system, it is not possible to say that perturbation above this wavelength would *necessarily* lead to instabilities but only that patchiness would occur under these conditions on grazing.

As an indication of the likely value, consider an ingestion rate which is half the maximum (i.e., $cP = 1$); then

$$L = 0(\sqrt{2}L_c)$$

Thus the combination of diffusion with the simplest type of ingestion relation does not greatly alter the order of magnitude of the critical wavelength. As with vertical variations, the horizontal variability is likely to decrease, but probably not eliminate, the need for threshold responses in simulation models. The general implications of patchiness are discussed in Steele (1975a).

4. Simulation of Vertical Distributions[1]

Any simulation, as distinct from a mathematical theory, is concerned with a particular situation. The aim in this section is not generality but a description of some of the consequences of the earlier discussion in one example of an ecosystem model. It is neither possible nor desirable to include all the aspects of zooplankton population structure and metabolism. This example concerns problems arising from variations in depth distribution of nutrients and phytoplankton and the way these affect the response of the zooplankton. It is an extension of a one-layer model (Steele, 1974a) to 12 layers with vertical exchange between the layers. Details are given in Steele and Henderson (1976). The changes with time in vertical distributions of a nutrient, nitrogen, and of phytoplankton carbon are necessary outputs from the model, but the emphasis here is on the herbivore component.

The first simplification is to assume that only one species of copepod is present, defined by an initial weight (WO) and a final adult weight (WN). The population is represented as six cohorts which are kept distinct by a second simplification; that recruitment of the next generation occurs after a fixed period of adult life. The initial and final weights used here (defined as $WO = 0.2$, $WN = 100$ μg C) are meant to approximate to a *Calanus* species. The period from start of adult life to recruitment, 20 days, is derived from the data of Marshall and Orr (1955) for *Calanus finmarchicus*, whose egg laying pattern approximates to the concept of a single brood.

The total depth (24 m) and mixing rates are intended to apply to shallower inshore areas (or large enclosed plastic columns for which this model was designed; Davies, Gamble, and Steele, 1975). The vertical distribution of maximum photosynthesis P_h (Fig. 6) is for rates with no nutrient limitation (from Steele, 1962). Nutrient limitation is introduced with the relation

$$\frac{N \cdot P_h}{1 + N}$$

That is, the half saturation constant (k_s in the Michaelis–Menten notation) is taken as 1 mg-at N/m^3. Depth distribution of phytoplankton is an important factor affecting zooplankton grazing and this involves sinking rate which, here, varies with ambient nitrogen concentration (Steele and Yentsch, 1960) using the simple relation

$$\text{Sinking (m/day)} = V(N^* - N), \qquad N < N^*$$
$$= 0, \qquad\qquad N \geqslant N^*$$

Nutrients are added to the deepest layer to simulate regeneration from the bottom (or addition at this depth in an enclosure). As described earlier zooplankton excretion also affects nutrient concentration in the model but this is not considered here; nor are the details of the nutrient–phytoplankton interactions required for the model.

To provide some illusions of reality the outputs to be described have specific time and space scales but from dimensional consideration the same relative distributions

---

[1] We thank Eric Henderson for assistance with the simulation of vertical structure.

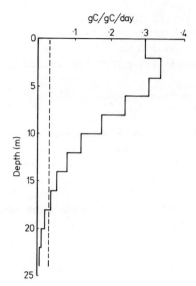

Fig. 6. The depth distribution of photosynthesis (g C/phytoplankton C/day) with the rate of respiration shown by a dashed line.

will occur on other scales. Thus if the present depth intervals were increased by a factor, λ, then with

$$\lambda \text{ (sinking)}, \qquad \lambda^2 \text{ (mixing)}, \qquad \lambda \text{ (zoop. biomass/m}^2)$$

N and P are invariant.

Similarly if all rate processes are changed by the same factor, the same outputs will occur on a different time scale. This would occur if a change in temperature affected *all* rate processes equally.

In the following description P is in units of mg C/m³ but numbers of animals, Z, are expressed as No's $\times 10^{-3}/\text{m}^2$. The time unit is days, and no diurnal effects are included.

The form used here for ingestion is

$$I = \frac{I_m (P - P_0)W^{0.7}}{\delta + P - P_0} \tag{22}$$

with $I_m = 2$, $\delta = 200$, and $P_0 = 50$ (similar to values in Steele, 1974a).

Growth is expressed as

$$[0.7 (1 - E)I - M_1]W^{0.7}$$

where it is assumed that 30 % of ingested food is voided as particulate feces and lost to the system. Of the food assimilated, a fraction E (= 0.4 here) is used for food processing (equivalent to $(M_2 + M_3)/0.7I$). There is also a basal metabolic rate M_1 (=0.1). The nutrient content associated with the E and M_1 terms is returned to the system.

Predation on each cohort is formulated generally as

$$\frac{G(Z - D_f)W}{G' + W} \tag{23}$$

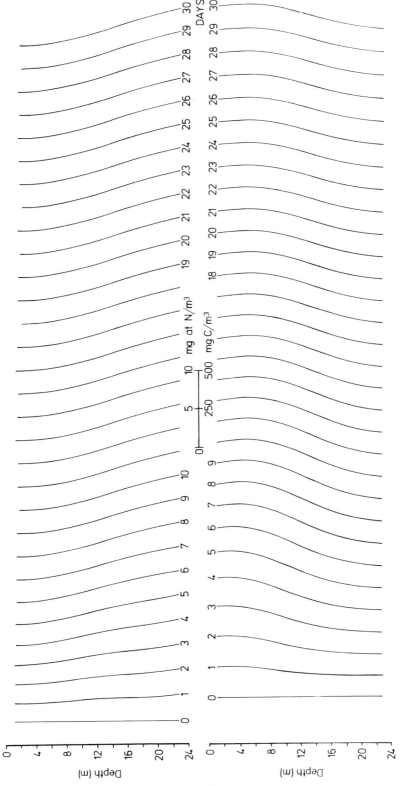

Fig. 7. Vertical profiles of N (as mg-at/m³) and P (as mg C/m³) at 1-day intervals for the first 30 days.

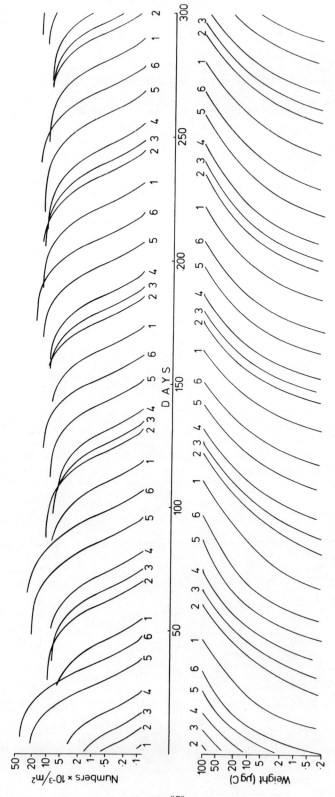

Fig. 8.   Numbers (Z) and weights (W) of six cohorts of a herbivorous copepod grazing on P.

By taking $G' = 10^3$ and $D_f = 0.5$ (to provide a very small threshold corresponding to an overwintering population level) predation is effectively proportional to the biomass in each cohort. Since little is known of predation patterns, G is chosen ($= 3$) to provide a reasonable output in the standard run. A test of the robustness of the model can be made by varying G; particularly by decreasing it so that the effects of overgrazing on the phytoplankton population can be simulated.

Recruitment of the population is derived from relations 17 and 18. For the standard run $X = 0.1$, $X' = 0$, providing a fixed rate of mortality and energy loss before recruitment. The effects of density dependence are introduced later.

Lastly, vertical migration can be assumed to have two possible patterns: (1) a purely mechanical scanning where the zooplankton spend equal time at all depths, and (2) a response to food such that time spent in any depth layer is proportional to the concentration of food there. For the standard run the first condition is used.

A. Standard Run

Given these relations, the vertical distributions of nutrient (nitrogen) and phytoplankton carbon, are obtained for a vertical eddy diffusivity of 0.5 cm²/sec and initial values of 3 mg-at N/m^3, 300 mg C/m^3, all constant with depth. The runs were made for 300 days but Fig. 7 shows only the first 30 days, which cover the initial bloom; thereafter there was little change in vertical structure. The changes in Z and W (Fig. 8) show the sequence of cohorts for the whole period. Lastly (Fig. 9) the changes in surface phytoplankton, P_s, and in total zooplankton biomass (BIOM $= \sum ZW$) show the general variations in populations with time.

In choosing values of the parameters, the aim was to obtain an output which showed relatively small fluctuations after the initial bloom and were reasonably close to levels found in inshore waters around Scotland. As before (Steele, 1974a) changes in the initial values of N and P affect the shape of the bloom but do not alter the later sequences. However, the initial cohort structure is very important and does affect output 100–200 days later (Steele and Henderson, 1976). The particular example to be considered here concerns the parameters in the ingestion term.

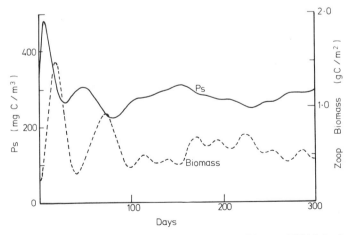

Fig. 9. Surface phytoplankton concentration, P_s, and zooplankton biomass, BIOM, for $I_m = 2$, $\delta = 200$, $P_0 = 50$.

B. Ingestion

The problem of choosing values for parameters can be illustrated by comparing the numbers used here with results obtained by Paffenhöfer (1971) for a *Calanus* species from the Pacific (Fig. 10). His experiments were made with unialgal foods at concentrations around 100 mg C/m^3 and at a temperature of 15°C. The formula for I used here is too low for these data, even taking $P_0 = 0$ as appropriate to feeding on unialgal cultures (Fig. 10). As discussed earlier, this could be rectified by increasing I_m and all other rates to give the same form of outputs but with shorter time scales. If, however, the data are fitted mainly by reducing δ ($I_m = 2.5$, $\delta = 100$) in equation 22, then a much more variable output is obtained (Fig. 11). The increased range in P_s and BIOM is associated with a greater variability in the growth rates of the zooplankton (Fig. 12). In particular, for any cohort there is a succession of life cycles with fast and slow growth and a similar period of two cycles in occurrence of high biomass and low phytoplankton. This periodicity of twice the life cycle is a common feature in other runs (not shown here). It emphasizes how the nature of the copepod life cycle may make as large a contribution to the variability in phytoplankton as physical factors.

C. Threshold Response

The other critical parameter in the ingestion relation is the threshold value taken here as $P_0 = 50$ mg C/m^3. In the earlier one-layer, one-cohort, model (Steele, 1974a)

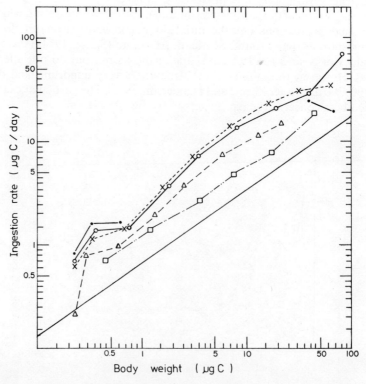

Fig. 10. Ingestion rates of *Calanus helgolandicus* as a function of body carbon (Paffenhöfer, 1971). The straight line is the theoretical relation for I with $I_m = 2$, $\delta = 200$, $P_0 = 0$.

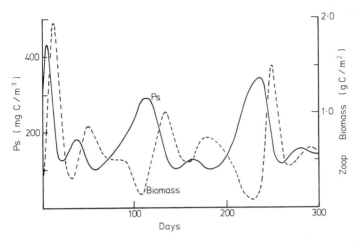

Fig. 11. P_s and BIOM with $I_m = 2.5$, $\delta = 100$, otherwise same parameters as in Fig. 9.

the output became extremely unrealistic with $P_0 = 0$. The same is true with six cohorts, if the model used here is converted into a single layer. One of the most interesting consequences of the 12-layer system is that if $P_0 = 0$ in the standard run, a variable but acceptable output is obtained (Fig. 13).

For the run with altered I_m and δ (Fig. 11) if $P_0 = 0$ the phytoplankton decline steadily after the bloom so that from 60 days on, the weights of individual zooplankton decrease through lack of food until the populations become "extinct." This confirms the qualitative conclusions of the earlier section that, for sufficiently large values of δ, the system will not be unstable.

D. Predation

The values used here for the two coefficients G, G' are arbitrary numbers chosen to obtain a reasonable response. If G is cut to one-third the value used in the standard run the system still survives but with a zooplankton growth rate about half the standard, and expected, rate, which provides three generations in 200 days. If in addition $P_0 = 0$, then the system breaks down completely after 100 days. Thus, again, the grazing threshold, although not essential in the presence of vertical variations, has a significant effect in permitting the system to absorb variations in other factors.

E. Recruitment

In the standard run a fixed fraction (0.1) of food energy available for egg production appeared as recruits. An alternative is to set an absolute upper limit of 20,000 recruits per cohort by taking $X = 0.4$, $X' = 0.1$ in relations 17 and 18. The effect of this can be illustrated by using these values in Fig. 11 ($I_m = 2.5$, $\delta = 100$, $P_0 = 50$) giving a much smoother response after the bloom (Fig. 14). If, however, this density dependence is applied with $P_0 = 0$, then the food goes to a very low level after 30 days and the animals lose weight, a similar result to that obtained without density dependent recruitment. Thus this factor, although it smooths the response, does not eliminate the effects of zero threshold.

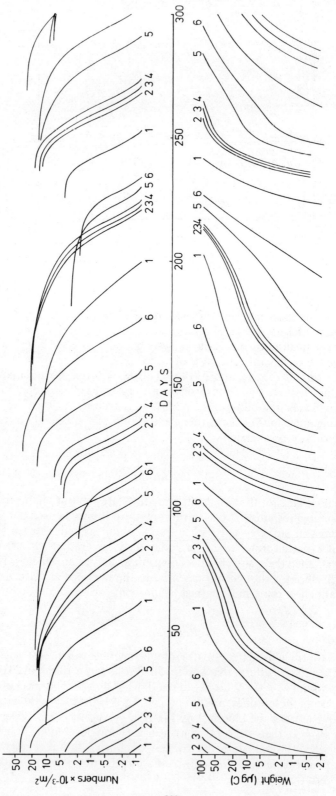

Fig. 12. Numbers (Z) and weights (W) for six cohorts corresponding to Fig. 11.

882

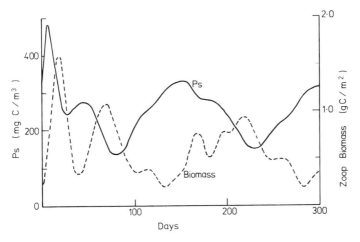

Fig. 13. P_s and IBOM for $I_m = 2$, $\delta = 200$, $P_0 = 0$.

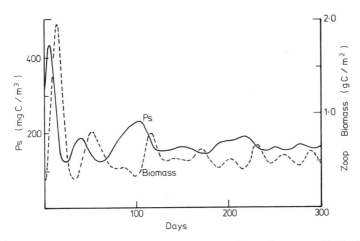

Fig. 14. P_s and BIOM with density-dependent recruitment (compare with Fig. 11).

F. Vertical Migration

By decreasing mixing rates to 0.1 cm²/sec a very marked midwater maximum can be obtained as a roughly steady state (Fig. 15). With other conditions the same as in the standard run, an index of the effect on P can be obtained from the range of P_{max}; see Table II. If the threshold is removed, the range is increased but the system does not break down showing that the earlier result holds for other vertical distributions. In these simulations it is assumed that the animals spend equal time in all layers. If, instead, time at any depth is proportional to food concentration, there is a marked effect on the vertical phytoplankton distribution (Table II) particularly with $P_0 = 0$. A comparison of the first and last runs (Fig. 15) shows how the whole character of the phytoplankton distribution can be changed depending on the interactions of different possible behavioral responses of the zooplankton to food.

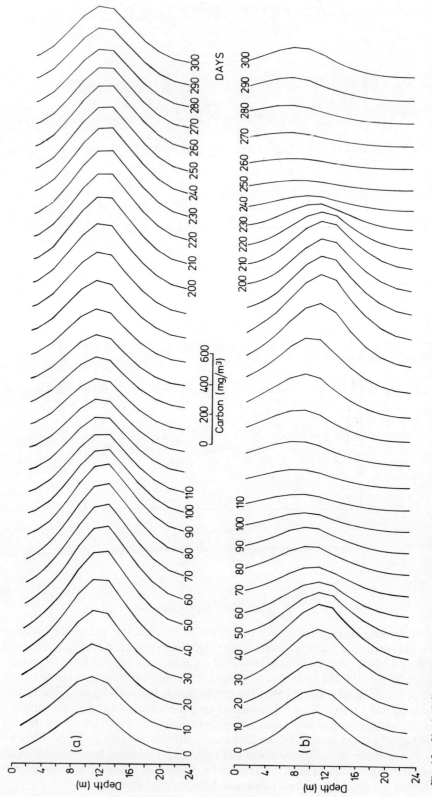

Fig. 15. Vertical distribution of P with low mixing rate, $I_m = 2$, $\delta = 200$, and (a) $P_0 = 0$, time at all depths; (b) $P_0 = 50$, uniform time at depth proportional to P.

884

TABLE II

Combined Effects of Vertical Migration and Grazing Threshold on Maximum Concentrations of Phytoplankton

Migration	P_0 Threshold (mg C/m^3)	Range of P_{max} (mg C/m^3)
Uniform	50	409–570
	0	312–539
Food dependent	50	212–550
	0	85–526

G. Stability

The last run illustrates a general problem, the practical use of the concept of stability. If this run is extended for the extremely long time period of 2000 days (Fig. 16) the increasing amplitude suggests that in some sense this run is unstable. The time period is unrealistic even for environments with little seasonal change. Thus stability is relevant only in the context of appropriate time scales. When the system breaks down completely within 100 days, it may be correct to describe the hypothetical regime portrayed by the simulation as unstable. Apart from this, however, there is the question of what degree of variation is acceptable. By a suitable choice of parameters concerned with ingestion, recruitment, or predation rates, it is possible to simulate systems which, after an initial perturbation, achieve fairly steady states relatively

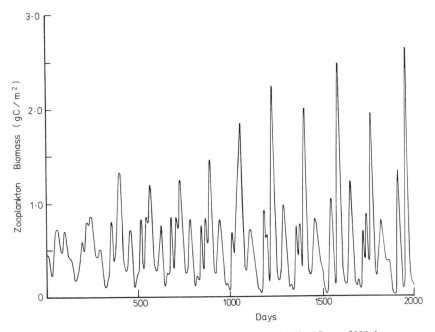

Fig. 16. Zooplankton biomass from the second run in Fig. 15 over 2000 days.

quickly. Other values of the parameters result in large-amplitude, long-period oscillations so that, with a constant physical environment, the system has a high degree of variability. Is there any reason to assume that the former is the more appropriate picture? In many natural systems there is normally a high degree of variability in such measures of plankton as chlorophyll or zooplankton biomass. For these systems we do not know how much of this variation is associated with the physical fluctuations, and how much is due to the inherent characteristics of population structure and to animal metabolic and behavioral processes. The simulation given here is intended to illustrate these problems and possible methods of investigating them. It also demonstrates the importance of spatial as well as temporal variations since the former can alter the nature of the interactions of plants and herbivores. The same general deductions are likely to be true for horizontal as well as vertical variability and, finally, these dimensions will need to be combined.

5. Conclusions

The aim of this chapter has been not to reach conclusions, but to describe techniques which may be of use for incorporating zooplankton into simulation models of ecosystems. The outstanding problem, however, is not theoretical but concerns the quality of field data used to test such simulations. This problem has been avoided here by restricting the discussion partly to the experimental basis for the components needed in any model, and partly to general questions of the structure of zooplankton models. Briefly, the problems of data in this context arise from the fact that the variability in space and time of zooplankton is usually so great that any model that has the right orders of magnitude in its outputs will fit the data. In consequence only gross differences between different ecosystems can be distinguished (Steele, 1975b). Thus even with models treating herbivores in some detail, the testing of these models may rest primarily upon the nutrient and phytoplankton levels, which can be measured with greater accuracy. Examples of this are given in the succeeding chapters dealing with special ecosystems.

The discussion here has tended to concern changes with time and with one spatial dimension, horizontal or vertical, for one species of copepod. This has resulted in emphasis on the problem of the effect of spatial heterogeneity on the stability, or at least the variability, of systems with single "populations" at each trophic level. The more detailed analysis of vertical migration suggests that vertical structure in the system, with certain limitations, may provide a functional response of the grazing component mathematically similar to that based on the purely biological concept of a threshold feeding concentration. It is possible that horizontal spatial variations could also provide the same result. Further, as a result of vertical velocity shear the vertical migration of the animals must be combined with this horizontal heterogeneity (Steele, 1975a), but we are not yet in a position to do this for a fully developed model of herbivorous populations.

The most important "dimension" in samples of natural communities is the species composition, but this remains the most intractable aspect for modeling trophic relations. From the earlier sections it is obvious that it is sufficiently difficult to provide functional relations for an "average" copepod. It will be much more difficult to identify the feeding and metabolic *differences* between species, and to relate these to the diversity of phytoplankton. Thus for conceptual as much as for technical reasons, modeling may have to deal with this dimension separately.

Correspondingly, methods of collection of data will have to attempt, somewhat arbitrarily, to separate these requirements. For the horizontal distributions, the theoretical structure of the problem (Steele, 1974c; Platt et al., Chapter 22 of this volume) suggests some form of spectral analysis. The practical difficulty is obtaining sufficiently fine resolution for the zooplankton, but this may be overcome by particle counters (Boyd, 1973). In the vertical, the use of large enclosed water columns, by removing the horizontal dimensions and by permitting detailed sampling of one population, may provide the required data. A first approximation to appropriate data for multispecies models may be possible if bodily size (volume or biomass) can be used as the most important dimension for both phytoplankton and animals (e.g., Petipa et al., 1970; Vinogradov et al., 1972).

List of Symbols

$a, b, c, g, h, \alpha, \beta, \gamma$	Arbitrary constants in various formulations
δ	Ingestion rate as I_m/δ for P slightly greater than P_0
H	Zooplankton biomass (herbivorous)
I	Ingestion rate of food by a herbivore
I_m	Maximum ingestion rate
J	Time from start of adult life to egg laying
K	Lateral diffusion coefficient
M_r	Rates of metabolic loss by zooplankton
N	Nutrient concentration
N_j	Number of copepods in stage j
P	Phytoplankton concentration (biomass)
p_j	Predation rate on stage j
P_m	Phytoplankton biomass in mixed layer
P_0	Threshold food concentration
P_s	Phytoplankton biomass at the surface
S	Egg (or naupliar) production by adults
S_j	Average weight of stage j
T_j	Average time spent in stage j
W	Weight of an individual copepod
WN	Weight of adult copepod
WO	Weight of juveniles at start of feeding (naupliar stage)
X, X'	Fraction of food assimilated by adults which is used for reproduction rather than for growth
Z	Number of copepods, depth
Z_b	Depth of bottom
Z_m	Depth of mixed layer
ZN	Number of adults
ZO	Number of juveniles

References

Adams, J. A. and J. H. Steele, 1966. Shipboard experiments on the feeding of *Calanus finmarchicus* (Gunnerus). In *Some Contemporary Studies in Marine Science.* H. Barnes, Ed. Allen & Unwin, London, pp. 19–35.

Ambler, J. W. and B. W. Frost, 1974. The feeding behavior of a predatory planktonic copepod, *Tortanus discaudatus. Limnol. Oceanogr.*, **19**, 446–451.

Anraku, M. and M. Omori, 1963. Preliminary survey of the relationship between the feeding habit and the structure of the mouthparts of marine copepods. *Limnol. Oceanogr.*, **8**, 116–126.

Berman, M. S. and S. Richman, 1974. The feeding behavior of *Daphnia pulex* from Lake Winnebago, Wisconsin. *Limnol. Oceanogr.*, **19**, 105–109.

Beverton, R. J. H. and S. J. Holt, 1957. On the dynamics of exploited fish populations. *Fish. Invest., Lond. Ser. 2*, **19**, 533 pp.

Blackburn, M., 1973. Regressions between biological oceanographic measurements in the eastern tropical Pacific and their significance to ecological efficiency. *Limnol. Oceanogr.*, **18**, 552–563.

Boyd, C. M., 1973. Small-scale spatial patterns of marine zooplankton examined by an electronic in situ zooplankton detecting device. *Neth. J. Sea Res.*, **7**, 103–111.

Champalbert, G., and R. Gaudy, 1972. Etude de la respiration chez copépodes de niveaux bathymétriques variés dans la région sud marocaine et canarienne. *Mar. Biol.*, **12**, 159–169.

Comita, G. W. and J. J. Comita, 1966. Egg production in *Tigriopus brevicornis*. In *Some Contemporary Studies in Marine Science*. H. Barnes, Ed. Allen & Unwin, London, pp. 171–185.

Conover, R. J., 1964. Food relations and nutrition of zooplankton. *Proc. Symp. Exp. Mar. Ecol., Occas. Publ. 2*. Graduate School of Oceanography, University of Rhode Island, 81–91.

Conover, R. J., 1966. Factors affecting the assimilation of organic matter by zooplankton and the question of superfluous feeding. *Limnol. Oceanogr.*, **11**, 346–354.

Conover, R. J. and V. Francis, 1973. The use of radioactive isotopes to measure the transfer of materials in aquatic food chains. *Mar. Biol.*, **18**, 272–284.

Conover, R. J., and C. M. Lalli, 1974. Feeding and growth in *Clione limacina* (Phipps), a pteropod mollusc. II. Assimilation, metabolism, and growth efficiency. *J. Exp. Mar. Biol. Ecol.*, **16**, 131–154.

Corkett, C. J. and I. A. McLaren, 1969. Egg production and oil storage by the copepod *Pseudocalanus* in the laboratory. *J. Exp. Mar. Biol. Ecol.*, **3**, 90–105.

Corner, E. D. S., R. N. Head, and C. C. Kilvington, 1972. On the nutrition and metabolism of zooplankton. VIII. The grazing of *Biddulphia* cells by *Calanus helgolandicus*. *J. Mar. Biol. Assoc. U.K.*, **52**, 847–862.

Cushing, D. H., 1959. On the nature of production in the sea *Fish. Invest. Lond. Ser. 2*, **22**, 1–40.

Cushing, D. H. and D. S. Tungate, 1963. Studies on a *Calanus* patch. I. The identification of a *Calanus* patch. *J. Mar. Biol. Assoc. U.K.*, **43**, 327–337.

Dagg, M. J., 1974. Loss of prey body contents during feeding by an aquatic predator. *Ecology*, **55**, 903–906.

Davies, J. M., J. C. Gamble, and J. H. Steele, 1975. Preliminary studies with a large plastic enclosure. In *Estuarine Research*, ed. L. Eugene Cronin. Academic Press Inc. New York, San Francisco, London, pp. 251–264.

DiToro, D. M., D. J. O'Connor, R. V. Thomann, and J. L. Mancini, 1975. Phytoplankton-zooplankton-nutrient interaction model for Western Lake Erie. *Systems Analysis and Simulation in Ecology*, B. C. Patten, Ed., Academic Press, New York, **3**, 423–474.

Edmondson, W. T. and G. G. Winberg, 1971. A manual on methods for the assessment of secondary productivity in fresh waters. *International Biological Program Handbook No. 17*. Blackwell, Oxford and Edinburgh, 358 pp.

Frost, B. W., 1972. Effects of size and concentration of food particles on the feeding behavior of the marine planktonic copepod *Calanus pacificus*. *Limnol. Oceanogr.*, **17**, 805–815.

Frost, B. W., 1974. Feeding processes at lower trophic levels in pelagic communities. In *The Biology of the Oceanic Pacific*, Proc. 33rd Ann. Biol. Coll. C. B. Miller, Ed. Oregon State University Press, Corvallis, pp. 59–77.

Frost, B. W., 1975. A threshold feeding behavior in *Calanus pacificus*. *Limnol. Oceanogr.*, **20**, 263–266.

Gauld, D. T., 1951. The grazing rate of planktonic copepods. *J. Mar. Biol. Assoc. U.K.*, **29**, 695–706.

Gilmer, R. W., 1974. Some aspects of feeding in thecosomatous pteropods. *J. Exp. Mar. Biol. Ecol.*, **15**, 127–144.

Hardy, A. C., 1924. The herring in relation to its animate environment. I. The food and feeding habits of the herring with special reference to the east coast of England. *Fish. Invest. Lond. Ser. 2*, **7** (3).

Harvey, H. W., 1937. Note on selective feeding by *Calanus*. *J. Mar. Biol. Assoc. U.K.*, **22**, 97–100.

Holling, C. S., 1965. The functional response of predators to prey density and its role in mimicry and population regulation. *Mem. ent. Soc. Can.*, **45**, 5–60.

Ikeda, T., 1970. Relationship between respiration rate and body size in marine plankton animals as a function of the temperature of habitat. *Bull. Fac. Fish. Hokkaido Univ.*, **21**, 91–112.

Isaacs, J. D., 1973. Potential trophic biomasses and trace-substance concentrations in unstructured marine food webs. *Mar. Biol.*, **22**, 97–104.

Ivlev, V. S., 1955. *Experimental Ecology of the Feeding of Fishes*. English trans. by D. Scott. Yale University Press, 302 pp.

Jacobs, J., 1974. Quantitative measurement of food selection. A modification of the forage ratio and Ivlev's electivity index. *Oecologia* (Berl.), **14**, 413–417.

Johannes, R. E., 1964. Uptake and release of phosphorus by a benthic marine amphipod. *Limnol. Oceanogr.*, **9**, 235–242.

Klyashtorin, L. B. and A. A. Yarzhombek, 1973. On energy losses for active movement of planktonic organisms. *Okeanol.*, **13**, 697–703 (English trans.).

Madin, L. P., 1974. Field observations on the feeding behavior of salps (Tunicata: Thaliacea). *Mar. Biol.*, **25**, 143–147.

Mann, K. H., 1969. The dynamics of aquatic ecosystems. *Adv. Ecol. Res.*, **6**, 1–81.

Marshall, S. M., 1973. Respiration and feeding in copepods. *Adv. Mar. Biol.*, **11**, 57–120.

Marshall, S. M. and A. P. Orr, 1955. *The Biology of a Marine Copepod.* Oliver and Boyd, Edinburgh, 188 pp.

McAllister, C. D., 1970. Zooplankton rations, phytoplankton mortality and the estimation of marine production. In *Marine Food Chains.* J. H. Steele, Ed. Oliver & Boyd, Edinburgh, pp. 419–457.

McLaren, I. A., 1963. Effects of temperature on growth of zooplankton and the adaptive value of vertical migration. *J. Fish. Res. Bd. Can.*, **20**, 685–727.

Mullin, M. M., 1963. Some factors affecting the feeding of marine copepods of the genus *Calanus*. *Limnol. Oceanogr.*, **8**, 239–250.

Mullin, M. M., 1966. Selective feeding by calanoid copepods from the Indian Ocean. In *Some Contemporary Studies in Marine Science.* H. Barnes, Ed. Allen & Unwin, London, pp. 545–554.

Mullin, M. M. and E. R. Brooks, 1970a. Growth and metabolism of two planktonic, marine copepods as influenced by temperature and type of food. In *Marine Food Chains.* J. H. Steele, Ed. Oliver & Boyd, Edinburgh, pp. 74–95.

Mullin, M. M. and E. R. Brooks, 1970b. The effect of concentration of food on body weight, cumulative ingestion, and rate of growth of the marine copepod *Calanus helgolandicus*. *Limnol. Oceanogr.*, **15**, 748–755.

Mullin, M. M. and E. R. Brooks, 1970c. Part VII. Production of the planktonic copepod, *Calanus helgolandicus*. In *The Ecology of the plankton off La Jolla, California, in the period April through September, 1967.* J. D. H. Strickland, Ed. *Bull. Scripps Inst. Oceanogr.*, **17**, 89–103.

Mullin, M. M., E. F. Stewart, and F. J. Fuglister, 1975. Ingestion by planktonic grazers as a function of concentration of food. *Limnol. Oceanogr.*, **20**, 259–262.

Nassogne, A., 1970. Influence of food organisms on the development and culture of pelagic copepods. *Helgoländer Wiss. Meeresunters.*, **20**, 333–345.

Paffenhöfer, G. A., 1971. Grazing and ingestion rates of nauplii, copepodids and adults of the marine planktonic copepod *Calanus helgolandicus*. *Mar. Biol.*, **11**, 286–298.

Paine, R. T., 1969. A note on trophic complexity and community stability. *Amer. Nat.*, **103**, 91–93.

Parker, R. A., 1973. Some problems associated with computer simulation of an ecological system. In *The Mathematical Theory of the Dynamics of Biological Populations.* M. S. Bartlett and R. W. Hiorns, Eds. Academic Press, London, pp. 269–288.

Parsons, T. R., R. J. LeBrasseur, J. D. Fulton, and O. D. Kennedy, 1969. Production studies in the Strait of Georgia. II. Secondary production under the Fraser River plume, February to May, 1967. *J. Exp. Mar. Biol. Ecol.*, **3**, 39–50.

Petipa, T. S., 1959. Feeding of the copepod, *Acartia clausi*. *Transl. Sevastopol Biol. Sta.*, **11**, 72–99 (English trans.).

Petipa, T. S., 1966. On the energy balance of *Calanus helgolandicus* (Claus) in the Black Sea. In *Physiology of Marine Animals.* Akad. Nauk. S.S.S.R., Oceanographical Comm., Scientific Publishing House, Moscow, pp. 60–81 (English trans.).

Petipa, T. S., E. V. Pavlova, and G. N. Mironov, 1970. The food web structure, utilization and transport of energy by trophic levels in the plankton communities. In *Marine Food Chains.* J. H. Steele, Ed. Oliver and Boyd, Edinburgh, pp. 142–167.

Petrusewicz, K. and A. Macfadyen, 1970. *Productivity of Terrestrial Animals—Principles and Methods.* International Biological Program Handbook No. 13. Blackwell, Oxford and Edinburgh, 190 pp.

Poulet, S. A., 1973. Grazing of *Pseudocalanus minutus* on naturally occurring particulate matter. *Limnol. Oceanogr.*, **18**, 564–573.

Poulet, S. A., 1974. Seasonal grazing of *Pseudocalanus minutus* on particles. *Mar. Biol.*, **25**, 109–123.

Reeve, M. R., 1963. Growth efficiency of *Artemia* under laboratory conditions. *Biol. Bull.*, **125**. 133–145.

Reeve, M. R., 1970. The biology of Chaetognatha I. Quantitative aspects of growth and egg production in *Sagitta hispida*. In *Marine Food Chains*. J. H. Steele, Ed. Oliver & Boyd, Edinburgh, pp. 168–189.

Reeve, M. R. and M. A. Walter, 1972. Conditions of culture, food size selection, and the effects of temperature and salinity on growth rate and generation time in *Sagitta hispida* Conant. *J. Exp. Mar. Biol. Ecol.*, **9**, 191–200.

Richman, S. and J. N. Rogers, 1969. The feeding of *Calanus helgolandicus* on synchronously growing populations of the marine diatom *Ditylum brightwellii*. *Limnol. Oceanogr.*, **14**, 701–709.

Riley, G. A., 1947. A theoretical analysis of the zooplankton population of Georges Bank. *J. Mar. Res.*, **6**, 104–113.

Riley, G. A., 1963. Theory of food-chain relations in the ocean. In *The Sea*. Vol. 2. M. N. Hill, Ed. Wiley-Interscience, New York, pp. 438–463.

Steele, J. H., 1962. Environmental control of photosynthesis in the sea. *Limnol. Oceanogr.*, **7**, 137–150.

Steele, J. H., 1974a. The structure of marine ecosystems. Harvard University Press, Cambridge, Mass., 128 pp.

Steele, J. H., 1974b. Stability of plankton ecosystems. In *Ecological Stability*. M. B. Usher and M. H. Williamson, Eds. Chapman & Hall, London, pp. 179–191.

Steele, J. H., 1974c. Spatial heterogeneity and population stability. *Nature*, **248** (5443), 83.

Steele, 1975a. Patchiness. In *The Ecology of the Sea*. D. H. Cushing and J. J. Walsh, Eds. Blackwell, London.

Steele, J. H. 1975b. The state of the art in biological modelling (A review of papers submitted to the conference). *Proc. NATO Sci. Comm. Conf. Modeling of Marine Systems*, J. C. J. Nihoul, Ed. Elsevier Oceanography Series, **10**, 207–216.

Steele, J. H. and E. Henderson, 1976. Simulation of vertical structure in a planktonic ecosystem. *Scot. Fish. Res. Rep.* (5) 27 pp.

Steele, J. H. and C. S. Yentsch, 1960. The vertical distribution of chlorophyll. *J. Mar. Biol. Assoc. U.K.*, **39**, 217–226.

Vinberg, G. G. and S. I. Anisimov, 1969. Investigations of the mathematical model of an aquatic ecosystem (from "Biological foundations of the fishing industry and regulations of marine fisheries"). *Fish. Res. Bd. Can. Trans. Ser. No. 1571*.

Vinogradov, M. E., V. V. Menshutkin, and E. A. Shushkina, 1972. On mathematical simulation of a pelagic ecosystem in tropical waters of the ocean. *Mar. Biol.*, **16**, 261–268.

Vlyman, W. J., 1970. Energy expenditure of swimming copepods. *Limnol. Oceanogr.*, **15**, 348–356.

Walsh, J. J., 1975. A spatial simulation model of the Peru upwelling ecosystem. *Deep-Sea Res.*, **22**, 201–236.

Ware, D. M., 1973. Risk of epibenthic prey to predation by rainbow trout (*Salmo gairdneri*). *J. Fish. Res. Bd. Can.*, **30**, 787–797.

Wilson, D. S., 1973. Food size selection among copepods. *Ecology*, **57**, 909–914.

23. THE MODELING OF OPEN-SEA ECOSYSTEMS

M. E. VINOGRADOV AND V. V. MENSHUTKIN

The functioning of communities is assured by relationships between populations, against a background of evolutionary adaptation of the organisms themselves to existence under the conditions of a given system, in which their morphological peculiarities, genetic characteristics, behavioral responses, etc. play an important role. However, the basic relationships in a community, those that integrate it as a whole and determine its structure and productivity, are trophic connections. Therefore the study of trophic relationships within a community, the evaluation of energy flows through a biological system and their utilization by various trophic groups, may be expected to yield the most essential information on the functioning of communities.

The energetic principle is especially effective in application to aquatic and, in particular, to pelagic communities. Owing to a considerable homogeneity of the biotope, trophic relationships occupy a foremost place in the regulation of aquatic ecosystems as entities. Any relationships not directly relating to feeding play but a secondary part in the pelagic ecosystems, unlike marine benthic and, especially, terrestrial ecosystems. Viewed from this standpoint pelagic ecosystems represent the simplest object for modeling and as yet the only one for which the construction of a comprehensive model may be attempted. Moreover, the abiotic conditions that directly affect the functioning of a pelagic community are easily determined and quantitatively evaluated, whereas their evaluation for benthic systems is far less accurate and sometimes unachievable.

1. Some Singularities of Open-Sea Ecosystems

A characteristic feature of pelagic communities is the decoupling between the zones of primary production of organic matter and the depths where the greater part of this production is consumed. The community inhabiting the waters of the producing zone includes autotrophs and consequently does not depend on an inflow of food, whereas the communities living at greater depths subsist on food (energy) coming from the surface community and are thus energy dependent.

The communities of different depths are united by a constant and fairly intensive exchange of individuals. Interzonal migrants, feeding in the surface layers, constitute up to 40% biomass of the whole bathypelagic (1000–3000 m) population in temperate–cold water regions; in tropical regions the values are somewhat lower. (Vinogradov, 1968; Vinogradov and Arashkevich, 1969; Araskhevich, 1972). Such species are components of both the surface and the deep-water communities. Therefore the community of the surface layers and the energy-dependent deep-water communities should be regarded as parts of a single larger community, encompassing the entire or nearly entire water column of the ocean, so that the entire water column itself, with the organisms inhabiting it, constitutes a single ecosystem.

The interrelationships between the communities of different depths are rather complicated and do not lend themselves to single-valued resolutions based on a priori considerations. This statement may be exemplified by the question raised by Banse (1964), who assumes that if active transfer by migrating organisms is prevailing in the transport of organic matter from surface layers into greater depths, then the quantity of deep-water plankton is relatively lower (per unit of production of organic matter in the surface community) in environments with balanced life cycles than in environments with unbalanced cycles.

Actually such a relationship may be nonexistent, if the trophic structure of deep-water communities under a balanced cycle differs substantially from that under an unbalanced cycle. Let us compare the subpolar (northwestern) and tropical (central) regions of the Pacific Ocean. The life cycle of surface communities is unbalanced in the first region and balanced in the second. In the tropics, in the waters below the producing zone, carnivorous macroplankton constitutes more than half the total mass of net plankton (up to 74%, Vinogradov, 1968). A considerable part of these animals perform diel feeding migrations to the subsurface layers, where they feed both on herbivores and on small predators (copepods, chaetognaths, etc). Thus they are to a considerable extent second-level predators. In subpolar regions where the share of carnivorous macroplankton does not exceed 12% even in the depths of its greatest abundance, the main mass of predators consists of small mesoplankton animals, chiefly chaetognaths, which feed mainly on interzonal filter-feeding copepods sinking during their diurnal migrations. Thus they are first-level predators.

It is obvious that in both regions the relative mass of second-level predators will be lower than the biomass of first-level predators even at an equal consumption of herbivorous zooplankton; consequently the relation assumed by Banse does not arise. This example illustrates the intricate complexity of the problems that face the researcher who attempts to evaluate quantitatively the energy flow in the trophic chains from the surface to the deep water.

Experimental observations of the energetic characteristics of deep-water populations are still hampered by substantial methodical difficulties and must be restricted to the examination of the communities of the surface trophic zone only. However, as these communities include practically the entire mass of phytoplankton and nearly half the mass of zooplankton, they are in fact responsible for the productivity of the ocean. The ecosystems of the open ocean occupy about 70% of the earth's surface and constitute the largest system of the biosphere.

Many works have been devoted to the study of pelagic ecosystems in the trophic layer of the temperate–cold regions of the ocean, and for these the greatest number of attempts at modeling the production cycle have been made (Patten, 1968; Riley, Stommel, and Bumpus, 1949; Steele, 1962; Vollenweider, 1965; Dugdale, 1967; Cushing, 1959, 1969). But the modeling of the functioning of the whole system is greatly impeded by large variations in seasonal production. The energetic balance of the system for a full production cycle requires year-round observations for the same particular community. To obtain such data for open-sea systems presents great technical difficulties.

In this respect the communities of open tropical waters present a great advantage; because they are far less subject to seasonal variations, the system may be considered to be in equilibrium at any given moment. The supply of nutrients from the deep-water layers depends on quasi-stationary rates of water ascent rather than on seasonal mixing, and the intensity of solar radiation varies but little during the year. Therefore a judgment on the energetic balance of the system may be formed from relatively short-term observations of one and the same community. The highly complex species structure of the tropical communities and their great diversity present no special difficulties in the evaluation of their functioning if based on energetic principles.

2. The Structural Characteristics of the System and Their Variability

The energetic approach to the study of the functioning of communities permits one to ignore, to a certain extent, their species composition, but in return it is very exacting in regard to information on their trophic and spatial structure. Special investigations are required to obtain their exact description.

A. Trophic Structure

Trophic groups may be recognized with any desired degree of detail, depending above all on the available amount of experimental data. However, if they are too numerous they will overload the model with superfluous detail. On the other hand, they must be sufficient to evaluate the major energy flows passing through the system.

As an example, in the modeling of communities in the tropical regions of the Pacific Ocean (Vinogradov et al., 1973) the following groups (elements) were recognized: phytoplankton; bacterioplankton; protozoans, nauplii; filter feeders smaller than 1 mm (such as *Oikopleura, Clausocalanus, Paracalanus, Acartia, Lucicutia*, small ostracods); filter feeders larger than 1 mm (such as *Undinula, Scolecithrix, Neocalanus*, young euphausiids; carnivores *Cyclopoida*; carnivores *Calanoida*; predatory *Chaetognatha* and *Polychaeta*; and the fry of plankton-feeding fishes (see also Sections 6–8).

B. Spatial Structure

The unevenness of the horizontal distribution of plankton may be traced on different scales (Cassie, 1959; Gitelzon et al., 1971; Cushing, 1962), but the laws governing the formation of the horizontal structure of zoo- and phytoplankton are not sufficiently known to attempt to construct a mathematical model of plankton distribution in an area with a horizontal scale of several tens and hundreds of meters. The vertical distribution of organisms in the water column is extremely uneven, too, being influenced both by abiotic and biotic patterns. Within the water column, layers very poor in plankton alternate with layers of high abundance. Even in oligotrophic tropical waters relatively thin quasi-stationary layers are encountered containing enough food organisms to satisfy the requirements of filter-feeding invertebrates. This accounts for a situation frequently observed in oligotrophic tropical waters, when zooplankton subsists on organisms whose average recorded biomass is less than 20–40 mg/m^3, that is, less than the threshold concentration at which filter feeders may compensate the loss of energy expended on metabolism (Jorgensen, 1962). Thus the very existence of communities in oligotrophic tropical regions of the ocean is possible only owing to the presence of layers of above-average concentrations of producers and consumers.

The basic abiotic factors determining the distribution of plankton in the surface zone are light, temperature, and salinity (and ensuing density of the water) and the concentration of nutrients. The distribution of certain microelements may play an important part, but their role in the system and distribution in the environment is still insufficiently known.

The decrease of light with depth proceeds gradually and may be easily described mathematically for all the parts of the spectrum. Its effect on the phytoplankton has been well studied.

The effects of changes in temperature, salinity, and density are most significant in the layers of their largest gradients, but as has been pointed out by Vinogradov, Gitelzon, and Sorokin (1970), the formation of layers of high phyto- and bacterio-plankton concentration cannot be explained merely by the phenomenon of the cells "hanging" in the zone of maximum density gradients. Cells are bound to aggregate in the zone of their neutral buoyancy, that is, to follow the absolute density value. The density gradient affects the thickness of the aggregation zone: the larger the gradient, the narrower the zone of cell aggregation. But the thickness of the zone is also affected by the range of variability of the cells themselves in regard to specific gravity, which they are able to change actively (Beklemishev et al., 1960). Therefore high cell concentrations occur often but not always in the layer of maximum density gradient (usually in the upper part of the thermocline).

For example, in oligotrophic tropical regions, the formation of a quasi-stationary thin layer of high phytoplankton bacterial and sometimes also zooplankton concentrations at depths between 60 and 120 m, near the thermocline or above it (the so-called "lower maximum") can be explained biologically. Phytoplankton (and, in consequence, bacterioplankton) develop in a comparatively narrow optimum layer, limited from below by lack of light and from above by lack of nutrient salts. This layer is characterized by intense photosynthesis of phytoplankton, depending on the dynamic equilibrium of these two limiting factors. This equilibrium is responsible for the depth of the layer, which may not coincide with the maximum density gradient. The advantageous rate of cell growth in this layer results in high cell concentrations, which hamper further accumulation unless washed away by turbulent mixing (Vinogradov, Gitelzon, and Sorokin, 1970).

The distribution of zooplankton is less related to variations of physical factors and it is not found associated with food concentrations, except for relatively brief feeding periods. Nevertheless zooplankton animals also tend to concentrate in great quantities within very narrow layers, which usually differ in ecologically similar species (Timonin, 1975; Zalkina, 1971, 1975; Rudjakov, 1971).

To obtain a picture approaching the actual vertical distribution of communities, zooplankton should be collected with nets in very narrow layers (preferably at 10-m intervals) whereas bottle samples for estimations of phytoplankton, microzoo-plankton, bacterioplankton, and detritus concentrations should be taken not at standard depths but in relation to the depths of their maximum concentrations. These depths are determined by lowering and hauling continuous-recording instruments, for such parameters as intensity of the bioluminescent field, chlorophyll fluorescence, water turbidity, and, of course, temperature, salinity, density, and oxygen concentration. Bottle samples collected at the extreme points of the curves of these parameters make it possible to record with a high degree of reliability all the major layers of maximum and minimum concentrations of the primary trophic links of the system.

3. Dynamics of the Ecosystem

By combining soundings of the bioluminescent field with simultaneous collection of bottle samples it became possible to reveal the existence of a quasi-stationary "two-maximum" structure of plankton concentration in the meso- and oligotrophic tropical regions (Vinogradov, Gitelzon, and Sorokin, 1970). The causes of the formation of a lower maximum were discussed above. In oligotrophic regions this "lower maximum" utilizes the flow of nutrients coming from below and, as it were, "cuts off" the overlying part of the community from the source of nutrients.

The upper maximum, situated at a depth of 10–15 m, probably owes its existence to the transport of nutrients by horizontal water flows from the zone of their most intensive ascent (chiefly from zones of divergences and upwellings). The nutrient salts brought into the surface layers are repeatedly used here in the production–destruction cycle involving phytoplankton, zooplankton, and bacterioplankton (Cushing, 1969).

The total content of nutrients in the surface layer decreases with distance from the zone of water ascent (with the "aging" of the upwelled waters) owing to their removal by animals migrating into greater depths and by sinking of phytoplankton and detritus. In this process, owing to the decrease in the quantity of phytoplankton and changes in its composition (Margalef, 1958) and also to the fact that the different trophic groupings of the community reach maturity and maximum abundance at

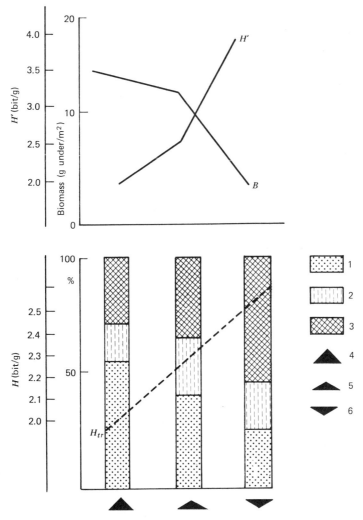

Fig. 1.   Structure differences of zooplankton communities in regions with different water regimes, Indian Ocean (after Timonin, 1971). B, Total mass of zooplankton; H', species diversity; H_{tr}, trophic structure diversity; 1, filter feeders (herbivores); 2, omnivores; 3, carnivores; 4, water of intensive divergence; 5, waters of weak divergence; 6, stable stratification and weak convergence.

different rates, the whole trophic structure of the community is subjected to successional changes. Simultaneously its biomass diminishes, species diversity increases,
and the stability of the system is enhanced (Fig. 1). The concept of successional changes
in a pelagic community differs somewhat from the classical "succession" of terrestrial
ecologists. Nevertheless the application of the term to developing aquatic systems is
fully justified and the special features of this succession have been specially examined
by many authors (Odum, 1963; Margalef, 1967, 1968; Timonin, 1971).

Figure 2 shows an idealized scheme of the turnover of nutrients and dissolved
organic matter, on which depends the development of pelagic communities in tropical
regions. Naturally the picture presented by the scheme may be modified by many

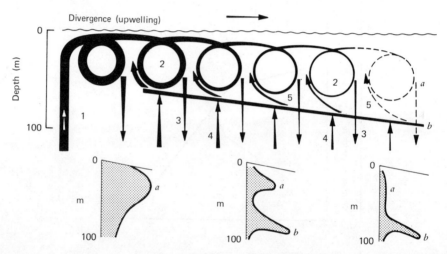

Fig. 2. Turnover of nutrients and dissolved organic matter (DOM) in a tropical pelagic community
(after Vinogradov, Gitelzon and Sorokin, 1970). 1, Ascent of nutrients and DOM in upwelling; 2, repeated
use of nutrients and DOM in the production–destruction cycle of a surface community; 3, loss of nutrients
with sinking organic remains and migrating organisms; 4, turbulent diffusion ascent of nutrients with
deep water and their retention in the layer of the maximum; 5, incorporation of nutrients carried up from
layer of lower maximum by migrating organisms into the production cycle of a surface community.
Below, changes in the vertical distribution of relative phytoplankton concentrations with distance from
the zone of upwelling (a, upper maximum; b, lower maximum).

factors but, if the scheme is true in its essence, the model should account for the
succession of the community from the moment of ascent of the water to the moment
of its sinking. In other words, the model should simulate the temporal change in a
community moving in space and observations of the changes should follow the flow
of water. Since long-term observations in a drift present great technical difficulties,
the task may be simplified by separate investigations of several states of the system
succeeding each other in time. In this manner data may be obtained on the changes
taking place in the structure of the community over the whole period of its existence.
Of course, no claim to equilibrium, or energetic balance, of the system can be made
for each transect. Near the zone of its formation the system accumulates energy,
farther "downstream" it spends it.

The concept that the formation of a pelagic community starts with the ascent to the surface of water rich in nutrients but nearly devoid of population, that only later does it begin to "bloom" with phytoplankton and still later become populated within zooplankton developing on the basis of phytoplankton, may be more or less fully realized only in regions of very intensive water ascent from relatively great depths. For instance, such a picture is characteristic of the Peruvian Upwelling, but in the East Pacific Equatorial Upwelling the water of the Cromwell Current that rises to the surface is already inhabited by a relatively mature community carried in from the west. In this case the whole community acquires a number of special features, inherent in communities in later stages of development, in the very region of its formation (Vinogradov, 1974).

4. Evaluation of Energy Flow Through the System of a Transect

As an example, Fig. 3 presents a scheme of energy flow (cal/m^3/day) through a system of pelagic organisms in the 0–150-m layer of a tropical region.

The trophic relations between the elements were derived from the analysis of gut contents, morphological study of buccal appendages, and experimental investigations of feeding under laboratory conditions and in situ, using ^{14}C-labeled food. Phytoplankton production was determined by the ^{14}C method, its biomass by direct counts (Koblentz–Mishke et al., 1971). The biomass of bacteria was determined by direct count, their production by the ^{14}C method from the value of heterotrophic CO_2 assimilation (Romanenko, 1964; Sorokin, 1971). The biomass of all animal groups was determined by direct count and measurement. The energetic (caloric) equivalent of body mass or the content of calories in the body of an individual of a definite size, was determined for pelagic animals of 42 systematic groups (Shushkina and Sokolova, 1972).

The average daily mean consumption, C, was estimated using the following expression:

$$C = H + P + R = \frac{P + R}{U}$$

where H = unassimilated food
P = rate of production
R = rate of metabolism (respiration)
U = efficiency of assimilation

The values R and U were determined directly during the cruise (Shushkina and Vilenkin, 1971; Shushkina and Pavlova, 1973; Petipa, Pavlova, and Sorokin, 1971). The rate of diel production was calculated from the ratio

$$P = R \frac{K_2}{1 - K_2}$$

where K_2 is the coefficient of assimilated food used up for growth. On the basis of experimental and published data a value of $K_2 = 0.5$ is assumed for protozoans and nauplii and $K_2 = 0.4$ for the other animals.

If experimental and calculated values are available on the metabolism and production rates of all the main plankton animals, then by arranging them into trophic and size groups, a scheme may be obtained of the flow of energy through the system investigated.

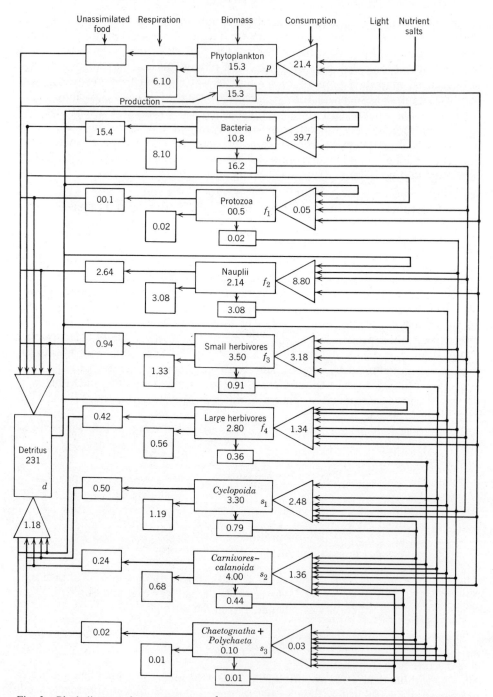

Fig. 3. Block diagram of energy flow (cal/m³/day) through a community in the 0–150-m layer in a meso-trophic tropical area (04°29′N, 142°30′E, *Vitjaz* St. N 6429, May 1971) (Vinogradov et al., 1973).

5. General Statement of the Problem of Modeling Pelagic Ecosystems

The construction and investigation of mathematical models of pelagic ecosystems has a fairly long and interesting history, well reflected in the review of Patten (1968). It must be said, however, that the overwhelming majority of such models describe only separate and for the most part initial stages of the production cycle. Substantial contributions to mathematical modeling of whole aquatic ecological systems have been made by Riley, Stommel and Bumpus (1949), Winberg and Anisimov (1966), and Lyapunov (1971). The work of Lyapunov provides the methodical foundations on which are built the models of marine ecosystems discussed in this chapter.

The approach to the modeling of complicated systems used by Lyapunov (Lyapunov and Yablonsky, 1963) is in its general features similar to the macroscopic method of Odum (1971). The system is presented in the form of an assemblage. These elements are linked together by connecting channels. The role of signals passing through these channels may be placed by portions of matter or information, and accordingly *material* and *informational* connections between the elements of the system may be recognized. The transmission of portions of matter from one element to another is inseparably associated with a transmission of energy. One of the basic principles of the construction of mathematical models of the energetics of ecological systems is the principle of conservation of matter and energy, which is interpreted in the form of a relationship for each living or inert element of the ecosystem.

Let us draw a distinction between two means of transfer from one element of the ecosystem to another. The first, which we call *flow*, is necessarily associated with some transformation of matter and energy, as exemplified by the consumption of nutrients in the process of formation of primary production, the eating of individuals of one trophic level by individuals of another level, or the decomposition of dead organic matter.

The second means of matter and energy transmission, which we term *transport*, is associated with active or passive displacements of elements within the water. As examples we have the displacements of elements by currents, vertical migrations of zooplankton, sinking of detritus, or penetration of light through the water column.

The whole pelagic ecosystem may be divided into *cells* in such a manner that the elements within a cell are connected only by flows, while the transmission of matter and energy between the cells is effected by transport.

A mathematical model of relations in an ecosystem can be constructed only on the condition that a certain degree of fullness has been achieved in the study of the simulated object. A concept must be formed of the distribution of matter and energy between the elements, of the laws governing the intensity of flows between the elements, of what enters the system, and of what (and in what quantities) leaves the system or is extracted from it.

In devising the model (Lyapunov, 1971) it was assumed that all the processes occur without time delays. Generally speaking, this is not always true, but with the passage to a discrete time step, lasting 24 hr or more, the assumption of elementary processes running without delays, for instance in regard to phytoplankton, is entirely justifiable.

Lyapunov's model was given in terms of differential equations. This is most suitable for the formulation of the problem, but it must be remembered that all the further results require a passage to finite difference equations expressed as programs for digital electronic computers.

The Lyapunov model contained merely six elements: light (I), nitrogen concentration in assimilable ionic form (n_N), concentration of assimilable phosphorus (n_P), biomass of phytoplankton (p), biomass of zooplankton (f), and concentration of detritus (d). The following assumptions were made for each element of the system: light is absorbed by water (a) and also by phytoplankton (a_1), zooplankton (a_2), and detritus (a_3). Nitrogen and phosphorus are expended in the formation of primary production and process of photosynthesis (coefficients h_N and h_P), and are released during the decomposition of detritus (v_N and v_P). Photosynthetic intensity is limited by light conditions and the concentration of nutrients.

$$P_p = \min\{lI, g_N h_N, g_P h_P\} \tag{1}$$

where l, g_N, and g_P are coefficients.

Phytoplankton consumption is assumed to follow Volterra (coefficient β). The effects of zooplankton reproduction, zooplankton consumption of detritus, the process of cannibalism (coefficients γ_1, γ_2, and γ_3) are accounted for in an analogous manner. Natural mortality of zooplankton (ε) and rate of vertical migration (ω'') are also taken into account.

It is assumed that detritus is formed from dying zooplankton and from zooplankton excreta, the quantity of which is proportional to the quantity of consumed food (coefficients $\delta_1, \delta_2, \delta_3$). Phytoplankton does not die off and is completely consumed by zooplankton; detritus decomposition (μ) and settling (ω_3) are taken into consideration. Dissolved nutrient compounds, phytoplankton, zooplankton, and detritus are transported in a vertical direction by turbulent diffusion, the intensity of which is described by the coefficient k. The depth axis z is assumed to be directed from the surface to the depth of the ocean.

The system of equations has the following form:

$$\frac{\partial I}{\partial z} = -I(a + a_1 p + a_2 f + a_3 d) \tag{2}$$

$$\frac{\partial n_N}{\partial t} = -h_N P_p p + v_N d + \frac{\partial}{\partial z}\left(k \frac{\partial n_N}{\partial z}\right) \tag{3}$$

$$\frac{\partial n_P}{\partial t} = -h_P P_p p + v_P d + \frac{\partial}{\partial z}\left(k \frac{\partial n_P}{\partial z}\right) \tag{4}$$

$$\frac{\partial p}{\partial t} = \alpha P_p p - \beta_P f + \frac{\partial}{\partial z}\left(k \frac{\partial p}{\partial z}\right) + \frac{\partial}{\partial z}(\omega_1 p) \tag{5}$$

$$\frac{\partial f}{\partial t} = \gamma_1 p f - \gamma_2 f^2 + \gamma_3 f d - \varepsilon f + \frac{\partial}{\partial z}\left(k \frac{\partial f}{\partial z}\right) - \frac{\partial}{\partial z}(\omega'' f) \tag{6}$$

$$\frac{\partial d}{\partial t} = \varepsilon f - \mu d + \delta_1 p f + \delta_2 f^2 + \delta_3 f d + \frac{\partial}{\partial z}\left(k \frac{\partial d}{\partial z}\right) - \frac{\partial}{\partial z}(\omega_3 d) \tag{7}$$

Figure 4 shows a block scheme of a cell of Lyapunov's model. The symbols given in the figure are preserved through the whole chapter and serve as a kind of language in the description of the structure of the models. Although the model developed by Lyapunov had not been carried out to its numerical calculation, it became the foundation around which were centered the efforts of a numerous and heterogeneous body of researchers.

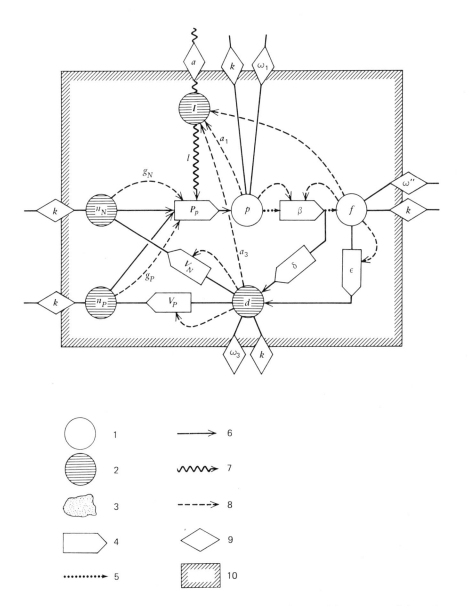

Fig. 4. Block scheme of a cell in the Lyapunov model. 1, Living element of the ecosystem; 2, inert element of the ecosystem; 3, group of elements in a cell; 4, flow of matter; 5, trophic connection; 6, transfer of matter; 7, solar radiation energy; 8, informative connection; 9, transport between cells; 10, cell of the ecosystem. I, Solar radiation; n_N, nitrogen concentration; n_P, phosphorus concentration; P_p, phytoplankton production; p, biomass of phytoplankton; f, biomass of zooplankton; d, detritus concentration; ω'', vertical migrations of the zooplankton; ω_1, phytoplankton settling; ω_3, detritus settling; k, turbulent diffusion. The other symbols correspond to the coefficients of equations 2–7 and are explained in the text.

6. A Model Simulating the Vertical Distribution of Elements

The results of investigations carried out during the forty-fourth and fiftieth cruises of the *Vityaz* compelled us to revise and refine many aspects of Lyapunov's model (Vinogradov, Menshutkin, and Shushkina, 1972; Vinogradov et al., 1973).

Because of the important part played by bacteria in the transmission of matter and energy and in the turnover of matter in pelagic ecosystems, a new structural element, bacterioplankton, was added to the model. This called forth the consideration of the transmission of energy from phytoplankton to bacteria through dissolved organic matter (DOM). It was assumed that 30 % of phytoplankton production consists of the release of dissolved organic matter. The intensity of bacterial proliferation is limited both by food concentration (detritus and dissolved organic matter) and by the maximum ratio of daily production to biomass, which for tropical regions is assumed to equal three (Sorokin, 1971).

Another modification of Lyapunov's model consisted in refusing to consider zooplankton as a homogeneous element, and recognizing in it (Fig. 5) such elements as protozoans (f_1), microzooplankton (f_2), small filter feeders (f_3), large filter feeders (f_4), carnivorous cyclopoids (s_1), carnivorous calanoids (s_2) and large predators; chaetognaths and polychaetes (s_3). Protozoans included naked heterotrophic flagellates, infusorians, and radiolarians. The nauplii of all copepods were assigned to microzooplankton. The systematic composition of other groups has been discussed above.

The numerous experimental data available on the dependence of the intensity of zooplankton feeding on food density led us to abandon the scheme of Volterra for the formula proposed by Ivlev (1957). According to this expression the relation between the real ration, C, and food concentration, B, can be represented in the form:

$$C = C_{max}\{1 - \exp\left[-\zeta(B - B_0)\right]\} \tag{8}$$

where C_{max} = maximum ration
$\quad\quad B_0$ = minimum prey density, below which consumption ceases
$\quad\quad \zeta$ = a constant

In establishing the values of feeding selectivity, when no direct experimental data were available, we made the hypothesis that the concentration of a given kind of food is proportional to its percentage in the real ration.

The rate of total metabolism, R, and of the coefficient of assimilated food used up for growth, K_2, were experimentally determined in different representatives of the plankton of the tropical Pacific Ocean. This permitted the determination of the productivity of zooplankton elements (f_i, s_i) from the expression discussed above.

$$P = R\frac{K_2}{I - K_2} \tag{9}$$

The functioning of phytoplankton in a pelagic community was substantially refined. In accordance with Voitov and Kopelevich (1971), an empirical relation was established between phytoplankton and detritus concentrations and the coefficient of light attenuation, which replaced the coefficients in equation 2 by

$$a = 0.01 + 0.001(p + d) \tag{10}$$

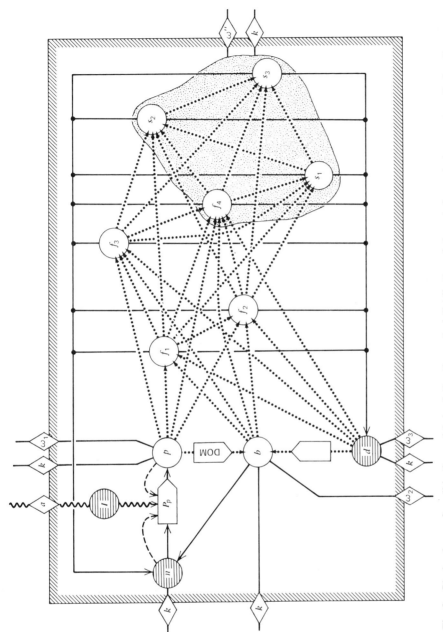

Fig. 5. Block scheme of a cell in the model simulating the vertical distribution of elements in a pelagic ecosystem. Designations same as in Fig. 4 plus the following: b, biomass of bacteria; f_1, biomass of protozoans; f_2, biomass of micro plankton; f_3, biomass of small herbivores; f_4, biomass of large herbivores; s_1, biomass of cyclopoids; s_2, biomass of carnivorous calanoids; s_3, biomass of chaetognaths and polychaets.

According to the data of Ryther (1956) the dependence of photosynthesis on the light flux I is presented in the form:

$$P'_p = P_{\max} \frac{I}{I_{opt}} \exp\left(1 - \frac{I}{I_{opt}}\right) \tag{11}$$

where I_{opt} is the light flux at which maximum photosynthesis $P_{\max}$ is achieved.

In one of the variants of the model relation 11 was replaced by the following expression, nearer to the empirical data

$$P_p = P'_p(1 - 10^{-0.1n})^{0.6}$$

which approximates the equation of Michaelis–Menten, used for analogous purposes by O'Connor, St John, and Di Toro (1968).

Besides the limitations imposed by light intensity and concentration of nutrients, the limitation of phytoplankton production by maximum rate of proliferation was taken into account. On the basis of experimental and published data the decrease in the maximum diel value of the P/B coefficient was assumed to change from 5 at the initiation of the community to 1 on the sixteenth day of its existence.

A three-layer model was used in the description of the hydrology. A high coefficient of turbulent diffusion was preset for the surface layer ($0 \leqslant z \leqslant z_1$); its value decreased sharply within the pycnocline ($z_1 \leqslant z \leqslant z_2$) and increased again only at greater depths ($z > z_2$). The values z_1 and z_2 determining the depth of occurrence and the thickness of the discontinuity layer were assumed in the model to depend on the time of development of the ecosystem, z_1, increasing from 10 to 120 m on the fiftieth day of its existence.

The rate of sinking of phytoplankton (ω_1), bacteria (ω_2), and detritus (ω_3) was assumed to depend on the viscosity of the water, determined by vertical temperature distribution, according to the above assumptions of changes with time in the depth of the discontinuity layer.

The vertical migrations of a part of the total zooplankton were simulated in the model not by direct transport of elements as suggested in Lyapunov's model but by transferring part of the food requirements of the elements (f_4, s_1, s_2, s_3) inhabiting the 0–50 m layer to the 50–150 m layer. The fraction transferred, K_m, increased with time from 0.02 to 0.1 which corresponds to the increase in the intensity of zooplankton migrations with the development of the community.

The spatial arrangement of the model is shown in Fig. 6. A water column is considered, extending from the surface to a depth of 200 m and divided into 20 elementary cells. The vertical connection between the cells is realized by light penetration (a), turbulent diffusion (k), and rates of phytoplankton (ω_1) and detritus (ω_3) sinking. The diel inflow of light energy from the surface to the ocean (I_0) and nutrient concentration at the depth of 200 m (n_{200}) were assumed to be constant.

The water column was assumed to be displaced horizontally by a constant current with velocities uniformly distributed along the vertical. Vertical water transport was effected by turbulent mixing.

With the accepted assumptions it becomes possible to substitute for the horizontal displacement of the water column containing the simulated ecosystem the time of existence of this ecosystem from some initial state. In our case the initial state corresponds to the moment of ascent of deep water in the upwelling area.

It was assumed that in the initial state ($t = 0$) all the elements of the simulated system are uniformly distributed along the vertical. An investigation of the model

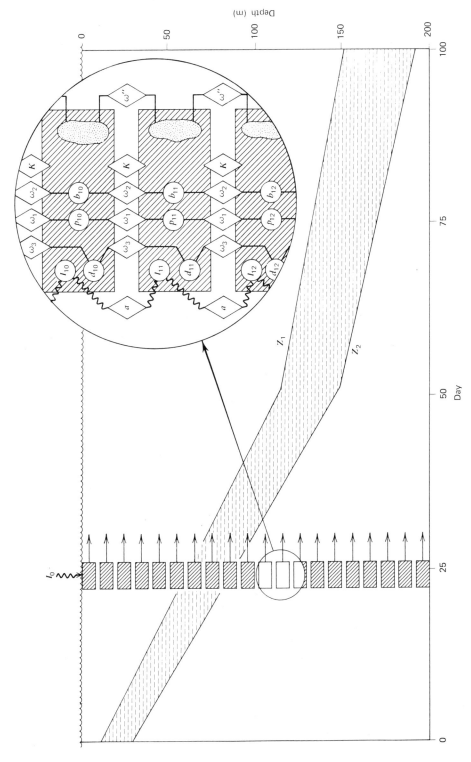

Fig. 6. Disposition of cells in the model simulating the vertical distribution of elements (horizontal shading represents the pycnocline area; designations in the inset circle same as in Figs. 4 and 5).

showed that the system is insensitive to the choice of the initial values of biomass, so that the arbitrariness of these initial values is in a certain measure justified ($n_N = 250$ mg/m^3; $b = 1$ cal/m^3; $p = 0.5$ cal/m^3; $f_1 \cdots f_4, s_1 \cdots s_3 = 0.01$–$0.5$ cal/m^3; $d = 0$).
The system of equations of the model is as follows:

$$\frac{dI}{dz} = -aI \tag{13}$$

$$\frac{\partial n}{\partial t} = -hPp + vd + \eta \sum_{i=p,b,f_1 \cdots f_4, s_1 \cdots s_3} Ri + k\frac{\partial^2 n}{\partial z^2} + \beta\frac{\partial n}{\partial z} \tag{14}$$

$$\frac{\partial p}{\partial t} = \alpha Pp - Rp - \mu p - \sum_{j=f_1 \cdots f_4} C_{pj} + k\frac{\partial^2 p}{\partial z^2} - \omega_1\frac{\partial p}{\partial z} \tag{15}$$

$$\frac{\partial b}{\partial t} = Pb - Rb - \mu b - \sum_{j=f_1 \cdots f_4} C_{bj} + k\frac{\partial^2 b}{\partial z^2} - \omega_2\frac{\partial b}{\partial z} \tag{16}$$

$$\frac{\partial X_i}{\partial t} = U \sum_{j=p,b,d,f_1 \cdots f_4, s_1, s_2} C_{ji} - Ri - \mu i X_i - \sum_{j=f_2 \cdots f_4, s_1 \cdots s_3} C_{ij} \tag{17}$$

$$\frac{\partial d}{\partial t} = \sum_{i=f_1 \cdots f_4, s_1 \cdots s_3} (H_i + \mu i X_i) - \sum_{i=f_1 \cdots f_4} C_{di} + k\frac{\partial^2 d}{\partial z^2} - \omega_3\frac{\partial d}{\partial z} \tag{18}$$

where $\eta =$ coefficient of nutrient release in the process of metabolism
$C_{ij} =$ specific ration of the j food consumer using the i source of food (the share of the i element in the food of the j element)
$\mu =$ coefficient of natural mortality
$X_i =$ biomass of the i element of zooplankton ($i = f_1, f_2, f_3, f_4, s_1, s_2, s_3$)
$H_i =$ unassimilated food of the i element of zooplankton

$$H_i = (1 - Ui) \sum_{j=p,b,d,f_1 \cdots f_4, s_1 \cdots s_3} C_{ji} \tag{19}$$

The other symbols have been explained above or are analogous to the values of Lyapunov's model (equations 2–7).

Figure 7 depicts the changes taking place in the biomasses of elements with time and, consequently, with distance from the zone of water ascent. Phytoplankton reaches its maximum development on the fifth to seventh day, bacterial on the tenth to twelfth day. In its absolute value the phytoplankton maximum (4500 cal/m^2) agrees with values observed in intensive tropical upwellings in the eastern part of the Pacific Ocean, or off the western coast of Africa.

Small herbivores develop more slowly, and large herbivores still more so. The biomass of the latter reaches its maximum only after 30 days. Nevertheless their joint pressure on phyto- and bacterioplankton, together with the slowing down of growth owing to the exhaustion of the nutrients, results in an abrupt decrease in the biomass of phyto- and bacterioplankton. The rate of development of carnivores is still slower than that of large herbivores. The biomass of the various groups of carnivores reaches its maximum only on the thirty-fifth to fiftieth day.

The system reaches a state near stationary on the fiftieth to sixtieth day of its existence. It is characterized by a low biomass of living elements and a balance between phytoplankton consumption and production. Such a mature state of the community, when its elements vary but little in time and, consequently, also in space, is inherent in oligotrophic ecosystems of tropical regions.

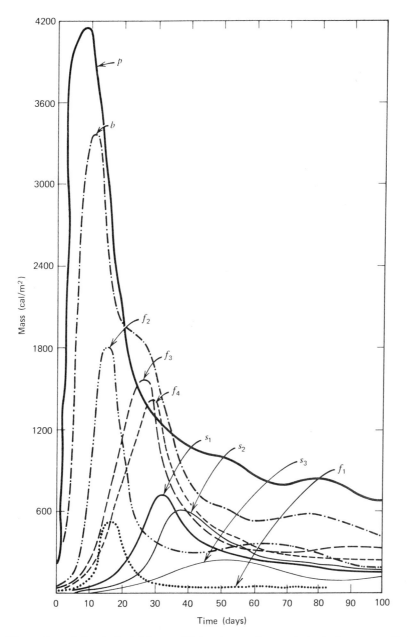

Fig. 7. Change through time in the total mass of living elements in a tropical pelagic ecosystem (0–150-m layer) (Vinogradov et al., 1973). p, phytoplankton; b, bacterioplankton; f_1, protozoans; f_2, microzooplankton; f_3 and f_4, small and large herbivores; s_1, *Cyclopoida*; s_2, carnivorous *Calanoida*; s_3, *Chaetognatha*.

In comparing model outputs with observations in the ocean we note a sequence of maxima of phytoplankton, herbivore, and carnivore biomasses with distance from the zones of water ascent, which has been remarked upon by many authors (Vinogradov and Voronina, 1961; Zalkina, 1971, Timonin, 1971). The overwhelming prevalence of herbivores in the communities of regions near upwelling areas is striking; carnivores constitute about half the biomass of total zooplankton in mature communities. The same relationships are observed in reality in the tropical regions of the ocean (Fig. 1) (Timonin, 1969; Gueredrat, 1972).

A comparison of the absolute values of biomasses derived from the model with observations under real conditions is given in Table I. Considering the roughness of the model, its agreement with the original may be regarded as acceptable. The actual quantity of bacteria proved to be considerably higher than that derived from the model, probably because the latter does not consider the growth of bacteria at the cost of consumed allochthonous dissolved organic matter.

The change in the vertical distributions of ecosystem elements with time is shown in Fig. 8. At an early period ($t = 5$), when the total quantity of phytoplankton is already approaching maximum, its biomass is uniformly high in the 10–50 m layer. All the other living elements of the ecosystem have more or less sharply expressed maxima. But after 10 days the stock of nutrients in the upper layer becomes almost entirely exhausted and only at a depth of 10–20 m is a maximum phytoplankton biomass still preserved. Deeper, at the upper boundary of the discontinuity layer, a second ("lower") maximum begins, supported by an inflow of nutrients through the thermocline.

TABLE I

Biomass (cal/m²) of the Elements of a System in the 0–150 m Layer

Elements	Ecosystem of Low Maturity (ages 30–40 days)		A More Mature Ecosystem (ages 60–80 days)	
	Model System 30–40 days	Actual System 30–40 days[a]	Model System 80 days	Actual System 60–80 days[b]
Phytoplankton (p)	1319	1092	2000	827	900
Bacteria (b)	1673	864	4100	564	2180
Nauplii (f_2)	394	303	321	300	?
Small herbivores (f_3)	1338	612	525	290	74
Large herbivores (f_4)	1416	726	420	252	164
$\sum f$	3184	1641	1266	842	
Carnivorous *Cyclopoida* (s_1)	624	491	495	203	236
Carnivorous *Calanidae* (s_2)	288	600	610	191	175
Carnivorous *Chaetognatha* + *Polychaeta* (s_3)	184	183	15	102	51
$\sum s$	1096	1274	1120	496	462

[a] "Vitjaz" station N 6429: 04°30′N, 142°30′E.
[b] "Vitjaz" station N 6493: 13°N, 140°E.

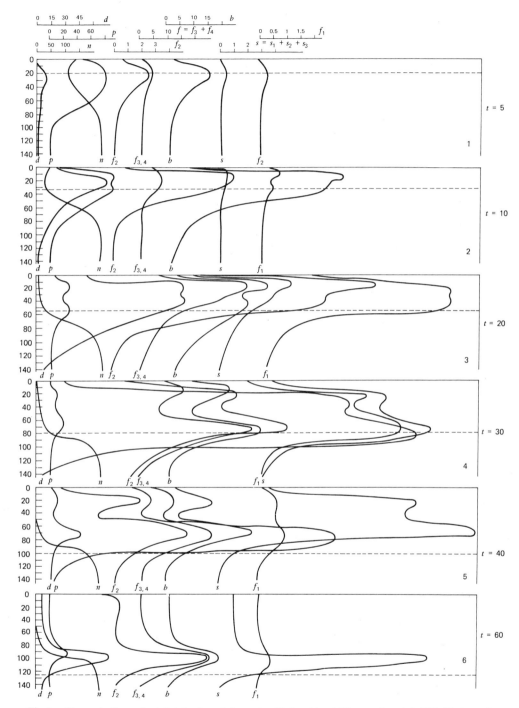

Fig. 8.   Change in the vertical distribution of elements with time (days) (Vinogradov et al., 173). Horizontal hatch line indicates the upper boundary of the pycnocline. Designations same as in Fig. 7; note that different scales are used for different elements.

As has been shown in Section 3, such a structure with two maxima at different depths is typical of the vertical distribution of nearly all the elements. It is most apparent on the twentieth to thirtieth day of the existence of the ecosystem.

With the depletion of the stock of nutrients in the surface layer, the vertical transport of nutrients from below the discontinuity layer becomes most important for the functioning of the ecosystem. The lower maximum now exceeds the upper one. With the shifting of the thermocline to a depth of 80–100 m and more, the illumination at its upper boundary is no longer sufficient for an intensive development of phytoplankton. A situation arises when the lower maximum "tears away" from the thermocline and its position becomes dependent on the intensity of illumination from above and the flow of nutrients from below (see Section 3 and Fig. 2).

According to the model a characteristic feature of oligotrophic and ultraoligotrophic regions is a nearly complete disappearance of the upper maxima of all elements as the community reaches a degree of maturity (sixtieth to eightieth day of existence). In fact, the upper maximum is absent only in phytoplankton, whereas in zooplankton even in the oligotrophic regions the upper maximum coexists with the lower maximum.

This model permits us to evaluate the effect of the assumed values of the parameters on the behavior of the system. The criterion used was the standard deviation of the simulated vertical distribution of living elements at $t = 40$ from the values observed at a "*Vitjaz*" station (04°29′N, 142°30′E)

$$\phi = \sqrt{\frac{1}{N} \sum_{z=0}^{N} (f_z - \hat{f}_z)^2}$$

where f_z = calculated biomass of herbivores
$\hat{f}_z$ = observed biomass of herbivores
N = number of sampled layers

The model was found to be most sensitive to the maximum P/B coefficient for phytoplankton, especially its decrease. Another parameter, extremely important for the functioning of a pelagic ecosystem, proved to be the coefficient of solar radiation absorption by water (a). The system is less sensitive to the coefficients of food assimilation, U, and values of metabolism, R.

The results show that many essential features of vertical and temporal distribution of the elements of pelagic ecosystems in tropical regions have been successfully reproduced and thus explained. There are, however, several important aspects that remain outside the range of the model. Thus the assumption of a constant horizontal velocity of currents in the 0–200 m layer is not in good agreement with the real picture observed in the ocean. The model does not account for the zone of upwelling where the vertical component of the vector of current velocities cannot be disregarded. Finally, the model does not include the last trophic links of the food chain: fishes, squids, and aquatic mammals, which are of great interest from the viewpoint of their utilization by man. In the following sections, therefore, we discuss some other possible approaches to the construction of mathematical models of pelagic ecosystems.

7. A Model Simulating the Areal Distribution of Elements

Another approximation to reality presents the simulation of changes in pelagic ecosystems associated with a given area rather than a particular water mass. This approach is especially effective when dealing with a complex system of surface currents and with plankton composed of different faunistic groupings.

The Japan Sea was chosen as a prototype for this model (Menshutkin, Vinogradov, and Shushkina, 1974). In this basin, the shallows occupy a very small area and the open part of the sea may be considered as a continental shelf pelagic ecosystem. The strong and inconstant flow of water through the Tsushima Strait creates considerable horizontal inhomogeneities which justify the choice of an areal model. Among the organisms passively carried with the currents through the Tsushima Strait are actively migrating nekton, entering the Japan Sea during the spring and summer and leaving it in winter.

The model used the ecological and faunistic characteristics of the Japan Sea plankton, and data on the trophic relationships between its major groups were obtained during cruise 52 of the R/V *Vitjaz* and on a number of nearshore expeditions (Pasternak and Shushkina, 1973; Shushkina, 1972).

The sea was divided into 32 squares (length of side 150 km). To each square corresponded an elementary cell of the ecosystem. The inhomogeneities in the horizontal distribution of abiotic and biotic elements within each cell were assumed to be insignificant. The transport of dissolved matter and particulate detritus, as well as of phyto- and zooplankton from one cell to another, was effected by currents. The scheme of currents (after Sisova, 1961) was an approximation determined by the size of the cells (Fig. 9). The seasonal changes in the velocity of the Tsushima current (after Nishimura, 1969) were assumed to be sinusoidal, with a maximum in September.

Temporal and latitudinal changes in solar radiation input on the surface of the sea were taken from Budyko (1956). The effective temperature, that is, the temperature at which the inhabitants of a given cell are living, was assumed to be a function of time and of the coordinates of the cell; the field of temperature was taken from Nishimura (1969).

Each cell of the model included 12 elements (Fig. 10). The group of inert elements was represented by the concentrations of nutrients, n, and detritus, d. Plankton was

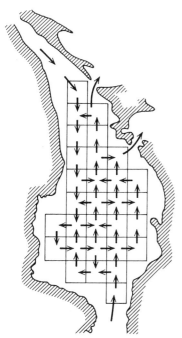

Fig. 9. Scheme of cell disposition in an ecosystem of the Japan Sea. Arrows show transport by currents.

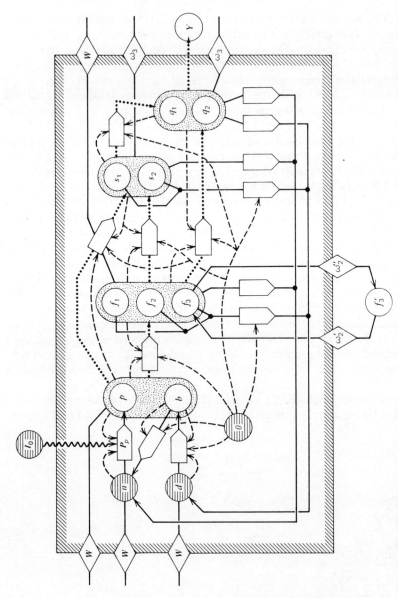

Fig. 10. Block scheme of a cell of a pelagic ecosystem in the Japan Sea. $I_\circ$, Solar radiation; n, nutrients concentration; p, phytoplankton biomass; b, bacteria; d, detritus; f_1, boreal herbivorous epizooplankton; f_2, warm-water herbivorous epizooplankton; f_3, interzonal herbivorous plankton; s_1, carnivores and omnivores of the boreal complex; s_2, carnivores and omnivores of the warm-water complex; q_1, fish; q_2, squid; θ, water temperature; W, transport by currents; Y, fish and squid catch; ω_3, active migrations of fish and squid; ω_2', ω_2'', seasonal vertical migrations of interzonal plankton.

912

divided into three groups of organisms: microplankton, mesoplankton I (herbivores), and mesoplankton II (carnivores and omnivores). Microplankton consisted of phytoplankton, p, and bacterioplankton, b. Mesoplankton I included boreal epiplankton, f_1 (*Psuedocalanus elongatus, Oithona similis, Acartia clausi*, appendicularia, larvae of bottom invertebrates, etc.); warm-water epiplankton, f_2 (*Paracalanus parvus, Clausocalanus pergens, Calanus tenuicornis, Mecynocera clausi*, etc.); and interzonal plankton, f_3 (*Calanus plumchrus, C. cristatus, C. finmarchicus, C. pacificus*, etc.). The seasonal vertical migrations of interzonal herbivores were accounted for and a special element, f_3, was recognized, the "wintering stock," that is, the interzonal plankton found in winter at depths greater than 200 m. It was assumed that the "wintering stock" is not transported by the system of surface currents.

Among the mesoplankton carnivores and omnivores (euphausiids, *Metridia*, jellyfish, etc.) two elements were recognized: species of the boreal complex, s_1, and species of the warm-water complex, s_2.

In contrast to the plankton elements passively transported from cell to cell by currents, the nekton elements (fish, q_1, and squid, q_2) are capable of active displacement. For want of a better hypothesis the principle was adopted that fish and squid search for maximum food concentrations in their own and neighboring cells. The rate of displacement was set at 10 km/day for fish and 15 km/day for squid.

A specific feature of the model under review, as compared with the model discussed in the preceding section, is that the elements of a cell are united into groups. A group comprises elements with similar sizes of individuals, types and spectra of feeding, and interactions with predators. In substance the organisms united in a group occupy the same broad ecological niche, but the responses to abiotic factors, energetic characteristics, and patterns of migrations remain specific for each element. This recognition of trophic groups greatly simplifies the model algorithms and is not inconsistent with the biological essence of the simulated process.

The time step of the model is 5 days. The runs of the model started from an arbitrary initial state with a uniform distribution of all the elements of the ecosystem in all the squares. After functioning 3 yr the system reached a near-stationary state, in which the difference in the states of the elements on the same date of a given year and the following year did not exceed 10%.

Figure 11 gives data on the monthly production of phyto- and zooplankton, bacteria, and nekton, averaged for the entire surface of the Japan Sea. Phytoplankton production shows two fairly well expressed maxima. The high value of the second maximum depends primarily on the increase in advection of nutrients through the Tsushima Strait during the autumn. According to the model the average annual production of phytoplankton in the Japan Sea is 1280 kcal/m^2, a value which seems highly probable, judging from the data obtained by Sorokin and Koblentz-Mishke (1958). It is difficult to estimate the reliability of the model values for bacterioplankton and zooplankton production as there are no observations for comparison. The production of higher trophic levels (fish and squid) can be compared with data on commercial catches (Moiseev, 1969). According to the model the annual catch of fish (30% of the standing stock, assuming 1 kcal/g of wet weight) in the Japan Sea amounts to 0.82×10^6 tons and the catch of squids to 0.54×10^6 tons, as compared with the actual catch of fish and squids of about 1×10^6 tons, and the potentially possible catch of 1.23×10^6 tons (Gulland, 1970).

Model output for the seasonal distribution of phytoplankton, zooplankton, and nekton biomass over the Japan Sea are shown in Fig. 12. Unfortunately, the data available on the actual spatial distributions of these biomasses are not sufficient to

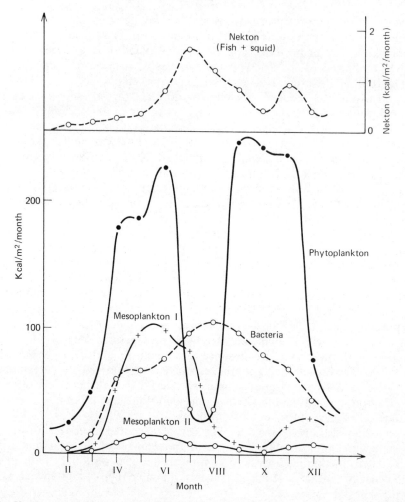

Fig. 11.   Yearly course of average monthly production of some elements in an ecosystem of the Japan Sea (Menshutkin, Vinogradov, and Shushkina, 1974).

compare them with the output of the model, so that the degree of reliability of the model still remains an open question. Nevertheless it may be stated that the picture of zooplankton distribution presented by the model is not inconsistent with the available, rather fragmentary, observations on the distribution of the biomass of zooplankton. The distribution of squid also agrees in its general features with observed data (Zuev and Nesis, 1971; anonymous, 1972).

Thus the model shows that a quantitative description of the seasonal course and distribution of the elements of a pelagic ecosystem over a complex sea area is potentially achievable. It must be kept in mind that models of this type are extremely sensitive to the completeness of hydrological data. Not only a general scheme of currents is needed, but also the values of mass transport in each square and the value of the coefficient of horizontal turbulent diffusion are required. These requirements impose definite restrictions in regard to the sphere of their application.

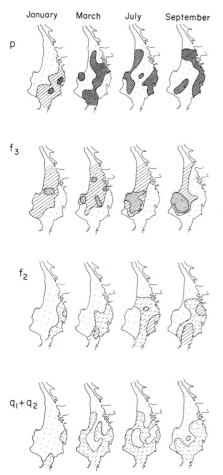

Fig. 12. Model output for the distribution of some elements of a pelagic ecosystem in the Japan Sea (Menshutkin, Vinogradov, and Shushkina, 1974). p, Phytoplankton, f_3, interzonal herbivores; f_2, herbivorous zooplankton (southern forms); q, nekton. 1, Biomass greater than 10 kcal/m^2; 2, 10–5 kcal/m^2; 3, 5–1 kcal/m^2; 4, 1–0.1 kcal/m^2. Points indicate the centers of cells.

8. A Model Simulating the Volumetric Distribution of Elements

The transition to three-dimensional models is a natural development of the models of vertical and areal distribution considered above. This transition is especially topical in dealing with zones of upwelling or oceanic regions with strong, deep countercurrents.

Thus, for instance, in the eastern part of the Equatorial Pacific Ocean the structure of the field of current is rather complex. To represent the situation by a horizontal model would be an unwarrantable simplification, but as we did not dispose of sufficient information on the field of currents and coefficients of turbulent diffusion in this part of the ocean we had to renounce the construction of a portrait model and content ourselves with an extremely simplified and approximate scheme having a qualitative rather than a quantitative character.

Figure 13 gives a block scheme of an elementary cell of the ecosystem and of the disposition of cells in space. Because of the qualitative character of the model the structure of the trophic chain of each cell was simplified as far as possible. It comprises

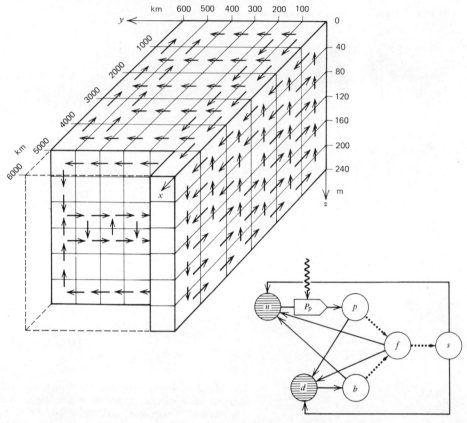

Fig. 13.   Block scheme of the cell of a three-dimensional model of a pelagic ecosystem (designations same as in Fig. 4) and the spatial disposition of the cells. Arrows show transport by currents.

nutrients, n, detritus, d, phytoplankton, p, bacterioplankton, b, herbivores, f, and carnivores, s. The interaction between the elements in an elementary cell is described in the same manner as in the model of Section 6.

Altogether the model contains 216 elementary cells forming a parallelopiped extending along the equator. The distance between the centers of the cells is 1000 km along axis x (longitude), 100 km along axis y (latitude), and 40 m along axis z (depth). The scheme of currents was chosen so as to imitate the surface equatorial current along axis x from the origin and the Cromwell Current gradually approaching the surface and directed in parallel to axis x but toward the origin. In the region directly adjoining the plane of the equator (plane xz) the current velocity has an important vertical component responsible for the ascent of water to the surface.

A surface current directed toward the origin, parallel to x for a distance of 500–600 km northward of the equator, simulates the Northern Equatorial Countercurrent, in which region is situated the zone of convergence. The model stipulates a backflow of surface waters from the equator (motion along y from the origin) and a corresponding deep counterflow (parallel to y toward the origin). Maximum vertical current velocity was set in the model at 5 m/day (0.6×10^{-3} m/sec) and maximum horizontal velocity at 50 km/day. The scheme of currents was selected in such a manner that the conservation of mass was satisfied for each cell. The model took into account the transport of elements by currents and the gravitational settling of detritus.

The initial state of the model was a uniform distribution of all elements through the whole volume. During the fortieth to fiftieth days of the run, the model reached a stable state when the inflow of nutrients from the lower surface of the parallelopiped was equal to the settling of detritus through the same surface.

Figure 14 shows the distribution of elements. A maximum concentration of phytoplankton in the region of most intensive deep-water ascent (at the origin) agrees with the concept of the role of upwelling in the formation of a plankton community. With progress westward along the equator the concentration of phytoplankton decreases and its maximum sinks from the surface to a depth of 80 m. This agrees with observations in the ocean and with the data obtained in the two-dimensional model discussed in Section 6.

A characteristic feature of the distribution of herbivores is the formation of two areas of concentration. The first is situated near the equator, to the west of the intensive upwelling; the second extends northward of the equator, and is characterized

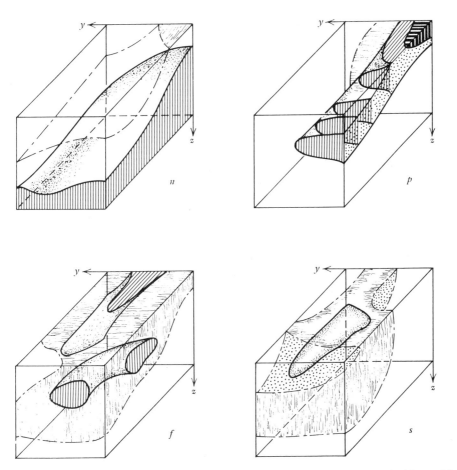

Fig. 14. Spatial distribution of the concentrations of nutrients (n), phytoplankton (p), herbivores (f), and carnivorous zooplankton (s). Higher concentrations of elements are indicated by denser hatching.

by a shift of the maximum of herbivore concentration to greater depths in the zone of convergence.

The region of predominance of carnivores forms a large band, extending from east to west for a distance of 500–600 km northward from the equator. Maximum concentration is observed at a depth of 60–120 m. At a stable state the carnivores are distributed through nearly the entire volume of water considered, down to 200 m. This type of distribution is not typical of herbivores or phytoplankton. It must be remembered that the model does not take into account the vertical migrations of zooplankton.

The comparison of the theoretical distribution of elements, with observations in the equatorial part of the eastern Pacific Ocean (Vinogradov and Voronina, 1963; Voronina, 1964; Blackburn et al., 1970) shows good agreement between distribution of phytoplankton and zooplankton in the model and in nature. The decrease in phytoplankton concentration in a westward direction along the equator is clearly expressed in the model despite the approximations made. The maximum of carnivorous zooplankton in the zone of convergence at 5°N also is reflected in the model, although another maximum of carnivores existing near the equator between 150 and 160°W is not recognized in the model. Perhaps this may be explained by an overestimation of the values of horizontal current velocities in the model, or by the fact that only carnivores with a typical size up to 10 mm were considered.

Although it is extremely schematic and yields data of a qualitative nature, the model demonstrates that it is possible to create spatial models of pelagic ecosystems. In our experience, however, in developing models of this type we were faced with many substantial difficulties. Firstly, the elucidation of the spatial vector field of current velocities even under stationary conditions presents a complex and not always soluble hydrographic problem. Secondly, processes of turbulent transport cannot be ignored as has been done in this model, as a first approximation. Moreover, new complexities arise with the consideration of horizontal and vertical turbulent transport. The consideration of the active displacements of zooplankton and nekton, especially important in three dimensions, creates implications that do not arise in one- or two-dimensional models.

9. Perspectives in Research on Open-Sea Ecosystems

The models of pelagic ecosystems described in this chapter are important not only from the viewpoint of the actual results obtained. No less important is the fact that the application of mathematical modeling in oceanographic research has substantially modified our approach to the collection and treatment of samples and to the organization of fieldwork.

If, indeed, a mathematical model of processes in the pelagic zone of the ocean is to be recognized as a valid generalization, then the need for a complex and comprehensive study of the object being modeled becomes imperative. Omissions or insufficient knowledge of processes referring to one of the elements, may depreciate efforts directed toward the study of other elements of the systems.

The method of mathematical modeling demands that data on the functioning of all the elements of the ecosystem be presented in a quantitative form relevant to the model, and this in itself contributes to the effectiveness of research. These requirements provide a practical foundation for an objective planning of scientific work using quantitative methods.

The use of mathematical models in describing the dynamics of pelagic ecosystems opens to oceanography possibilities for predicting the changes that may occur in ecosystems subjected to various abiotic, biotic, and, particularly, anthropogenic effects.

In the near future, the results obtained will permit the construction of (a) models of ecosystems in the productive layer of cold-water regions with a well-developed seasonal cycle of production; (b) models of ecosystems in oceanic gyres covering considerable areas of the ocean; and (c) models taking into account the effect of commercial exploitation and anthropogenic pollution of the ocean.

Still greater difficulties probably attend the modeling of communities for the entire water column, including the populations of the deepest layers of the ocean.

The creation of adequate models of ecosystems for the different regions of the ocean will probably necessitate not only widening present investigations but also devising instruments for continuous, automated collection of information on plankton populations and environmental parameters, as well as devices for quantitative sampling of macroplankton and nekton. Methods of experimental investigations on feeding, production, and energetic balance in the different elements of a community will have to be perfect and the spatial distribution of populations will have to be investigated in greater detail. The major deterrent is our insufficient knowledge, not so much of the biological objects themselves as of their hydrological background and its variability. Without concrete data on the system of currents and absolute values of water transport in any given region, without information about their variability, no three-dimensional model of ecosystems may be constructed with any approximation to reality. But precisely such models are needed as criteria in the elaboration of definite measures for controlling oceanic communities or recommendations for fishery regulation.

The practical application of the present models is hindered also by the impossibility of evaluating their adequacy in regard to the final trophic links of the system: the pelagic fish exploited by the fisheries. Existing methods do not permit us to assess the total biomass (not only the aggregations but also the dispersed parts of these populations, and their distribution in different regions of the ocean). We do not doubt, however, that these impediments will be overcome during the next few years.

It is now obvious that no further important increase in the amount of animal protein obtained from the ocean, urgently needed by the ever-growing humanity, can be achieved by a mere increase of fishing fleets. There is a pressing need for a transition from "free hunting" to a rational management of the biological productivity of the ocean, to the same type of economy that is practiced on land. A way out of the difficulties presented by the complexity of marine systems as objects of exploitation, and by the techniques of their investigation, is to develop the cybernetics of marine biology (Monin et al., 1974), the use of mathematical models for predicting the behavior of systems subject to purposeful modifications of one or another of their parameters.

References

Anonymous 1972. Studies of Japanese common squid (*Todorodes pacificus*). *Bull. Tech. Counc. Agric. Fish. Res.*, **57**.

Arashkevich, E. G., 1972. Vertical distribution of different trophic groups of copepods in the boreal and tropical regions of the Pacific ocean. *Oceanology*, **12**, 315–325.

Banse, K., 1964. On the vertical distribution of zooplankton in the sea. In *Progress in Oceanography*. Vol. 2. Pergamon Press, New York, pp. 127–212.

Beklemishev, C. W., M. N. Petrikova, and H. J. Semina, 1961. On the cause of the buoyancy of plankton diatoms. *Tr. Inst. Oceanol.*, **51**, 33–36.

Blackburn, M., R. M. Laurus, R. W. Owen, and B. Zeitschel, 1970. Seasonal and areal changes in standing stocks of phytoplankton, zooplankton and micronecton in the eastern tropical Pacific. *Mar. Biol.*, **7**, 14–31.

Cassie, M. R., 1959. An experimental study of factors inducing of aggregation in marine plankton. *N.Z. J. Sci.*, **2**, 339–365.

Cushing, D. H., 1959. On the nature of production in the sea. *Fish. Invest. Lond.*, Ser. 2, **22**, 40 pp.

Cushing, D. H., 1962. Patchiness. *Rapp. P.-V. Reun. Cons. Perm. Int. Explor. Mer*, **153**, 152–164.

Cushing, D. H., 1969. Models of the productive cycle in the sea. In *Morning Review Lectures of the Second International Oceanographic Congress.* UNESCO, pp. 103–115.

Dugdale, R. C., 1967. Nutrient limitation in the sea: dynamics, identification and significance. *Limnol. Oceanogr.*, **12**, 685–695.

Gitelzon, I. I., L. A. Levin, A. P. Shevyrnogov, R. N. Utyushev, and A. S. Artemkin, 1971. Bathyphotometric sounding of the pelagic zone of the ocean and its use for the studies of spatial structure of a biocoenossis. In *Functioning of Pelagic Communities in the Tropical Regions of the Ocean.* M. E. Vinogradov, ed. Nauka, Moscow, pp. 50–64.

Gulland, J. A., ed., 1970. The fish resources of the ocean. *FAO Fish Tech. Pap. 97.*

Ivlev, V. A., 1955. Experimental ecology of nutrition of fishes. Pishchepromisdat, Moscow, p. 252. (English transl. by D. Scott, Yale University Press, New Haven, 1961.)

Jorgensen, C. B., 1962. The food of filter feeding organisms. *Rapp. P.-V. Reun. Cons. Perm. Int. Explor. Mer*, **152**, 99–105.

Koblentz-Mishke, O. I., A. M. Tsvetkova, M. M. Gromov, and L. I. Paramonova, 1971. Primary production and chlorophyll "a" in the western Pacific. In *Functioning of Pelagic Communities in the Tropical Regions of the Ocean.* M. E. Vinogradov, ed. Nauka, Moscow, pp. 70–79.

Lyapunov, A. A., 1971. On the construction of a mathematical model of balance correlations in the ecosystem of the tropical ocean. In *Functioning of Pelagic Communities in the Tropical Regions of the Ocean.* M. E. Vinogradov, ed. Nauka, Moscow, pp. 13–24.

Lyapunov, A. A. and S. V. Yablonsky, 1963. Theoretical problems of cybernetics. *Probl. Cybernet.*, **9**.

Margalef, R., 1958. Temporal succession and spatial heterogeneity in natural phytoplankton. In *Perspectives in Marine Biology.* University of California Press, Berkeley, pp. 323–349.

Margalef, R., 1967. Some concepts relative to the organization of plankton. *Oceanogr. Mar. Biol. Ann. Rev.*, **5**, 257–289.

Margalef, R., 1968. *Perspectives in Ecological Theory.* University of Chicago Press, 111 pp.

Menshutkin, V. V., M. E. Vinogradov, and E. A. Shushkina, 1974. A mathematical model of the pelagic ecosystem of the sea of Japan. *Oceanology*, **14**, 880–887.

Moiseev, P. A., 1969. Living resources of the World Ocean. Pishepromisdat, Moscow, pp. 3–339.

Monin, A. S., V. M. Kamenkovich, and V. G. Kort, 1974. The variability of the world ocean. Gidrometeoisdat, Leningrad, 262 pp.

Nishimura, S., 1969. The zoogeographical aspects of the Japan Sea. Part. V. *Publ. Seto Mar. Biol. Lab., Kyoto Univ.*, **17**, 67–124.

O'Connor, D. J., J. P. St John, and D. M. Di Toro, 1968. Water quality analysis of the delaware river estuary. *J. Sanit. Eng. Div.*, **94**, 1225–1252.

Odum, E. P., 1963. *Ecology.* Holt, Rinehart & Winston, New York.

Odum, H. T., 1971. *Environment, Power, and Society.* Wiley-Interscience, 331 pp.

Pasternak, F. A. and E. A. Shushkina, 1973. Biological investigations in the 52nd cruise of r/v "Vitjaz." *Oceanology*, **13**, 737–740.

Patten, B. C., 1968. Mathematical models of plankton production. *Int. Rev. Ges. Hydrobiol.*, **53** (3).

Petipa, T. A., E. V. Pavlova, and I. I. Sorokin, 1971. Radiocarbon studies of feeding of the mass plankton forms in the tropical Pacific. In *Functioning of Pelagic Communities in the Tropical Regions of the Ocean.* M. E. Vinogradov, ed. Nauka, Moscow, pp. 123–141.

Riley, G. A., H. Stommel, and D. F. Bumpus, 1949. Quantitative ecology of the plankton of the western north Atlantic. *Bull. Bingham Oceanogr. Coll.*, **12**.

Romanenko, V. I., 1964. Heterotroph CO_2 assimilation by bacterial flora of water. *Microbiology*, **33**, 679–683.

Rudjakov, J. A., 1971. Details of the horizontal distribution and diel vertical migrations of Cypridina sinuosa (Müller) (Crustacea, Ostracoda) in the Western Equatorial Pacific. In *Functioning of Pelagic Communities in the Tropical Regions of the Ocean.* M. E. Vinogradov, ed. Nauka, Moscow, pp. 213–227.

Ryther, J. H., 1956. Photosynthesis in the ocean as a function of light intensity. *Limnol. Oceanogr.*, **1**, 61–70.

Shushkina, E. A., 1972. Intensities of production and the use of assimilated food for growth of the Japan sea mysids. *Oceanology*, **12**, 326–337.

Shushkina, E. A. and B. Y. Vilenkin, 1971. Respiration rate of planktonic crustaceans from the tropical Pacific. In *Functioning of Pelagic Communities in the Tropical Regions of the Ocean*. M. E. Vinogradov, ed. Nauka, Moscow, pp. 117–171.

Shushkina, E. A. and E. P. Pavlova, 1973. Rates of metabolism and production of zooplankton in the Equatorial Pacific. *Oceanology*, **13**, 339–345.

Shushkina, E. A., Yu. A. Kislyakov, and F. A. Pasternak, 1974. Combination of the radiocarbon method with mathematical simulation for the estimation of productivity of the marine zooplankton. *Oceanology*, **14**, 319–326.

Shushkina, E. A. and I. A. Sokolova, 1972. Caloric equivalents of the body mass of the tropical organisms from the pelagic part of the ocean. *Oceanology*, **12**, 860–867.

Sisova, I. V., 1961. Currents of the Sea of Japan. In *Basic Features of Geology and Hydrology of the Sea of Japan*. Publishing House of the USSR Academy of Sciences, Moscow.

Sorokin, Y. I., 1971. On bacterial numbers and production in the water column on the Central Pacific. *Oceanology*, **11**, 105–116.

Sorokin, Y. I. and O. I. Koblentz-Mishke, 1958. Primary production in the Japan Sea and the Pacific near Japan in spring 1957. *Dokl. Akad. Nauk SSR*, **122**, 1018–1020.

Steele, J. H., 1962. Environmental control of photosynthesis in the sea. *Limnol. Oceanogr.*, **7**, 137–150.

Sushchenya, L. M., 1972. Qualitative regularities in crustacean metabolism. Kiev, *Naukova Dumka*, 3–196.

Timonin, A. G., 1969. Structure of the pelagic communities. The quantitative relationship between different trophic groups of plankton in the frontal zones of the tropical ocean. *Oceanology*, **9**, 846–856.

Timonin, A. G., 1971. The structure of plankton communities of the Indian Ocean. *Mar. Biol.*, **9**, 281–289.

Timonin, A. G., 1975. Vertical microdistribution of zooplankton in the tropical western Pacific. *Tr. Inst. Oceanol.*, **102**, 245–259.

Vinogradov, M. E., 1968. *Vertical Distribution of the Oceanic Zooplankton*. Nauka, Moscow, 320 pp.

Vinogradov, M. E., 1974. The 17th cruise of the r/v "Akademik Kurchatov" (Investigation of the plankton communities in the east Pacific upwelling regions). *Oceanology*, **14**, 941–946.

Vinogradov, M. E. and E. G. Arashkevich, 1969. The vertical distribution of interzonal copepod-filtrators and their role in the communities of different depths in the North-west Pacific. *Oceanology*, **9**, 488–499.

Vinogradov, M. E., I. I. Gitelzon, and Y. I. Sorokin, 1970. The vertical structure of a pelagic community in the tropical ocean. *Mar. Biol.*, **6**, 187–194.

Vinogradov, M. E., V. V. Menshutkin, and E. A. Shushkina, 1972. On mathematical simulation of a pelagic ecosystem in tropical waters of the ocean. *Mar. Biol.*, **16**, 261–268.

Vinogradov, M. E., V. F. Krapivin, V. V. Menshutkin, B. S. Fleishman, and E. A. Shushkina, 1973. Mathematical simulation of functioning of the pelagic ecosystem in the tropical ocean (based on the materials of 50th cruise of the r/v "Vitjaz"). *Oceanology*, **13**, 852–866.

Vinogradov, M. E. and N. M. Voronina, 1963. Quantitative distribution of plankton in the upper layers of the Pacific equatorial currents. I. *Tr. Inst. Oceanol.*, **71**, 22–59.

Vinogradov, M. E. and N. M. Voronina, 1964. Some peculiarities of the plankton distribution in the Pacific and Indian Ocean's Equatorial currents. *Okeanol. Issled.*, **13**, 128–136.

Voitov, V. I. and O. V. Kopelevich, 1971. Optical studies in the western Pacific. In *Functioning of Pelagic Communities in the Tropical Regions of the Ocean*. Nauka, Moscow, pp. 25–34.

Vollenweider, R. A., 1965. Calculation models of photosynthesis depth curves and some implications regarding day rate estimates in primary production. *Mem. Inst. Ital. Idrobiol.*, **18**, Suppl.

Voronina, N. M., 1964. The distribution of macroplankton in the waters of equatorial currents of the Pacific ocean. *Oceanology*, **4**, 884–895.

Winberg, G. G. and S. I. Anisimov, 1966. A mathematical model of an aquatic ecosystem. In *Photosynthesizing System of High Productivity*. Nauka, Moscow, pp. 213–223.

Zalkina, A. V., 1971. Vertical distribution and diurnal vertical migration of species of Cyclopoida (Copepoda) in the western equatorial Pacific. In *Functioning of Pelagic Communities in the Tropical Regions of the Ocean*. M. E. Vinogradov, ed. Nauka, Moscow, pp. 204–212.

Zalkina, A. V., 1975. Vertical distribution and diel migrations of Cyclopoida (Copepoda) in the northern equatorial current and in the Sulu Sea. *Tr. Inst. Oceanol.*, **102**, 260–279.

Zuev, G. V. and K. N. Nesis, 1971. Squids (biology and fishery). Pischepromizdat, Moscow, 360 pp.

24. A BIOLOGICAL SKETCHBOOK FOR AN EASTERN BOUNDARY CURRENT

J. J. WALSH

I should have been a pair of ragged claws
Scuttling across the floors of silent seas.

—T. S. ELIOT

1. Introduction

Mathematical models are formal representation of one's understanding of the phenomenon under study. Such formal representation consists of a conscious abstraction of the processes considered to be important, with the wisdom of one's selection of these processes demonstrated in the validity of the model. One could imagine oneself to be a copepod or a crab and ask what factors and processes are important in the life history and in the survival of this anthropomorphic orgainsm. In a search for both completeness and generality, however, a list of "essential" variables might ensue which could be as complex as the "real" world. An appropriate mathematical translation of this list, that is, a model, might then be nearly impossible to construct, too costly to run, and as hard to comprehend as the "real" world.

An alternative approach, initially lacking in generality and perhaps biological realism, is to pose a specific question or problem that might justify the use of systems analysis in place of direct observation. If such a question is properly posed and non-trivial, and if the solution neither presently exists nor can be obtained by simple experimentation, simulation models may provide an answer through proper manipulation of available data sets. At the very least, a number of critical field and laboratory experiments will then be identified in this iterative process (Walsh, 1972). Models of varying complexity are presented elsewhere in this volume and they can be considered as quasi-solutions to questions of varying complexity about the interactions of the ocean. The purpose of this chapter is to discuss the use of simulation models at the ecosystem level of complexity within the eastern boundary currents.

On a relative basis, understanding of physical subsets of oceanographic data does not depend on a multitude of interacting factors, as does consideration of biological subsystems, which are forced by the physical and chemical variables. A large number of measurements on a few physical properties such as wind, density, pressure, and current velocity have led to a long history of physical theory involving linear approximations and analytical solutions, which have generated an impressive literature on both the general physics of the oceans and the eastern boundary currents (see preceding chapters). Early descriptive and theoretical work on the physical processes of upwelling have been summarized by Wooster and Reid (1963) and Smith (1968), and O'Brien (1975) has discussed the more recent advances in theory.

Biological oceanography, however, has been plagued by problems of serious non-linearity, by only a few measurements of an order-of-magnitude larger number of variables, and by a zeal for the exception rather than the general principle. A decade ago, only a rough description of geographical gradients in marine productivity (Ryther, 1963) and a glimmer of food-chain theory (Riley, 1963) were possible, with

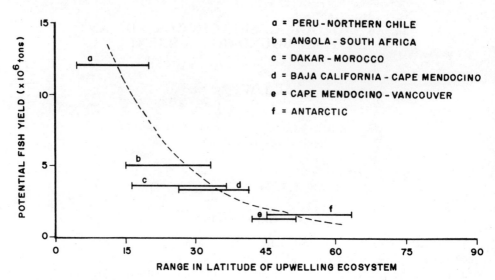

Fig. 1. The potential fish yield of the world's major upwelling regions as a function of latitude (after Walsh, 1974).

almost no ecological analyses focused directly on the eastern boundary currents, except as areas of high fish yield. Recent international programs on upwelling eco-systems have led to large data sets with a more detailed description of the biological dynamics of these systems (Cushing, 1971a; Dugdale, 1972; Walsh, 1972). The advent of large digital computers has also led to increased activity in biological theory, with a growing awareness of the utility of mathematical analysis in biological ocean-ography (Steele, 1974).

Older questions of why upwelling regions are the most productive of the marine pelagic ecosystems (Ryther, 1969) can now, perhaps, be more cogently rephrased (Walsh, 1974) as to why some of the eastern boundary currents have lower potential yield than others (Fig. 1). After more than 10 yr of conducting ecological research on these systems, it is not clear that I or anybody else have the answers to these simple questions, of course, but a start has been made to unravel the temporal and spatial scales of nonlinear interactions that lead to the observed distribution of biological variables in the sea (Walsh, 1976a; Walsh et al., 1976). Moreover, the collaborative efforts of a large number of the oceanographic community have provided input and feedback to a series of simulation models on upwelling ecosystems that my students

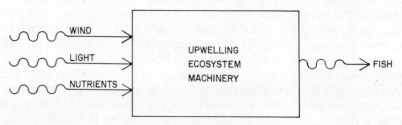

Fig. 2. A conceptual black-box model of time dependent interactions within upwelling ecosystems.

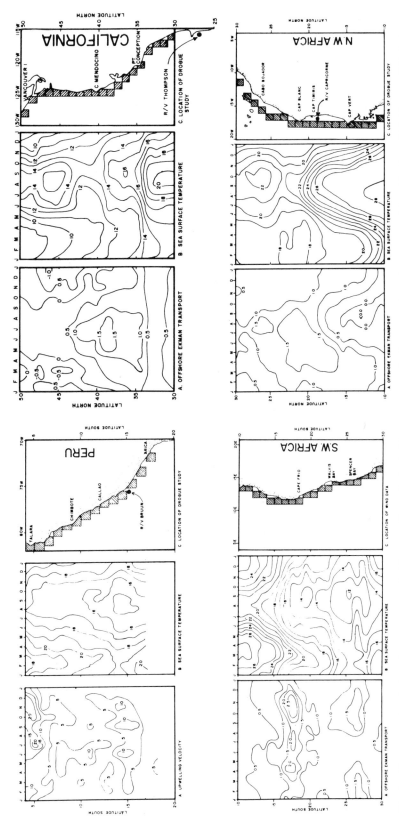

Fig. 3. The seasonal variation of temperature and either upwelling velocity or Ekman transport in the eastern boundary currents (after Maeda and Kishimoto, 1970; Bakun, McLain, and Mayo, 1974; and Wooster, Bakun, and McLain, 1976); the locations of drogue studies, used for validation of the models, are indicated off Peru, Baja California, and Northwest Africa.

925

and I have built over the last 6 yr, and although all individuals cannot be acknowledged, it is stressed that these results stem from gestalt research. A comparative review of these models may thus shed light on answers to the above questions as well as provide a partial index of our present understanding of marine ecosystem dynamics.

A simple, conceptual black-box model (Fig. 2) of oscillations in the forcing functions of the biological community within the euphotic zone of the eastern boundary currents might suffice, for example, to consider the impact of El Niños on the harvest of clupeids (Idyll, 1973). One can further imagine a multivariate analysis that would yield the ultimate in predictability

$$\text{Fish} = a_0 + a_1 \,(\text{wind}) + a_2 \,(\text{light})$$

with appropriate tuning; that is, Smayda (1966) has attempted to predict phytoplankton production in the Gulf of Panama from previous wind velocities during coastal upwelling, and Walsh (1971) has tried to predict phytoplankton distribution across the ecotone of the austral oceanic upwelling system from a series of physical and chemical variables. One could assemble data on the time dependence of wind stress, or Ekman transport, in the eastern boundary currents (Fig. 3), read in average radiation values (Kimball, 1928) to a computer, and accumulate reams of computer printout with the above algebraic paradigm. Unfortunately, none of this exercise would be causal, it would not directly lead to answers of the above questions, and it is of marginal help in providing understanding of systems that involve nonlinear interaction, such as thresholds, prey switching, patchiness, and species succession. The above regression equation might even suggest, for example, that increased wind leads to increased fish, when in fact the opposite evidently occurs (Walsh, 1975c, 1976a).

Increased sophistication of a model requires, however, greater resolution of the data sets and subsequent loss of generality with specification of time and space scales, that is, the important frequency domains of both biological and physical variability (Fig. 4). If one wishes to compare simulation output with field data, additional logic must be built in the model to allow for specification of properties such as seasonal transients, vertical structure, horizontal patchiness, bottom topography, and trophic level shifts of the "real" system. One-dimensional steady-state models are the simplest to build (Walsh and Dugdale, 1971) with increasing technical problems as one proceeds to two- and three-dimensional models involving time-dependent processes.

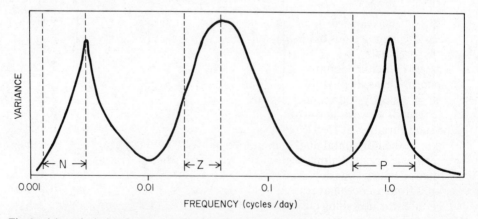

Fig. 4. A hypothetical variance spectrum of physical forcing functions in relation to life cycles of nekton (n), zooplankton (z), and phytoplankton (p) within the eastern boundary currents (after Walsh et al., 1976).

This chapter first considers implications of two different seasonal cases of a three-dimensional, periodically steady-state model off Peru, with its application to Baja California and Northwest Africa data. The dominant modes of seasonal, diel, and broader event frequencies of marine systems (Fig. 4) are then examined with a series of seasonally time-dependent, two-dimensional models of the Antarctic and Sargasso Seas, of daily time-dependent, one-dimensional models of Alta and Baja California, and of a stochastic, two-dimensional model of fish schools, foraging in an idealized upwelling ecosystem.

> And should I then presume?
> And how should I begin?

> —idem

2. Steady-State Dynamics

A. Peru

Consideration of the biological state variables, the network of fluxes, and most importantly, the matrix of forcing functions of an aquatic ecosystem model defines its spatial and temporal boundaries. Although the Peru ecosystem appears to have the least seasonal variability of upwelling in comparison with the other eastern boundary currents (Fig. 3), a definite seasonal cycle of weak and strong upwelling occurs (Meada and Kishimoto, 1970) with respective congregation and dispersal of the anchoveta stocks (Paulik, 1971). Further, the Peru shelf at 15°S is less than half the width of the comparable austral upwelling system in the Atlantic (Fig. 5) with a narrower zonal gradient of nutrients that may partially reflect differences in the predominant cross-stream circulation patterns (Fig. 6). Size-selective grazing stress and increased ecological efficiencies during these periods of weak upwelling (Walsh, 1972, 1976a), together with a one-cell circulation and no phytoplankton sinking loss within a nearshore convergence zone (Fig. 6c), may lead to higher productivity, greater nutrient utilization, and larger terminal yield off Peru (Fig. 1) than in the other eastern boundary currents. A final complication of these forcing functions is the longshore presence of capes (Yoshida, 1967) and canyons (Shaffer, 1974; Walsh et al., 1974) which may lead to local intensification of the vertical velocity field (Fig. 7), invalidating assumptions of longshore homogeneity and net local balance of onshore and offshore flow.

With this specific habitat in mind, the hourly flux of nitrogen, carbon, silica, and phosphorus through phytoplankton, zooplankton, anchoveta, and detritus is considered in a steady-state model over 0–50 m depth, from 0 to 50 km off the coast, during periods of weak and strong upwelling that lead to plumes and parallel bands of the state variables at 15°S (Fig. 3), off the Peru coast (Walsh, 1975c). The grid mesh is 10 km in the horizontal and 10 m in the vertical, with a time step of 1 hr, and the flow regime is oriented along the 100 km axis of a plume at a 30° angle to the coast. Although an hourly time dependence is introduced with periodic functions, the model's spin up is from an ocean at rest, and the daily time invariant state occurs in less than 10 days after the upwelling circulation is imposed. Unless otherwise specified, a right-hand cartesian coordinate system, with x positive eastward, y positive northward, and z positive upward, is used in this chapter.

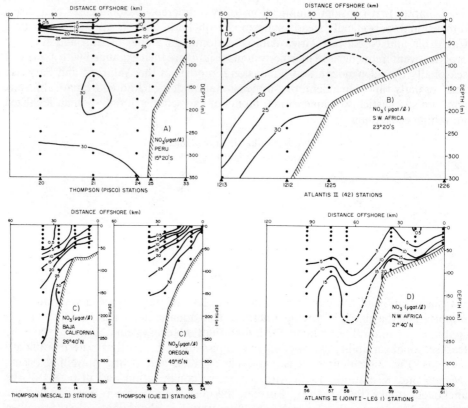

Fig. 5. A comparison of shelf width and zonal gradients of nitrate in the eastern boundary currents (after Walsh, 1976a).

The assumptions and numerical solutions of this model's nonlinear partial differential equations are discussed in much detail by Walsh (1975b, c, 1976b), Walsh and Dugdale (1971), and Walsh and Howe (1976) and only the form and a brief description of the terms are given below. Additional discussion of these types of submodels can be found in other biological chapters of this book as well. The following state equations were solved simultaneously in finite difference form (Morishima, Bass, and Walsh, 1974) of the Peru steady-state model:

Nitrate:

$$\frac{\partial NO_3}{\partial t} = -\frac{\partial uNO_3}{\partial x} - \frac{\partial wNO_3}{\partial z} + \frac{K_y\partial^2NO_3}{\partial y^2} - \frac{V_mNO_3P_n}{K_t + NO_3}$$

where $V_m = (0.11 - 0.02\ NH_3)\sin 0.2618t$

Recycled nitrogen:

$$\frac{\partial NH_3}{\partial t} = -\frac{\partial uNH_3}{\partial x} - \frac{\partial wNH_3}{\partial z} + \frac{K_y\partial^2NH_3}{\partial y^2} - \frac{V_nNH_3P_n}{K_t + NH_3}$$

$$+ (0.67)\frac{G_z(P_n - P_0)Z_n}{P_k + (P_n - P_0)} + (0.67)\frac{G_f(P_n - P_a)F_n}{P_f + (P_n - P_a)}$$

where $V_n = 0.11\sin 0.2618t$.

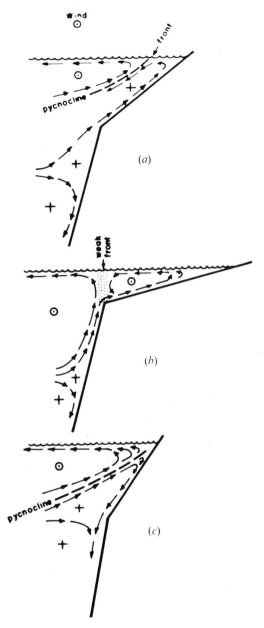

Fig. 6. Alternative cross-stream circulation patterns for upwelling ecosystems: (*a*) two-cell, possibly observed off Oregon; (*b*) two-cell, possibly observed off Northwest Africa; and (*c*) one-cell, possibly observed off Peru: + indicates poleward flow and ⊙ equatorward flow—note the relative widths of continental shelf (after Hagen, 1974; and after Mooers, et al., 1976).

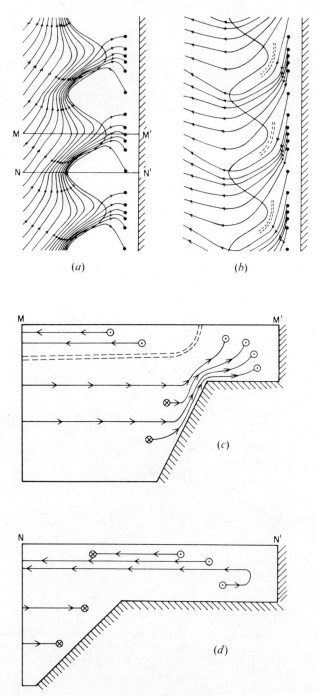

Fig. 7. A possible consequence of a series of submarine ridges and canyons in deformation of the long-shore flow off Northwest Africa; (a) the longshore flow in the subsurface, poleward compensation layer; (b) the longshore flow in the surface equatorward layer; (c) the cross-stream flow through a canyon head (MM'); and (d) the cross-stream flow through a dividing ridge (NN') between canyons. The solid lines are mass transport streamlines, the dotted lines represent possible fronts, the black dots are loci of upwelling; ⊗ indicates poleward flow and ⊙ equatorward flow (after Shaffer, 1974).

Phosphate:

$$\frac{\partial PO_4}{\partial t} = - \frac{\partial u PO_4}{\partial x} - \frac{\partial w PO_4}{\partial z} + \frac{K_y \partial^2 PO_4}{\partial y^2} - \frac{V_p(PO_4)(0.067 P_n)}{K_p + PO_4}$$

$$+ (0.13) \frac{G_z(P_n - P_0)Z_n}{P_k + (P_n - P_0)} + (0.13) \frac{G_f(P_n - P_a)F_n}{P_f + (P_n - P_a)}$$

where $V_p = 0.11 \sin 0.2618t$.

Silicate:

$$\frac{\partial SiO_4}{\partial t} = - \frac{\partial u SiO_4}{\partial x} - \frac{\partial w SiO_4}{\partial z} + \frac{K_y \partial^2 SiO_4}{\partial y^2} - \frac{V_s(SiO_4)(0.67 P_n)}{K_s + SiO_4}$$

where $V_s = 0.11 \sin 0.2618t$.

Phytoplankton carbon:

$$\frac{\partial P_c}{\partial t} = - \frac{\partial u P_c}{\partial x} - \frac{\partial w P_c}{\partial z} + \frac{K_y \partial^2 P_c}{\partial y^2} - \frac{G_z(P_c - 5P_0)(5Z_n)}{5P_k + (P_c + 5P_0)}$$

$$- \frac{G_f(P_c - 5P_a)(5F_n)}{5P_f + (P_c - 5P_a)} + \frac{V_a\{\exp [1 - (I_s/I_m)] - \exp [1 + (I_a/I_m)]\}(P_c)}{rz}$$

where $I_a = I_t (0.1309 \sin 0.2618t)$

$I_s = I_a e^{-rz}$

$r = 0.04 + 0.0021 P_c + 0.021(P_c)^{0.67}$

Phytoplankton nitrogen:

$$\frac{\partial P_n}{\partial t} = - \frac{\partial u P_n}{\partial x} - \frac{\partial w P_n}{\partial z} + \frac{K_y \partial^2 P_n}{\partial y^2} - \frac{G_z(P_n - P_0)Z_n}{P_k + (P_n - P_0)}$$

$$- \frac{G_f(P_n - P_a)F_n}{P_f + (P_n - P_a)} + VP_n$$

where V is the minimum of

$$\left\{ \frac{V_m NO_3}{K_t + NO_3} + \frac{V_n NH_3}{K_t + NH_3}; \frac{V_s SiO_4}{K_s + SiO_4}; \frac{V_p PO_4}{K_p + PO_4}; \frac{V_n I_z}{K_I + I_z} \right\}$$

with $I_z = I_0 e^{-jz}$

$j = 0.16 + 0.0053 P_n + 0.039(P_n)^{0.67}$

Zooplankton nitrogen:

$$\frac{\partial Z_n}{\partial t} = (1.00 - 0.67 - 0.13 - 0.20) \frac{G_z(P_n - P_0)Z_n}{P_k + (P_n - P_0)}$$

where $G_z = 0.03 \cos (0.2618t + 1.571)$ if $z < 30$ m

        $= 0.03 \sin 0.2618t$ if $z > 30$ m

Nekton nitrogen:

$$\frac{\partial F_n}{\partial t} = (1.00 - 0.67 - 0.13 - 0.20) \frac{G_f(P_n - P_a)F_n}{P_f + (P_n - P_a)}$$

where $G_f = 0.008 \cos (0.2618t + 1.571)$ if $z < 30$ m

         $= 0.008 \sin 0.2618t$ if $z > 30$ m

TABLE I

Values of Parameters in a Series of Simulation Models of Marine Ecosystems

Parameter	System					
	Peru	NW Africa	Baja California	Alta California	Southern Ocean	Sargasso Sea
Vertical velocity (w) cm/sec	10^{-2}	10^{-2}	10^{-2}	—	10^{-4}–10^{-5}	10^{-3}–10^{-4}
Downstream velocity (u) cm/sec	25	25	25	—	4	1
Eddy coefficient (K_y) cm^2/sec	10^6	0	1–5×10^6	—	10^7	10^6
Maximum nutrient uptake (V_i) day^{-1}	1.0	1.0	1.0	2.0	1.0	0.5
N half saturation (K_t) μg-at/l	1.5	1.5	1.5	1.0	1.0	0.5
Si half saturation (K_s) μg-at/l	0.75	0.75	0	1.0	—	—
PO$_4$ half saturation (K_p) μg-at/l	0.25	0.25	0.25	—	—	—
Zooplankton grazing rate (G_z) day^{-1}	0.24	0.24	0.24	—	0.10	0.43
Nekton grazing rate (G_n) day^{-1}	0.06	0.06	0.06	—	0.01	0.06
Z_n grazing threshold (P_0) μg-at/l	0.5	0.5	0.5	—	0.1	0.16
F_n grazing threshold (P_a) μg-at/l	3.0	3.0	3.0	—	0	0.02

Detrital nitrogen:

$$\frac{\partial D_n}{\partial t} = -\frac{\partial u D_n}{\partial x} - \frac{\partial w D_n}{\partial z} + 0.2 \frac{G_z(P_n - P_0)Z_n}{P_k + (P_n - P_0)}$$

$$+ 0.2 \frac{G_f(P_n - P_a)F_n}{P_f + (P_n - P_a)} - w_s \frac{\partial D_n}{\partial z}$$

The values of the parameters of this and other models discussed in this chapter are given in Table I.

The spatially varying downstream and vertical flow field (u, w) and the lateral mixing coefficient (K_y) are specified for each season of the model (Walsh, 1975c) in accordance with assumptions of the changes in physical structure during the austral winter and autumn. Diel variation of nutrient utilization in the hyperbolic Michaelis–Menten terms, of photosynthesis in the carbon budget, and of diurnal migration of the herbivores in the grazing terms are parameterized by periodic functions of the specific uptake rates, the incident radiation, and the specific grazing rates, that is, V_m, V_n, V_s, V_p, I_t, G_z, and G_f. Light regulation at depth is simulated in the shelf-shading, exponential I_z and I_s terms of the phytoplankton nitrogen and carbon equations. Suppression of nitrogen uptake (V_m) by concentrations of reduced nitrogen and hourly alternative regulation of phytoplankton growth by nitrogen, silica, phosphorus, or light are incorporated in the model with FORTRAN If statements. Thresholds of prey concentration (P_0, P_a) are specified in the nonlinear grazing terms, which are also used for differential recycling of nitrogen, phosphorus, and silica. Finally, detrital accumulation is assumed to occur through production and sinking (w_s) of fecal pellets with no mineralization or coprophagy within the time domain of the model. Note that zooplankton and nekton are not dynamic variables in this formulation, that is, $\partial Zn/\partial t = \partial Fn/\partial t = 0$.

The steady-state predictions of surface nutrients and phytoplankton along the axis of a plume in the autumn season of the model agree fairly well in one dimension (x) when compared with independent data from a drogue study (Fig. 8); the second dimension (y) of this surface model has been parameterized in the lateral boundary assumptions. These simulation results imply that advection and mixing are most important inshore, while silica limitation is suggested offshore, and assumptions of synopticity cannot be made for phytoplankton populations that take 2–3 days to drift from one end of the model to the other (Walsh, 1975c). The predicted phytoplankton atomic C/N ratio is close to the observed particulate ratio, implying that gains and losses to the phytoplankton population have been properly estimated, that nutrient limitation is not operative inshore of the upwelling ecotone (Walsh and Howe, 1976), and that protein yield of individual organisms might not be increased through resource management.

The two-dimensional x–z extension of the above solution (i.e., the third dimension is still parameterized in the lateral mixing assumption) does not match the rest of the independent data set (Fig. 9) as well as the surface sector. A priori, one might have required a better fit if additional parameters were used to simulate the vertical structure; that is, the degrees of freedom would have been reduced; the same number of parameters are used in both sections of the model, however, and the sample size is larger in the x–z case. No formal attempt has been made to distribute error within the model, nor to consider goodness of fit criteria, and the observed data are means over appropriate depth intervals of the drogue stations nearest to each grid point; one student, Tom Rawson, is presently considering these oceanographic sampling

PERU

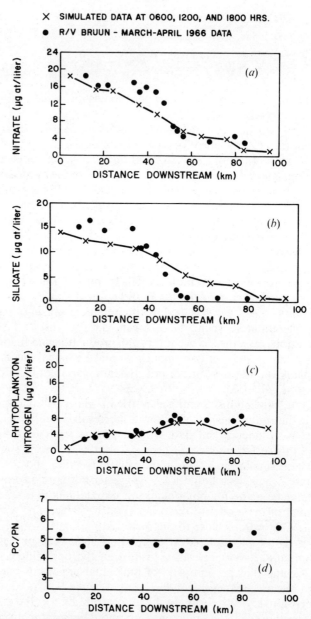

Fig. 8. A comparison of observations (●) and simulated results (×) within the surface layer (0–10 m) of a steady-state Peru model: (a) nitrate, (b) silicate; (c) phytoplankton and particulate nitrogen: (d) carbon nitrogen ratio (after Walsh, 1975c; and after Walsh and Howe, 1976).

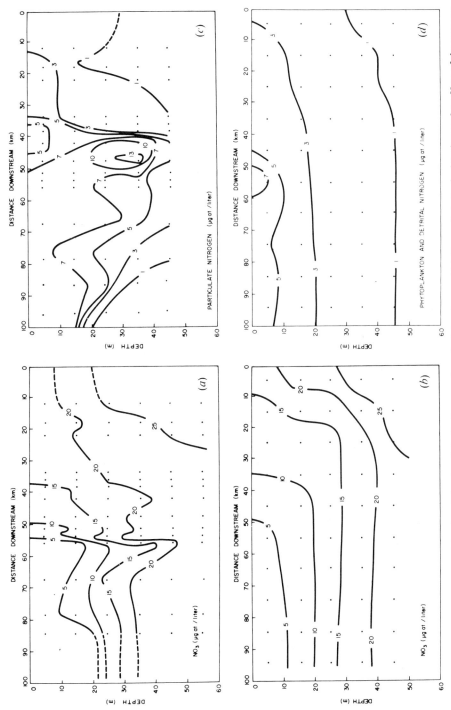

Fig. 9. A comparison of observed (*a, c*) and simulated (*b, d*) nitrate and particulate nitrogen (detritus and phytoplankton) over 0 to 50 m of the autumn water column off Peru (after Walsh, 1975c).

problems of linking models and the "real" world. The subsurface sector of the autumn model severely underestimates the amount of particulate nitrogen in the aphotic zone, nevertheless, implying that the regeneration, detritus, and vertical migration submodels are inadequate.

Fewer data place smaller constraints on the model, of course, and the winter x–z case appeared to match the meager June observations (Fig. 10). These winter results

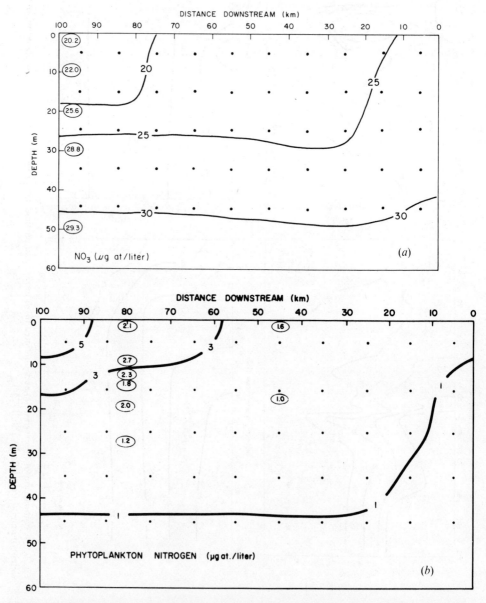

Fig. 10. A comparison of observed (circled values) and simulated nitrate (a) and measured particulate (circled values) and computed phytoplankton nitrogen (b) over 0 to 50 m of the winter water column off Peru (after Walsh, 1975c).

were obtained by doubling the velocity at each grid point to simulate stronger winds, whereas the lateral gradients were assumed to be zero in this homogeneous longshore situation; the offshore boundary conditions were also adjusted to those of winter rather than autumn.

Comparison of the seasonal results of the Peru $x-z$ model suggested that the food chain might be lengthened in periods of strong upwelling, with anchoveta switching from a phytophagous to zoophagous niche (Walsh 1975c, 1976b). It has been estimated that zooplankton herbivores crop only 25% of the Peru winter productivity (Beers et al., 1971) in contrast to 80% utilization of the autumn productivity by both zoo-plankton and anchoveta herbivores (Walsh, 1975c), with a terminal yield to anchoveta of only 1% of the productivity in strong upwelling and 10% in weak upwelling (Parsons, 1976). The simulation analysis also suggested that one mechanism to explain a possible shift in the Peru trophic structure might be an effective algal threshold of the anchoveta, with the nekton finding large concentrations of phyto-plankton in the plumes of weak upwelling, but few in the dispersed longshore structure of strong upwelling; these implications are discussed in more detail below.

B. Baja California

The generality of the Peruvian upwelling ecosystem model was considered (Walsh, 1976b) by applying it to analyses of data from other eastern boundary currents. The upwelling systems of the northern hemisphere appear to have greater seasonal intermittency of the wind-forcing function than those of the austral ones (Fig. 3), with alternation from heterogeneous plume to homogeneous longshore structure in a matter of weeks (Fig. 11) rather than months. Within the northern hemisphere, the narrower zonal nutrient gradients of the Pacific upwelling regions (Fig. 5) reflect their higher daily productivity (Huntsman and Barber, 1976), and perhaps narrower shelves and circulation patterns (Figs. 6a, 6c), in contrast to the Atlantic systems of similar latitude, but of wider shelf area and two-cell structure (Fig. 6b). The yields to fish are about the same off Baja California and Northwest Africa (Fig. 1), however, while the zooplankton biomass is also about the same; the additional primary productivity off Baja California may be possibly shunted to the red crabs, *Pleuroncodes planipes* (Walsh, 1976b).

Lateral mixing off Baja California at 27°N (Fig. 3) was considered in the model above 20 m, but not below 20 m, to simulate the combined effects of an upstream canyon as a point source within a relatively strong, homogeneous longshore upwelling regime (Walsh et al., 1974); the northwest African case considers the separate effects of a canyon as a line source of upwelled nutrients. The advective and biological assumptions over 0–50 km of this Baja California model were similar to the above Peru case (Table I), but dinoflagellates were considered to be the dominant phyto-plankton, rather than diatoms, whereas red crabs were assumed to be the main nektonic organisms instead of anchovy. Finally, the dinoflagellates were allowed to take up nutrients deeper in the water column by decreasing the extinction coefficient of the model, while their diel migration was simulated through vertical integration of dinoflagellate nitrogen over varying depth intervals at different times of the day.

The steady-state predictions of the one-dimensional case of the Baja California model agree fairly well with independent data (Fig. 12) from a 1972 drogue study off Punta San Hipolito, Baja California (Walsh et al., 1974). The model suggests that the red crabs were not phytophagous during the dinoflagellate bloom, however, for a combination of nocturnal upward migration and the feeding threshold of the red

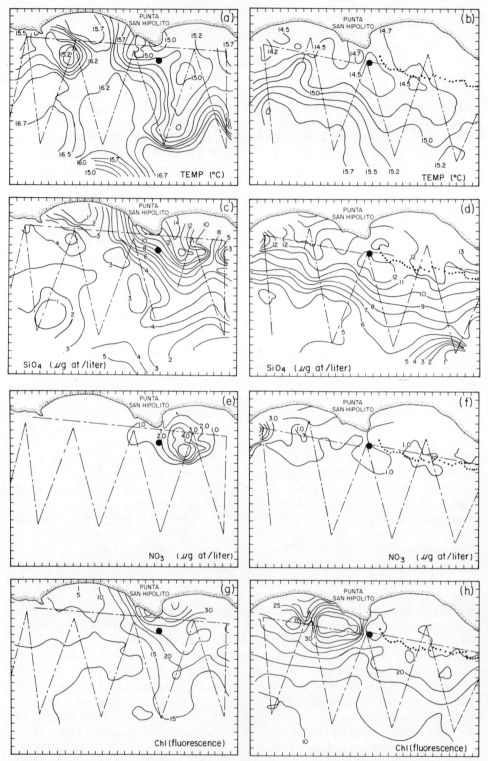

Fig. 11. The temporal variability at 3 m of plume and longshore band structure of temperature, nutrients, and phytoplankton off Baja California at the beginning (*a, c, e, g*) and end (*b, d, f, h*) of a 2-week study of the spin up of upwelling (after Walsh et al., 1974); the smaller dots refer to the drogue trajectory of Figs. 3 and 12, and the large dot is the site of the time series in Fig. 23.

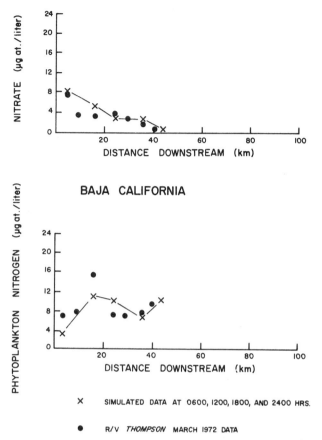

Fig. 12. A comparison of observation (●) and simulated results (×) within the surface layer (0–10 m) of a steady-state Baja California model for phytoplankton, particulate nitrogen, and dissolved nitrate (after Walsh, 1976b).

crabs in the Baja California model leads to little simulated grazing on the dino-flagellates. As a result of their own migration pattern in the model, the algal concentrations were dispersed at night with their numbers below the assumed grazing threshold when the crabs were present in the upper waters, whereas the dinoflagellates were concentrated above the grazing threshold in the daytime when the crabs were located in the lower part of the water column.

The model's artifice of spatial separation of predator and prey through a grazing threshold of phytoplankton biomass may really represent a more general one of size in the eastern boundary currents (Walsh, 1976a); in particular, the red crabs are thought to require prey particles above 25 μ (Longhurst, Lorenzen, and Thomas, 1967) and the mean size of the dinoflagellates in 1972 was 30–40 μ. A second cruise to the same study area in 1973 suggested (Walsh et al., 1976), moreover, that the red crabs were phytophagous later in the upwelling season when diatoms had replaced the dinoflagellates as dominants of the phytoplankton community. The mean red crab biomass of the two cruises was 5 g wet wt/m^3, or ~ 6 μg-at PN/l with a conversion of 10% dry wt = PN (Walsh et al., 1974, 1976), and is of the same order as the simulated biomass (10 μg-at PN/l), suggesting that either the threshold or the grazing rate is the critical parameter in the grazing stress term of the state equations. We are presently building a time-dependent model of the Baja California system (see below) to examine the consequence of prey switching by these facultative herbivores.

C. Northwest Africa

Irrespective of possible differences in the circulation patterns of upwelling off Baja California and Northwest Africa, the higher daily productivity in the Pacific system, yet the same potential fish yield in both eastern boundary currents, suggested that different trophodynamics might be operative in the two systems. The epibenthic red crabs may have replaced clupeids in the phytophagous nekton niche off Baja California (Walsh, 1976b), with the adult fish supported, instead, by a two-step phytoplankton–zooplankton food chain off Alta and Baja California (Longhurst, 1971). The low terminal yield and low primary production off Northwest Africa (~ 1–2 g C/m^2/day instead of ~ 5–6 g C/m^2/day off Baja California or Peru) suggested that there might not be any phytophagous shunts of energy flow to nekton in this system; the Peru steady-state model was thus applied to a drogue study at 19°N (Fig. 3) to test this hypothesis off Northwest Africa.

The Peru and Baja California validation data were gathered, 100 and 60 km, respectively, downstream of the heads of submarine canyons which might be expected to generate point sources of upwelling (Fig. 7) as parameterized in the lateral mixing term over 0–50 m in the Peru autumn model and over 0–20 m in the Baja California model. Drogues were also set south of a large canyon off Cape Timiris, Mauretania, in 1972 and 1973, and drift studies were conducted after the seasonal onset of upwelling in this region (Herbland, Le Borgne, and Voituriez, 1973, 1974). The remarkable similarity of the surface longshore structure of nutrients within and between these successive drogue studies off Northwest Africa (Fig. 13) suggested that this canyon might be treated as an initial east–west line source of nutrients with little additional north–south nutrient input during the course of the drogue observations. Since all these steady-state models are oriented along the axes of drogue studies, the only change from the Peru assumptions in the Northwest Africa model (Table I) was to eliminate the lateral mixing term at all depths to simulate a uniform north–south change of properties, that is, no east–west gradients landward of the 50 m isobath. The grazing stress was assumed to be the same off Northwest Africa as off Peru, and the phytoplankton were again considered to be diatomaceous.

The nutrient predictions of the one-dimensional case of the Northwest Africa model agree with the 1972 drogue observations (Fig. 14), but the phytoplankton standing stock is underestimated in the downstream sector of the simulation. As perhaps might have been expected, this result suggests that, notwithstanding the large grazing threshold (P_a), the high grazing stress of the Peru model cannot be supported in the Northwest Africa ecosystem. In fact, few clupeids were observed during the 1972 drogue study and it was thought that mesozooplankton were the major herbivores; that is, the assumed biomass in the grazing stress term of the state equation was too large for this case. If the intermittency of upwelling off Northwest Africa is comparable to seasonal pulses of temperate systems (Cushing, 1971b) with spatial translation of time lags in a classical phytoplankton–zooplankton–herring type food chain, such an herbivore population of weakly motile zooplankton would be expected to be located downstream of a point source, subject to variability of the Ekman transport (Wickett, 1967). Large populations of zooplankton were, in fact, found 180 km downstream of the Cape Timiris canyon and 60 km downstream of the high phytoplankton popula-

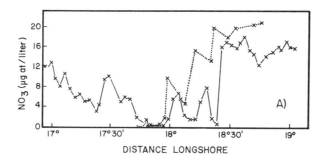

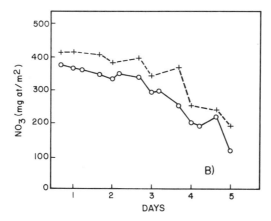

Fig. 13.   The weekly and yearly variability of the longshore nitrate distribution off Northwest Africa (after Herbland et al., 1974, 1975). (*A*) The solid line is surface nitrate between St. Louis and Cape Timiris on March 26, 1972, and the dashed line is surface nitrate (the validation data of Fig. 14) from the March 30–April 7, 1972 drogue study, south of Cape Timiris; (*B*) the dashed line is the integrated nitrate over the water column during each day of the 1972 study, and the solid line is the integrated nitrate during the 1973 drogue study.

tions in both years. Such a high phytoplankton biomass, upstream of the herbivores and unpredicted by the model, can be attributed to little grazing by weakly motile organisms, unable to stay near the source of upwelling, that is, in contrast to Peru and Baja California nekton.

Consideration of the downstream trajectories of zooplankton in a longshore homogeneous case has been formalized by the SCOR Working Group 36 on upwelling during their 1974 meeting in Kiel: assuming a two-cell, cross-stream circulation (Fig. 6*b*) with a $\pm u$ of 4 cm/sec, a $\pm v$ of 25 cm/sec, and a $\pm w$ of 2×10^{-2} cm/sec, that is, about the same inshore circulation pattern as in the above steady-state models, it would take ~ 15 days for a zooplankter to circle through the helix of the inshore cell, while traveling ~ 190 km longshore (Fig. 15). *Calanus helgolandicus* adults take about 18 days to develop in the laboratory at 15°C (Paffenhöfer, 1970), and if the origin of the helix were a point source in a heterogeneous longshore case, that is, an offshore closure of Shaffer's (1974) three-dimensional hypothesis (Fig. 7), the adult zooplankton populations would be displaced downstream about the same distance from a point source as that observed in the above two drogue studies at 19°N.

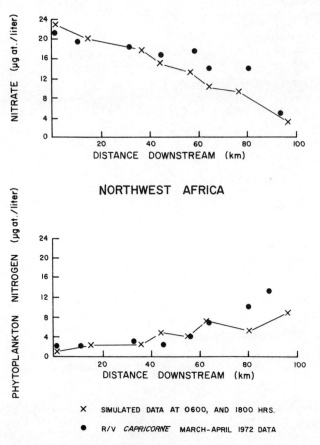

NORTHWEST AFRICA

× SIMULATED DATA AT 0600, AND 1800 HRS.

● R/V *CAPRICORNE* MARCH–APRIL 1972 DATA

Fig. 14. A comparison of the observations (●) and simulated results (×) within the surface layer (0–10 m) of a steady-state Northwest Africa model for phytoplankton, particulate nitrogen, and dissolved nitrate (after Walsh, 1976b).

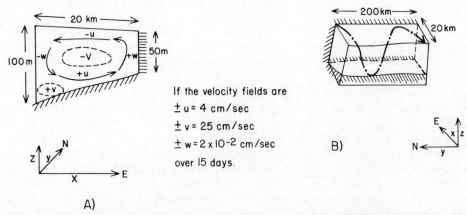

If the velocity fields are
± u = 4 cm/sec
± v = 25 cm/sec
± w = 2 × 10⁻² cm/sec
over 15 days.

Fig. 15. The downstream helix (*B*) of a hypothetical trajectory of a zooplankton population within a two-cell, cross-stream circulation pattern (*A*) of an eastern eastern boundary current, which is described in a right-hand cartesian coordinate system with a $\pm u, v, w$ of respectively 4, 25, 2×10^{-2} cm/sec over a 15-day period.

942

From such helix calculations, moreover, one would expect to see a uniform distribution of poorly motile zooplankton over the shelf in any two-cell case with a homogeneous longshore dimension greater than 200 km. Our recent 1974 study at 21°40'N off Northwest Africa suggests that during periods of strong winds along a coast with little irregularity in botton topography for over 300 km, between Cape Blanc and Cape Barbas, the smaller components of the zooplankton were indeed found evenly distributed landward of the shelf break (Fig. 16), with over 70% of the zooplankton wet weight in the <500 μ size fraction. The larger, more motile zooplankton were found seaward of the shelf break, where the shallowness of the shelf may no longer interrupt their diel migratory patterns (Longhurst, 1975); as off Peru (Beers et al.,

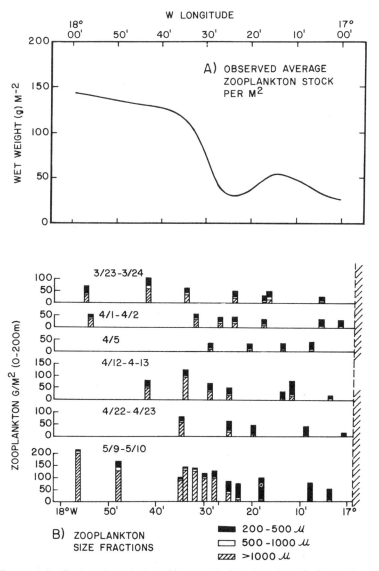

Fig. 16. The zonal distribution of zooplankton biomass (A), size (B), and metabolism (C) in a cross-shelf transect (D) at 21°40'N, off Northwest Africa (after R. Clutter and M. Blackburn, personal communications).

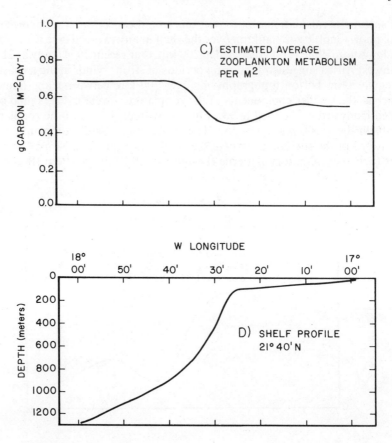

Fig. 16.   (*Continued*).

1971), only 25% of the zooplankton wet weight consisted of the $< 500\ \mu$ size fraction in water deeper than 100 m. The dominance of microzooplankton on the wide, shallow Northwest African shelf suggests the fauna may not be typical of the other eastern boundary currents (Walsh, 1976a). Irrespective of the composition of the shelf plankton, however, these zooplankton observations support the model's conclusions in suggesting that an extra step of poorly motile herbivores may exist in the Northwest African clupeid food chain; in this respect, it is interesting to note that total zooplankton biomass off Peru may be less than those off either Northwest Africa or Baja California (Walsh, 1976b).

Examination of steady-state models of three eastern boundary currents suggests that the observed range in primary production, ecological efficiency, and terminal yield may be a function of intermittency, shelf width, bottom topography, and species replacement, that is, the biological response to varying composition of the dominant frequencies of the physical forcing functions (Fig. 4) in structuring the physical variability of each habitat. If the seasonal intermittency of upwelling is low, if a one-cell system, without a convergence front, is present on a narrow shelf, and if the bottom topography leads to surface concentration of prey organisms, the primary productivity appears to be transferred directly to clupeids off Peru; if the intermittency is high, if a one-cell system is present on a shelf of intermediate width, and if the bottom topography is appropriate for plume formation, the primary productivity may be transferred both directly to a facultative, phytophagous benthos (red crabs) and, in a lengthened food chain, to clupeids off Baja California; finally, if the intermittency is high, and if a two-cell system is present on a wide shelf, the clupeid food chain seems to be lengthened off Northwest Africa in either a homogeneous or a heterogeneous longshore case. Such seasonal "snapshots" of geographically separated and temporally varying ecosystems would be more convincing, of course, if the reputed causal relationships could be shown to hold in time-dependent studies of individual systems as well as in these steady-state parameterizations.

> And indeed there will be time
> To wonder, "Do I dare?" and, "Do I dare?"
> Time to turn back and descend the stair

> —idem

3. Seasonal Time Dependency

A. The Southern Ocean

Analysis of the time-dependent and spatial nature of upwelling ecosystems has been extended beyond the above steady-state hypotheses with a simulation model of the seasonal cycle of production in the Southern Ocean. This analysis was developed by two of my former students, Drs. N. Kachel and L. Conway, as a continuation of a class project in marine systems analysis; they hope to continue these studies and I present only some of our preliminary results. Their model of nutrients, phytoplankton, euphausiids, and whales was assumed to be directed along 90°W, between the Antarctic Divergence at 67°S and the Antarctic Convergence at 59–60°S. The sparse grid mesh consisted of two 75-m layers over the upper 150-m water column and three boxes of 220 km length between 67 and 61°S; that is, the convergence area was not considered in the model. The time step was 1 day, and the time-dependent forcing functions (Fig. 17) of light availability, wind stress, and whale predation on euphausiids were updated each month; only the northern time-dependent input at 62°S is shown, because the form of the other forcing functions at 64 and 66°S was the same, with progressively less light and wind in the more southern areas, and the 3-month feeding migration of baleen whales was assumed to be the same in each area during the austral summer.

Without the hourly periodic functions, similar equations to those for nonconservative variables of the coastal models (see above) were solved numerically for the seasonal distribution of nitrate, ammonia, phytoplankton nitrogen, and zooplankton nitrogen, south of the Antarctic Convergence, with an IBM 1130 simulation package (Walsh and Bass, 1971) that agrees quite well (Walsh, 1975b) with the more complex CDC 6400 program (Morishima, Bass, and Walsh, 1974) used for the steady-state models. East–west gradients associated with the Antarctic Circumpolar Current ($\sim$20 cm/sec) were assumed to be negligible; the northerly Ekman drift (v) was considered to vary spatially with the wind stress, and a range of values from 0.7 to 4.2 cm/sec were computed, with the larger velocities estimated for both the northern and winter sectors of the model; a range of upwelling rates (w) from 1×10^{-4} to 1×10^{-5} cm/sec were used, with the smallest values in both the northern and summer

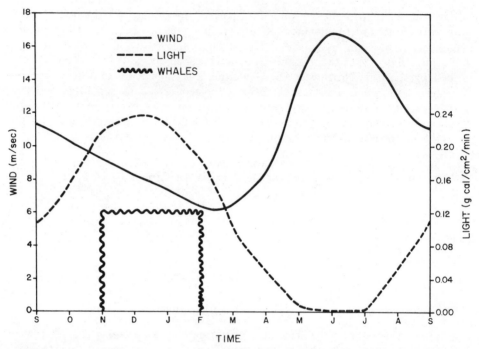

Fig. 17.   The time-dependent forcing functions of light, wind, and predation for a simulation model of the
Southern Ocean.

sectors of the model. The horizontal and vertical eddy coefficients (K_y, K_z) were
estimated to be 10^7 and 1 cm^2/sec.

The biological and chemical assumptions were similar to the previous models
(Table I) with a bottom boundary condition of 30 μg-at NO_3/l; a Michaelis–Menten
uptake of NO_3 and NH_3; light inhibition of phytoplankton growth; and the same
maximum phytoplankton division rate of 1.4 doublings/day. Excretory return of
NH_3 to the inorganic nitrogen pool was also considered; a grazing threshold of 0.1
μg-at N/l of phytoplankton was assumed for a 1-yr cohort of *Euphausia superba* with
a specific grazing rate of 0.1/day, that is, half that of the zooplankton in the coastal
models; finally, the zooplankton of the Southern Ocean were considered as a dynamic
variable, with losses to molting, predatory whales, seals, and birds.

Since the computation interval of these simulations was 1 day, neither diurnal
metazoan migration nor phytoplankton periodicity were considered in the analysis.
The model's logic does suggest, however, that the simulated seasonal, surface drift
of the zooplankton north to the Antarctic Convergence (Mackintosh, 1937) would
take place in about 165 days at the fastest rate of drift of Antarctic Surface Water from
67 to 61°S within the model, allowing sufficient time for their southerly return
migration, within rising Antarctic Circumpolar Water, during the rest of the year.
Additional logic of the model considers the seasonal, feeding migration of the baleen
whales by imposing a 3-month predation stress on the euphausiids, without specification of the predator biomass similar to Steele's (1974) approach in a recent model of
the North Sea ecosystem.

I have superimposed the available seasonal observations on the results of one of
Kachel and Conway's simulation runs in Fig. 18; only their data for the upper 75 m

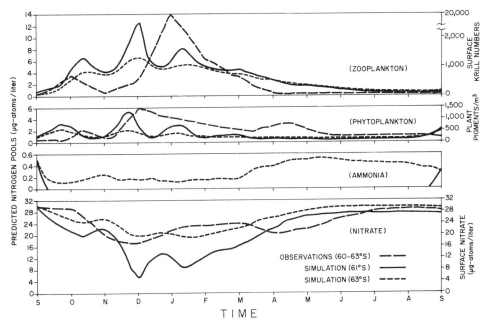

Fig. 18.   A comparison of observations and the simulated nitrogen pools of nutrients, phytoplankton, and zooplankton during a year in the Southern Ocean.

at 61 and 63°S are shown for the sake of clarity; the model's results at 65°S show similar patterns in continuing the increasing latitudinal trends of higher nutrients and lower phytoplankton between the middle and northern sectors of the model. The observed nitrate data are from Cruises 7 to 27 of the USNS *Eltanin* (Jacobs and Amos, 1967; Jacobs, 1965, 1966) for each month of the austral year at stations taken between 60 and 63°S, from 35°W to 158°E, except for the September observation made at 58°40'S. I know of no routine ammonia determinations in the Southern Ocean. The phytoplankton data are from Hart's (1942) northernmost observations, for he defined this sector as extending 5.5° south of the convergence, which is about 60°S at 90°W (Walsh, 1971), and should thus be applicable to at least the 61°S predictions of the model. The zooplankton data (Marr, 1962) are the number of adult euphausiids observed in the surface Weddell Current from 30 to 60°W, between the Antarctic Convergence and the ice pack at ∼65°S.

The spatial, light dependency of the Antarctic upwelling ecosystem (Walsh, 1969) can be seen in the model's results between 61 and 63°S. As Hart (1942) suggested, light becomes more limiting the farther south one proceeds, and the model predicts more nitrate and less phytoplankton and zooplankton in the middle sector than in the northern sector. The simulated time-dependent trends of the phytoplankton and zooplankton spring blooms appear to follow the observed ones as well, with a decline in both plant and animal biomass at the onset of the austral winter.

The absolute values of the model's predictions are also in reasonable agreement with observations: the simulated nitrate at 63°S is fairly close to the *Eltanin* data; if it is assumed that little detritus was present and the PN : Chl *a* conversion were 1 : 1, a maximum of 6 μg-at PN/l in the model implies a chlorophyll biomass of 6 mg Chl *a*/m^3 during the austral summer, which is of the right order of magnitude (Walsh,

1969); finally, if one assumes that there is 7 mg nitrogen within each euphausiid (Mauchline and Fisher, 1969), the model's seasonal zooplankton maxima at 61 and 63°S are 12 and 24 times the estimated mean concentration of 1 animal/m³ (Mauchline and Fisher, 1969), in comparison with the seasonal peak of observed euphausiids, which appears to be ~ 10 times the mean.

The dynamics of the Southern Ocean model suggest an interesting comparison of control mechanisms between coastal and oceanic upwelling ecosystems (Walsh, 1976a). The prediction of low nitrate at 61°S in Fig. 18 may reflect too small an upwelling rate (10^{-5} cm/sec) in this northerly sector of the model, because neither the predicted phytoplankton nor zooplankton biomass appears to be excessive in comparison with the observed data; that is, the predicted nitrate value is still far above the half-saturation constant (Table I), and it is thus dubious that overutilization of nitrate occurred in this area of the model. The closer fit of the model's nitrate at 63°S to the *Eltanin* observations implies, therefore, that the vertical velocity of 1×10^{-4} cm/sec in the middle sector may be a better estimate of upwelling in the Southern Ocean. This value, however, is still two orders of magnitude less than that of the above models of coastal upwelling (Table I).

The low rate of input of nutrients over a wide area, coupled with poor seasonal availability of light (Fig. 17) and subsequent meager yearly biological production, might account for at least an order-of-magnitude larger horizontal nutrient gradient in the Southern Ocean (Fig. 19) than off Peru (Fig. 5), despite the much higher up-welling rates of the coastal systems. Examination of the nutrient gradient in equatorial upwelling indicates, however (see Fig. 22 on page 953) that light limitation is not the only factor in this paradox of nutrient input, utilization, and horizontal gradients. For example, a more efficient grazing stress during the spring blooms of the Southern

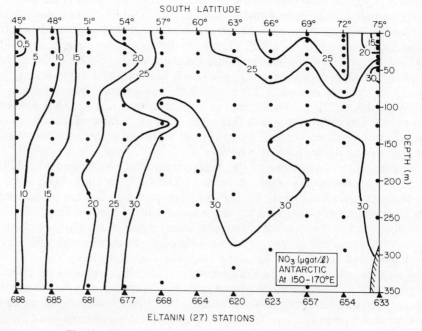

Fig. 19. The gradient of nitrate across the Antarctic Divergence.

Ocean might also account for differences in the horizontal nutrient gradients between coastal and oceanic upwelling systems.

Such an efficient grazing stress cannot be generated with either the specific grazing rate or herbivore biomass terms of the state equation, however. The specific grazing rate of zooplankton in the coastal upwelling models was twice that of the Southern Ocean model, in agreement with Q_{10} temperature arguments (Taniguchi, 1973), whereas the herbivorous zooplankton of the coastal regime tend to be smaller (~ 5–25×10^2 μ) short-lived (weeks) calanoid copepods with higher metabolic costs (Fig. 16) in contrast to the larger (25–50×10^3 μ), long-lived (years) euphausiids of the Southern Ocean. It is interesting to note, moreover, that the range in maximal biomass (6–12 μg-at PN/l) of the austral model's micronekton is close to that (10 μg-at PN/l) used in the coastal models for anchoveta off Peru and for red crabs off Baja California.

As one solution to this possible nutrient anomaly in both high and low latitudes, I have, in fact, speculated (Walsh, 1976a) that the relative predictability of seasonal events in the offshore divergences may have led to the evolution of shorter time lags between phytoplankton and zooplankton blooms, with subsequently less phytoplankton biomass and wider nutrient gradients of the Southern Ocean. The seasonal migratory habits of the Southern Ocean euphausiids may allow them to anticipate the spring outburst of phytoplankton; this is simulated in the model with a low grazing threshold (0.1 μg-at PN/l). A higher grazing threshold (3.0 μg-at PN/l) was used in the coastal upwelling models, of course, which leads to their higher phytoplankton standing stocks. The order-of-magnitude higher coastal threshold may reflect a greater intrinsic patchiness, that is, spatial translation of longer time lags, within the ocean's boundary layer in contrast to its interior, that is, the Southern Ocean. Such theory is also in accordance with long-term current meter records that suggest more of the physical variability is at the high-frequency bands on the continental shelves than in the deep oceans (Walsh, 1976a).

There has been much speculation about the effects of overfishing by the whaling industry in the Southern Ocean, and it seems appropriate to comment on implications of this austral model. For example, it has been suggested (Steele, 1974) that the drastic decline of whale stocks over the last 50 yr may have had little effect on changes in euphausiid biomass, although few data are available to contradict or support this hypothesis. Steele (1974) assumed that the most sensitive control link of marine ecosystems is at the plant–herbivore interface rather than at the herbivore–carnivore interface. Little loss of euphausiids to whales, seals, birds, or molting was included in this particular simulation run (Fig. 18), yet neither drastic increases in zooplankton biomass nor declines in phytoplankton biomass occurred, supporting Steele's (1974) hypothesis. The euphausiids are responding seasonally in the model to the declining light instead of predation, and thus directly to the decreasing supply of phytoplankton as winter approaches. These results are preliminary, of course, and additional numerical analyses (see below) were conducted to further consider the implications of the relative importance of patchiness at the plant–herbivore and herbivore–carnivore interfaces.

B. The Sargasso Sea

Our studies of the synergistic interaction of nutrient limitation, light regulation, grazing control, and frequency composition of the physical forcing functions were continued in a separate time-dependent study of the seasonal cycle of production in the Sargasso Sea. Two other former students, Drs. T. Whitledge and L. Codispoti, performed this simulation analysis of the daily nitrogen flux within an oligotrophic ecosystem near station S (32°N, 65°W), 15 mi southeast of Bermuda. They also intend to continue this work, and again I only report some of our preliminary results of a model extending east–west, along 35°N, from the Gulf Stream at 75°W to the interior of the North Central Atlantic Gyre at 45°W. Because of weak horizontal nutrient gradients, both within the Sargasso Sea and between it and the Gulf Stream (Stefansson and Atkinson, 1971), an irregular x–y grid mesh of 2500×500 km^2 and 250×500 km^2 rectangles, respectively, was adopted for the Sargasso Sea and for both the Gulf Stream and the Atlantic Gyre boundaries, with two 50-m layers over the upper 100-m water column of each sector. The time step was 1 day, as in the Southern Ocean model, but the only time-dependent input was the seasonal overturn of the water column (Fig. 20) during January, February, and March with time-invariant forcing by light and predation. Finally, equations similar to those of the state variables in the above models of the coastal and oceanic divergences were solved numerically for the seasonal distribution of nitrate, ammonia, phytoplankton nitrogen, and zooplankton nitrogen in the Sargasso Sea and Gulf Stream with the same IBM 1130 simulation package (Walsh and Bass, 1971).

The Sargasso Sea is thought to be a convergence area (Montgomery, 1959), and our salinity calculations of the water's residence time suggested that a negative vertical velocity of 5×10^{-4} cm/sec and 2.5×10^{-4} cm/sec might approximate sinking at the bottom of the euphotic zone (100 m) in the Sargasso Sea and the western gyre; upwelling was assumed to occur at $\sim 5 \times 10^{-5}$ cm/sec at the bottom of the euphotic zone within the eastern Gulf Stream. Considerations of the source of 18°C water in the study area (Worthington, 1959) suggested that the southeasterly flow from the Gulf Stream might have a speed of ~ 0.7 cm/sec and the northwesterly flow from the South Atlantic might be ~ 1.4 cm/sec to balance the mass transport of downwelled water within the model. Finally, winter convection was simulated with a special vertical mixing term, equivalent to a net upward velocity of $\sim 10^{-3}$ cm/sec, while the rest of the turbulent motion was parameterized with vertical and horizontal eddy coefficients assumed to be 1 and 10^6 cm^2/sec, leading to a net upward input equivalent to $\sim 10^{-5}$ cm/sec during the rest of the year (Fig. 20).

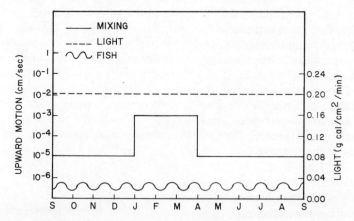

Fig. 20. The time-dependent forcing functions of light, mixing, and predation for a simulation model of the Sargasso Sea.

The biological terms of the state equations were the same as for the preceding models, but different values of the parameters were selected to represent the structure and organisms of a subtropical, oligotrophic ecosystem (Table I). For example, in contrast to a bottom boundary condition of 30 μg-at NO_3/l for the eutrophic models, a value of 2.0 μg-at NO_3/l was chosen to represent impoverished nutrient conditions below the euphotic zone, whereas a V_n and K_t of 0.5/day and 0.5 μg-at N/l were assumed for the Sargasso Sea phytoplankton instead of the previous 1.0/day and 1.0 μg-at N/l; that is, these algae of the simulated subtropics would have the competitive advantage (Dugdale, 1967) in conditions of low nutrients.

The same concepts of ammonia inhibition, light regulation, and a grazing threshold were employed with a value of 0.1 μg-at N/l for P_0 in the nonlinear grazing term, but the zooplankton grazing rate was assumed to be 0.4/day, that is, twice that of the zooplankton in the coastal models and four times that of the Southern Ocean. As in the Antarctic case, the zooplankton of the Sargasso Sea model were a dynamic variable, with the same 60% assimilation efficiency and similar losses to defecation, excretion, and predation, although this second model involved no consideration of molting, reproductive products, or cohort structure. Predacious fish were an explicit variable, with a formulation similar to the nekton grazing term in the coastal upwelling models.

I have superimposed seasonal observations from station S on one of Whitledge and Codispoti's simulation runs in Fig. 21; only their results for the upper 50 m of the Sargasso Sea are shown for comparison with the output of the Peru, Baja California, Northwest Africa, and Southern Ocean models. The observed mean nitrate over 0–100 m is from April 1958 to March 1959 (Menzel and Ryther, 1960); the mean chlorophyll over 0–50 m is from June 1958 to May 1959 (Menzel and Ryther, 1960); the integrated zooplankton biomass over 0–500 m is from September 1958 to September 1959, because there are suggestions of spurious data in the 1957–1958 time series (Menzel and Ryther, 1961); and the monthly mean ammonia at 50 m is from July 1960 to June 1961 (Menzel and Spaeth, 1962).

The seasonal dependency of the Sargasso Sea productivity on the winter convective overturn is reproduced in the model, with the right order of magnitude of nitrate, phytoplankton, and zooplankton standing crops (Fig. 21). If a Chl a : PN ratio of 1 : 1 is again assumed, the model's predictions are close to the seasonal chlorophyll observations, and also approximate a range of 0.3–1.5 μg-at PN/l measured from September 1962 to January 1963 at station S in the Sargasso Sea (Dugdale and Goering, 1967). If one assumes that 1–5 mg dry wt/m^3 is a reasonable estimate of the seasonal range of zooplankton biomass over the upper 100 m (Menzel and Ryther, 1961), the equivalent nitrogen biomass is 0.007–0.04 μg-at PN/l (with a conversion factor of 1 mg dry wt/m^3 = 0.007 μg-at PN/l), in agreement with most of the model's solution. The predicted ammonia is too low, and if all of the measured ammonia were actually ammonia (Walsh, 1975a), part of the mismatch of reduced nitrogen between simulation and observation could be explained by input from rainfall in the "real" world (Menzel and Spaeth, 1962) and not in the model. If the nitrogen K_t is really of the order of 0.2 μg-at N/l for oligotrophic organisms (MacIsaac and Dugdale, 1969), however, it is hard to rationalize ambient ammonia concentrations > 0.5 μg-at N/l throughout the water column at station S.

The serious mismatch of the Sargasso Sea model's ammonia predictions with measured values, nevertheless, as well as the poor phase relationship of this model's zooplankton response to winter overturn (the nitrate match would be improved if I had selected a simulation run with convective mixing started in January instead of

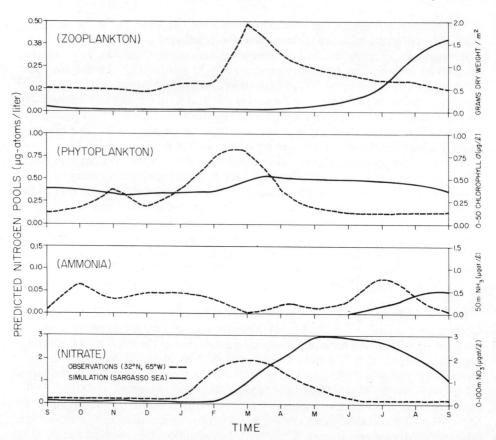

Fig. 21.   A comparison of observations and the simulated nitrogen pools of nutrients, phytoplankton, and zooplankton during a year in the Sargasso Sea.

February), may shed additional light on the dynamics of coastal upwelling ecosystems. The zooplankton grazing rate of the Sargasso Sea case was 0.4/day, or four times the Southern Ocean model, which is reasonable in light of Q_{10} temperature relationships for smaller organisms of subtropical oceanic waters (Taniguchi, 1973); but the oligotrophic grazing threshold was about the same as in the Antarctic case and it may have been too large. A smaller P_0 of the Sargasso Sea model would have both reflected the lower phytoplankton biomass of the subtropics, and led to a larger grazing flux, cropping more phytoplankton during a shorter time in the model, which in turn would have generated both a faster growth of the zooplankton population and a higher rate of herbivore excretion, that is, a better zooplankton phase relationship and increased stocks of ammonia in the model.

   A smaller, more appropriate grazing threshold of, perhaps, 0.01 μg-at PN/l for the Sargasso Sea model would then imply less patchiness or even shorter time lags at the phytoplankton–herbivore interface within the interior of the tropical ocean, as I have, indeed, argued (Walsh, 1976a) on the basis of the horizontal nitrate gradients across equatorial divergences (Fig. 22). The upwelling rate ($\sim 10^{-3}$ cm/sec) of the Pacific equatorial divergence is at least an order of magnitude (Knauss, 1966) less than coastal upwelling, and the potential growth rates of the phytoplankton may be about the

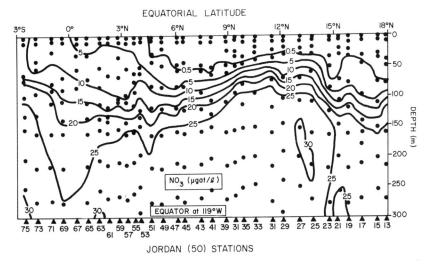

Fig. 22. The gradient of nitrate across the Equatorial Divergence.

same in the two systems (Walsh, 1976a); yet the low latitude nutrient gradient is of almost the same order of magnitude as in the Southern Ocean (Fig. 19).

I have argued that the microzooplankton of these areas may be a very important component of the grazing stress, immediately cropping most of the phytoplankton production in the subtropical interior of the ocean (Sheldon, Sutcliffe, and Prakash, 1973), and thus leading to a low phytoplankton biomass and wide nutrient gradients. These smaller, herbivorous organisms (< 100–$200 \, \mu$) contribute to only 25% of the grazing stress in the Californian eastern boundary current in contrast to 73% across the equatorial divergence (Beers and Stewart, 1971), whereas the biomass of the small herbivores has been predicted to be greater than that of the large herbivores in a recent simulation model of equatorial upwelling (Vinogradov, Menshutkin, and Shushkina, 1972).

The possible similarity of function of the plant–herbivore interface between the Southern Ocean and the Sargasso Sea is not continued in consideration of the subtropical herbivore–carnivore interface, perhaps reflecting the two-orders-of-magnitude difference in size of dominant herbivores between the two systems. The fish biomass of this particular run (Fig. 21) of the Sargasso Sea model was very small (0.01 μg-at PN/l), similar to concept to the lack of predators in the Southern Ocean case (Fig. 18). The predicted zooplankton of the Sargasso Sea were an order of magnitude higher than the observed biomass after the winter overturn, however, suggesting that a predator control mechanism was needed in the model.

The above simulation results imply that small herbivores of the subtropical interior have small predators (Sheldon, Sutcliffe, and Prakash, 1973), which may experience less patchiness in their prey distribution than the larger carnivores find in the Southern Ocean or the larger herbivores in the eastern boundary currents, because of a possible gradient in seasonal intermittency among the three systems. For example, an equivalent size herbivore, with a life cycle of 1 month, would experience more environmental variability relative to its life history within the high-frequency regimes of the continental shelves than in the low-frequency regimes of the deep ocean (Walsh, 1976a). The small primary carnivores of this hypothetical Sargasso Sea might then support larger secondary and tertiary carnivores, thus corresponding to observations of higher diversity of zooplankton carnivores within the ocean's interior in comparison with the more variable, high-frequency shelf areas (Grice and Hart, 1962). Finally, these speculations of low thresholds in the subtropics suggest that the transfer efficiency is greater at both the plant–herbivore and herbivore–carnivore interfaces of the Sargasso Sea model, supporting Blackburn's (1973) conclusions of greater ecological efficiency in oligotrophic ecosystems; the greater patchiness at the herbivore–carnivore interface of the Southern Ocean and at the plant–herbivore interface of the eastern boundary currents may lead to less ecological efficiency within these eutrophic systems.

> Do I dare
> Disturb the universe?
> In a minute there is time
> For decisions and revisions which a minute will reverse

—idem

4. Daily Transients

A. Baja California

One of the dominant modes of the wind and current forcing functions of the eastern boundary currents appears to be transient responses of 1–2 days after the onset of equatorward winds (Smith, 1974), with total duration of each upwelling event on the order of 1–2 weeks between relaxations of the system. To assess the implications of higher-frequency phenomena for the species composition and productivity of upwelling ecosystems (Walsh et al., 1976; Wang and Walsh, 1976), we have initiated a time-dependent simulation model of the Baja California system. In contrast to our previously discussed time-dependent studies, however, the time step was 12 min in this analysis instead of 1 day, and upwelling events of 7–14 days duration were considered instead of the above seasonal phenomena on the time scale of months.

Two cruises were made to an area near Punta San Hipolito at 27°N off Baja California (Walsh et al., 1974; Walsh et al., 1976) to generate time series of wind, current, nutrient, phytoplankton, and herbivore data in order to document the transient stages of coastal upwelling; Fig. 23 is an example of the daily variability in the wind and nutrient fields during 1973. A z–t model over 0–50 m depth was then constructed to analyze this data set at the site of the 1972–1973 time series (see Walsh et al., 1974, 1976), 5 km off the coast, with the assumption of no longshore gradient in concentrations of the state variables. A surface and bottom Ekman layer is also assumed with a vertical velocity (w) of $\sim 1 \times 10^{-2}$ cm/sec and both a zonal ($\partial u/\partial z$) and a southward longshore ($\partial v/\partial z$) shear of $\sim 10^{-3}$/sec over 50 m during the spin-up phase of the model; the decay of upwelling, after relaxation of the winds, is simulated with a similar zonal shear, but now of onshore flow at the surface and offshore at the bottom, with a northward flow that is constant with depth, and with a downwelling velocity of 2×10^{-2}–2×10^{-3} cm/sec. The horizontal eddy coefficients (K_x, K_y) vary over the water column from 1×10^6 to 6×10^6 cm^2/sec and the vertical eddy coefficient (K_z) from 1 to 55 cm^2/sec with the highest values of mixing in the surface layers of the model.

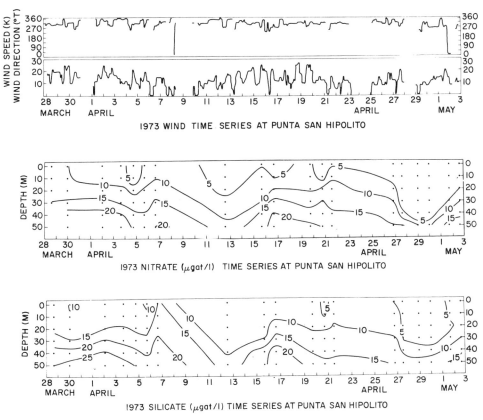

Fig. 23.   A 1973 time series of daily variability in wind and nutrient structure off Baja California (after Walsh et al., 1976).

As mentioned in the discussion of the steady-state model of the Baja California system, a dinoflagellate bloom of *Gonyaulax polyedra* was observed during the 1972 cruise; these organisms have no known silicon requirement and silicon then appeared to be a quasi-conservative parameter with a high correlation with the density field in contrast to the nitrate distribution (Wang and Walsh, 1976). Diatoms were dominant during the first part of the 1973 cruise, however, and ^{29}Si uptake studies then demonstrated a large flux of silicon to the phytoplankton. We thus elected to test our preliminary physical assumptions about the time response of the Baja California system to the onset and decay of upwelling by examining the distribution of the quasiconservative silicate in 1972 and of the nonconservative silicate in 1973.

No biological processes are operative in the present version of the model in which the spin up of upwelling in 1972 is reproduced (Fig. 24). Two spin downs or relaxations are reproduced with respect to the 1973 data set as well, suggesting that our first approximation of the time response of the system may be correct. A time-dependent circulation submodel is not included in the present model, however, and it must be remembered that these results stem from the response of the model to the imposition of a steady spin up or spin down circulation pattern. The short response time (on the order of 2–5 days) of the model's silicate distribution to these two circulation modes suggests that our preliminary fit of the simulated and observed silicate may mainly

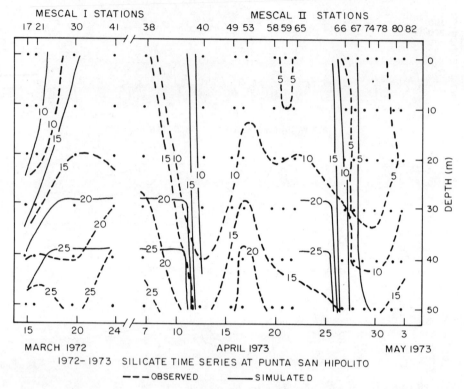

Fig. 24. A comparison of observations and the simulated spin up and spin down of upwelling off Baja California in 1972 and 1973.

reflect the barotropic response of the system, particularly after relaxation of the wind; we are presently incorporating a time varying circulation submodel (Hamilton and Rattray, 1976) that computes the velocity at each grid point from the primitive equations of motion which may then allow us to separate the baroclinic and barotropic responses. Biological questions of phytoplankton species competition and size in relation to a fluctuating environment are now being explored with this model; the actual response of an algal population to a varying ambient nutrient supply is considered in the following model.

B. *Alta California*

The short-term response of phytoplankton to a pulse of nutrients has been studied (Howe and Walsh, in preparation) in another time-dependent study of simulated algal populations in the upwelling ecosystem off Alta California. An attempt has been made to estimate the reliability of the biological submodels by considering a situation in which advection, diffusion, and, perhaps, grazing are not operative. Eppley et al. (1971) collected phytoplankton populations from oligotrophic surface waters of the California Current, filtered these through a net of 183-μ mesh to remove most of the zooplankton, enriched these shipboard collections with nutrients, and monitored their physiological and chemical properties, that is, an artificial, shipboard analogue of the seasonal onset of upwelling off Alta California. We have attempted to simulate this set of experiments in a model with a time step of 1 hr over 5 days, involving nutrient injections at time zero, after ~ 1 day, and after ~ 3 days; see Fig. 25A for the time dependence of the nitrate forcing function.

Our model considers their second experiment from July 11 to 15 1970 under nitrate limitation with presumably saturating values of light, other nutrients, trace metals, and vitamins. Advection and diffusion are not considered in this analysis. Diel periodicity of nutrient uptake, with a maximum division rate of 2.0/day (Table I), is assumed as in the previous steady-state models. Additional consideration is given, however, to chlorophyll synthesis, cell biomass, and the amount of particulate nitrogen and carbon per cell (i.e., the internal pool) with hourly computation of the particulate carbon:chlorophyll a and particulate carbon:particulate nitrogen ratios. Chlorophyll per cell is assumed to increase under light (carbon) limitation

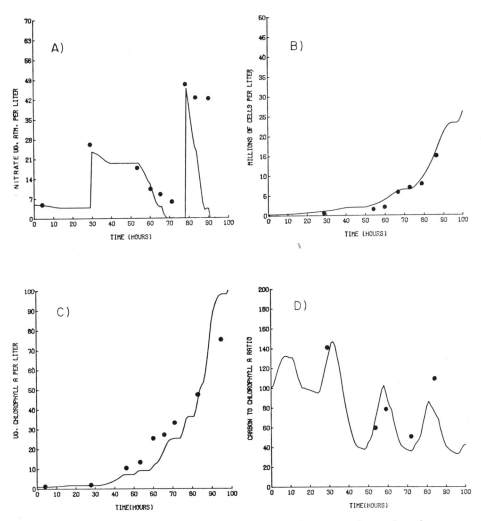

Fig. 25. A comparison of observations (●) and the simulated response of a number of parameters (B, cell number; C, chlorophyll biomass; D, carbon:chlorophyll a ratio) of phytoplankton populations to pulses of nutrients (A) in an analogue of upwelling off Alta California.

and to decrease under nitrogen limitation; more detail is given by Howe and Walsh (in preparation). Grazing and thus excretory recycling of nutrients are not considered in this simulation, while physiological death is considered insignificant and the population is allowed to settle at an exponentially decreasing rate between sampling intervals.

The response of the phytoplankton population to variation of the ambient nutrient regime (Fig. 25A) can be seen in the time history of the cell concentration (Fig. 25B), the chlorophyll biomass (Fig. 25C), and the carbon:chlorophyll a ratio (Fig. 25D). Up until 80 hr in the model, the simulated state variables appear to follow the observations fairly well. For example, nitrate depletion, 2 days after the second inoculum of 27 μg-at NO_3/l, leads to lower cell and chlorophyll biomass with a higher carbon:chlorophyll a ratio in the model; that is, protein synthesis decreased under nitrogen limitation while photosynthesis continues. After nitrate addition at 5, 30, and 80 hr, the simulated population increased with a smaller carbon:chlorophyll a ratio.

The match of the simulated and observed variables up to 80 hr in this shipboard experiment suggests that our hypothesis of poor grazing control in most of the eastern boundary currents may be correct, that is, the model has zero grazing stress, which is equivalent to a 3-day time lag before initiation of any grazing loss, when cell division would have led to larger particles capable of utilization by nekton in the "real" world (Walsh, 1976a). Ammonia concentrations increased at least fivefold during this time, however, suggesting that a population of microzooplankton might have developed during this period, although there is no mention of this in their paper (Eppley et al., 1971). Such a time-dependent grazing hypothesis would account for the higher phytoplankton biomass and lower nitrate of the model, in comparison with observations, after 80 hr. An alternate hypothesis is vitamin limitation (Eppley et al., 1971), which was not considered in the model, and may account for the lower plant biomass and higher nutrients within the observed data set. We have been sufficiently encouraged with the results of these short-term, time-dependent submodels, nevertheless, to begin incorporation of these features in a spatial, time-dependent model of the Northwest African ecosystem. Explicit time dependence and spatial translation of these time lags raise additional problems, however.

> Would it have been worth while,
> To have bitten off the matter with a smile,
> To have squeezed the universe into a ball
> To roll it toward some overwhelming question,

> —idem

5. Patchiness

Grazing thresholds have been invoked as the balm for most model problems in this chapter, when in fact there is some question as to whether their existence can be demonstrated in the laboratory (Corner, Head, and Kilvington, 1972; Frost, 1972). Moreover, if patchiness of prey could be treated in a full, three-dimensional time-dependent model, the need for the parameterization of prey availability in grazing thresholds of one- and two-dimensional models (Steele, 1974; Walsh, 1975c) may disappear. Patchiness itself may also be a non-problem because one must be very careful in specification of the scales and processes that are supposed to bring about the "patchy" distribution of the quantity under consideration.

For example, there has been a renaissance of interest (Steele, 1976; Okubo, 1974; Wroblewski, O'Brien, and Platt, 1975) in Kierstead's and Slobodkin's (1953) theory of critical patch size in which the growth of a phytoplankton population is counterbalanced by diffusion losses. Additional consideration has been given to the loss term of grazing (Wroblewski and O'Brien, 1975), but all the recent studies appear to be concerned with how the patch is maintained, given a particular growth rate and mixing coefficient. The critical length scale is $l = 4.8 \sqrt{K_y/V_{max}}$ (Steele, 1976) and if values of 10^6 cm^2/sec and 1/day (Table I) are selected, a length of ~ 0.5 km is obtained; that is, the phytoplankton patch has to be at least this large, in theory, to avoid dissipation by turbulent mixing. Unfortunately phytoplankton are single-cell organisms, usually $< 100\,\mu$ long, and the really critical question is not how a 0.5×10^9 μ patch is maintained, but how it is formed in the first place.

Patches or plumes of phytoplankton on a length scale of 10–20 km have certainly been observed in the eastern boundary currents (Walsh, 1972; Walsh et al., 1974), and discontinuous production of zooplankton cohorts, that is, relaxation of grazing stress, has been advocated as a formation mechanism for observed algal patches in the North Sea (Steele, 1974). It is possible that a combination of geographical discontinuities (canyons, capes, and submarine ridges leading to local intensification of upwelling), of weak winds, and of *poor* grazing control leads to the patchy structures of nutrients and phytoplankton in upwelling ecosystems. For example, plumes of

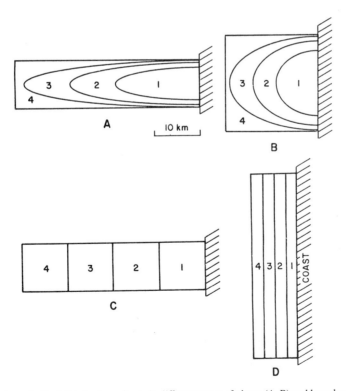

Fig. 26.   The hypothetical distribution of prey in different types of plume (A, B) and longshore band (C, D) structures of the eastern boundary currents.

these variables have been observed (Medriprod, 1973) off Cape Corbiero, Northwest Africa (21°40′N) under weak winds (∼ 1–3 m/sec) during 1972, in contrast to no plume structure in the same area under strong winds (∼ 10 m/sec) during 1974, yet the zoo-plankton biomass appears to be similar from year to year (Howe and Walsh, in preparation).

We have explored (Wirick and Walsh, in preparation) the role of grazing in the formation of patchiness within a stochastic model of nekton foraging in an idealized upwelling ecosystem. Our preliminary model assumes different constant shapes of prey distribution (Fig. 26) and inquires where the fish are likely to be found and which structure (plume or longshore band) might provide the optimal habitat. In our present analysis, the phytoplankton population is held constant; that is, there is no consideration of nutrient or phytoplankton dynamics, and the total fish population is also time invariant without consideration of their metabolic requirements. This submodel expands the spatial resolution of the feeding behavior of anchoveta in the previous Peru steady-state model, but does not contain a grazing threshold or the feedback loops of excretion and diurnal migration; advection and diffusion are also not explicitly considered in this analysis.

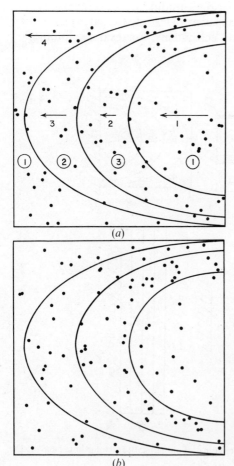

(a)

(b)

Fig. 27. A Monte Carlo simulation of fish schools (●) foraging in an upwelling ecosystem at time zero (a) and after 2½ days (b); the length of the arrows in the first panel is an indication of the flight length, or inverse time spent, in each subdivision of resource (indicated by the circled integer) within the plume habitat, that is, the smallest jumps occur, and thus the most time is spent, in the area 2 of highest resource ③.

A total of 100 fish schools is allowed to move through an x–y 20 × 20 km space of reflecting boundaries with a continuous grid mesh and a time step of 2 hr over a $2\frac{1}{2}$-day period. The consumer's searching mode is modeled with the concept of random flights (Pearson, 1906) in which the organism moves a specified distance, that is, flight length, in a random direction. In our first analysis, the direction was chosen with a random number generator, that is, a Monte Carlo simulation, and the flight length was an inverse function of the available resource with a maximum fish speed of 14 cm/sec at low food concentrations. For example, after 12 iterations, more than 40% of the fish schools had moved to the area with the highest resource of case B (Fig. 27) and a quasi-steady state appears to have been obtained in 2.5 days.

We then refined the searching algorithm with a Markov process (Frank, 1964; Riffenburgh, 1969) in which the transition probabilities from one subdivision of the resource field to another (sections 1–4 of Fig. 26) are specified as an algebraic function of the flight length and the geometric shape of each subdivision rather than with the random number generator; the flight length formulation was the same and the time step was reduced to 20 min. Such analytical expression of this stochastic process appears to fit fairly well the Monte Carlo simulation of case B (Fig. 28) and it also allows us to tractably examine the role of patchiness in nekton distribution.

We defined an index of exploitation of a particular habitat (cases A–D of Fig. 26) to be the expected value of the ration a consumer might obtain while foraging for one time step in a given habitat; at any one time interval of the model, this expectation is the sum of the products of the probability that a fish school is in a subdivision of the resource field and that a ration will be obtained while in this subdivision, that is, the ration is a single hyperbolic function of the ambient resource. At time zero, the expected ration is 44.6 mg C/day in all habitats (Fig. 29) and the steady-state range is 33.7 to 60.6, with the lower value obtained if all fish schools were in subdivisions 1 or 4 (Fig. 26) and the higher value if all the consumers remained in subdivision 2 where the resource is most abundant.

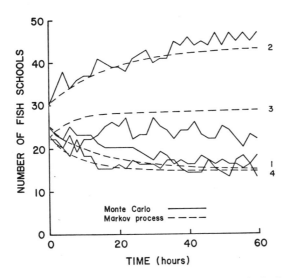

Fig. 28.   A comparison of a Markov Process and a Monte Carlo simulation in the description of stochastic behavior of fish schools within resource areas 1–4 of the upwelling habitat.

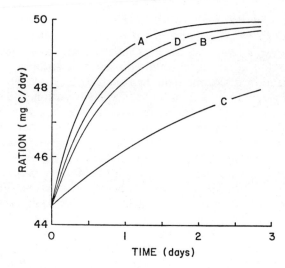

Fig. 29.   An index of exploitation by nekton of various habitats (plumes and longshore bands) of the eastern boundary currents.

   Our preliminary results thus far (Fig. 29) would tend to suggest that our hypo-thetical anchoveta might obtain approximately the same ration in either a plume or longshore band structure; the patchiness of their prey would then have little effect on fish ingestion or distribution. With dynamic coupling to the phytoplankton field, that is, explicit grazing and excretion, our stochastic submodel might suggest a different location of the fish schools, because of probably a new location of the highest algal biomass as a result of the dynamic balance of phytoplankton gains and losses in the x–y space, but the results of our patchiness study would probably still hold under the present assumptions.
   The present submodel assumes that the same type of food is found under the same subdivisional area in each geometric case or habitat, however, and, as previously suggested from the steady-state model, the change from plume to band structure may be accompanied by changes in prey size, in prey species, and in overall trophic structure. Incorporation of different yields of the ration from different types of resources within cases A–D perhaps would lead to quite different values of the exploi-tation index. We are now coupling this submodel with the other previous time-dependent submodels of circulation and lower trophic levels to explore this hypothesis of prey switching between plume and longshore band structures.
   These present stochastic simulations imply, nevertheless, that rather than the cause of patchiness, the consumer may be simply responding to the biological consequences of the heterogeneous physical forcing functions. One might further speculate that detached plumes, or patches, may represent either a temporary decay of upwelling (Wang and Walsh, 1976), a focused, periodic forcing function such as internal waves in a canyon (Shepard, Marshall, and McLoughlin, 1974) or large-scale wave phenom-ena (Gill and Clarke, 1974), and that, irrespective of grazing pressure, these types of intermittency at the 1–2 day time scale (Fig. 4) of the phytoplankton domain could lead to boluses of water 20–40 km apart in a steady 25 cm/sec longshore current. As we begin to consider the spatial and temporal properties of the eastern boundary

ANTCR

currents in more detail, this question of the strength of the nekton coupling to the rest of the system becomes very important.

> We have lingered in the chambers of the Sea
> By Sea-girls wreathed with seaweed red and brown
> Till human voices wake us, and we drown.

— idem

6. Prospectus

It is admittedly difficult to distinguish natural fluctuations of clupeid stocks from man-induced alterations (Soutar and Isaacs, 1974), but the record of man's exploitation of these resources appears to be all too familiar (Murphy, 1973). Most of the world's previous clupeid fisheries now yield less than 10% of their maximum harvest (Table II), and these declines might be attributed to overfishing. One could argue, however, that the most serious declines of clupeid stocks have occurred in nonupwelling systems without the continual input of nutrients to maintain large yields of nekton and further that the sardines have been replaced by anchovy in the California Current.

Unfortunately the most well studied and supposedly well managed clupeid fishery off Peru has begun to show signs of fishing stress as well (Murphy, 1973). Coincident with the 1972 El Niño (Wooster and Guillen, 1974), the yield of anchoveta from this ecosystem dropped precipitously (Table III). It is possible that the present collapse of the Peru fishery over the last three years (1972–1974) is linked to a temporary recruitment failure in relation to changed hydrographic conditions, but it is also possible that this anchoveta stock might not recover, with the energy flow diverted instead to the Chilean sardine (Longhurst, 1971) or the demersal hake populations (Cushing, personal communication); for example, the heavy fishing pressure on the North Sea herring might have led to the large demersal catches of more than 2×10^6 tons in this area during 1971 (Gulland, 1976).

There is the haunting possibility that nekton of the eastern boundary currents are poorly coupled to the lower trophic levels in the sense that, once the clupeid stocks are fished below their carrying capacity, the release of grazing stress on a patchy resource is not a sufficiently strong feedback to allow rapid increase of prey and thus

TABLE II

Yields from Clupeoid Resources that Have Collapsed (after Murphy, 1973)

Stock	Present Yield (1970) (thousands of tons)	Maximum Yield (thousands of tons)
Hokkaido–Sakhalin herring	0	800
Atlanto–Scandian herring	21	1723
Downs herring	0	98
Pacific sardine	0	791
Japanese sardine	21	1590
South African sardine	81	452
Total	104	4454

TABLE III

Yield of Anchoveta from the
Peru Upwelling Ecosystem
(after Jordan, 1974)

Year	Fish Catch (thousands of tons)
1959	1,909
1960	2,944
1961	4,580
1962	6,275
1963	6,423
1964	8,863
1965	7,233
1966	8,530
1967	9,825
1968	10,263
1969	8,960
1970	12,277
1971	10,282
1972	4,449
1973	2,035
1974[a]	3,692

[a] January to November 26th.

rapid recovery of the predator. A corollary hypothesis might be that with the removal of the fish, in situ chelators or other organic growth substances (Barber, Dugdale, MacIsaac, and Smith, 1971) in the form of excretory products may no longer be available to stimulate phytoplankton growth as in the previously conditioned water within areas of dense schools of anchoveta. Such a nonlinear interaction might lead to poor productivity at a time when such a hypothetically recovering stock would have the most critical demand for high availability of resources.

One might ask how simulation models can be used to answer this evidently new question of not why upwelling areas are so rich, nor why their potential yield varies (Fig. 1), but why the upwelling systems may now be so poor. In the discussions of this chapter, I have demonstrated that we have had some beginning success in coupling physical, chemical, and biological submodels to describe the steady-state and transient processes of upwelling ecosystems. Further, the oceanographic community may soon be able to develop elegant black-box models (Fig. 2) of El Niño phenomena in which future Peruvian anchoveta oscillations may be predicted from indices (Quinn, 1974) of large-scale atmospheric (Namais, 1973) and oceanic (Wyrtki, 1973) perturbations. From this future work, we will still not know how the nekton are coupled to circulation and lower trophic level interactions, however, and an expanded knowledge of the important biological frequency domains (Fig. 4) must accompany any future studies of transients in the physical forcing functions of the eastern boundary currents.

The event time scale, with its associated spatial patchiness, is probably one of the least understood oceanographic phenomena, and, analogous to numerical weather prediction, may represent a serious impediment to future simulation models of upwelling ecosystems. Numerical models are isomorphs of available data sets, however, and as we begin to obtain coupled time series of both physical and biological data (e.g., Fig. 23 of our 1973 Baja California study) in which physical events and the biological response are embedded, we can begin to assess the importance of all scales of variability in the eastern boundary currents. Our present state of knowledge suggests, nevertheless, that simulation models will be of most utility in data analysis and experimental design rather than in prognostications; ecosystem models and their submodels of various processes are very efficient in pinpointing which critical data are missing and which scales must be studied in the ocean.

An additional hope for understanding the dynamics of marine ecosystems is the subtle shift in scientific emphasis from individual to group endeavors in which cross-disciplinary information exchanges occur, as evidenced in the rapidly growing literature of Ph.D. dissertations on all facets of the eastern boundary currents.

Acknowledgments

In addition to ideas in this review from my other colleagues in the Coastal Upwelling Ecosystems Analysis (CUEA) program and from O. Poppell, I am indebted to recent unpublished work of D. Barton, M. Brodersen, L. Castiglione, L. Codispoti, L. Conway, C. Davis, A. Griffiths, P. Hamilton, P. Harrison, T. Hopkins, S. Howe, H. Hurlburt, A. Huyer, N. Kachel, C. Mooers, L. Pietrafesa, D. Pillsbury, T. Rawson, G. Shaffer, R. Strickland, D. Thompson, R. Trumble, D. Wang, T. Whitledge, C. Wirick, and J. Wroblewski. Finally, I would like to acknowledge that this multi-disciplinary research has been funded by the U.S. Antarctic Research Program (USARP), the U.S. International Biological Program (IBP), and the U.S. International Decade of Ocean Exploration Program (IDOE), all of the National Science Foundation.

References

Bakun, A., D. R. McLain, and F. V. Mayo, 1974. The mean annual cycle of coastal upwelling off western North America as observed from surface measurements. *Fish. Bull.*, **72**, 843–844.

Barber, R. T., R. C. Dugdale, J. J. MacIsaac, and R. L. Smith, 1971. Variations in phytoplankton growth associated with the source and conditioning of upwelling water. *Invest. Pesq.*, **35**, 171–193.

Beers, J. R., M. R. Stevenson, R. W. Eppley, and E. R. Brooks, 1971. Plankton populations and upwelling off the coast of Peru, June 1969. *Fish. Bull.*, **69**, 859–876.

Beers, J. R. and G. L. Stewart, 1971. Micro-zooplankters in the plankton communities of the upper waters of the eastern tropical Pacific. *Deep-Sea Res.*, **18**, 861–865.

Blackburn, M., 1973. Regression between biological oceanographic measurements in the eastern tropical Pacific and their significance to ecological efficiency. *Limnol. Oceanogr.*, **18** (4), 552–563.

Corner, E. D. S., R. N. Head, and C. C. Kilvington, 1972. On the nutrition and metabolism of zooplankton. 8. The grazing of *Biddulphia* cells by *Calanus helgolandicus*. *J. Mar. Biol. Assoc. U.K.*, **52**, 847–861.

Cushing, D. H., 1971a. Upwelling and the production of fish. *Adv. Mar. Biol.*, **9**, 255–334.

Cushing, D. H., 1971b. A comparison of production in temperate seas and the upwelling areas. *Trans. Roy. Soc. S. Afr.*, **40**, 17–33.

Dugdale, R. C., 1967. Nutrient limitation in the sea: dynamics, identification, and significance. *Limnol. Oceanogr.*, **12**, 685–695.

Dugdale, R. C., 1972. Chemical oceanography and primary productivity in upwelling regions. *Geoforum.*, **11**, 47–62.

Dugdale, R. C. and J. J. Goering, 1967. Uptake of new and regenerated forms of nitrogen in primary productivity. *Limnol. Oceanogr.*, **12**, 196–206.

Eppley, R. W., A. F. Carlucci, O. Holm-Hansen, D. Kiefer, J. J. McCarthy, E. Venrick, and P. M. Williams, 1971. Phytoplankton growth and composition in shipboard cultures supplied with nitrate, ammonia, or urea as the nitrogen source. *Limnol. Oceanogr.*, **16**, 741–751.

Frank, P. W., 1964. On home range of limpets. *Am. Nat.*, **98**, 99–104.

Frost, B. W., 1972. Effects of size and concentration of food particles on the feeding behavior of the marine planktonic copepod *Calanus pacificus*. *Limnol. Oceanogr.*, **17**, 805–815.

Gill, A. E. and A. J. Clarke, 1974. Wind-induced upwelling, coastal currents, and sea-level changes. *Deep-Sea Res.*, **21**, 325–346.

Grice, G. D. and A. D. Hart, 1962. The abundance, seasonal occurrence, and distribution of the epizooplankton between New York and Bermuda. *Ecol. Monogr.*, **32**, 287–309.

Gulland, J. A., 1976. Production and catches of fish in the sea. In *Ecology of the Seas*. D. H. Cushing and J. J. Walsh, eds. Blackwell, Oxford, pp. 283–314.

Hagen, E., 1974. Ein einfaches Schema der Entwicklung von Kaltwasserauftriebszellen vor der Nordwestafrikanische Küste. *Beitr. Meereskd.*, **33**, 115–126.

Hamilton, P. and M. Rattray, 1976. A numerical model of the depth-dependent, wind-driven upwelling circulation on a continental shelf. *J. Phys. Oceanogr.*, in press.

Hart, T. J., 1942. Phytoplankton periodicity in Antarctic surface waters. *Discovery Rep.*, **12**, 261–365.

Herbland, A., R. LeBorgne, and B. Voituriez, 1973. Production primaire, secondaire, et regeneration des sels nutritifs dans l'upwelling de Mauretanie. *Doc. Sci. Cent. Rech. Oceanogr. Abidjan*, **4**, 1–75.

Herbland, A., R. LeBorgne, and B. Voituriez, 1974. Main physical and biological features of the Cape Timiris area (Northwest Africa). *C. U. E. A. Newsl.*, **3** (5), 10–13.

Huntsman, S. A. and R. T. Barber, 1976. Primary production off Northwest Africa: the relationship to wind and nutrient conditions. *Deep-Sea Res.*, in press.

Idyll, C. P., 1973. The anchovy crisis. *Sci. Am.*, **228**, 22–29.

Jacobs, S. S., 1965. Physical and chemical oceanographic observations in the Southern Oceans. U.S.N.S. *Eltanin* Cruises 7–15, 1963–1964. *Tech. Rep. 1-cu-1-65*. Columbia University, 321 pp.

Jacobs, S. S., 1966. Physical and chemical oceanographic observations in the Southern Ocean. U.S.N.S. *Eltanin* Cruises 16–21, 1965. *Tech. Rep. 1-cu-1-66*. Columbia University, 128 pp.

Jacobs, S. S. and A. F. Amos, 1967. Physical and chemical oceanographic observations in the Southern Oceans. U.S.N.S. *Eltanin* Cruises 22–27. 1966–1967. *Tech. Rep.* 1-cu-k-67. Columbia University, 287 pp.

Jordan, R., 1974. Biology of the anchoveta. Part I: Status of the present knowledge. *Proc. Workshop on the "El Niño Phenomenon, Guayaquil, Ecuador, 4–12 December 1974*, 30 pp.

Kierstead, H. and L. B. Slobodkin, 1953. The size of water masses containing plankton blooms. *J. Mar. Res.*, **12**, 141–147.

Kimball, H. H., 1928. Amount of solar radiation that reaches the surface of the earth on the land and on the sea and the methods by which it is measured. *Mon. Weather Rev.*, **56**, 393–398.

Knauss, J. A., 1966. Further measurements and observations on the Cromwell Current. *J. Mar. Res.*, **24**, 205–240.

Longhurst, A. R., 1971. The clupeoid resources of tropical seas. *Oceanogr. Mar. Biol. Annu. Rev.*, **9**, 349–385.

Longhurst, A. R., 1976. Vertical migration in the sea. In *Ecology of the seas*. D. H. Cushing and J. J. Walsh, eds. Blackwell, Oxford, pp. 116–137.

Longhurst, A. R., C. J. Lorenzen, and W. H. Thomas, 1967. The role of pelagic crabs in the grazing of phytoplankton off Baja California. *Ecology*, **48**, 190–200.

MacKintosh, M. A., 1937. The seasonal circulation of the Antarctic macroplankton. *Discovery Rep.*, **16**, 367–412.

MacIsaac, J. J. and R. C. Dugdale, 1969. The kinetics of nitrate and ammonia uptake by natural populations of marine phytoplankton. *Deep-Sea Res.*, **16**, 45–57.

Maeda, S. and R. Kishimoto, 1970. Upwelling off the coast of Peru. *J. Ocean. Soc. Japan*, **26**, 300–309.

Marr, J. W. S., 1962. The natural history and geography of the Antarctic Krill (*Euphausia superba dana*). *Discovery Rep.*, **32**, 33–464.

Mauchline, J. and L. R. Fisher, 1969. The biology of euphausiids. *Adv. Mar. Biol.*, **7**, 1–454.

Medriprod, 1973. Donnees physiques, chemiques, et biologiques. Campagne "CINECA-CHARCOT II," Programme de la RCP-247 du C. N. R. S.

Menzel, D. W. and J. H. Ryther, 1960. The annual cycle of primary production in the Sargasso Sea off Bermuda. *Deep-Sea Res.*, **6**, 351–367.

Menzel, D. W. and J. H. Ryther, 1961. Zooplankton in the Sargasso Sea off Bermuda and its relation to organic production. *J. Cons.*, **26**, 250–258.

Menzel, D. W. and J. P. Spaeth, 1962. Occurrence of ammonia in Sargasso Sea waters and in rain water at Bermuda. *Limnol. Oceanogr.*, **7** (2), 159–162.

Montgomery, R. B., 1959. Salinity and the residence time of subtropical oceanic surface water. In *The Atmosphere and the Sea in Motion*. B. Bolin, ed. Oxford University Press, pp. 143–146.

Mooers, C. N. K., C. A. Collins, and R. L. Smith, 1976. The dynamic structure of the frontal zone in the coastal upwelling region off Oregon, *J. Phys. Oceanogr.*, **6**, 3–21.

Morishima, D. L., P. B. Bass, and J. J. Walsh, 1974. AUGUR, a three-dimensional simulation program for non-linear analysis of aquatic ecosystems. *C. U. E. A. Tech. Rep.*, **7**, 239 pp.

Murphy, G. I., 1973. Clupeoid fishes under exploitation with special reference to the Peruvian anchovy. *Univ. Hawaii Tech. Rep.*, **30**, 73 pp.

Namais, J., 1973. Response of the equatorial countercurrent to the subtropical atmosphere. *Science*, **181**, 1244–1245.

O'Brien, J. J., 1975. Models of coastal upwelling. In *Proc. Symp. Numerical Models of Ocean Circulation*. National Academy of Sciences, Washington, D.C., pp. 204–215.

Okubo, A., 1974. Diffusion-induced instability in model ecosystems: another possible explanation of patchiness. *Chesapeake Bay Inst. Tech. Rep.*, **86**, 17 pp.

Paffenhöfer, G. A., 1970. Cultivation of *Calanus helgolandicus* under controlled conditions. *Helgol. Wiss. Meeresunters.*, **20**, 346–259.

Parsons, T. R., 1976. The structure of life in the sea. In *Ecology of the Seas*. D. H. Cushing and J. J. Walsh, eds. Blackwell, Oxford, pp. 81–97.

Paulik, G. J., 1971. The anchoveta fishery of Peru. *Center Quant. Stud. Pap. 13*. University of Washington, Seattle, 1–79 pp.

Pearson, K., 1906. A mathematical theory of random migration. *Drapers Co. Res. Mem., Biometric Ser. No. 3*.

Quinn, W. H., 1974. Monitoring and predicting El Niño invasions. *J. Appl. Meteor.*, **13**, 825–830.

Riley, G. A., 1963. Theory of food chain relations in the ocean. In *The Sea*. Vol. 2, M. N. Hill, ed. Wiley-Interscience, New York, pp. 438–463.

Riffenburgh, R. H., 1969. A stochastic model of interpopulation dynamics in marine ecology. *J. Fish. Res. Bd. Can.*, **26**, 2843–2879.

Ryther, J. H., 1963. Geographical variations in productivity. In *The Sea*. Vol. 2. M. N. Hill, ed. Wiley-Interscience, New York, pp. 347–380.

Ryther, J. H., 1969. Photosynthesis and fish production in the sea. *Science*, **166**, 72–76.

Shaffer, G., 1974. On the Northwest African coastal upwelling system. Ph.D. thesis. Kiel University, 178 pp.

Sheldon, R. W., W. H. Sutcliffe, and A. Prakash, 1973. The production of particles in the surface waters of the ocean with particular reference to the Sargasso Sea. *Limnol. Oceanogr.*, **18**, 719–733.

Shepard, F. P., N. F. Marshall, and P. A. McLouglin, 1974. Currents in submarine canyons. *Deep-Sea Res.*, **21**, 691–706.

Smayda, T. J., 1966. A quantitative analysis of the phytoplankton of the Gulf of Panama. III. General ecological conditions and the plankton dynamics at 8°45′ N, 79°23′ W from November 1954 to May 1957. *Inter-Amer. Trop. Tuna Comm. Bull.*, **11**, 355–612.

Smith, R. L., 1968. Upwelling. *Ann. Rev. Ocean. Mar. Biol.*, **6**, 11–47.

Smith, R. L., 1974. A description of current, wind, and sea level variations during coastal upwelling off the Oregon coast, July–August, 1972. *J. Geophys. Res.*, **79**, 435–443.

Soutar, A. and J. D. Isaacs, 1974. Abundance of pelagic fish during the 19th and 20th centuries as recorded in anaerobic sediment off the Californias. *Fish. Bull.*, **72**, 257–273.

Steele, J. H., 1974. The structure of marine ecosystems. Harvard University Press, Cambridge, Mass. 128 pp.

Steele, J. H., 1976. Patchiness in the Sea. In *Ecology of the Seas*. D. H. Cushing and J. J. Walsh, eds. Blackwell, Oxford, pp. 98–115.

Stefansson, U. and L. P. Atkinson, 1971. Nutrient-density relationships in the western North Atlantic between Cape Lookout and Bermuda. *Limnol. Oceanogr.*, **16**, 51–59.

Taniguchi, A., 1973. Phytoplankton-zooplankton relationships in the western Pacific Ocean and Adjacent seas. *Mar. Biol.*, **21**, 115–121.

Vinogradov, M. E., V. V. Menshutkin, and E. A. Shushkina, 1972. On mathematical simulation of a pelagic ecosystem in tropical waters of the ocean. *Mar. Biol.*, **16** (4), 261–268.

Walsh, J. J., 1969. Vertical distribution of antarctic phytoplankton. II. A comparison of phytoplankton standing crops in the Southern Ocean with that of the Florida Strait. *Limnol. Oceanogr.*, **14** (1), 86–94.

Walsh, J. J., 1971. Relative importance of habitat variables in predicting the distribution of phytoplankton at the ecotone of the Antarctic Upwelling Ecosystem. *Ecol. Monogr.*, **41**, 291–309.

Walsh, J. J., 1972. Implications of a systems approach to oceanography. *Science*, **176**, 969–975.

Walsh, J. J., 1974. Primary production in the sea. *Proc. First Intern. Congr. Ecol.* PUDOC, Wageningen, pp. 150–154.

Walsh, J. J., 1975a. Utility of systems models: a consideration of some possible feedback loops of the Peru upwelling ecosystem. In *Proc. Second Intern. Estuar. Res. Conf.* L. E. Cronin, ed. Academic Press, New York, pp. 617–633.

Walsh, J. J., 1975b. A biological interface of numerical models and the real world—an elegy for E. J. Ferguson Wood. In *Proc. Symp. Numerical Models of Ocean Circulation.* National Academy of Sciences, Washington, D.C., pp. 5–9.

Walsh, J. J., 1975c. A spatial simulation model of the Peru upwelling ecosystem. *Deep-Sea Res.*, **22**, 201–236.

Walsh, J. J., 1976a. Herbivory as a factor in patterns of nutrient utilization in the sea. *Limnol. Oceanogr.*, **21**, 1–13.

Walsh, J. J., 1976b. Modelled processes in the sea. In *Ecology of the Seas.* D. H. Cushing and J. J. Walsh, eds. Blackwell, Oxford, pp. 338–407.

Walsh, J. J. and P. B. Bass, 1971. OCEANS—a seagoing simulation program. A user's guide to the University of Washington's IBM 1130 spatial version of COMSYS 1. *Spec. Rep. 48.* University of Washington, 98 pp.

Walsh, J. J. and R. C. Dugdale, 1971. A simulation model of the nitrogen flow in the Peruvian upwelling system. *Invest. Pesq.*, **35**, 309–330.

Walsh, J. J. and S. O. Howe, 1976. Protein from the Sea: a comparison of the simulated nitrogen and carbon productivity in the Peru upwelling ecosystem. In *Systems Analysis and Simulation in Ecology.* Vol. 4. B. C. Patten, ed. Academic Press, New York, pp. 47–61.

Walsh, J. J., J. C. Kelley, T. E. Whitledge, J. J. MacIsaac, and S. A. Huntsman. 1974. Spin-up of the Baja California upwelling ecosystem. *Limnol. Oceanogr.*, **19**, 553–572.

Walsh, J. J., J. C. Kelley, T. E. Whitledge, S. A. Huntsman, and R. D. Pillsbury, 1976. Further transition states of the Baja California upwelling ecosystem. *Limnol. Oceanogr.*, in press.

Wang, D. P. and J. J. Walsh, 1976. Objective analysis of the upwelling ecosystem off Baja California. *J. Mar. Res.*, **34**, 43–60.

Wickett, W. P., 1967. Ekman transport and zooplankton concentration in the North Pacific Ocean. *J. Fish. Res. Bd. Can.*, **24**, 581–594.

Wooster, W. S. and J. Reid, 1963. Eastern boundary currents. In *The Sea.* Vol. 2, M. N. Hill, ed. Wiley-Interscience, New York, pp. 253–280.

Wooster, W. S. and O. Guillen, 1974. Characteristics of El Niño in 1972. *J. Mar. Res.*, **32**, 387–404.

Wooster, W. S., A. Bakun, and D. R. McLain, 1976. The seasonal upwelling cycle along the eastern boundary of the North Atlantic. *J. Mar. Res.*, **34**, 131–141.

Worthington, L. V., 1959. The 18° water in the Sargasso Sea. *Deep-Sea Res.*, **5**, 297–305.

Wroblewski, J. S., J. J. O'Brien, and T. Platt, 1975. On the physical and biological scales of phytoplankton patchiness in the ocean. *Proc. Sixth Liege Coll. Ocean Hydrodyn., Mem. Soc. Roy. Sci. Liege*, **7**, 43–57.

Wyrtki, K., 1973. Teleconnections in the equatorial Pacific Ocean. *Science*, **180**, 66–68.

Yoshida, K., 1967. Circulation in the eastern tropical ocean with special reference to upwelling and undercurrents. *Jap. J. Geophys.*, **4**, 1–75.

25. ESTUARINE PHYTOPLANKTON BIOMASS MODELS—VERIFICATION ANALYSES AND PRELIMINARY APPLICATIONS

Dominic M. Di Toro, Robert V. Thomann, Donald J. O'Connor, and John L. Mancini

1. Introduction

The quantitative analysis of water quality problems has become an important component of comprehensive water quality management and planning for estuaries. Decisions relating, for example, to the type and degree of treatment of industrial and municipal discharges to estuarine waters are usually conditioned by some type of quantitative assessment of the response of the estuary to these actions. Rational water quality management requires such an assessment for it is only through these calculations that it is possible to estimate the benefits of the proposed actions. The quantitative framework or model, as it has become known, relates the concentrations of certain substances in the estuary to the quantities introduced into the estuary either from the upstream river flow or from direct sources within the estuary proper.

The complexity of the model depends to a large extent on the time and space scales for which the analysis is required. Thus temporal steady-state models in one spatial dimension are an order of magnitude less complicated than seasonally varying two-dimensional models. The other complicating feature is the number of interacting dependent variables involved, and the degree of their interaction. Models of conservative quantities, the classical example being chloride ion, are straightforward when compared to models which involve bacterially mediated reactions such as the oxidation of reduced carbonaceous and nitrogenous substances. These, in turn, are less involved than models which attempt to describe biological reactions and transformations in some detail.

The subject of this chapter is a model in the latter category—a model that attempts to relate the growth of phytoplankton biomass to the concentrations of available nutrients on the one hand and the activity of zooplankton predation on the other. In addition, the effects of temperature and available solar radiation are incorporated into the framework as well as the bacterially mediated mineralizations of organic detritus which provide additional sources of nutrients. This framework is a direct descendant of previous models of estuarine water quality (O'Connor and Thomann, 1971), with their emphasis on dissolved oxygen distributions and of models of phytoplankton biomass in coastal waters (Riley et al., 1949), which were developed more from a scientific point of view and less from a problem-solving orientation. Thus the treatment of the physical transport phenomena is virtually identical to that used for the analysis of salinity intrusion in estuaries. The kinetics employed for the biological interactions are elaborations of those used in the original phytoplankton biomass models, which stress the fundamental interactions between temperature, light, nutrient concentrations, and zooplankton predation. More empirical structures are used to describe the mineralization reactions in a fashion analogous to the estuarine water quality models for dissolved oxygen. Because the emphasis is toward establishing the causal relations between the nutrients discharged to the estuary and the resulting phytoplankton biomass, a heavy stress is placed on a direct incorporation of the rates of these discharges to the estuary.

The model that results is a synthesis of the currently available quantitative methods for describing transport phenomena in estuaries and the biological reactions that are known to be important. Certain reactions are relatively well understood, for example, growth of phytoplankton biomass, whereas other reactions are less well defined, such as mineralization and zooplankton predation. Thus the model is not uniformly based on solid experimentally determined fact. This is a common situation in engineering analysis. The justification for the procedure is the extent to which the calculations agree with observed data, and the extent to which the calculations are sensitive to the parts of the model that are not firmly grounded. Since these models are not direct expressions of fundamental scientific fact even in their most refined form, their use is justifiable only after a detailed and thorough verification analysis.

The review presented here deals with the application of essentially similar models to the estuary of San Francisco Bay, and the Potomac Estuary. In each case the emphasis is on the verification and the procedure used to arrive at what is judged to be an acceptable verification. Examples of applications to questions involving water quality management issues are also presented to illustrate the utility of such calculations within the larger context of managing and protecting the water quality of estuaries.

The underlying point of view in these formulations is the distinction between kinetics, the mathematical expressions which relate rates of microorganism growths and death to concentrations of nutrients, temperature, and, for phytoplankton, light; and the specific setting which governs these concentrations, temperature, etc. If this point of view is adopted, it is clear that experiments in the laboratory and in situ bottle experiments should conform reasonably well to the calculated behavior using the same kinetics. Further, and of primary importance, the prototype behavior should also conform. The features that severely modify the prototype behavior are the mass transport effects and the external sources of nutrients. But these are, in a sense, incidental if the kinetics are the same in both cases. The San Francisco Bay estuary investigation is presented to highlight this point of view. The Potomac Estuary model is presented to highlight the application of such calculations to practical water quality questions.

2. Modeling Framework

The basis of the model to be discussed is the principle of conservation of mass, which requires that there be a strict accounting of the mass of nutrients, phytoplankton, etc., either introduced into the estuary or created within it. The mathematical statement of this principle is the equation of continuity which gives a relationship at a point $\mathbf{r}$ and time t between the mass flux ($g/m^2/day$) of a substance $\mathbf{j}$, which is due to the physical transport phenomena of the estuary; the sources and sinks ($mg/l/day$) of the substance within the estuary S, which are due to both internal sources such as biological reactions and external inputs; and the time rate of change of the concentration, $c(\mathbf{r}, t)$:

$$\frac{\partial c}{\partial t} + \nabla \cdot \mathbf{j} = S(c, \mathbf{r}, t) \tag{1}$$

The equation requires that if there exists a source of mass at a location $\mathbf{r}$, the mass being introduced must be accounted for either by being transported away from that location ($\nabla \cdot \mathbf{j} = \sum_i \partial j_i / \partial x_i$ is the sum of the gradients of the mass flux in the three coordinate directions and therefore is the rate of mass transport away from the point) or by an increase of the concentration with time. This equation must apply to all the variables of the model; in fact it is the basis of the detailed equation for each variable.

The equation as it stands is too general to be used directly: what is required is a specification for the mass flux, $\mathbf{j}$, which reflects the mass transport phenomena in the estuary, and the sources and sinks for each substance, which involve the reaction kinetics of the substances being considered.

A. Transport Phenomena

Transport phenomena in estuaries have received considerable attention and there exist quite a number of techniques for describing the mass flux of a substance in an estuary. The most detailed of these involves calculating or measuring the velocity variations throughout the tidal cycle and using these velocities, $\mathbf{v}(\mathbf{r}, t)$, directly to specify the mass flux:

$$\mathbf{j} = \mathbf{v}c \tag{2}$$

For two-dimensional tidal embayments with no significant vertical circulation such a procedure is feasible. However, the time and space scales for such a calculation are small relative to the tidal period and a typical tidal excursion, respectively. For larger time and space scales the mass flux is represented as the sum of two components: the flux due to the net-over-tidal cycle velocity field $\langle \mathbf{v} \rangle$, and the flux due to the mixing motions of the estuary.

$$\mathbf{j} = \langle \mathbf{v} \rangle c - E\nabla c \tag{3}$$

where E is the diagonal matrix of dispersion coefficients which parameterize the mixing.

This is the representation that will be used for the phytoplankton biomass model since the time scales for these phenomena are long relative to the tidal period.

B. Implementation

The numerical solution of sets of conservation equations requires that the estuary be divided into finite volumes. For the variables of concern this corresponds to dividing the estuary into a series of segments or cells which are chosen so that the assumption of spatial homogeneity within each segment is reasonable.

A typical conservation of mass equation used in the model for concentration c_{ij} of substance i in segment j has the general form:

$$V_j \frac{dc_{ij}}{dt} = \sum_k Q_{kj}c_{ik} + \sum_k E'_{kj}(c_{ik} - c_{ij}) + S_{ij} \tag{4}$$

V_j is the segment volume, S_{ij} is the net source of substance i in segment j; E'_{kj} is the bulk rate of transport of c_{ik} into and c_{ij} out of segment j for all segments k adjacent to segment j, and Q_{kj} is the net advective flow rate between segments k and j. Numerical integration of these equations gives the seasonal distribution of the concentrations in each of the spatial segments of the model.

C. Kinetics

The kinetic structure employed for this model is comprised of reaction kinetic equations for the relevant interactions in the estuary. The 11 variables considered and their interactions are illustrated in Fig. 1. The cyclical structure of the pathways is

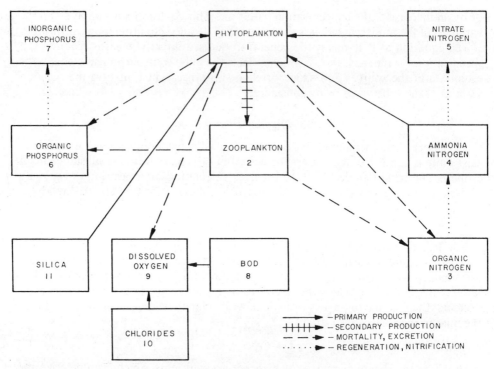

Fig. 1.   Kinetic interactions of the variables.

apparent: the primary production which converts inorganic nutrients to the phytoplankton; the secondary production of zooplankton accomplished by their grazing
on phytoplankton; the mortality and excretion pathways which release organic
material in detrital and soluble form; and the regeneration pathways which convert
organic forms into inorganic forms that are then available for the primary production
pathway. A discussion of each of the kinetic equations is presented below.

The Phytoplankton System

The kinetic source terms for the phytoplankton biomass equation relate the rate of
production, S_P, of biomass, measured in this model as chlorophyll *a* concentration, to
the rates of growth and death of the population. Thus if P is the chlorophyll *a*
concentration, then

$$S_P = (G_P - D_P)P \tag{5}$$

where G_P is the phytoplankton growth rate and D_P is the phytoplankton death rate.

The formulation of the phytoplankton growth rate for a depth-averaged model is
based on the following reasoning (Di Toro et al., 1971): At optimal conditions of
light availability and nutrient concentration the growth rate of population $K_1(T)$ is
dependent on temperature only. The effect of nonoptimal light intensity is to reduce
this growth rate. If I_s is the optimal or saturating light intensity, then it has been
proposed (Steele, 1965) that the reduction in growth rate due to an intensity I is
given by

$$F(I) = \frac{I}{I_s} \exp\left[-\frac{I}{I_s} + 1 \right] \tag{6}$$

In the natural environment the light available at any depth $I(z)$, varies inversely with depth according to the equation:

$$I(z) = I_0 \exp(-K_e z) \tag{7}$$

where z is the depth (positive downward), I_0 is the surface light intensity, and K_e is the extinction coefficient. In addition, the surface light intensity varies throughout the day. For the time scale of this model, however, it is adequately represented by a constant I_a, the mean daily solar radiation, which is incident for a fraction of a day, f, the photoperiod.

For a model that is depth averaged and with a time scale on the order of 1 week, it is appropriate to use a depth averaged time-averaged growth rate reduction factor, r, due to nonoptimal light. The result, using equations 6 and 7, is (Di Toro et al., 1971)

$$r = \frac{ef}{K_e H} \left[\exp(-\alpha_1) - \exp(-\alpha_0) \right] \tag{8}$$

where

$$\alpha_1 = \frac{I_a}{I_s} \exp(-K_e H) \tag{9}$$

$$\alpha_0 = \frac{I_a}{I_s} \tag{10}$$

and H = depth of the segment, f = photoperiod, K_e = extinction coefficient, I_s = optimal light intensity, and I_a = mean daily light intensity.

The effect of nonoptimal nutrient concentrations is to further reduce the growth rate. The form of the reduction factor chosen is the same as that adopted by Monod for bacterial growth, namely, the Michaelis–Menten expression: $N/(K_{mN} + N)$, where N is the nutrient concentration and K_{mN} is the Michaelis or half-saturation constant for that nutrient (Eppley et al., 1969). The two nutrients considered are total inorganic nitrogen: $c_N(NH_3 + NO_3$, assuming NO_2 concentrations are negligible) and ortho-phophorus: $c_p(PO_4)$. Based on an analysis (Di Toro et al., 1971) of a set of laboratory experiments (Ketchum, 1939), and for lack of a better assumption, it is assumed that the growth rate reduction due to low nutrient concentrations is expressible as a product of two Michaelis–Menten expressions: $c_N c_p/(K_{mN} + c_N)(K_{mp} + c_p)$ where K_{mN} and K_{mp} are the half-saturation constants for total inorganic nitrogen and ortho-phosphorus, respectively. The growth rate expression is then assumed to be the product of these reduction factors:

$$G_P = K_1(T)r \frac{c_N}{K_{mN} + c_N} \frac{c_p}{K_{mp} + c_p} \tag{11}$$

where $K_1(T)$ is the temperature-dependent saturated growth rate.

The formulation of the phytoplankton death rate follows a previous analysis (Di Toro et al., 1971). It is assumed that the phytoplankton biomass is reduced by its endogenous respiration, which is temperature dependent, and by the grazing of the zooplankton population, which is proportional to the zooplankton biomass concentration Z. Thus the death rate expression is given by

$$D_P = K_2(T) + C_g(T)Z \tag{12}$$

where $K_2(T)$ = temperature-dependent endogenous respiration rate constant, and $C_g(T)$ = temperature-dependent grazing rate of the zooplankton biomass. Equations

11 and 12 specify the growth and death rates of the phytoplankton and, therefore, also specify the behavior of the phytoplankton population's interaction with temperature, light, extinction coefficient, depth, nutrient concentrations, and zooplankton predation.

The Zooplankton System

The zooplankton kinetic source term is analogous in form to that of the phytoplankton (equation 5):

$$S_Z = (G_Z - D_Z)Z \tag{13}$$

where G_Z and D_Z are the growth and death rates of the zooplankton population whose biomass concentration, Z, is expressed as its organic carbon concentration. If it is assumed that there is sufficient phytoplankton biomass to provide the food source for the zooplankton that affect the phytoplankton (i.e., the herbivorous zooplankton which are the zooplankton of concern) then their growth rate is directly related to their grazing of the phytoplankton and can be formulated as (Di Toro et al., 1971):

$$G_Z = a_1 a_{ZP} \frac{K_{mP}}{P + K_{mP}} C_g(T)P \tag{14}$$

where a_1 = the assimilation efficiency of zooplankton, K_{mP} = the half-saturation constant for the phytoplankton biomass grazed, and a_{ZP} = the carbon/chlorophyll ratio of the phytoplankton population. The formulation of the zooplankton death rate presents somewhat of a problem, because in addition to their endogenous respiration rate the zooplankton are being preyed upon by the upper levels of the food chain. In order to simplify the model framework it is necessary to introduce this effect empirically as an additional death rate constant. Thus the zooplankton death rate is expressed as

$$D_Z = K_3(T) + K_4 \tag{15}$$

where $K_3(T)$ = the temperature-dependent endogenous respiration rate, K_4 = empirical mortality constant.

The Nitrogen System

The major components of the nitrogen system included in this model are nonliving organic nitrogen, c_3; ammonia nitrogen, c_4; and nitrate nitrogen, c_5. In natural waters there is a stepwise transformation, mediated by bacteria, of the organic nitrogen to ammonia nitrogen which itself is subsequently transformed to nitrite and then to nitrate nitrogen. The first of these steps can be an important source of inorganic nitrogen for phytoplankton growth, which is the reason for its inclusion in this model, whereas the second step, nitrification, can have important consequences in the dissolved oxygen balance. The kinetics of these transformations are assumed to be the first-order reactions with temperature-dependent rate coefficients (Thomann et al., 1970).

Two sources of detrital organic nitrogen are considered: (1) the organic nitrogen produced by phytoplankton and zooplankton endogenous respiration (the assumption being that only organic forms of nitrogen result from this process), and (2) the organic nitrogen equivalent of the grazed but not metabolized phytoplankton excreted by the zooplankton. The sink of organic nitrogen included in the formulation is the transformation of organic nitrogen to ammonia nitrogen.

TABLE I

A. The Nitrogen System, Sources ($+$) and Sinks ($-$)

Process	c_3 (Organic N)	c_4 (NH_3-N)	c_5 (NO_3-N)
Organic nitrogen– ammonia transformation	$-K_{34}(T)c_3$	$+K_{34}(T)c_3$	
Nitrification		$-K_{45}(T)c_4$	$+K_{45}(T)c_4$
Phytoplankton uptake		$-\alpha a_{NP} G_P P$	$-(1-\alpha)a_{NP} G_P P$
Phytoplankton endogenous respiration	$+a_{NP} K_2(T)P$		
Zooplankton endogenous respiration	$+a_{NZ} K_3(T)Z$		
Zooplankton excretion	$+(a_{NP} C_g(T)ZP - a_{NZ} G_Z Z)$		

B. The Phosphorus System, Sources ($+$) and Sinks ($-$)

	c_6 (Organic P)	c_7 (PO_4-P)
Organic phosphorus– orthophosphorus transformation	$-K_{67}(T)c_6$	$+K_{67}(T)c_7$
Phytoplankton uptake		$-a_{pP} G_P P$
Phytoplankton endogenous respiration	$+a_{pP} K_2(T)P$	
Zooplankton endogenous respiration	$+a_{pP} K_3(T)Z$	
Zooplankton excretion	$+(a_{pP} C_g(T)ZP - a_{pZ} G_Z Z)$	

The primary kinetic source of inorganic nitrogen is via the organic nitrogen transformation into ammonia nitrogen. It is assumed that there are no direct kinetic pathways from organic nitrogen to nitrate nitrogen. The primary sink of the inorganic nitrogen forms is the phytoplankton uptake. The algebraic forms used for these kinetic interactions are shown in Table IA. The source for each variable is the sum of the expressions in the appropriate column. In order to conform with the suspicion that ammonia nitrogen is preferentially used by phytoplankton, a preference coefficient is introduced:

$$\alpha = \left(\frac{c_5}{K_{mN} + c_5}\right)\left(\frac{c_4}{K_{mN} + c_4}\right) + \left(\frac{K_{mN}}{K_{mN} + c_5}\right)\left(\frac{c_4}{c_4 + c_5}\right) \qquad (16)$$

If $\alpha \to 1$ ammonia, (c_4) is used; if $\alpha \to 0$ nitrate, (c_5) is used. The expression is best understood by considering first the case for abundant nitrate ($c_5 \gg K_{mN}$). The first term of the sum dominates and ammonia is utilized until it declines to the vicinity of the Michaelis constant when nitrate starts to be used. As nitrate declines the second term of the sum dominates and the preference is proportional to the ratio of ammonia to total inorganic nitrogen so that the forms are utilized in proportion to their abundance.

The Phosphorus System

The formulation of the phosphorus conservation of mass equations is somewhat simpler than the nitrogen equations because only two forms of phosphorus are considered: nonliving organic phosphorus, c_6; and orthophosphorus, c_7. The mechanisms that are included parallel those for the nitrogen systems with the exception of nitrification, for which there appears to be no phosphorus counterpart. The sources and sinks that result are shown in Table 1B.

The Silica System

The silica system contains just one sink of silica—the uptake by phytoplankton. As with the nitrogen and phosphorus system, the source equation has the form:

$$S_{Si} = -a_{SiP} G_P P \tag{17}$$

The Dissolved Oxygen System

The treatment of the dissolved oxygen (DO) system is entirely classical with the single exception that the oxygen produced and consumed by photosynthetic production and respiration is calculated rather than externally specified. The decay of carbonaceous organic material (BOD), assumed to be first order with respect to BOD concentration, utilizes dissolved oxygen. DO is supplied via atmospheric reaeration. The reaction is first order with respect to DO deficit with reaction rate

$$K_a(20) = \sqrt{\frac{D_L|v|}{H^3}} + \frac{\beta W}{H} \tag{18}$$

where D_L is the molecular diffusivity of oxygen, $|v|$ is the average tidal velocity magnitude (as distinct from the net tidal velocity), H is the mean tide depth, W is the average wind velocity, and β is an empirical constant. Nitrification also consumes oxygen, which constitutes an additional sink as are phytoplankton and zooplankton respiration. Primary production is the second source. The algebraic forms are shown in Table II.

D. Temperature Dependence

The reaction rates for the kinetics described in the preceding section are all dependent on temperature, an effect of critical importance in the seasonal development of phytoplankton and zooplankton biomass, as well as to the rate of supply of

TABLE II

The Dissolved Oxygen System Sources (+) and Sinks (−)

Process	c_9 (Dissolved Oxygen)	c_8 (BOD)
BOD oxidation	$-K_d(T)c_8$	$-K_d(T)c_8$
DO reaeration	$K_a(T)(c_s - c_9)$	
Nitrification	$-a_{ON} K_{45}(T)c_4$	
Phytoplankton primary production	$a_{OP} G_P P$	
Phytoplankton respiration	$-a_{OP} K_2(T)P$	
Zooplankton respiration	$-a_{OZ} K_3(T)Z$	

nutrients via regeneration. In the earliest models of dissolved oxygen distribution in streams, the temperature dependence is specified, following Arrhenius, using

$$\frac{d}{dT} \ln K(T) = -\frac{E}{RT^2} \tag{19}$$

where T is temperature in degrees Kelvin, R is the universal gas constant, and E is the activation energy. To a very good approximation this leads to an exponential relationship

$$K(T) = K(20)\theta^{(T-20)} \tag{20}$$

where T is now temperature in degrees centigrade and $K(20)$ is the reaction rate at 20°C. This relationship is also consistent with the use of Q_{10}, the ratio of $K(20)$ to $K(10)$, since clearly

$$Q_{10} = \theta^{10} \tag{21}$$

The applicability of exponential temperature dependence to phytoplankton growth has been suggested (Eppley, 1972) and as shown in Fig. 2, the dependence is evident in the endogenous respiration as well as zooplankton grazing and endogenous respiration rate constants. Thus exponential temperature dependence appears to be justified within the temperature range relevant to the applications to be discussed.

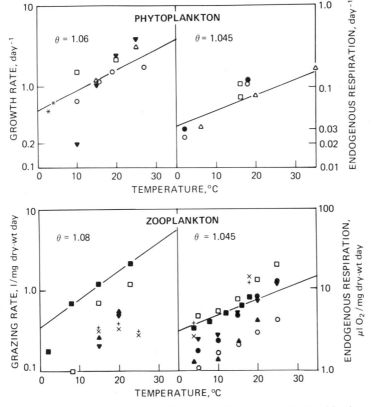

Fig. 2. Temperature dependence of phytoplankton and zooplankton kinetics.

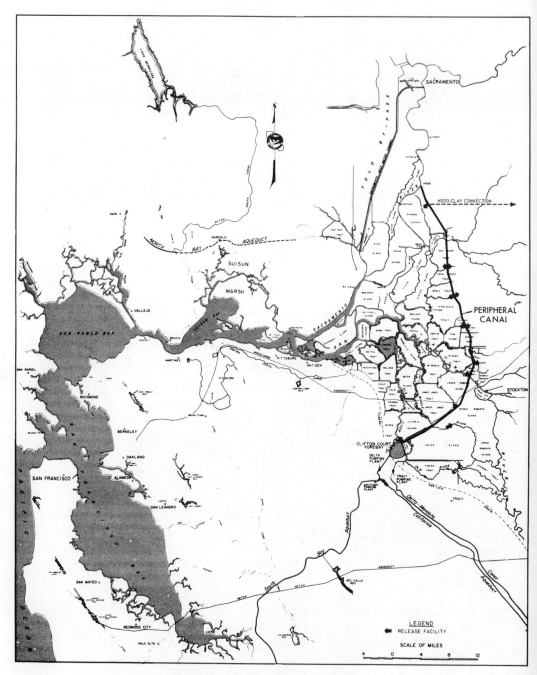

Fig. 3.   Location map—San Francisco Bay.

3. Kinetic and Prototype Verifications—a Comparison

The equations of the preceding sections are an attempt to formulate the interactions between nutrients, phytoplankton biomass, and zooplankton biomass in a way that is entirely analogous to the methods of physical chemistry. Following this analogy it should also be true that these equations apply to situations where the macroscopic mass transport mechanisms are quite dissimilar. The division of phenomena into mass transport on the one hand and kinetic on the other implies a degree of independence between the two. Of course there are situations for which the microscopic mass transport of molecular diffusion controls the kinetic rates. However, the independence advocated here is between the macroscopic advective flow and dispersive mixing in a particular estuary and the kinetics of the microbiology. The degree to which this is the case has been investigated in a preliminary way, for a phytoplankton model of the estuary of San Francisco Bay. The results to be discussed are from the latest (Hydroscience, 1974) in a sequence of progressively more complex models (Di Toro et al., 1971; Hydroscience, 1972), which were more highly detailed in both their spatial and kinetic structures.

As shown in Fig. 3, the confluence of the Sacramento and San Joaquin Rivers occurs near Chipps Island and an estuary is formed in the Suisun Bay region of the San Francisco Bay complex. This is the region of application of the calculations to be discussed. Owing to increases in population and demand for water, changes are expected to occur in both nutrients discharged to the region and in incoming freshwater flow. The purpose of the model development is to analyze the probable results of these expected changes and to suggest methods of mitigating any adverse conditions that may result.

A. Kinetic Verification

It became apparent during the development of the model that some independent checks on the various coefficients and formulations would be desirable. Thus a set of laboratory experiments were designed to investigate, primarily, the nutrient uptake kinetics of the phytoplankton since the calculations indicated that this was a significant feature of the model.

The laboratory experiments were conducted using samples of the natural population from the San Francisco Bay estuary in the region between Antioch and Big Break (refer to Fig. 7 for the locations). The intent of the experiments was to investigate the response of the population to additions of nutrients, and they lend themselves to a direct comparison with theoretical calculations using the kinetic equations of the previous sections.

The data were obtained as part of the Algal Growth Potential (AGP) studies conducted by the U.S. Bureau of Reclamation (Ball, 1972). Samples were taken at four stations in the vicinity of Antioch during September 28 and 29, and October 26 and 27, 1971. The laboratory procedure was to spike the samples with the following levels of nutrients:

(I) 0.2 mg/l Nitrate (N)
(II) 0.5 mg/l Nitrate (N)
(III) 0.5 mg/l Nitrate and 0.1 mg/l Phosphate (P)

and monitor the chlorophyll a, nitrate, orthophosphate, and dissolved silica concentrations daily for 4–5 days under controlled laboratory conditions of constant light

(400 fc) and temperature (24°C ± 1°). Owing to the similarity of the resulting data for the four stations, averages of the four stations for October are utilized in the analyses. Comparisons of the control with the various levels of nutrient additions are presented in Figs. 4–6.

The behavior of nitrate was as expected. In all cases, the production of chlorophyll was accompanied by a simultaneous decrease in nitrate with the peaks in chlorophyll dependent upon the magnitude of nitrate spiking only. Nitrate was eventually depleted below the limit of detection (0.01 mg/l) for all experiments analyzed.

However, phosphorus metabolism is more complex. In the controls, phosphate concentrations were depleted to the detection limit of 0.01 mg/l without chlorophyll increasing. Also in the samples spiked with nitrate only, phosphate was again depleted to below the detection limit. Although the peak chlorophyll concentrations substantially increased with nitrate spiking, the net decrease in phosphate was essentially the same. In addition, comparisons of the samples spiked with 0.5 mg/l nitrate and 0.1 mg/l phosphate with those spiked with only 0.5 mg/l nitrate, shows that although

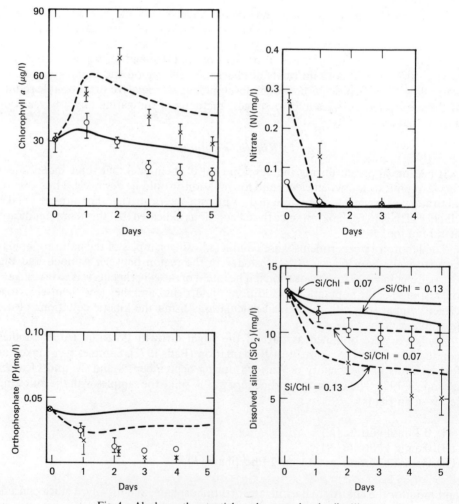

Fig. 4. Algal growth potential results control and spike (1).

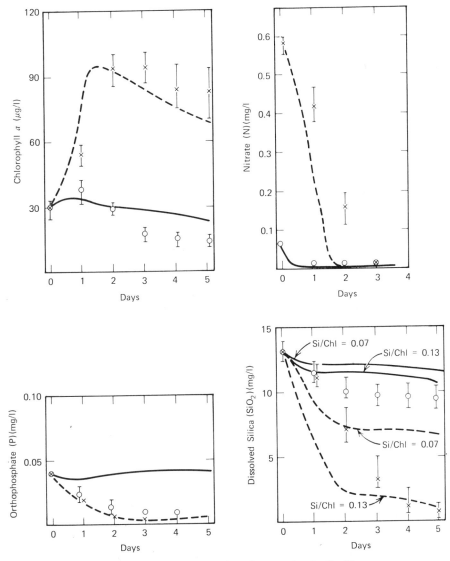

Fig. 5.   Algal growth potential results control and spike (2).

the peak chlorophyll concentrations were essentially unaltered, phosphate concentrations were again depleted to at least the detection limit. This indicates a variable phosphorus stoichiometry for the phytoplankton, a well-known phenomena in short-term growth experiments.

Silica metabolism was reasonably straightforward during chlorophyll increase. However silica uptake continued for about two days following the peak concentration in chlorophyll. The addition of 0.5 mg/l nitrate was, in all cases, sufficient to nearly deplete the silica.

The kinetic modeling results are also presented in Figs. 4–6. A consistent set of constants, presented subsequently in Table III, is used for all verifications; the initial conditions of the various constituents are the only differences between calculations.

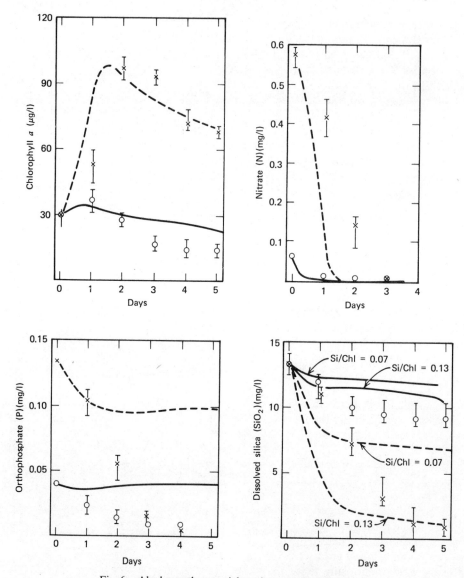

Fig. 6.   Algal growth potential results control and spike (3).

The inclusion of the dissolved organic nutrients and ammonia in the model were necessary for proper evaluation. Since measurements of these constituents were not conducted in the laboratory, estimates based on field measurements made at the same time were used for initial condition values.

In general, the model verifications with respect to chlorophyll and nitrate are quite satisfactory. Peak chlorophyll concentrations are approximated in almost all cases. A nitrogen to chlorophyll ratio of 7 mg N/mg chlorophyll *a* proved appropriate. The phytoplankton death rate used resulted in chlorophyll concentrations calculated after the peak to be somewhat higher than those observed for the control and 0.2 mg/l nitrate spiked samples. Attempts to simulate these samples by increasing the death rate resulted in poor correlation for the other spiked samples.

A variable stoichiometry with respect to phosphorus is indicated by the data. This phenomenon is often referred to as "luxury" uptake. Since the model assumes a constant stoichiometry, correct simulation of all cases is impossible.

Therefore, it was decided that the samples spiked with only 0.5 mg/l nitrate would be verified on the basis that these samples represented the conservative condition of the highest algal crop attained with the least available phosphorus for these experiments. With a phosphorus to chlorophyll ratio of 0.47 mg PO_4-P/mg chlorophyll a, a satisfactory phosphorus simulation of the 0.5 mg/l nitrate spiked samples results. The observed variations of the phosphorus to chlorophyll ratio may not be characteristic of the estuarine environment over long time scales. Under these conditions, the stoichiometry of the phytoplankton probably attains a mean value with minor variations. As a result, the inclusion of variable stoichiometry in the model without further detailed experimentation is considered unjustified at this time.

The uptake of dissolved silica is species dependent and, as such, cannot be properly simulated by the model if a mixed phytoplankton population exists. Qualitative descriptions of the AGP experiments indicated a shift in predominance from diatoms to green algae, with high levels of nitrate spiking. A dissolved silica to chlorophyll ratio of 70 μg SiO_2/μg chlorophyll, was utilized in the simulations presented.

This value is appropriate for the prototype simulation. The results for a ratio of 130 are also shown. This value appears to be more suitable for the AGP experiments. This difference is attributed to a species-dependent effect.

B. Prototype Verification

The comparison between prototype calculations and observations are discussed in this section. To apply the kinetics to the estuary itself it is necessary to specify the transport mechanisms—both the net advective flow and the dispersive mixing coefficients.

For an estuary with complex geometry it is a significant advantage to be able to calculate the net advective flows from a tidal hydrodynamic model. Such a model has been developed for the whole of San Francisco Bay (Water Resources Engineers, 1968) and its use has aided considerably the specification of the seasonal transport regime in the estuary.

In addition to the net advective flows, the mixing due to tidal motions, density currents, and velocity gradients must be accounted for by the dispersion coefficients. The method of choosing these coefficients is largely trial and error, the criterion being to achieve agreement between observed and computed chloride concentrations.

In addition to the transport regime, the external sources of nutrients are necessary for the calculations. The collection of complete nutrient mass discharge information is a tedious and costly enterprise. However, it is essential to have such information for a model based on mass conservation equations. Also it is necessary to check the reliability of the information, once collected.

If it is assumed that no significant removal of phytoplankton or other particulate forms of the nutrients occur by settling, or that no significant regeneration occurs from the sediments, then the concentrations of total nitrogen, total phosphorus, and total silica are variables unaffected by kinetic transformations. Thus a verification based on the observed quantity of these variables is an additional check on the transport regime and in the case of nitrogen and phosphorus the quantities of internal discharges of these materials.

Such comparisons have been made, (Hydroscience, Inc., 1974) and they indicate that the above assumptions are reasonable. This appears to be a uniquely estuarine situation since it is clear that phytoplankton sinking is of primary importance in both freshwater and marine settings.

There are some additional details that must be taken into account to reflect the specific situation in the prototype. The most important of these is the light regime to which the phytoplankton are exposed. The incident solar radiation and photoperiod, together with the water temperature variation and wind velocity are required. The extinction coefficient, the reciprocal of which is the depth at which the light is 37% of the surface intensity, must also be specified.

The difficulty is that the phytoplankton population themselves contribute a substantial fraction of the extinction coefficient, owing to their absorption of light (Riley, 1956) (this effect is often called self-shading). To establish the extinction coefficients for the model, it is necessary to subtract the phytoplankton component from the observed extinction coefficient and then to recompute it as the calculation proceeds. Thus in a sense the extinction coefficient becomes another variable that must be verified since it is computed as the phytoplankton population develops.

Chlorophyll

The stations chosen for comparison of data and computation are shown in Fig. 7. The boundary condition at Rio Vista is included to illustrate the difference between its value and the values calculated for the model interior. The stations are Sherman Lake, a shallow productive region which exchanges with the Sacramento River; Honker Bay, a shallow productive bay in the center of the Delta; Grizzly Bay, a larger bay further west which exchanges with Suisun Bay; and two stations in the San Joaquin, Big Break and Antioch, the locations from which the samples were taken for the kinetic experiments. Port Chicago, a deep-channel section opposite Honker Bay, and Benicia, the most westward segment of the model, are included.

Computed chlorophyll concentrations are compared to the observed data for 1970 in Fig. 8. Calculated concentrations are quite small during the early part of the year until the temperature and light conditions are suitable for phytoplankton growth at which time, approximately $t = 120$, the phytoplankton population initiates the spring growing period. Peak concentrations calculated for the summer range from 25

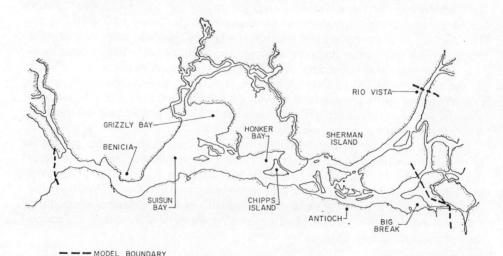

– – – MODEL BOUNDARY

Fig. 7.   Study area—estuary study area.

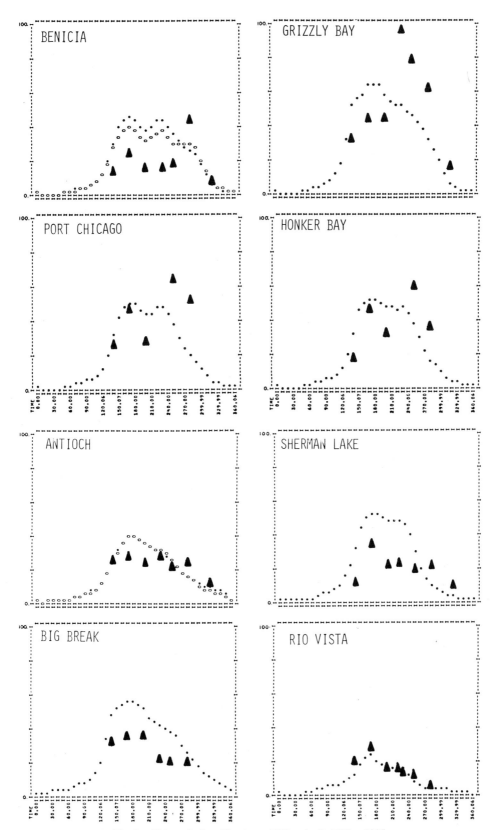

Fig. 8. Chlorophyll verification—1970 scale: 0–100/μg Chl/l.

985

µg/l in the Sacramento River to 70 µg/l in the shallow productive region of Grizzly Bay. The computations in some cases overestimate the chlorophyll concentrations and in some cases underestimate them. The kinetic constants, specifically the growth rate, are chosen to reflect the substantial growth observed.

The major difficulty with the calculation is also evident in Fig. 8, namely, the inability to reproduce the fall maximum. This is thought to be related to species-dependent effects. A calculation that included these effects would almost parallel that presented with the exception that the biomass would be partitioned into functional groupings with different kinetic properties.

The verification for the 1966 conditions is shown in Fig. 9. Calculated and observed total chlorophyll are shown for six stations. The comparison is quite good, indicating that the model can reproduce the population behavior under somewhat different conditions of flow, nutrient, and light penetration reflecting both the changes in boundary conditions for the 2 years and the variation in turbidity.

It is interesting to note that for both years the October population biomass is approximately 20 µg/l and is both calculated and observed to be declining. However, as the AGP experiments show, if this population is exposed to new and more favorable conditions, it will respond and produce a substantial increase in biomass. This behavior is also calculated from the kinetics, indicating that the formulation corresponds to the observations even for this rather substantial perturbation.

Primary Production

Measurements of primary production, using the light bottle–dark bottle technique, were made at seven different times during 1970. In situ primary production measurements, a direct measurement of the rate of phytoplankton oxygen production, are also a direct measurement of the kinetic growth rate term in the conservation equation in oxygen units. Therefore, they provide a valuable check on the computation since the primary production is computed internally to predict phytoplankton growth. Both net and gross areal primary production are compared to computed values in Figs. 10 and 11. During the early part of the year, at $t = 106$, the net primary production is computed and observed to be approximately zero, indicating that no phytoplankton population, on balance, is being produced. The gross production is also quite small, less than 1 g $O_2/m^2/day$, as indicated in Fig. 10. Thirty days later ($t = 138$) positive net primary production is beginning to occur in the shallow bays as indicated by both the computed and observed values. Main channel primary production in both the Sacramento and San Joaquin Rivers is still approximately zero. The gross primary production has increased to approximately 2 g $O_2/m^2/day$ as indicated. This is due primarily to the increase in light and temperature over 30 days which separate the two data sets. Thirty days later, at $t = 167$, the net primary production is now substantial in the shallow areas and is beginning to increase at the downstream main channel stations as indicated by both the computation and observation. Slightly negative net primary production is both observed and computed at the upstream San Joaquin stations. Gross primary production is increasing to approximately 6 g $O_2/m^2/day$ in the downstream regions and 4 g $O_2/m^2/day$ in upstream regions. Essentially, the same picture is repeated at day $t = 196$. The time from day 120 to 180 corresponds to the maximum growth period during the year. Thirty days later at day $t = 225$ a decrease occurs in the primary production in the shallow areas owing essentially to an increase in light extinction coefficient values. However, substantial production is still occurring in the main channel region. Gross primary production for that period in the main channel is quite substantial whereas

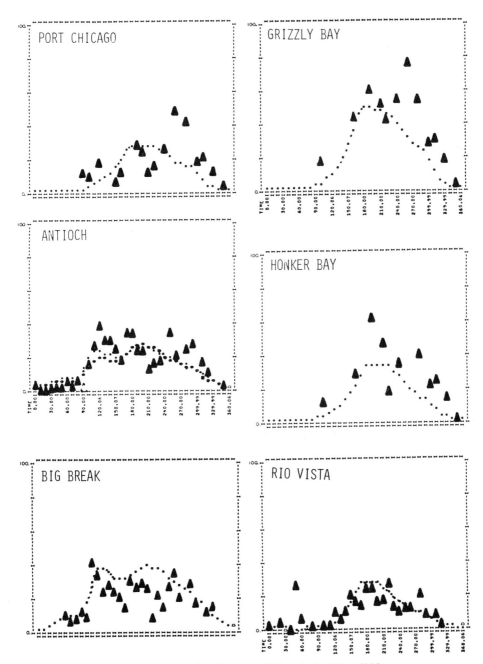

Fig. 9. Chlorophyll verification—1970 scale: 0–100 μg Chl/l.

987

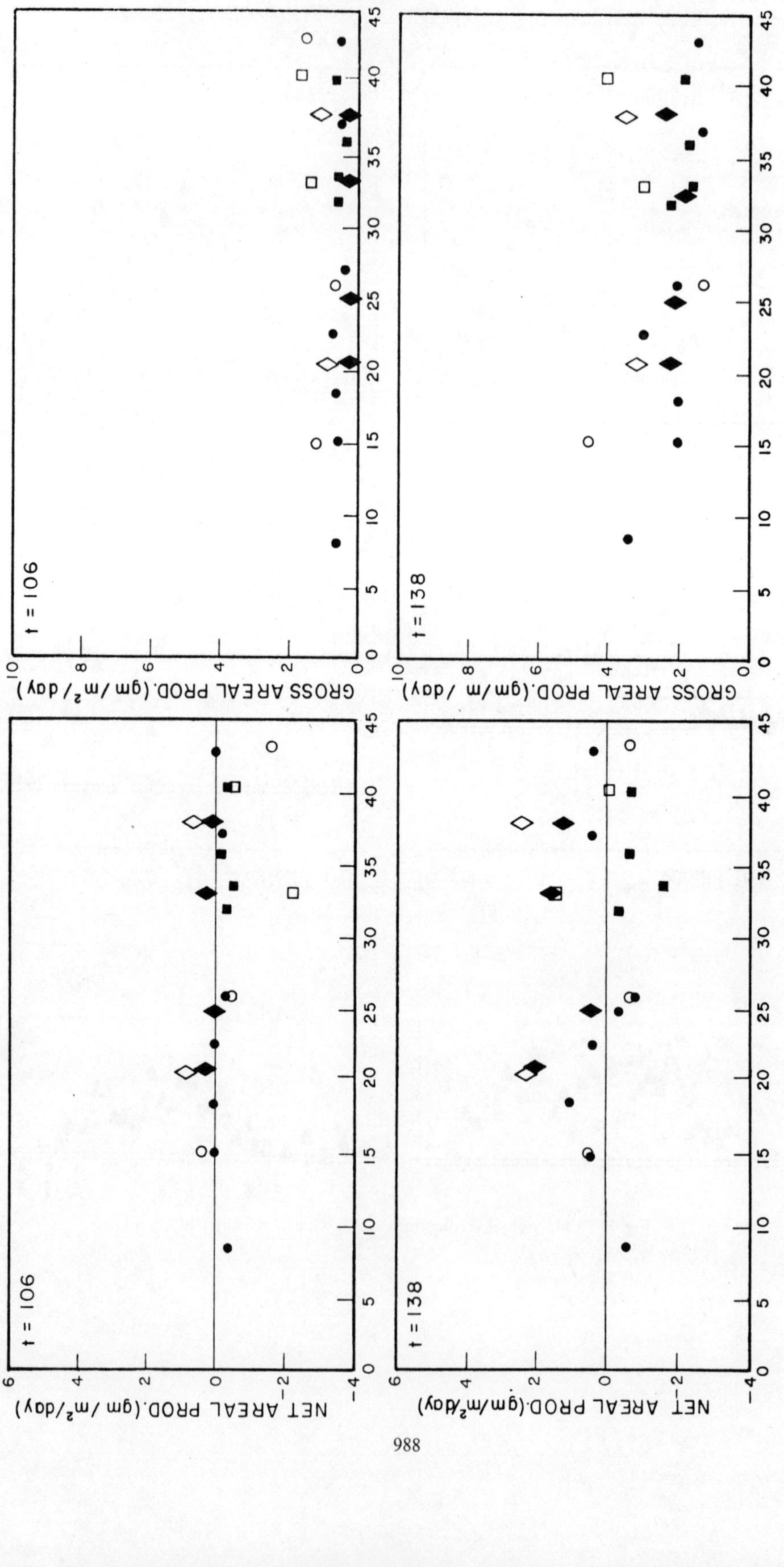

t = 106
t = 138
NET AREAL PROD. (gm/m²/day)
GROSS AREAL PROD. (gm/m²/day)

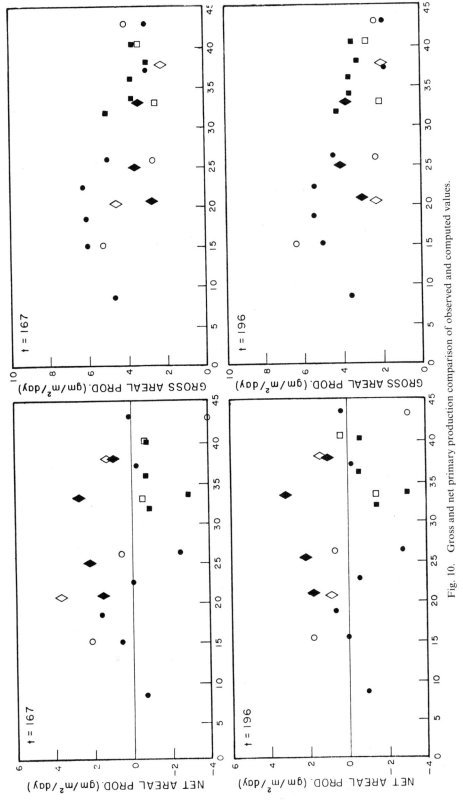

Fig. 10.   Gross and net primary production comparison of observed and computed values.

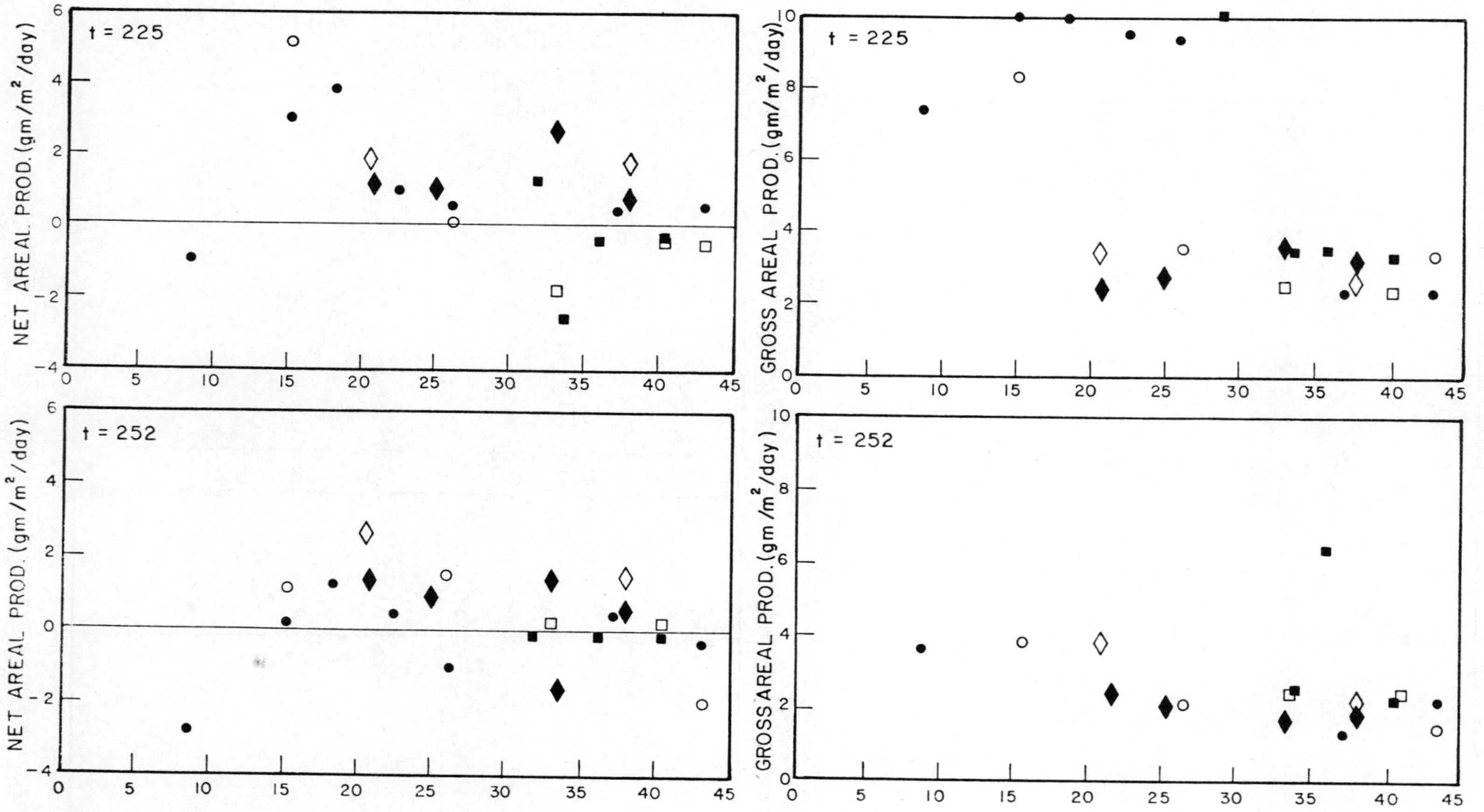
t = 225
NET AREAL PROD. (gm/m^2/day)
GROSS AREAL PROD. (gm/m^2/day)
t = 252
NET AREAL PROD. (gm/m^2/day)
GROSS AREAL PROD. (gm/m^2/day)

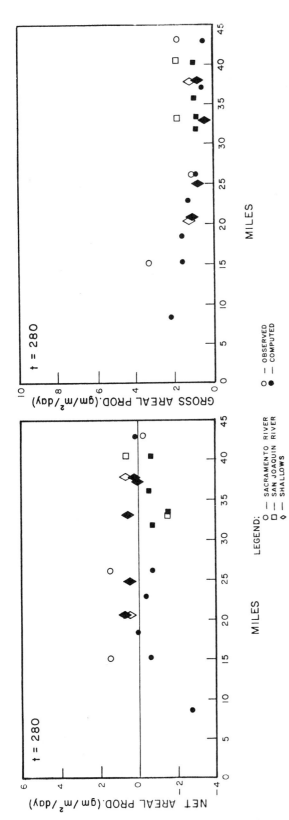

Fig. 11. Gross and net primary production comparison of observed and computed values.

in the shallow regions and the San Joaquin, production has been reduced substantially. At $t = 250$ the net and gross primary production both are decreasing and by $t = 280$ the values are substantially zero for net primary production and less than 2 g O_2/m^2/day for gross primary production indicating that the population is stationary or declining.

Considering the variability of the experimental data, due to the nature of the experiments required to obtain the data, the comparison between computed and observed primary production is regarded as quite satisfactory.

Primary production measurements can be viewed as in situ kinetic experiments of short duration from which a rate of oxygen evolution is measured and a carbon or chlorophyll fixation rate can be deduced, assuming a fixed stoichiometric ratio. The model calculations for the estuary continually calculate the gross $(a_{OP}G_P P)$ and net $[a_{OP}(G_P - D_P)P]$ primary production; a_{OP} being the oxygen to chlorophyll stoichiometric coefficient. Thus the comparison is direct, and is, in a sense, a comparison of the result of an in situ kinetic experiment with the model calculations.

Nitrogen

Three forms of nitrogen are considered in the calculations: organic nitrogen, ammonia nitrogen, and nitrate nitrogen. The calculations indicate that nitrogen is the primary nutrient which reduces the phytoplankton growth during the summer months. As mentioned previously, the computed total nitrogen distribution when nitrogen is considered a conservative variable correctly reproduces the observed total nitrogen calculations. However, the distribution of nitrogen among the three species is an important prerequisite for calculating the phytoplankton distribution, since organic nitrogen is not in a form that is readily available to the phytoplankton.

Organic nitrogen is contributed primarily by the upstream Sacramento and San Joaquin sources and the internal discharges.

The concentration is calculated to remain essentially constant with a slight decrease during the early period owing to the conversion of organic nitrogen to ammonia nitrogen. Then the organic nitrogen concentration begins to increase primarily owing to the formation of organic nitrogen as phytoplankton. Since the prototype measurements do not discriminate between phytoplankton nitrogen and nonliving organic nitrogen, the calculated lines are the sum of computed nonliving organic nitrogen and the nitrogen equivalent of the computed phytoplankton. The comparison between calculations and observed data is shown in Fig. 12 for 1970. Observed and computed organic nitrogen concentrations average 0.4–0.8 mg N/l, which represents a substantial quantity of nitrogen. Since this form is unavailable to the phytoplankton directly and must be transformed to inorganic nitrogen by bacterial action, its effect on phytoplankton growth is indirect.

Ammonia nitrogen concentrations for both years are quite low in the entire estuary, averaging less than 50 μg N/l. Nevertheless, comparisons between observed ammonia nitrogen and calculated concentrations are quite satisfactory. The comparisons are presented in Fig. 13 for 1970. The primary hypothesis employed in these calculations is that the phytoplankton have a preference for ammonia nitrogen as the nitrogen form utilized for synthesis. When ammonia nitrogen concentrations reach levels comparable to the nitrogen Michaelis constant (25. μg N/l), the form of nitrogen assimilated changes to nitrate nitrogen.

Nitrate nitrogen concentrations are compared to observed data for 1970 in Fig. 14. As can be seen observed nitrate nitrogen concentrations average 0.3 mg N/1 during the early part of the year. They then decline sharply until almost no nitrate nitrogen is measured. Again, the comparison is quite satisfactory. The shallow areas are

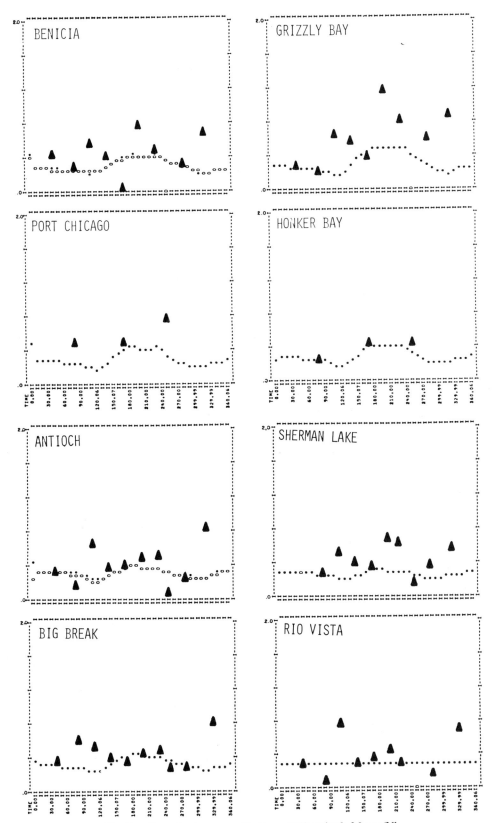

Fig. 12. Organic nitrogen verification—1970 scale: 0–2.0 mg /Nl.

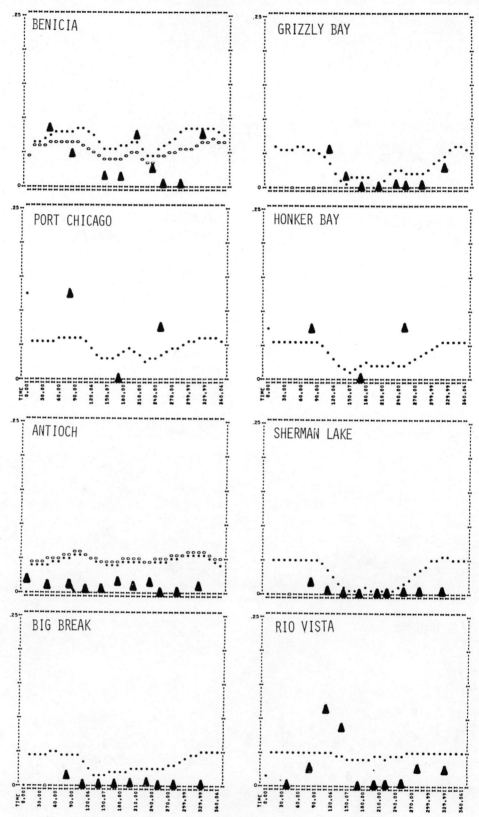

Fig. 13. Ammonia–nitrogen verification—1970 scale: 0–0.25 mg N/l.

994

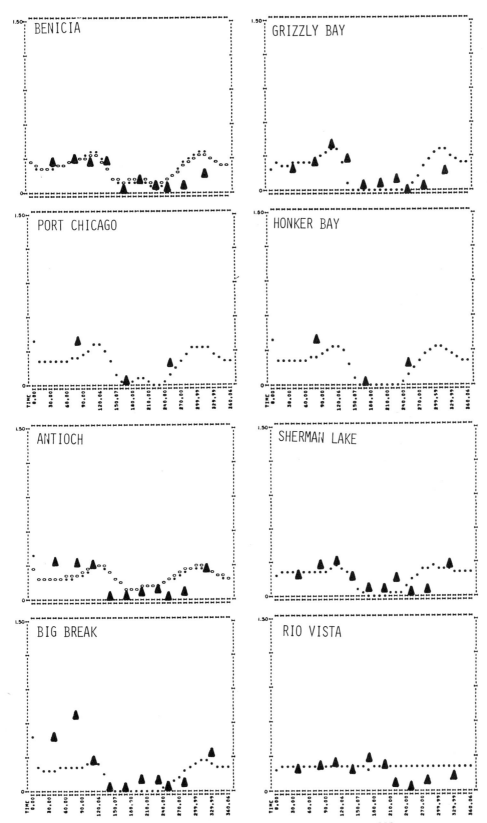

Fig. 14.　Nitrate–nitrogen verification—1970 scale: 0–1.5 mg N/l.

995

observed to be substantially depleted of nitrate nitrogen; this agrees with the calculated distributions. Some nitrate nitrogen is calculated to be present in the deeper
areas, which appears to be born out by the observations. Thus not all the inorganic
nitrogen available for phytoplankton growth is utilized during the growing period.
The other factor that controls phytoplankton growth is the available light which is
such that vigorous growth is not possible in the main channel regions. Thus a substantial quantity of inorganic nitrogen remains unassimilated through the year.
This has important implications for projected conditions. The nitrogen concentrations for the AGP experiments are similar to the October concentrations in these
figures: a substantial quantity of organic nitrogen; almost no ammonia nitrogen;
and very little nitrate nitrogen. The control for the experiments (Fig. 4) indicated
that for these conditions the population would start to decline after the initial small
quantity of available inorganic nitrogen is consumed. However the prototype
behavior is quite different. The reason is that a continual supply of nutrients is available to the population via the external sources. What is significant is that both types of
phytoplankton population behavior are predicted by model calculations.

Phosphorus

Together with nitrogen, phosphorus is required for phytoplankton growth and a
shortage of this nutrient can slow and eventually stop their growth. The forms of
phosphorus considered are organic and inorganic; the former is unavailable directly
for phytoplankton growth but is transformed by bacterial action to inorganic or
orthophosphorus, which can then be assimilated.

The concentration of organic phosphorus is approximately 0.1 mg PO_4-P/l
through the estuary. Its seasonal variation is approximately comparable to the
organic nitrogen, showing an increase during the summer months due to the increase
in phytoplankton-associated organic phosphorus.

Inorganic phosphorus concentrations also average 0.1 mg PO_4-P/l throughout
the year. A seasonal depletion is neither observed nor calculated (Hydroscience, Inc.,
1974) and the quantity of phosphorus available is at all times sufficient so that the
phytoplankton growth is not significantly affected by these concentration levels.

Thus although the kinetic experiments showed a depletion of phosphorus with no
equivalent phytoplankton biomass increase, the prototype shows no similar behavior
and the prototype calculations are in agreement with this observation. It appears
that this short-term luxury uptake effect does not influence the prototype behavior.
However, this is a case for which kinetic and prototype behavior is different.

Silica

Substantial quantities of dissolved silica are assimilated by the resident phytoplankton population during the growing period, indicating that a substantial proportion of the population are diatoms. Since the other groups of algae do not require
silica for their growth, silica is not considered a limiting nutrient in the sense that
growth rates of the biomass are reduced owing to the presence of low silica concentrations. However, the sink due to silica uptake is included in a mass balance
equation so that silica distributions can be calculated and compared to observations.
As shown in Fig. 15, the calculations compare quite well with the observations. A
depletion occurs in the shallow downstream regions and since there is no comparable
oceanic source of silica, the concentration at Benicia is quite low during the summer.

As can be seen the October silica concentrations of Antioch have recovered from
the summer depression and are nearly 15 mg SiO_2/l. The depletion observed in the
AGP experiments corresponds to a silica/chlorophyll ratio of 130 whereas the prototype comparison shown is calculated with a ratio of 70. This indicates that the

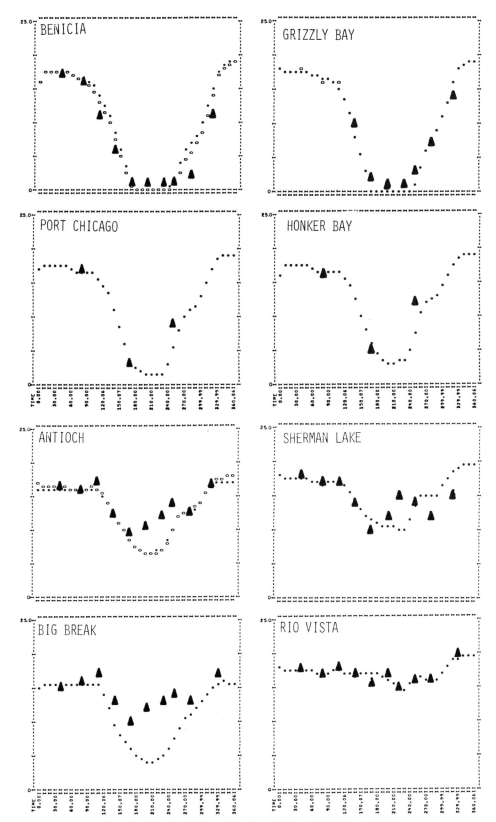

Fig. 15.   Dissolved silica verification—1970 scale: 0–25 mg SiO_2/l.

TABLE III
Kinetic Parameters

Notation	Description	Reported Range[a]	Value Used
K_1	Saturated growth rate of phytoplankton at 20°C	1.0–3.0	2 5/day
θ_1	Temperature correction for phytoplankton growth		1.06
I_s	Saturation light intensity	150–550	300 ly/day
K_{mN}	Michaelis (half-saturation) constant for total inorganic nitrogen	0.0015–0.15	0.025 mg N/l
K_{mp}	Michaelis constant for orthophosphorus	0.006–0.025	0.005 mg P/l
K_{mP}	Michaelis constant for phytoplankton chlorophyll grazing		60 μg Chl/l
K_2	Endogenous respiration rate of phytoplankton at 20°C	0.05–0.15	0.1/day
θ_2	Temperature correction for phytoplankton endogenous respiration		1.045
K_{34}	Organic nitrogen to ammonia hydrolysis rate at 20°C	0.1–0.2	0.08/day
θ_{34}	Temperature correction for K_{34}		1.045
K_{45}	Ammonia to nitrate nitrification rate at 20°C		0.3/day
θ_{45}	Temperature correction for K_{45}		1.045
K_{67}	Organic phosphorus to orthophosphorus conversion rate at 20°C		$0.03 \dfrac{\text{mg PO}_4\text{-P}}{\text{mg org P-day}}$
θ_{67}	Temperature correction for K_{67}		1.045
K_{MSi}	Michaelis constant for silica	<1.0	0.0 mg SiO$_2$/l
a_{ZP}	Carbon to chlorophyll ratio	0.01–0.1	$0.05 \dfrac{\text{mg C}}{\mu\text{g Chl}}$
a_{NP}	Nitrogen to chlorophyll ratio	0.0027–0.0091	$0.007 \dfrac{\text{mg N}}{\mu\text{g Chl}}$
a_{pP}	Phosphorus to chlorophyll ratio		$0.47 \dfrac{\mu\text{g PO}_4\text{-P}}{\mu\text{g Chl}}$
a_{OP}	Oxygen to chlorophyll ratio		$0.133 \dfrac{\text{mg O}_2}{\mu\text{g Chl}}$
a_{SiP}	Silica to chlorophyll ratio		$0.07 \dfrac{\text{mg SiO}_2}{\mu\text{g Chl}}$

[a] Di Toro et al., 1971; Thomann et al., 1970.

October population has a higher fraction of diatoms as part of the biomass than the summer population for which a ratio of 70. is indicated. This is in agreement with the observed composition of the population.

C. Discussion

The prototype comparisons indicate that the same relationships which describe the kinetic experiments in the laboratory are directly applicable to the estuary itself. The result that inorganic nitrogen is the critical nutrient is clear from both sets of calculations. The fact that there is sufficient inorganic phosphorus is amply demonstrated. The verification of primary production data also indicates that bottle experiments and prototype behavior are governed by essentially similar relationships.

The major differences observed are the luxury uptake of phosphorus and the different silica/biomass ratio in the bottle experiments, when compared to prototype calculations. The former appears to be of little consequence under present conditions of high phosphorus concentrations, and the silica stoichiometry is related to the seasonally changing ratio of diatoms to other forms in the estuary.

From a practical point of view the objective of such an intensive verification procedure is to provide credibility for the theoretical structure, to check its implementation as a computer program, and to estimate the parameters of the model which are functions of that part of the hydrodynamics and biology not encompassed by the theory. It is interesting to note that the situation is roughly parallel with respect to both the transport and the kinetics. The hydrodynamic theory used provides net tidal flows; the observed data is used to obtain the dispersion coefficients. For the biology, theory provides the forms of the equations and experiments provide the ranges of certain parameters. The observed data is used to estimate the remaining parameters.

The final parameter values used are listed in Table III. It should again be pointed out that these values were used for the AGP calculations, the 1970 verifications, and the 1966 verifications. Thus they are consistent with that body of data. It is essentially this consistency which is required for a model as a minimum prerequisite to its use in making projections of future situations.

4. Verification Analysis–The Upper Potomac Estuary

The Potomac Estuary has been the subject of investigation and analysis over many decades, most intensively within the past several years. This work has been motivated in large measure by the location of Washington, D.C. along its shores and by a long history of poor water quality. Several Enforcement Conferences have been held on the water quality of the Potomac Estuary with the aim of establishing required effluent controls to achieve specified water quality objectives. Important recommendations which evolved from the 1969 conference included waste load allocations which restrict the mass discharge of oxygen-demanding material, and nitrogen and phosphorus residuals. These recommendations relied to some degree on the application of mathematical modeling to key water quality constituents such as dissolved oxygen and total phosphorus.

The purpose of this section is to present the verification of a preliminary model of the dynamic behavior of phytoplankton in the Upper Potomac Estuary. This preliminary phytoplankton model is intended, therefore, to shed further light on the water quality changes that can be expected when a waste reduction program is completed for the Potomac Estuary. Primary emphasis here is on assessing in a preliminary way the efficacy of nutrient removal programs, especially in the Washington, D.C., area. In addition, the difficulty of making meaningful decisions regarding nutrient removal in the face of a rapidly expanding modeling technology is discussed.

The area of interest for the model is centered in the upper 40-mi reach of the Potomac Estuary, although as indicated below, the model extends geographically over the entire 114-mi length of the estuary from Little Falls to Chesapeake Bay. The upper reach of 30–40 mi is generally unaffected by the incursion of salts from Chesapeake Bay. The Potomac is tidal in the vicinity of Washington, D.C., and is several hundred feet wide with a shipping channel of minimum depth of 24 ft maintained to Washington. The tidal portion averages about 18 ft and is characterized by numerous coves and embayments along its length with average depths significantly less than the main estuary. Presently the major waste source in the upper reach of the Potomac is the effluent from the Washington, D.C., secondary treatment plant.

Water quality problems include low values of dissolved oxygen in the vicinity of the District of Columbia and high concentrations of phytoplankton, especially of *Anacystis.*

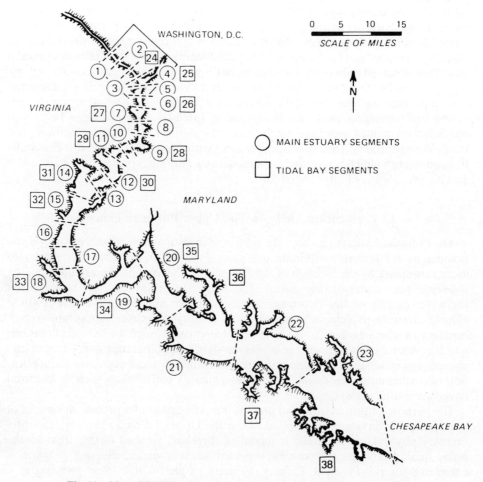

Fig. 16.   Map of Potomac Estuary showing longitudinal and lateral segments.

A. Model Geometry and Kinetics

The basic model builds on previous efforts of phytoplankton dynamics (Di Toro et al., 1971) and nitrification effects in estuaries (Thomann et al., 1970) and specific modeling applications of nutrients in the Potomac (Jaworski et al., 1969a, b, 1971; Aalto et al., 1970). The model, therefore, extends previous efforts on the Potomac to incorporate explicity the space–time variability of chlorophyll *a* as a water quality parameter indicative of a eutrophied environment.

Spatially, the estuary is divided into 23 longitudinal segments using depths, cross-sectional areas, and segment volumes given by earlier modeling work in the Potomac. Figure 16 shows the location of these main estuary segments. As discussed more fully below, it was found necessary to include the effects of the tidal bays on main channel quality. The tidal flat areas have been cited as an important phenomenon in the Potomac River as far back as 1916 (Phelps, 1944). Accordingly, an additional 15 spatial segments were incorporated in the model to reflect lateral effects of the shallow-water areas. These tidal bay segments are also shown in Fig. 16. A total of 38 spatial segments were therefore used to represent the Potomac with primary emphasis on segments 1–15, the Upper Estuary. To simplify the analysis, waste loads were in-putted into segments 5–7, which accounts for the major portion of direct discharge load. No attempt was made in this preliminary model to input urban or suburban runoff or overflows from combined sewers.

The model incorporates interactions between nine variables which are space- and time-dependent (Fig. 17). The principle features are similar to those of the Suisun Bay model discussed previously.

The nine system equations for each of 38 spatial segments result in a total of 342 simultaneous nonlinear ordinary differential equations. The entire set was programmed for solution by a CDC 6600 computer. For a 1-yr simulation at integration time steps of about 0.1 day, normal central processing times were about 6 min.

B. Verification and Sensitivity Analysis—1968 Data

Two data periods were available for testing the validity and degree of applicability of the preliminary model. The procedure followed in the verification was to tune the model to the first data set collected in 1968. A forecasting situation was then set up for 1969. The validated model for 1968 was used to estimate "independently" the 1969 data. The only new input data used for 1969 were the river inflow with associated water quality, water temperature, and waste loads. All coefficients determined from the 1968 validation were used without modification for the 1969 verification. This latter verification provides a basis for determining the degree to which the model will simulate future conditions under different waste removal policies.

Figure 18 shows the data periods and the flow and temperature regimes used for the 1968 and 1969 computations. During the 1968 data period, Fig. 18 indicates a significant flow transient in June 1968 with a lesser transient in September. The piecewise linear approximations to the time variable flow were determined from the available data; they do not account for the day to day variability in inflow. The emphasis was on describing the seasonal variation rather than on shorter-term fluctuations.

Only summary plots of observed data were available for verification of the 1968 data (Jaworski et al., 1969b). Figures 19 and 20 show a variety of results from the validation analyses of the May to November 1968 data. As indicated previously, the system equations are all time variable; the results in Figs. 19 and 20 show the average spatial profile during August 1968, a period of low flow just prior to the late summer flow transient that year.

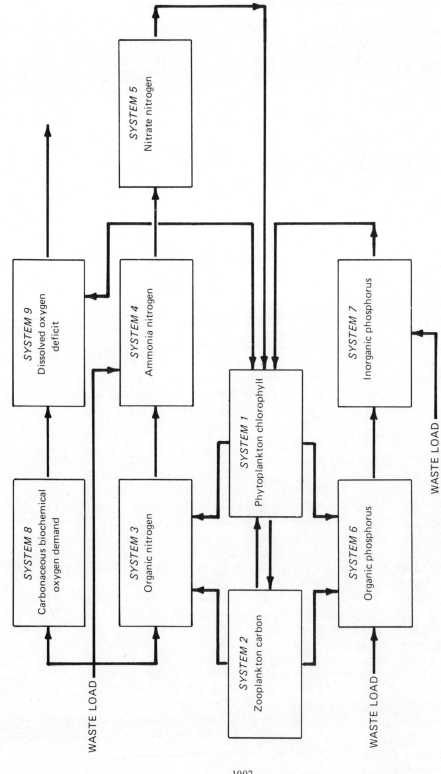

Fig. 17.   Interactions of nine systems used in preliminary phytoplankton model.

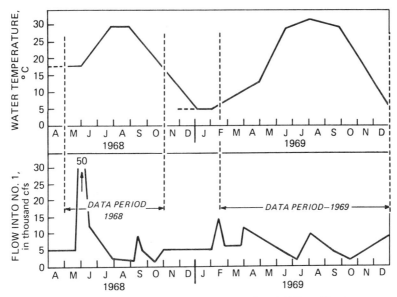

Fig. 18.   Temperature and flow regimes used for 1968 and 1969 verification runs.

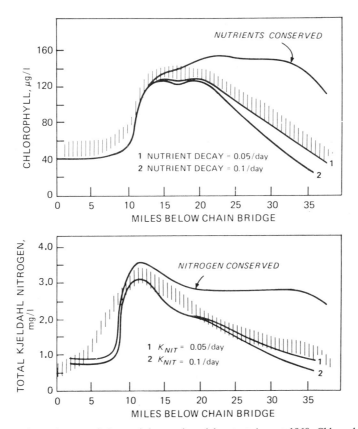

Fig. 19.   Comparison of range of observed data and model output August 1968. Chlorophyll *a* (μg/l); total Kjeldahl nitrogen (mg/l).

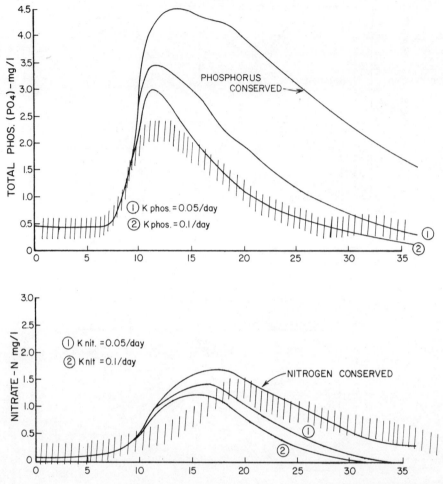

Fig. 20. Top, total phosphorus-PO$_4$ (mg/l) comparison; bottom, nitrate–nitrogen comparison, August 1968.

The three curves of model output in the figures represent three conditions on the nutrients: nutrients are totally conserved in the model; and nutrients are decayed out of the system at rates of 0.05/day and 0.1/day. As shown, the agreement is good when both the organic nitrogen and total phosphorus are allowed to decay at rates of 0.05/day or 0.1/day but there is poor agreement when the nutrients are totally conserved in the system. This illustrates one of the steps in the tuning of the model using the 1968 data. It was obvious from the early runs of the model for which the nutrients were allowed to recycle entirely within the model that the results were not desirable below about Milepoint 15. The observed data on all variables shown in Figs. 19 and 20 decreased more rapidly than the computations. Accordingly, the hypothesis was adopted that incoming sediment load permitted sorption of phosphorus and nitrogen with subsequent settling to the bottom of the estuary and out of the domain of the modeling framework. This hypothesis is consistent with that given by Jaworski et al. (1971) in determining phosphorus balances in the estuary.

In addition, the determination of model parameters using the 1968 data indicated that zooplankton grazing of the phytoplankton was probably not a significant factor in the Potomac. This is supported by the fact that the phytoplankton of concern in the Washington, D.C., area down to about segment 15 are *Anacystis*, a blue-green form that is toxic to zooplankton. Several runs were made to show the sensitivity of the various systems to perturbations on the zooplankton system.

Figure 21*b* shows the effect of zooplankton grazing on the phytoplankton chlorophyll a concentration in segment 9. The time period extends from March to December 1968, with an additional 60 days to show the general trend after December 1968. As indicated in Fig. 21*b* with extensive zooplankton grazing, phytoplankton populations are almost completely depleted in September and maximum concentrations of only 60 μg/l are computed as compared to observed values of about 130 μg/l. It is interesting

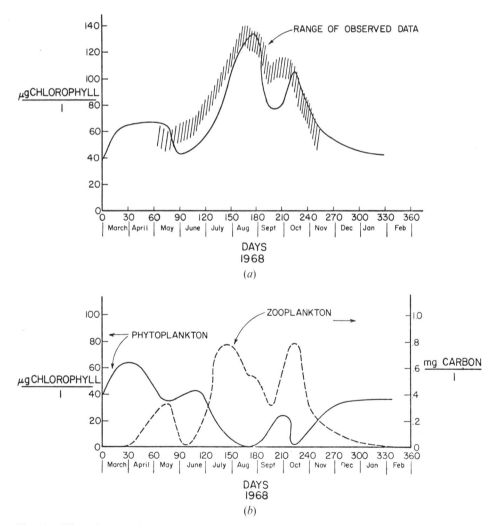

Fig. 21. Effect of zooplankton grazing on phytoplankton in segment 9: (*a*) no zooplankton grazing; (*b*) zooplankton grazing at 0.42 1/mg C-day, 1968 flow regime.

to note that the sharp drop in zooplankton at day 90 is the effect of flow transient at that time and as a consequence phytoplankton populations remain high since the predators are flushed past the segment.

Figure 21*a* with no zooplankton grazing is considerably closer to the observed data range as shown. The transient drop around September 1968 is due to a flow increase (Fig. 18).

The effect of zooplankton predation on the nutrient forms is interesting (Fig. 22). The ammonia nitrogen is quite insensitive to the zooplankton effect under the conditions run. The drop in ammonia concentration at day 90 and day 180 is due to flow transients during that time (Fig. 18). With the cycling of organic nitrogen to ammonia nitrogen there is a sufficient feed forward through the ammonia system even when the phytoplankton population is high.

The effect on the nitrate concentration is marked (top of Fig. 22). Under extensive zooplankton grazing, the utilization of nitrate is decreased and nitrate behaves approximately as a conservative variable responding only to changes in the flow regime while being fed by the nitrification system. As a consequence, NO_3 values build up to 3 mg/l under extensive zooplankton grazing—considerably greater than observed (Fig. 20). With no zooplankton grazing, phytoplankton populations increase and therefore utilize nitrate nitrogen and reduce the values to the 1–1.5 mg/l level.

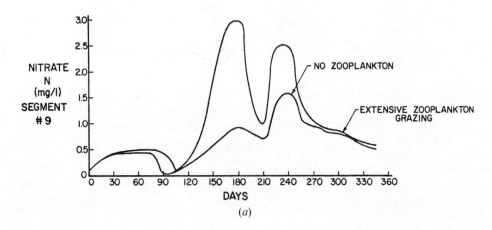

(*a*)

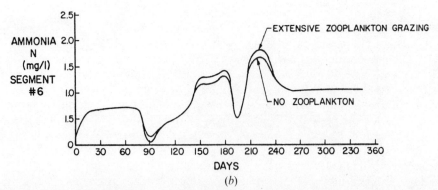

(*b*)

Fig. 22.   Effect of zooplankton grazing on nitrate–nitrogen (*a*) in segment 9 and ammonia–nitrogen (*b*) in segment 6, 1968 flow regime.

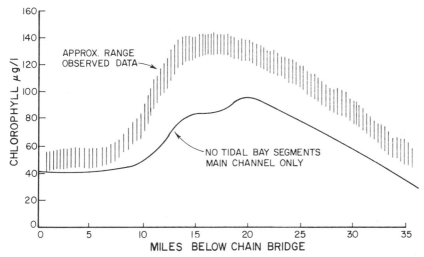

Fig. 23.   Sensitivity run—no tidal bay segments, August 1968 profile and 1968 conditions.

This is the range observed during 1968. All these runs indicate that even with minimal grazing rates of the zooplankton on the phytoplankton, the phytoplankton population is never computed to grow to the levels observed in 1968. It was concluded that the predatory effect of zooplankton on the phytoplankton was minimal at least for the upper end of the estuary below Washington, D.C.

A sensitivity run was made to indicate the effects of including the tidal bay segments (Fig. 23). Observed chlorophyll concentrations in the main channel appear to be significantly influenced by the growth of phytoplankton in the shallower side-channel areas. With average depths of about 5 ft in some of the tidal bay segments, significant biomass is computed for these areas. This biomass is then tidally exchanged with the main channel flow and contributes to the observed biomass at that location. It is hypothesized then that the growth of phytoplankton in the shallow areas is significant (even without direct discharges to embayments) and can account for as much as 40 μg/l chlorophyll in the observed main channel data.

C. Verification Analysis—1969 Data

The parameters determined from the 1968 verification analysis are listed in Table IV. This set of parameters was used to verify independently the 1969 data under the flow and temperature regime for that year. As shown in Fig. 18, freshwater inflow conditions were generally unsteady throughout 1969. There was a gradual decrease in the flow from April to mid-July, at which point the flow increased markedly.

Figures 24 and 25 show the comparison between the computed output and some observed data for the period at the end of the gradual decline in river inflow, from June 30, 1969 to July 15, 1969. Three data surveys were conducted during this period. It should be recalled that the computed output for 1969 was generated directly using the parameters from the 1968 runs; the parameters were not adjusted for the 1969 runs. Only flow, temperature, and incoming concentrations into segment 1 based on observed 1969 data were used in the 1969 computations. It should be noted, however, that boundary conditions at segment 1 do influence results for about the first 15 mi of

TABLE IV

Parameters used in Verifications of 1968 and 1969 Potomac River Data
Preliminary Phytoplankton Model

Notation	Description	Unit	Value Used
$K_1(T)$	Saturated growth rate[a]	1/day-°C	0.1
I_s	Saturating light intensity	ly/day	300.0
K_{mN}	Michaelis constant–nitrogen	mg N/1	0.025
K_{mp}	Michaelis constant–phosphorus	mg PO_4-P/l	0.005
K_{mP}	Michaelis constant–chlorophyll	μg Chl/l	60.0
$K_2(T)$	Phytoplankton endogenous respiration rate[a]	1/day-°C	0.005
a_{ZP}	Carbon–chlorophyll ratio	mg C/μg Chl	0.05
a_{NP}	Nitrogen–chlorophyll ratio	mg N/μg Chl	0.01
a_{pP}	Phosphorus–chlorophyll ratio	mg PO_4-P/μg Chl	0.001
K_{34}	Org N-NH_3 hydrolysis rate[a]	1/day-°C	0.007
K_{33}	Decay of organic nitrogen	1/day	0.10
K_{45}	NH_3-NO_3 nitrification[a]	1/day-°C	0.01
K_{67}	Org P-Inorg P conversion rate[a]	1/day-°C	0.007
K_{77}	Decay of org P	1/day	0.10
K_{66}	Decay of inorg P	1/day	0.10
K_d	BOD decay coefficient[a]	1/day-°C	0.01

[a] The temperature dependence used for these rates was linear with respect to temperature in °C.

the estuary. Incoming chlorophyll concentrations were not measured during 1969; the first station for which data were available was at Key Bridge, Mile 3.3. Boundary chlorophyll values were inputted to approximately duplicate the order of magnitude of observed data at that station. The values ranged from 4 μg/l in the winter to a high of 55 μg/l in September 1969.

Figure 24 shows the comparison between chlorophyll values as observed and the computed values for the beginning and end of the survey period. The June 30 values varied markedly, especially in the vicinity of Mile 15 where a low value of 30 μg/1 was observed, an apparently abnormal value for this location at that particular time of the year. The general shape of the spatial profile as computed is good and approximately reproduced the spatial behavior of the observed data. Peak values as computed during the period tend to be higher than observed, although the average of the two computed lines of June 30 and July 15 is reasonably close to the average of the observed data and differs in the peak region by about 20 μg/l and by 5–10 μg/l downstream of the peak region.

The total Kjeldahl nitrogen verification is also shown in Fig. 24 and agreement is very good with the exception of the July 8 survey, where there was no observed increase in nitrogen in the vicinity of the major discharges. There is no readily apparent explanation for this discrepancy.

Figure 25 shows the comparison between the spatial profile of total phosphorus (as PO_4) observed during the surveys of June 30, July 8, and July 15, 1969, and the computed profiles for the time period bracketing those surveys. The agreement is quite good although in the downstream direction, computed values decline more rapidly than observed values. This is probably a consequence of the simple first-order

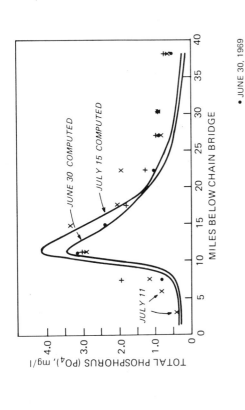

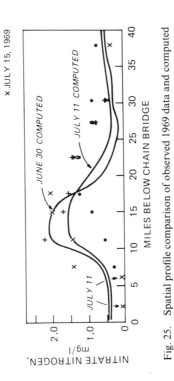

Fig. 25. Spatial profile comparison of observed 1969 data and computed values for total phosphorus (top) and nitrate–nitrogen (bottom).

Fig. 24. Spatial profile comparison of observed 1969 data and computed values for chlorophyll *a* (top) and total Kjeldahl nitrogen (bottom).

kinetics that were used and of the decay of both organic and inorganic forms. A second-order assumption (Jaworski et al., 1971) would give better results. Computed values in general are within 0.5–1.0 mg/l of observed values of PO_4.

For the nitrate profiles (Fig. 25), the agreement is fair for the July 8 and July 15 surveys. Agreement was not obtained for the June 30 survey. During that survey the data showed a considerable downstream shift in the observed nitrate which was not duplicated by the computed values. This could be due to low dissolved oxygen values which would delay the onset of nitrification but DO data were not available to confirm this hypothesis. There is a more rapid decline in computed values of nitrate nitrogen in the vicinity of Mile 20 than was observed. This discrepancy may be due to the simple nitrogen preference structure used in the model. A more detailed analysis of phytoplankton preference for differing forms of nitrogen appears to be warranted to provide better agreement.

In summary, the spatial agreement between observed and computed data for 1969 conditions is good. The general shapes of the spatial profiles are obtained and approximate quantitative agreement is obtained. Several areas remain to be explored, notably the model structure of the different nitrogen forms where anomalous results were occasionally obtained.

Comparisons between the temporal variation in chlorophyll *a* at four stations show (Fig. 26) the observed data are scattered but with a general peak in July 1969. The order and timing of this peak for both Mile 12.1 and Mile 18.3 is properly duplicated by the model. The rapid drop in chlorophyll concentration at the end of July is attributed to an increase in freshwater inflow at that time. The decrease is also successfully duplicated by the model. The model approximates the subsequent fall bloom of phytoplankton at Mile 18.3, but at Mile 12.1 the model calculations are somewhat higher than the observed data.

Figure 27 shows the comparison at two stations further downstream. At both stations a winter bloom of phytoplankton occurred which was not modeled in this work. At Hallowing Point, Mile 26.9, the spring growth pattern and subsequent decrease is adequately modeled. In the fall of 1969 at Hallowing Point, however, data indicated a significant increase in phytoplankton (maximum levels of 445 μg/l chlorophyll) which was not duplicated by the phytoplankton model. It is not clear from the data whether the high values in the fall of 1969 at Mile 26.9 are a surface phenomenon and therefore beyond the scope of the model formulation or whether the high values extended throughout the water column.

For Milepoint 38, Possum Point, similar comments can be made. The apparent winter bloom of phytoplankton is not captured by the model since the growth parameters for the phytoplankton biomass reflect a warm-water population. The model does reasonably well until about August 1969 after which a wide scatter of chlorophyll values was observed. Peak values during August to October are not obtained by the model although the model does approximate conditions again during November and December.

In general, the phytoplankton model as formulated herein, provided a reasonable approximation to 1969 conditions using a 1968 tuning of the model. As an independent check, recognizing that no changes were made to the model from the results of the 1968 analysis, the verification for 1969 provides an added measure of credibility to the overall model structure. Spatial profiles during July 1969 verified well to about 40 mi downstream from Chain Bridge. Dynamic variability in phytoplankton was verified well throughout the first 20 mi and then only approximately for the remaining 20 miles. Transient blooms in the late winter and early fall of 1969 were not duplicated by the model.

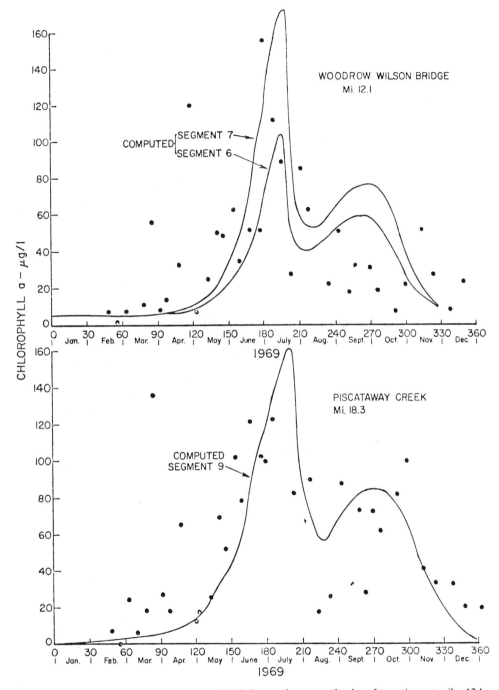

Fig. 26. Temporal comparison of observed 1969 data and computed values for stations at miles 12.1 and 18.3.

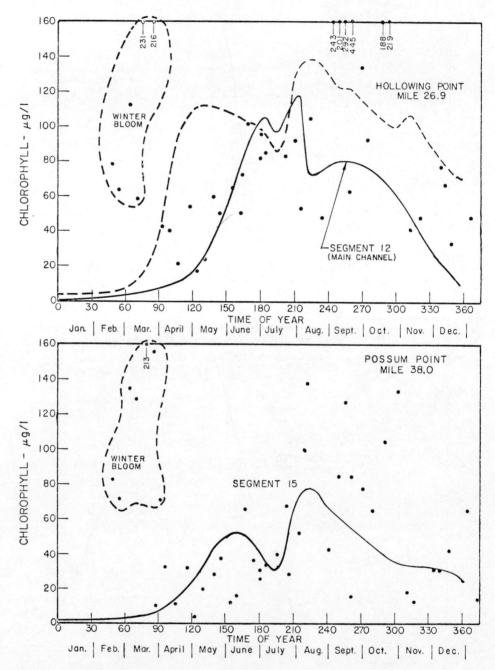

Fig. 27. Temporal comparison of observed 1969 data and computed values for stations at miles 26.9 and 38.0.

5. Preliminary Model Application—Potomac Estuary

As noted previously, the 1969 water pollution enforcement conference resulted in recommendation for nutrient removal including 96% removal of phosphorus and 85% removal of total nitrogen from discharges to the Potomac Estuary. It was assumed that the nutrient removal program would improve phytoplankton conditions in the estuary to objectives of 25–50 μg chlorophyll/l. The estimated capital cost at that time was $360 million for the improvements at Blue Plains (Perry, 1975) and by 1975, the cost had risen to $488 million, of which fully $100 million was estimated as the cost of the denitrification facilities. Increasing concern was being expressed over the relative need for the removal of total nitrogen, especially when a revised estimate of the operation and maintenance costs indicated that the denitrification facilities would cost about $16.6 million/yr.

The concern was heightened by the increase in the state of the art of phytoplankton modeling. Fig. 28 illustrates the development of models for the Potomac Estuary. In a period of about 5 yr, the modeling structure for the Potomac has changed rapidly and has included progressively more detail. It should be noted that a detailed model of nitrogen in the estuary was not available in 1969, the time at which decisions were made regarding nitrogen removal. The inclusion of increasing degrees of complexity improved the understanding of the behavior of the Potomac Estuary nutrient and phytoplankton interactions. However, the decision-making apparatus was not able to respond until several years later. The application of the Potomac phytoplankton model, therefore, was of particular relevance and provided a partial basis for decisions reached in 1975 regarding the efficiency of nitrogen removal.

To illustrate the application of the preliminary model, two simulations were prepared. It should be stressed that this model is largely exploratory in nature and the simulations discussed below are not to be considered as quantitative projections. At best, the simulations indicate general qualitative trends. Both simulations used a 90% removal policy of present raw waste loads. For the first simulation the 1969 flow regime was used. Untreated nitrogen, phosphorus and carbonaceous waste loads were reduced by 90%, and incoming boundary values into segment 1 were the same as those used in the 1969 verification analysis. For the second simulation, median flow conditions were used with reduced waste inputs as in the first simulation. In addition several boundary conditions on chlorophyll were explored in the median flow simulation. For both simulations it was assumed that all organic and ammonia nitrogen was converted to nitrate nitrogen at the treatment plants and all phosphorus residual load is inorganic phosphorus. The summary of the waste loads is given in Table V.

TABLE V
Direct Discharge Waste Loads

	Waste Input for Verification (lb/day)	Waste Input For Simulation (lb/day)
Total Nitrogen (N)	46,500	7,070
Total Phosphorus (P)	20,300	2,620
Carbonaceous BOD	151,200	10,000

Development of Nutrient Models for Potomac Estuary

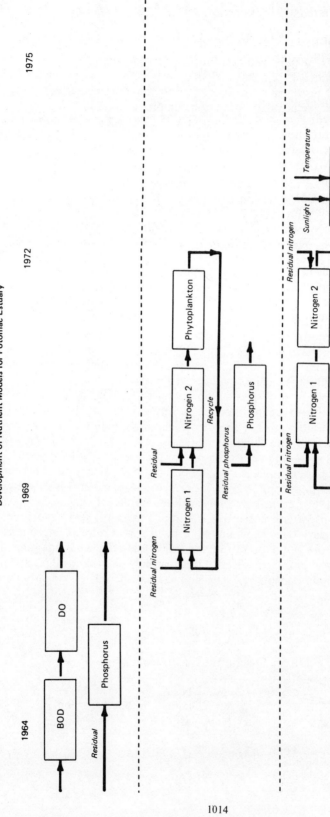

Fig. 28. Development of nutrient models for Potomac estuary, 1964, 1969, 1972, and 1975.

The 90% removal of the untreated waste loads results in about a 15% residual discharge of the 1968–1969 discharged loads. Although the residual loads used in the simulations are in excess of the allocations established for the Potomac (Jaworski et al., 1971), the level is considered a practical achievable level on a sustained basis. Figure 29 shows the spatial chlorophyll profile for June 30–July 15, 1969 and the 1969 simulation with 90% reduction of nutrient waste discharge. Even after waste reduction, maximum concentrations of phytoplankton chlorophyll exceed 100 μg/l or four times the objective of 25 μg/l. The effect of the removal program is significant, however, from about Mile 15 to Mile 40. It should be recalled that no distributed sources of nutrient were included which would increase the nutrients available for growth.

The results of the 1969 simulation tended to indicate that phosphorus was more limiting than nitrogen which for all channel regions was above limiting concentrations. From about segment 2 downstream, the phytoplankton appear to be increasingly limited by phosphorus almost uniformly throughout the year. Also, concentrations of inorganic phosphorus in the vicinity of Washington, D.C., range from 0.05 to 0.1 mg/l P, significantly above the Michaelis constant, indicating that in that immediate area, phosphorus is not limiting.

Flows for the median flow simulation ranged from 13,000 cfs during the winter to a high of 20,500 cfs in the spring, decreasing to 4,200 cfs in the later summer and then increasing again to 9,500 cfs in December. Incoming boundary concentrations, as noted previously, have an important relative impact on the estuary, especially after waste loads are reduced by treatment. For the nutrients, median concentrations as given in Jaworski et al., (1971) were used. Kjeldahl nitrogen of 0.95 mg/l and nitrate nitrogen levels ranging from 0.1 to 0.9 mg/l were inputted as boundary values. Organic phosphorus of 0.05 mg/l and an inorganic phosphorus concentration of 0.05 mg/l were also inputted as constant values throughout the year into segment 1. The

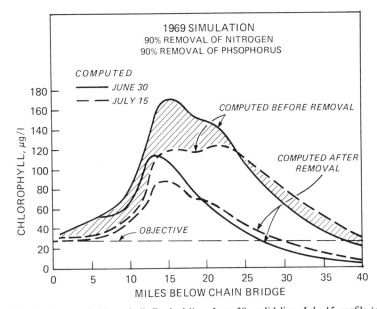

Fig. 29. 1969 simulation of chlorophyll. Dashed line, June 30; solid line, July 15 profile (compare to Fig. 17).

effect of incoming phytoplankton chlorophyll was examined under median flows for two cases: (*a*) chlorophyll *a* concentrations of 1 μg/l entering the estuary from upriver and (*b*) chlorophyll *a* boundary concentration of 25 μg/l.

Figure 30*a* shows the temporal variation in chlorophyll for the main channel segment 9 and its associated Piscataway embayment, using median flows. For the case of 1 μg/l chlorophyll entering the estuary (Fig. 30*a*), the effect of the reduced

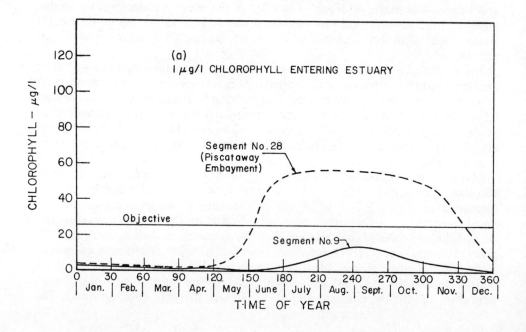

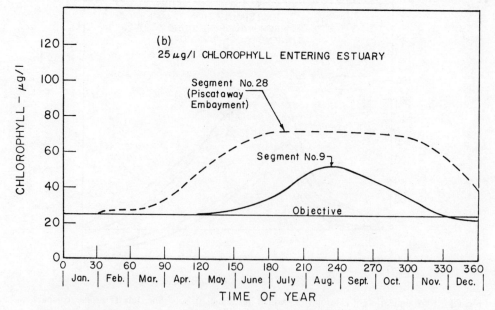

Fig. 30.   Temporal variation in chlorophyll *a* at segments 9 and 28, median flow simulation.

waste discharge on the main channel is minor and results in a late summer increase to about 14 $\mu g/l$ chlorophyll. In the embayment section, however, chlorophyll concentrations rise to about 55 $\mu g/l$ for a period of about 120 days. It should be noted that the level of 50 $\mu g/l$ is considerably less than the concentrations of 150–200 $\mu g/l$ before any nutrient reduction. The incoming concentration of 1 $\mu g/l$ chlorophyll is unrealistic, however, and one would normally expect higher incoming concentrations.

Figure 30*b* shows the effect of constraining the boundary chlorophyll concentration at a level equal to the objective of 25 $\mu g/l$. Main channel chlorophyll levels rise to more than 50 $\mu g/l$ in August and in the embayment, concentrations rise to over 70 $\mu g/l$ again for a period of about 4 months.

These simulations permit the following general observations. Achievement of the objective of 25 $\mu g/l$ chlorophyll in the estuary may not be possible, primarily because of the effect of discharges from the Upper Potomac into the estuary. A level of 50 $\mu g/l$ may be possible in the main channel although higher concentrations would still be expected to occur in some embayments. Under a median flow regime (4,000 cfs during the growing season), maximum concentrations of 50 $\mu g/l$ chlorophyll arc calculated for the main channel and 75 $\mu g/l$ in some embayments. Additional sources of nutrients such as local drainage and urban runoff have not been included in the model. Therefore, even if reductions greater than the 90% used in these applications could be achieved, such reductions may be offset by other distributed nutrient sources. On the basis of the preliminary model, the state-of-the-art nutrient reduction (90% or better) in the Potomac may provide reductions in average chlorophyll levels by about 60% under median flow conditions. Again, the simulations should be considered as indicative of general trends only and not as a predictive certainty.

The transport through the estuary assumes an important role in the phytoplankton dynamics of the Upper Potomac Estuary. The nutrients and phytoplankton associated with the incoming river flow are particularly important. Unlike other problems such as dissolved oxygen, critical conditions for phytoplankton growth may occur during nondrought flow conditions. This is due primarily to the addition of nutrients during high flows. The verification analyses indicate that, at present, phosphorus and nitrogen are probably sorbed from the estuary and enter the estuarine sediments. The preliminary model does not incorporate any recycle of nutrients from the sediments, although the model does indicate that phosphorus and nitrogen are not conserved in the water column and losses to the sediment can be considerable. These losses could alter the effect of incoming river concentrations.

Simulations with the preliminary model indicate that under median flows and a 90% reduction of untreated raw nutrient loads, chlorophyll concentrations in the main channel may rise to about 50 $\mu g/l$ with embayment values greater than 70 $\mu g/l$. The concentration during 1969 flow conditions after 90% removal of nitrogen and phosphorus may rise to greater than 100 μg chlorophyll/l in the main channel. The simulations also tend to indicate that phosphorus is more limiting to phytoplankton growth than nitrogen. However, it should be stressed that the simulations are intended only to show possible trends under future loading conditions and are not to be interpreted in any detailed quantitative fashion. Models of the type presented herein require additional evaluation and testing on a continuing basis.

Nevertheless, partly as a result of the continuing development of models for the Potomac Estuary and a continuing increase in the understanding of the system, a decision was made by the EPA to postpone for 2 yrs construction of denitrification facilities at Blue Plains. The two main factors in this decision were the high cost of the facilities and the increasing evidence that nitrogen removal may only be marginally effective in reducing phytoplankton growth. Because of the long-range consequences of any further decisions on nitrogen removal, further refinement of the model is mandatory before any definite statements can be made of the effects of nutrient reduction on phytoplankton chlorophyll in the Upper Potomac Estuary.

6. Summary and Discussion

A quantitative framework for the analysis of the seasonal development of phytoplankton biomass distribution in estuaries is presented in this chapter. The method is based on the equations of conservation of mass which combine the effects of estuarine mass transport, direct discharges of nutrients to the estuary, ocean and river boundary conditions, the kinetics of the phytoplankton and zooplankton populations, and nutrient recycling. Each of these components is susceptible to an independent quantitative assessment: mass transport is related to the distribution of salinity; direct discharges and boundary conditions are measured directly; and at least a part of the kinetics can be examined using laboratory or in situ experiments. These components can then be combined into the model of the estuary itself.

It is important that the structure of the equations and the coefficients used have a basis in experimental fact. The formulation of the kinetics of the model is a deliberate attempt to incorporate the major features of phytoplankton biomass behavior as observed in the laboratory and the field for the seasonal time scale of interest. The coefficients used for the kinetic expressions must be within the ranges observed. Site-specific experiments, such as the AGP results for the San Francisco Bay Estuary, are a stringent check on the realism of the formulation. As shown in Tables III and IV the magnitudes of the kinetic coefficients for both the San Francisco Bay and Potomac Estuaries are quite similar. This indicates that the kinetic structure is not necessarily estuary specific but appears to be applicable in other settings.

The degree to which the calculations agree with observation is an important criterion which governs their acceptability for use in engineering decisions. At least one independent check, a calculation for a set of data that was not used for model calibration, is a prerequisite for a claim of some predictive capability for the model. However, it is even more important that the calculations agree with observations of variables, as many as are relevant, and that the agreement is simultaneous. Thus the behavior of all the dependent variables must be checked against observations, and in addition, direct measurements of rates, such as primary production, add confidence to the reliability of the calculations.

The effort devoted to verification of the models for the San Francisco Bay Estuary and the Potomac Estuary are the result of a conviction that only with a solidly based verification can reasonable projections be attempted; even with such verifications, if the projections involve large perturbations, which change the interactions substantially, there is always the possibility that effects are occurring for which the verification was insensitive. Thus if the results of a set of projections indicate that new phenomena are occurring, a change in the limiting nutrient, for example, then it is reasonable to require that kinetic experiments be designed to confirm the behavior as calculated.

It is apparent that the major features of the regional development of phytoplankton biomass in estuaries can be understood in terms of the equations of the model. A number of important interactions are basic: the productivity of shallow estuarine regions and the interplay of mass transport of both plankton and nutrients between regions of negative and positive primary production is characteristic of both estuaries considered. The notion of an absolutely limiting nutrient, a nutrient which is completely depleted—the candidates are nitrogen in the San Francisco Bay Estuary and phosphorus in the Potomac Estuary—is too simple a characterization. It appears to be necessary to use the product Michaelis–Menten kinetics for phosphorus and nitrogen, a formulation that is supported by the AGP experiments. More complex short time scale interactions appear to be irrelevent when averaged over the seasonal time scale. The excess of phosphorus in the San Francisco Bay Estuary and the excess of nitrogen in the Potomac Estuary are correctly accounted for by the kinetics; that is, the growth rate is not affected by the excess nutrient. The other extreme, observed nutrient depletion, is also correctly incorporated. Population production continues to occur even if the measured concentrations of the nutrient are small because it is being supplied either by direct discharge or by the recycle of the organic forms. This is the purpose of a model, to balance correctly the many phenomena that contribute to phytoplankton biomass development and decline and to do it quantitatively in a way that is beyond the capabilities of qualitative assessments.

The extent to which the model presented in this chapter succeeds is still a matter of judgment since the calculations are not a priori in the sense of classical physics. The difficulty of arriving at a completely determined theory is apparent. However, it is our judgment that the model at its present level of development can be used for evaluating certain questions within the scope of the framework if the level of verification is adequate to the question. The calculations are critical, however, since they bound the possibilities and reduce the number of questions that must be addressed in the final analysis. It is for this reason that calculations of this sort are becoming a fundamental part of the rational management of the water quality of estuaries.

Acknowledgment

The investigation of the San Francisco Bay Estuary was sponsored by a contract between the Department of Water Resources, State of California and Hydroscience, Inc. The participation of Gerald Cox, Austin Nelson, Fred Bachman, and Harlan Proctor is acknowledged. Also the participation of Harold Chadwick of the Department of Fish and Game, and James Arthur and Douglas Ball of the U.S. Bureau of Reclamation is appreciated. The efforts of James Fitzpatrick, Joseph A. Nusser, and Henry Salas of Hydroscience are gratefully acknowledged.

The investigation of the Potomac Estuary was sponsored in part by a grant from the Environmental Protection Agency and by assistance provided through the research program of Hydroscience, Inc. The participation of Richard Winfield, Manhattan College, is appreciated.

References

Aalto, J. A., L. S. Clark, and K. D. Feigner, 1970. Current water quality conditions and investigations in the upper Potomac River tidal system. Tech. Rept. 41, Ches. Tech. Sup. Lab. FWQA, Annapolis, Md.

Ball, Douglas, 1972. Personal communication. U.S. Bureau of Reclamation, Sacramento, Calif.

Di Toro, D. M., D. J. O'Connor, and R. V. Thomann, 1971. A dynamic model of the phytoplankton population in the Sacramento–San Joaquin Delta. In *Nonequilibrium Systems in Natural Water Chemistry*, Adv. Chem. Ser 106. American Chemical Society, Washington, D.C., pp. 131–180.

Eppley, R. W., 1972. Temperature and phytoplankton growth in the sea. *Fish. Bull.*, **70** (4), 1063–1085.

Eppley, R. W., J. N. Rogers, and J. J. McCarthy, 1969. Half saturation constants for uptake of nitrate and ammonium by marine phytoplankton. *Limnol. Oceanogr.*, **14** (6), 912–920.

Hydroscience, Inc., 1972. *Mathematical Model of Phytoplankton Population Dynamics in the Sacramento–San Joaquin Bay Delta, Preliminary Results.* Prepared for California Department of Water Resources, Sacramento, Calif.

Hydroscience, Inc., 1974. *Western Delta and Suisan Bay Phytoplankton Model—Verifications and Projections.* Prepared for California Department of Water Resources, Sacramento, Calif.

Interagency Ecological Study Program for the Sacramento–San Joaquin Estuary, 1970–1973. Annual Reports California Department of Fish and Game; California Department of Water Resources; U.S. Fish and Wildlife Service; U.S. Bureau of Reclamation.

Jaworski, N. A., et al., 1969a. Nutrients in the Potomac River Basin. *Tech Rept. 9, Ches. Tech. Sup. Lab.* FWPCA, Dept of Interior, 40 pp.

Jaworski, N. A., L. S. Clark, and K. D. Feigner, 1971. A water resource water supply study of the Potomac Estuary. *Tech. Rept. 35, Ches. Tech. Sup. Lab.* EPA, Annapolis, Md.

Jaworski, N. A., D. W. Lear, and J. A. Aalto, 1969b. A technical assessment of current water quality conditions and factors affecting water quality in the upper Potomac Estuary. *Tech. Rept. 5, Ches. Tech. Sup. Lab.* FWQA, Annapolis, Md.

Ketchum, B. H., 1939. The absorption of phosphate and nitrate by illuminated cultures of Nitzschia closterium. *Am. J. Bot.*, June issue, 26.

O'Connor, D. J. and R. V. Thomann, 1971. Water quality models: chemical, physical, and biological constituents. Chapter 3 in *Estuarine Modeling: An Assessment.* EPA Water Pollution Control Res. Ser. 16070 DZV 02/71.

Perry, R. R., 1975. Statement for the ICPRB conference on the status of water quality improvement efforts in the Washington Metropolitan Area of the Potomac River Basin, District of Columbia. Presented at ICPRB Conference, Great Falls, Md., Feb. 25, 1975. 6pp.

Phelps, E. B., 1944. *Stream Sanitation.* Wiley, New York, pp. 148–149.

Riley, G. A., 1956. Oceanography of Long Island Sound 1952–54 II. Physical oceanography. *Bull. Bingham Oceanogr. Coll.*, **15**, 15–46.

Riley, G. A., H. Stommel, and D. F. Bumpus, 1949. Quantitative ecology of the plankton of the Western North Atlantic. *Bull. Bingham Oceanogr. Coll.*, **12** (3), 1–169.

Steele, J. H., 1965. Notes on some theoretical problems in production ecology: In *Primary Production in Aquatic Environments.* C. R. Goldman, ed. Mem. Inst. Idrobiol. University of California Press, Berkeley, pp. 383–398.

Thomann, R. V., D. J. O'Connor, and D. M. Di Toro, 1970. *Modeling of the Nitrogen and Algal Cycles in Estuaries.* Fifth International Water Pollution Conference, San Francisco.

Water Resources Engineers. *Hydrologic-Water Quality Model, Development and Testing Task III San Francisco Bay—Delta Water Quality Control Program.* Prepared for California Department of Water Resources.

AUTHOR INDEX

Numbers in *italics* indicate the pages on which the full references appear.

SUBJECT INDEX